A-Level Chemistry

The Complete Course for AQA

Like a heavy fraction of crude oil, A-Level Chemistry is difficult to crack. But don't worry — this CGP book will help you put in a high-octane performance all the way through the AQA course!

It's packed with clear explanations, exam practice, advice on maths skills and practical investigations... everything you need, basically.

It even includes a free Online Edition to read on your PC, Mac or tablet.

How to get your free Online Edition

Go to **cgpbooks.co.uk/extras** and enter this code...

2023 0109 9161 7878

This code will only work once. If someone has used this book before you, they may have already claimed the Online Edition.

Contents

Introduction

Practical Skills

PRACTICAL SKILLS

Unit 1

Section 1: Atomic Structure

Section 2: Amount of Substance

Section 3: Bonding

Section 4: Energetics

Section 5: Kinetics

Section 6: Equilibria and Redox Reactions

Unit 2

Section 1: Periodicity

Section 2: Group 2 and Group 7 Elements

Unit 3

Section 1: Introduction to Organic Chemistry

Section 2: Alkanes and Halogenoalkanes

Section 3: Alkenes and Alcohols

Section 4: Organic Analysis

Unit 4

Section 1: Thermodynamics

Section 2: Rate Equations and K_p

Section 3: Electrode Potentials and Cells

Section 4: Acids, Bases and pH

Unit 5

Section 1: Period 3 Elements

Section 2: Transition Metals

Unit 6

Section 1: Isomerism and Carbonyl Compounds

Section 2: Aromatic Compounds and Amines

Section 3: Polymers

Section 4: Amino Acids, Proteins and DNA

Section 5: Further Synthesis and Analysis

Maths Skills

MATHS SKILLS

Exam Help

EXAM HELP

Reference

How to use this book

Learning Objectives

- These tell you exactly what you need to learn, or be able to do, for your exams.
- There's a specification reference at the bottom that links to the AQA specification.

Tips

These are here to help you understand the theory.

Exam Tips

There are tips throughout the book to help with all sorts of things to do with answering exam questions.

Learning Objectives:

- Know the importance of the conditions when measuring the electrode potential, *E*.
- Understand how cells are used to measure electrode potentials by reference to the standard hydrogen electrode.
- Know that standard electrode potential, $E^{\ominus}$, refers to conditions of 298 K, 100 kPa and 1.00 mol dm^{-3} solution of ions.

Specification Reference 3.1.11.1

Tip: See the earlier page on Le Chatelier's Principle (pages 133-135) for more about how changes in temperature, pressure and concentration affect the position of equilibria.

Tip: A platinum electrode is needed because you can't have a gas electrode.

Figure 2: A nugget of platinum. Platinum is used as the electrode in the standard hydrogen electrode.

Exam Tip
Make sure you learn the standard conditions — you might well be asked about them in the exam.

2. Standard Electrode Potentials

Electrode potentials are influenced by things like temperature and pressure. So if you want to compare electrode potentials, they need to be standardised. This is done using a standard hydrogen electrode.

Factors affecting the electrode potential

Half-cell reactions are reversible. So just like any other reversible reaction, the equilibrium position is affected by changes in temperature, pressure and concentration. Changing the equilibrium position changes the cell potential. To get around this, **standard conditions** are used to measure electrode potentials — using these conditions means you always get the same value for the electrode potential and you can compare values for different cells.

The standard hydrogen electrode

You measure the electrode potential of a half-cell against a **standard hydrogen electrode**. In the standard hydrogen electrode, hydrogen gas is bubbled through a solution of aqueous H^+ ions. A platinum electrode is used as a platform for the oxidation/reduction reactions — see Figure 1.

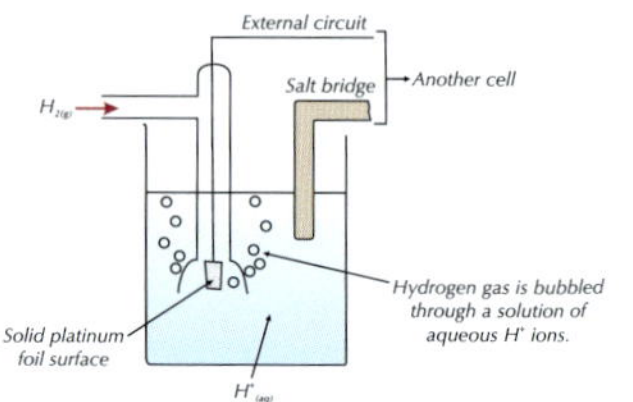

External circuit
Another cell
Salt bridge
$H_{2(g)}$
Hydrogen gas is bubbled through a solution of aqueous H^+ ions.
Solid platinum foil surface
$H^+_{(aq)}$

Figure 1: The standard hydrogen electrode

When measuring electrode potentials using the standard hydrogen electrode it's important that everything is done under standard conditions:

1. Any solutions of ions must have a concentration of 1.00 mol dm^{-3}.
2. The temperature must be 298 K (25 °C).
3. The pressure must be 100 kPa.

Measuring standard electrode potentials

The standard electrode potential of a half-cell is the voltage measured under standard conditions when the half-cell is connected to a standard hydrogen electrode — see Figure 3 (next page).

328 Unit 4: Section 3 Electrode Potentials and Cells

Indicators

REQUIRED PRACTICAL 9

When you use an indicator, you need it to change colour exactly at the end point of your titration. So you need to pick one that changes colour over a narrow pH range that lies entirely on the vertical part of the pH curve. So for the titration shown in Figure 3 (below) you'd want an indicator that changed colour somewhere between pH 8 and pH 11:

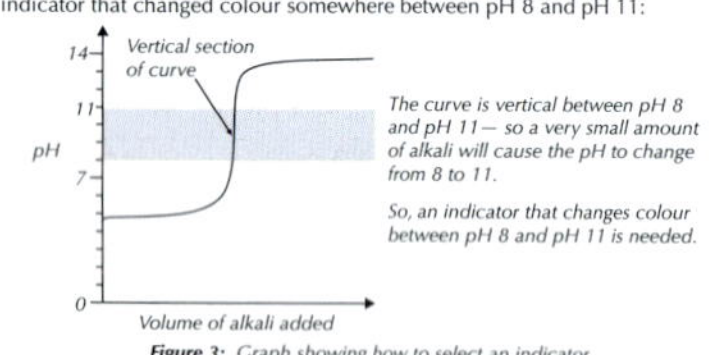

Vertical section of curve
14
11
7
0
pH
The curve is vertical between pH 8 and pH 11 — so a very small amount of alkali will cause the pH to change from 8 to 11.
So, an indicator that changes colour between pH 8 and pH 11 is needed.
Volume of alkali added

Figure 3: Graph showing how to select an indicator.

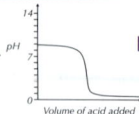

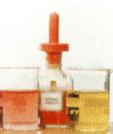

Figure 4: The red to yellow colour change of methyl orange.

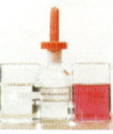

Figure 5: The colourless to pink colour change of phenolphthalein

Methyl orange and **phenolphthalein** are indicators that are often used for acid–base titrations. They each change colour over a different pH range:

Name of indicator	Colour at low pH	Approx. pH of colour change	Colour at high pH
Methyl orange	red	3.1 – 4.4	yellow
Phenolphthalein	colourless	8.3 – 10	pink

- For a strong acid/strong base titration, you can use either of these indicators — there's a rapid pH change over the range for both indicators.
- For a strong acid/weak base only methyl orange will do. The pH changes rapidly across the range for methyl orange, but not for phenolphthalein.
- For a weak acid/strong base, phenolphthalein is the stuff to use. The pH changes rapidly over phenolphthalein's range, but not over methyl orange's.
- For weak acid/weak base titrations there's no sharp pH change, so no indicator will work — you should just use a pH meter.

You need to be able to use a pH curve to select an appropriate indicator:

Example

The graph to the right shows the pH curve produced when a strong acid is added to a weak base. From the table below, select an indicator that you could use for this titration.

Indicator	pH range
Bromophenol blue	3.0 – 4.6
Litmus	5.0 – 8.0
Cresol purple	7.6 – 9.2

pH 14, 7, 0 — Volume of acid added

The graph shows that the vertical part of the pH curve is between about pH 2 and pH 6. So you need an indicator with a pH range between 2 and 6. The only indicator that changes colour within this range is bromophenol blue. So in this example, bromophenol blue is the right indicator to choose.

Exam Tip
In your exam, you'll usually be given a table of indicators to choose from, like in this example — so don't worry, you don't have to learn all these indicators and their pH ranges.

352 Unit 4: Section 4 Acids, Bases and pH

Required Practicals

- As part of your course, you'll be expected to do a set of Required Practicals. You'll be tested on your knowledge of them in your exams too.
- Information about these practicals is marked with a Required Practical stamp.

Practical Skills

- There are some key practical skills you'll not only need to use in your Required Practicals, but you could be tested on in the exams too.
- There's a Practical Skills section at the front of this book with loads of information on how to plan experiments and analyse data.

Examples

These are here to help you understand the theory.

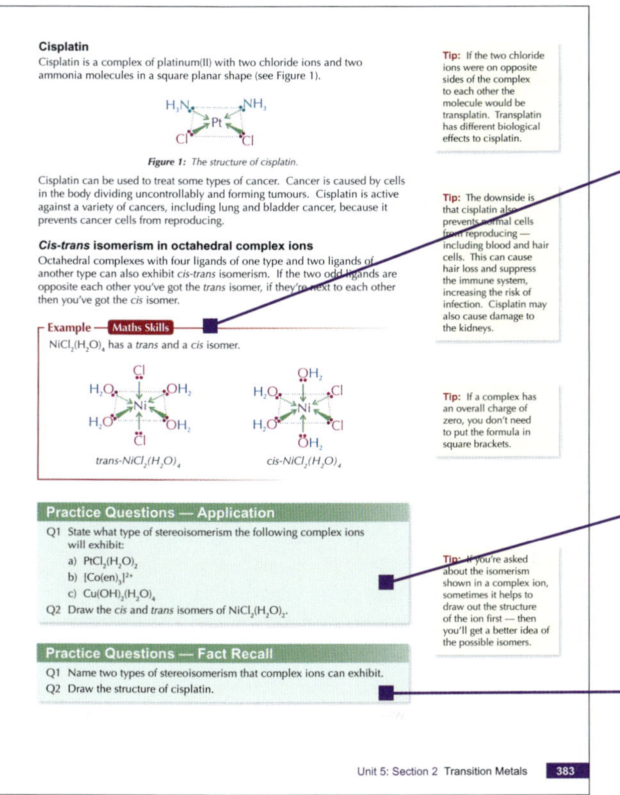

Cisplatin
Cisplatin is a complex of platinum(II) with two chloride ions and two ammonia molecules in a square planar shape (see Figure 1).

Figure 1: The structure of cisplatin.

Cisplatin can be used to treat some types of cancer. Cancer is caused by cells in the body dividing uncontrollably and forming tumours. Cisplatin is active against a variety of cancers, including lung and bladder cancer, because it prevents cancer cells from reproducing.

Cis-trans **isomerism in octahedral complex ions**
Octahedral complexes with four ligands of one type and two ligands of another type can also exhibit *cis-trans* isomerism. If the two odd ligands are opposite each other you've got the *trans* isomer, if they're next to each other then you've got the *cis* isomer.

Example — Maths Skills
$NiCl_2(H_2O)_4$ has a *trans* and a *cis* isomer.

trans-$NiCl_2(H_2O)_4$ *cis*-$NiCl_2(H_2O)_4$

Practice Questions — Application
Q1 State what type of stereoisomerism the following complex ions will exhibit:
a) $PtCl_2(H_2O)_2$
b) $[Co(en)_3]^{2+}$
c) $Cu(OH)_2(H_2O)_4$
Q2 Draw the *cis* and *trans* isomers of $NiCl_2(H_2O)_2$.

Practice Questions — Fact Recall
Q1 Name two types of stereoisomerism that complex ions can exhibit.
Q2 Draw the structure of cisplatin.

Tip: If the two chloride ions were on opposite sides of the complex to each other the molecule would be transplatin. Transplatin has different biological effects to cisplatin.

Tip: The downside is that cisplatin also prevents normal cells from reproducing — including blood and hair cells. This can cause hair loss and suppress the immune system, increasing the risk of infection. Cisplatin may also cause damage to the kidneys.

Tip: If a complex has an overall charge of zero, you don't need to put the formula in square brackets.

Tip: If you're asked about the isomerism shown in a complex ion, sometimes it helps to draw out the structure of the ion first — then you'll get a better idea of the possible isomers.

Unit 5: Section 2 Transition Metals 383

Maths Skills

There's a range of maths skills you could be expected to apply in your exams.

- Examples that show these maths skills in action are marked up like this.
- There's also a Maths Skills section at the back of the book.

Practice Questions — Application

- Annoyingly, the examiners expect you to be able to apply your knowledge to new situations — these questions are here to give you plenty of practice at doing this.
- All the answers are in the back of the book (including any calculation workings).

Practice Questions — Fact Recall

- There are a lot of facts you need to learn — these questions are here to test that you know them.
- All the answers are in the back of the book.

Exam-style Questions

- Practising exam-style questions is really important — you'll find some at the end of each section.
- They're the same style as the ones you'll get in the real exams — some will test your knowledge and understanding and some will test that you can apply your knowledge.
- All the answers are in the back of the book, along with a mark scheme to show you how you get the marks.

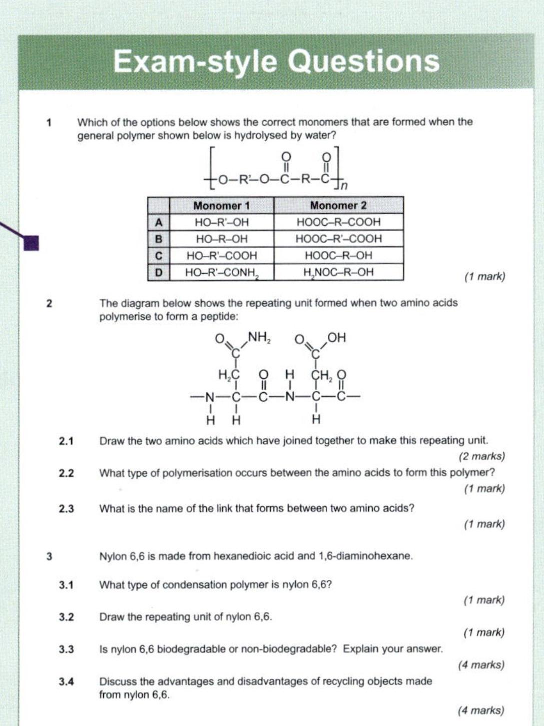

Exam-style Questions

1 Which of the options below shows the correct monomers that are formed when the general polymer shown below is hydrolysed by water?

	Monomer 1	Monomer 2
A	HO–R'–OH	HOOC–R–COOH
B	HO–R–OH	HOOC–R'–COOH
C	HO–R'–COOH	HOOC–R–OH
D	HO–R'–CONH$_2$	H$_2$NOC–R–OH

(1 mark)

2 The diagram below shows the repeating unit formed when two amino acids polymerise to form a peptide:

2.1 Draw the two amino acids which have joined together to make this repeating unit. (2 marks)

2.2 What type of polymerisation occurs between the amino acids to form this polymer? (1 mark)

2.3 What is the name of the link that forms between two amino acids? (1 mark)

3 Nylon 6,6 is made from hexanedioic acid and 1,6-diaminohexane.

3.1 What type of condensation polymer is nylon 6,6? (1 mark)

3.2 Draw the repeating unit of nylon 6,6. (1 mark)

3.3 Is nylon 6,6 biodegradable or non-biodegradable? Explain your answer. (4 marks)

3.4 Discuss the advantages and disadvantages of recycling objects made from nylon 6,6. (4 marks)

Unit 6: Section 3 Polymers 475

Exam Help

There's a section at the back of the book stuffed full of things to help with your exams.

Glossary

There's a glossary at the back of the book full of all the definitions you need to know for your exams, plus loads of other useful words.

Published by CGP

Editors:
Katie Braid, Emma Clayton, Mary Falkner, Katherine Faudemer, Gordon Henderson, Emily Howe, Paul Jordin, Rachel Kordan, Sophie Scott, Ben Train.

Contributors:
Mike Bossart, Robert Clarke, Ian H. Davis, John Duffy, Emma Grimwood, Lucy Muncaster, Derek Swain, Paul Warren.

ISBN: 978 1 78908 047 6

With thanks to Glenn Rogers and Karen Wells for the proofreading.
With thanks to Laura Jakubowski for the copyright research.
AQA Specification material is reproduced by permission of AQA.

Printed by Elanders Ltd, Newcastle upon Tyne.
Clipart from Corel®

Practical Skills

1. Planning Experiments

You'll get to do loads of practicals in class during this course, which is great. Unfortunately, you can also get tested on how to carry out experiments in your exams. The first thing you need is a good plan...

Initial Planning

Scientists solve problems by suggesting answers and then doing experiments that test their ideas to see if the evidence supports them. Being able to plan experiments that will give you valid and accurate results (see page 15) is a very important part of this process. Here's how you go about it:

First, you have to identify what question you are trying to answer. This is the **aim** of your experiment. It could be something like 'the aim of this experiment is to determine how temperature affects the rate of this reaction'. You can then make a **prediction** — a specific, testable statement about what will happen in the experiment, based on observation, experience or a **hypothesis** (a suggested explanation for a fact or observation).

Tip: Science experiments are all about identifying problems and using the knowledge you already have to solve them.

The next step is to identify what the independent and dependent **variables** (see below) are for your experiment. You can then decide what data you need to collect and how to collect it. This includes working out what equipment would be appropriate to use (see pages 2 and 3) — it needs to be the right size and have the right level of sensitivity.

Next you write out a detailed **method** for your experiment. As part of this, you'll need to do a **risk assessment** and plan any safety precautions. This ensures that you minimise the risk of any dangers there might be in your experiment, such as harmful chemicals you might be using. Your method will also need to include details of what steps you are going to take to control any other variables which may affect your results.

Finally, it's time to carry out the experiment — by following your method you can gather evidence to address the aim of your experiment.

Variables

You probably know this all off by heart, but it's easy to get mixed up sometimes. So here's a quick recap. A **variable** is a quantity that has the potential to change, e.g. temperature, mass or volume. There are two types of variable commonly referred to in experiments:

Independent variable — the thing that you change in an experiment.

Dependent variable — the thing that you measure in an experiment.

Tip: When drawing graphs, the dependent variable should go on the *y*-axis, the independent variable on the *x*-axis.

As well as the independent and dependent variables, you need to think of all the other variables that could affect the result of the experiment and plan ways to keep each of them the same. These other variables are called **control variables**.

Tip: There's lots more about what electrochemical cells are and how you set them up on pages 323-324.

Example

You could investigate the effect of changing the Cu^{2+} ion concentration in a $Cu^{2+}_{(aq)}/Cu_{(s)}$ half-cell on the EMF of the cell using the apparatus in Figure 1, below:

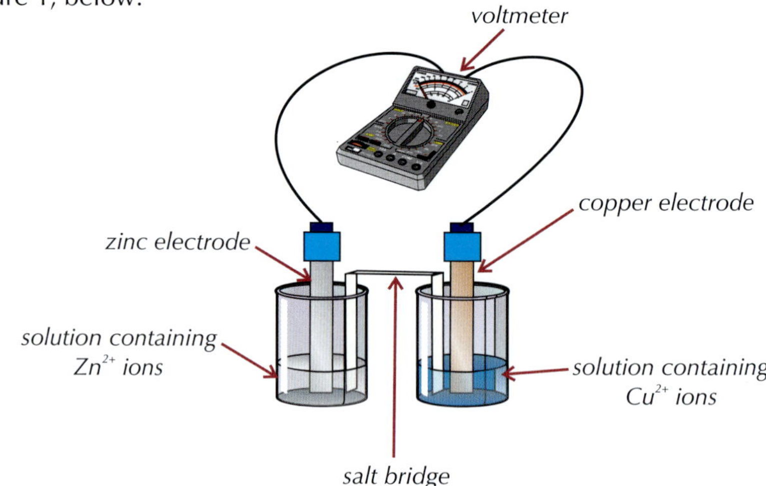

Figure 1: *Apparatus for measuring the EMF of an electrochemical cell.*

- The independent variable is the concentration of Cu^{2+} ions in the electrolyte of the $Cu^{2+}_{(aq)}/Cu_{(s)}$ half-cell.
- The dependent variable is the voltage (EMF) of the cell.
- For this experiment to be a fair test, all other variables must be kept the same each time. This includes the concentration of Zn^{2+} ions in the $Zn^{2+}_{(aq)}/Zn_{(s)}$ half cell, the volume of both electrolyte solutions, the pressure and the temperature. It also means using the same equipment (such as the wires and voltmeter) each time.

Tip: In a fair test, you only change one variable between experiments. This means you can directly compare the results from each experiment to see how they depend on the variable you've been changing.

Risks, hazards and ethical considerations

When you plan an experiment, you need to think about how you're going to make sure that you work safely. This involves carrying out a risk assessment, where you identify all the hazards that might be involved in your experiment and come up with ways to reduce the risks that these hazards pose. There's more detail about risk assessments on pages 4 and 5.

You need to make sure you work ethically, too. This is most important if there are other people or animals involved. You have to put their welfare first.

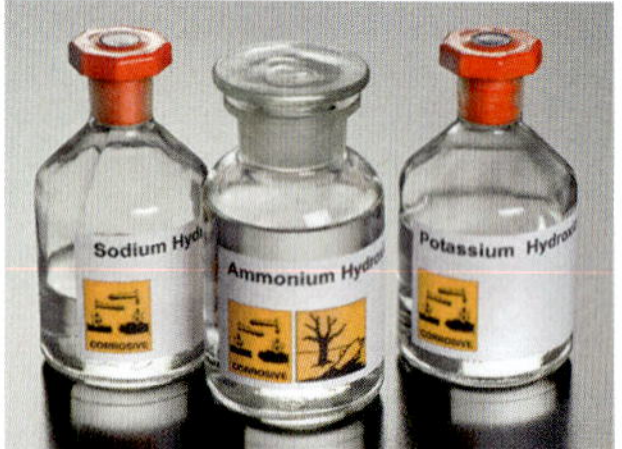

Figure 2: *Hazard symbols on bottles of chemicals can tell you if a chemical is dangerous, for example if it is flammable or toxic.*

Choosing equipment

When you're planning an experiment, you should always plan to use equipment that is appropriate for the experiment you're doing.

Example

If you want to measure the amount of gas produced in a reaction, you need to make sure you use apparatus which will collect the gas, without letting any escape.

The equipment that you use needs to be the right size for your experiment too.

Example

If you're using a gas syringe to measure the volume of a gas produced by a reaction, it needs to be big enough to collect all the gas, or the plunger will be pushed out of the end. You might need to do some rough calculations to work out what size of equipment to use.

The equipment also needs to have the right level of sensitivity.

Example

If you want to measure out 3.8 g of a substance, you need a balance that measures to the nearest tenth of a gram, not the nearest gram.

If you want to measure out 6 cm^3 of a solution, you need to use a measuring cylinder that has a scale marked off in steps of 1 cm^3, not one that only has markings every 10 cm^3.

If you want to measure very small changes in pH, then you need to use a pH meter, which can measure pH to several decimal places, rather than using indicator paper.

Tip: If you want to measure out a solution really accurately (e.g. 20.0 cm^3 of solution) you'll need to use a burette or a pipette.

Tip: There's more about using pH meters and other methods to measure the pH of a solution on page 6.

Methods

When you come to write out your method, you need to keep all of the things on the preceding pages in mind.

The method must be clear and detailed enough for anyone to follow — it's important that other people can recreate your experiment and get the same results. Make sure your method includes:

- All the substances needed and what quantity of each to use.
- How to control variables.
- The exact apparatus needed (a diagram is often helpful to show the set-up).
- Any safety precautions that should be taken.
- What data to collect and how to collect it.

Evaluating experiment plans

If you ever need to evaluate the plan for someone else's experiment, you need to think about the same sorts of things that you would if you were designing the experiment yourself:

- Does the experiment actually test what it sets out to test?
- Is the method clear enough for someone else to follow?
- Apart from the independent and dependent variables, is everything else going to be properly controlled?
- Are the apparatus and techniques appropriate for what's being measured? Will they be used correctly?
- Are enough repeated measurements going to be taken?
- Is the experiment going to be conducted safely?

Exam Tip
This is the sort of thing you could be asked to do in an exam as a written test of your practical skills — either to evaluate an experiment plan, or to spot any issues with a small bit of it, e.g. the safety precautions.

2. Working Safely

When you do an experiment, you need to carry out a risk assessment and work safely at all times. This reduces the risk of anyone being hurt.

Tip: A hazard is anything that has the potential to cause harm or damage. The risk associated with that hazard is the probability of someone (or something) being harmed if they are exposed to the hazard.

Risks and hazards

Many chemistry experiments have risks associated with them. These can include risks associated with the equipment you're using (such as the risk of burning from an electric heater) as well as risks associated with chemicals.

When you're planning an experiment, you need to identify all the hazards and what the risk is from each hazard. This includes working out how likely it is that something could go wrong, and how serious it would be if it did. You then need to think of ways to reduce these risks. This procedure is called **a risk assessment**.

Tip: The CLEAPSS® website has a database with details of the potential harm lots of the hazardous chemicals you're likely to come across could cause. It also has student safety sheets, and your school or college may have CLEAPSS® Hazcards® you can use. These are all good sources of information if you're writing a risk assessment.

Any hazardous chemicals you use should come with a list of dangers associated with them — this can be often be found on the bottle they come in, or you can look the information up on a Material Safety Data Sheet.

Tip: This isn't a comprehensive list of all the types of hazard you may encounter — just some of the more common ones.

Hazard word	Potential harm	Examples of how to reduce the risk
Corrosive	May cause chemical burns to tissues such as skin and eyes.	Use as little of the substance as possible. If the chemical is a solution, use it in low concentrations. If the chemical is a gas, carry out the experiment in a fume cupboard. Wear a lab coat, goggles and gloves when handling the chemical.
Irritant	May cause inflammation and discomfort.	Use as little of the substance as possible. If the chemical is a solution, use it in low concentrations. If the chemical is a gas, carry out the experiment in a fume cupboard. Wear a lab coat, goggles and gloves when handling the chemical.
Flammable	May catch fire.	Keep the chemical away from any naked flames. If you have to heat it, use a water bath, an electric heater or a sand bath instead of a Bunsen burner.
Toxic	May cause illness or even death.	Use as little of the substance as possible. If the chemical is a solution, use it in low concentrations. If the chemical is a gas, carry out the experiment in a fume cupboard. Wear a lab coat, goggles and gloves when handling the chemical.
Oxidising	Reacts to form oxygen, so other substances may burn more easily in its presence.	Use as little of the substance as possible. Use it in solution rather than as a solid powder. Use in clean, well-ventilated areas. Keep away from combustible materials, such as wood or paper.

Figure 2: *Some common types of hazard and how to reduce the risks they present.*

Figure 1: *The hazard symbols for flammable (top left), oxidising (top middle), toxic (top right), corrosive (bottom left) and irritant (bottom right).*

Example

Imagine that you have been asked to synthesise a small amount of the ester propyl ethanoate using the following procedure:

- Add some propan-1-ol and concentrated ethanoic acid to a test tube containing concentrated sulfuric acid.
- Gently heat the mixture for 1 minute.
- Cool the mixture.
- Pour the mixture into a solution of sodium carbonate and mix to neutralise any remaining acid.

Before carrying out the experiment, you would need to do a risk assessment to identify any steps you should take to ensure that you were working safely.

The experiment involves hazards in both the procedure and the chemicals used. The hazards associated with the chemicals in the experiment are summarised in the table below.

Hazard	Type of hazard	Ways to reduce the risk
Propan-1-ol	Flammable	Keep away from naked flames — carry out the experiment using a water bath to heat the reaction mixture.
	Irritant	Use as little propan-1-ol as possible. Wear a lab coat, goggles and gloves when handling it.
Concentrated ethanoic acid	Corrosive	Use as little concentrated ethanoic acid as possible. Wear a lab coat, goggles and gloves when handling it.
Concentrated sulfuric acid	Corrosive	Use as little concentrated sulfuric acid as possible. Wear a lab coat, goggles and gloves when handling it.
Sodium carbonate solution	Irritant	Use as little sodium carbonate solution as possible, in a low concentration. Wear a lab coat, goggles and gloves when handling it.

Figure 3: *Chemical hazards in an experiment to synthesise propyl ethanoate.*

The main hazard in the procedure comes from the need to heat the reaction mixture. To reduce the risk of being burnt by any equipment, you shouldn't touch the apparatus during or soon after the heating step. You should use tongs to remove the test tube from the heating apparatus and allow it to cool completely before continuing with the experiment.

You could now rewrite the procedure, taking into account the steps that you should take in order to reduce the risks in this experiment:

- Wear a lab coat, goggles and gloves throughout the procedure.
- Add 1 cm^3 propan-1-ol and 1 cm^3 concentrated ethanoic acid to a test-tube containing 0.5 cm^3 concentrated sulfuric acid.
- Gently heat the mixture in a water bath for 1 minute.
- Remove the test tube from the water bath using tongs, and leave it in a test tube rack on a heatproof mat to cool.
- Pour the mixture into 5 cm^3 of sodium carbonate solution and mix.

Tip: Esters are organic compounds made by reacting carboxylic acids with alcohols. There's more about how to synthesise them and how they react on pages 426-429.

Tip: The hazard symbols in Figure 1 are quite new, so you may come across the old-style ones on bottles in your laboratory. The symbols are very similar, but they're in orange squares instead of red and white diamonds (see Figure 4).

Figure 4: *A bottle of ethanoic acid with a 'corrosive' hazard symbol.*

Tip: The risk associated with a solution might change depending on how concentrated it is. For example, whilst concentrated sulfuric acid is corrosive, dilute sulfuric acid is an irritant.

3. Practical Techniques

This section is an introduction to some of the techniques you'll be expected to know for A-level Chemistry. You'll have met many of them before, and others will be covered in more detail as they crop up throughout the book.

Tip: Make sure you do a risk assessment before you carry out any of the techniques in this topic.

Measuring pH

The pH of a solution is a measure of how acidic or basic it is. There's more than one way of measuring this — you can use pH charts, pH meters or pH probes.

pH charts

Universal indicator is a substance that changes colour according to the pH of the solution that it is added to. A **pH chart** tells you what colour an indicator will be at different pHs.

The pH chart for universal indicator is shown in Figure 1. So, if you add a few drops of Universal indicator to a solution, you can compare the resulting colour to a pH chart and determine the pH of the solution.

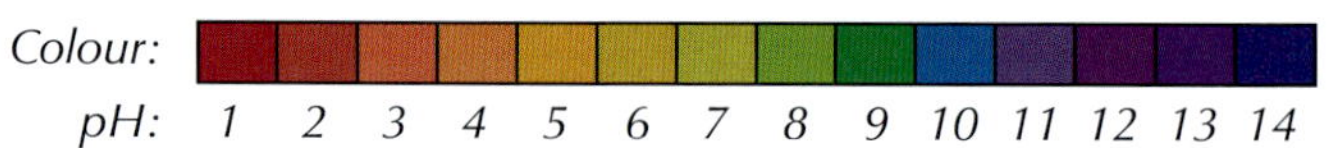

Figure 1: *The pH chart of Universal indicator.*

pH charts are good for giving a rough value of the pH of the solution, but since the colour of a solution is a qualitative observation (it's based on opinion), the data won't be precise.

pH meters

A **pH meter** is an electronic gadget that can be used to give a precise value for the pH of a solution. A pH meter is made up of a probe attached to a digital display. The probe is placed in the solution, and the pH reading is shown on the display.

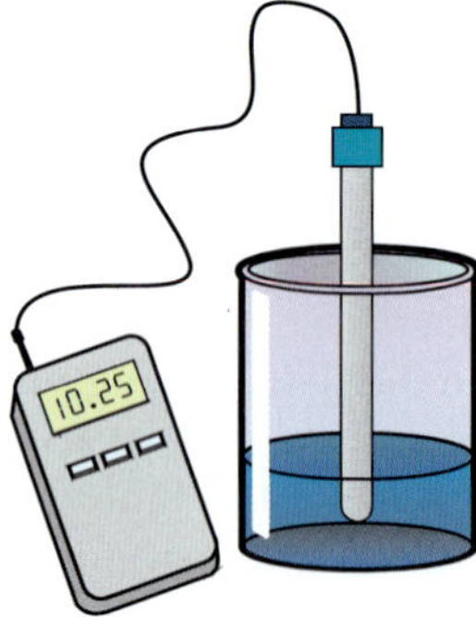

Figure 2: *A pH meter.*

pH probe on a data logger

A pH probe can be attached to something called a **data logger**. A data logger is a device that will record the readings from the pH probe at set intervals and store them so you can look back at them later. This makes them useful for experiments where you're measuring how pH changes, as the data logger will record the results, meaning you don't have to write them down separately. You might have to connect the data logger to a computer or tablet to collect the results from it. You may also need to print the results off once the experiment is complete, so you can keep a record of them in your lab book.

Heating substances

Many chemical substances are flammable, so are at risk of catching fire if they're heated with a naked flame, e.g. with a Bunsen burner. Luckily, there are a number of other techniques you can use to heat a reaction mixture.

Electric heaters

Electric heaters are often made up of a metal plate that can be heated to a specified temperature. The reaction vessel is placed on top of the hot plate (see Figure 2). The mixture is only heated from below, so you'll usually have to stir the reaction mixture to make sure it's heated evenly.

Water baths

A **water bath** is a container filled with water that can be heated to a specific temperature (see Figure 3). You place the reaction vessel in the water bath. The level of the water outside the vessel should be just above the level of the reaction mixture inside the vessel. The mixture will then be warmed to the same temperature as the water.

As the reaction mixture is surrounded by water, the heating is very even. However, water boils at 100 °C, so you can't use a water bath to heat the reaction mixture to a higher temperature than this.

Sand baths

Sand baths are similar to water baths, but the container is filled with sand. Sand can be heated to a much higher temperature than water, so sand baths are useful for heating reaction mixtures to temperatures above 100 °C.

Heating under reflux

You heat a reaction mixture under **reflux** when you're making volatile organic compounds (substances that evaporate easily) from compounds that need to be heated in order to react. The reaction mixture is heated in a flask that's fitted with a condenser, so that any materials that evaporate will condense and drip back into the mixture. This stops you losing any of the reactants or products from the mixture during the reaction. Figure 4 shows how reflux apparatus would be set up.

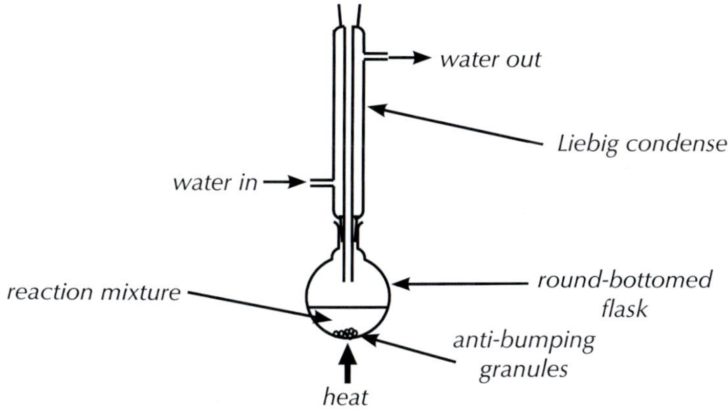

Figure 4: *A scientific diagram of reflux apparatus.*

Tip: Organic compounds are generally flammable, so you might use these methods for heating them in synthesis (see page 494).

Figure 2: *An electric heater.*

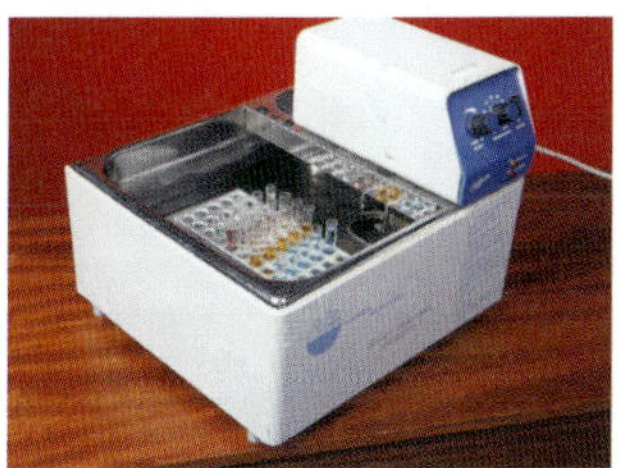

Figure 3: *A water bath.*

Tip: Organic compounds are just molecules that contain carbon.

Tip: Once you've synthesised a product under reflux, you might need to purify it. There are lots of techniques for doing this on pages 437-439.

Tip: You also need to know how to synthesise and purify a compound using distillation. See page 438 for more on this.

Filtration

Filtration is used to separate a solid from a liquid.
There are two types of filtration you need to know about:

1. Filtration under reduced pressure

Filtration under reduced pressure is normally used when you want to keep the solid and discard the liquid (filtrate). You use a piece of equipment called a Büchner funnel — a flat-bottomed funnel with holes in the base. The Büchner funnel is attached to the top of a side-arm flask, which is secured with a stand-and-clamp to stop it from falling over. The side-arm flask is attached to a vacuum line, which, when it is switched on, causes the flask to be under reduced pressure — this helps to suck the liquid through the filter.

Here's how you carry out the filtration:

- Place a piece of filter paper, slightly smaller than the diameter of the funnel, on the bottom of the Büchner funnel so that it lies flat and covers all the holes.
- Wet the paper with a little solvent, so that it sticks to the bottom of the funnel, and doesn't slip around when you pour in your mixture.
- Turn the vacuum on, and then pour your mixture into the funnel. As the flask is under reduced pressure, the liquid is sucked through the funnel into the flask, leaving the solid behind.
- Rinse the solid with a little of the solvent that your mixture was in. This will wash off any of the original liquid from the mixture that stayed on your crystals (and also any soluble impurities), leaving you with a more pure solid.
- Disconnect the vacuum line from the side-arm flask and then turn off the vacuum.
- The solid will be a bit wet from the solvent, so leave it to dry completely.

Tip: Filtration is often the first step in purifying an organic compound, either to remove liquid impurities if your product is solid, or solid impurities if your compound is a liquid. It's also an important step in recrystallisation (page 439), to remove the solute from the recrystallised product.

Figure 5: *The apparatus used to carry out a filtration under reduced pressure.*

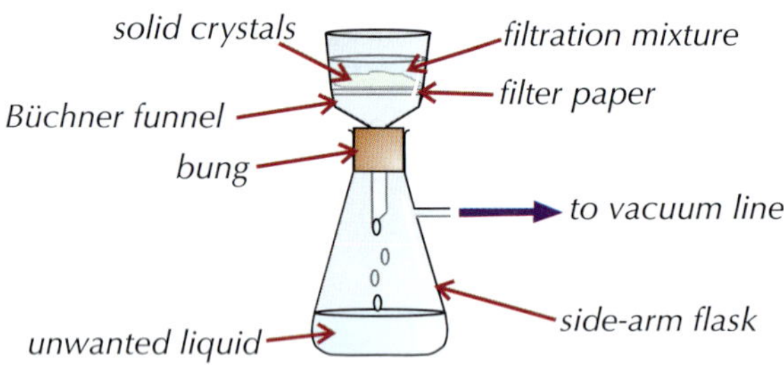

Figure 6: *Apparatus for filtration under reduced pressure.*

2. Filtration using fluted filter paper

Filtration using fluted filter paper is normally used when you want to keep the liquid (the filtrate) and discard the solid. For example, it can be used to remove the solid drying agent from the organic layer of a liquid that has been purified by separation (see page 437). A piece of fluted filter paper is placed in a funnel that feeds into a conical flask. The mixture to be separated is poured gently into the filter paper. The solution will pass through the filter paper into the conical flask, and the solid will be trapped. A pure sample of the solvent present in the solution can be used to rinse the solid left in the filter paper. This makes sure that all the soluble material has passed through the filter paper and collected in the conical flask.

Tip: If you're doing a filtration using fluted filter paper, take care not to fill the funnel above the top of the filter paper when you're pouring in your mixture.

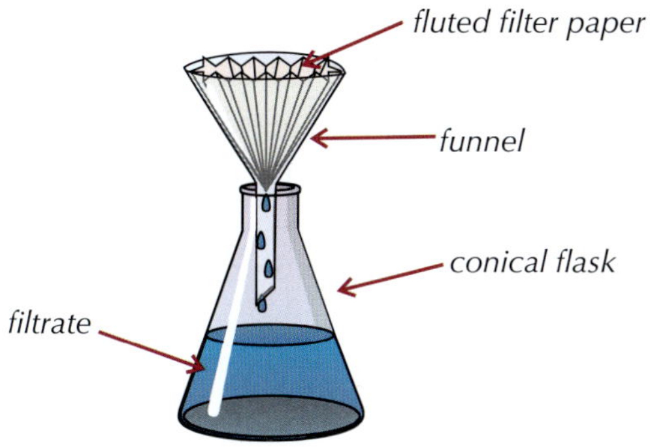

Figure 7: *Filtration using fluted filter paper.*

Tip: Fluted filter paper is a cone with a concertinaed edge. The concertina gives the paper a greater surface area than if the paper were smooth, so the solution is filtered faster.

Filter paper normally comes as a flat disc. To make fluted filter paper, follow the steps in Figure 8.

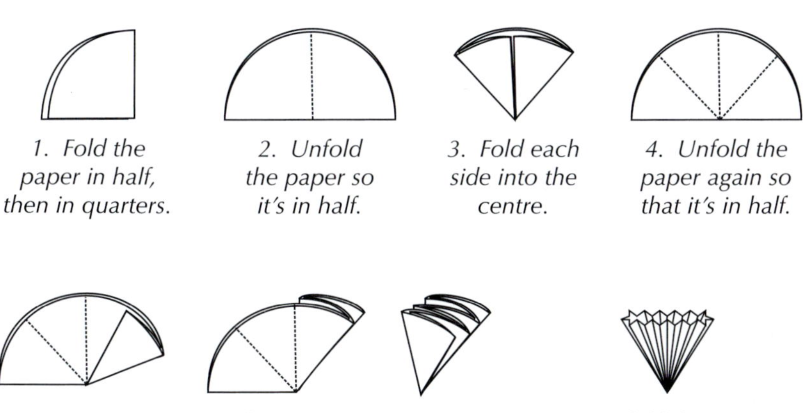

1. Fold the paper in half, then in quarters.
2. Unfold the paper so it's in half.
3. Fold each side into the centre.
4. Unfold the paper again so that it's in half.
5. Concertina the paper, using the folds as guidelines.
6. Unfold the paper, and make it into a cone.

Figure 8: *Making fluted filter paper.*

4. Data

When you're doing an experiment you need to think about how you're going to record your results and process the data that you've collected.

Figure 1: *An acid-carbonate reaction. The number of bubbles produced is discrete data, but the volume of gas produced is continuous data.*

Types of data

Experiments always involve some sort of measurement to provide data. There are different types of data — and you need to know what they are.

1. Discrete data

You get **discrete data** by counting — a discrete variable can only have certain values on a scale. E.g. the number of bubbles produced in a reaction would be discrete (see Figure 1). You can't have 1.25 bubbles. That'd be daft. Shoe size is another good example of a discrete variable.

2. Continuous data

A **continuous variable** can have any value on a scale. For example, the volume of gas produced or the mass of products from a reaction. You can never measure the exact value of a continuous variable.

3. Categoric data

A **categoric variable** has values that can be sorted into categories. For example, the colours of solutions might be blue, red and green (see Figure 2). Or types of material might be wood, steel, glass.

Figure 2: *Different coloured solutions. Colour is a type of categoric data.*

4. Ordered (ordinal) data

Ordered data is similar to categoric, but the categories can be put in order. For example, if you classified reactions as 'slow', 'fairly fast' and 'very fast' you'd have ordered data.

Tables of data

It's a good idea to set up a table to record the results of your experiment. Make sure that you include enough rows and columns to record all the data you need. You might also need to include a column for processing your data (e.g. working out an average).

Each column should have a heading so that you know what's going to be recorded where. In the column heading, you should include the units — this is to avoid having to write them out repeatedly in every cell.

[KI] / mol dm^{-3}	Duration of clock reaction / s				Mean duration of clock reaction / s
	Run 1	Run 2	Run 3	Run 4	
0.045	32	34	33	33	
0.033	50	50	48	48	
0.021	72	71	89	70	

Figure 3: *A table of results.*

Calculating means

A mean is just an **average** of your repeated results. Calculating an average helps to balance out any random errors in your data (see page 17).

To calculate the mean result for a data point, first remove any anomalous results. Then add up all the other measurements from each repeat and divide by the number of (non-anomalous) measurements.

Tip: Have a look at page 13 for more on anomalous results.

Example — Maths Skills

The average duration of the clock reaction at each concentration of potassium iodide point for the table in Figure 3 is:

- When [KI] = 0.045 mol dm^{-3}: (32 + 34 + 33 + 33) ÷ 4 = **33 s**
- When [KI] = 0.033 mol dm^{-3}: (50 + 50 + 48 + 48) ÷ 4 = **49 s**
- When [KI] = 0.021 mol dm^{-3}: The result for Run 3 is anomalously high, so ignore it when calculating the mean. (72 + 71 + 70) ÷ 3 = **71 s**

Tip: There's more about dealing with data (e.g. using the right number of significant figures and converting units) in the Maths Skills section on pages 519 to 531.

Types of charts and graphs

You'll often be expected to make a graph of your results. Graphs make your data easier to understand — so long as you choose the right type.

Tip: Use simple scales when you draw graphs — this'll make it easier to plot points.

Bar charts

You should use a bar chart when one of your data sets is categoric or ordered data, like in Figure 4.

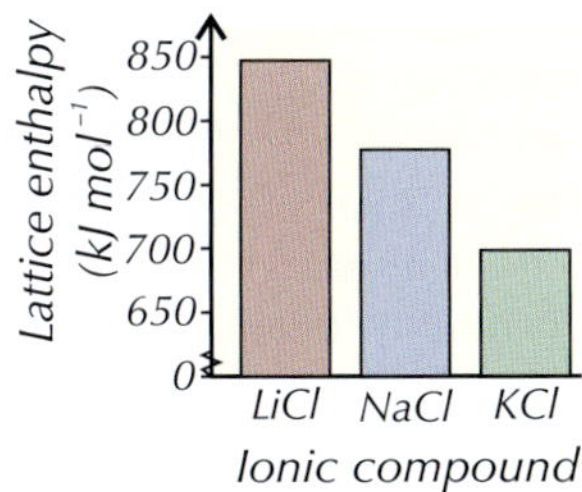

Figure 4: *Bar chart to show the lattice enthalpies of various chloride compounds.*

Exam Tip

Whatever type of graph you make, you'll only get full marks if you:

1. Choose a sensible scale — don't do a tiny graph in the corner of the paper.
2. Label both axes — including units.
3. Plot your points accurately — using a sharp pencil.

Pie charts

Pie charts are normally used to display categoric data, like in Figure 5.

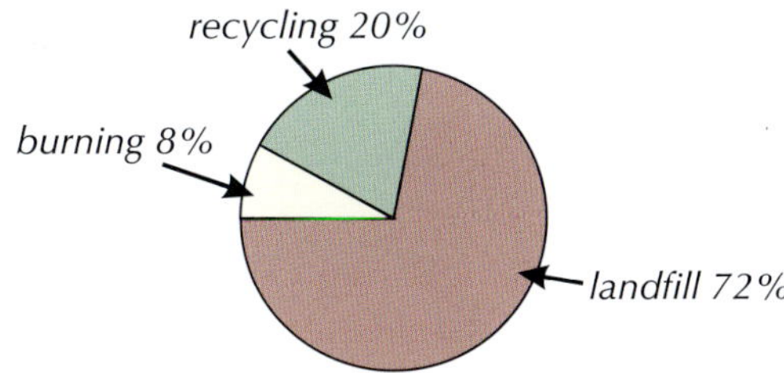

Figure 5: *Pie chart to show how plastic waste is disposed of by a country.*

Scatter graphs

Scatter graphs, like the graphs shown in Figures 6 and 7, are great for showing how two sets of continuous data are related (or correlated — see page 13).

Don't try to join up all the points on a scatter graph — you should draw a line of best fit to show the trend of the data instead. A line of best fit can be a straight line, like the example in Figure 6...

Tip: A line of best fit should have about half of the points above it and half of the points below. You should ignore any anomalous results, like the one circled in Figure 6 — there's more about anomalous results coming up on page 13.

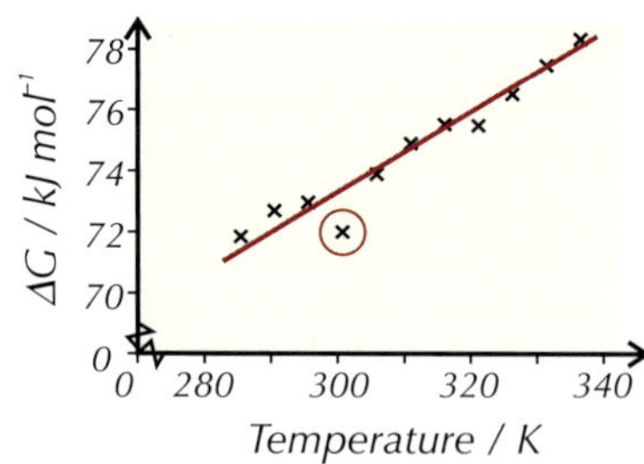

Figure 6: *Scatter graph showing the relationship between ΔG and temperature for a reaction.*

...or a curve, like the example in Figure 7.

Exam Tip
If you're asked to sketch a graph in an exam, you only need to show the labelled axes and the general shape of the line. You don't need to plot points and draw the line accurately for a sketch graph.

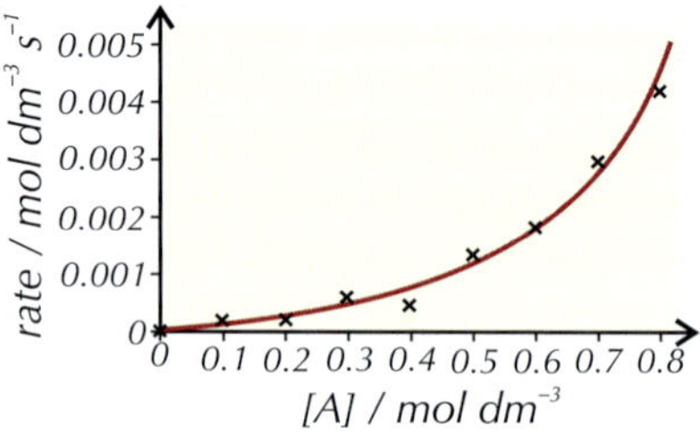

Figure 7: *Scatter graph to show how the rate of a reaction changes with the concentration of one of the reactants.*

5. Analysing Results

Once you've got your results nicely presented you can start to draw a conclusion. But be careful — you may have a graph showing a lovely correlation, but that doesn't always tell you as much as you might think.

Anomalous results

Anomalous results are ones that don't fit in with the other values — this means they are likely to be wrong. They're often caused by experimental errors, e.g. if a drop in a titration is too big and shoots past the end point, or if a syringe plunger gets stuck whilst collecting gas produced in a reaction. When looking at results in tables or graphs, you need to check if there are any anomalies — you ignore these results when calculating means or drawing a line of best fit.

Tip: Even though you ignore anomalies when you're calculating means and drawing lines of best fit, if possible you should try to find an explanation for why they happened (and include it in your write-up).

Examples — Maths Skills

The table below shows the titre volume of a number of titrations.

Titration Number	1	2	3	4
Titre Volume (cm^3)	15.20	15.30	15.70	15.25

Titre 3 isn't concordant with (doesn't match) the other results, so you need to ignore that one and just use the other three.

$$\text{mean} = \frac{15.20 + 15.30 + 15.25}{3} = 15.25 \text{ cm}^3$$

The graph below shows how the pH of a reaction mixture changes as an alkali is added.

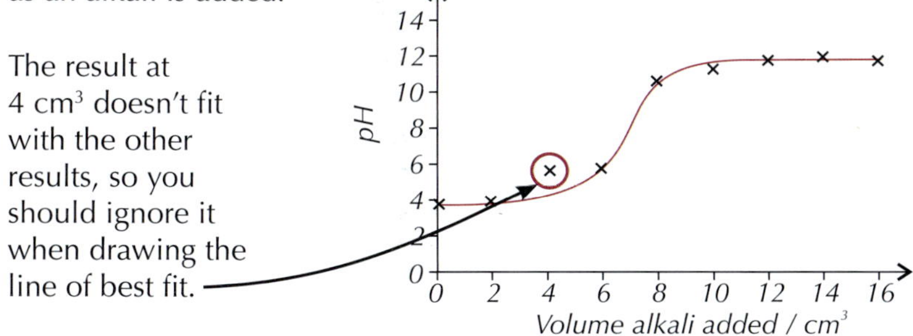

Tip: In titrations, saying that your results are concordant means that they're all within 0.1 cm^3 of each other.

Tip: Saying that one result '<u>isn't concordant</u>' with the other results is just another way of saying that it's an anomalous result.

Scatter graphs and correlation

Correlation describes the relationship between two variables — usually the independent one and the dependent one. Data can show positive correlation, negative correlation or no correlation. A scatter graph can show you how two variables are correlated:

Positive correlation
As one variable increases the other also increases.

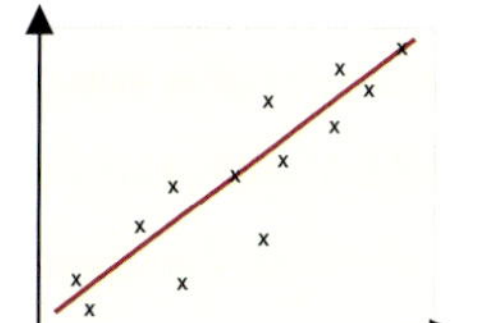

Negative correlation
As one variable increases the other decreases.

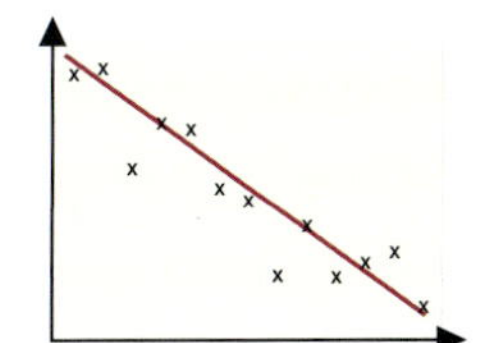

No correlation
There is no relationship between the variables.

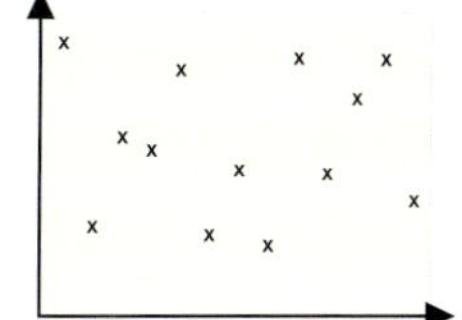

Tip: If an experiment really does confirm that changing one variable causes another to change, we say there's a causal link between them.

Tip: Watch out for bias too — for instance, a bottled water company might point these studies out to people without mentioning any of the doubts.

Correlation and cause

Ideally, only two quantities would ever change in any experiment — everything else would remain constant. But in experiments or studies outside the lab, you can't usually control all the variables. So even if two variables are correlated, the change in one may not be causing the change in the other. Both changes might be caused by a third variable.

Example

Some studies have found a correlation between drinking chlorinated tap water and the risk of developing certain cancers. So some people argue that this means water shouldn't have chlorine added. But it's hard to control all the variables between people who drink tap water and people who don't. It could be many lifestyle factors. Or, the cancer risk could be affected by something else in tap water — or by whatever the non-tap water drinkers drink instead.

Drawing conclusions

The data should always support the conclusion. This may sound obvious, but it's easy to jump to conclusions. Conclusions have to be specific — not make sweeping generalisations.

Example

The rate of an enzyme-controlled reaction was measured at 10 °C, 20 °C, 30 °C, 40 °C, 50 °C and 60 °C. All other variables were kept constant, and the results are shown in Figure 1.

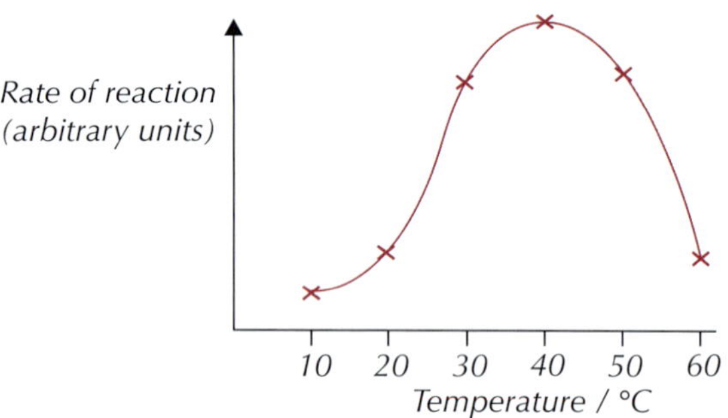

Figure 1: *Graph to show the effect of temperature on the rate of an enzyme-controlled reaction.*

A science magazine concluded from this data that this enzyme works best at 40 °C. The data doesn't support this exact claim. The enzyme could work best at 42 °C or 47 °C, but you can't tell from the data because increases of 10 °C at a time were used. The rate of reaction at in-between temperatures wasn't measured. All you know is that it's faster at 40 °C than at any of the other temperatures tested.

Also, the experiment only gives information about this particular enzyme-controlled reaction. You can't conclude that all enzyme-controlled reactions happen faster at a particular temperature — only this one. And you can't say for sure that doing the experiment at, say, a different constant pressure wouldn't give a different optimum temperature.

Tip: Whoever funded the research (e.g. a chemical manufacturer) may have some influence on what conclusions are drawn from the results, but scientists have a responsibility to make sure that the conclusions they draw are supported by the data.

6. Evaluations and Errors

Now that you've drawn some conclusions from your experiment, you can consider how accurate those conclusions could be.

Evaluations

There are a few terms that you need to understand. They'll be useful when you're evaluating how convincing your results are.

1. Valid results

Valid results answer the original question. For example, if you haven't controlled all the variables your results won't be valid, because you won't be testing just the thing you wanted to.

2. Accurate results

Accurate results are those that are really close to the true answer.

3. Precise results

The smaller the amount of spread of your data around the mean (see page 11), the more **precise** it is. Calculating a mean (average) result from your repeats will increase the precision of your result, because it helps to reduce the effect of random errors on the answer (see page 17).

4. Repeatable results

Your results are **repeatable** if you get the same results when you repeat the experiment using the same method and the same equipment. You really need to repeat your readings at least three times to demonstrate that your results really are repeatable.

Tip: Results that fulfil all of these criteria are sometimes referred to as being reliable.

5. Reproducible results

Your results are **reproducible** if other people get the same results when they repeat your experiment.

Uncertainty

Any measurements you make will have **uncertainties** (or **errors**) in them, due to the limits of the sensitivity of the equipment.

Uncertainties are usually written with a ± sign, e.g. ±0.05 cm. The ± sign tells you the actual value of the measurement is likely to lie somewhere between your reading minus the uncertainty value and your reading plus the uncertainty value.

The uncertainty will be different for different pieces of equipment.

Example

Pieces of equipment such as pipettes, volumetric flasks and thermometers will have uncertainties that depend on how well made they are.

The manufacturers provide these uncertainty values — they're usually written on the equipment somewhere (see Figure 1).

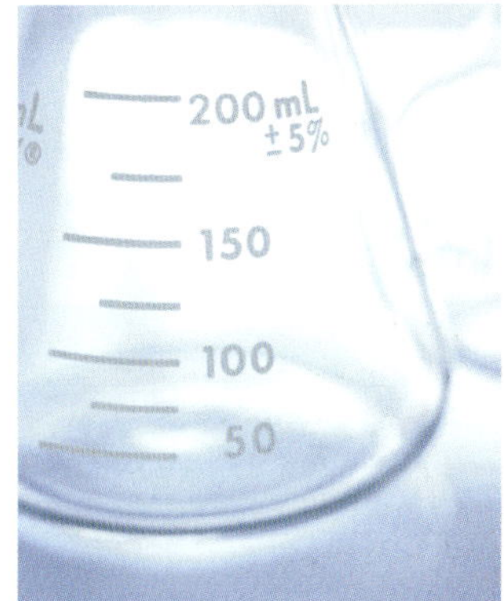

Figure 1: *A conical flask where the graduations have an uncertainty of ±5%.*

For any piece of equipment you use that has a **scale**, the uncertainty will be **half** the **smallest increment** the equipment can measure, in either direction.

Example

The scale on a 50 cm^3 burette usually has marks every 0.1 cm^3. You should be able to tell which mark the level's closest to, so any reading you take won't be more than 0.05 cm^3 out.
So the uncertainty on your burette readings is ±0.05 cm^3.

The level in this burette is between the 44.9 cm^3 and 45.0 cm^3 marks. It's closer to 45.0 — so the level is between 44.95 and 45.0. So a reading of 45.0 cm^3 can't have an uncertainty of more than 0.05 cm^3.

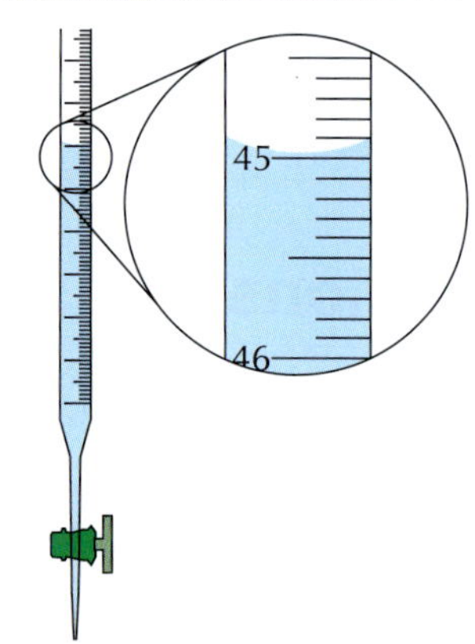

Figure 2: *Reading a volume from a burette.*

Tip: When you combine uncertainties, the overall uncertainty should be given to the same number of decimal places as the measurement.

If you use two readings to work out a measurement, you'll need to combine their uncertainties. For example, if you find a temperature change by subtracting a final temperature from an initial temperature, the uncertainty for the temperature change is the uncertainty for both readings added together.

Uncertainty on a mean

The uncertainty on a **mean** is equal to half the range of the measured values:

$$\text{uncertainty on a mean} = \frac{\text{largest measurement} - \text{smallest measurement}}{2}$$

Percentage uncertainty

Tip: The same formula works for finding the percentage uncertainty on a <u>mean</u> value — just divide the uncertainty on the mean by the mean and multiply the result by 100.

If you know the **uncertainty** (or **error**) in a reading that you've taken, you can use it to calculate the percentage uncertainty in your measurement:

$$\text{percentage uncertainty} = \frac{\text{uncertainty}}{\text{reading}} \times 100$$

Examples — Maths Skills

The EMF of an electrochemical cell is measured as 0.42 V. The voltmeter has an uncertainty of ±0.01 V. Find the percentage uncertainty on the EMF.

Percentage uncertainty is $\frac{0.01}{0.42} \times 100 = \mathbf{2.4}$ **%**.

In a titration a burette with an uncertainty of ±0.05 cm^3 is used. The initial reading on the burette is 0.0 cm^3. The final reading is 21.2 cm^3. Calculate the percentage uncertainty on the titre value.

The titre value is 21.2 – 0.0 = 21.2 cm^3. The uncertainty on each burette reading is ±0.05 cm^3. Two readings have been combined to find the titre value, so the total uncertainty is 0.05 + 0.05 = ±0.1 cm^3.

So, percentage uncertainty on the titre value

$= \frac{0.1}{21.2} \times 100 = 0.472 = \mathbf{0.5\%}$

Tip: Percentage uncertainty means you can directly compare the degree of uncertainty in different measurements as a proportion of the total measurement.

Percentage uncertainty is useful because it tells you how significant the uncertainty in a reading is in comparison to its size — e.g. an uncertainty of ±0.1 g is more significant when you weigh out 0.2 g of a solid than when you weigh out 100.0 g.

You can reduce uncertainties in your measurements by using the most sensitive equipment available. There's not much you can do about this at school or college though — you're stuck with whatever's available.

But there are other ways to lower the uncertainty in experiments. The larger the reading you take with a piece of equipment, the smaller the percentage uncertainty on that reading will be. Here's a quick example:

Example

If you measure out 2 g of a solid using a mass balance with an uncertainty of ±0.05 g cm^3 then the percentage uncertainty is (0.05 ÷ 2) × 100 = 2.5%.

But if you measure 4 g of the solid using the same mass balance the percentage uncertainty is (0.05 ÷ 4) × 100 = 1.25%. You've just halved the percentage uncertainty.

So you can reduce the percentage uncertainty of this experiment by using a greater mass of solid.

You can apply the same principle to other measurements too. For example, if you measure out a small volume of liquid, the percentage uncertainty will be larger than if you measured out a larger volume using the same measuring cylinder.

Types of error

Errors in your results can come from a variety of different sources.

Random errors

Random errors cause readings to be spread about the true value due to the results varying in an unpredictable way. You get random error in all measurements and no matter how hard you try, you can't correct them. The tiny errors you make when you read a burette are random — you have to estimate the level when it's between two marks, so sometimes your figure will be above the real one and sometimes below.

Repeating an experiment and finding the mean of your results helps to deal with random errors. The results that are a bit high will be cancelled out by the ones that are a bit low. So your results will be more precise.

Tip: In some cases, errors from human judgement have more of an effect than equipment uncertainty. For example, if you have a stopwatch that measures to two decimal places, your ability to start and stop it precisely will probably introduce more error than the error on the watch itself.

Systematic errors

Systematic errors cause each reading to be different to the true value by the same amount, i.e. they shift all of your results. They may be caused by the set-up or the equipment you're using. If the 10.00 cm^3 pipette you're using to measure out a sample for titration actually only measures 9.95 cm^3, your sample will be 0.05 cm^3 too small every time you repeat the experiment.

Repeating your results won't get rid of any systematic errors, so your results won't get more accurate. The best way to get rid of systematic errors is to carefully calibrate any equipment you're using, if possible.

Tip: 'Calibration' is when you mark a scale on a measuring instrument or check the scale by measuring a known value.

7. The Practical Endorsement

Alongside your A-level exams, you have to do a separate 'Practical Endorsement'. This assesses practical skills that can't be tested in a written exam.

What is the Practical Endorsement?

The Practical Endorsement is assessed slightly differently to the rest of your course. Unlike the exams, you don't get a mark for the Practical Endorsement — you just have to get a pass grade.

In order to pass the Practical Endorsement, you have to carry out at least twelve practical experiments and demonstrate that you can:

- use a range of specified apparatus, e.g. you must be able to use melting point apparatus (see page 439),
- carry out a range of specified practical techniques, e.g. you must be able to purify a solid by recrystallisation (see page 439).

Exam Tip
You could also get asked questions about the Required Practicals in your written exams.

The twelve practicals that you do are most likely to be the twelve Required Practicals that form part of the AQA A-level Chemistry course — these cover all the techniques you need to be able to demonstrate for your Practical Endorsement. You may carry out other practicals as well, or instead, which could also count towards your Practical Endorsement. You'll do the practicals in class, and your teacher will assess you as you're doing them.

You'll need to keep a record of all your assessed practical activities. This book contains information about all of the twelve Required Practicals that you'll need to do during your course.

Tip: Throughout this book, information about the Required Practicals and opportunities to carry out Required Practicals are marked up with a big stamp, like this one:

Assessment of the Practical Endorsement

When assessing your practical work, your teacher will be checking that you're able to do five things:

1. Follow written methods and instructions

Make sure you read any instructions you are given fully before you start work and that you understand what you're about to do. This will help you to avoid missing out any important steps or using the wrong piece of apparatus.

2. Use apparatus and investigative methods correctly

You'll need to demonstrate that you can select appropriate apparatus for the experiment that you're doing and decide what measurements you need to take with it (see pages 2 and 3 for more). Once you've selected your apparatus and set it up correctly, you'll also need to show that you use it to take accurate measurements.

Tip: You won't necessarily need to demonstrate that you can do all five of these things every time you carry out a practical.

You'll also need to show that you can identify which variables need to be controlled in an experiment and work out how best to control them (see page 1).

3. Use apparatus and materials safely

This means being able to carry out a risk assessment (see pages 4 and 5) to identify the hazards in your experiment and what you can do to reduce the risks associated with those hazards. You'll also have to show that you can work safely in the lab, using appropriate safety equipment to reduce the risks you've identified, and that you adapt your method as you go along if necessary.

4. Make observations and record results

You need to show that you can collect data that's valid, accurate and precise. When you record the data, e.g. in a table, you need to make sure you do so to an appropriate level of precision and you include things like the units. There's more on recording data on page 10.

5. Carry out supporting research and write reports using appropriate references

You need to show that you can write up an investigation properly. As well as reporting your results and drawing a conclusion about your findings, you'll need to describe the method you used and any safety precautions you took. You'll also need to show that you can use computer software to process your data. This might mean drawing a graph of your results using the computer, or using a computer programme to help you quickly and accurately carry out calculations on large amounts of data.

You'll need to write up any research you've done too (e.g. to help you with planning a method or to draw your conclusions) and properly cite any sources that you've used.

Research, references and citations

You can use books or the Internet to carry out research, but there a few things you'll need to bear in mind:

- Not all the information you find on the Internet will be true. It's hard to know where information comes from on forums, blogs and websites that can be edited by the general public, so you should avoid using these. Websites of organisations such as the Royal Society of Chemistry and the National Institute of Standards and Technology (NIST) provide lots of information that comes from reliable scientific sources. Scientific papers and textbooks are also good sources of reliable information.
- It may sound obvious, but when you're using the information that you've found during your research, you can't just copy it down word for word. Any data you're looking up should be copied accurately, but you should rewrite everything else in your own words.
- When you've used information from a source, you need to cite the reference properly. Citations allow someone else to go back and find the source of your information. This means they can check your information and see you're not making things up out of thin air. Citations also mean you've properly credited other people's data that you've used in your work. A citation for a particular piece of information may include the title of the book, paper or website where you found the information, the author and/or the publisher of the document and the date the document was published.

Tip: If you're unsure whether the information on a website is true or not, try and find the same piece of information in a different place. The more sources you can find for the information, the more likely it is to be correct.

Tip: There are lots of slightly different ways of referencing sources, but the important thing is that it's clear where you found the information.

1. The Atom

Atoms are the basis of all of chemistry. You learned about them at GCSE and they're here again. They're super important.

Learning Objectives:
- Understand that an atom consists of a nucleus containing protons and neutrons surrounded by electrons.
- Be able to describe the properties of protons, neutrons and electrons in terms of relative charge and relative mass.
- Know the meaning of mass number (*A*) and atomic (proton) number (*Z*).
- Be able to determine the number of protons, neutrons and electrons in atoms and ions using mass number, atomic number and charge.
- Be able to explain the existence of isotopes.

Specification Reference 3.1.1.1, 3.1.1.2

The structure of the atom

All elements are made of **atoms**. Atoms are made up of 3 types of particle — **protons**, **neutrons** and **electrons**. Figure 1 shows how they are arranged in the atom.

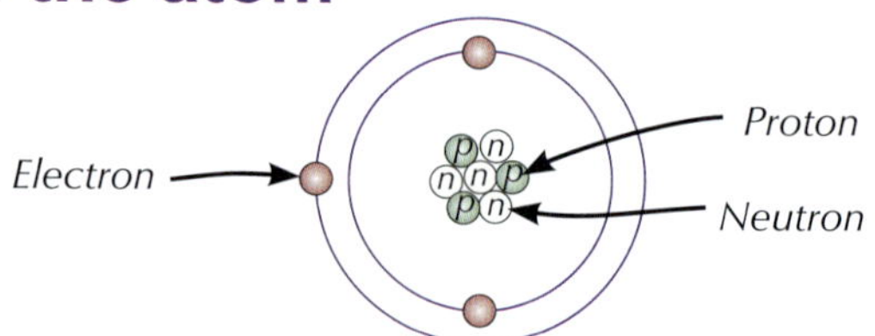

Figure 1: *The atom.*

Electrons have –1 charge. They whizz around the nucleus in orbitals, which take up most of the volume of the atom. Most of the mass of the atom is concentrated in the **nucleus**, although the diameter of the nucleus is rather titchy compared to the whole atom. The nucleus is where you find the protons and neutrons. The mass and charge of these subatomic particles is really small, so relative mass and relative charge are used instead. Figure 2 shows the relative masses and charges of protons, neutrons and electrons.

Subatomic particle	*Relative mass*	*Relative charge*
Proton	*1*	*+1*
Neutron	*1*	*0*
Electron, e^-	$\frac{1}{2000}$	*–1*

Figure 2: *Relative masses and charges of subatomic particles.*

Tip: The mass of an electron is negligible compared to a proton or a neutron — so you can usually ignore it.

Mass number and atomic number

You can figure out the number of protons, neutrons and electrons in an atom from the nuclear symbol.

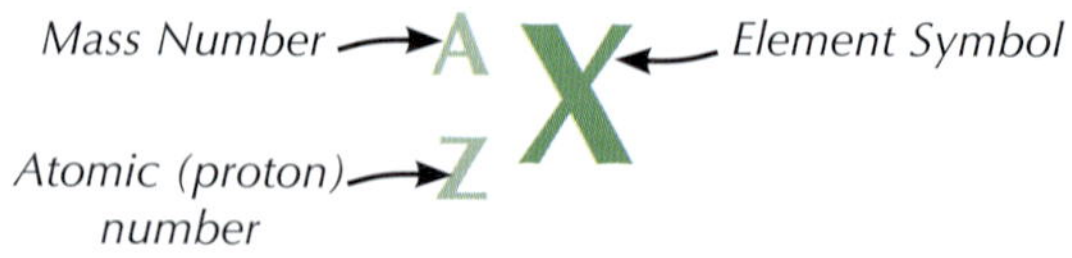

Figure 3: *Nuclear symbol.*

Mass number

This is the total number of protons and neutrons in the nucleus of an atom.

Atomic (proton) number

This is the number of protons in the nucleus of an atom — it identifies the element. All atoms of the same element have the same number of protons. Sometimes the atomic number is left out of the nuclear symbol, e.g. ^{7}Li. You don't really need it because the element's symbol tells you its value.

Tip: You can find the symbols and atomic numbers for each element using the periodic table. The other number in the periodic table isn't the mass number though — it's the relative atomic mass, which is a bit different. (See page 25 for more on relative atomic mass.)

Atoms

For neutral atoms, which have no overall charge, the number of electrons is the same as the number of protons. The number of neutrons is just mass number minus atomic number (top minus bottom in the nuclear symbol). Figure 4 shows some examples.

Nuclear symbol	Atomic number, Z	Mass number, A	Protons	Electrons	Neutrons
${}^{7}_{3}Li$	3	7	3	3	7 – 3 = 4
${}^{79}_{35}Br$	35	79	35	35	79 – 35 = 44
${}^{24}_{12}Mg$	12	24	12	12	24 – 12 = 12

Figure 4: *Calculating the number of neutrons in atoms.*

Ions

Atoms form **ions** by gaining or losing electrons. Ions have different numbers of protons and electrons — negative ions have more electrons than protons, and positive ions have fewer electrons than protons. For example:

Examples

Br^- is a negative ion

The negative charge means that there's 1 more electron than there are protons. Br has 35 protons (see table above), so Br^- must have 36 electrons. The overall charge = + 35 – 36 = –1.

Mg^{2+} is a positive ion

The 2+ charge means that there's 2 fewer electrons than there are protons. Mg has 12 protons (see table above), so Mg^{2+} must have 10 electrons. The overall charge = +12 – 10 = +2.

Tip: Ions are easy to spot because they've always got a $^+$ or a $^-$ next to them. If they've got a $^+$ it means they've lost electrons; if it's a $^-$ then they've gained electrons. If there's a number next to the sign it means more than one electron has been lost or gained. For example, $^{3+}$ means 3 electrons have been lost, and $^{2-}$ means that 2 have been gained.

Isotopes

Isotopes of an element are atoms with the same number of protons but different numbers of neutrons.

Examples

Chlorine-35 and chlorine-37 are examples of isotopes. They have different mass numbers, so they have different numbers of neutrons. Their atomic numbers are the same — both isotopes have 17 protons and 17 electrons.

Chlorine-35: ${}^{35}_{17}Cl$

35 – 17 = 18 neutrons

Chlorine-37: ${}^{37}_{17}Cl$

37 – 17 = 20 neutrons

Here's another example — naturally occurring magnesium consists of 3 isotopes.

${}^{24}Mg$ (79%)	${}^{25}Mg$ (10%)	${}^{26}Mg$ (11%)
12 protons	12 protons	12 protons
12 neutrons	13 neutrons	14 neutrons
12 electrons	12 electrons	12 electrons

Figure 5: *Subatomic particles in Mg isotopes.*

Tip: You can show isotopes in different ways. For example, the isotope of magnesium with 12 neutrons can be shown as:

Magnesium-24,

${}^{24}Mg$ or ${}^{24}_{12}Mg$

The number and arrangement of the electrons decides the chemical properties of an element. Isotopes have the same configuration of electrons, so they have the same chemical properties. Isotopes of an element do have slightly different physical properties though, e.g. different densities and rates of diffusion. This is because physical properties tend to depend more on the mass of the atom.

Exam Tip
In your exams you'll probably have to look at the periodic table to find the nuclear symbol of an element — you won't usually be given it in the question like this.

Practice Questions — Application

Q1 Aluminium has the nuclear symbol: $^{27}_{13}Al$

a) How many protons does an atom of aluminium have?
b) How many electrons does an atom of aluminium have?
c) How many neutrons does an atom of aluminium have?

Q2 A potassium atom has 19 electrons and 20 neutrons.

a) How many protons does a potassium ion have?
b) What is the mass number of a potassium atom?
c) Write the nuclear symbol for potassium.
d) Potassium can form ions with a charge of 1+.
How many electrons does one of these potassium ions have?

Q3 Calcium has the nuclear symbol: $^{40}_{20}Ca$
It forms Ca^{2+} ions.

a) How many electrons does a Ca^{2+} ion have?
b) How many neutrons does a Ca^{2+} ion have?

Q4 Isotope X has 41 protons and 52 neutrons.
Identify which element isotope X is an isotope of.
Write the nuclear symbol of isotope X.

Q5 This question relates to the atoms or ions A to D:

A $^{16}_{8}O^{2-}$ B $^{17}_{7}N$ C $^{20}_{10}Ne$ D $^{18}_{8}O$

Identify the similarity for each of the following pairs.

a) A and C.
b) A and D.
c) B and C.
d) B and D.
e) Which two of the atoms or ions are isotopes of each other?
Explain your reasoning.

Tip: Here we mean similarities in the numbers of protons, neutrons or electrons between the two atoms or ions.

Practice Questions — Fact Recall

Q1 Name the three types of particle found in an atom.
Q2 State where in the atom each of these particles would be found.
Q3 Give the relative masses of these particles.
Q4 What is a mass number?
Q5 What is an atomic number?
Q6 How can you work out the number of neutrons an atom has?
Q7 What are isotopes?
Q8 Why do isotopes have the same chemical properties?
Q9 Explain why isotopes can have different physical properties.

2. Atomic Models

Models of the atom are useful for understanding loads of ideas in chemistry. But the accepted model of the atom has changed throughout history.

Learning Objective:

- Appreciate that knowledge and understanding of atomic structure has evolved over time.

Specification Reference 3.1.1.1

Dalton's and Thomson's models

The **model** of the atom you need to know (the one on page 20) is one of the currently accepted ones. But in the past, completely different models were accepted, because they fitted the evidence available at the time. As scientists did more experiments, new evidence was found and the models were modified to fit it.

At the start of the 19th century John Dalton described atoms as solid spheres (see Figure 1) and said that different spheres made up the different elements.

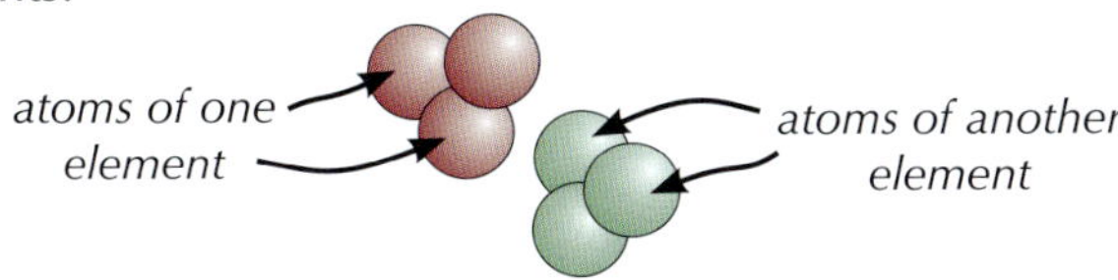
atoms of one element
atoms of another element

Figure 1: *Dalton's model of the atom.*

Tip: A model is a simplified description of a real-life situation. Scientists use models to help make complicated real-world science easier to explain or understand.

In 1897 J. J. Thomson concluded from his experiments that an atom must contain even smaller, negatively charged particles — electrons. The 'solid sphere' idea of atomic structure had to be changed. The new model was known as the 'plum pudding model' — see Figure 2.

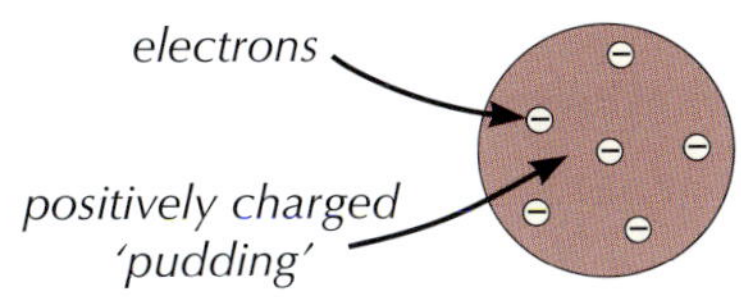
electrons
positively charged 'pudding'

Figure 2: *Thomson's model of the atom.*

Rutherford's model

In 1909 Ernest Rutherford and his students Hans Geiger and Ernest Marsden conducted their famous gold foil experiment. They fired alpha particles (which are positively charged) at a very thin sheet of gold. From the plum pudding model, they were expecting most of the alpha particles to be deflected slightly by the positive 'pudding' that made up most of an atom. In fact, most of the alpha particles passed straight through the gold atoms and a very small number were deflected backwards. So the plum pudding model couldn't be right.

So Rutherford came up with an idea that could explain this new evidence — the nuclear model of the atom (see Figure 4). In this, there's a tiny, positively charged nucleus at the centre, surrounded by a 'cloud' of negative electrons. Most of the atom is empty space.

Figure 3: *Thomson and Rutherford worked together at Cambridge University.*

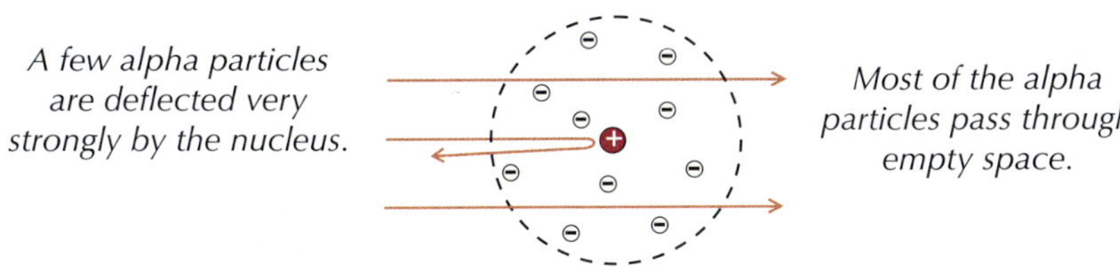
A few alpha particles are deflected very strongly by the nucleus.
Most of the alpha particles pass through empty space.

Figure 4: *Rutherford's model of the atom.*

Bohr's model

Scientists realised that electrons in a 'cloud' around the nucleus of an atom, as Rutherford described, would quickly spiral down into the nucleus, causing the atom to collapse. Niels Bohr proposed a new model of the atom with four basic principles:

- Electrons only exist in fixed orbits (shells) and not anywhere in between.
- Each shell has a fixed energy.
- When an electron moves between shells electromagnetic radiation is emitted or absorbed.
- Because the energy of shells is fixed, the radiation will have a fixed frequency.

Figure 6: *Rutherford and Bohr worked together at the University of Manchester.*

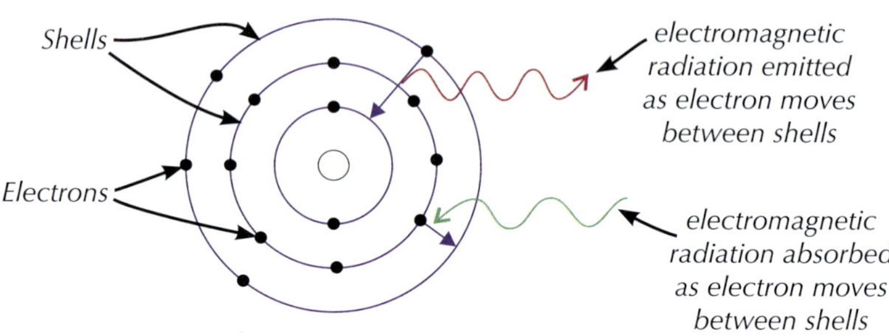

Figure 5: *Bohr's model of the atom.*

The frequencies of radiation emitted and absorbed by atoms were already known from experiments. The Bohr model fitted these observations.

Other atomic models

Scientists later discovered that not all the electrons in a shell had the same energy. This meant that the Bohr model wasn't quite right, so they refined it to include sub-shells (see page 33). The refined model fitted observations even better than Bohr's original model. We now know that the refined Bohr model is not perfect either — but it's still widely used to describe atoms because it is simple and explains many observations from experiments, like bonding and ionisation energy trends.

The most accurate model we have today is based on quantum mechanics. The quantum model explains some observations that can't be accounted for by the Bohr model, but it's a lot harder to get your head round and visualise. Scientists use whichever model is most relevant to whatever they're investigating.

Practice Questions — Fact Recall

Q1 Describe how J. J. Thomson's model of the atom was different from Dalton's model.

Q2 Explain how Rutherford's gold foil experiment provided evidence that Thomson's model was wrong.

Q3 Describe Rutherford's model of the atom.

Q4 Describe the main features of Bohr's model of the atom.

3. Relative Mass

The actual mass of an atom is very, very tiny. Don't worry about exactly how tiny for now, but it's far too small to weigh. So you usually talk about the mass of an atom compared to the mass of one carbon atom instead — this is its relative mass. You need to know about relative atomic mass, relative isotopic mass and relative molecular mass.

Learning Objectives:

- Be able to define relative atomic mass (A_r) and relative molecular mass (M_r) in terms of ^{12}C.
- Understand that the term relative formula mass is used for ionic compounds.

Specification Reference 3.1.2.1

Relative atomic mass

The relative atomic mass, A_r, is the average mass of an atom of an element on a scale where an atom of carbon-12 is exactly 12. The relative atomic mass of each element is shown in the periodic table (see Figure 1).

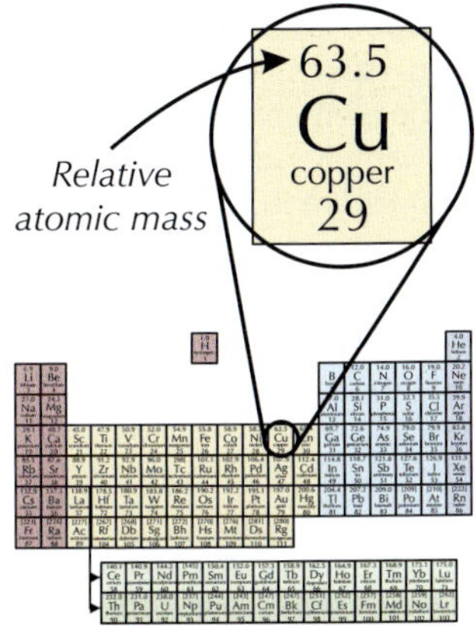

Figure 1: Location of relative atomic masses on the periodic table.

Relative isotopic mass

Relative isotopic mass is the mass of an atom of an isotope of an element on a scale where an atom of carbon-12 is exactly 12.

Calculating relative atomic mass

Relative isotopic mass is usually a whole number (at A-level anyway). Relative atomic mass is an average, so it's not usually a whole number.

Example — Maths Skills

A natural sample of chlorine contains a mixture of ^{35}Cl and ^{37}Cl, whose relative isotopic masses are 35 and 37. 75% of the sample is ^{35}Cl and 25% is ^{37}Cl. You need to take these percentages into account when you work out the relative atomic mass of chlorine.

$$\text{Relative atomic mass} = \frac{(35 \times 75) + (37 \times 25)}{100}$$

isotopic masses × percentages (numerator); *total percentage (75% ^{35}Cl + 25% ^{37}Cl)* (denominator)

Relative atomic mass = 35.5

Tip: You could turn the percentages into decimals and multiply instead. For the chlorine example, 75% = 0.75 and 25% = 0.25, so the relative atomic mass is (35 × 0.75) + (37 × 0.25) = 35.5

Relative molecular mass

The relative molecular mass, M_r, is the average mass of a molecule on a scale where an atom of carbon-12 is exactly 12. To find the M_r, just add up the relative atomic mass values of all the atoms in the molecule.

Example — Maths Skills

Calculating the relative molecular mass of C_2H_6O.

In one molecule of C_2H_6O there are 2 atoms of carbon, 6 of hydrogen and 1 of oxygen. The relative atomic masses (A_r) of each atom are shown in Figure 2.

$$M_r \text{ of } C_2H_6O = (2 \times 12.0) + (6 \times 1.0) + (1 \times 16.0) = 46.0$$

2 C atoms, *A_r of C*, *6 H atoms*, *A_r of H*, *1 O atom*, *A_r of O*

Calculating the relative molecular mass of C_4H_{10}.

In one molecule of C_4H_{10} there are 4 atoms of carbon and 10 of hydrogen.

$$M_r \text{ of } C_4H_{10} = (4 \times 12.0) + (10 \times 1.0) = 58.0$$

Atom	A_r
Carbon (C)	12.0
Hydrogen (H)	1.0
Oxygen (O)	16.0
Calcium (Ca)	40.1
Fluorine (F)	19.0

Figure 2: Table of relative atomic masses.

> **Tip:** Relative molecular mass and relative formula mass are basically the same thing — it's just that ionic compounds aren't made of molecules so they can't have a molecular mass. You work them out the same way though.

> **Tip:** The relative atomic masses (A_r) of all of the elements in these examples are shown in Figure 2.

Relative formula mass

Relative formula mass is the average mass of a formula unit on a scale where an atom of carbon-12 is exactly 12. It's used for compounds that are ionic (or giant covalent, such as SiO_2). To find the relative formula mass, just add up the relative atomic masses (A_r) of all the ions in the formula unit.

Examples — Maths Skills

Calculating the relative formula mass of CaF_2.

The formula unit of CaF_2 contains one Ca^{2+} ion and two F^- ions.

A_r of Ca (there's only one calcium ion)

$$M_r \text{ of } CaF_2 = 40.1 + (2 \times 19.0) = 78.1$$

2 ions of F^- *A_r of F*

Calculating the relative formula mass of $CaCO_3$.

In the $CaCO_3$ formula unit, there is one Ca^{2+} ion and one CO_3^{2-} ion. The CO_3^{2-} ion contains 1 carbon atom and 3 oxygen atoms, so the A_r values of all these atoms need to be included in the calculation.

$$M_r \text{ of } CaCO_3 = 40.1 + 12.0 + (3 \times 16.0) = 100.1$$

Practice Questions — Application

Use the periodic table to help you answer Questions 1-3.

Q1 Find the relative atomic mass of the following elements:

a) Rubidium

b) Mercury

c) Zinc

Q2 Find the relative molecular mass of the following compounds:

a) NH_3

b) CO_2

c) $C_2H_4O_6N_2$

Q3 Find the relative formula mass of the following compounds:

a) $CaCl_2$

b) $MgSO_4$

c) NaOH

Q4 A sample of tungsten is 0.1% ^{180}W, 26.5% ^{182}W, 14.3% ^{183}W, 30.7% ^{184}W and 28.4% ^{186}W. Calculate the A_r of tungsten.

Q5 A sample of zirconium is 51.5% ^{90}Zr, 11.2% ^{91}Zr, 17.1% ^{92}Zr, 17.4% ^{94}Zr and 2.8% ^{96}Zr. Calculate the A_r of zirconium.

> **Exam Tip**
> It's really important that you can calculate relative molecular mass (and relative formula mass). It crops up in loads of different calculations, so you need to be really confident that you can do it correctly.

Practice Questions — Fact Recall

Q1 What is relative atomic mass?

Q2 What is relative molecular mass?

Q3 What is relative formula mass?

4. The Mass Spectrometer

A mass spectrometer is a machine used to analyse elements or compounds. You need to know how a time of flight (TOF) mass spectrometer works.

Learning Objectives:

- Know that the mass spectrometer gives accurate information about relative isotopic mass and also about the relative abundance of isotopes.
- Understand the principles of a simple time of flight (TOF) mass spectrometer, limited to ionisation, acceleration to give all ions constant kinetic energy, ion drift, ion detection and data analysis.
- Be able to interpret simple mass spectra of elements.
- Know that mass spectrometry can be used to identify elements.

Specification Reference 3.1.1.2

How a mass spectrometer works

A mass spectrometer can give you information about the relative atomic mass of an element and the relative abundance of its isotopes, or the relative molecular mass of a molecule if you use it to analyse a compound.

There are four things that happen when a sample is squirted into a time of flight (TOF) mass spectrometer:

1. Ionisation

The sample needs to be ionised before it enters the mass spectrometer. Two ways of doing this are:

Electrospray ionisation, — in this method the sample is dissolved in a solvent and pushed through a small nozzle at high pressure. A high voltage is applied to it, causing each particle to gain an H^+ ion. The solvent is then removed, leaving a gas made up of positive ions.

Electron impact ionisation — in this method, the sample is vaporised and an 'electron gun' is used to fire high energy electrons at it. This knocks one electron off one each particle, so they become +1 ions.

2. Acceleration

The positive ions are accelerated by an electric field. The electric field gives the same kinetic energy to all the ions. The lighter ions experience a greater acceleration — they're given as much energy as the heavier ions, but they're lighter, so they accelerate more.

3. Ion drift

Next, the ions enter a region with no electric field. They drift through it at the same speed as they left the electric field. So the lighter ions will be drifting at higher speeds.

4. Detection

Because lighter ions travel through the drift region at higher speeds, they reach the detector in less time than heavier ions. The detector detects the current created when the ions hit it and records how long they took to pass through the spectrometer. This data is then used to calculate the mass/charge values needed to produce a mass spectrum (see next page).

Tip: The particles need to be ionised when they're put into the mass spectrometer, otherwise they couldn't be accelerated by the electric field or detected by the ion detector.

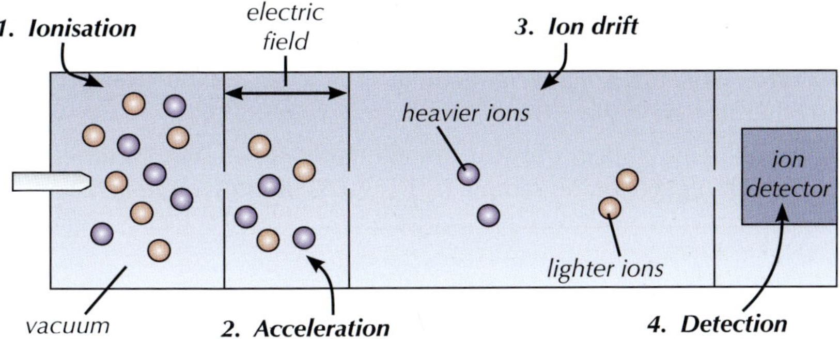

Figure 2: *Diagram showing how a TOF mass spectrometer works.*

Figure 1: *Electrospray ionisation apparatus.*

Interpreting a mass spectrum

A **mass spectrum** is a type of chart produced by a mass spectrometer. It shows information about the sample that was passed through the mass spectrometer.

If the sample is an element, each line will represent a different isotope of the element (see Figure 3). The *y*-axis gives the abundance of ions, often as a percentage. For an element, the height of each peak gives the **relative isotopic abundance** (the relative amount of each isotope present in a sample). The *x*-axis units are given as a 'mass/charge' ratio.

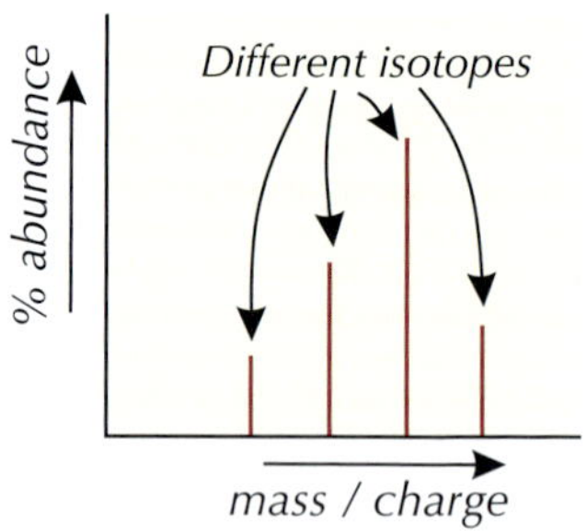

Figure 3: *The mass spectrum of an element*

Tip: Mass/charge is often shown as *m*/*z*.

The spectrum in Figure 3 was produced using electron impact ionisation. One electron has been knocked off each particle to turn them into +1 ions — so the mass/charge ratio of each peak is the same as the relative mass of that isotope. (If electrospray ionisation had been used instead, a H^+ ion would have been added to each particle to form +1 ions — so the mass/charge ratio of each peak would be one unit greater than the relative mass of each isotope.) All of the spectra shown in this topic have been produced using electron impact ionisation.

Example — Maths Skills

The mass spectrum produced when a sample of lithium is passed through a mass spectrometer is shown below.

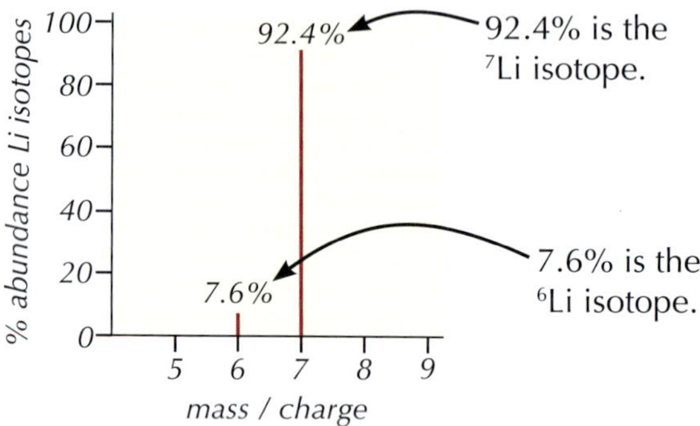

There are two peaks, so the sample contains two isotopes of lithium.

One peak has a mass/charge ratio of 6, so the relative isotopic mass of one isotope is 6. (Since the lithium ions produced in a mass spectrometer have a charge of 1+ (Li^+) and $6 \div 1 = 6$.)

The other peak has a mass/charge ratio of 7, so the relative isotopic mass of the other isotope is 7.

So, from the mass spectrum you can tell that there are two isotopes present, ^{6}Li and ^{7}Li. 92.4% of the sample is ^{7}Li. 7.6% of the sample is ^{6}Li.

Tip: The mass spectrometer produces ions by removing an electron, so the ions it produces are always positive — even for substances like chlorine which normally form negative ions.

Practice Question — Application

Q1 The mass spectrum for a sample of copper is shown below.

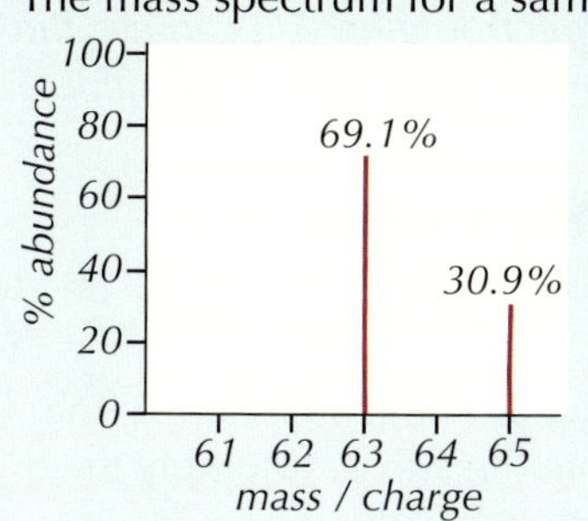

a) How many isotopes of copper were in the sample?

b) Give the relative isotopic mass of each isotope.

c) Give the relative isotopic abundance of each isotope.

Identifying elements

Mass spectrometry can be used to identify elements. Elements with different isotopes produce more than one line in a mass spectrum because the isotopes have different masses. This produces characteristic patterns which can be used as 'fingerprints' to identify certain elements.

Tip: Many elements only have one stable isotope. They can still be identified in a mass spectrum by looking for a line at their relative atomic mass.

Example

Magnesium has three isotopes with the percentage abundances shown in the table below. If a sample being analysed contains magnesium, this isotopic distribution will show up in the mass spectrum.

Mg isotopes	*% Abundance*
^{24}Mg	79
^{25}Mg	10
^{26}Mg	11

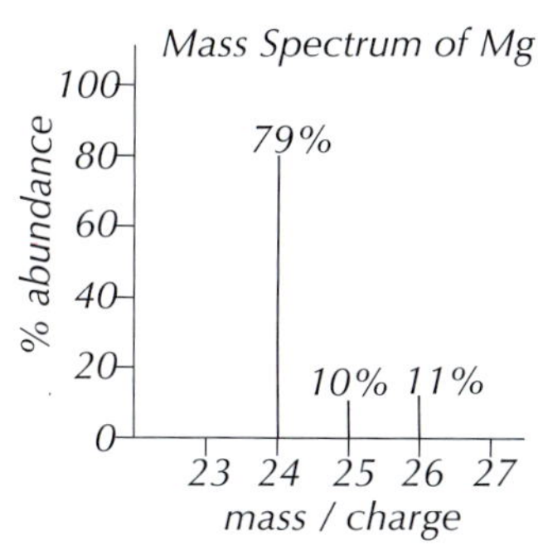

Figure 4: The Phoenix Mars lander probe. The Phoenix lander had a mass spectrometer that was used to analyse dust on the surface of Mars.

Practice Question — Application

Q2 The table below shows the relative abundance of the stable isotopes of three elements.

Magnesium		*Silicon*		*Indium*	
^{24}Mg	79.0%	^{28}Si	92.2%	^{113}In	4.3%
^{25}Mg	10.0%	^{29}Si	4.7%	^{115}In	95.7%
^{26}Mg	11.0%	^{30}Si	3.1%		

The mass spectrum for a sample of an unknown element is shown on the right.

For each element listed in the table above, explain why the mass spectrum suggests that the unknown substance is unlikely to be that element.

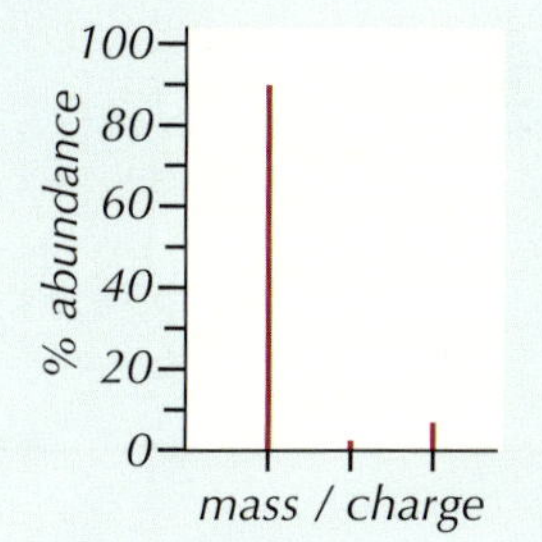

5. Using Mass Spectra

Learning Objectives:

- Be able to calculate relative atomic mass from isotopic abundance, limited to mononuclear ions.
- Know that mass spectrometry can be used to determine relative molecular mass.

Specification Reference 3.1.1.2

Once you've analysed a sample in a mass spectrometer, you can use the mass spectrum that the sample produces to find out about what's in it.

Calculating relative atomic mass

You need to know how to calculate the relative atomic mass (A_r) of an element from a mass spectrum.

- Step 1: For each peak, read the % relative isotopic abundance from the *y*-axis and the relative isotopic mass from the *x*-axis. Multiply them together to get the total relative mass for each isotope.
- Step 2: Add up these totals.
- Step 3: Divide by 100 (since percentages were used).

Tip: Remember — the relative isotopic mass for each isotope is the same as its mass/charge value.

Example — Maths Skills

Here's how to calculate A_r for magnesium, using the mass spectrum below:

Step 1:

Total mass of 1st isotope: 79 × 24 = 1896

Total mass of 2nd isotope: 10 × 25 = 250

Total mass of 3rd isotope: 11 × 26 = 286

Step 2:

1896 + 250 + 286 = 2432

Step 3:

2432 ÷ 100 = 24.32

So A_r (Mg) = 24.3 (to 3 s.f.)

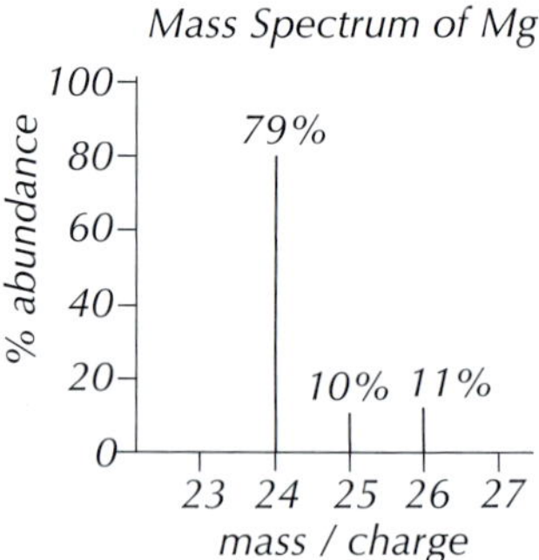

If the relative abundance is not given as a percentage, the total abundance may not add up to 100. In this case, don't panic. Just do steps 1 and 2 as above, but then divide by the total relative abundance instead of 100.

Example — Maths Skills

Here's how to calculate A_r for neon, using the mass spectrum below:

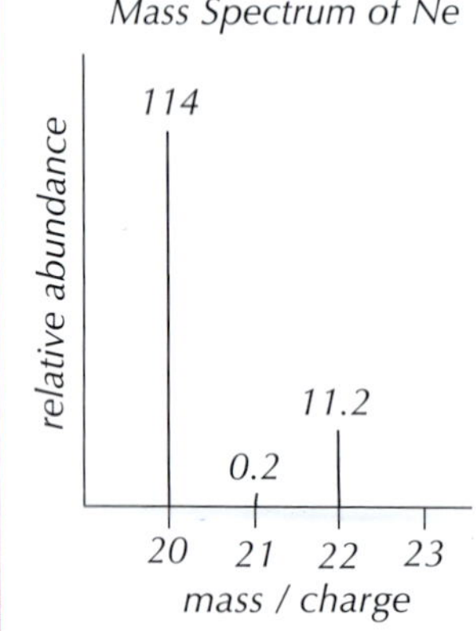

Step 1:

Total mass of 1st isotope: 114 × 20 = 2280

Total mass of 2nd isotope: 0.2 × 21 = 4.2

Total mass of 3rd isotope: 11.2 × 22 = 246.4

Step 2:

2280 + 4.2 + 246.4 = 2530.6

Step 3:

2530.6 ÷ (114 + 0.2 + 11.2) = 20.18...

So A_r (Ne) = 20.2 (to 3 s.f.)

total relative abundance

Tip: To find the total relative abundance, you just add up the relative abundances of all the isotopes in the sample.

Practice Questions — Application

Q1 Use the mass spectrum below to calculate the A_r of copper.

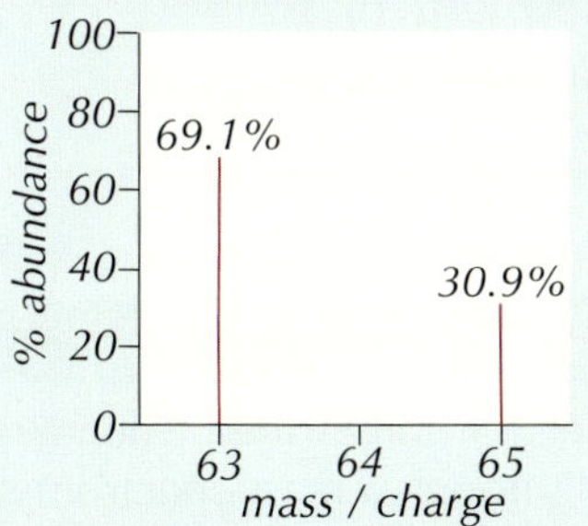

Q2 Use the mass spectrum below to calculate the A_r of boron.

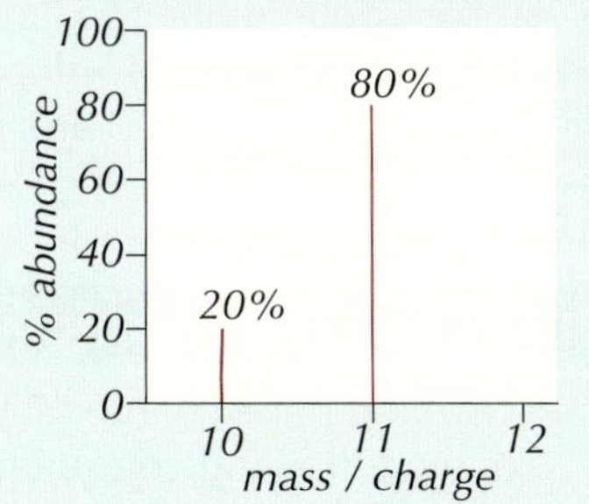

Q3 Use the mass spectrum below to calculate the A_r of lithium.

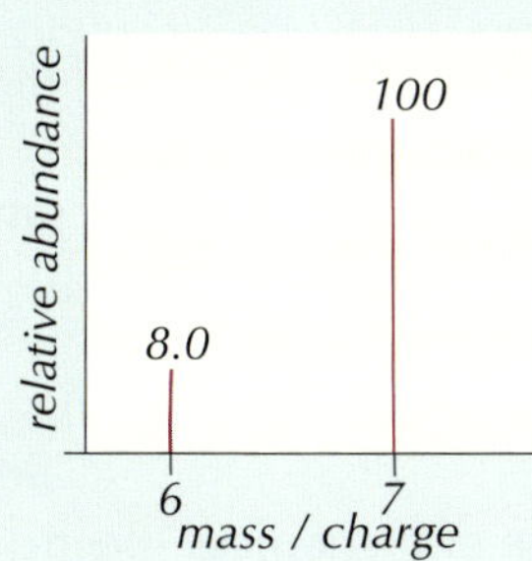

Q4 Use the mass spectrum below to calculate the A_r of gallium.

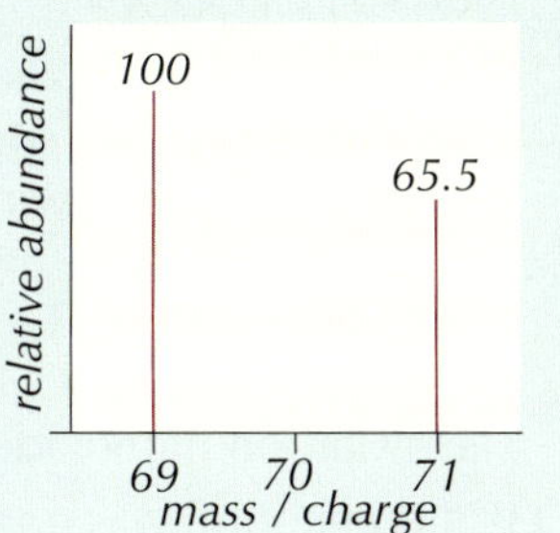

Exam Tip
Remember to check whether the *y*-axis shows relative abundance or percentage abundance before you start your calculation.

Calculating relative molecular mass

You can also get a mass spectrum for a molecular sample. A molecular ion, $M^+_{(g)}$, is formed when 1 electron is removed from the molecule. This gives a peak in the spectrum with a mass/charge ratio equal to the M_r of the molecule. This can be used to help identify an unknown compound. There's more about using mass spectra to identify compounds on page 254.

Tip: If a compound contains any atoms with more than one common isotope (e.g. chlorine), individual molecules will have different masses — so you'll get more than one molecular ion peak in its mass spectrum. But it's unlikely that this will come up in your exams, unless the examiners are feeling really mean.

Example

A sample of an unknown alcohol was analysed in a mass spectrometer. The mass/charge ratio of its molecular ion was 46.0.

The M_r values of the first few alcohols are:

Alcohol	M_r
Methanol CH_3OH	$12.0 + (4 \times 1.0) + 16.0 = 32.0$
Ethanol C_2H_5OH	$(2 \times 12.0) + (6 \times 1.0) + 16.0 = 46.0$
Propanol C_3H_7OH	$(3 \times 12.0) + (8 \times 1.0) + 16.0 = 60.0$
Butanol C_4H_9OH	$(4 \times 12.0) + (10 \times 1.0) + 16.0 = 74.0$

The mass/charge ratio of the molecular ion on a mass spectrum is equal to the M_r value of the compound being analysed.

So the unknown alcohol must be ethanol ($M_r = 46.0$).

Tip: Some types of mass spectrometer produce spectra with many smaller peaks, representing fragment ions created when the molecular ion breaks up. You don't need to know about these fragmentation patterns for this course.

If you have a mixture of compounds with different M_r values, you'll get a peak for the molecular ion of each one.

Example

A sample containing a mixture of ethanol (M_r = 46.0) and butanol (M_r = 74.0) will produce a mass spectrum with peaks at m/z = 46.0 and m/z = 74.0.

Tip: For all these questions, you can assume that the molecular ions had a charge of +1.

Practice Questions — Application

Q1 Give the mass/charge ratio of the molecular ion formed when each of the following compounds is passed through a mass spectrometer.

a) C_3H_6O

b) $CH_3CH_2CHCH_2$

c) CH_3CH_2COOH

Q2 A sample of an unknown gas was passed through a time of flight mass spectrometer. The spectrum produced had a peak with m/z = 28. Which of the following cannot be the gas contained in the sample?

A Nitrogen (N_2)
B Ethene (CH_2CH_2)
C Methylamine (CH_3NH_2)
D Carbon monoxide (CO)

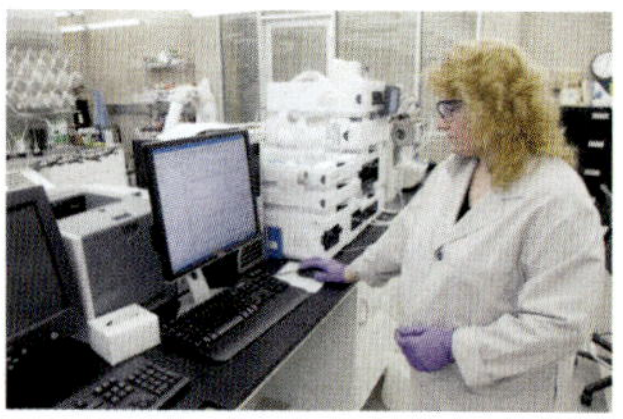

Figure 1: A scientist using a mass spectrometer to analyse a sample in a research lab.

Q3 A compound was analysed using a time of flight mass spectrometer. Its molecular ion produced a peak with m/z = 34 on the mass spectrum. Which of the following compounds could be responsible for this peak?

A Ethane (CH_3CH_3)
B Fluoromethane (CH_3F)
C Carbon dioxide (CO_2)
D Hydrogen cyanide (HCN)

Q4 A list of compounds and a mass spectrum are shown below. A mixture is known to contain a combination of some of the compounds listed. The mixture was analysed in a time of flight mass spectrometer and produced the peaks shown on the mass spectrum.

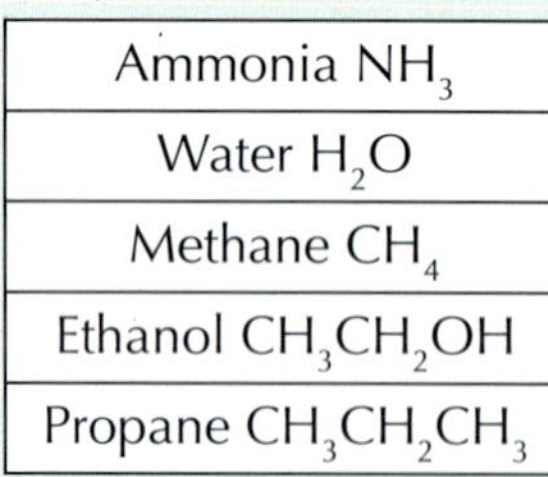

Ammonia NH_3
Water H_2O
Methane CH_4
Ethanol CH_3CH_2OH
Propane $CH_3CH_2CH_3$

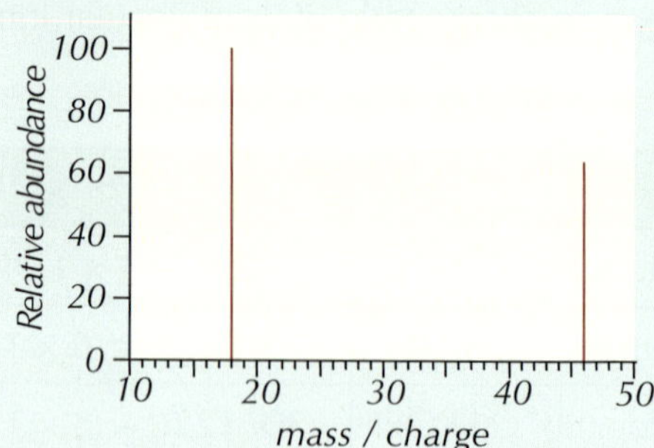

Which of the compounds listed must make up the mixture?

6. Electronic Structure

Electronic structure is all about how electrons are arranged in atoms.

Learning Objective:
- Know the electron configurations of atoms and ions up to $Z = 36$ in terms of shells and sub-shells (orbitals) s, p and d.

Specification Reference 3.1.1.3

Electron shells

In the currently accepted model of the atom, electrons have fixed energies. They move around the nucleus in certain regions of the atom called **shells** or **energy levels**. Each shell is given a number called the principal quantum number. The further a shell is from the nucleus, the higher its energy and the larger its principal quantum number — see Figure 1.

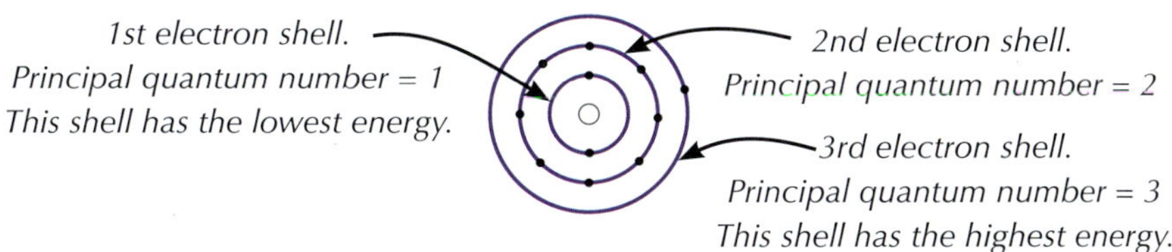
1st electron shell.
Principal quantum number = 1
This shell has the lowest energy.
2nd electron shell.
Principal quantum number = 2
3rd electron shell.
Principal quantum number = 3
This shell has the highest energy.

Figure 1: *A sodium atom.*

Experiments show that not all the electrons in a shell have exactly the same energy. The atomic model explains this — shells are divided up into **sub-shells**. Different electron shells have different numbers of sub-shells, which each have a different energy. Sub-shells can be s sub-shells, p sub-shells, d sub-shells or f sub-shells.

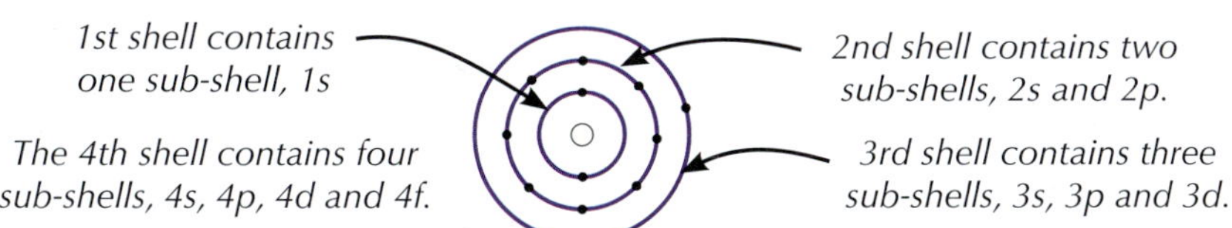
1st shell contains one sub-shell, 1s
The 4th shell contains four sub-shells, 4s, 4p, 4d and 4f.
2nd shell contains two sub-shells, 2s and 2p.
3rd shell contains three sub-shells, 3s, 3p and 3d.

Figure 2: *A sodium atom.*

Tip: Don't get confused by notation like 2s or 4f. The letter shows what type of sub-shell it is, the number shows what shell it's in. So 3p means a p sub-shell in the 3rd electron shell.

The sub-shells have different numbers of **orbitals** which can each hold up to 2 electrons. The table on the right shows the number of orbitals in each sub-shell. You can use it to work out the number of electrons that each shell can hold.

Sub-shell	Number of orbitals	Maximum electrons
s	1	$1 \times 2 = 2$
p	3	$3 \times 2 = 6$
d	5	$5 \times 2 = 10$
f	7	$7 \times 2 = 14$

Exam Tip
The f subshell has been included here so you know all the details of the first four shells. But in your exams you won't be asked to give the electron configuration of any atom that goes past the d subshell.

Example

The third shell contains 3 sub-shells: 3s, 3p and 3d.
- An s sub-shell contains 1 orbital, so can hold 2 electrons (1×2).
- A p sub-shell contains 3 orbitals, so can hold 6 electrons (3×2).
- A d sub-shell contains 5 orbitals, so can hold 10 electrons (5×2).

So the total number of electrons the third shell can hold is $2 + 6 + 10 = 18$

The table on the right shows the number of electrons that the first four electron shells can hold.

Shell	Sub-shells	Total number of electrons
1st	1s	$2 = 2$
2nd	2s 2p	$2 + (3 \times 2) = 8$
3rd	3s 3p 3d	$2 + (3 \times 2) + (5 \times 2) = 18$
4th	4s 4p 4d 4f	$2 + (3 \times 2) + (5 \times 2) + (7 \times 2) = 32$

Exam Tip
Make sure you learn how many electrons each shell and sub-shell can hold — you won't get far with electronic structures if you don't know these numbers.

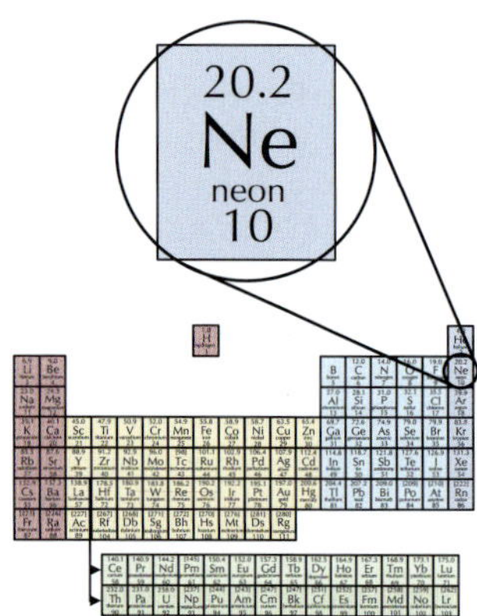

Figure 3: Neon has the atomic number 10, so an atom of neon must have 10 electrons.

Tip: Sub-shell notation is the 'standard' way of showing electron configurations. The other ways can be useful for showing the information in a different way, but if you're just asked to give 'the electron configuration' in an exam question, it will be the sub-shell notation they want.

Showing electron configurations

The number of electrons that an atom or ion has, and how they are arranged, is called its **electron configuration**. Electron configurations can be shown in different ways. For example, an atom of neon has 10 electrons — two electrons are in the 1s sub-shell, two are in the 2s sub-shell and six are in the 2p sub-shell. You can show this electron configuration in three ways...

1. Sub-shell notation

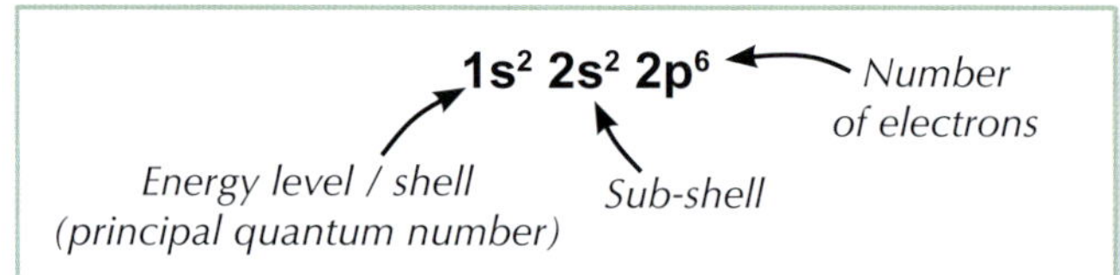

2. Arrows in boxes

Each of the boxes represents one orbital. Each of the arrows represents one electron. The up and down arrows represent the electrons spinning in opposite directions. Two electrons can only occupy the same orbital if they have opposite spin.

1s 2s 2p

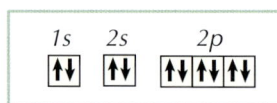

3. Energy level diagrams

These show the energy of the electrons in different orbitals, as well as the number of electrons and their arrangement.

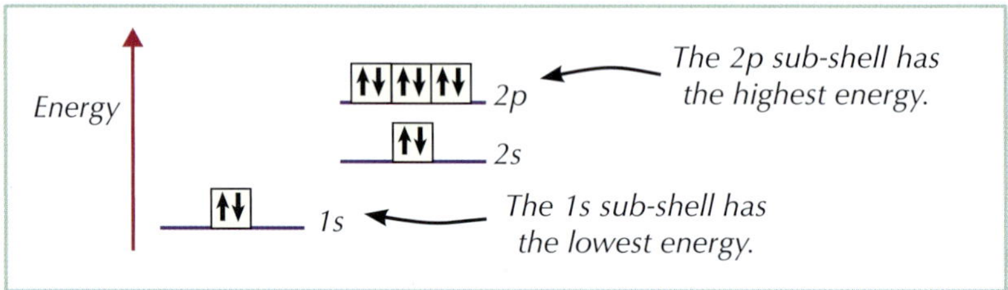

Working out electron configurations

You can figure out most electron configurations pretty easily, so long as you know a few simple rules:

Rule 1

Electrons fill up the lowest energy sub-shells first.

Example

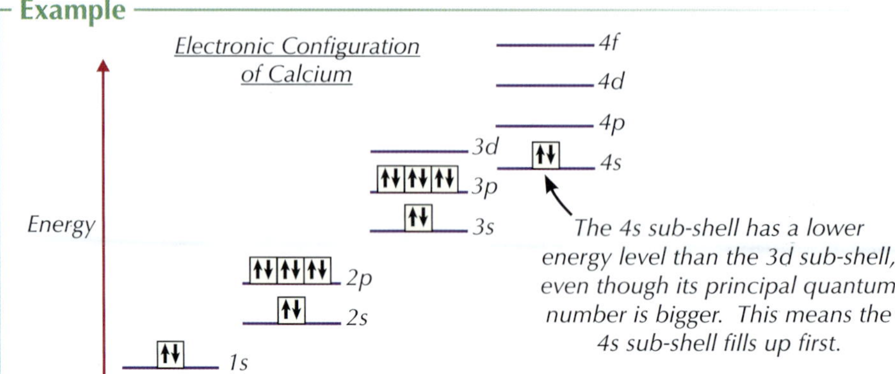

Rule 2

Electrons fill orbitals in a sub-shell singly before they start sharing.

Examples

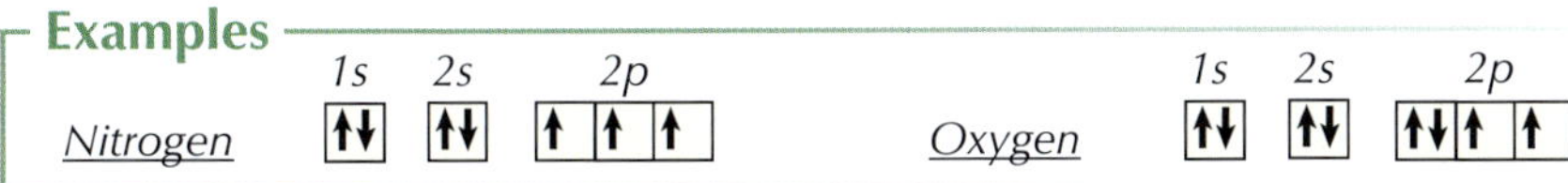

Rule 3

For the configuration of ions from the s and p blocks of the periodic table, just add or remove the electrons to or from the highest energy occupied sub-shell.

Examples

Mg atom: $1s^2\ 2s^2\ 2p^6\ 3s^2$ Cl atom: $1s^2\ 2s^2\ 2p^6\ 3s^2\ 3p^5$

Mg^{2+} ion: $1s^2\ 2s^2\ 2p^6$ Cl^- ion: $1s^2\ 2s^2\ 2p^6\ 3s^2\ 3p^6$

Tip: Elements with their outer electrons in an s sub-shell are called s block elements. Elements with their outer electrons in a p sub-shell are called p block elements. The configurations of d block elements (transition metals) are covered on the next page.

Shortened electron configurations

Noble gas symbols in square brackets, such as [Ar], are sometimes used as shorthand in electron configurations. E.g. calcium ($1s^2\ 2s^2\ 2p^6\ 3s^2\ 3p^6\ 4s^2$) can be written as $[Ar]4s^2$, where $[Ar] = 1s^2\ 2s^2\ 2p^6\ 3s^2\ 3p^6$.

Tip: Writing electron configurations using noble gas symbols can save time. Just make sure you've got your head around sub-shell notation before you start to use it — otherwise you're likely to get confused.

Practice Questions — Application

Q1 Use sub-shell notation to show the full electron configurations of the elements listed below.

a) Lithium

b) Titanium

c) Gallium

d) Nitrogen

Q2 Draw arrows in boxes to show the electron configurations of the elements listed below.

a) Calcium

b) Nickel

c) Sodium

d) Oxygen

Q3 Draw energy level diagrams to show the electron configurations of the elements listed below.

a) Magnesium

b) Argon

c) Carbon

d) Arsenic

Q4 Use sub-shell notation to show the full electron configurations of the ions listed below.

a) Na^+

b) O^{2-}

c) Al^{3+}

d) S^{2-}

Q5 Which elements have the electron configurations given below?

a) $[Ar]3d^{10}\ 4s^2\ 4p^5$

b) $[Ne]3s^2\ 3p^3$

c) $[Ar]3d^3\ 4s^2$

Exam Tip

If a question asks you to give the full configuration, that means 'don't use the noble gas shorthand or you'll lose marks'.

Exam Tip
Chromium and copper don't follow the normal rules for electron configurations. Make sure you learn these two exceptions — they're really important.

Tip: Some people prefer to write the 4s sub-shell before the 3d subshell, because it's lower in energy. Either way is fine — for example, you could write the electron configuration of copper like this instead:

$1s^2 2s^2 2p^6 3s^2 3p^6 4s^1 3d^{10}$

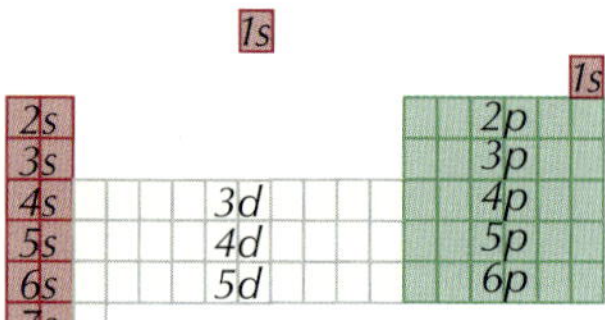

Figure 4: Elements can be classified as s block elements, p block elements or d block elements depending on where they are in the periodic table and which sub-shell their outer electrons are in.

Tip: Group 0 elements are know as the noble gases (or the inert gases).

Electron configuration of transition metals

Chromium (Cr) and copper (Cu) are badly behaved. They donate one of their 4s electrons to the 3d sub-shell. It's because they're happier with a more stable full or half-full d sub-shell.

So, the electron configuration of a Cr atom is: $1s^2 2s^2 2p^6 3s^2 3p^6 3d^5 4s^1$ (not ending in $3d^4 4s^2$ as you'd expect).

And the electron configuration of a Cu atom is: $1s^2 2s^2 2p^6 3s^2 3p^6 3d^{10} 4s^1$ (rather than finishing with $3d^9 4s^2$).

Here's another weird thing about transition metals — when they become ions, they lose their 4s electrons before their 3d electrons.

Example

The electron configuration of an Fe atom is: $1s^2 2s^2 2p^6 3s^2 3p^6 3d^6 4s^2$

To become an Fe^{3+} ion it loses three electrons — two 4s electrons and one 3d electron. So the electron configuration of an Fe^{3+} ion is : $1s^2 2s^2 2p^6 3s^2 3p^6 3d^5$

Electronic structure and chemical properties

The number of outer shell electrons decides the chemical properties of an element. You can use the periodic table to help you work them out.

- The s block elements (Groups 1 and 2) have 1 or 2 outer shell electrons. These are easily lost to form positive ions with an inert gas configuration. E.g. Na — $1s^2 2s^2 2p^6 3s^1 \rightarrow$ Na^+ — $1s^2 2s^2 2p^6$ (the electron configuration of neon).
- The elements in Groups 5, 6 and 7 (in the p block) can gain 1, 2 or 3 electrons to form negative ions with an inert gas configuration. E.g. O — $1s^2 2s^2 2p^4 \rightarrow$ O^{2-} — $1s^2 2s^2 2p^6$. Groups 4 to 7 can also share electrons when they form covalent bonds.
- Group 0 elements have completely filled s and p sub-shells and don't need to gain, lose or share electrons — their full sub-shells make them inert.

Practice Questions — Application

Q1 Give the full electron configuration of Cr^{2+} using sub-shell notation.

Q2 Give the full electron configuration of Ni^{2+} using sub-shell notation.

Q3 Give the full electron configuration of V^{3+} using sub-shell notation.

Practice Questions — Fact Recall

Q1 How many orbitals does a p sub-shell contain?

Q2 How many electrons can a p sub-shell hold?

Q3 How many electrons can the 3rd electron shell hold in total?

Q4 What does "electron configuration" mean?

Q5 The electron configuration shown here is wrong. Explain why.

1s	2s	2p		
↑↓	↑↓	↑↓		

Q6 What is the electron configuration of a chromium atom?

Q7 The electron configuration of copper is $1s^2 2s^2 2p^6 3s^2 3p^6 3d^{10} 4s^1$. Why isn't it $1s^2 2s^2 2p^6 3s^2 3p^6 3d^9 4s^2$?

Q8 Describe the ions that Group 5, 6 and 7 elements form.

7. Ionisation Energies

Electrons might be tiny, but moving them around can be pretty hard work.

Ionisation

When electrons have been removed from an atom or molecule, it's been ionised. The energy you need to remove the first electron is called the **first ionisation energy**.

The first ionisation energy is the energy needed to remove 1 electron from each atom in 1 mole of gaseous atoms to form 1 mole of gaseous 1+ ions.

You can write equations for this process — here's the equation for the first ionisation of oxygen:

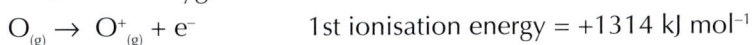

$O_{(g)} \rightarrow O^{+}_{(g)} + e^{-}$ 1st ionisation energy = +1314 kJ mol^{-1}

Here are a few rather important points about ionisation energies:

- You must use the gas state symbol, (g), because ionisation energies are measured for gaseous atoms.
- Always refer to 1 mole of atoms, as stated in the definition, rather than to a single atom.
- The lower the ionisation energy, the easier it is to form a positive ion.

Factors affecting ionisation energy

A high **ionisation energy** means there's a high attraction between the electron and the nucleus, so more energy is needed to remove the electron. There are three things that can affect ionisation energy:

1 Nuclear charge

The more protons there are in the nucleus, the more positively charged the nucleus is and the stronger the attraction for the electrons.

2 Distance from nucleus

Attraction falls off very rapidly with distance. An electron close to the nucleus will be much more strongly attracted than one further away.

3 Shielding

As the number of electrons between the outer electrons and the nucleus increases, the outer electrons feel less attraction to the nucleus. This lessening of the pull of the nucleus thanks to the inner electron shells is called shielding.

Example

There are only two electrons between the nucleus and the outer electron in a lithium atom.

There are ten electrons between the nucleus and the outer electron in a sodium atom — the shielding effect is greater.

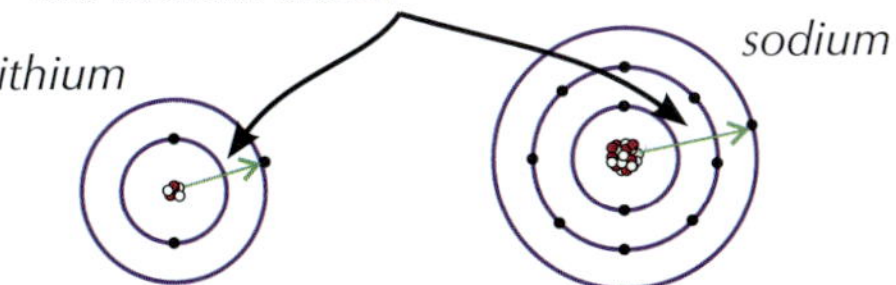
The distance between the nucleus and the electron being removed is greater in the sodium atom.
lithium
sodium

Figure 1: *A lithium atom and a sodium atom.*

This means that lithium has a higher first ionisation energy (+519 kJ mol^{-1}) than sodium (+496 kJ mol^{-1}). (The shielding and distance from the nucleus have a bigger effect than the increased nuclear charge in this example.)

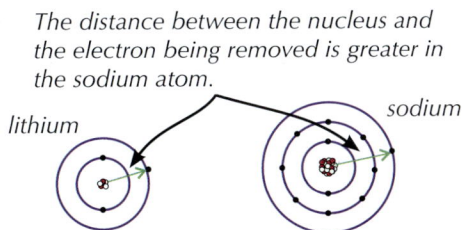

Learning Objectives:

- Know the meaning of the term ionisation energy.
- Be able to define first ionisation energy.
- Know how to write equations for first and successive ionisation energies.
- Explain how first and successive ionisation energies in Period 3 (Na–Ar) and in Group 2 (Be–Ba) give evidence for electron configuration in sub-shells and shells.

Specification Reference 3.1.1.3

Tip: You might see 'ionisation energy' referred to as 'ionisation enthalpy' instead.

Tip: You have to put energy in to remove an electron from an atom or molecule, so ionisation (in this sense) is always an endothermic process (see page 108).

Tip: You can only really see the effect of nuclear charge on ionisation energy if you compare atoms with outer electrons the same distance from the nucleus and with equal shielding effects. That only really happens with elements that are in the same period of the periodic table.

Tip: The third ionisation energy is the energy needed to remove an electron from each ion in 1 mole of 2+ ions. Third ionisation energies are greater than second ionisation energies — there's a greater attraction between a 2+ ion and its electrons than between a 1+ ion and its electrons.

Second ionisation energy

The second ionisation energy is the energy needed to remove an electron from each ion in 1 mole of gaseous 1+ ions. For example:

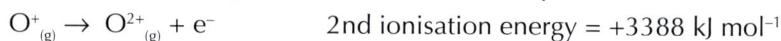

$$O^{+}_{(g)} \rightarrow O^{2+}_{(g)} + e^{-} \qquad \text{2nd ionisation energy} = +3388 \text{ kJ mol}^{-1}$$

Just like first ionisation energy, the value of second ionisation energy depends on nuclear charge, the distance of the electron from the nucleus and the shielding effect of inner electrons. Second ionisation energies are greater than first ionisation energies because the electron is being removed from a positive ion (and not an atom), which will require more energy. The electron configuration of the atom will also play a role in how much larger the second ionisation energy is than the first.

Example

The first electron removed from a lithium atom comes from the second shell ($2s^1$). The second electron removed comes from the first shell ($1s^2$). So the second electron to be removed is closer to the nucleus and experiences a stronger nuclear attraction than the first electron to be removed. The second electron will also not experience shielding from any inner electron shells, unlike the first electron. This means that the second ionisation energy of lithium is much higher than the first.

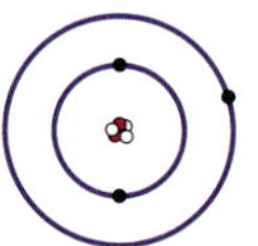

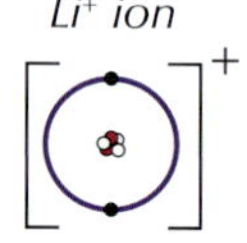

Figure 2: *The 2nd electron removed from lithium is closer to the nucleus than the 1st electron, so the attraction between it and the nucleus is greater.*

Successive ionisation energies

You can remove all the electrons from an atom, leaving only the nucleus. Each time you remove an electron, there's a successive ionisation energy.

You need to be able to write equations for any successive ionisation.

The general equation for the nth ionisation is:

$$X^{(n-1)+}_{(g)} \rightarrow X^{n+}_{(g)} + e^{-}$$

Example

The equation for the fifth ionisation of oxygen is: $O^{4+}_{(g)} \rightarrow O^{5+}_{(g)} + e^{-}$

Figure 3: *The first five Group 2 elements (left to right: beryllium, magnesium, calcium, strontium, barium).*

Ionisation trends down Group 2

First ionisation energy decreases down Group 2.

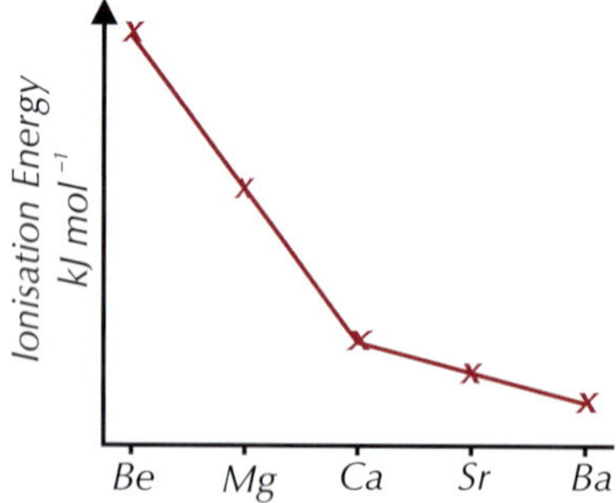

Figure 4: *First ionisation energies of Group 2.*

This provides evidence that electron shells really exist. If each element going down Group 2 has one more electron shell than the one above, the extra shell will shield the outer electrons from the attraction of the nucleus. Also, the extra shell means that the outer electrons will be further from the nucleus, so the nucleus's attraction will be reduced. It makes sense that both of these factors make it easier to remove outer electrons, resulting in lower ionisation energies.

Ionisation trends across periods

The graph below shows the first ionisation energies of the elements in Period 3.

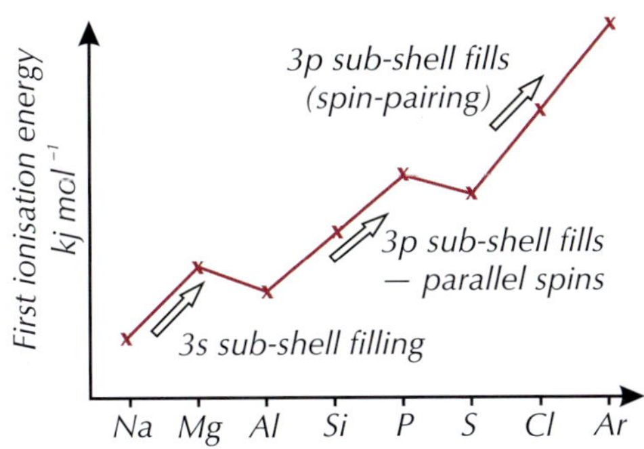

Figure 5: *First ionisation energies of Period 3.*

As you move across a period, the general trend is for the ionisation energies to increase — i.e. it gets harder to remove the outer electrons. This can be explained by the fact that the number of protons is increasing, which means a stronger nuclear attraction.

All the extra electrons are at roughly the same energy level, even if the outer electrons are in different orbital types. This means there's generally little extra shielding effect or extra distance to lessen the attraction from the nucleus. But, there are small drops between Groups 2 and 3, and 5 and 6. Tell me more I hear you cry. Well, alright then...

The drop between Groups 2 and 3 shows sub-shell structure.

Example

Mg $1s^2 2s^2 2p^6 3s^2$ 1st ionisation energy = +738 kJ mol^{-1}

Al $1s^2 2s^2 2p^6 3s^2 3p^1$ 1st ionisation energy = +578 kJ mol^{-1}

- Aluminium's outer electron is in a 3p orbital rather than a 3s. The 3p orbital has a slightly higher energy than the 3s orbital, so the electron is, on average, to be found further from the nucleus.
- The 3p orbital has additional shielding provided by the 3s electrons.

These two factors together are strong enough to override the effect of the increased nuclear charge, resulting in the ionisation energy dropping slightly. This pattern in ionisation energies provides evidence for the theory of electron sub-shells.

The drop between Groups 5 and 6 is due to electron repulsion.

Example

P $1s^2 2s^2 2p^6 3s^2 3p^3$ 1st ionisation energy = +1012 kJ mol^{-1}

S $1s^2 2s^2 2p^6 3s^2 3p^4$ 1st ionisation energy = +1000 kJ mol^{-1}

The shielding is identical in the phosphorus and sulfur atoms, and the electron is being removed from an identical orbital.

In phosphorus's case, the electron is being removed from a singly-occupied orbital. But in sulfur, the electron is being removed from an orbital containing two electrons.

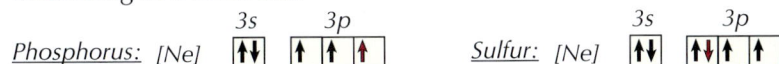

Phosphorus: *[Ne]* 3s ↑↓ 3p ↑ ↑ ↑ *Sulfur:* *[Ne]* 3s 3p

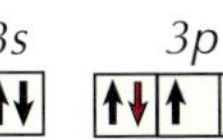

The repulsion between two electrons in an orbital means that electrons are easier to remove from shared orbitals. It's yet more evidence for the electronic structure model.

Exam Tip

You don't just need to be able to state the ionisation energy trends down Group 2 and across Period 3 — make sure you can explain them too. You could also be asked to explain why the ionisation energy for one element is higher or lower than it is for another element.

Tip: Writing out or drawing the electron configurations of elements can help you work out why their ionisation energies are what they are. For example, drawing

will show you that the next electron being removed is paired, so there will be repulsion.

Ionisation energies and shell structure

If you know the successive ionisation energies of an element you can work out the number of electrons in each shell of the atom and which group the element is in. A graph of successive ionisation energies provides evidence for the shell structure of atoms (see Figure 6).

Tip: The *y*-axis of this graph has a log (logarithmic) scale. Log scales go up in powers of a number (e.g. 1, 10, 100, etc.) rather than in units (1, 2, 3, etc.). Log scales are often used for graphs like this because ionisation energy values have such a huge range.

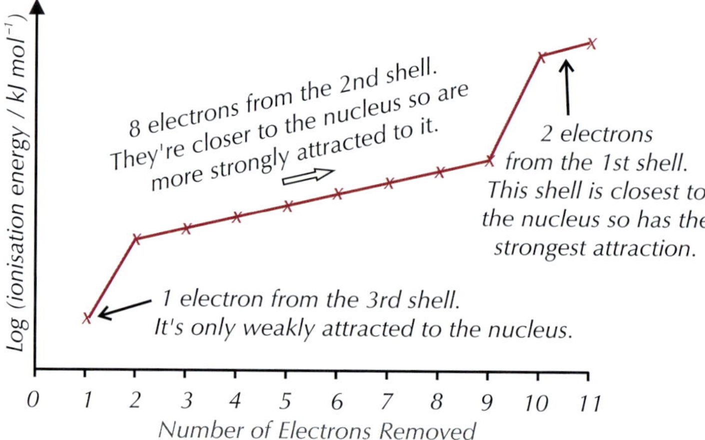

Figure 6: *Successive ionisation energies of sodium.*

Within each shell, successive ionisation energies increase. This is because electrons are being removed from an increasingly positive ion — there's less repulsion amongst the remaining electrons, so they're held more strongly by the nucleus. The big jumps in ionisation energy happen when a new shell is broken into — an electron is being removed from a shell closer to the nucleus.

Graphs like the one in Figure 7 can tell you which group of the periodic table an element belongs to. Just count how many electrons are removed before the first big jump to find the group number.

Example

In Figure 6, one electron is removed before the first big jump — sodium is in Group 1.

These graphs can be used to predict the electronic structure of an element. Working from right to left, count how many points there are before each big jump to find how many electrons are in each shell, starting with the first.

Example

Working from right to left in Figure 6, the graph has 2 points on the right-hand side, then a jump, then 8 points, a jump, and 1 final point. Sodium has 2 electrons in the first shell, 8 in the second and 1 in the third.

Practice Questions — Application

Q1 a) Write an equation for the first ionisation of chlorine.

b) Write an equation for the second ionisation of chlorine.

Q2 Sketch a graph showing the trend of the first ionisation energies of the elements in Period 2.

Q3 The first ionisation energy of nitrogen is +1402 kJ mol^{-1}.
The first ionisation energy of oxygen is +1314 kJ mol^{-1}.
Explain why oxygen has a lower first ionisation energy than nitrogen.

Q4 The first ionisation energy of beryllium is +900 kJ mol^{-1}.
The first ionisation energy of boron is +801 kJ mol^{-1}.
Explain why boron has a lower first ionisation energy than beryllium.

Q5 The first ionisation energy of lithium is +519 kJ mol^{-1}.
The second ionisation energy is +7298 kJ mol^{-1} and the third ionisation energy is +11815 kJ mol^{-1}.

Explain why the difference between the first and second ionisation energies is much greater than the difference between the second and third ionisation energies.

Q6 The graph below shows the successive ionisation energies of an element.

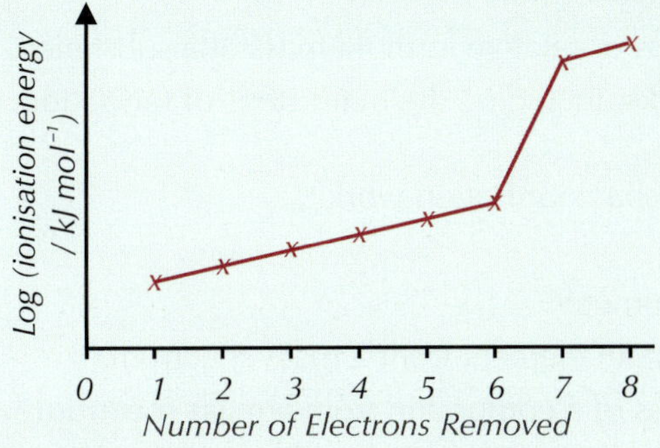

a) Which group is this element in?

b) State the number of electrons it has in each shell.

c) Name the element.

Q7 a) Sketch a graph showing the pattern of the successive ionisation energies of magnesium.

b) Explain the shape of the graph.

Exam Tip
If you get a question like this in the exam asking you to compare ionisation energies, make sure you remember the three factors that affect ionisation energy. Work out how the factors differ between each element, and you should have a very good idea of the answer.

Practice Questions — Fact Recall

Q1 Define first ionisation energy.

Q2 How does the number of protons in an atom affect its first ionisation energy?

Q3 Give two other factors that affect first ionisation energy.

Q4 Define second ionisation energy.

Q5 Explain why successive ionisation energies increase within each shell of an atom.

Q6 What happens to the first ionisation energy of elements in Group 2 as you go down the group?

Q7 What is the general trend in first ionisation energy across a period?

Exam Tip
Make sure you learn the definition of first ionisation energy — if it comes up in an exam question it'll be a nice easy way to get marks.

Section Summary

Make sure you know...

- The structure of the atom — including the location, relative masses and relative charges of protons, neutrons and electrons.
- What nuclear symbols, mass numbers (*A*) and atomic (proton) numbers (*Z*) are.
- What atoms, ions and isotopes are.
- How to calculate the number of protons, neutrons and electrons in an atom or ion using its mass number, proton number and charge.
- How the accepted model of the structure of the atom has changed over time.
- What relative atomic mass (A_r) is.
- What relative isotopic mass is.
- How to calculate relative atomic mass.
- What relative molecular mass (M_r) and relative formula mass are.
- How to calculate the relative molecular mass or relative formula mass of a substance.
- How a time of flight mass spectrometer works, including the main steps of ionisation, acceleration, ion drift and detection.
- How electrospray ionisation and electron impact ionisation work.
- How to interpret a mass spectrum.
- How to identify elements using mass spectrometry.
- How to calculate the relative atomic mass of an element from a mass spectrum.
- How to calculate the relative molecular mass of a compound from a mass spectrum.
- How electrons are arranged in energy levels (shells) within an atom.
- That each electron shell is divided into sub-shells.
- That sub-shells can be s, p, d or f sub-shells.
- That each sub-shell is divided into orbitals.
- How many electrons each orbital and sub-shell can hold.
- How to show electron configurations using sub-shell notation, arrows in boxes and energy level diagrams.
- How to work out electron configurations of the first 36 elements of the periodic table and their ions (including the electron configurations of chromium and copper).
- What ionisation energy and first ionisation energy are.
- The factors that affect ionisation energy.
- How to write equations to show ionisation energies.
- The trend in ionisation energy down Group 2, and the reasons for this trend.
- How ionisation energy changes across Period 3, and the reasons for these changes.
- How the successive ionisation energies of an element are related to its electron shell structure.
- How ionisation energies give evidence for electron arrangement in shells and sub-shells.

Exam-style Questions

1 An element, Z, has the electron configuration $1s^2\ 2s^2\ 2p^6\ 3s^2\ 3p^2$.
A sample of element Z is passed through a mass spectrometer.

1.1 Identify element Z.

(1 mark)

1.2 State the block of the periodic table that element Z belongs to.

(1 mark)

1.3 Give the full electron configuration of the particle formed when a single electron is removed from an atom of element Z.

(1 mark)

The mass spectrum produced when element Z is passed through the mass spectrometer is shown below.

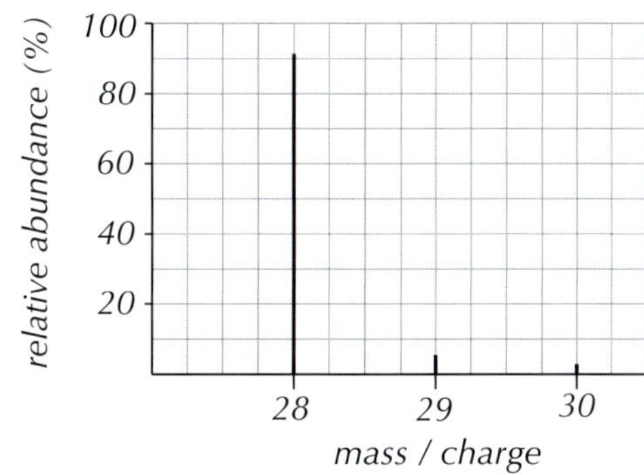

1.4 State how many different isotopes of element Z were present in the sample.

(1 mark)

1.5 Give the nuclear symbol of the isotope in the sample with the highest relative isotopic abundance.

(2 marks)

1.6 State whether you would expect the isotopes of element Z to have the same chemical properties or different chemical properties. Explain your answer.

(1 mark)

2 Compound Q is analysed in a time of flight mass spectrometer. A sample of compound Q in solution is injected into the mass spectrometer at high pressure. A high voltage is then applied to remove electrons from atoms of the sample.

2.1 Give the name of this process.

(1 mark)

2.2 Explain how a time of flight mass spectrometer separates charged particles with different mass/charge ratios.

(4 marks)

3 The table below shows the first ionisation energies of some of the elements in Period 3.

Element	Na	Mg	Al	Si	P	S	Cl
First ionisation energy / kJ mol^{-1}	+496	+738	+578		+1012	+1000	+1251

3.1 State the general trend in first ionisation energy across Period 3.

(1 mark)

3.2 Write an equation for the first ionisation of sodium.

(1 mark)

3.3 Explain why the first ionisation energy of aluminium is lower than the first ionisation energy of magnesium.

(3 marks)

3.4 In which of the following ranges would you expect to find the first ionisation energy of silicon?

A +300 kJ mol^{-1} to +600 kJ mol^{-1}

B +600 kJ mol^{-1} to +1000 kJ mol^{-1}

C +1000 kJ mol^{-1} to +1200 kJ mol^{-1}

D +1200 kJ mol^{-1} to +1600 kJ mol^{-1}

(1 mark)

3.5 Use your understanding of the electronic structures of aluminium, silicon and phosphorus to explain your answer to question 3.4.

(4 marks)

3.6 Explain why the first ionisation energy of sulfur is lower than the first ionisation energy of phosphorus.

(2 marks)

4 Vanadium and copper are both transition metals.

4.1 Give the full electron configuration of vanadium.

(1 mark)

4.2 When vanadium is ionised it can form V^{2+} and V^{3+} ions. Give the full electron configurations of both of these ions.

(2 marks)

4.3 Give the full electron configuration of copper and explain why it deviates from the normal rules of electron configuration.

(2 marks)

4.4 Identify the transition metal with the electron configuration $1s^2\ 2s^2\ 2p^6\ 3s^2\ 3p^6\ 3d^6\ 4s^2$.

(1 mark)

4.5 A transition metal is ionised to form a 2+ ion. The ion has the electron configuration $1s^2\ 2s^2\ 2p^6\ 3s^2\ 3p^6\ 3d^5$. Identify the transition metal.

(1 mark)

5 The relative isotopic masses and relative abundances of element X were identified using mass spectrometry. The isotopes and their relative abundances are shown in the table below.

Isotope	Relative abundance
^{20}X	90.48
^{21}X	0.27
^{22}X	9.25

5.1 Calculate the relative atomic mass of element X, and identify the element.

(3 marks)

5.2 State the number of protons, neutrons and electrons in an atom of each isotope of element X.

(3 marks)

Isotopes of an element, such as the isotopes of element X, have the same proton number but different mass numbers.

5.3 Define the term **mass number (*A*)**.

(1 mark)

5.4 Define the term **proton number (*Z*)**.

(1 mark)

The first ionisation energy of element X is +2080.7 kJ mol^{-1}. Atoms of another element, element A, contain 24 more electrons than atoms of element X.

5.5 Define the term **first ionisation energy**.

(2 marks)

5.6 Using your understanding of the factors that affect ionisation energy, state how you would expect the first ionisation energy of element A to be different from element X. Explain your answer.

(2 marks)

5.7 There is a large jump in the trend of the successive ionisation energies of element X between the removal of the eight and ninth electrons.

Use your understanding of the factors that affect ionisation energy to explain why this is the case.

(3 marks)

Learning Objectives:

- Understand the concept of a mole as applied to electrons, atoms, molecules, ions, formulas and equations.
- Understand the concept of the Avogadro constant and use it to carry out calculations.
- Be able to carry out calculations using mass of substance, M_r and amount in moles.
- Understand the concentration of a substance in solution, measured in mol dm^{-3}.
- Be able to carry out calculations using concentration, volume and amount of substance in a solution.

Specification Reference 3.1.2.2

Figure 1: *1 mole of carbon.*

Tip: The Avogadro constant's a huge number so it's always written in standard form. There's more about standard form on pages 519-520.

1. The Mole

Amount of substance is a really important idea in chemistry. It's all about working out exactly how much of a chemical you have and what amount of it is reacting with other chemicals. Then you can use that information in all sorts of calculations to do with things like mass, concentration and volume.

What is a mole?

Amount of substance is measured using a unit called the **mole** (mol for short) and given the symbol *n*. One mole is roughly 6.02×10^{23} particles (the Avogadro constant). It doesn't matter what the particles are. They can be atoms, molecules, electrons, ions, penguins — anything.

Examples

1 mole of carbon (C) contains 6.02×10^{23} atoms.

1 mole of methane (CH_4) contains 6.02×10^{23} molecules.

1 mole of sodium ions (Na^+) contains 6.02×10^{23} ions.

1 mole of electrons contains 6.02×10^{23} electrons.

The Avogadro constant

You can use the **Avogadro constant** to convert between number of particles and number of moles. Just remember this formula:

Number of particles = Number of moles × Avogadro's constant

Examples — Maths Skills

How many atoms are in 0.450 moles of pure iron?

Number of atoms = $0.450 \times (6.02 \times 10^{23}) = 2.71 \times 10^{23}$

How many ions are in 0.724 moles of calcium ions?

Number of ions = $0.724 \times (6.02 \times 10^{23}) = 4.36 \times 10^{23}$

You can rearrange the formula and use it to work out the number of moles:

Example — Maths Skills

How many moles are in 1.14×10^{24} molecules of NH_3?

Rearrange the formula to find number of moles (divide both sides by the Avogadro constant):

number of moles = number of particles ÷ Avogadro's constant

Number of moles of NH_3 = $(1.14 \times 10^{24}) \div (6.02 \times 10^{23}) = 1.89$ moles

Calculations with moles

1 mole of any substance has a mass that's the same as its relative molecular mass (M_r) in grams.

Example

What is the mass of 1 mole of water?

M_r of H_2O = (1.0 × 2) + 16.0 = 18.0, so 1 mole of water has a mass of 18.0 g.

This means that you can work out how many moles of a substance you have from the mass of the substance and its relative molecular mass (M_r). Here's the formula you need:

$$\text{Number of moles} = \frac{\text{mass of substance}}{M_r}$$

Exam Tip
This formula crops up in all sorts of chemistry calculations — you'll need to know it off by heart for your exams. You might see it with 'molar mass' instead of M_r — molar mass just means 'the mass of 1 mole of a substance', so its value for a compound is the same as the M_r.

Examples — Maths Skills

How many moles of aluminium oxide are present in 5.10 g of Al_2O_3?

M_r of Al_2O_3 = (2 × 27.0) + (3 × 16.0) = 102

Number of moles of Al_2O_3 = $\frac{5.10}{102}$ = 0.0500 moles

How many moles of calcium bromide are present in 39.98 g of $CaBr_2$?

M_r of $CaBr_2$ = 40.1 + (2 × 79.9) = 199.9

Number of moles of $CaBr_2$ = $\frac{39.98}{199.9}$ = 0.200 moles

Exam Tip
Make sure you give your final answer to an appropriate number of significant figures. See page 520 for more about this.

You can also rearrange the formula and use it to work out either the mass of a substance or its relative molecular mass:

Examples — Maths Skills

What is the mass of 2 moles of NaF?

Rearrange the formula to find mass (multiply both sides by relative molecular mass):

$$\text{mass of substance} = \text{number of moles} \times M_r$$

M_r of NaF = 23.0 + 19.0 = 42.0

Mass of 2 moles of NaF = 2 × 42.0 = 84.0 g

0.0500 moles of a solid weighs 2.60 g. Find its relative molecular mass.

Rearrange the formula:

$$M_r = \text{mass} \div \text{number of moles}$$

So, relative molecular mass = 2.60 ÷ 0.0500 = 52.0.

Tip: If it helps you to remember how to rearrange the equation, you could use this formula triangle:

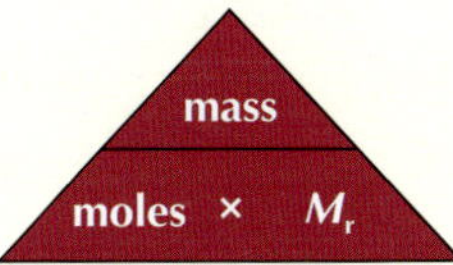

Just cover the thing you want to calculate to find the right formula.

(For example, if you cover mass it will tell you that to calculate mass you multiply moles by M_r.)

Practice Questions — Application

Q1 How many molecules are in 0.360 moles of H_2O?

Q2 How many ions are in 0.0550 moles of magnesium ions?

Q3 1.5 moles of a mystery compound weighs 66 g. Find its relative molecular mass.

Q4 How many moles of sodium nitrate are present in 212.5 g of $NaNO_3$?

Q5 How many moles of zinc chloride are present in 15.5 g of $ZnCl_2$?

Q6 What is the mass of 2 moles of NaCl?

Tip: 1 dm^3 is the same as 1000 cm^3 or 1 litre.

Exam Tip
You need to know all of the formulas in this section by heart — so look out for the formula boxes, and learn them.

Tip: This is another formula that you can stick in a formula triangle if it helps you:

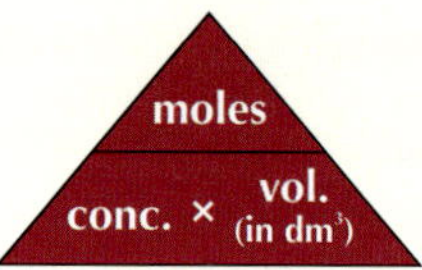

Exam Tip
Remember to watch out for the units in this type of calculation. Double check whether the volume you've been given is in cm^3 or dm^3.

Moles and concentration

The concentration of a solution is how many moles are dissolved per 1 dm^3 of solution. The units are mol dm^{-3}.

Here's the formula to find the number of moles.

$$\text{Number of moles} = \frac{\text{Concentration} \times \text{Volume (in cm}^3\text{)}}{1000}$$

$$\text{Or: Number of moles} = \text{Concentration} \times \text{Volume (in dm}^3\text{)}$$

You need to be able to use these formulas to do calculations in the exam.

Examples — Maths Skills

How many moles of lithium chloride are present in 25 cm^3 of a 1.2 mol dm^{-3} solution of LiCl?

$$\text{Number of moles} = \frac{\text{concentration} \times \text{volume (in cm}^3\text{)}}{1000}$$

$$= \frac{1.2 \times 25}{1000} = 0.030 \text{ moles}$$

A solution of $FeCl_3$ contains 0.2 moles of iron(III) chloride in 0.4 dm^3. What is the concentration of the solution?

Rearrange the formula to find concentration (divide both sides by volume):

$$\text{concentration} = \frac{\text{number of moles}}{\text{volume (in dm}^3\text{)}}$$

$$= \frac{0.2}{0.4} = 0.5 \text{ mol dm}^{-3}$$

A 0.50 mol dm^{-3} solution of zinc sulfate contains 0.080 moles of $ZnSO_4$. What volume does the solution occupy?

Rearrange the formula to find volume (divide both sides by volume):

$$\text{volume (in dm}^3\text{)} = \frac{\text{number of moles}}{\text{concentration}}$$

$$= \frac{0.080}{0.50} = 0.16 \text{ dm}^3$$

You might be asked to combine a concentration calculation with a moles and mass calculation. This just means using both formulas, one after the other.

Example — Maths Skills

What mass of sodium hydroxide needs to be dissolved in water to give 50.0 cm^3 of solution with a concentration of 2.00 mol dm^{-3}?

First look at the question and see what information it gives you. You've got concentration and volume — so you can work out number of moles.

$$\text{Number of moles} = \frac{2.00 \times 50.0}{1000} = 0.100 \text{ moles of NaOH}$$

Then you can use this to work out the mass using the equation number of moles = mass ÷ M_r

M_r of NaOH = 23.0 + 16.0 + 1.0 = 40.0

Mass = number of moles × M_r = 0.100 × 40.0 = 4.00 g

Practice Questions — Application

Q1 How many moles of potassium phosphate are present in 50 cm^3 of a 2 mol dm^{-3} solution?

Q2 How many moles of sodium chloride are present in 0.5 dm^3 of a 0.08 mol dm^{-3} solution?

Q3 How many moles of silver nitrate are present in 30 cm^3 of a 0.70 mol dm^{-3} solution?

Q4 A solution contains 0.25 moles of copper bromide in 0.50 dm^3. What is the concentration of the solution?

Q5 A solution contains 0.080 moles of lithium chloride in 0.75 dm^3. What is the concentration of the solution?

Q6 A solution contains 0.10 moles of magnesium sulfate in 36 cm^3. What is the concentration of the solution?

Q7 A solution of calcium chloride contains 0.46 moles of $CaCl_2$. The concentration of the solution is 1.8 mol dm^{-3}. What volume, in dm^3, does the solution occupy?

Q8 A solution of copper sulfate contains 0.010 moles of $CuSO_4$. The concentration of the solution is 0.55 mol dm^{-3}. What volume, in dm^3, does the solution occupy?

Q9 The molecular formula of sodium oxide is Na_2O. What mass of sodium oxide would you have to dissolve in 75 cm^3 of water to make a 0.80 mol dm^{-3} solution?

Q10 The molecular formula of cobalt(II) bromide is $CoBr_2$. What mass of cobalt(II) bromide would you have to dissolve in 30 cm^3 of water to make a 0.50 mol dm^{-3} solution?

Q11 A solution is made by dissolving 4.08 g of a compound in 100 cm^3 of pure water. The solution has a concentration of 1.20 mol dm^{-3}. What is the relative molecular mass of the compound?

Figure 2: *Preparing a solution of copper sulfate.*

Practice Questions — Fact Recall

Q1 a) How many particles are there in a mole?

b) What's the name for this special number?

Q2 What's the formula that links number of particles and number of moles?

Q3 What's the formula that links relative molecular mass and number of moles?

Q4 How many cm^3 are there in one dm^3?

Q5 Give an example of the units that you could use to describe concentration.

Q6 What's the formula that links number of moles and concentration? (Write it out twice, once using each volume measurement.)

Learning Objectives:

- Be able to recall the ideal gas equation $pV = nRT$ with the variables in SI units.
- Be able to use the ideal gas equation in calculations.

Specification Reference 3.1.2.3

Tip: All of these units are SI units. That means they're part of an agreed system of measurements used by scientists all over the world.

Exam Tip
Make sure you've learnt the ideal gas equation off by heart, because you won't be given it in the exam. You will be given the value of the gas constant if you need it though. (In fact, if you're given the value of the gas constant in a question, it's a good clue that you need to use the ideal gas equation.)

Tip: There's more about converting units in the Maths Skills section on pages 526-527.

Tip: There's no need to worry about units for number of moles (it doesn't have any units) or R (it always has the same units).

2. Gases and the Mole

Lots of things in chemistry relate back to how many moles of a substance you have. And that includes how much volume a gas takes up.

The ideal gas equation

In the real world (and AQA exam questions), it's not always room temperature and pressure. The **ideal gas equation** lets you find the number of moles in a certain volume at any temperature and pressure:

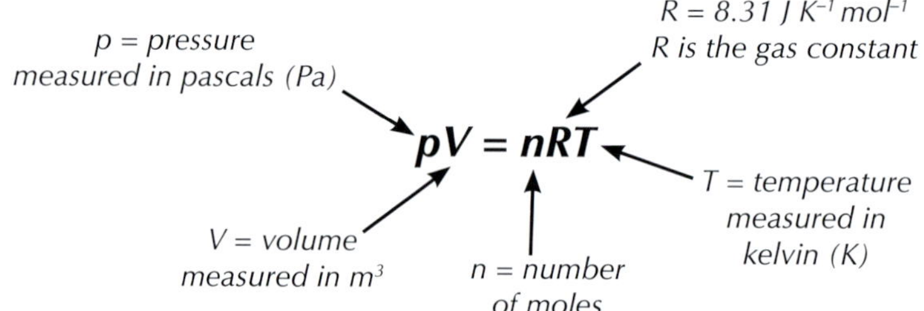

You could be asked to find any of the values in the equation. Its the same old idea — just rearrange the equation and put in the numbers you know.

Examples — Maths Skills

How many moles are there in $0.0600\ m^3$ of hydrogen gas, at 283 K and 50 000 Pa?

Rearrange the equation to find number of moles (divide both sides by RT):

$$n = \frac{pV}{RT} = \frac{50000 \times 0.0600}{8.31 \times 283} = 1.28 \text{ moles (3 s.f.)}$$

At what pressure would 0.400 moles of argon gas occupy $0.0100\ m^3$ at 298 K?

Rearrange the equation to find pressure (divide both sides by V):

$$p = \frac{nRT}{V} = \frac{0.400 \times 8.31 \times 298}{0.0100} = 99\ 100 \text{ Pa (3 s.f.)}$$

If you're given the values in different units from the ones used in the ideal gas equation you'll need to convert them to the right units first.

- You might be given pressure in kPa (kilopascals). To convert from kPa to Pa you multiply by 1000 (e.g. 2 kPa = 2000 Pa).
- You might be given temperature in °C. To convert from °C to K you add 273 (e.g. 25 °C = 298 K).
- You might be given volume in cm^3 or dm^3. To convert from cm^3 to m^3 you multiply by 10^{-6}. To convert from dm^3 to m^3 you multiply by 10^{-3}. ($1\ m^3 = 1 \times 10^6\ cm^3 = 1 \times 10^3\ dm^3$)

Example — Maths Skills

What volume does 2.00 moles of argon gas occupy at 27.0 °C and 100 kPa?

First put all the values you have into the right units:

T = 27.0 °C = (27.0 + 273) K = 300 K

p = 100 kPa = 100 000 Pa

Now rearrange the equation to find volume (divide both sides by pressure):

$$V = \frac{nRT}{p} = \frac{2.00 \times 8.31 \times 300}{100\ 000} = 0.0499\ m^3 \text{ (3 s.f.)}$$

You might be asked to combine an ideal gas equation calculation with another type of calculation.

Examples — Maths Skills

At a temperature of 60.0 °C and a pressure of 250 kPa, a gas occupied a volume of 1100 cm^3 and had a mass of 1.60 g. Find its relative molecular mass.

You've been given temperature, pressure and volume, so you need to find the number of moles:

$$n = \frac{pV}{RT} = \frac{(2.50 \times 10^5) \times (1.1 \times 10^{-3})}{8.31 \times 333} = 0.993\text{... moles}$$

Now you've got the number of moles, you can calculate relative molecular mass using the formula M_r = mass ÷ number of moles.

M_r = mass ÷ number of moles = 1.60 ÷ 0.993... = 16.1...

So the relative molecular mass is 16.1 (3 s.f.).

Tip: There's a 10^5 and a 10^{-3} in this formula because the numbers have been put into standard form. It's easier to write very big or very small numbers in standard form — see page 519 for more.

Tip: This is just the formula from page 47 again, but it's been rearranged to make M_r the subject. There's more about rearranging formulas on page 525.

Practice Questions — Application

Q1 How many moles are there in 0.040 m^3 of oxygen gas at a temperature of 350 K and a pressure of 70 000 Pa?

Q2 What volume would 0.65 moles of carbon dioxide gas occupy at a temperature of 280 K and a pressure of 100 000 Pa?

Q3 How many moles are there in 0.55 dm^3 of nitrogen gas at a temperature of 35 °C and a pressure of 90 000 Pa?

Q4 At a pressure of 110 000 Pa, 0.0500 moles of hydrogen gas occupied a volume of 1200 cm^3. What was the temperature in °C?

Q5 What volume, in m^3, would 0.75 moles of helium gas occupy at a temperature of 22 °C and a pressure of 75 kPa?

Q6 At a temperature of 300 K and a pressure of 80 kPa a gas had a volume of 1.5 dm^3 and a mass of 2.6 g. Find its relative molecular mass.

Q7 A student had a sample of neon gas, Ne. They heated it to 44 °C. At this temperature the gas had a volume of 0.00300 m^3. If the pressure was 100 kPa, what was the mass of the neon gas?

Practice Question — Fact Recall

Q1 Write out the ideal gas equation. Say what the terms mean and give the standard units that each is measured in.

Figure 1: *Neon gas lighting up restaurant signs.*

Learning Objectives:

- Be able to balance equations for unfamiliar reactions when reactants and products are specified.
- Be able to write balanced full and ionic equations for reactions studied.

Specification Reference 3.1.2.5

Exam Tip
Being able to balance equations is a key skill in Chemistry. Whenever you write an equation in an exam it should be balanced (whether the question specifically asks you to balance it or not).

Tip: You can use ½ to balance equations, but you should only use it for diatomic molecules like O_2, H_2 or Cl_2. (For example, you can't really have $\frac{1}{2}H_2O$. That would mean splitting an oxygen atom.)

Tip: Always do a quick check to make sure your final equation balances.

3. Chemical Equations

Writing and balancing equations is one of those topics that gets everywhere in chemistry, so here's your chance to make sure you've got your head round it.

How to balance equations

Balanced equations have the same number of each atom on both sides. If you've got an equation that isn't balanced, you can add more atoms to balance it, but only by adding whole reactants or products. You do this by changing the number in front of a reactant or product — you never mess with formulas (e.g. you can change H_2O to $2H_2O$, but never to H_4O).

Examples — Maths Skills

Balance the equation $H_2SO_4 + NaOH \rightarrow Na_2SO_4 + H_2O$.

First you need to count how many of each atom you have on each side.

$$H_2SO_4 + NaOH \rightarrow Na_2SO_4 + H_2O$$

Left		Right	
H = 3	*Na = 1*	*H = 2*	*Na = 2*
O = 5	*S = 1*	*O = 5*	*S = 1*

The left side needs 2 Na's, so try changing NaOH to 2NaOH:

$$H_2SO_4 + 2NaOH \rightarrow Na_2SO_4 + H_2O$$

Left		Right	
H = 4	*Na = 2*	*H = 2*	*Na = 2*
O = 6	*S = 1*	*O = 5*	*S = 1*

Now the right side needs 4 H's, so try changing H_2O to $2H_2O$:

$$H_2SO_4 + 2NaOH \rightarrow Na_2SO_4 + 2H_2O$$

Left		Right	
H = 4	*Na = 2*	*H = 4*	*Na = 2*
O = 6	*S = 1*	*O = 6*	*S = 1*

Both sides have the same number of each atom — the equation is balanced.

Balance the equation $C_2H_6 + O_2 \rightarrow CO_2 + H_2O$.

First work out how many of each atom you have on each side.

$$C_2H_6 + O_2 \rightarrow CO_2 + H_2O$$

Left		Right	
C = 2	*H = 6*	*C = 1*	*H = 2*
	O = 2		*O = 3*

The right side needs 2 C's, so try $2CO_2$. It also needs 6 H's, so try $3H_2O$.

$$C_2H_6 + O_2 \rightarrow 2CO_2 + 3H_2O$$

Left		Right	
C = 2	*H = 6*	*C = 2*	*H = 6*
	O = 2		*O = 7*

The left side needs 7 O's, so try $3\frac{1}{2}O_2$

$$C_2H_6 + 3\tfrac{1}{2}O_2 \rightarrow 2CO_2 + 3H_2O$$

Left		Right	
C = 2	*H = 6*	*C = 2*	*H = 6*
	O = 7		*O = 7*

This balances the equation.

Ionic equations

You can write an **ionic equation** for any reaction involving ions that happens in solution. In ionic equations, only the reacting particles (and the products they form) are included.

To write an ionic equation, you start by writing a full, balanced equation for the reaction. Then you split any dissolved ionic species up into ions. Finally, you take out any ions that appear on both sides of the equation.

Examples — Maths Skills

Write an ionic equation for the reaction between sodium hydroxide and nitric acid: $\mathbf{HNO_{3\,(aq)} + NaOH_{(aq)} \rightarrow NaNO_{3\,(aq)} + H_2O_{(l)}}$

First, check the full equation is balanced. This one is — there are the same numbers of each type of atom on each side of the equation.

The ionic substances from the equation will dissolve, breaking up into ions in solution. So you can rewrite the equation to show all the ions that are in the reaction mixture:

$$H^+ + NO_3^- + Na^+ + OH^- \rightarrow Na^+ + NO_3^- + H_2O$$

To get from this to the ionic equation, just cross out any ions that appear on both sides — in this case, that's the sodium ions and the nitrate ions.

$$H^+ + \cancel{NO_3^-} + \cancel{Na^+} + OH^- \rightarrow \cancel{Na^+} + \cancel{NO_3^-} + H_2O$$

So the ionic equation for this reaction is: $\mathbf{H^+ + OH^- \rightarrow H_2O}$

Once you've written the ionic equation, check that the charges are balanced. In this example, the net charge on the left-hand side is +1 + (–1) = 0, and the net charge on the right-hand side is 0 — so the charges balance.

Write an ionic equation for this reaction:

$$\mathbf{Na_3PO_{4\,(aq)} + CaCl_{2\,(aq)} \rightarrow NaCl_{(aq)} + Ca_3(PO_4)_{2\,(s)}}$$

This time the full equation isn't balanced. So the first thing to do is to balance it. (Look back at the last page for more on how to do this.)

$$2Na_3PO_4 + 3CaCl_2 \rightarrow 6NaCl + Ca_3(PO_4)_2$$

Now split up everything that dissolves into its constituent ions. Remember to make sure you've included the correct number of atoms of each type from the full, balanced equation.

$$(2 \times 3)Na^+ + 2PO_4^{3-} + 3Ca^{2+} + (3 \times 2)Cl^- \rightarrow 6Na^+ + 6Cl^- + Ca_3(PO_4)_2$$

$$6Na^+ + 2PO_4^{3-} + 3Ca^{2+} + 6Cl^- \rightarrow 6Na^+ + 6Cl^- + Ca_3(PO_4)_2$$

Finally, cross out the ions that appear on both sides of the equation.

$$\cancel{6Na^+} + 2PO_4^{3-} + 3Ca^{2+} + \cancel{6Cl^-} \rightarrow \cancel{6Na^+} + \cancel{6Cl^-} + Ca_3(PO_4)_2$$

This leaves you with the ionic equation: $\mathbf{2PO_4^{3-} + 3Ca^{2+} \rightarrow Ca_3(PO_4)_2}$

Tip: The state symbols tell you that HNO_3, NaOH and $NaNO_3$ are in solution, but the water is a liquid. (There's more about state symbols on page 56.)

Tip: Leave anything that isn't an ion in solution (like the water) as it is.

Tip: An ion that's present in the reaction mixture but doesn't get involved in the reaction is called a spectator ion.

Tip: As long as you started off with a full, balanced equation, the charges should balance.

Tip: The state symbols in the full equation show that the $Ca_3(PO_4)_2$ is solid. You don't split it up into ions, because it isn't in solution.

Tip: The net charge on the left-hand side is (2 × –3) + (3 × +2) = 0. This is the same as the net charge on the right-hand side, so your equation balances.

Practice Questions — Application

Q1 Balance these equations:

a) $Mg + HCl \rightarrow MgCl_2 + H_2$

b) $S_8 + F_2 \rightarrow SF_6$

c) $Ca(OH)_2 + H_2SO_4 \rightarrow CaSO_4 + H_2O$

d) $Na_2CO_3 + HCl \rightarrow NaCl + CO_2 + H_2O$

e) $C_4H_{10} + O_2 \rightarrow CO_2 + H_2O$

Q2 Write ionic equations for these reactions:

a) $Fe_{(s)} + CuSO_{4\,(aq)} \rightarrow FeSO_{4(aq)} + Cu_{(s)}$

b) $BaCl_{2\,(aq)} + Na_2SO_{4(aq)} \rightarrow NaCl_{(aq)} + BaSO_{4\,(s)}$

c) $Na_2CO_{3\,(aq)} + HNO_{3\,(aq)} \rightarrow NaNO_{3\,(aq)} + H_2O_{(l)} + CO_{2\,(g)}$

Tip: If you're not sure what the charges on the ions of different elements are, or how to work them out, look at pages 76 to 77.

4. Equations and Calculations

Learning Objectives:

- Be able to calculate masses from balanced equations.
- Be able to calculate volumes of gases from balanced equations.

Specification Reference 3.1.2.5

Once you've made sure that an equation is balanced, you can use it to calculate all sorts of things — like how much product a reaction will make...

Calculating masses

You can use the balanced equation for a reaction to work out how much product you will get from a certain mass of reactant.

Here are the steps to follow:

1. Write out the balanced equation for the reaction.
2. Work out how many moles of the reactant you have.
3. Use the **molar ratio** from the balanced equation to work out the number of moles of product that will be formed from this much reactant.
4. Calculate the mass of that many moles of product.

Tip: The ratio of the moles of each reactant and product in a balanced chemical equation is called the molar ratio. The big numbers in front of the species in a balanced equation tell you what the molar ratio is.

Here's a nice juicy example to help you get to grips with the method:

Example — Maths Skills

Find the mass of iron(III) oxide produced when 28.0 g of iron is burnt in air.

1. Write out the balanced equation: $2Fe_{(s)} + 1\frac{1}{2}O_{2\,(g)} \rightarrow Fe_2O_{3\,(s)}$
2. Work out how many moles of iron you have:
   A_r of Fe = 55.8
   Moles = mass ÷ A_r = 28.0 ÷ 55.8 = 0.502... moles of iron
3. The molar ratio of Fe : Fe_2O_3 is 2 : 1. This means that for every 2 moles of Fe that you have, you will produce 1 mole of Fe_2O_3.
   But you only have 0.502... moles of Fe here.
   So you will produce: 0.502... ÷ 2 = 0.251... moles of Fe_2O_3
4. Now find the mass of 0.251... moles of Fe_2O_3:
   M_r of Fe_2O_3 = (2 × 55.8) + (3 × 16) = 159.6
   Mass = moles × M_r = 0.25 × 159.6 = 40.0 g of iron(III) oxide (3 s.f.)

Tip: Look — it's that moles = mass ÷ M_r formula yet again...

You can use similar steps to work out how much of a reactant you had at the start of a reaction when you're given a certain mass of product:

Tip: Reactants are the chemicals you start with that get used up during a reaction. Products are the chemicals that are formed during a reaction.

Example — Maths Skills

Hydrogen gas can react with nitrogen gas to give ammonia (NH_3). Calculate the mass of hydrogen needed to produce 6.8 g of ammonia.

1. $N_{2\,(g)} + 3H_{2\,(g)} \rightarrow 2NH_{3\,(g)}$
2. M_r of NH_3 = 14.0 + (3 × 1.0) = 17.0
   Moles = mass ÷ M_r = 6.8 ÷ 17.0 = 0.4 moles of NH_3
3. From the equation: the molar ratio of NH_3 : H_2 is 2 : 3.
   So to make 0.4 moles of NH_3, you must need to start with (0.4 ÷ 2) × 3 = 0.6 moles of H_2
4. M_r of H_2 = 2 × 1.0 = 2.0
   Mass = moles × M_r = 0.6 × 2.0 = 1.2 g of hydrogen

Practice Questions — Application

Q1 3.4 g of zinc is dissolved in hydrochloric acid, producing zinc chloride ($ZnCl_2$) and hydrogen gas.

a) Write a balanced equation for this reaction.

b) Calculate the number of moles of zinc in 3.4 g.

c) How many moles of zinc chloride does the reaction produce?

d) What mass of zinc chloride does the reaction produce?

Q2 A student burns some ethene gas (C_2H_4) in oxygen, producing carbon dioxide gas and 15 g of water.

a) Write a balanced equation for this reaction.

b) Calculate the number of moles of water in 15 g.

c) How many moles of ethene did the student begin with?

d) What mass of ethene did the student begin with?

Q3 Calculate the mass of barium carbonate ($BaCO_3$) produced if 4.58 g of barium chloride ($BaCl_2$) is reacted with sodium carbonate (Na_2CO_3).

Figure 1: Zinc dissolving in hydrochloric acid.

Calculating gas volumes

It's handy to be able to work out how much gas a reaction will produce, so that you can use large enough apparatus. Or else there might be a large bang. The first three steps of this method are the same as the method on the last page. Once you've found the number of moles of product, the final step is to put that number into the ideal gas equation (see page 50).

Examples — Maths Skills

What volume of hydrogen gas, in m³, is produced when 15.0 g of sodium is reacted with excess water at a temperature of 25.0 °C and a pressure of 100 kPa? The gas constant is 8.31 J K⁻¹ mol⁻¹.

1. $2Na_{(s)} + 2H_2O_{(l)} \rightarrow 2NaOH_{(aq)} + H_{2\,(g)}$
2. A_r of Na = 23.0
   number of moles = mass ÷ A_r
   = 15.0 ÷ 23.0 = 0.652... moles of sodium
3. From the equation: the molar ratio of Na : H_2 is 2 : 1.
   So 0.652... moles of Na produces (0.652... ÷ 2) = 0.326... moles of H_2.
4. Volume = $\frac{nRT}{p} = \frac{0.326... \times 8.31 \times 298}{100\,000}$ = 0.00808 m³ (3 s.f.)

Tip: 'Excess water' just means that all of the sodium will react.

What volume of carbon dioxide, in dm³, is produced when 10.0 g of calcium carbonate reacts with excess hydrochloric acid at a temperature of 25.0 °C and a pressure of 100 kPa? The gas constant is 8.31 J K⁻¹ mol⁻¹.

1. $CaCO_{3\,(s)} + 2HCl_{(aq)} \rightarrow CaCl_{2\,(aq)} + CO_{2\,(g)} + H_2O_{(l)}$
2. M_r of $CaCO_3$ = 40.1 + 12.0 + (3 × 16.0) = 100.1
   number of moles = mass ÷ M_r
   = 10.0 ÷ 100.1 = 0.0999... moles of calcium carbonate
3. From the equation: the molar ratio of $CaCO_3$: CO_2 is 1 : 1.
   So 0.0999... moles of $CaCO_3$ produces 0.0999... moles of CO_2.
4. Volume = $\frac{nRT}{p} = \frac{0.0999... \times 8.31 \times 298}{100\,000}$
   = 0.00247... m³ = 2.47 dm³ (3 s.f.)

Exam Tip
Sometimes, exam questions will tell you what units to give your answer in — if they do, make sure you follow those instructions, or you could lose marks.

State symbols

State symbols are put after each compound in an equation. They tell you what state of matter things are in:

s = solid, l = liquid, g = gas, aq = aqueous (solution in water).

Exam Tip
Make sure you include state symbols for the equations you write in the exam.

Example

$$CaCO_{3\,(s)} + 2HCl_{(aq)} \rightarrow CaCl_{2\,(aq)} + H_2O_{(l)} + CO_{2\,(g)}$$

solid *aqueous* *aqueous* *liquid* *gas*

Practice Questions — Application

Q1 Give the state symbols that you would use in an equation to show the state of the following substances.

a) a solution of magnesium chloride in water

b) a piece of magnesium metal

c) a measured amount of water at room temperature and pressure

d) a solution of sodium nitrate in water

e) ethane gas

f) copper oxide powder

Q2 9.00 g of water is split apart to give hydrogen gas and oxygen gas.

a) Write a balanced equation for this reaction.

b) Calculate the number of moles of water in 9.00 g.

c) How many moles of oxygen gas does the reaction produce?

d) What volume of oxygen gas will the reaction produce at 298 K and 100 000 Pa?

Q3 7.0 g of zinc sulfide (ZnS) is burnt in oxygen. This produces solid zinc oxide (ZnO) and sulfur dioxide gas (SO_2).

a) Write a balanced equation for this reaction.

b) Calculate the number of moles of zinc sulfide in 7.0 g.

c) How many moles of sulfur dioxide gas does the reaction produce?

d) What volume of sulfur dioxide gas will the reaction produce at 298 K and 100 000 Pa?

Q4 A sample of hexane gas (C_6H_{14}) is cracked to give butane gas (C_4H_{10}) and ethene gas (C_2H_4). The mass of butane produced is 3.0 g.

a) Write a balanced equation for this reaction.

b) Calculate the number of moles of butane in 3.0 g.

c) How many moles of hexane gas were present in the sample?

d) What volume would this many moles of hexane gas occupy at a temperature of 308 K and a pressure of 100 000 Pa?

Q5 Magnesium metal will react with steam to produce solid magnesium oxide and hydrogen gas. Calculate the volume of steam needed to create 10 g of MgO at 100 °C and 101 325 Pa.

5. Titrations

You can do a titration to find the concentration of an acid or an alkali. But first you'll have to prepare a standard solution to use in your titration...

Neutralisation

When an acid reacts with an alkali you get a salt and water. This is called a neutralisation reaction.

Example

$$H_2SO_{4\,(aq)} + 2NaOH_{(aq)} \rightarrow Na_2SO_{4\,(aq)} + 2H_2O_{(l)}$$

acid *alkali* *salt* *water*

Titrations involve neutralisation reactions to work out the concentration of an acidic or alkaline solution.

REQUIRED PRACTICAL 1

Making a standard solution

Before you do a titration, you might have to make up a **standard solution** to use. A standard solution is any solution that you know the exact concentration of. Making a standard solution involves dissolving a known amount of solid in a known amount of water to create a known concentration.

Example

Make 250 cm³ of a 2.00 mol dm⁻³ solution of sodium hydroxide.

1. First work out how many moles of sodium hydroxide you need using the formula:
   moles = concentration × volume
   = 2.00 mol dm^{-3} × 0.250 dm^3 = 0.500 moles
2. Now work out how many grams of sodium hydroxide you need using the formula: mass = moles × M_r
   M_r of NaOH = 23.0 + 16.0 + 1.0 = 40.0
   Mass = 0.500 × 40.0 = 20.0 g
3. Place a weighing bottle on a digital balance. Weigh out the required mass of solid approximately and tip it into a beaker.
4. Weigh the weighing bottle (which may still contain traces of the solid). Subtract the mass of the bottle from the mass of the bottle and the solid together to find the precise mass of solid you have weighed out.
5. Add distilled water to the beaker and stir until all the sodium hydroxide has dissolved.
6. Tip the solution into a 250 cm^3 volumetric flask — use a funnel to make sure it all goes in.
7. Rinse the beaker, stirring rod and funnel with distilled water and add that to the flask too. This makes sure there's no solute clinging to the beaker or rod.
8. Now top the flask up to the correct volume with more distilled water. Make sure the bottom of the meniscus reaches the line (see Figure 1). When you get close to the line add the water drop by drop — if you go over the line you'll have to start all over again.
9. Stopper the flask and turn it upside down a few times to make sure it's mixed well.
10. Now calculate the exact concentration of your standard solution.

Learning Objective:

- Be able to make up a volumetric solution and carry out a simple acid–base titration (Required Practical 1).
- Be able to use balanced equations to calculate concentrations and volumes for reactions in solutions.

Specification Reference 3.1.2.5

Tip: Standard solutions can also be called volumetric solutions.

Tip: If you're doing this experiment, make sure you carry out a risk assessment before you start.

Figure 1: *The meniscus of a liquid in a volumetric flask.*

Performing titrations

REQUIRED PRACTICAL 1

Titrations allow you to find out exactly how much acid is needed to neutralise a measured quantity of alkali (or the other way round). You can use this data to work out the concentration of the alkali.

Start off by using a pipette to measure out a set volume of the solution that you want to know the concentration of. Put it in a flask. Add a few drops of an appropriate indicator (see below) to the flask. Then fill a burette (see Figure 3) with a standard solution of the acid — remember, that means you know its exact concentration. Use a funnel to carefully pour the acid into the burette. Always do this below eye level to avoid any acid splashing on to your face or eyes. (You should wear safety glasses too.) Now you're ready to titrate.

Tip: Make sure you are aware of any safety issues involved in the experiment before carrying it out.

Figure 2: *A student doing a titration. She is adding acid from the burette to the alkali in the flask. The alkali looks pink because it contains phenolphthalein.*

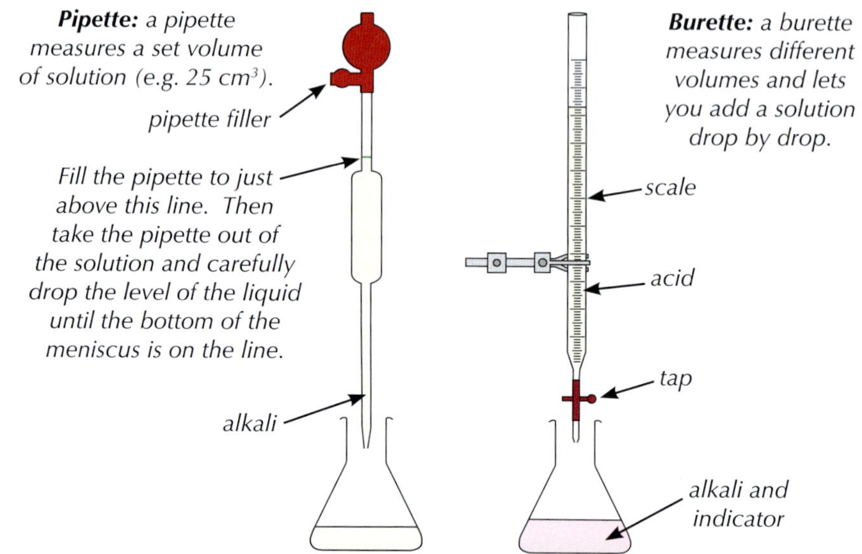

Pipette: a pipette measures a set volume of solution (e.g. 25 cm³).
pipette filler
Fill the pipette to just above this line. Then take the pipette out of the solution and carefully drop the level of the liquid until the bottom of the meniscus is on the line.
alkali
Burette: a burette measures different volumes and lets you add a solution drop by drop.
scale
acid
tap
alkali and indicator

Figure 3: *The apparatus needed for a titration.*

First do a rough titration to get an idea where the end point (the point where the alkali is neutralised and the indicator changes colour) is. Take an initial reading of how much acid is in the burette. Then gradually add the acid to the alkali, giving the flask a regular swirl. When the colour changes, take a final reading.

Then do an accurate titration. Take an initial reading, then run the acid in to within 2 cm^3 of the end point. When you get to this stage, add it dropwise. If you don't see exactly when the colour changes you'll overshoot and your result won't be accurate. Find the amount of acid used to neutralise the alkali by subtracting the final reading from the initial reading. This is called the **titre.**

Repeat the titration a few times, until you have at least three results that are concordant (for titrations, this means three results within 0.1 cm^3 of each other). Use the results from each repeat to calculate the mean volume of acid used. Remember to leave out any anomalous results when calculating your mean — anomalous results can distort your answer.

Tip: There's loads of stuff about how to handle the data you get from experiments like this on pages 10-17. Have a look if you want to know more about means, anomalous results and error.

Indicators

In titrations, **indicators** that change colour quickly over a very small pH range are used so you know exactly when the reaction has ended. The main two indicators used for acid/alkali titrations are methyl orange, which is red in acids and yellow in alkalis, and phenolphthalein, which is colourless in acids and pink in alkalis. It's a good idea to stand your flask on a white tile when you're titrating — it'll make it easier to see exactly when the end point is.

Tip: Universal indicator is no good here — its colour change is too gradual.

Calculating concentrations

You need to be able to use the results of a titration to calculate the concentration of acids and alkalis. There's more on concentration calculations on page 48.

Examples — Maths Skills

In a titration experiment, 25.0 cm³ of 0.500 mol dm⁻³ HCl neutralised 35.0 cm³ of NaOH solution. Calculate the concentration of the sodium hydroxide solution in mol dm⁻³.

First write a balanced equation and decide what you know and what you need to know:

$$HCl + NaOH \rightarrow NaCl + H_2O$$

	HCl	NaOH
Volume:	*25.0 cm³*	*35.0 cm³*
Concentration:	*0.500 mol dm⁻³*	*?*

You know the volume and concentration of the HCl, so first work out how many moles of HCl you have:

$$\text{Number of moles HCl} = \frac{\text{concentration} \times \text{volume}(\text{cm}^3)}{1000}$$

$$= \frac{0.500 \times 25.0}{1000} = 0.0125 \text{ moles}$$

From the equation, you know 1 mole of HCl neutralises 1 mole of NaOH.

So 0.0125 moles of HCl must neutralise 0.0125 moles of NaOH.

Now it's a doddle to work out the concentration of NaOH.

$$\text{Concentration of NaOH} = \frac{\text{moles of NaOH} \times 1000}{\text{volume}(\text{cm}^3)}$$

$$= \frac{0.0125 \times 1000}{35.0} = 0.357 \text{ mol dm}^{-3} \text{ (3 s.f)}$$

Here's an example where it's an alkali being added to an acid instead.

40.0 cm³ of 0.250 mol dm⁻³ KOH was used to neutralise 22.0 cm³ of HNO_3 solution. Calculate the concentration of the nitric acid in mol dm⁻³.

Write out the balanced equation and the information that you have:

$$HNO_3 + KOH \rightarrow KNO_3 + H_2O$$

	HNO_3	KOH
Volume:	*22.0 cm³*	*40.0 cm³*
Concentration:	*?*	*0.250 mol dm⁻³*

You know the volume and concentration of the KOH, so now work out how many moles of KOH you have:

$$\text{Number of moles KOH} = \frac{\text{concentration} \times \text{volume}(\text{cm}^3)}{1000}$$

$$= \frac{0.250 \times 40.0}{1000} = 0.0100 \text{ moles}$$

From the equation, you know 1 mole of KOH neutralises 1 mole of HNO_3.

So 0.0100 moles of KOH must neutralise 0.0100 moles of HNO_3.

$$\text{Concentration of } HNO_3 = \frac{\text{moles of } HNO_3 \times 1000}{\text{volume}(\text{cm}^3)}$$

$$= \frac{0.0100 \times 1000}{22.0} = 0.455 \text{ mol dm}^{-3} \text{ (3 s.f)}$$

Exam Tip

Keep a close eye on the units in questions like these. You don't want to lose marks just because you didn't check whether you needed the formula for dm³ or cm³ before you started.

Practice Questions — Application

Q1 28 cm^3 of 0.75 mol dm^{-3} hydrochloric acid (HCl) was used to neutralise 40 cm^3 of potassium hydroxide (KOH) solution.

a) Write a balanced equation for this reaction.

b) Calculate the number of moles of HCl used to neutralise the solution.

c) How many moles of KOH were neutralised by the HCl?

d) What was the concentration of the KOH solution?

Q2 15.3 cm^3 of 1.5 mol dm^{-3} sodium hydroxide (NaOH) was used to neutralise 35 cm^3 of nitric acid (HNO_3).

a) Write a balanced equation for this reaction.

b) Calculate the number of moles of NaOH used to neutralise the nitric acid.

c) How many moles of HNO_3 were neutralised by the NaOH?

d) What was the concentration of the HNO_3 solution?

Q3 12 cm^3 of 0.50 mol dm^{-3} HCl solution was used to neutralise 24 cm^3 of KOH solution. What was the concentration of the KOH solution?

Exam Tip
Don't forget to think about numbers of significant figures when doing any calculations. See page 520 for more on this.

Calculating volumes

You can use a similar method to find the volume of acid or alkali that you need to neutralise a solution. You'll need to use this formula again:
number of moles = (concentration × volume (cm^3)) ÷ 1000
but this time rearrange it to find the volume:

$$\text{volume (cm}^3\text{)} = \frac{\text{number of moles} \times 1000}{\text{concentration}}$$

Examples — Maths Skills

20.4 cm^3 of a 0.500 mol dm^{-3} solution of sodium carbonate reacts with 1.50 mol dm^{-3} nitric acid. Calculate the volume of nitric acid required to neutralise the sodium carbonate.

Like before, first write a balanced equation for the reaction and decide what you know and what you want to know:

	Na_2CO_3 +	$2HNO_3 \rightarrow 2NaNO_3 + H_2O + CO_2$
Volume:	*20.4 cm^3*	*?*
Concentration:	*0.500 mol dm^{-3}*	*1.50 mol dm^{-3}*

Now work out how many moles of Na_2CO_3 you've got:

$$\text{Number of moles } Na_2CO_3 = \frac{\text{concentration} \times \text{volume (cm}^3\text{)}}{1000}$$

$$= \frac{0.500 \times 20.4}{1000} = 0.0102 \text{ moles}$$

1 mole of Na_2CO_3 neutralises 2 moles of HNO_3, so 0.0102 moles of Na_2CO_3 neutralises 0.0204 moles of HNO_3.

Now you know the number of moles of HNO_3 and the concentration, you can work out the volume:

$$\text{Volume of } HNO_3 = \frac{\text{number of moles} \times 1000}{\text{concentration}}$$

$$= \frac{0.0204 \times 1000}{1.50} = 13.6 \text{ cm}^3$$

And here's an example where you're finding the volume of alkali used.

Examples — Maths Skills

18.2 cm³ of a 0.800 mol dm⁻³ solution HCl reacts with 0.300 mol dm⁻³ KOH. Calculate the volume of KOH required to neutralise the HCl.

Write out the balanced equation and the information that you have:

$$HCl + KOH \rightarrow KCl + H_2O$$

	HCl	KOH
Volume:	*18.2 cm³*	*?*
Concentration:	*0.800 mol dm⁻³*	*0.300 mol dm⁻³*

Now work out how many moles of HCl you've got:

$$\text{Number of moles HCl} = \frac{\text{concentration} \times \text{volume (cm}^3)}{1000}$$

$$= \frac{0.800 \times 18.2}{1000} = 0.0145\text{... moles}$$

1 mole of HCl neutralises 1 moles of KOH, so 0.0145... moles of HCl neutralises 0.0145... moles of KOH.

Now use this to work out the volume:

$$\text{Volume of KOH} = \frac{\text{number of moles} \times 1000}{\text{concentration}}$$

$$= \frac{0.0145... \times 1000}{0.300} = 48.53... = 48.5 \text{ cm}^3 \text{ (to 3 s.f.)}$$

Figure 4: *A titration where an alkali is being added to an acid. The indicator in the flask is phenolphthalein, so the solution starts clear and turns pink at the endpoint.*

Practice Questions — Application

Q1 18.8 cm³ of a 0.20 mol dm⁻³ solution of nitric acid (HNO_3) reacts with 0.45 mol dm⁻³ potassium hydroxide (KOH) solution.

a) Write a balanced equation for this reaction.

b) Calculate the number of moles of HNO_3 present in the acid added.

c) How many moles of KOH were in the sample of the alkali?

d) What volume of KOH was required to neutralise the HNO_3 solution?

Q2 37.3 cm³ of a 0.420 mol dm⁻³ solution of potassium hydroxide (KOH) reacts with 1.10 mol dm⁻³ ethanoic acid (CH_3COOH) solution.

a) Write a balanced equation for this reaction.

b) Calculate the number of moles of KOH present in the alkali added.

c) How many moles of CH_3COOH were in the sample of the acid?

d) What volume of CH_3COOH was required to neutralise the KOH solution?

Q3 14 cm³ of a 1.5 mol dm⁻³ NaOH solution reacts with 0.60 mol dm⁻³ H_2SO_4 solution. What volume of H_2SO_4 was required to neutralise the NaOH solution?

Practice Questions — Fact Recall

Q1 What is a standard solution?

Q2 Name the piece of equipment that you would use to measure out a set volume of alkali or acid for a titration.

Q3 Name the piece of equipment that you would use to add liquid drop by drop to the flask during a titration.

Q4 What is the 'end point' of a titration?

6. Formulas

Now for a few pages about chemical formulas. A formula tells you what atoms are in a compound. Useful, I think you'll agree.

Learning Objectives:

- Know that the empirical formula is the simplest whole number ratio of atoms of each element in a compound.
- Know that the molecular formula is the actual number of atoms of each element in a compound.
- Understand the relationship between empirical and molecular formulas.
- Be able to calculate a molecular formula from the empirical formula and relative molecular mass.
- Be able to calculate empirical formulas from data giving percentage by mass or composition by mass.

Specification Reference 3.1.2.4

Empirical and molecular formulas

You need to know what's what with empirical and molecular formulas. The **empirical formula** gives the smallest whole number ratio of atoms of each element present in a compound. The **molecular formula** gives the actual numbers of atoms in a molecule. The molecular formula is made up of a whole number of empirical units.

Example

This molecule is butane:

```
   H  H  H  H
   |  |  |  |
H—C—C—C—C—H
   |  |  |  |
   H  H  H  H
```

A butane molecule contains 4 carbon (C) atoms and 10 hydrogen (H) atoms. So its molecular formula is C_4H_{10}.

Butane's empirical formula is C_2H_5. This means that the ratio of carbon atoms to hydrogen atoms in the molecule is 2:5. That's as much as you can simplify it.

Tip: There's more about the molecular an empirical formulas of organic compounds on pages 180-181.

If you know the empirical formula and the relative molecular mass of a compound, you can calculate its molecular formula. Just follow these steps:

1. Find the empirical mass (that's just the relative mass of the empirical formula).
2. Divide the relative molecular mass by the empirical mass. This tells you how many multiples of the empirical formula are in the molecular formula.
3. Multiply the empirical formula by that number to find the molecular formula.

Here are a couple of examples to show you how it works.

Example — Maths Skills

A molecule has an empirical formula of $C_4H_3O_2$, and a relative molecular mass of 166. Work out its molecular formula.

1. Find the empirical mass — add up by the relative atomic mass values of all the atoms in the empirical formula.

$$\text{empirical mass} = (4 \times 12.0) + (3 \times 1.0) + (2 \times 16.0) = 83.0$$

(4 C atoms × A_r of C) + (3 H atoms × A_r of H) + (2 O atoms × A_r of O)

2. Divide the relative molecular mass by the empirical mass. The relative molecular mass is 166, so there are (166 ÷ 83.0) = 2 empirical units in the molecule.
3. The molecular formula is the empirical formula × 2, so the molecular formula = $C_8H_6O_4$.

Example — Maths Skills

The empirical formula of glucose is CH_2O. Its relative molecular mass is 180. Find its molecular formula.

1. Find the empirical mass of glucose.
   empirical mass = $(1 \times 12.0) + (2 \times 1.0) + (1 \times 16.0) = 30.0$
2. Divide the relative molecular mass by the empirical mass. The relative molecular mass is 180, so there are $(180 \div 30.0) = 6$ empirical units in the molecule.
3. Molecular formula = $C_6H_{12}O_6$

Practice Questions — Application

Q1 A molecule has the empirical formula C_4H_9, and a relative molecular mass of 171. Find its molecular formula.

Q2 A molecule has the empirical formula $C_3H_5O_2$, and a relative molecular mass of 146. Find its molecular formula.

Q3 A molecule has the empirical formula C_2H_6O, and a relative molecular mass of 46. Find its molecular formula.

Q4 A molecule has the empirical formula $C_4H_6Cl_2O$, and a relative molecular mass of 423. Find its molecular formula.

Calculating empirical formulas

You need to know how to work out empirical formulas from the percentages of the different elements. Follow these steps each time:

1. Assume you've got 100 g of the compound — you can turn the percentages straight into masses. Then you can work out how many moles of each element are in 100 g of the compound.
2. Divide each number of moles by the smallest number of moles you found in step 1. This gives you the ratio of the elements in the compound.
3. Apply the numbers from the ratio to the formula.

Example — Maths Skills

A compound is found to have percentage composition 56.5% potassium, 8.70% carbon and 34.8% oxygen by mass.
Calculate its empirical formula.

1. If you had 100 g of the compound you would have 56.5 g of potassium, 8.70 g of carbon and 34.8 g of oxygen. Use the formula moles = mass ÷ A_r to work out how many moles of each element that is.

   K: $\frac{56.5}{39.1} = 1.45$ moles C: $\frac{8.70}{12.0} = 0.725$ moles O: $\frac{34.8}{16.0} = 2.18$ moles

2. Divide each number of moles by the smallest number (0.725 here).

   K: $\frac{1.45}{0.725} = 2.0$ C: $\frac{0.725}{0.725} = 1.0$ O: $\frac{2.18}{0.725} = 3.0$

   This tells you that the ratio of K : C : O in the molecule is 2 : 1 : 3.

3. So you know the empirical formula's got to be K_2CO_3.

Tip: The formula for the number of moles is a bit different here — when you're dealing with elements, you need to use A_r instead of M_r.

Exam Tip
Make sure you write down all your working for calculation questions. You'll be more likely to spot any mistakes and if you do go wrong you might get some marks for the working.

Sometimes you might only be given the percentage of some of the elements in the compound. Then you'll have to work out the percentages of the others.

Examples — Maths Skills

An oxide of nitrogen contains 26.0% by mass of nitrogen. Calculate its empirical formula.

1. The compound only contains nitrogen and oxygen, so if it is 26.0% N it must be 100 – 26.0 = 74.0% O. So if you had 100 g of the compound you would have 26.0 g of nitrogen and 74.0 g of oxygen.

   N: $\frac{26.0}{14.0}$ = 1.86 moles   O: $\frac{74.0}{16.0}$ = 4.63 moles

2. Divide each number of moles by 1.86.

   N: $\frac{1.86}{1.86}$ = 1.0   O: $\frac{4.63}{1.86}$ = 2.5

   This tells you that the ratio of N : O in the molecule is 1 : 2.5.

3. All the numbers in an empirical formula have to be whole numbers, so you need to multiply the ratio by 2 to put it into its simplest whole number form: 2 × (1 : 2.5) = 2 : 5.

   So the empirical formula is N_2O_5.

You need to be able to work out empirical formulas from experimental results too. In a way this is easier than by percentages, as you already know the masses, however you have to think a bit more about what is actually happening in the experiment.

Tip: When hydrocarbons undergo complete combustion, the only products are carbon dioxide and water. If they undergo incomplete combustion (combustion in limited oxygen) then other products such as carbon monoxide and soot may form. See page 206 for more on this.

Examples — Maths Skills

When a hydrocarbon is burnt in excess oxygen, 4.40 g of carbon dioxide and 1.80 g of water are made. What is the empirical formula of the hydrocarbon?

1. First work out how many moles of the products you have:

   No. of moles of CO_2 = $\frac{\text{mass}}{M_r} = \frac{4.40}{12.0 + (16.0 \times 2)} = \frac{4.40}{44.0}$ = 0.100 moles

   No. of moles of H_2O = $\frac{\text{mass}}{M_r} = \frac{1.80}{(2 \times 1.0) + 16.0} = \frac{1.80}{18.0}$ = 0.100 moles

2. All the hydrogen and carbon in the carbon dioxide and water must have come from the hydrocarbon. This means we can use the number of moles of carbon dioxide and water to work out how many moles of hydrogen and carbon atoms were in the hydrocarbon:

   1 mole of CO_2 contains 1 mole of carbon atoms, so the original hydrocarbon must have contained 0.100 moles of carbon atoms.

   1 mole of H_2O contains 2 moles of hydrogen atoms, so the original hydrocarbon must have contained 0.200 moles of hydrogen atoms.

3. Divide each number of moles by the smallest number (0.100).

   C: $\frac{0.100}{0.100}$ = 1   H: $\frac{0.200}{0.100}$ = 2

   This tells you that the ratio of C : H in the hydrocarbon is 1 : 2.

4. So the empirical formula must be CH_2.

Example — Maths Skills

2.50 g of copper is heated to form 3.13 g of copper oxide. What is the empirical formula of the oxide?

1. First work out the mass of oxygen which reacted with the copper to form the copper oxide: 3.13 – 2.50 = 0.630 g of oxygen.
2. Then work out the number of moles of each reactant:

   $$\text{No. of moles of Cu} = \frac{\text{mass}}{A_r} = \frac{2.50}{63.5} = 0.0394 \text{ moles}$$

   $$\text{No. of moles of O} = \frac{\text{mass}}{A_r} = \frac{0.630}{16.0} = 0.0394 \text{ moles}$$

3. Divide each number of moles by the smallest number (0.0394).

   $$\text{Cu: } \frac{0.0394}{0.0394} = 1 \qquad \text{O: } \frac{0.0394}{0.0394} = 1$$

   This tells use that the ratio of Cu : O in the oxide is 1 : 1.
4. So the empirical formula must be CuO.

Tip: Here you don't really need to divide each number of moles by the smallest number — you can tell the ratio will be 1 : 1 as the number of moles of copper is the same as the number of moles of oxygen.

Practice Questions — Application

Q1 A compound is found to have percentage composition 5.9% hydrogen and 94.1% oxygen by mass. Find its empirical formula.

Q2 A compound is found to have percentage composition 20.2% aluminium and 79.8% chlorine by mass. Find its empirical formula.

Q3 A compound is found to have percentage composition 8.5% carbon, 1.4% hydrogen and 90.1% iodine by mass. Find its empirical formula.

Q4 A compound is found to have percentage composition 50.1% copper, 16.3% phosphorus and 33.6% oxygen by mass. Find its empirical formula.

Q5 A compound containing only vanadium and chlorine is found to be 32.3% vanadium by mass. Find its empirical formula.

Q6 An oxide of chromium contains 31.58% by mass of oxygen. Find its empirical formula.

Q7 2.00 g of phosphorus was burnt to form 4.58 g of phosphorus oxide. Find the empirical formula of the phosphorus oxide.

Q8 0.503 g of silver reacts with chlorine to form 0.669 g of silver chloride. Find the empirical formula of the silver chloride.

Q9 A hydrocarbon is burnt to form 9.70 g of CO_2 and 7.92 g of H_2O. Find the empirical formula of the hydrocarbon.

Tip: You should add up the percentages each time there's a question like this to make sure they add up to 100% and you haven't missed out any elements.

Practice Questions — Fact Recall

Q1 What information does the empirical formula of a compound give you?

Q2 What information does the molecular formula of a compound give you?

7. Chemical Yield

Learning Objective:

- Be able to calculate percentage yields from balanced equations.

Specification Reference 3.1.2.5

If you're making a chemical (in a lab or a factory), it helps to know how much of it you can expect to get. In real life you'll never manage to make exactly that much — but percentage yield can give you an idea of how close you got.

Calculating theoretical yield

The **theoretical yield** is the mass of product that should be formed in a chemical reaction. It assumes no chemicals are 'lost' in the process.
You can use the masses of reactants and a balanced equation to calculate the theoretical yield for a reaction. It's a bit like calculating reacting masses (see page 54) — here are the steps you have to go through:

1. Work out how many moles of the reactant you have.
2. Use the equation to work out how many moles of product you would expect that much reactant to make.
3. Calculate the mass of that many moles of product — and that's the theoretical yield.

Tip: 'Hydrated' means that the crystals have a bit of water left in them. That's what the dot in $(NH_4)_2Fe(SO_4)_2{\cdot}6H_2O$ means — for every mole of $(NH_4)_2Fe(SO_4)_2$ that the salt contains, it also contains 6 moles of water.

Example — Maths Skills

1.40 g of iron filings react with ammonia and sulfuric acid to make hydrated ammonium iron(II) sulfate. The balanced equation for the reaction is:

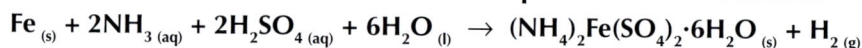

$$Fe_{(s)} + 2NH_{3\,(aq)} + 2H_2SO_{4\,(aq)} + 6H_2O_{(l)} \rightarrow (NH_4)_2Fe(SO_4)_2{\cdot}6H_2O_{(s)} + H_{2\,(g)}$$

Calculate the theoretical yield of this reaction.

1. Work out how many moles of iron you have:
   M_r of Fe = 55.8
   Number of moles Fe = mass ÷ M_r = 1.40 ÷ 55.8 = 0.0251
2. Work out how many moles of product you would expect to make:
   From the equation, you know that 1 mole of Fe produces 1 mole of $(NH_4)_2Fe(SO_4)_2{\cdot}6H_2O$ so 0.0251 moles of Fe will produce 0.0251 moles of $(NH_4)_2Fe(SO_4)_2{\cdot}6H_2O$.
3. Now calculate the mass of that many moles of product:
   M_r of $(NH_4)_2Fe(SO_4)_2{\cdot}6H_2O$ = $[2 \times (14.0 + (4 \times 1.0))] + 55.8$
   $+ [2 \times (32.1 + (4 \times 16.0))] + [6 \times ((2 \times 1.00) + 16.0)]$
   $= 392.0$

   Theoretical yield = number of moles × M_r = 0.0251 × 392.0 = 9.84 g.

Figure 1: A student pouring and filtering a solution. Some chemicals will be left on the glassware and some will be left on the filter paper.

Calculating percentage yield

For any reaction, the actual mass of product (the actual yield) will always be less than the theoretical yield. There are many reasons for this. For example, sometimes not all the starting chemicals react fully. And some chemicals are always 'lost', e.g. some solution gets left on filter paper, is lost during transfers between containers, or forms other products you don't want in side reactions.

Once you've found the theoretical yield and the actual yield, you can work out the **percentage yield**.

$$\text{Percentage Yield} = \frac{\text{Actual Yield}}{\text{Theoretical Yield}} \times 100$$

Examples — Maths Skills

In the ammonium iron(II) sulfate example on the previous page, the theoretical yield was 9.84 g. Say you weighed the hydrated ammonium iron(II) sulfate crystals produced and found the actual yield was 5.2 g. Calculate the percentage yield of this reaction.

Then you just have to plug the numbers into the formula:

$$\text{Percentage yield} = \frac{\text{actual yield}}{\text{theoretical yield}} \times 100$$

$$= (5.2 \div 9.84) \times 100 = 53\%$$

Here's another example:

In an experiment 5.00 g of copper was heated in air to produce copper oxide. The theoretical yield of this reaction was 6.26 g. When the copper oxide was weighed it was found to have a mass of 5.23 g.

Calculate the percentage yield of this reaction.

All you need to do here is put the right numbers into the formula:

$$\text{Percentage yield} = \frac{\text{actual yield}}{\text{theoretical yield}} \times 100$$

$$= (5.23 \div 6.26) \times 100 = 83.5\%$$

Exam Tip

This is a percentage yield, so it can never be more than 100%. If your answer is bigger than 100%, check the working for mistakes.

If you're not given the theoretical yield, you need to work this out first before you can calculate the percentage yield.

Example — Maths Skills

0.475 g of CH_3Br reacts with excess NaOH in the following reaction:

$$CH_3Br + NaOH \rightarrow CH_3OH + NaBr$$

0.153 g of CH_3OH is produced. What is the percentage yield?

1. Work out how many moles of CH_3Br you have:
   M_r of $CH_3Br = 12.0 + (3 \times 1.0) + 79.9 = 94.9$
   Number of moles CH_3Br = mass $\div M_r$
   $= 0.475 \div 94.9 = 0.00501...$ moles.
2. Work out how many moles of product you would expect to make:
   From the equation, you know that 1 mole of CH_3Br produces 1 mole of CH_3OH so 0.00501... moles of CH_3Br will produce 0.00501... moles of CH_3OH.
3. Now calculate the mass of that many moles of product:
   M_r of $CH_3OH = 12.0 + (4 \times 1.0) + 16.0 = 32.0$
   Theoretical yield = number of moles $\times M_r$
   $= 0.00501... \times 32.0 = 0.160...$ g.
4. Now put these numbers into the percentage yield formula:

$$\text{Percentage yield} = \frac{\text{actual yield}}{\text{theoretical yield}} \times 100$$

$$= (0.153... \div 0.160...) \times 100 = 95.5\%$$

Practice Questions — Application

Q1 The theoretical yield of a reaction used in an experiment was 3.24 g. The actual yield was 1.76 g. Calculate the percentage yield of the reaction.

Q2 In an experiment sodium metal was reacted with chlorine gas to produce sodium chloride. The theoretical yield of this reaction was 6.1 g. The sodium chloride produced had a mass of 3.7 g. Calculate the percentage yield of this reaction.

Q3 Hydrogen reacts with oxygen to produce 138 g of water. The theoretical yield of this reaction was 143 g. Calculate the percentage yield of this reaction.

Q4 3.0 g of iron filings are burnt in air to give iron oxide (Fe_2O_3):

$$4Fe_{(s)} + 3O_{2\,(g)} \rightarrow 2Fe_2O_{3\,(s)}$$

a) How many moles of iron are there in 3.0 g of metal?

b) Calculate the theoretical yield of iron oxide for this reaction.

c) Calculate the percentage yield if 3.6 g of Fe_2O_3 is made.

Q5 Aluminium metal can be extracted from aluminium oxide by electrolysis. The balanced equation for this reaction is:

$$2Al_2O_{3\,(l)} \rightarrow 4Al_{(l)} + 3O_{2\,(g)}$$

What mass of aluminium would you expect to get from 1000 g of Al_2O_3?

Q6 4.70 g of sodium hydroxide is dissolved in water. This solution is reacted with an excess of sulfuric acid to make sodium sulfate:

$$2NaOH_{(aq)} + H_2SO_{4\,(g)} \rightarrow Na_2SO_{4\,(aq)} + H_2O_{(l)}$$

The sodium sulfate is allowed to crystallise. The dry crystals have a mass of 6.04 g. Calculate the percentage yield of this reaction.

Q7 In an experiment 40.0 g of magnesium was reacted with an excess of nitric acid. The balanced equation for this reaction is:

$$Mg_{(s)} + 2HNO_{3\,(aq)} \rightarrow Mg(NO_3)_{2\,(aq)} + H_{2\,(g)}$$

If 1.70 g of hydrogen was produced, what was the percentage yield of the reaction?

Exam Tip
If you're asked to calculate yields in an exam, make sure you've written down a balanced equation to work from.

Practice Questions — Fact Recall

Q1 What is meant by the 'theoretical yield' of a reaction?

Q2 Write down the formula for percentage yield.

8. Atom Economy

Atom economy is one way to work out how efficient a reaction is. Efficient reactions are better for the environment and save the chemical industry money.

Learning Objectives:

- Know the economic, ethical and environmental advantages for society and for industry of developing chemical processes with a high atom economy.
- Know the formula for calculating percentage atom economy.
- Be able to calculate percentage atom economies from balanced equations.

Specification Reference 3.1.2.5

What is atom economy?

In the previous topic you met the idea of percentage yield. Percentage yield can give you useful information about how wasteful a process is. It's based on how much of the product is lost because of things like reactions not completing or losses during collection and purification.

But percentage yield doesn't measure how wasteful the reaction itself is. A reaction that has a 100% yield could still be very wasteful if a lot of the atoms from the reactants wind up in by-products rather than the desired product. **Atom economy** is a measure of the proportion of reactant atoms that become part of the desired product (rather than by-products) in the balanced chemical equation.

Advantages of high atom economy

Companies in the chemical industry will often choose to use reactions with high atom economies. Reactions with high atom economies have environmental, economic and ethical benefits.

Economic advantages

A company using a process that has a high atom economy will make more efficient use of its raw materials. A process with low atom economy will involve using large quantities of raw materials to make relatively small amounts of the desired product and lots of by-products. It's a waste of money if a high proportion of the reactant chemicals they buy end up as useless by-products.

Using a process with a high atom economy also means that the company will have less waste to deal with. This means they will spend less on separating the desired product from the waste products. Any waste products that are produced need to be disposed of safely, which increases costs further.

Figure 1: *Tablets of the painkiller ibuprofen. Ibuprofen was originally made using a reaction with a 40% atom economy. Now a new way of making it with a 77% atom economy is used. This produces much less waste.*

Environmental and ethical advantages

Processes that use fewer raw materials and produce less waste are better for the environment as well as for business. Many raw materials are in limited supply, so it makes sense to use them efficiently so they last as long as possible. Producing less waste is better for the environment as waste chemicals are often harmful to the environment. It can be difficult to dispose of them in a way that minimises their harmful effects.

Using high atom economy processes tends to be more **sustainable** than using low atom economy processes. Doing something sustainably means using up as little of the Earth's resources as you can and not putting loads of damaging chemicals into the environment — in other words not messing things up for the future.

It can also be good for society if chemical companies can find easier, cheaper ways to mass produce medicines and other useful chemicals, as this may mean the products can be sold for lower prices and be made available to more people.

Calculating atom economy

Atom economy is calculated using this formula:

$$\% \text{ atom economy} = \frac{\text{molecular mass of desired product}}{\text{sum of molecular masses of all reactants}} \times 100$$

Tip: Any reaction where there's only one product will have a 100% atom economy.

To calculate the atom economy for a reaction, you just need to add up the molecular masses of the reactants, find the molecular mass of the product you're interested in and put them both into the formula.

Example — Maths Skills

Bromomethane is reacted with sodium hydroxide to make methanol:

$$CH_3Br + NaOH \rightarrow CH_3OH + NaBr$$

Calculate the percentage atom economy for this reaction.

First, calculate the total mass of the reactants — add up the relative molecular masses of everything on the left side of the balanced equation:

Total mass = (12.0 + (3 × 1.0) + 79.9) + (23.0 + 16.0 + 1.0) = 134.9

Then find the mass of the desired product — that's the methanol:

Mass of desired product = 12.0 + (3 × 1.0) + 16.0 + 1.0 = 32.0

Now you can find the % atom economy:

$$\% \text{ atom economy} = \frac{\text{molecular mass of desired product}}{\text{sum of molecular masses of all reactants}} \times 100$$

$$= \frac{32.0}{134.9} \times 100 = 23.7\%$$

Exam Tip
You should always calculate atom economy from a balanced equation, so the first thing you do when you start a question like this is to make sure you've got one written down.

When you calculate the masses, you should use the number of moles of each compound that is in the balanced equation (e.g. the mass of '$2H_2$' should be 2 × (2 × 1.0) = 4.0). Here's a quick example:

Example — Maths Skills

Ethanol can be produced by fermenting glucose, $C_6H_{12}O_6$:

$$C_6H_{12}O_6 \rightarrow 2C_2H_5OH + 2CO_2$$

Calculate the percentage atom economy for this reaction.

Calculate the total mass of the reactants (1 mole of glucose):

Total mass of reactants = (6 × 12.0) + (12 × 1.0) + (6 × 16.0) = 180.0

Then find the mass of the desired product (2 moles of ethanol):

Mass of desired product = 2 × ((12.0 × 2) + (5 × 1.0) + 16.0 + 1.0) = 92.0

$$\text{So, } \% \text{ atom economy} = \frac{\text{molecular mass of desired product}}{\text{sum of molecular masses of all reactants}} \times 100$$

$$= \frac{92.0}{180.0} \times 100 = 51.1\%$$

Practice Questions — Application

Q1 Chlorine gas can react with excess methane to make chloromethane:

$$CH_4 + Cl_2 \rightarrow CH_3Cl + HCl$$

a) Find the total molecular mass of the reactants in this reaction.

b) Find the molecular mass of the chloromethane produced.

c) Calculate the percentage atom economy of this reaction.

d) A company wants to use this reaction to make chloromethane, despite its low atom economy. Suggest one way that they could increase their profit and reduce the waste they produce.

Q2 Aluminium chloride can be produced using this reaction:

$$2Al + 3Cl_2 \rightarrow 2AlCl_3$$

Calculate the percentage atom economy of this reaction.

Q3 Pure iron can be produced from iron oxide in the blast furnace:

$$2Fe_2O_3 + 3C \rightarrow 4Fe + 3CO_2$$

Calculate the percentage atom economy of this reaction.

Q4 In industry, ammonia (NH_3) is usually produced using this reaction:

Reaction 1: $N_2 + 3H_2 \rightarrow 2NH_3$

It can also be made using this reaction:

Reaction 2: $2NH_4Cl + Ca(OH)_2 \rightarrow CaCl_2 + 2NH_3 + 2H_2O$

a) Calculate the percentage atom economy of both reactions.

b) Give one reason why reaction 1 is used to produce ammonia industrially rather than reaction 2.

Exam Tip

To save time on calculations, make sure you check how many products there are in the reaction first. If there's only one, you already know the atom economy will be 100%.

Practice Questions — Fact Recall

Q1 What is meant by the 'atom economy' of a reaction?

Q2 What are the economic and environmental advantages of using a process with a high atom economy?

Q3 Write down the formula for calculating % atom economy.

Section Summary

Make sure you know...

- What a mole is and how many particles there are in one.
- What the Avogadro constant represents.
- How to calculate the number of particles in a substance using the Avogadro constant.
- How to find the number of moles, mass or relative molecular mass of a substance using the equation 'number of moles = mass of substance ÷ M_r'.
- How to find the number of moles of a substance in a solution, or its concentration or volume, using the concentration equation.
- What the ideal gas equation is.
- The standard units of all the values in the ideal gas equation.
- How to convert units of temperature, pressure and volume into the correct units for the ideal gas equation.
- How to use the ideal gas equation to calculate the pressure, volume, number of moles or temperature of a gas.
- How to write and balance full and ionic equations for reactions.
- How to calculate the mass of a reactant or a product from a balanced equation.
- How to calculate the volume of gas produced by a reaction from a balance equation.
- The four state symbols used in equations.
- How to make up a standard solution.
- How to perform an accurate titration.
- How to calculate the concentration of an acid or alkali from the results of a titration.
- How to calculate the volume of acid needed to neutralise an alkali (and vice versa).
- What an empirical formula and a molecular formula are.
- How to find the molecular formula of a compound from its empirical formula and molecular mass.
- How to find the empirical formula of a compound from its percentage or composition by mass.
- What the ethical, economic and environmental advantages of high atom economy processes are.
- How to calculate the percentage yield of a reaction.
- How to calculate the atom economy of a reaction.

Exam-style Questions

1 Which of these contains the largest number of particles?

A 1 mole of C_2H_6

B 3 moles of N

C 2 moles of CO_2

D 2 moles of Ag^+ *(1 mark)*

2 5.00 g of CO reacts with excess O_2 to produce 6.40 g of CO_2.
What is the percentage yield of CO_2?

$$2CO_{(g)} + O_{2\,(g)} \rightarrow 2CO_{2\,(g)}$$

A 72.3%

B 79.4%

C 81.5%

D 83.5% *(1 mark)*

3 Chlorine gas (Cl_2) is used in water treatment and in the production of plastics.
It is usually produced by the electrolysis of sodium chloride solution:

$$2NaCl_{(aq)} + 2H_2O_{(l)} \rightarrow Cl_{2\,(g)} + H_{2\,(g)} + 2NaOH_{(aq)}$$

3.1 20.0 g of NaCl was dissolved in an excess of water, and the resulting solution was electrolysed.
Calculate the amount, in moles, of NaCl in 20.0 g.

(2 marks)

3.2 Calculate the amount, in moles, of Cl_2 gas that would be produced from a 20.0 g sample of NaCl.

(1 mark)

3.3 In another experiment a different sample of sodium chloride solution was electrolysed to produce 0.65 moles of chlorine gas.
Calculate the volume in m^3 that this gas would occupy at 330 K and 98 kPa.
(The gas constant, $R = 8.31\ J\ K^{-1}\ mol^{-1}$.)

(3 marks)

3.4 Calculate the percentage atom economy for the formation of Cl_2 gas from the electrolysis of NaCl solution.

(2 marks)

3.5 This reaction has a low atom economy. There are other ways of producing chlorine that have a higher atom economy. Given that this is the case, suggest why most chemical companies still use this reaction to produce chlorine.

(1 mark)

4 Phosphoric acid (H_3PO_4) is made by dissolving oxides of phosphorus in water.

4.1 An oxide of phosphorus contains 43.6% oxygen by mass.
Find the empirical formula of this oxide.

(3 marks)

4.2 Given that the molecular mass of the oxide in **4.1** is 220,
find its molecular formula.

(2 marks)

4.3 Diammonium phosphate $(NH_4)_2HPO_4$ can be used as a fertiliser.
It can be made from ammonia and phosphoric acid according to this equation:

$$2NH_{3\,(g)} + H_3PO_{4\,(aq)} \rightarrow (NH_4)_2HPO_{4\,(s)}$$

Calculate the amount, in moles, of NH_3 in 2.50 g of ammonia gas.

(2 marks)

4.4 Calculate the amount, in moles, of $(NH_4)_2HPO_4$ that would be produced from 2.50 g of ammonia gas. (Assume the phosphoric acid was present in excess.)

(1 mark)

4.5 Calculate the mass of $(NH_4)_2HPO_4$ that would be
produced from 2.50 g of ammonia gas.

(2 marks)

5 A 3.40 g piece of calcium metal was burned in air to make calcium oxide.
Here is the balanced equation for this reaction:

$$2Ca_{\,(s)} + O_{2\,(g)} \rightarrow 2CaO_{\,(s)}$$

5.1 State the atom economy of this reaction.
Explain your answer.

(2 marks)

5.2 Calculate the amount, in moles, of Ca in 3.40 g of calcium metal.

(2 marks)

5.3 Calculate the maximum amount, in moles, of CaO that could
be produced by burning the piece of calcium metal.

(1 mark)

5.4 Calculate the maximum mass of CaO that could be produced
by burning this piece of calcium metal.

(2 marks)

5.5 The actual mass of CaO produced was 3.70 g.
Calculate the percentage yield of CaO.

(1 mark)

6 Octane is a hydrocarbon with the molecular formula C_8H_{18}.
It can be used as a fuel in cars, and is a liquid at room temperature.

6.1 Octane burns completely in air to give water and carbon dioxide.
Write a balanced equation for this reaction, including state symbols.

(2 marks)

6.2 A sample of octane is burnt in air. The amount of carbon dioxide produced occupies a volume of 0.020 m^3 at 308 K and 101 000 Pa.

Calculate the number of moles of CO_2 present in 0.020 m^3 of carbon dioxide at this temperature and pressure. (The gas constant, R = 8.31 J K^{-1} mol^{-1})

(2 marks)

6.3 How many moles of octane were burnt to produce the amount of CO_2 in **6.2**?

(1 mark)

6.4 A different hydrocarbon contains 85.7% carbon by mass.
Find the empirical formula of this hydrocarbon.

(3 marks)

7 A 20 cm^3 sample of hydrochloric acid was titrated with 0.50 mol dm^{-3} potassium hydroxide in order to determine its concentration.

The acid and the base reacted according to the following equation:

$$HCl_{(aq)} + KOH_{(aq)} \rightarrow KCl_{(aq)} + H_2O_{(l)}$$

7.1 Calculate the mass of pure solid potassium hydroxide that you would need to dissolve in 150 cm^3 of water to make a 0.50 mol dm^{-3} solution.
Give your answer to two significant figures.

(3 marks)

7.2 The results of the titration experiment are shown in the table below.

Titration	1	2	3	4
Volume of KOH solution added (cm^3)	26.00	26.05	28.30	26.00

Use this data to calculate the concentration of the hydrochloric acid solution that was used in the titration.

(4 marks)

7.3 Hydrogen chloride gas ($HCl_{(g)}$) is a hydrogen halide.
Hydrochloric acid ($HCl_{(aq)}$) can be made by dissolving this gas in water.

A different acid was made by dissolving 3.07 g of a mystery hydrogen halide, HX, in pure water. This produced a 0.20 dm^3 sample of the acid.

This sample of acid was then titrated with KOH and found to have a concentration of 0.12 mol dm^{-3}.

Find the relative atomic mass of the halogen, X, and suggest its identity.

(6 marks)

Learning Objectives:

- Be able to predict the charge on a simple ion using the position of the element in the Periodic Table.
- Know the formulas of compound ions, e.g. sulfate, hydroxide, nitrate, carbonate and ammonium.
- Know that ionic bonding involves electrostatic attraction between oppositely charged ions in a lattice.
- Be able to construct formulas for ionic compounds.
- Know that the ionic structure is one of the four types of crystal structure.
- Know that the structure of sodium chloride is an example of an ionic crystal structure.
- Be able to draw diagrams to represent the structure of giant ionic lattices.
- Be able to relate the melting point and conductivity of ionic compounds to their structure and bonding.

Specification Reference 3.1.3.1, 3.1.3.4

Tip: The notation, e.g. '2, 8, 7' shows the electron configuration of chlorine. There's more about how to work out electron configurations on page 34.

1. Ionic Bonding

When atoms join together, you get a compound. There are two main types of bonding in compounds — ionic and covalent. First up is ionic bonding.

Ions

Ions are formed when electrons are transferred from one atom to another. The simplest ions are single atoms which have either lost or gained 1, 2 or 3 electrons so that they've got a full outer shell.

Examples

The sodium ion

A sodium atom (Na) loses 1 electron to form a sodium ion (Na^+) — see Figure 1.

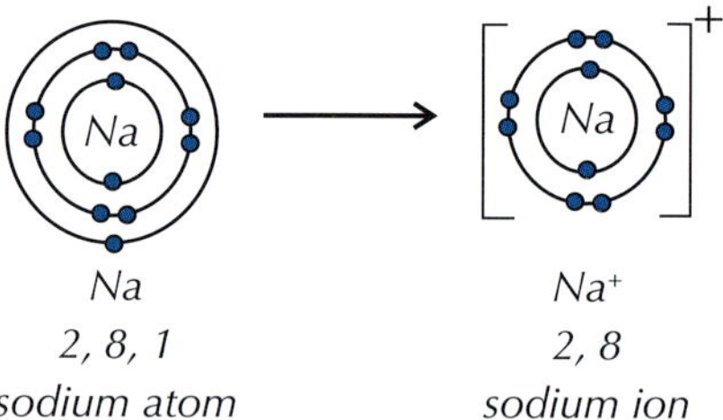

Figure 1: *Formation of a sodium ion.*

This can be shown by the equation: $\mathbf{Na \rightarrow Na^+ + e^-}$.

The chloride ion

A chlorine atom (Cl) gains 1 electron to form a chloride ion (Cl^-) — see Figure 2.

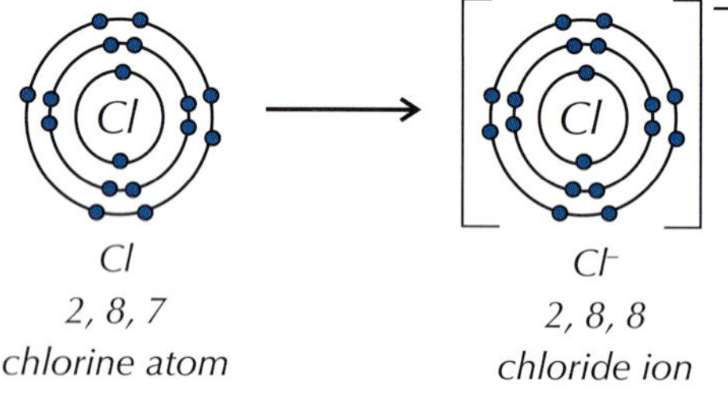

Figure 2: *Formation of a chloride ion.*

This can be shown by the equation: $\mathbf{Cl + e^- \rightarrow Cl^-}$.

Other examples:

A magnesium atom (Mg) loses 2 electrons to form a magnesium ion (Mg^{2+}), shown by the equation: $\mathbf{Mg \rightarrow Mg^{2+} + 2e^-}$.

An oxygen atom (O) gains 2 electrons to form an oxide ion (O^{2-}), shown by the equation: $\mathbf{O + 2e^- \rightarrow O^{2-}}$.

You don't have to remember what ion each element forms — nope, for many of them you just look at the periodic table. Elements in the same group all have the same number of outer electrons. So they have to lose or gain the same number to get the full outer shell that they're aiming for. And this means that they form ions with the same charges. Figure 3 shows the ions formed by the elements in different groups.

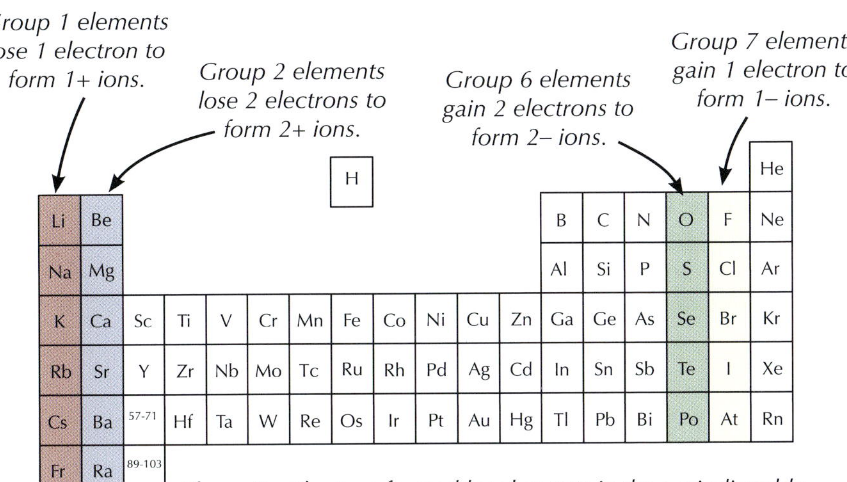

Figure 3: *The ions formed by elements in the periodic table.*

Tip: Metals tend to form ions with a charge that is the same as their group number. Non-metals tend to form ions with a charge that is their group number minus eight.

Exam Tip
Make sure you know how to find the charges on different ions from the periodic table — you never know when it'll come in handy in the exam.

There are lots of ions that are made up of groups of atoms with an overall charge. These are called **compound ions**. Figure 4 gives the formulas of some common compound ions that you need to know for your exam.

Compound ion	Ionic formula
Ammonium	NH_4^+
Carbonate	CO_3^{2-}
Hydroxide	OH^-
Nitrate	NO_3^-
Sulfate	SO_4^{2-}

Figure 4: *The names and formulas of some common compound ions.*

Tip: Positive ions can also be called 'cations', and negative ions can be called 'anions'.

Ionic compounds

Electrostatic attraction holds positive and negative ions together — it's very strong. When atoms are held together in a lattice like this, it's called ionic bonding. When oppositely charged ions come together and form ionic bonds you get an ionic compound.

Examples

Sodium chloride

The formula of sodium chloride is NaCl. Each sodium atom loses an electron to form an Na^+ ion, and each chlorine atom gains an electron to form a Cl^- ion — see Figure 6.

So sodium chloride is made up of Na^+ ions and Cl^- ions held together by electrostatic attraction in a 1:1 ratio. The single positive charge on the Na^+ ion balances the single negative charge on the Cl^- ion so the compound is neutral overall.

Figure 5: *The reaction between sodium and chlorine to form sodium chloride.*

Tip: The diagram on the right is a 'dot-and-cross' diagram of sodium chloride. The dots represent the electrons from sodium, and the crosses show the electrons that come from chlorine. Dot-and-cross diagrams are a great way of showing how ions form, as they clearly show where the electrons move from and to.

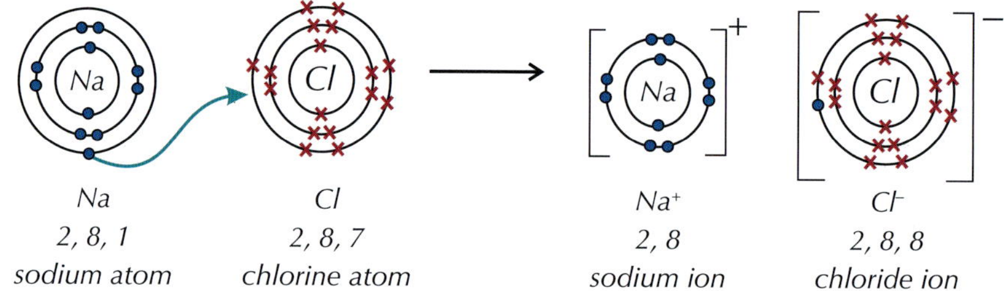

Figure 6: *Formation of sodium chloride from a sodium atom and a chlorine atom.*

Magnesium oxide

Magnesium oxide, MgO, is another example of an ionic compound. The formation of magnesium oxide involves the transfer of two electrons — each magnesium atom loses two electrons to form an Mg^{2+} ion, whilst each oxygen atom gains two electrons to form an O^{2-} ion — see Figure 7.

So as you can see from the formula, magnesium oxide is made up of Mg^{2+} ions and O^{2-} ions in a 1:1 ratio.

Tip: The numbers outside the brackets tells you the charge on each ion in the compound.

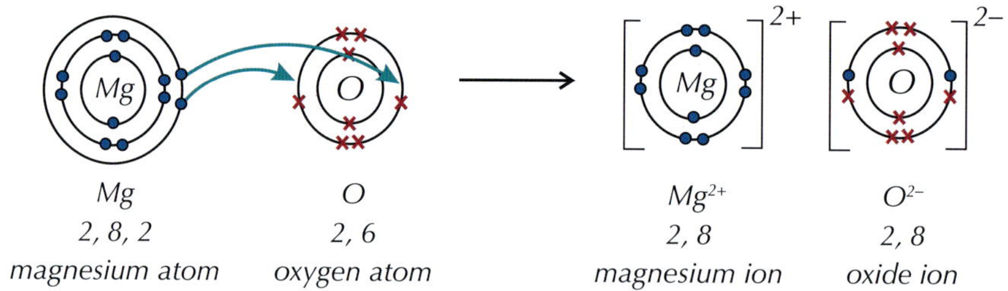

Figure 7: *Formation of magnesium oxide from a magnesium atom and an oxygen atom.*

Magnesium chloride

Magnesium chloride, $MgCl_2$, is yet another ionic compound. When it forms, each magnesium atom loses two electrons to form an Mg^{2+} ion, whilst each chlorine atom gains one electron to form a Cl^- ion — see Figure 8.

So each magnesium atom donates its two outer electrons to two chlorine atoms, resulting in the formula being $MgCl_2$.

Tip: These diagrams only show the ions given in the ionic formula, but don't forget that ionic compounds are actually made up of loads of these basic units.

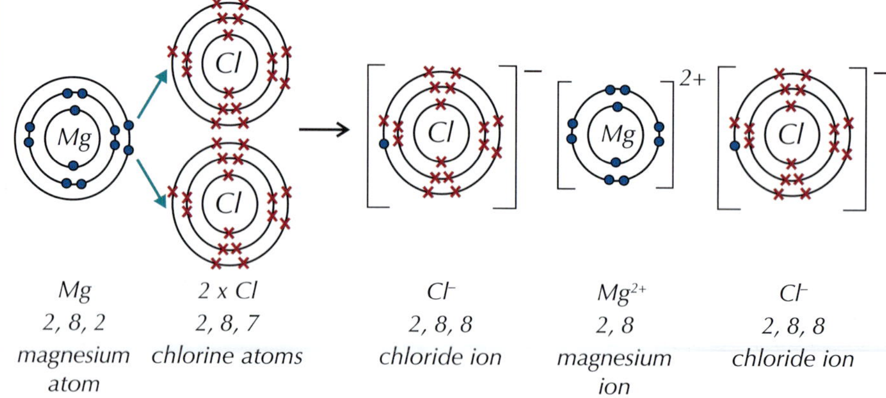

Figure 8: *Formation of magnesium chloride from a magnesium atom and two chlorine atoms.*

The formula of a compound tells you what ions the compound has in it. The positive charges in the compound balance the negative charges exactly — so the total overall charge is zero. This is a dead handy way of working out the formula of an ionic compound, if you know what ions it contains.

Examples

Sodium nitrate

Sodium nitrate contains Na^+ (+1) and NO_3^- (–1) ions.

The charges are balanced with one of each ion, so the formula of sodium nitrate is $NaNO_3$.

Calcium chloride

Calcium chloride contains Ca^{2+} (+2) and Cl^- (–1) ions.

Chloride ions only have a –1 charge so two of them are needed to balance out the +2 charge of a calcium ion. This gives the formula $CaCl_2$.

Exam Tip

If you're trying to work out the formula of an ionic compound in the exam, always make sure your charges balance.

Giant ionic lattices

Ionic crystals are **giant lattices** of ions. A lattice is just a regular structure. The structure's called 'giant' because it's made up of the same basic unit repeated over and over again. In sodium chloride, the Na^+ and Cl^- ions are packed together. Sodium chloride is an example of a compound with an ionic crystal structure. The sodium chloride lattice is cube shaped (see Figure 10).

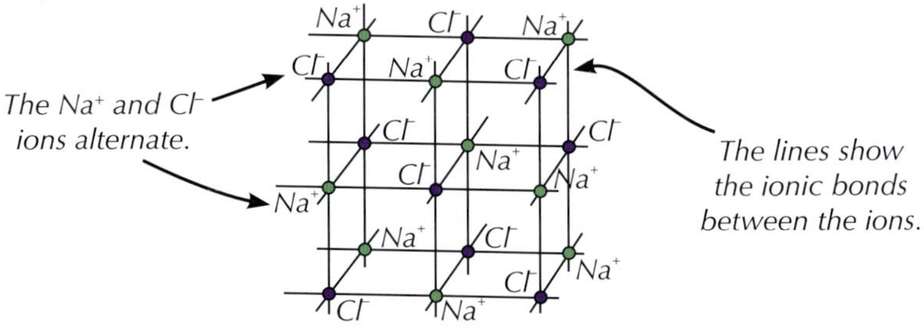
The Na⁺ and Cl⁻ ions alternate.
The lines show the ionic bonds between the ions.
Na⁺
Cl⁻

Figure 10: *The structure of sodium chloride.*

Figure 9: *Crystals of table salt (sodium chloride).*

Tip: Ionic lattices are a type of crystal structure — a regular arrangement of atoms in a solid (or liquid) lattice.

Different ionic compounds have different shaped structures, but they're all still giant lattices.

Behaviour of ionic compounds

The structure of ionic compounds decides their physical properties — things like their electrical conductivity, melting point and solubility.

Electrical conductivity

Ionic compounds conduct electricity when they're molten or dissolved — but not when they're solid. The ions in a liquid are free to move (and they carry a charge). In a solid they're fixed in position by the strong ionic bonds.

Melting point

Ionic compounds have high melting points. The giant ionic lattices are held together by strong electrostatic forces. It takes loads of energy to overcome these forces, so melting points are very high (801 °C for sodium chloride).

Solubility

Ionic compounds tend to dissolve in water. Water molecules are polar — part of the molecule has a small negative charge, and the other bits have small positive charges (see pages 92-93). The water molecules pull the ions away from the lattice and cause it to dissolve.

Exam Tip

Make sure you learn these — you might get asked in the exam what properties you'd expect a specific ionic compound to have. It doesn't matter if you've never heard of the compound before — if you know it's an ionic compound you know it'll have these properties.

Exam Tip
In the exam you'll be given a copy of the periodic table on the data sheet. You can use this to work out the charges of simple ions (see page 77).

Tip: Remember — even though ionic compounds contain charged particles, they are always neutral overall.

Practice Questions — Application

Q1 Use the periodic table to give the charge on the following ions:
a) bromide b) potassium c) beryllium

Q2 Calcium reacts with iodine to form an ionic compound.
a) What is the charge on a calcium ion?
b) What is the charge on an iodide ion?
c) Give the formula of calcium iodide.

Q3 Fluorine forms ionic bonds with lithium.
a) Give the formula of the compound formed.
b) Describe the formation of an ionic bond between fluorine and lithium atoms.

Q4 Magnesium sulfate is an ionic compound. With reference to its bonding, explain whether you would expect magnesium sulfate to have a high or a low melting point.

Exam Tip
If you're asked to give the formula of an ion, remember to give its charge as well as its formula, or you might lose marks in the exam.

Practice Questions — Fact Recall

Q1 Give the formulas of the following compound ions:
a) sulfate
b) ammonium

Q2 What effect does electrostatic attraction have on oppositely charged ions?

Q3 Explain what an ionic lattice is.

Q4 Draw the structure of sodium chloride, showing at least 12 ions.

Q5 Explain why ionic compounds conduct electricity when molten.

Q6 Magnesium oxide is an ionic compound. Apart from electrical conductivity when molten or dissolved, describe two physical properties you would expect magnesium oxide to have.

2. Covalent Bonding

Ionic bonding done — now it's on to covalent bonding.

Molecules

Molecules form when two or more atoms bond together — it doesn't matter if the atoms are the same or different. Chlorine gas (Cl_2), carbon monoxide (CO), water (H_2O) and ethanol (C_2H_5OH) are all molecules. Molecules are held together by strong covalent bonds. Covalent bonds can be single, double or triple bonds. In Chemistry, you'll often see covalent bonds represented as lines.

Single bonds

In covalent bonding, two atoms share electrons, so they've both got full outer shells of electrons. A single covalent bond contains a shared pair of electrons. Both the positive nuclei are attracted electrostatically to the shared electrons.

Examples

Two iodine atoms (I) bond covalently to form a molecule of iodine (I_2) — see Figure 1. These diagrams just show the electrons in the outer shells. (The electrons are really all the same, but dots and crosses are used to make it obvious which atoms the electrons originally came from.)

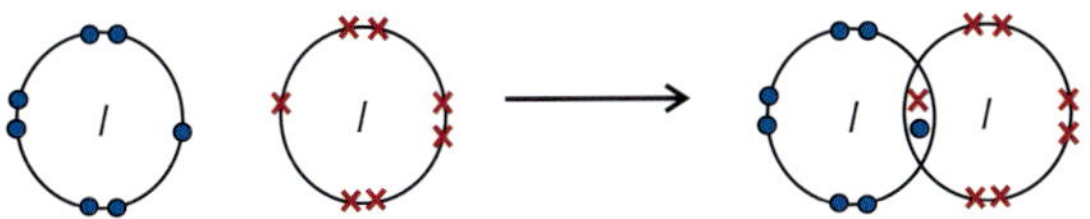

Figure 1: *Formation of a molecule of iodine.*

An iodine molecule can also be drawn as:

I—I

The diagrams below show other examples of covalent molecules.

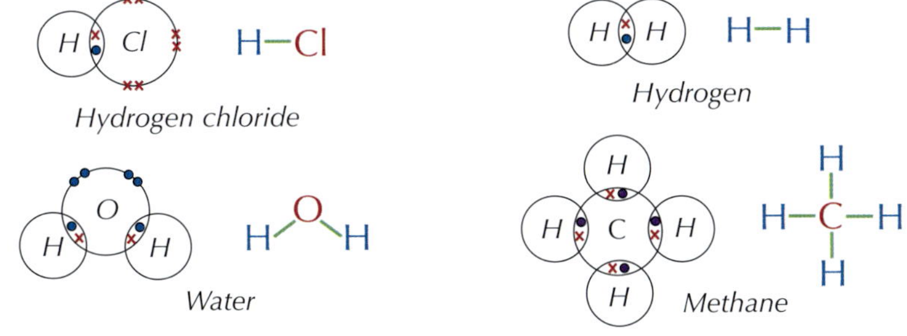

Double and triple bonds

Atoms in covalent molecules don't just form single bonds — double or even triple covalent bonds can form too. These are shown using multiple lines.

Examples

Double bonds

One carbon atom (C) can bond to two oxygen atoms (O). Each oxygen atom shares two pairs of electrons with the carbon atom. So, each molecule of carbon dioxide (CO_2) contains two double bonds.

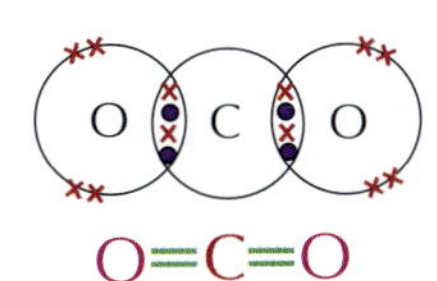

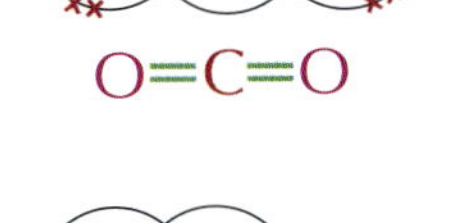

Triple bonds

When a molecule of nitrogen (N_2) forms, the nitrogen atoms share three pairs of electrons. So, each molecule of nitrogen contains one triple bond.

N≡N

Learning Objectives:

- Know that a single covalent bond contains a shared pair of electrons.
- Know that multiple bonds contain multiple shared pairs of electrons.
- Be able to represent covalent bonds using a line.
- Know that the macromolecular (giant covalent) structure is one of the four types of crystal structure.
- Know that the structures formed in diamond and graphite are examples of macromolecular crystal structures.
- Be able to draw diagrams to represent macromolecular structures.
- Be able to relate the melting point and conductivity of macromolecular compounds to their structure and bonding.
- Know that a co-ordinate (dative covalent) bond contains a shared pair of electrons with both electrons supplied by one atom.
- Be able to represent co-ordinate bonds using arrows.

Specification Reference 3.1.3.2, 3.1.3.4

Tip: The molecules shown on the previous page are all simple covalent compounds.

Simple covalent compounds

Compounds that are made up of lots of individual molecules are called **simple covalent compounds**. The atoms in the molecules are held together by strong covalent bonds, but the molecules within the simple covalent compound are held together by much weaker forces called **intermolecular forces** (see page 94). It's the intermolecular forces, rather than the covalent bonds within the molecules, that determine the properties of simple covalent compounds. In general, they have low melting and boiling points and are electrical insulators (see page 98).

Giant covalent structures

Giant covalent structures are type of crystal structure. They have a huge network of covalently bonded atoms. (They're sometimes called **macromolecular** structures.)

Carbon atoms can form this type of structure because they can each form four strong, covalent bonds. There are two types of giant covalent carbon structure you need to know about — graphite and diamond.

Tip: 'Delocalised' means an electron isn't attached to a particular atom — it can move around between atoms.

Graphite

The carbon atoms in graphite are arranged in sheets of flat hexagons covalently bonded with three bonds each (see Figure 2). The fourth outer electron of each carbon atom is delocalised. The sheets of hexagons are bonded together by weak van der Waals forces (see page 94).

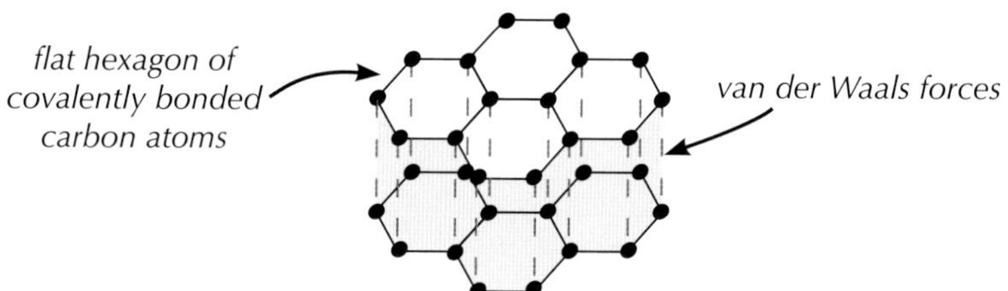

Figure 2: *The structure of graphite.*

Graphite's structure means it has certain properties:

- The weak bonds between the layers in graphite are easily broken, so the sheets can slide over each other — graphite feels slippery and is used as a dry lubricant and in pencils.
- The delocalised electrons in graphite are free to move along the sheets, so an electric current can flow.
- The layers are quite far apart compared to the length of the covalent bonds, so graphite has a low density and is used to make strong, lightweight sports equipment.
- Because of the strong covalent bonds in the hexagon sheets, graphite has a very high melting point (it sublimes at over 3900 K).
- Graphite is insoluble in any solvent.
  The covalent bonds in the sheets are too difficult to break.

Tip: 'Sublimes' means it changes straight from a solid to a gas, skipping out the liquid stage.

Tip: Tetrahedral is a molecular shape — see pages 86-90 for more on the shapes of molecules.

Diamond

Diamond is also made up of carbon atoms. Each carbon atom is covalently bonded to four other carbon atoms (see Figure 3). The atoms arrange themselves in a tetrahedral shape — its crystal lattice structure.

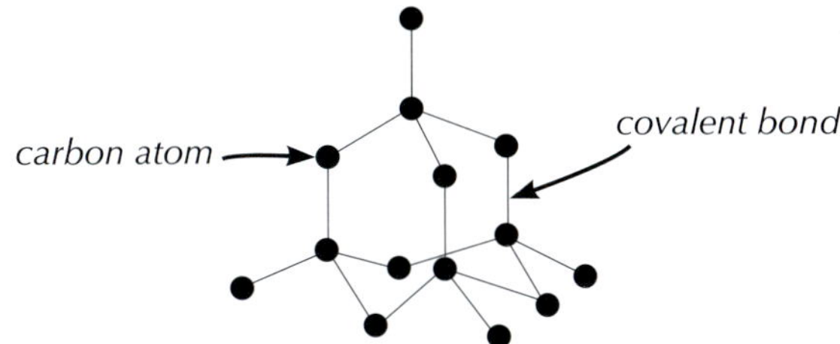

Figure 3: *The structure of diamond.*

Because of its strong covalent bonds:

- Diamond has a very high melting point — it actually sublimes at over 3800 K.
- Diamond is extremely hard — it's used in diamond-tipped drills and saws.
- Vibrations travel easily through the stiff lattice, so it's a good thermal conductor.
- It can't conduct electricity — all the outer electrons are held in localised bonds.
- Like graphite, diamond won't dissolve in any solvent.

Tip: You can 'cut' diamond to form gemstones (see Figure 4). Its structure makes it refract light, which is why it sparkles.

Figure 4: *A cut and polished diamond.*

Co-ordinate (dative covalent) bonds

In a normal single covalent bond, atoms share a pair of electrons — with one electron coming from each atom. In a **co-ordinate bond**, also known as a **dative covalent bond**, one of the atoms provides both of the shared electrons.

Example

The ammonium ion

The ammonium ion (NH_4^+) is formed by co-ordinate bonding. It forms when the nitrogen atom in an ammonia molecule donates a pair of electrons to a proton (H^+) — see Figure 5.

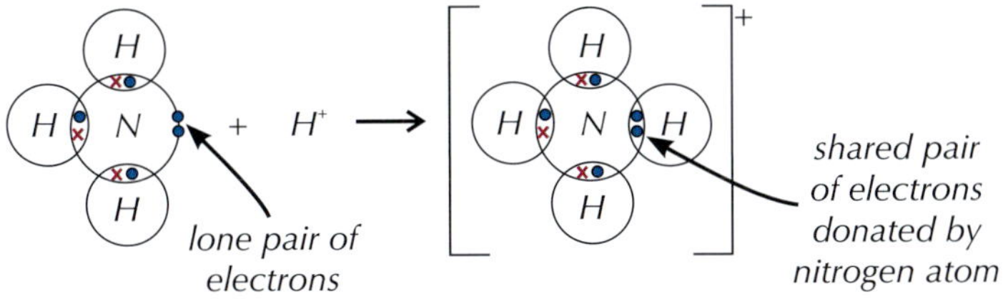

Figure 5: *Co-ordinate bonding in NH_4^+.*

Co-ordinate bonding can also be shown in diagrams by an arrow, pointing away from the 'donor' atom (see Figure 6).

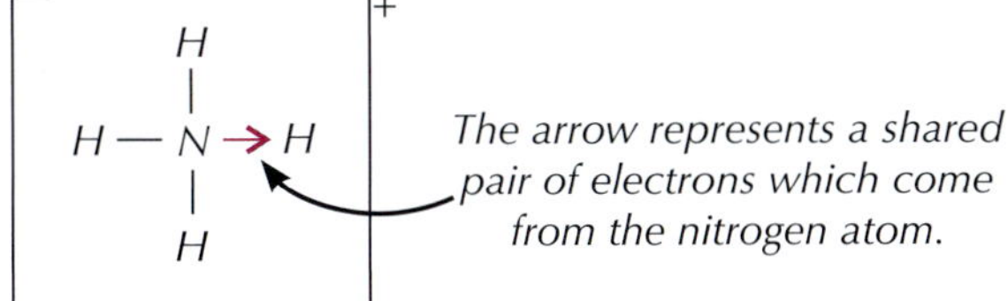

H
H — N → H
H
+
The arrow represents a shared pair of electrons which come from the nitrogen atom.

Figure 6: *An alternative way of showing co-ordinate bonding in NH_4^+.*

Co-ordinate bonds form when one of the atoms in the bond has a lone pair of electrons, and the other doesn't have any electrons available to share.

Tip: Once a co-ordinate bond has formed, it's no different to a normal covalent bond.

Tip: A lone pair of electrons is a pair of electrons in the outer shell of an atom that isn't involved in bonding with other atoms.

Tip: There's more about working out how many electrons surround an atom on page 86.

Example

The hydroxonium ion

The hydroxonium ion (H_3O^+), is formed when H_2O reacts with H^+. Hydrogen ions have no electrons so can only receive electrons, not donate them. However, a water molecule contains an oxygen atom and two hydrogen atoms. Oxygen has six electrons in its outer shell, which is then filled by sharing one electron from each hydrogen atom. This makes eight electrons in total. Only four of these electrons take part in covalent bonding with H atoms so the oxygen is left with two lone pairs of electrons.

So, oxygen donates one lone pair of electrons to the hydrogen ion, forming a dative bond (see Figure 7).

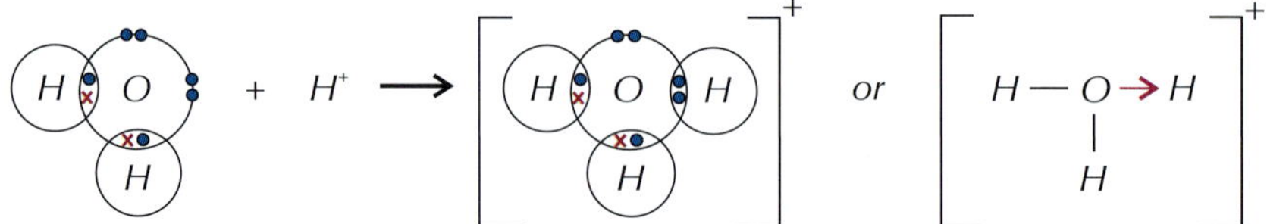

Figure 7: *Dative bonding in H_3O^+.*

Practice Questions — Application

Q1 PH_3 is a covalent compound that reacts with H^+ to form PH_4^+.

a) How many electrons are in the outer shell of a phosphorus atom?

b) How many electrons are in the outer shell of phosphorus in a PH_3 molecule?

c) Name the donor atom in PH_4^+.

d) Draw a diagram to show the formation of PH_4^+.

Q2 SiO_2 is a giant covalent compound with a structure similar to diamond. Use your knowledge of structure and bonding to predict and explain the properties of SiO_2, limited to its melting point, electrical conductivity and solubility.

Practice Questions — Fact Recall

Q1 How is a covalent bond formed?

Q2 Draw a 'dot-and-cross' diagram to show two iodine atoms coming together to form an iodine molecule.

Q3 Explain what a triple covalent bond is.

Q4 Explain why graphite is used as a lubricant.

Q5 Describe the structure of diamond.

Q6 What is a co-ordinate bond?

Q7 In a diagram showing the bonding in a molecule, what does an arrow show?

3. Charge Clouds

Electrons can be found whizzing around nuclei in charge clouds.

Charge clouds

Molecules and molecular ions come in loads of different shapes. The shape depends on the number of pairs of electrons in the outer shell of the central atom. Pairs of electrons can be shared in a covalent bond or can be unshared. Shared electrons are called bonding pairs, unshared electrons are called **lone pairs** or non-bonding pairs.

Bonding pairs and lone pairs of electrons exist as charge clouds. A charge cloud is an area where you have a big chance of finding an electron. The electrons don't stay still — they whizz around inside the charge cloud.

Example

In ammonia, the outer shell of nitrogen has four pairs of electrons. These electrons can be shown in a 'dot-and-cross' diagram or as charge clouds:

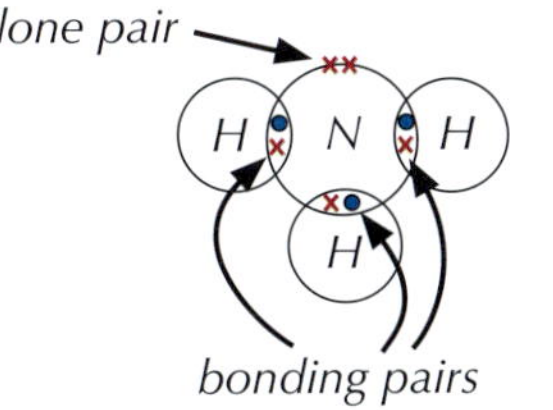

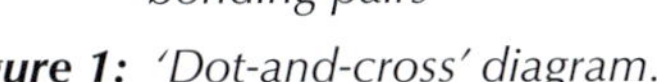

Figure 1: *'Dot-and-cross' diagram.*

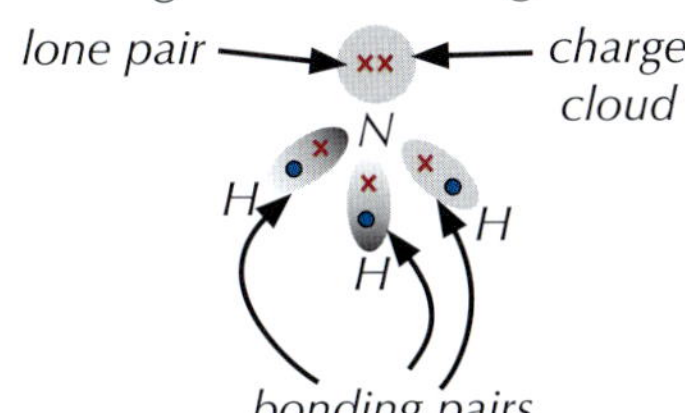

Figure 2: *Ammonia's charge clouds.*

Electron pair repulsion

Electrons are all negatively charged, so charge clouds repel each other until they're as far apart as possible. This sounds simple, but the shape of a charge cloud affects how much it repels other charge clouds. Lone-pair charge clouds repel more than bonding-pair charge clouds, so bond angles are often reduced because bonding pairs are pushed together by lone-pair repulsion. This is known by the snappy name '**Valence Shell Electron Pair Repulsion Theory**'.

Example

The central atom in methane, ammonia and water each has four pairs of electrons in its outer shell, but their bond angles are different:

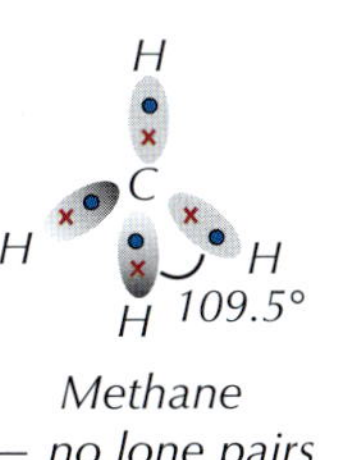

Methane — no lone pairs

The lone pair repels the bonding pairs.

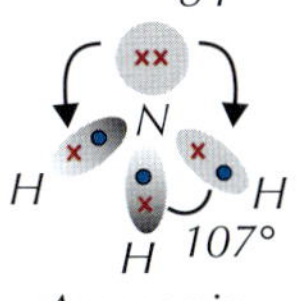

Ammonia — 1 lone pair

2 lone pairs reduce the bond angle even more.

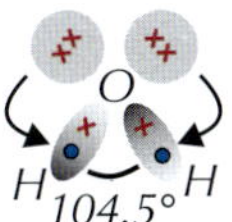

Water— 2 lone pairs

So lone pair/lone pair angles are the biggest, lone pair/bonding pair angles the second biggest and bonding pair/bonding pair angles are the smallest.

Practice Questions — Fact Recall

Q1 What is a lone pair of electrons?

Q2 Briefly describe a charge cloud.

Q3 Describe valence shell electron pair repulsion theory.

Learning Objectives:

- Be able to describe bonding pairs and lone (non-bonding) pairs of electrons as charge clouds that repel each other.
- Know that pairs of electrons in the outer shell of atoms arrange themselves as far apart as possible to minimise repulsion.
- Know that lone pair-lone pair repulsion is greater than lone pair-bond pair repulsion, which is greater than bond pair-bond pair repulsion.
- Understand the effect of electron pair repulsion on the bond angles in molecules.

Specification Reference 3.1.3.5

Tip: There's lots more coming up about bond angles and shapes of molecules in the next few pages.

4. Shapes of Molecules

Learning Objectives:
- Be able to explain the shapes of, and bond angles in, simple molecules and ions with up to six electron pairs (including lone pairs of electrons) surrounding the central atom.
- Understand the effect of electron pair repulsion on the bond angles in molecules.

Specification Reference 3.1.3.5

There's a lot of variation in molecular shape and you need to understand how to work out the shape of any molecule or molecular ion. Don't worry though, the next few pages have lots of advice to help you along.

Drawing shapes of molecules

It can be tricky to draw molecules showing their shapes, because you're trying to show a 3D shape on a 2D page. Usually you do it is by using different types of lines to show which way the bonds are pointing. In a molecule diagram, use wedges to show a bond pointing towards you, and a broken (or dotted) line to show a bond pointing away from you (see Figure 1).

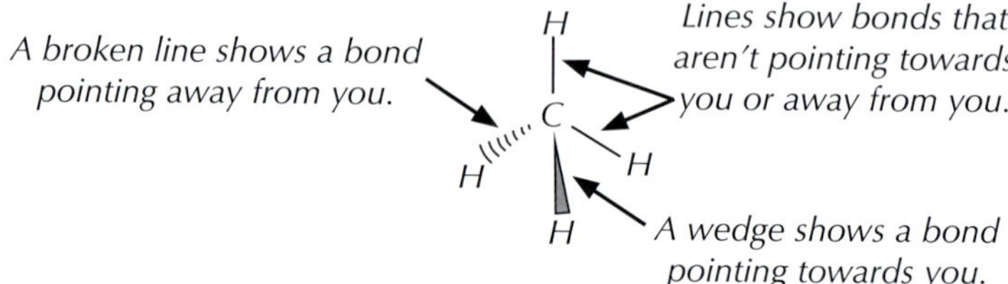
A broken line shows a bond pointing away from you.
H
C
H
H
H
Lines show bonds that aren't pointing towards you or away from you.
A wedge shows a bond pointing towards you.

Figure 1: *A molecular diagram showing the shape of methane.*

Tip: It might help you to think of wedges as bonds that stick out of the page, broken lines as bonds that point behind the page and straight lines as bonds that are flat against the page.

Finding the number of electron pairs

To work out the shape of a molecule or an ion you need to know how many lone pairs and how many bonding pairs of electrons are on the central atom. To find that out, you just follow these steps:

1. Find the central atom — it's the one all the other atoms are bonded to.
2. Work out how many electrons are in the outer shell of the central atom. This will be the same as its group number in the periodic table.
3. Add 1 electron for every atom that the central atom is bonded to. (You can work this out from the formula of the molecule or ion.)
4. If you're looking at an ion, you need to take its charge into account — add 1 electron for each negative charge or subtract 1 for each positive charge.
5. Add up all the electrons. Divide by 2 to find the number of electron pairs.
6. Compare the number of electron pairs to the number of bonds to find the number of lone pairs and the number of bonding pairs on the central atom.

Figure 2: *A molecular model of a molecule of methane.*

Example

Carbon tetrafluoride, CF_4

1. The central atom in this molecule is carbon.
2. Carbon's in group 4. It has 4 electrons in its outer shell.
3. The carbon atom is bonded to 4 fluorine atoms.
4. CF_4 isn't an ion.
5. There are $4 + 4 = 8$ electrons in the outer shell of the carbon atom, which is $8 \div 2 = 4$ electron pairs.
6. 4 pairs of electrons are involved in bonding the fluorine atoms to the carbon so there must be 4 bonding pairs of electrons. That's all the electrons, so there are no lone pairs.

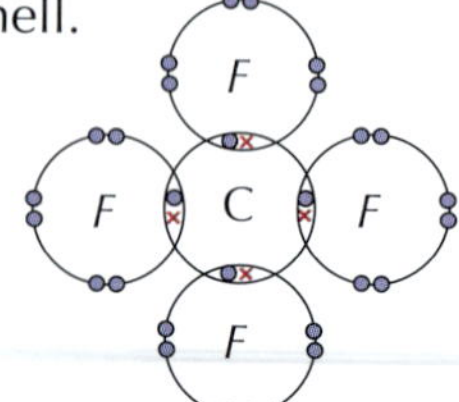
F
F
C
F
F

Figure 3: *Dot-and-cross diagram for CF_4.*

Tip: The exception to the rule that the number of electrons in the outer shell of an atom being the same as its group number is group 0. Group 0 elements have 8 electrons in their outer shells (apart from helium, which has 2).

Example

Phosphorus trihydride, PH_3

1. The central atom in this molecule is phosphorus.
2. Phosphorus is in group 5, so it has five electrons in its outer shell.
3. Phosphorus has formed 3 bonds with hydrogen atoms.
4. PH_3 isn't an ion.
5. There are 5 + 3 = 8 electrons in the outer shell of the phosphorus atom, so there are 8 ÷ 2 = 4 electron pairs.
6. 3 electron pairs are involved in bonding with the hydrogen atoms (bonding pairs), so there must also be 1 lone pair of electrons on the phosphorus atom.

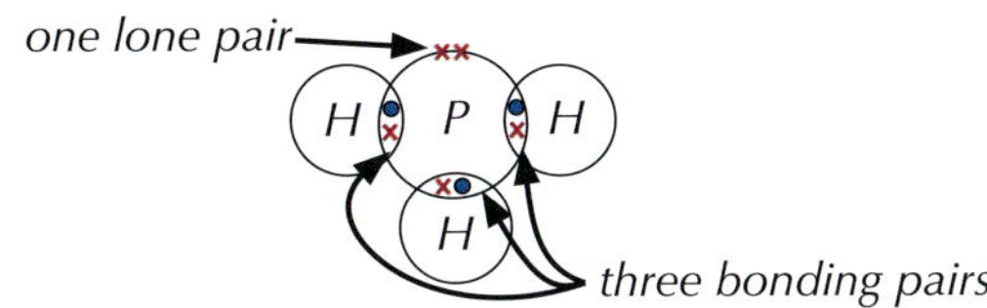

***Figure 4:** Dot-and-cross diagram for PH_3.*

Tip: A hydrogen atom will form one covalent bond with another atom to complete its outer shell.

Molecular shapes

Once you know how many bonding pairs and how many lone pairs are on the central atom, you can work out the shape of the molecule. Here's a run down of the different shapes that you might come across (and their bond angles).

Tip: You need to know how to calculate the number of electron pairs before you try and learn this bit. Make sure you've got the previous page firmly in your head before you go any further.

Central atoms with two electron pairs

Molecules with two electron pairs have a bond angle of 180° and have a **linear** shape. This is because the pairs of bonding electrons want to be as far away from each other as possible.

Example — Maths Skills

Beryllium chloride, $BeCl_2$

In $BeCl_2$, the central beryllium atom has two bonding pairs of electrons and no lone pairs. So the bond angle in $BeCl_2$ is 180° and it has a linear shape.

180°
Cl—Be—Cl

Exam Tip
You'll be expected to know the bond angles for different shapes of molecules in your exam — so make sure you learn them.

Central atoms with three electron pairs

Molecules that have three electron pairs around the central atom don't always have the same shape — the shape depends on the combination of bonding pairs and lone pairs of electrons.

If there are three bonding pairs of electrons the repulsion of the charge clouds is the same between each pair and so the bond angles are all 120°. The shape of the molecule is called **trigonal planar**.

Example — Maths Skills

Boron trifluoride, BF_3

The central boron atom has three bonding pairs and no lone pairs, so the bond angle in BF_3 is 120° and it has a **trigonal planar** shape.

F F
B 120°
F

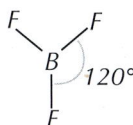

Tip: As a general rule, each lone pair decreases the bond angle by about 2.5° (compared to a molecule with the same number of electron pairs where they're all bonding pairs).

If you have a molecule with two bonding pairs of electrons and one lone pair of electrons, you'll get a non-linear or 'bent' molecule (see below for more on bent molecules). The bond angle will be a bit less than 120°.

Central atoms with four electron pairs

If there are four pairs of bonding electrons and no lone pairs on a central atom, all the bond angles are 109.5° — the charge clouds all repel each other equally. The shape of the molecule is **tetrahedral**.

Tip: The bond between nitrogen and hydrogen shown by the arrow is a co-ordinate bond — both of the electrons in the bond come from the nitrogen atom.

Example — Maths Skills

Ammonium ion, NH_4^+

In NH_4^+, the central nitrogen atom has four bonding pairs of electrons and no lone pairs, so the shape of NH_4^+ is **tetrahedral**.

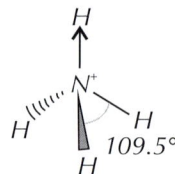

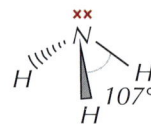

If there are three bonding pairs of electrons and one lone pair, the lone-pair/bonding-pair repulsion will be greater than the bonding-pair/bonding-pair repulsion and so the angles between the atoms will change. There'll be smaller bond angles between the bonding pairs of electrons and larger bond angles between the lone pair and the bonding pairs. The bond angle is 107° and the shape of the molecules is **trigonal pyramidal**.

Tip: Take a look back at page 85 for why more lone pairs means smaller bond angles.

Example — Maths Skills

Ammonia, NH_3

In NH_3, the central nitrogen atom has three bonding pairs of electrons and a lone pair so the shape of NH_3 is **trigonal pyramidal**.

Exam Tip

If you're drawing the shape of a molecule in an exam, make sure that you show the lone pairs as well as the atoms.

If there are two bonding pairs of electrons and two lone pairs of electrons the lone-pair/lone-pair repulsion will squish the bond angle even further. The bond angle will be around 104.5° and the shape of the molecules is **bent** (or **non-linear**).

Example — Maths Skills

Water, H_2O

In H_2O, the central oxygen atom has two bonding pairs shared with hydrogen atoms and two lone pairs so the shape of H_2O is **bent** (or **non-linear**).

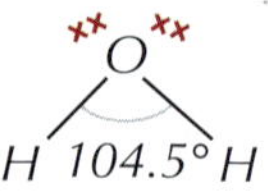

Figure 5: A molecular model of water, showing it to have a bent (non-linear) shape.

Central atoms with five electron pairs

Some central atoms can 'expand the octet' — which just means that they can have more than eight bonding electrons in their outer shells.

A molecule with five bonding pairs will be **trigonal bipyramidal**. Repulsion between the bonding pairs means that three of the atoms will form a trigonal planar shape with bond angles of 120° and the other two atoms will be at 90° to them.

Example — Maths Skills

Phosphorus pentachloride, PCl_5

In PCl_5, the central phosphorus atom has five bonding pairs and no lone pairs, so it has a **trigonal bipyramidal** shape.

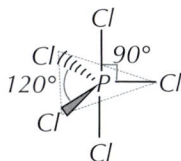

If there are four bonding pairs and one lone pair of electrons, the molecule forms a **seesaw** shape. The lone pair is always positioned where one of the trigonal planar atoms would be in a trigonal bipyramidal molecule.

Tip: If you look at the shapes of the molecules where five pairs of electrons surround the central atom, you can see how they're all based on a trigonal bipyramidal shape. Some just have the bonds replaced by lone pairs of electrons. Similarly, molecules with six pairs of electrons are based on an octahedral shape, and molecules with four pairs of electrons are based on a tetrahedral shape.

Example — Maths Skills

Sulfur tetrafluoride, SF_4

In SF_4, the central sulfur atom has four bonding pairs and one lone pair so it has a **seesaw** shape.

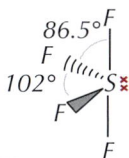

If there are three bonding pairs and two lone pairs of electrons, the molecule will be **T-shaped**.

Example — Maths Skills

Chlorine trifluoride, ClF_3

In ClF_3, the central chlorine atom has three bonding pairs and two lone pairs so it has a **T-shape**.

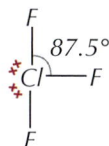

Central atoms with six electron pairs

A molecule with six bonding pairs will be **octahedral**. All of the bond angles in the molecule will be 90°.

Example — Maths Skills

Sulfur hexafluoride, SF_6

In SF_6, the central sulfur atom has six bonding pairs and no lone pairs, making its shape **octahedral**.

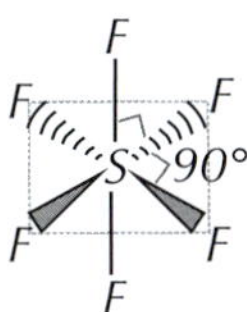

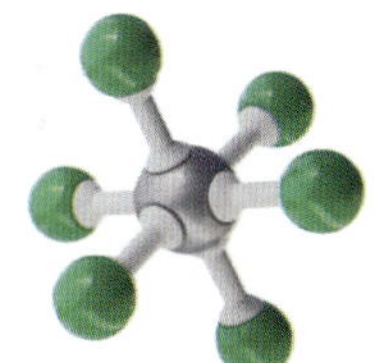

Figure 6: A molecular model of sulfur hexafluoride.

If there are five bonding pairs and one lone pair, the molecule forms a **square pyramidal** structure. (Molecules with this shape are very rare.)

Example — Maths Skills

Chlorine pentafluoride, ClF_5

In ClF_5, the central chlorine has five bonding pairs and one lone pair of electrons, making its shape **square pyramidal**.

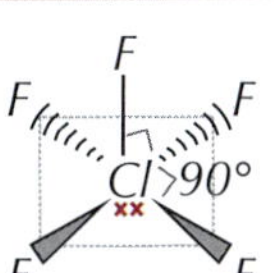

If there are four bonding pairs and two lone pairs of electrons, the molecule will be **square planar**.

Example — Maths Skills

Xenon tetrafluoride, XeF_4

In XeF_4, the central xenon atom has four bonding pairs and two lone pairs of electrons — its shape is **square planar**.

Finding the shape of an unfamiliar molecule

In the exam you could be asked to draw the shape of a molecule you've never met before. Don't panic, just take it step by step. Work out how many electron pairs the molecule has, then work out how many of those are lone pairs. Decide what the bond angles are in the molecule, then draw and label it neatly.

Example

Predict the shape of the BF_4^- ion.

First, follow the steps on page 86 to find the number of electron pairs:

1. The central atom is boron.
2. Boron is in Group 3, so it has 3 electrons in its outer shell.
3. The boron atom is bonded to 4 fluorine atoms.
4. BF_3 is an ion with a 1– charge, so you need to add one extra electron to account for that.
5. That means there are 3 + 4 + 1 = 8 electrons in the outer shell of the central boron atom. That's 8 ÷ 2 = 4 pairs.
6. The boron atom has 4 electron pairs around it, and has made 4 bonds. So it has 4 bonding pairs and no lone pairs.

A molecule with 4 bonding pairs and no lone pairs will have a tetrahedral shape, with bond angles of 109.5°. Now all you have to do is draw it.

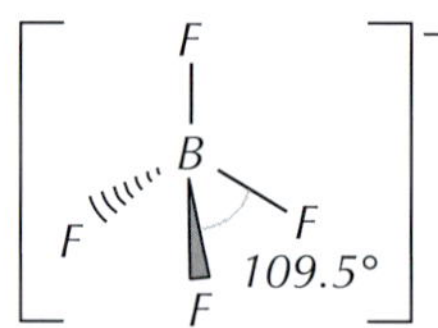

Figure 7: The shape of the BF_4^- ion.

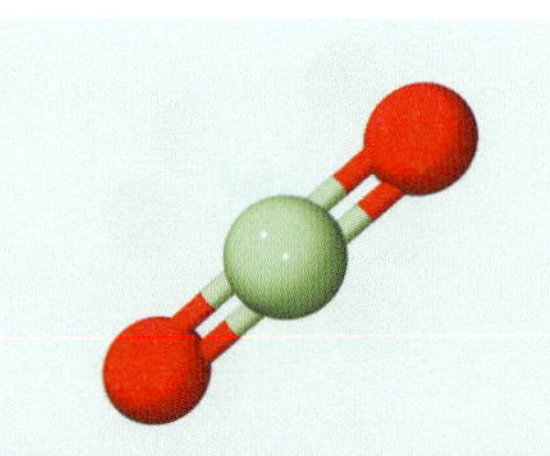

Figure 8: A molecular model of carbon dioxide, showing it to be a linear molecule with a bond angle of 180°.

Tip: If the central atom forms multiple bonds (i.e. double or triple bonds), then treat each multiple bond as a single electron pair when you're working out the shape.

Awkward molecules

There are some special cases where molecules don't follow the normal rules. If a molecule has multiple bonds, you treat each multiple bond as if it was one single bond when you're working out the shape (even though there's usually slightly more repulsion between double bonds).

Examples

Carbon dioxide, CO_2

180°
O=C=O

Carbon has four bonding pairs of electrons (found in two carbon-oxygen double bonds) and no lone pairs. Double bonds can be treated as one bond, so you can say that there are two bonds and no lone pairs — CO_2 will be **linear**.

Sulfur dioxide, SO_2

xx
S
O O
120°

The extra electron density in the double bonds cancels out the extra repulsion from the lone pair, so you get 120° angles.

Sulfur has four bonding pairs of electrons (found in two sulfur-oxygen double bonds) and one lone pair. Double bonds can be treated as one bond so you can say there that are two bonds and one lone pair — SO_2 will be **bent** (non-linear).

Practice Questions — Application

Q1 a) How many electron pairs are on the central atom of an H_2S molecule?

b) How many lone pairs does a molecule of H_2S have?

c) Draw the shape of an H_2S molecule.

d) Name the shape of an H_2S molecule.

e) Give the bond angle between bonding pairs in H_2S.

Q2 a) How many electron pairs are on the central atom of an H_3O^+ molecule?

b) How many lone pairs does a molecule of H_3O^+ have?

c) Draw and name the shape of H_3O^+.

d) Give the bond angle between bonding pairs in H_3O^+.

Q3 a) Draw and name the shape of a molecule of AsH_3.

b) Give the bonding pair/bonding pair bond angle in AsH_3.

Q4 Draw and name the shape of a molecule of CCl_2F_2.

Practice Questions — Fact Recall

Q1 What is the bond angle between electron pairs in a trigonal planar molecule?

Q2 How many electron pairs are on the central atom in a tetrahedral molecule?

Q3 Name the shape that a molecule will have if it has four bonding pairs and one lone pair on its central atom.

Exam Tip
Make sure you practise drawing the different shapes that molecules can have — it'll make it easier if you have to in the exam.

Learning Objectives:

- Know that electronegativity is the power of an atom to attract the pair of electrons in a covalent bond.
- Know that the electron distribution in a covalent bond between elements with different electronegativities will be unsymmetrical and that this produces a polar covalent bond and may cause a molecule to have a permanent dipole.
- Be able to use partial charges to show that a bond is polar.
- Be able to explain why some molecules with polar bonds do not have a permanent dipole.

Specification Reference 3.1.3.6

Exam Tip
You don't need to learn the electronegativity values — if you need them you'll be given them in the exam.

Tip: The atoms in polar bonds generally have an electronegativity difference that is greater than 0.4.

5. Polarisation

Polarisation of bonds occurs because of the nature of different atomic nuclei — some are just more attractive than others.

Electronegativity

The ability to attract the bonding electrons in a covalent bond is called **electronegativity**. Electronegativity is measured on the Pauling Scale. A higher number means an element is better able to attract the bonding electrons. Fluorine is the most electronegative element. Oxygen, nitrogen and chlorine are also very strongly electronegative — see Figure 1.

Element	*H*	*C*	*N*	*Cl*	*O*	*F*
Electronegativity (Pauling Scale)	*2.20*	*2.55*	*3.04*	*3.16*	*3.44*	*3.98*

Figure 1: *The electronegativity of different elements.*

Polar and non-polar bonds

The covalent bonds in diatomic gases (e.g. H_2, Cl_2) are non-polar because the atoms have equal electronegativities and so the electrons are equally attracted to both nuclei (see Figure 2). Some elements, like carbon and hydrogen, have pretty similar electronegativities, so bonds between them are essentially non-polar.

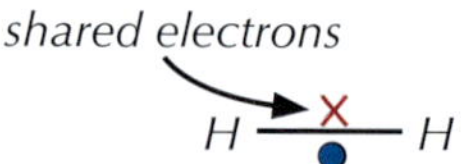
shared electrons
H
H

Figure 2: *A non-polar covalent bond in a hydrogen molecule.*

In a covalent bond between two atoms of different electronegativities, the bonding electrons are pulled towards the more electronegative atom. This makes the bond polar (see Figure 3). The greater the difference in electronegativity, the more polar the bond.

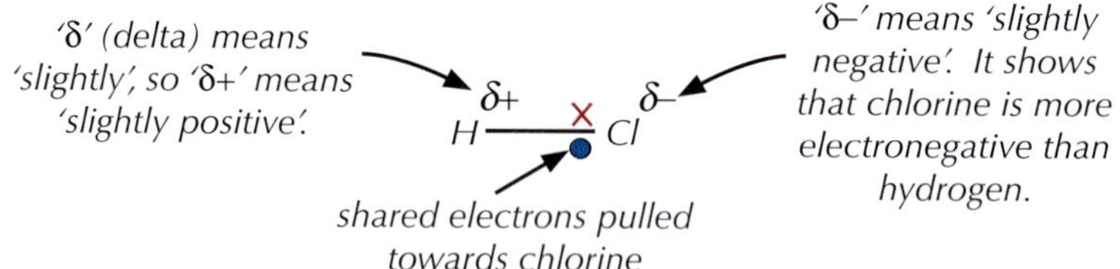
'δ' (delta) means 'slightly', so 'δ+' means 'slightly positive'.
δ+
δ–
H
Cl
'δ–' means 'slightly negative'. It shows that chlorine is more electronegative than hydrogen.
shared electrons pulled towards chlorine

Figure 3: *A polar covalent bond in a hydrogen chloride molecule.*

In a polar bond, the difference in electronegativity between the two atoms causes a **dipole**. A dipole is a difference in charge between the two atoms caused by a shift in electron density in the bond.

Polar molecules

If charge is distributed unevenly over a whole molecule, then the molecule will have a **permanent dipole**. Molecules that have a permanent dipole are called **polar molecules**. Whether or not a molecule is polar depends on whether it has any polar bonds, and its overall shape.

In simple molecules, such as hydrogen chloride, the one polar bond means charge is distributed unevenly across the whole molecule, so it has a permanent dipole (see Figure 4).

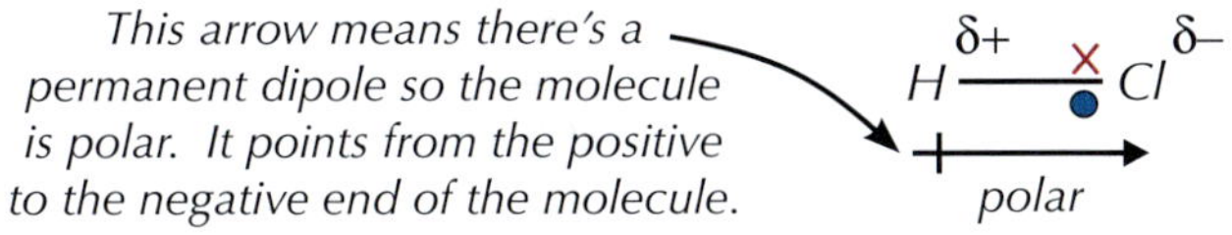

This arrow means there's a permanent dipole so the molecule is polar. It points from the positive to the negative end of the molecule.

δ+ H — Cl δ–

polar

Figure 4: *The permanent dipole in a molecule of hydrogen chloride.*

More complicated molecules might have several polar bonds. The shape of the molecule will decide whether or not it has an overall permanent dipole. If the polar bonds are arranged symmetrically so that the dipoles cancel each other out, such as in carbon dioxide, then the molecule has no permanent dipole and is non-polar — see Figure 5.

The two polar C=O bonds exactly cancel each other out, so the molecule has no permanent dipole moment.

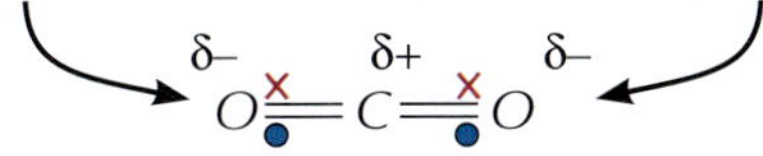

Figure 5: *A molecule of carbon dioxide has no permanent dipole moment.*

If the polar bonds are arranged so that they all point in roughly the same direction, and they don't cancel each other out, then charge will be arranged unevenly across the whole molecule. This results in a polar molecule — the molecule has a permanent dipole (see Figure 6).

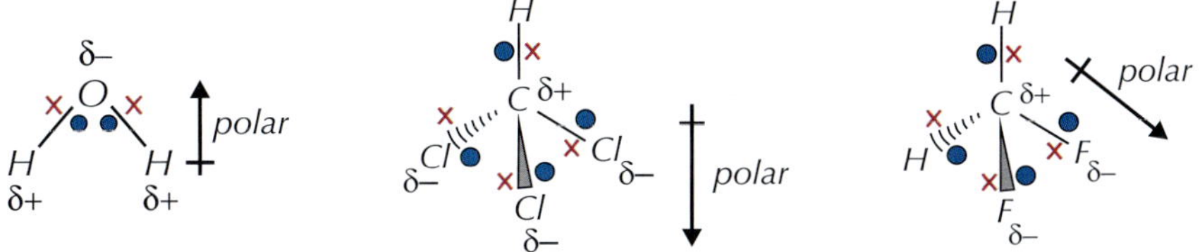

Figure 6: *The permanent dipoles in some polar molecules.*

Tip: It's really, really important that you get your head around the relationship between electronegativity, polarisation and dipoles. Differences in the electronegativity of atoms cause bonds to become polarised, which results in a dipole — a difference in charge between the two atoms. If these dipoles don't cancel out then the molecule will be polar (it will have a permanent dipole).

Practice Questions — Application

Q1 Draw the shapes of the following molecules.
Then predict whether or not they have a permanent dipole:

a) BCl_3 (The B–Cl bonds are polar.)

b) CH_2Cl_2

c) PF_3 (The P–F bonds are polar.)

Q2 Given that the Pauling electronegativities of silicon and chlorine are 1.90 and 3.16 respectively, explain whether silicon tetrachloride ($SiCl_4$) will possess polar bonds and/or have a permanent dipole.

Tip: Have a look back at pages 86-90 for the rules on predicting the shapes of molecules.

Practice Questions — Fact Recall

Q1 Chlorine is more electronegative than hydrogen.
Explain what this means.

Q2 Explain why the H–F bond is polarised.

Q3 What is a dipole?

6. Intermolecular Forces

Molecules don't just exist independently — they can interact with each other. And you need to know how they interact.

Learning Objectives:

- Know that intermolecular forces exist between molecules, including induced dipole–dipole forces (van der Waals forces), permanent dipole–dipole forces and hydrogen bonds.
- Be able to draw diagrams of molecular crystal structures.
- Be able to explain how the melting and boiling points of molecular substances are influenced by the strength of these intermolecular forces.
- Explain the importance of hydrogen bonds in the low density of ice and the anomalous boiling points of compounds.
- Be able to explain the existence of intermolecular forces between familiar and unfamiliar molecules.
- Know that the molecular lattice structure is one of the four types of crystal structure.
- Know that the structures found in iodine and ice are examples of molecular crystal structures.
- Be able to relate the melting point and conductivity of molecular compounds to their structure and bonding.

Specification Reference 3.1.3.4, 3.1.3.7

Tip: Temporary dipoles are also called induced dipoles.

What are intermolecular forces?

Intermolecular forces are forces between molecules. They're much weaker than covalent, ionic or metallic bonds. There are three types you need to know about: induced dipole-dipole (or van der Waals) forces, permanent dipole-dipole forces and hydrogen bonding (this is the strongest type).

Van der Waals forces

Van der Waals forces cause all atoms and molecules to be attracted to each other. Electrons in charge clouds are always moving really quickly. At any particular moment, the electrons in an atom are likely to be more to one side than the other. At this moment, the atom would have a temporary dipole. This dipole can cause another temporary dipole in the opposite direction on a neighbouring atom (see Figure 1). The two dipoles are then attracted to each other. The second dipole can cause yet another dipole in a third atom. It's kind of like the domino effect. Because the electrons are constantly moving, the dipoles are being created and destroyed all the time. Even though the dipoles keep changing, the overall effect is for the atoms to be attracted to each other.

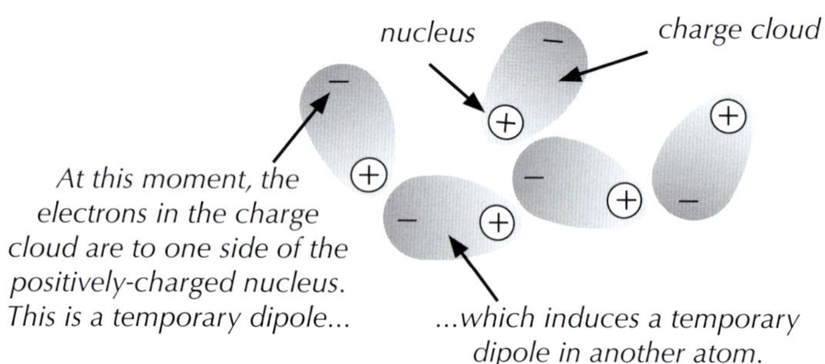
nucleus
charge cloud
At this moment, the electrons in the charge cloud are to one side of the positively-charged nucleus. This is a temporary dipole...
...which induces a temporary dipole in another atom.

Figure 1: *Temporary dipoles in a liquid resulting in van der Waals forces.*

Example

Van der Waals forces are responsible for holding iodine molecules together in a lattice. Iodine atoms are held together in pairs by strong covalent bonds to form molecules of I_2 (see Figure 2). But the molecules are then held together in a molecular lattice arrangement by weak van der Waals forces (see Figure 3). Molecular lattices are a type of crystal structure.

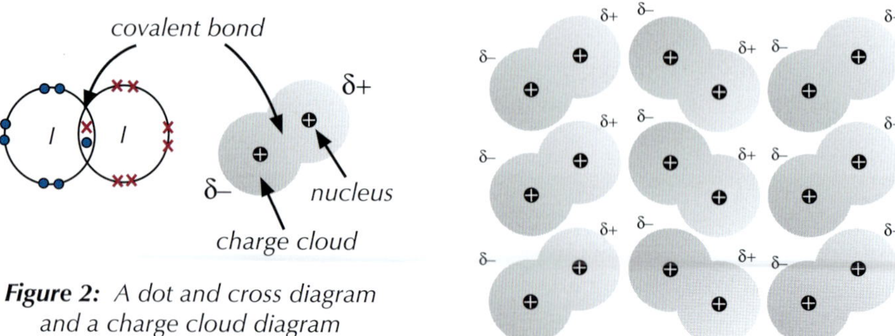
covalent bond
I
I
δ+
δ–
nucleus
charge cloud
δ+
δ–

Figure 2: *A dot and cross diagram and a charge cloud diagram showing a molecule of iodine.*

Figure 3: *Lattice of iodine molecules held together by van der Waals forces.*

Not all van der Waals forces are the same strength — larger molecules have larger electron clouds, meaning stronger van der Waals forces.

The shape of molecules also affects the strength of van der Waals forces. Long, straight molecules can lie closer together than branched ones — the closer together two molecules are, the stronger the forces between them.

When you boil a liquid, you need to overcome the intermolecular forces, so that the particles can escape from the liquid surface. It stands to reason that you need more energy to overcome stronger intermolecular forces, so liquids with stronger Van der Waals forces will have higher boiling points. Van der Waals forces affect other physical properties, such as melting point and viscosity too.

Tip: As well as induced dipole-dipole forces, van der Waals forces can also be called London forces or dispersion forces.

Tip: Remember — there are van der Waals forces between the molecules in every chemical.

Examples

As you go down the group of noble gases, the number of electrons increases. So the van der Waals forces increase, and so do the boiling points (see Figure 4).

Tip: The noble gases are all in Group 0 in the periodic table.

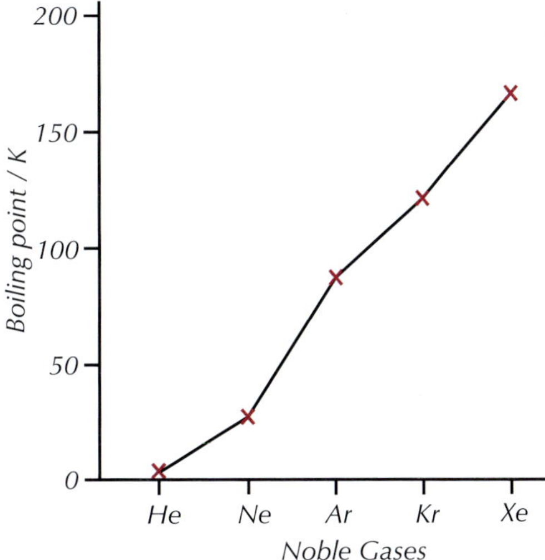

Figure 4: *Graph showing the boiling points of noble gases.*

As the alkane chains get longer, the number of electrons in the molecules increases. The area over which the van der Waals forces can act also increases. This means the van der Waals forces are stronger, and so the boiling points increase (see Figure 5).

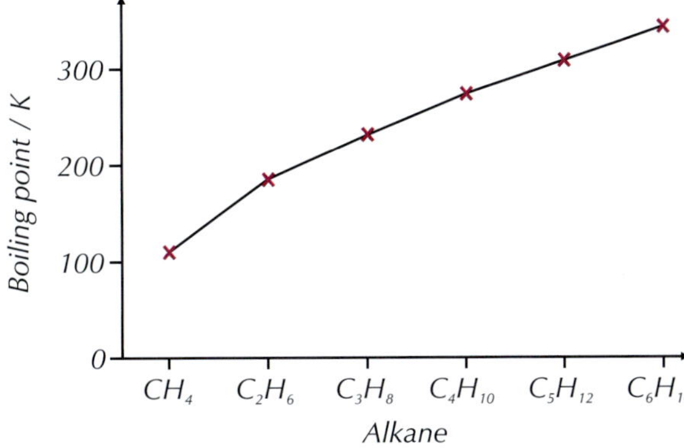

Figure 5: *Graph showing the boiling points of straight-chain alkanes*

Permanent dipole-dipole forces

In a substance made up of molecules that have permanent dipoles, there will be weak electrostatic forces of attraction between the δ+ and δ– charges on neighbouring molecules. These are called permanent dipole-dipole forces.

Example

Hydrogen chloride gas has polar molecules due to the difference in electronegativity of hydrogen and chlorine.

$\overset{\delta+}{H}$—$\overset{\delta-}{Cl}$ • • • $\overset{\delta+}{H}$—$\overset{\delta-}{Cl}$ • • • $\overset{\delta+}{H}$—$\overset{\delta-}{Cl}$

The molecules have weak electrostatic forces between them because of the shift in electron density.

Figure 6: *Permanent dipole-dipole forces in hydrogen chloride gas.*

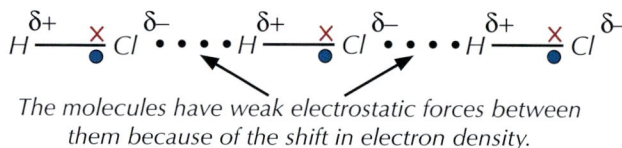

Exam Tip
When you're drawing dipoles in the exam, make sure you include the δ+ and δ– symbols to show the charges.

If you put an electrostatically charged rod next to a jet of a polar liquid, like water, the liquid will move towards the rod. It's because polar liquids contain molecules with permanent dipoles. It doesn't matter if the rod is positively or negatively charged. The polar molecules in the liquid can turn around so the oppositely charged end is attracted towards the rod (see Figures 7 and 8).

The more polar the liquid, the stronger the electrostatic attraction between the rod and the jet, so the greater the deflection will be. By contrast, liquids made up of non-polar molecules, such as hexane, will not be affected at all when placed near a charged rod.

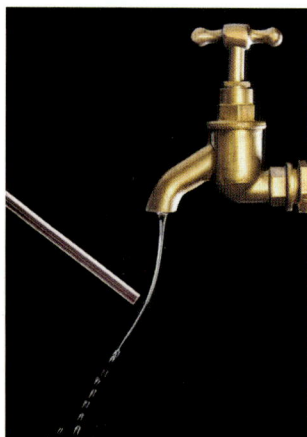

Figure 7: *A charged glass rod bends water.*

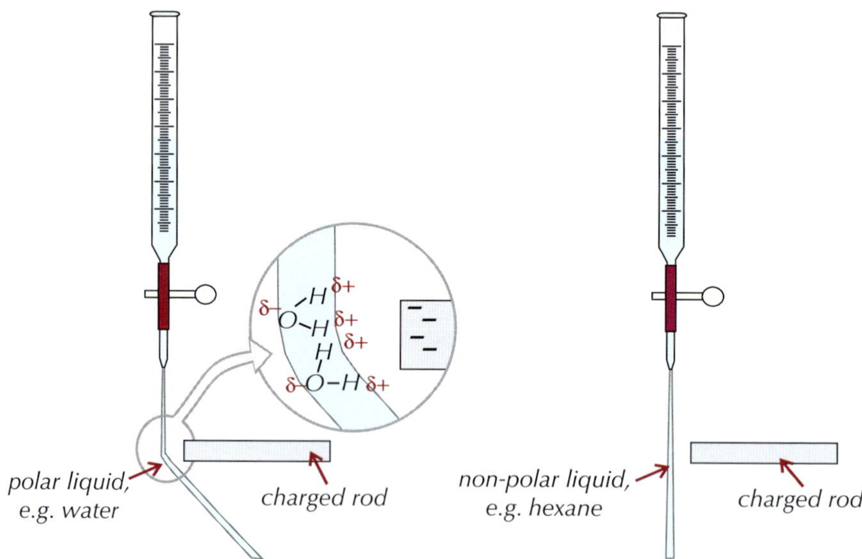

Figure 8: *A charged glass rod bends polar liquids such as water but has no effect on non-polar liquids such as hexane.*

Hydrogen bonding

Hydrogen bonding is the strongest intermolecular force. It only happens when hydrogen is covalently bonded to fluorine, nitrogen or oxygen. Fluorine, nitrogen and oxygen are very electronegative, so they draw the bonding electrons away from the hydrogen atom.

Exam Tip
Hydrogen bonding is a special case scenario — it only happen in specific molecules. In the exam, you could be asked to compare intermolecular forces in different substances, so you'll need to know the different intermolecular forces and their relative strengths. Don't forget that not every molecule with hydrogen in it makes hydrogen bonds.

The bond is so polarised, and hydrogen has such a high charge density because it's so small, that the hydrogen atoms form weak bonds with lone pairs of electrons on the fluorine, nitrogen or oxygen atoms of other molecules. Molecules which have hydrogen bonding are usually organic, containing -OH or -NH groups.

Tip: Charge density is just a measure of how much positive or negative charge there is in a certain volume.

Examples

Water and ammonia both have hydrogen bonding (see Figures 9 and 10).

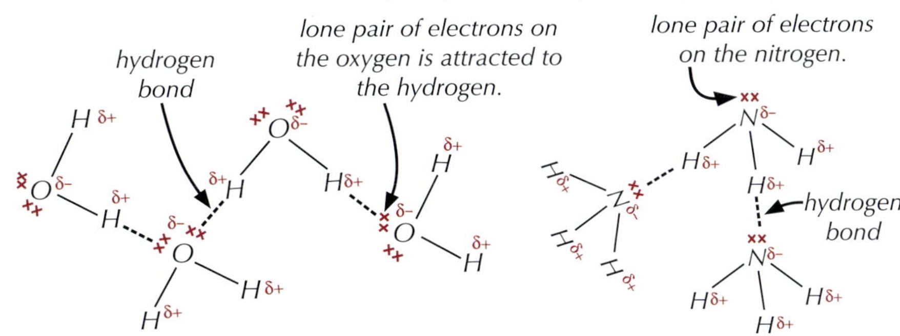

Figure 9: *Hydrogen bonding in water.* **Figure 10:** *Hydrogen bonding in ammonia.*

Hydrogen bonding has a huge effect on the properties of substances. Substances with hydrogen bonds have higher boiling and melting points than other similar molecules because of the extra energy needed to break the hydrogen bonds. This is the case with water which has a much higher boiling point than the other group 6 hydrides — see Figure 11.

Tip: The melting points of compounds that form hydrogen bonds are higher than other simple covalent compounds but still much lower than ionic or giant covalent substances.

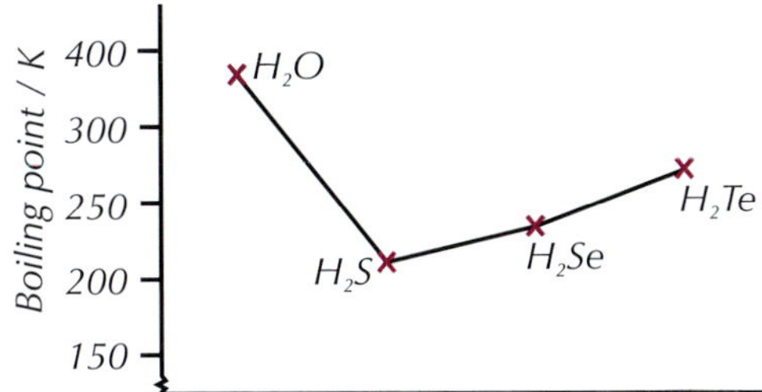

Figure 11: *Graph showing the boiling points of group 6 hydrides.*

The effects of hydrogen bonding are very apparent in other physical properties of water as well. As liquid water cools to form ice, the molecules make more hydrogen bonds and arrange themselves into a regular lattice structure. Since hydrogen bonds are relatively long, the average distance between H_2O molecules is greater in ice than in liquid water — so ice is less dense than liquid water (see Figures 12 and 13). This is unusual — most substances are more dense as solids than they are as liquids.

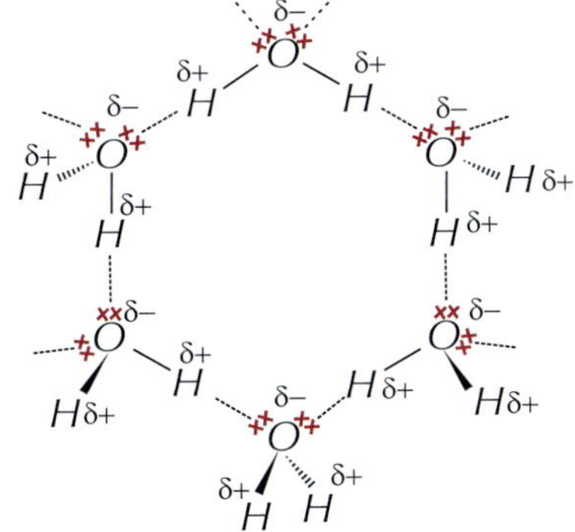

Figure 13: *Lattice of water molecules in ice held together by hydrogen bonds.*

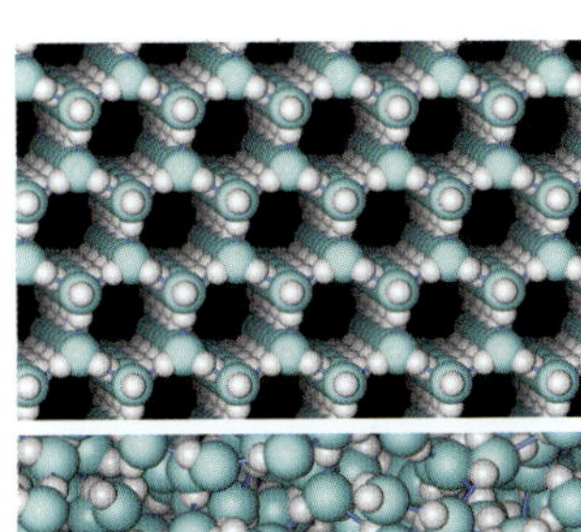

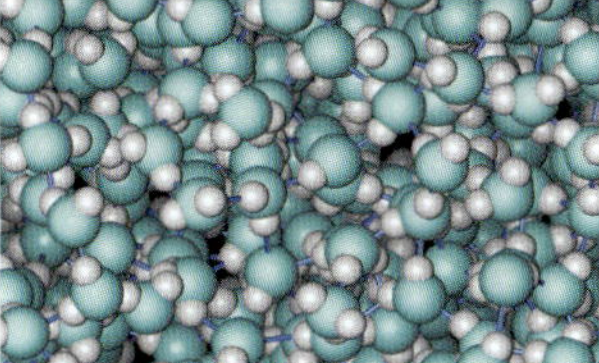

Figure 12: *The structures of ice (top) and liquid water (bottom).*

Behaviour of simple covalent compounds

Simple covalent compounds have strong covalent bonds within molecules (see pages 81-82), but weak forces between the molecules. Their physical properties, such as electrical conductivity, melting point and solubility are determined by the bonding in the compound.

Electrical conductivity

Simple covalent compounds don't conduct electricity because there are no free ions or electrons to carry the charge.

Melting point

Simple covalent compounds have low melting points because the weak forces between molecules are easily broken.

Solubility

Some simple covalent compounds dissolve in water depending on how polarised the molecules are (see pages 92-93 for more on polarisation).

Tip: When you melt or boil a simple covalent compound, the weak intermolecular forces are broken. The strong covalent bonds between the atoms in a molecule stay the same.

Trends in melting and boiling points

In general, the main factor that determines the melting or boiling point of a substance will be the strength of the induced dipole-dipole forces (unless the molecule can form hydrogen bonds).

Example

As you go down the Group 7 hydrides from HCl to HI, there are two competing factors that could affect the overall strength of the intermolecular bonds, and so the boiling points:

- The polarity of the molecules decreases, so the strength of the permanent dipole-dipole interactions decreases.
- The number of electrons in the molecules increases, so the strength of the induced dipole-dipole interactions increases.

As you can see from Figure 14, the boiling points of the Group 7 hydrides increase from HCl to HI. So the increasing strength of the induced dipole-dipole interactions has a greater effect on the boiling point than the decreasing strength of the permanent dipole-dipole interactions.

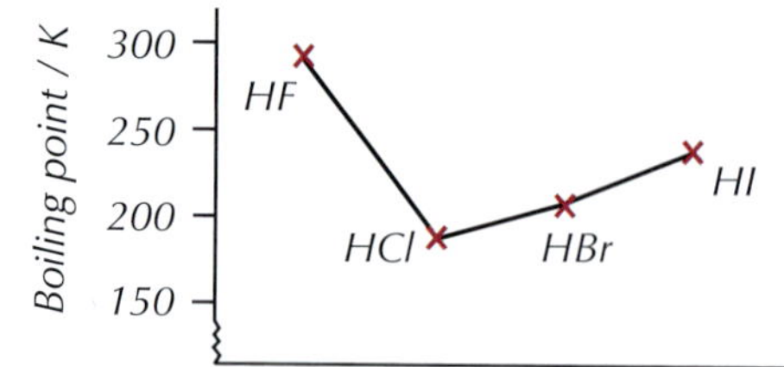

Figure 14: *Boiling points of the Group 7 hydrides.*

Tip: The boiling points of the Group 7 hydrides are still very low — except for HF, they're all gases at room temperature.

Tip: HF has the highest boiling point of the Group 7 hydrides because it can form hydrogen bonds.

If you have two molecules with a similar number of electrons, then the strength of their induced dipole-dipole interactions will be similar. So if one of the substances has molecules that are more polar than the other, it will have stronger permanent dipole-dipole interactions and so a higher boiling point.

Practice Questions — Application

Q1 What intermolecular force(s) exist(s) in H_2?

Q2 Chlorine (Cl_2) is a simple covalent molecule.

a) Explain why chlorine has a very low melting point.

b) Would you expect chlorine to conduct electricity? Explain your answer.

Q3 Look at the information in the table below.

Compound	Melting point / °C
Decane, $C_{10}H_{22}$	–30.5
Methane, CH_4	–182

Explain why the melting point of decane is higher than the melting point of methane.

Q4 The table below shows the electronegativity values of some elements

Element	H	C	Cl	O	F
Electronegativity (Pauling Scale)	2.20	2.55	3.16	3.44	3.98

a) Use the table above to explain why there are hydrogen bonds between H_2O molecules but not between HCl molecules.

b) Identify one other element from the table that would form hydrogen bonds when covalently bonded to hydrogen.

c) Name one other element from the table that would not form hydrogen bonds when covalently bonded to hydrogen.

Q5 Hydrogen has an electronegativity value of 2.20 on the Pauling scale, nitrogen has a value of 3.04 and phosphorous has a value of 2.19.

a) The boiling point of NH_3 is –33 °C and the boiling point of PH_3 is –88 °C. Explain why the boiling point of PH_3 is lower.

b) Arsenic (As) has an electronegativity value of 2.18. Would you expect the boiling point of AsH_3 to be higher or lower than that of NH_3?

Exam Tip
It's pretty common to be asked to identify and compare the intermolecular forces in different substances in the exam. You'll also need to know the effects that the intermolecular forces have on the properties of the substances, so make sure you know all this stuff inside out.

Practice Questions — Fact Recall

Q1 Describe the bonding within and between iodine molecules.

Q2 What are permanent dipole-dipole forces?

Q3 Name the strongest type of intermolecular force in water.

Q4 a) What is the strongest intermolecular force in ammonia?

b) Draw a diagram to show this intermolecular force between two ammonia molecules.

Q5 Explain why ice is less dense than liquid water.

Exam Tip
If you're asked to draw a diagram to show hydrogen bonding you'll need to include the lone pairs of electrons on the electronegative atom (O, N or F). You may also have to show the partial charges — the $\delta+$ goes on the H atom and the $\delta-$ goes on the electronegative atom.

7. Metallic Bonding

You might remember metallic bonding from GCSE, but there's more to know...

Learning Objectives:

- Know that metallic bonding involves attraction between delocalised electrons and positive ions arranged in a lattice.
- Know that metallic lattices are one of the four types of crystal structure.
- Know that the structure found in magnesium is an example of a metallic lattice.
- Be able to draw diagrams to represent the structure of metallic lattices.
- Be able to relate the melting point and conductivity of metallic compounds to their structure and bonding.

Specification Reference 3.1.3.3, 3.1.3.4

Metallic bonding

Metal elements exist as **giant metallic lattice structures**. The outermost shell of electrons of a metal atom is delocalised — the electrons are free to move about the metal. This leaves a positive metal ion, e.g. Na^+, Mg^{2+}, Al^{3+}. The positive metal ions are attracted to the delocalised negative electrons. They form a lattice of closely packed positive ions in a sea of delocalised electrons — this is **metallic bonding** (see Figure 1).

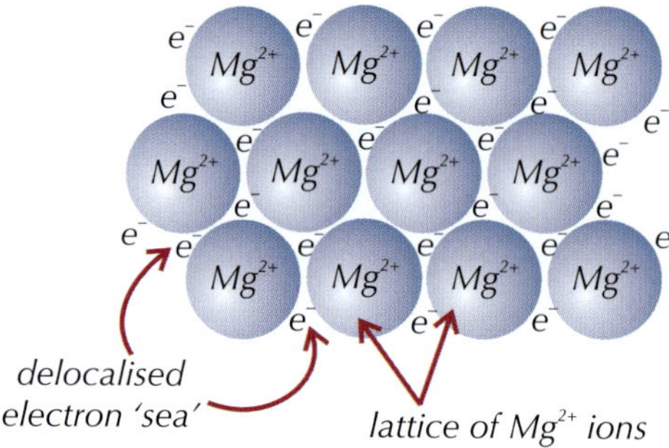

Figure 1: *Metallic bonding in magnesium.*

Metallic bonding explains the properties of metals.

Melting point

Metals have high melting points because of the strong electrostatic attraction between the positive metal ions and the delocalised sea of electrons. The number of delocalised electrons per atom affects the melting point. The more there are, the stronger the bonding will be and the higher the melting point. Mg^{2+} has two delocalised electrons per atom, so it's got a higher melting point than Na^+, which only has one. The size of the metal ion and the lattice structure also affect the melting point.

Ability to be shaped

As there are no bonds holding specific ions together, the metal ions can slide over each other when the structure is pulled, so metals are malleable (can be shaped) and ductile (can be drawn into a wire, see Figure 2).

Figure 2: *Copper drawn into a wire.*

Conductivity

The delocalised electrons can pass kinetic energy to each other, making metals good thermal conductors. Metals are good electrical conductors because the delocalised electrons can move and carry a charge.

Solubility

Metals are insoluble, except in liquid metals, because of the strength of the metallic bonds.

Practice Questions — Fact Recall

Q1 Describe the structure of magnesium.

Q2 What type of bonding can be found in magnesium?

Q3 Explain the following:

a) Copper can be drawn into wires.

b) Copper is a good thermal conductor.

8. Properties of Materials

You've covered loads of different types of bonds and intermolecular forces. Here are the last few bits to compare how these bonds and forces affect the properties of materials.

Learning Objectives:

- Be able to explain the energy changes associated with changes of state.
- Be able to relate the melting point and conductivity of materials to the type of structure and the bonding present.

Specification Reference 3.1.3.4

Solids, liquids and gases

A typical solid has its particles very close together. This gives it a high density and makes it incompressible. The particles vibrate about a fixed point and can't move about freely. A typical liquid has a similar density to a solid and is virtually incompressible. The particles move about freely and randomly within the liquid, allowing it to flow. In gases, the particles have loads more energy and are much further apart. So the density is generally pretty low and it's very compressible. The particles move about freely, with not a lot of attraction between them, so they'll quickly diffuse to fill a container (see Figures 1 and 2).

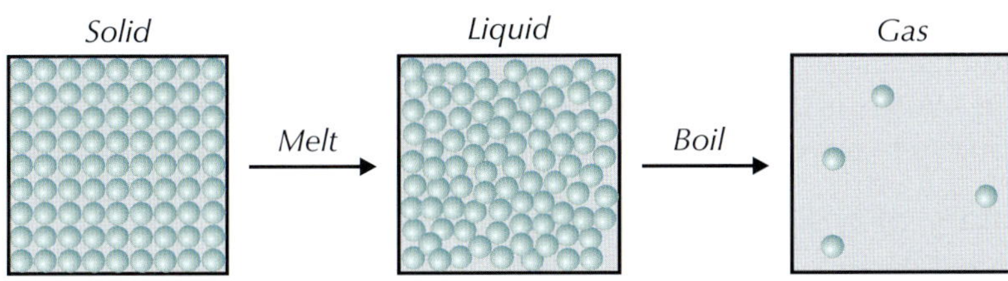

Figure 1: *The arrangement of particles in solids, liquids and gases.*

Figure 2: *Bromine liquid changing state to become a gas and diffusing to fill the container.*

In order to change from a solid to a liquid or a liquid to a gas, you need to break the forces that are holding the particles together. To do this you need to give the particles more energy, e.g. by heating them.

Melting and boiling covalent substances

In simple covalent substances, the covalent bonds don't break during melting and boiling. To melt or boil simple covalent substances you only have to overcome the weak intermolecular forces that hold the molecules together. You don't need to break the strong covalent bonds that hold the atoms together within the molecules. That's why simple covalent compounds have relatively low melting and boiling points.

By contrast, to melt or boil a giant covalent substance you do need to break the covalent bonds holding the atoms together. That's why giant covalent compounds have very high melting and boiling points.

Examples

Chlorine, Cl_2, is a simple covalent substance. To melt or boil chlorine, all you have to do is break the weak Van der Waals forces that hold the molecules together. Because of this, chlorine has a melting point of –101 °C and a boiling point of –34 °C — it's a gas at room temperature and pressure.

Bromine, Br_2, is also a simple covalent substance with low melting and boiling points. But bromine has slightly larger molecules than chlorine, which gives it slightly stronger Van der Waals forces. So bromine has a melting point of –7 °C and a boiling point of 59 °C — it's a liquid at room temperature and pressure.

Diamond is a giant covalent substance. To turn it into a liquid or a gas, you have to break the covalent bonds between carbon atoms. Diamond never really melts, but sublimes (goes straight from solid to gas) at over 3600 °C.

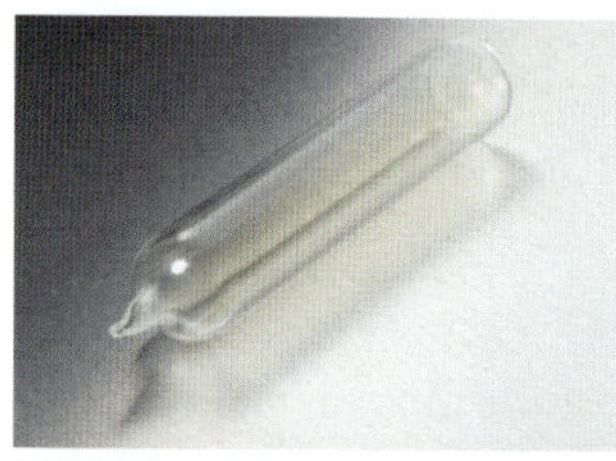

Figure 4: Chlorine (top) is a simple covalent compound and a gas at room temperature. Cobalt chloride (bottom) is an ionic compound and is a solid at room temperature.

Tip: A solvent is a liquid that a solid will dissolve in to form a solution.

Physical properties of materials

The particles that make up a substance, and the type of bonding that exists between them, will affect the physical properties of a material.

Melting and boiling points

The melting and boiling points of a substance are determined by the strength of the attraction between its particles. For example, ionic compounds have much higher boiling and melting points than simple covalent substances — the strong electrostatic attraction between the ions requires a lot more energy to break than the weak intermolecular forces between molecules (see Figure 4).

Electrical conductivity

A substance will only conduct electricity if it contains charged particles that are free to move, such as the delocalised electrons in a metal.

Solubility

How soluble a substance is in water depends on the type of particles that it contains. Water is a polar solvent, so substances that are polar or charged will dissolve in it well, whereas non-polar or uncharged substances won't.

Summary of typical properties

The table in Figure 5 is a summary of the typical physical properties of the types of materials you've met in this section. You should make sure you know it like the back of your hand.

Bonding	Melting/ boiling points	Typical state under standard conditions	Does it conduct electricity?	Is it soluble in water?
Ionic	High	Solid	Not as a solid (ions are held in place), but it will when molten or dissolved (ions are free to move).	Yes
Simple covalent (molecular)	Low	May be solid (e.g. I_2) but usually liquid or gas.	No	Depends on how polar the molecules are.
Giant covalent	High	Solid	No (except graphite)	No
Metallic	High	Solid	Yes (delocalised electrons)	No

Figure 5: A summary of the physical properties of different types of compound.

Exam Tip
In the exam, make sure you can explain why each material has these properties, as well as what the properties are.

Tip: Make sure you do a risk assessment before you carry out any experiments in class.

If you have an unknown compound and you want to predict the type of structure it has, you can carry out experiments to work out its physical properties. For example, you could test its electrical conductivity as a solid and a liquid, or investigate whether it has a high or low boiling point. From its physical properties, you can use the table above to predict what structure the compound is likely to have.

Example

A student is trying to predict the structure of an unknown compound called substance X, which is a solid at room temperature. He carries out the following tests:

- First, he tests the conductivity of the compound when it is solid and finds that, when solid, it is an electrical insulator.
- Next, he tries to dissolve a sample of the compound in water and finds that it is soluble.
- Finally, he tests the conductivity of the solution of substance X and finds that it now conducts electricity.

From this information the student predicts that substance X is an ionic compound. This is because it doesn't conduct electricity when it's solid, but does conduct electricity once dissolved. It's also soluble in water.

It's also clear that substance X isn't metallic because it doesn't conduct electricity when it's solid, and it definitely isn't giant covalent or simple covalent because it does conduct electricity when dissolved.

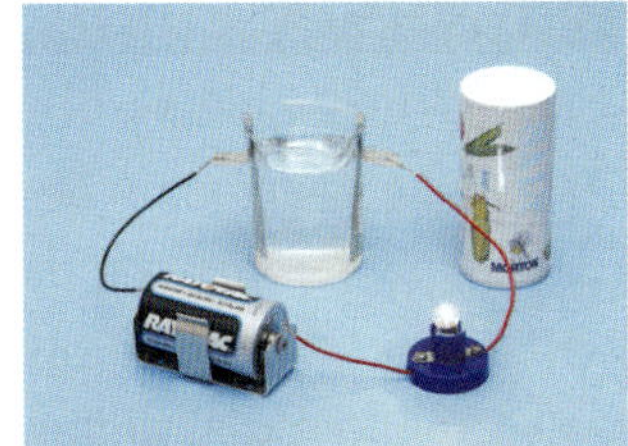

Figure 6: *An experiment to test the conductivity of a solution.*

Tip: If you're trying to work out the structure of a compound, you have to plan your experiment carefully to prove that it must be one structure and can't be any of the others. If the student in the example had just tested the electrical conductivity of the solid compound and then the solubility, he couldn't have ruled out the possibility of it being a polar, simple covalent compound.

Practice Questions — Application

Q1 The melting point of silicon dioxide is 1610 °C. It is insoluble in water and doesn't conduct electricity.

a) Suggest what type of structure silicon dioxide has.

b) Explain why silicon dioxide has a high melting point.

Q2 A student has a sample of a solid substance. She performs a series of tests on the sample and determines that it will not conduct electricity when solid, will not dissolve in water and has a low melting point.

Predict the structure of the substance.

Practice Questions — Fact Recall

Q1 Describe what happens to the particles in a substance when a solid changes to a liquid.

Q2 Explain why simple covalent substances have lower melting points than giant covalent substances.

Q3 Give three typical physical properties of metallic compounds.

Section Summary

Make sure you know...

- That ions form when electrons are transferred from one atom to another.
- How to predict the charge of a simple ion based on the Group of the Periodic Table that it is in.
- That compound ions are made up of groups of two or more atoms.
- The formulas of the compound ions sulfate, hydroxide, nitrate, carbonate and ammonium.
- That electrostatic attraction holds ions together and that this is called ionic bonding.
- How to work out the formulas of neutral ionic compounds.
- That ions form crystals that are giant ionic lattices.
- The structure of sodium chloride.
- How the structure of ionic compounds decides their physical properties — their electrical conductivity, melting point and solubility.
- That covalent bonds form when atoms share pairs of electrons.
- How single and multiple (double and triple) covalent bonds form between atoms.
- What a giant covalent (macromolecular) structure is.
- The structures of graphite and diamond and how the structures determine their properties.
- That co-ordinate (dative covalent) bonds form when one atom donates both shared electrons in a bond.
- That covalent bonds can be represented using a line and co-ordinate bonds can be represented using an arrow that points away from the donor atom.
- That charge clouds represent the areas where bonding pairs and lone pairs of electrons are most likely to be found.
- That charge clouds repel each other and arrange so that they are as far apart as possible.
- That Valence Shell Electron Pair Repulsion Theory states that lone-pair/lone-pair bond angles are the biggest, lone-pair/bonding-pair bond angles are the second biggest and bonding-pair/bonding-pair bond angles are the smallest.
- How to predict the shapes of molecules with up to six pairs of electrons surrounding the central atom, including their bond angles and shape names.
- That electronegativity is the ability to attract the bonding electrons in a covalent bond.
- How differences in electronegativities between bonding atoms causes polarisation.
- The difference between polar and non-polar bonds.
- That a molecule with polar bonds may have a permanent dipole, depending on its shape.
- What van der Waals forces and permanent dipole-dipole forces are, and what causes them.
- That simple covalent molecules can form molecular lattices.
- The structure of solid iodine.
- How hydrogen bonds form and their effect on the properties of compounds.
- The structure of ice, and how the formation of hydrogen bonds explains its unusually low density.
- How the structure of simple covalent compounds decides their physical properties — their electrical conductivity, melting point and solubility.
- What metallic bonding is and how to recognise giant metallic lattice structures.
- The structure of magnesium.
- How the structure of metals decides their physical properties — their conductivity, melting point, ability to be shaped and solubility.
- The energy changes that take place when a substance changes state.
- How to predict the structure of a compound based on its physical properties.

Exam-style Questions

1 Solid aluminium is an electrical conductor, used in overhead cables.
It has a melting point of 660 °C and is insoluble in water.
Predict the structure of aluminium.

A giant covalent

B giant ionic lattice

C metallic

D simple covalent

(1 mark)

2 Which of the following statements about hydrogen selenide, H_2Se, is correct?

A Hydrogen selenide is a linear molecule.

B Hydrogen selenide has a higher boiling point than hydrogen sulfide (H_2S).

C There is one lone pair and two bonding pairs around the central Se atom in H_2Se.

D Hydrogen selenide is able to form hydrogen bonds.

(1 mark)

3 Germanium is in the same group of the periodic table as carbon.
Germanium reacts with hydrogen to form the compound germane, GeH_4.

3.1 Name the type of bonding between germanium and hydrogen in germane, and describe how the bonds are formed.

(2 marks)

3.2 Draw the shape of a molecule of GeH_4, labelling the bond angles, and name the shape.

(3 marks)

3.3 State whether or not you would expect germane to conduct electricity. Explain your answer.

(2 marks)

3.4 Germanium also combines with chlorine to form germanium dichloride, $GeCl_2$. Draw and name the shape of a molecule of $GeCl_2$.

(2 marks)

3.5 Suggest a value for the bond angle in $GeCl_2$ and explain your answer.

(2 marks)

4 The Group 5 elements include nitrogen, phosphorus, arsenic and antimony. They can form covalent bonds with hydrogen.

The graph below shows the boiling points of some Group 5 hydrides.

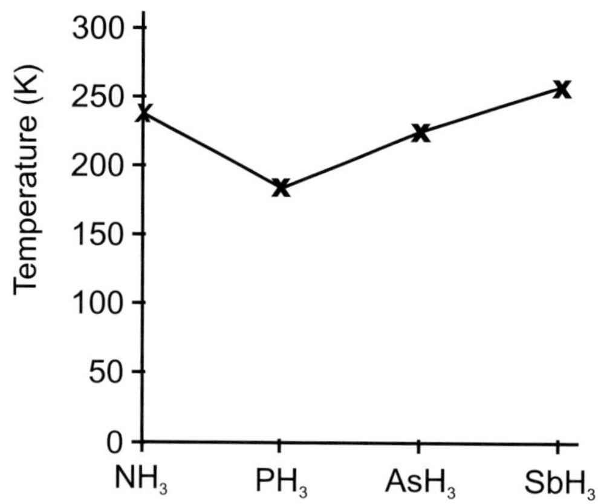

4.1 Explain the trend in boiling points shown by the graph for PH_3, AsH_3 and SbH_3.

(2 marks)

4.2 Name the strongest type of intermolecular force found in NH_3.

(1 mark)

4.3 NH_3 reacts with H^+ to form an NH_4^+ ion. Name the type of bond that forms in this reaction and explain how it is formed.

(3 marks)

4.4 Explain why the bond angle in NH_3 is smaller than in NH_4^+.

(3 marks)

5 Sodium chloride (NaCl) is an ionic compound formed from sodium metal and chlorine gas (Cl_2).

5.1 Name and describe the structure and bonding in sodium metal.

(3 marks)

5.2 Explain how this bonding structure allows sodium to be easily shaped.

(1 mark)

5.3 State what is meant by the term ionic bond.

(1 mark)

5.4 Describe the structure of sodium chloride.

(2 marks)

6 The table below shows the electronegativities of some elements.

Element	C	H	Cl	O	F
Electronegativity (Pauling Scale)	2.55	2.20	3.16	3.44	3.98

6.1 Define the term electronegativity and explain how electronegativity can give rise to permanent dipole-dipole interactions.

(4 marks)

6.2 Use the information in the table to name all the intermolecular forces present in each of the following compounds:

HCl

CH_4

H_2O

(3 marks)

6.3 Draw a diagram to show the strongest intermolecular forces between HF molecules. Include partial charges and all lone pairs.

(3 marks)

6.4 Explain why the only forces between Cl_2 molecules are van der Waals forces.

(1 mark)

7 Carbon can form lots of different structures and can combine with other elements to form lots of different compounds.

7.1 Graphite and diamond contain only carbon atoms. Both substances have the same type of structure. Name this structure.

(1 mark)

7.2 Describe the structures and bonding of graphite and diamond

(4 marks)

7.3 Explain why graphite can conduct electricity but diamond cannot.

(1 mark)

7.4 State the type of intermolecular forces found between molecules of methane, CH_4.

Compare the strength of the intermolecular forces between molecules of CH_4 to the intermolecular forces between molecules of C_3H_8.

(2 marks)

7.5 Explain why the boiling point of diamond is much higher than the boiling point of methane.

(2 marks)

Learning Objectives:

- Understand that enthalpy change (ΔH) is the heat energy change measured under conditions of constant pressure.
- Know that standard enthalpy changes refer to standard conditions, i.e. 100 kPa and a stated temperature (e.g. $\Delta H_{298}^{\ominus}$).
- Know that reactions can be endothermic or exothermic.

Specification Reference 3.1.4.1

1. Enthalpy

When chemical reactions happen, there'll be a change in energy. The souped-up chemistry term for this is enthalpy change.

Enthalpy notation

Enthalpy change, ΔH (delta H), is the heat energy transferred in a reaction at constant pressure. The units of ΔH are kJ mol^{-1}. You write $\Delta H^{\ominus}$ to show that the reactants and products were in their standard states and that the measurements were made under **standard conditions**. Standard conditions are 100 kPa (about 1 atm) pressure and a stated temperature (e.g. $\Delta H^{\ominus}_{298}$). In this book, all the enthalpy changes are measured at 298 K (25 °C). Sometimes the notation will also include a letter to signify whether the enthalpy change is for a reaction (r), combustion (c), or the formation of a new compound (f). See page 111 for more on this notation.

Tip: $\Delta_c H^{\ominus}_{298}$ is the notation for the enthalpy change of a combustion under standard conditions (a pressure of 100 kPa with all substances in their standard states) and at a temperature of 298 K.

Exothermic reactions

Exothermic reactions give out energy to their surroundings, so the temperature in the reaction usually goes up. The products of the reaction end up with less energy than the reactants. This means that the enthalpy change for the reaction, ΔH, will be negative.

Examples

Oxidation is usually exothermic. Here are two examples:

The combustion of a fuel like methane:

$$CH_{4(g)} + 2O_{2(g)} \rightarrow CO_{2(g)} + 2H_2O_{(l)} \qquad \Delta_c H^{\ominus}_{298} = -890 \text{ kJ mol}^{-1}$$

ΔH is negative so the reaction is **exothermic**.

The oxidation of carbohydrates, like glucose, in respiration is exothermic.

Endothermic reactions

Endothermic reactions take in energy from their surroundings, so the temperature in the reaction usually falls. This means that the products of the reaction have more energy than the reactants, so the enthalpy change for the reaction, ΔH, is positive.

Examples

The thermal decomposition of calcium carbonate is endothermic.

$$CaCO_{3(s)} \rightarrow CaO_{(s)} + CO_{2(g)} \qquad \Delta_r H^{\ominus}_{298} = +178 \text{ kJ mol}^{-1}$$

ΔH is positive so the reaction is **endothermic**.

The main reactions of photosynthesis are also endothermic — sunlight supplies the energy.

Figure 1: Photosynthesis in plants is endothermic.

Practice Questions — Fact Recall

Q1 Give the notation for an enthalpy change under standard conditions, at a temperature of 298 K.

Q2 Describe the difference between exothermic and endothermic reactions.

2. Bond Enthalpies

Reactions involve breaking and making bonds. The enthalpy change for a reaction depends on which bonds are broken and which are made.

Learning Objectives:

- Define, understand and use the terms bond enthalpy and mean bond enthalpy.
- Be able to use mean bond enthalpies to calculate an approximate value of ΔH for reactions in the gaseous phase.
- Be able to explain why values from mean bond enthalpy calculations differ from those determined using Hess's Law.
- Be able to recall the definition of standard enthalpies of combustion ($\Delta_c H^{\ominus}$) and formation ($\Delta_f H^{\ominus}$).

Specification Reference 3.1.4.1, 3.1.4.4

What are bond enthalpies?

Atoms in molecules are held together by strong covalent bonds. It takes energy to break bonds and energy is given out when new bonds form. **Bond enthalpy** is the energy needed to break a bond. Bond enthalpies have specific values that differ depending on which atoms are attached on either side of the bond.

Breaking and making bonds

When reactions happen, reactant bonds are broken and product bonds are formed. You need energy to break bonds, so bond breaking is endothermic (ΔH is positive). Stronger bonds take more energy to break. Energy is released when bonds are formed, so this is exothermic (ΔH is negative). Stronger bonds release more energy when they form. The enthalpy change for a reaction is the overall effect of these two changes. If you need more energy to break bonds than is released when bonds are made, ΔH is positive. If it's less, ΔH is negative.

Example — Maths Skills

Nitrogen reacts with hydrogen to form ammonia (NH_3) in this reaction:

$$N_2 + 3H_2 \rightarrow 2NH_3$$

The energy needed to break all the bonds in N_2 and H_2 = 2253 kJ mol^{-1}.

The energy released when forming the bonds in NH_3 = 2346 kJ mol^{-1}.

The amount of energy released is bigger than the amount needed, so the reaction is exothermic, and ΔH is negative.

Tip: You can look up the mean (average) bond enthalpies for different bonds in a data book, or calculate the mean bond enthalpies from given data. In an exam you'll be given any bond enthalpies you need.

Mean bond enthalpies

Bond enthalpy is the energy needed to break a bond. The energy required to break a certain type of bond can change depending on where it is, e.g. it takes different amounts of energy to break the C-C bond in methane, ethane, ethene, etc. So in calculations, you use **mean bond enthalpy** — that's the average energy needed to break a certain type of bond, over a range of compounds. It is often given in data tables and used in calculations.

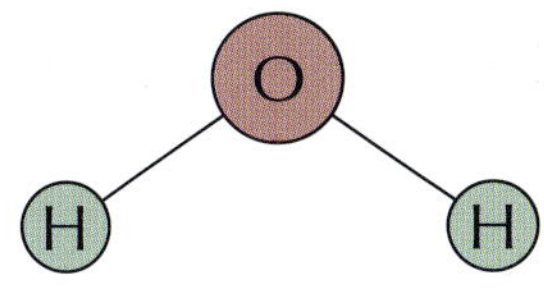

***Figure 1:** The bonds in a water molecule.*

Example — Maths Skills

Water (H_2O) has got two O–H bonds (see Figure 1). You'd think it'd take the same amount of energy to break them both, but it doesn't.

The first bond, $H–OH_{(g)}$: E(H–OH) = +492 kJ mol^{-1}

The second bond, $H–O_{(g)}$: E(H–O) = +428 kJ mol^{-1}

(OH^- is a bit easier to break apart because of the extra electron repulsion.)

So, the mean bond enthalpy for O–H bonds in water is:

$$\frac{492 + 428}{2} = +460 \text{ kJ mol}^{-1}.$$

The data book says the bond enthalpy for O–H is +464 kJ mol^{-1}. It's a bit different than the one calculated above because it's the average for a much bigger range of molecules, not just water. For example, it includes the O–H bonds in alcohols and carboxylic acids too.

Tip: Breaking bonds is always an endothermic process, so mean bond enthalpies are always positive.

Calculating enthalpy changes

In any chemical reaction, energy is absorbed to break bonds and given out during bond formation. The difference between the energy absorbed and released is the overall enthalpy change of reaction:

Enthalpy change of reaction = Total energy absorbed – Total energy released

- To calculate the overall enthalpy change for a reaction, first calculate the total energy needed to break the bonds in the reactants. You'll usually be given the average bond enthalpies for each type of bond, so just multiply each value by the number of each bond present. This total will be the total energy absorbed in the reaction.
- To find the total energy released by the reaction, calculate the total energy needed to form all the new bonds in the products. Use the average bond enthalpies to do this.
- The overall enthalpy change for the reaction can then be found by subtracting the total energy released from the total energy absorbed.

Tip: Draw sketches to show the bonds present in the reactants and products to make sure you include them all in your calculations.

Bond	Bond Enthalpy (Mean value except where stated)
N≡N	945 kJ mol^{-1}
H–H	436 kJ mol^{-1}
N–H	391 kJ mol^{-1}
O=O	498 kJ mol^{-1}
O–H	464 kJ mol^{-1}

Figure 2: *Table of bond enthalpies.*

Examples — Maths Skills

Calculate the overall enthalpy change for the following reaction:

$$N_{2(g)} + 3H_{2(g)} \rightarrow 2NH_{3(g)}$$

Use the mean bond enthalpy values shown in Figure 2.

You might find it helpful to draw a sketch of the molecules in the reaction (it doesn't have to be 3D, it's just to help you to see all the bonds):

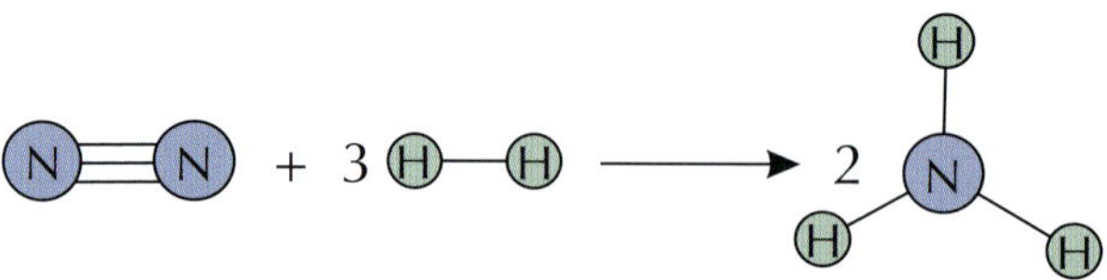

Bonds broken: 1 × N≡N bond broken = 1 × 945 = 945 kJ mol^{-1}
3 × H–H bonds broken = 3 × 436 = 1308 kJ mol^{-1}
Total Energy Absorbed = 945 + 1308 = 2253 kJ mol^{-1}

Bonds formed: 6 × N–H bonds formed = 6 × 391 = 2346 kJ mol^{-1}
Total Energy Released = 2346 kJ mol^{-1}

Now you just subtract 'total energy released' from 'total energy absorbed':
Enthalpy change of reaction = 2253 – 2346 = –93 kJ mol^{-1}.

Tip: Calculating this enthalpy change using Hess's Law gives a value of –92.4 kJ mol^{-1}. This is slightly different because Hess's Law uses precise data, whilst this calculation uses mean bond enthalpy values. See pages 116-118 for more on Hess's Law.

Calculate the overall enthalpy change for the following reaction:

$$H_{2(g)} + \tfrac{1}{2}O_{2(g)} \rightarrow H_2O_{(g)}$$

Use the mean bond enthalpy values shown in Figure 2.

The molecules present are shown below:

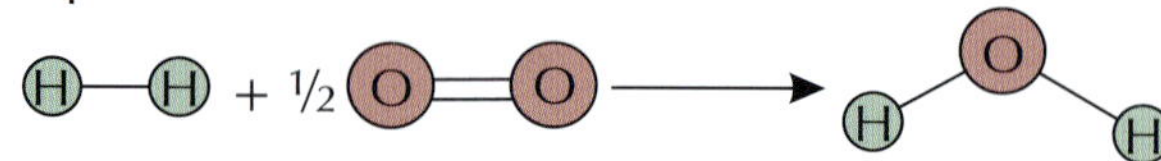

Bonds broken: 1 × H–H bond broken = 1 × 436 = 436 kJ mol^{-1}
½ × O=O bonds broken = ½ × 498 = 249 kJ mol^{-1}
Total Energy Absorbed = 436 + 249 = 685 kJ mol^{-1}

Bonds formed: 2 × O–H bonds formed = 2 × 464 = 928 kJ mol^{-1}
Total Energy Released = 928 kJ mol^{-1}

Enthalpy change of reaction = 685 – 928 = –243 kJ mol^{-1}.

Tip: If you can't remember which value to subtract from which, just take the smaller number from the bigger one then add the sign at the end — positive if 'bonds broken' was the bigger number (endothermic), negative if 'bonds formed' was bigger (exothermic).

The different types of ΔH

Standard enthalpy change of formation

Standard enthalpy change of formation, $\Delta_f H^\circ$, is the enthalpy change when 1 mole of a compound is formed from its elements in their standard states under standard conditions, e.g $2C_{(s)} + 3H_{2(g)} + \frac{1}{2}O_{2(g)} \rightarrow C_2H_5OH_{(l)}$

Standard enthalpy change of combustion

Standard enthalpy change of combustion, $\Delta_c H^\circ$, is the enthalpy change when 1 mole of a substance is completely burned in oxygen under standard conditions with all reactants and products in their standard states.

Standard enthalpy change of reaction

Standard enthalpy change of reaction, $\Delta_r H^\circ$, is the enthalpy change when a reaction occurs in the molar quantities shown in the chemical equation, under standard conditions with all reactants and products in their standard states.

Exam Tip
There's quite a bit of maths in this section but you need to learn all these definitions too — you can be asked to write them down in an exam, and it's often the thing that people forget or get wrong.

Tip: $\Delta_c H^\circ$ is sometimes written as ΔH_c° — the same goes for $\Delta_f H^\circ$ and $\Delta_r H^\circ$.

Practice Questions — Application

Q1 Use the mean bond enthalpies shown in Figure 3 to calculate the enthalpy changes for the following reactions:

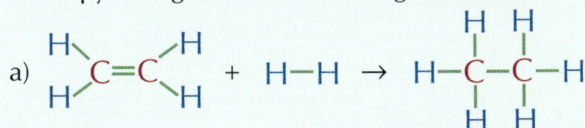

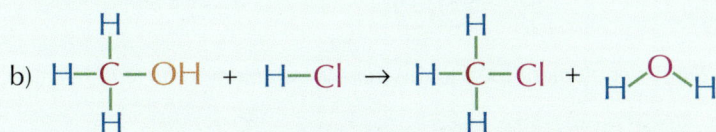

a) $H_2C{=}CH_2 + H{-}H \rightarrow H_3C{-}CH_3$

b) $H{-}CH_2{-}OH + H{-}Cl \rightarrow H{-}CH_2{-}Cl + H{-}O{-}H$

c) $C_3H_8 + 5O_2 \rightarrow 3CO_2 + 4H_2O$

d) $C_2H_5Cl + NH_3 \rightarrow C_2H_5NH_2 + HCl$

Q2 Calculate the enthalpy change for the complete combustion of ethene (C_2H_4) using the bond enthalpies given in Figure 3. (The products of complete combustion are CO_2 and H_2O.)

Q3 Calculate the enthalpy change for the formation of hydrogen chloride ($HCl_{(g)}$) from hydrogen ($H_{2(g)}$) and chlorine ($Cl_{2(g)}$) using the bond enthalpies given in Figure 3.

Q4 The enthalpy change for the following reaction is –181 kJ mol^{-1}:

$$2NO_{(g)} \rightarrow N_{2(g)} + O_{2(g)}$$

Use this value for $\Delta_r H$, along with the data in Figure 3, to estimate a value for the mean bond enthalpy for the bond between nitrogen and oxygen in NO.

Tip: You can ignore any bonds that don't actually change during the reaction. Just work out which bonds actually break and which new bonds form.

Bond	Bond Enthalpy (Mean value except where stated)
N≡N	945 kJ mol^{-1}
H–H	436 kJ mol^{-1}
N–H	391 kJ mol^{-1}
O=O	498 kJ mol^{-1}
O–H	464 kJ mol^{-1}
C–H	413 kJ mol^{-1}
C=C	612 kJ mol^{-1}
C–C	347 kJ mol^{-1}
C–O	358 kJ mol^{-1}
C–Cl	346 kJ mol^{-1}
C=O (in CO_2)	805 kJ mol^{-1}
C–N	286 kJ mol^{-1}
H–Cl	432 kJ mol^{-1}
Cl–Cl	243 kJ mol^{-1}

Figure 3: Table of bond enthalpies.

Practice Questions — Fact Recall

Q1 What notation is used for the standard enthalpy change of:

a) formation?

b) combustion?

Q2 Define the 'standard enthalpy change of formation'.

3. Measuring Enthalpy Changes

A lot of the data we have on enthalpy changes has come from someone, somewhere, measuring the enthalpy change of a reaction in a lab.

Learning Objectives:

- Know how to measure an enthalpy change (Required Practical 2).
- Know that the heat change, q, in a reaction is given by the equation $q = mc\Delta T$, where m is the mass of the substance that has a specific heat capacity, c, and ΔT is the change in temperature.
- Be able to use the equation $q = mc\Delta T$ in related calculations, including calculating the molar enthalpy change for a reaction.

Specification Reference 3.1.4.2

Tip: Always carry out a risk assessment and put in place any safety precautions that you might need before starting an experiment.

Measuring an enthalpy change in the lab

REQUIRED PRACTICAL 2

To find the enthalpy change for a reaction, you only need to know three things — the number of moles of the stuff that's reacting, the change in temperature, and how much stuff you're heating. Experiments that measure the heat given out by reactions are called **calorimetry** experiments. How you go about doing an experiment like this depends on what type of reaction it is.

For reactions that happen in solution (see page 114), you just put the reactants in a container and use a thermometer to measure the temperature of the mixture at regular intervals (see Figure 1). It's best to use a polystyrene beaker to reduce the amount of heat lost or gained through the sides.

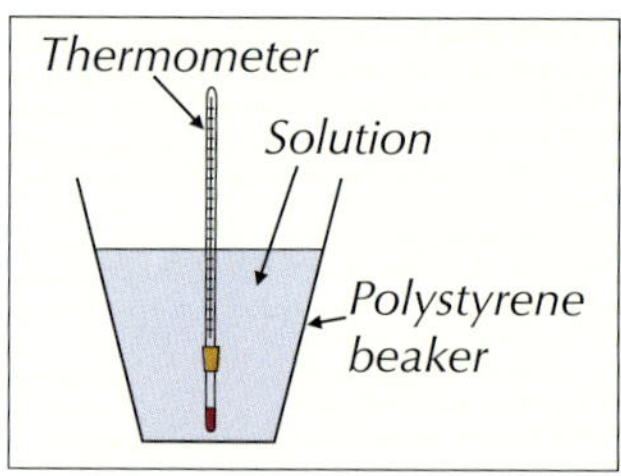

Figure 1: *Simple equipment used to measure the enthalpy change of reaction.*

Calorimetry and combustion reactions

You can find out how much energy is given out by a combustion reaction by measuring the temperature change it causes as it burns. To find the enthalpy of combustion of a flammable liquid, you burn it in a calorimeter (see Figure 2).

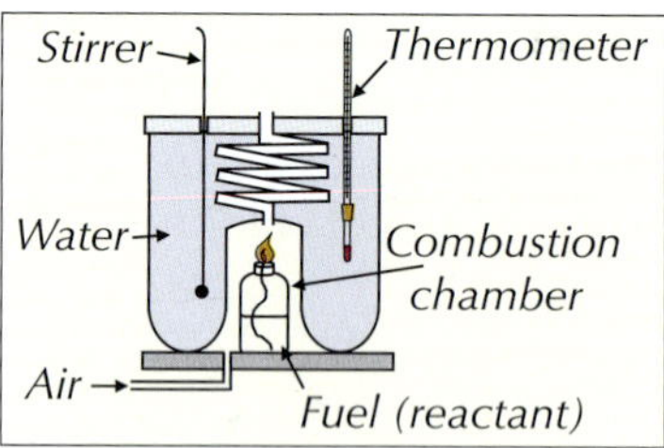

Figure 2: *A calorimeter used to measure the enthalpy change of combustion.*

Figure 3: *A bomb calorimeter. This is a very accurate piece of equipment that works on the same principle as the one shown in the diagram in Figure 2.*

As the fuel burns, it heats the water. You can work out the heat energy that has been absorbed by the water if you know the mass of the water, the temperature change (ΔT), and the specific heat capacity of water ($= 4.18 \text{ J g}^{-1} \text{ K}^{-1}$) — see the next page for the details of how to do this.

Ideally, all the heat given out by the burning fuel would be absorbed by the water, allowing you to work out the enthalpy change of combustion exactly. In practice you always lose some heat to the surroundings, however well your calorimeter is insulated. This makes it hard to get an accurate result.

Also, when you burn a fuel, some of the combustion that takes place may be incomplete, meaning that less energy is given out. Flammable liquids are often quite volatile too, so you may lose some of the fuel to evaporation.

Calculating an accurate temperature change

The most obvious way of finding the temperature change in a calorimetry experiment is to subtract the starting temperature from the highest temperature you recorded. But that won't give you a very accurate value, because of the heat lost from the calorimeter to the surroundings. Instead, you can use a graph of your results to find a more accurate value. Here's what you do:

1. During the experiment, record the temperature at regular intervals, beginning a couple of minutes before you start the reaction.
2. Plot a graph of your results.
3. Draw two lines of best fit: one going through the points from before the reaction started and one going through the points from after it started.
4. Extend both lines so that they both pass the time when the reaction started.
5. The distance between the two lines at the time the reaction started (before any heat was lost) is the accurate temperature change (ΔT) for the reaction.

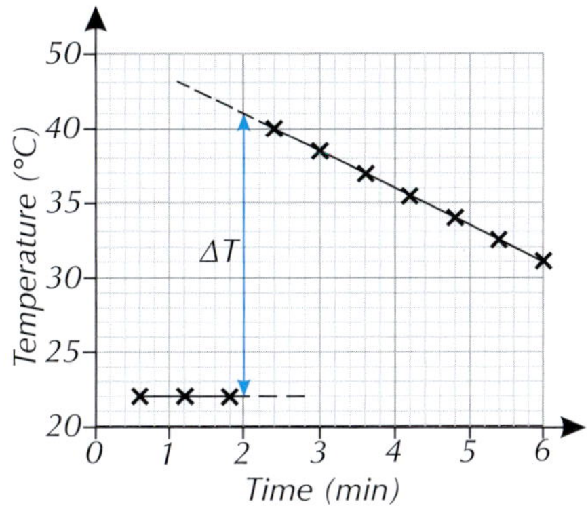

***Figure 4:** A graph being used to find the temperature change for a reaction beginning at time = 2 minutes.*

Using the equation $q = mc\Delta T$

The equation used to calculate the enthalpy change of a reaction is:

$$q = mc\Delta T$$

***q** = heat lost or gained (in J). This is the same as the enthalpy change if the pressure is constant.*

***m** = mass (in g) of solution in the polystyrene beaker (or mass of water in the calorimeter).*

***c** = specific heat capacity of the solution / water (4.18 J $g^{-1}K^{-1}$).*

***ΔT** = the change in temperature of the solution / water (in Kelvin).*

Tip: The specific heat capacity of a solution is the amount of heat energy it takes to raise the temperature of 1 g of the solution by 1 K.

Calculating the standard enthalpy change of combustion

To calculate the standard enthalpy change of combustion, $\Delta_c H^\circ$, using data from a laboratory experiment, follow these steps:

Step 1: Calculate the amount of heat lost or gained during the combustion using $q = mc\Delta T$ and your measured or given values of m and ΔT. You'll then need to change the units of q from joules to kilojoules, because standard enthalpies of combustion are always given in units of kJ mol^{-1}.

Step 2: Calculate the number of moles of fuel that caused this enthalpy change, from the mass that reacted. Use the equation:

$$n = \frac{mass}{M_r}$$

n is the number of moles of fuel burned.
M is the fuel's relative molecular mass.

Step 3: Calculate the standard enthalpy change of combustion, $\Delta_c H^\circ$ (in kJ mol^{-1}), using the actual heat change for the reaction, q (in kJ), and the number of moles of fuel that burned, n. Use the equation:

$$\Delta_c H^\circ = \frac{q}{n}$$

Tip: ΔH° is the standard enthalpy change of a reaction carried out at 100 kPa with all reactants and products in their standard states (see page 108). If the experiment was carried out under different conditions, this method wouldn't give you the value for ΔH°.

Example — Maths Skills

In a laboratory experiment, 1.16 g of an organic liquid fuel was completely burned in oxygen. The heat formed during this combustion raised the temperature of 100 g of water from 295.3 K to 357.8 K. Calculate the standard enthalpy of combustion, $\Delta_c H^\circ$, of the fuel. Its M_r is 58.

Step 1: Calculate the amount of heat given out by the fuel using $q = mc\Delta T$. Remember that m is the mass of water, not the mass of fuel.

$q = mc\Delta T$

$q = 100 \times 4.18 \times (357.8 - 295.3) = 26\,125$ J

Change the amount of heat from J to kJ: $q = 26.125$ kJ.

Step 2: Find out how many moles of fuel produced this heat:

$$n = \frac{\text{mass}}{M_r} = \frac{1.16\text{ g}}{58\text{ g mol}^{-1}} = 0.0200 \text{ moles of fuel.}$$

Step 3: The standard enthalpy of combustion involves 1 mole of fuel.

So $\Delta_c H^\circ = \frac{q}{n} = \frac{-26.125\text{ kJ}}{0.0200\text{ mol}} \approx$ **-1310 kJ mol^{-1}** (to 3 s.f.)

(q is negative here because combustion is an exothermic reaction.)

Tip: –26.125 kJ is the enthalpy change of this reaction. That's the amount of energy given out when this mass of this fuel is burned. –1310 kJ mol^{-1} is the molar enthalpy change of this reaction. That's the amount of energy that would be given out when 1 mole of this fuel was burned.

Measuring enthalpy changes in solution

The same formula can also be used to calculate the enthalpy change for a reaction that happens in solution, such as neutralisation, dissolution (dissolving) or displacement. Here's an example:

Measuring an enthalpy of neutralisation

REQUIRED PRACTICAL 2

The enthalpy of neutralisation is the energy change when one mole of water is formed by the reaction of an acid and an alkali.

To find the enthalpy change for a neutralisation reaction, add a known volume of acid to an insulated container (e.g. a polystyrene cup) and measure the temperature. Then add a known volume of alkali and record the temperature at regular intervals. Stir the solution to make sure it's evenly heated. When you've worked out the temperature change (using the method from page 113), you can work out the heat energy given out by the reaction using the formula from the previous page.

Tip: Make sure that you carry out all necessary safety precautions whilst doing any experiment.

Example

50 cm^3 of 1.0 mol dm^{-3} sodium hydroxide was added to 50 cm^3 of 1.0 mol dm^{-3} hydrochloric acid. The temperature rose by 6.9 °C. Calculate the enthalpy change of neutralisation for this reaction. The equation for the reaction is: $HCl + NaOH \rightarrow NaCl + H_2O$

First calculate the heat energy given out by the reaction:

- You can assume that all solutions have the same density as water. Since 1 cm^3 of water has a mass of 1 g, each cm^3 of solution you use will have a mass of 1g. So $m = 50 + 50 = 100$ g.
- You can also usually assume that all solutions have the same specific heat capacity as water (4.18 kJ g^{-1} K^{-1}).
- ΔT is the change in temperature in K, which is equal to the change in temperature in °C — so $\Delta T = 6.9$.

$q = mc\Delta T = 100 \times 4.18 \times 6.9 = 2884.2$ J $= 2.8842$ kJ.

Tip: Don't forget that if you're mixing two solutions you need to include the masses of both when you're finding m.

Now calculate the enthalpy change of neutralisation — to do this you need to divide the heat energy change by the number of moles of H_2O produced.

- First work out how many moles of acid you started off with
  moles = concentration (mol dm^{-3}) × volume (dm^3)
  = 1.0 × (50 ÷ 1000) = 1.0 × 0.050 = 0.05
- From the balanced equation, the ratio between HCl and H_2O is 1:1 — so if 0.050 moles of HCl reacted, 0.050 moles of H_2O was produced.
- So enthalpy of neutralisation = = $\frac{q}{n} = \frac{-2.8842}{0.050}$ = **–58 kJ mol⁻¹** (to 2 s.f.)

(q is negative because this is an exothermic reaction.)

Tip: You could work out how many moles of alkali you started with here instead — it's fine to use either.

Measuring the enthalpy change of a dissolution reaction works in a very similar way. You just add a known mass of the solid that you're dissolving to a known volume of water in the polystyrene cup and stir, recording the temperature regularly. Once you've found the temperature change, you can use it to work out the energy change for the reaction.

If you're asked to find the molar enthalpy change for this type of reaction, you need to find the enthalpy change per mole of solute dissolved (this is the **enthalpy of solution**). This means that you need to divide the heat energy change for the reaction by the number of moles of solid that you started off with.

Displacement reactions often mean mixing a solid (such as a metal) with a salt solution. Some displacement reactions involve mixing two solutions though, so you can just use whichever method fits your reactants.

Tip: Dissolution can be exothermic or endothermic, so make sure that the sign of your final answer is right. If the temperature increases, your answer should be negative. If it decreases, your answer should be positive.

Tip: If you're finding the molar enthalpy change for a displacement reaction, you might be told which substance to find the number of moles of and divide by. (For example "Find the enthalpy change per mole of copper sulfate used".) If you're not told, use the number of moles of the reactant that isn't present in excess.

Practice Questions — Application

Q1 0.0500 mol of a compound dissolves in water, causing the temperature of the solution to increase from 298 K to 301 K. The total mass of the solution is 220 g. Calculate the enthalpy change for the reaction in kJ mol^{-1}. Assume c = 4.18 J g^{-1} K^{-1}.

Q2 A calorimeter, containing 200 g of water (c = 4.18 J g^{-1} K^{-1}), was used to measure the enthalpy change of combustion of pentane ($C_5H_{12(l)}$, M_r = 72). 0.500 g of pentane was burnt. The temperature of the water increased by 29.0 K.

a) Calculate the enthalpy change of combustion of pentane. Give your answer in kJ mol^{-1}.

b) Suggest one reason why this value may be different to the standard enthalpy change of combustion of pentane given in a data book.

Q3 The standard enthalpy of combustion of octane ($C_8H_{18(l)}$, M_r = 114) is –5470 kJ mol^{-1}. Some octane was burnt in a calorimeter containing 300 g of water (c = 4.18 J g^{-1} K^{-1}). The temperature of the water went up by 55 K. Calculate an estimate of the mass of octane burnt.

Practice Questions — Fact Recall

Q1 What three things need to be measured in order to calculate the enthalpy change for a reaction in a laboratory?

Q2 Sketch and label a calorimeter that could be used in the lab to measure the enthalpy change of a combustion.

Learning Objective:

- Understand Hess's Law.
- Be able to use Hess's Law to perform calculations, including the calculation of enthalpy changes for reactions from enthalpies of combustion or from enthalpies of formation.

Specification Reference 3.1.4.3

4. Hess's Law

For some reactions, there is no easy way to measure enthalpy changes in the lab. For these, we can use Hess's Law.

What is Hess's Law?

Hess's Law says that:

> The total enthalpy change for a reaction is independent of the route taken.

This law is handy for working out enthalpy changes that you can't find directly by doing an experiment — for example, the enthalpy change of the reaction that breaks down NO_2 into N_2 and O_2. We can call this reaction 'route 1'. But we can also think of the reaction as NO_2 breaking down into NO and O_2, and then reacting further to form N_2 and O_2. This longer route, with an intermediate step, can be called 'route 2' (see Figure 1).

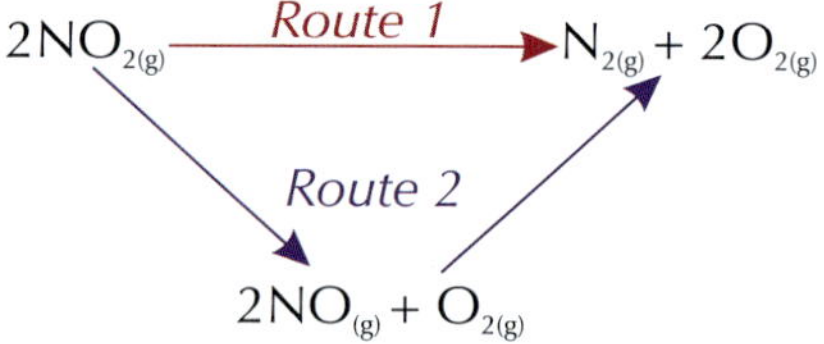

Figure 1: *Two possible routes for the formation of nitrogen and oxygen from nitrogen dioxide.*

Hess's Law says that the total enthalpy change for route 1 is the same as for route 2. So if you know the enthalpy changes for the stages of route 2, you can calculate the enthalpy change for route 1, as shown in the example below.

Example — Maths Skills

Use Hess's Law to calculate the enthalpy change, $\Delta_r H^\circ$, for route 1 of the reaction shown below.

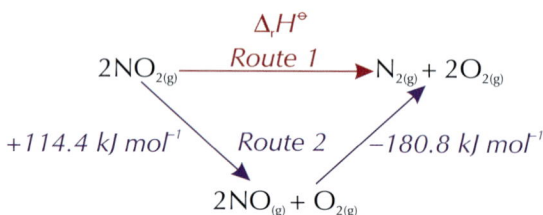

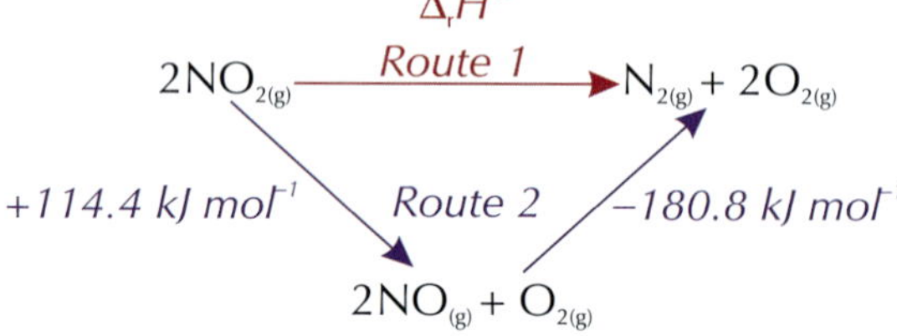

The total enthalpy change for route 1 is the same as the total enthalpy change for route 2. So the enthalpy change for route 1 is the sum of the steps in route 2:

$\Delta_r H^\circ = 114.4 \text{ kJ mol}^{-1} + (-180.8 \text{ kJ mol}^{-1}) = -66.4 \text{ kJ mol}^{-1}$.

Using enthalpies of formation

Enthalpy changes of formation are useful for calculating enthalpy changes you can't find directly. You need to know $\Delta_f H^\circ$ for all the reactants and products that are compounds. The value of $\Delta_f H^\circ$ for elements is zero — the element's being formed from the element, so there's no change in enthalpy. The standard enthalpy changes are all measured at 298 K.

Exam Tip
You'll be given all the information you need in your exam — you don't need to memorise any enthalpy values.

Example — Maths Skills

Calculate $\Delta_r H^\circ$ for this reaction using the enthalpies of formation in Figure 2:

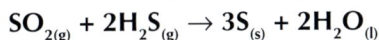

$$SO_{2(g)} + 2H_2S_{(g)} \rightarrow 3S_{(s)} + 2H_2O_{(l)}$$

Compound	$\Delta_f H^\circ$
$SO_{2(g)}$	–297 kJ mol^{-1}
$H_2S_{(g)}$	–20.6 kJ mol^{-1}
$H_2O_{(l)}$	–286 kJ mol^{-1}

Figure 2: Table of enthalpies of formation for three compounds.

- Write under the reaction a list of all the elements present in the reaction, balanced in their correct molar quantities, as shown below:

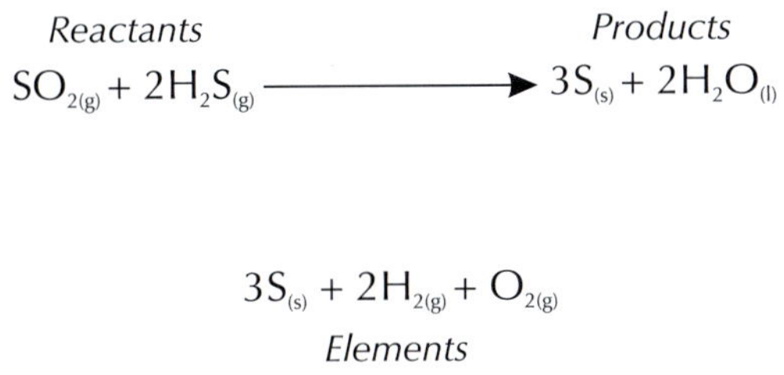

Reactants $SO_{2(g)} + 2H_2S_{(g)} \longrightarrow$ *Products* $3S_{(s)} + 2H_2O_{(l)}$

$3S_{(s)} + 2H_{2(g)} + O_{2(g)}$
Elements

- Enthalpies of formation ($\Delta_f H^\circ$) tell you the enthalpy change going from the elements to the compounds. The enthalpy change of reaction ($\Delta_r H^\circ$) is the enthalpy change going from the reactants to the products. Draw and label arrows to show this on your diagram:

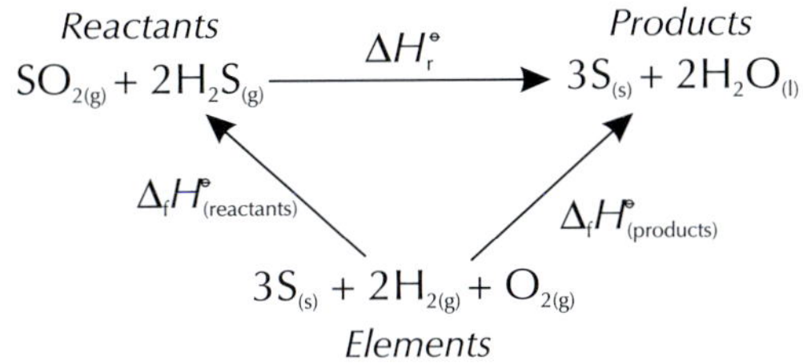

Reactants $SO_{2(g)} + 2H_2S_{(g)} \xrightarrow{\Delta H_r^\circ}$ *Products* $3S_{(s)} + 2H_2O_{(l)}$

$\Delta_f H^\circ_{(reactants)}$ (arrow from elements to reactants); $\Delta_f H^\circ_{(products)}$ (arrow from elements to products)

$3S_{(s)} + 2H_{2(g)} + O_{2(g)}$
Elements

- The calculation is often simpler if you keep the arrows end to end, so make both routes go from the elements to the products. Route 1 gets there via the reactants (and includes $\Delta_r H^\circ$), whilst route 2 gets there directly. Label the enthalpy changes along each arrow, as shown below. There are 2 moles of H_2O and 2 moles of H_2S, so their enthalpies of formation will need to be multiplied by 2. $\Delta_f H^\circ$ of sulfur is zero because it's an element, but you can still label it on the diagram.

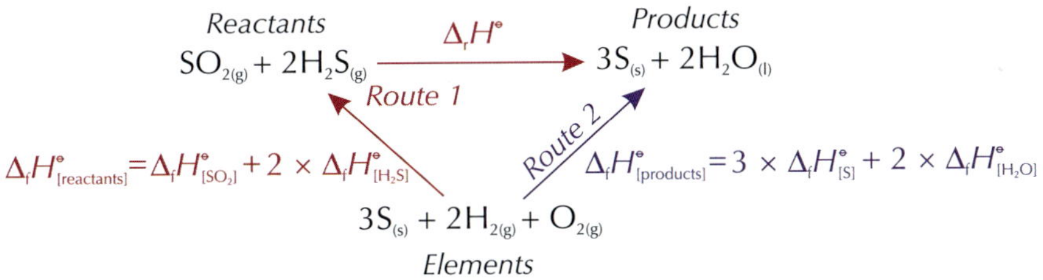

Tip: You don't have to pick a route that follows the direction of the arrows. If your route goes against an arrow you can just change the signs (so negative enthalpies become positive and positive enthalpies become negative).

Reactants $SO_{2(g)} + 2H_2S_{(g)} \xrightarrow{\Delta_r H^\circ}$ *Products* $3S_{(s)} + 2H_2O_{(l)}$

Route 1: $\Delta_f H^\circ_{[reactants]} = \Delta_f H^\circ_{[SO_2]} + 2 \times \Delta_f H^\circ_{[H_2S]}$

Route 2: $\Delta_f H^\circ_{[products]} = 3 \times \Delta_f H^\circ_{[S]} + 2 \times \Delta_f H^\circ_{[H_2O]}$

$3S_{(s)} + 2H_{2(g)} + O_{2(g)}$
Elements

- Use Hess's Law, Route 1 = Route 2, and plug the numbers from Figure 2 into the equation:

$\Delta_f H^\circ[SO_2] + 2\Delta_f H^\circ[H_2S] + \Delta_r H^\circ = 3\Delta_f H^\circ[S] + 2\Delta_f H^\circ[H_2O]$

$-297 + (2 \times -20.6) + \Delta_r H^\circ = (3 \times 0) + (2 \times -286)$

$\Delta_r H^\circ = (3 \times 0) + (2 \times -286) - [-297 + (2 \times -20.6)] = -233.8$ kJ mol^{-1}.

Using enthalpies of combustion

You can use a similar method to find an enthalpy change from enthalpy changes of combustion, instead of using enthalpy changes of formation.

Example — Maths Skills

Calculate $\Delta_f H^\circ$ of ethanol using the enthalpies of combustion in Figure 3.

- The desired reaction in this case is the formation of ethanol from its elements, so write out the balanced equation:

Reactants *Product*

$$2C_{(s)} + 3H_{2(g)} + \frac{1}{2}O_{2(g)} \longrightarrow C_2H_5OH_{(l)}$$

Substance	$\Delta_c H^\circ$
$C_{(s)}$	–394 kJ mol^{-1}
$H_{2(g)}$	–286 kJ mol^{-1}
$C_2H_5OH_{(l)}$	–1367 kJ mol^{-1}

Figure 3: *Table of enthalpies of combustion for three substances.*

Tip: The products of a complete combustion are carbon dioxide (CO_2) and water (H_2O).

- Figure 3 tells you the enthalpy change when each of the 'reactants' and 'products' is burned in oxygen. Add these combustion reactions to your diagram, making sure they are balanced, as shown below:

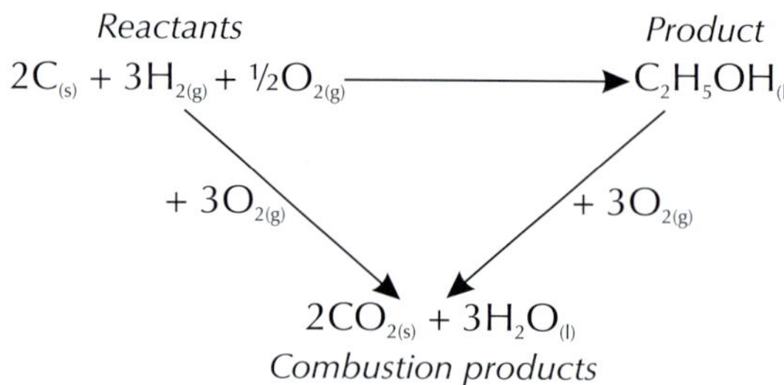

- Choose which reactions will form which route. Label the diagram with the enthalpy changes along each arrow as before (taking into account molar quantities):

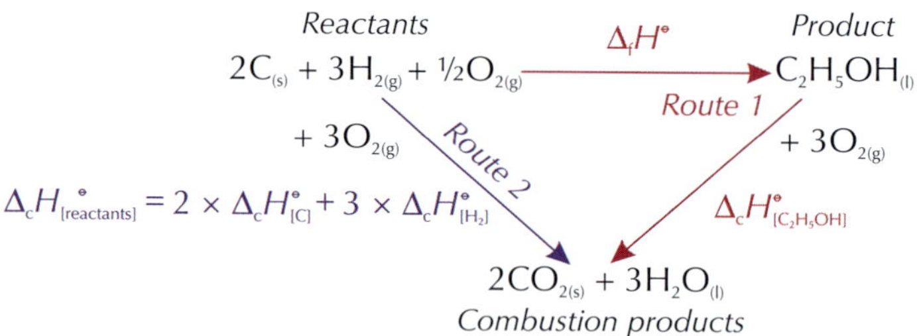

- Use Hess's Law as follows: Route 1 = Route 2

$\Delta_f H^\circ[\text{ethanol}] + \Delta_c H^\circ[C_2H_5OH] = 2\Delta_c H^\circ[C] + 3\Delta_c H^\circ[H_2]$

$\Delta_f H^\circ[\text{ethanol}] + (-1367) = (2 \times -394) + (3 \times -286)$

$\Delta_f H^\circ[\text{ethanol}] = -788 + -858 - (-1367)$

$\Delta_f H^\circ[\text{ethanol}] = -279$ kJ mol^{-1}.

Tip: You can ignore the enthalpy change of combustion of oxygen in these calculations. Oxygen doesn't have an enthalpy change of combustion — you can't burn 1 mole of oxygen in oxygen.

Practice Questions — Application

Q1 Calculate $\Delta_r H^\circ$ for the following reactions using Hess's Law, and the enthalpies of formation given in Figure 4:

a) $CaCO_{3(s)} \rightarrow CaO_{(s)} + CO_{2(g)}$

b) $N_2O_{5(s)} \rightarrow 4NO_{2(g)} + O_{2(g)}$

Q2 Calculate $\Delta_f H^\circ$ for the following organic compounds using Hess's Law, and the enthalpies of combustion given in Figure 5 and below:

a) propan-1-ol ($CH_3CH_2CH_2OH$): $\Delta_c H^\circ = -2021$ kJ mol^{-1}.

b) ethane-1,2-diol (CH_2OHCH_2OH): $\Delta_c H^\circ = -1180$ kJ mol^{-1}.

c) butan-2-one ($CH_3COCH_2CH_3$): $\Delta_c H^\circ = -2442$ kJ mol^{-1}.

Compound	$\Delta_f H^\circ$
$CO_{2(g)}$	–394 kJ mol^{-1}
$CaO_{(s)}$	–635 kJ mol^{-1}
$CaCO_{3(s)}$	–1207 kJ mol^{-1}
$N_2O_{5(s)}$	–41 kJ mol^{-1}
$NO_{2(g)}$	33 kJ mol^{-1}

Figure 4: *Table of enthalpies of formation for five compounds.*

Element	$\Delta_c H^\circ$
$C_{(s)}$	–394 kJ mol^{-1}
$H_{2(g)}$	–286 kJ mol^{-1}

Figure 5: *Enthalpies of combustion for carbon and hydrogen.*

Section Summary

Make sure you know...

- That enthalpy change, ΔH (in kJ mol^{-1}), is the heat energy transferred in a reaction at constant pressure.
- That ΔH° is the enthalpy change for a reaction where the reactants and products are in their standard states and the measurements are made at 100 kPa pressure and a stated temperature (usually 298 K).
- That exothermic reactions give out energy, so ΔH is negative.
- That endothermic reactions absorb energy, so ΔH is positive.
- That mean bond enthalpies tell us the average energy (per mole) required to break the bond between two atoms.
- How to use mean bond enthalpies to calculate enthalpy changes for reactions, using the equation: Enthalpy change of reaction = Total energy absorbed – Total energy released.
- That $\Delta_r H^\circ$ is the enthalpy change when a reaction occurs in the molar quantities shown in the chemical equation, under standard conditions with all reactants and products in their standard states.
- That $\Delta_f H^\circ$ is the enthalpy change when 1 mole of a compound is formed from its elements in their standard states under standard conditions.
- That $\Delta_c H^\circ$ is the enthalpy change when 1 mole of a substance is completely burned in oxygen under standard conditions.
- How to measure an enthalpy change.
- How to calculate the heat lost or gained (q) by a reaction using the equation $q = mc\Delta T$, where m is the mass of the reaction mixture, c is its specific heat capacity, and ΔT is the temperature change due to the reaction.
- How to calculate the enthalpy change of combustion and the enthalpy change of reaction given values for q and n, the number of moles reacted.
- That Hess's Law states that: The total enthalpy change for a reaction is independent of the route taken.
- How to use Hess's Law to calculate enthalpy changes for reactions from enthalpies of formation.
- How to use Hess's Law to calculate enthalpy changes for reactions from enthalpies of combustion.

Exam-style Questions

1 A scientist is conducting the following reaction:

$$CH_3COOH_{(l)} + C_2H_5OH_{(l)} \rightarrow CH_3COOC_2H_{5(l)} + H_2O_{(l)}$$

ethanoic acid *ethanol* *ethyl ethanoate* *water*

She adds 0.0500 mol of ethanoic acid to an excess of ethanol in a polystyrene beaker. She records the temperature of the mixture at regular intervals and uses her data to find the temperature change associated with the reaction.

The solution in the beaker has a total mass of 50.0 g, and a specific heat capacity of 2.46 J g^{-1} K^{-1}.

1.1 The temperature rises by 1.00 °C.
Calculate the enthalpy change due to the reaction.

(3 marks)

1.2 State two conditions necessary for the enthalpy change calculated in part 1.1 to be the standard enthalpy change of reaction.

(2 marks)

2 The table below shows the standard enthalpy change of combustion, $\Delta_c H^\circ$, for carbon, hydrogen and octane ($C_8H_{18(l)}$). The standard enthalpy of formation of octane can be calculated from this data using Hess's Law.

	$\Delta_c H^\circ$
$C_{(s)}$	–394 kJ mol^{-1}
$H_{2(g)}$	–286 kJ mol^{-1}
$C_8H_{18(l)}$	–5470 kJ mol^{-1}

2.1 State Hess's Law.

(1 mark)

2.2 Write out a balanced chemical equation for the complete combustion of octane.

(1 mark)

2.3 Use your answers to **2.1** and **2.2**, and the data in the table above, to calculate the standard enthalpy change of formation of octane, $\Delta_f H^\circ$.

(3 marks)

2.4 State whether the formation of octane is exothermic or endothermic.
Explain your answer.

(2 marks)

3 The structure of but-1-ene is shown below.

$$H-\overset{\overset{H}{|}}{\underset{\underset{H}{|}}{C}}-\overset{\overset{H}{|}}{\underset{\underset{H}{|}}{C}}-\overset{\overset{H}{|}}{C}=C\begin{matrix}\nearrow H\\ \searrow H\end{matrix}$$

But-1-ene will burn completely in oxygen to produce CO_2 and H_2O.

The table below shows bond enthalpies for the bonds present in the reactants and products of this combustion reaction.

Bond	Bond Enthalpy (Mean value except where stated)
C–H	413 kJ mol^{-1}
C=C	612 kJ mol^{-1}
C–C	347 kJ mol^{-1}
O=O	498 kJ mol^{-1}
C=O (in CO_2)	805 kJ mol^{-1}
O–H	464 kJ mol^{-1}

These bond enthalpies can be used to calculate the standard enthalpy change of combustion for but-1-ene.

3.1 Define the term 'standard enthalpy of combustion'.

(3 marks)

3.2 Use the data in the table to calculate a value for the standard enthalpy change of combustion for but-1-ene.

(3 marks)

3.3 The standard enthalpy change of combustion for but-1-ene calculated from the mean bond enthalpies is different to the value given in the data book. Explain why.

(1 mark)

4 Potassium hydroxide reacts with sulfuric acid in the following way:

$$2KOH_{(s)} + H_2SO_{4(l)} \rightarrow K_2SO_{4(s)} + 2H_2O_{(l)}$$

The table below shows the standard enthalpies of formation of each of the reactants and products in this reaction.

	$\Delta_f H^{\ominus}$
$KOH_{(s)}$	–425 kJ mol^{-1}
$H_2SO_{4(l)}$	–814 kJ mol^{-1}
$K_2SO_{4(s)}$	–1438 kJ mol^{-1}
$H_2O_{(l)}$	–286 kJ mol^{-1}

4.1 Define the term 'standard enthalpy of formation'.

(3 marks)

4.2 Use the data in the table to calculate a value for the standard enthalpy change of the reaction, $\Delta_r H^{\ominus}$.

(3 marks)

Learning Objectives:

- Understand that reactions can only occur when collisions take place between particles having sufficient energy and that this is called the activation energy.
- Be able to define the term activation energy.
- Be able to explain why most collisions do not lead to a reaction.
- Understand the Maxwell–Boltzmann distribution of molecular energies in gases.
- Be able to draw and interpret distribution curves for different temperatures.
- Understand the qualitative effect of temperature changes on the rate of reaction.
- Use the Maxwell-Boltzmann distribution to explain why a small temperature increase can lead to a large increase in rate.
- Understand the qualitative effect of changes in concentration, or a change in the pressure of a gas, on collision frequency.
- Be able to explain how a change in concentration or pressure influences the rate of a reaction.

Specification Reference 3.1.5.1, 3.1.5.2, 3.1.5.3, 3.1.5.4

1. Reaction Rates

The rate of a reaction is the change in the amount of a reactant or product over time — it describes how fast a reaction's happening. But if particles are going to react, they have to meet first. That's where collision theory comes in...

Collision theory and activation energy

Particles in liquids and gases are always moving and colliding with each other. They don't react every time though — only when the conditions are right. **Collision theory** says that a reaction won't take place between two particles unless they collide in the right direction (they need to be facing each other the right way) and they collide with at least a certain minimum amount of kinetic (movement) energy.

The minimum amount of kinetic energy particles need to react is called the **activation energy**. This much energy is needed to break the bonds within reactant particles to start the reaction. Reactions with low activation energies often happen pretty easily. But reactions with high activation energies don't. You need to give the particles extra energy by heating them. You can show the activation energy of a reaction on an enthalpy profile diagram like the one shown below in Figure 1.

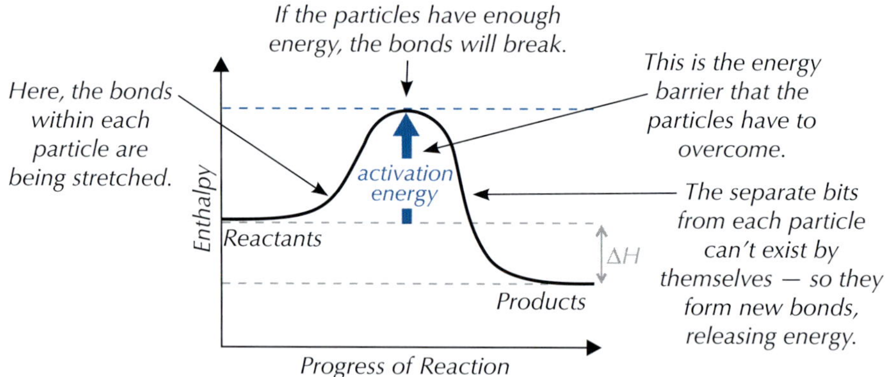

If the particles have enough energy, the bonds will break.
Here, the bonds within each particle are being stretched.
This is the energy barrier that the particles have to overcome.
activation energy
The separate bits from each particle can't exist by themselves — so they form new bonds, releasing energy.
Enthalpy
Reactants
ΔH
Products
Progress of Reaction

Figure 1: *An enthalpy profile diagram.*

Maxwell-Boltzmann distributions

Imagine looking down on Oxford Street when it's teeming with people. You'll see some people ambling along slowly, some hurrying quickly, but most of them will be walking with a moderate speed. It's the same with the molecules in a gas. Some don't have much kinetic energy and move slowly. Others have loads of kinetic energy and whizz along. But most molecules are somewhere in between.

If you plot a graph of the numbers of molecules in a gas with different kinetic energies you get a **Maxwell-Boltzmann distribution**. The Maxwell-Boltzmann distribution is a theoretical model that has been developed to explain scientific observations. It looks like this:

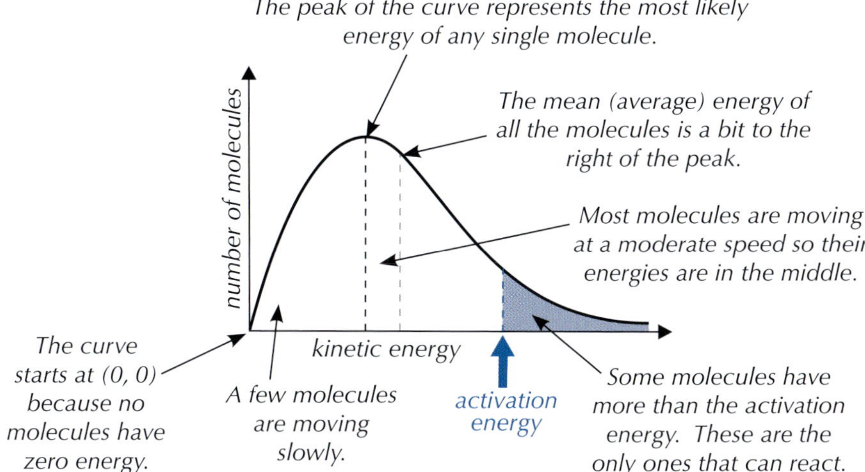

The peak of the curve represents the most likely energy of any single molecule.
The mean (average) energy of all the molecules is a bit to the right of the peak.
Most molecules are moving at a moderate speed so their energies are in the middle.
number of molecules
kinetic energy
The curve starts at (0, 0) because no molecules have zero energy.
A few molecules are moving slowly.
activation energy
Some molecules have more than the activation energy. These are the only ones that can react.

Figure 2: *A Maxwell-Boltzmann distribution curve showing the different kinetic energies of molecules in a gas.*

The area under a Maxwell-Boltzmann distribution curve is equal to the total number of molecules.

Figure 3: *James Clerk Maxwell, the Scottish physicist who studied the motion of gas molecules.*

Figure 4: *Ludwig Boltzmann, the Austrian physicist who developed Maxwell's ideas on the energy distribution of gas molecules.*

The effect of temperature on reaction rate

If you increase the temperature of a gas, the molecules will on average have more kinetic energy and will move faster. So, a greater proportion of molecules will have at least the activation energy and be able to react. This changes the shape of the Maxwell-Boltzmann distribution curve — it pushes it over to the right (see Figure 5). The total number of molecules is still the same, which means the area under each curve must be the same.

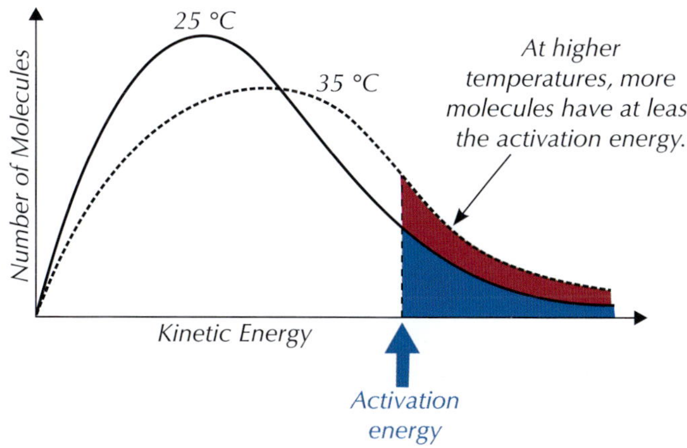

Figure 5: *Two Maxwell-Boltzmann distribution curves for a gas at different temperatures. Increasing the temperature of the gas shifts the distribution of the kinetic energies of the molecules.*

Because the molecules are flying about faster, they'll collide more often. This is another reason why increasing the temperature makes a reaction faster. So, small temperature increases can lead to large increases in reaction rate.

Exam Tip

You need to be able to draw distribution curves for different temperatures so remember — if the temperature **i**ncreases the curve moves to the r**i**ght, if it d**e**creases the curve moves to the l**e**ft.

The effect of concentration on reaction rate

If you increase the concentration of reactants in a solution, the particles will on average be closer together. If they're closer, collisions are more frequent. If collisions are more frequent, they'll be more chances for particles to react.

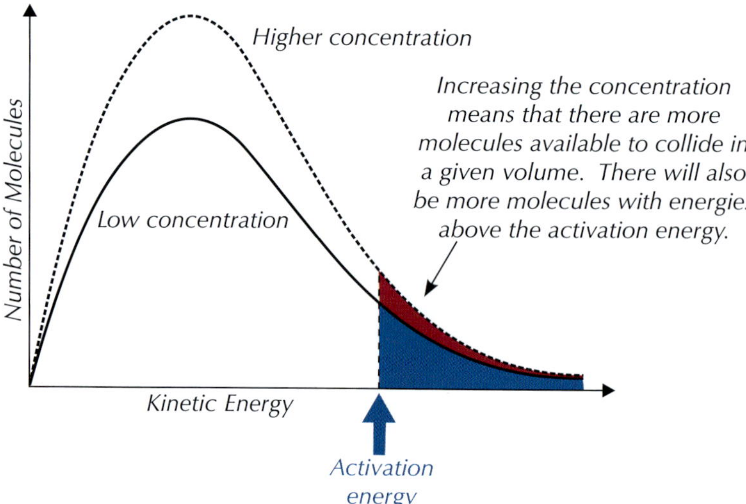

Higher concentration
Increasing the concentration means that there are more molecules available to collide in a given volume. There will also be more molecules with energies above the activation energy.
Number of Molecules
Low concentration
Kinetic Energy
Activation energy

Figure 6: *Two Maxwell-Boltzmann distribution curves for a solution at different concentrations.*

Exam Tip
Exam questions about reaction rates sometimes come with a lot of extra detail about someone doing an experiment. Don't let that throw you though — just remember, if the temperature, concentration or pressure increases, the rate of reaction increases too (or vice versa).

If the reaction involves gases, increasing the pressure of the gases works in just the same way. Raising the pressure pushes all of the gas particles closer together, increasing the number in a given volume. This increases the frequency of collisions, which increases the reaction rate.

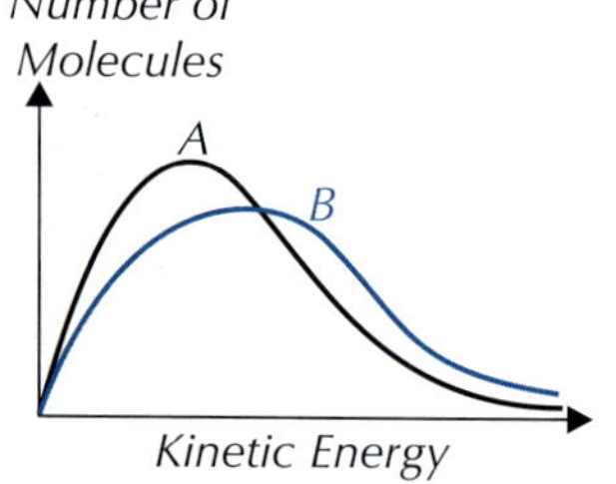

Number of Molecules
A
B
Kinetic Energy

Figure 7: *Two Maxwell-Boltzmann distribution curves for a volume of gas at different temperatures.*

Practice Question — Application

Q1 The two Maxwell-Boltzmann distribution curves shown in Figure 7 are for the same volume of the same gas. Which curve, A or B, is for the gas at a higher temperature? Explain your answer.

Practice Questions — Fact Recall

Q1 What conditions are required for a collision between two particles to result in a reaction?

Q2 What does the term 'activation energy' mean?

Q3 Explain why a small increase in temperature can lead to a large increase in reaction rate.

Q4 Describe and explain the effect that increasing the concentration of a solution has on the rate of a reaction involving that solution.

2. Catalysts

Sometimes you need to speed up a reaction, but you can't (or don't want to) increase the temperature, concentration or pressure any further. That's where catalysts come in.

Learning Objectives:

- Understand that a catalyst is a substance that increases the rate of a chemical reaction without being changed in chemical composition or amount.
- Understand that catalysts work by providing an alternative reaction route of lower activation energy.
- Explain, using a Maxwell-Boltzmann distribution, how a catalyst increases the rate of a reaction involving a gas.

Specification Reference 3.1.5.5

What is a catalyst?

You can use **catalysts** to make chemical reactions happen faster. A catalyst increases the rate of a reaction by providing an alternative reaction pathway with a lower activation energy. The catalyst is chemically unchanged at the end of the reaction.

Catalysts are great. They don't get used up in reactions, so you only need a tiny bit of catalyst to catalyse a huge amount of stuff. They do take part in reactions, but they're remade at the end. Catalysts are very fussy about which reactions they catalyse. Many will usually only work on a single reaction. Catalysts save heaps of money in industrial processes.

Example

The Haber-Bosch process uses an iron catalyst to increase the rate of forming ammonia from nitrogen and hydrogen in the following reaction:

$$N_{2(g)} + 3H_{2(g)} \rightleftharpoons 2NH_{3(g)}$$

This reaction has a very high activation energy, due to a very strong N≡N bond in N_2. For the reaction rate to be high enough to make ammonia in any great quantity, the temperature and pressure would have to be extremely high — too high to be practical or profitable.

In reality, the reaction is performed with the use of an iron catalyst, which increases the reaction rate at a workable temperature and pressure (around 400-500 °C and 20 MPa).

The nitrogen and hydrogen molecules bind to the surface of the catalyst. This makes it easier to break the bonds at lower energies, and so the activation energy of the reaction decreases. The broken nitrogen and hydrogen molecules then form ammonia molecules, and break away from the surface of the catalyst.

Figure 1: Fritz Haber, the German chemist who developed the process of ammonia production with Carl Bosch.

Tip: The reaction used to make ammonia is a reversible reaction — there's more on reversible reactions on pages 133-136.

How do catalysts work?

If you look at an enthalpy profile (see Figure 2) alongside a Maxwell-Boltzmann distribution (Figure 3, on the next page), you can see why catalysts work.

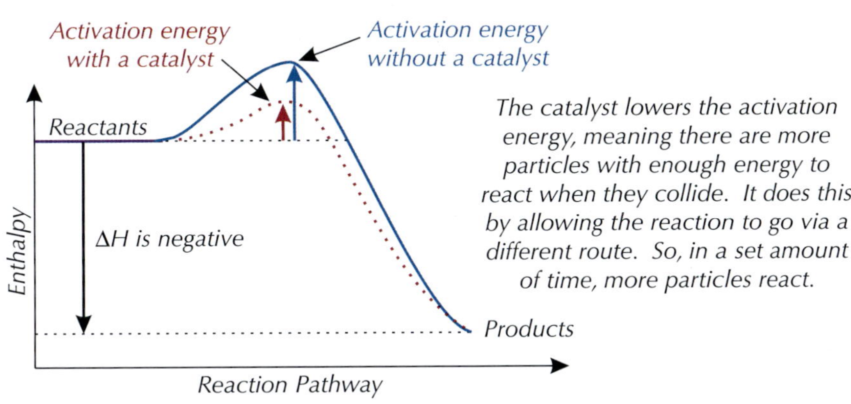
Activation energy with a catalyst
Activation energy without a catalyst
Reactants
Enthalpy
ΔH is negative
Products
Reaction Pathway
The catalyst lowers the activation energy, meaning there are more particles with enough energy to react when they collide. It does this by allowing the reaction to go via a different route. So, in a set amount of time, more particles react.

Figure 2: *Enthalpy profile diagram for a reaction with and without a catalyst.*

Tip: Catalysts speed up the reaction in a different way to increasing temperature, concentration or pressure. These things all change the energy distribution but the addition of a catalyst does not.

With a catalyst present, the molecules still have the same amount of energy, so the Maxwell-Boltzmann distribution curve is unchanged. But because the catalyst lowers the activation energy, more of the molecules have energies above this threshold and are able to react, as shown in Figure 3.

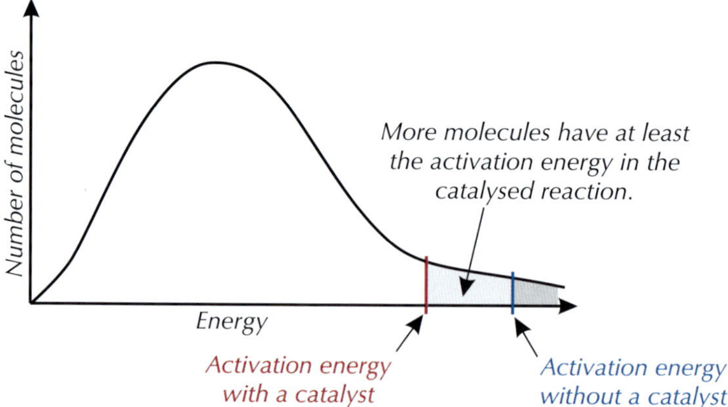

Figure 3: *A Maxwell-Boltzmann distribution curve for a reaction with and without a catalyst.*

Practice Question — Application

Tip: 'Uncatalysed' just means that no catalyst has been used.

Q1 The Maxwell-Boltzmann distribution curve shown below is for an uncatalysed chemical reaction to produce 'Product X'.

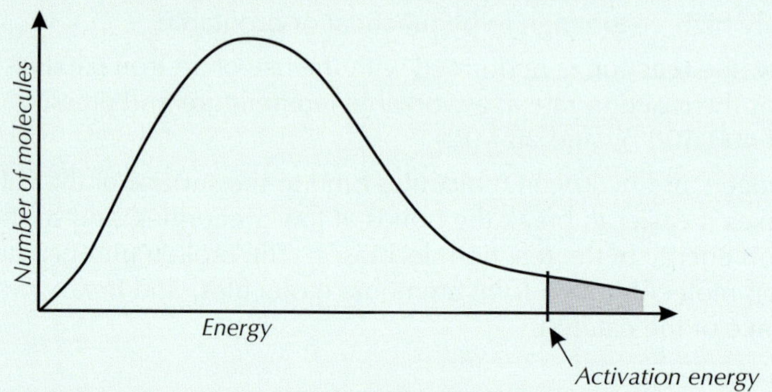

A company wants to produce 'Product X' on a large scale. They are considering using a catalyst.

a) Draw a sketch to show how the addition of a catalyst would affect the Maxwell-Boltzmann distribution curve for the reaction.

b) The uncatalysed reaction will only take place at temperatures above 1000 °C. Suggest how adding a catalyst would improve the industrial process.

Practice Questions — Fact Recall

Q1 What is a catalyst?

Q2 Explain how a catalyst can speed up the rate of a reaction.

3. Measuring Reaction Rates

Understanding all about the rates of chemical reactions is a really important part of chemistry. You need to know how to measure reaction rate as well.

Learning Objectives:
- Understand the meaning of the term rate of reaction.
- Investigate how the rate of a reaction changes with temperature (Required Practical 3).

Specification Reference 3.1.5.3

Calculating reaction rates

Rate of reaction is the change in the amount of a reactant or product over time. The units of reaction rate will be 'change you're measuring ÷ unit of time' (e.g. $g\ s^{-1}$ or $cm^3\ s^{-1}$). Here's a simple formula for finding the rate of a reaction:

$$\text{rate of reaction} = \frac{\text{amount of reactant used or product formed}}{\text{time}}$$

Measuring the rate of a reaction

If you want to find the rate of a reaction, you need to be able to follow the reaction as it's occurring. Although there are quite a few ways to follow reactions, not every method works for every reaction. You've got to pick a property that changes as the reaction goes on. Here are a few examples:

REQUIRED PRACTICAL 3

Tip: Carrying out experiments can be hazardous so you should always do a risk assessment before beginning.

Time taken for a precipitate to form

You can use this method when the product's a precipitate that clouds a solution.

Example

When you mix colourless sodium thiosulfate solution and colourless hydrochloric acid solution, a yellow precipitate of sulfur is formed:

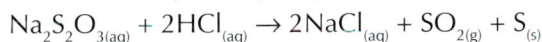

$$Na_2S_2O_{3(aq)} + 2HCl_{(aq)} \rightarrow 2NaCl_{(aq)} + SO_{2(g)} + S_{(s)}$$

You can stand a conical flask on top of a white tile with a black mark on it. Then you add fixed volumes of the reactant solutions to the flask and start a stopwatch. Look through the solutions to observe the mark on the tile. As the precipitate forms, the mark will become harder to see clearly. Stop the timer when the mark is no longer visible. The reading on the timer is recorded as the time taken for the precipitate to form.

Stopwatch
0 1:08 50
Sodium thiosulfate and hydrochloric acid solution.
A mark is made on a tile underneath the reaction vessel, which is visible through the initial reaction mixture.
A yellow sulfur precipitate forms which clouds the solution.

Figure 1: *Experimental setup for measuring reaction rate by monitoring the time taken for a precipitate to form.*

You can repeat this reaction for solutions at different temperatures to investigate how temperature affects reaction rate. Use a water bath to gently heat both solutions to the desired temperature before mixing them. The volumes and concentrations of the solutions must be kept the same each time. The results should show that the higher the temperature, the less time it takes for the mark to disappear and the faster the rate of the reaction gets.

Tip: If the same person uses the same mark each time you can compare the reaction rate, because roughly the same amount of precipitate will have been formed when the mark is obscured. But this method is subjective — different people might not agree on exactly when the mark has 'disappeared'.

Tip: In this experiment, temperature is the independent variable because you're changing it to see what happens. Time is the dependent variable because that's what you're measuring. See page 1 for more about this.

Change in mass

When the product is a gas, its rate of formation can be measured using a mass balance.

Example

This would work for the reaction between hydrochloric acid and calcium carbonate in which carbon dioxide gas is given off.

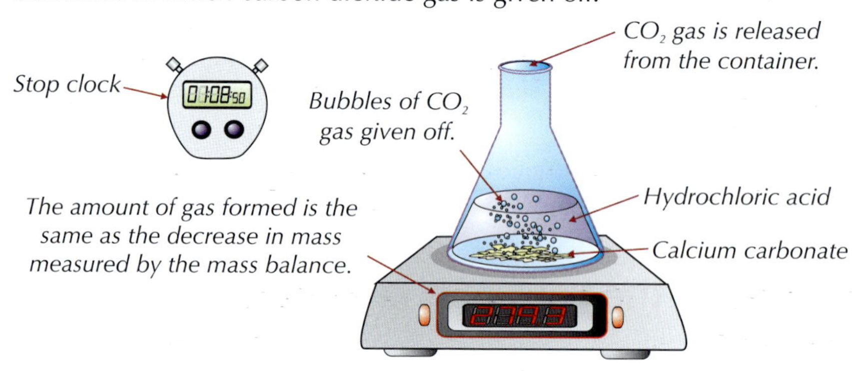

CO_2 gas is released from the container.
Stop clock
Bubbles of CO_2 gas given off.
The amount of gas formed is the same as the decrease in mass measured by the mass balance.
Hydrochloric acid
Calcium carbonate

Figure 2: *Experimental setup for measuring the reaction rate by monitoring the change in mass of a reaction mixture.*

Time (s)	Mass (g)
0	280.0
30	279.3
60	278.7
90	278.2
120	278.0
150	277.9
180	277.8
210	277.7
240	277.7
270	277.7

Figure 3: *Example results for the experiment carried out in Figure 2.*

When the reaction starts, start a stop clock or timer, then read off the mass at regular time intervals. Make a table with a column for 'time' and a column for 'mass' and fill it in as the reaction goes on (see Figure 3). You'll know the reaction is finished when the reading on the mass balance stops decreasing. This method is very accurate and easy to use but does release gas into the room, which could be dangerous if the gas is toxic or flammable. So it's best to carry out the experiment in a fume cupboard.

You can repeat this reaction for acids at different temperatures to investigate how temperature affects reaction rate. All other experimental variables must be kept the same. The results should show that the higher the temperature, the faster the mass decreases and the faster the reaction rate gets.

Tip: These methods are known as continuous monitoring methods because you can monitor the reaction rate from the beginning of the reaction to the end.

Gas volume

If a gas is given off during a reaction, you can measure reaction rate by collecting it in a gas syringe and recording how much you've got at regular time intervals.

Example

This would work for the reaction between magnesium and an acid in which hydrogen gas is given off.

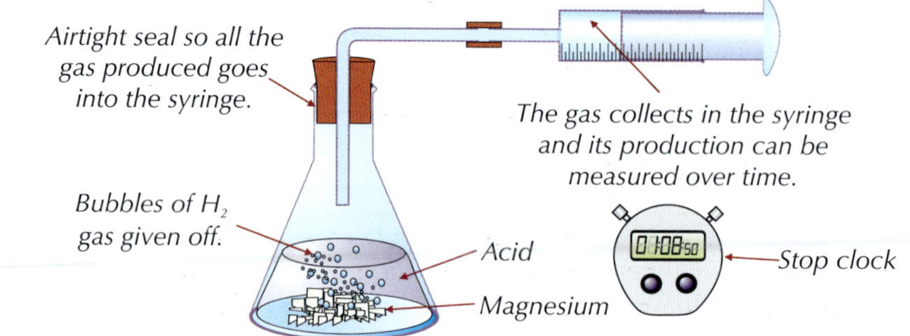

Airtight seal so all the gas produced goes into the syringe.
The gas collects in the syringe and its production can be measured over time.
Bubbles of H_2 gas given off.
Acid
Magnesium
Stop clock

Figure 4: *Experimental setup for measuring the reaction rate by monitoring the volume of gas given off by a reaction mixture.*

Again, start a stop clock or timer when the reaction starts, then read off the volume of gas in the gas syringe at regular time intervals (see Figure 5 for a sample results table). You know that the reaction has finished when the gas volume stops increasing.

This method is accurate because gas syringes usually give volumes to the nearest 0.1 cm^3. Because no gas escapes, you can use this method for reactions that produce toxic or flammable gases (although you should still do reactions like these in a fume cupboard to be safe). Vigorous reactions can blow the plunger out of the syringe, so you should do a rough calculation of how much gas you expect the reaction to produce before you begin. Then you can use an appropriate size of gas syringe.

The reaction can be repeated with the acid at different temperatures to investigate the effect on reaction rate. The results should show that the higher the temperature, the faster the gas is produced, and therefore the faster the rate of reaction.

Time (s)	Volume (cm^3)
0	0.0
30	6.1
60	10.9
90	15.2
120	17.7
150	20.0
180	21.1
210	21.9
240	22.0
270	22.0

Figure 5: *Example results for the experiment carried out in Figure 4.*

Practice Question — Application

Q1 A student is asked to measure the rate of the reaction below.

$$A_{(aq)} \rightarrow B_{(l)} + C_{(g)}$$

a) Suggest an experimental method the student could use to measure the rate of reaction.

b) How would the student know when the reaction has finished, using the method in your answer to part a)?

c) The student carries out the reaction at eight different temperatures and measures how long it takes the reaction to finish. The results are presented in the table in Figure 6. Make an estimate of how long it takes the reaction to finish at 45 °C.

Temp (°C)	Time (s)
20	230
25	228
30	225
35	219
40	208
45	
50	160
55	117

Figure 6: *Results of the experiment carried out in Q1 c).*

Practice Questions — Fact Recall

Q1 What is meant by the term ‘rate of reaction’?

Q2 Outline a method which could be used to measure the rate of a reaction in which one of the products is a precipitate.

Section Summary

Make sure you know...

- That most collisions between particles don't result in a reaction.
- That collision theory says that a collision will only result in a reaction if it's in the right direction and has at least a certain minimum amount of kinetic energy.
- That the minimum amount of kinetic energy required for a reaction is called the activation energy.
- That the Maxwell-Boltzmann distribution describes the spread of energies of the molecules in a gas.
- How to draw and interpret Maxwell-Boltzmann distribution curves for gases at different temperatures.
- That even a small increase in temperature can increase the reaction rate, by increasing the number of molecules with energies above the activation energy. This will increase the frequency of collisions and also mean that more particles will react when they collide.
- That increasing the concentration of reactants (or the pressure if they're gases) will increase the reaction rate, because the molecules will be closer together and so collisions will be more frequent.
- That a catalyst is a substance that increases the rate of a reaction by providing an alternative pathway with a lower activation energy, and is chemically unchanged at the end of the reaction.
- That the shape of the Maxwell-Boltzmann distribution curve doesn't change for a catalysed reaction, only the position of the activation energy on the curve.
- That the rate of a reaction is the change in the amount of reactants or products per unit time.
- That various experimental methods can be used to investigate how the rate of a reaction changes with temperature.

Exam-style Questions

Questions 1 and 2 are about the two Maxwell-Boltzmann energy distributions for molecules of a gas shown below.

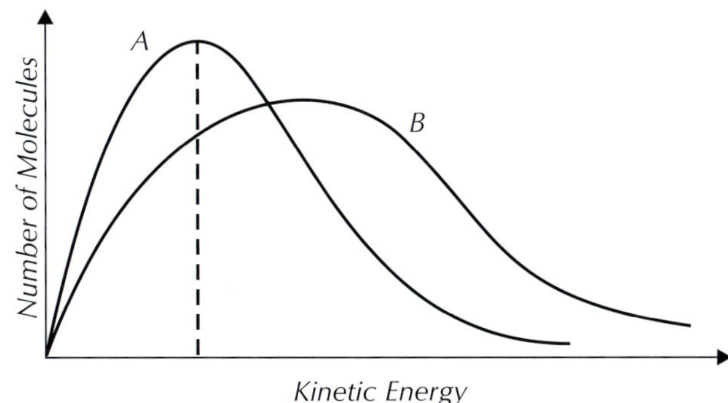

1 What does the dotted line on curve A represent?

A The total number of molecules in the distribution.

B The most likely energy of molecules in the distribution.

C The enthalpy change of the reaction.

D The mean energy of the molecules in the distribution. *(1 mark)*

2 Distributions A and B are for the same volume of the same gas.
Which of these could distribution B represent?

A Distribution A at lower concentration.

B Distribution A at lower temperature.

C Distribution A at higher temperature.

D Distribution A at higher concentration. *(1 mark)*

3 1.5 g of magnesium is added to a flask containing 40 cm^3 1.0 mol dm^3 hydrochloric acid and the volume of hydrogen gas given off over time is measured using a gas syringe. The experiment is then repeated at different temperatures, using an identical set-up, to see what effect this has on the rate of reaction.
Which of the following variables is the independent variable in the experiment?

A Temperature of the hydrochloric acid.

B Volume of hydrochloric acid.

C Mass of magnesium added.

D Volume of hydrogen gas produced. *(1 mark)*

4 Which of the following is a unit that could be used to show the rate of a reaction?

A $g\ l^{-1}$

B $kJ\ mol^{-1}$

C $mol\ dm^{-3}$

D $g\ s^{-1}$

(1 mark)

5 The Maxwell-Boltzmann distribution curve for a gas is shown below.

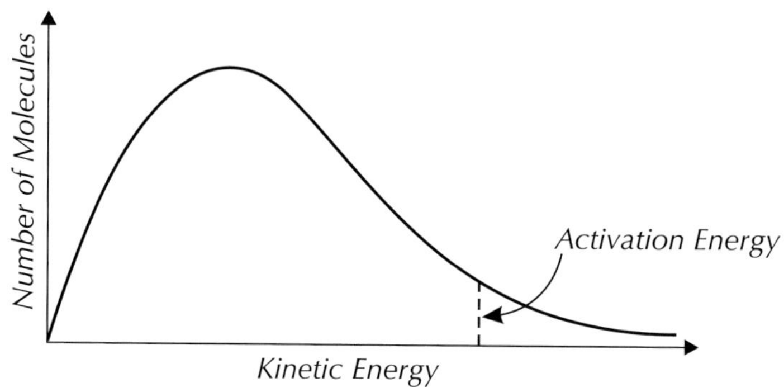

5.1 What is meant by the term 'activation energy'?

(1 mark)

5.2 Draw another curve on the axes above that would show the same volume of the same gas at a higher temperature.

(1 mark)

5.3 Define the term catalyst.

(1 mark)

5.4 How does the addition of a catalyst to a reaction affect the shape of a Maxwell-Boltzmann distribution curve?

Explain your answer.

(2 marks)

5.5 The reactants in a particular reaction are heated.

Explain the effect that this will have on the rate of the reaction.

(3 marks)

5.6 For a particular reaction, the best yield is obtained by keeping the pressure as low as possible.

Explain the effect that lowering the pressure will have on the rate of the reaction.

(3 marks)

1. Reversible Reactions

We usually think of a reaction as a one-way process to make products from reactants. In reality though, many reactions are reversible.

Dynamic equilibrium

Lots of chemical reactions are reversible — they go both ways. To show a reaction's reversible, you stick in a $\rightleftharpoons$.

Example

Hydrogen reacts with iodine to give hydrogen iodide: $\mathbf{H_{2(g)} + I_{2(g)} \rightleftharpoons 2HI_{(g)}}$

This reaction can go in either direction,
forwards: $H_{2(g)} + I_{2(g)} \rightarrow 2HI_{(g)}$ ...or backwards: $2HI_{(g)} \rightarrow H_{2(g)} + I_{2(g)}$

As the reactants get used up, the forward reaction slows down — and as more product is formed, the reverse reaction speeds up. After a while, the forward reaction will be going at exactly the same rate as the backward reaction. The concentration of reactants and products won't be changing any more, so it'll seem like nothing's happening. It's a bit like you're digging a hole, while someone else is filling it in at exactly the same speed. This is called a **dynamic equilibrium**. A dynamic equilibrium can only happen in a **closed system** (this just means nothing can get in or out) which is at a constant temperature.

Le Chatelier's principle

If you change the concentration, pressure or temperature of a reversible reaction, you're going to alter the position of equilibrium. This just means you'll end up with different amounts of reactants and products at equilibrium. If the position of equilibrium moves to the left, the backwards reaction is faster than the forwards reaction, and so you'll get more reactants.

Example

If the position of equilibrium in the reaction $H_{2(g)} + I_{2(g)} \rightleftharpoons 2HI_{(g)}$ shifts to the left, the backwards reaction is fastest, so more H_2 and I_2 are produced:

$$\mathbf{2HI_{(g)} \rightarrow H_{2(g)} + I_{2(g)}}$$

If the position of equilibrium moves to the right, the forwards reaction is faster than the backwards reaction, and so you'll get more products.

Example

If the position of equilibrium in the reaction $H_{2(g)} + I_{2(g)} \rightleftharpoons 2HI_{(g)}$ shifts to the right, the forwards reaction is fastest, so more HI is produced:

$$\mathbf{H_{2(g)} + I_{2(g)} \rightarrow 2HI_{(g)}}$$

Le Chatelier's principle tells you how the position of equilibrium will change if a condition changes:

> If a reaction at equilibrium is subjected to a change in concentration, pressure or temperature, the position of equilibrium will move to counteract the change.

Learning Objectives:

- Know that many chemical reactions are reversible.
- Know that in a reversible reaction at equilibrium, forward and reverse reactions proceed at equal rates.
- Know that in a reversible reaction at equilibrium, the concentrations of reactants and products remain constant.
- Know and be able to use Le Chatelier's principle to predict qualitatively the effects of changes in temperature, pressure and concentration on the position of equilibrium in homogeneous reactions.
- Know that a catalyst does not affect the position of equilibrium.

Specification Reference 3.1.6.1

Tip: $H_{2(g)} + I_{2(g)} \rightarrow 2HI_{(g)}$ is an example of a homogeneous reaction — the reactants and products are all in the same state (in this case they're all gases).

Tip: Le Chatelier's principle only applies to homogeneous equilibria — i.e. when every species in the reaction is in the same physical state (e.g. all liquid or all gas).

So, basically, if you raise the temperature, the position of equilibrium will shift to try to cool things down. And, if you raise the pressure or concentration, the position of equilibrium will shift to try to reduce it again. Catalysts have no effect on the position of equilibrium. They can't increase yield — but they do mean equilibrium is reached faster.

Using Le Chatelier's principle

Changing concentration

If you increase the concentration of a reactant, the equilibrium tries to get rid of the extra reactant. It does this by making more product. So the equilibrium's shifted to the right. If you increase the concentration of the product, the equilibrium tries to remove the extra product. This makes the reverse reaction go faster — so the equilibrium shifts to the left. Decreasing the concentrations has the opposite effect.

Examples

Sulfur dioxide reacts with oxygen to produce sulfur trioxide:

$$2SO_{2(g)} + O_{2(g)} \rightleftharpoons 2SO_{3(g)}$$

If you increase the concentration of SO_2 or O_2, the equilibrium tries to get rid of it by making more SO_3, so the equilibrium shifts to the right. If you increase the concentration of SO_3, the equilibrium shifts to the left to make the backwards reaction faster to get rid of the extra SO_3.

In the Haber process, nitrogen reacts with hydrogen to produce ammonia:

$$N_{2(g)} + 3H_{2(g)} \rightleftharpoons 2NH_{3(g)}$$

If you increase the concentration of N_2 or H_2, the equilibrium shifts to the right and you'll make more NH_3. If you increase the concentration of NH_3, the equilibrium shifts to the left and you'll make more N_2 and H_2.

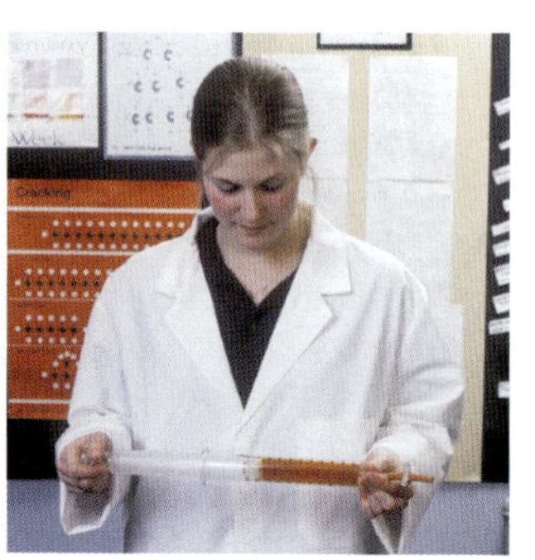

Figure 1a: Equilibrium between $NO_{2(g)}$ (brown) and $N_2O_{4(l)}$ (colourless).

Figure 1b: When pressure is applied the colour changes because the equilibrium shifts in favour of $N_2O_{4(l)}$.

Changing pressure

Changing the pressure only affects equilibria involving gases. Increasing the pressure shifts the equilibrium to the side with fewer gas molecules. This reduces the pressure. Decreasing the pressure shifts the equilibrium to the side with more gas molecules. This raises the pressure again.

Examples

When sulfur dioxide reacts with oxygen you get sulfur trioxide:

$$2SO_{2(g)} + O_{2(g)} \rightleftharpoons 2SO_{3(g)}$$

There are 3 moles on the left, but only 2 on the right. So, an increase in pressure shifts the equilibrium to the right, making more SO_3 and reducing the pressure. Decreasing the pressure favours the backwards reaction, so the equilibrium shifts to the left and more SO_2 and O_2 will be made to increase the pressure.

Methane reacts with water to produce carbon monoxide and hydrogen:

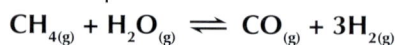

$$CH_{4(g)} + H_2O_{(g)} \rightleftharpoons CO_{(g)} + 3H_{2(g)}$$

There are 2 moles on the left and 4 on the right. So for this reaction, an increase in pressure shifts the equilibrium to the left, making more CH_4 and H_2O. Decreasing the pressure shifts the equilibrium to the right to make more CO and H_2. This reaction is used in industry to produce hydrogen. It is best performed at a low pressure to favour the forwards reaction so that more H_2 is produced.

Changing temperature

Increasing the temperature means adding heat. The equilibrium shifts in the endothermic (positive ΔH) direction to absorb this heat. Decreasing the temperature removes heat. The equilibrium shifts in the exothermic (negative ΔH) direction to try to replace the heat. If the forward reaction's endothermic, the reverse reaction will be exothermic, and vice versa.

Tip: An enthalpy change given with a reversible reaction always refers to the forwards reaction, unless you're told otherwise.

Examples

This reaction's exothermic in the forward direction, which means it is endothermic in the backward direction.

Exothermic →

$$2SO_{2(g)} + O_{2(g)} \rightleftharpoons 2SO_{3(g)} \qquad \Delta H = -197 \text{ kJ mol}^{-1}$$

← *Endothermic*

If you increase the temperature, the equilibrium shifts to the left (the endothermic direction) to absorb the extra heat. This means more SO_2 and O_2 are produced.

If you decrease the temperature, the equilibrium shifts to the right (the exothermic direction) to produce more heat. This means more SO_3 is produced.

This reaction's endothermic in the forward direction (and so exothermic in the backward direction).

Endothermic →

$$N_2O_{4(g)} \rightleftharpoons 2NO_{2(g)} \qquad \Delta H = +57.2 \text{ kJ mol}^{-1}$$

← *Exothermic*

Increasing the temperature will shift the equilibrium to the right, producing more NO_2.

Decreasing the temperature shifts the equilibrium to the left, producing more N_2O_4.

Exam Tip
In an exam question, make it clear exactly how the equilibrium shift opposes a temperature change — i.e. by removing or producing heat.

Tip: A lot of questions in this section ask about the effect of increasing temperature, pressure and concentration on the position of equilibrium. These questions can look quite similar to the ones on reaction rate from the last section, so make sure you're really clear what you're being asked about.

Following equilibrium reactions

There are some reactions you can do in the lab to follow equilibrium shifts. The reaction of $[Cu(H_2O)_6]^{2+}$ with concentrated hydrochloric acid (HCl) is one of these reactions.

When $[Cu(H_2O)_6]^{2+}$ reacts with hydrochloric acid, a copper chloride complex, $[CuCl_4]^{2-}$, forms (see Figure 2). This reaction is a reversible reaction, so at any point there will be a mixture of $[Cu(H_2O)_6]^{2+}$ and $[CuCl_4]^{2-}$ present in the reaction container.

$[Cu(H_2O)_6]^{2+}$ is a light blue colour, while $[CuCl_4]^{2-}$ is a greeny-yellow. You can therefore monitor the equilibrium position of this reaction by noting what colour the solution is. If the solution is blue, then the position of equilibrium must lie to the left and there'll be more reactants than products. But, if the solution's greeny-yellow, the equilibrium position must lie to the right and there'll be more of the products.

Tip: Before carrying out any reactions in the lab, you should do a thorough risk assessment and take any relevant safety precautions.

$$[Cu(H_2O)_6]^{2+}_{(aq)} + 4Cl^-_{(aq)} \rightleftharpoons [CuCl_4]^{2-}_{(aq)} + 6H_2O_{(l)}$$

blue copper aqua complex (under $[Cu(H_2O)_6]^{2+}$) — *greeny-yellow copper chloride complex* (under $[CuCl_4]^{2-}$)

Figure 2: *The reaction between $[Cu(H_2O)_6]^{2+}$ and concentrated HCl.*

Figure 3: *A solution of $[Cu(H_2O)_6]^{2+}$ (left). Upon addition of an excess of HCl, the $[CuCl_4]^{2-}$ complex forms which turns the solution green (right).*

Changing the concentration

If you have a test tube containing $[Cu(H_2O)_6]^{2+}$, you'll see it's a light blue colour. If you slowly add concentrated HCl, you'll notice the solution turn from light blue to a bluey-green as the equilibrium in Figure 2 is established. The more HCl you add, the more green the solution goes as more and more of the $[CuCl_4]^{2-}$ complex forms — this is because, by adding HCl, you're increasing the concentration of Cl^- in the solution. The equilibrium shifts to the right to try and remove the excess Cl^- ions from the solution, and more greeny-yellow $[CuCl_4]^{2-}$ forms (see Figure 3).

You can push the equilibrium back to the right by adding distilled water to the reaction container. The equilibrium position moves to try and mop up all the extra H_2O molecules you're adding to the solution by forming more $[Cu(H_2O)_6]^{2+}$, which turns the solution blue again.

Changing the temperature

The forward reaction of the equilibrium is endothermic (it takes in heat). So, if you heat a sample containing an equilibrium mixture of $[Cu(H_2O)_6]^{2+}$ and $[Cu(Cl)_4]^{2-}$, the equilibrium will move to the right to try and absorb the extra heat. This means more of the product forms, and you'll see the solution turn green as more $[CuCl_4]^{2-}$ is made.

The opposite happens if you cool the mixture down. Since the reverse reaction is exothermic, the equilibrium will move to the left to favour the reverse reaction to try and make up for the loss of heat. This means more of the $[Cu(H_2O)_6]^{2+}$ complex will form, and the solution will turn more blue.

Practice Questions — Application

Q1 An industrial process uses the following reversible reaction:

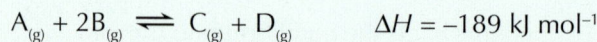

$$A_{(g)} + 2B_{(g)} \rightleftharpoons C_{(g)} + D_{(g)} \qquad \Delta H = -189 \text{ kJ mol}^{-1}$$

a) Explain the effect of increasing the concentration of A on the position of equilibrium.

b) Explain the effect of increasing the pressure on the position of equilibrium.

c) Explain the effect of increasing the temperature on the position of equilibrium.

d) Briefly outline the best reaction conditions (in terms of high or low concentration, pressure and temperature) to maximise the production of product D.

Q2 What will be the effect of increasing the pressure on the position of equilibrium of the following reaction? Explain your answer.

$$H_{2(g)} + I_{2(g)} \rightleftharpoons 2HI_{(g)}$$

Practice Questions — Fact Recall

Q1 What does it mean if a reaction is at dynamic equilibrium?

Q2 What conditions are needed for dynamic equilibrium to be established?

Q3 What is Le Chatelier's Principle?

Q4 How does the addition of a catalyst affect the position of equilibrium in a reversible reaction?

Exam Tip

If you're asked to define something like 'dynamic equilibrium' in an exam, look at the number of marks allocated for the answer. If there's more than one, you'll need more than one point in your definition.

2. Industrial Processes

Le Chatelier's principle can be applied to lots of industrial processes — like the production of ethanol and methanol.

Learning Objective:
- Be able to explain why, for a reversible reaction used in an industrial process, a compromise of temperature and pressure may be used.

Specification Reference 3.1.6.1

Compromise conditions in industry

Companies have to think about how much it costs to run a reaction and how much money they can make from it. This means they have a few factors to think about when they're choosing the best conditions for a reaction.

Example

Ethanol can be made via a reversible reaction between ethene and steam:

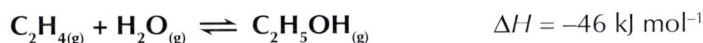

$$C_2H_{4(g)} + H_2O_{(g)} \rightleftharpoons C_2H_5OH_{(g)} \qquad \Delta H = -46 \text{ kJ mol}^{-1}$$

The industrial conditions for the reaction are:
- a pressure of 60-70 atmospheres,
- a temperature of 300 °C,
- a phosphoric acid catalyst.

Because it's an exothermic reaction, lower temperatures favour the forward reaction. This means that at lower temperatures more ethene and steam are converted to ethanol — you get a better **yield**. But lower temperatures mean a slower rate of reaction. You'd be daft to try to get a really high yield of ethanol if it's going to take you 10 years. So the 300 °C is a compromise between maximum yield and a faster reaction.

Higher pressure favours the forward reaction, since it moves the reaction to the side with fewer molecules of gas, so a pressure of 60-70 atmospheres is used. Increasing the pressure also increases the rate of reaction. Cranking up the pressure as high as you can sounds like a great idea, but high pressures are expensive to produce. You need stronger pipes and containers to withstand high pressure. So the 60-70 atmospheres is a compromise between maximum yield and minimum expense.

Only a small proportion of the ethene reacts each time the gases pass through the reactor. To save money and raw materials, the unreacted ethene is separated from the ethanol and recycled back into the reactor. Thanks to this, around 95% of the ethene is eventually converted to ethanol.

Tip: The yield is the amount of product you get from a reaction. Increasing the reaction rate will give you a higher yield in a given time, but you need to shift the equilibrium to increase the maximum yield.

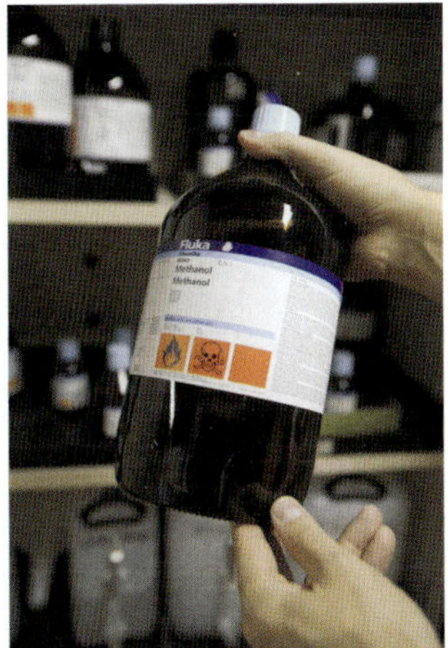

Figure 1: *Methanol has many uses. For example, it is widely used as a solvent.*

Example

Methanol is also made industrially in a reversible reaction:

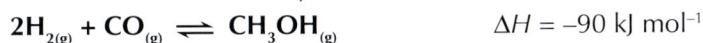

$$2H_{2(g)} + CO_{(g)} \rightleftharpoons CH_3OH_{(g)} \qquad \Delta H = -90 \text{ kJ mol}^{-1}$$

Just like ethanol production, the conditions are a compromise between keeping costs low and yield high. The conditions for this reaction are:
- a pressure of 50-100 atmospheres,
- a temperature of 250 °C,
- a catalyst of a mixture of copper, zinc oxide and aluminium oxide.

Tip: As with ethanol production, low temperatures favour the forward (exothermic) reaction, so the temperature is kept as low as possible without reducing the reaction rate too much.

Practice Question — Fact Recall

Q1 In the production of ethanol from ethene and steam, using a low temperature and a high pressure favours the forward reaction and gives a better yield. Given that this is the case, explain why a moderate temperature and pressure are used when making ethanol industrially.

Learning Objectives:

- Know that the equilibrium constant, K_c, is deduced from the equation for a reversible reaction.
- Know that the concentration, in mol dm^{-3}, of a species X involved in the expression for K_c is represented by [X].
- Be able to construct an expression for K_c for a homogeneous system in equilibrium.
- Be able to calculate a value for K_c from the equilibrium concentrations for a homogeneous system at constant temperature.
- Be able to perform calculations involving K_c.

Specification Reference 3.1.6.2

Exam Tip
In exam questions on K_c you'll often be told the temperature. You don't need this value for your calculation — it's just so you know the temperature is constant.

Exam Tip
Always work out the units for K_c — there may well be a mark for the units in the exam.

Tip: When you're working out the units of K_c, if you end up having to multiply mol dm^{-3} by itself, you just add the powers. For example (mol dm^{-3})(mol dm^{-3}) = mol^2 dm^{-6}.

3. The Equilibrium Constant

You don't just need to know what dynamic equilibrium means — you'll also need to be able to describe it, using some mathsy bits...

K_c, the equilibrium constant

If you know the molar concentration of each substance at equilibrium, you can work out the **equilibrium constant, K_c**. This is a ratio worked out from the concentrations of the products and reactants after equilibrium is reached. Your value of K_c will only be true for that particular temperature. Before you can calculate K_c, you have to write an expression for it. Here's how:

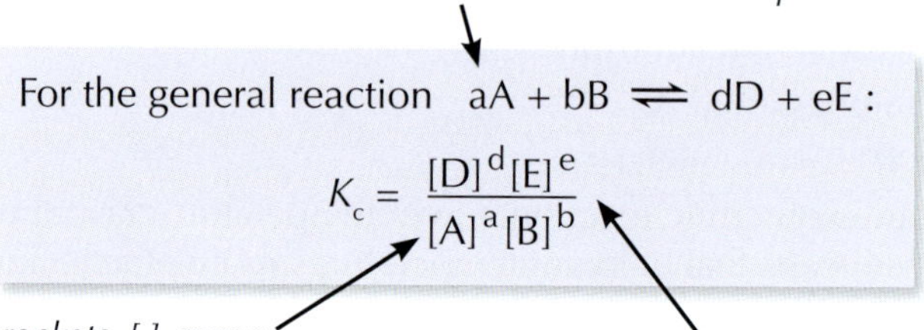

The lower–case letters a, b, d and e are the number of moles of each substance in the equation.
For the general reaction aA + bB ⇌ dD + eE :
$K_c = \frac{[D]^d[E]^e}{[A]^a[B]^b}$
The square brackets, [ ], mean concentration in mol dm^{-3}.
The products go on the top line and the reactants go on the bottom line.

Example

For the reaction $H_{2(g)} + I_{2(g)} \rightleftharpoons 2HI_{(g)}$ there are two reactants (H_2 and I_2) and one product (HI). There's one mole of each of the reactants and two moles of the product. So the expression for K_c is:

$$K_c = \frac{[HI]^2}{[H_2]^1[I_2]^1} = \frac{[HI]^2}{[H_2][I_2]}$$

Calculating K_c

If you know the equilibrium concentrations, just bung them in your expression. Then with a bit of help from the old calculator, you can work out the value for K_c. The units are a bit trickier though — they vary, so you have to work them out after each calculation.

Example — Maths Skills

For the hydrogen iodide example above, the equilibrium concentrations are: [HI] = 0.80 mol dm^{-3}, [H_2] = 0.10 mol dm^{-3} and [I_2] = 0.10 mol dm^{-3} at 640 K. What is the equilibrium constant for this reaction at 640 K?

Just stick the concentrations into the expression for K_c:

$$K_c = \frac{[HI]^2}{[H_2][I_2]} = \frac{0.80^2}{0.10 \times 0.10} = 64$$

To work out the units of K_c, put the units in the expression instead of the numbers:

$$\text{Units of } K_c = \frac{(\text{mol dm}^{-3})^2}{(\text{mol dm}^{-3})(\text{mol dm}^{-3})} = \frac{\cancel{(\text{mol dm}^{-3})}\cancel{(\text{mol dm}^{-3})}}{\cancel{(\text{mol dm}^{-3})}\cancel{(\text{mol dm}^{-3})}}$$

The concentration units cancel, so there are no units and K_c is just **64**.

You might have to figure out some of the equilibrium concentrations before you can find K_c. To do this follow these steps:

Step 1: Find out how many moles of each reactant and product there are at equilibrium. You'll usually be given the number of moles at equilibrium for one of the reactants. You can then use the balanced reaction equation to work out the number of moles of all the others.

Step 2: Calculate the molar concentrations of each reactant and product by dividing each number of moles by the volume of the reaction. You'll be told the volume in the question but you may have to convert it into different units. To work out molar concentrations you need the volume to be in dm^3.

Once you've done this you're ready to substitute your values into the expression for K_c and calculate it.

Tip: The molar concentration is just the concentration in mol dm^{-3}.

Example — Maths Skills

0.20 moles of phosphorus(V) chloride decomposes at 600 K in a vessel of 5.0 dm^3. The equilibrium mixture is found to contain 0.080 moles of chlorine. Write the expression for K_c and calculate its value, including units.

$$PCl_{5(g)} \rightleftharpoons PCl_{3(g)} + Cl_{2(g)}$$

1. Find out how many moles of PCl_5 and PCl_3 there are at equilibrium:
   - The equation tells you that when 1 mole of PCl_5 decomposes, 1 mole of PCl_3 and 1 mole of Cl_2 are formed.
   - So if 0.080 moles of chlorine are produced at equilibrium, then there will be 0.080 moles of PCl_3 as well.
   - 0.080 mol of PCl_5 must have decomposed, so there will be 0.12 moles left (0.20 – 0.080 = 0.12).
2. Divide each number of moles by the volume of the flask to give the molar concentrations:

$$[PCl_3] = [Cl_2] = \frac{0.080}{5.0} = 0.016 \text{ mol dm}^{-3}$$

$$[PCl_5] = \frac{0.12}{5.0} = 0.024 \text{ mol dm}^{-3}$$

3. Put the concentrations in the expression for K_c and calculate it:

$$K_c = \frac{[PCl_3][Cl_2]}{[PCl_5]} = \frac{[0.016][0.016]}{[0.024]} = 0.011$$

4. Now find the units of K_c:

$$\text{Units of } K_c = \frac{(\text{mol dm}^{-3})(\text{mol dm}^{-3})}{\text{mol dm}^{-3}} = \frac{(\text{mol dm}^{-3})\cancel{(\text{mol dm}^{-3})}}{\cancel{\text{mol dm}^{-3}}} = \text{mol dm}^{-3}$$

So K_c = 0.011 mol dm^{-3}

Exam Tip
With long wordy questions like this it can help to circle important bits of information before you start — e.g. any concentrations or volumes you're likely to need for the calculation.

Exam Tip
You may be asked to calculate the molar amounts of some substances in an earlier part of the question — if this happens you can reuse your answers to find K_c. Handy.

Exam Tip
When you're writing expressions for K_c make sure you use [square brackets]. If you use (rounded brackets) you won't get the marks.

Tip: If you're finding the units of K_c and you end up with units left on their own on the bottom of the fraction, you can simplify them by swapping the signs of the powers. For example:

$$\frac{\cancel{\text{mol dm}^{-3}}}{\cancel{(\text{mol dm}^{-3})}(\text{mol dm}^{-3})} = \text{mol}^{-1}\text{ dm}^3.$$

Using K_c

If you know the value of K_c you can use it to find unknown equilibrium concentrations. Here's how you do it:

Step 1: Put all the values you know into the expression for K_c.

Step 2: Rearrange the equation and solve it to find the unknown values.

Example — Maths Skills

When ethanoic acid was allowed to reach equilibrium with ethanol at 25 °C, it was found that the equilibrium mixture contained 2.0 mol dm^{-3} ethanoic acid and 3.5 mol dm^{-3} ethanol. The K_c of the equilibrium is 4.0 at 25 °C. What are the concentrations of the other components?

$$CH_3COOH_{(l)} + C_2H_5OH_{(l)} \rightleftharpoons CH_3COOC_2H_{5(l)} + H_2O_{(l)}$$

1. Put all the values you know in the K_c expression:

$$K_c = \frac{[CH_3COOC_2H_5][H_2O]}{[CH_3COOH][C_2H_5OH]} \quad \text{so} \quad 4.0 = \frac{[CH_3COOC_2H_5][H_2O]}{2.0 \times 3.5}$$

2. Rearranging this gives:

$$[CH_3COOC_2H_5][H_2O] = 4.0 \times 2.0 \times 3.5 = 28$$

3. From the equation, you know that $[CH_3COOC_2H_5] = [H_2O]$, so:

$$[CH_3COOC_2H_5] = [H_2O] = \sqrt{28} = 5.3 \text{ mol dm}^{-3}$$

The concentration of $CH_3COOC_2H_5$ and H_2O is 5.3 mol dm^{-3}.

Tip: The units of concentration should always be mol dm^{-3}. If your answer doesn't give you this then go back and check your calculation to see where you've gone wrong.

Practice Questions — Application

Q1 The following equilibrium exists under certain conditions:

$$C_2H_4 + H_2O \rightleftharpoons C_2H_5OH$$

a) Write out the expression for K_c for this reaction.

5.00 moles of C_2H_5OH was placed in a container and allowed to reach equilibrium. At a certain temperature and pressure the equilibrium mixture was found to contain 1.85 moles of C_2H_4, and have a total volume of 15.0 dm^3.

b) Determine the number of moles of each substance at equilibrium.

c) Calculate the molar concentrations (in mol dm^{-3}) of all the reagents at equilibrium.

d) Calculate K_c for this equilibrium.

e) At a different temperature and pressure the equilibrium constant (K_c) for this reaction is 3.8 and the equilibrium mixture contained 0.80 mol dm^{-3} C_2H_5OH. Determine the equilibrium concentrations of C_2H_4 and H_2O under these conditions.

Tip: Don't forget — you need to work out the units of K_c too.

Q2 Under certain conditions the following equilibrium is established:

$$2SO_2 + O_2 \rightleftharpoons 2SO_3$$

a) Write out an expression for K_c for this reaction.

At a certain temperature the equilibrium concentrations for the three reagents were found to be:

SO_2 = 0.250 mol dm^{-3} , O_2 = 0.180 mol dm^{-3} , SO_3 = 0.360 mol dm^{-3}

b) Calculate K_c for this equilibrium.

c) If all other conditions (including the concentrations of O_2 and SO_3) were to stay the same, what would the equilibrium concentration of SO_2 have to be for K_c to be 15?

4. Factors Affecting the Equilibrium Constant

By tweaking some of the conditions of a system, you can change the position of the equilibrium. However, not all conditions have an effect...

Learning Objectives:

- Be able to predict the qualitative effects of changes of temperature on the value of K_c.
- Know that the value of the equilibrium constant is not affected either by changes in concentration or addition of a catalyst.

Specification Reference 3.1.6.2

Changing the temperature

If you increase the temperature, you add heat. The equilibrium shifts in the **endothermic** (positive ΔH) direction to absorb the heat. Decreasing the temperature removes heat energy. The equilibrium shifts in the **exothermic** (negative ΔH) direction to try to replace the heat. If the forward reaction's endothermic, the reverse reaction will be exothermic, and vice versa.
If the change means more product is formed, K_c will rise.
If it means less product is formed, then K_c will decrease.

Examples

The reaction below is exothermic in the forward direction:

Exothermic →

$$2SO_{2(g)} + O_{2(g)} \rightleftharpoons 2SO_{3(g)} \qquad \Delta H = -197 \text{ kJ mol}^{-1}$$

← *Endothermic*

If you increase the temperature, the equilibrium shifts to the left (in the endothermic direction) to absorb some of the extra heat energy. This means that less product's formed so the concentration of product ($[SO_3]$) will be less.

$$K_c = \frac{[SO_3]^2}{[SO_2]^2[O_2]}$$

As $[SO_3]$ will be a smaller value, and $[SO_2]$ and $[O_2]$ will have higher values, K_c will be lower.

The reaction below is endothermic in the forward direction:

Endothermic →

$$2CH_{4(g)} \rightleftharpoons 3H_{2(g)} + C_2H_{2(g)} \qquad \Delta H = +377 \text{ kJ mol}^{-1}$$

← *Exothermic*

This time increasing the temperature shifts the equilibrium to the right. This means more product's formed, so K_c increases.

$$K_c = \frac{[H_2]^3[C_2H_2]}{[CH_4]^2}$$

As $[H_2]$ and $[C_2H_2]$ will have higher values and $[CH_4]$ will have a smaller value, K_c will be higher.

Tip: The ΔH values given for reversible reactions show the ΔH of the forward reaction.

Tip: If the temperature decreases the opposite will happen — for reactions that are exothermic in the forward direction K_c will increase and for reactions that are endothermic in the forward direction K_c will decrease.

Changing the concentration

The value of the equilibrium constant, K_c, is fixed at a given temperature. So if the concentration of one thing in the equilibrium mixture changes then the concentrations of the others must change to keep the value of K_c the same.

Example

$$CH_3COOH_{(l)} + C_2H_5OH_{(l)} \rightleftharpoons CH_3COOC_2H_{5(l)} + H_2O_{(l)}$$

If you increase the concentration of CH_3COOH then the equilibrium will move to the right to get rid of some of the extra CH_3COOH — so more $CH_3COOC_2H_5$ and H_2O are produced. This keeps the equilibrium constant the same.

Tip: Remember — saying that the equilibrium moves to the right is just another way of saying more of the products form. If the equilibrium moves to the left, the opposite happens — more products are converted into reactants.

Adding a catalyst

Catalysts have no effect on the position of equilibrium or on the value of K_c. This is because a catalyst will increase the rate of both the forward and backward reactions by the same amount. As a result, the equilibrium position will be the same as the uncatalysed reaction, but equilibrium will be reached faster. So catalysts can't increase yield (the amount of product produced) — but they do decrease the time taken to reach equilibrium.

Exam Tip
Don't be thrown if a reaction you're given in an exam question about K_c has a catalyst. Catalysts don't affect K_c.

Practice Questions — Application

Q1 The following equilibrium is established under certain conditions:

$$2CHClF_{2(g)} \rightleftharpoons C_2F_{4(g)} + 2HCl_{(g)} \qquad \Delta H = +128 \text{ kJ mol}^{-1}$$

State and explain how you would expect the following to affect the value of K_c for this equilibrium:

a) Increasing the concentration of C_2F_4.

b) Increasing the temperature.

c) Adding a catalyst

Q2 The value of K_c for the following equilibrium increases if the temperature is decreased:

$$2SO_{2(g)} + O_{2(g)} \rightleftharpoons 2SO_{3(g)}$$

Is the forward reaction endothermic or exothermic? Explain your answer.

Exam Tip
Writing out the expression for K_c can really help solve questions like these. Remember — if the numbers on the top of the expression for K_c increase, then K_c will generally increase. If the numbers on the bottom of the expression for K_c increase, then K_c will generally decrease.

Practice Questions — Fact Recall

Q1 How will increasing the temperature affect K_c if:

a) the forward reaction is endothermic?

b) the forward reaction is exothermic?

Q2 Explain why changing the concentration of a reagent does not affect K_c.

Q3 How does adding a catalyst affect the equilibrium constant?

5. Redox Reactions

This'll probably ring a bell from GCSE, but don't go thinking you know it all already — there's plenty to learn about redox reactions.

Learning Objectives:

- Know that oxidation is the process of electron loss.
- Know that oxidising agents are electron acceptors.
- Know that reduction is the process of electron gain.
- Know that reducing agents are electron donors.
- Know the rules for assigning the oxidation state of an element in a compound or ion from the formula.
- Be able to work out the oxidation state of an element in a compound or ion from the formula.

Specification Reference 3.1.7

What are redox reactions?

A loss of electrons is called **oxidation**. A gain in electrons is called **reduction**. Reduction and oxidation happen simultaneously — hence the term "redox" reaction. An **oxidising agent** accepts electrons and gets reduced. A **reducing agent** donates electrons and gets oxidised (see Figure 1).

Tip: Now's your chance to learn the most famous memory aid thingy in the world...
OIL RIG
Oxidation Is Loss
Reduction Is Gain
(of electrons)

Example

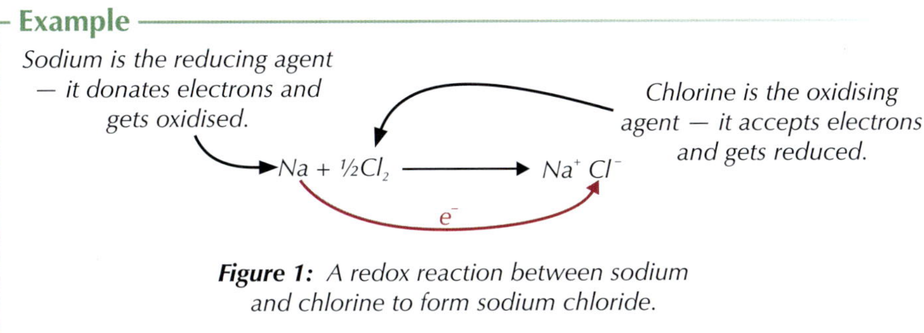

Sodium is the reducing agent — it donates electrons and gets oxidised.
Chlorine is the oxidising agent — it accepts electrons and gets reduced.
Na + ½Cl₂ → Na⁺ Cl⁻
e⁻

Figure 1: *A redox reaction between sodium and chlorine to form sodium chloride.*

Oxidation states

The **oxidation state** of an element tells you the total number of electrons it has donated or accepted. Oxidation states are also called oxidation numbers. There are lots of rules for working out oxidation states. Take a deep breath...

Uncombined elements have an oxidation state of 0. Elements just bonded to identical atoms also have an oxidation state of 0.

Examples

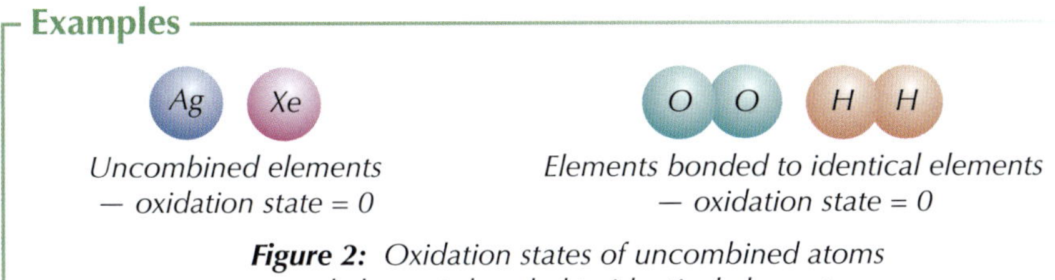

Figure 2: *Oxidation states of uncombined atoms and elements bonded to identical elements.*

The oxidation state of a simple monatomic ion is the same as its charge.

Examples

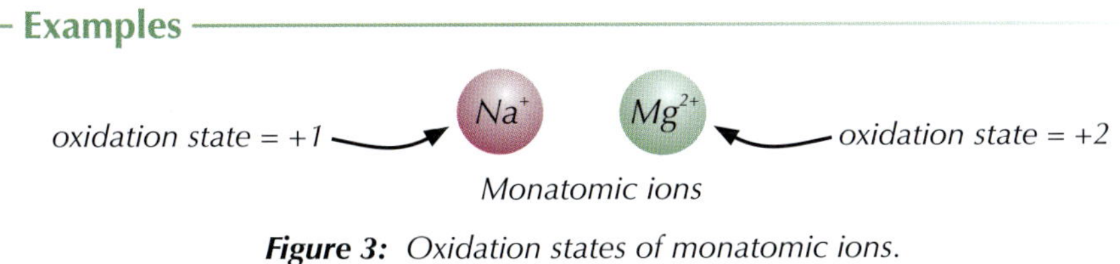

Figure 3: *Oxidation states of monatomic ions.*

In compounds or compound ions, each of the constituent atoms has an oxidation state of its own and the sum of the oxidation states equals the overall oxidation state (see Figure 4). This overall oxidation state is equal to the overall charge on the ion.

Within a compound ion, the most electronegative element has a negative oxidation state (equal to its ionic charge). Other elements have more positive oxidation states.

Tip: Take a look back at page 92 for more about electronegativity.

Tip: Combined oxygen usually has an oxidation state of –2. There are exceptions for combined oxygen — in OF_2 it's +2, and in O_2F_2 it's +1. And don't forget O_2, where it's 0.

Example

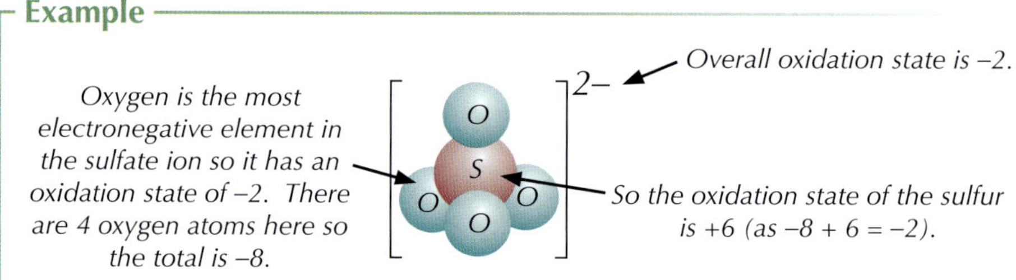

Overall oxidation state is –2.
Oxygen is the most electronegative element in the sulfate ion so it has an oxidation state of –2. There are 4 oxygen atoms here so the total is –8.
So the oxidation state of the sulfur is +6 (as –8 + 6 = –2).
O
S
O
O
O
2–

Figure 4: *Oxidation states of elements in the SO_4^{2-} ion.*

The sum of the oxidation states for a neutral compound is 0 (see Figure 5). If the compound is made up of more than one element, each element will have its own oxidation number.

Example

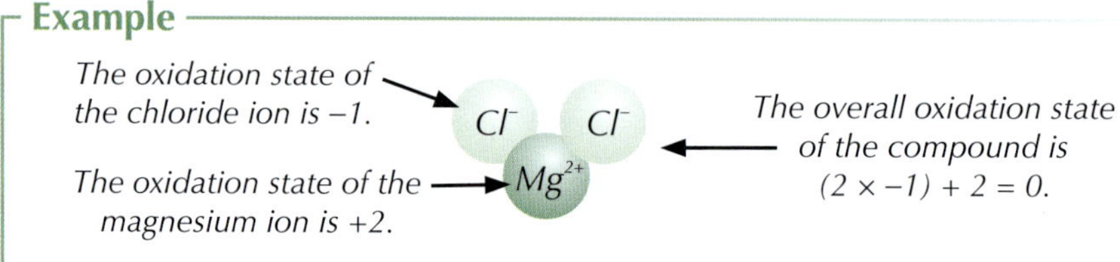

The oxidation state of the chloride ion is –1.
The oxidation state of the magnesium ion is +2.
Cl⁻
Cl⁻
Mg²⁺
The overall oxidation state of the compound is (2 × –1) + 2 = 0.

Figure 5: *Oxidation states of elements in magnesium chloride ($MgCl_2$).*

Combined oxygen is nearly always –2, except in peroxides, where it's –1 (see Figure 6). Combined hydrogen is +1, except in metal hydrides where it is –1 (see Figure 7) and H_2, where it's 0.

Examples

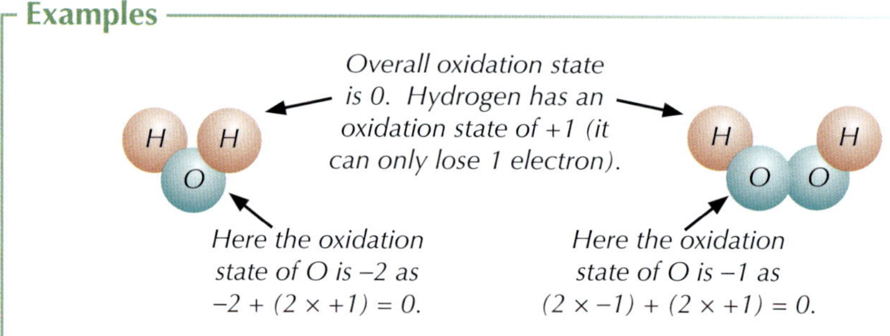

Overall oxidation state is 0. Hydrogen has an oxidation state of +1 (it can only lose 1 electron).
H
H
O
H
O
O
H
Here the oxidation state of O is –2 as –2 + (2 × +1) = 0.
Here the oxidation state of O is –1 as (2 × –1) + (2 × +1) = 0.

Figure 6: *Oxidation states of hydrogen and oxygen in water (H_2O) and hydrogen peroxide (H_2O_2).*

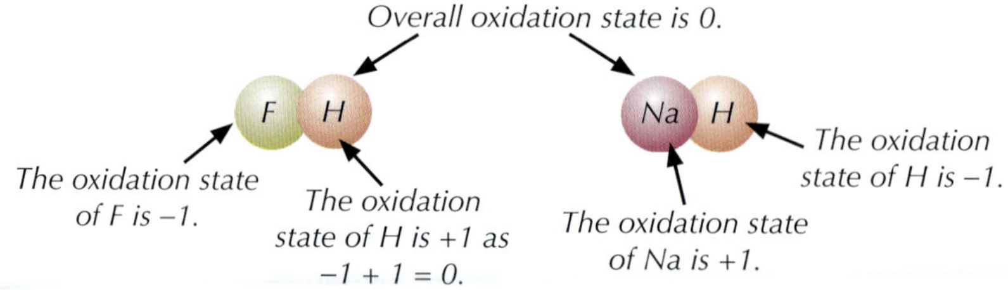

Figure 7: *Oxidation states of hydrogen in hydrogen fluoride (HF) and sodium hydride (NaH).*

Exam Tip
You may get a question asking you to work out the oxidation state of one element in a compound. Just follow all the rules and you'll be fine.

Finding oxidation states

You can work out oxidation states from formulas or systematic names.

Finding oxidation states from formulas

In your exam, you may get a question asking you to work out the oxidation state of one element in a compound. To do this you just have to follow all the rules on the previous two pages and you'll be fine.

Examples

Find the oxidation state of Zn in $Zn(OH)_2$.

- $Zn(OH)_2$ is neutral (it has no charge), so its overall oxidation state is 0.
- Oxygen's oxidation state is usually –2, and hydrogen's is usually +1.
- So the oxidation state of the $(OH)_2$ bit of the molecule is $2 \times (-2 + 1) = -2$.
- So the oxidation state of Zn in $Zn(OH)_2$ is $0 - (-2) = +2$.

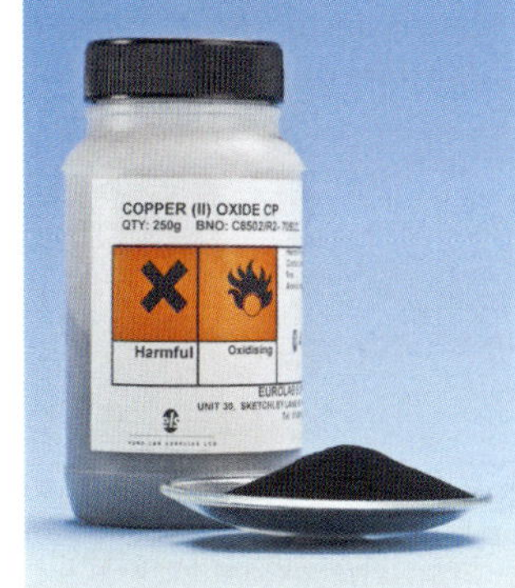

Figure 8: A bottle of copper(II) oxide. The Roman numerals show that the copper has an oxidation number of +2.

Finding oxidation states from systematic names

If an element can have multiple oxidation states (or isn't in its 'normal' oxidation state) its oxidation state is sometimes shown using Roman numerals, e.g. (I) = +1, (II) = +2, (III) = +3 and so on. The Roman numerals are written after the name of the element they correspond to.

Examples

In iron(II) sulfate, iron has an oxidation state of +2. Formula = $FeSO_4$

In iron(III) sulfate, iron has an oxidation state of +3. Formula = $Fe_2(SO_4)_3$

This is particularly useful when looking at -ate ions. Ions with names ending in -ate (e.g. sulfate, nitrate, carbonate) contain oxygen, as well as another element. For example, sulfates contain sulfur and oxygen, nitrates contain nitrogen and oxygen... and so on. But sometimes the 'other' element in the ion can exist with different oxidation states, and so form different '-ate ions'. You can use the systematic name to work out the formula of the ion.

Examples

In sulfate(VI) ions the sulfur has oxidation state +6. This is the SO_4^{2-} ion.

In sulfate(IV) ions, the sulfur has oxidation state +4. This is the SO_3^{2-} ion.

In nitrate(III), nitrogen has an oxidation state of +3. This is the NO_2^- ion.

Tip: The oxidation state in sulfate ions applies to the sulfur, not the oxygen, because oxygen always has an oxidation state of –2 in -ate ions.

Practice Questions — Application

Q1 Give the oxidation states of the following ions.

a) Na^+

b) F^-

c) Ca^{2+}

Q2 Give the overall oxidation states of the following ions.

a) OH^-

b) CO_3^{2-}

c) NO_3^-

Tip: Several ions have widely used common names that are different from their correct systematic names. E.g. the sulfate(IV) ion (SO_3^{2-}) is often called the sulfite ion.

Exam Tip
It's really important to remember the basic rules for working out oxidation states, as well as the exceptions — it's the kind of thing that could easily trip you up in exam questions.

Q3 Work out the oxidation states of all the elements in the following compounds and compound ions.

a) HCl b) SO_2

c) CO_3^{2-} d) ClO_4^-

e) Cu_2O f) HSO_4^-

Q4 Work out the oxidation states of carbon in the following.

a) CO b) CO_2

c) CCl_4 d) C

e) $CaCO_3$

Q5 Work out the oxidation states of phosphorus in the following.

a) P_4 b) PH_3

c) PO_4^{3-} d) P_2F_4

e) PBr_5 f) P_2H_4

Q6 Look at the reaction below.

$$Cu_{(s)} + H_2SO_{4(aq)} \rightarrow CuSO_{4(aq)} + H_{2(g)}$$

Give the oxidation states at the beginning and end of the reaction for the following elements:

a) Cu

b) S

c) H

d) O

Tip: The oxidation state of an atom doesn't always change when it reacts, so don't be alarmed if your answer is the same for the beginning and end of the reaction.

Practice Questions — Fact Recall

Q1 What is oxidation?

Q2 What is reduction?

Q3 Describe the role of an oxidising agent in a redox reaction.

Q4 Describe the role of a reducing agent in a redox reaction.

Q5 Give the oxidation state of an element bonded to an identical atom.

Q6 What is the sum of the oxidation states for a neutral compound?

Q7 What is the oxidation state of oxygen in a peroxide?

Q8 Give the oxidation state of hydrogen in a metal hydride.

6. Redox Equations

In redox reactions, oxidation and reduction go on simultaneously. You can write separate equations to show the two things happening, or you can package them up into one nice, neat redox equation.

Learning Objectives:

- Be able to write half-equations identifying the oxidation and reduction processes in redox reactions.
- Be able to combine half-equations to give an overall redox equation.

Specification Reference 3.1.7

Half-equations and redox equations

Ionic **half-equations** show oxidation or reduction (see Figure 1). The electrons are shown in a half-equation so that the charges balance.

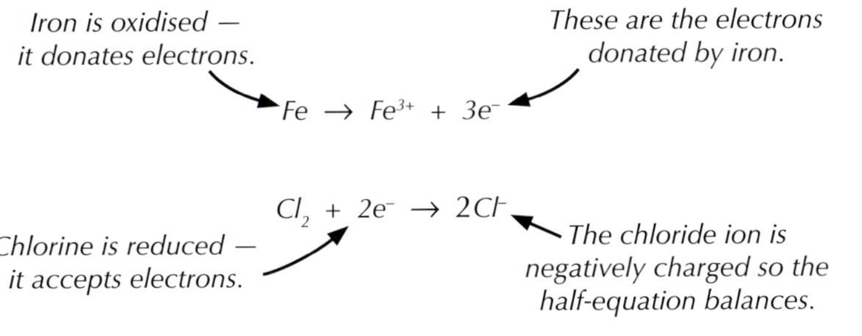

Iron is oxidised — it donates electrons.
These are the electrons donated by iron.
$Fe \rightarrow Fe^{3+} + 3e^-$
$Cl_2 + 2e^- \rightarrow 2Cl^-$
Chlorine is reduced — it accepts electrons.
The chloride ion is negatively charged so the half-equation balances.

Figure 1: *Half-equations showing the oxidation of iron and the reduction of chlorine.*

Tip: Remember — the charges as well as the number of atoms must be balanced in a balanced equation. See pages 52-53 for more on balancing equations.

You can combine half-equations for oxidising and reducing agents to make full equations for redox reactions. Just make sure both half-equations have the same number of electrons in, stick them together and cancel out the electrons.

Example

Iron reacts with chlorine to form iron(III) chloride ($FeCl_3$).

You can see the half-equations for the oxidation of iron and the reduction of chlorine in Figure 1. Oxidising iron produces three electrons — but to reduce chlorine you only need two.

If you multiply the iron equation by two and the chlorine equation by three then they'll both have six electrons in:

$$Fe \rightarrow Fe^{3+} + 3e^- \xrightarrow{\times 2} 2Fe \rightarrow 2Fe^{3+} + 6e^-$$

$$Cl_2 + 2e^- \rightarrow 2Cl^- \xrightarrow{\times 3} 3Cl_2 + 6e^- \rightarrow 6Cl^-$$

Now you can combine them. If you stick them together you get:

$$2Fe + 3Cl_2 + 6e^- \rightarrow 2Fe^{3+} + 6e^- + 6Cl^-$$

But the electrons on each side cancel out:

$$2Fe + 3Cl_2 + \cancel{6e^-} \rightarrow 2Fe^{3+} + \cancel{6e^-} + 6Cl^-$$

So the full redox equation for this reaction is:

$$2Fe + 3Cl_2 \rightarrow 2FeCl_3$$

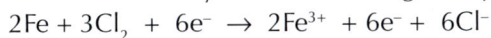

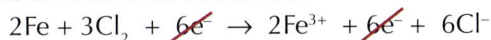

Tip: In any redox reaction the number of electrons released by the oxidation reaction must be the same as the number used up by the reduction reaction.

You can also work out the half-equations for a given equation — just make sure the atoms and charges balance.

Example

Magnesium burns in oxygen to form magnesium oxide:

$$2Mg + O_2 \rightarrow 2MgO$$

Write half-equations for the oxidation and reduction reactions that are part of this process.

Tip: Take your time balancing half-equations and always double check them — it's easy to get confused and end up with electrons all over the place.

Tip: The only things you're allowed to <u>add</u> to half-equations to balance them are electrons, H^+ ions and water.

Exam Tip
If you're told that it's an acid solution, that's a really big hint that you need to have H^+ ions in the equation somewhere.

Tip: If you come across other reactions like this with complicated oxidising or reducing agents, just follow the same steps and you'll be fine.

Start with the half-equation of oxygen being reduced to O^{2-}: $O_2 \rightarrow 2O^{2-}$

Now balance the charges by adding some electrons in: $\mathbf{O_2 + 4e^- \rightarrow 2O^{2-}}$

Then do the same for the half-equation of magnesium being oxidised to Mg^{2+}: $2Mg \rightarrow 2Mg^{2+}$.

Balance it by adding in the electrons: $\mathbf{2Mg \rightarrow 2Mg^{2+} + 4e^-}$

You can check your answer by making sure all the electrons cancel out:

$$2Mg + O_2 + \cancel{4e^-} \rightarrow 2MgO + \cancel{4e^-}$$

Sometimes you might have to write half-equations for a more complicated reaction where the oxidising or reducing agent contains oxygen or hydrogen. If so, you might need to add in H_2O and H^+ ions to balance the equation.

Example

Write a half-equation for the conversion of manganate(VII) ions (MnO_4^-) in acid solution into Mn^{2+} ions.

Start by writing out the basic reaction:

$$MnO_4^- \rightarrow Mn^{2+}$$

Add some H_2O to the right side to balance the oxygen in MnO_4^-:

$$MnO_4^- \rightarrow Mn^{2+} + 4H_2O$$

Next, add H^+ ions to the left side to balance the hydrogen:

$$MnO_4^- + 8H^+ \rightarrow Mn^{2+} + 4H_2O$$

Finally, add electrons in to balance the charges:

$$MnO_4^- + 8H^+ + 5e^- \rightarrow Mn^{2+} + 4H_2O$$

Practice Questions — Application

Q1 Combine the 2 half-equations below to give the full redox equation for the displacement of silver by zinc:

$Zn \rightarrow Zn^{2+} + 2e^-$ and $Ag^+ + e^- \rightarrow Ag$

Q2 Write the oxidation and reduction half-equations for this reaction:

$$Ca + Cl_2 \rightarrow CaCl_2$$

Q3 Balance this half-equation: $NO_3^- + H^+ + e^- \rightarrow N_2 + H_2O$

Q4 Write a half-equation for the reduction of $Cr_2O_7^{2-}$ ions in acid solution to Cr^{3+} ions.

Q5 H_2SO_4 can act as an oxidising agent. Give the half-equation for the reduction of H_2SO_4 to H_2S and water.

Practice Questions — Fact Recall

Q1 What does an ionic half-equation show?

Q2 Which of these (**A**, **B** or **C**) is a half-equation?

A $2Mg + O_2 \rightarrow 2MgO$

B $Fe \rightarrow Fe^{3+} + 3e^-$

C $2Mg + O_2 + 4e^- \rightarrow 2MgO + 4e^-$

Section Summary

Make sure you know...

- That reversible reactions have both a forwards and a backwards reaction.
- That reversible reactions can reach dynamic equilibrium when the concentrations of reactants and products stay constant and the forwards and backwards reactions have the same reaction rate.
- That dynamic equilibrium can only be reached in a closed system which is at a constant temperature.
- That Le Chatelier's principle states that "if a reaction at equilibrium is subjected to a change in concentration, pressure or temperature, the equilibrium will move to help counteract the change."
- That a catalyst does not affect the position of equilibrium in a reversible reaction.
- That increasing the concentration of a reactant shifts the equilibrium to remove the extra reactant.
- That increasing the pressure shifts the equilibrium in favour of the reaction that produces the fewest moles of gas, in order to reduce the pressure.
- That increasing the temperature shifts the equilibrium in favour of the endothermic reaction, to remove the excess heat. (Low temperatures favour exothermic reactions.)
- Why, in industrial processes, there is a compromise in the temperatures and pressures used, in terms of rate, equilibrium and production costs.
- That K_c is the equilibrium constant.
- How to derive expressions for the equilibrium constant, K_c.
- How to calculate K_c and its units from the equilibrium concentrations and molar ratios for a reaction.
- How to use K_c to find unknown equilibrium concentrations for a reaction.
- That increasing the temperature of exothermic reactions will decrease K_c.
- That increasing the temperature of endothermic reactions will increase K_c.
- That changing the concentrations of reactants or adding a catalyst has no effect on the value of K_c.
- That loss of electrons is called oxidation and gain of electrons is called reduction.
- That oxidising agents are electron acceptors and reducing agents are electron donors.
- What redox reactions are.
- The rules for assigning the oxidation states of common atoms and ions.
- How to work out the oxidation state of an element in a compound or ion.
- How to write half-equations for the oxidation and reduction parts of a redox reaction.
- How to combine half-equations to make a full redox equation.

Exam-style Questions

1 Which of the following statements about redox is true?

A An oxidising agent loses electrons.

B Hydrogen always has an oxidation state of +1.

C Reduction is gain of electrons.

D The manganese in a manganate(VI) ion has an oxidation state of –6.

(1 mark)

2 Four different units are listed below.

1 $mol^2\ dm^{-6}$ **2** $mol^2\ dm^{-1}$ **3** $mol^3\ dm^{-1}$ **4** $mol\ dm^{-3}$

Which of the units listed above could be the units for the equilibrium constant, K_c?

A 4 only

B 1, 2 and 4 only

C 1, 2, 3 and 4

D 1 and 4 only

(1 mark)

3 Silver nitrate reacts with copper to form copper nitrate and silver.
The reaction is shown below.

$$2AgNO_3 + Cu \rightarrow Cu(NO_3)_2 + 2Ag$$

Which of the following options shows the correct half-equations for this reaction?

A $2Ag^+ + 2e^- \rightarrow 2Ag$ and $Cu \rightarrow Cu^{2+} + 2e^-$

B $Ag^{6+} + 6e^- \rightarrow 2Ag$ and $Cu \rightarrow Cu^{6+} + 6e^-$

C $Ag^+ \rightarrow Ag + e^-$ and $NO_3^- + e^- \rightarrow NO_3^{2-}$

D $2Ag^+ + e^- \rightarrow 2Ag$ and $Cu \rightarrow Cu^{2+} + e^-$

(1 mark)

4 What is the oxidation state of N in N_2O_5?

A +2

B +5

C –2

D –10

(1 mark)

5 A chemical factory produces ethanol (C_2H_5OH) from ethene (C_2H_4) and steam (H_2O) using the following reversible reaction:

$$C_2H_{4(g)} + H_2O_{(g)} \rightleftharpoons C_2H_5OH_{(g)} \qquad \Delta H = -46 \text{ kJ mol}^{-1}$$

The reaction is carried out under the following conditions:

Pressure = 60 atm

Temperature = 300 °C

Catalyst = phosphoric acid

5.1 Without the phosphoric acid catalyst the rate of reaction is so slow that dynamic equilibrium takes a very long time to occur. Describe what it means for a reaction to be at dynamic equilibrium.

(2 marks)

5.2 The process conditions for the reaction were chosen with consideration of Le Chatelier's principle. State Le Chatelier's principle.

(1 mark)

5.3 Explain why the pressure chosen for the process is a compromise.

(3 marks)

5.4 A leak in one of the pipes reduces the amount of H_2O in the reaction mixture. Explain the effect this has on the maximum yield of ethanol.

(3 marks)

6 Ammonia (NH_3) is produced industrially using the Haber-Bosch process. It uses the following reaction between nitrogen and hydrogen:

$$N_{2(g)} + 3H_{2(g)} \rightleftharpoons 2NH_{3(g)} \qquad \Delta H = -92 \text{ kJ mol}^{-1}$$

The reaction is usually carried out under the following conditions:

Pressure = 200 atm

Temperature = 400 °C – 500 °C

Catalyst = iron

6.1 What effect does the iron catalyst have on the position of equilibrium?

(1 mark)

The reaction needs to be carried out at a reasonably high temperature in order to keep the reaction rate high. As well as affecting the rate, increasing the temperature also affects the position of equilibrium for the reaction.

6.2 State whether this reaction is endothermic or exothermic.

(1 mark)

6.3 Explain the effect of increasing the temperature on the position of equilibrium.

(3 marks)

6.4 Other than changing the temperature, suggest two ways to shift the position of equilibrium in order to get an increased yield of ammonia from the reaction.

(2 marks)

7 The following equilibrium establishes at temperature X:

$$CH_{4(g)} + 2H_2O_{(g)} \rightleftharpoons CO_{2(g)} + 4H_{2(g)} \qquad \Delta H = +165 \text{ kJ mol}^{-1}$$

At equilibrium the mixture was found to contain 0.0800 mol dm^{-3} CH_4, 0.320 mol dm^{-3} H_2O, 0.200 mol dm^{-3} CO_2 and 0.280 mol dm^{-3} H_2.

7.1 Write an expression for K_c for this equilibrium.

(1 mark)

7.2 Calculate the value of K_c at temperature X, and give its units.

(3 marks)

At a different temperature, Y, the value of K_c was found to be 0.0800 and the equilibrium concentrations were as follows:

Gas	CH_4	H_2O	CO_2	H_2
Concentration (mol dm^{-3})	?	0.560	0.420	0.480

7.3 Calculate the equilibrium concentration of CH_4 at this temperature.

(2 marks)

7.4 At another temperature, Z, the value of K_c was found to be 1.2×10^{-3}. Suggest whether temperature Z is higher or lower than temperature Y. Explain your answer.

(3 marks)

7.5 State how the value of K_c would change if a catalyst was added to the reaction. Explain your answer.

(2 marks)

8 Aluminium chloride, ($AlCl_3$), is a useful ionic substance, used as a catalyst in Friedel-Crafts reactions. It is synthesised in industry by reacting aluminium with chlorine (Cl_2) at high temperatures.

8.1 Write the ionic half-equations for this reaction.

(2 marks)

8.2 Identify the oxidising agent in the reaction between chlorine and aluminium.

(1 mark)

8.3 Aluminium is also used to form the ionic compound, $Al_2(SO_4)_3$. What is the oxidation state of sulfur in $Al_2(SO_4)_3$?

(1 mark)

1. The Periodic Table

You'll remember from GCSE that the periodic table isn't just arranged how it is by chance. There are well-thought-out reasons behind it, and you can find out lots of stuff from it, not least about the numbers of electron shells and electrons each element has. Read on...

Learning Objective:
- Know that an element is classified as s, p, d or f block according to its position in the periodic table, which is determined by its proton number.

Specification Reference 3.2.1.1

How is the periodic table arranged?

Dmitri Mendeleev developed the modern periodic table in the 1800s. Although there have been changes since then, the basic idea is still the same. The periodic table is arranged into periods (rows) and groups (columns), by atomic (proton) number.

Figure 1: Dmitri Mendeleev (1834-1907) was a Russian chemist who developed the periodic table.

Period	Group 1	Group 2											Group 3	Group 4	Group 5	Group 6	Group 7	Group 0
								1 H 1										4 He 2
2	7 Li 3	9 Be 4											11 B 5	12 C 6	14 N 7	16 O 8	19 F 9	20 Ne 10
3	23 Na 11	24 Mg 12											27 Al 13	28 Si 14	31 P 15	32 S 16	35.5 Cl 17	40 Ar 18
4	39 K 19	40 Ca 20	45 Sc 21	48 Ti 22	51 V 23	52 Cr 24	55 Mn 25	56 Fe 26	59 Co 27	59 Ni 28	63.5 Cu 29	65 Zn 30	70 Ga 31	73 Ge 32	75 As 33	79 Se 34	80 Br 35	84 Kr 36
5	86 Rb 37	88 Sr 38	89 Y 39	91 Zr 40	93 Nb 41	96 Mo 42	99 Tc 43	101 Ru 44	103 Rh 45	106 Pd 46	108 Ag 47	112 Cd 48	115 In 49	119 Sn 50	122 Sb 51	128 Te 52	127 I 53	131 Xe 54
6	133 Cs 55	137 Ba 56	57-71	179 Hf 72	181 Ta 73	184 W 74	186 Re 75	190 Os 76	192 Ir 77	195 Pt 78	197 Au 79	201 Hg 80	204 Tl 81	207 Pb 82	209 Bi 83	210 Po 84	210 At 85	222 Rn 86
7	223 Fr 87	226 Ra 88	89-103															

Figure 2: The periodic table.

The period and group of an element gives you information about the number of electrons and electron shells that an element has.

Exam Tip

You'll be given a periodic table in your exam, so you'll always have a copy to refer to to answer exam questions. It's a good idea to be familiar with how it works before you go into the exam though.

Elements and periods

All the elements within a period have the same number of electron shells (if you don't worry about s and p sub-shells).

Example

The elements in Period 2 have 2 electron shells.

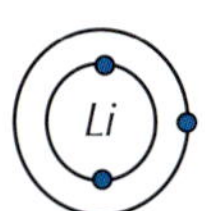

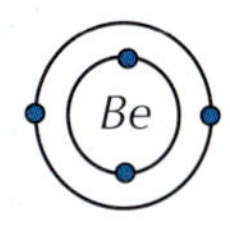

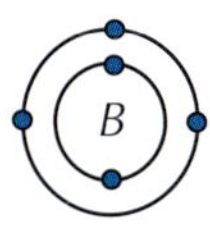

Figure 3: Atoms of the first three elements in Period 2.

Elements and groups

All the elements within a group have the same number of electrons in their outer shell — so they have similar properties (see Figure 5). The group number tells you the number of electrons in the outer shell.

Figure 5: *The Group 1 elements potassium, sodium and lithium all react strongly with water.*

Examples

Group 1 elements have 1 electron in their outer shell (see Figure 4), Group 4 elements have 4 electrons and so on.

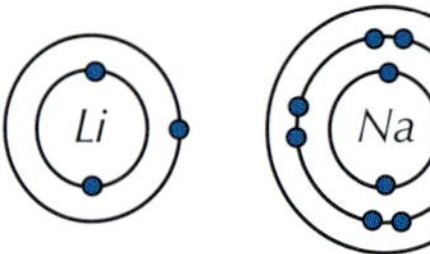

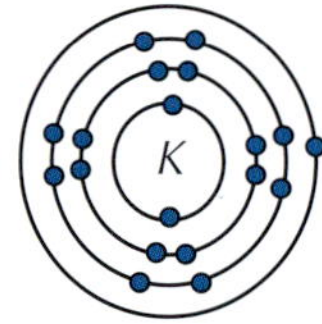

Figure 4: *Atoms of elements in Group 1.*

The exception to this rule is Group 0. All Group 0 elements have eight electrons in their outer shell (apart from helium, which has two), giving them full outer shells.

Electron configurations

The periodic table can be split into an s block, d block, p block and f block (see Figure 6). Doing this shows you which sub-shells all the electrons go into.

Exam Tip

You won't get a periodic table split up like this in the exam, so you need to remember where the different blocks are.

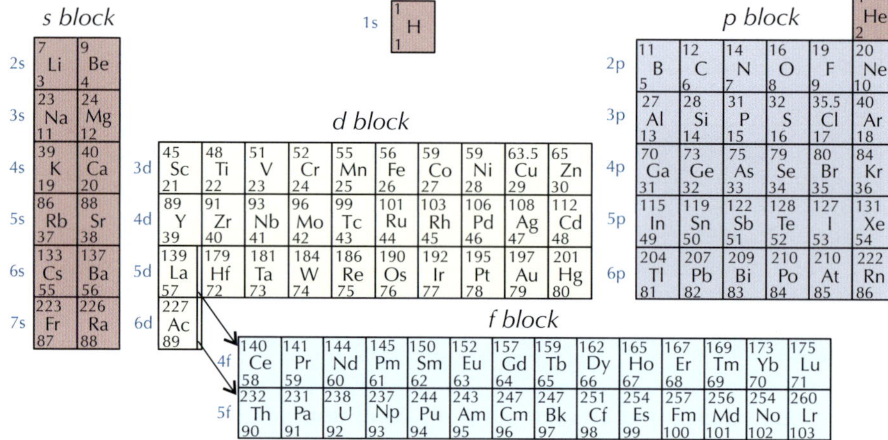

Figure 6: *The periodic table showing the s block, p block, d block and f block.*

Tip: If you can't remember what sub-shells are, don't know what the electron configuration of an element shows, or can't quite get your head around what all this $1s^2\ 2s^2\ 2p^6$ business is, then have a look back at pages 33-35 — it's all explained in lots of lovely detail there.

When you've got the periodic table labelled with the shells and sub-shells, it's pretty easy to read off the electron structure of any element by starting at the top and working your way across and down until you get to your element.

Example

To work out the electron configuration of phosphorus (P), you can use the periodic table to see that it's in Group 5 and Period 3. Starting with Period 1, the electron configuration of a full shell is $1s^2$. For Period 2 it's $2s^2\ 2p^6$. However, phosphorus' outer shell is only partially filled — it's got 5 outer electrons in the configuration $3s^2\ 3p^3$.

So: Period 1 — $1s^2$

Period 2 — $2s^2\ 2p^6$

Period 3 — $3s^2\ 3p^3$

The full electron structure of phosphorus is: $1s^2\ 2s^2\ 2p^6\ 3s^2\ 3p^3$.

Practice Questions — Application

Q1 Give the number of electron shells that atoms of the following elements have:

a) Sulfur, S
b) Beryllium, Be
c) Bromine, Br
d) Neon, Ne
e) Rubidium, Rb

Q2 How many electrons are in the outer shell of atoms of the following elements?

a) Selenium, Se
b) Potassium, K
c) Fluorine, F
d) Aluminium, Al
e) Strontium, Sr

Q3 Work out the electron configurations of the following elements:

a) Sodium, Na
b) Calcium, Ca
c) Chlorine, Cl
d) Arsenic, As
e) Vanadium, V
f) Scandium, Sc

Tip: It doesn't really matter how you work out the electron configurations of elements, whether you use the rules on pages 33-35 or whether you read them off the periodic table like on these pages. The important thing is that you get them right — so find a method you're happy with and stick with it.

Tip: Don't forget that for d-block elements, the sub-shells aren't written in the order that they're filled.

Practice Questions — Fact Recall

Q1 How is the periodic table arranged?

Q2 Look at the diagram below.

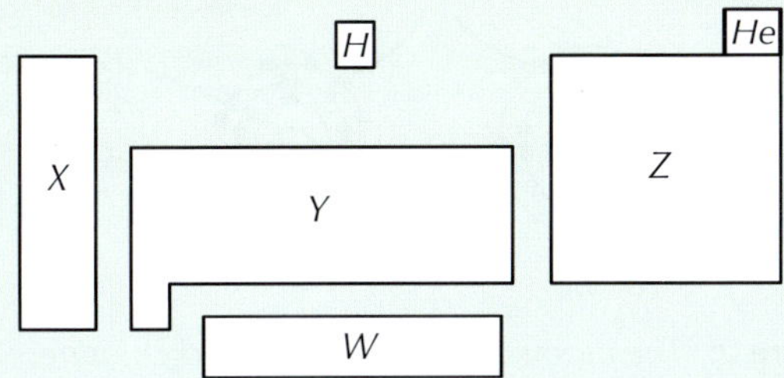

a) Which letter(s) (W, X, Y or Z) represent(s) the p block of the periodic table?
b) Which letter(s) (W, X, Y or Z) represent(s) the d block of the periodic table?
c) Which letter(s) (W X, Y or Z) represent(s) the s block of the periodic table?

2. Periodicity

Learning Objectives:

- Know the trends in atomic radius, first ionisation energy and melting point of the elements Na–Ar.
- Understand and explain the reasons for these trends in terms of the structure of and bonding in the elements.

Specification Reference 3.2.1.2

Periodicity is an important idea in chemistry. It's all to do with the trends in physical and chemical properties of elements across the periodic table — things like atomic radius, melting point, boiling point, and ionisation energy.

Atomic radius

Atomic radius decreases across a period (see Figure 1). As the number of protons increases, the positive charge of the nucleus increases. This means electrons are pulled closer to the nucleus, making the atomic radius smaller (see Figure 2). The extra electrons that the elements gain across a period are added to the outer energy level, so they don't really provide any extra shielding effect (shielding is mainly provided by the electrons in the inner shells).

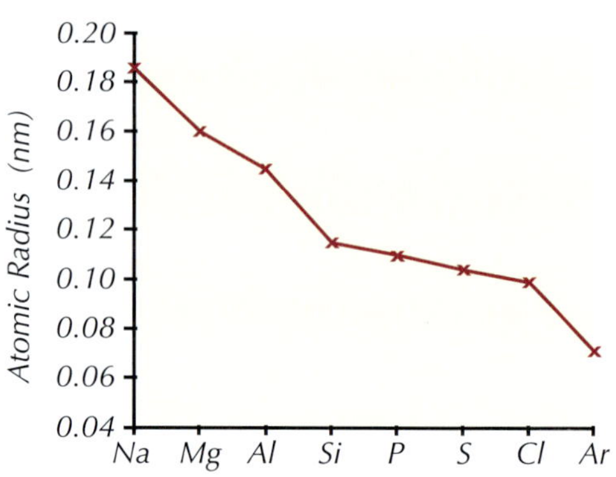

Figure 1: *The atomic radii of the Period 3 elements.*

Tip: Shielding is when the inner electrons effectively 'screen' the outer electrons from the pull of the nucleus. Look back at page 37 for more on shielding.

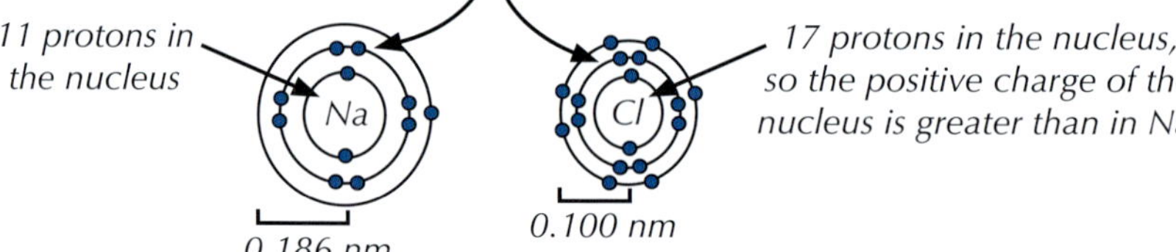

Figure 2: *The atomic radii of sodium and chlorine.*

Figure 3: *Chlorine (top) has lower melting and boiling points than silicon (bottom). So chlorine is a gas at r.t.p., whilst silicon is a solid.*

Melting points

If you look at how the melting points change across Period 3, the trend isn't immediately obvious. The melting points increase from sodium to silicon, but then generally decrease from silicon to argon (see Figure 4).

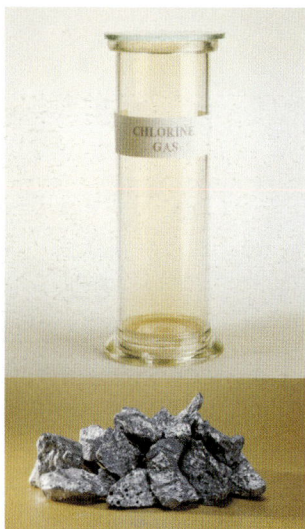

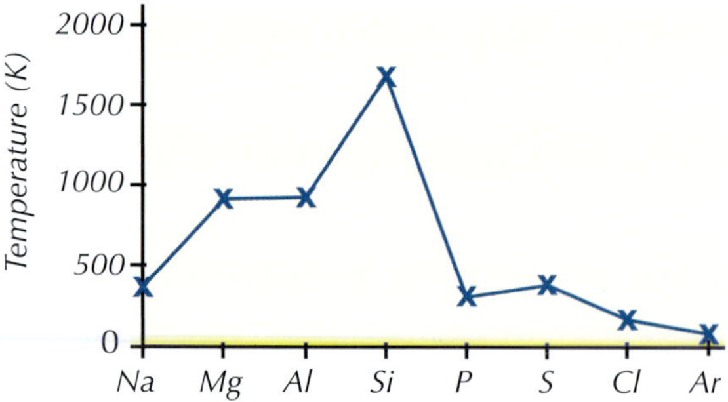

Figure 4: *Melting points of Period 3 elements.*

Sodium, magnesium and aluminium

Sodium, magnesium and aluminium are metals. Their melting points increase across the period because the metal-metal bonds get stronger. The bonds get stronger because, as you go across the period, the metal ions have an increasing positive charge, an increasing number of delocalised electrons and a decreasing radius.

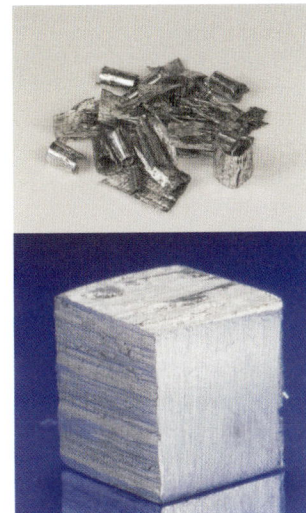

Figure 5: *Magnesium (top) and aluminium (bottom).*

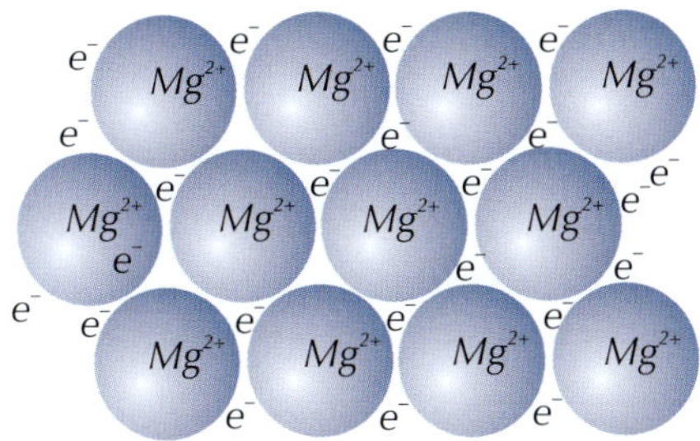

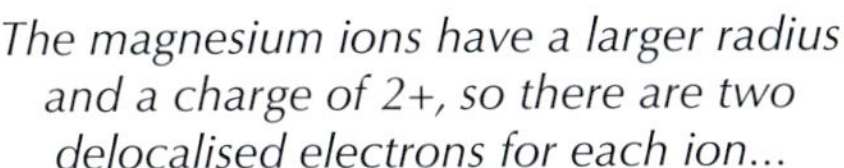

The magnesium ions have a larger radius and a charge of 2+, so there are two delocalised electrons for each ion...

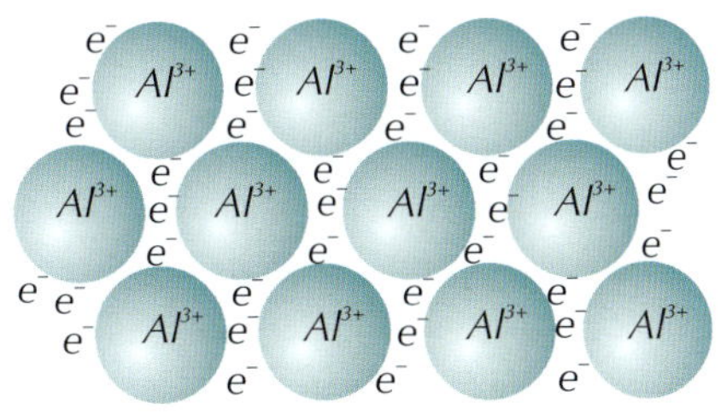

...whereas the aluminium ions have a smaller radius and a charge of 3+, so there are three delocalised electrons for each ion.

Figure 6: *The structures of magnesium and aluminium.*

Silicon

Silicon is **macromolecular**, with a tetrahedral structure — strong covalent bonds link all its atoms together (see Figure 7). A lot of energy is needed to break these bonds, so silicon has a high melting point.

Tip: The structure of silicon should look familiar — it's similar to diamond (page 83). There's a good reason for this — carbon and silicon are both in Group 4, so have the same number of electrons in their outer shell.

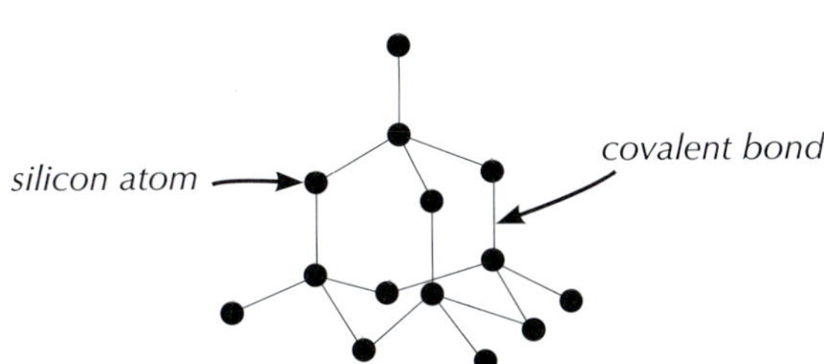

Figure 7: *The structure of silicon.*

Phosphorus, sulfur, chlorine and argon

Phosphorus (P_4), sulfur (S_8) and chlorine (Cl_2) are all molecular substances. Their melting points depend upon the strength of the van der Waals forces (see pages 94-95) between the molecules. Van der Waals forces are weak and easily overcome, so these elements have low melting points. More atoms in a molecule mean stronger van der Waals forces. Sulfur is the biggest molecule (S_8 — see Figure 9), so it's got a higher melting point than phosphorus or chlorine. Argon has a very low melting point because it exists as individual atoms (it's monatomic), resulting in very weak van der Waals forces.

Figure 8: *Sulfur (yellow powder) and phosphorus (stored under water) are solids at room temperature, whereas chlorine is a gas (see Figure 4).*

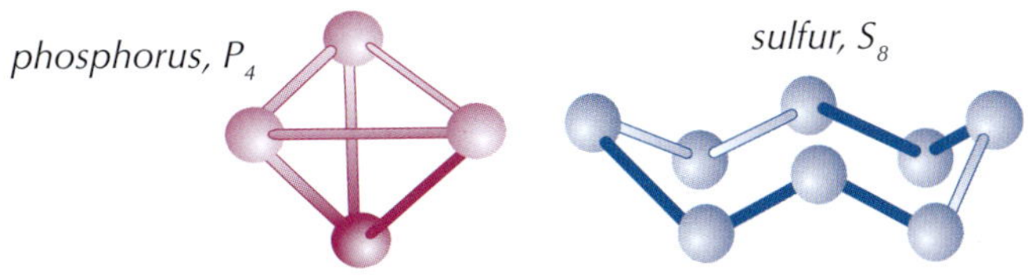

Figure 9: *The structures of phosphorus and sulfur.*

Exam Tip
There's no need to memorise the graphs on these pages, but you do need to know their shapes and be able to explain them.

Tip: There are a few blips in the trend for first ionisation energy — check back to page 39 for more details.

First ionisation energy

The first ionisation energy is the energy needed to remove 1 electron from each atom in 1 mole of gaseous atoms to form 1 mole of gaseous 1+ ions. There's a general increase in the first ionisation energy as you go across Period 3 (see Figure 10). This is because of the increasing attraction between the outer shell electrons and the nucleus, due to the number of protons increasing.

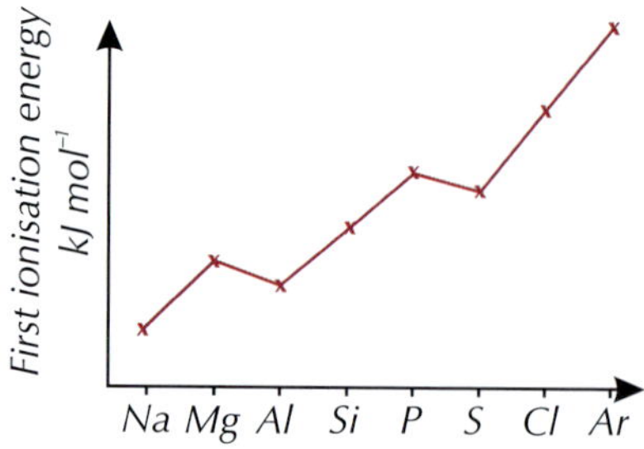

Figure 10: *The first ionisation energies of the Period 3 elements.*

Exam Tip
If you're asked about a specific trend, it might help you to roughly sketch out the shape of the relevant graph. That way you'll easily be able to see how the values for the elements compare to each other.

Practice Questions — Application

Q1 a) Explain why the atomic radius of aluminium is larger than the atomic radius of sulfur.

b) Name a Period 3 element with a larger atomic radius than aluminium.

Q2 Explain why the first ionisation energy of sulfur is higher than the first ionisation energy of aluminium.

Q3 The melting point of silicon is 1414 °C and the melting point of phosphorus is 44 °C.

a) Explain why the melting point of phosphorus is lower than the melting point of silicon.

b) Name a Period 3 element with a lower melting point than phosphorus.

Practice Questions — Fact Recall

Q1 Describe the trend in atomic radius across Period 3 of the periodic table.

Q2 Describe the trend in melting points across Period 3 of the periodic table.

Q3 Describe the general trend in first ionisation energy across Period 3 of the periodic table.

Section Summary

Make sure you know...

- That elements are classed as s block, p block, d block or f block depending on their position in the periodic table, which is determined by their proton number.
- The trends in atomic radius, melting point and first ionisation energy across Period 3.
- The reasons for the trends in atomic radius, melting point and first ionisation energy across Period 3.

Exam-style Questions

1 Which of the following determines the order that elements are arranged in in the periodic table?

A atomic mass

B proton number

C number of electrons in outer shell

D atomic radius

(1 mark)

2 Which of the following statements about sulfur is **not** correct?

A Its atomic radius is smaller than that of magnesium.

B Its electron configuration is $1s^2 2s^2 2p^6 3s^2 3p^4$.

C It's in the s block of the periodic table.

D Its first ionisation energy is higher than that of sodium.

(1 mark)

3 Which of the following elements has the lowest first ionisation energy?

A Cl

B Si

C P

D Ar

(1 mark)

4 Sodium, magnesium and aluminium are all metals in Period 3 of the periodic table.

4.1 Explain why the atomic radius of aluminium is smaller than that of sodium.

(2 marks)

4.2 Explain why the melting point of magnesium is greater than that of sodium.

(3 marks)

4.3 Argon is another element in Period 3. Explain why the melting point of argon is lower than the melting points of sodium, magnesium and aluminium.

(2 marks)

1. Group 2 — The Alkaline Earth Metals

Learning Objective:
- Know the trends in atomic radius, first ionisation energy and melting point of the elements Mg – Ba.
- Be able to explain the trends in atomic radius and first ionisation energy.
- Be able to explain the melting point of the elements in terms of their structure and bonding.

Specification Reference 3.2.2

The alkaline earth metals are in the s block of the periodic table. You have to know the trends in their properties as you go down Group 2 — in atomic radius, ionisation energy and melting points.

Atomic radius

As you go down a group in the periodic table, the atomic radius gets larger. This is because extra electron shells are added as you go down the group (see Figures 2 and 3).

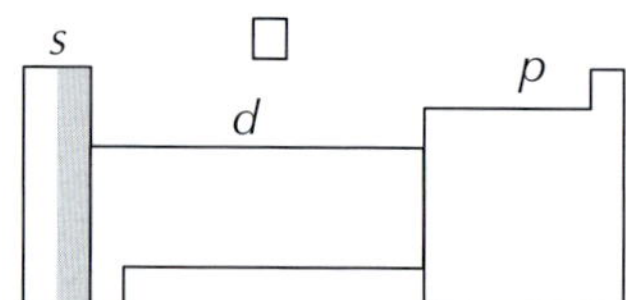

Figure 1: *The s, p and d blocks of the periodic table (see page 36 for more on this). Group 2 is highlighted in grey.*

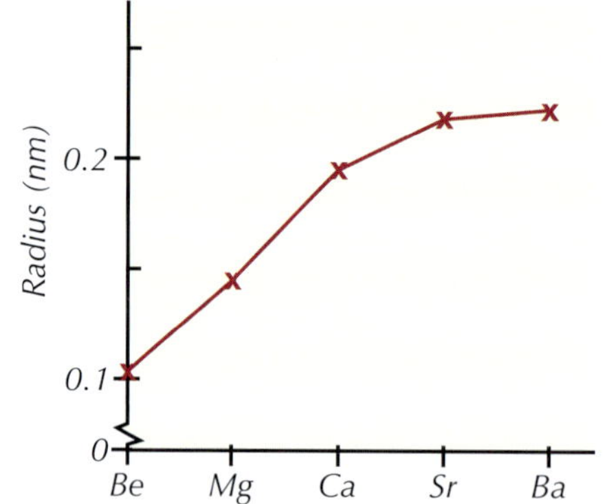

Figure 2: *Atomic radii of the first five elements in Group 2.*

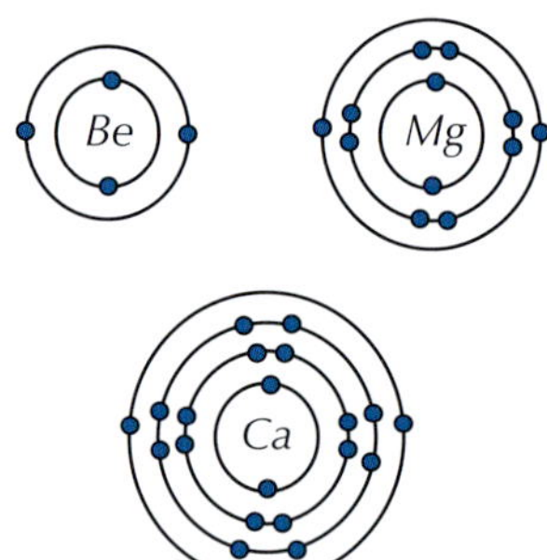

Figure 3: *Electron configurations of the first three elements in Group 2.*

First ionisation energy

Group 2 elements all have two electrons in their outer shell (s^2). They can lose their two outer electrons to form 2+ ions. Their ions then have every atom's dream electronic structure — that of a noble gas (see Figure 4).

Element	Atom	Ion
Be	$1s^2\,2s^2$	$1s^2$
Mg	$1s^2\,2s^2\,2p^6\,3s^2$	$1s^2\,2s^2\,2p^6$
Ca	$1s^2\,2s^2\,2p^6\,3s^2\,3p^6\,4s^2$	$1s^2\,2s^2\,2p^6\,3s^2\,3p^6$

Figure 4: *Electronic structures of Group 2 atoms and ions.*

Tip: See pages 37-40 for more on ionisation energies.

First ionisation energy decreases down the group (see Figure 5). This is because each element down Group 2 has an extra electron shell compared to the one above. The extra inner shells shield the outer electrons from the attraction of the nucleus. Also, the extra shell means that the outer electrons are further away from the nucleus, which greatly reduces the nucleus's attraction. Both of these factors make it easier to remove outer electrons, resulting in a lower first ionisation energy. The positive charge of the nucleus does increase as you go down a group (due to the extra protons), but this effect is overridden by the effect of the extra shells.

Group 2 element	1st ionisation energy / kJ mol^{-1}
Be	900
Mg	738
Ca	590
Sr	550
Ba	503

Figure 5: *First ionisation energies of Group 2 elements.*

Example

The first ionisation energy of calcium is lower than the first ionisation energy of magnesium (see Figure 6).

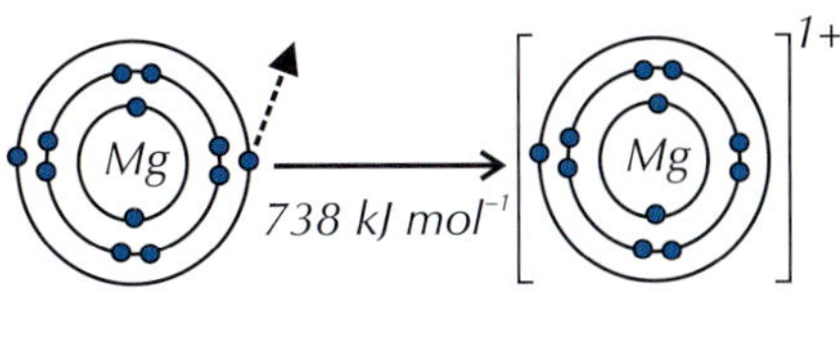

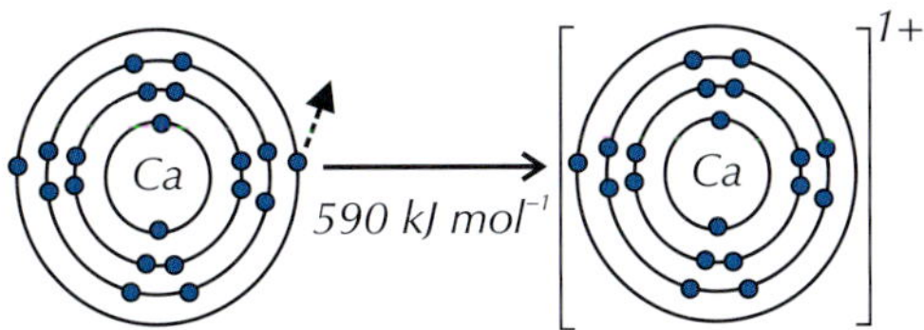

Figure 6: *First ionisation energy of magnesium and calcium.*

A magnesium atom has three electron shells, whereas a calcium atom has four electron shells. This means that the outer shell electrons are further from the nucleus in calcium than in magnesium. Also, a calcium atom has 18 electrons in inner shells, compared to only 10 in a magnesium atom. This means that shielding is greater in calcium atoms. So, less energy is needed to remove an electron from calcium than from magnesium.

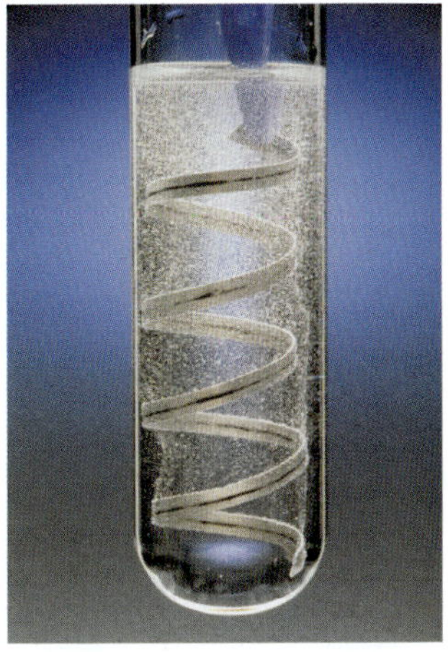

Figure 7: *Magnesium ribbon reacting with hydrochloric acid.*

Reactivity

When Group 2 elements react they lose electrons, forming positive ions. The easier it is to lose electrons (i.e. the lower the first ionisation energy), the more reactive the element, so reactivity increases down the group (see Figures 7 and 8).

Figure 8: *Calcium reacting with hydrochloric acid.*

Melting point

Melting points generally decrease down the group (see Figure 9). The Group 2 elements have typical metallic structures, with positive ions in a crystal structure surrounded by delocalised electrons from the outer electron shells. Going down the group the metal ions get bigger. But the number of delocalised electrons per atom doesn't change (it's always 2) and neither does the charge on the ion (it's always +2). The larger the ionic radius, the further away the delocalised electrons are from the positive nuclei. So it takes less energy to break the bonds, which means the melting points generally decrease as you go down the group. However, there's a big 'blip' at magnesium, because the crystal structure (the arrangement of the metallic ions) changes.

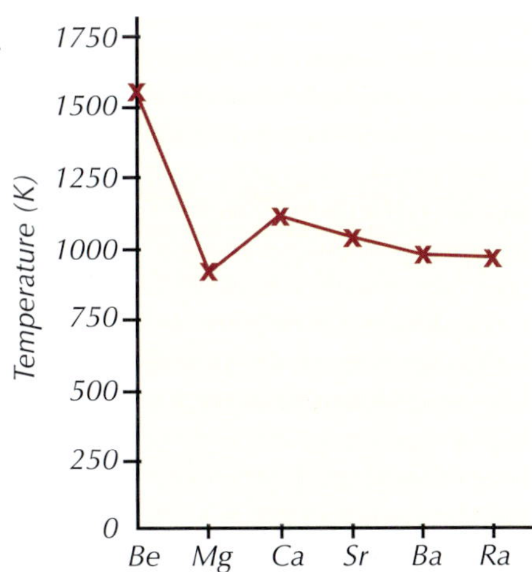

Figure 9: *Melting points of Group 2 elements.*

Exam Tip

There's no need to memorise the exact values on this graph, or the one on page 160, but you do need to know their shapes and be able to explain them.

Exam Tip
There's no need to memorise these structures and numbers, but you do have to understand how atomic radius affects melting point.

Example

The atomic radii increase from calcium to strontium to barium, but there are still only two delocalised electrons per ion (see Figure 10). So melting point decreases.

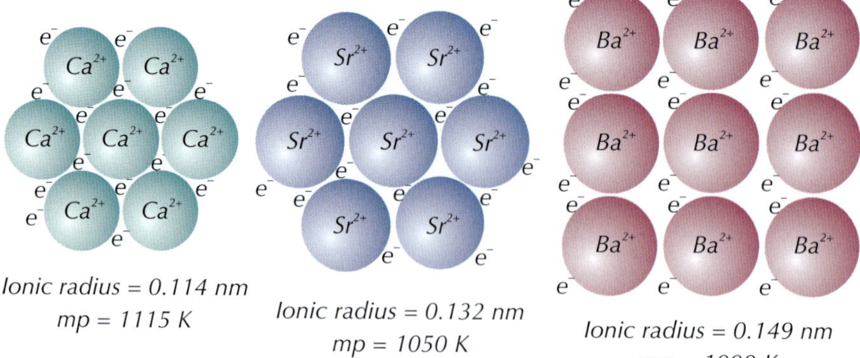

Figure 10: *Comparison of calcium, strontium and barium crystals. ('mp' means melting point.)*

Practice Questions — Application

Q1 The higher the ionisation energy of a group 2 element, the less readily it will react. Calcium and strontium react with dilute hydrochloric acid. Which reaction you would expect to occur more rapidly?

Q2 The table below shows the atomic radii of three elements from Group 2.

Element	Atomic radius/nm
X	0.105
Y	0.200
Z	0.145

a) Which element you would expect to have the highest first ionisation energy?

b) Which element you would expect to have the lowest melting point?

Practice Questions — Fact Recall

Q1 Describe the trend in atomic radius in Group 2.

Q2 Explain the trend in first ionisation energy in Group 2.

Q3 a) Why is the melting point of barium higher than the melting point of radium?

b) At which element in Period 2 is there an anomaly in the trend of melting points? Explain why.

2. Group 2 Compounds

Here's a bit more about the alkaline earth metals and their compounds to follow on from what you learned in the previous topic. It's a Group 2 bonus...

Learning Objectives:

- Know the reactions of the elements Mg – Ba with water.
- Know the relative solubilities of the hydroxides of the elements Mg – Ba in water and that $Mg(OH)_2$ is sparingly soluble.
- Know the relative solubilities of the sulfates of the elements Mg – Ba in water and that $BaSO_4$ is insoluble.
- Be able to explain why $BaCl_2$ solution is used to test for sulfate ions and why it is acidified.
- Know the use of $BaSO_4$ in medicine.
- Know the use of magnesium in the extraction of titanium from $TiCl_4$.
- Know the use of CaO or $CaCO_3$ to remove SO_2 from flue gases.
- Know the use of $Mg(OH)_2$ in medicine and of $Ca(OH)_2$ in agriculture.
- Be able to carry out a simple test-tube reaction to identify SO_4^{2-} ions (Required Practical 4).

Specification Reference 3.2.2

Reactions with water

When Group 2 elements react, they are oxidised from a state of 0 to +2, forming M^{2+} ions.

$$M \rightarrow M^{2+} + 2e^-$$

Oxidation state: $0 \rightarrow +2$

Example

$$\mathbf{Ca \rightarrow Ca^{2+} + 2e^-}$$

Oxidation state: $0 \rightarrow +2$

The Group 2 metals react with water to give a metal hydroxide and hydrogen.

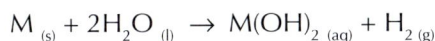

$$M_{(s)} + 2H_2O_{(l)} \rightarrow M(OH)_{2\,(aq)} + H_{2\,(g)}$$

Oxidation state: $0 \rightarrow +2$

Example

Calcium reacts with water to form calcium hydroxide and hydrogen.

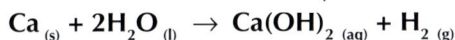

$$\mathbf{Ca_{(s)} + 2H_2O_{(l)} \rightarrow Ca(OH)_{2\,(aq)} + H_{2\,(g)}}$$

Oxidation state: $0 \rightarrow +2$

The elements react more readily down the group because the ionisation energies decrease (see Figure 1).

Group 2 element	*1st ionisation energy / kJ mol⁻¹*	*Rate of reactivity with water*
Be	*900*	*doesn't react*
Mg	*738*	*VERY slow*
Ca	*590*	*steady*
Sr	*550*	*fairly quick*
Ba	*503*	*rapid*

Figure 1: *Comparison of first ionisation energies and reactivity with water for Group 2 elements.*

Solubility of compounds

The solubility of Group 2 compounds depends on the anion (negative ion) in the compound. Generally, compounds of Group 2 elements that contain singly charged negative ions (e.g. OH^-) increase in solubility down the group, whereas compounds that contain doubly charged negative ions (e.g. SO_4^{2-}) decrease in solubility down the group (see Figures 2 and 3).

Group 2 element	*hydroxide (OH^-)*	*sulfate (SO_4^{2-})*
magnesium	*least soluble* ↓	*most soluble* ↑
calcium		
strontium		
barium	*most soluble*	*least soluble*

Figure 2: *Solubility of Group 2 anions.*

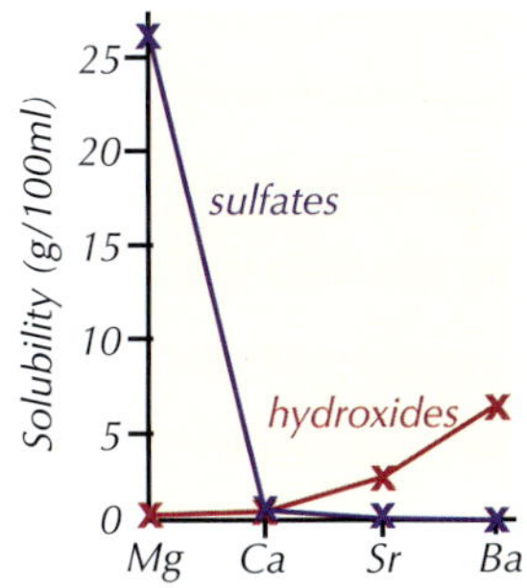

Figure 3: *Solubilities of Group 2 compounds at room temperature and pressure.*

Tip: Group 2 hydroxides and sulfates are all white when solid.

Tip: Remember to carry out any necessary safety precautions before doing an experiment.

Tip: You need to acidify the barium chloride with hydrochloric acid to get rid of any lurking sulfites or carbonates, which will also produce a white precipitate. (You can't use sulfuric acid, because that would add extra sulfate ions.)

Compounds like magnesium hydroxide, $Mg(OH)_2$, which have very low solubilities are said to be sparingly soluble.

REQUIRED PRACTICAL 4

Most sulfates are soluble in water, but barium sulfate ($BaSO_4$) is insoluble. The test for sulfate ions makes use of this property. If you add dilute hydrochloric acid and then barium chloride ($BaCl_2$) solution to a solution containing sulfate ions, then a white precipitate of barium sulfate is formed (see Figure 4).

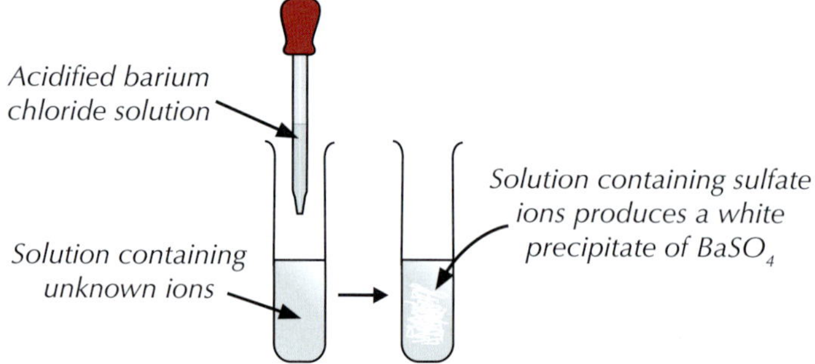
Acidified barium chloride solution
Solution containing unknown ions
Solution containing sulfate ions produces a white precipitate of $BaSO_4$

Figure 4: *The test for identifying sulfate ions in solution.*

Barium meals

The fact that barium sulfate is insoluble is also useful in medicine. X-rays are great for finding broken bones, but they pass straight through soft tissue — so soft tissues, like the digestive system, don't show up on conventional X-ray pictures. Barium sulfate is opaque to X-rays — they won't pass through it. It's used in 'barium meals' to help diagnose problems with the oesophagus, stomach or intestines. A patient swallows the barium meal, which is a suspension of barium sulfate. The barium sulfate coats the tissues, making them show up on the X-rays, showing the structure of the organs (see Figure 5). You couldn't use other barium compounds for this because solutions containing barium ions are poisonous — barium sulfate is insoluble so forms a suspension rather than a solution.

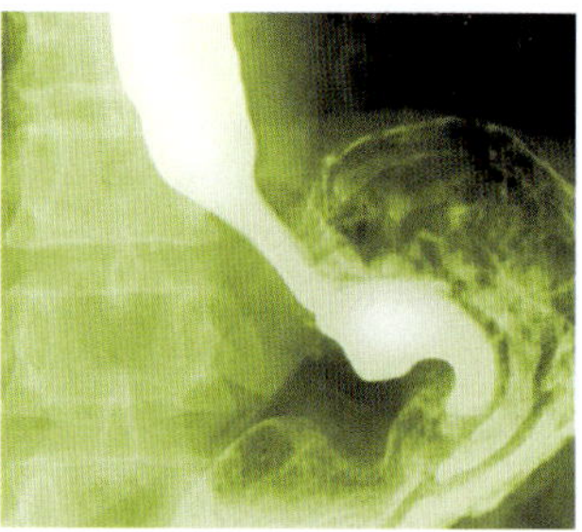

Figure 5: *X-ray showing the oesophagus and stomach following a barium meal.*

Extraction of titanium

Magnesium is used as part of the process of extracting titanium from its ore. The main titanium ore, titanium(IV) oxide (TiO_2), is first converted to titanium(IV) chloride ($TiCl_4$) by heating it with carbon in a stream of chlorine gas. The titanium chloride is then purified by fractional distillation, before being reduced by magnesium in a furnace at almost 1000 °C.

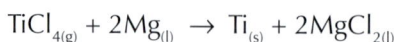

$$TiCl_{4(g)} + 2Mg_{(l)} \rightarrow Ti_{(s)} + 2MgCl_{2(l)}$$

Tip: Magnesium (Mg) is the reducing agent in this reaction.

Removal of sulfur dioxide from flue gases

Burning fossil fuels to produce electricity also produces sulfur dioxide, which pollutes the atmosphere. The acidic sulfur dioxide can be removed from flue gases by reacting with an alkali — this process is called wet scrubbing. Calcium oxide (lime, CaO) and calcium carbonate (limestone, $CaCO_3$) can both be used for this. A slurry is made by mixing the calcium oxide or calcium carbonate with water. It's then sprayed onto the flue gases. The sulfur dioxide reacts with the alkaline slurry and produces a solid waste product, calcium sulfite.

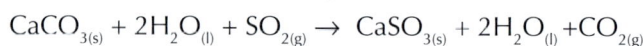

$$CaO_{(s)} + 2H_2O_{(l)} + SO_{2(g)} \rightarrow CaSO_{3(s)} + 2H_2O_{(l)}$$

$$CaCO_{3(s)} + 2H_2O_{(l)} + SO_{2(g)} \rightarrow CaSO_{3(s)} + 2H_2O_{(l)} + CO_{2(g)}$$

Tip: Flue gases are the gases emitted from industrial exhausts and chimneys.

Other uses of Group 2 compounds

Group 2 elements are known as the alkaline earth metals, and many of their common compounds are used for neutralising acids. Calcium hydroxide (slaked lime, $Ca(OH)_2$) is used in agriculture to neutralise acidic soils. Magnesium hydroxide ($Mg(OH)_2$) is used in some indigestion tablets as an antacid (a substance that neutralises excess stomach acid).

Figure 6: A tractor spreading slaked lime on a field with acidic soil.

Practice Questions — Application

Q1 One Group 2 element has a first ionisation energy of 550 kJ mol^{-1} and another has a first ionisation energy of 738 kJ mol^{-1}. Explain which element you would expect to react most rapidly with water.

Q2 Acidified barium chloride is added to a solution. No precipitate forms. What does this result show?

Tip: Remember that for Group 2 elements, a lower ionisation energy means they are more reactive. See page 161 for more.

Practice Questions — Fact Recall

Q1 Describe the trend in reactivity of Group 2 elements with water.

Q2 How does the solubility of hydroxides change down Group 2?

Q3 How does the solubility of sulfates change down Group 2?

Q4 Why is barium chloride solution acidified before it is used to test for sulfate ions?

Q5 What is a 'barium meal'?

Q6 Describe how magnesium is used in the extraction of titanium from TiO_2.

Q7 Suggest two Group 2 compounds are used to remove sulfur dioxide from flue gases?

Q8 Give the chemical name of a Group 2 hydroxide that is used in agriculture and say what it's used for.

3. Group 7 — The Halogens

The halogens are highly-reactive non-metals found in Group 7 of the periodic table. You need to know about their properties and trends — oh, and just how much we rely on chlorine to give us nice clean water.

Learning Objectives:

- Be able to explain the trend in electronegativity of the halogens.
- Be able to explain the trend in the boiling point of the halogens in terms of their structure and bonding.
- Know the trend in oxidising ability of the halogens down the group, including displacement reactions of halide ions in aqueous solution.
- Know the reaction of chlorine with cold, dilute, aqueous NaOH and uses of the solution formed.
- Know the reaction of chlorine with water to form chloride ions and chlorate(I) ions.
- Know the reaction of chlorine with water to form chloride ions and oxygen.
- Understand the use of chlorine in water treatment.
- Appreciate that the benefits to health of water treatment by chlorine outweigh its toxic effects.
- Appreciate that society assesses the advantages and disadvantages when deciding if chemicals should be added to water supplies.

Specification Reference 3.2.3.1, 3.2.3.2

Properties of halogens

The table below gives some of the main properties of the first four halogens, at room temperature.

Halogen	Formula	Colour	Physical state	Electron configuration of atom
fluorine	F_2	pale yellow	gas	$1s^2\ 2s^2\ 2p^5$
chlorine	Cl_2	green	gas	$1s^2\ 2s^2\ 2p^6\ 3s^2\ 3p^5$
bromine	Br_2	red-brown	liquid	$1s^2\ 2s^2\ 2p^6\ 3s^2\ 3p^6\ 3d^{10}\ 4s^2\ 4p^5$
iodine	I_2	grey	solid	$1s^2\ 2s^2\ 2p^6\ 3s^2\ 3p^6\ 3d^{10}\ 4s^2\ 4p^6\ 4d^{10}\ 5s^2\ 5p^5$

Boiling points

The boiling points of the halogens increase down the group. This is due to the increasing strength of the van der Waals forces as the size and relative mass of the molecules increases. This trend is shown in the changes of physical state from fluorine (gas) to iodine (solid).

Electronegativity

Electronegativity decreases down the group. Electronegativity, remember, is the tendency of an atom to attract a bonding pair of electrons. The halogens are all highly electronegative elements. But larger atoms attract electrons less than smaller ones. This is because their outer electrons are further from the nucleus and are more shielded, because they have more inner electrons.

Displacement reactions

When the halogens react, they gain an electron. This means they are oxidising agents. They get less reactive down the group, because the atoms become larger and the outer shell gets further from the nucleus. So the halogens become less oxidising down the group.

The relative oxidising strengths of the halogens can be seen in their displacement reactions with the halide ions. A halogen will displace a halide from solution if the halide is below it in the periodic table (e.g. chlorine can displace bromide ions, but chloride ions are displaced by fluorine). You can see this if you add a few drops of an aqueous halogen to a solution containing halide ions. A colour change is seen if there's a reaction:

	Potassium chloride solution $KCl_{(aq)}$ (colourless)	Potassium bromide solution $KBr_{(aq)}$ (colourless)	Potassium iodide solution $KI_{(aq)}$ (colourless)
Chlorine water $Cl_{2(aq)}$ (colourless)	no reaction	orange solution (Br_2) formed	brown solution (I_2) formed
Bromine water $Br_{2(aq)}$ (orange)	no reaction	no reaction	brown solution (I_2) formed
Iodine solution $I_{2(aq)}$ (brown)	no reaction	no reaction	no reaction

Tip: 'Halogen' is used to describe the atom (X) or molecule (X_2), but 'halide' describes the negative ion (X^-).

These displacement reactions can be used to help identify which halogen (or halide) is present in a solution. Halide ions are colourless in solution, but when the halogen is displaced it shows a distinctive colour, e.g. when bromide ions come out of solution to form bromine the colour changes from colourless to orange.

Tip: You don't need to know about fluorine here (because it will oxidise the water instead of forming a solution).

Examples

Chlorine

If you add chlorine to a solution containing bromide ions (e.g. potassium bromide), it will displace the bromide ions — and there will be a colour change.

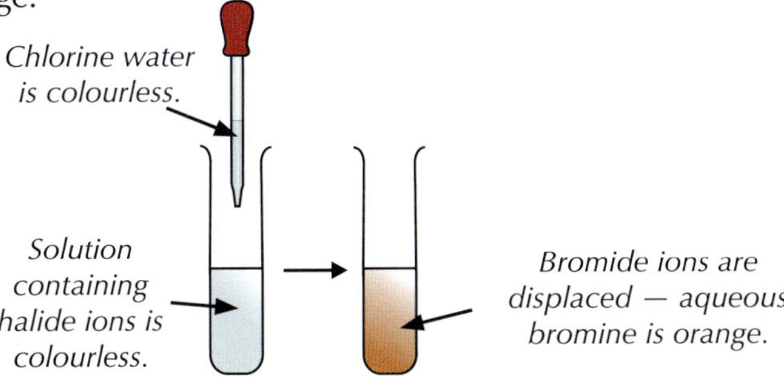

The equation for this reaction is: $Cl_{2(aq)} + 2KBr_{(aq)} \rightarrow 2KCl_{(aq)} + Br_{2(aq)}$

It can also be written as an ionic equation: $Cl_{2(aq)} + 2Br^{-}_{(aq)} \rightarrow 2Cl^{-}_{(aq)} + Br_{2(aq)}$

If you add chlorine to a solution of potassium iodide ions, it will displace the iodide ions. This time the colour change will be from colourless to brown.

The equation for this reaction is: $Cl_{2(aq)} + 2KI_{(aq)} \rightarrow 2KCl_{(aq)} + I_{2(aq)}$

The ionic equation is: $Cl_{2(aq)} + 2I^{-}_{(aq)} \rightarrow 2Cl^{-}_{(aq)} + I_{2(aq)}$

Bromine

If you add bromine to a solution of potassium iodide, it will displace the iodide ions — and there will be a colour change.

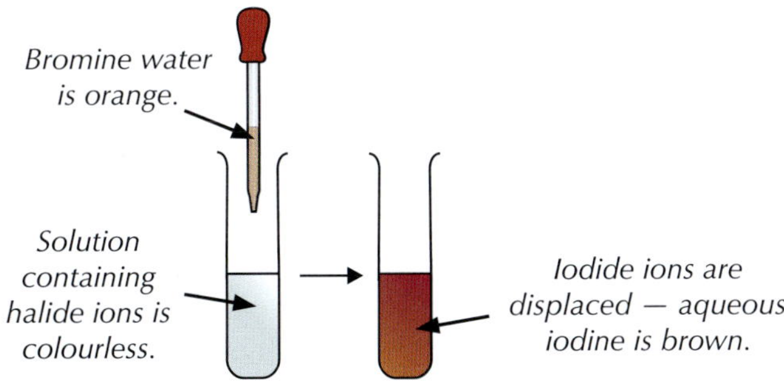

Bromine water is orange.
Solution containing halide ions is colourless.
Iodide ions are displaced — aqueous iodine is brown.

The equation for this reaction is: $Br_{2(aq)} + 2KI_{(aq)} \rightarrow 2KBr_{(aq)} + I_{2(aq)}$

The ionic equation is: $Br_{2(aq)} + 2I^{-}_{(aq)} \rightarrow 2Br^{-}_{(aq)} + I_{2(aq)}$

There's no reaction if you add bromine water to a solution of chloride ions. Chlorine is above bromine in Group 7 so is more reactive and can't be displaced by it.

Iodine

Iodine is below chlorine and bromine in Group 7, so it's less reactive than them and won't displace either halide.

Figure 1: *Green chlorine gas.*

Tip: Carry out any safety precautions before doing these tests.

Figure 2: *Red-brown bromine liquid.*

Figure 3: *Grey iodine crystals.*

Tip: Take a look at pages 143-145 for more on oxidation states and how to work them out.

Making bleach

If you mix chlorine gas with cold, dilute sodium hydroxide solution at room temperature, you get sodium chlorate(I) solution($NaClO_{(aq)}$). This just happens to be common household bleach (which kills bacteria). In this reaction chlorine is both oxidised and reduced. This is called **disproportionation**.

$$2NaOH_{(aq)} + Cl_{2\,(g)} \rightarrow NaClO_{(aq)} + NaCl_{(aq)} + H_2O_{(l)}$$

Chlorine is bonded to chlorine so its oxidation state is 0. (arrow to Cl_2)

ClO^- is the chlorate(I) ion. Chlorine's oxidation state is +1 in this ion. (arrow to NaClO)

Here, chlorine's oxidation state is −1. (arrow to NaCl)

The sodium chlorate(I) solution (bleach) has loads of uses — it's used in water treatment, to bleach paper and textiles... and it's good for cleaning toilets, too. Handy...

Chlorine and water

When you mix chlorine with water, it undergoes disproportionation. You end up with a mixture of chloride ions and chlorate(I) ions.

$$Cl_{2(g)} + H_2O_{(l)} \rightleftharpoons 2H^+_{(aq)} + Cl^-_{(aq)} + ClO^-_{(aq)}$$

Chlorine's oxidation state is 0.

Chloride's oxidation state is −1.

In chlorate(I) ions chlorine's oxidation state is +1.

Figure 4: *Chlorine is used to treat tap water in the UK.*

In sunlight, chlorine can also decompose water to form chloride ions and oxygen.

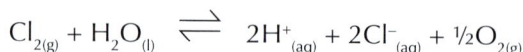

$$Cl_{2(g)} + H_2O_{(l)} \rightleftharpoons 2H^+_{(aq)} + 2Cl^-_{(aq)} + \tfrac{1}{2}O_{2(g)}$$

Water treatment

Chlorate(I) ions kill bacteria. So, adding chlorine (or a compound containing chlorate(I) ions) to water can make it safe to drink or swim in. On the downside, chlorine is toxic.

In the UK our drinking water is treated to make it safe. Chlorine is an important part of water treatment. It kills disease-causing microorganisms (and some chlorine persists in the water and prevents reinfection further down the supply). It also prevents the growth of algae, eliminating bad tastes and smells, and removes discolouration caused by organic compounds.

However, there are risks from using chlorine to treat water. Chlorine gas is very harmful if it's breathed in — it irritates the respiratory system. Liquid chlorine on the skin or eyes causes severe chemical burns. Accidents involving chlorine could be really serious, even fatal.

Water contains a variety of organic compounds, e.g. from the decomposition of plants. Chlorine reacts with these compounds to form chlorinated hydrocarbons, e.g. chloromethane (CH_3Cl), and many of these chlorinated hydrocarbons are carcinogenic (cancer-causing). However, this increased cancer risk is small compared to the risks from untreated water — a cholera epidemic, say, could kill thousands of people. We have to weigh up these risks and benefits when making decisions about whether we should add chemicals to drinking water supplies.

Practice Questions — Application

Q1 Three test tubes, A, B and C, contain different halide solutions. Several drops of chlorine water are added to each test tube and the following colour changes are observed.

Tube A — colourless to orange

Tube B — no colour change

Tube C — colourless to brown

a) Suggest the halide ion present in each solution.

b) The test is repeated, but iodine solution ($I_{2(aq)}$) is added to the test tubes instead of chlorine water. Explain how the results would be different.

Q2 Chlorine gas is mixed with sodium hydroxide solution. The solution is tested with litmus paper, which turns white. Explain why.

Practice Questions — Fact Recall

Q1 Describe the trend in the boiling points of the halogens.

Q2 Name the most electronegative halogen.

Q3 Which halide ions are displaced by reaction with chlorine water?

Q4 a) Describe the colour change when bromine water is added to potassium iodide solution.

b) Give the full equation for this reaction.

c) Give the ionic equation for this reaction.

Q5 a) Name three products of the reaction between sodium hydroxide and chlorine.

b) Give the balanced equation for this reaction.

Q6 Describe the reactions that occur when chlorine is mixed with water, in and out of sunlight.

Q7 a) Explain why chlorine is used to treat water.

b) Describe the disadvantages of using chlorine to treat water.

Figure 5: *The distinctive 'swimming pool smell' is due to the chlorine in the water.*

4. Halide Ions

Halides ions are the 1– ions formed by the halogens. The different halide ions react slightly differently, so telling them apart is easier than you might think.

Learning Objectives:

- Know the trend in reducing ability of the halide ions, including the reactions of solid sodium halides with concentrated sulfuric acid.
- Understand the use of acidified silver nitrate solution to identify and distinguish between halide ions.
- Be able to explain why silver nitrate solution is used to identify halide ions, why the silver nitrate solution is acidified and why ammonia solution is added.
- Know the trend in solubility of the silver halides in ammonia.
- Be able to carry out simple test-tube reactions to identify halide ions (Required Practical 4).

Specification Reference 3.2.3.1

Halide ion formation and oxidation

You'll remember from your chemistry basics that the elements in Group 7 form ions by gaining one electron. They end up as 1– ions with a full outer shell. For example:

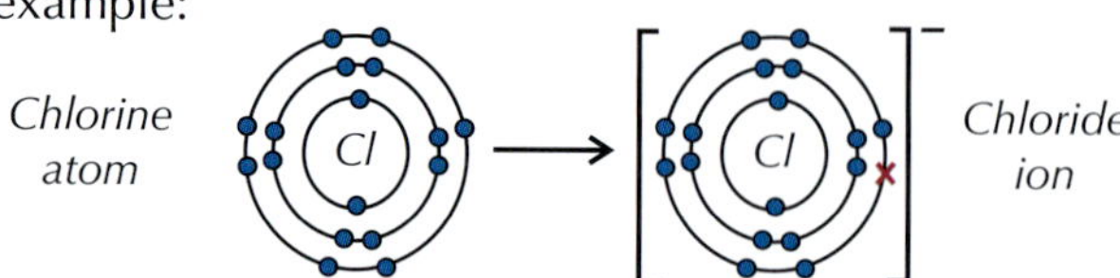

***Figure 1:** Chlorine gains an electron to form an ion.*

When a halide ion takes part in a **redox reaction**, it reduces something and is oxidised itself. To reduce something, the halide ion needs to lose an electron from its outer shell — think OIL RIG (see page 143).

The reducing power of halides

How easy it is for a halide ion to lose an electron depends on the attraction between the nucleus and the outer electrons. As you go down the group, the attraction gets weaker because the ions get bigger, so the electrons are further away from the positive nucleus. There are extra inner electron shells too, so there's a greater shielding effect (see Figure 2). Therefore, the reducing power of the halides increases down the group.

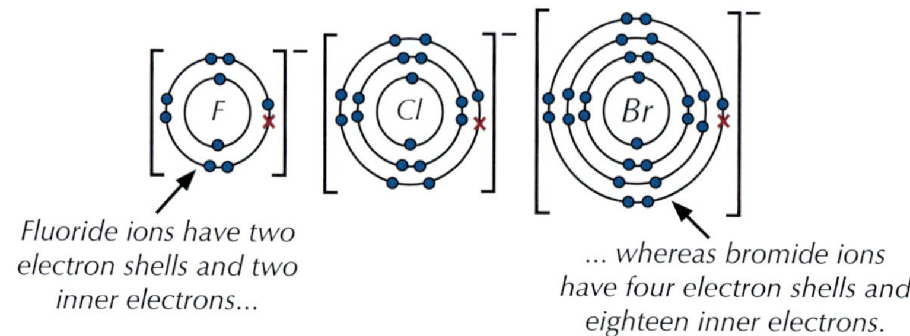

F
Cl
Br
Fluoride ions have two electron shells and two inner electrons...
... whereas bromide ions have four electron shells and eighteen inner electrons.

***Figure 2:** Electron shells and shielding in ions of the first three halogens.*

Tip: You already know one good example of halide ions as reducing agents — it's the good old halogen / halide displacement reaction (see pages 166-167). For example:
$Cl_2 + 2Br^- \rightarrow 2Cl^- + Br_2$

Reactions with sulfuric acid

All the halides react with concentrated sulfuric acid to give a hydrogen halide as a product to start with.

$$NaX + H_2SO_4 \rightarrow NaHSO_4 + HX$$

A sodium halide with the halogen labelled as 'X'. — *A hydrogen halide is produced.*

Tip: In chemistry, X is often used to stand for 'any halogen'.

But what happens next depends on which halide you've got. Some halide ions are strong enough reducing agents that they can reduce the sulfuric acid to water and sulfur dioxide. Overall the reaction is:

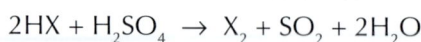

$$2HX + H_2SO_4 \rightarrow X_2 + SO_2 + 2H_2O$$

The half equations are:

$$2X^-_{(g)} \rightarrow X_{2(s)} + 2e^-$$ ← *The halide is oxidised.*

$$H_2SO_4 + 2H^+ + 2e^- \rightarrow SO_2 + 2H_2O$$ ← *The sulfuric acid is reduced.*

Iodide is such a strong reducing agent that it can reduce the SO_2 to H_2S or S.

Exam Tip
In an exam, you might have to construct half-equations, given the main reactants and products.

Examples

Reaction of NaF or NaCl with H_2SO_4

$$NaF_{(s)} + H_2SO_{4(l)} \rightarrow NaHSO_{4(s)} + HF_{(g)}$$

$$NaCl_{(s)} + H_2SO_{4(l)} \rightarrow NaHSO_{4(s)} + HCl_{(g)}$$

Hydrogen fluoride (HF) or hydrogen chloride gas (HCl) is formed. You'll see misty fumes as the gas comes into contact with moisture in the air. But HF and HCl aren't strong enough reducing agents to reduce the sulfuric acid, so the reaction stops there. It's not a redox reaction — the oxidation states of the halide and sulfur stay the same (–1 and +6).

Reaction of NaBr with H_2SO_4

$$NaBr_{(s)} + H_2SO_{4(l)} \rightarrow NaHSO_{4(s)} + HBr_{(g)}$$

The first reaction gives misty fumes of hydrogen bromide gas (HBr). But the HBr is a stronger reducing agent than HCl and reacts with the H_2SO_4 in a redox reaction.

$$2HBr_{(g)} + H_2SO_{4(l)} \rightarrow Br_{2(g)} + SO_{2(g)} + 2H_2O_{(l)}$$

Oxidation state of S: +6 → +4 *reduction*
Oxidation state of Br: –1 → 0 *oxidation*

The reaction produces choking fumes of SO_2 and orange fumes of Br_2.

Reaction of NaI with H_2SO_4

$$NaI_{(s)} + H_2SO_{4(l)} \rightarrow NaHSO_{4(s)} + HI_{(g)}$$

Same initial reaction giving HI gas. The HI then reduces H_2SO_4, as above.

$$2HI_{(g)} + H_2SO_{4(l)} \rightarrow I_{2(s)} + SO_{2(g)} + 2H_2O_{(l)}$$

Oxidation state of S: +6 → +4 *reduction*
Oxidation state of I: –1 → 0 *oxidation*

But HI (being the strongest reducing agent) keeps going and reduces the SO_2 to H_2S.

$$6HI_{(g)} + SO_{2(g)} \rightarrow H_2S_{(g)} + 3I_{2(s)} + 2H_2O_{(l)}$$

Oxidation state of S: +4 → –2 *reduction*
Oxidation state of I: –1 → 0 *oxidation*

The reaction produces fumes of H_2S and solid iodine.

Tip: This may seem like an awful lot of information at first glance. Don't worry though — just learn the principles and keep referring back to the equations on the previous page. It will really help if you can learn the general pattern of the reactions — you can always work out oxidation states if you need to (see pages 143-145).

Tip: Make sure that you have taken all necessary safety precautions if you're carrying out these experiments.

Tip: When iodine is produced, you'll see a grey solid and/or a purple gas.

Tip: This is no one's favourite reaction — H_2S is toxic and smells of bad eggs.

Testing for halides

REQUIRED PRACTICAL 4

The halogens are pretty distinctive to look at (see pages 166 and 167). Unfortunately, the same can't be said of halide solutions, which are colourless. You can test for halides using the **silver nitrate test** — it's easy. First you add dilute nitric acid to remove ions which might interfere with the test. Then you just add a few drops of silver nitrate solution ($AgNO_{3(aq)}$). A precipitate is formed (of the silver halide).

$$Ag^+_{(aq)} + X^-_{(aq)} \rightarrow AgX_{(s)} \quad \text{...where X is Cl, Br or I}$$

Tip: You can't use hydrochloric acid instead of nitric acid because the silver nitrate would just react with the chloride ions from the HCl — and that would mess up your results completely.

The colour of the precipitate identifies the halide (see Figures 3 and 4).

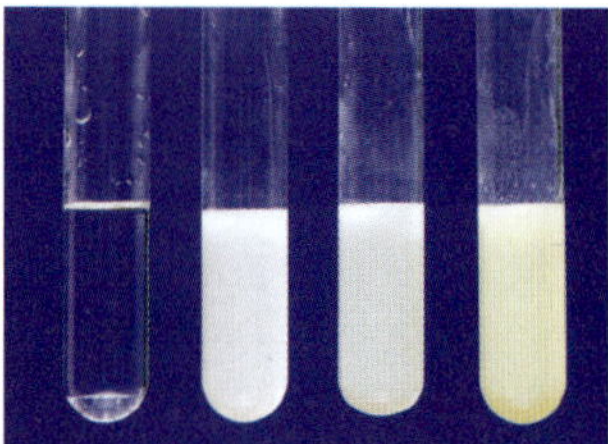

Figure 3: Results of silver nitrate tests for solutions containing (L-R) fluoride, chloride, bromide and iodide ions.

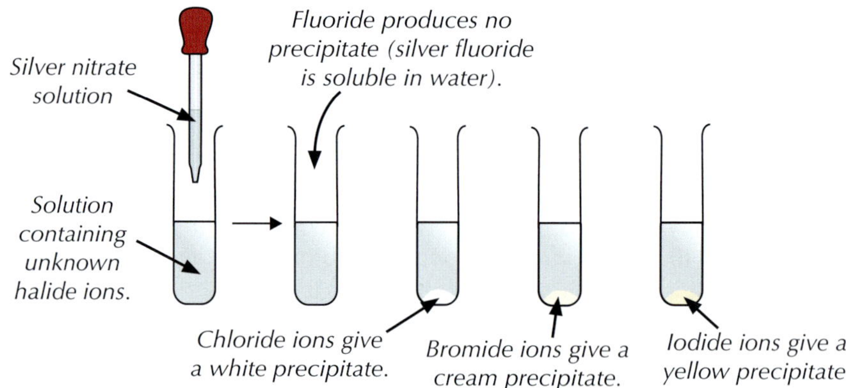

Silver nitrate solution
Fluoride produces no precipitate (silver fluoride is soluble in water).
Solution containing unknown halide ions.
Chloride ions give a white precipitate.
Bromide ions give a cream precipitate.
Iodide ions give a yellow precipitate.

Figure 4: *The silver nitrate test for identifying an unknown halide ion in solution.*

Then to be extra sure, you can test your results by adding ammonia solution. Each silver halide has a different solubility in ammonia (see Figures 5 and 6).

Halide	Result	
Chloride Cl^-	precipitate dissolves in dilute $NH_{3(aq)}$	most soluble
Bromide Br^-	precipitate dissolves in conc. $NH_{3(aq)}$	↓
Iodide I^-	precipitate insoluble in conc. $NH_{3(aq)}$	least soluble

Figure 5: *Solubility of silver halide precipitates in ammonia.*

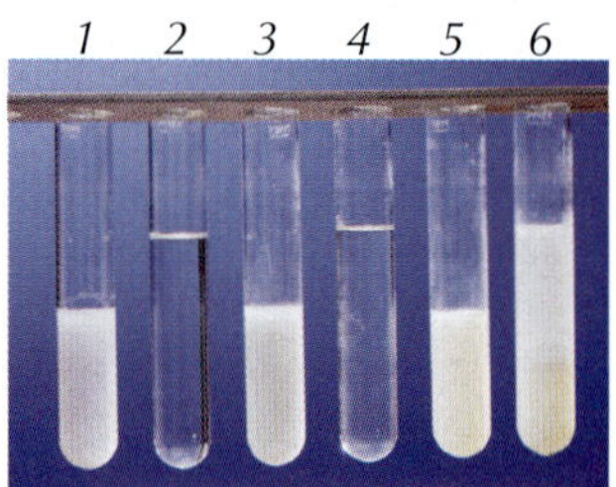

1 2 3 4 5 6

Figure 6: The chloride, bromide and iodide test tubes from Figure 3 (1, 3 and 5), and the same tubes with dilute $NH_{3(aq)}$ (2) or concentrated $NH_{3(aq)}$ (4 and 5)added.

Practice Questions — Application

Q1 Sunil carries out a reaction between solid sodium bromide and concentrated sulfuric acid. Then he does the same reaction, replacing sodium bromide with sodium chloride. He predicts that the only gaseous product of both reactions will be a hydrogen halide. Explain whether Sunil's prediction is correct.

Q2 An experiment is carried out to identify the halide ions in three different solutions. The results are shown in the table below.

Sample	Colour of precipitate following addition of silver nitrate	Effect of adding concentrated NH_3 solution to the precipitate
A	yellow	no change
B	no precipitate	no change
C	cream	precipitate dissolves

Identify the halide ion in each sample.

Practice Questions — Fact Recall

Q1 Explain why the reducing power of the halide ions increases as you go down the group.

Q2 Write the equation(s) for the reactions that occur when sulfuric acid is mixed with:

a) sodium fluoride, b) sodium iodide.

Q3 a) Describe a test that could be used to distinguish between solutions of fluoride ions and chloride ions.

b) Describe how you could use ammonia solution to confirm the result for the chloride ion.

5. Tests for Ions

REQUIRED PRACTICAL 4

You have to be able to carry out tests to find out which ions are in a solution.

Learning Objective:

- Be able to carry out simple test-tube reactions to identify cations (Group 2, NH_4^+) and anions (Group 7 (halide ions), OH^-, CO_3^{2-}, SO_4^{2-}) (Required Practical 4).

Specification Reference 3.2.3.2

Identifying positive ions

Positive ions (or cations) include things like the ions of Group 2 metals and ammonium ions. Here are some chemical tests that help to identify them:

Tests for Group 2 ions

Compounds of some Group 2 metals burn with characteristic colours. You can identify them using a **flame test**. First you dip a nichrome wire loop in concentrated hydrochloric acid (to clean it) and then dip it into the unknown compound. Hold the loop in the clear blue part of a Bunsen burner flame and observe the colour change in the flame (see Figures 1a and b).

Metal ion	*Flame colour*
Calcium, Ca^{2+}	*brick red*
Strontium, Sr^{2+}	*red*
Barium, Ba^{2+}	*pale green*

Figure 1a: *The colours of the flames when different metal ions are burnt.*

Figure 1b: *Results of flame tests for solutions containing calcium (top), strontium (bottom left) and barium (bottom right).*

You can also use dilute sodium hydroxide solution (NaOH) to help you identify Group 2 ions. Add the NaOH dropwise to a test tube containing the metal ion solution and observe the precipitate that forms (if there is one). Keep adding the NaOH until it is in excess and record any changes that you see. Figure 2 shows the results for each ion.

Metal ion	*With OH^-*	*With excess OH^-*
Magnesium, Mg^{2+}	*slight white precipitate*	*white precipitate*
Calcium, Ca^{2+}	*slight white precipitate*	*slight white precipitate*
Strontium, Sr^{2+}	*slight white precipitate*	*slight white precipitate*
Barium, Ba^{2+}	*no change*	*no change*

Figure 2: *The observed results when NaOH is added to metal ions in solution.*

Test for ammonium ions

Ammonia gas (NH_3) is alkaline — so you can test for it using a damp piece of red litmus paper. The litmus paper needs to be damp so the ammonia gas can dissolve. If there's ammonia present, the paper will turn blue. If you add hydroxide ions to (OH^-) a solution containing ammonium ions (NH_4^+), they will react to produce ammonia gas and water, like this:

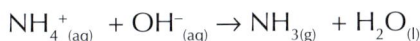

$$NH_4^+{}_{(aq)} + OH^-{}_{(aq)} \rightarrow NH_{3(g)} + H_2O_{(l)}$$

You can use this reaction to test whether a substance contains ammonium ions (NH_4^+). Add some dilute sodium hydroxide solution to your mystery substance in a test tube and gently heat the mixture. If there's ammonia given off, ammonium ions must be present (see Figure 3).

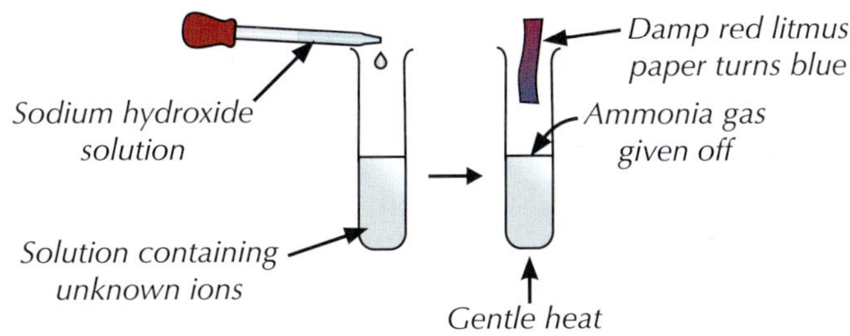

Figure 3: *The test for identifying ammonium ions in solution.*

Tip: Before carrying out any of the tests on these pages, you need to think about any safety issues that might be involved.

Identifying negative ions

Negative ions (or anions) include things like sulfate ions, hydroxide ions, halides and carbonate ions. Here are the chemical tests for these ions:

Tip: The hydrochloric acid is added to get rid of any traces of carbonate and sulfite ions before you do the test. (These would also produce a precipitate, so they'd confuse the results.) You can't use sulfuric acid, because you don't want to add any sulfate ions.

Test for sulfate ions

You've already met this test back on page 164, but here's a quick reminder in case you've forgotten. To identify a sulfate ion (SO_4^{2-}), you add a little dilute hydrochloric acid, followed by barium chloride solution, $BaCl_{2(aq)}$. If a white precipitate of barium sulfate forms, it means the original compound contained a sulfate.

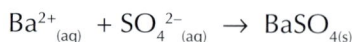

$$Ba^{2+}_{(aq)} + SO_4^{2-}{}_{(aq)} \rightarrow BaSO_{4(s)}$$

Test for hydroxide ions

Hydroxide ions make solutions alkaline. So if you think a solution might contain hydroxide ions, you can use a pH indicator to test it. For example, if you dip a piece of red litmus paper into the solution and hydroxide ions are present, the paper will turn blue.

Test for halide ions

To test for chloride (Cl^-), bromide (Br^-) or iodide (I^-) ions, you just add dilute nitric acid (HNO_3), followed by silver nitrate solution ($AgNO_3$).

Tip: There's more info about this test on pages 171-172.

- A chloride gives a white precipitate of silver chloride.
- A bromide gives a cream precipitate of silver bromide.
- An iodide gives a yellow precipitate of silver iodide.

These precipitates can look a bit similar, so you might have to add ammonia solution to tell them apart. Silver chloride will dissolve in dilute ammonia solution. Silver bromide will only dissolve in concentrated ammonia solution. Silver iodide won't dissolve in either.

Test for carbonate ions

You can test to see if a solution contains carbonate ions (CO_3^{2-}) by adding an acid. When you add dilute hydrochloric acid, a solution containing carbonate ions will fizz. This is because the carbonate ions react with the hydrogen ions in the acid to give carbon dioxide:

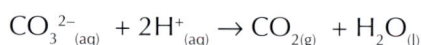

$$CO_3^{2-}{}_{(aq)} + 2H^+_{(aq)} \rightarrow CO_{2(g)} + H_2O_{(l)}$$

You can test for carbon dioxide using limewater. Carbon dioxide turns limewater cloudy — just bubble the gas through a test tube of limewater and watch what happens. If the limewater goes cloudy, your solution contains carbonate ions (see Figures 4 and 5).

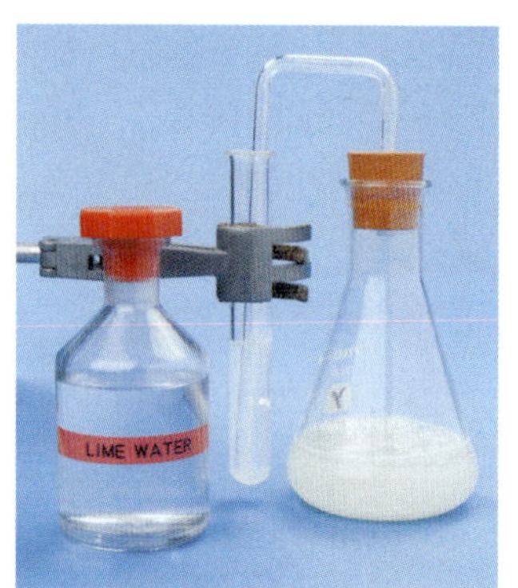

Figure 4: Testing for carbonate ions with hydrochloric acid and limewater.

Tip: You need to put a bung on the flask so that the carbon dioxide gas doesn't escape.

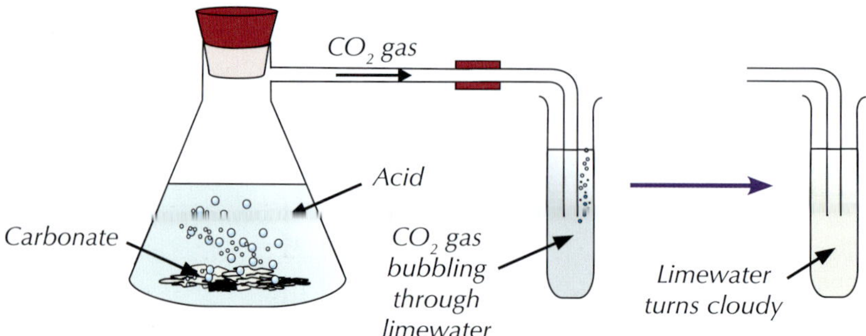

Figure 5: The test for identifying carbonate ions in solution.

Practice Questions — Fact Recall

Q1 Write the colour that a blue Bunsen burner flame would turn in the presence of:
a) barium ions, b) calcium ions, c) strontium ions.

Q2 What solution do you need to add to an unknown substance when testing for ammonium ions?

Q3 Describe a test that could be used to determine the presence of sulfate ions in a solution.

Q4 Describe a test you could do to confirm whether hydroxide ions are present in a solution.

Q5 Which two solutions do you need to add when testing for halide ions in an unknown solution?

Q6 Describe how you can use limewater to test for carbonate ions.

Exam Tip
Make sure you've absolutely definitely memorised which test is for which kind of ion — you don't want to mix them up if they come up in your exams.

Section Summary

Make sure you know...

- Why atomic radius increases as you go down Group 2.
- Why first ionisation energy decreases as you go down Group 2.
- How and why melting point changes as you go down Group 2.
- The reactions of Group 2 metals with water.
- How the solubilities of Group 2 hydroxides and sulfates vary.
- What the barium chloride test for sulfate ions is and what a positive test result is.
- How barium sulfate and magnesium hydroxide are used in medicine.
- How magnesium is used in the extraction of titanium from its ore.
- How calcium oxide and calcium carbonate are used to remove sulfur dioxide from flue gases.
- How calcium hydroxide is used in agriculture.
- Why the electronegativity of the halogens decreases as you go down the group.
- Why the boiling points of the halogens increase as you go down the group.
- That the oxidising ability of the halogens decreases down the group, and how this can be demonstrated by halide ion displacement reactions.
- The products of the reaction between chlorine and cold dilute sodium hydroxide and the uses of the solution formed.
- How chlorine reacts with water.
- Why chlorine is used in water treatment, even though it can be toxic.
- Why the reducing ability of the halide ions increases down the group.
- The reactions between concentrated sulfuric acid and the sodium halides.
- What the silver nitrate test for halides is and why the silver nitrate solution is acidified.
- The results of the silver nitrate test for solutions containing fluoride, chloride, bromide and iodide ions.
- The trend in the solubility of silver halides in ammonia and why this is useful.
- How to carry out tests to identify certain positive and negative ions in an unknown solution.

Exam-style Questions

1 To an unknown solution, X, a few drops of hydrochloric acid are added, followed by some barium chloride solution. A white precipitate forms. When a drop of solution X on a nichrome wire loop was placed in a blue Bunsen burner flame, the flame turned brick red. What is the identity of solution X?

A Strontium sulfate

B Calcium chloride

C Barium hydroxide

D Calcium sulfate

(1 mark)

2 Which of the following describes a simple test for ammonium ions?

A Add dilute nitric acid to the solution, followed by silver nitrate solution.

B Add some dilute sodium hydroxide solution, gently heat the mixture and test any gas that's released with damp red litmus paper.

C Add dilute hydrochloric acid and test any gas formed with limewater.

D Dip a piece of red litmus paper into the solution.

(1 mark)

3 After the 2010 earthquake in Haiti, there was an outbreak of cholera which has since killed thousands of people. The spread of the disease has partly been blamed on a lack of water treatment. It is possible to use chlorine to treat water.

3.1 Write an equation for the reaction between chlorine and water.

(1 mark)

3.2 Why does adding chlorine to water help to stop the spread of diseases such as cholera?

(2 marks)

3.3 Explain why some people do not support the addition of chlorine to water supplies.

(1 mark)

3.4 Using chlorine has some disadvantages.
Suggest why we continue to treat water with it.

(1 mark)

4 Chlorine and bromine are halogens.

4.1 The halogens have different boiling points.

State whether chlorine has a higher or lower boiling point than bromine and explain why.

(3 marks)

4.2 Chlorine water and bromine water are added to solutions **A** and **B**. Each solution contains a potassium halide. The table below shows the results.

	Solution A	Solution B
Chlorine water	solution turns orange	solution turns brown
Bromine water	no change	solution turns brown

Identify solution **A** and explain your reasoning.

(3 marks)

4.3 Identify solution **B** and write an ionic equation for its reaction with chlorine water.

(2 marks)

4.4 Chlorine can undergo the following reaction:

$$2NaOH_{(aq)} + Cl_{2\,(g)} \rightarrow NaClO_{(aq)} + NaCl_{(aq)} + H_2O_{(l)}$$

Give one use for the NaClO formed.

(1 mark)

4.5 Silver nitrate can be used to identify halide ions in solution.

Explain why dilute nitric acid (HNO_3) is added to the solution.

(1 mark)

4.6 State what you would observe if bromide ions are present in the solution being tested in **4.5**.

(1 mark)

4.7 Explain how ammonia solution can be used to confirm the result that you gave in **4.6**.

(1 mark)

5 Samples of three of the alkaline earth metals, strontium, calcium and magnesium, are placed in jars labelled **D**, **E** and **F**. Some information about the three metals is shown in the table below.

	D	**E**	**F**
Atomic radius	0.16 nm	0.19 nm	0.215 nm
1st ionisation energy	738 kJ mol^{-1}	590 kJ mol^{-1}	550 kJ mol^{-1}
Melting point	923 K	1115 K	1050 K

5.1 Which of the elements in the table is magnesium?

(1 mark)

5.2 Explain why the first ionisation energy of **D** is higher than the first ionisation energy of **E**.

(3 marks)

5.3 Using the information in the table, explain how the reactivity of metal **F** with water will compare with **E**.

(2 marks)

5.4 Barium has a lower melting point than metal **F**. Explain why this is the case.

(3 marks)

5.5 State the solubility of magnesium hydroxide in water.

(1 mark)

5.6 Describe how magnesium hydroxide is used in medicine.

(2 marks)

5.7 Magnesium can be used to reduce titanium(IV) chloride to pure titanium.

Write the equation for this reaction.

(1 mark)

Unit 3 Section 1: Introduction to Organic Chemistry

Organic Chemistry

Organic chemistry is just the study of carbon-containing compounds.

The basics

There are a few basic concepts in organic chemistry that you'll need to get your head around before you study organic chemistry in more detail.

Formulas

Picturing molecules can be pretty difficult when you can't see them all around you. We can use the elemental symbols from the periodic table to help visualise molecules. For example, a molecule of methane is one carbon atom attached to four hydrogen atoms. You could show this by giving its molecular formula, CH_4, or you could draw its displayed formula (see Figure 1). There's more on formulas on pages 180-182.

```
    H
    |
H — C — H
    |
    H
```

Figure 1: *The displayed formula of methane.*

This isn't exactly what methane looks like, but visualising it like this lets us compare it to other molecules and means we can predict its properties and how it might react with other molecules. Molecular models can also be used to represent molecules (see Figure 2).

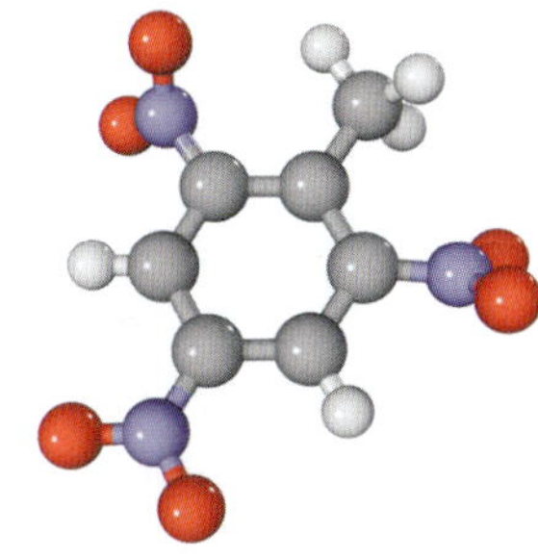

Figure 2: *A molecular model of a TNT molecule. Each grey sphere represents a carbon atom, each white sphere represents a hydrogen atom, the purple ones represent nitrogen atoms and the red ones oxygen atoms.*

Functional groups

The functional group of a molecule is the group of atoms that's responsible for its characteristic reactions — it's where all the interesting stuff happens. They're usually pretty easy to spot because they're the bits which aren't just hydrogen and carbon atoms (e.g. bromine atoms, oxygen atoms, etc.). You'll come across a few different functional groups in this course — here are a few examples...

Functional groups of...	*alcohols*	*carboxylic acids*	*alkenes*
	—OH	—C(=O)—OH	>C=C<

For now, remember that carbon atoms have four bonds, hydrogen atoms have one bond and oxygen atoms have two bonds joining them to other atoms.

Nomenclature

There are thousands, if not millions, of known organic compounds and it would be pretty silly if we didn't have an easy way to describe them. That's where **nomenclature** comes in. Don't be put off by the long name — all it means is naming molecules using specific rules. These rules (known as the IUPAC system for naming organic compounds) allow scientists to discuss organic chemistry safe in the knowledge that they're all talking about the same molecules. It means that some molecules end up with really long and complicated looking names (e.g. 1,2-dichloro-3-methylbutane), but once you know the rules it's easy to work out what they all mean. There's more about nomenclature on pages 188-191.

So there you go. That's pretty much all you need to know to get started. You'd better get on — next up is formulas...

1. Formulas

Organic compounds can be represented in lots of different ways, using different types of formulas. You need to be familiar with what these formulas show and how to switch between them.

Learning Objectives:

- Know that organic compounds can be represented by: an empirical formula, a molecular formula, a general formula, a structural formula, a displayed formula and a skeletal formula.
- Know the characteristics of a homologous series.
- Be able to draw structural, displayed and skeletal formulas for given organic compounds.

Specification Reference 3.3.1.1

Types of formula

Molecular formulas

A molecular formula gives the actual number of atoms of each element in a molecule.

Examples

Ethane has the molecular formula C_2H_6 — each molecule is made up of 2 carbon atoms and 6 hydrogen atoms.

Pentene has the molecular formula C_5H_{10} — each molecule is made up of 5 carbon atoms and 10 hydrogen atoms.

1,4-dibromobutane has the molecular formula $C_4H_8Br_2$ — each molecule is made up of 4 carbon atoms, 8 hydrogen atoms and 2 bromine atoms.

1,3-dichloropropane has the molecular formula $C_3H_6Cl_2$ — each molecule is made up of 3 carbon atoms, 6 hydrogen atoms and 2 chlorine atoms.

Tip: Don't worry for now if you're not sure exactly how all the molecules in this topic get their names. The system for naming organic molecules is covered in full over the next couple of topics.

Structural formulas

A structural formula shows the atoms carbon by carbon, with the attached hydrogens and functional groups.

Examples

Ethane has the structural formula CH_3CH_3.

Pent-1-ene has the structural formula $CH_3CH_2CH_2CHCH_2$.

1,4-dibromobutane has the structural formula $BrCH_2CH_2CH_2CH_2Br$.

1,3-dichloropropane has the structural formula $ClCH_2CH_2CH_2Cl$.

Displayed formulas

A displayed formula shows how all the atoms are arranged, and all the bonds between them.

Examples

Displayed formula of ethane:

```
  H H
  | |
H-C-C-H
  | |
  H H
```

Displayed formula of pent-1-ene:

```
  H H H H
  | | | |    H
H-C-C-C-C=C<
  | | |      H
  H H H
```

Displayed formula of 1,4-dibromobutane:

```
   H H H H
   | | | |
Br-C-C-C-C-Br
   | | | |
   H H H H
```

Displayed formula of 1,3-dichloropropane:

```
   H H Cl
   | | |
Cl-C-C-C-H
   | | |
   H H H
```

Empirical formulas

An empirical formula gives the simplest whole number ratio of atoms of each element in a compound. To find the empirical formula you have to find the highest number that will go into each number in the molecular formula, then divide by it. For example, if the molecular formula is $C_6H_8Cl_4$, the highest number that goes into each number is 2, so divide the molecular formula by 2 and get the empirical formula $C_3H_4Cl_2$. Sometimes the empirical formula will be the same as the molecular formula. This happens when you can't divide all the numbers in the molecular formula by the same number and still end up with whole numbers of atoms.

Tip: If there's just one atom of something in a formula then you know you've got an empirical formula — it can't be simplified any further.

$C_4H_{10}O$ is an empirical formula — there's only 1 oxygen atom.

Examples

Name	Molecular Formula	Divide by...	Empirical Formula
Ethane	C_2H_6	2	CH_3
Pentene	C_5H_{10}	5	CH_2
1,4-dichlorobutane	$C_4H_8Cl_2$	2	C_2H_4Cl
1,3-dichloropropane	$C_3H_6Cl_2$		$C_3H_6Cl_2$
1,2,3-trichloroheptane	$C_7H_{13}Cl_3$		$C_7H_{13}Cl_3$

In the last two examples in the table, the molecular formula is the same as the empirical formula — there's nothing you can divide all the numbers in the molecular formula by and still get whole numbers.

Exam Tip
You need to make sure you know which type of formula is which. You won't get any marks for writing a structural formula when the examiner wants a molecular one.

General formulas and homologous series

A **general formula** is an algebraic formula that can describe any member of a family of compounds. Organic chemistry is more about groups of similar chemicals than individual compounds. These groups are called **homologous series**. A homologous series is a family of compounds that have the same functional group and general formula. Consecutive members of a homologous series differ by $-CH_2-$.

Tip: There's a lot more about the homologous series you need to know on pages 184-187.

Example

The simplest homologous series is the alkanes. They're straight-chain molecules that contain only carbon and hydrogen atoms. There are always twice as many hydrogen atoms as carbon atoms, plus two more. So the general formula for alkanes is C_nH_{2n+2}. You can use this formula to work out how many hydrogen atoms there are in any alkane if you know the number of carbon atoms. For example...

If an alkane has 1 carbon atom, n = 1.
This means the alkane will have $(2 \times 1) + 2 = 4$ hydrogen atoms.
So the molecular formula of this alkane would be CH_4 (you don't need to write the 1 in C_1).

If an alkane has 5 carbon atoms, n = 5.
This means the alkane will have $(2 \times 5) + 2 = 12$ hydrogen atoms.
So its molecular formula would be C_5H_{12}.

If an alkane has 15 carbon atoms, n = 15.
This means the alkane will have $(2 \times 15) + 2 = 32$ hydrogen atoms.
So its molecular formula is $C_{15}H_{32}$.

Figure 1: *Molecular model of methane — the alkane where n = 1.*

Skeletal formulas

A skeletal formula shows the bonds of the carbon skeleton only, with any functional groups. The hydrogen and carbon atoms that are part of the main carbon chain aren't shown. This is handy for drawing large complicated structures, like cyclic hydrocarbons. The carbon atoms are found at each junction between bonds and at the end of bonds (unless there's already a functional group there). Each carbon atom has enough hydrogen atoms attached to make the total number of bonds from the carbon up to four.

Exam Tip
Skeletal formulas are notoriously tricky to draw, so make sure that you get plenty of practice using them. Then if they come up in an exam question it'll be a breeze (relatively speaking).

Examples

Displayed and skeletal formulas of 1,5-difluoropentane:

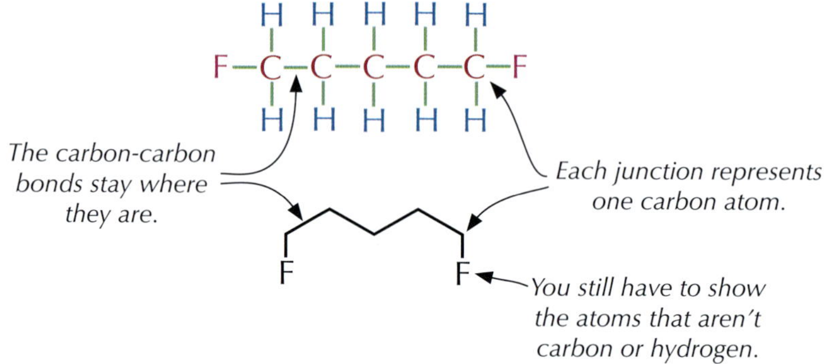
H H H H H
F–C–C–C–C–C–F
H H H H H
The carbon-carbon bonds stay where they are.
Each junction represents one carbon atom.
F
F
You still have to show the atoms that aren't carbon or hydrogen.

Skeletal formula of hex-1-ene:

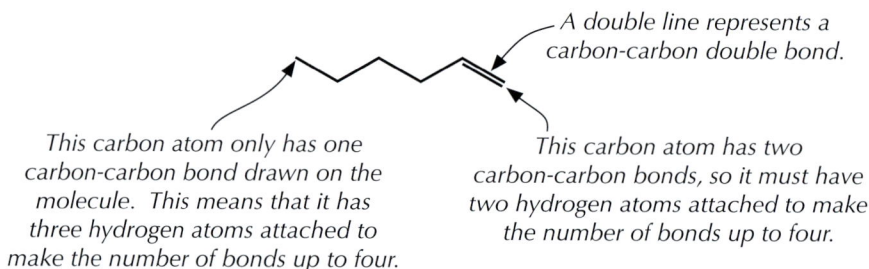

A double line represents a carbon-carbon double bond.

This carbon atom only has one carbon-carbon bond drawn on the molecule. This means that it has three hydrogen atoms attached to make the number of bonds up to four.

This carbon atom has two carbon-carbon bonds, so it must have two hydrogen atoms attached to make the number of bonds up to four.

Tip: You have to draw skeletal formulas as zig-zag lines, otherwise you can't tell where one bond ends and the next begins.

Practice Questions — Application

Q1 2-bromopropane has the structural formula $CH_3CHBrCH_3$. Draw the displayed formula of 2-bromopropane.

Q2 Here is the structure of 3-ethyl-2-methylpentane.

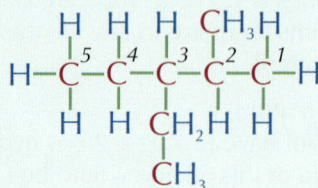

H H H CH_3 H
H–C(5)–C(4)–C(3)–C(2)–C(1)–H
H H CH_2 H H
CH_3

Write down the molecular formula for 3-ethyl-2-methylpentane.

Q3 Write down the empirical formula of the following compounds:

a) C_2H_4

b) $C_8H_{14}Br_2$

c) $C_9H_{17}Cl_3$

Tip: Drawing a displayed formula from a structural formula is dead easy — just draw it out exactly as it's written:

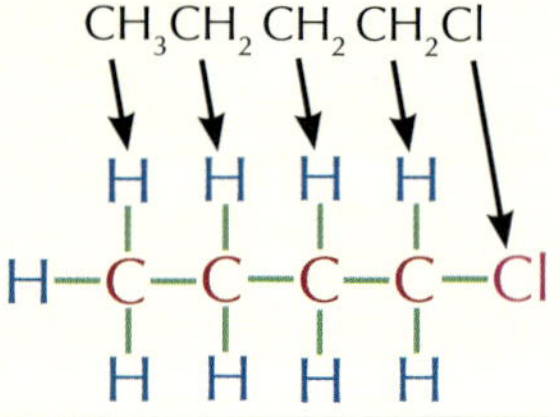
$CH_3CH_2CH_2CH_2Cl$
H H H H
H–C–C–C–C–Cl
H H H H

Q4 Alkenes have the general formula C_nH_{2n}.

a) Butene is an alkene with 4 carbon atoms. Write the molecular formula of butene.

b) Heptene is an alkene with 7 carbon atoms. How many hydrogen atoms does it contain?

Q5 1,2-dibromopropane has the structural formula $CH_3CHBrCH_2Br$.

a) Write down the molecular formula of 1,2-dibromopropane.

b) Draw the displayed formula of 1,2-dibromopropane.

c) Write the empirical formula of 1,2-dibromopropane.

Q6 Here is the displayed formula of pent-1-ene.

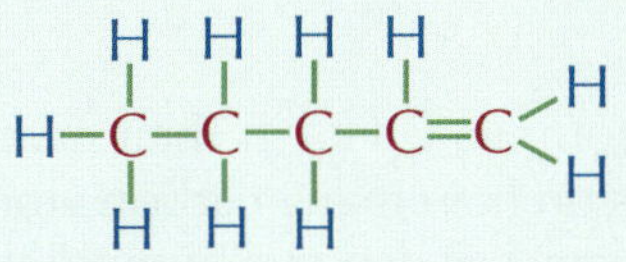

```
    H  H  H  H
    |  |  |  |     H
 H—C—C—C—C=C<
    |  |  |        H
    H  H  H
```

a) Write down the molecular formula of pent-1-ene.

b) Write down the structural formula of pent-1-ene.

c) What is the empirical formula of pent-1-ene?

Q7 Draw skeletal formulas of the molecules below:

a) 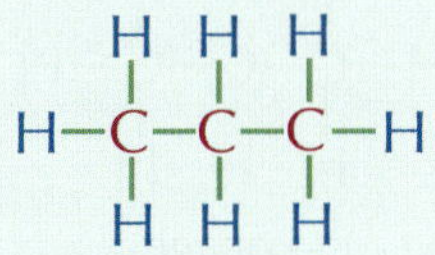
```
    H  H  H
    |  |  |
 H—C—C—C—H
    |  |  |
    H  H  H
```

b)
```
    H CH3 H  H  H
    |  |  |  |  |
 H—C—C—C—C—C—H
    |  |  |  |  |
    H  H OH  H  H
```

c)
```
      CH3
       |
    H CH2 H  H  H  H
    |  |  |  |  |  |
 H—C—C—C—C—C—C—H
    |  |  |  |  |  |
    H  H CH2 H  H  H
          |
         CH3
```

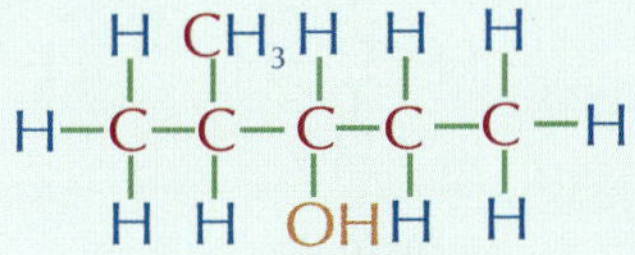

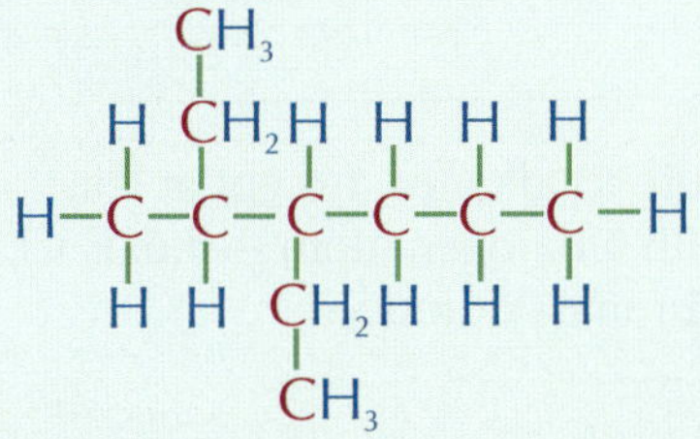

d) 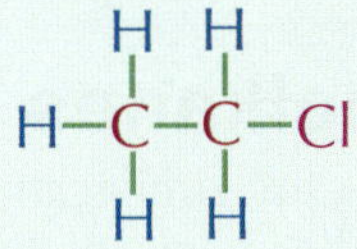
```
    H  H
    |  |
 H—C—C—Cl
    |  |
    H  H
```

Q8 Give structural formulas for the molecules below:

a)

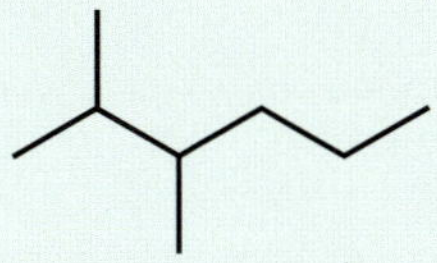

b)

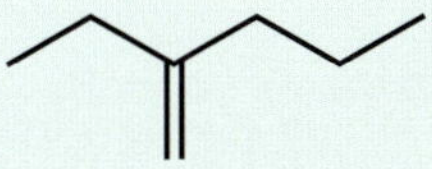

c) 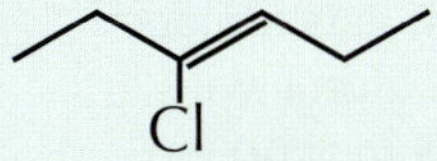 Cl

d) 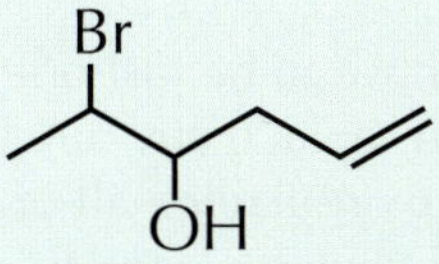 Br OH

Exam Tip
It's really important to double-check your answers for questions like these. It's so easy to miscount — and you need to make sure you collect all the easy marks in the exam.

Tip: It doesn't matter whether you show atoms bonding above, below or to the side of a carbon atom — they all mean the same thing. It's which atoms they're bonding to that's the important thing.

Tip: When you're writing the molecular formulas of branched molecules, you put the branch in brackets. E.g. methylpropane...

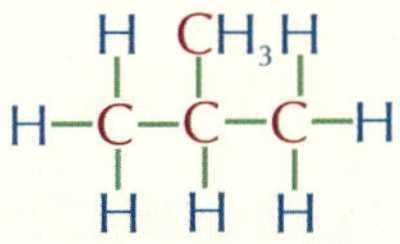

```
    H CH3 H
    |  |  |
 H—C—C—C—H
    |  |  |
    H  H  H
```

... has the structural formula:

$CH_3CH(CH_3)CH_3$

Practice Questions — Fact Recall

Q1 What is a molecular formula?

Q2 What does a displayed formula show?

Q3 How do you work out the empirical formula of a compound?

Q4 What is a homologous series?

2. Functional Groups

Learning Objective:

- Be able to apply IUPAC rules for nomenclature to name and draw the structure of organic compounds limited to chains and rings with up to six carbon atoms each.

Specification References 3.3.1.1

The functional groups are the parts of a molecule that define it. There are loads of different functional groups — all the ones you need to know about are described over the next few pages.

Homologous series

A homologous series is a series of molecules that all have the same functional group and the same general formula. If you know what homologous series a molecule is a part of, you can predict some of the ways that it will behave or react with other molecules. Below you're going to see a whole load of different homologous series, starting with the most basic of them — alkanes.

Alkanes

Alkanes have the general formula C_nH_{2n+2}. Their names all end in -ane, and the stem of the name depends on how many carbon atoms there are in the chain (see Figure 1). Naming alkanes is covered in more detail on pages 188-189). They've only got carbon and hydrogen atoms, so they're **hydrocarbons**. Every carbon atom in an alkane has four single bonds with other atoms. It's impossible for carbon to make more than four bonds, so alkanes are **saturated**.

No. of Carbon Atoms	Stem
1	*meth-*
2	*eth-*
3	*prop-*
4	*but-*
5	*pent-*
6	*hex-*

Figure 1: *Table showing the stems of organic compounds for up to six carbons.*

Examples

```
    H           H   H             H   H   H
    |           |   |             |   |   |
H — C — H   H — C — C — H     H — C — C — C — H
    |           |   |             |   |   |
    H           H   H             H   H   H
```

methane *ethane* *propane*

Halogenoalkanes

Halogenoalkanes are similar in structure to alkanes, except at least one of the hydrogen atoms has been replaced with a halogen atom — i.e. F, Cl, Br or I. They have the prefix fluoro-, chloro-, bromo- or iodo-.

Examples

By replacing one hydrogen atom in methane with a chlorine atom, you get chloromethane.

```
    Cl
    |
H — C — H
    |
    H
```

chloromethane, CH_3Cl

It doesn't have to be just one hydrogen atom that gets replaced — you can replace any number of the hydrogen atoms with halogen. By replacing all of the hydrogen atoms in propane with fluorine atoms, you get octafluoropropane.

```
    F   F   F
    |   |   |
F — C — C — C — F
    |   |   |
    F   F   F
```

octafluoropropane, C_3F_8

Figure 2: *C_3F_8 could potentially be used to create a habitable atmosphere on Mars.*

Tip: The 'octa-' tells you that there are eight fluorine atoms (see page 189 for more on this).

Cycloalkanes

Cycloalkanes have a ring of carbon atoms with two hydrogens attached to each carbon. Cycloalkanes have two fewer hydrogens than other alkanes (assuming they have only one ring), so cycloalkanes have a different general formula from that of normal alkanes (C_nH_{2n}), but they are still saturated. The molecules in this homologous series have the prefix cyclo- and the suffix -ane.

Examples

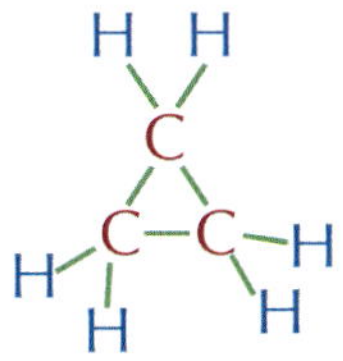

cyclopropane, C_3H_6

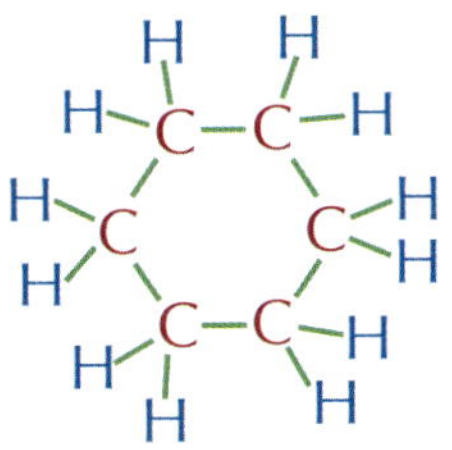

cyclohexane, C_6H_{12}

The smallest number of carbons you need to make a ring is three, so cyclopropane is the smallest cycloalkane.

Branched alkanes

A branched alkane is an alkane that doesn't have all the carbon atoms in one straight chain. They will have a main chain of carbons (whichever chain is the longest) and one or more carbons coming off this main chain. These branches are called alkyl groups. Branched alkanes have the same general formula as straight-chained alkanes.

Side Chain Length	Suffix
1	*methyl-*
2	*ethyl-*
3	*propyl-*
4	*butyl-*
5	*pentyl-*
6	*hexyl-*

Figure 3: *Table showing the suffixes for alkyl groups up to six carbons long.*

Examples

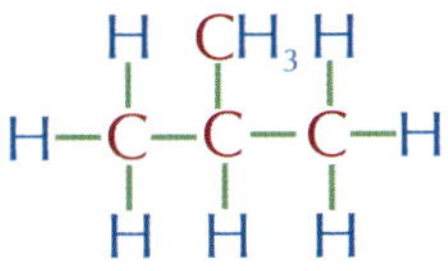

methylpropane, C_4H_{10}

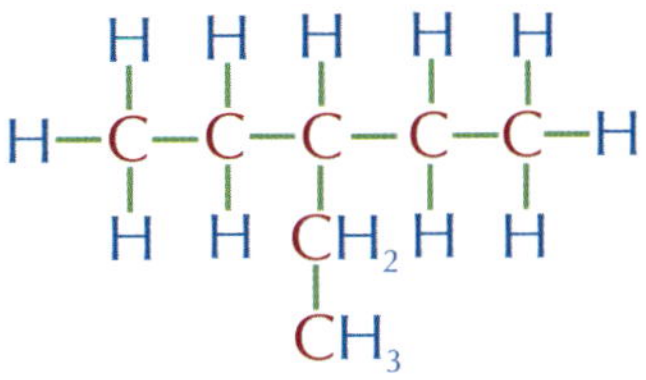

3-ethylpentane, C_7H_{16}

The alkyl group in the example on the left is just one carbon long, so it's called 'methyl-'. The alkyl group in the example on the right is two carbons long, so it's called 'ethyl-'.

Tip: The '3' in 3-ethylpentane means that the branch is on the third carbon along the main chain (see page 188 for more details).

Alkenes

An alkene is a hydrocarbon with a carbon-carbon double bond. They have the general formula C_nH_{2n}. The carbons on either end of the double bond are only bonded to three atoms each, rather than the maximum of four. This means that they could form another bond, so they are unsaturated. This makes alkenes fairly reactive (see page 222 for more on this).

Tip: The general formula C_nH_{2n} only applies to alkenes with exactly one C=C bond. The general formula's different for alkenes with two or more C=C bonds.

Examples

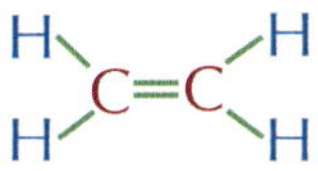

ethene

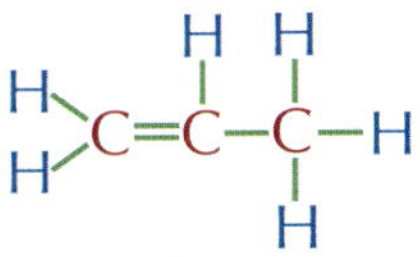

propene

Figure 4: *Ethanol (C_2H_5OH) has been produced by humans for thousands of years.*

Alcohols

Alcohols are organic molecules that contain the –OH, or hydroxyl, functional group. They have the suffix -ol, and the general formula $C_nH_{2n+1}OH$. Alcohols can be reacted to give alkenes, and vice versa. These reactions are discussed in more detail on pages 235 and 238.

Examples

```
    H                  H   H   H   H
    |                  |   |   |   |
H—C—OH            H—C—C—C—C—H
    |                  |   |   |   |
    H                  H   H  OH   H
```

methanol   *butan-2-ol*

Aldehydes

In aldehydes, one of the end carbons has a double bond to an oxygen atom and a single bond to a hydrogen atom, like this:

```
  O
  ||
—C—H
```

The suffix for aldehydes is -al. The general formula is written as R–CHO, where R is just an alkyl group or an H atom.

Tip: Anything with the functional group C=O is called a carbonyl. Aldehydes, ketones and carboxylic acids are all carbonyls.

Examples

The alkyl group can be a straight chain or a branched chain.

```
   H   H   O              H₃C  H   O
   |   |   ||               |   |   ||
H—C—C—C—H           H—C—C—C—H
   |   |                    |   |
   H   H                    H  CH₃
```

propanal   *2-methylbutanal*

Ketones

Like aldehydes, ketones also contain the C=O bond, except it isn't one of the end carbons. So the general structure of a ketone is this:

```
    O
    ||
R—C—R′
```

Here, R and R′ are alkyl groups, which may or may not be the same. The general formula is written R–CO–R′.

Tip: In the general formula for ketones, R or R′ can't represent an H atom, otherwise it would be an aldehyde.

Examples

In propanone, the alkyl groups R and R′ are the same.
In pentan-2-one, they're different.

```
   H   O   H              H   H   H   O   H
   |   ||  |              |   |   |   ||  |
H—C—C—C—H           H—C—C—C—C—C—H
   |       |              |   |   |       |
   H       H              H   H   H       H
```

propanone   *pentan-2-one*

Tip: Aldehydes, ketones and carboxylic acids can all be produced by oxidising alcohols — this is covered in detail on pages 242-245.

Carboxylic acids

Carboxylic acids all contain the carboxyl functional group:

```
  O
  ||
—C—OH
```

Its suffix is -oic acid. The general formula is written R–COOH, where R is an alkyl group.

Examples

The alkyl group R can also represent a hydrogen atom, like in methanoic acid. This is the simplest carboxylic acid.

```
  O
  ||
H-C-OH
```

methanoic acid

```
  H H O
  | | ||
H-C-C-C-OH
  | |
  H CH3
```

2-methylpropanoic acid

Figure 5: *Methanoic acid occurs naturally in the venom of some ants.*

Practice Questions — Fact Recall

Q1 Give the general formula for an alkane.

Q2 What is the stem for a carbon chain containing six carbon atoms?

Q3 Describe a cycloalkane.

Q4 What is the name for a carbon side chain containing two carbon atoms?

Q5 Which two homologous series have the general formula C_nH_{2n}?

Q6 What does the 'R' represent in the general formula R–CHO?

Q7 What is the difference between an aldehyde and a ketone?

3. Nomenclature

Nomenclature is just a fancy word for naming organic compounds. You have to follow a strict set of rules for naming, but it's dead handy — this way anyone anywhere can know what compound you're talking about.

Learning Objective:

- Be able to apply IUPAC rules for nomenclature to name and draw the structure of organic compounds limited to chains and rings with up to six carbon atoms each.

Specification Reference 3.3.1.1

Naming alkanes

The IUPAC system for naming organic compounds is the agreed international language of chemistry. Years ago, organic compounds were given whatever names people fancied, such as acetic acid and ethylene. But these names caused confusion between different countries.

The IUPAC system means scientific ideas can be communicated across the globe more effectively. So it's easier for scientists to get on with testing each other's work, and either support or dispute new theories.

You need to be able to name straight-chain and branched alkanes using the IUPAC system for naming organic compounds.

Exam Tip
Always double check that you've spelled the IUPAC names of compounds correctly in exams — if you don't spell them right, you won't get the marks.

Straight-chain alkanes

There are two parts to the name of a straight-chain alkane. The first part — the stem — states how many carbon atoms there are in the molecule (see Figure 1 on page 184). The second part is always "-ane". It's the "-ane" bit that lets people know it's an alkane.

Tip: These stems come up again and again in chemistry so it's really important that you know all of them.

Example

```
    H  H  H  H  H
    |  |  |  |  |
 H—C—C—C—C—C—H
    |  |  |  |  |
    H  H  H  H  H
```

pentane

There are 5 carbon atoms, so the stem is 'pent-' — the alkane is called pentane.

***Figure 1:** August Kekulé was one of the first scientists to recognise the need for a way to systematically name molecules.*

Branched alkanes

Branched alkanes have side chains. These are the carbon atoms that aren't part of the longest continuous chain. To name branched alkanes you first need to count how many carbon atoms are in the longest chain and work out the stem (just like you would for a straight-chain alkane). Once you've done that you can name the side chains.

The side chains are named according to how many carbon atoms they have (see Figure 3 on page 185) and which carbon atom they are attached to. If there's more than one side chain in a molecule, you place them in alphabetical order. So but- groups come before eth- groups, which come before meth- groups.

Examples

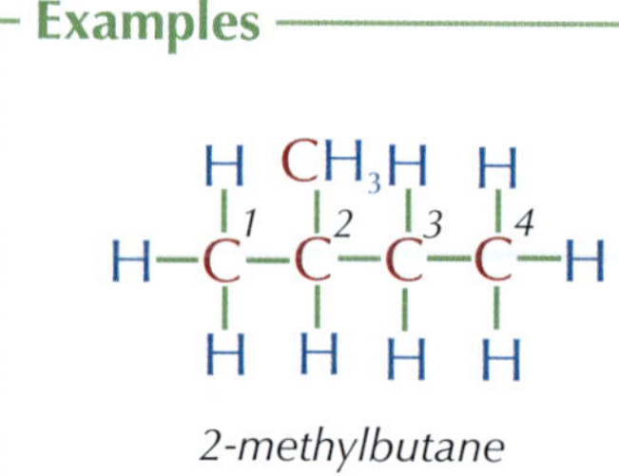

2-methylbutane

The longest continuous carbon chain is 4 carbon atoms, so the stem is butane.

There's one side chain, which has one carbon atom, so it's a methyl group.

It's joined to the main carbon chain at the 2nd carbon atom, so it's a 2-methyl group.

The alkane is called 2-methylbutane.

The longest continuous carbon chain is 5 carbon atoms, so the stem is pentane.

There are two side chains.

One side chain is a methyl group joined to the 2nd carbon atom: 2-methyl-.

The other is an ethyl group (2 carbons) joined to the 3rd carbon atom: 3-ethyl-.

Side chains go in alphabetical order, so the alkane is 3-ethyl-2-methylpentane.

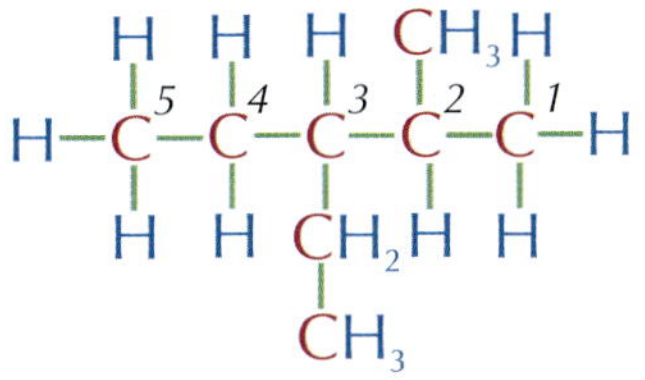

3-ethyl-2-methylpentane

If there are two or more side chains of the same type then you add a prefix of di- for two, tri- for three, etc. (You should ignore these prefixes when you're putting the other prefixes in alphabetical order.)

Example

The longest carbon chain is 5 atoms long, so the stem is pentane.

There's an ethyl group on the 3rd carbon atom: 3-ethyl-.

There are methyl groups on the 2nd and the 4th carbon atoms: 2,4-dimethyl-.

The alkane is called 3-ethyl-2,4-dimethylpentane.

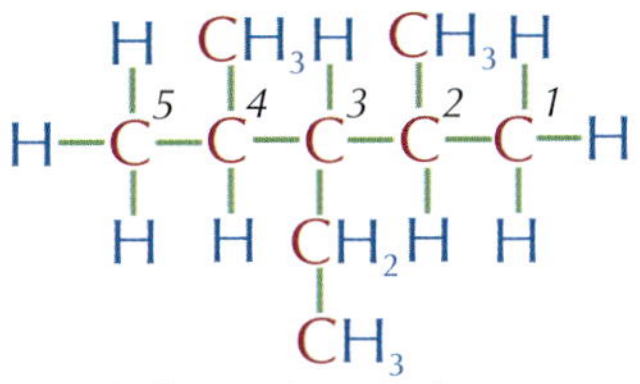

3-ethyl-2,4-dimethylpentane

Cycloalkanes

Cycloalkanes have the same name as their straight-chain alkane equivalent, but with cyclo- attached to the front. If the cycloalkane has an alkyl group attached, then just add the alkyl prefix (you don't need a number). If there's more than one alkyl group, then make the numbers as low as possible, and the alkyl that's first alphabetically goes on the 1-carbon.

Example

The carbon ring is 5 atoms long, so it's a cyclopentane.

There's a methyl group and an ethyl group. Ethyl comes first alphabetically: 1-ethyl.

Depending on which way round the ring you count, the methyl is on the 3rd or 4th carbon. Make the numbers as low as possible: 3-methyl.

So the molecule is 1-ethyl-3-methyl-cyclopentane.

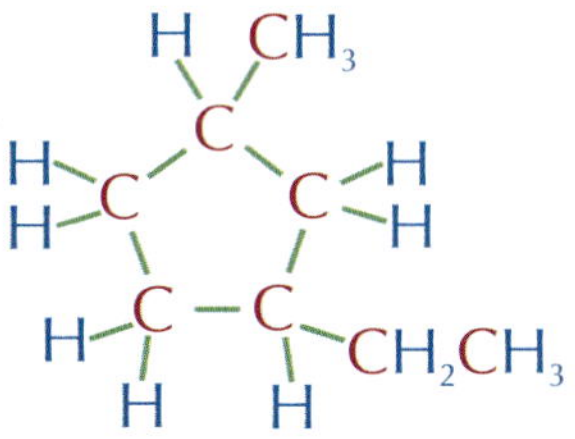

Naming other functional groups

Once you know how to name alkanes, the other functional groups follow nicely. The stem comes from the longest carbon chain that contains the functional group. Prefixes and suffixes come from the functional groups as well as any alkyl side chains. If you need to use a number to show the position of the functional group, give the carbon which the functional group is on the lowest number possible.

Tip: Always number the longest continuous carbon chain so that the name contains the lowest numbers possible. For example, you could number this chain:

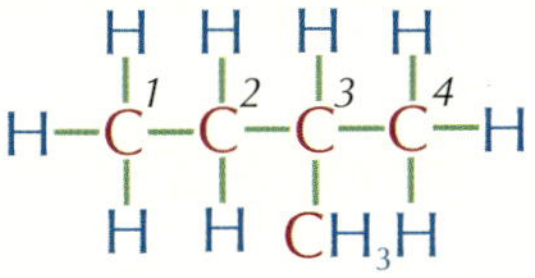

which would make it 3-methylbutane. But you should actually number it in the opposite direction to get 2-methylbutane.

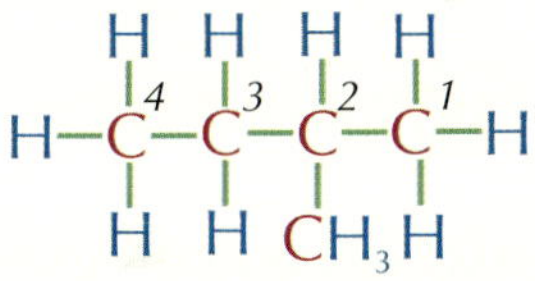

Tip: Be careful — the longest carbon chain may not be in a straight line:

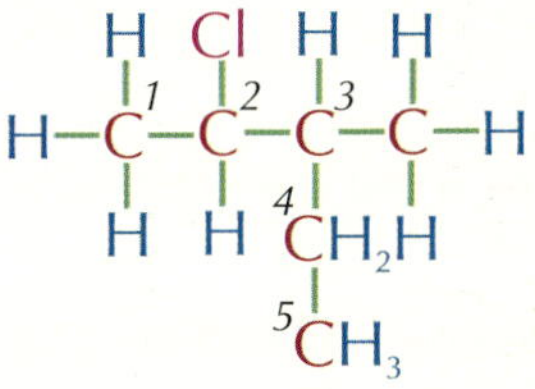

Tip: When you're naming molecules commas are put between numbers (for example 2,2) and dashes are put between numbers and letters (for example 2-methyl).

Examples

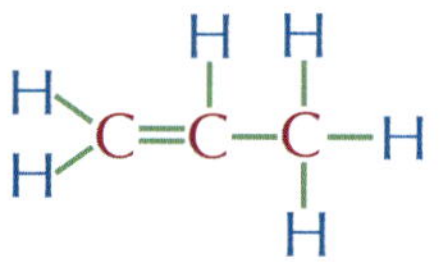

propene

There are three carbon atoms, so the stem is prop-. It's an alkene, so the suffix is -ene.

So this is propene. The double bond must always be on the first carbon in propene, so you don't need a number in the name.

This is an alkene with a straight chain of four carbon atoms — so it's butene.

The double bond is between the first and second carbon, so you say it's on the 1st carbon. So the name of the molecule is but-1-ene.

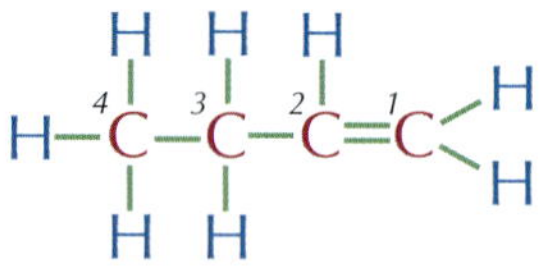

but-1-ene

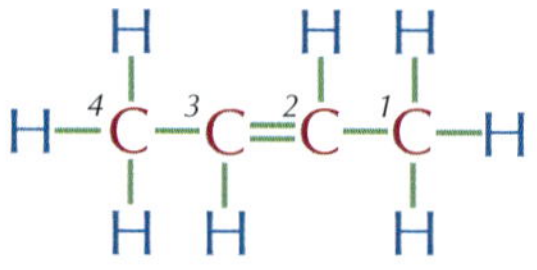

but-2-ene

This is also butene, but the double bond is between the second and third carbon atoms.

So the name of the molecule is but-2-ene.

If there's more than one functional group you have to work out which one has the highest priority (see Figure 2) — this is the main functional group. The stem of the name then comes from the longest carbon chain containing the main functional group, and you number the carbons so that the main functional group has the lowest number possible.

Tip: You might see molecules with several 'other functional groups', but you won't be asked to name them.

Halogens	Alkyl groups	Alkenes	Other functional groups
Lowest priority	→	→	*Highest priority*

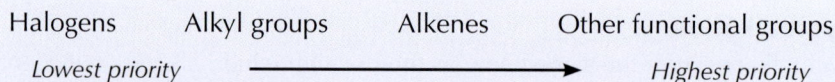

Figure 2: *The order of priority of different functional groups.*

Examples

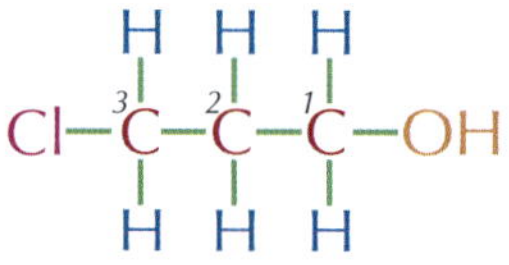

3-chloropropan-1-ol

This molecule has just one carbon chain, which is 3 atoms long, so the stem is prop-.

It's got two functional groups, Cl and OH, so it needs a chloro- prefix and an -ol suffix.

The alcohol group has a higher priority than the chlorine, so number the carbons to give the alcohol the lowest number possible. So this is 3-chloropropan-1-ol (not 1-chloropropan-3-ol).

The highest priority functional group on this molecule is the CHO group, so it's an aldehyde, and the suffix is -al. The longest carbon chain containing the CHO group is 4 atoms long, so the stem is but-.

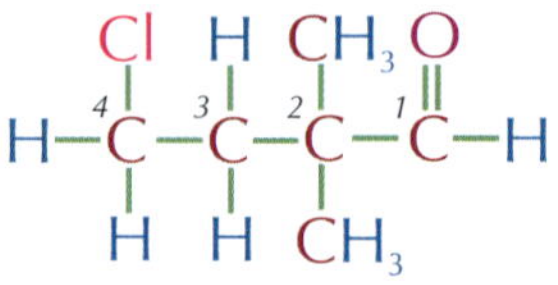

4-chloro-2,2-dimethylbutanal

Number the carbons in this chain starting from the carbonyl group, as it has the highest priority.

Both methyl groups are on the 2nd carbon: 2,2-dimethyl.

The chlorine atom is on the 4th carbon: 4-chloro.

Chloro- comes before methyl- alphabetically.

So the molecule is 4-chloro-2,2-dimethylbutanal.

Tip: If there are no double or triple carbon-carbon bonds in an aldehyde, you need to write '-an-' between the stem and the suffix. This is also true for alcohols, ketones and carboxylic acids.

```
      H  H  O  H
H  5 4|  3| 2|| 1|
  >C=C--C--C--C--H
H         |     |
          H     H
```

pent-4-en-2-one

The highest priority functional group here is the ketone. The longest carbon chain containing the ketone is 5 atoms long, so the stem is pent-.

It's a ketone, so the suffix is -one. It's also an alkene, so it ends '-enone' instead of '-anone'.

Number the carbons in the longest chain so that the carbonyl group has as low a number as possible (i.e. start on the right): -2-one.

The C=C bond is between the 4th and 5th carbons: -4-ene.

This molecule is called pent-4-en-2-one.

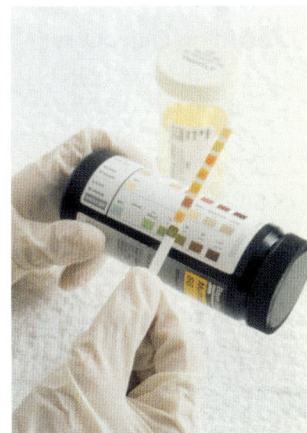

Figure 3: *Ketones can be used in tests for diabetes.*

Practice Questions — Application

Q1 Name the following branched alkanes:

a)
```
    H  H  H  H
    |  |  |  |
 H--C--C--C--C--H
    |  |  |  |
    H  H CH2 H
          |
         CH3
```

b)
```
         CH3
          |
    H  H CH2 H
    |  |  |  |
 H--C--C--C--C--H
    |  |  |  |
    H  H CH2 H
          |
         CH3
```

c)
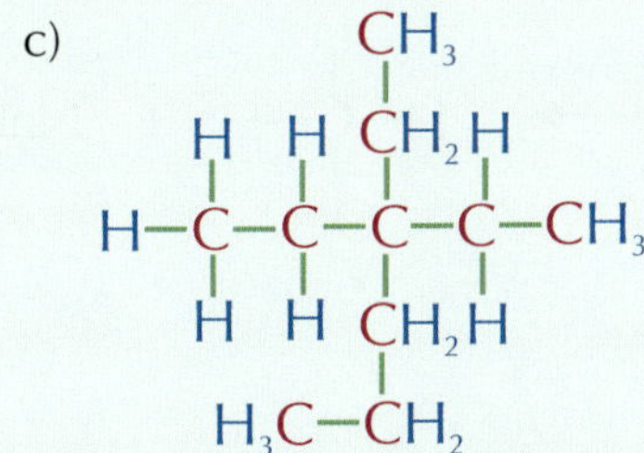

d)
```
         CH3
          |
    H CH3 CH2 H
    |  |  |  |
 H--C--C--C--C--CH3
    |  |  |  |
    H  H CH2 H
          |
       H3C-CH2
```

Q2 Name the following molecules:

a)
```
    H  H CH3 H  O
    |  |  |  |  ||
 H--C--C--C--C--C--OH
    |  |  |  |
    H CH3 H CH3
```

b)
```
    H  H  O  H  H
    |  |  ||  |  |
 H--C--C--C--C--C--H
    |  |     |  |
    H  H   H3C  H
```

Q3 Draw the structure of each of the following molecules:

a) 2-methylhexane

b) 3-bromobutan-2-ol

c) 3-ethylpentan-2-one

d) 1,3-difluorobut-2-ene

Exam Tip

Make sure you know the order of priority of the functional groups.

Practice Questions — Fact Recall

Q1 If there are two methyl- side chains, what prefix should you add to methyl- when naming the molecule?

Q2 Put the following functional groups in order of priority, from highest to lowest: Alkene Alcohol Halogen Alkyl

4. Mechanisms

Learning Objectives:

- Know that reactions of organic compounds can be explained using mechanisms.
- Know that the formation of a covalent bond is shown by a curly arrow that starts from a lone electron pair or from another covalent bond.
- Know that the breaking of a covalent bond is shown by a curly arrow starting from the bond.
- Be able to outline mechanisms by drawing the structures of the species involved and curly arrows to represent the movement of electron pairs.

Specification Reference 3.3.1.2

It's all very well knowing the outcome of a reaction, but it can also be useful to know how a reaction happens.

What are mechanisms?

Mechanisms break reactions down into a sequence of stages. Reaction mechanism diagrams show how molecules react together by using curly arrows to show which bonds are made or broken.

Curly arrows

In order to make or break a bond in a reaction, electrons have to move around. A curly arrow shows where a pair of electrons goes during a reaction. They look like this:

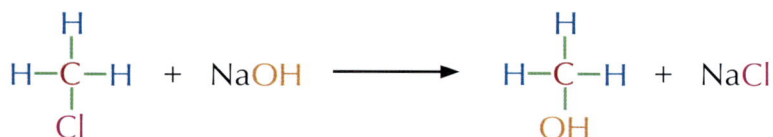

The arrow starts at the bond or lone pair where the electrons are at the beginning of the step.

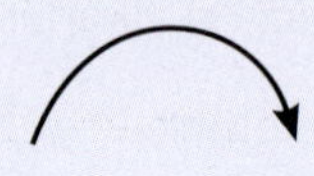

The arrow points to where the new bond is formed at the end of the step.

Example

Draw a reaction mechanism to show how chloromethane reacts with sodium hydroxide to form methanol and sodium chloride.

Reaction:

$$CH_3Cl + NaOH \rightarrow CH_3OH + NaCl$$

Mechanism:

1. The C–Cl bond is polar. The $C^{\delta+}$ attracts a lone pair of electrons from the OH^- ion.

3. A new bond forms between the C and the OH^- ion, making an alcohol...

2. The OH^- ion attacks the slightly positive carbon atom.

4. ... and the C–Cl bond breaks. Both the electrons from the bond are taken by the Cl.

NaOH and NaCl are both ionic compounds that are dissociated in a solution. This means that Na^+ doesn't get involved in the reaction — so you don't need to include it in the mechanism.

Tip: Free radical mechanisms, which are a different type of mechanism, are covered on pages 208-210.

Tip: There are lots of mechanisms coming up in the next two sections, so if it all seems a bit strange now, don't worry — before long you'll have seen loads of them.

Practice Questions — Fact Recall

Q1 What does a curly arrow point from and to in a reaction mechanism diagram?

Q2 In a reaction between chloroethane and aqueous potassium hydroxide, why would potassium not feature in the mechanism?

5. Isomers

You can put the same atoms together in different ways to make completely different molecules. Two molecules that have the same molecular formula but are put together in a different way are isomers of each other.

Structural isomers

Structural isomers have the same molecular formula, but a different structural formula (i.e the atoms are connected in different ways). There are three types of structural isomers — chain isomers, position isomers and functional group isomers.

1. Chain isomers

Chain isomers have the same functional groups but different arrangements of the carbon skeleton. Some are straight chains and others are branched in different ways.

Examples

There are two chain isomers of C_4H_{10}. The diagrams below show the straight-chain isomer butane and the branched-chain isomer methylpropane.

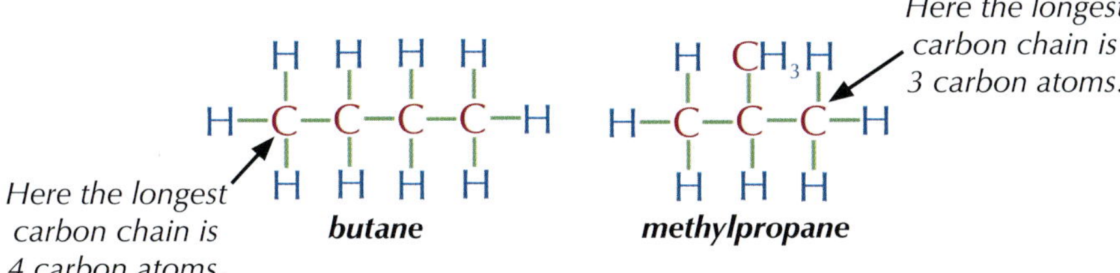
Here the longest carbon chain is 4 carbon atoms.
Here the longest carbon chain is 3 carbon atoms.

butane ***methylpropane***

There are two chain isomers of C_3H_7COOH. The diagrams below show the straight-chain isomer butanoic acid and the branched-chain isomer methylpropanoic acid.

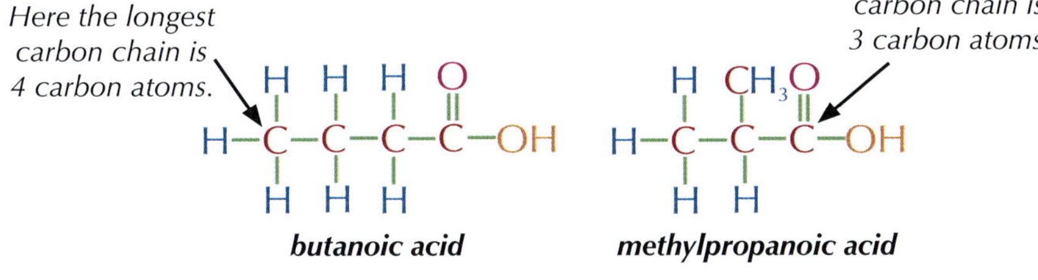
Here the longest carbon chain is 4 carbon atoms.
Here the longest carbon chain is 3 carbon atoms.

butanoic acid ***methylpropanoic acid***

2. Position isomers

Position isomers have the same skeleton and the same atoms or groups of atoms attached. The difference is that the atoms or groups of atoms are attached to different carbon atoms.

Examples

There are two position isomers of C_4H_9Cl. The chlorine atom is attached to different carbon atoms in each isomer.

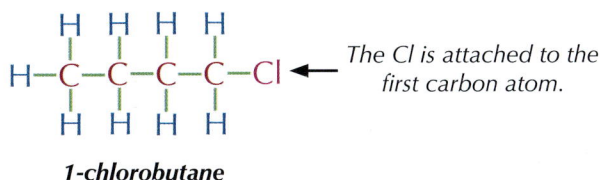

The Cl is attached to the first carbon atom.

1-chlorobutane

Learning Objectives:

- Know and understand the meaning of the term structural isomerism.
- Be able to draw the structures of chain, position and functional group isomers.

Specification Reference 3.3.1.3

Tip: When a functional group can only go in one place on a molecule (e.g. 2-methylpropane), you don't need to write the number (e.g. methylpropane).

Exam Tip
You don't always have to draw all of the bonds when you're drawing a molecule — writing CH_3 next to a bond is just as good as drawing out the carbon atom, three bonds and three hydrogen atoms. But if you're asked for a displayed formula you <u>must</u> draw out all of the bonds to get the marks.

Tip: When you're drawing isomers, always number the carbons. It makes it easier to see what the longest chain is and where side chains and atoms are attached:

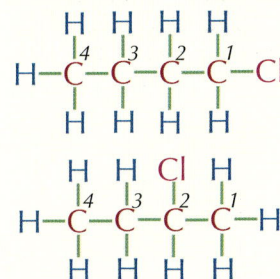

Tip: If the chlorine atom was attached to the second carbon atom from the left, it would still be the same molecule — just drawn the other way round. It would still be 2-chlorobutane.

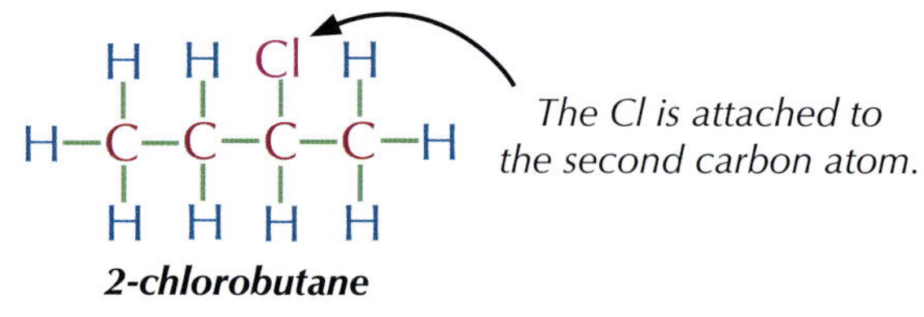
The Cl is attached to the second carbon atom.

3. Functional group isomers

Functional group isomers have the same atoms arranged into different functional groups.

Example

The formulas below show two functional group isomers of C_6H_{12}.

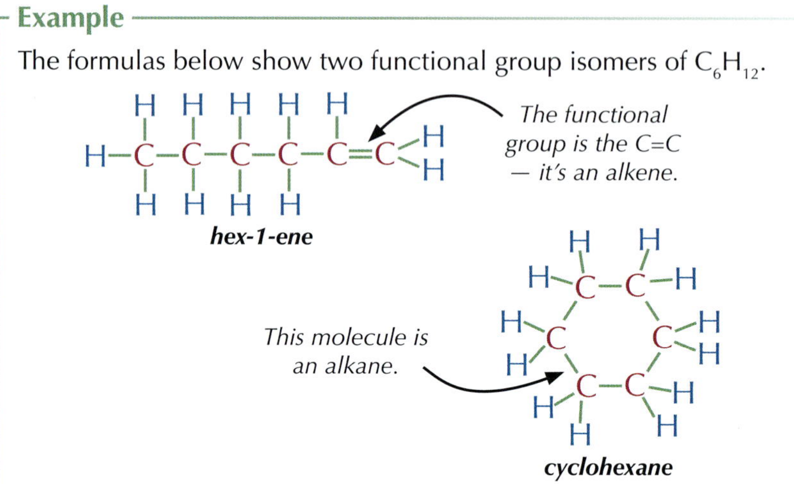
The functional group is the C=C — it's an alkene.
hex-1-ene
This molecule is an alkane.
cyclohexane

Exam Tip

To avoid mistakes when you're identifying isomers in an exam, draw the molecule so the longest carbon chain goes left to right across the page. This will make it easier to see the isomers.

Identifying isomers

Atoms can rotate as much as they like around single C–C bonds. Remember this when you work out structural isomers — sometimes what looks like an isomer, isn't.

Examples

There are only two position isomers of C_3H_7Br — 1-bromopropane and 2-bromopropane.

1-bromopropane

The Br is always on the first carbon atom.

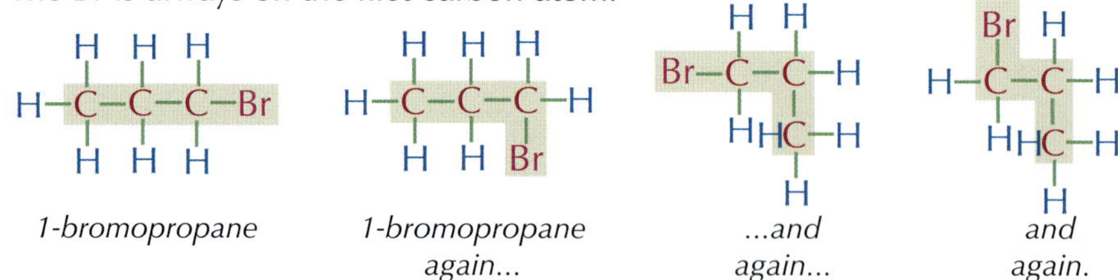

1-bromopropane   *1-bromopropane again...*   *...and again...*   *and again.*

All these molecules are the same — they're just drawn differently.

2-bromopropane

The Br is always on the second carbon atom.

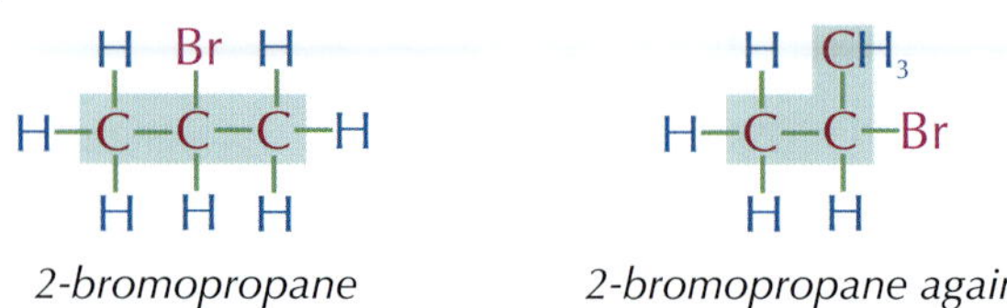

2-bromopropane   *2-bromopropane again*

Tip: In propane, the Br can only really go on the first or second carbon atom. If it was on the "third" it would be the same as being on the first again because you start counting from whichever end the Br is on.

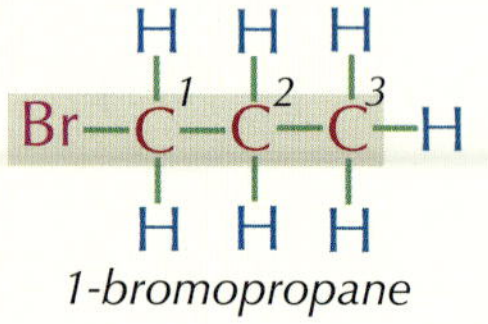

1-bromopropane

Practice Questions — Application

Q1 Here is an isomer of 2-chloro-2-methylpropane.

```
      CH3
      |
H3C—C—CH3
      |
      Cl
```

Draw the other position isomer of chloro-2-methylpropane.

Q2 Draw all the chain isomers of C_5H_{12}.

Q3 Here is the displayed formula of propanal.

```
   H  H  O
   |  |  ||
H—C—C—C—H
   |  |
   H  H
```

Propanal has the functional group C(=O)—H.
Draw an isomer of propanal with a carbonyl group C=O.

Q4 Here is the displayed formula of 1-chlorohexane.

```
   H  H  H  H  H  H
   |  |  |  |  |  |
H—C—C—C—C—C—C—Cl
   |  |  |  |  |  |
   H  H  H  H  H  H
```

a) For each of the molecules (i–iv), say whether they are isomers of 1-chlorohexane or not.

```
i)     H  Cl H   H              ii)    H  Cl CH3 H
       |  |  |   |                     |  |  |   |
    H—C—C—C—C—H                    H—C—C—C—C—H
       |  |  |   |                     |  |  |   |
       H  H  CH2 H                     H  H  CH3 H
             |
             CH3

iii)   H  H  H   H              iv)    H  H  Cl H  H
       |  |  |   |                     |  |  |  |  |
    H—C—C—C—C—H                    H—C—C—C—C—C—H
       |  |  |   |                     |  |  |  |  |
       H  H  CH2 H                     H  H  H  H  H
             |
             CH3
```

b) State the types of isomerism shown in part a).

Exam Tip
Remember, isomers have the same atoms arranged in a different way — you can do this type of question simply by counting the number of carbon, hydrogen and chlorine atoms.

Practice Questions — Fact Recall

Q1 What is a chain isomer?

Q2 What is a position isomer?

Q3 What is a functional group isomer?

6. *E*/*Z* Isomers

Structural isomers aren't the only isomers you need to know about. You also need to know about E/Z isomerism, which is a type of stereoisomerism.

Learning Objectives:

- Know and understand the meaning of the term stereoisomerism.
- Understand that *E*/*Z*-isomerism is a form of stereoisomerism and that it occurs as a result of restricted rotation about the planar carbon-carbon double bond.
- Be able to draw the structural formulas of *E* and *Z* isomers.
- Know and be able to apply the Cahn-Ingold-Prelog priority rules to *E* and *Z* isomers.

Specification Reference 3.3.1.3

Stereoisomers

Stereoisomers have the same structural formula, but their atoms are arranged differently in space. One type of stereoisomerism is *E*/*Z* isomerism, which you see in molecules with C=C double bonds. Before getting to that, you need to know a bit more about the structure of a C=C double bond. Read on...

Planar double bonds and restricted rotation

Carbon atoms in a C=C double bond and the atoms bonded to these carbons all lie in the same plane (they're planar). Because of the way they're arranged, they're said to be trigonal planar — the atoms attached to each double-bonded carbon form the corners of an imaginary equilateral triangle:

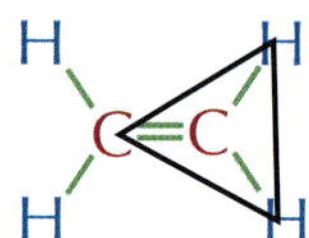

The bond angles in the planar unit are all 120°.

Ethene (C_2H_4) is totally planar. In larger alkenes only the C=C unit is planar.

Example

This molecule is but-1-ene. The carbon-carbon double bond section of the molecule is planar. The section of the molecule that only contains single bonds is non-planar.

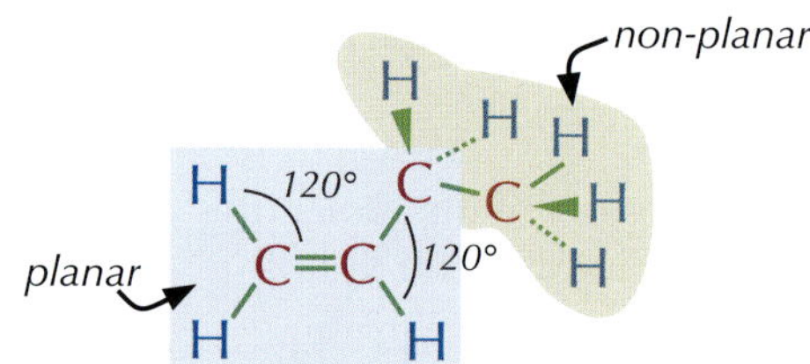

Another important thing about C=C double bonds is that atoms can't rotate around them like they can around single bonds. In fact, double bonds are fairly rigid — they don't bend much. Things can still rotate about any single bonds in the molecule though.

Tip: Don't get confused by all those crazy green lines on the diagram. The dotted ones mean that the bond is pointing into the paper and the wedgy ones mean that the bond is pointing out of the paper. There's more on this on page 86.

Example

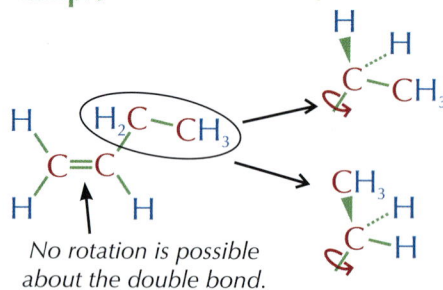

No rotation is possible about the double bond.

In this molecule of but-1-ene:

- The C=C double bond can't rotate.
- But the C–C single bonds can rotate.

E/*Z* isomerism in alkenes

The restricted rotation around the C=C double bond in alkenes causes a type of stereoisomerism called *E*/*Z* isomerism. If both double-bond carbons have two different atoms or groups attached to them, the arrangement of the groups around the double bond becomes important — you get two stereoisomers. One of these isomers is called the **'E-isomer'** and the other is the **'Z-isomer'**.

The simplest cases are when each carbon in the double bond has the same two 'different groups' attached. The *E*-isomer is the one that has the matching groups across the double bond from each other. The *Z*-isomer is the one with the matching groups both above or both below the double bond:

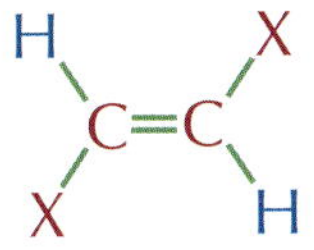

H X
C=C
X H
***E*-isomer**

H H
C=C
X X
***Z*-isomer**

In these diagrams X represents any group larger than a single H atom.

Tip: *E* stands for 'entgegen', a German word meaning 'opposite'. *Z* stands for 'zusammen', the German for 'together'.

Example

The double-bonded carbon atoms in but-2-ene (C_4H_8) each have an H and a CH_3 group attached.

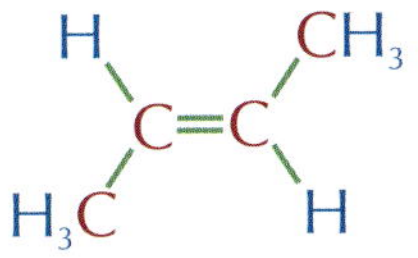

H CH_3
C=C
H_3C H
E-isomer

When the CH_3 groups are across the double bond then it's the *E*-isomer.

This molecule is *E*-but-2-ene.

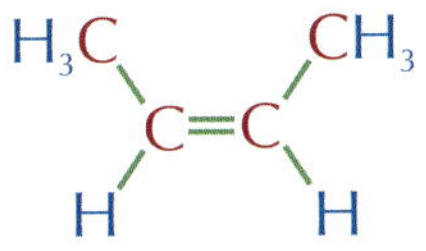

H_3C CH_3
C=C
H H
Z-isomer

When the CH_3 groups are both above or both below the double bond then it's the *Z*-isomer.

This molecule is *Z*-but-2-ene.

Tip: If all this isomer stuff is a bit confusing, try to get your hands on some molecular models and have a go making the isomers yourself — it should make everything a bit clearer.

In unbranched hydrocarbon chains, it is fairly straightforward to identify which isomer is which. The *E*-isomer is the one that has the carbon groups across the double bond from each other. The *Z*-isomer is the one which has the carbon groups both above or both below the double bond.

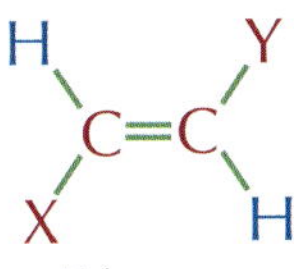

***E*-isomer**

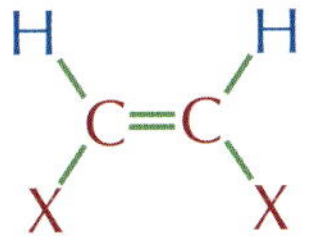

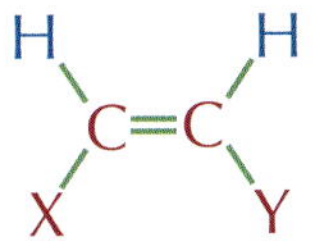

H H
C=C
X Y
***Z*-isomer**

In these diagrams, X and Y are carbon chains.

Example

In pent-2-ene (C_5H_{10}) one of the double-bonded carbon atoms has an H and a CH_3 group attached to it. The other has an H and a CH_2CH_3 group attached.

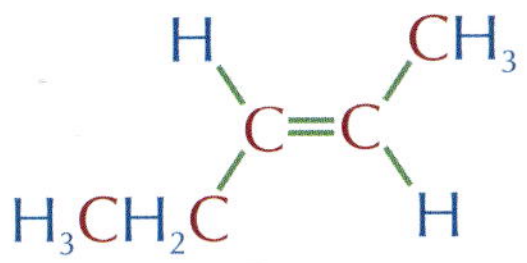

H CH_3
C=C
H_3CH_2C H
E-isomer

The high priority groups (CH_3 and CH_2CH_3) are across the double bond, so it's the *E*-isomer. This molecule is *E*-pent-2-ene.

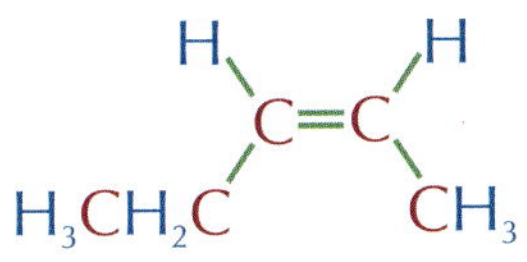

H H
C=C
H_3CH_2C CH_3
Z-isomer

The high priority groups are both below the double bond, so it's the *Z*-isomer.

This molecule is *Z*-pent-2-ene.

Cahn-Ingold-Prelog Priority Rules

A molecule that has a C=C bond surrounded by four different groups still has an E- and a Z-isomer — it's just harder to work out which is which. Fortunately, you can solve this problem using the Cahn-Ingold-Prelog (CIP) priority rules:

- Start by assigning a priority to the two atoms attached to each side of the double bond. To do this, you look at the atoms that are directly bonded to each of the C=C carbon atoms.
- The atom with the higher atomic number on each carbon is given the higher priority.
- If the atoms directly bonded to each carbon are the same, then you look at the next atom in the groups to work out which has the higher priority.
- To work out which isomer you have, just look at how the two highest priority groups are arranged. If they're positioned across the double bond from each other, you have the *E*-isomer. If they're both above or below the double bond, you have the *Z*-isomer.

Figure 1: *Sir Christopher Ingold contributed to the nomenclature of stereoisomers.*

Examples

A stereoisomer of 1-bromo-1-chloro-2-fluoro-ethene

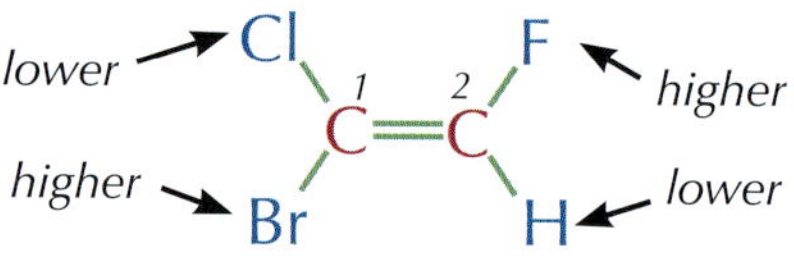

- The atoms directly attached to carbon-1 are bromine and chlorine. Bromine has an atomic number of 35 and chlorine has an atomic number of 17. So bromine is the higher priority group.
- The atoms directly attached to carbon-2 are fluorine and hydrogen. Fluorine has an atomic number of 9 and hydrogen has an atomic number of 1. So fluorine is the higher priority group.
- The two higher priority groups (Br and F) are positioned across the double bond from one another — so this is ***E*-1-bromo-1-chloro-2-fluoroethene**.

A stereoisomer of 1-bromo-1-chloro-2-methylbut-1-ene

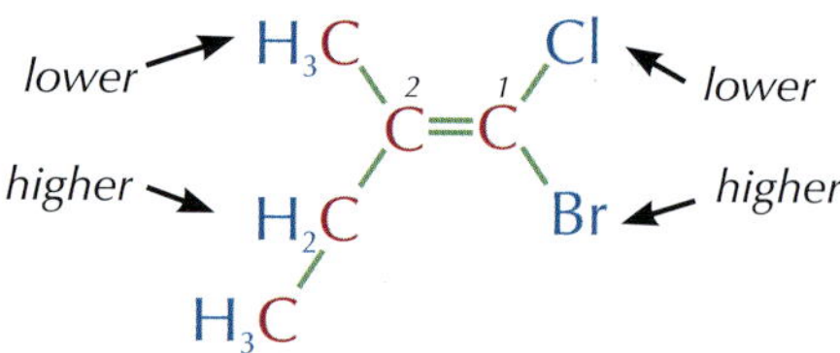

- The atoms attached to carbon-1 are bromine and chlorine. Bromine has the higher atomic number, so it is the higher priority group.
- The atoms attached to carbon-2 are both carbons, so you need to go further along the chain to work out the priority. The methyl carbon is attached to hydrogen (atomic number = 1), but the first ethyl carbon is attached to another carbon (atomic number = 6). So the ethyl group has higher priority.
- Both higher priority groups are below the double bond — so this molecule is ***Z*-1-bromo-1-chloro-2-methylbut-1-ene**.

Tip: If a C=C bond is surrounded by three different groups, then the E/Z label doesn't necessarily describe the positions of the two matching groups. You still need to use the priorities of the atoms to work out whether it's an E or Z isomer.

Practice Questions — Application

Q1 State whether the following molecules are E-isomers or Z-isomers.

a)

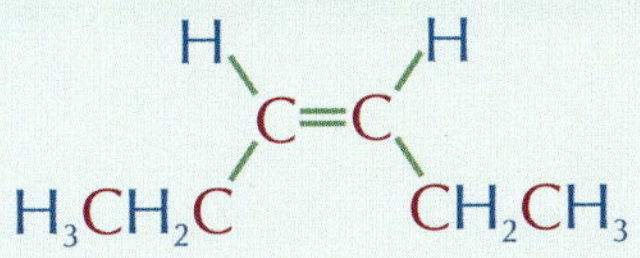

```
        H       H
         \     /
          C = C
         /     \
   H3CH2C       CH2CH3
```

b)

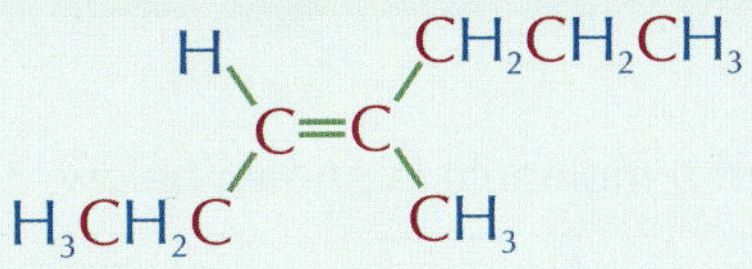

```
        H       CH2CH2CH3
         \     /
          C = C
         /     \
   H3CH2C       CH3
```

Q2 Draw the two stereoisomers of 3,4-dimethylhex-3-ene. Label the E-isomer and the Z-isomer. The structure of 3,4-dimethylhex-3-ene is shown below.

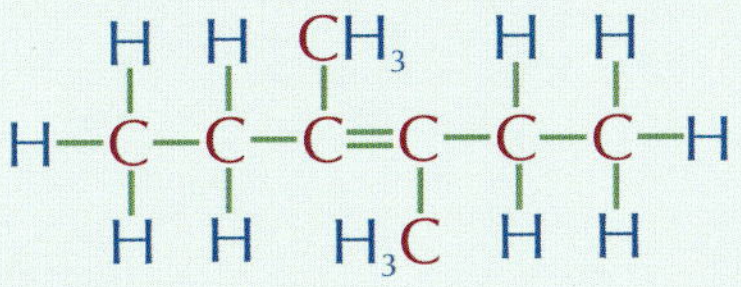

```
    H  H  CH3     H  H
    |  |  |       |  |
 H—C—C—C=C—C—C—H
    |  |      |   |  |
    H  H    H3C   H  H
```

Practice Questions — Fact Recall

Q1 What is a stereoisomer?

Q2 a) What is a Z-isomer?

b) What is an E-isomer?

Q3 Each of the following pairs are joined to the same carbon in a carbon-carbon double bond. Use the CIP rules to decide which one has the higher priority:

a) CH_3 , H

b) CH_2CH_3 , Cl

c) $CH_2CH_2CH_3$, OH

d) $CH_2CH_2CH_2CH_2COOH$, $CH_2CH_2CH_2CH_2CH_2OH$

Tip: Remember — if the first atom in both chains is the same, you need to look at the next one (and so on...).

Section Summary

Make sure you know...

- What molecular formulas, structural formulas, displayed formulas, empirical formulas, general formulas, homologous series and skeletal formulas are. You also need to be able to use all of the different types of formulas.
- What alkanes are.
- How to name straight-chain and branched alkanes with up to 6 carbon atoms in the longest chain.
- How to draw reaction mechanisms using curly arrows.
- What structural isomerism is.
- How to draw the structures of chain, position and functional group isomers.
- What stereoisomerism is, and that it occurs as a result of carbon-carbon double bonds.
- That *E*/*Z*-isomerism is a form of stereoisomerism.
- How to identify *E*-isomers and *Z*-isomers.
- How to apply the Cahn-Ingold-Prelog priority rules to naming isomers.

Exam-style Questions

1 The skeletal formula of a molecule is shown below.

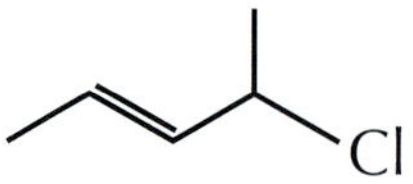

Cl

How many carbon atoms are in this molecule?

A 4

B 5

C 6

D 7

(1 mark)

2 Which of the options below represents a different molecule to the other three?

A

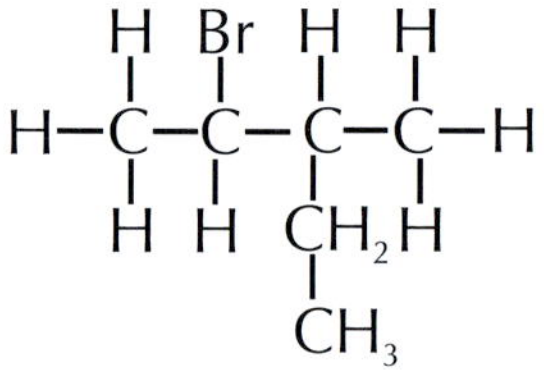

```
   H  Br  H   H
   |  |   |   |
H—C—C—C—C—H
   |  |   |   |
   H  H  CH2  H
          |
         CH3
```

B $CH_3CHBrCH(CH_3)CH_2CH_3$

C 2-bromo-3-methylpentane

D $C_6H_{12}Br$

(1 mark)

3 Which of the skeletal formulas below represents a molecule that has E/Z-isomerism?

A OH

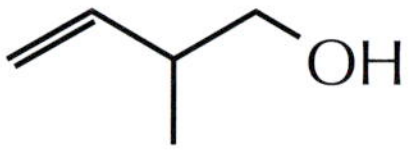

B

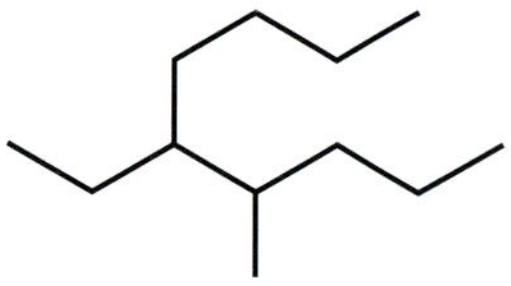

C

D

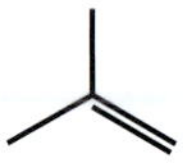

(1 mark)

4 The diagram below shows the displayed formula of molecule **A**.

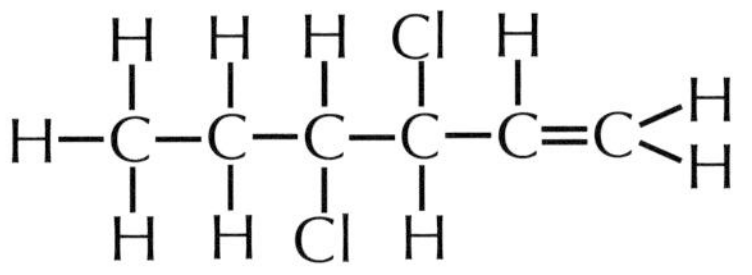

```
     H  H  H  Cl H
     |  |  |  |  |         H
H -  C -C -C -C -C = C <
     |  |  |  |            H
     H  H  Cl H
```

4.1 Name molecule **A**.

(1 mark)

4.2 Write down the molecular formula of molecule **A**.

(1 mark)

4.3 Give the empirical formula of molecule **A**.

(1 mark)

The diagram below shows a structural isomer of molecule **A** — molecule **B**.

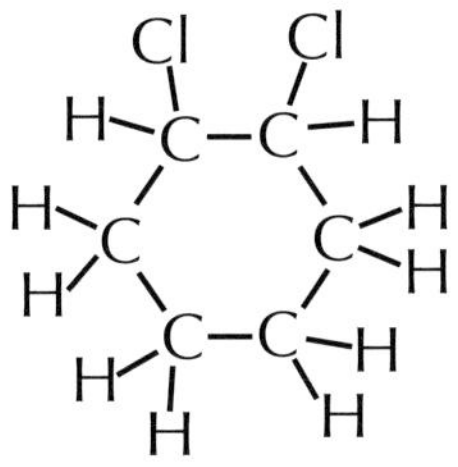

```
        Cl    Cl
        |    /
   H -  C -  C - H
       /      \
H \   C        C - H
H /             \ H
       \       /
   H -  C  -  C - H
        |      \
        H       H
```

4.4 Identify what type of structural isomer molecule **B** is.

(1 mark)

4.5 Draw a position isomer of molecule **B**.

(1 mark)

The diagram below shows another isomer of molecule **A** — molecule **C**.

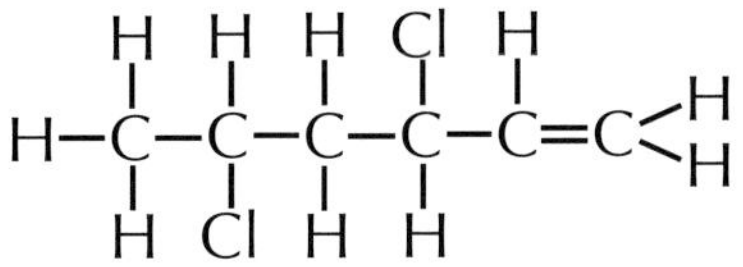

```
     H  H  H  Cl H
     |  |  |  |  |         H
H -  C -C -C -C -C = C <
     |  |  |  |            H
     H  Cl H  H
```

4.6 Identify what type of structural isomer of **A** molecule **C** is.

(1 mark)

4.7 Does molecule **C** show *E*/*Z*-isomerism? Explain your answer.

(2 marks)

5 Heptane is an alkane which is found in crude oil.

5.1 Give the general formula for an alkane.

(1 mark)

5.2 Below is the displayed formula for heptane.

$$\begin{array}{ccccccccccccccccc}
 & & H & & H & & H & & H & & H & & H & & H & & \\
 & & | & & | & & | & & | & & | & & | & & | & & \\
H & - & C & - & C & - & C & - & C & - & C & - & C & - & C & - & H \\
 & & | & & | & & | & & | & & | & & | & & | & & \\
 & & H & & H & & H & & H & & H & & H & & H & &
\end{array}$$

Draw the displayed formula of a chain isomer of heptane.

(1 mark)

Lots of other alkanes can be found in crude oil.
Name the following alkanes.

5.3

$$\begin{array}{ccccccccccccc}
 & & H & & H & & H & & H & & H & & \\
 & & | & & | & & | & & | & & | & & \\
H & - & C & - & C & - & C & - & C & - & C & - & H \\
 & & | & & | & & | & & | & & | & & \\
 & & H & & H & & H & & H & & H & &
\end{array}$$

(1 mark)

5.4

$$\begin{array}{ccccccccccc}
 & & H & & CH_3 & & H & & H & & \\
 & & | & & | & & | & & | & & \\
H & - & C & - & C & - & C & - & C & - & H \\
 & & | & & | & & | & & | & & \\
 & & H & & H & & CH_3 & & H & &
\end{array}$$

(1 mark)

6 A student has been given a sample of the alkene 3-methylpent-2-ene.
The structure of 3-methylpent-2-ene is shown below.

$$\begin{array}{ccccccccccccc}
 & & H & & H & & & & H & & H & & \\
 & & | & & | & & & & | & & | & & \\
H & - & C & - & C & - & C & = & C & - & C & - & H \\
 & & | & & | & & | & & & & | & & \\
 & & H & & H & & CH_3 & & & & H & &
\end{array}$$

3-methylpent-2-ene has a number of different stereoisomers.

6.1 Define the term stereoisomer.

(1 mark)

6.2 Draw two stereoisomers of 3-methylpent-2-ene.

(2 marks)

6.3 Name the two stereoisomers you have drawn in **6.2**.

(2 marks)

1. Alkanes and Petroleum

Petroleum is just a fancy word for crude oil — the sticky black stuff they get out of the ground with oil wells. It's a mixture that is mostly made up of alkanes. They range from small alkanes, like pentane, to massive alkanes with more than 50 carbons.

Learning Objectives:

- Know that alkanes are saturated hydrocarbons.
- Know that petroleum is a mixture consisting mainly of alkane hydrocarbons that can be separated by fractional distillation.
- Be able to explain the economic reasons for cracking alkanes.
- Know that cracking involves breaking C–C bonds in alkanes.
- Know that thermal cracking takes place at high pressure and high temperature and produces a high percentage of alkenes.
- Know that catalytic cracking takes place at a slight pressure, high temperature and in the presence of a zeolite catalyst and is used mainly to produce motor fuels and aromatic hydrocarbons.

Specification Reference 3.3.2.1, 3.3.2.2

Alkanes

You've already met alkanes on page 184, but just to remind you — they're **saturated** hydrocarbons. This means they only contain carbon and hydrogen atoms, and each of their carbon atoms forms four single bonds (the most they can make).

Fractional distillation

Crude oil isn't very useful as it is, but you can separate it out into more useful bits (or fractions) by **fractional distillation**. Here's how fractional distillation works — don't try this at home.

- First, the crude oil is vaporised at about 350 °C.
- The vaporised crude oil goes into the bottom of the fractionating column and rises up through the trays.
- The largest hydrocarbons don't vaporise at all, because their boiling points are too high — they just run to the bottom and form a gooey residue.
- As the crude oil vapour goes up the fractionating column, it gets cooler, creating a temperature gradient.
- Because boiling points of alkanes increase as the molecules get bigger, each fraction condenses at a different temperature. The fractions are drawn off at different levels in the column.
- The hydrocarbons with the lowest boiling points don't condense. They're drawn off as gases at the top of the column.

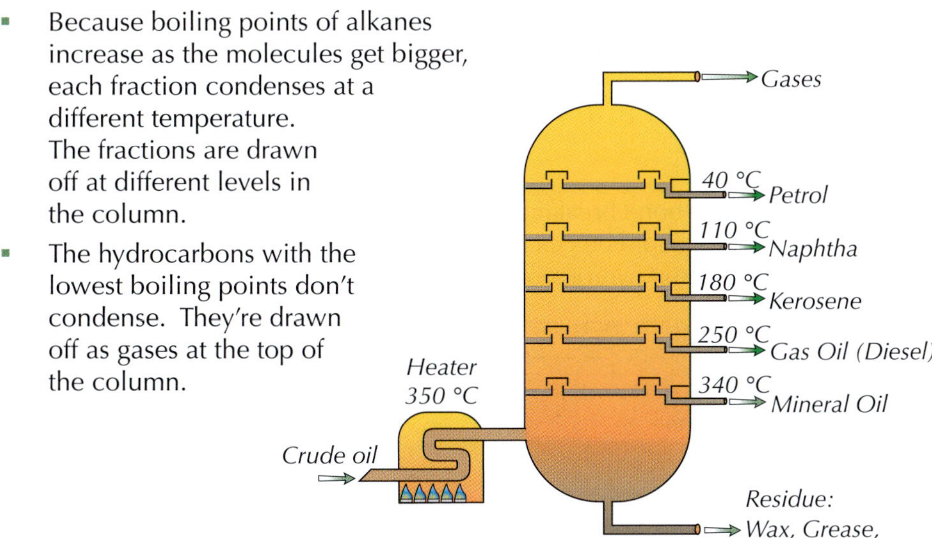

Figure 1: *A fractionating column.*

Tip: You might do fractional distillation in the lab, but if you do you'll use a safer crude oil substitute instead. This could involve using high temperatures, so remember to carry out a risk assessment first.

Uses of crude oil fractions

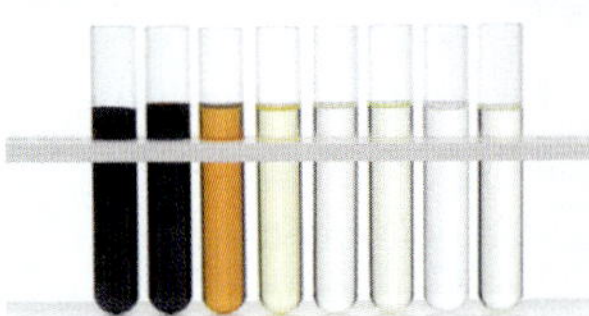

Figure 2: Fractions produced from fractional distillation of crude oil. The fractions are arranged in order of boiling point, with higher boiling points towards the left.

Fraction	Carbon Chain	Uses
Gases	C_1 - C_4	Liquefied Petroleum Gas (LPG), camping gas
Petrol (gasoline)	C_5 - C_{12}	petrol
Naphtha	C_7 - C_{14}	processed to make petrochemicals
Kerosene (paraffin)	C_{11} - C_{15}	jet fuel, petrochemicals, central heating fuel
Gas oil (diesel)	C_{15} - C_{19}	diesel fuel, central heating fuel
Mineral Oil (lubricating)	C_{20} - C_{30}	lubricating oil
Fuel Oil	C_{30} - C_{40}	ships, power stations,
Wax, grease	C_{40} - C_{50}	candles, lubrication
Bitumen	C_{50+}	roofing, road surfacing

Figure 3: A table showing the products of fractional distillation of crude oil and their uses.

Cracking hydrocarbons

Most people want loads of light fractions, like petrol and naphtha. They don't want so much of the heavier stuff like bitumen though. Stuff that's in high demand is much more valuable than the stuff that isn't.

To meet this demand, the less popular heavier fractions are cracked. **Cracking** is breaking long-chain alkanes into smaller hydrocarbons (which can include alkenes). It involves breaking the C–C bonds.

Exam Tip
In the exam you might have to write equations for cracking. The marks are pretty simple to pick up, as long as you check that you have the same number of carbons and hydrogens on each side of the equation.

Example

Decane could be cracked into smaller hydrocarbons like this:

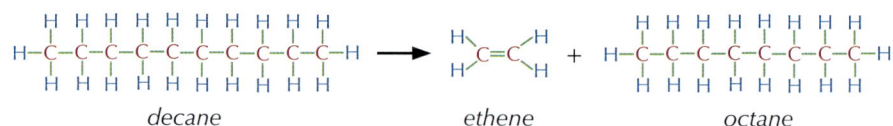

decane → *ethene* + *octane*

But, because the bond breaking in cracking is random, this isn't the only way that decane could be cracked — it could be cracked to produce different short chain hydrocarbons. For example...

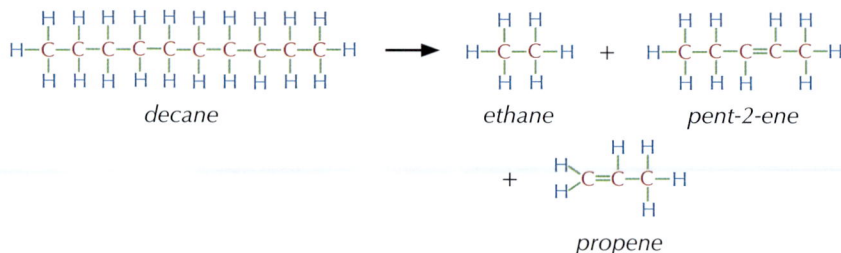

decane → *ethane* + *pent-2-ene* + *propene*

Types of cracking

There are two types of cracking you need to know about — thermal cracking and catalytic cracking.

Thermal cracking

It takes place at high temperature (up to 1000 °C) and high pressure (up to 70 atm). It produces a lot of alkenes. These alkenes are used to make heaps of valuable products, like polymers. A good example is poly(ethene), which is made from ethene (have a look at pages 229-231 for more on polymers).

Catalytic cracking

Catalytic cracking uses something called a zeolite catalyst (hydrated aluminosilicate), at a slight pressure and high temperature (about 500 °C). This mostly produces aromatic hydrocarbons and the alkanes needed to produce motor fuels.

Using a catalyst cuts costs, because the reaction can be done at a low pressure and a lower temperature. The catalyst also speeds up the rate of reaction, saving time (and time is money).

Tip: Aromatic compounds contain substituted benzene rings. Benzene rings look like this:

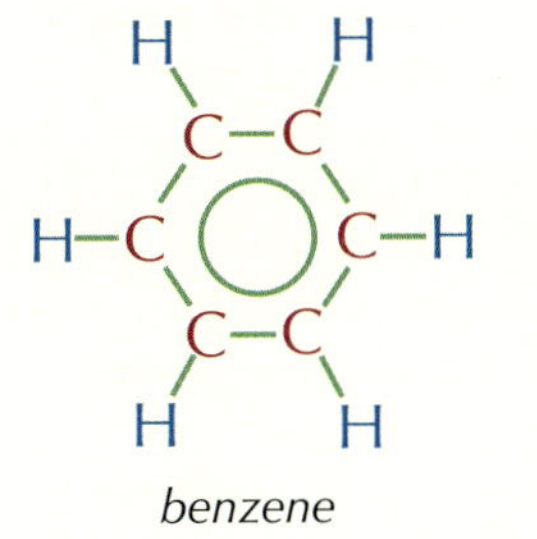

benzene

Practice Questions — Fact Recall

Q1 What is petroleum?

Q2 Why are alkanes described as saturated hydrocarbons?

Q3 Fractional distillation separates hydrocarbons. What property are they separated by?

Q4 Explain why some fractions are drawn off higher up the fractional distillation column than others.

Q5 What is cracking?

Q6 Why do we crack heavier petroleum fractions?

Q7 Why does using a catalyst for catalytic cracking cut costs?

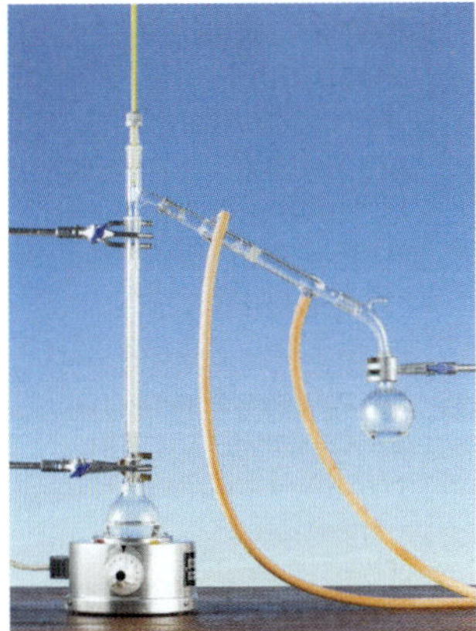

Figure 4: *Laboratory fractional distillation apparatus.*

2. Alkanes as Fuels

Alkanes are found in fossil fuels. Alkanes make great fuels — burning just a small amount of methane releases a humongous amount of energy. They're burnt in power stations, central heating systems and, of course, to power car engines. Unfortunately, burning them can release pollutants.

Learning Objectives:

- Know that alkanes are used as fuels.
- Know that the combustion of alkanes and other organic compounds can be complete or incomplete.
- Know that the internal combustion engine produces a number of pollutants including NO_x, CO, carbon and unburned hydrocarbons.
- Know that gaseous pollutants from internal combustion engines can be removed using catalytic converters.
- Know that the combustion of hydrocarbons containing sulfur leads to sulfur dioxide that causes air pollution.
- Be able to explain why sulfur dioxide can be removed from flue gases using calcium oxide or calcium carbonate.

Specification Reference 3.3.2.3

Combustion

Complete combustion

If you burn (oxidise) alkanes (and other hydrocarbons) with plenty of oxygen, you get only carbon dioxide and water. This is **complete combustion.**

Example

If you burn a molecule of propane in plenty of oxygen, you get three molecules of carbon dioxide and four water molecules.
The equation for this is shown below.

$$C_3H_{8(g)} + 5O_{2(g)} \rightarrow 3CO_{2(g)} + 4H_2O_{(g)}$$

Incomplete combustion

If there's not enough oxygen around when you burn a hydrocarbon, you get **incomplete combustion** happening instead. This produces particulate carbon (soot) and carbon monoxide gas instead of, or as well as, carbon dioxide.

Example

If you burn propane in a limited supply of oxygen you'll produce carbon monoxide and carbon as well as carbon dioxide and water.

$$C_3H_{8(g)} + 3\tfrac{1}{2}O_{2(g)} \rightarrow 3CO_{(g)} + 4H_2O_{(g)}$$

This is bad news because carbon monoxide gas is poisonous. Carbon monoxide molecules bind to the same sites on haemoglobin molecules in red blood cells as oxygen molecules. So oxygen can't be carried around the body. Luckily, carbon monoxide can be removed from exhaust gases by catalytic converters on cars.

Soot is also thought to cause breathing problems, and it can build up in engines, meaning they don't work properly.

Tip: Not all incomplete combustion reactions will produce carbon dioxide, carbon monoxide, particulate carbon and water. Some, for example, will only produce water and carbon monoxide. Just make sure your equation balances and you should be fine.

Pollution from burning fuels

Unburnt hydrocarbons and oxides of nitrogen

Nitrogen oxides are a series of toxic and poisonous molecules which have the general formula NO_x. Nitrogen monoxide is produced when the high pressure and temperature in a car engine cause the nitrogen and oxygen atoms from the air to react together. Nitrogen monoxide can react further to produce nitrogen dioxide — the equations for these reactions are shown below.

$$N_{2(g)} + O_{2(g)} \rightarrow 2NO_{(g)}$$

$$2NO_{(g)} + O_{2(g)} \rightarrow 2NO_{2(g)}$$

Engines don't burn all the fuel molecules. Some of these come out as unburnt hydrocarbons. These hydrocarbons react with nitrogen oxides in the presence of sunlight to form ground-level ozone (O_3), which is a major component of smog. Ground-level ozone irritates people's eyes, aggravates respiratory problems and even causes lung damage (ozone isn't nice stuff, unless it is high up in the atmosphere as part of the ozone layer).

The three main pollutants from vehicle exhausts are nitrogen oxides, unburnt hydrocarbons and carbon monoxide. Catalytic converters on cars remove these pollutants from the exhaust by the following reactions:

$$C_3H_{8(g)} + 5O_{2(g)} \rightarrow 3CO_{2(g)} + 4H_2O_{(g)}$$

This equation is just an equation for the complete combustion of a hydrocarbon.

$$2NO_{(g)} \rightarrow N_{2(g)} + O_{2(g)}$$

$$2NO_{(g)} + 2CO_{(g)} \rightarrow N_{2(g)} + 2CO_{2(g)}$$

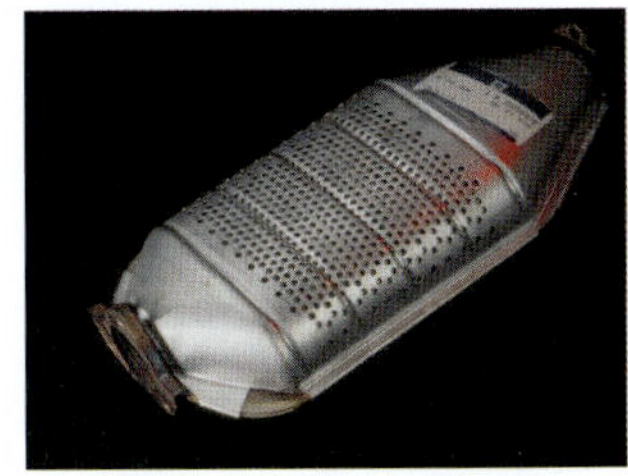

Figure 1: *A catalytic converter.*

Tip: Catalytic converters are designed to have very large surface areas to make sure that the harmful pollutants have the best chance of being turned into less harmful chemicals.

Sulfur dioxide

Some fossil fuels contain sulfur. When they are burnt in, for example, car engines and power stations, the sulfur reacts to form sulfur dioxide gas (SO_2). If sulfur dioxide gets into the atmosphere, it dissolves in the moisture and is converted into sulfuric acid. This is what causes acid rain. The same process occurs when nitrogen dioxide escapes into the atmosphere — nitric acid is produced. Acid rain destroys trees and vegetation, as well as corroding buildings and statues and killing fish in lakes.

Fortunately, sulfur dioxide can be removed from power station flue gases before it gets into the atmosphere — you've already come across this process on page 164. Powdered calcium carbonate (limestone) or calcium oxide is mixed with water to make an alkaline slurry. When the flue gases mix with the alkaline slurry, the acidic sulfur dioxide gas reacts with the calcium compounds to form a harmless salt (calcium sulfite).

$$CaO_{(s)} + SO_{2(g)} \rightarrow CaSO_{3(s)}$$

Figure 2: *Trees killed by acid rain.*

Global warming

Burning fossil fuels produces carbon dioxide. Carbon dioxide is a greenhouse gas. Greenhouse gases in our atmosphere are very good at absorbing infrared energy (heat). They emit some of the energy they absorb back towards the Earth, keeping it warm. This is called the greenhouse effect. Most scientists agree that by increasing the amount of carbon dioxide in our atmosphere, we are making the Earth warmer. This process is known as global warming.

Tip: There's more about global warming and the greenhouse effect on page 258.

Practice Questions — Application

Q1 Write the equation for the complete combustion of pentane (C_5H_{12}).

Q2 Write the equation for the incomplete combustion of pentane (C_5H_{12}) to produce carbon monoxide and water only.

Exam Tip
You could be asked to write equations for the complete or incomplete combustion of other alkanes. Make sure you get lots of practice writing out different equations.

Practice Questions — Fact Recall

Q1 Give three pollutants produced by vehicle exhausts.

Q2 What can be used to remove pollutants from vehicle exhausts?

Q3 Explain how acid rain is caused by burning fossil fuels containing sulfur.

Q4 Describe one method for removing sulfur dioxide from flue gases.

Learning Objectives:

- Know that the unpaired electron in a free radical is represented by a dot.
- Know the reaction of methane with chlorine and be able to explain this reaction as a free radical substitution mechanism involving initiation, propagation and termination steps.
- Be able to write balanced equations for the steps in a free radical mechanism.
- Know that ozone is beneficial because it absorbs UV radiation.
- Know that chlorine atoms are formed in the upper atmosphere when UV radiation causes CCl bonds in chlorofluorocarbons (CFCs) to break.
- Understand that chlorine atoms catalyse the decomposition of ozone and contribute to the hole in the ozone layer.
- Be able to use equations to explain how chlorine atoms catalyse decomposition of ozone.
- Appreciate that results of research by different groups in the scientific community provided evidence for legislation to ban the use of CFCs as solvents and refrigerants.
- Know that chemists have now developed chlorine-free alternatives to CFCs.

Specification Reference 3.3.1.2, 3.3.2.4, 3.3.3.3

3. Synthesis of Chloroalkanes

Chloroalkanes are alkanes with one or more hydrogen atoms substituted by a chlorine atom. They are pretty important to chemists, so it's important to understand how they're made. That's where the synthesis part comes in — a synthesis is just a step-wise method detailing how to create a chemical.

Photochemical reactions

Halogens react with alkanes in photochemical reactions to form halogenoalkanes (see page 211 for more on halogenoalkanes).

Photochemical reactions are reactions started by ultraviolet (UV) light. A hydrogen atom is substituted (replaced) by chlorine or bromine. This is a **free radical** substitution reaction. A free radical is a particle with an unpaired electron. Free radicals form when a covalent bond splits equally, giving one electron to each species. The unpaired electron makes them very reactive. You can show something's a free radical in a mechanism by putting a dot next to it, like this: Cl• or $\bullet CH_3$. The dot represents the unpaired electron.

Synthesis of chloromethane

A mixture of methane and chlorine will not react on its own but when exposed to UV light it reacts with a bit of a bang to form chloromethane. The overall equation for this reaction is shown below.

$$CH_4 + Cl_2 \xrightarrow{UV} CH_3Cl + HCl$$

A reaction mechanism shows each step in the synthesis of a chemical. The reaction mechanism for the synthesis of chloromethane by a photochemical reaction has three stages — initiation, propagation and termination.

Initiation

In the initiation step, free radicals are produced. Sunlight provides enough energy to break some of the Cl–Cl bonds — this is photodissociation.

$$Cl_2 \xrightarrow{UV} 2Cl\bullet$$

The bond splits equally and each atom gets to keep one electron. The atom becomes a highly reactive free radical, Cl•, because of its unpaired electron.

Propagation

During propagation, free radicals are used up and created in a chain reaction. First, Cl• attacks a methane molecule:

$$Cl\bullet + CH_4 \rightarrow \bullet CH_3 + HCl$$

The new methyl free radical, $CH_3\bullet$, can then attack another Cl_2 molecule:

$$\bullet CH_3 + Cl_2 \rightarrow CH_3Cl + Cl\bullet$$

The new Cl• can attack another CH_4 molecule, and so on, until all the Cl_2 or CH_4 molecules are used up.

Substitutions

If the chlorine's in excess, the hydrogen atoms on methane will eventually be replaced by chlorine atoms. This means you'll get dichloromethane CH_2Cl_2, trichloromethane $CHCl_3$, and tetrachloromethane CCl_4.

$$CH_4 + Cl_2 \rightarrow CH_3Cl + HCl$$

$$CH_3Cl + Cl_2 \rightarrow CH_2Cl_2 + HCl$$

$$CH_2Cl_2 + Cl_2 \rightarrow CHCl_3 + HCl$$

$$CHCl_3 + Cl_2 \rightarrow CCl_4 + HCl$$

But if the methane's in excess, then the chlorine will be used up quickly and the product will mostly be chloromethane.

$$CH_4 + Cl_2 \rightarrow CH_3Cl + HCl$$

Termination

In the termination step, free radicals are mopped up. If two free radicals join together, they make a stable molecule — the two unpaired electrons form a covalent bond. This terminates the chain reaction. Here are three possible termination reactions from the synthesis of chloromethane:

$$\bullet CH_3 + Cl\bullet \rightarrow CH_3Cl$$

$$\bullet CH_3 + CH_3\bullet \rightarrow C_2H_6$$

$$Cl\bullet + Cl\bullet \rightarrow Cl_2$$

Tip: Some of the products formed in the termination step will be trace impurities in the final sample.

Exam Tip
When you're writing radical equations you need to make sure that there's the same number of radicals on each side of the equation or that two radicals are combining to create a non-radical.

Chlorofluorocarbons

Chlorofluorocarbons (CFCs) are halogenoalkane molecules where all of the hydrogen atoms have been replaced by chlorine and fluorine atoms.

Examples

```
     Cl                F
     |                 |
Cl — C — Cl        F — C — F
     |                 |
     F                 Cl
```

trichlorofluoromethane   *chlorotrifluoromethane*

Tip: There's more on halogenoalkanes later on (see page 211).

Chlorofluorocarbons and the ozone layer

Ozone (O_3) in the upper atmosphere acts as a chemical sunscreen. It absorbs a lot of ultraviolet radiation from the Sun, stopping it from reaching us. Ultraviolet radiation can cause sunburn or even skin cancer. Ozone's formed naturally when an oxygen molecule is broken down into two free radicals by ultraviolet radiation:

$$O_2 + h\nu \rightarrow O\bullet + O\bullet$$

The free radicals attack other oxygen molecules forming ozone:

$$O_2 + O\bullet \rightarrow O_3$$

Tip: The $h\nu$ in the first equation is the notation for a quantum of UV radiation — a photon. It just means that the reaction needs to be initiated by electromagnetic radiation.

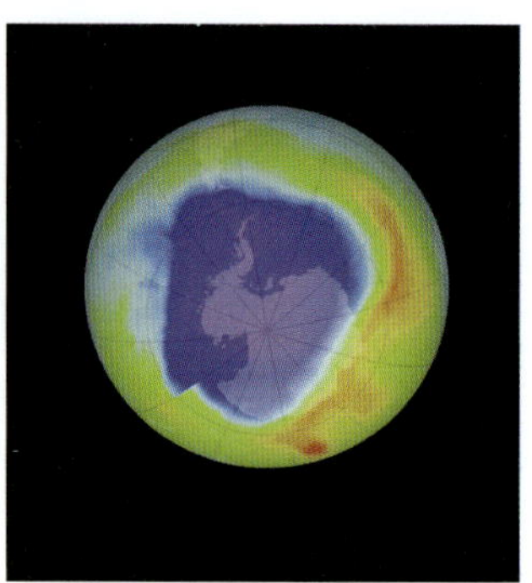

Figure 1: *Satellite image of the ozone layer over Antarctica in 2013. The 'hole' is shown by the blue area.*

You've probably heard of how the ozone layer's being destroyed by CFCs, right. Well, here's what's happening. Chlorine free radicals, Cl•, are formed in the upper atmosphere when the C–Cl bonds in CFCs are broken down by ultraviolet radiation, like this:

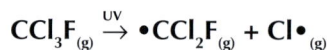

$$\mathbf{CCl_3F_{(g)} \xrightarrow{UV} \bullet CCl_2F_{(g)} + Cl\bullet_{(g)}}$$

These free radicals are **catalysts**. They react with ozone to form an **intermediate** (ClO•), and an oxygen molecule.

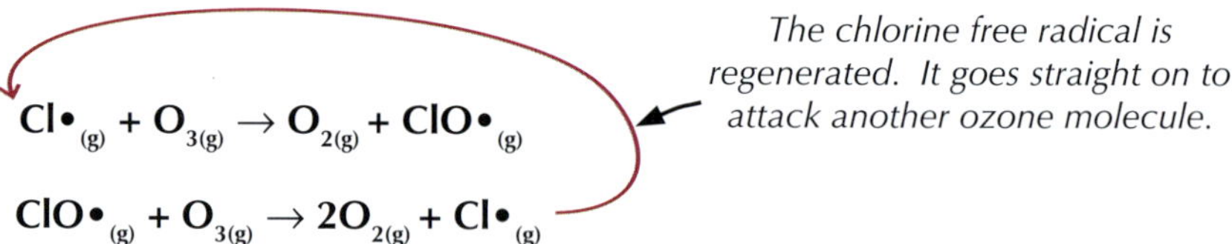

The chlorine free radical is regenerated. It goes straight on to attack another ozone molecule.

$$\mathbf{Cl\bullet_{(g)} + O_{3(g)} \rightarrow O_{2(g)} + ClO\bullet_{(g)}}$$

$$\mathbf{ClO\bullet_{(g)} + O_{3(g)} \rightarrow 2O_{2(g)} + Cl\bullet_{(g)}}$$

Because the Cl• free radical is regenerated, it only takes one little chlorine free radical to destroy loads of ozone molecules. So, the overall reaction is...

$$\mathbf{2O_{3(g)} \rightarrow 3O_{2(g)}}$$

... and Cl• is the catalyst.

Environmental problems of CFCs

CFCs are pretty unreactive, non-flammable and non-toxic. They used to be used in fire extinguishers, as propellants in aerosols and as the coolant gas in fridges. They were also added to foam plastics to make insulation and packaging materials.

In the 1970s, research by several different scientific groups demonstrated that CFCs were causing damage to the ozone layer. The advantages of CFCs couldn't outweigh the environmental problems they were causing, so they were banned.

Chemists have developed safer alternatives to CFCs which contain no chlorine. HCFCs (hydrochlorofluorocarbons) and HFCs (hydrofluorocarbons) are less dangerous than CFCs, so they're being used as temporary alternatives until safer products are developed. Most aerosols now have been replaced by pump spray systems, or use nitrogen as the propellant. Many industrial fridges use ammonia or hydrocarbons as the coolant gas and carbon dioxide is used to make foamed polymers.

Practice Questions — Fact Recall

Q1 What is a photochemical reaction?

Q2 Write an equation for the initiation step of the synthesis of chloromethane by a photochemical reaction.

Q3 Describe what takes place in the termination step of a free radical substitution mechanism.

Q4 Why is ozone in the upper atmosphere beneficial?

Q5 Describe how chlorine free radicals are formed in the upper atmosphere.

Q6 Write two equations to show how chlorine atoms catalyse the decomposition of ozone.

Q7 Explain why CFCs were banned from use as solvents and refrigerants.

4. Halogenoalkanes

Halogenoalkanes pop up a lot in chemistry so it's important that you know exactly what they are and how they react.

Learning Objective:

- Explain that halogenoalkanes contain polar bonds.

Specification Reference 3.3.3.1

What are halogenoalkanes?

A **halogenoalkane** is an alkane with at least one halogen atom in place of a hydrogen atom.

Examples

```
    Cl              H  H  H               Cl H
    |               |  |  |               |  |
H—C—Cl          H—C—C—C—H           Cl—C—C—Br
    |               |  |  |               |  |
    H               H  I  H               H  H
```

dichloromethane *2-iodopropane* *2-bromo-1,1-dichloroethane*

Tip: There's more about how to name halogenoalkanes on page 184.

Polarity of halogenoalkanes

Halogens are generally much more electronegative than carbon. So, most carbon-halogen bonds are **polar**.

Example

$-\overset{|}{\underset{|}{C}}^{\delta+}-Br^{\delta-}$

The bromine atom is more electronegative than the carbon atom and so withdraws electron density from the carbon atom. This leaves the carbon atom with a partial positive charge and the bromine atom with a partial negative charge.

Tip: Don't worry if you see halogenoalkanes called haloalkanes. It's a government conspiracy to confuse you.

The δ+ carbon doesn't have enough electrons. This means it can be attacked by a **nucleophile**. A nucleophile's an electron-pair donor. It donates an electron pair to somewhere without enough electrons.

Examples

Here are some examples of nucleophiles that will react with halogenoalkanes.

$:\bar{C}N$ — *cyanide ion*

$:NH_3$ — *ammonia*

$:\bar{O}H$ — *hydroxide ion*

The pairs of dots represent lone pairs of electrons.

Tip: Nucleophiles are often negative ions, and because they form by gaining electrons, they have extra electrons that they can donate. They don't have to be ions though. For example, NH_3 is a nucleophile — it's got a non-bonding pair of electrons that it can donate.

There are several examples of reactions where nucleophiles react with halogenoalkanes coming up on the next few pages.

Practice Questions — Fact Recall

Q1 What is a halogenoalkane?

Q2 Explain why halogen-carbon bonds are polar.

Q3 What is a nucleophile?

Q4 Give two examples of nucleophiles that will react with halogenoalkanes.

Q5 What does a pair of dots represent on a nucleophile?

Learning Objectives:

- Understand that halogenoalkanes undergo nucleophilic substitution reactions with the nucleophiles OH^-, CN^- and NH_3.
- Be able to outline the nucleophilic substitution mechanisms of the reactions of halogenalkanes with the nucleophiles OH^-, CN^- and NH_3.
- Be able to explain why the carbon–halogen bond enthalpy influences the rate of reaction.

Specification Reference 3.3.3.1

5. Nucleophilic Substitution

Nucleophilic substitution is a reaction where one functional group is substituted for another. You need to know about the nucleophilic substitution reactions of halogenoalkane molecules.

Nucleophilic substitution reactions

As you saw on page 192, **mechanisms** are diagrams that show how a reaction works. They show how the bonds in molecules are made and broken, how the electrons are transferred and how you get from the reactants to the products.

You need to know the mechanism for the **nucleophilic substitution** of halogenoalkanes. In a nucleophilic substitution reaction, a nucleophile attacks a polar molecule, kicks out a functional group and settles itself down in its place. The general equation for the nucleophilic substitution of a halogenoalkane is:

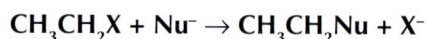

$$CH_3CH_2X + Nu^- \rightarrow CH_3CH_2Nu + X^-$$

And here's how it all works:

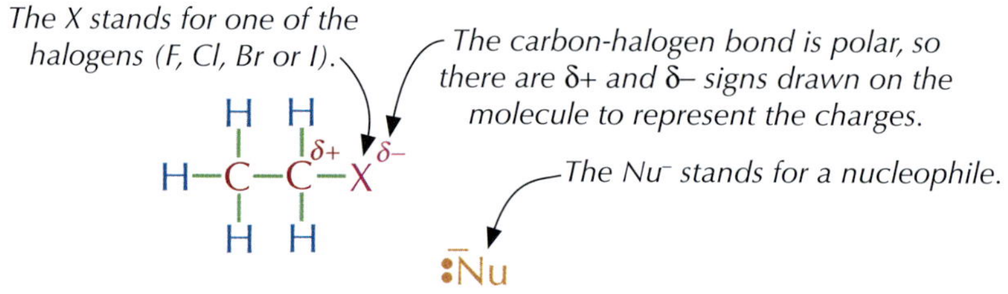

The X stands for one of the halogens (F, Cl, Br or I).

The carbon-halogen bond is polar, so there are δ+ and δ– signs drawn on the molecule to represent the charges.

The Nu^- stands for a nucleophile.

The lone pair of electrons on the nucleophile attacks the slightly positive charge on the carbon — this is shown by a black curly arrow. In mechanisms, curly arrows always show the movement of an electron pair. The lone pair of electrons creates a new bond between the nucleophile and the carbon.

Exam Tip
You must draw the curly arrows coming from the lone pair of electrons or from the bond. If you don't, you won't get the marks for the mechanism in the exam.

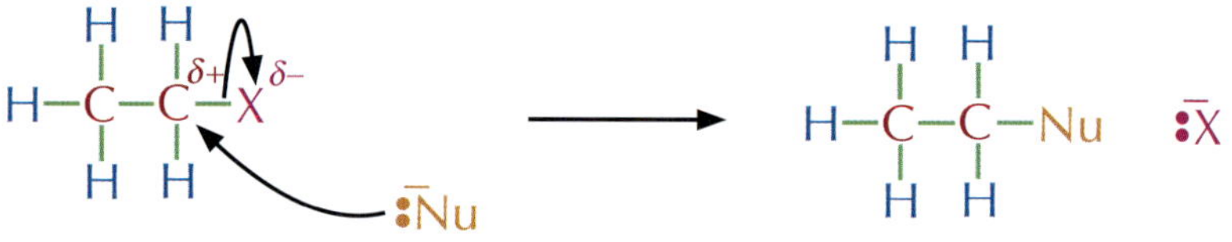

The carbon can only be bonded to four other atoms so the addition of the nucleophile breaks the bond between the carbon and the halogen — this is shown by another curly arrow. The pair of electrons from the carbon-halogen bond are taken by the halogen and become a lone pair.

Tip: This reaction is sometimes called hydrolysis, because the same reaction will happen (albeit very slowly) with water. Hydrolysis means splitting a molecule apart by reacting it with water.

Reaction of halogenoalkanes with OH^-

Halogenoalkanes will react with hydroxides to produce alcohols.

Example

Bromoethane can be changed to ethanol in a nucleophilic substitution reaction. You have to use warm aqueous sodium or potassium hydroxide or it won't work.

In the equation for the reaction, R represents an alkyl group. X stands for one of the halogens (F, Cl, Br or I). As it's a nucleophilic substitution reaction, the nucleophile (^-OH) kicks out the halogen (X) from the R–X molecule and takes its place. So the overall reaction can just be written as:

$$\textbf{R–X} + {}^-\textbf{OH} \rightarrow \textbf{ROH} + \textbf{X}^-$$

Here's how it happens:

1. The C–Br bond is polar. The $C^{\delta+}$ attracts a lone pair of electrons from the OH^- ion.
3. A new bond forms between the C and the OH^- ion, making an alcohol...
4. ... and the C–Br bond breaks. Both the electrons from the bond are taken by the Br.
H H
H–C–C–Br
H H
δ+
δ–
:OH
H H
H–C–C–OH
H H
:Br
2. The OH^- ion acts as a nucleophile, attacking the slightly positive carbon atom.

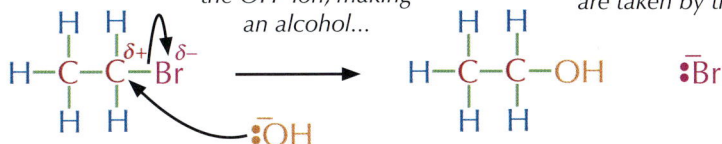

Tip: When you're drawing mechanisms make sure the charges balance. That way you'll know if you've managed to lose or gain electrons along the way. In this example, the left hand side of the equation has one negative charge on the hydroxide nucleophile and the right hand side has one negative charge from a bromide ion — so it's balanced.

Making nitriles from halogenoalkanes

Nitriles have CN groups. The carbon atom and nitrogen atom are held together with a triple bond.

Examples

Nitriles are derived from hydrogen cyanide:

H–C≡N

hydrogen cyanide

Here is ethanenitrile:

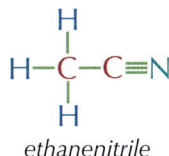

ethanenitrile

Tip: Hydrogen cyanide could also be called methanenitrile. Hydrogen cyanide is the molecule's old name and methane nitrile is its name assigned by the IUPAC naming conventions. Either name is correct.

If you warm a halogenoalkane with ethanolic potassium cyanide (that's just potassium cyanide dissolved in ethanol), you get a nitrile. It's yet another nucleophilic substitution reaction — the cyanide ion, CN^-, is the nucleophile.

Example

Reacting bromoethane with potassium cyanide under reflux will produce propanenitrile and potassium bromide.

$$\mathbf{CH_3CH_2Br + KCN \rightarrow CH_3CH_2CN + KBr}$$

The ethanolic potassium cyanide dissociates to form a K^+ ion and a CN^- ion. It's the CN^- ion that acts as the nucleophile in the reaction.

$$\mathbf{KCN \rightarrow K^+ + CN^-}$$

The reaction mechanism follows the same pattern as above — the lone pair of electrons on the CN^- ion attacks the δ+ carbon, the C–Br bond breaks and the bromine leaves.

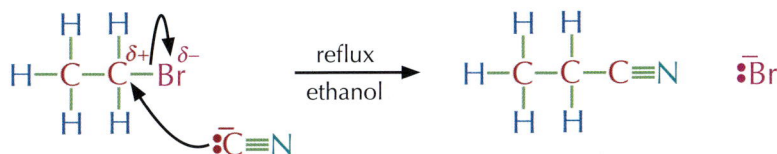

Tip: When you heat a liquid mixture, eventually it will start to boil and some of the mixture will be lost as vapour. When you heat under reflux, you use equipment that stops the vapour escaping from the reaction mixture. There's more about heating under reflux on page 244.

You have to use ethanol as a solvent here instead of water. If you used water, it could act as a competing nucleophile and you'd get some alcohol product.

Making amines from halogenoalkanes

An amine has the structure R_3N. The R groups can be hydrogens or another group. In amines, the nitrogen always has a lone pair (shown as a pair of dots next to the nitrogen atom).

Tip: Amines are derivatives of the ammonia molecule, shown below:

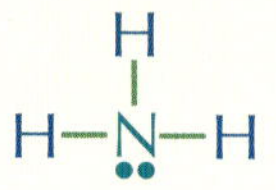

Examples

The molecules below are both amines.

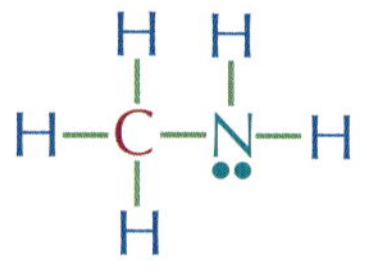

methylamine 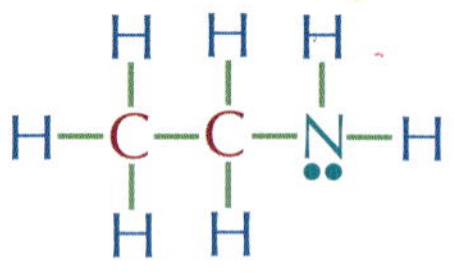　　*ethylamine*

If you warm a halogenoalkane with excess ethanolic ammonia (ammonia dissolved in ethanol) in a sealed tube, the ammonia swaps places with the halogen to form an amine — yes, it's another one of those nucleophilic substitution reactions.

Exam Tip

When you're drawing amines in the exam, it's a good idea to draw the lone pair of electrons in so that you don't forget that they're there. And if the amine's part of a mechanism you'll have to draw them in to get all the marks.

Example

In this reaction bromoethane is reacting with ammonia to form ethylamine.

- The first step is the same as in the mechanism on the previous page, except this time the nucleophile is NH_3. The nitrogen atom donates its lone pair of electrons to the carbon atom to create a bond.
- The nitrogen atom was neutral to begin with, so this means the nitrogen is left with a positive charge.

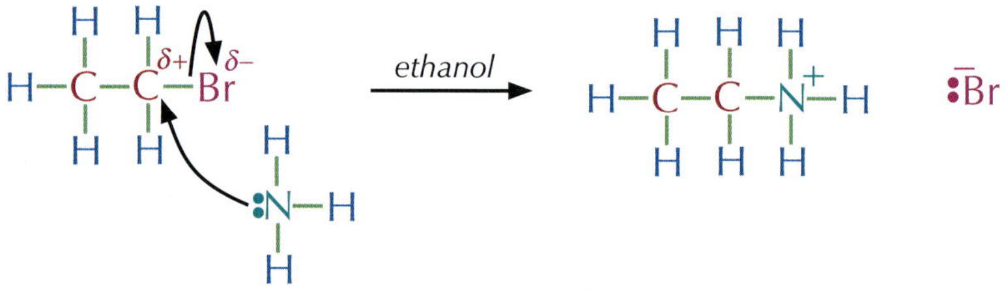

$\delta+$　$\delta-$　*ethanol*　$:\bar{B}r$

- In the second step, a second ammonia molecule removes a hydrogen from the NH_3 group to form an ammonium ion (NH_4^+) and an amine.
- The ammonia molecule donates its lone pair of electrons to the hydrogen to form a bond, so the nitrogen atom in the ammonium ion now has a positive charge, and the amine has no charge.

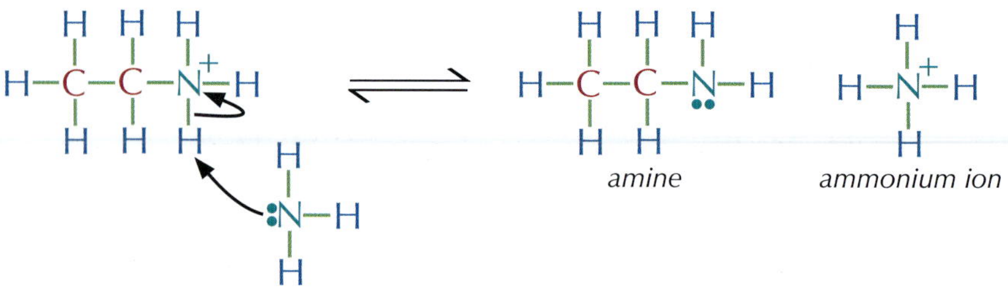

Exam Tip

This example shows ammonia reacting with bromoethane, but in the exam you could be given a question that involves a different halogenoalkane. Don't panic though — the mechanism is exactly the same.

amine　*ammonium ion*

The ammonium ion formed can react with the bromide ion to form ammonium bromide. Ammonium bromide is held together by an ionic bond (see page 77).

$$\left[\begin{array}{c} H \\ | \\ H-N-H \\ | \\ H \end{array} \right]^+ \; :\bar{Br}$$

So the overall reaction is:

$$H-\underset{H}{\overset{H}{C}}-\underset{H}{\overset{H}{C}}-Br \; + \; 2\,:NH_3 \xrightarrow{ethanol} H-\underset{H}{\overset{H}{C}}-\underset{H}{\overset{H}{C}}-\underset{\cdot\cdot}{\overset{H}{N}}-H \; + \; NH_4Br$$

The amine group in the product still has a lone pair of electrons. This means that it can also act as a nucleophile — so it may react with halogenoalkane molecules itself, giving a mixture of products.

Exam Tip
You could be asked for the mechanism for this reaction, or for the overall reaction equation, so make sure that you know them both.

Reactivity of halogenoalkanes

The carbon-halogen bond strength (or enthalpy) decides reactivity. For a reaction to occur the carbon-halogen bond needs to break. The C–F bond is the strongest — it has the highest bond enthalpy. So fluoroalkanes undergo nucleophilic substitution reactions more slowly than other halogenoalkanes. The C–I bond has the lowest bond enthalpy, so it's easier to break. This means that iodoalkanes are substituted more quickly.

bond	bond enthalpy kJ mol^{-1}
C–F	467
C–Cl	346
C–Br	290
C–I	228

Faster substitution as bond enthalpy decreases (the bonds are getting weaker).

Figure 1: *Carbon-halogen bond enthalpies.*

Tip: If you've got a molecule with more than one halogen in it, the halogen with the lowest bond enthalpy will get replaced first.

Summary of nucleophilic substitution

- Nucleophilic substitution reactions can occur between a halogenoalkane and a nucleophile.
- The nucleophile attacks the δ+ carbon atom, which breaks the carbon-halogen bond.
- One new bond is formed (between the nucleophile and the δ+ carbon atom) and one bond is broken (the carbon-halogen bond).
- When you're drawing mechanisms for nucleophilic substitution reactions, it's important to draw the curly arrows coming from the electrons and going to an atom. The electrons can come from either a bond or from a lone pair on an atom or ion.

Exam Tip
In the exam you could be asked to draw a mechanism for a nucleophilic substitution reaction you haven't seen before. You should practise drawing mechanisms for lots of different reactions before the exam so that you get into the swing of it.

- Make sure the charges are balanced at every stage of a mechanism — if you start with a negative charge you should end up with one too.
- And finally, it doesn't matter which nucleophile you use (^-CN, ^-OH or NH_3), the mechanism for nucleophilic substitution of a halogenoalkane is always the same.

Practice Questions — Application

Q1 Draw the mechanism for the reaction of 1-chlorobutane with warm ethanolic potassium cyanide. The molecule 1-chlorobutane is shown below.

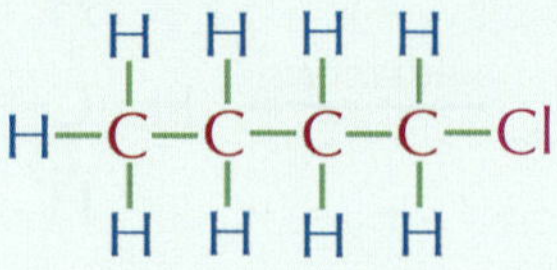

```
   H   H   H   H
   |   |   |   |
H— C — C — C — C —Cl
   |   |   |   |
   H   H   H   H
```

Q2 Which of the following reactions would be quickest? Explain your answer.

A: $CH_3CH_2Cl + H_2O \rightarrow CH_3CH_2OH + HCl$

B: $CH_3CH_2Br + H_2O \rightarrow CH_3CH_2OH + HBr$

C: $CH_3CH_2I + H_2O \rightarrow CH_3CH_2OH + HI$

Q3 Draw the mechanism for the reaction of iodopropane with ammonia. The reaction is done in a sealed tube in warm ethanol. The reactants are shown below.

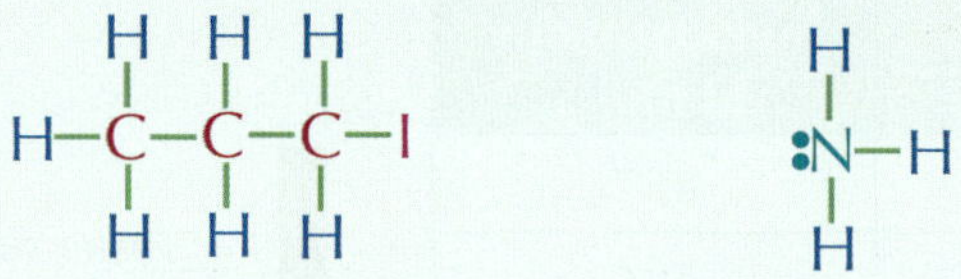

```
   H   H   H             H
   |   |   |             |
H— C — C — C —I        :N—H
   |   |   |             |
   H   H   H             H
```

Q4 Draw the mechanism for the hydrolysis of chloroethane by warm aqueous sodium hydroxide.

Practice Questions — Fact Recall

Q1 What happens in a nucleophilic substitution reaction?

Q2 What chemical would you react with bromoethane to get ethanol?

Q3 Name the nucleophile that is present when bromoethane reacts with ethanolic potassium cyanide under reflux.

Q4 Under what reaction conditions do you react bromoethane with ammonia to form ethylamine?

Q5 Explain why fluoroalkanes are substituted more slowly than other halogenoalkanes.

6. Elimination Reactions

In an elimination reaction, a small group of atoms breaks away from a larger molecule. This small group is not replaced by anything else (whereas it would be in a substitution reaction).

Learning Objectives:

- Understand concurrent substitution and elimination reactions of a halogenoalkane and be able to explain the role of the reagent as both nucleophile and base.
- Be able to outline the mechanisms of concurrent substitution and elimination reactions.

Specification Reference 3.3.3.2

Halogen elimination from a halogenoalkane

If you warm a halogenoalkane with hydroxide ions dissolved in ethanol instead of water, an elimination reaction happens and you end up with an alkene.

Example

Reacting 2-bromopropane with potassium hydroxide dissolved in warm ethanol under **reflux** produces an elimination reaction and forms propene, water and potassium bromide. Here's the equation for this reaction.

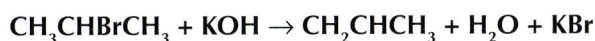

$$CH_3CHBrCH_3 + KOH \rightarrow CH_2CHCH_3 + H_2O + KBr$$

In the reaction, H and Br are eliminated from neighbouring carbon atoms in $CH_3CHBrCH_3$ to leave CH_2CHCH_3. Here's how the reaction works:

1. OH^- acts as a base and takes a proton, H^+, from the carbon on the left (making water).

2. The left carbon now has a spare pair of electrons, so it forms a double bond with the middle carbon.

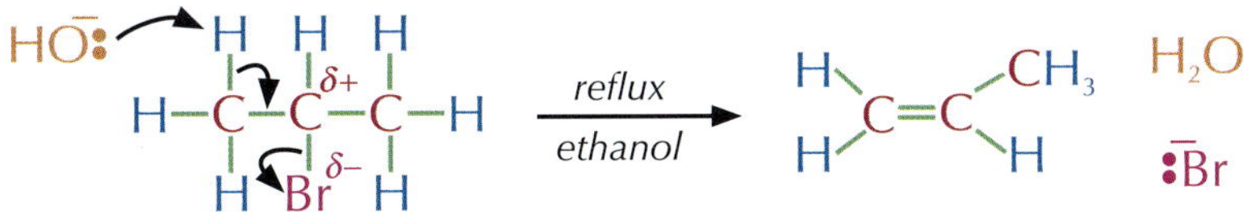

HO:⁻ → H–C(H)(H)–C(H)(δ+)(Br δ−)–C(H)(H)–H —*reflux*, *ethanol*→ $H_2C{=}CH(CH_3)$ + H_2O + :Br^-

3. To form the double bond, the middle carbon has to let go of the Br, which drops off as a Br^- ion.

Tip: If you used a longer, unsymmetrical secondary alkene, you'd get two products, because the double bond could go on either side of the carbon with the bromine attached to it. For example, 2-bromobutane forms but-1-ene and but-2-ene.

Nucleophilic substitution vs elimination

You can influence which type of reaction will happen most by changing the conditions. By reacting a halogenoalkane with water under reflux, the molecule will predominantly undergo nucleophilic substitution to form an alcohol. You'll still get a bit of elimination to form an alkene but not a lot.

Example

Reacting bromoethane with water under reflux will produce ethanol.

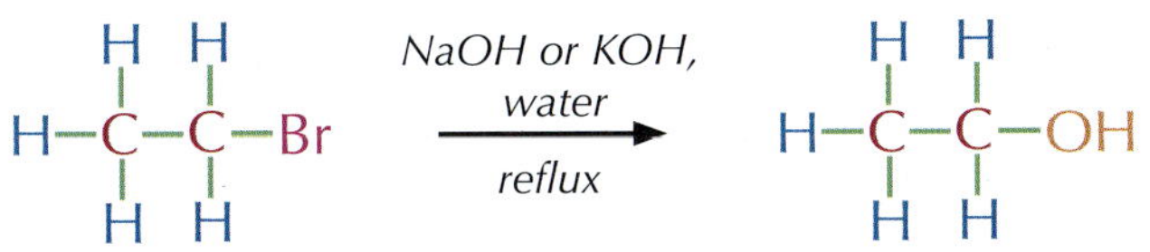

H–C(H)(H)–C(H)(H)–Br —*NaOH or KOH, water*, *reflux*→ H–C(H)(H)–C(H)(H)–OH

This is because under aqueous conditions, the OH^- acts as a nucleophile — it donates an electron pair to the δ+ carbon atom.

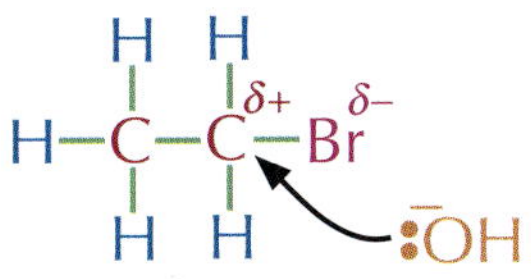

H–C(H)(H)–C(H)(H)(δ+)–Br(δ−) ← :OH^-

Here the OH^- nucleophile is attacking the δ+ carbon atom.

Exam Tip

It's really important that you know the difference between nucleophilic substitution and elimination. So make sure you know both really well and you can describe the differences between them. You don't want to get them muddled in the exam.

By reacting a halogenoalkane with ethanol under reflux, the molecule will predominantly undergo elimination to form an alkene.

Example

Reacting bromoethane with ethanol under reflux will produce ethene.

$$CH_3CH_2Br \xrightarrow[\text{reflux}]{\text{NaOH or KOH, ethanol}} CH_2{=}CH_2$$

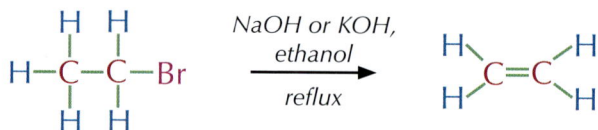

This is because the OH⁻ acts as a base — it removes a hydrogen atom from the halogenoalkane.

Here the OH⁻ is acting as a base and pulling the hydrogen off the halogenoalkane.

Figure 1: *Two students carrying out an organic reaction under reflux, to stop the vapour escaping from the reaction mixture.*

If you use a mixture of water and ethanol as the solvent, both reactions will happen and you'll get a mixture of the two products.

Practice Question — Application

Q1 Draw the mechanism for the main reaction of 1-bromopropane with ethanol under reflux. A molecule of 1-bromopropane is shown below.

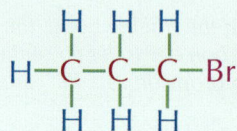

$$CH_3CH_2CH_2Br$$

Practice Questions — Fact Recall

Q1 Briefly describe an elimination reaction.

Q2 a) Reacting bromoethane with water under reflux will produce mainly ethanol. What type of reagent does the ⁻OH act as in this reaction?

b) Name the type of reaction that is occurring.

Q3 a) Reacting bromoethane with ethanol under reflux will produce mainly ethene. What type of reagent does the ⁻OH act as in this reaction?

b) Name the type of reaction that is occurring.

Section Summary

Make sure you know...

- That alkanes are saturated hydrocarbons.
- That petroleum is a mixture, mostly consisting of alkanes.
- How petroleum can be separated by fractional distillation.
- That cracking breaks long-chain alkanes into smaller hydrocarbons.
- That cracking is a valuable process as smaller hydrocarbons are in much higher demand than long-chain hydrocarbons.
- The conditions needed for thermal cracking and catalytic cracking and what the products are.
- That alkanes are used as fuels.
- What complete combustion and incomplete combustion are.
- What pollutants are produced by internal combustion engines and how they are removed from exhaust gases.
- How air pollution is created.
- How sulfur dioxide can be removed from flue gases.
- That the unpaired electron in a free radical is represented by a dot.
- The reaction mechanism of methane with chlorine as a free radical substitution reaction involving initiation, propagation and termination steps.
- That ozone, which is formed naturally in the upper atmosphere, provides protection from the Sun's UV rays.
- That chlorine atoms are formed in the upper atmosphere when energy from ultraviolet radiation causes C–Cl bonds in chlorofluorocarbons (CFCs) to break.
- How chlorine atoms catalyse the decomposition of ozone and have helped form a hole in the ozone layer.
- That legislation to ban the use of CFCs was supported by chemists, and that alternative chlorine-free compounds have been developed to replace CFCs.
- That halogenoalkanes contain polar bonds and are susceptible to nucleophilic attack
- The mechanism of nucleophilic substitution in primary halogenoalkanes.
- That the carbon–halogen bond enthalpy influences the rate of reaction of reactions where carbon-halogen bonds break.
- That halogenoalkanes can undergo nucleophilic substitution and elimination reactions where the reagent can act as either a nucleophile or a base.
- The mechanisms of concurrent nucleophilic substitution and elimination reactions in halogenoalkanes.

Exam-style Questions

1 Bromoethane reacts with aqueous potassium hydroxide to form ethanol.
Which of the following statements about this reaction is **not** correct?

A The OH^- ions act as nucleophiles.

B Iodoethane will react more quickly than bromoethane in the same reaction.

C The reaction is an elimination reaction.

D The reaction needs to take place under reflux.

(1 mark)

2 Chloroethane reacts with ethanolic potassium cyanide under reflux.
The following steps describe the mechanism of the reaction.

1) A new bond forms between the C and the CN^- ion, making a nitrile.
2) The C–Cl bond breaks and both the electrons from the bond are taken by the Cl.
3) The $C^{\delta+}$ attracts a lone pair of electrons from the CN^- ion.
4) The CN^- ion acts as a nucleophile, attacking the slightly positive carbon atom.

Which is the correct order of steps?

A 1, 3, 2, 4

B 3, 4, 1, 2

C 3, 2, 4, 1

D 4, 3, 2, 1

(1 mark)

3 Chlorine and methane react in ultraviolet light to form the halogenoalkane, chloromethane. This reaction is initiated by the breaking of the Cl-Cl bond.

3.1 Write two equations to show the propagation steps for this reaction.

(2 marks)

3.2 Write an equation of a termination reaction showing the formation of chloromethane.

(1 mark)

3.3 When chlorofluorocarbons enter the upper atmosphere, ultraviolet light breaks the C-Cl bonds to form chlorine free radicals. Use equations to show how this leads to the decomposition of ozone, and explain why this is a problem.

(4 marks)

4 Heptane is an alkane which is found in petroleum.

4.1 Fractional distillation can be used to separate out heptane.

Explain how fractional distillation allows the fractions in petroleum to be separated.

(3 marks)

4.2 Petroleum fractions can be burnt to generate energy.

Write an equation for the incomplete combustion of heptane.

(2 marks)

4.3 Explain why the incomplete combustion of heptane can be dangerous.

(3 marks)

4.4 Heptane can be cracked to form smaller chain hydrocarbons.

Explain why it is necessary to crack heptane into smaller chain hydrocarbons.

(1 mark)

4.5 Write an equation for the cracking of heptane into smaller chain hydrocarbons.

(1 mark)

4.6 In industry, a catalyst can be used to crack heptane.
Explain why a catalyst is used.

(2 marks)

5 Ethanol is a simple alcohol that can be used as a fuel, in alcoholic beverages, as a solvent and also as a starting molecule for many organic synthesis reactions. Its structure is shown below.

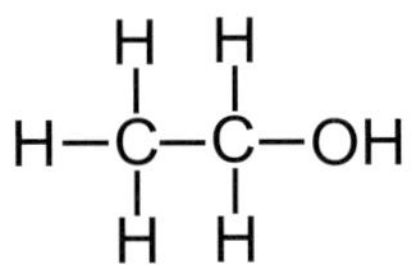

```
    H   H
    |   |
H — C — C — OH
    |   |
    H   H
```

5.1 Ethanol can be created by a nucleophilic substitution reaction of bromoethane with aqueous sodium hydroxide.

Write the equation for this reaction and draw out the mechanism by which it proceeds.

(2 marks)

5.2 Ethanol can also be produced by reaction of iodoethane with aqueous sodium hydroxide.

Which of the reactions would proceed more quickly?
Explain your answer.

(3 marks)

5.3 Sodium hydroxide can be dissolved in warm ethanol, instead of water.

How will this affect the type of reaction that occurs with bromoethane?
Explain your answer.

(2 marks)

1. Alkenes

Alkenes might sound very similar to alkanes, but they're a whole different breed of organic compound...

Learning Objectives:
- Know that alkenes are unsaturated hydrocarbons.
- Know that bonding in alkenes involves a double covalent bond, a centre of high electron density.
- Understand the use of bromine to test for unsaturation.
- Understand and be able to outline the mechanism of electrophilic addition reactions of alkenes with Br_2 and H_2SO_4.

Specification Reference 3.3.4.1, 3.3.4.2

What is an alkene?

Alkenes have the general formula C_nH_{2n}. They're just made of carbon and hydrogen atoms, so they're hydrocarbons. Alkene molecules all have at least one C=C double covalent bond. Molecules with C=C double bonds are **unsaturated** because they can make more bonds with extra atoms in **addition reactions**. Because there are two pairs of electrons in the C=C double bond, it has a really high electron density. This makes alkenes pretty reactive.

Examples

Here are a few pretty diagrams of alkenes:

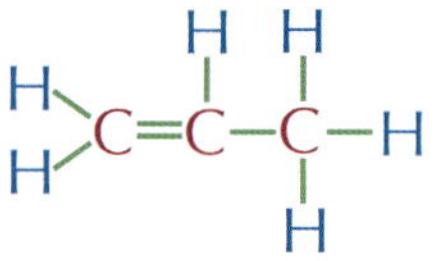

propene, CH_2CHCH_3

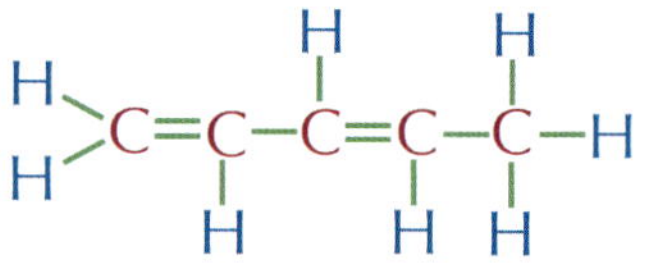

penta-1,3-diene, $CH_2CHCHCHCH_3$

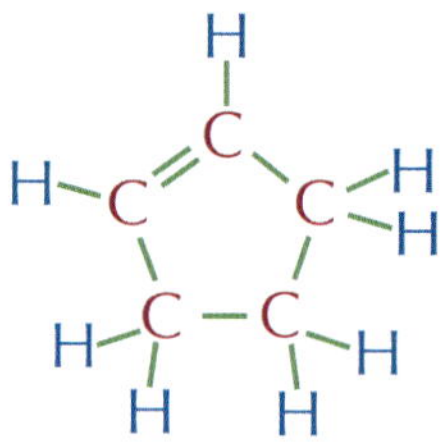

cyclopentene, C_5H_8

A cyclic alkene has two fewer hydrogen atoms than an open-chain alkene. Carbons can only have four bonds — a double bond means that the carbons can make one less bond with a hydrogen.

Tip: Alkenes with more than one double bond have fewer hydrogen atoms than the general formula suggests.

Electrophilic addition reactions

Electrophilic addition reactions aren't too complicated. The double bond in an alkene opens up and atoms are added to the carbon atoms. Electrophilic addition reactions happen because the double bond has got plenty of electrons and is easily attacked by **electrophiles**. Electrophiles are electron-pair acceptors — they're usually a bit short of electrons, so they're attracted to areas where there's lots of electrons about.

Tip: C=C double bonds are nucleophilic — they're attracted to places that don't have enough electrons.

Examples

Here are a few examples of electrophiles.

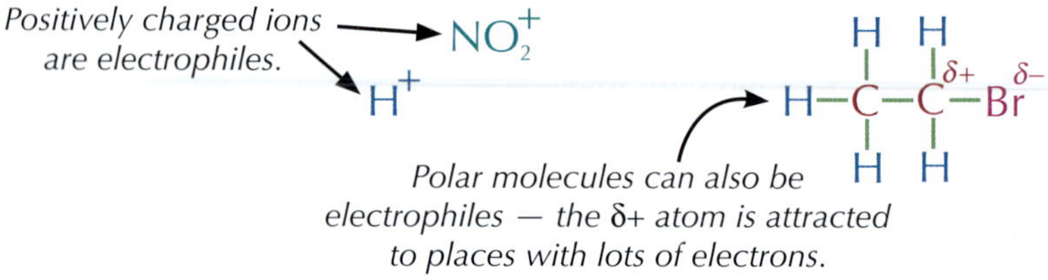

Positively charged ions are electrophiles. → NO_2^+, H^+

Polar molecules can also be electrophiles — the δ+ atom is attracted to places with lots of electrons. → $CH_3CH_2^{\delta+}Br^{\delta-}$

Tip: Just as nucleophiles are often negatively charged ions (see page 211), electrophiles are often positively charged ions. Positive ions lose electrons as they form, so they're ready and waiting to accept any electrons that come their way.

Mechanism

You need to know how to draw mechanisms for electrophilic addition reactions with alkenes. Here's the general equation for this type of reaction, using ethene and an electrophile, X–Y.

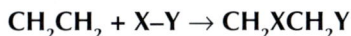

$$\mathbf{CH_2CH_2 + X{-}Y \rightarrow CH_2XCH_2Y}$$

And here's the mechanism for this reaction:

1. The C=C double bond repels the electrons in X–Y, which polarises the X–Y bond (or the bond could already be polar, as in HBr — see page 93).

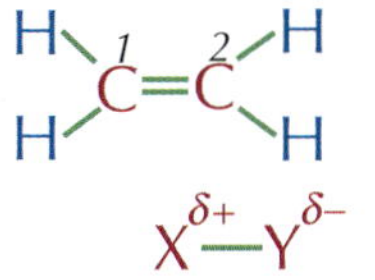

2. Two electrons from the C=C double bond attack the δ+ X atom creating a new bond between carbon 1 and the X atom. The X–Y bond breaks and the electrons from the bond are taken by the Y atom to form a negative ion with a lone pair of electrons. Carbon 2 left with a positive charge (since when the double bond broke carbon 1 took the electrons to form a bond with the X atom) so you now have a **carbocation intermediate**.

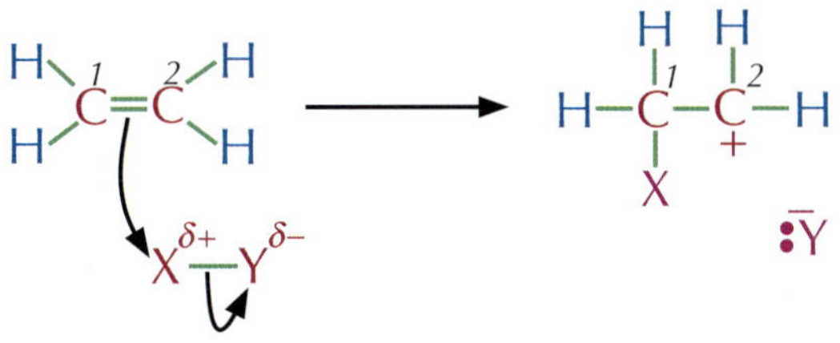

3. The Y^- ion then acts as a nucleophile, attacking the positively charged carbocation, donating its lone pair of electrons and forming a new bond with carbon 2.

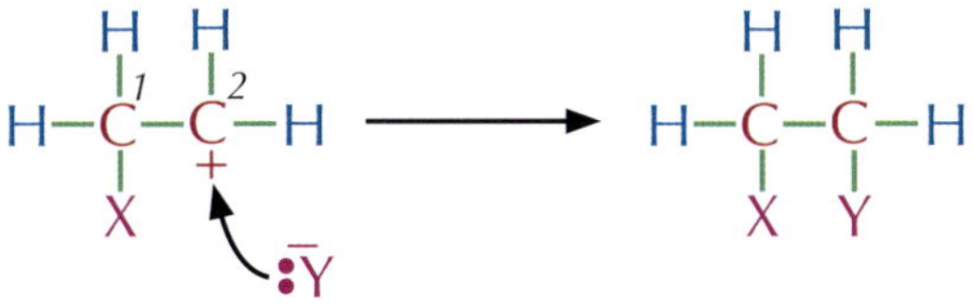

So overall, the X–Y molecule has been added to the alkene across the double bond to form a saturated compound.

Tip: Reactions of other alkenes look similar, but with longer carbon chains.

Tip: A carbocation is an organic ion containing a positively charged carbon atom.

Tip: An intermediate is a short-lived, reactive species that forms in the middle of a reaction mechanism — you can't easily isolate them from the reaction mixture.

Tip: If you're drawing this mechanism for a reaction involving an unsymmetrical alkene, they'll be two different carbocations that could form during step 2. The more stable carbocation will be more likely to form, so you need to make sure you draw the right one. Don't panic though — there's lots about the stability of carbocations coming up on page 227.

Example

Heating ethene with water in the presence of concentrated sulfuric acid produces ethanol. The overall equation for this reaction is shown below.

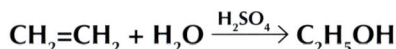

$$\mathbf{CH_2{=}CH_2 + H_2O \xrightarrow{H_2SO_4} C_2H_5OH}$$

You have to do the reaction in two steps. First, concentrated sulfuric acid reacts with ethene in an electrophilic addition reaction. This forms ethyl hydrogen sulfate.

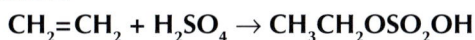

$$\mathbf{CH_2{=}CH_2 + H_2SO_4 \rightarrow CH_3CH_2OSO_2OH}$$

If you then add cold water and warm the product, it's hydrolysed to form ethanol.

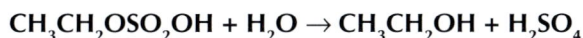

$$\mathbf{CH_3CH_2OSO_2OH + H_2O \rightarrow CH_3CH_2OH + H_2SO_4}$$

The sulfuric acid isn't used up — it acts as a catalyst.

Mechanism

In the first step of this reaction, the carbon-carbon double bond attacks a δ+ hydrogen atom on the sulfuric acid molecule. A new bond is formed between one of the carbons and the hydrogen, and the electrons from the O–H bond are taken by the oxygen atom to form a lone pair. The second carbon is left with a positive charge because it has lost the electron from the double bond.

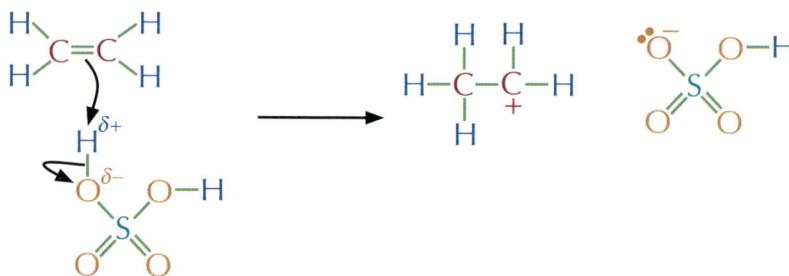

The negative ion created in the first step then acts as a nucleophile and attacks the carbocation creating a new intermediate.

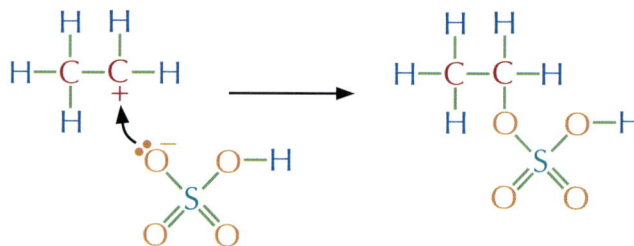

Once this stage of the reaction is over, water can be added to produce sulfuric acid and ethanol.

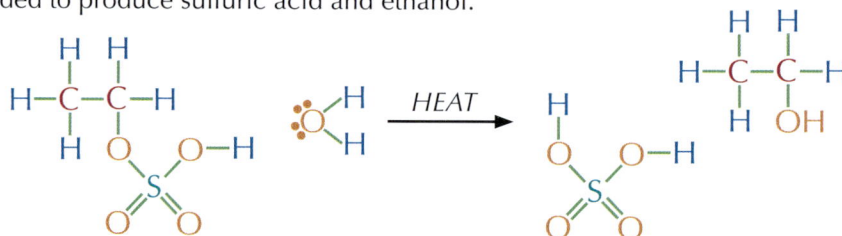

Exam Tip
You could be asked about this reaction in your exams, but you should only be asked to draw the mechanism for the first part (i.e. the reaction of an alkene with H_2SO_4 — the bit before the reaction with water). Read the question carefully, so you're sure you know what it's asking for.

Testing for unsaturation

When you shake an alkene with orange bromine water, the solution quickly turns from orange to colourless (see Figure 1 and 2). Bromine is added across the double bond to form a colourless dibromoalkane — this happens by electrophilic addition.

***Figure 1:** Bromine water test. The test tube on the right contains hex-1-ene, a compound with a C=C that has reacted with the bromine water. The one on the left contains hexane, a saturated substance that doesn't react with bromine water.*

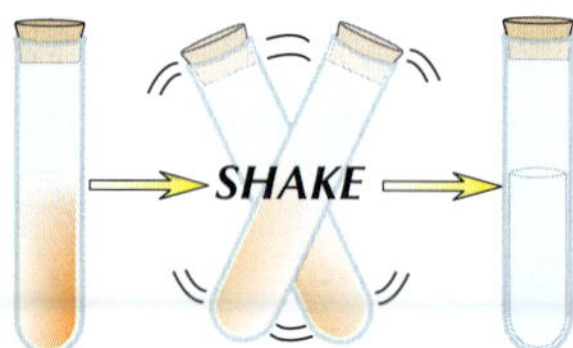

***Figure 2:** Adding bromine water to a solution containing a carbon-carbon double bond turns the bromine water colourless.*

Tip: Because they turn it from orange to colourless, alkenes are sometimes described as 'decolourising' the bromine water.

Example

When you shake ethene orange bromine water, the solution turns from orange to colourless. Here's the equation for this reaction:

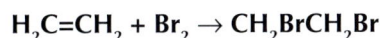

$$H_2C{=}CH_2 + Br_2 \rightarrow CH_2BrCH_2Br$$

Here's the mechanism...

The double bond repels the electrons in Br_2, polarising Br–Br. This is called an induced dipole.

A pair of electrons in the double bond attracts the $Br^{\delta+}$ and forms a bond with it. This repels electrons in the Br–Br bond further, until it breaks.

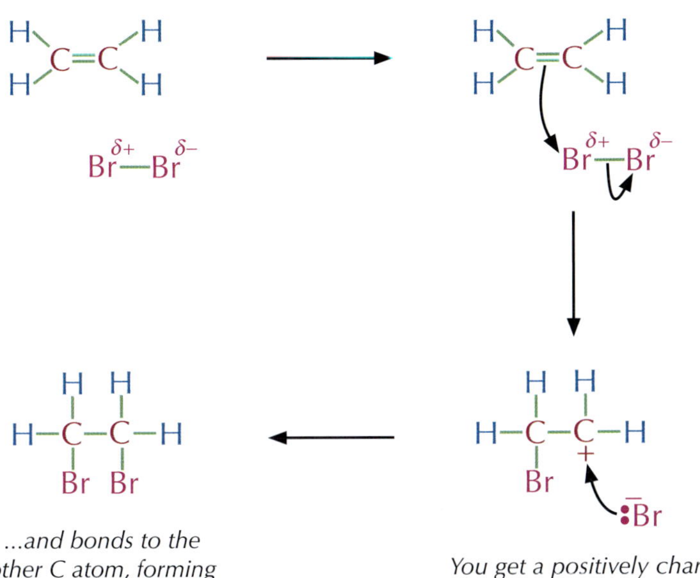

...and bonds to the other C atom, forming 1,2-dibromoethane.

You get a positively charged carbocation intermediate. The Br^- now zooms over...

Exam Tip
Make sure the first curly arrow comes from the carbon-carbon double bond.

Practice Question — Application

Q1 Draw the mechanism for the reaction of but-2-ene with bromine. The structures of the reactants are shown below.

```
    H H H H
    | | | |
  H-C-C=C-C-H        Br—Br
    |     |
    H     H
```

Tip: Remember to be careful where your curly arrows start and finish — they should come from bonds or lone pairs, and go to atoms.

Practice Questions — Fact Recall

Q1 Give the general formula for an alkene.

Q2 Explain why alkenes are unsaturated.

Q3 Explain why alkenes can undergo electrophilic addition reactions.

Q4 What is an electrophile?

Q5 Give two examples of electrophiles.

Q6 What can you use bromine water to test for?

Learning Objectives:

- Understand and be able to outline the mechanism of electrophilic addition reactions of alkenes with HBr.
- Explain the formation of major and minor products in addition reactions of unsymmetrical alkenes by reference to the relative stabilities of primary, secondary and tertiary carbocation intermediates.

Specification Reference 3.3.4.2

2. Reactions of Alkenes

Sometimes a chemical reaction has more than one product — you've got to be able to decide which product is more likely to form. Don't panic, it's not just a wild stab in the dark — there are some handy rules to help you out.

Reactions with hydrogen halides

Alkenes undergo electrophilic addition reactions with hydrogen halides to form halogenoalkanes.

Example

This is the reaction between ethene and hydrogen bromide, to form bromoethane.

$$C_2H_4 + HBr \rightarrow C_2H_5Br$$

It's an electrophilic addition reaction so the mechanism follows the pattern that you saw on page 223.

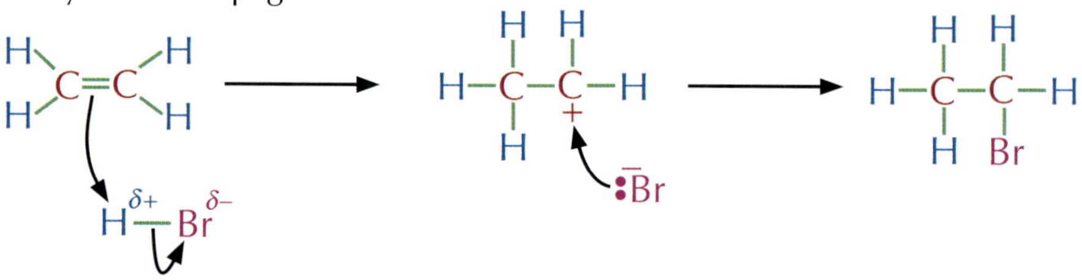

H H H H H H H H H Br δ+ δ– Br Br +

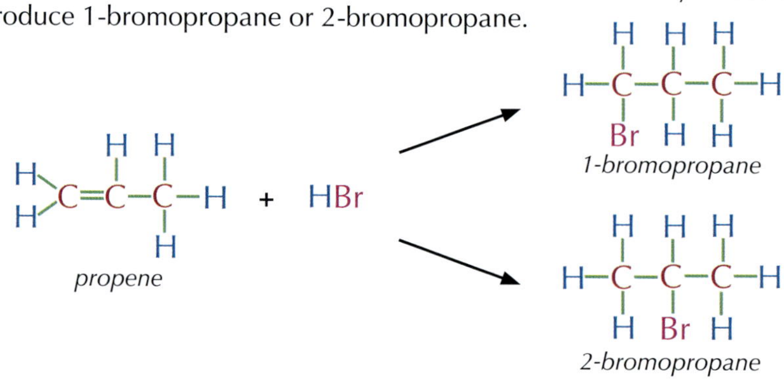

Exam Tip
Other alkenes react in a similar way with HBr. Don't be put off if they give you a different alkene in the exam — the mechanism works in exactly the same way.

Addition of hydrogen halides to unsymmetrical alkenes

If the hydrogen halide adds to an unsymmetrical alkene, there are two possible products.

Example

If you add hydrogen bromide to propene, the bromine atom could add to either the first carbon or the second carbon. This means you could produce 1-bromopropane or 2-bromopropane.

propene + HBr

1-bromopropane

2-bromopropane

Tip: "Primary carbocation" can also be written as 1° carbocation — the 1° stands for primary. Secondary carbocation can be written as 2° carbocation, and tertiary as 3° carbocation.

The amount of each product formed depends on how stable the carbocation formed in the middle of the reaction is. This is known as the carbocation intermediate. The three possible carbocations are:

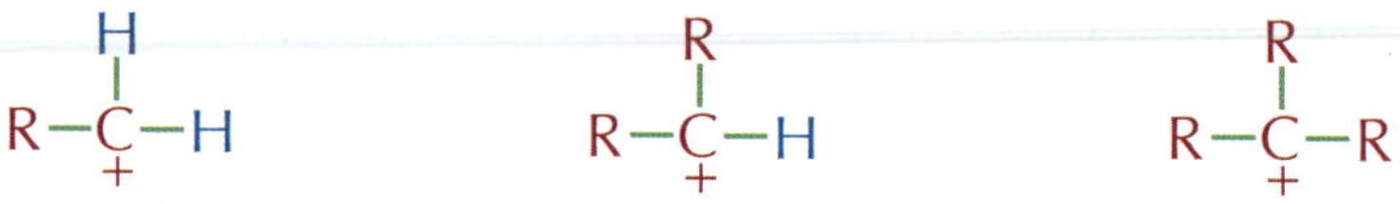

primary carbocation *secondary carbocation* *tertiary carbocation*

R is an alkyl group — an alkane with a hydrogen removed, e.g. $-CH_3$.

Carbocations with more alkyl groups are more stable because the alkyl groups feed electrons towards the positive charge.

You can show that an alkyl group is donating electrons by drawing an arrow on the bond that points to where the electrons are donated.

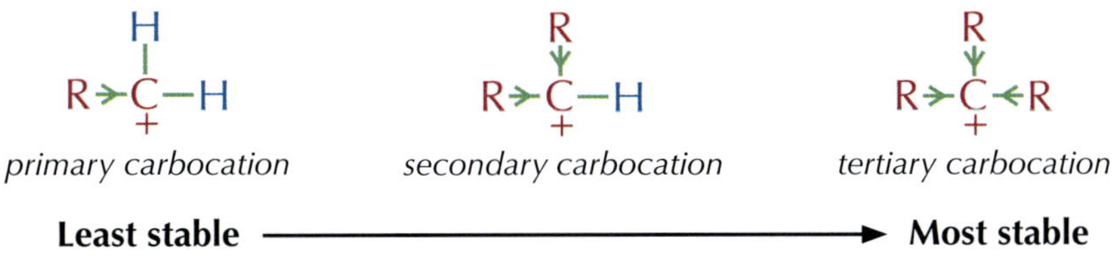

primary carbocation *secondary carbocation* *tertiary carbocation*

Least stable → **Most stable**

More stable carbocations are much more likely to form than less stable ones. This means that there will be more of the product formed via the more stable carbocation than there is via the less stable carbocation.

Examples

Here's how hydrogen bromide reacts with propene:

$H_2C{=}CHCH_3 + HBr \rightarrow CH_3CHBrCH_3$ *2-bromopropane major product*

$H_2C{=}CHCH_3 + HBr \rightarrow CH_2BrCH_2CH_3$ *1-bromopropane minor product*

The secondary carbocation's more stable because it's got two alkyl groups. This carbocation will form most of the time.

The primary carbocation's less stable as it's only got one alkyl group. It forms less often.

2-bromopropane major product

1-bromopropane minor product

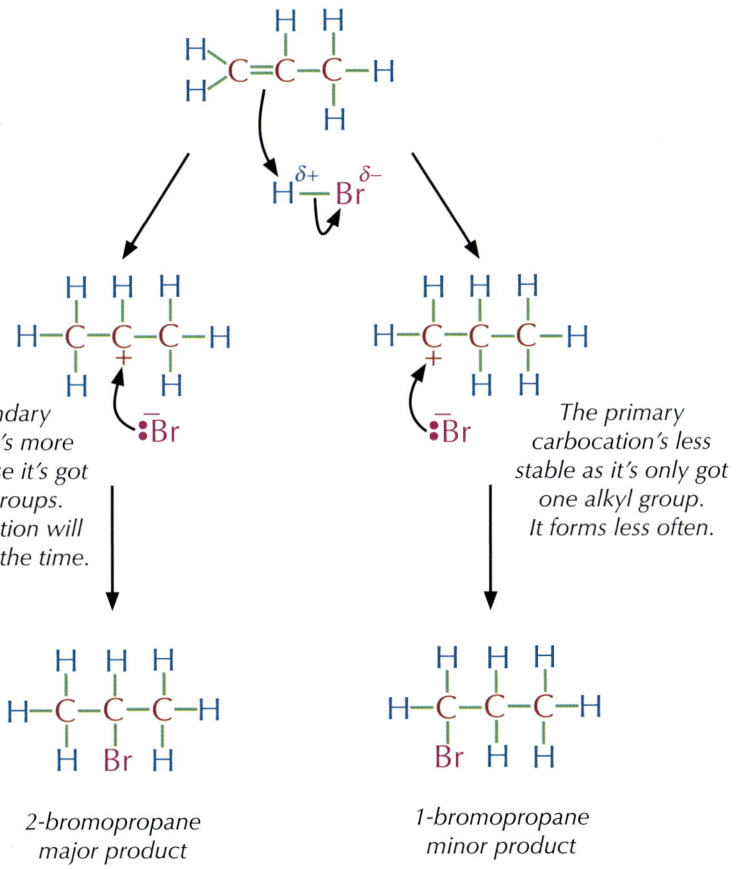

Tip: The alkyl groups don't give up their electrons to the carbon atom — they just move some of their negatively charged electrons nearer to it, which helps to stabilise the positive charge.

Tip: The product that there's most of is called the major product, or the Markovnikov product — after the Russian scientist, Vladimir Markovnikov, who came up with the theory that the product formed most often is formed via a more stable carbocation.

Tip: If you were adding Br_2 to propene instead, they'd only be one possible product (1,2-dibromopropane) — but the reaction would still be more likely to go by the route with a secondary carbocation intermediate.

Exam Tip
It makes it easier to see what the main product is if you draw out all the possible carbocations each time — then there's less chance of making a mistake.

Here's how hydrogen bromide reacts with 2-methylbut-2-ene:

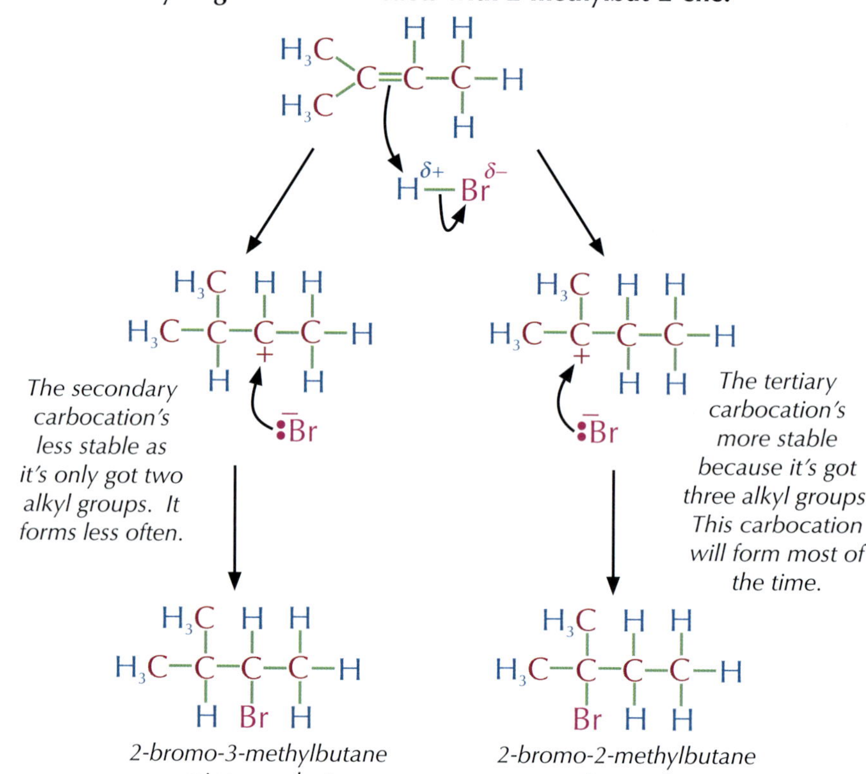

δ+
δ–
H–Br
:Br
The secondary carbocation's less stable as it's only got two alkyl groups. It forms less often.
The tertiary carbocation's more stable because it's got three alkyl groups. This carbocation will form most of the time.
2-bromo-3-methylbutane
minor product
2-bromo-2-methylbutane
major product

Tip: The bigger the difference in stability of the carbocations, the more of the major product you'll get at the end.

Practice Questions — Application

Q1 a) Draw the mechanism for the reaction between but-2-ene and hydrogen bromide. The reactants for this reaction are shown below.

```
  H     H  H
  |     |  |
H-C-C=C-C-H          HBr
  | |     |
  H H     H
```

b) Write the overall equation for this reaction.

Q2 Hydrogen bromide reacts with but-1-ene to form either 1-bromobutane or 2-bromobutane. Explain why 2-bromobutane is the major product of the reaction. The structures of 1-bromobutane and 2-bromobutane are shown below.

```
  H H H H              H H Br H
  | | | |              | | |  |
H-C-C-C-C-Br         H-C-C-C--C-H
  | | | |              | | |  |
  H H H H              H H H  H
```

Practice Questions — Fact Recall

Q1 What could you react with ethene in order to produce bromoethane?

Q2 Place the following carbocations in order of stability from most to least stable: secondary, primary and tertiary.

3. Addition Polymers

There is a way of joining up lots of alkene molecules to make all sorts of different materials — these are called polymers.

Polymers

Polymers are long chain molecules formed when lots of small molecules, called **monomers**, join together.

Polymers can be natural, e.g. DNA, or synthetic (man-made), e.g. polythene. We use them for all sorts of things in everyday life, including in plastic bags, rain coats, non-stick pans and car tyres. People have used natural polymers to make things like fabrics and jewellery for many years.

During the 19th century, researchers managed to synthesise artificial polymers, such as artificial silk and hard rubber. Further developments in polymers came during the 20th century, when materials such as nylon and Kevlar® were developed. Polymers are still being developed today — scientists are still looking to develop materials that are cheaper to produce or that perform their functions better than current materials.

Addition polymers

The double bonds in alkenes can open up and join together to make long chains called polymers. It's kind of like they're holding hands in a big line. The individual, small alkenes are called monomers. This is called **addition polymerisation**.

Example

Poly(ethene) is made by the addition polymerisation of ethene.

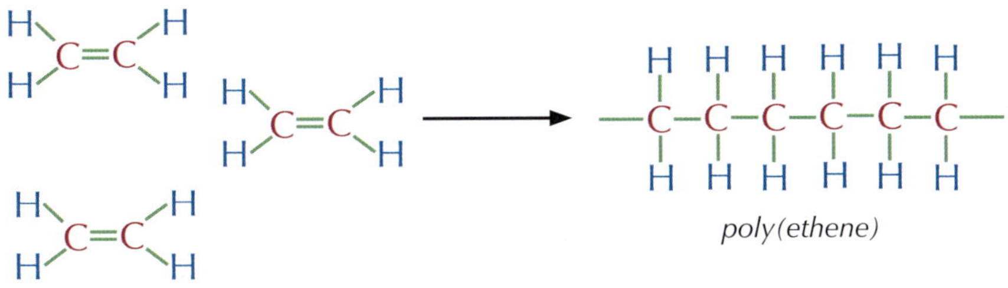

ethene monomers

poly(ethene)

Addition polymerisation reactions can be written like this...

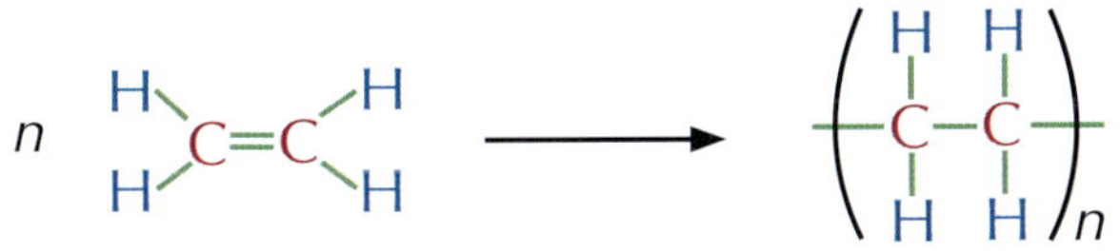

monomer

polymer

...where the *n* stands for the number of repeating units in the polymer.

You can also use **substituted alkenes** as monomers in addition polymerisation. A substituted alkene is just an alkene where one of the hydrogen atoms has been swapped for another atom or group. For example, if you swap one of the hydrogen atoms in ethene for a chlorine atom, you get chloroethene. Polymerising chloroethene makes poly(chloroethene) — see page 231.

Learning Objectives:

- Appreciate that knowledge and understanding of the production and properties of polymers has developed over time.
- Know that addition polymers are formed from alkenes and substituted alkenes.
- Be able to draw the repeating unit from a monomer structure, and the repeating unit or monomer from a section of the polymer chain.
- Know IUPAC rules for naming addition polymers.
- Be able to explain why addition polymers are unreactive.
- Be able to explain the nature of intermolecular forces between molecules of polyalkenes.
- Know typical uses of poly(chloroethene) and how its properties can be modified using a plasticiser.

Specification Reference 3.3.4.3

Exam Tip

If you're asked for the formula of a polymer, you need to include the brackets and the n. If you're asked for the repeating unit, you should just draw the bit inside the brackets.

Tip: Remember to show the 'trailing bonds' at the ends of the molecule. This shows that the polymer continues.

To find the monomer used to form an addition polymer, take the repeating unit, remove unnecessary side bonds and add a double bond.

Example

To find the monomer used to make the polymer below you first need to look for the repeating unit.

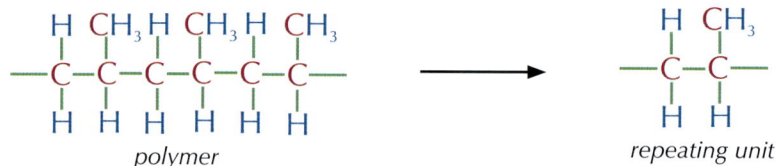

```
  H CH3 H CH3 H CH3                     H CH3
  |  |  |  |  |  |                      |  |
—C—C—C—C—C—C—        ———————▶        —C—C—
  |  |  |  |  |  |                      |  |
  H  H  H  H  H  H                      H  H
      polymer                        repeating unit
```

Then replace the horizontal carbon-carbon bond with a double bond and remove the unnecessary side bonds to find the monomer.

```
   H CH3                     H       CH3
   |  |                       \     /
—C—C—      ———————▶         C=C
   |  |                       /     \
   H  H                      H       H
repeating unit            monomer — propene
```

IUPAC nomenclature

Naming addition polymers is fairly straightforward once you've found the monomer used to form the addition polymer — they have the form **poly(X)**, where X is the name of the monomer. For example, if the monomer is but-2-ene, the polymer it forms will be called poly(but-2-ene).

If the name of the monomer doesn't have a number in it, then you can write it without the brackets. So in the example above, the addition polymer is called poly(propene), or just polypropene.

Properties of polymers

Alkene monomers are unsaturated (they contain one or more double covalent bonds), but once they form polymers they become saturated (there are only single bonds in the carbon chain). The main carbon chain of polyalkenes is also usually non-polar. These factors result in addition polymers being very unreactive — polyalkenes are chemically inert.

Tip: The carbon chain of a polyalkene can be polar in cases where the monomer contains electronegative atoms — the main example of this is poly(chloroethene) (see next page), which contains chlorine atoms.

The monomers within a polymer chain have strong covalent bonds. However, the intermolecular forces between polymer chains are much weaker, which affects the properties of the polymer. Longer chains with fewer branches have stronger intermolecular forces, making these polymer materials stronger and more rigid.

Examples

Polyethene

A polymer with few or no branches, such as polyethene, can pack closely together (see Figure 1). The polymer chains are attracted to each other by van der Waals forces. This makes a strong, rigid material.

Tip: Look back at page 94 for more information on van der Waals forces.

Figure 1: *Skeletal diagram of packed polyethene chains.*

Polystyrene

Poly(phenylethene), more commonly known as polystyrene, is formed from phenylethene.

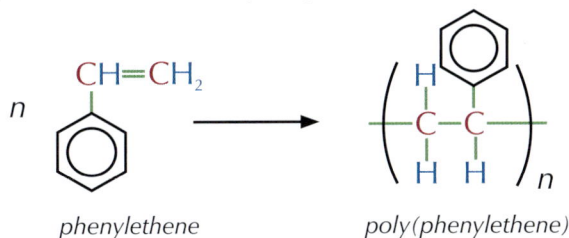

$$n\ C_6H_5CH{=}CH_2 \longrightarrow \left(-CH_2-CH(C_6H_5)-\right)_n$$

phenylethene — *poly(phenylethene)*

The benzene ring coming off the main carbon chain of poly(phenylethene) is a large branch. This makes it difficult for the chains to pack closely together, so they only form weak van der Waals forces. This makes poly(phenylethene) a more flexible material than polyethene.

Poly(chloroethene)

Poly(chloroethene), also known as polyvinyl chloride, or **PVC**, is an addition polymer formed from chloroethene monomers.

$$n\ H_2C{=}CHCl \longrightarrow \left(-CH_2-CHCl-\right)_n$$

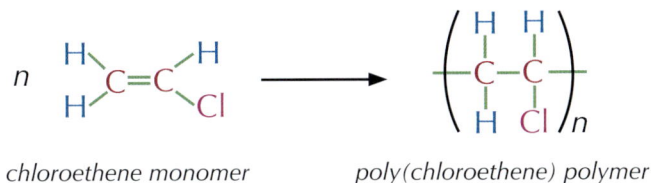

chloroethene monomer — *poly(chloroethene) polymer*

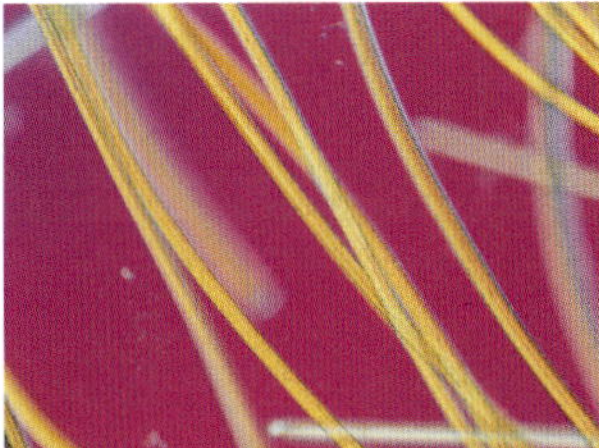

***Figure 2:** Polyvinyl chloride (PVC) fibres. PVC is an addition polymer.*

The covalent bonds between the chlorine and the carbon atoms are polar, with chlorine being more electronegative. The δ– charges on the chlorine atoms and the δ+ charges on the carbon atoms mean that there are permanent dipole-dipole forces between the polymer chains (see Figure 3). This makes PVC a hard but brittle material. It is used to make drain pipes and window frames.

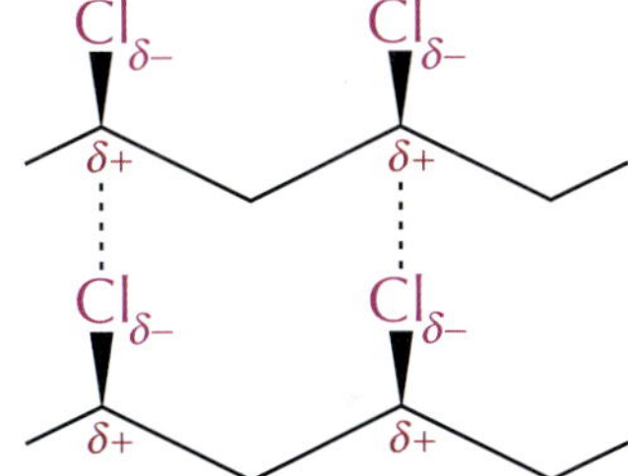

***Figure 3:** Skeletal formula showing the permanent dipole-dipole forces in PVC.*

Plasticisers

You can add chemicals, such as **plasticisers**, to polymers to modify their properties. Adding a plasticiser makes a polymer bendier. The plasticiser molecules get between the polymer chains and push them apart. This reduces the strength of the intermolecular forces between the chains — so the chains can slide around more, making them more flexible.

Plasticised PVC is much more flexible than rigid PVC. It's used to make electrical cable insulation, flooring tiles and clothing.

Practice Questions — Application

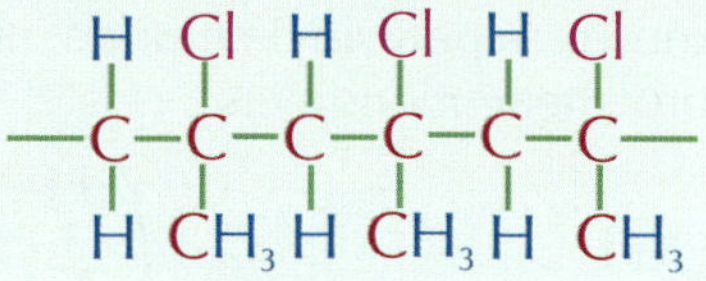

Exam Tip
Make sure you know how to convert between the monomer structure, the repeating unit and the polymer structure.

Q1 a) Write a reaction for the addition polymerisation of the monomer fluoroethene. The structure of fluoroethene is shown below.

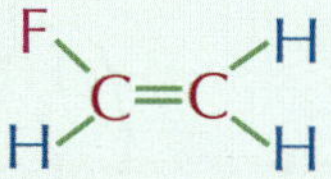

```
F       H
 \     /
  C = C
 /     \
H       H
```

b) Name the polymer that is formed in the reaction.

Q2 Draw the structure of the monomer used to form the polymer shown below.

```
   H  Cl H  Cl H  Cl
   |  |  |  |  |  |
 — C— C— C— C— C— C —
   |  |  |  |  |  |
   H  H  H  H  H  H
```

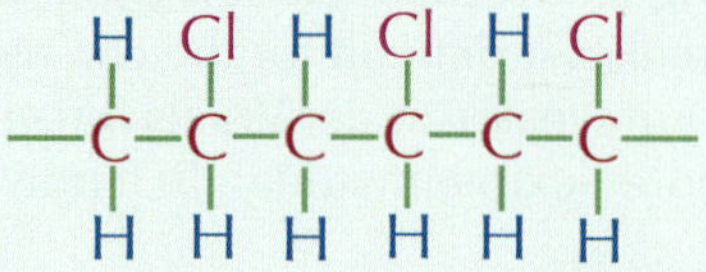

Q3 Draw the repeating unit of the polymer shown below.

```
   H  Cl  H  Cl  H  Cl
   |  |   |  |   |  |
 — C— C — C— C — C— C —
   |  |   |  |   |  |
   H  CH3 H  CH3 H  CH3
```

Practice Questions — Fact Recall

Q1 Briefly describe addition polymerisation.

Q2 Why are poly(alkenes) unreactive?

Q3 a) Describe and explain the intermolecular forces that exist between the polymer chains in unplasticised poly(chloroethene).

b) Give two uses of rigid poly(chloroethene).

Tip: Plasticisers are just chemicals that make a polymer more flexible.

Q4 a) Briefly describe how plasticisers work.

b) Give two uses of flexible poly(chloroethene).

4. Alcohols

These pages on alcohols are pretty important — this topic is packed full of real life applications of chemistry. But before we get onto all that fun stuff here's a bit more nomenclature for you to learn.

Learning Objectives:

- Be able to apply IUPAC rules for nomenclature to name alcohols limited to chains with up to 6 carbon atoms.
- Understand that alcohols can be classified as primary, secondary or tertiary.

Specification Reference 3.3.1.1, 3.3.5.2

Nomenclature of alcohols

The alcohol homologous series has the general formula $C_nH_{2n+1}OH$. Alcohols are named using the same IUPAC naming rules found on pages 188-191, but the suffix -ol is added in place of the -e on the end of the name. You also need to indicate which carbon atom the alcohol functional group is attached to — the carbon number(s) comes before the -ol suffix. If there are two –OH groups the molecule is a -diol and if there are three it's a -triol.

Examples

The longest continuous carbon chain is 2 carbon atoms, so the stem is ethane.

There's one –OH attached to the carbon chain so the suffix is -ol.

There are two carbon atoms it could be attached to, but they are equivalent (they'd both be labelled carbon atom 1) so there's no need to put a number.

So, the alcohol is called ethanol.

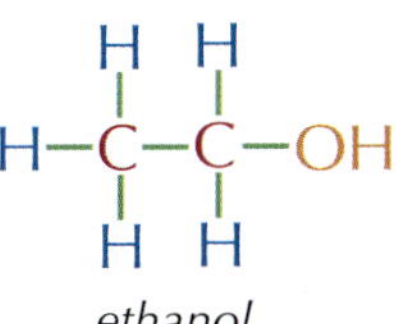

ethanol

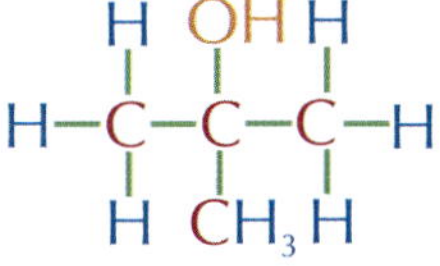

2-methylpropan-2-ol

The longest continuous carbon chain is 3 carbon atoms, so the stem is propane.

There's one –OH attached to the carbon chain so the suffix is -ol.

It's attached to the second carbon so there's a 2 before the -ol.

There's also a methyl group attached to the second carbon so there's also a 2-methyl- prefix.

The alcohol is called 2-methylpropan-2-ol.

The longest continuous carbon chain is 2 carbon atoms, so the stem is ethane.

There are two –OH groups attached to the carbon chain so the suffix is -diol.

There's one –OH attached to each carbon atom so there's a 1,2- before the -diol.

So, the alcohol is called ethane-1,2-diol.

ethane-1,2-diol

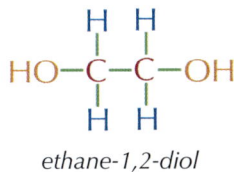

Tip: When you're naming alcohols with only one -OH group, you lose the 'e' at the end of the alkane stem, as usual — but when you're naming diols and triols, you keep the 'e' For example:

- butanol (not butan**e**ol),
- hexan**e**-2,4-diol (not hexan-2,4-diol).

Primary, secondary and tertiary alcohols

An alcohol is primary, secondary or tertiary, depending on which carbon atom the hydroxyl group –OH is bonded to. **Primary alcohols** are given the notation 1° and the –OH group is attached to a carbon with one alkyl group attached (see Figure 1). **Secondary alcohols** are given the notation 2° and the –OH group is attached to a carbon with two alkyl groups attached. **Tertiary alcohols** are given the notation 3° (you can see where I'm going with this) and the –OH group is attached to a carbon with three alkyl groups attached.

Tip: Remember that an alkyl group is an alkane with a hydrogen removed, for example $-CH_3$ or CH_3CH_2-.

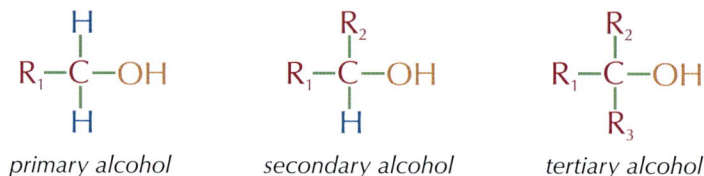

primary alcohol　　secondary alcohol　　tertiary alcohol

Figure 1: *Diagrams of 1°, 2° and 3° alcohols. R = alkyl group.*

Examples

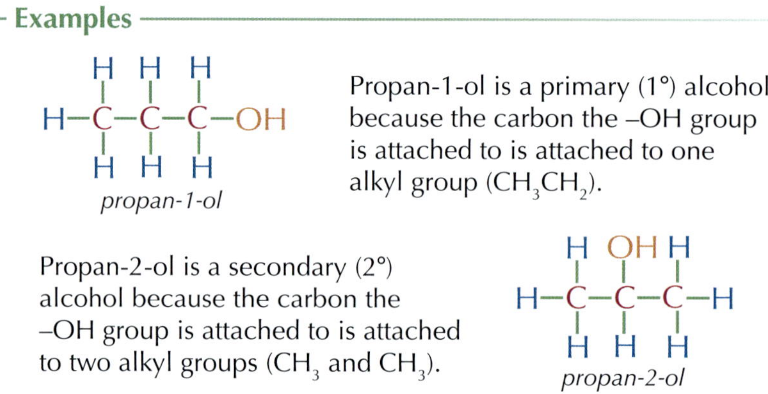

Propan-1-ol is a primary (1°) alcohol because the carbon the –OH group is attached to is attached to one alkyl group (CH_3CH_2).

Propan-2-ol is a secondary (2°) alcohol because the carbon the –OH group is attached to is attached to two alkyl groups (CH_3 and CH_3).

Practice Questions — Application

Q1 Name the following alcohols.

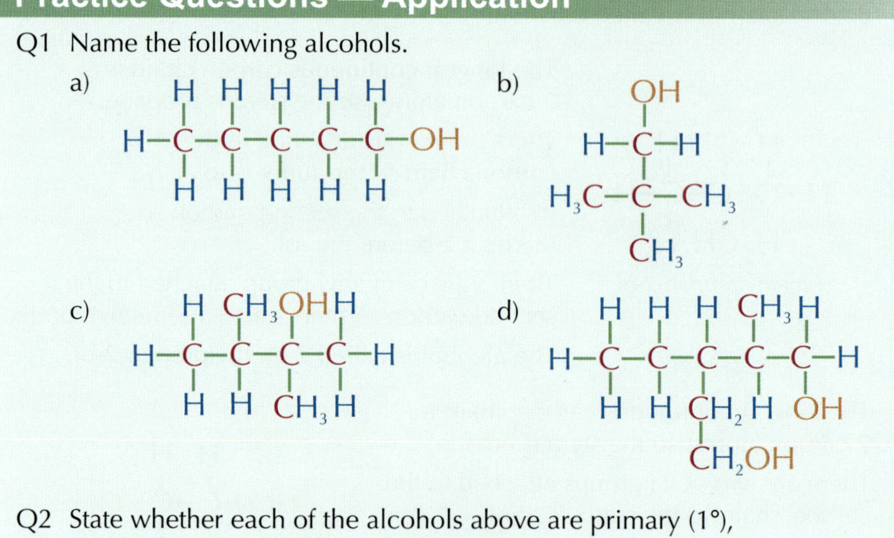

Tip: Remember — name alcohols so that the names include the lowest numbers possible.

Q2 State whether each of the alcohols above are primary (1°), secondary (2°) or tertiary (3°) alcohols.

Practice Questions — Fact Recall

Q1 Give the general formula for the alcohol homologous series.

Q2 What is a secondary alcohol?

5. Dehydrating Alcohols

Making ethene from ethanol means kicking out a molecule of water, so it's called 'dehydration'. The opposite reaction is in the next topic (on page 238).

How to dehydrate an alcohol

You can make alkenes by eliminating water from alcohols in a dehydration reaction (i.e. elimination of water).

$$C_nH_{2n+1}OH \rightarrow C_nH_{2n} + H_2O$$

This reaction allows you to produce alkenes from renewable resources — you can produce ethanol by fermentation of glucose, which you can get from plants. This is important, because it means that you can produce polymers (poly(ethene), for example) without needing oil.

One of the main industrial uses for alkenes is as the starting material for polymers. Here's how you can make ethene from ethanol:

Example

Water can be eliminated from ethanol in a dehydration reaction. Ethanol is heated with a concentrated sulfuric acid catalyst.

$$C_2H_5OH \xrightarrow{H_2SO_4} CH_2{=}CH_2 + H_2O$$

The product is usually in a mixture with water, acid and reactant in it, so the alkene has to be separated out. Here is the mechanism:

1. A lone pair of electrons from the oxygen bonds to an H^+ from the acid. The alcohol is protonated, giving the oxygen a positive charge.

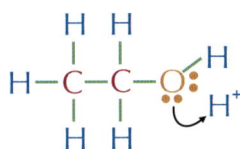

2. The positively charged oxygen pulls electrons away from the carbon. An H_2O molecule leaves, creating an unstable carbocation intermediate.

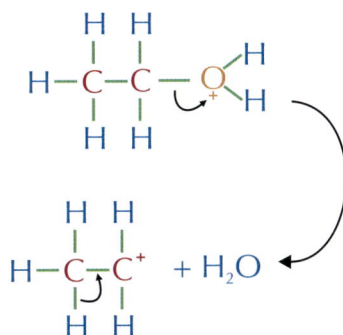

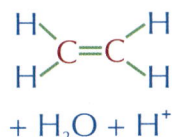

3. The carbocation loses an H^+...

4. ...and the alkene is formed.

$$CH_3CH_2OH + H^+ \rightarrow CH_3CH_2OH_2^+ \rightarrow CH_3CH_2^+ + H_2O \rightarrow CH_2{=}CH_2 + H_2O + H^+$$

Dehydration of longer, unsymmetrical alcohols results in more than one product, because the double bond can go on either side of the carbon that had the OH group on it.

Example

$$CH_3CH(OH)CH_2CH_3 \xrightarrow{+ H_2SO_4} CH_3CH{=}CHCH_3 \text{ and } CH_3CH_2CH{=}CH_2$$

Butan-2-ol → *But-2-ene* and *But-1-ene*

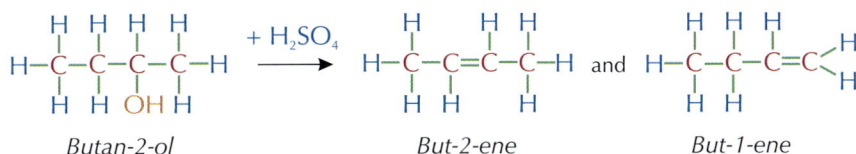

Learning Objectives:

- Know that alkenes can be formed from alcohols by acid-catalysed elimination reactions.
- Know that alkenes produced by this method can be used to produce addition polymers without using monomers derived from crude oil.
- Be able to outline the mechanism for the elimination of water from alcohols.
- Know how to distil a product from a reaction (Required Practical 5).

Specification Reference 3.3.5.3

Tip: In an elimination reaction, a small group of atoms breaks away from a larger molecule. It's not replaced by anything else.

Tip: Phosphoric acid can also be used as a catalyst in this reaction.

Tip: The acid catalysed elimination of water from ethanol is the reverse of the acid catalysed hydration of ethene — see page 238.

Tip: Butan-2-ol actually forms three products, because but-2-ene exists in two isomer forms — see pages 193-197 for more on isomerism.

Purifying the product of a reaction

REQUIRED PRACTICAL 5

The products of organic reactions are often impure — so you definitely need to know how to get rid of any unwanted by-products or leftover reactants from the reaction mixture.

In the dehydration reaction of alcohols to form alkenes, the mixture at the end contains the product, the reactant, acid, water and other impurities. To get a pure alkene, you need a way to separate it from the other substances.

Distillation is a technique which uses the fact that different chemicals have different boiling points, to separate them. You need to know how to distil a product from a reaction — there's a description of how to do this in the example below.

Sometimes you'll need to perform a series of steps to collect and purify the product of a reaction. A good example of this is preparing an alkene from an alcohol.

Tip: Remember to consider any safety precautions that you might need to take before you start doing an experiment like this.

Example

Here's how to produce cyclohexene from cyclohexanol and separate and purify the product.

Stage 1 — reaction and first distillation

1) Add concentrated H_2SO_4 and H_3PO_4 to a round-bottom flask containing cyclohexanol. Mix the solution by swirling the flask and add 2-3 carborundum boiling chips (these make the mixture boil more calmly).
2) Connect the flask to the rest of the distillation apparatus, including a thermometer, condenser and a cooled collection flask. Figure 1, below, shows how the distillation apparatus should be set up.

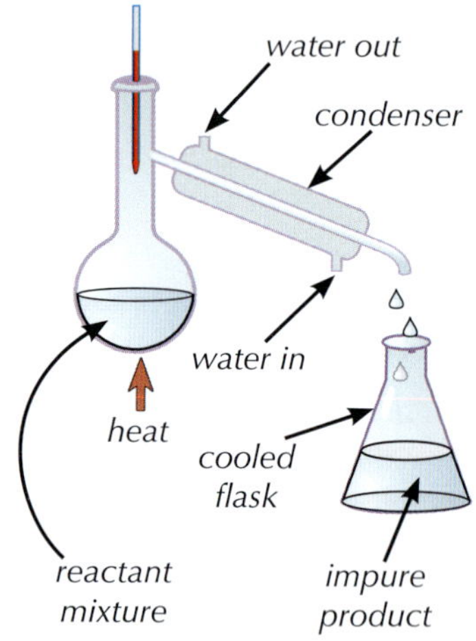

Figure 1: *Set-up of distillation apparatus.*

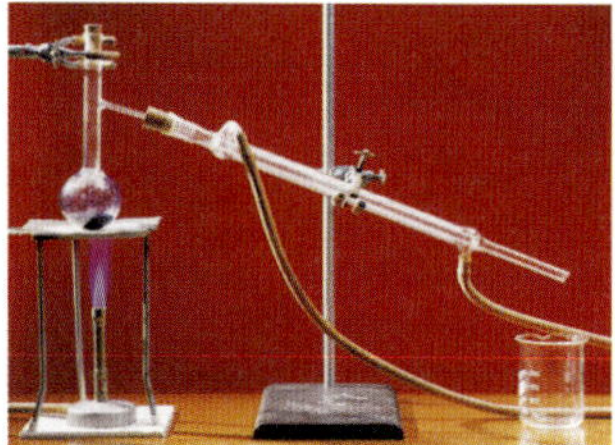

Figure 2: *Photograph of distillation apparatus.*

Tip: You should use a heating method without a flame here because the reactants and the products are flammable.

3) Gently heat the mixture in the flask to around 83 °C (the boiling point of cyclohexene) using a water bath or electric heater.
4) Chemicals with boiling points up to 83 °C will evaporate. The warm gas will rise out of the flask and into the condenser
5) The condenser has cold water running through the outside. The cooler temperatures turn the gas back into a liquid.
6) The liquid product can then be collected in a cooled flask.

Stage 2 — separation

1) The product collected after the first distillation will still contain some impurities which need to be removed.
2) Transfer the product mixture to a separating funnel (see Figure 3, below) and add water to dissolve water soluble impurities and create an aqueous solution.
3) Allow the mixture to settle into layers. Drain off the aqueous layer at the bottom, leaving behind the impure cyclohexene.

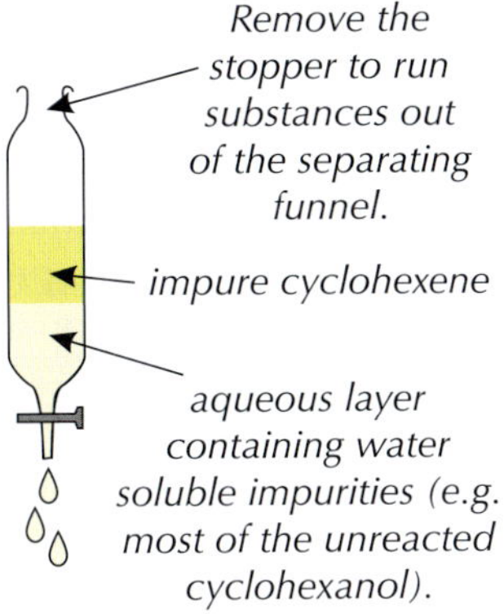
Remove the stopper to run substances out of the separating funnel.
impure cyclohexene
aqueous layer containing water soluble impurities (e.g. most of the unreacted cyclohexanol).

Figure 3: *Set-up of separation apparatus.*

Stage 3 — purification

1) Drain off the impure cyclohexene into a round-bottomed flask.
2) Add anhydrous $CaCl_2$ (a drying agent — it removes any remaining traces of water) and stopper the flask. Let the mixture dry for at least 20 minutes with occasional swirling.
3) The cyclohexene may still have small amounts of impurities so distil the mixture one last time. This time, only collect the product that is released when the mixture is at around 83 °C — this will be the pure cyclohexene.

Tip: The water in the aqueous layer is polar and the cyclohexene is non-polar, so the two liquids won't mix — like oil and water. This comes in really handy when you're trying to separate them...

Tip: In some cases, the desired product might be the denser bottom layer, with impurities in the top layer. In this case you'd drain off and collect the bottom layer, leaving the top layer behind.

Tip: The cyclohexene will be fairly pure by the end of the second purification step, so you may not be asked to do the second distillation.

Practice Question — Application

Q1 a) Write an equation for the dehydration of propan-1-ol to form propene.

b) Draw out a the mechanism for this reaction.

Practice Questions — Fact Recall

Q1 Write out the general equation for the dehydration of an alcohol to form an alkene.

Q2 Draw and label a diagram to show how you would set up distillation apparatus.

Q3 Imagine you are making cyclohexene from cyclohexanol. You heat the alcohol with acid in distillation apparatus, collect the resulting product and separate it using a separating funnel. You are left with impure cyclohexene. Describe how you would purify it.

6. Ethanol Production

Learning Objectives:

- Know that alcohols are produced industrially by hydration of alkenes in the presence of an acid catalyst.
- Be able to outline the mechanism for the formation of an alcohol by the reaction of an alkene with steam in the presence of an acid catalyst.
- Be able to describe how ethanol is produced industrially by fermentation of glucose.
- Be able to justify the conditions used in the production of ethanol by fermentation of glucose.
- Know that ethanol produced industrially by fermentation is separated by fractional distillation and can then be used as a biofuel.
- Be able to explain the meaning of the term biofuel.
- Be able to write equations to support the statement that ethanol produced by fermentation is a carbon neutral fuel, and give reasons why this statement is not valid.
- Be able to discuss the environmental (including ethical) issues linked to decision making about biofuel use.

Specification Reference 3.3.5.1

There are two main methods that are used to produce ethanol and other alcohols industrially — hydrating alkenes and fermenting sugars. You need to know the details of both methods, so here they come...

Hydrating alkenes

The standard industrial method for producing alcohols is to hydrate an alkene using steam in the presence of an acid catalyst. Here's the general equation for this type of reaction:

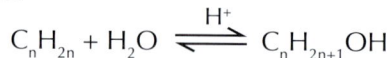

$$C_nH_{2n} + H_2O \xrightleftharpoons{H^+} C_nH_{2n+1}OH$$

Steam hydration of ethene is used industrially to produce ethanol. Ethene can be hydrated by steam at 300 °C and a pressure of 60 atm. It needs a solid phosphoric(V) acid catalyst. Here's the equation for the industrial hydration of ethene:

$$\mathbf{CH_2{=}CH_{2(g)} + H_2O_{(g)} \xrightleftharpoons[300\ °C,\ 60\ atm]{H_3PO_4} CH_3CH_2OH_{(g)}}$$

The reaction's reversible and the reaction yield is low — only about 5%. (This sounds rubbish, but you can recycle the unreacted ethene gas, making the overall yield a much more profitable 95%.)

Mechanism

Here's the mechanism for the reaction of ethene with steam (it's the reverse of the dehydration mechanism from page 235):

1. *A pair of electrons from the double bond bonds to an H^+ from the acid.*

2. *A lone pair of electrons from a water molecule bonds to the carbocation.*

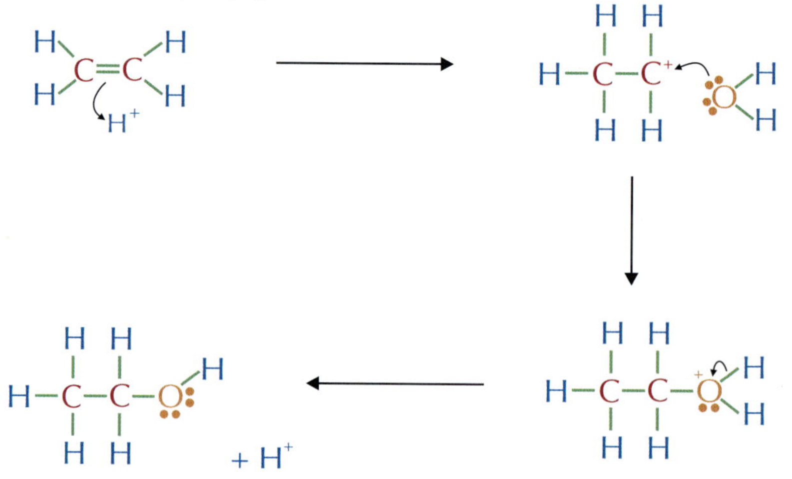

3. *The water loses an H^+...*

4. *...and the alcohol is formed.*

Make sure you're familiar with how this mechanism works — you could be asked to draw the mechanism for the reaction of steam with other alcohols too.

(This reaction is similar to the reaction of ethene with sulfuric acid (see pages 223-224), but because the reaction conditions are different the mechanism's slightly different too.)

Fermentation of glucose

At the moment most industrial ethanol is produced by steam hydration of ethene with a phosphoric acid catalyst (see page 238). The ethene comes from cracking heavy fractions of crude oil. But in the future, when crude oil supplies start running out, petrochemicals like ethene will be expensive — so producing ethanol by fermentation will become much more important...

Industrial production of ethanol by fermentation

Fermentation is an exothermic process, carried out by yeast in anaerobic conditions (without oxygen). Here's the equation for the reaction.

$$C_6H_{12}O_{6(aq)} \xrightarrow[\text{yeast}]{30\text{-}40\,°C} 2C_2H_5OH_{(aq)} + 2CO_{2(g)}$$

Exam Tip
When you're writing out this equation make sure you always include the conditions above and below the arrow.

Yeast produces enzymes which convert glucose ($C_6H_{12}O_6$) into ethanol and carbon dioxide. The enzyme works at an optimum (ideal) temperature of 30-40 °C. If it's too cold, the reaction is slow — if it's too hot, the enzyme is denatured (damaged). Figure 2 shows how the rate of reaction of fermentation is affected by temperature.

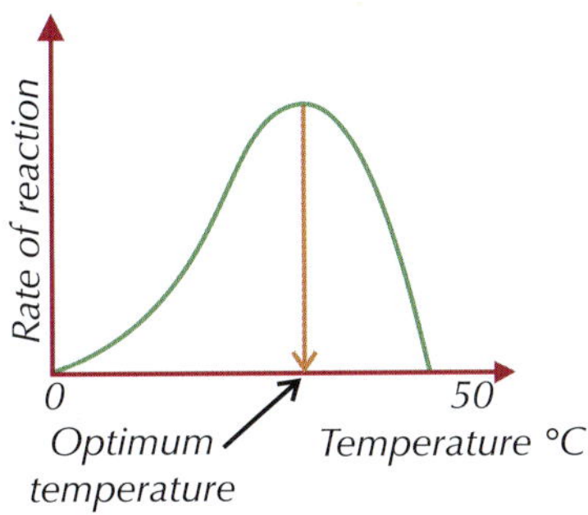

Figure 2: *Graph to show the effect of temperature on fermentation.*

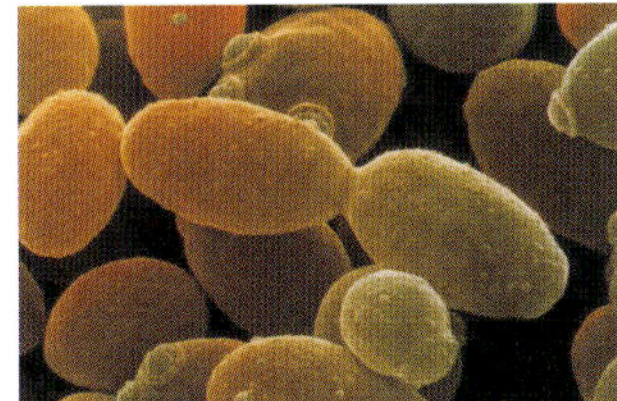

Figure 1: *A scanning electron micrograph (SEM) of yeast cells.*

When the solution reaches about 15% ethanol, the yeast dies. Fractional distillation is used to increase the concentration of the ethanol. Fermentation is low-tech — it uses cheap equipment and renewable resources. But the fractional distillation step that is needed to purify the ethanol produced using this method takes extra time and money.

Figure 3: *Vat for the fermentation of yeast.*

Comparison of ethanol production methods

Figure 4 shows a quick summary of the advantages and disadvantages of the two main industrial methods of making ethanol.

	Hydration of Ethene	Fermentation
Rate of Reaction	Very fast	Very slow
Quality of Product	Pure	Very impure — needs further processing.
Raw Material	Ethene from oil — a finite resource.	Sugars — a renewable resource.
Process/Costs	Continuous process, so expensive equipment needed, but low labour costs.	Batch process, so cheap equipment needed, but high labour costs.

Figure 4: *Table comparing methods of ethanol production.*

Issues with biofuels

A **biofuel** is a fuel that's made from biological material that's recently died. For example, sugars from sugar cane can be fermented to produce ethanol, which can be added to petrol. Ethanol produced in this way is a biofuel (sometimes called bioethanol)

You need to know about both the advantages and the disadvantages of using biofuels.

Figure 5: *A maize field that grows crops for the production of biofuel.*

Advantages of biofuel use

One of the big advantages of using biofuels instead of fuels that come from crude oil is that biofuels are renewable energy sources. Unlike fossil fuels, biofuels won't run out. This makes them more sustainable.

Like conventional fuels, biofuels do produce carbon dioxide when they're burnt. But burning a biofuel only releases the same amount of carbon dioxide that the crop plant took in as it was growing. So most biofuels are considered to be **carbon neutral** (although this is not quite true — see below).

Disadvantages of biofuel use

The main ethical problem with biofuel production is known as the "food vs. fuel" debate. When you use land used to grow crops for fuel, that land can't be used to grow food. If countries start using land to grow biofuel crops instead of food, they may be unable to feed everyone in the country.

There are also environmental problems with the production of biofuels. In some places, trees may be cut down in order to create more land to grow crops for biofuels. Deforestation destroys habitats and removes trees, which are very efficient at taking carbon dioxide out of the air themselves. The trees that are cut down are often burnt, releasing more carbon dioxide.

Fertilisers are often added to soils in order to increase biofuel crop production. Fertilisers can pollute waterways, and some fertilisers also release nitrous oxide, which is a greenhouse gas (see page 207).

There are also some practical problems with switching from fossil fuels to biofuels. For example, most current car engines would be unable to run on fuels with high ethanol concentrations without being modified.

Tip: Carbon dioxide is a greenhouse gas — See page 207 for more details.

Is ethanol a carbon neutral biofuel?

Just like burning the hydrocarbons from fossil fuels, burning ethanol produces carbon dioxide (CO_2). But the plants that are grown to produce bioethanol take in carbon dioxide from the atmosphere as they grow. When you burn the fuel produced from the plants, you only release the same amount of carbon dioxide that the plant took in in the first place — this is described as being carbon neutral. So bioethanol is sometimes thought of as a carbon neutral fuel. Here are the chemical equations to support that argument:

1. During photosynthesis, plants use carbon dioxide from the atmosphere to produce glucose.

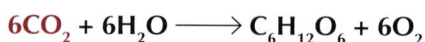

$$6CO_2 + 6H_2O \longrightarrow C_6H_{12}O_6 + 6O_2$$

6 moles of carbon dioxide are taken from the atmosphere to produce 1 mole of glucose.

2. In the fermentation process, glucose is converted into ethanol.

$$C_6H_{12}O_6 \longrightarrow 2C_2H_5OH + 2CO_2$$

2 moles of carbon dioxide are released into the atmosphere when 1 mole of glucose is converted to 2 moles of ethanol.

3. When ethanol is burned, carbon dioxide and water are produced.

$$2C_2H_5OH + 6O_2 \longrightarrow 4CO_2 + 6H_2O$$

4 moles of carbon dioxide are released into the atmosphere when 2 moles of ethanol are burned completely.

If you combine these three equations, you'll see that exactly 6 moles of CO_2 are taken in and exactly 6 moles of CO_2 are given out.

However, fossil fuels will need to be burned to power the machinery used to make fertilisers for the crops and the machinery used to harvest the crops. Refining and transporting the bioethanol also uses energy. It's usually fossil fuels that are being burnt in order to produce the energy needed to carry out these processes, and burning that fuel will produce carbon dioxide. So bioethanol isn't a completely carbon neutral fuel.

Tip: If these equations are combined, all of the species on the left and right sides completely cancel each other out. Overall, CO_2 isn't taken from or put into the atmosphere.

Practice Question — Application

Q1 Ethene reacts with steam in the presence of an acid catalyst. Draw the mechanism for this reaction.

Practice Questions — Fact Recall

Q1 Write down the equation for the production of ethanol by fermentation.

Q2 Why is it necessary for the reaction to take place between 30 °C and 40 °C?

Q3 What is a biofuel?

Q4 a) Write down three equations which support the statement "bioethanol is a carbon neutral fuel".

b) Explain why bioethanol is not a carbon neutral fuel.

Q5 a) Give one advantage of using biofuels rather than fossil fuels.

b) Give three disadvantages of using biofuels.

Learning Objectives:

- Be able to apply IUPAC rules for nomenclature to naming aldehydes, ketones and carboxylic acids limited to chains with up to 6 carbon atoms.
- Know that primary alcohols can be oxidised to aldehydes, which can be further oxidised to carboxylic acids.
- Know that secondary alcohols can be oxidised to ketones.
- Know that tertiary alcohols are not easily oxidised.
- Know that acidified potassium dichromate(VI) is a suitable oxidising agent.
- Be able to write equations for these oxidation reactions, showing the oxidant as [O].
- Be able to explain how the method used to oxidise a primary alcohol determines whether an aldehyde or carboxylic acid is obtained.
- Be able to use chemical tests to distinguish between aldehydes and ketones including Fehling's solution and Tollens' reagent.

Specification Reference 3.3.1.1, 3.3.5.2

Tip: A molecule has to contain at least 3 carbon atoms for it to be a ketone.

7. Oxidising Alcohols

Oxidising an alcohol creates a carbon-oxygen double bond. Substances that contain these carbon-oxygen double bonds are known as carbonyl compounds — they're great fun. Honest.

The basics

The simple way to oxidise alcohols is to burn them. But you don't get the most exciting products by doing this. If you want to end up with something more interesting, you need a more sophisticated way of oxidising. You can use the oxidising agent acidified potassium dichromate(VI), $K_2Cr_2O_7$, to mildly oxidise 1° and 2° alcohols. In the reaction the orange dichromate(VI) ion, $Cr_2O_7^{2-}$, is reduced to the green chromium(III) ion, Cr^{3+}. Primary alcohols are oxidised to aldehydes and then to carboxylic acids. Secondary alcohols are oxidised to ketones only. Tertiary alcohols aren't oxidised.

Aldehydes, ketones and carboxylic acids

Aldehydes

Aldehydes and ketones are carbonyl compounds — they have the functional group C=O. Their general formula is $C_nH_{2n}O$. Aldehydes have a hydrogen and one alkyl group attached to the **carbonyl** carbon atom...

CH_3CH_2CHO

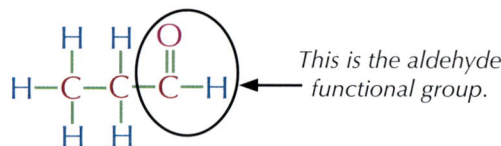

This is the aldehyde functional group.

Aldehydes have the suffix -al. You don't have to say which carbon the functional group is on — it's always on carbon-1. Naming aldehydes follows very similar rules to the naming of alcohols (see page 233).

Examples

CH_3CH_2CHO

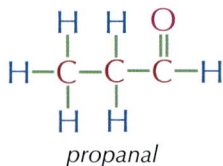

propanal

The longest continuous carbon chain is 3 carbon atoms, so the stem is propane.

So, the aldehyde is called propanal.

The longest continuous carbon chain is 4 carbon atoms, so the stem is butane.

There's a methyl group attached to the second carbon atom so there's a 2-methyl- prefix.

So, the aldehyde is called 2-methylbutanal.

$CH_3CH_2CH(CH_3)CHO$

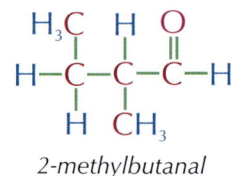

2-methylbutanal

Ketones

Ketones have two alkyl groups attached to the carbonyl carbon atom.

CH_3COCH_3

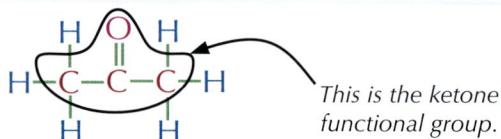

This is the ketone functional group.

The suffix for ketones is -one. For ketones with five or more carbons, you need to say which carbon the functional group is on.

Examples

```
   H  O  H
   |  ‖  |
H—C—C—C—H
   |     |
   H     H
```

propanone

The longest continuous carbon chain is 3 carbon atoms, so the stem is propane. So, the ketone is called propanone.

The longest continuous carbon chain is 5 carbon atoms, so the stem is pentane. The carbonyl is found on the second carbon atom. So, the ketone is called pentan-2-one.

```
   H  H  H  O  H
   |  |  |  ‖  |
H—C—C—C—C—C—H
   |  |  |     |
   H  H  H     H
```

pentan-2-one

> **Tip:** Remember that a suffix comes at the end of a name and a prefix comes at the beginning.

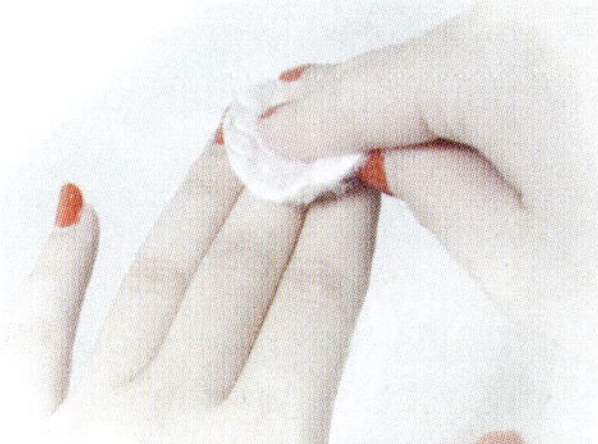

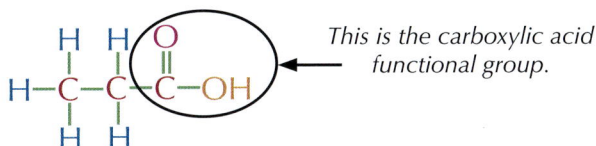

Figure 1: *The ketone propanone (also known as acetone) is commonly used as a nail varnish remover.*

Carboxylic acids

Carboxylic acids have a COOH group at the end of their carbon chain. Their general formula is $C_nH_{2n+1}COOH$.

```
   H  H  O
   |  |  ‖
H—C—C—C—OH
   |  |
   H  H
```

This is the carboxylic acid functional group.

The suffix for carboxylic acids is -oic. You also add the word 'acid' to the end of the name.

> **Tip:** The carboxylic acid functional group is usually written COOH and not CO_2H to show that the two oxygens are different — one is part of a carbonyl group (CO) and the other is part of an OH group.

Examples

```
   H  H  O
   |  |  ‖
H—C—C—C—OH
   |  |
   H  H
```

propanoic acid

The longest continuous carbon chain is 3 carbon atoms, so the stem is propane. So, the carboxylic acid is called propanoic acid.

The longest continuous carbon chain is 3 carbon atoms, so the stem is propane. There's a COOH group at each end of the carbon chain so it has a -dioic acid suffix. So, the carboxylic acid is called propanedioic acid.

```
    O  H  O
    ‖  |  ‖
HO—C—C—C—OH
       |
       H
```

propanedioic acid

> **Exam Tip**
> Don't get all these functional groups mixed up in the exam — you could lose easy marks. The only way to learn them is practice, practice and more practice.

Oxidation of primary alcohols

A primary alcohol is first oxidised to an aldehyde. This aldehyde can then be oxidised to a carboxylic acid. You can use the notation [O] to represent an oxidising agent. This means you can write equations like this:

$$\underset{\text{primary alcohol}}{R{-}CH_2{-}OH} + [O] \longrightarrow \underset{\text{aldehyde}}{R{-}\overset{\overset{\displaystyle O}{\|}}{C}{-}H} + H_2O$$

$$\underset{\text{aldehyde}}{R{-}\overset{\overset{\displaystyle O}{\|}}{C}{-}H} + [O] \xrightarrow{\text{reflux}} \underset{\text{carboxylic acid}}{R{-}\overset{\overset{\displaystyle O}{\|}}{C}{-}OH}$$

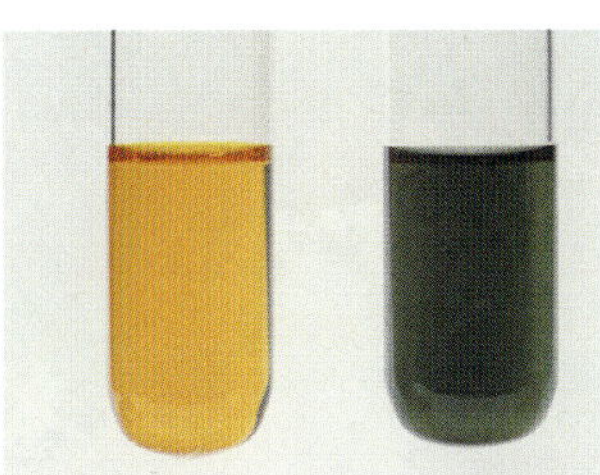

Figure 2: *Alcohol oxidation. The acidified potassium dichromate(VI) ion turns from orange to green when it oxidises an alcohol.*

You can control how far the alcohol is oxidised by controlling the reaction conditions.

Oxidising primary alcohols to aldehydes

Gently heating ethanol with potassium dichromate(VI) solution and sulfuric acid in a test tube should produce "apple" smelling ethanal (an aldehyde).

Tip: Make sure you balance the oxidising agents when you're writing these equations.

$$\begin{array}{c}\text{H}\;\;\text{H}\\ |\;\;\;\;|\\ \text{H}-\text{C}-\text{C}-\text{OH}\\ |\;\;\;\;|\\ \text{H}\;\;\text{H}\end{array} + [\text{O}] \xrightarrow[\text{distillation}]{H_2SO_4} \begin{array}{c}\text{H}\;\;\text{O}\\ |\;\;\;\;\|\\ \text{H}-\text{C}-\text{C}-\text{H}\\ |\;\;\;\;\;\;\\ \text{H}\;\;\;\;\;\;\end{array} + H_2O$$

ethanol → *ethanal*

However, it's really tricky to control the amount of heat and the aldehyde is usually oxidised to form "vinegar" smelling ethanoic acid.

Tip: As usual, carry out any necessary safety precautions whilst doing these experiments — here, you're using acid and you're heating the mixture so safety equipment should be used and care taken.

To get just the aldehyde, you need to get it out of the oxidising solution as soon as it forms.

You can do this by gently heating excess alcohol with a controlled amount of oxidising agent in distillation apparatus (see Figure 3). The aldehyde (which boils at a lower temperature than the alcohol) is distilled off immediately.

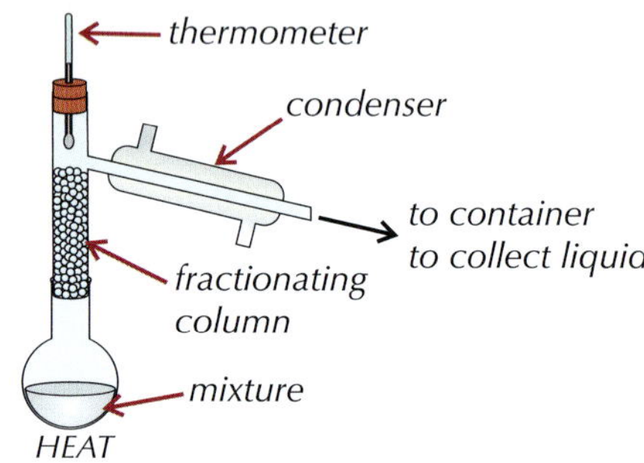

Figure 3: *Distillation apparatus.*

Exam Tip
It's really important that you learn the conditions required to produce both aldehydes and carboxylic acids. It can be pretty easy to mix them up but make sure you don't — you'll be expected to know them for the exam.

Oxidising primary alcohols to carboxylic acids

To produce the carboxylic acid, the alcohol has to be vigorously oxidised.

$$\begin{array}{c}\text{H}\;\;\text{H}\\ |\;\;\;\;|\\ \text{H}-\text{C}-\text{C}-\text{OH}\\ |\;\;\;\;|\\ \text{H}\;\;\text{H}\end{array} + 2[\text{O}] \xrightarrow{\text{reflux}} \begin{array}{c}\text{H}\;\;\text{O}\\ |\;\;\;\;\|\\ \text{H}-\text{C}-\text{C}-\text{OH}\\ |\;\;\;\;\;\;\\ \text{H}\;\;\;\;\;\;\end{array} + H_2O$$

ethanol → *ethanoic acid*

The alcohol is mixed with excess oxidising agent and heated under reflux (see Figures 4 and 5). Heating under reflux means you can increase the temperature of an organic reaction to boiling without losing volatile solvents, reactants or products. Any vapourised compounds are cooled, condense and drip back into the reaction mixture. So the aldehyde stays in the reaction mixture and is oxidised to carboxylic acid.

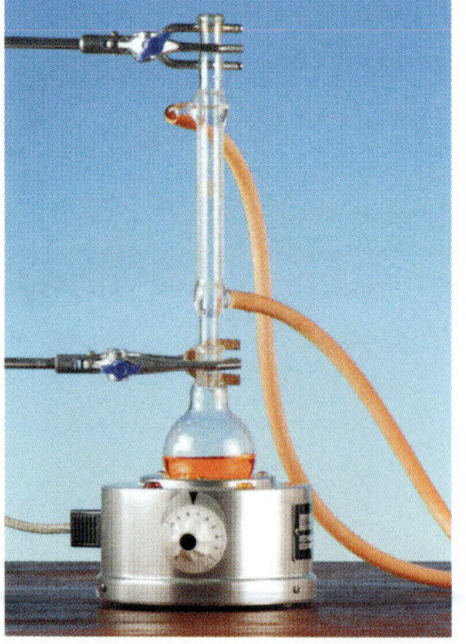

Figure 4: *Refluxing apparatus.*

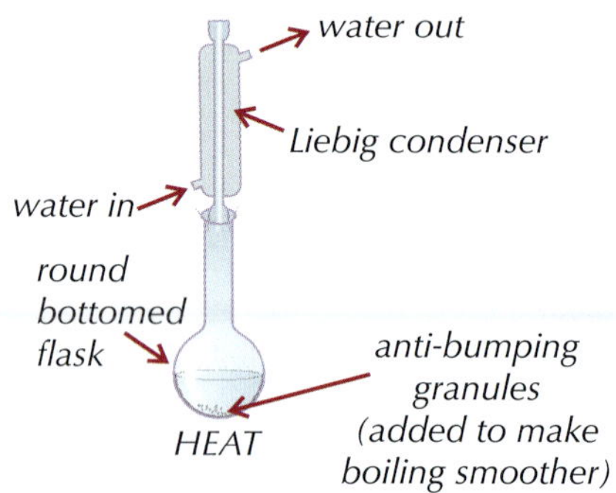

Figure 5: *Refluxing apparatus.*

Oxidation of secondary alcohols

Refluxing a secondary alcohol with acidified dichromate(VI) will produce a ketone.

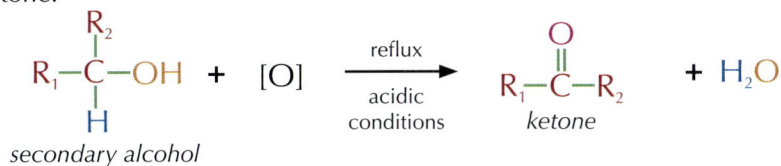

$$R_1-\underset{\underset{H}{|}}{\overset{\overset{R_2}{|}}{C}}-OH \;+\; [O] \xrightarrow[\text{acidic conditions}]{\text{reflux}} R_1-\overset{\overset{O}{||}}{C}-R_2 \;+\; H_2O$$

secondary alcohol → *ketone*

Example

$$H-\underset{\underset{H}{|}}{\overset{\overset{H}{|}}{C}}-\underset{\underset{H}{|}}{\overset{\overset{OH}{|}}{C}}-\underset{\underset{H}{|}}{\overset{\overset{H}{|}}{C}}-H \;+\; [O] \xrightarrow[\text{acidic conditions}]{\text{reflux}} H-\underset{\underset{H}{|}}{\overset{\overset{H}{|}}{C}}-\overset{\overset{O}{||}}{C}-\underset{\underset{H}{|}}{\overset{\overset{H}{|}}{C}}-H \;+\; H_2O$$

propan-2-ol → *propanone*

Ketones can't be oxidised easily, so even prolonged refluxing won't produce anything more.

Oxidation of tertiary alcohols

Tertiary alcohols don't react with acidified potassium dichromate(VI) at all — the solution stays orange. The only way to oxidise tertiary alcohols is by burning them.

Testing for aldehydes and ketones

Aldehydes and ketones can be distinguished using oxidising agents — aldehydes are easily oxidised but ketones aren't. Fehling's solution and Benedict's solution are both deep blue Cu^{2+} complexes (alkaline solutions of copper(II) sulfate), which reduce to a brick-red Cu_2O precipitate when warmed with an aldehyde, but stay blue with a ketone (see Figure 6). Tollens' reagent is a colourless $[Ag(NH_3)_2]^+$ complex — it's reduced to silver when warmed with an aldehyde, but not with a ketone. The silver will coat the inside of the apparatus to form a silver mirror (see Figure 7).

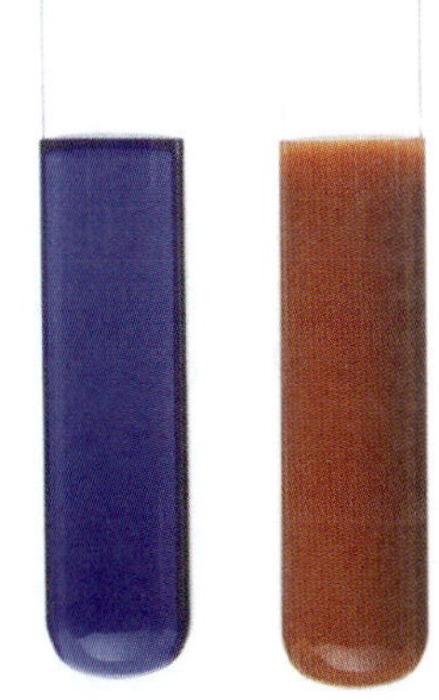

Figure 6: *Fehling's solution. The test-tube on the left shows the unreacted Fehling's solution. The test-tube on the right shows the result of the reaction of Fehling's solution with an aldehyde.*

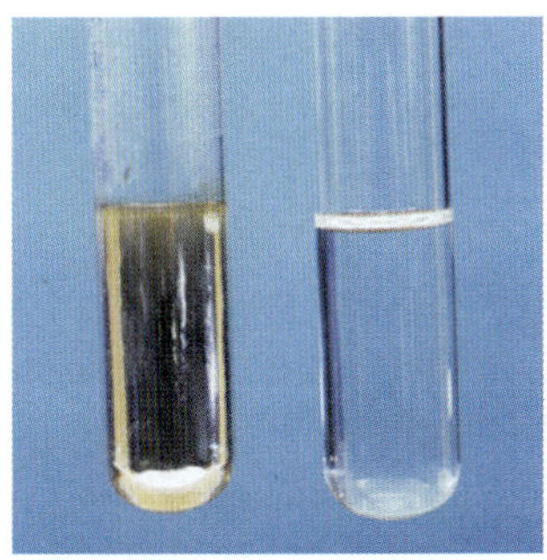

Figure 7: *The test-tube on the left shows the result of warming Tollens' reagent with an aldehyde. The test-tube on the right shows the result of warming Tollens' reagent with an ketone.*

Practice Question — Application

Q1 Draw the structures of the organic products of the following reactions.

a) A reaction between butan-2-ol and acidified potassium dichromate(VI) under reflux.

b) A reaction between butan-1-ol and acidified potassium dichromate(VI) using distillation apparatus.

c) A reaction between butan-1-ol and acidified potassium dichromate(VI) under reflux.

Practice Questions — Fact Recall

Q1 What are the functional groups of aldehydes, ketones and carboxylic acids?

Q2 Write a general equation for the reaction of a primary alcohol with an oxidising agent under reflux.

Q3 Name two reagents you could use to distinguish between an aldehyde and a ketone.

Section Summary

Make sure you know...

- That alkenes are unsaturated hydrocarbons which contain a double covalent bond.
- That the double bond in an alkene is a centre of high electron density.
- The mechanism of electrophilic addition reactions of alkenes with HBr, H_2SO_4 and Br_2.
- That bromine water can be used to test for unsaturation.
- How to predict the products of addition to unsymmetrical alkenes.
- How addition polymers are formed from alkenes.
- How to recognise and draw the repeating unit in a polyalkene.
- How to draw the structure of a monomer when given a section of the polymer.
- How to name addition polymers.
- That addition polymers are unreactive.
- The nature of intermolecular forces between molecules of polyalkenes.
- Some typical uses of poly(chloroethene) and how its properties can be modified using a plasticiser.
- How to name alcohols.
- That alcohols can be primary, secondary or tertiary.
- That alkenes can be formed from alcohols by acid catalysed elimination reactions and that this method provides a possible route to polymers without using monomers derived from oil.
- The mechanism for the elimination of water from alcohols.
- That alcohols are produced industrially by hydration of alkenes in the presence of an acid catalyst.
- The typical conditions for the industrial production of ethanol from ethene.
- The mechanism for the formation of an alcohol by the reaction of an alkene with steam in the presence of an acid catalyst.
- How ethanol is produced industrially by fermentation, including the conditions for the reaction.
- What a biofuel is and what carbon neutral means.
- Equations to show that ethanol, produced by fermentation, can be thought of as a carbon-neutral biofuel, but that this statement is invalid.
- The environmental and ethical issues linked to decision-making about biofuel use.
- How to name aldehydes, ketones and carboxylic acids.
- That primary and secondary alcohols can be oxidised to aldehydes, carboxylic acids and ketones by using an oxidising agent such as acidified potassium dichromate(VI).
- How to use Fehling's and Benedict's solution or Tollens' reagent to distinguish between aldehydes and ketones.

Exam-style Questions

1 Which substance is not involved in the reaction during the dehydration of propanol using a phosphoric acid catalyst?

A propene

B propane

C phosphoric acid

D water

(1 mark)

2 What type of reaction produces bromoethane from ethene and hydrogen bromide?

A nucleophilic addition

B electrophilic substitution

C electrophilic addition

D nucleophilic substitution

(1 mark)

3 Ethanol is a simple alcohol that can be used as a fuel, in alcoholic beverages, as a solvent and also as a starting molecule for many organic synthesis reactions. Its structure is shown below.

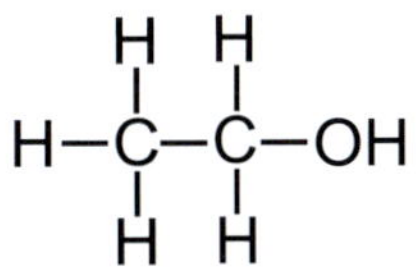

```
    H   H
    |   |
H — C — C — OH
    |   |
    H   H
```

A fermentation reaction can be used to produce ethanol.

3.1 Write down the equation for this reaction and state the conditions needed for the reaction to occur.

(2 marks)

3.2 Explain why production of ethanol by fermentation may be important in the future.

(1 mark)

3.3 Industrially, ethanol is also produced by steam hydration. Describe the conditions required for this reaction.

(2 marks)

3.4 Draw the mechanism for the production of ethanol by steam hydration.

(4 marks)

Ethanol can be used as a carbon neutral biofuel.

3.5 Define the term biofuel.

(1 mark)

3.6 By using equations, show that ethanol made by fermentation is a carbon neutral fuel.

(4 marks)

3.7 Explain why your answer to **3.6** is not a valid conclusion.

(2 marks)

3.8 Outline two environmental issues associated with biofuels.

(2 marks)

3.9 Outline one ethical issue associated with biofuels.

(1 mark)

4 A student has been given a sample of the alkene 3-methylpent-2-ene. The structure of 3-methylpent-2-ene is shown below.

```
    H  H     H  H
    |  |     |  |
H — C— C— C= C— C—H
    |  |  |     |
    H  H  CH3   H
```

4.1 Describe how the student could prove that the sample is unsaturated.

(2 marks)

The student reacts 3-methylpent-2-ene with hydrogen bromide (HBr).

4.2 Write the equation for the reaction.

(1 mark)

4.3 Draw the structure of the major product of the reaction.

(1 mark)

4.4 Use your knowledge of carbocation stability to explain why the structure given in **4.3** is the major product of this reaction.

(3 marks)

4.5 3-methylpent-2-ene can be polymerised using an addition polymerisation reaction. Draw the repeating unit of the polymer formed in this reaction.

(1 mark)

5 Poly(tetrafluoroethene) is a polymer, more commonly known as Teflon®. Teflon® is formed by addition polymerisation from its monomer.

5.1 What is a monomer?

(1 mark)

5.2 Give the IUPAC name of the monomer of Teflon®.

(1 mark)

5.3 Explain why poly(tetrafluoroethene) is unreactive.

(2 marks)

5.4 A plasticiser is sometimes added to a polymer.
How does a plasticiser affect the properties of a polymer?

(1 mark)

6 The structure of an isomer of butanol is shown below.

```
    H  H  H  H
    |  |  |  |
 H—C—C—C—C—H
    |  |  |  |
    H  H  H  OH
```

6.1 Give the IUPAC name of this isomer.

(1 mark)

The isomer is oxidised to produce butanal.

6.2 Name a suitable oxidising agent for this reaction.

(1 mark)

6.3 Describe the method used to carry out the oxidation.

(3 marks)

6.4 Give the equation for the reaction.
You may use [O] to represent the oxidising agent.

(1 mark)

6.5 A student is given the oxidised product, but she is not told whether it is an aldehyde or a ketone. She must carry out a test to determine what the oxidised product is. Describe two tests that she could carry out, and the expected outcome of each.

(4 marks)

6.6 Draw the structure of an isomer of butanol that will not oxidise.

(1 mark)

Learning Objectives:
- Be able to identify functional groups using reactions in the specification.
- Know the tests for alcohols, aldehydes, alkenes and carboxylic acids (Required Practical 6).

Specification Reference 3.3.6.1

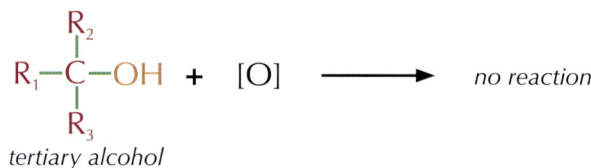

Tip: Primary alcohols would be oxidised to carboxylic acids if the aldehyde product was not distilled off when it formed.

Tip: Look back at pages 243-245 for more on these reactions.

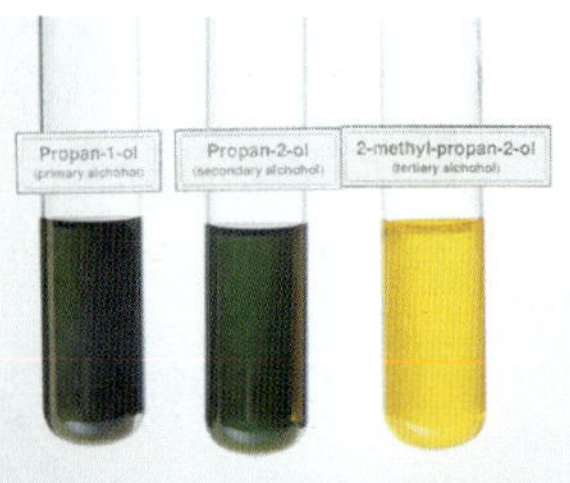

Figure 1: With a primary or secondary alcohol, the colour changes from orange to green as Cr(VI) changes to Cr(III). With a tertiary alcohol there is no reaction, so it stays orange.

Tip: If you're doing any of the tests shown on pages 250-252 in the lab, make sure you're aware of any hazards or safety issues involved.

1. Tests for Functional Groups

REQUIRED PRACTICAL 6

Organic compounds can be a bit tricky to identify. But handily there are a few simple tests that you can do that will help you tell them apart.

Testing for primary, secondary and tertiary alcohols

Potassium dichromate(VI) is an oxidising agent. It can oxidise primary and secondary alcohols to form aldehydes and ketones respectively (tertiary alcohols can't be oxidised). You can use the notation [O] to represent an oxidising agent when you write out a reaction.

$$\underset{\text{primary alcohol}}{R{-}CH_2{-}OH} + [O] \xrightarrow{\text{distil}} \underset{\text{aldehyde}}{R{-}C({=}O){-}H} + H_2O$$

$$\underset{\text{secondary alcohol}}{R_1{-}C(R_2)(H){-}OH} + [O] \xrightarrow[\text{acidic conditions}]{\text{reflux}} \underset{\text{ketone}}{R_1{-}C({=}O){-}R_2} + H_2O$$

$$\underset{\text{tertiary alcohol}}{R_1{-}C(R_2)(R_3){-}OH} + [O] \longrightarrow \text{no reaction}$$

As primary and secondary alcohols are oxidised, as in the reactions above, potassium dichromate(VI) is reduced. This is accompanied by a colour change (see Figure 1):

$$Cr_2O_7^{2-} + 6e^- + 14H^+ \rightarrow 2Cr^{3+} + 7H_2O$$

orange dichromate(VI) ions are reduced ($Cr_2O_7^{2-}$) — *green chromium(III) ions are formed* ($2Cr^{3+}$)

You can use this colour change to test for the presence of primary, secondary or tertiary alcohols. Here's the test you'll need to carry out:

1. Add 10 drops of the alcohol to 2 cm^3 of acidified potassium dichromate solution in a test tube.
2. Warm the mixture gently in a hot water bath.
3. Watch for a colour change — with primary and secondary alcohols, the orange solution slowly turns green as an aldehyde or ketone forms. No colour change is seen with tertiary alcohols.

This test is useful, but there is a problem. This test doesn't help to work out whether an alcohol is a primary or a secondary alcohol — it gives the same result for both. There are a few more tests you can carry out, however, to tell whether an unknown alcohol is primary or secondary, using the oxidised versions of the primary or secondary alcohol:

- If you oxidise an alcohol under reflux and it tests positive for being a carboxylic acid, then it's a primary alcohol.
- If you oxidise an alcohol under distillation conditions and it tests positive for being an aldehyde, then it's a primary alcohol.
- If you oxidise an alcohol under reflux (or distillation) and it tests positive for being a ketone, then it's a secondary alcohol.

The tests for aldehydes, ketones and carboxylic acids are covered on the next two pages.

Tip: You can also test for alcohols using sodium metal. If you add a small piece of sodium to a pure alcohol, it will fizz as it gives off H_2 gas. There are lots of other chemicals which react with sodium like this though, so it may not be conclusive. Also, sodium reacts violently with acids, so you have to be very sure that the solution you're testing isn't acidic first.

Testing for aldehydes and ketones

As you saw on pages 243-245, aldehydes are easily oxidised to carboxylic acids, whereas it's not easy to oxidise ketones further. You can use this property to tell the difference between them.

Fehling's and Benedict's solution

Fehling's solution is a blue solution of complexed copper(II) ions dissolved in sodium hydroxide. **Benedict's solution** is effectively the same as Fehling's solution except the copper(II) ions are dissolved in sodium carbonate instead of sodium hydroxide. To test an aldehyde or ketone with Benedict's or Fehling's solution, follow this method:

1. Add 2 cm^3 of Fehling's or Benedict's solution to a test tube. (Whichever one you use, it should be a clear blue solution.)
2. Add 5 drops of the aldehyde or ketone to the test tube.
3. Put the test tube in a hot water bath to warm it for a few minutes.

Benedict's and Fehling's solutions will both reduce to a brick red Cu_2O precipitate when warmed with an aldehyde. No reaction happens with ketones, so the solution will stay blue (see Figure 2).

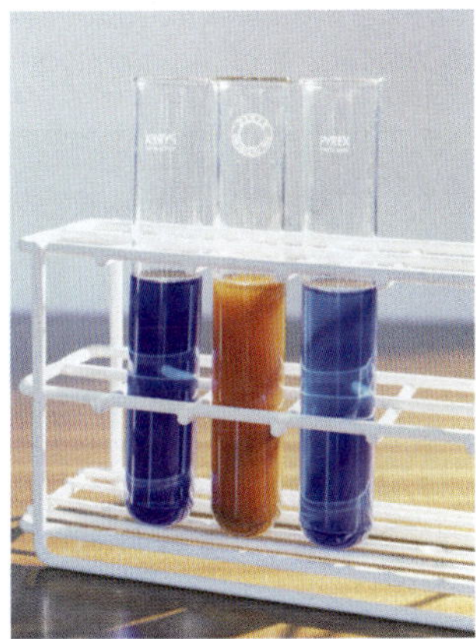

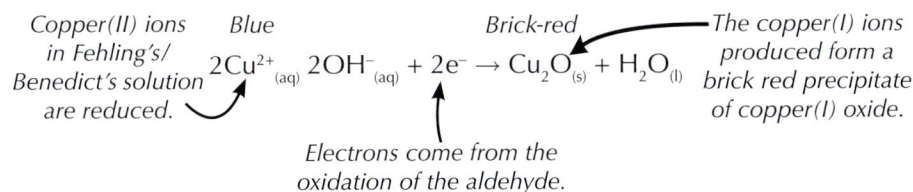

Figure 2: From left to right, these test tubes show: Fehling's solution, the brick-red precipitate made by reacting Fehling's solution with an aldehyde, and the result of reacting Fehling's solution with an ketone (i.e. no change).

$$2Cu^{2+}_{(aq)}\ 2OH^-_{(aq)} + 2e^- \rightarrow Cu_2O_{(s)} + H_2O_{(l)}$$

Copper(II) ions in Fehling's/Benedict's solution are reduced. (Blue)

Brick-red — *The copper(I) ions produced form a brick red precipitate of copper(I) oxide.*

Electrons come from the oxidation of the aldehyde.

Tip: Don't panic — you don't need to know this equation for what happens to the ions in the oxidising agent for your exams.

Tollens' reagent

Tollens' reagent contains a $[Ag(NH_3)_2]^+$ complex. It can be made using the following method.

1. Put 2 cm^3 of 0.10 mol dm^{-3} silver nitrate solution in a test tube.
2. Add a few drops of dilute sodium hydroxide solution. A light brown precipitate should form.
3. Add drops of dilute ammonia solution until the brown precipitate dissolves completely — this solution is Tollens' reagent.

Tip: You don't need to learn this equation for what happens to the silver ions here.

Figure 3: *The test-tube on the left contains unreacted Tollens' reagent. The one on the right shows the result of its reaction with an aldehyde.*

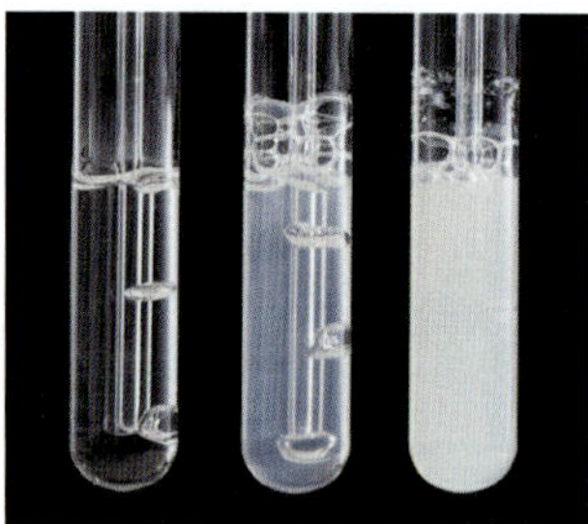

Figure 4: *Limewater turns from clear to cloudy when carbon dioxide is bubbled through it.*

Tip: When carbon dioxide is bubbled through limewater, a calcium carbonate precipitate is formed. It's this that makes the limewater turn cloudy.

Figure 5: *Bromine water turns colourless when shaken with alkenes.*

The silver ions in Tollens' reagent are reduced to silver metal when warmed with an aldehyde, but not with a ketone. The silver will coat the inside of the apparatus to form a silver mirror (see Figure 3).

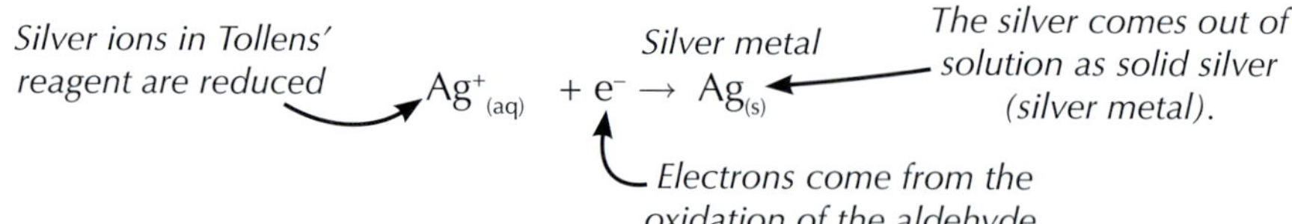
Silver ions in Tollens' reagent are reduced
Silver metal
The silver comes out of solution as solid silver (silver metal).
$Ag^+_{(aq)} + e^- \rightarrow Ag_{(s)}$
Electrons come from the oxidation of the aldehyde.

This doesn't always produce a lovely even, shiny silver coating, so don't worry if you end up with a silvery-grey precipitate — that's a positive result too.

Aldehydes and ketones are flammable so you have to take great care when heating them (such as in these tests). Aldehydes and ketones should always be warmed using a water bath, rather than using a Bunsen burner, to prevent them from catching alight.

Testing for carboxylic acids

Carboxylic acids are formed by oxidising aldehydes or primary alcohols — there's more about how this happens on page 244.

If you've got a sample of a substance that you think might be a carboxylic acid, you can test it using the following method:

1. Add 2 cm^3 of the solution that you want to test to a test tube.
2. Add 1 small spatula of solid sodium carbonate (or 2 cm^3 of sodium carbonate solution).
3. If the solution begins to fizz, bubble the gas that it produces through some limewater in a second test tube.

If the solution tested contains a carboxylic acid, carbon dioxide gas will be produced. When carbon dioxide is bubbled through limewater, the limewater turns cloudy.

Be careful though — this test will give a positive result with any acid, so you can only use it to distinguish between organic compounds when you already know that one of them is a carboxylic acid.

Testing for alkenes

Testing for alkenes involves testing for the presence of unsaturation. In the case of alkenes, this is testing for double bonds.

1. Add 2 cm^3 of the solution that you want to test to a test tube.
2. Add 2 cm^3 of bromine water to the test tube.
3. Shake the test tube.

SHAKE

Figure 6: *Shaking alkenes with bromine water causes orange bromine water to turn colourless.*

If an alkene is present, the bromine water will turn from orange to colourless.

Practice Questions — Application

Q1 A student has an unknown organic compound. It contains only carbon, oxygen and hydrogen atoms. He adds it to a test tube containing a solution of Tollens' reagent. He warms the test tube for a while, and silver metal starts to coat the inside of the test tube. What homologous series does his compound belong to?

Q2 A student has two compounds, A and B. She knows one is a ketone. She adds the two compounds, A and B, to separate test tubes containing Benedict's solution. A red precipitate forms in the test tube containing compound A, but nothing happens with compound B. Which compound, A or B, is the ketone?

Q3 A student is investigating the behaviour of primary alcohols.

a) She firstly adds a primary alcohol to a solution containing potassium dichromate(VI) ions and refluxes the mixture. What colour change would you expect her to see?

b) She then adds a spatula of calcium carbonate to the resulting solution. The solution starts to fizz. The student bubbles the resulting gas through a test tube containing limewater. What would you expect the student to observe?

Q4 A student has three test tubes. Each test tube contains one of three carbonyl compounds: propanal, propanoic acid or propanone. Outline a series of tests the student could carry out to identify which test tube contains which compound. You should include details of any reagents, conditions and expected observations.

Q5 a) State why warming samples of a substance with acidified potassium dichromate(VI) solution is not enough on its own to distinguish whether it is a primary alcohol or a secondary alcohol.

b) Describe an experiment that you could perform to distinguish between samples of a primary alcohol and a secondary alcohol.

Practice Questions — Fact Recall

Q1 What colour precipitate forms when Fehling's solution is mixed with an aldehyde?

Q2 Describe the procedure for making Tollens' reagent.

Q3 What observation would you make when an alkene was shaken with bromine water?

Tip: This topic will be so much easier if you know the chemistry of alcohols and carbonyl compounds inside out. Have a look back at pages 233-245 for all this chemistry.

2. Mass Spectrometry

Learning Objective:

- Know that mass spectrometry can be used to determine the molecular formula of a compound.
- Be able to use precise atomic masses and the precise molecular mass to determine the molecular formula of a compound.

Specification Reference 3.3.6.2

An analytical technique is a method of analysing a substance to learn more about it. This topic deals with one specific analytic technique — mass spectrometry. Mass spectrometry uses the mass of a compound to identify it.

Finding relative molecular masses

You saw on pages 27-32 how mass spectrometry can be used to find relative isotopic masses, the abundance of different isotopes of an element, and the relative molecular mass, M_r, of a compound. Remember — the M_r of a compound is given by the molecular ion peak on the spectrum. The mass/charge (m/z) value of the molecular ion peak is equal to the M_r of the molecule.

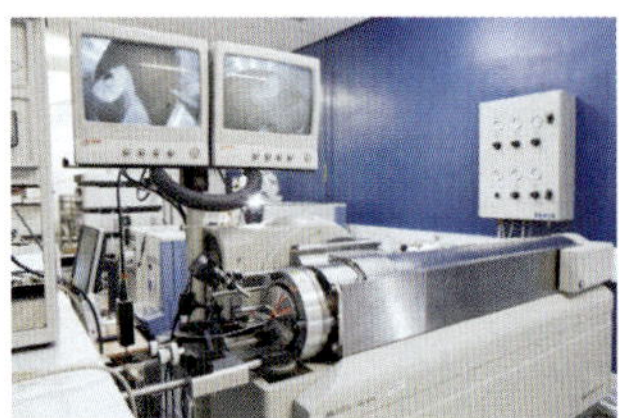

Figure 1: *Mass spectrometer.*

Example

The mass spectrum of a straight chain alkane contains a molecular ion peak with m/z = 72. Identify the compound.

1. The m/z value of the molecular ion peak is 72, so the M_r of the compound must be 72.
2. If you calculate the molecular masses of the first few straight-chain alkanes, you'll find that the one with a molecular mass of 72 is pentane (C_5H_{12}): M_r of pentane = (5 × 12.0) + (12 × 1.0) = 72.0
3. So the compound must be pentane.

High resolution mass spectrometry

High resolution mass spectrometers can measure atomic and molecular masses extremely accurately (to several decimal places). This can be useful for identifying compounds that appear to have the same M_r when they're rounded to the nearest whole number.

For example, propane (C_3H_8) and ethanal (CH_3CHO) both have an M_r of 44 to the nearest whole number. But on a high resolution mass spectrum, propane has a molecular ion peak with m/z = 44.0624 and ethanal has a molecular ion peak with m/z = 44.0302.

Tip: On a normal (low resolution) mass spectrum, all of these molecules would show up as having an m/z of 98.

Example

On a high resolution mass spectrum, a compound has a molecular ion peak with m/z = 98.0336. What is its molecular formula?

A $C_5H_{10}N_2$ B $C_6H_{10}O$ C C_7H_{14} D $C_5H_6O_2$

Use these precise atomic masses to work out your answer:

^{1}H — 1.0078 ^{12}C — 12.0000 ^{14}N — 14.0031 ^{16}O — 15.9949

Work out the precise molecular mass of each compound:

$C_5H_{10}N_2$: M_r = (5 × 12.0000) + (10 × 1.0078) + (2 × 14.0031) = 98.0842

$C_6H_{10}O$: M_r = (6 × 12.0000) + (10 × 1.0078) + 15.9949 = 98.0729

C_7H_{14} : M_r = (7 × 12.0000) + (14 × 1.0078) = 98.1092

$C_5H_6O_2$: M_r = (5 × 12.0000) + (6 × 1.0078) + (2 × 15.9949) = 98.0336

So the answer is **D**, $C_5H_6O_2$.

Practice Questions — Application

Q1 A student runs low resolution mass spectrometry on a sample of a compound, which he knows is either C_3H_6O, C_3H_8N or C_4H_{10}.

a) Why might low resolution mass spectrometry not be able to distinguish between these molecules?

b) The sample is injected into a high resolution mass spectrometer. A molecular ion peak is seen at m/z = 58.0655.
What is the molecular formula of the compound?
Use these precise atomic masses to work out your answer:
^{1}H: 1.0078 ^{12}C: 12.0000
^{14}N: 14.0031 ^{16}O: 15.9949

Q2 A mixture of two compounds was analysed in a high resolution time of flight mass spectrometer. The sample produced the two molecular ion peaks shown on the spectrum below.

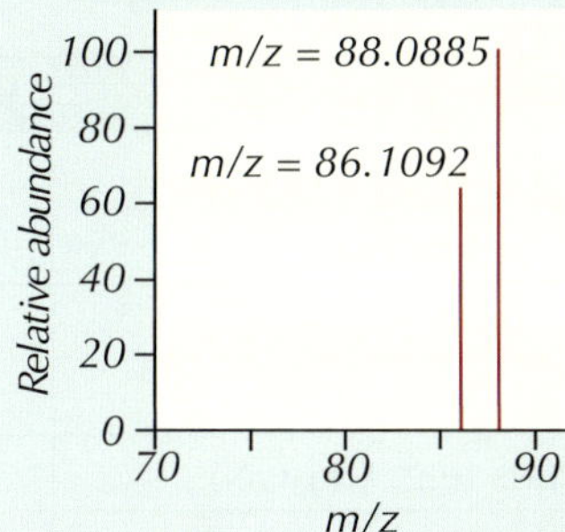

The mixture consists of two of the compounds listed below. Using the precise atomic masses given in Q1, work out which two of the compounds make up the mixture.

Molecule	Formula
butanoic acid	$CH_3CH_2CH_2COOH$
pentan-1-ol	$CH_3CH_2CH_2CH_2CH_2OH$
pentan-3-one	$CH_3CH_2COCH_2CH_3$
hexane	$CH_3CH_2CH_2CH_2CH_2CH_3$

Learning Objectives:

- Know that bonds in a molecule absorb infrared radiation at characteristic wavenumbers.
- Know that 'fingerprinting' allows identification of a particular molecule by comparison of spectra.
- Be able to use infrared spectra and the Chemistry Data Sheet or Booklet to identify particular bonds, and therefore functional groups, and also to identify impurities.
- Understand the link between absorption of infrared radiation by bonds in CO_2, methane, water vapour and global warming.

Specification Reference 3.3.6.3

3. Infrared Spectroscopy

Infrared spectroscopy is another analytical technique. It uses the fact that bonds in different functional groups absorb different frequencies of infrared light. We can use an infrared spectrum of a molecule to identify its functional groups.

The basics

In **infrared (IR) spectroscopy**, a beam of IR radiation is passed through a sample of a chemical. The IR radiation is absorbed by the covalent bonds in the molecules, increasing their vibrational energy. Bonds between different atoms absorb different frequencies of IR radiation. Bonds in different places in a molecule absorb different frequencies too — so the O–H bond in an alcohol and the O–H bond in a carboxylic acid absorb different frequencies. Figure 1 shows what frequencies different bonds absorb — you don't need to learn this data, but you do need to understand how to use it. Wavenumber is the measure used for the frequency (it's just 1/wavelength).

Bond	Where it's found	Wavenumber (cm^{-1})
N–H (amines)	amines (e.g. CH_3NH_2)	3300 - 3500
O–H (alcohols)	alcohols	3230 - 3550
C–H	most organic molecules	2850 - 3300
O–H (acids)	carboxylic acids	2500 - 3000
C≡N	nitriles (e.g. CH_3CN)	2220 - 2260
C=O	aldehydes, ketones, carboxylic acids, esters	1680 - 1750
C=C	alkenes	1620 - 1680
C–O	alcohols, carboxylic acids	1000 - 1300
C–C	most organic molecules	750 - 1100

Figure 1: *Table showing the absorption of different bonds.*

Exam Tip
You'll get a table like the one in Figure 1 on the Data Sheet in your exam. So there's no need to memorise all those numbers — yay.

An infrared spectrometer produces a graph that shows you what frequencies of radiation the bond in the molecules are absorbing. So you can use it to identify the functional groups in a molecule. The peaks show you where radiation is being absorbed — the 'peaks' on IR spectra are upside-down.

Tip: Most organic molecules will have loads of C–H bonds in them so the region at ~3000 cm^{-1} on an IR spectrum isn't always very useful.

Examples — Maths Skills

The structure of ethanal is shown on the right:

This is the infrared spectrum of ethanal:

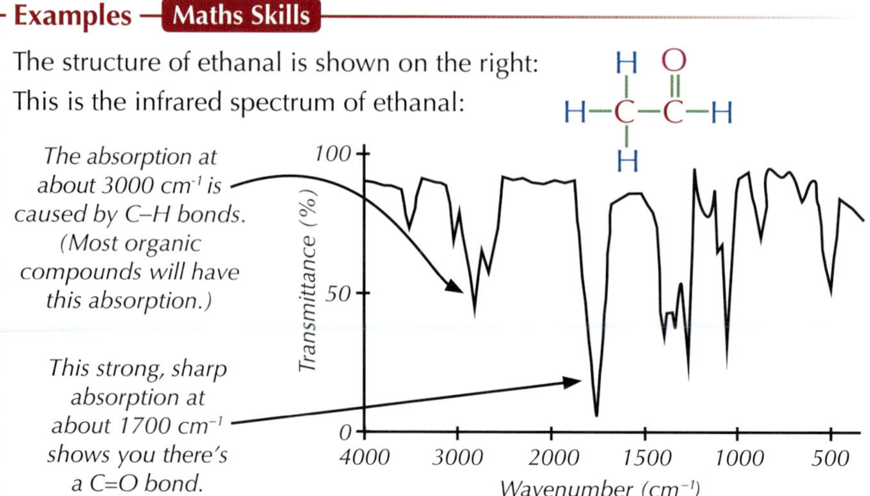
H O
H–C–C–H
H
The absorption at about 3000 cm⁻¹ is caused by C–H bonds. (Most organic compounds will have this absorption.)
This strong, sharp absorption at about 1700 cm⁻¹ shows you there's a C=O bond.
Transmittance (%)
100
50
0
4000
3000
2000
1500
1000
500
Wavenumber (cm⁻¹)

Here is the structure and infrared spectrum of ethylamine.

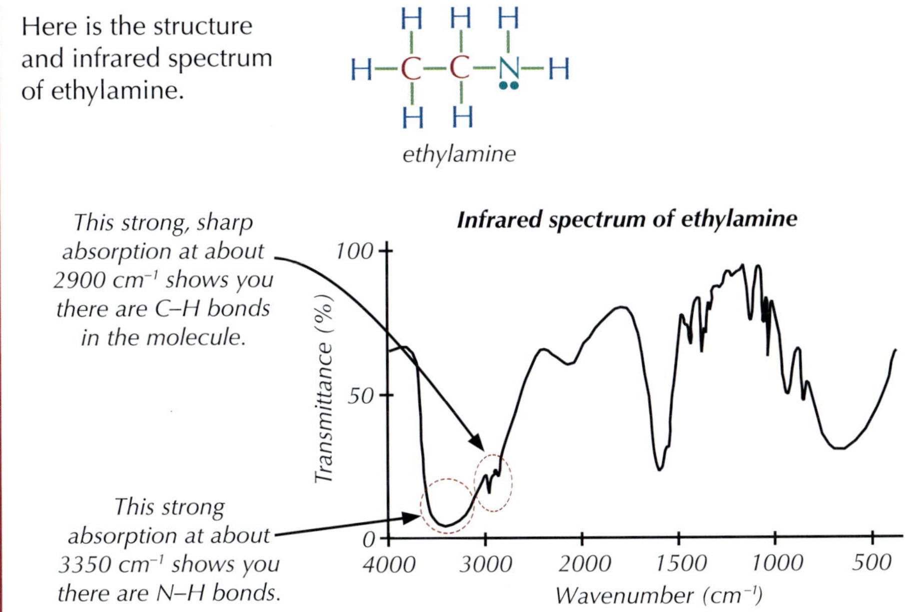
H H H
H–C–C–N–H
H H
ethylamine
Infrared spectrum of ethylamine
This strong, sharp absorption at about 2900 cm⁻¹ shows you there are C–H bonds in the molecule.
This strong absorption at about 3350 cm⁻¹ shows you there are N–H bonds.
Transmittance (%)
100
50
0
4000
3000
2000
1500
1000
500
Wavenumber (cm⁻¹)

Tip: When you're reading an infrared spectrum, always double check the scale. The wavenumbers increase from right to left — don't get caught out.

The fingerprint region

The region between 1000 cm^{-1} and 1550 cm^{-1} on the spectrum is called the **fingerprint region**. It's unique to a particular compound. You can check this region of an unknown compound's IR spectrum against those of known compounds. If it matches one of them, you know what the molecule is.

Tip: There are computer databases that store the infrared spectra of thousands of pure organic compounds. So you can do this checking and comparing relatively quickly using a computer.

Example — Maths Skills

Here is the structure and the infrared spectrum of ethanoic acid.

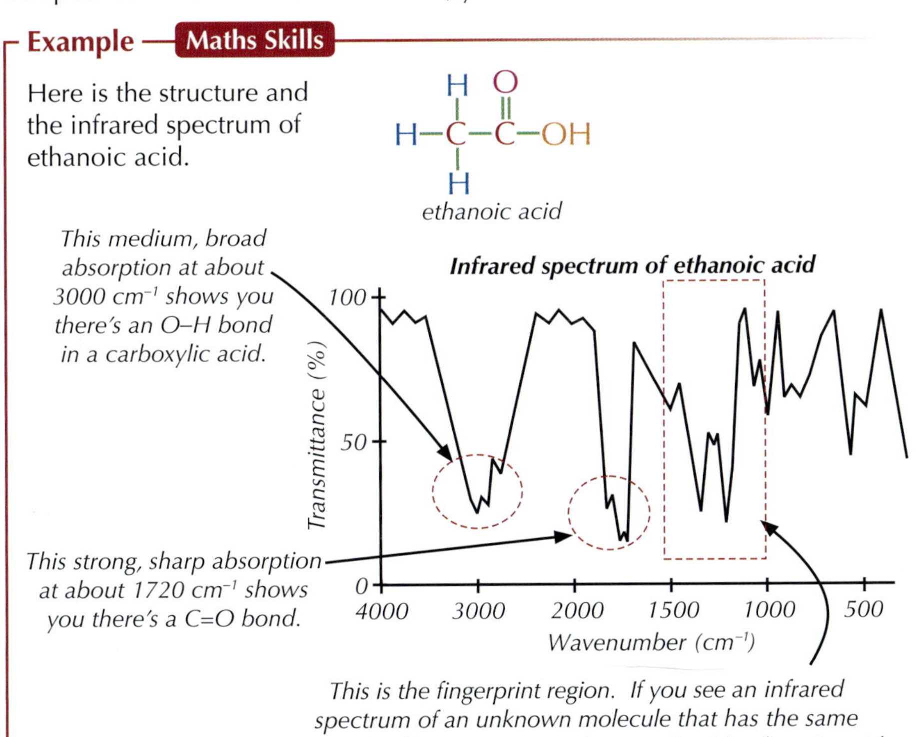
H O
H–C–C–OH
H
ethanoic acid
Infrared spectrum of ethanoic acid
This medium, broad absorption at about 3000 cm⁻¹ shows you there's an O–H bond in a carboxylic acid.
This strong, sharp absorption at about 1720 cm⁻¹ shows you there's a C=O bond.
Transmittance (%)
100
50
0
4000
3000
2000
1500
1000
500
Wavenumber (cm⁻¹)
This is the fingerprint region. If you see an infrared spectrum of an unknown molecule that has the same pattern in this area, you can be sure that it's ethanoic acid.

Infrared spectroscopy can also be used to assess how pure a compound is and identify impurities — impurities produce extra peaks in the fingerprint region.

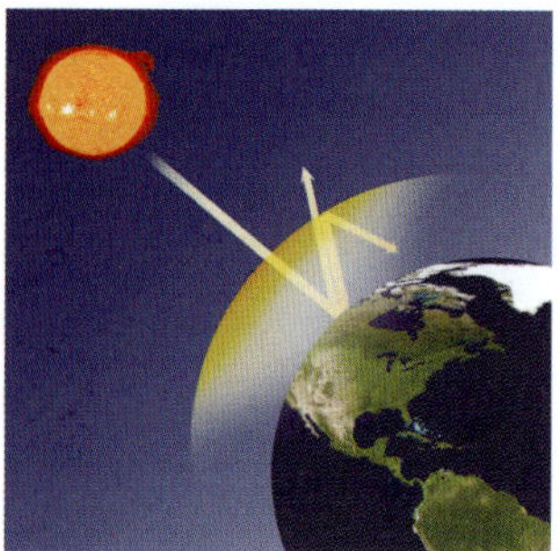

Figure 2: *The greenhouse effect.*

Tip: You can use the table on page 256 to help you out with these questions.

Infrared absorption and global warming

The Sun emits mainly UV/visible radiation which is absorbed by the Earth's surface and re-emitted as IR radiation. Molecules of greenhouse gases, like carbon dioxide, methane and water vapour, have bonds that are really good at absorbing infrared energy — so if the amounts of these gases in the atmosphere increase, more IR radiation is absorbed which leads to global warming. The more IR radiation a molecule absorbs, the more effective they are as greenhouse gases.

Practice Questions — Application

Q1 The spectrum below is the infrared spectrum of a carboxylic acid with $M_r = 74$.

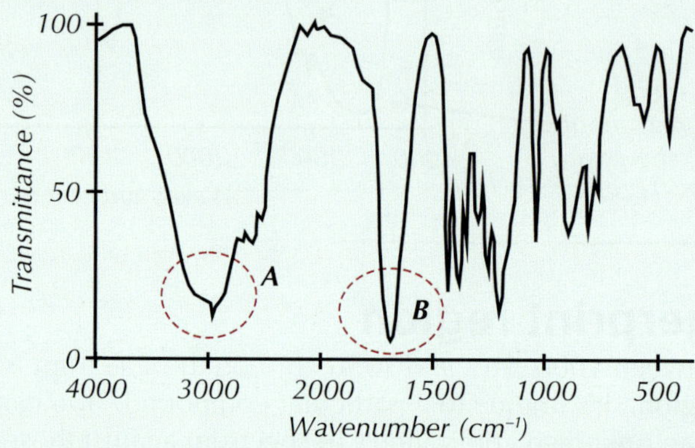

a) Identify the bonds that create the peaks marked **A** and **B** in the diagram.

b) Draw the displayed formula of the molecule.

Q2 The spectrum below shows the infrared spectrum for an unknown molecule. Use the spectrum to identify one important bond that can be found in the molecule.

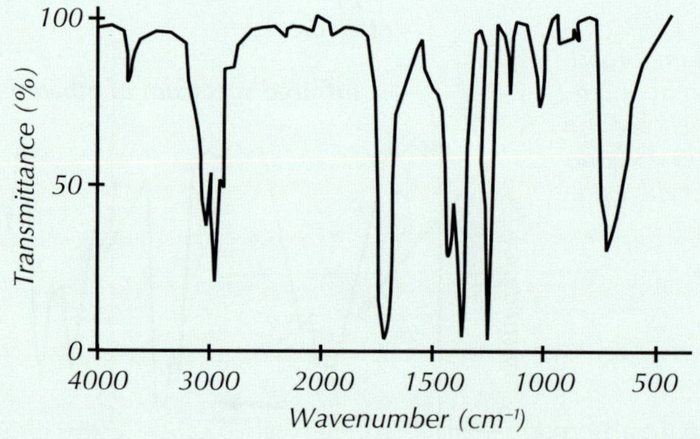

Practice Questions — Fact Recall

Q1 Give a brief explanation of how an infrared spectrum is created.

Q2 Over what frequency range is the fingerprint region of an IR spectrum found?

Section Summary

Make sure you know...

- How to use potassium dichromate(VI) to distinguish between primary and secondary and tertiary alcohols.
- How to use Fehling's (or Benedict's) solution to distinguish between aldehydes and ketones.
- How to make Tollens' reagent and use it to distinguish between aldehydes and ketones.
- How to use calcium carbonate as a test for carboxylic acids.
- How to use bromine water to test for alkenes.
- How to use mass spectrometry to identify the molecular mass of a compound.
- How to use mass spectrometry to determine the molecular formula of a compound.
- That high resolution mass spectrometry can be used to identify compounds with similar M_r values.
- How to use precise atomic masses and precise molecular masses to determine the molecular formula of a compound.
- That certain functional groups absorb infrared radiation at characteristic frequencies.
- That the 'fingerprint region' of an infrared spectrum allows identification of a molecule by comparison of spectra.
- How to use infrared spectra to identify particular functional groups and to identify impurities.
- That greenhouse gases, such as CO_2, water and methane, absorb infrared radiation and so increasing the amounts of them in the atmosphere can cause global warming.

Exam-style Questions

1 A scientist has synthesised two molecules — molecule **A** and molecule **B**. Both of the molecules were synthesised by reacting 1-bromopropane with OH^- ions. The structure of 1-bromopropane is shown below.

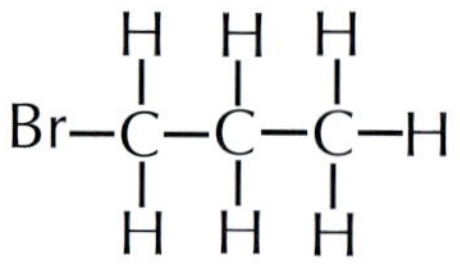

```
      H   H   H
      |   |   |
Br—C—C—C—H
      |   |   |
      H   H   H
```

The infrared spectra of the molecules are shown below.

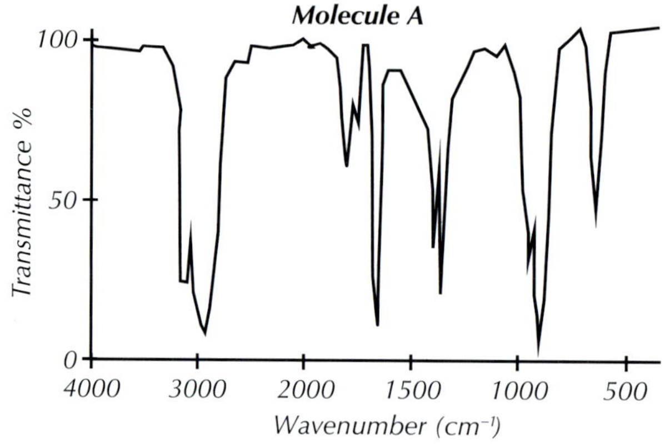

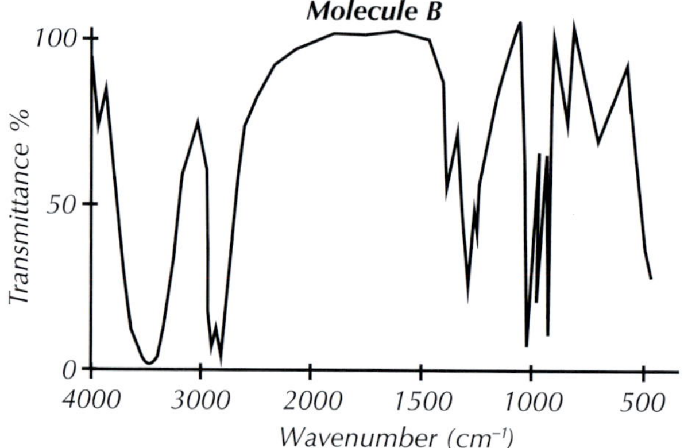

1.1 Neither molecule **A** nor molecule **B** contains halogen atoms. Use the table on page 256 to help you predict the structures of molecule **A** and molecule **B**. Explain your reasoning.

(4 marks)

1.2 Name molecule **A** and molecule **B**.

(2 marks)

1.3 Give the reagents and conditions that are needed to produce each molecule from 1-bromopropane.

(2 marks)

2 A student is investigating the nature of various greenhouse gases using infrared spectroscopy. The graphs below show the spectra of water vapour, carbon dioxide and methane.

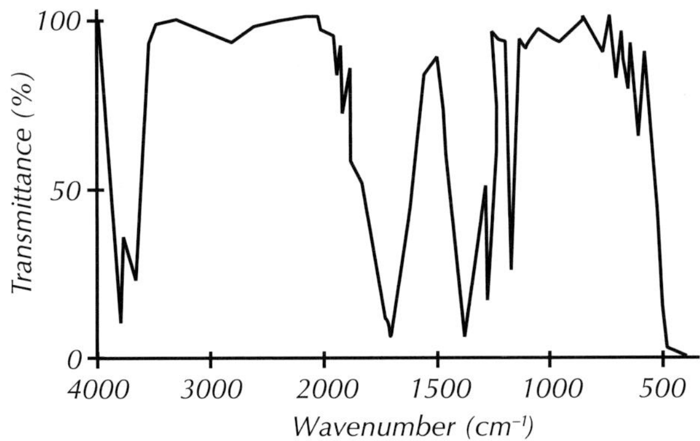

Spectrum of carbon dioxide

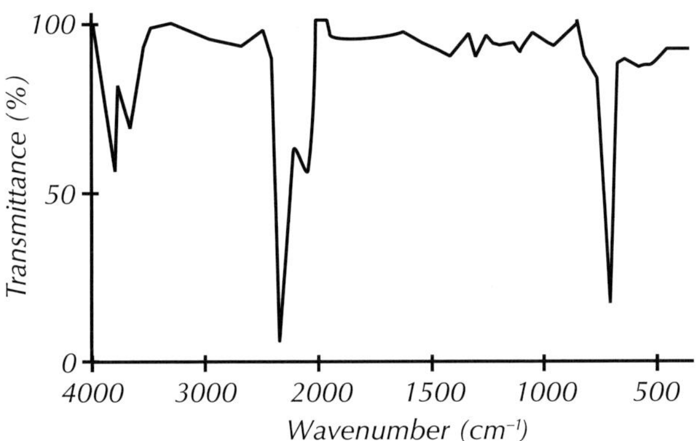

Spectrum of methane

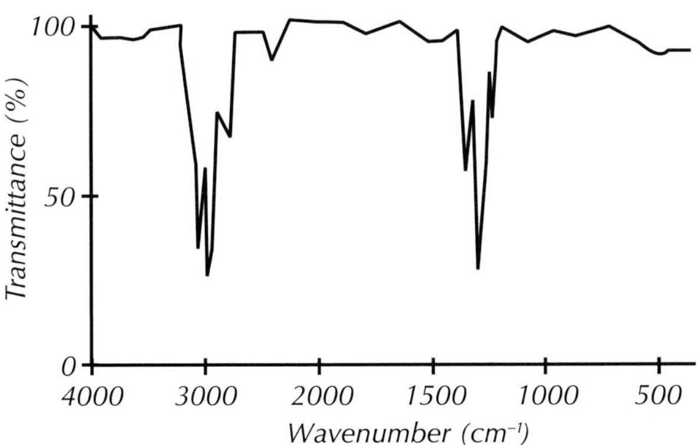

Predict, with reasoning, whether water vapour, carbon dioxide or methane is the most effective greenhouse gas.

(2 marks)

3 A student has three organic compounds — butanal, propanoic acid and ethanol. The substances are each in a separate test tube, labelled **A**, **B** and **C**.

The student carries out a series of tests to work out which compound is contained in each test tube.

Her results are summarised in the table below.

	Test		
	Addition to potassium dichromate(VI) solution and warmed	Addition of a spatula of calcium carbonate	Addition to Tollens' reagent and warmed
Compound A	Solution turns from orange to green	No reaction	No reaction
Compound B	Solution turns from orange to green	No reaction	Silver mirror coats the inside of the test tube
Compound C	No reaction	Fizzing — gas produced turns limewater turns cloudy	No reaction

With reasoning, correctly assign the identity of compounds **A**, **B** and **C**.

(3 marks)

4 A student has two molecules — molecule **X** and molecule **Y**. She knows each molecule is either propanal or propanone.

4.1 Draw out the displayed formulae of propanal.

(1 mark)

4.2 Draw out the displayed formula of propanone.

(1 mark)

4.3 Explain why the infrared spectroscopy would not be a suitable technique to use to distinguish between molecule **X** and molecule **Y**.

(1 mark)

4.4 Explain why high resolution mass spectrometry would not be a suitable technique to use to distinguish between molecule **X** and molecule **Y**.

(1 mark)

4.5 Outline a simple test that the student could use to identify molecule **X** and molecule **Y**.

You should include the names of any reagents used, any specific safety precautions that should be taken and how the test results would show the identity of each molecule.

(3 marks)

5 Three organic compounds, **A**, **B** and **C** were analysed using high resolution mass spectrometry and infrared spectroscopy. None of the compounds contain any atoms other than hydrogen, carbon and oxygen.

The tables below show the *m/z* value of the molecular ion peak in the mass spectrum of each compound and the precise atomic masses of hydrogen, carbon and oxygen.

Compound	*m/z*
A	72.0573
B	72.0936
C	70.0780

Atom	Atomic mass
^{1}H	1.0078
^{12}C	12.0000
^{16}O	15.9949

Each of the infrared spectra below matches one of compounds **A**, **B** and **C**.

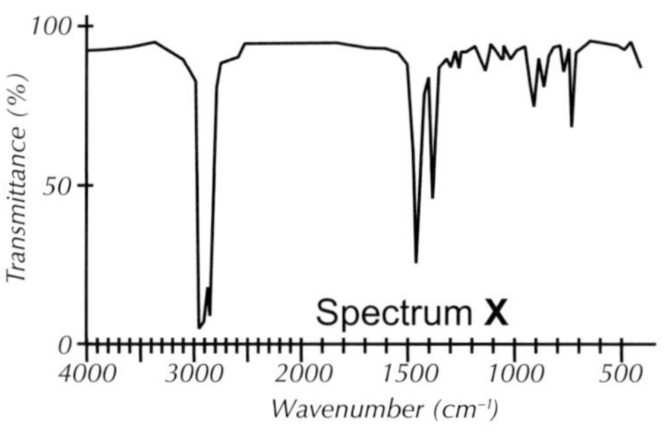

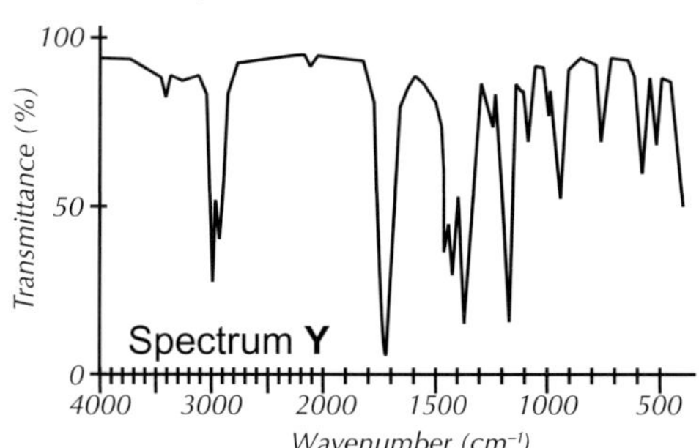

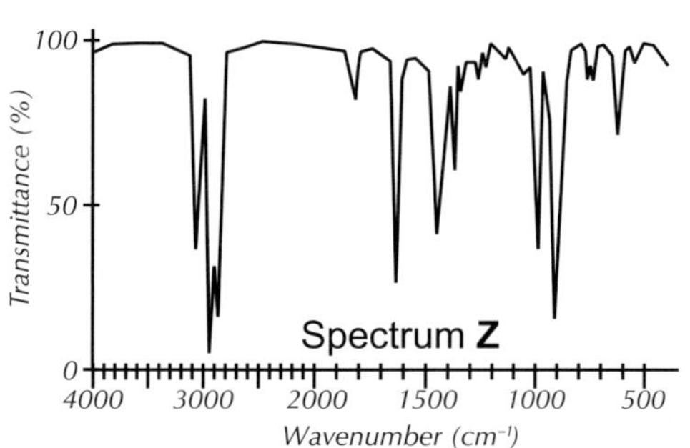

5.1 Using your knowledge of organic chemistry and analytical techniques, along with data given above and in the table on page 256, find the molecular formula of each of compounds **A**, **B** and **C**, and to match each of the infrared spectra above to one of the three compounds. Explain your reasoning.

(8 marks)

5.2 Given that none of compounds **A**, **B** and **C** contain branched carbon chains, suggest one possible identity for each of the three compounds. (There may be more than one possible answer for each.)

(3 marks)

5.3 What feature of infrared spectra would allow you to check whether your suggestions in **5.2** were correct?

(1 mark)

1. Enthalpy Changes

Learning Objective:

- Be able to define the terms enthalpy of formation, ionisation energy, enthalpy of atomisation, bond enthalpy, electron affinity, lattice enthalpy (defined as either lattice dissociation or lattice formation) and enthalpy of hydration.

Specification Reference 3.1.8.1

Some of this stuff may ring a few bells from the energetics section in Unit 1 (pages 108-119). Make sure you understand all the definitions in this topic because they're really important for the rest of the section.

The basics

Enthalpy notation

Enthalpy change, ΔH (delta H), is the heat energy transferred in a reaction at constant pressure. The units of ΔH are kJ mol^{-1}. You write ΔH° to show that the substances were in their standard states and that the measurements were made under **standard conditions**. Standard conditions are 100 kPa (about 1 atm) pressure and a stated temperature (e.g. ΔH°_{298}). In this book, all the enthalpy changes are measured at 298 K (25 °C). Sometimes the notation will also include a subscript to signify whether the enthalpy change is for a reaction (r), for combustion (c), or for the formation of a new compound (f).

Exothermic and endothermic reactions

Exothermic reactions have a negative ΔH value, because heat energy is given out (the chemicals lose energy). **Endothermic reactions** have a positive ΔH value, because heat energy is absorbed (the chemicals gain energy).

Tip: You covered enthalpy notation on page 108 so if you need a reminder, have a flick back now.

What is lattice enthalpy?

Ions in lattices are held together by ionic bonds. It takes energy to break apart the bonds, and energy is given out when new bonds form. The energy needed to break a bond between two ions is the same amount of energy that is given out when that bond is formed. These 'lattice enthalpies' have specific values that differ depending on the ions involved (see Figure 1). So, lattice enthalpy is a measure of ionic bond strength. It can be defined in two ways.

Lattice formation enthalpy: the enthalpy change when 1 mole of a solid ionic compound is formed from its gaseous ions.

Compound	*Lattice formation enthalpy (kJ mol^{-1})*
LiCl	*–826*
NaCl	*–787*
KCl	*–701*
RbCl	*–692*

Figure 1: *Lattice formation enthalpies of chloride compounds.*

Example

Formation of sodium chloride:

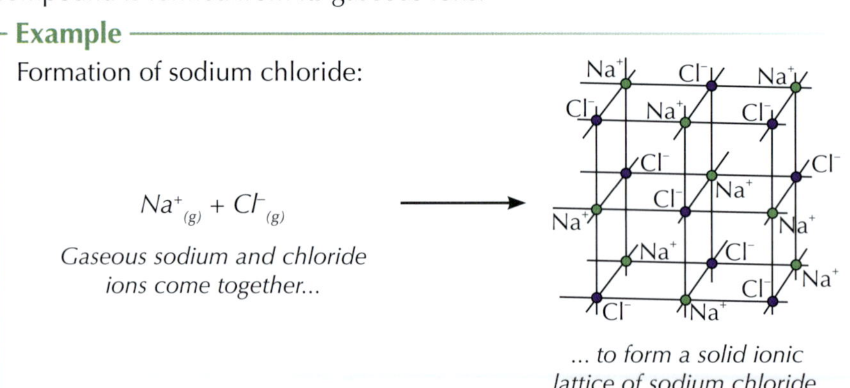

Na+(g) + Cl–(g)
Gaseous sodium and chloride ions come together...
Na+
Cl–
... to form a solid ionic lattice of sodium chloride.

Lattice formation enthalpy, $\Delta H_{lattice}$, for this reaction is –787 kJ mol^{-1}.
The negative ΔH value shows that lattice formation is an exothermic process.

Lattice dissociation enthalpy: the enthalpy change when 1 mole of a solid ionic compound is completely dissociated into its gaseous ions.

Example

Dissociation of sodium chloride:

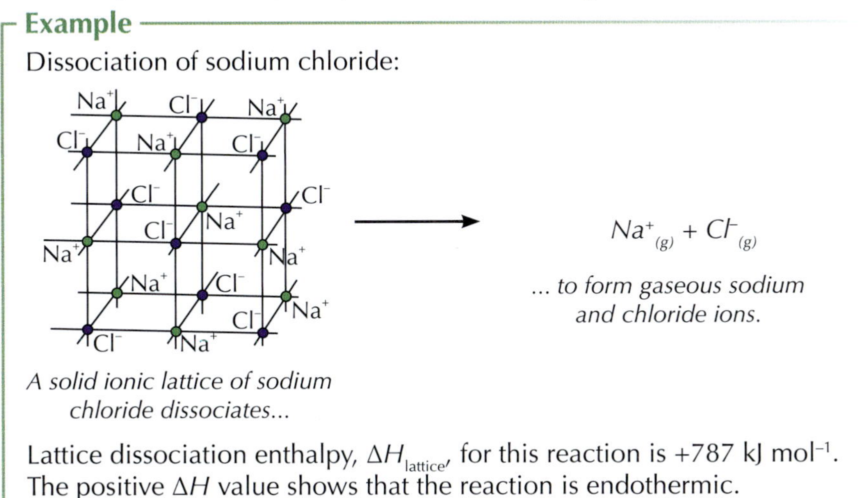

$Na^+_{(g)} + Cl^-_{(g)}$
... to form gaseous sodium and chloride ions.

A solid ionic lattice of sodium chloride dissociates...

Lattice dissociation enthalpy, $\Delta H_{lattice}$, for this reaction is +787 kJ mol^{-1}. The positive ΔH value shows that the reaction is endothermic.

Exam Tip

In the exam, make sure you use the term 'enthalpy' and not 'energy' otherwise you'll lose a mark.

Tip: Notice that the lattice formation enthalpy and the lattice dissociation enthalpy are exactly the same size — the only difference is that one's negative and the other one's positive.

Determining lattice enthalpy

Lattice enthalpy can't be measured directly. Instead, you have to combine the enthalpies from a number of other processes to work out the lattice enthalpy — see Figure 2.

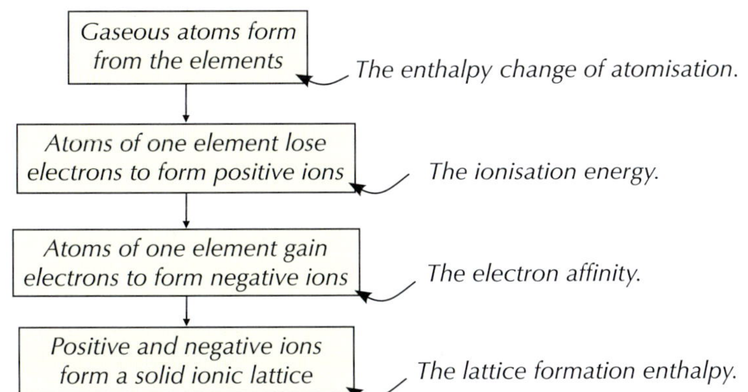

Gaseous atoms form from the elements
The enthalpy change of atomisation.
Atoms of one element lose electrons to form positive ions
The ionisation energy.
Atoms of one element gain electrons to form negative ions
The electron affinity.
Positive and negative ions form a solid ionic lattice
The lattice formation enthalpy.

Figure 2: *The processes which lead to the formation of a lattice.*

Tip: Although lattice enthalpy can't be measured directly, it can be calculated using a Born-Haber cycle — all will be revealed on page 267.

Example

The processes that lead to the formation of sodium chloride:

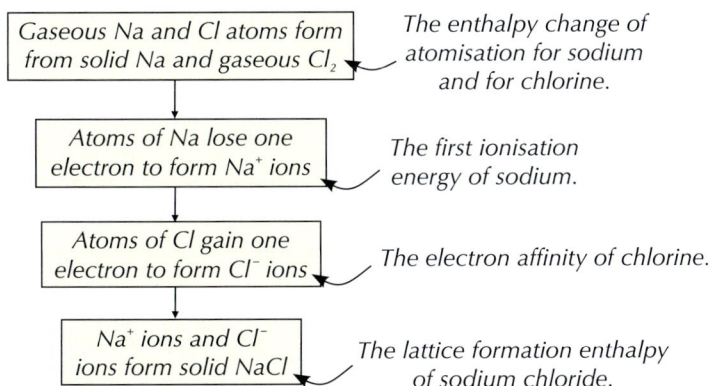

Gaseous Na and Cl atoms form from solid Na and gaseous Cl_2 → *The enthalpy change of atomisation for sodium and for chlorine.*

↓

Atoms of Na lose one electron to form Na^+ ions → *The first ionisation energy of sodium.*

↓

Atoms of Cl gain one electron to form Cl^- ions → *The electron affinity of chlorine.*

↓

Na^+ ions and Cl^- ions form solid NaCl → *The lattice formation enthalpy of sodium chloride.*

Figure 3: *A 3D model of a sodium chloride lattice.*

Unfortunately, each specific type of enthalpy change has its own definition and you need to learn them all.

- **Enthalpy change of formation**, ΔH_f, is the enthalpy change when 1 mole of a compound is formed from its elements in their standard states, e.g. $Ca_{(s)} + Cl_{2(g)} \rightarrow CaCl_{2(s)}$
- The **bond dissociation enthalpy**, ΔH_{diss}, is the enthalpy change when all the bonds of the same type in 1 mole of gaseous molecules are broken, e.g. $Cl_{2(g)} \rightarrow 2Cl_{(g)}$
- **Enthalpy change of atomisation of an element**, ΔH_{at}, is the enthalpy change when 1 mole of gaseous atoms is formed from an element in its standard state, e.g. $\frac{1}{2}Cl_{2(g)} \rightarrow Cl_{(g)}$
- **Enthalpy change of atomisation of a compound**, ΔH_{at}, is the enthalpy change when 1 mole of a compound in its standard state is converted to gaseous atoms, e.g. $NaCl_{(s)} \rightarrow Na_{(g)} + Cl_{(g)}$
- The **first ionisation energy**, ΔH_{ie1}, is the enthalpy change when 1 mole of gaseous 1+ ions is formed from 1 mole of gaseous atoms, e.g. $Mg_{(g)} \rightarrow Mg^+_{(g)} + e^-$
- The **second ionisation energy**, ΔH_{ie2}, is the enthalpy change when 1 mole of gaseous 2+ ions is formed from 1 mole of gaseous 1+ ions, e.g. $Mg^+_{(g)} \rightarrow Mg^{2+}_{(g)} + e^-$
- **First electron affinity**, ΔH_{ea1}, is the enthalpy change when 1 mole of gaseous 1– ions is made from 1 mole of gaseous atoms, e.g. $O_{(g)} + e^- \rightarrow O^-_{(g)}$
- **Second electron affinity**, ΔH_{ea2}, is the enthalpy change when 1 mole of gaseous 2– ions is made from 1 mole of gaseous 1– ions, e.g. $O^-_{(g)} + e^- \rightarrow O^{2-}_{(g)}$
- The **enthalpy change of hydration**, ΔH_{hyd}, is the enthalpy change when 1 mole of aqueous ions is formed from gaseous ions, e.g. $Na^+_{(g)} \rightarrow Na^+_{(aq)}$
- The **enthalpy change of solution**, $\Delta H_{solution}$, is the enthalpy change when 1 mole of an ionic substance dissolves in enough solvent to form an infinitely dilute solution.

Exam Tip
You could be asked to give any of these definitions in the exam, so make sure you've learned them all word for word.

Tip: These terms will crop up a lot in this section so don't move on until you understand what each one means — you might want to fold this page over so you can find it again if you need to.

Tip: An infinitely dilute solution is one which contains so much solvent that when you add more liquid, there is no change in concentration.

Practice Questions — Fact Recall

Q1 What is meant by the term enthalpy change?

Q2 Give the symbol for enthalpy change.

Q3 Give two different definitions for lattice enthalpy.

Q4 Define the following:
  a) enthalpy change of formation,
  b) second electron affinity,
  c) enthalpy change of hydration.

Q5 Name the changes in enthalpy defined below:
  a) The enthalpy change when 1 mole of gaseous atoms is formed from an element in its standard state.
  b) The enthalpy change when 1 mole of gaseous 1+ ions is formed from 1 mole of gaseous atoms.

Q6 Write the symbol for the following enthalpy changes:
  a) The enthalpy change of hydration.
  b) The bond dissociation enthalpy.

2. Born-Haber Cycles

Get those rulers and pencils out — it's time to draw some Born-Haber cycles.

Calculating lattice enthalpies

You can't measure a lattice enthalpy directly, so you have to use a **Born-Haber cycle** to figure out what the enthalpy change would be if you took another, less direct, route. There are two routes you can follow to get from the elements in their standard states to the ionic lattice. On the diagram below, the blue arrow shows the direct route and the red arrows show the indirect route.

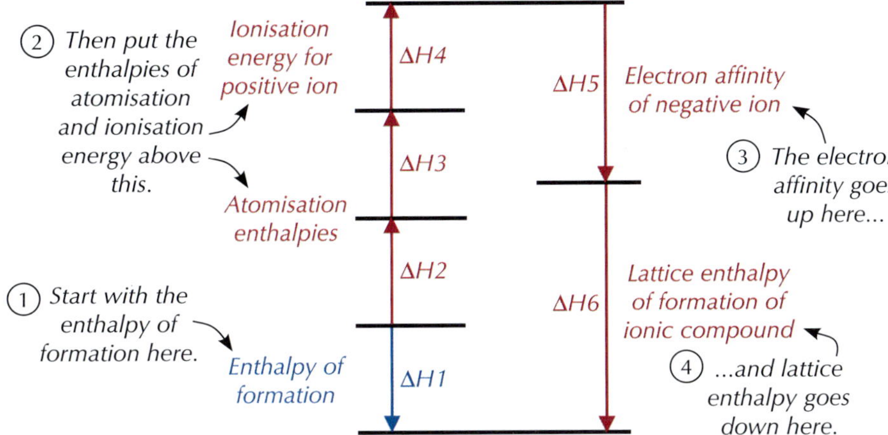

The enthalpy change for each route is the same. This is **Hess's law**:

> The total enthalpy change of a reaction is always the same, no matter which route is taken.

So, to calculate the lattice enthalpy change of formation ($\Delta H6$), you use an alternative route around the diagram:

$$\Delta H6 = -\Delta H5 - \Delta H4 - \Delta H3 - \Delta H2 + \Delta H1$$

Notice that you add in a minus sign if you go the wrong way along an arrow.

Example

To calculate the lattice enthalpy of NaCl:

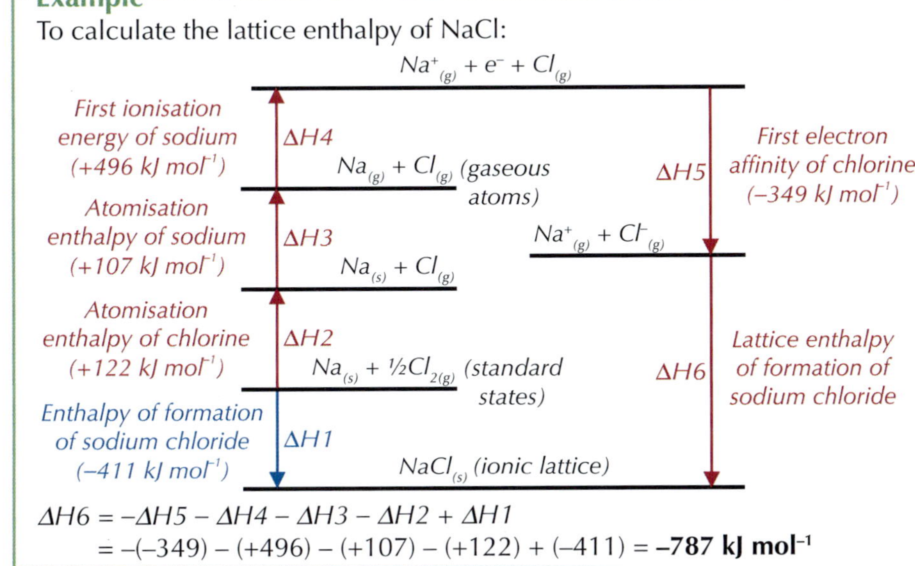

$\Delta H6 = -\Delta H5 - \Delta H4 - \Delta H3 - \Delta H2 + \Delta H1$

$= -(-349) - (+496) - (+107) - (+122) + (-411) = \mathbf{-787\ kJ\ mol^{-1}}$

Learning Objectives:

- Be able to construct Born-Haber cycles to calculate lattice enthalpies and other enthalpy changes.
- Be able to compare lattice enthalpies from Born-Haber cycles with those from calculations based on a perfect ionic model to provide evidence for covalent character in ionic compounds.

Specification Reference 3.1.8.1

Figure 1: *German physicist Max Born who, along with Fritz Haber, developed Born-Haber cycles.*

Tip: You can also get the equation above by re-arranging this equation:
$\Delta H1 = \Delta H2 + \Delta H3 + \Delta H4 + \Delta H5 + \Delta H6$.

Tip: The Haber who helped develop Born-Haber cycles is the same Haber who invented the Haber process for synthesising ammonia (see page 399).

Exam Tip
Make sure you understand what's going on with these Group 2 elements so you can handle whatever compound they throw at you in the exam.

Born-Haber cycles for compounds containing Group 2 elements have an extra step compared to the one on the previous page. Here's a Born-Haber cycle for a compound containing a Group 2 element:

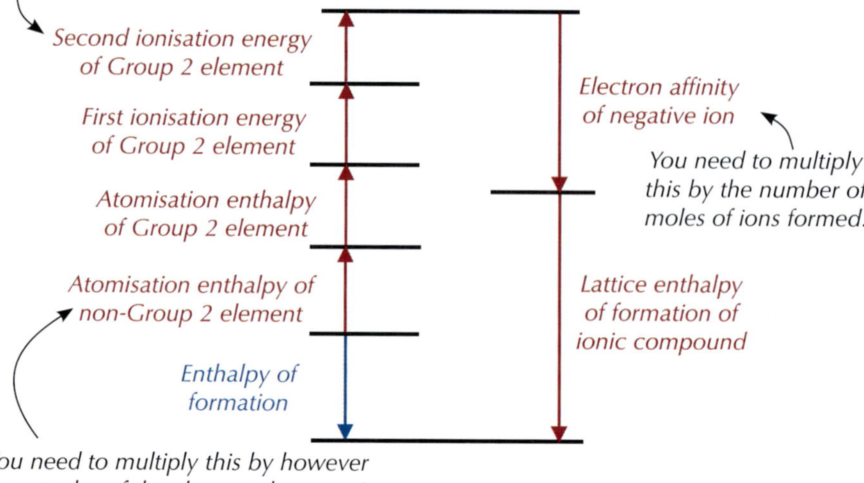

Tip: Remember, enthalpy change of formation, ΔH_f, is the enthalpy change when <u>one mole of a compound</u> is formed from its elements in their standard states under standard conditions.

Example

Magnesium can react with chlorine to form magnesium chloride. Here's how you calculate the lattice enthalpy of magnesium chloride ($MgCl_2$).

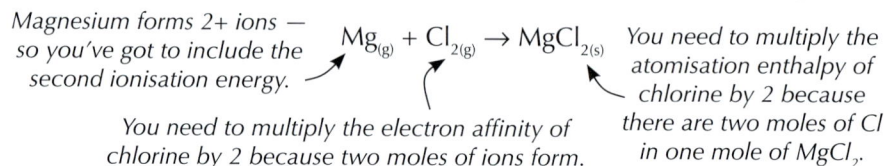

$$Mg_{(g)} + Cl_{2(g)} \rightarrow MgCl_{2(s)}$$

Magnesium forms 2+ ions — so you've got to include the second ionisation energy.

You need to multiply the electron affinity of chlorine by 2 because two moles of ions form.

You need to multiply the atomisation enthalpy of chlorine by 2 because there are two moles of Cl in one mole of $MgCl_2$.

Here's the Born-Haber cycle:

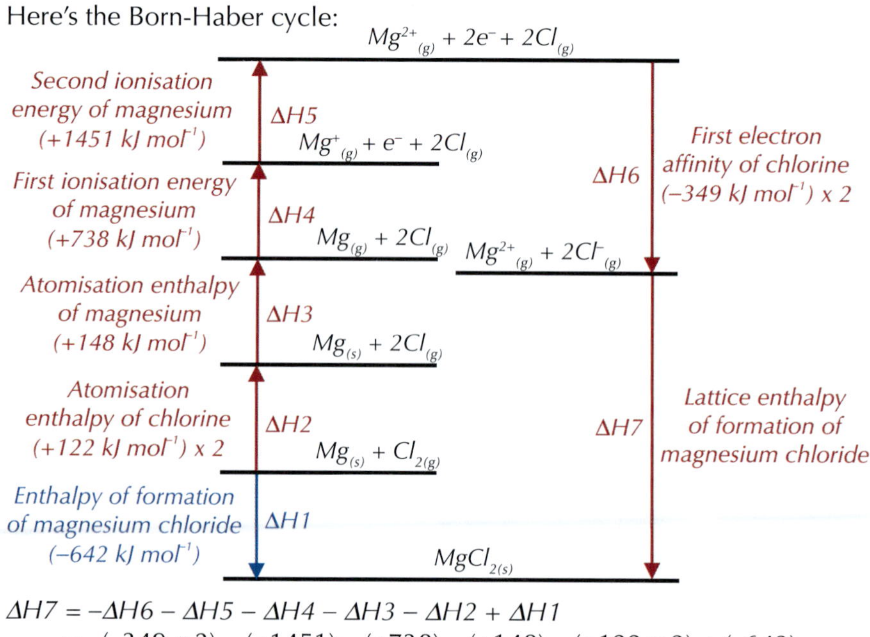

$\Delta H7 = -\Delta H6 - \Delta H5 - \Delta H4 - \Delta H3 - \Delta H2 + \Delta H1$
$= -(-349 \times 2) - (+1451) - (+738) - (+148) - (+122 \times 2) + (-642)$
$= \mathbf{-2525\ kJ\ mol^{-1}}$

Figure 2: *Dry magnesium chloride.*

Tip: State symbols are really important here. If you're not sure what symbols should go where, double-check the definitions on page 266.

Calculating other enthalpy changes

You could be asked to calculate any of the enthalpy or energy values used in Born-Haber cycles. Don't worry though, you just construct the diagram and calculate the unknown value in exactly the same way as for lattice enthalpy.

Example

Magnesium can react with oxygen to form magnesium oxide.

$$Mg_{(g)} + \frac{1}{2}O_{2(g)} \rightarrow MgO_{(s)}$$

Magnesium forms 2+ ions — so you've got to include the second ionisation energy.

Oxygen forms 2– ions — so you've got to include the second electron affinity.

Here's how to calculate the atomisation enthalpy of magnesium using the Born-Haber cycle for MgO:

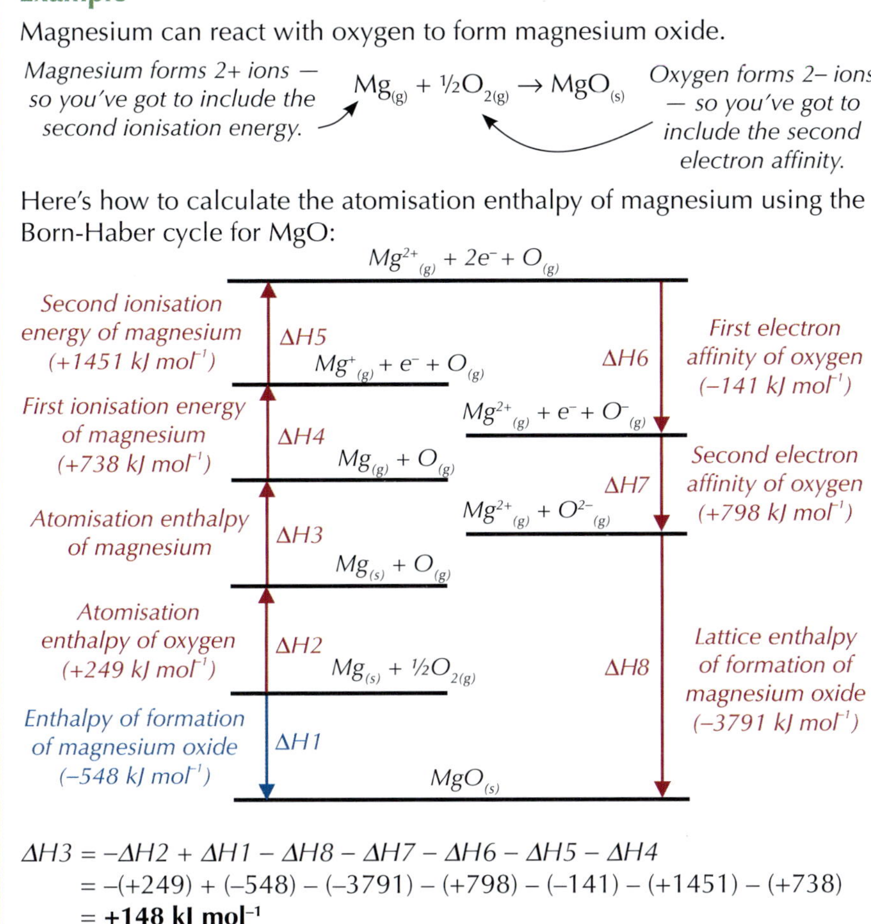

$\Delta H3 = -\Delta H2 + \Delta H1 - \Delta H8 - \Delta H7 - \Delta H6 - \Delta H5 - \Delta H4$

$= -(+249) + (-548) - (-3791) - (+798) - (-141) - (+1451) - (+738)$

$=$ **+148 kJ mol⁻¹**

Tip: The unknown value you need to find out is in a different position in the cycle to the previous examples. You work it out in the same way as before though. In this example, work your way round the arrows from $[Mg_{(s)} + O_{(g)}]$ to $[Mg_{(g)} + O_{(g)}]$ by the indirect route.

Theoretical and experimental enthalpies

You can work out a theoretical lattice enthalpy by doing some calculations based on the purely ionic model of a lattice. The **purely ionic model of a lattice** assumes that all the ions are spherical, and have their charge evenly distributed around them. But the experimental lattice enthalpy from the Born-Haber cycle is usually different. This is evidence that ionic compounds usually have some covalent character. The positive and negative ions in a lattice aren't usually exactly spherical (see Figure 3). Positive ions polarise neighbouring negative ions to different extents, and the more polarisation there is, the more covalent the bonding will be.

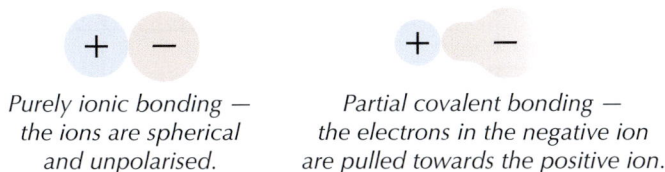

Purely ionic bonding — the ions are spherical and unpolarised.

Partial covalent bonding — the electrons in the negative ion are pulled towards the positive ion.

Figure 3: *Purely ionic and partial covalent bonding.*

Tip: The purely ionic model of a lattice is also known as the perfect ionic model.

Tip: Although lattice enthalpy cannot be directly measured, the value obtained from the Born-Haber cycle is the actual, true value of the lattice enthalpy because all of the other values which have been used to calculate it are real experimental values for that substance.

If the experimental and theoretical lattice enthalpies for a compound are very different, it shows that the compound has a lot of covalent character.

Example

Here are both lattice enthalpy values for some magnesium halides.

Compound	Lattice Enthalpy of Formation (kJ mol^{-1})	
	From experimental values in Born-Haber cycle	From theory
Magnesium chloride	*–2526*	*–2326*
Magnesium bromide	*–2440*	*–2079*
Magnesium iodide	*–2327*	*–1944*

The experimental lattice energies are more exothermic than the theoretical values by a fair bit. This tells you that the bonding is stronger than the calculations from the ionic model predict. The difference shows that the ionic bonds in the magnesium halides are quite strongly polarised and so they have quite a lot of covalent character.

Exam Tip
Don't worry — you don't have to memorise any of these enthalpy values. You'll always be given them in the question if you need them.

If the experimental and theoretical lattice enthalpies for a compound are very similar, it shows that the compound has very little covalent character.

Example

Here are some more lattice energies, for sodium halides this time:

Compound	Lattice Enthalpy of Formation (kJ mol^{-1})	
	From experimental values in Born-Haber cycle	From theory
Sodium chloride	*–787*	*–766*
Sodium bromide	*–742*	*–731*
Sodium iodide	*–698*	*–686*

The experimental and theoretical values are a pretty close match — so you can say that these compounds fit the 'purely ionic' model very well. This indicates that the structure of the lattice for these compounds is quite close to being purely ionic. There's almost no polarisation so they don't have much covalent character.

Tip: If you're struggling to remember which way round it goes, just think — lots of difference means lots of covalent character, little difference means little covalent character.

Practice Questions — Application

Q1 Complete the following Born-Haber cycle for the formation of $CaBr_2$.

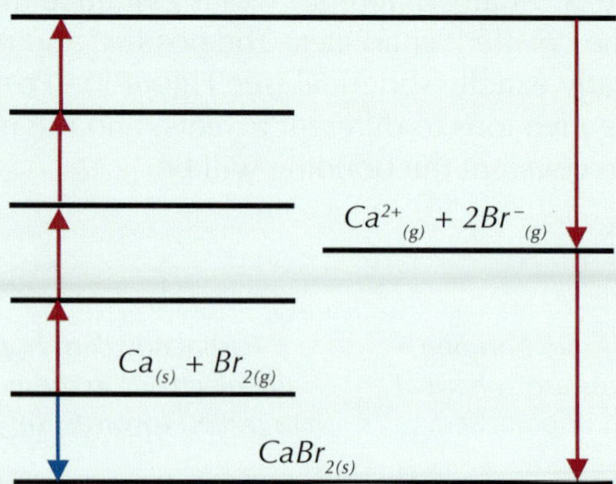

Exam Tip
If you're asked to complete a Born-Haber cycle, it just means fill in the black bits on the diagram — you don't have to write out what each stage is unless it specifically asks you to.

Tip: Calcium is a Group 2 element.

Q2 Look at the table below.

Enthalpy change	$\Delta H^{\ominus}$ / kJ mol^{-1}
Enthalpy change of atomisation of chlorine	+122
Enthalpy change of atomisation of lithium	+159
First ionisation energy of lithium	+520
First electron affinity of chlorine	–349
Lattice enthalpy of formation of lithium chloride	–861

a) Draw the Born-Haber cycle for the formation of LiCl.

b) Use Hess's law to calculate the enthalpy of formation of LiCl.

Q3 Look at the table below.

Enthalpy change	$\Delta H^{\ominus}$ / kJ mol^{-1}
Enthalpy change of atomisation of potassium	+89
Enthalpy change of atomisation of fluorine	+79
First ionisation energy of potassium	+419
First electron affinity of fluorine	–328
Enthalpy of formation of potassium fluoride	–563

a) Draw the Born-Haber cycle for the formation of KF.

b) Use Hess's law to calculate the lattice enthalpy of formation of KF.

c) Explain why theoretical lattice enthalpies are often different from experimental values.

Exam Tip

Always double-check that you've got the plus and minus signs right when you're doing calculations using a Born-Haber cycle. It's really easy to make a mistake and get it wrong in the exam — and you will lose precious marks.

Exam Tip

When completing Born-Haber cycles in the exam, make sure you get the enthalpy changes in the right order. Atomisation always comes before ionisation, and they always come before electron affinity.

Learning Objective:

- Be able to use enthalpy cycles to calculate enthalpies of solution for ionic compounds from lattice enthalpies and enthalpies of hydration.

Specification Reference 3.1.8.1

Tip: Take a look back at pages 92-93 if you need a reminder about electronegativity and the δ+ and δ– charges on polar molecules such as water: $\overset{\delta-}{O}-H^{\delta+}$ (with the second H also $\delta+$)

3. Enthalpies of Solution

Lattice enthalpies aren't the only enthalpies you can calculate using Hess's law. You can work out enthalpies of solution too. You'll need to remind yourself of the definitions for lattice dissociation enthalpy (page 265) and enthalpy changes of solution and hydration (page 266) before you start.

Dissolving ionic lattices

When a solid ionic lattice dissolves in water these two things happen:

- The bonds between the ions break to give gaseous ions — this is endothermic. This enthalpy change is the **lattice enthalpy of dissociation**.
- Bonds between the ions and the water are made — this is exothermic. The enthalpy change here is called the **enthalpy change of hydration**.

Water can form bonds with the ions because it is a polar molecule. Oxygen is more electronegative than hydrogen, so it draws the bonding electrons toward itself, creating a dipole. Consequently, positive ions form weak bonds with the partial negative charge on the oxygen atom and negative ions form weak bonds with the partial positive charge on the hydrogen atoms (see Figure 1).

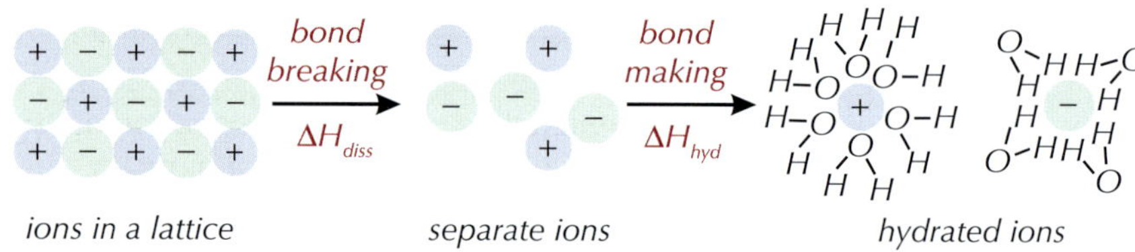

Figure 1: *A solid ionic lattice dissolving in water.*

The enthalpy change of solution is the overall effect on the enthalpy of bond breaking and bond making.

Calculating enthalpy change of solution

You can work out the enthalpy change of solution using an enthalpy cycle. You just need to know the lattice dissociation enthalpy of the compound and the enthalpies of hydration of the ions. Here's how to draw an enthalpy cycle for calculating the enthalpy change of solution:

(1) *Put the ionic lattice and the dissolved ions on the top — connect them by the enthalpy change of solution. This is the direct route.*

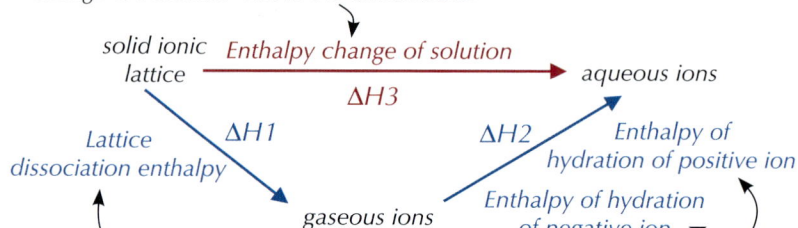

solid ionic lattice → *Enthalpy change of solution* ($\Delta H3$) → aqueous ions

solid ionic lattice → *Lattice dissociation enthalpy* ($\Delta H1$) → gaseous ions

gaseous ions → $\Delta H2$ (*Enthalpy of hydration of positive ion*, *Enthalpy of hydration of negative ion*) → aqueous ions

(2) *Connect the ionic lattice to the gaseous ions by the lattice enthalpy of dissociation. This will be a positive number. If you're given a negative value for lattice enthalpy, it'll be the lattice enthalpy of formation. It's the reverse of this that you want (see page 265).*

(3) *Connect the gaseous ions to the dissolved ions by the hydration enthalpies of each ion. This completes the indirect route.*

Figure 2: *Copper(II) sulfate ($CuSO_4$) dissolved in water.*

Examples

Here's the enthalpy cycle for working out the enthalpy change of solution for sodium chloride.

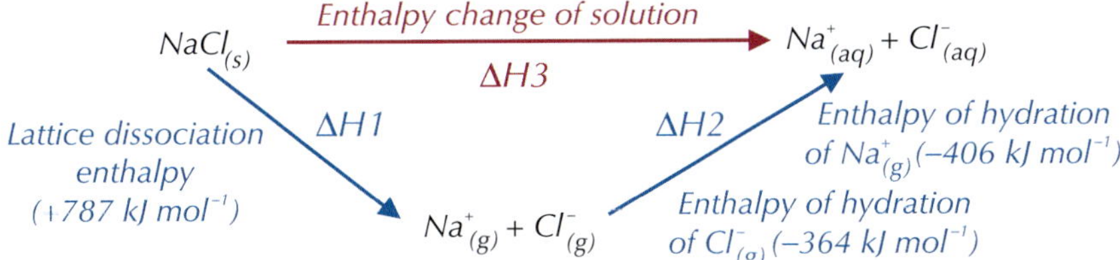

From Hess's law: $\Delta H3 = \Delta H1 + \Delta H2$
$= +787 + (-406 + -364) = \mathbf{+17\ kJ\ mol^{-1}}$

And here's another example. This enthalpy cycle is for working out the enthalpy change of solution for silver chloride.

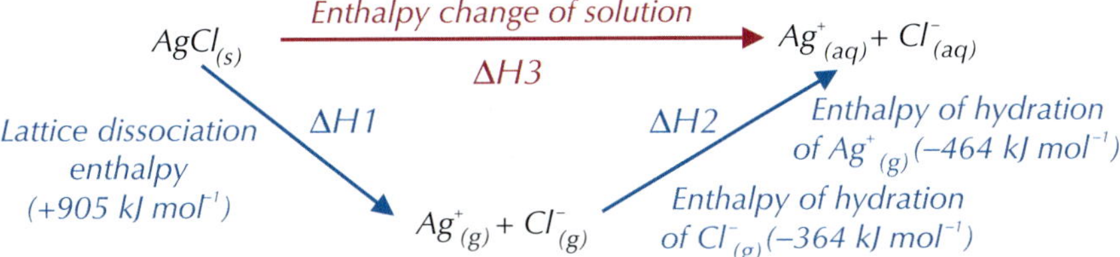

From Hess's law: $\Delta H3 = \Delta H1 + \Delta H2$
$= +905 + (-464 + -364) = \mathbf{+77\ kJ\ mol^{-1}}$

Tip: Take a look back at page 267 for more on Hess's law.

Tip: The enthalpy change of solution for silver chloride is much more endothermic than the enthalpy change of solution for sodium chloride. Take a look at page 274 to see why changes in entropy mean that sodium chloride is soluble in water but silver chloride isn't.

Practice Questions — Application

Q1 The cycle below shows the enthalpy change of solution for LiCl.

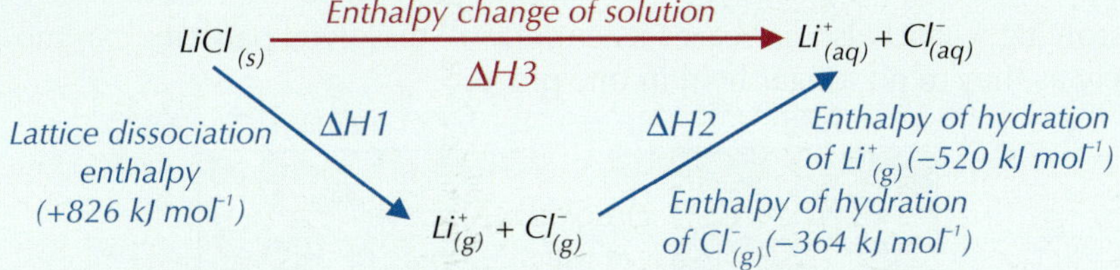

Calculate the enthalpy change of solution for LiCl.

Q2 a) Draw a cycle to show the enthalpy change of solution for sodium bromide. Use the following values:

Lattice dissociation enthalpy of NaBr = +747 kJ mol^{-1}
Enthalpy of hydration of Na^+ = –406 kJ mol^{-1}
Enthalpy of hydration of Br^- = –336 kJ mol^{-1}

b) Calculate the enthalpy change of solution for sodium bromide.

Exam Tip
In the exam, there's no need to draw the cycle out unless the question specifically asks for it. However, if you find it easier to calculate values from a cycle, it's a good idea to draw one out anyway.

Practice Question — Fact Recall

Q1 a) What two processes happen when a solid ionic lattice dissolves in water?

b) State whether each of the processes named in part a) is exothermic or endothermic.

4. Entropy

Left alone, things generally tend towards disorder — that's entropy.

Learning Objectives:

- Understand the concept of increasing disorder (entropy change ΔS).
- Be able to illustrate ΔS in terms of physical change and chemical change.
- Be able to calculate entropy changes from absolute entropy values.

Specification Reference 3.1.8.2

What is entropy?

Entropy tells you how much disorder there is. It's a measure of the number of ways that particles can be arranged and the number of ways that the energy can be shared out between the particles. Substances really like disorder, so the particles move to try to increase the entropy. Entropy is represented by the symbol S. There are a few things that affect entropy, such as:

Physical state

Physical state affects entropy. You have to go back to the good old solid-liquid-gas particle explanation thingy to understand this. Solid particles just wobble about a fixed point — there's hardly any disorder, so they have the lowest entropy. Gas particles whizz around wherever they like. They've got the most disordered arrangements of particles, so they have the highest entropy.

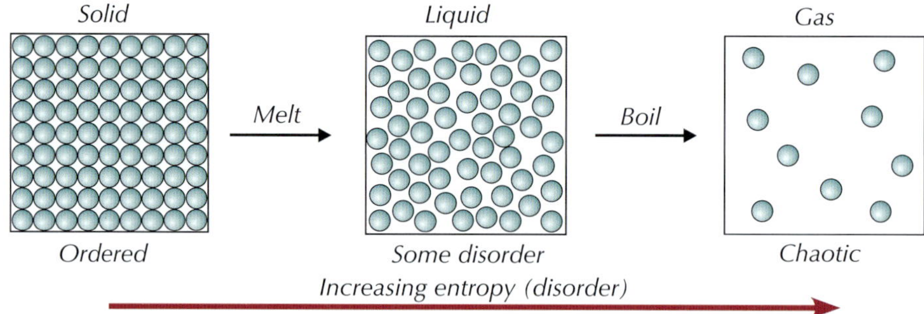

Figure 1: *Melting ice. When ice melts, its entropy increases.*

Dissolution

Dissolving a solid also increases its entropy — dissolved particles can move freely as they're no longer held in one place:

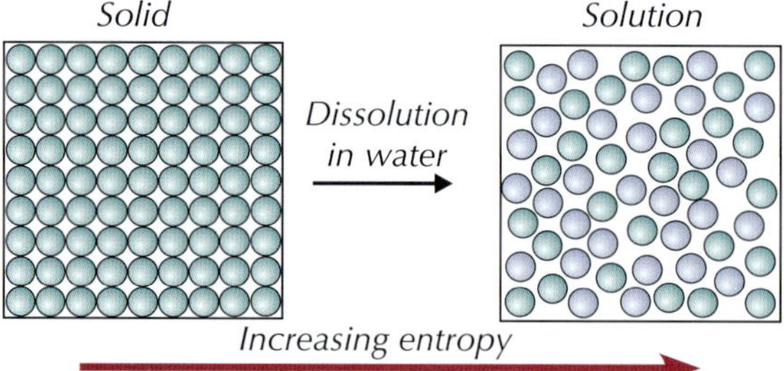

Number of particles

More particles means more entropy. It makes sense — the more particles you've got, the more ways they and their energy can be arranged. So in a reaction like $N_2O_{4(g)} \rightarrow 2NO_{2(g)}$, entropy increases because the number of moles increases:

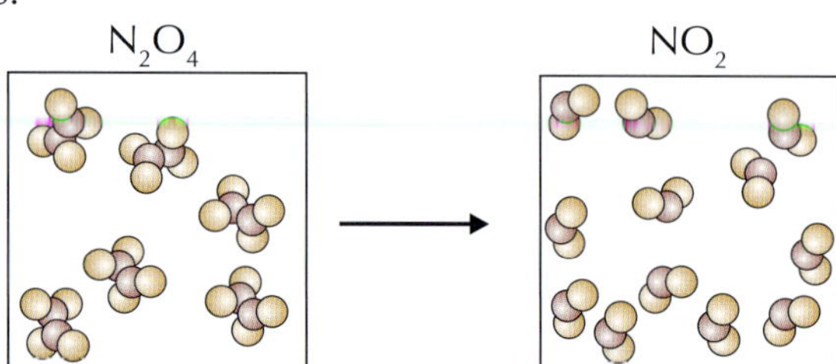

Examples

Evaporation of water

When water evaporates, there is an increase in entropy because there's a change in physical state from liquid to gas.

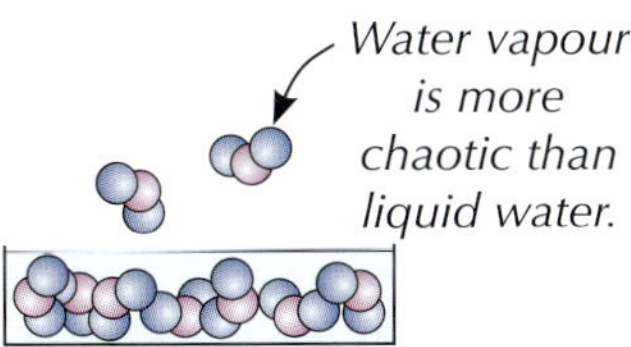
Water vapour
is more
chaotic than
liquid water.

Reaction of $NaHCO_3$ and HCl

When sodium hydrogen carbonate reacts with hydrochloric acid there is an increase in entropy.

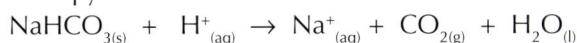

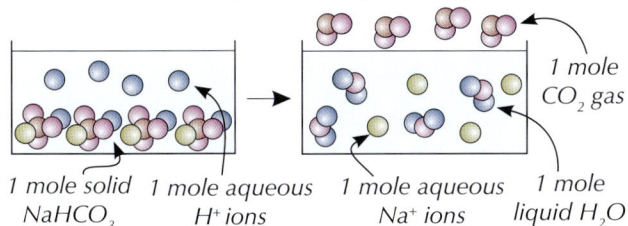

$$NaHCO_{3(s)} + H^+_{(aq)} \rightarrow Na^+_{(aq)} + CO_{2(g)} + H_2O_{(l)}$$

1 mole CO_2 gas

1 mole solid $NaHCO_3$ — 1 mole aqueous H^+ ions — 1 mole aqueous Na^+ ions — 1 mole liquid H_2O

There are more particles in the products — and gases and liquids have more entropy than solids too.

Figure 2: The evaporation of water, to form clouds, brings about an increase in entropy.

Calculating entropy changes

Most reactions won't happen unless the entropy change is positive. During a reaction, there's an entropy change (ΔS) between the reactants and products.

This is just the difference between the entropies of the reactants and products.

$$\Delta S = S_{products} - S_{reactants}$$

Exam Tip

You need to be able to calculate entropy changes in your exam so make sure you learn this equation.

Example — Maths Skills

Calculate the entropy change for the reaction of ammonia and hydrogen chloride under standard conditions.

$$NH_{3(g)} + HCl_{(g)} \rightarrow NH_4Cl_{(s)}$$

$S^\ominus[NH_{3(g)}] = 192.3 \text{ J K}^{-1}\text{ mol}^{-1}$

$S^\ominus[HCl_{(g)}] = 186.8 \text{ J K}^{-1}\text{ mol}^{-1}$

$S^\ominus[NH_4Cl_{(s)}] = 94.6 \text{ J K}^{-1}\text{ mol}^{-1}$

$\Delta S = S_{products} - S_{reactants} = 94.6 - (192.3 + 186.8)$

$= \mathbf{-284.5 \text{ J K}^{-1}\text{ mol}^{-1}}$

This reaction has a negative change in entropy. It's not surprising, as 2 moles of gas have combined to form 1 mole of solid.

Tip: $S^\ominus$ is the standard entropy. The $S^\ominus$ value of a substance is the entropy of one mole of that substance, measured under standard conditions.

Tip: When you calculate an entropy change, don't forget to give the units — they're always J K^{-1} mol^{-1}.

Substance	Standard entropy of substance, S° ($J\ K^{-1}\ mol^{-1}$)
$CH_{4(g)}$	186
$O_{2(g)}$	205
$CO_{2(g)}$	214
$H_2O_{(l)}$	69.9
$SO_{2(g)}$	248
$H_2S_{(g)}$	206
$S_{(s)}$	31.6

Figure 3: *Standard entropy values for different substances.*

Practice Questions — Application

Q1 Solid sodium hydroxide is added to aqueous hydrogen chloride. The reaction produces sodium chloride solution. The solution is heated to produce solid sodium chloride and water vapour.

Describe the entropy changes that take place during these processes.

Q2 Using the data from Figure 3, work out the entropy change for this reaction under standard conditions:

$$CH_{4(g)} + 2O_{2(g)} \rightarrow CO_{2(g)} + 2H_2O_{(l)}$$

Q3 Using the data in Figure 3 work out the entropy change for this reaction under standard conditions:

$$SO_{2(g)} + 2H_2S_{(g)} \rightarrow 3S_{(s)} + 2H_2O_{(l)}$$

Practice Questions — Fact Recall

Q1 a) What is entropy?

b) Give the symbol for entropy change.

Q2 Explain how the following affect the entropy of a system:

a) a substance changing from a liquid to a gas.

b) a solid dissolving in water.

c) a reaction that results in an increased number of gaseous particles.

5. Free-Energy Change

Everyone likes free things, so I can almost guarantee you'll like free-energy.

Learning Objectives:

- Understand that ΔH, whilst important, is not sufficient to explain feasible change and that ΔS accounts for this deficiency.
- Understand that the balance between entropy and enthalpy determines the feasibility of a reaction; know that this is given by the relationship $\Delta G = \Delta H - T\Delta S$.
- Be able to use this equation to determine how ΔG varies with temperature.
- Understand that for a reaction to be feasible, the value of ΔG must be zero or negative.
- Be able to use this relationship to determine the temperature at which a reaction is feasible.

Specification Reference 3.1.8.2

What is free-energy change?

Free-energy change, ΔG, is a measure used to predict whether a reaction is **feasible**. A feasible reaction is one that, once started, will carry on to completion, without any energy being supplied to it. If ΔG is negative or equal to zero, then the reaction is feasible. Free-energy change takes into account the changes in enthalpy and entropy in the system. And of course, there's a formula for it:

$$\Delta G = \Delta H - T\Delta S$$

Free-energy change (in J mol⁻¹) — ΔG; *Enthalpy change (in J mol⁻¹)* — ΔH; *Temperature (in K)* — T; *Entropy change (in J K⁻¹ mol⁻¹)* — ΔS

A negative ΔG doesn't guarantee a reaction will happen or tell you about its rate. Even if ΔG shows that a reaction is theoretically feasible, it might have a really high activation energy or be so slow that you wouldn't notice it happening at all. Here's an example of calculating free-energy change.

Example — Maths Skills

Calculate the free-energy change for the following reaction at 298 K.

$MgCO_{3(s)} \rightarrow MgO_{(s)} + CO_{2(g)}$ $\Delta H^{\ominus} = +117 \text{ kJ mol}^{-1}$

$\Delta S^{\ominus} = +175 \text{ J K}^{-1} \text{ mol}^{-1}$

$\Delta G = \Delta H - T\Delta S = +117 \times 10^3 - [298 \times (+175)]$

$= \mathbf{+64\ 900\ J\ mol^{-1}}$ (3 s.f.)

ΔG is positive — so the reaction isn't feasible at this temperature.

Effect of temperature

If a reaction is exothermic (negative ΔH) and has a positive entropy change, then ΔG is always negative since $\Delta G = \Delta H - T\Delta S$. These reactions are feasible at any temperature.

If a reaction is endothermic (positive ΔH) and has a negative entropy change, then ΔG is always positive. These reactions are not feasible at any temperature. But for other combinations, temperature has an effect.

If ΔH is positive (endothermic) and ΔS is positive then the reaction will only be feasible above a certain temperature.

Example

The decomposition of calcium carbonate is endothermic but results in an increase in entropy (the number of molecules increases and CO_2 is a gas).

$$CaCO_{3(s)} \rightarrow CaO_{(s)} + CO_{2(g)}$$

The reaction only occurs when $CaCO_3$ is heated — it isn't feasible at 298 K.

Figure 1: Calcium carbonate ($CaCO_3$) being heated to produce CaO and CO_2.

If ΔH is negative (exothermic) and ΔS is negative then the reaction will only be feasible below a certain temperature.

Example

The process of turning water from a liquid to a solid is exothermic but results in a decrease in entropy (a solid is more ordered than a liquid). So it will only occur at certain temperatures (i.e. at 0 °C or below).

Tip: You might see free-energy referred to as 'Gibbs free-energy'. They're the same thing.

Tip: The units of ΔH and ΔS must be the same. So, if you're given a value for ΔH in kJ, multiply it by 10^3 to get in J. Or if you want ΔG in kJ mol^{-1}, divide ΔS by 10^3 instead.

Tip: Don't forget — the value of ΔG has to be less than or equal to zero for a reaction to be feasible.

Here are some examples of how to decide if a reaction is feasible at different temperatures.

Examples — Maths Skills

Reaction 1: $\Delta H = +10$ kJ mol^{-1}, $\Delta S = +10$ J K^{-1} mol^{-1}

At 300 K:

$$\Delta G = \Delta H - T\Delta S$$
$$= +10 \times 10^3 - (300 \times +10)$$
$$= \mathbf{+7000\ J\ mol^{-1}}$$

At 1500 K:

$$\Delta G = \Delta H - T\Delta S$$
$$= +10 \times 10^3 - (1500 \times +10)$$
$$= \mathbf{-5000\ J\ mol^{-1}}$$

So this reaction is feasible at 1500 K, but not at 300 K.

Reaction 2: $\Delta H = -10$ kJ mol^{-1}, $\Delta S = -10$ J K^{-1} mol^{-1}

At 300 K:

$$\Delta G = \Delta H - T\Delta S$$
$$= -10 \times 10^3 - (300 \times -10)$$
$$= \mathbf{-7000\ J\ mol^{-1}}$$

At 1500 K:

$$\Delta G = \Delta H - T\Delta S$$
$$= -10 \times 10^3 - (1500 \times -10)$$
$$= \mathbf{+5000\ J\ mol^{-1}}$$

So this reaction is feasible at 300 K, but not at 1500 K.

When ΔG is zero, a reaction is *just* feasible. You can find the temperature when ΔG is zero by rearranging the free-energy equation.

$\Delta G = \Delta H - T\Delta S$, so when $\Delta G = 0$, $T\Delta S = \Delta H$. So:

$$T = \frac{\Delta H}{\Delta S}$$

temperature at which a reaction becomes feasible (in K) → T

enthalpy change (in J mol⁻¹) → ΔH

entropy change (in J K⁻¹ mol⁻¹) → ΔS

Substance	Standard entropy of substance, S° ($J\ K^{-1}\ mol^{-1}$)
$WO_{3(s)}$	76.0
$H_{2(g)}$	65.0
$W_{(s)}$	33.0
$H_2O_{(g)}$	189.0

Figure 2: *Standard entropy values for different substances.*

Example — Maths Skills

Tungsten, W, can be extracted from its ore, WO_3, by reduction using hydrogen.

$$WO_{3(s)} + 3H_{2(g)} \rightarrow W_{(s)} + 3H_2O_{(g)} \qquad \Delta H^\circ = +117\ kJ\ mol^{-1}$$

Use the data in Figure 2 to find the minimum temperature at which the reaction becomes feasible.

First, convert the enthalpy change, ΔH, to joules per mole:

$$\Delta H = 117 \times 10^3\ J\ mol^{-1} = 117\ 000\ J\ mol^{-1}$$

Then find the entropy change, ΔS:

$$\Delta S = S_{products} - S_{reactants} = [33.0 + (3 \times 189.0)] - [76.0 + (3 \times 65.0)]$$
$$= +329\ J\ K^{-1}\ mol^{-1}$$

Then divide ΔH by ΔS to find the temperature at which the reaction just becomes feasible:

$$T = \frac{\Delta H}{\Delta S} = \frac{117\ 000}{329} = \mathbf{356\ K}$$

Tip: For more on the ΔS formula, see page 275.

Free-energy graphs

If you're given a graph of free-energy (ΔG) versus temperature (T), you can use it to work out the enthalpy change (ΔH) and entropy change (ΔS) for the reaction. The ΔG vs T graph is a straight line so can be described using the equation of a straight line, $y = mx + c$. You can rearrange the free-energy equation to get it into the same form as the equation of a straight line:

$$\Delta G = \Delta H - T\Delta S$$
$$\Delta G = (-\Delta S \times T) + \Delta H$$
$$y \;\; = (m \times x) + c$$

So on a graph where ΔG is the y-axis and T is the x-axis, the gradient of the line (m) is equal to $-\Delta S$ and the y-intercept (c) is equal to ΔH.

Tip: There are some tips about drawing and using straight line graphs on page 530.

Tip: When you're working out the gradient of a straight line, pick two points on the line where it's easy to read off an x and y value. This usually means points that lie on grid lines.

Example — Maths Skills

A graph showing the free-energy change (ΔG) versus temperature for a reaction is shown below. Use the graph to determine the enthalpy change and the entropy change for this reaction.

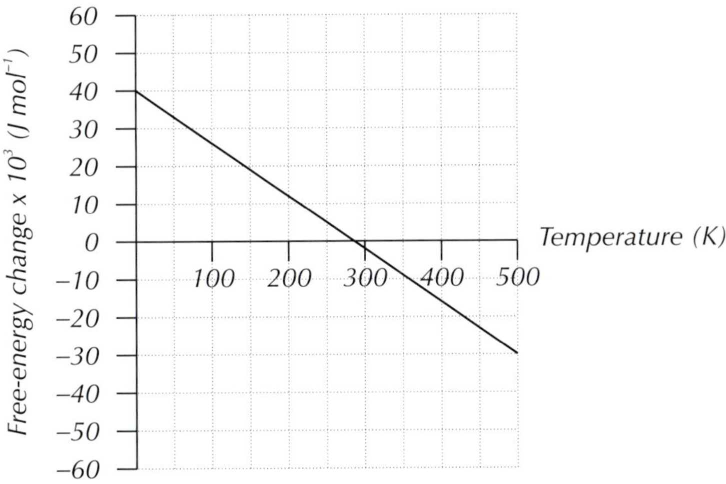

Figure 3: *Graph showing free-energy change versus temperature for a reaction.*

By rearranging the equation $\Delta G = \Delta H - T\Delta S$ into the form of the equation of a straight line, we know that:

$\Delta H = y$-intercept (c)

$-\Delta S =$ gradient (m)

The y-intercept of the line is at 40 000 J mol^{-1}:

$\Delta H =$ **+40 kJ mol^{-1}**

The gradient of a straight line is equal to the change in the y-axis divided by the change in the x-axis:

$$m = \Delta y \div \Delta x = (-30\,000 - 40\,000) \div (500 - 0)$$
$$= -70\,000 \div 500 = -140$$

The entropy change is the same size as the gradient but with the opposite sign:

$-\Delta S =$ gradient $(m) = -140$

$\Delta S =$ **+140 J K^{-1} mol^{-1}**

Tip: ΔG vs T graphs for different reactions can look a little different. This one has a positive y-intercept and a negative gradient. Others might have a negative y-intercept or a positive gradient. Any combination is possible but it will always be a straight line so treat them all in the same way.

Tip: This graph is for a reaction with a positive enthalpy change and a positive entropy change. In this situation, the reaction is unfeasible at low temperatures and only becomes feasible above a certain temperature. You can see this from the graph. The free-energy change only becomes negative at higher temperatures. The temperature at which this reaction becomes feasible is 286 K, which is the x-intercept of the graph.

Substance	Standard entropy of substance, S° ($J\ K^{-1}\ mol^{-1}$)
$Al_2O_{3(s)}$	51.0
$Mg_{(s)}$	32.5
$Al_{(s)}$	28.3
$MgO_{(s)}$	27.0

Figure 4: *Standard entropy values for different substances.*

Tip: Be careful with the scale on the vertical axis here. You need to multiply the values on the graph by 10^3 to get the actual ΔG values in $J\ mol^{-1}$.

Practice Questions — Application

Q1 Using the data from Figure 4 for this reaction under standard conditions: $Al_2O_{3(s)} + 3Mg_{(s)} \rightarrow 2Al_{(s)} + 3MgO_{(s)}$ $\Delta H^\circ = -130\ kJ\ mol^{-1}$

a) calculate ΔS.

b) calculate ΔG.

c) explain whether the reaction is feasible at 298K.

Q2 Calculate the temperature at which this reaction becomes feasible:

$$CaCO_{3(s)} \rightarrow CaO_{(s)} + CO_{2(g)}$$

$\Delta H^\circ = +178.0\ kJ\ mol^{-1}$ $\Delta S = +165.0\ J\ K^{-1}\ mol^{-1}$

Q3 Using the graph below of free-energy change versus temperature for a reaction:

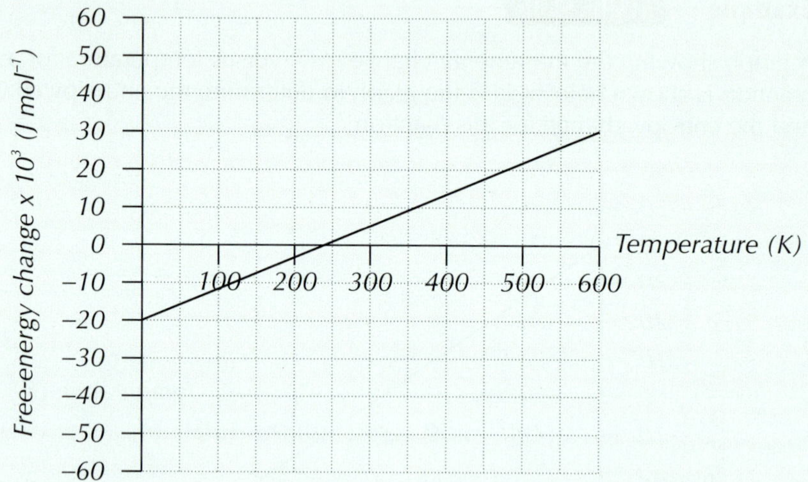

a) determine ΔH.

b) determine ΔS.

c) explain whether the reaction is feasible at 400K.

Practice Questions — Fact Recall

Q1 a) What is free-energy change?

b) Give the symbol for free-energy change.

c) What are the units of free-energy change?

Q2 Give the formula needed to work out free-energy.

Q3 A reaction is endothermic, has a negative entropy change and so has a positive value for free-energy. Is this reaction feasible?

Q4 Give the formula that you'd use to calculate the temperature at which a reaction becomes feasible.

Section Summary

Make sure you know...

- That enthalpy change is the heat energy transferred in a reaction at constant pressure, and the symbol for it is ΔH.
- That exothermic reactions have negative ΔH values and endothermic reactions have positive ΔH values.
- What lattice enthalpy is and the two different ways that it can be defined.
- The definitions of enthalpy change of formation, bond dissociation enthalpy, enthalpy change of atomisation for elements and for compounds, ionisation energy, electron affinity, enthalpy change of hydration and enthalpy change of solution.
- What a Born-Haber cycle is and how to use one to calculate enthalpy changes from experimental data.
- Why theoretical lattice enthalpies are often different from experimental values.
- How to determine the amount of covalent character in an ionic lattice by comparing its theoretical and experimental lattice enthalpy values.
- The different enthalpy changes that take place when a solid ionic lattice dissolves.
- How to calculate the enthalpy change of solution for an ionic compound from lattice dissociation enthalpy and enthalpy of hydration values.
- That entropy, S, is a measure of disorder.
- How entropy is affected by physical state, dissolution and number of particles.
- Why an increase in entropy may (or may not) mean a reaction is feasible.
- How to calculate the total entropy change for a reaction.
- That it's the balance between entropy and enthalpy that determines the feasibility of a reaction.
- What free-energy change is and how to calculate it using $\Delta G = \Delta H - T\Delta S$.
- That the feasibility of some reactions depends on temperature, and how to calculate the temperature at which a reaction becomes feasible.
- How to determine enthalpy change and entropy change from a graph of free-energy change versus temperature.

Exam-style Questions

1 A graph of free-energy change versus temperature for a certain reaction is shown below.

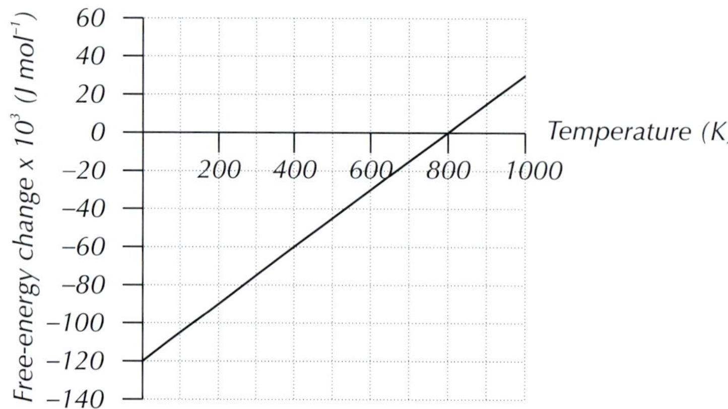

Which statement about this reaction is true?

A The enthalpy change is positive and the entropy change is positive.

B The enthalpy change is positive and the entropy change is negative.

C The enthalpy change is negative and the entropy change is positive.

D The enthalpy change is negative and the entropy change is negative.

(1 mark)

2 Strontium fluoride is a solid ionic lattice which dissolves in water and can be used as an optical coating for lenses.

2.1 Draw a Born-Haber cycle for the enthalpy change of solution of $SrF_{2(s)}$. Label each enthalpy change.

(2 marks)

2.2 Calculate the enthalpy change of solution for SrF_2, using the data in the table below.

(2 marks)

Enthalpy change	ΔH° / kJ mol^{-1}
Lattice dissociation enthalpy of $SrF_{2(s)}$	+2492
Enthalpy change of hydration of Sr^{2+} ions	−1480
Enthalpy change of hydration of F^- ions	−506

3 Rubidium chloride is an ionic compound that dissolves easily in water and can be used as a cell marker in laboratories. The table below shows thermodynamic data for rubidium chloride.

Enthalpy change	***ΔH° / kJ mol^{-1}***
Enthalpy change of atomisation of rubidium	+81
First ionisation energy of rubidium	+403
Enthalpy change of hydration of Rb^+ ions	−296
Enthalpy change of atomisation of chlorine	+122
First electron affinity of chlorine	−349
Enthalpy change of hydration of Cl^- ions	−364
Enthalpy change of formation of rubidium chloride	−435

3.1 Define the term enthalpy change of atomisation of an element.

(1 mark)

3.2 Complete the Born-Haber cycle for the formation of rubidium chloride by filling in the blank lines. You should include chemical symbols and state symbols.

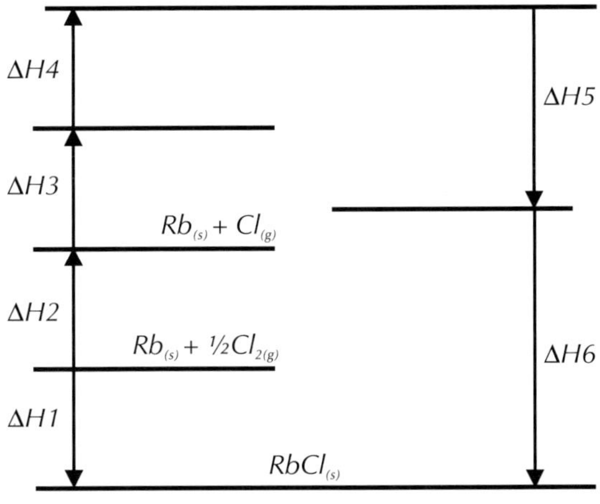

(3 marks)

3.3 Use the data in the table to calculate the lattice enthalpy change of formation of rubidium chloride.

(3 marks)

3.4 The theoretical value for the lattice enthalpy change of formation of rubidium chloride is very close to the experimental value calculated using a Born-Haber cycle. What does this tell you about the bonding character of rubidium chloride?

(2 marks)

3.5 State the value of the lattice dissociation enthalpy of rubidium chloride.

(1 mark)

3.6 Name the enthalpy change for rubidium chloride which is equal to the sum of its lattice dissociation enthalpy, the enthalpy change of hydration of Rb^+ ions and the enthalpy change of hydration of Cl^- ions.

(1 mark)

3.7 Calculate the enthalpy change you named in **3.6**
(If you have not been able to answer **3.5** use the value of +300 kJ mol^{-1}. This value is incorrect.)

(1 mark)

4 When methane undergoes incomplete combustion, carbon monoxide and water are produced as shown:

$$2CH_{4\,(g)} + 3O_{2\,(g)} \rightarrow 2CO_{(g)} + 4H_2O_{(g)} \qquad \Delta H^{\circ} = -1034 \text{ kJ mol}^{-1}$$

The standard entropies for the products and reactants are given in the table below.

	$CH_{4\,(g)}$	$O_{2\,(g)}$	$CO_{(g)}$	$H_2O_{(g)}$
S° / J K^{-1} mol^{-1}	+186	+205	+198	+189

4.1 Use the balanced equation to explain whether you would expect entropy to decrease or increase during this reaction.

(2 marks)

4.2 The water vapour produced in the reaction condenses to form liquid water. Explain how the entropy of the system would change during this process.

(1 mark)

4.3 Calculate the entropy change for this reaction.

(3 marks)

4.4 Give the equation for calculating the free-energy change ΔG.

(1 mark)

4.5 Use the enthalpy change of reaction and your answer to **4.3** to explain why this reaction is feasible at any temperature. (If you have not been able to answer **4.3** use +611 kJ mol^{-1} for the entropy change. This value is incorrect.)

(2 marks)

5 Manganese can be extracted from its ore manganese(IV) oxide, MnO_2, by reduction using carbon at 1473 K. The reaction is shown in the equation below.

$$MnO_{2(s)} + C_{(s)} \rightarrow Mn_{(s)} + CO_{2(g)} \qquad \Delta H^{\circ} = +127 \text{ kJ mol}^{-1}$$

	$MnO_{2(s)}$	$C_{(s)}$	$Mn_{(s)}$	$CO_{2(g)}$
S° / J K^{-1}mol^{-1}	53.0	5.70	32.0	214

5.1 Calculate the free-energy change for the extraction of manganese from its ore.

(4 marks)

5.2 Explain how the free-energy change of a reaction relates to the feasibility of that reaction.

(1 mark)

5.3 Give the equation which links the temperature at which a reaction becomes feasible, T, the enthalpy change for the reaction, ΔH, and the entropy change of the system, ΔS.

(1 mark)

5.4 Calculate the temperature at which the reduction of manganese oxide using carbon becomes feasible.

(2 marks)

The standard enthalpy of change formation of MnO_2 is –521 kJ mol^{-1}.

5.5 Define the term standard enthalpy change of formation.

(1 mark)

5.6 State whether the formation of manganese(IV) oxide from manganese and oxygen will be endothermic or exothermic.

(1 mark)

1. Monitoring Reactions

Learning Objective:
- Be able to measure the rate of reaction by a continuous monitoring method (Required Practical 7).

Specification Reference 3.1.9.2

Being able to work out the rate of a chemical reaction is a really important part of chemistry. You'll have come across reaction rates before, but now it's time to cover things in a bit more detail.

Reaction rates

The reaction rate is the change in the amount of reactants or products per unit time (normally per second). If the reactants are in solution, the rate will be change in concentration per second and the units will usually be mol dm^{-3} s^{-1}.

Measuring the progress of a reaction

If you want to find the rate of a reaction, you need to be able to follow the reaction as it's occurring. You can follow a reaction all the way through to its end by recording the amount of product (or reactant) you have at regular time intervals — this is called **continuous monitoring**.

The results can be used to work out how the rate changes over time. Although there are quite a few ways to follow reactions, not every method works for every reaction. You've got to pick a property that changes as the reaction goes on. Here are a few examples:

Tip: Before carrying out any experiment, you should do a thorough risk assessment. You should then take any necessary safety precautions to minimise the risks associated with the experiment.

Gas volume

If a gas is given off, you could collect it in a gas syringe and record how much you've got at regular time intervals (see Figure 1).

Tip: You need to make sure that your system is airtight. If any gas is allowed to escape, then the measurements you take will be lower than the true values. This means that you won't get accurate results.

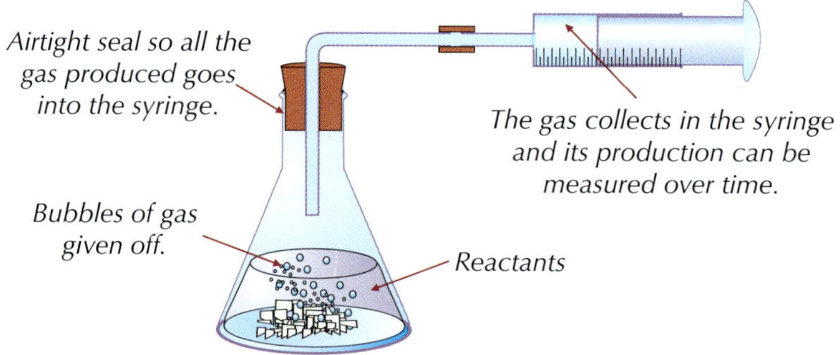

Airtight seal so all the gas produced goes into the syringe.
The gas collects in the syringe and its production can be measured over time.
Bubbles of gas given off.
Reactants

Figure 1: *The experimental set-up for a reaction where the rate of reaction is monitored by collecting a gas.*

You can find the rate of reaction in terms of concentration per unit of time by drawing a concentration-time graph (see pages 291-292). But before you do that, you need to find the concentration of a reactant or product at each time point. This is often quite hard to measure, so you'll usually have to collect some other data, and then convert that into concentration data.

In experiments where you're collecting gases, you'll need to use the ideal gas equation (see page 50) to work out how many moles of gas you've got, then use the molar ratio to work out the concentration of an aqueous reactant.

Tip: The more regularly you take readings, the more clearly you'll be able to see the trend in the rate of reaction against time.

Example — Maths Skills

A scientist is following the reaction between 500 cm³ 2.00 mol dm^{-3} hydrochloric acid and 25 g of solid sodium carbonate, by monitoring the volume of gas produced at r.t.p.. She takes measurements at 15 second intervals. Her results are recorded in the table below. Calculate the concentration of HCl at each time point.

Time / seconds	*0*	*15*	*30*	*45*	*60*	*75*
Volume of gas produced / cm³	*0.0*	*140*	*282*	*423*	*565*	*682*

1. First, it's a good idea to write out the chemical equation of the reaction, so you can see exactly what reaction you're dealing with.

$$2HCl_{(aq)} + Na_2CO_{3(s)} \rightarrow 2NaCl_{(aq)} + CO_{2(g)} + H_2O_{(l)}$$

2. From the reaction above, you can see that the gas produced is carbon dioxide. So, now work out how many moles of carbon dioxide have been produced at each time point, using the ideal gas equation. For example, using the data for the time interval at 15 seconds:

moles of CO_2 = $pV \div RT$ = (100 000 × 1.4 × 10^{-4}) ÷ (8.31 × 298)
= **5.65... × 10^{-3} moles**

3. Now, work out how many moles of HCl you had initially, using the equation: concentration = moles ÷ volume. You're told in the question that the concentration of HCl is 2.0 mol dm^{-3}, and you've got 500 cm³ of it.

2.0 = moles ÷ (500 ÷ 1000)

moles = 2 × 0.5 = **1.0 mole**

4. From the balanced equation, you can see that for every one mole of CO_2 produced, two moles of HCl are used up. This means you can work out the number of moles of HCl lost at each time point. So, at 15 seconds:

number of moles of HCl = initial moles of HCl – (2 × moles of CO_2)
= 1 – (2 × 5.65... × 10^{-3}) = **0.988... moles**

5. Given that the volume of solution doesn't change throughout the experiment, you can work out the concentration of HCl at each time point. So...

Concentration of HCl at 15 seconds = 0.988... ÷ (500 ÷ 1000)
= **1.98 mol dm^{-3}**

6. Repeating steps 2. to 5. with each data point gives you concentration/time data. You can use this to plot a concentration-time graph (see page 291), which you can use to work out the rate of reaction.

Time / seconds	*0*	*15*	*30*	*45*	*60*	*75*
Concentration of HCl / mol dm^{-3}	*2.00*	*1.98*	*1.95*	*1.93*	*1.91*	*1.89*

Tip: At r.t.p, T = 298 K and p = 100 kPa. You'll need these values later...

Tip: The ideal gas equation is $pV = nRT$. You need to use it to find out the number of moles (n). You'll need to rearrange the equation to be in terms of n. This is: $n = pV \div RT$. Make sure everything's in the right units too. Look at page 50 if you need a reminder about this.

Tip: Remember to read the question thoroughly — it may tell you vital information that you'll need to answer the question.

Tip: You couldn't work out the concentration of sodium carbonate as it's a solid — hydrochloric acid is aqueous so its concentration is easy to work out, using the information given in the question.

Tip: To make sure you really understand what's going on in this example, try working through steps 2. to 5. for each time interval — make sure you can get all the answers shown in this table.

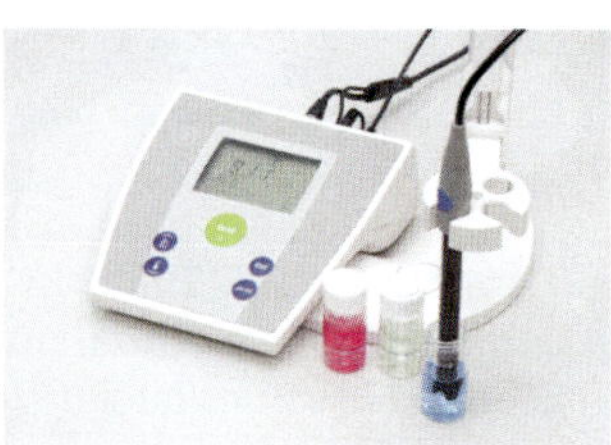

Figure 2: *A pH meter is an electronic probe that you can use to accurately measure pH.*

Tip: Make sure you don't mix up colorimeters (which measure absorbance) and calorimeters (which measure enthalpy changes).

Changes in pH

If the reaction produces or uses up H^+ ions, the pH of the solution will change. So you could measure the pH of the solution at regular intervals and calculate the concentration of H^+ (see page 343).

Example

In the reaction between propanone and iodine, H^+ ions are produced, so you can monitor the progress of the reaction by monitoring the decrease in pH.

Colour change

Sometimes you can track the colour change of a reaction using a gadget called a colorimeter. A colorimeter measures the absorbance (the amount of light of particular wavelengths absorbed) of the solution. The more concentrated the colour of the solution, the higher the absorbance is.

Example

In the reaction between propanone and iodine, the brown colour fades.

$$\underset{\text{colourless}}{CH_3COCH_{3(aq)}} + \underset{\text{brown}}{I_{2(aq)}} \rightarrow \underbrace{CH_3COCH_2I_{(aq)} + H^+_{(aq)} + I^-_{(aq)}}_{\text{colourless}}$$

You collect and convert absorbance data into concentration like this:

1. Plot a **calibration curve** — a graph of known concentrations of I_2 plotted against absorbance (you'll need to collect this data yourself by measuring the absorbances of some standard solutions of I_2 of various concentrations).

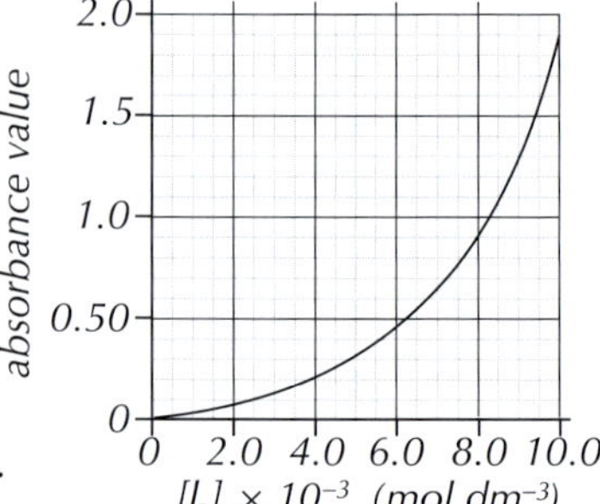

2. During the experiment, take a small sample from your reaction solution at regular intervals and read the absorbance.
3. Use your calibration curve to convert the absorbance at each time point into a concentration.
4. So here for example, if the absorbance value at time *t* seconds was 1.3, then the concentration of iodine at this time would be 9.0×10^{-3} mol dm^{-3}.

Figure 3: *A student using a colorimeter as part of a colorimetry experiment. A digital display shows the absorbance value of a sample at any given moment.*

Tip: The 'loss of mass' method and the 'gas volume' method will both tend to work on the same reactions — they both rely on the production of a gas.

Loss of mass

If a gas is given off, the system will lose mass. You can measure this at regular intervals with a balance, as shown in Figure 4.

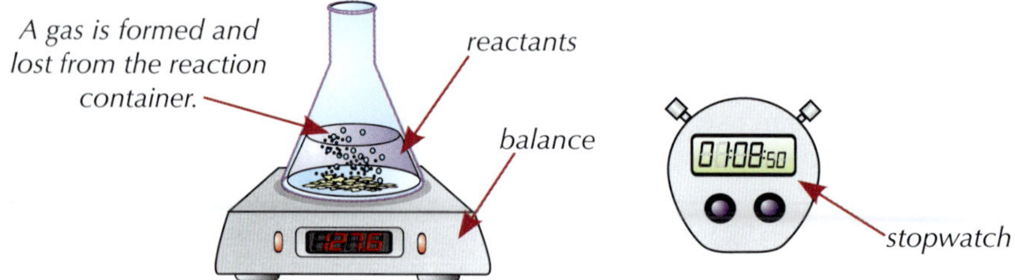

A gas is formed and lost from the reaction container.
reactants
balance
stopwatch

Figure 4: *The experimental set-up for a reaction where the rate of reaction is monitored by measuring the mass lost from a reaction vessel.*

You can use mole calculations to work out how much gas you've lost, and therefore how many moles of reactants are left.

Example — Maths Skills

A student places 5.2 g of zinc in a conical flask containing 225 cm^3 of 1.0 mol dm^{-3} hydrochloric acid. The initial mass of the reaction vessel and all its contents is 389 g. He monitors the loss in mass of the container over the course of a minute. His results are in the table below. Calculate the concentration of HCl at each time point.

Time / seconds	*0*	*10*	*20*	*30*	*40*	*50*	*60*
Mass of reaction vessel / g	*389.00*	*388.97*	*388.95*	*388.93*	*388.91*	*388.89*	*388.87*

1. Start by working out the mass lost at each time interval. This just involves subtracting the mass of the vessel at each time from its initial mass.

Time / seconds	*0*	*10*	*20*	*30*	*40*	*50*	*60*
Mass lost / g	*0*	*0.03*	*0.05*	*0.07*	*0.09*	*0.11*	*0.13*

2. Now write out the chemical equation for the reaction.

$$2HCl_{(aq)} + Zn_{(s)} \rightarrow ZnCl_{2(aq)} + H_{2(g)}$$

3. From the reaction above, you can see that the only gas produced is hydrogen, so all the loss in mass is due to the production of H_2. You can work out how many moles of hydrogen are lost at each time interval, using the equation: moles = mass ÷ M_r. So, after 10 seconds:

   moles of H_2 = 0.03 ÷ (2 × 1)
   = **0.015 moles**

4. Now, work out how many moles of HCl you had initially using the equation: concentration = moles ÷ volume. This will allow you to work out the concentration of HCl at each time point.

   1.0 = moles ÷ (225 ÷ 1000)

   moles = 1.0 × 0.225 = **0.225 moles**

5. From the balanced equation, you can see that for every one mole of H_2 produced, two moles of HCl are used up. So, using the data value at 10 seconds:

   number of moles of HCl = initial moles of HCl – (2 × moles of H_2)
   = 0.225 – (2 × 0.015) = **0.195 moles**

6. So, given that the volume of solution doesn't change throughout the experiment, you can work out the concentration of HCl at each time point. For example...

   Concentration HCl at 10 seconds = 0.195 ÷ (225 ÷ 1000)
   = **0.87 mol dm^{-3}**

Tip: As you can see, this method takes quite a long time. The initial rates method (see pages 298-300) and clock reactions (see pages 302-303) are often used instead of continuous monitoring, to cut down on the amount of maths you need to do.

Tip: Make sure your equation at this point is balanced — if it's not, it could throw off all your calculations.

Tip: Dividing the volume by 1000 at this stage converts the volume from cm^3 into dm^3.

Tip: These mole calculations were covered earlier. Look back at pages 47-48 if you need a reminder on how to do them.

7. Repeating this with each data point gives you concentration/time data. You can use this to plot a concentration-time graph (see page 291), which you can use to work out the rate of reaction.

Time / seconds	*0*	*10*	*20*	*30*	*40*	*50*	*60*
Concentration HCl / mol dm^{-3}	*1.0*	*0.87*	*0.78*	*0.69*	*0.60*	*0.51*	*0.42*

Practice Questions — Application

Q1 Suggest a technique you could use to follow the progress of each of the following reactions.

a) The reaction between chromate(VI) ions (CrO_4^{2-}) and an acid to form dichromate(VI) ions ($Cr_2O_7^{2-}$) and water.

b) The reaction between zinc metal and hydrochloric acid.

c) The reaction between thiosulfate ($S_2O_3^{2-}$) and acid to produce sulfur dioxide gas, water and a yellow precipitate of solid sulfur.

Tip: Think carefully about the reactants and products present in each of these reactions. It may help you to write out a chemical equation.

Q2 A student is monitoring the rate of a reaction between an acid and a metal by collecting gas in a gas syringe. How would her calculated rate be affected if there was a leak in the system?

Q3 A student is using a balance to monitor the loss of mass over the course of the following reaction between zinc sulfide and 250 cm^3 of 1.50 mol dm^{-3} hydrochloric acid.

$$ZnS_{(s)} + 2HCl_{(aq)} \rightarrow ZnCl_{2(aq)} + H_2S_{(g)}$$

She records the mass of the reaction vessel at various time intervals. Calculate the concentration of HCl at each time interval shown below.

Time / seconds	*0*	*10*	*20*	*30*	*40*	*50*	*60*
Mass of vessel / g	*411.02*	*410.42*	*409.84*	*409.25*	*408.65*	*408.08*	*407.51*

Q4 A student is following the reaction between 100 cm^3 1.00 mol dm^{-3} sulfuric acid and 15 g of solid sodium carbonate, by monitoring the volume of gas produced at r.t.p.. She takes measurements at 15 second intervals. Her results are recorded in the table below. Calculate the concentration of H_2SO_4 at each time point.

Tip: You'll need the ideal gas equation for this question. It's $pV = nRT$.

Time / seconds	*0*	*15*	*30*	*45*	*60*	*75*
Volume of gas produced / cm^3	*0.00*	*3.20*	*6.51*	*9.94*	*13.0*	*15.9*

Practice Questions — Fact Recall

Q1 What is meant by the term 'continuous monitoring'?

Q2 Draw and label the set-up you would use to measure the volume of gas produced over the course of a reaction.

Q3 What does a colorimeter measure?

Q4 Describe how you could follow the rate of reaction for a reaction where a gas is lost from the reaction container.

2. Reaction Rates and Graphs

Once you've converted the data you've collected from continuously monitoring a reaction into concentration data, you can plot a concentration-time graph. You can use the gradient of the line of this graph to work out the reaction rate.

Learning Objective:

- Be able to use concentration-time graphs to deduce the rate of reaction.

Specification Reference 3.1.9.2

Finding the rate from a straight-line graph

If you draw a graph of the amount of reactant or product against time for a reaction, the rate at any point is given by the gradient at that point on the graph. You can work out the gradient at any point using the equation:

gradient = change in y ÷ change in x

Example — Maths Skills

The graph below shows data collected from an experiment where the decomposition of a reactant X was monitored over time.
Work out the rate of the reaction.

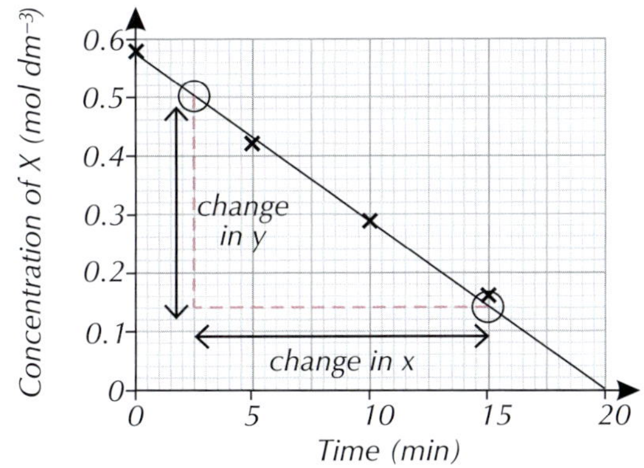

1. To find the rate of this reaction, you can use the gradient of the line. Because this is a straight-line graph, the gradient is the same at all points on the line.
2. Try to pick two points on the line that are easy to read — ideally these are points where the line passes exactly through the corners of the grid squares.
3. Now use the equation, gradient = change in y ÷ change in x, to work out the gradient of the graph.
   - change in y = 0.14 – 0.50 = –0.36 mol dm^{-3}
   - change in x = 15 – 2.5 = 12.5 minutes
   - gradient = –0.36 ÷ 12.5 = –0.029 mol dm^{-3} min^{-1}
4. The gradient is negative, but you can ignore the sign when you use it for the rate. So the rate of reaction is 0.029 mol dm^{-3} min^{-1}.

Exam Tip
In the exam, you might have to draw a line of best fit going through the data points on your graph before you can start working out the gradient and rate — so make sure you know how to draw one (see page 530).

Tip: Straight-line graphs are pretty simple — the gradient is the same across the entire data set, so the reaction rate is the same too. This means you only have to work out a single rate as it's always the same.

Tip: To help you work out the changes in x and y, you should draw a vertical line down from one point, and a horizontal line across from the other to make a triangle.

Tip: If you're not sure whether you've calculated the gradient of a straight-line correctly, you can check your answer by working out the gradient at another point too. If your answer is right, they should be the same.

Finding the rate from a curve

If the graph's a curve, you have to draw a tangent to the curve and find the gradient of that. The gradient, and so the rate, at a particular point in the reaction is the same as the gradient of the tangent at that point.
A tangent is a line that just touches a curve and has the same gradient as the curve does at that point. Figure 1 shows how a graph of the concentration of a reactant against time might look.

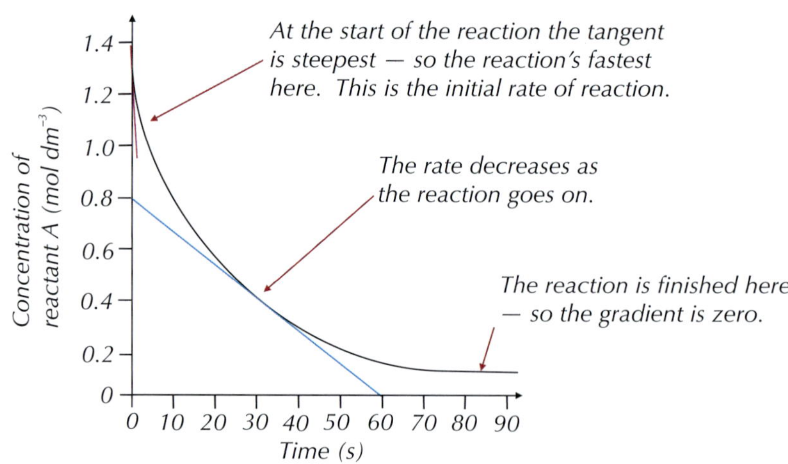

Figure 1: *Graph showing the concentration of a reactant against time.*

Tip: Concentration-time graphs pop up loads throughout this section, so make sure you really understand how to use them to work out rates before moving on to the next topic.

Example — Maths Skills

The graph below shows how the concentration of a solution changes over time. Calculate the rate of the reaction at 3 minutes.

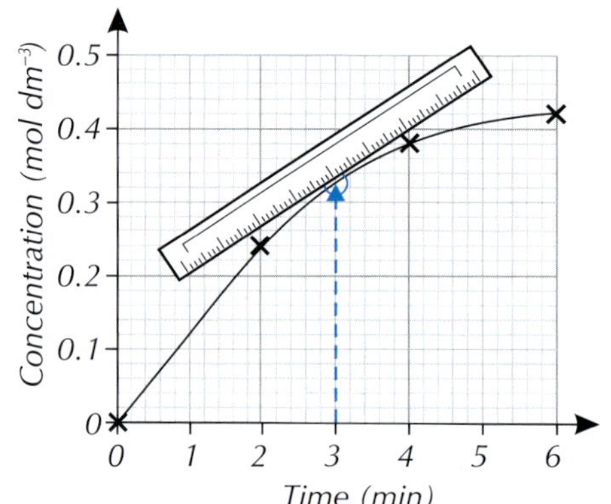

1. Find the point on the curve that you need to look at. The question asks you to find the rate of reaction at 3 minutes, so find 3 on the *x*-axis and go up to the curve from there.
2. Place a ruler at that point so that it's just touching the curve. Position the ruler so that you can see the whole curve.
3. Adjust the ruler until the space between the ruler and the curve is equal on both sides of the point.
4. Draw a line along the ruler to make the tangent. Extend the line right across the graph — it'll help to make your gradient calculation easier as you'll have more points to choose from.

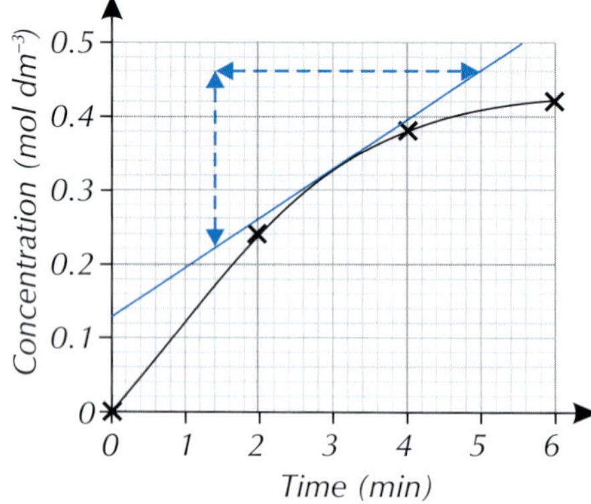

5. Calculate the gradient of the tangent to find the rate:
   - gradient = change in y ÷ change in x
     = (0.46 – 0.22) ÷ (5.0 – 1.4)
     = 0.24 mol dm^{-3} ÷ 3.6 mins = 0.067 mol dm^{-3} min^{-1}
   - So, the rate of reaction at 3 mins was 0.067 mol dm^{-3} min^{-1}.

Exam Tip
Drawing a tangent is pretty much impossible without a ruler — so make sure you remember to bring one into your exam.

Tip: In many cases, it's useful to measure the gradient at various points along the curve. This gives you a more detailed idea of how the rate changes over time.

Tip: Working out the gradient of a tangent is exactly the same as working out the gradient of a straight line, so look back at the last page for more details on how to do this.

Practice Questions — Application

Q1 The graph below shows the results of a reaction between a metal and hydrochloric acid, where the mass of the reaction vessel was measured throughout the reaction.

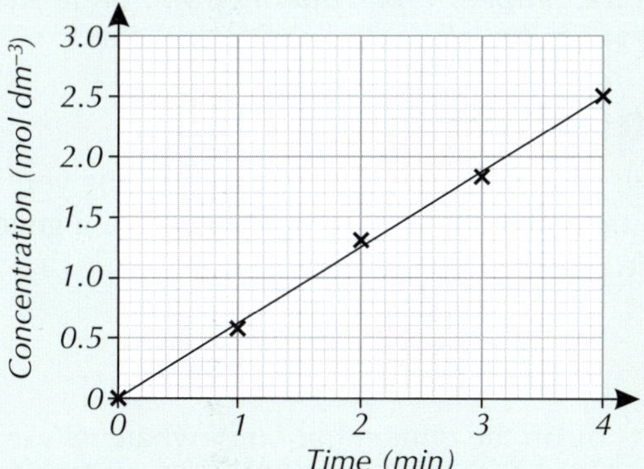

a) Calculate the rate of reaction.

b) How does the rate of reaction vary over the 4 minutes?

Q2 A student is measuring the volume of carbon dioxide gas formed by a reaction between limestone and hydrochloric acid. Her results are shown in the following table.

Time / minutes	Concentration of acid / mol dm^{-3}
0	5.0
1	3.1
2	2.6
3	2.5
4	2.4
5	2.4

a) Plot the data in the table on a graph.

b) Draw a line of best fit on the graph drawn in part a).

c) Calculate the rate of the reaction at 2 minutes.

Q3 The following graph shows the data collected from a pH experiment.

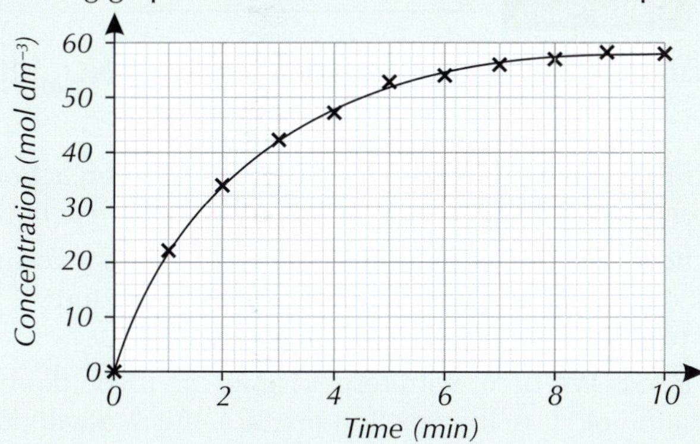

a) Calculate the rate of the reaction at: i) 2 minutes, ii) 6 minutes.

b) A student says that the rate of the reaction is 0 mol dm^{-3} min^{-1}. Comment on this statement.

Exam Tip
Always remember to include your units when you're working out the rate. It could cost you valuable marks in the exam if you miss them out...

Tip: There's a lot to remember when it comes to drawing graphs. You've got to think about the size of the graph, what goes on which axis, the scale, not to mention making sure you plot the data points correctly. Look at page 530 for some tips on how to draw graphs.

3. Rate Equations

Rate equations tell us a bit more about what affects the rate. You'll have to get your head around something called the rate constant, and you'll meet reaction orders too. It's not the simplest topic, but once you understand the theory, the maths really isn't too hard at all.

Learning Objectives:

- Know that the rate of a chemical reaction is related to the concentration of reactants by a rate equation of the form: Rate = $k[A]^m[B]^n$ where m and n are the orders of reaction with respect to reactants A and B, and k is the rate constant.
- Be able to define the terms 'order of reaction' and 'rate constant'.
- Know that the rate equation is an experimentally determined relationship.
- Be able to derive the rate equation for a reaction from the orders with respect to each of the reactants.
- Be able to perform calculations using the rate equation.

Specification Reference 3.1.9.1, 3.1.9.2

The rate equation

Rate equations look ghastly, but all they're really telling you is how the rate is affected by the concentrations of reactants. For a general reaction: A + B → C + D, the rate equation is:

$$\text{Rate} = k[A]^m[B]^n$$

The square brackets mean the concentration of whatever's inside them. So [A] means 'the concentration of A' and [B] means 'the concentration of B'. The units of rate are usually mol dm^{-3} s^{-1}. m and n are the **orders of reaction** and k is the **rate constant** (but more on these below...).

Reaction orders

m and n are the orders of the reaction with respect to reactant A and reactant B. m tells you how the concentration of reactant A affects the rate and n tells you the same for reactant B.

- If [A] changes and the rate stays the same, the order of reaction with respect to A is 0 (or zero order with respect to A). So if [A] doubles, the rate will stay the same. If [A] triples, the rate will stay the same.
- If the rate is proportional to [A], then the order of reaction with respect to A is 1 (or first order with respect to A). So if [A] doubles, the rate will double. If [A] triples, the rate will triple.
- If the rate is proportional to $[A]^2$, then the order of reaction with respect to A is 2 (or second order with respect to A). So if [A] doubles, the rate will be $2^2 = 4$ times faster. If [A] triples, the rate will be $3^2 = 9$ times faster.

The overall order of the reaction is $m + n$. So, if m was 2 and n was 1 the overall order of the reaction would be 2 + 1 = 3.

Exam Tip
m and *n* could take any value but in the exam you'll only have to deal with rate equations where *m* and *n* are 0, 1 or 2.

Example — Maths Skills

The rate equation of the following reaction is rate = $k[NO_{(g)}]^2[H_{2(g)}]^1$.

$$2H_{2(g)} + 2NO_{(g)} \rightarrow 2H_2O_{(l)} + N_{2(g)}$$

- From this rate equation, you can see that the rate of reaction depends on the concentrations of H_2 and NO present in the reaction mixture.
- The reaction is second order with respect to NO, so if you double the concentration of NO in the reaction mixture, then the reaction rate will be four times faster.
- The reaction is first order with respect to H_2, so if you double the concentration of H_2 in the reaction mixture, the rate will double too.
- The overall order of the reaction is 2 + 1 = 3.

Tip: In this example there are only two reactants but rate equations can be written for equations with any number of reactants. E.g. if there were three reactants the equation would be:

Rate = $k[A]^m[B]^n[C]^x$

You can only find orders of reactions, and so rate equations, by doing experiments, such as the ones shown on pages 286-288. You can't tell what the rate equation will be from the balanced chemical equation.

The rate constant

k is the rate constant — it's a constant that links the concentration of reactants to the rate of a reaction. The bigger it is, the faster the reaction. The rate constant is always the same for a certain reaction at a particular temperature — but if you increase the temperature, the rate constant rises too. When you increase the temperature of a reaction, the rate of reaction increases. The concentrations of the reactants and the orders of reaction haven't changed though. So it must be an increase in the value of k which has increased the rate of reaction.

Tip: The link between the rate constant and temperature is explained in more detail by the Arrhenius equation. It's covered on pages 308-312.

Writing rate equations

You need to know how to write rate equations for reactions and use them to calculate the rate constant. This example shows you how:

Figure 1: *Students measuring the rate of a reaction which produces a gas as a product.*

Example — Maths Skills

The chemical equation below shows the acid-catalysed reaction between propanone and iodine.

$$CH_3COCH_{3(aq)} + I_{2(aq)} \xrightarrow{H^+_{(aq)}} CH_3COCH_2I_{(aq)} + H^+_{(aq)} + I^-_{(aq)}$$

a) This reaction is first order with respect to propanone, first order with respect to $H^+_{(aq)}$ and zero order with respect to iodine. Write the rate equation for this reaction.

In this example there are three things that you need to think about — the orders of reaction of propanone (CH_3COCH_3), iodine (I_2) and hydrogen ions (H^+). (Even though H^+ is a catalyst, rather than a reactant, it can still be in the rate equation because it affects the rate of reaction.) So the rate equation will be in the form Rate = $k[A]^m[B]^n[C]^x$.

You're told the reaction orders with respect to each reactant in the question so just use that information to construct the rate equation:

$$\text{Rate} = k[CH_3COCH_3]^1[H^+]^1[I_2]^0$$

But $[X]^1$ is usually written as $[X]$, and $[X]^0$ equals 1 so is usually left out of the rate equation. So you can simplify the rate equation to:

$$\text{Rate} = k[CH_3COCH_3][H^+]$$

This rate equation shows that the rate of reaction is proportional to the concentrations of propanone and H^+. So doubling the concentration of either propanone or H^+ will double the rate of the reaction. The overall order of this reaction is 2 — it's a second order reaction.

Tip: When simplifying rate equations think about the index laws from maths — the same rules apply here. Anything to the power of zero is one (e.g. $[X]^0 = 1$) and anything to the power of one doesn't change (e.g. $[X]^1 = [X]$). Look at pages 528-529 for more about this.

b) At a certain temperature, k was found to be 520 $mol^{-1}\,dm^3\,s^{-1}$ when $[CH_3COCH_3] = [I_2] = [H^+] = 1.50 \times 10^{-3}$ $mol\,dm^{-3}$. Calculate the rate at this temperature.

Calculating the rate is as simple as popping in the values the question into the rate equation you worked out in part a).

$$\text{Rate} = k[CH_3COCH_3][H^+] = 520 \times 1.50 \times 10^{-3} \times 1.50 \times 10^{-3} = 1.17 \times 10^{-3}$$

You can find the units for the rate by putting the other units into the rate equation.

The units for the rate are $\cancel{mol^{-1}}\,\cancel{dm^3}\,s^{-1} \times \cancel{mol}\,\cancel{dm^{-3}} \times mol\,dm^{-3} = mol\,dm^{-3}\,s^{-1}$.

So the answer is: rate = $\mathbf{1.17 \times 10^{-3}\ mol\,dm^{-3}\,s^{-1}}$

Tip: Look at the Maths Skills section if you need more information on how to work out the units for the rate constant (page 527).

Calculating the rate constant

If you know the orders of a reaction, you can use the rate equation and experimental data to work out the rate constant, k. The units of k vary, so you'll need to work them out too.

Tip: If anything doesn't feature in the rate equation, it just means that its concentration doesn't affect the rate of reaction.

Example — Maths Skills

The reaction below is second order with respect to NO and zero order with respect to CO and O_2.

$$NO_{(g)} + CO_{(g)} + O_{2(g)} \rightarrow NO_{2(g)} + CO_{2(g)}$$

At a certain temperature, the rate is 1.76×10^{-3} mol dm^{-3} s^{-1}, when $[NO_{(g)}] = [CO_{(g)}] = [O_{2(g)}] = 2.00 \times 10^{-3}$ mol dm^{-3}. Find the value of the rate constant, k, at this temperature.

- To answer this question you first need to write out the rate equation:

$$\text{Rate} = k[NO]^2[CO]^0[O_2]^0$$
$$= k[NO]^2$$

- Next insert the concentration and the rate, which were given to you in the question:

$$\text{Rate} = k[NO]^2$$
$$1.76 \times 10^{-3} = k \times (2.00 \times 10^{-3})^2$$

- Rearrange the equation and calculate the value of k:

$$k = \frac{1.76 \times 10^{-3}}{(2.00 \times 10^{-3})^2} = 440$$

- Find the units of k by putting the other units in the rate equation:

$$\text{Rate} = k[NO]^2 \quad \text{so} \quad \text{mol dm}^{-3}\text{ s}^{-1} = k \times (\text{mol dm}^{-3})^2$$

- Rearrange the equation to get k:

$$k = \frac{\text{mol dm}^{-3}\text{ s}^{-1}}{(\text{mol dm}^{-3})^2}$$

- Cancel out units wherever possible. In this example you can cancel out a mol dm^{-3} from the top and bottom lines of the fraction:

$$k = \frac{\cancel{\text{mol dm}^{-3}}\text{ s}^{-1}}{\cancel{(\text{mol dm}^{-3})}(\text{mol dm}^{-3})} = \frac{\text{s}^{-1}}{\text{mol dm}^{-3}}$$

- Get rid of the fraction using index laws:

$$k = \frac{\text{s}^{-1}}{\text{mol dm}^{-3}} = \text{mol}^{-1}\text{ dm}^{3}\text{ s}^{-1}$$

- So the answer is:

$$k = 440 \text{ mol}^{-1}\text{ dm}^{3}\text{ s}^{-1}$$

Exam Tip
If you're asked to calculate the rate constant in an exam, make sure you show your working out. That way you might get some marks even if your final answer is wrong.

Exam Tip
If you get stuck in a calculation, re-read the question to check that you've not missed any information — it's dead easy to miss stuff the first time round.

Tip: You need to know how to work out what units your answer is in — it crops up a lot in this section so it's important that you understand how to do it. Have a look at page 526 for more on units.

Tip: Remember — $1/x^n = x^{-n}$. So, for example, if you've got a mol^2 dm^{-6} on the bottom of the fraction it becomes mol^{-2} dm^6, if you've got mol^{-3} dm^9 it becomes mol^3 dm^{-9}, and so on. See page 527 for more...

Practice Questions — Application

Q1 A student is trying to determine the rate equation for the reaction:

$$2Fe^{3+}_{(aq)} + 2I^-_{(aq)} \rightarrow 2Fe^{2+}_{(aq)} + I_{2(aq)}$$

He says, "because the equation involves the reaction of two moles of iron(III) ions and 2 moles of iodide ions, the rate equation must be rate = $k[Fe^{3+}]^2[I^-]^2$".

a) Do you agree with the student? Explain your answer.

b) According to the student's rate equation, how would halving the concentration of iodide ions affect the rate of reaction?

Q2 The following reaction occurs under certain conditions:

$$2H_{2(g)} + 2NO_{(g)} \rightarrow 2H_2O_{(g)} + N_{2(g)}$$

This reaction is first order with respect to H_2 and second order with respect to NO.

a) Construct the rate equation for this reaction.

b) What is the overall order of this reaction?

c) How would you expect the rate of reaction to change if the concentration of NO were doubled?

d) At a certain temperature, k was found to be 221 $mol^{-2}\ dm^6\ s^{-1}$ when $[H_2]$ = [NO] = 1.54×10^{-3} mol dm^{-3}. Calculate the rate at this temperature.

Q3 The reaction below is second order with respect to NO and first order with respect to Cl_2:

$$2NO_{(g)} + Cl_{2(g)} \rightarrow 2NOCl_{(g)}$$

The rate of reaction is 5.85×10^{-6} mol $dm^{-3}\ s^{-1}$ at 50 °C, when the concentration of both NO and Cl_2 is 0.400 mol dm^{-3}.

a) Write the rate equation for this reaction.

b) Calculate the value of the rate constant (k) for this reaction at 50 °C.

c) How would you expect the value of k to change if the temperature were increased?

d) Calculate the expected rate of reaction if 0.500 mol dm^{-3} NO were mixed with 0.200 mol dm^{-3} Cl_2 at 50 °C.

Exam Tip
When answering calculation questions like these always remember to give the units for your answer.

Exam Tip
In the exam, you should give your answers to the same number of significant figures as used in the question. Here, the values in the question are given to 3 s.f. so you should give your answer to 3 s.f. as well (see page 520).

Practice Questions — Fact Recall

Q1 What does the rate equation tell you?

Q2 Write out the rate equation for the following, general reaction:

$$A + B \rightarrow C + D$$

Q3 In the rate equation Rate = $k[X]^a[Y]^b$, what do the following represent:

a) k?

b) [Y]?

c) a?

Q4 How do you work out the units of the rate constant?

Tip: This isn't the last you'll see of rate equations. Oh no — there's lots more on reaction orders to come, and the rate equation is really useful for working out reaction mechanisms (see page 306). So make sure you really understand the basics covered on the last few pages, and that you can answer all the questions before moving on to the next topic.

4. The Initial Rates Method

Rate equations are really useful, but to work out the rate equation you need to know the orders of a reaction. These can be determined experimentally by monitoring the initial rate of a reaction as you change the concentration of one of the reactants.

Learning Objectives:

- Be able to use initial concentration-time data to deduce the initial rate of a reaction.
- Be able to use rate-concentration data to deduce the order (0, 1 or 2) with respect to a reactant.

Specification Reference 3.1.9.2

Finding the initial rate

The initial rate of a reaction is the rate right at the start of the reaction. You can find this from a concentration–time graph. You can collect data for a concentration–time graph by setting up a reaction and then monitoring the amount of a reactant or product over time (see pages 286-288 for lots of ways to do this). The amount of reactant decreases with time because it gets used up in the reaction. Once the reaction is complete, plotting the concentration of reactant against time gives a concentration–time graph.

Once you have a concentration–time graph, you can find the initial rate of reaction by calculating the gradient of the curve at time = 0. Just draw a tangent to the curve at time = 0 and draw in horizontal and vertical lines to make a triangle with the tangent as its longest side. Label the horizontal line x and the vertical line y. The gradient of the line is the change in y divided by the change in x and this represents the initial rate of reaction — see Figure 1.

Tip: Graphs and gradients are covered in a bit more detail on pages 530-531.

Tip: A tangent is a line that just touches a curve and has the same gradient as the curve does at that point.

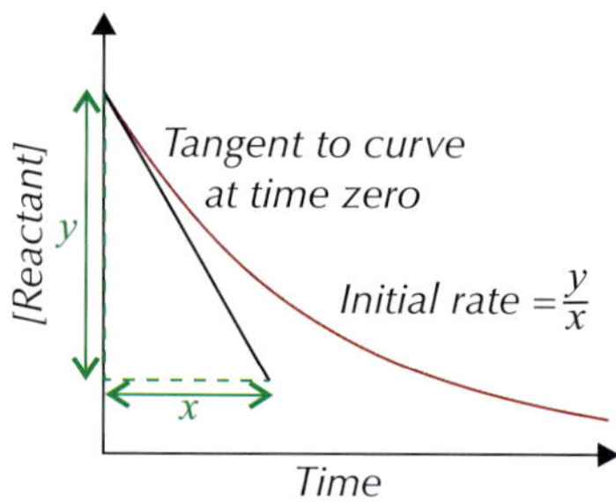

Figure 1: *Graph showing how to calculate the initial rate of reaction.*

Tip: Look back at pages 286-288 for different ways to measure the rate of reaction using continuous monitoring.

Example — Maths Skills

The data on the graph below was obtained by continuously monitoring the concentration of H⁺ ions over the course of a reaction between a metal and an acid. Calculate the initial rate of the reaction.

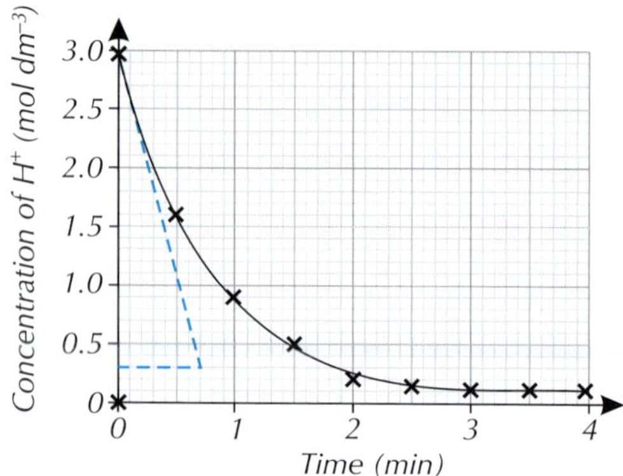

- To work out the initial rate, you need to work out the gradient of the curve right at the start of the reaction.

- Draw a tangent to the curve at time = 0.
- Then you just need to work out the gradient of the tangent using exactly the same method as the one you saw, back on page 292.

  gradient = change in y ÷ change in x

  = (0.3 – 3.0) ÷ (0.7 – 0.0)

  = –2.7 mol dm^{-3} ÷ 0.7 mins = –3.857... mol dm^{-3} min^{-1}
- So the initial rate of reaction was 3.9 mol dm^{-3} min^{-1}.

Tip: Remember, you can ignore the sign of the gradient when using it to work out the rate.

Exam Tip
Missing off the units from your answer is a sure-fire way of losing precious marks. Don't let this trip you up...

Using the initial rates method

You can use a process called the **initial rates method** to work out the order of reaction for each reactant. Here's how it works:

- Carry out the reaction and monitor its progress. Unlike with continuous monitoring, you don't have to monitor the reaction all the way through to the end — you just need to collect enough data to allow you to work out an initial rate. Use this data to draw a concentration-time graph.
- Repeat the experiment using a different initial concentration of one of the reactants. Keep the concentrations of other reactants and other conditions the same. Draw another concentration-time graph.
- Use your graphs to calculate the initial rate for each experiment using the method above.
- Now look at how the different initial concentrations affect the initial rate — use this to work out the order for that reactant.
- Repeat the process for each reactant (different reactants may have different orders).

The examples below and on the next page show you how to deduce the orders of reaction, with respect to each reactant, from experimental concentration and initial rates data. Once you know the orders, you can work out the rate equation.

Tip: When deciding how much data to collect, you could wait for a set amount of product to form or for a set amount of reactant to be used up.

Tip: Look back at page 289 for how to use experimental data, from continuous monitoring techniques, to calculate the concentration of reactants.

Example — Maths Skills

The table below shows the results of a series of initial rate experiments for the reaction $2NO_{(g)} + Cl_{2(g)} \rightarrow 2NOCl_{(g)}$. Write down the rate equation for the reaction. The temperature remained constant throughout.

Experiment number	[NO] / mol dm^{-3}	[Cl_2] / mol dm^{-3}	Initial rate / mol dm^{-3} s^{-1}
1	0.125	0.125	1.79×10^{-7}
2	0.250	0.125	7.16×10^{-7}
3	0.250	0.250	1.43×10^{-6}

- Look at experiments 1 and 2 — when [NO] doubles (and the concentration of Cl_2 stays constant) the rate is four times faster, and $4 = 2^2$. So the reaction is second order with respect to NO.
- Look at experiments 2 and 3 — when [Cl_2] doubles (but the concentration of NO stays constant), the rate is two times faster, and $2 = 2^1$. So the reaction is first order with respect to Cl_2.

So the rate equation is: Rate = $k[NO]^2[Cl_2]$

Tip: Look back at page 294 if you need a reminder about how to link a change in concentration and a change in rate to the reaction order with respect to a certain reactant.

Tip: You can also work out reaction orders by looking at the shapes of concentration-time graphs and rate-concentration graphs. This is covered later in this section.

In the exam you could get a question where more than one concentration changes. These are a bit trickier. The example below shows you what to do.

Exam Tip
You would usually only change one concentration at a time but you could get a question where more than one concentration changes — don't get confused, the principle is the same.

Example — Maths Skills

The table below shows the results of a series of initial rate experiments for the reaction $NO_{(g)} + CO_{(g)} + O_{2(g)} \rightarrow NO_{2(g)} + CO_{2(g)}$. The experiments were carried out at a constant temperature. Write down the rate equation for the reaction and use it to calculate *k* at this temperature.

Experiment number	[NO] / mol dm^{-3}	[CO] / mol dm^{-3}	[O_2] / mol dm^{-3}	Initial rate / mol dm^{-3} s^{-1}
1	2.0×10^{-2}	1.0×10^{-2}	1.0×10^{-2}	0.17
2	6.0×10^{-2}	1.0×10^{-2}	1.0×10^{-2}	1.53
3	2.0×10^{-2}	2.0×10^{-2}	1.0×10^{-2}	0.17
4	4.0×10^{-2}	1.0×10^{-2}	2.0×10^{-2}	0.68

- Look at experiments 1 and 2 — when [NO] triples (and all the other concentrations stay constant) the rate is nine times faster, and $9 = 3^2$. So the reaction is second order with respect to NO.
- Look at experiments 1 and 3 — when [CO] doubles (but all the other concentrations stay constant), the rate stays the same. So the reaction is zero order with respect to CO.
- Look at experiments 1 and 4 — the rate of experiment 4 is four times faster than experiment 1. The reaction is second order with respect to [NO], so the rate will quadruple when you double [NO]. But in experiment 4, [O_2] has also been doubled. As doubling [O_2] hasn't had any additional effect on the rate, the reaction must be zero order with respect to O_2.

Tip: It doesn't matter which experiment you use to calculate *k*. The value of *k* should be the same for all of them because they have all been done at the same temperature.

Now that you know the order with respect to each reactant you can write the rate equation, Rate = $k[NO]^2$.

You can then calculate *k* at this temperature by putting the concentrations and the initial rate from one of the experiments into the rate equation.

So, using experiment 1: $0.17 = k(2.0 \times 10^{-2})^2$

$k = 0.17 \div (2.0 \times 10^{-2})^2 = 425$

Tip: You might not always be given the rate of the reaction at different concentrations. Instead, you could be given the time taken for the reaction to reach a certain point. The time taken is proportional to the rate — that means you can use 1 ÷ time as a measure of the rate.

To work out the units of *k*, substitute the units of each variable into the equation for *k*.

$$k = \frac{\cancel{\text{mol dm}^{-3}}\ \text{s}^{-1}}{\cancel{\text{mol dm}^{-3}} \times \text{mol dm}^{-3}}$$

And get rid of the fraction using index laws:

$$\frac{\text{s}^{-1}}{\text{mol dm}^{-3}} = \text{mol}^{-1}\ \text{dm}^{3}\ \text{s}^{-1}$$

Meaning the rate constant, $k = 425$ mol^{-1} dm^3 s^{-1}.

Practice Questions — Application

Q1 The data shown on the graph below was obtained experimentally from the decomposition of SO_2Cl_2.
Work out the initial rate of the reaction.

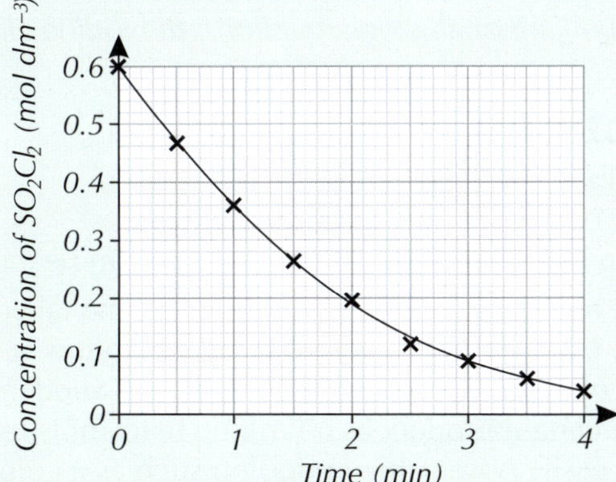

Q2 A student is carrying out a reaction to try and work out the rate equation of the following reaction:

$$I_2 + 2S_2O_3^{2-} \rightarrow 2I^- + S_4O_6^{2-}$$

His results are in the table below.

Trial number	$[I_2]$ /mol dm^{-3}	$[S_2O_3^{2-}]$ / mol dm^{-3}	Reaction time / s
1	0.040	0.040	312
2	0.080	0.040	156
3	0.040	0.020	624

Work out the rate equation for this reaction.

Q3 The table below shows the results of a series of initial rate experiments for the reaction: $2A + B + C \rightarrow AB + AC$

Experiment number	[A] / mol dm^{-3}	[B] / mol dm^{-3}	[C] / mol dm^{-3}	Initial rate / mol dm^{-3} s^{-1}
1	1.2	1.2	1.2	0.25
2	1.2	2.4	1.2	1.00
3	1.2	2.4	3.6	3.00
4	0.6	2.4	1.2	0.50

a) Write down the rate equation for this reaction.

b) Calculate the rate constant (*k*).

Tip: The reaction in question 2 is an example of a clock reaction. They're covered in more detail in the next topic.

Tip: Remember — the rate is proportional to 1/time, so you can use 1/time as a measure of the rate.

Exam Tip
It can be a good idea to check the value of your rate constant by making sure you get the same value for it when you use the data from each experiment.

Practice Questions — Fact Recall

Q1 How do you find out the initial rate of a reaction from a concentration-time graph?

Q2 Briefly outline how to carry out the initial rates method to work out the order of reaction with respect to each reactant.

5. Clock Reactions

Learning Objective:

- Be able to measure the rate of reaction by using an initial rates method, such as a clock reaction (Required Practical 7).

Specification Reference 3.1.9.2

Clock reactions are an example of an initial rates method. They involve measuring how long it takes for a certain amount of product to form. These pages cover everything you'll need to know about clock reactions, and you'll even get to meet the king of all clock reactions, the iodine clock reaction.

Clock reactions

REQUIRED PRACTICAL 7

The method described in the last topic for working out initial rates is a bit of a faff — lots of measuring and drawing graphs. In clock reactions, the initial rate can be easily estimated.

In a clock reaction, you measure how the time taken for a set amount of product to form changes as you vary the concentration of one of the reactants. As part of a clock reaction, there will be a sudden increase in the concentration of a certain product as a limiting reactant is used up. There's usually an easily observable endpoint, such as a colour change, to tell you when the desired amount of product has formed. The quicker the clock reaction finishes, the faster the initial rate of the reaction. You need to make the following assumptions:

- The concentration of each reactant doesn't change significantly over the time period of your clock reaction.
- The temperature stays constant.
- When the endpoint is seen, the reaction has not proceeded too far.

As long as these assumptions are reasonable for your experiment, you can assume that the rate of reaction stays constant during the time period of your measurement. So the rate of your clock reaction will be a good estimate for the initial rate of your reaction.

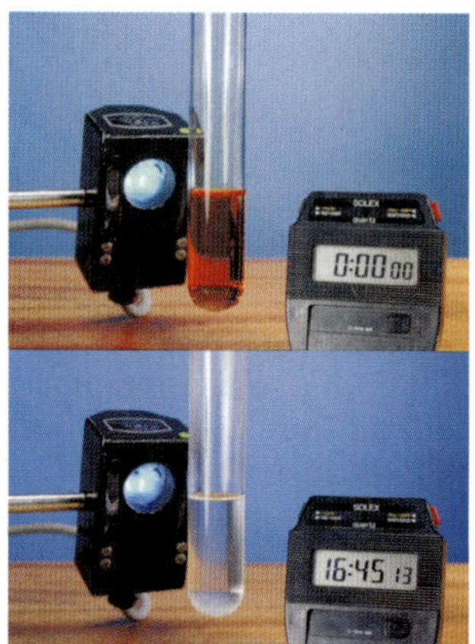

***Figure 1:** Using a colorimeter to measure a colour change. In clock reactions, there is often a sudden, clear colour change, so a colorimeter isn't always needed.*

The iodine clock reaction

The most famous clock reaction is the iodine clock reaction.

In an iodine clock reaction, the reaction you're monitoring is:

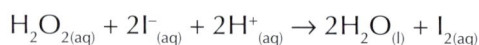

$$H_2O_{2(aq)} + 2I^-_{(aq)} + 2H^+_{(aq)} \rightarrow 2H_2O_{(l)} + I_{2(aq)}$$

- A small amount of sodium thiosulfate solution and starch are added to an excess of hydrogen peroxide and iodide ions in acid solution. (Starch is used as an indicator — it turns blue-black in the presence of iodine.)
- The sodium thiosulfate that is added to the reaction mixture reacts instantaneously with any iodine that forms:

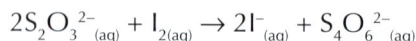

$$2S_2O_3^{2-}{}_{(aq)} + I_{2(aq)} \rightarrow 2I^-_{(aq)} + S_4O_6^{2-}{}_{(aq)}$$

- To begin with, all the iodine that forms in the first reaction is used up straight away in the second reaction. But once all the sodium thiosulfate is used up, any more iodine that forms will stay in solution, so the starch indicator will suddenly turn the solution blue-black. This is the end of the clock reaction.
- Varying iodide or hydrogen peroxide concentration while keeping the others constant will give different times for the colour change.

Tip: These are ionic equations — they only show the reacting particles. You met them on pages 52-53.

Tip: When carrying out clock reactions, the data you collect will be in units of time, rather than units of rate. You can use 1/time as a measure of the rate.

You may have to carry out an iodine-clock reaction in the lab. Here's the method of how to do it...

1. Rinse a clean 25.0 cm^3 pipette with sulfuric acid. Then, use this pipette to transfer 25.0 cm^3 sulfuric acid, of known concentration (e.g. 0.25 mol dm^{-3}), to a clean 250 cm^3 beaker. This beaker is your reaction vessel.
2. Using a clean pipette or measuring cylinder, add 20.0 cm^3 of distilled water to the beaker containing the sulfuric acid.
3. Using a dropping pipette, add a few drops (or about 1 cm^3) of starch solution to the same beaker.
4. Measure out 5.0 cm^3 of potassium iodide solution of a known concentration (e.g. 0.1 mol dm^{-3}), using either a 5.0 cm^3 pipette or a burette, rinsed with potassium iodide solution. Transfer this volume to the reaction vessel.
5. Next, using a pipette rinsed with sodium thiosulfate solution, or a clean measuring cylinder, add 5.0 cm^3 sodium thiosulfate to the reaction vessel. Swirl the contents of the beaker so all the solutions are evenly mixed.
6. Finally, rinse a 10.0 cm^3 pipette with hydrogen peroxide solution. Then, use the pipette to transfer 10.0 cm^3 hydrogen peroxide solution to the reaction vessel and simultaneously start a stop watch. Stir the contents of the beaker throughout the reaction, using a glass rod.
7. Stop the stop watch when the contents of the beaker turn from colourless to blue-black, this marks the endpoint. Record this time in a results table, along with the quantities of sulfuric acid, water, potassium iodide and sodium thiosulfate solutions you used in that experiment.
8. Repeat the experiment varying the volume of potassium iodide solution. Use varying amounts of distilled water in each experiment so the overall volume of the reaction mixture remains constant.

You can then use this data to work out the reaction order, with respect to potassium iodide, using the method on pages 299-300. Varying the concentration of other reagents allows you to find the order of reaction with respect to them too.

Tip: Before you start this experiment, you'll need standard solutions of potassium iodide, hydrogen peroxide and sodium thiosulfate. If you're not given these, look back at page 57 for some information on how to make up standard solutions from solids and liquids.

Tip: Make sure you carry out a full risk assessment before doing any experiments. This will help prevent any nasty accidents.

Tip: You need to vary the volumes of KI solution and H_2O in each run. So, you could do 5 runs using, e.g:

Vol. $KI_{(aq)}$	Vol. H_2O
5 cm^3	20 cm^3
10 cm^3	15 cm^3
15 cm^3	10 cm^3
20 cm^3	5 cm^3
25 cm^3	0 cm^3

Practice Questions — Fact Recall

Q1 State what is meant by a clock reaction.

Q2 What assumptions do you have to make when carrying out a clock reaction?

Q3 What colour change is seen in an iodine clock reaction, when starch is used as an indicator?

6. Rate-concentration Graphs

Learning Objective:

- Be able to use rate-concentration graphs to deduce the order (0, 1 or 2) with respect to a reactant.

Specification Reference 3.1.9.2

You can work out the order of reaction, with respect to a certain reactant, from the shape of its concentration-time graph, and its rate-concentration graph. Read on to find out more...

Working out reaction orders from graphs

You can work out the reaction order with respect to a particular reactant using a rate-concentration graph. You can make a rate-concentration graph using a concentration-time graph. Here's how:

- Find the gradient (which is the rate) at various points along the concentration-time graph (see page 291 for a reminder of how to do this). This gives you a set of points for the rate-concentration graph.
- Then just plot the points and join them up with a line or smooth curve.

The shape of the rate-concentration graph tells you the order:

Zero order

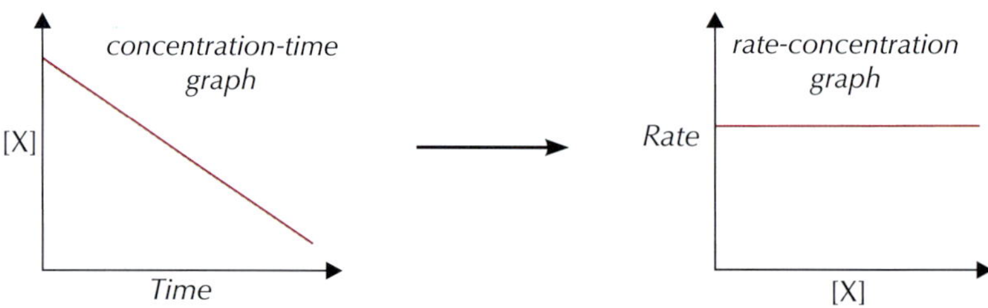

A horizontal line on a rate-concentration graph means that changing the concentration doesn't change the rate, so the reaction is zero order.

Tip: If something is in [square brackets] it's a concentration. So [X] means the concentration of reactant X.

First order

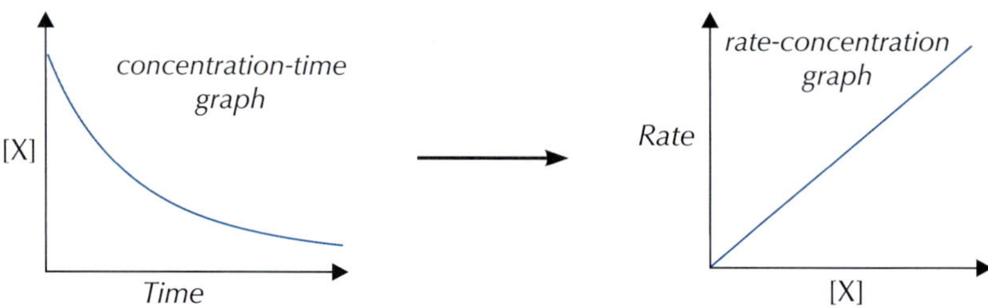

If the rate-concentration graph is a straight line through the origin, then the reaction is first order. The rate is proportional to [X]. For example, if the concentration of X triples, the rate will triple. If it halves, the rate will halve.

Second order

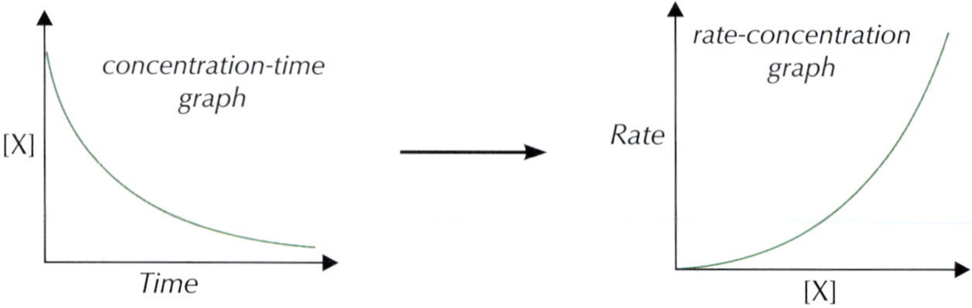

A curved rate-concentration graph means that the reaction is second order. The rate is proportional to $[X]^2$. For example, if the concentration of X triples, the rate will be nine times as fast (because $3^2 = 9$).

Tip: In theory, a curved rate-concentration graph could mean a higher order than 2. But you won't be asked about any with a higher order than 2 in the exam — so if you see a curve, always say order 2.

Practice Questions — Application

Q1 Below are the rate-concentration graphs produced as the concentrations of three different reactants (A, B and C) changed during a reaction.

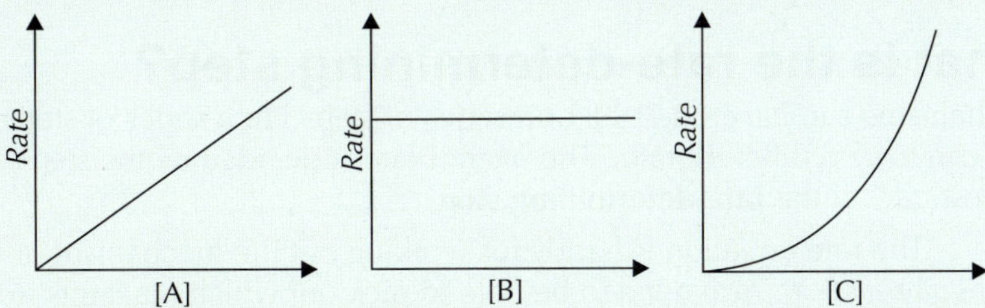

a) Deduce the reaction orders with respect to reactants A, B and C.

b) State how you would expect the rate of reaction to change if the concentration of:

(i) reactant A was halved.

(ii) reactant B was tripled.

(iii) reactant C was doubled.

Q2 The following reaction occurs between nitrogen monoxide (NO) and oxygen (O_2): $2NO_{(g)} + O_{2(g)} \rightarrow 2NO_{2(g)}$

This reaction is second order with respect to NO and first order with respect to O_2.

a) Sketch a graph showing the rate of reaction against the concentration of NO.

b) State how you would expect the rate of reaction to change if the concentration of O_2 was doubled.

Practice Questions — Fact Recall

Q1 Describe how you would produce a rate-concentration graph from a concentration-time graph.

Q2 What is the gradient of the straight-line rate-concentration graph produced by a reactant that is zero order?

Q3 Sketch the shape of the rate-concentration graph for a reactant that is first order.

Exam Tip

It's easy to get concentration-time graphs and rate-concentration graphs confused. But make sure you learn the different shapes of each graph, for zero, first and second order reactions. They may well come up in your exam...

Tip: There's more about orders on page 294, so look back there if you need more information.

Tip: If you're asked to sketch a graph, you don't need to plot any actual data. Just make sure you show the correct shape, label the axes and give any other information that the question asks for.

Learning Objectives:

- Know that the orders of reactions with respect to reactants can be used to provide information about the mechanism of a reaction.
- Be able to use the orders with respect to reactants to provide information about the rate-determining/limiting step of a reaction.

Specification Reference 3.1.9.2

7. The Rate-determining Step

Reaction mechanisms show step by step how a chemical reaction takes place. The most important step is the rate-determining step.

What is the rate-determining step?

Mechanisms can have one step or a series of steps. In a series of steps, each step can have a different rate. The overall rate is decided by the step with the slowest rate — the **rate-determining step**.

The rate equation is handy for working out the mechanism of a chemical reaction. You need to be able to pick out which reactants from the chemical equation are involved in the rate-determining step. This isn't too hard — if a reactant appears in the rate equation, it must affect the rate. So this reactant, or something derived from it, must be in the rate-determining step. If a reactant doesn't appear in the rate equation, then it won't be involved in the rate-determining step and neither will anything derived from it.

An important point to remember about rate-determining steps and mechanisms is that the rate-determining step doesn't have to be the first step in a mechanism. Also, the reaction mechanism can't usually be predicted from just the chemical equation.

Tip: The rate-determining step is also known as the rate-limiting step.

Orders of reaction and the rate-determining step

The order of a reaction with respect to a reactant shows the number of molecules of that reactant that are involved in the rate-determining step. So, if a reaction's second order with respect to X, there'll be two molecules of X in the rate-determining step.

Example

The mechanism for the reaction between chlorine free radicals (Cl•) and ozone (O_3) consists of two steps:

$Cl\bullet_{(g)} + O_{3(g)} \rightarrow ClO\bullet_{(g)} + O_{2(g)}$ *This step is slow — it's the rate determining step.*

$ClO\bullet_{(g)} + O_{3(g)} \rightarrow Cl\bullet_{(g)} + 2O_{2(g)}$ *This step is fast.*

Cl• and O_3 are both in the rate determining step so must both be in the rate equation. So, the rate equation will be:

$$\text{Rate} = k[Cl\bullet]^m[O_3]^n$$

There's only one Cl• and one O_3 molecule in the rate-determining step, so the orders, m and n, are both 1. So the rate equation is:

$$\text{Rate} = k[Cl\bullet][O_3]$$

Tip: Look back at pages 209-210 for more about this reaction between ozone and chlorine radicals.

Tip: See pages 294-297 for more on rate equations.

Tip: Catalysts can appear in rate equations, so they can be in rate-determining steps too.

Reaction mechanisms

If you know which reactants are in the rate-determining step, you can work out the reaction mechanism.

Examples

2–bromo–2–methylpropane can react with the nucleophile OH^- to give 2–methyl propan–2–ol and bromide ions (Br^-).

There are two possible mechanisms for this reaction. Here's one...

$$\begin{array}{c} CH_3 \\ | \\ H_3C-C-Br \\ | \\ CH_3 \end{array} + OH^- \longrightarrow \begin{array}{c} CH_3 \\ | \\ H_3C-C-OH \\ | \\ CH_3 \end{array} + Br^-$$

... and here's the other one:

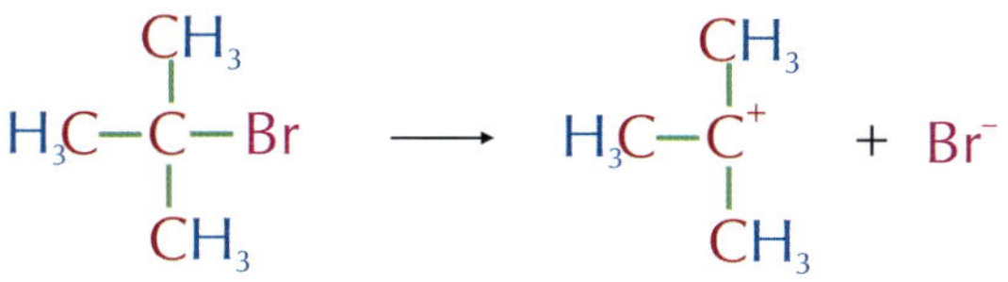

$$\begin{array}{c} CH_3 \\ | \\ H_3C-C-Br \\ | \\ CH_3 \end{array} \longrightarrow \begin{array}{c} CH_3 \\ | \\ H_3C-C^+ \\ | \\ CH_3 \end{array} + Br^-$$

This step is slow — it's the rate-determining step.

$$\begin{array}{c} CH_3 \\ | \\ H_3C-C^+ \\ | \\ CH_3 \end{array} + OH^- \longrightarrow \begin{array}{c} CH_3 \\ | \\ H_3C-C-OH \\ | \\ CH_3 \end{array}$$

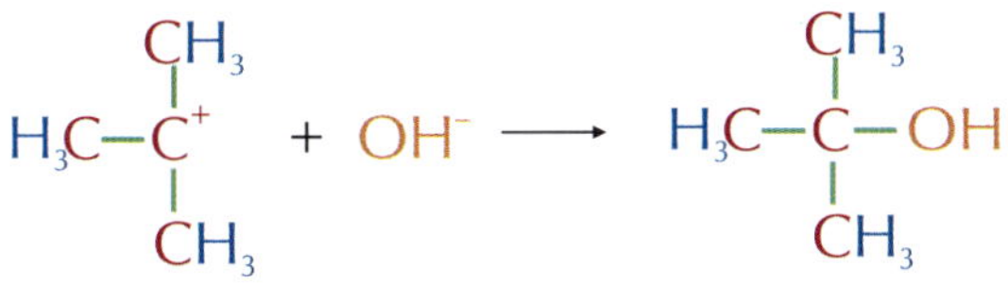

This step is fast.

The actual rate equation was worked out using rate experiments. It is:

$$\text{Rate} = k[(CH_3)_3CBr]$$

OH^- isn't in the rate equation, so it can't be involved in the rate-determining step. The second mechanism is correct because OH^- isn't in the rate-determining step.

Exam Tip
In the exam, you could be given reactions with three or even four steps, but don't be put off — you work it out in exactly the same way.

Nitrogen monoxide can react with oxygen to produce nitrogen dioxide:

$$2NO + O_2 \rightarrow 2NO_2$$

The reaction mechanism for this reaction is made up of two steps:

Step 1 — $NO + NO \rightarrow N_2O_2$
Step 2 — $N_2O_2 + O_2 \rightarrow 2NO_2$

The rate equation for this reaction is:

$$\text{Rate} = k[NO]^2[O_2]$$

So you know that the rate-determining step must involve 2 molecules of NO and 1 molecule of O_2. Neither step 1 nor step 2 contains all the molecules you'd expect from the rate equation. But in step 2 there's an intermediate molecule, N_2O_2, that's derived from 2 molecules of NO. So step 2 is the rate-determining step.

Practice Questions — Application

Q1 The reaction $NO_2 + CO \rightarrow NO + CO_2$ has a two-step mechanism:

Step 1: $2NO_2 \rightarrow NO + NO_3$
Step 2: $NO_3 + CO \rightarrow NO_2 + CO_2$

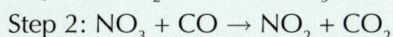

The rate equation for this reaction is rate = $k[NO_2]^2$.

a) What is the rate-determining step? Explain your answer.

b) A one-step mechanism was also proposed for this reaction. How can you tell that this reaction isn't a one-step mechanism?

Q2 The rate-determining step for a reaction between three reactants (A, B and C) is $2A + B + C \rightarrow X + Y$. Predict the rate equation for this reaction.

Tip: Working out mechanisms is a little bit of trial and error. You have to keep going until you find a mechanism that both fits the rate equation, and is likely to happen in real life.

Practice Questions — Fact Recall

Q1 What is the rate-determining step of a chemical reaction?

Q2 A reaction is second order with respect to oxygen. How many molecules of oxygen are involved in the rate-determining step?

8. The Arrhenius Equation

You met the rate constant (k) back on page 294, and you saw briefly how it increases with temperature. Now it's time to cover this idea in a bit more detail, with the help of a really yucky looking equation.

Learning Objectives:

- Know that the rate constant, k, varies with temperature as shown by the equation: $k = Ae^{\frac{-E_a}{RT}}$ where A is a constant, known as the Arrhenius constant, E_a is the activation energy, and T is the temperature in K.
- Be able to explain the qualitative effect of changes in temperature on the rate constant k.
- Be able to perform calculations using the equation $k = Ae^{\frac{-E_a}{RT}}$.
- Understand that the equation $k = Ae^{\frac{-E_a}{RT}}$ can be rearranged into the form $\ln k = -E_a/RT + \ln A$ and know how to use this rearranged equation with experimental data to plot a straight-line graph with a slope $-E_a/R$.

Specification Reference 3.1.9.1

What is the Arrhenius equation?

The **Arrhenius equation** (that nasty-looking thing in the purple box) shows how the rate constant (k) varies with temperature (T) and activation energy (E_a, the minimum amount of kinetic energy particles need to react).

This is probably the worst equation you're going to meet. Luckily, it'll be given to you in your exams if you need it, so you don't have to learn it off by heart. But you do need to know what all the different bits mean, and how it works.

Here it is:

$$k = Ae^{\frac{-E_a}{RT}}$$

Where:

- k = rate constant
- E_a = activation energy (J mol^{-1})
- T = temperature (K)
- R = gas constant (8.31 J K^{-1} mol^{-1})
- A = the Arrhenius constant (another constant)

Tip: In the Arrhenius equation, $e^{-E_a/RT}$ is known as the exponential.

The rate constant and activation energy

The Arrhenius equation is a really useful tool for helping us to work out what happens to the rate constant if various conditions, such as the temperature or activation energy, change.

As the activation energy, E_a, gets bigger, k gets smaller. You can test this out by trying different numbers for E_a in the equation... ahh go on, have a go.

Examples — Maths Skills

If you increase the activation energy, then this fraction becomes a larger, negative number...

$$k = Ae^{\frac{-E_a}{RT}}$$

...this results in the value of the exponential becoming smaller, and so k also gets smaller.

If you decrease the activation energy, then this fraction becomes a smaller, less negative number...

$$k = Ae^{\frac{-E_a}{RT}}$$

...this results in the value of the exponential becoming larger, and so k also increases.

Tip: There are two ways to answer questions about how activation energy affects the rate. You can answer using the Arrhenius equation, or using kinetics (see Unit 1: Section 5).

So, a large E_a will mean a slow rate. This makes sense when you think about it. If a reaction has a high activation energy, then not many of the reactant particles will have enough energy to react. So only a few of the collisions will result in the reaction actually happening, and the rate will be slow.

The rate constant and temperature

The equation also shows that as the temperature rises, k increases. (You can test this out by trying different numbers for T as well. Will the fun never cease?)

Examples — Maths Skills

If you increase the temperature, then this fraction becomes a smaller, less negative number...

$$k = Ae^{\frac{-E_a}{RT}}$$

...this results in the value of the exponential increasing, and so k increases.

If you decrease the temperature, then this fraction becomes a larger, more negative number...

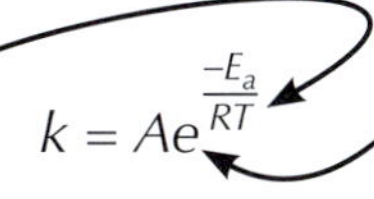

$$k = Ae^{\frac{-E_a}{RT}}$$

...this results in the value of the exponential becoming smaller, and so k also decreases.

The temperature dependence makes sense too. Higher temperatures mean reactant particles move around faster and with more energy so they're more likely to collide and more likely to collide with at least the activation energy, so the reaction rate is higher.

Using the Arrhenius equation

You might be given four of the five values in the Arrhenius equation and asked to use the equation to find the value of the fifth. All you need to do is rearrange the equation to make the unknown the subject, and then substitute the values you know into the equation. Here are some examples to show you how it's done...

Examples — Maths Skills

The decomposition of N_2O_5 at 308 K has a rate constant of 1.35×10^{-4} s^{-1}. The Arrhenius constant for this reaction is 4.79×10^{13} s^{-1}. Calculate the activation energy of this reaction. (R = 8.31 J K^{-1} mol^{-1})

- First it's a good idea to get the Arrhenius equation into a simpler form so it's easier to use. That means getting rid of the nasty exponential bit — so you need to take the natural log (ln) of everything in the equation:

$$k = Ae^{\frac{-E_a}{RT}} \longrightarrow \ln(k) = \ln(Ae^{\frac{-E_a}{RT}})$$
$$= \ln(A) + \ln(e^{\frac{-E_a}{RT}})$$
$$= \ln(A) - \frac{E_a}{RT}$$

- Now re-arrange the equation to get E_a on the left hand side:

$$\frac{E_a}{RT} = \ln(A) - \ln(k)$$

- And another quick re-arrangement to get E_a on its own:

$$E_a = (\ln(A) - \ln(k)) \times RT$$

- Now you can just pop the numbers from the question into this formula:

$$E_a = (\ln(4.79 \times 10^{13}) - \ln(1.35 \times 10^{-4})) \times (8.31 \times 308)$$
$$= (31.5... - (-8.91...)) \times (8.31 \times 308) = 103\,429.54 \text{ J mol}^{-1}$$
$$= 103 \text{ kJ mol}^{-1} \text{ (3 s.f.)}$$

Tip: If you're asked what happens to a variable or constant in the Arrhenius equation when something changes, just use this method to work out what will happen. The only thing that can't change is R, the gas constant — this is a defined physical constant that can't change. Though k and A are known as constants, they're only constant for a given reaction under certain conditions.

Figure 1: *Svante August Arrhenius, the Swedish chemist after whom the Arrhenius equation is named.*

Tip: 'e' and 'ln' are inverses, so you can use them to 'get rid' of each other.

Tip: This example uses the log law that states: ln(ab) = ln(a) + ln(b), but you can look at pages 528-529 for lots more on how to use logs.

Tip: There are buttons on your calculator for 'e' and for 'ln'. They make life much easier when it comes to dealing with calculations involving the Arrhenius equation.

Tip: The Arrhenius constant is temperature-dependent, so it won't always have the same value for the same reaction.

Tip: Rearranging equations can be quite tricky. Always check your answer at the end. It's a good idea to make sure all the variables that were in the starting equation have ended up in the final one, and that you've not lost or gained any along the way.

The same reaction is set up at a different temperature, and the rate constant is found to have a value of 2.4×10^{-3} s^{-1}. Given that the Arrhenius constant for this reaction is 6.22×10^{13} s^{-1}, but all other conditions remain the same, calculate the temperature at which this reaction occurs.

- First of all, rearrange your equation so it's in terms of your unknown. In this case, it's *T*.

$$k = Ae^{\frac{-E_a}{RT}} \longrightarrow \ln(k) = \ln(A) - \frac{E_a}{RT}$$ *First take the natural log (ln)...*

$$\frac{E_a}{RT} = \ln(A) - \ln(k)$$ *...then subtract ln(k) from, and add E_a/RT to, both sides...*

$$\frac{E_a}{R} = T(\ln(A) - \ln(k))$$ *...then multiply by T...*

$$T = \frac{E_a}{R(\ln(A) - \ln(k))}$$ *...and finally divide by (ln(A) – ln(k)).*

- Then just go about substituting everything you know into your new equation (the activation energy will be your answer from the last question).

$$T = \frac{103\,429.54}{8.31 \times (\ln(6.22 \times 10^{13}) - \ln(2.4 \times 10^{-3}))} = 329 \text{ K (3 s.f.)}$$

Tip: The logarithmic version of the Arrhenius equation, $\ln(k) = -\frac{E_a}{RT} + \ln(A)$ is in the form, $y = mx + c$, the equation for a straight-line graph.

Arrhenius plots

Putting the **Arrhenius equation** into logarithmic form generally makes it a bit easier to use. By creating a graph of ln *k* against 1/*T*, known as an Arrhenius plot, you can use the relationship in the logarithmic form of the Arrhenius equation to find out certain variables. For example, the line of best fit on an Arrhenius plot will be a straight line with a gradient of $-E_a/R$, and the *y*-intercept will be equal to ln *A*.

Tip: The units of the *x*-axis are K^{-1}. The *y*-axis doesn't have any units — no natural logarithms have units.

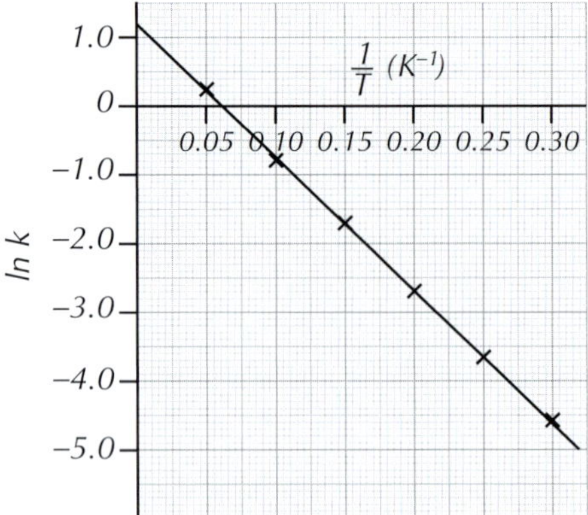

Figure 2: *An example of an Arrhenius plot. The value of $-E_a/R$ is given by the gradient, and ln A is equal to the y-intercept.*

Tip: Values of 1/*T* can often be quite small, and it can get confusing writing out lots of 0s all the time. Sometimes axes are written with units in standard form. If this is the case, you'll need to watch out for the units on the *x*-axis — they might show that there are actually some extra decimal places to put in.

To gather the data for an Arrhenius plot, you'll need to carry out the same experiment a few times, each time at a different temperature, and work out the rate at the same point for each one.

Example — Maths Skills

The rate constants for a reaction at a variety of temperatures are shown in the table below. Use the data to create an Arrhenius plot, and therefore work out the activation energy and the Arrhenius constant.

T / K	k / s^{-1}	1/T / K^{-1}	ln k
150	4.59×10^{-48}	0.0067	–109
200	2.44×10^{-36}	0.0050	–82.0
250	5.90×10^{-29}	0.0040	–65.0
300	9.60×10^{-24}	0.0033	–53.0
350	1.05×10^{-20}	0.0029	–46.0
400	4.25×10^{-18}	0.0025	–40.0

- If you're given data for T and k, and asked to plot an Arrhenius plot, the first thing you need to do is to work out the values of $1/T$ and ln k for each T and k value respectively.
- Now, plot your values of $1/T$ and ln k on a suitable set of axes, and draw a line of best fit through the data set.

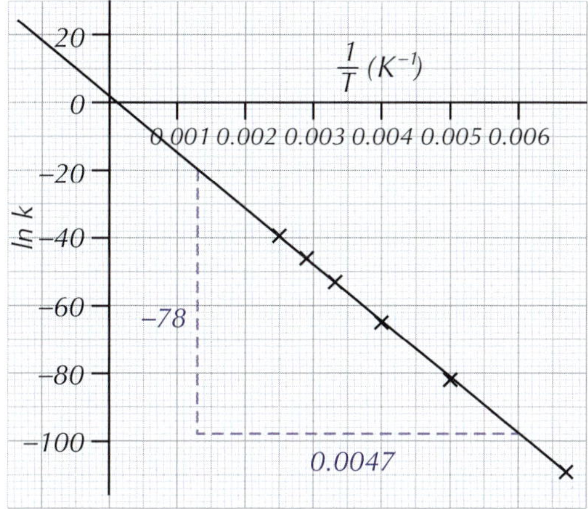

- To find the value of E_a, find the gradient of the straight-line.

  gradient = change in y ÷ change in x

  = –78 ÷ 0.0047 = –16 595.7...
- You know that gradient = $-E_a/R$, so just substitute in the value of R, and solve to find the activation energy.

  $-E_a/R = -16\,595.7...$

  $-E_a = -16\,595.7... \times 8.31$

  $E_a = 137\,910.6...$ J mol^{-1} = **140 kJ mol^{-1}** (2 s.f.)
- To find out A, substitute your value of E_a back into the equation $k = Ae^{\frac{-E_a}{RT}}$, along with a value from the table of T and k.

  So, using the data point at 150 K: $A = \frac{k}{e^{-E_a/RT}}$

  $A = 4.59 \times 10^{-48} \div e^{(-137\,910.6... \div (8.31 \times 150))}$

  $A =$ **5.1 s^{-1}** (2 s.f.)

Tip: If you need to work out the values of $1/T$ and ln k, draw out a table with four columns, like the one on the left. This makes sure you don't miss out any values, and you don't get your T's and k's muddled up.

Tip: When you're drawing your line of best fit through your data set, if you can, extend it so it goes through the y-axis. This can sometimes help when you're working out the Arrhenius constant.

Tip: Look back at pages 291-292 if you need a reminder on how to calculate the gradient of a straight line.

Tip: If the x-axis on your Arrhenius plot extends to 0, you can work out A by looking at the y-intercept — this is just where the line of best fit crosses the y-axis, and is equal to ln (A).
So in this example:
ln (A) = 2.0
$A = e^{2.0} = 7.4$ s^{-1} (2 s.f.)

Tip: The units of A are the same as the units of k (which you're given in the question).

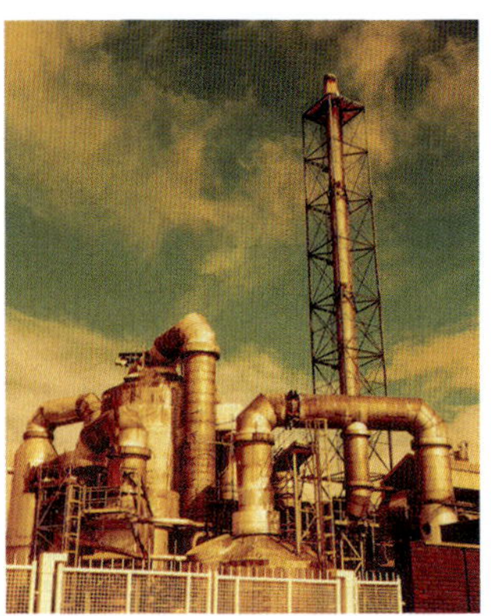

Figure 2: *A sulfuric acid plant in Cleveland.*

Practice Questions — Application

Q1 In part of the industrial manufacture of sulfuric acid, SO_2 is converted into SO_3. In the Contact process, this step is catalysed by a vanadium(V) oxide catalyst.
The rate constant for the reaction can be expressed by the Arrhenius equation:

$$k = Ae^{\frac{-E_a}{RT}}$$

a) With reference to the Arrhenius equation, would you expect the rate constant of the catalysed reaction to be greater, smaller or the same as the uncatalysed reaction?

b) A manufacturer tries decreasing the temperature at which they run the Contact process to save money. With reference to the Arrhenius equation, explain how will this affect the rate constant.

Q2 The activation energy of the following reaction is 180 kJ mol^{-1}.

$$2NH_{3(g)} \rightarrow N_{2(g)} + 3H_{2(g)}$$

a) Given that the rate constant of this reaction is 1.10×10^{-5} mol^{-1} dm^3 s^{-1} at 600 K, calculate the Arrhenius constant for this reaction.

b) A catalyst is added to the reaction vessel which increases the rate constant to 2.02×10^{-5} mol^{-1} dm^3 s^{-1}. Given that all other variables stay the same, calculate the activation energy of the catalysed reaction.

Q3 The rate of reaction at different temperatures are recorded below.

T (K)	*k*
555	3.50×10^{-7}
630	3.00×10^{-5}
665	2.20×10^{-4}
700	1.15×10^{-3}
790	3.94×10^{-2}

a) Construct an Arrhenius plot for the data above.

b) Work out the activation energy of this reaction.

c) Work out the Arrhenius constant for this reaction.

Tip: Be careful when working out the Arrhenius constant. The value of the *y*-intercept is when *x* = 0. But sometimes you'll find your axes don't meet at (0, 0). If this is the case, then you can't use the *y*-intercept to work out the value of ln A (unless you redraw your axes) — you'll need to use the substitution method instead.

Tip: When you're using real experimental data, you may find your calculated value of *A* varies a little bit, depending on which data point you use, but as long as it's within the same order of magnitude, you can be reasonably happy you've got the right answer.

Practice Questions — Fact Recall

Q1 Write out the Arrhenius equation, defining all the variables in it.

Q2 What happens to the rate of a reaction when the temperature is increased?

Q3 What is the logarithmic form of the Arrhenius equation?

Q4 What should go on the *x*- and *y*- axes on an Arrhenius plot?

9. Gas Equilibria

When we're dealing with equilibrium reactions involving gases, it's easier to think in terms of pressure than in terms of concentrations. A gaseous equilibrium system will have a total pressure, but there will also be individual pressures for each of the individual gases. These are called partial pressures.

Learning Objectives:

- Be able to derive partial pressures from mole fraction and total pressure.
- Know that the equilibrium constant, K_p, is deduced from the equation for a reversible reaction occurring in the gas phase.
- Know that K_p is the equilibrium constant calculated from partial pressures for a system at constant temperature.
- Be able to construct an expression for K_p for a homogeneous system in equilibrium.
- Be able to perform calculations involving K_p.

Specification Reference 3.1.10

Partial pressures

You've met equilibrium reactions before — they're just reactions that can go in both a forwards direction, and a reverse direction.

Example

The reaction between hydrogen and iodine vapour is an example of a gas equilibrium reaction.

$$H_{2(g)} + I_{2(g)} \rightleftharpoons 2HI_{(g)}$$

The forwards reaction is: $H_{2(g)} + I_{2(g)} \rightarrow 2HI_{(g)}$

And the backwards reaction is: $2HI_{(g)} \rightarrow H_{2(g)} + I_{2(g)}$

In a mixture of gases, each individual gas exerts its own pressure — this is called its partial pressure. So, in the example above, hydrogen, iodine and hydrogen iodide gases will all exert a pressure on the system. The size of the pressure exerted by each gas doesn't have to be the same. If gases are present in equal quantities, their partial pressures will be the same.

The total pressure of a gas mixture is the sum of all the partial pressures of the individual gases. Partial pressures are often written as p_X — this just means the partial pressure of X.

Example

So, in the reaction between hydrogen and iodine vapour to make hydrogen iodide, the total pressure can be expressed as:

Total pressure = $p_{H_2} + p_{I_2} + p_{HI}$

Tip: Look back at pages 133-142 for a reminder on reversible reactions and equilibria.

You can put this fact into use when dealing with pressure calculations, such as the one below.

Example — Maths Skills

When 3.00 moles of the gas PCl_5 is heated, it decomposes into PCl_3 and Cl_2:

$$PCl_{5(g)} \rightleftharpoons PCl_{3(g)} + Cl_{2(g)}$$

In a sealed vessel at 500 K, the equilibrium mixture contains chlorine with a partial pressure of 263 kPa. If the total pressure of the mixture is 714 kPa, what is the partial pressure of PCl_5?

From the equation you know that PCl_3 and Cl_2 are produced in equal amounts, so the partial pressures of these two gases are the same at equilibrium — they're both 263 kPa.

Total pressure = $p_{PCl_5} + p_{PCl_3} + p_{Cl_2}$

$$714 = p_{PCl_5} + 263 + 263$$

So the partial pressure of PCl_5 = 714 – 263 – 263 = 188 kPa

Figure 1: John Dalton, the English chemist, who came up with the theory that the total pressure of a mixture of gases is equal to the sum of the partial pressures of the gases in the mixture. This theory is also known as Dalton's Law.

Mole fractions

A '**mole fraction**' is just the proportion of a gas mixture that is a made up of particular gas. So if you have four moles of gas in total and two of them are gas A, the mole fraction of gas A is ½. There are two formulas you need to know.

Exam Tip
You won't be given these equations in the exam — make sure you learn them off by heart so you can handle any K_p calculations the examiners could throw at you.

$$\text{Mole fraction of a gas in a mixture} = \frac{\text{number of moles of gas}}{\text{total number of moles of gas in the mixture}}$$

$$\text{Partial pressure of a gas in a mixture} = \text{mole fraction of gas} \times \text{total pressure of the mixture}$$

Here are some examples to show you how to use these equations.

Tip: You need to use a balanced chemical equation when working out mole fractions — you have to take into account the ratios of the number of moles of each gas when working out the mole fraction of a particular reactant or product.

Tip: Lots of gases made up of only one element are diatomic, e.g. N_2, O_2, Cl_2, H_2 etc. So, make sure you remember to divide by the M_r of the whole gas molecule, rather than just the A_r of the element when you're doing calculations with moles.

Tip: Try not to round your intermediate answers. Rounding too early could lead to an error in your final answer.

Examples — Maths Skills

When 2.00 mol of PCl_5 is heated in a sealed vessel, the equilibrium mixture contains 1.25 mol of chlorine. What is the mole fraction of PCl_5?

- First, write out a chemical equation so you can see what's happening:
  $$PCl_{5(g)} \rightleftharpoons PCl_{3(g)} + Cl_{2(g)}$$
- From the equation, PCl_3 and Cl_2 are produced in equal amounts, so there'll be 1.25 moles of PCl_3 too.
- 1.25 moles of PCl_5 must have decomposed so, (2.00 – 1.25) = 0.75 moles of PCl_5 must be left at equilibrium.
- This means that the total number of moles of gas at equilibrium = 0.75 + 1.25 + 1.25 = 3.25.
- So the mole fraction of $PCl_5 = \frac{0.75}{3.25} = 0.2307...$
- So the mole fraction of PCl_5 is 0.231 (to 3 s.f.)

A mixture of 8.02 g of nitrogen and 2.42 g of oxygen is contained inside a sealed vessel. Given that the total pressure of the mixture is 921 kPa, what is the partial pressure of the oxygen?

- First, work out the number of moles of each gas present in the vessel:

  moles of N_2 = 8.02 ÷ (14 × 2) = 0.286...

  moles of O_2 = 2.42 ÷ (16 × 2) = 0.0756...
- Add the values of each of these together to find out the total number of moles present in the mixture:

  0.286... + 0.0756... = 0.362...
- The question asks for the partial pressure of oxygen, so you need to work out the mole fraction of oxygen, using the trusty equation, mole fraction = no. moles of gas ÷ total moles of gas in the system

  = 0.0756... ÷ 0.362... = 0.208...

- Finally, substitute the mole fraction of oxygen, along with the total pressure that you were given in the question, into the equation for partial pressure:

  partial pressure = mole fraction × total pressure

  = 0.208... × 921 = 192 kPa (3 s.f.)

The gas equilibrium constant, K_p

K_p is the equilibrium constant for a reversible reaction where some, or all, of the reactants and products are gases. The expression for K_p is just like the one for K_c, except you use partial pressures instead of concentrations.

For the equilibrium $aA_{(g)} + bB_{(g)} \rightleftharpoons dD_{(g)} + eE_{(g)}$

$$K_p = \frac{(p_D)^d (p_E)^e}{(p_A)^a (p_B)^b}$$

Example

Write an expression for K_p for the following reaction.

$$O_{3(g)} + H_2O_{(g)} \rightleftharpoons H_{2(g)} + 2O_{2(g)}$$

There are two moles of O_2 in the balanced equation, so the partial pressure of O_2 needs to be raised to the power of 2.

$$K_p = \frac{(p_{O_2})^2 \times p_{H_2}}{p_{O_3} \times p_{H_2O}}$$

The reactants go on the bottom of the expression.

The products go on the top of the expression.

To calculate K_p you put the partial pressures in the expression. Then you work out the units like you did for K_c.

Example — Maths Skills

Calculate K_p for the decomposition of PCl_5 gas at 500 K. The partial pressures of each gas are: p_{PCl_5} = 188 kPa, p_{PCl_3} = 263 kPa, p_{Cl_2} = 263 kPa

- Start by writing out a balanced chemical equation:

$$PCl_{5(g)} \rightleftharpoons PCl_{3(g)} + Cl_{2(g)}$$

- Now, write an expression for K_p. Remember, the reactants go on the bottom of the expression, and the products go on the top. Then just substitute in all the partial pressure values you were given in the question.

$$K_p = \frac{p_{Cl_2} \times p_{PCl_3}}{p_{PCl_5}} = \frac{263 \times 263}{188} = 368$$

- To find the units, substitute all the partial pressure units into your expression for K_p, then cancel down as much as you can.

$$K_p = \frac{\cancel{kPa} \times kPa}{\cancel{kPa}} = kPa \qquad \text{So } K_p = 368 \text{ kPa}$$

Tip: K_c and K_p work in pretty much exactly the same way — K_p is just specifically for gases. Look back at pages 138-142 for more on K_c.

Tip: Solid, liquid and aqueous substances don't appear in the expression for K_p. They also don't contribute to the total pressure of a system (see page 313).

Tip: Make sure your chemical equation is balanced — otherwise you'll get your expression for K_p wrong.

Tip: K_p is temperature dependent (see page 317). So, a particular value of K_p is only valid at a given temperature.

Tip: Being able to write balanced chemical equations is a really important skill in chemistry. It will help you solve loads of questions. Look back at pages 52-53 for more details.

Tip: Remember — moles = mass ÷ M_r.

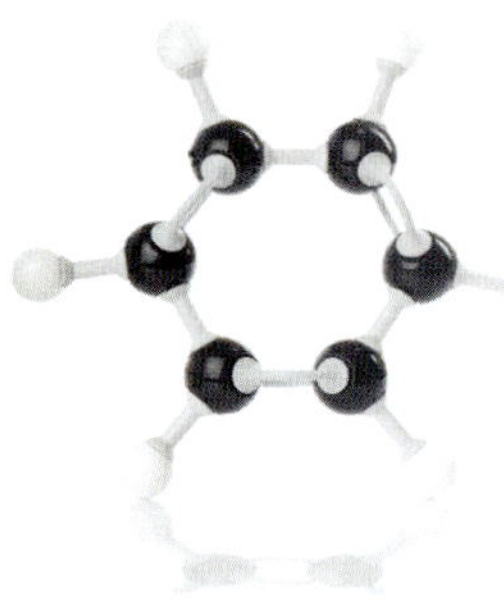

Figure 2: *A molecular model of benzene (C_6H_6). In the hydrogenation of benzene, hydrogen is added to the aromatic ring to create a saturated alkane — cyclohexane.*

Practice Questions — Application

Q1 Ammonia is reacted with oxygen. The products are nitrogen gas and water vapour. The reaction is a reversible reaction.

a) Write out a balanced chemical equation for this reaction.

b) Write out an expression for the total pressure of the system.

c) p_{oxygen} = 85 kPa, p_{ammonia} = 42 kPa, p_{nitrogen} = 21 kPa, and $p_{\text{water vapour}}$ = 12 kPa. Calculate the total pressure of the system.

d) What is the mole fraction of water vapour at the partial pressures given in c)?

Q2 A sealed container holds 4.00 g of helium gas (He) and 2.81 g of oxygen gas. Given that the total pressure of the container is 8.12 kPa, what is the partial pressure of oxygen?

Q3 The following equation shows the reaction between benzene and hydrogen to create cyclohexane:

$$C_6H_{6(g)} + 3H_{2(g)} \rightleftharpoons C_6H_{12(g)}$$

a) Write an equation for K_p for the reaction above.

b) At equilibrium, the partial pressures of the three gases are all equal. Given that $K_p = 4.80 \times 10^{-13}$ kPa^{-3}, what is the partial pressure of hydrogen?

Q4 When 24.32 mol of hydrogen fluoride is heated in a sealed vessel, the equilibrium mixture contains 9.340 mol of fluorine gas.

a) If the total pressure of the mixture is 2313 kPa, what is the partial pressure of hydrogen?

b) Calculate the gas equilibrium constant for this reaction.

Practice Questions — Fact Recall

Q1 What is meant by the term 'partial pressure'?

Q2 What equation links the mole fraction of a gas, the number of moles of that gas and the total number of moles of gas in a mixture?

Q3 Write an expression for K_p for the equation: $aA_{(g)} + bB_{(g)} \rightleftharpoons dD_{(g)} + eE_{(g)}$.

10. Changing Gas Equilibria

It's time to see how changing temperature, pressure or adding a catalyst affects equilibrium in gases. You'll need to use Le Chatelier's principle, which you met on page 133.

Learning Objectives:

- Predict the qualitative effects of changes in temperature and pressure on the position of equilibrium.
- Predict the qualitative effects of changes in temperature on the value of K_p.
- Understand that, whilst a catalyst can affect the rate of attainment of an equilibrium, it does not affect the value of the equilibrium constant.

Specification Reference 3.1.10

Le Chatelier's Principle

You met Le Chatelier's principle earlier in the course. It's a theory that states that, if a reversible reaction at equilibrium is subjected to a change in temperature, pressure or concentration, then the equilibrium position shifts to counteract the change. The effects of Le Chatelier's principle apply to gas equilibria too. Like K_c, K_p is affected by changes in temperature but not by changes in pressure.

Temperature

If you increase the temperature, the equilibrium shifts in the endothermic (positive ΔH) direction. If you decrease the temperature, the equilibrium shifts in the exothermic (negative ΔH) direction. Changing the temperature of a system will change K_p, depending on which way the equilibrium shifts.

Example

The reaction below is exothermic in the forward direction.

$$2SO_{2(g)} + O_{2(g)} \rightleftharpoons 2SO_{3(g)} \quad \Delta H = -197 \text{ kJ mol}^{-1}$$

- If the temperature is increased, the equilibrium shifts to the left to try and absorb the extra heat and counteract the change. This means that less product is formed.

$$K_p = \frac{(p_{SO_3})^2}{(p_{SO_2})^2 \times p_{O_2}}$$

If there's less product, the partial pressure of SO_3 will be reduced...

... and the partial pressures of SO_2 and O_2 will increase.

So, as the temperature increases, p_{SO_3} will decrease and p_{SO_2} and p_{O_2} will increase, so K_p will decrease.

- If the temperature is decreased, the equilibrium shifts to the right to try and make up for the decrease in heat. This means more product is formed, so K_p will increase.

Tip: Exothermic reactions release heat and endothermic reactions absorb heat. If a reversible reaction is exothermic in one direction, it will always be endothermic in the other direction.

Pressure

Le Chatelier's principle states that if you increase the pressure, the equilibrium shifts to the side with fewer moles of gas, and if you decrease the pressure, the equilibrium shifts to the side with more moles of gas. However, these changes don't affect K_p. Instead, if you change the pressure, the position of equilibrium shifts in such a way the partial pressures of reactants and products at the new equilibrium position keep K_p constant.

Tip: If you change the pressure of a system, there's only a shift in equilibrium position of reactions where there is an unequal number of moles of gas on each side of the reaction.

Example

The equilibrium of the reaction below shifts in response to changes in pressure.

$$2SO_{2(g)} + O_{2(g)} \rightleftharpoons 2SO_{3(g)} \quad \Delta H = -197 \text{ kJ mol}^{-1}$$

- If the pressure is increased, the equilibrium shifts to the right, the side with fewer moles of gas, to counteract the change. This means that the partial pressure of SO_3 increases.

Tip: You'll need to make sure you're using a balanced chemical equation when looking at how the number of moles changes as the reactants change to products.

$$K_p = \frac{(p_{SO_3})^2}{(p_{SO_2})^2 \times p_{O_2}}$$

The partial pressure of SO_3 will increase, so this value gets bigger...

...and the increase in the overall pressure cause the partial pressures of SO_2 and O_2 to increase, so that K_p remains constant.

- Similarly, if the pressure is decreased, the equilibrium shifts to the left to try and make up for the drop in pressure — so at equilibrium, there's more SO_2 and O_2, and less SO_3. But the partial pressures of all of the gases in the mixture change so as to keep K_p the same.

Figure 1: *Sulfur burning in oxygen to produce sulfur dioxide (SO_2).*

Catalysts

As with K_c, catalysts have no effect on the position of equilibrium or on the value of K_p. This is because a catalyst will increase the rate of both the forward and backward reactions by the same amount. As a result, the equilibrium position will be the same as the uncatalysed reaction, but equilibrium will be reached faster. So catalysts can't increase the amount of product produced, but they do decrease the time taken to reach equilibrium.

Tip: To work out how K_p is affected by changes in temperature, it's often helpful to write out the expression for K_p. You can then see how increasing amounts of products or reactants affect K_p much more easily.

Practice Questions — Application

Q1 The reaction between hydrogen and iodine to produce hydrogen iodide is a reversible reaction.

$$H_{2(g)} + I_{2(g)} \rightleftharpoons 2HI_{(g)}$$

A student says, "altering the pressure of a reaction affects the equilibrium position. Therefore, changing the pressure of this reaction would move equilibrium." Comment on the statement.

Q2 Nitrogen dioxide reacts to produce dinitrogen tetroxide as part of a reversible reaction. The forward reaction is exothermic.

$$2NO_{2(g)} \rightleftharpoons N_2O_{4(g)}$$

a) What will happen to the equilibrium position if there is a decrease in pressure? Explain your answer.

b) A student is investigating the effect on K_p of changing the temperature of the system. How would K_p change if the student decreased the temperature of the system?

Practice Questions — Fact Recall

Q1 What does Le Chatelier's principle state?

Q2 How would an increase in temperature affect the equilibrium position of a reaction that is exothermic in the forward direction?

Q3 What happens to K_p when there is a change in pressure for a reaction where there are a different number of moles of gas in the reactants and products?

Q4 How do catalysts affect the position of equilibrium and the value of K_p?

Section Summary

Make sure you know...

- That the reaction rate is the change in the amount of reactants or products per unit time.
- That continuous monitoring involves following a reaction through to completion, by monitoring the amount of product or reactant you have at regular intervals.
- That measuring gas volumes, loss of mass, changes in pH or colour are all ways to follow a reaction.
- How to find the rate of a reaction from a straight-line concentration-time graph by calculating the gradient.
- How to find the rate of a reaction at a given time from a curved concentration-time graph, by calculating the gradient of a tangent to the curve.
- That rate equations take the general form rate = $k[A]^m[B]^n$ where m and n are the orders of reaction with respect to A and B respectively.
- What is meant by the term 'order of reaction', and how the concentration of reactants with order 0, 1 and 2, affects the rate of reaction.
- That the overall order is the sum of the individual orders of reaction in a rate equation.
- That rate equations and orders of reactions can only be determined experimentally.
- Know that the rate constant of a reaction is a measure of how fast the reaction is going.
- How to write rate equations for reactions, given appropriate information.
- How to calculate the rate constant, its units, and related values, given information about the rate, and the concentration and relative orders of reactants in a reaction.
- How to find the initial rate of a reaction from a concentration-time graph.
- How to use the initial rates method to work out the orders of reaction, with respect to reactants in a chemical reaction, using concentration and rates.
- How clock reactions work.
- What the shapes of zero, first and second order concentration-time and rate-concentration graphs look like.
- That the rate-determining step of a reaction is the slowest step in a reaction mechanism.
- How to determine the rate-determining step from a rate equation.
- How to use the rate equation to suggest a possible reaction mechanism for a given chemical reaction.
- The variables and constants that make up the Arrhenius equation.
- How to use the Arrhenius equation to explain how the rate varies when the activation energy and temperature of a system are altered.
- How to use the Arrhenius equation to find the rate and related quantities.
- How to construct and use an Arrhenius plot to work out the activation energy and Arrhenius constant.
- That the partial pressure of a gas is how much pressure it exerts on a system, and that the total pressure of a gaseous system is equal to the sum of the partial pressures of its constituent gases.
- How to use partial pressures to work out the total pressure of a system.
- How to work out the mole fraction of a gas, and how to use mole fractions in calculations involving pressure.
- How to work out the expression for the gas equilibrium constant, K_p, for a given chemical reaction involving gases.
- How to calculate K_p for a chemical reaction, given the partial pressures of the constituent gases.
- How K_p is affected by changes in temperature, but not by changes in pressure or the addition of a catalyst.
- The effect of changes in temperature and pressure on the position of a gas equilibrium.

Exam-style Questions

1 What is the overall order of the following reaction?

$$3ClO^-_{(aq)} \rightarrow ClO_3^-{}_{(aq)} + 2Cl^-_{(aq)} \qquad \text{rate} = k[ClO^-]^2$$

A 1

B 2

C 3

D 4

(1 mark)

2 Which of these techniques would **not** produce data that would be suitable for working out the rate of the reaction?

A Measuring the loss of mass from a reaction container, every 10 seconds, until the reaction is complete.

B Measuring the mass of a reaction container, containing a reaction which involves the production of a gas, at the start and end of a reaction.

C Recording the absorbance of a sample of a reaction mixture, that involves the formation of a coloured solution, every 20 seconds.

D Measuring the pH of a reaction mixture, where H^+ ions are produced, at regular intervals throughout a reaction.

(1 mark)

3 Increasing the value of which of the following variables will result in a decrease in rate?

A The Arrhenius constant.

B The temperature.

C The activation energy.

D The pressure.

(1 mark)

4 Calculate the partial pressure of water vapour in the following reaction, given that p_{CO} = 109 kPa, p_{CO_2} = 623 kPa, p_{H_2} = 468 kPa and K_p = 31.4.

$$CO_{(g)} + H_2O_{(g)} \rightleftharpoons CO_{2(g)} + H_{2(g)}$$

A 77.7 kPa

B 83.9 kPa

C 840 kPa

D 85.2 kPa

(1 mark)

5 A student is investigating the rate of a reaction between a sample of zinc metal and hydrochloric acid. She measures the mass lost from a reaction container at regular intervals until the reaction is complete. She converts the mass loss into concentration data, with respect to HCl.

5.1 Suggest another way that the student could monitor this reaction.

(1 mark)

5.2 The data that the student recorded is shown in the table below.
Use the data to plot a graph.

Time / min	0	1	2	3	4	5	6	7	8
Concentration of HCl / mol dm^{-3}	5.9	5.1	4.5	4.0	3.6	3.3	3.1	2.9	2.8

(4 marks)

5.3 Calculate the initial rate of the reaction.

(2 marks)

6 The Arrhenius equation, shown below, describes the relationship between the rate constant, the activation energy and the temperature.

$$k = Ae^{\frac{-E_a}{RT}}$$

6.1 Find the activation energy of a reaction that has a rate constant of 1.28×10^{-5} mol^{-1} dm^3 s^{-1} at 600 K. The Arrhenius constant is 2.21 mol^{-1} dm^3 s^{-1} and the gas constant is 8.31 J K^{-1} mol^{-1}.

(2 marks)

6.2 When a catalyst is added to a reaction, the rate constant of the reaction increases. With reference to the Arrhenius equation, explain why this is.

(2 marks)

6.3 A student is carrying out an investigation into the Arrhenius equation for a certain reaction. He calculates the rate constant for the reaction at various temperatures. His results are shown in the table below.

Temperature / K	k / mol dm^{-3} s^{-1}
286	7.49×10^{-5}
294	2.03×10^{-4}
299	3.71×10^{-4}
313	1.50×10^{-3}
322	3.52×10^{-3}
334	1.11×10^{-2}

Calculate the activation energy and the Arrhenius constant for this reaction.

(8 marks)

7 The Haber Process, as shown below, is used in the manufacture of ammonia gas.

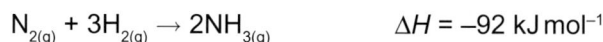

$$N_{2(g)} + 3H_{2(g)} \rightarrow 2NH_{3(g)} \qquad \Delta H = -92 \text{ kJ mol}^{-1}$$

7.1 In a certain reaction container, p_{H_2} = 19 kPa, p_{NH_3} = 8920 kPa and K_p = 52.7 kPa^{-2}. Calculate the partial pressure of nitrogen in this reaction.

(2 marks)

7.2 An iron catalyst is used when ammonia is manufactured on an industrial scale. Explain how the presence of a catalyst affects the value of K_p.

(1 mark)

7.3 The temperature of the reaction container in **7.1** is increased. Explain how this affects the position of equilibrium and the value of K_p.

(2 marks)

8 Under certain conditions the following reaction occurs between nitrogen monoxide (NO) and hydrogen (H_2):

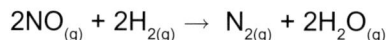

$$2NO_{(g)} + 2H_{2(g)} \rightarrow N_{2(g)} + 2H_2O_{(g)}$$

The table below shows the results of a series of initial rate experiments for this reaction.

Experiment	[NO] / mol dm^{-3}	[H_2] / mol dm^{-3}	Initial rate / mol dm^{-3} s^{-1}
1	3.0×10^{-3}	6.0×10^{-3}	4.50×10^{-3}
2	3.0×10^{-3}	3.0×10^{-3}	2.25×10^{-3}
3	6.0×10^{-3}	1.5×10^{-3}	4.50×10^{-3}

8.1 Determine the orders of reaction with respect to NO and H_2.

(2 marks)

8.2 Write out the rate equation for this reaction.

(1 mark)

8.3 Using the data above, calculate the rate constant (k) for this reaction and give its units.

(3 marks)

8.4 What would the rate of reaction be if 4.5×10^{-3} mol dm^{-3} NO were mixed with 2.5×10^{-3} mol dm^{-3} H_2 under the same conditions as the experiment above?

(2 marks)

8.5 Using the rate equation, explain why a one-step mechanism is not possible for this reaction.

(2 marks)

8.6 By writing reaction equations, suggest a possible two step mechanism for this reaction. The first step is the slowest step.

(2 marks)

1. Electrode Potentials

In redox reactions, electrons move from one atom to another. When electrons move, you get electricity. So redox reactions can be used to make electricity.

REQUIRED PRACTICAL 8

Electrochemical cells

Electrochemical cells can be made from two different metals dipped in salt solutions of their own ions and connected by a wire (the external circuit). There are always two reactions within an electrochemical cell — one's an oxidation and one's a reduction — so it's a **redox** process.

Example

The diagram below shows an electrochemical cell made using copper and zinc.

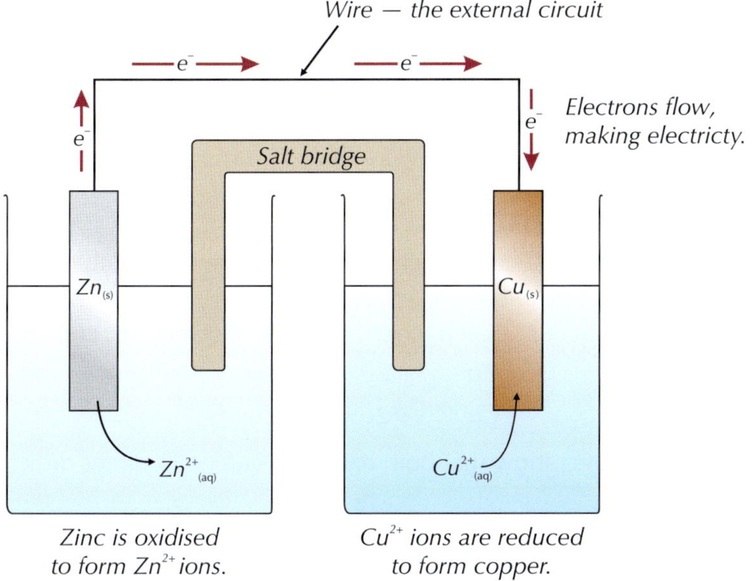

A copper electrode is dipped in a solution of Cu^{2+} ions and a zinc electrode is dipped in a solution of Zn^{2+} ions. Zinc loses electrons more easily than copper. So in the **half-cell** on the left, zinc (from the zinc electrode) is oxidised to form $Zn^{2+}_{(aq)}$ ions. This releases electrons into the external circuit. In the other half-cell, the same number of electrons are taken from the external circuit, reducing the Cu^{2+} ions to copper atoms.

The solutions are connected by a **salt bridge**, e.g. a strip of filter paper soaked in $KNO_{3(aq)}$. This allows ions to flow between the half-cells and balance out the charges — it completes the circuit.

Electrons flow through the wire from the most reactive metal to the least. You can put a voltmeter in the external circuit to measure the voltage between the two half-cells. This is the **cell potential** or **EMF**, E_{cell}.

Learning Objectives:

- Be able to write and apply the conventional representation of cells.
- Know the IUPAC convention for writing half-equations for electrode reactions.
- Be able to calculate the EMF of a cell.
- Know how to measure the EMF of electrochemical cells (Required Practical 8).

Specification Reference 3.1.11.1

Tip: Always be careful when carrying out experiments involving electricity. Carry out proper risk assessment before you start.

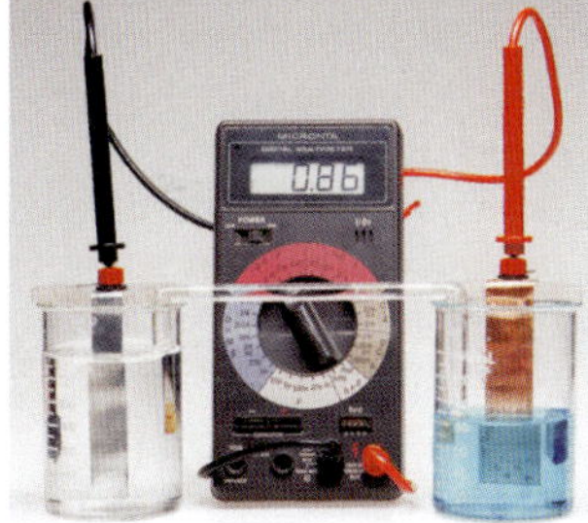

Figure 1: *A zinc/copper electrochemical cell.*

Tip: When you set up a cell, you need to clean the surfaces of the metal electrodes before you begin. Rub them with emery paper and then clean off any grease or oil using propanone.

Tip: EMF stands for electromotive force.

Tip: An electrode has to be a solid that conducts electricity. If a half-cell doesn't contain anything like this, you can use something else (like platinum) as an electrode.

Tip: Inert means it won't react with anything. (The electrode has to be inert or it could react with the solution it's in.)

Tip: The electrode the electrons flow to is the positive electrode. The electrode the electrons flow from is the negative electrode.

You can also have half-cells involving solutions of two aqueous ions of the same element.

Example

You can make an electrochemical half-cell using solutions of Fe^{2+} and Fe^{3+} ions. A platinum electrode is dipped into the solution.

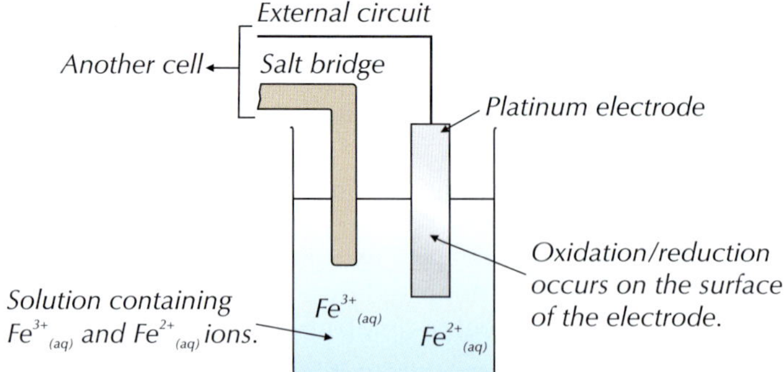

Platinum is used as the electrode because it is inert and conducts electricity. The conversion from Fe^{2+} to Fe^{3+}, or vice versa, happens on the surface of the electrode.

The direction of the conversion depends on the other half-cell in the circuit. If the other cell contains a metal that is less reactive than iron, then Fe^{2+} will be oxidised to Fe^{3+} at the electrode. But if the other cell contains a more reactive metal, Fe^{3+} will be reduced to Fe^{2+} at the electrode.

Electrode potentials

The reactions that occur at each electrode in a cell are reversible.

Tip: These reactions are called half-reactions and, even though they're reversible, they're always written with the reduction reaction going in the forward direction, with the electrons being added on the left-hand side.

Example

The reactions that occur at each electrode in the zinc/copper cell are:

$$Zn^{2+}_{(aq)} + 2e^- \rightleftharpoons Zn_{(s)}$$

$$Cu^{2+}_{(aq)} + 2e^- \rightleftharpoons Cu_{(s)}$$

The reversible arrows show that both reactions can go in either direction.

Which direction each reaction goes in depends on how easily each metal loses electrons (i.e. how easily it's oxidised). How easily a metal is oxidised is measured using **electrode potentials**. A metal that's easily oxidised has a very negative electrode potential, while one that's harder to oxidise has a less negative (or positive) electrode potential.

Figure 2: *Analogue or digital voltmeters can be used to measure electrode potentials.*

Tip: There's more on how electrode potentials are measured coming up.

Example

The table below shows the electrode potentials for the copper and zinc half-cells:

Half-cell	Electrode potential / V
$Zn^{2+}_{(aq)}/Zn_{(s)}$	–0.76
$Cu^{2+}_{(aq)}/Cu_{(s)}$	+0.34

The zinc half-cell has a more negative electrode potential, so in a zinc/copper cell, zinc is oxidised (the reaction goes backwards), while copper is reduced (the reaction goes forwards).

Drawing electrochemical cells

It's a bit of a faff drawing pictures of electrochemical cells. There's a shorthand way of representing them though. You just take the oxidised and reduced forms of the species from each of the half-cells and arrange them in a row (see Figure 3). This is called the 'conventional representation'.

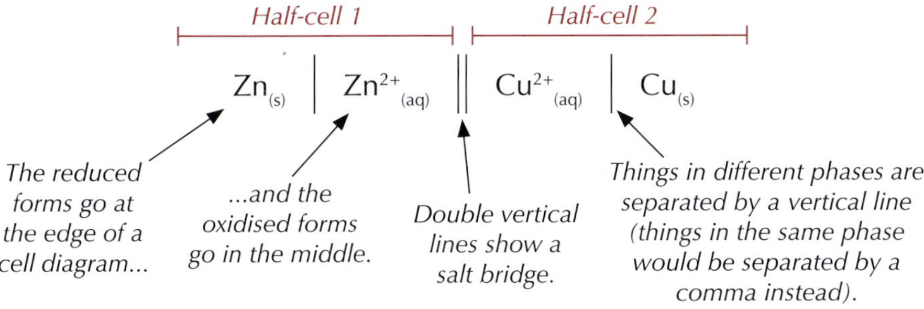

Half-cell 1
Half-cell 2
Zn(s) | Zn2+(aq) || Cu2+(aq) | Cu(s)
The reduced forms go at the edge of a cell diagram...
...and the oxidised forms go in the middle.
Double vertical lines show a salt bridge.
Things in different phases are separated by a vertical line (things in the same phase would be separated by a comma instead).

Figure 3: *The conventional representation of a zinc/copper cell.*

You always put the half-cell that has the more negative electrode potential on the left. In Figure 3, the zinc half-cell is on the left because it has an electrode potential of –0.76 V, while the potential of the copper half-cell is +0.34 V.

The zinc/copper electrochemical cell is quite simple. Some cells are a bit trickier to draw, e.g. they might have platinum electrodes or involve more reagents. If you need to draw a cell diagram for any cell, just follow these steps and you'll be fine:

1. Use the electrode potentials to work out which half-cell goes on the left and which goes on the right.
2. Write out the left-hand half-equation as an oxidation reaction. Then write out the right-hand half-equation as a reduction reaction.
3. Write out the reactants and products of the oxidation reaction, followed by the reactants and products of the reduction reaction.
4. Add in a salt bridge by drawing two vertical lines between the half-cells.
5. Draw a vertical line between any reagents that are in different phases. Put a comma between any reagents that are in the same phase.
6. If either of your half-cells has a separate electrode (like a platinum electrode, Pt), put that on the outside and separate it from the half-cell with a vertical line.

Tip: When you're talking about electrochemical cells, the term 'phase' just means whether a substance is a solid, a liquid (which includes aqueous solutions) or a gas.

Tip: Remember, the one with the more negative electrode potential goes on the left.

Tip: If you write out the half-equations like this, you'll always end up with the reduced forms on the outside and the oxidised forms in the middle.

Tip: You might see some cells which are a bit more complicated and have some more reagents going around. Don't worry though, you won't need to draw these reagents in representations of cells.

Example

The two half-equations below show the reactions taking place in an electrochemical cell:

$Sn^{2+}_{(aq)} + 2e^- \rightarrow Sn_{(s)}$ electrode potential = –0.14 V

$Fe^{3+}_{(aq)} + e^- \rightarrow Fe^{2+}_{(aq)}$ electrode potential = +0.77 V

A platinum rod is used as the electrode in the iron half-cell. Draw the conventional representation of this cell.

1. The Sn^{2+}/Sn reaction has a more negative electrode potential so it will go on the left. The Fe^{3+}/Fe^{2+} reaction will go on the right.
2. Oxidation: $Sn_{(s)} \rightarrow Sn^{2+}_{(aq)} + 2e^-$ *(left)*
   Reduction: $Fe^{3+}_{(aq)} + e^- \rightarrow Fe^{2+}_{(aq)}$ *(right)*

3. Write out the reactants and products of the oxidation equation, followed by the reactants and products of the reduction equation:

$$Sn_{(s)} \quad Sn^{2+}_{(aq)} \quad Fe^{3+}_{(aq)} \quad Fe^{2+}_{(aq)}$$

4. Put in the salt bridge between the half-cells:

Half-cell 1 | *Half-cell 2*

$$Sn_{(s)} \; Sn^{2+}_{(aq)} \,||\, Fe^{3+}_{(aq)} \; Fe^{2+}_{(aq)}$$

5. Add in the phase boundaries:

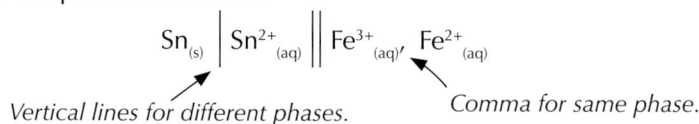

$$Sn_{(s)} \,|\, Sn^{2+}_{(aq)} \,||\, Fe^{3+}_{(aq)}, Fe^{2+}_{(aq)}$$

Vertical lines for different phases. *Comma for same phase.*

6. Finally, add in the platinum electrode that's part of the right-hand cell:

$$Sn_{(s)} \,|\, Sn^{2+}_{(aq)} \,||\, Fe^{3+}_{(aq)}, Fe^{2+}_{(aq)} \,|\, Pt$$

Tip: The platinum electrode goes on the left if it's for the half-cell on the left. It goes on the right if it's for the half-cell on the right.

Writing half-equations from cell diagrams

You won't always be asked to draw a cell diagram from the half-equations. Sometimes, you might be given a cell diagram and asked to write half-equations for one or both of the electrodes.

Example

A conventional representation of an aluminium/lead cell is given below:

$$Al_{(s)} \,|\, Al^{3+}_{(aq)} \,||\, Pb^{2+}_{(aq)} \,|\, Pb_{(s)}$$

Write half-equations for the reactions occurring at the positive and negative electrodes of this cell.

The positive electrode is the one on the right-hand side (Pb^{2+}/Pb) and the negative electrode is the one on the left-hand side (Al/Al^{3+}).

Reduction always happens at the positive electrode, so the half-equation for the reaction occurring at the positive electrode is:

$$Pb^{2+}_{(aq)} + 2e^- \rightleftharpoons Pb_{(s)}$$

Oxidation always happens at the negative electrode, so the half-equation for the reaction occurring at the negative electrode will be:

$$Al_{(s)} \rightleftharpoons Al^{3+}_{(aq)} + 3e^-$$

Tip: Each half-equation should just look like one half of the cell diagram (with the arrow and the electrons put back in).

Calculating the cell potential

$E^\ominus$ is the symbol for standard electrode potential. If you follow the conventions for drawing electrochemical cells, you can calculate the standard cell potential ($E^\ominus_{cell}$) (see page 323) using this formula:

$$E^\ominus_{cell} = E^\ominus_{\text{right-hand side}} - E^\ominus_{\text{left-hand side}}$$

The cell potential will always be a positive voltage, because the more negative $E^\ominus$ value is being subtracted from the more positive $E^\ominus$ value.

Tip: If you don't have a cell diagram you can still use this formula. You just work out which half-cell would go on the left — it's always the one with the more negative electrode potential.

Example

For the Zn/Cu cell: $E^\ominus_{cell} = E^\ominus_{\text{right-hand side}} - E^\ominus_{\text{left-hand side}}$

$= +0.34 - (-0.76)$

$= +1.10\text{ V}$

Tip: See page 328 for more on how electrode potentials for half-cells are measured.

Example

The cell below has an EMF of +2.00 V:

$$Al_{(s)} \mid Al^{3+}_{(aq)} \parallel Cu^{2+}_{(aq)} \mid Cu_{(s)}$$

The standard electrode potential of the copper half-cell is +0.34 V. Calculate the standard electrode potential of the aluminium half-cell.

$$E^{\ominus}_{cell} = E^{\ominus}_{\text{right-hand side}} - E^{\ominus}_{\text{left-hand side}}$$
$$E^{\ominus}_{\text{left-hand side}} = E^{\ominus}_{\text{right-hand side}} - E^{\ominus}_{cell}$$
$$= 0.34 - 2.00$$
$$= -1.66 \text{ V}$$

Exam Tip
Because the Al/Al^{3+} half-cell is on the left-hand side, you know it should have a more negative electrode potential than the Cu^{2+}/Cu half-cell. You can use this idea to check your answers in the exam.

Practice Questions — Application

Q1 An electrochemical cell containing a calcium half-cell and a silver half-cell was set up.

Half-cell	Electrode potential / V
$Ca^{2+}_{(aq)}/Ca_{(s)}$	–2.87
$Ag^{+}_{(aq)}/Ag_{(s)}$	+0.80

a) (i) Write a half-equation for the reaction at the positive electrode.
(ii) Write a half-equation for the reaction at the negative electrode.

b) Draw a conventional representation of this cell.

c) Calculate the cell potential.

Q2 The conventional representation of an iron/thallium electrochemical cell is shown below:

$$Pt \mid Fe^{2+}_{(aq)}, Fe^{3+}_{(aq)} \parallel Tl^{3+}_{(aq)}, Tl^{+}_{(aq)} \mid Pt$$

a) Write half-equations for the reactions occurring at the positive and negative electrodes.

b) Calculate the electrode potential of the Tl^{3+}/Tl^{+} electrode given that the overall cell potential is +0.48 V and the Fe^{3+}/Fe^{2+} electrode potential is +0.77 V.

Q3 In an electrochemical cell with a $Mg^{2+}_{(aq)}/Mg_{(s)}$ half-cell and a $Fe^{3+}_{(aq)}/Fe^{2+}_{(aq)}$ half-cell (using a Pt electrode), oxidation occurs in the $Mg^{2+}_{(aq)}/Mg_{(s)}$ half-cell and reduction occurs in the $Fe^{3+}_{(aq)}/Fe^{2+}_{(aq)}$ half-cell.

a) Which half-cell has the more negative electrode potential?

b) Draw the conventional representation of this cell.

Exam Tip
You don't have to memorise values for electrode potentials. You'll always be told them in the question.

Exam Tip
If you're drawing a cell diagram in the exam, use the state symbols you're given in the question to work out whether things should be separated by vertical lines or commas.

Tip: You don't need to include the platinum electrode in the half-equations because it's not actually involved in the redox reaction.

Practice Questions — Fact Recall

Q1 a) Suggest a metal that could be used for the electrode in half-cells involving solutions of two aqueous ions of the same element.

b) Why is this type of electrode suitable?

Q2 Does oxidation or reduction occur in the half-cell with the more positive electrode potential?

Q3 Give two important conventions that you have to follow when you're drawing electrochemical cells using the conventional representation.

2. Standard Electrode Potentials

Electrode potentials are influenced by things like temperature and pressure. So if you want to compare electrode potentials, they need to be standardised. This is done using a standard hydrogen electrode.

Learning Objectives:

- Know the importance of the conditions when measuring the electrode potential, *E*.
- Understand how cells are used to measure electrode potentials by reference to the standard hydrogen electrode.
- Know that standard electrode potential, $E^{\ominus}$, refers to conditions of 298 K, 100 kPa and 1.00 mol dm^{-3} solution of ions.

Specification Reference 3.1.11.1

Factors affecting the electrode potential

Half-cell reactions are reversible. So just like any other reversible reaction, the equilibrium position is affected by changes in temperature, pressure and concentration. Changing the equilibrium position changes the cell potential. To get around this, **standard conditions** are used to measure electrode potentials — using these conditions means you always get the same value for the electrode potential and you can compare values for different cells.

Tip: See the earlier page on Le Chatelier's Principle (pages 133-135) for more about how changes in temperature, pressure and concentration affect the position of equilibria.

The standard hydrogen electrode

You measure the electrode potential of a half-cell against a **standard hydrogen electrode**. In the standard hydrogen electrode, hydrogen gas is bubbled through a solution of aqueous H^+ ions. A platinum electrode is used as a platform for the oxidation/reduction reactions — see Figure 1.

Tip: A platinum electrode is needed because you can't have a gas electrode.

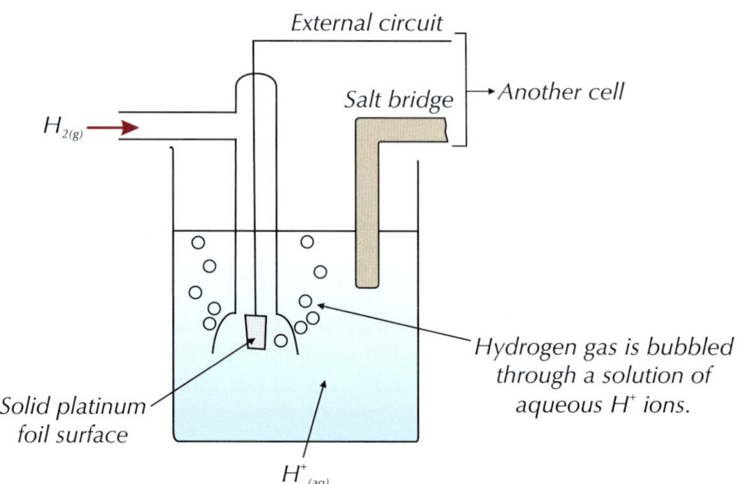

Figure 1: *The standard hydrogen electrode*

Figure 2: *A nugget of platinum. Platinum is used as the electrode in the standard hydrogen electrode.*

When measuring electrode potentials using the standard hydrogen electrode it's important that everything is done under standard conditions:

1. Any solutions of ions must have a concentration of 1.00 mol dm^{-3}.
2. The temperature must be 298 K (25 °C).
3. The pressure must be 100 kPa.

Exam Tip
Make sure you learn the standard conditions — you might well be asked about them in the exam.

Measuring standard electrode potentials

The standard electrode potential of a half-cell is the voltage measured under standard conditions when the half-cell is connected to a standard hydrogen electrode — see Figure 3 (next page).

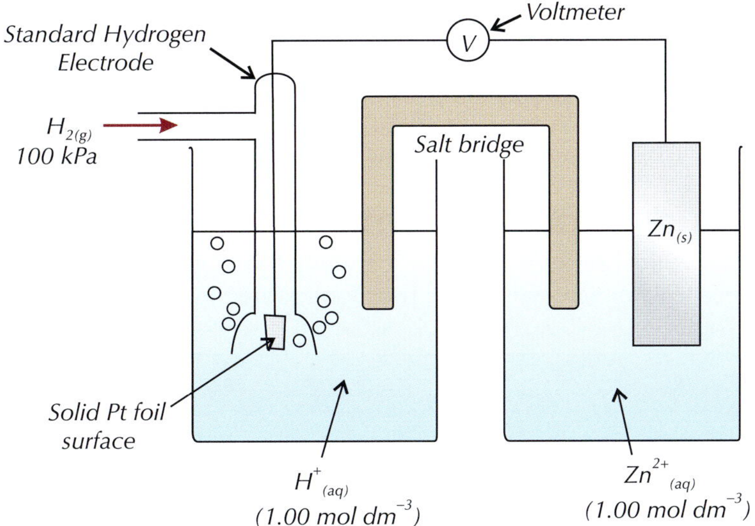

Figure 3: *A zinc electrode connected to a standard hydrogen electrode. This system could be used to measure the standard electrode potential of the zinc electrode.*

The electrochemical cell in Figure 3 can be written as:

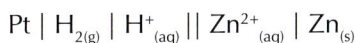

$$Pt \mid H_{2(g)} \mid H^+_{(aq)} \mid\mid Zn^{2+}_{(aq)} \mid Zn_{(s)}$$

The standard hydrogen electrode is always shown on the left — it doesn't matter whether or not the other half-cell has a more positive value.

Standard hydrogen electrodes can be used to calculate standard electrode potentials because the standard hydrogen electrode half-cell has an electrode potential of 0.00 V. The whole cell potential is:

$$E^{\ominus}_{cell} = E^{\ominus}_{\text{right-hand side}} - E^{\ominus}_{\text{left-hand side}}$$

The cell on the left-hand side is the standard hydrogen electrode so:

$$E^{\ominus}_{\text{left-hand side}} = 0.00\ \text{V}$$

So the voltage reading will be equal to $E^{\ominus}_{\text{right-hand side}}$. This reading could be positive or negative, depending which way the electrons flow.

Tip: Notice how the H_2 gas is at a pressure of 100 kPa and the H^+ and Zn^{2+} solutions both have concentrations of 1.00 mol dm^{-3}. These are the standard conditions.

Tip: If you've forgotten what cell potential is, have a read of page 323.

Tip: The standard hydrogen electrode has an electrode potential of 0.00 V by definition — it's that because scientists say it is.

Practice Question — Application

Q1 When a $Pb^{2+}_{(aq)}/Pb_{(s)}$ half-cell was connected to a standard hydrogen electrode via a voltmeter, the reading on the voltmeter was –0.13 V.

a) Draw the conventional representation of this cell.

b) What is the standard electrode potential of the Pb^{2+}/Pb half-cell?

c) Is oxidation or reduction occurring in the Pb^{2+}/Pb half-cell?

Tip: Remember, oxidation happens in the half-cell with the more negative electrode potential — see page 324 for more on this.

Practice Questions — Fact Recall

Q1 Give three factors that can influence electrode potentials.

Q2 Describe how a standard hydrogen electrode is set up and give the standard conditions used when measuring electrode potentials.

Q3 a) What is the electrode potential of the standard hydrogen electrode?

b) Why is it this value?

Q4 Define the term "standard electrode potential".

Learning Objectives:

- Know that standard electrode potentials can be listed as an electrochemical series.
- Be able to use $E^\ominus$ values to predict the direction of simple redox reactions.

Specification Reference 3.1.11.1

3. Electrochemical Series

The standard electrode potentials of different reactions are different (unsurprisingly). If you write a list of electrode potentials in order, you get an electrochemical series, which you can use to predict the direction of a reaction.

What is an electrochemical series?

An **electrochemical series** is basically a big long list of electrode potentials for different electrochemical half-cells. They look something like this:

Half-reaction	$E^\ominus$ / V
$Mg^{2+}_{(aq)} + 2e^- \rightleftharpoons Mg_{(s)}$	–2.38
$Al^{3+}_{(aq)} + 3e^- \rightleftharpoons Al_{(s)}$	–1.66
$Zn^{2+}_{(aq)} + 2e^- \rightleftharpoons Zn_{(s)}$	–0.76
$Ni^{2+}_{(aq)} + 2e^- \rightleftharpoons Ni_{(s)}$	–0.25
$2H^+_{(aq)} + 2e^- \rightleftharpoons H_{2(g)}$	0.00
$Sn^{4+}_{(aq)} + 2e^- \rightleftharpoons Sn^{2+}_{(aq)}$	+0.15
$Cu^{2+}_{(aq)} + 2e^- \rightleftharpoons Cu_{(s)}$	+0.34
$Fe^{3+}_{(aq)} + e^- \rightleftharpoons Fe^{2+}_{(aq)}$	+0.77
$Ag^+_{(aq)} + e^- \rightleftharpoons Ag_{(s)}$	+0.80
$Br_{2(aq)} + 2e^- \rightleftharpoons 2Br^-_{(aq)}$	+1.07
$Cr_2O_7^{2-}{}_{(aq)} + 14H^+_{(aq)} + 6e^- \rightleftharpoons 2Cr^{3+}_{(aq)} + 7H_2O_{(l)}$	+1.33
$Cl_{2(aq)} + 2e^- \rightleftharpoons 2Cl^-_{(aq)}$	+1.36
$MnO_4^-{}_{(aq)} + 8H^+_{(aq)} + 5e^- \rightleftharpoons Mn^{2+}_{(aq)} + 4H_2O_{(l)}$	+1.52

Figure 1: *A table showing an electrochemical series.*

The electrode potentials are written in order, starting from the most negative and going down to the most positive. The half-equations are always written as reduction reactions — but the reactions are reversible and can go the opposite way. When two half-equations are put together in an electrochemical cell, the one with the more negative electrode potential goes in the direction of oxidation (backwards) and the one with the more positive electrode potential goes in the direction of reduction (forwards).

Exam Tip
Don't panic — you don't have to memorise any of these values. You'll always be told them in the question if you need to use them.

Tip: If you can't remember which half-reaction in a cell goes backwards and which goes forwards, think NO P.R. — the more Negative electrode potential will go in the Oxidation direction and the more Positive electrode potential will go in the Reduction direction.

Calculating cell potentials

You can use the information in an electrochemical series to calculate the standard cell potential, or EMF, when two half-cells are connected together. All you have to do is work out which half-reaction is going in the direction of oxidation and which half-reaction is going in the direction of reduction. Then just substitute the $E^{\ominus}$ values into this equation:

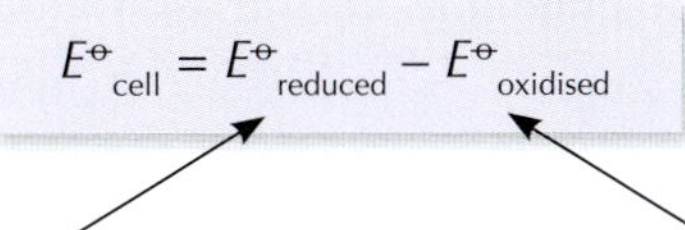

$$E^{\ominus}_{cell} = E^{\ominus}_{reduced} - E^{\ominus}_{oxidised}$$

This is the standard electrode potential of the half-cell which goes in the direction of reduction (the one with the more positive electrode potential).

This is the standard electrode potential of the half-cell which goes in the direction of oxidation (the one with the more negative electrode potential).

Example

Calculate the EMF of an Mg/Ag electrochemical cell using the two redox reaction equations shown below:

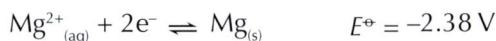

$$Mg^{2+}_{(aq)} + 2e^- \rightleftharpoons Mg_{(s)} \qquad E^{\ominus} = -2.38\,V$$

$$Ag^{+}_{(aq)} + e^- \rightleftharpoons Ag_{(s)} \qquad E^{\ominus} = +0.80\,V$$

The Mg/Mg^{2+} half-cell has the more negative electrode potential, so this half-reaction will go in the direction of oxidation. The Ag/Ag^+ half-cell has the more positive electrode potential and so will go in the direction of reduction.

$$E^{\ominus}_{cell} = E^{\ominus}_{reduced} - E^{\ominus}_{oxidised} = 0.80 - (-2.38) = +3.18\,V$$

Tip: Don't forget — cell potential is the voltage between two half-cells. See page 323 for more.

Predicting the direction of reactions

You can use electrode potentials to predict whether a redox reaction will happen and to show which direction it will go in. Just follow these steps:

1. Find the two half-equations for the redox reaction, and write them both out as reduction reactions.
2. Use an electrochemical series to work out which half-equation has the more negative electrode potential.
3. Write out the half-equation with the more negative electrode potential going in the backwards direction (oxidation) and the half-equation with the more positive electrode potential going in the forwards direction (reduction).
4. Combine the two half-equations and write out a full redox equation.

This is the feasible direction of the reaction and will give a positive overall $E^{\ominus}$ value. The reaction will not happen the other way round.

Tip: Once you know these steps, you can apply them to predict the direction of any redox reaction.

Tip: These values for the electrode potentials came from the electrochemical series on page 330.

Tip: Have a look back at pages 147-148 if you're unsure how to combine half-equations.

Example

Work out the direction of the reaction when a Zn/Zn^{2+} half-cell is connected to a Cu^{2+}/Cu half-cell.

1. Write down the two half-equations for the redox reaction as reduction reactions:

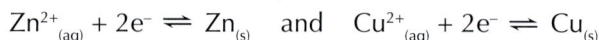

$$Zn^{2+}_{(aq)} + 2e^- \rightleftharpoons Zn_{(s)} \quad \text{and} \quad Cu^{2+}_{(aq)} + 2e^- \rightleftharpoons Cu_{(s)}$$

2. Find out the electrode potentials for the two half-equations:

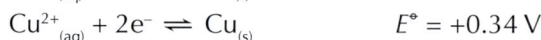

$$Zn^{2+}_{(aq)} + 2e^- \rightleftharpoons Zn_{(s)} \qquad E^\circ = -0.76\text{ V}$$
$$Cu^{2+}_{(aq)} + 2e^- \rightleftharpoons Cu_{(s)} \qquad E^\circ = +0.34\text{ V}$$

3. The zinc half-reaction has the more negative electrode potential, so write it out going backwards and the copper half-reaction going forwards.

$$Zn_{(s)} \rightarrow Zn^{2+}_{(aq)} + 2e^-$$
$$Cu^{2+}_{(aq)} + 2e^- \rightarrow Cu_{(s)}$$

4. Write out the combined full equation:

$$Zn_{(s)} + Cu^{2+}_{(aq)} \rightarrow Zn^{2+}_{(aq)} + Cu_{(s)}$$

This full equation is the outcome of the reaction. Zinc metal will reduce copper(II) ions and copper(II) ions will oxidise zinc metal. The overall reaction has a cell potential of +1.10 V, so is feasible.

$$E^\circ_{cell} = E^\circ_{reduced} - E^\circ_{oxidised} = 0.34 - (-0.76) = +1.10\text{ V}$$

Zinc(II) ions will not oxidise copper metal as this would give an electrode potential of –1.10 V.

You could be given a reaction and asked whether it is feasible or not. Answer the question in a similar way to the one above. Work out the EMF of the cell. If the EMF has a positive value, the reaction is feasible. If the EMF has a negative value, the reaction is not feasible.

Example

Work out whether the following redox reaction is feasible or not:

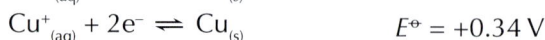

$$\mathbf{Cu_{(s)} + Ni^{2+}_{(aq)} \rightarrow Cu^{2+}_{(aq)} + Ni_{(s)}}$$

$$Ni^{2+}_{(aq)} + 2e^- \rightleftharpoons Ni_{(s)} \qquad E^\circ = -0.25\text{ V}$$
$$Cu^{+}_{(aq)} + 2e^- \rightleftharpoons Cu_{(s)} \qquad E^\circ = +0.34\text{ V}$$

In the full-equation above, Ni is reduced and Cu is oxidised.
Work out the EMF of the cell:

$$E^\circ_{cell} = E^\circ_{reduced} - E^\circ_{oxidised} = -0.25 - (+0.34) = -0.59\text{ V}$$

The EMF of the cell has a negative value, so the reaction is not feasible.

Practice Questions — Application

For these questions use the electrochemical series on page 330.

Q1 Calculate the EMF for the following reactions:

a) $Al_{(s)} + 3Ag^{+}_{(aq)} \rightarrow Al^{3+}_{(aq)} + 3Ag_{(s)}$

b) $Cu_{(s)} + Cl_{2(aq)} \rightarrow Cu^{2+}_{(aq)} + 2Cl^{-}_{(aq)}$

Q2 Predict the direction of the redox reactions of the following half-cells:

a) $Mg^{2+}_{(aq)}/Mg_{(s)}$ and $Ni^{2+}_{(aq)}/Ni_{(s)}$

b) $2Br^{-}_{(aq)}/Br_{2(aq)}$ and $Fe^{3+}_{(aq)}/Fe^{2+}_{(aq)}$

Q3 Predict whether or not these reactions will happen:

a) $Sn^{2+}_{(aq)} + Cu^{2+}_{(aq)} \rightarrow Sn^{4+}_{(aq)} + Cu_{(s)}$

b) $Sn^{2+}_{(aq)} + Zn^{2+}_{(aq)} \rightarrow Sn^{4+}_{(aq)} + Zn_{(s)}$

Q4 Will Ag^{+} ions react with Sn^{2+} ions in solution? Explain your answer.

Exam Tip
Sometimes you have to multiply one of the half-equations by something to make the full equation balance — but you don't need to multiply its $E^\ominus$ value when you calculate the EMF of the cell.

Exam Tip
In the exam, don't assume you'll always be given a reaction that will happen — you're just as likely to be given one that won't.

Practice Questions — Fact Recall

Q1 What is an electrochemical series?

Q2 In what direction are half-equations written in electrochemical series?

Q3 A half-reaction has a very positive electrode potential.
Is it more likely to go in the direction of oxidation or reduction?

4. Electrochemical Cells

Learning Objectives:

- Appreciate that electrochemical cells can be used as a commercial source of electrical energy.
- Appreciate that cells can be non-rechargeable (irreversible), rechargeable or fuel cells.
- Be able to describe the simplified electrode reactions in a lithium cell. Positive electrode: $Li^+ + CoO_2 + e^- \rightarrow Li^+[CoO_2]^-$. Negative electrode: $Li \rightarrow Li^+ + e^-$.
- Be able to use given electrode data to deduce the reactions occurring in non-rechargeable and rechargeable cells and to deduce the EMF of a cell.
- Understand the electrode reactions in an alkaline hydrogen-oxygen fuel cell and appreciate that a fuel cell does not need to be electrically recharged.
- Explain how the electrode reactions in a cell can be used to generate an electric current.
- Appreciate the benefits and risks to society associated with the use of these cells.

Specification Reference 3.1.11.2

The last few topics told you all about electrochemical cells. This one's all about what electrochemical cells are actually used for.

Batteries

The batteries that we use to power things like watches, digital cameras and mobile phones are all types of electrochemical cell. Some types of cell are rechargeable while others can only be used until they run out.

Non-rechargeable cells

Non-rechargeable cells use irreversible reactions. A common type of non-rechargeable cell is a dry cell alkaline battery. You'll probably find some of these in the TV remote control, torch or smoke alarms in your house. They're useful for gadgets that don't use a lot of power or are only used for short periods of time.

Example

Zinc-carbon dry cell batteries have a zinc negative electrode and a mixture of manganese dioxide and carbon for a positive electrode. In between the electrodes is a paste of ammonium chloride, which acts as an electrolyte.

The half-equations are:

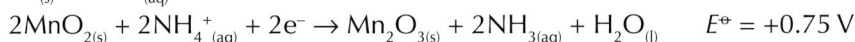

$$Zn_{(s)} \rightarrow Zn^{2+}_{(aq)} + 2e^- \qquad E^\ominus = -0.76\text{ V}$$

$$2MnO_{2(s)} + 2NH_4{}^+_{(aq)} + 2e^- \rightarrow Mn_2O_{3(s)} + 2NH_{3(aq)} + H_2O_{(l)} \qquad E^\ominus = +0.75\text{ V}$$

The EMF of this type of cell is:

$$E^\ominus_{cell} = E^\ominus_{\text{right-hand side}} - E^\ominus_{\text{left-hand side}} = +0.75 - (-0.76) = +1.51\text{ V}$$

The half-equations have non-reversible arrows because it is not practical to reverse them in a battery. They can be made to run backwards under the right conditions, but trying to do this in a battery can make it leak or explode. This is because the zinc electrode forms the casing of the battery, so becomes thinner as the zinc is oxidised.

Rechargeable cells

Rechargeable cells use reversible reactions. Rechargeable batteries are found in loads of devices, such as mobile phones, laptops and cars.

Example

Lithium cells are used in phones and laptops. One type of lithium cell is made up of a lithium cobalt oxide ($LiCoO_2$) electrode and a graphite electrode. The electrolyte is a lithium salt in an organic solvent.

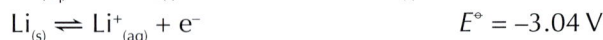

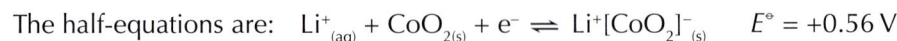

The half-equations are:

$$Li^+_{(aq)} + CoO_{2(s)} + e^- \rightleftharpoons Li^+[CoO_2]^-_{(s)} \qquad E^\ominus = +0.56\text{ V}$$

$$Li_{(s)} \rightleftharpoons Li^+_{(aq)} + e^- \qquad E^\ominus = -3.04\text{ V}$$

So the reactions which happen when the battery supplies power are:

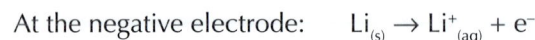

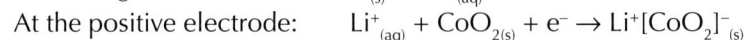

At the negative electrode: $Li_{(s)} \rightarrow Li^+_{(aq)} + e^-$

At the positive electrode: $Li^+_{(aq)} + CoO_{2(s)} + e^- \rightarrow Li^+[CoO_2]^-_{(s)}$

The EMF of this type of cell is:

$$E^\ominus_{cell} = E^\ominus_{\text{right-hand side}} - E^\ominus_{\text{left-hand side}} = +0.56 - (-3.04) = +3.60\text{ V}$$

Exam Tip

Learn these lithium cell half-equations — you could be asked to give them in the exam.

Two other types of rechargeable battery are NiCad (nickel-cadmium) and lead-acid. To recharge batteries, a current is supplied to force electrons to flow in the opposite direction around the circuit and reverse the reactions. This is possible because none of the substances in a rechargeable battery escape or are used up. The reactions that take place in non-rechargeable batteries are difficult or impossible to reverse in this way.

Figure 1: *A lithium battery being recharged.*

Fuel cells

In most cells the chemicals that generate the electricity are contained in the electrodes and the electrolyte that form the cell. In a **fuel cell** the chemicals are stored separately outside the cell and fed in when electricity is required.

Example

One example is the alkaline hydrogen-oxygen fuel cell, which can be used to power electric vehicles. Here's how this type of fuel cell works:

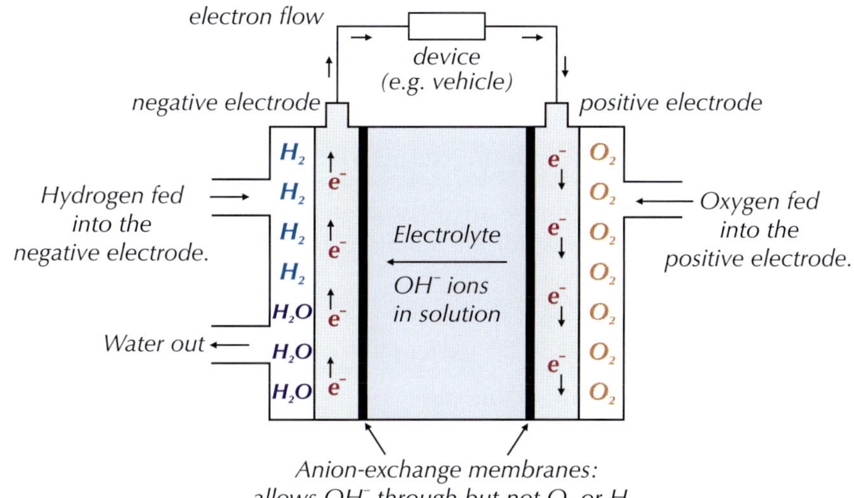

Figure 2: *A bus powered by a hydrogen-oxygen fuel cell.*

Hydrogen and oxygen gases are fed into two separate platinum-containing electrodes. These electrodes are usually made by coating a porous ceramic material with a thin layer of platinum. This is cheaper than using solid platinum rods and it provides a larger surface area so the reactions go faster.

The electrodes are separated by an anion-exchange membrane that allows anions (OH^- ions) and water to pass through it, but stops hydrogen and oxygen gas passing through it. The electrolyte is an aqueous alkaline (KOH) solution. Hydrogen is fed to the negative electrode.

The reaction that occurs at the negative electrode is:

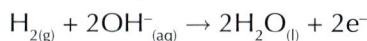

$$H_{2(g)} + 2OH^-_{(aq)} \rightarrow 2H_2O_{(l)} + 2e^-$$

The electrons flow from the negative electrode through an external circuit to the positive electrode. The OH^- ions pass through the anion-exchange membrane towards the negative electrode. Oxygen is fed to the positive electrode.

The reaction that occurs at the positive electrode is:

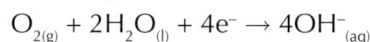

$$O_{2(g)} + 2H_2O_{(l)} + 4e^- \rightarrow 4OH^-_{(aq)}$$

The overall effect is that H_2 and O_2 react to make water:

$$2H_{2(g)} + O_{2(g)} \rightarrow 2H_2O_{(l)}$$

Tip: An electrolyte is a substance that contains free ions and can conduct electricity.

Tip: These equations are for fuel cells working in alkaline conditions. That means the electrolyte is a base (e.g. KOH).

You can make fuel cells with acidic electrolytes (e.g. H_3PO_4) too. Then the electrode reactions look like this instead —
Negative electrode:
$H_2 \rightarrow 2H^+ + 2e^-$
Positive electrode:
$O_2 + 4H^+ + 4e^- \rightarrow 2H_2O$

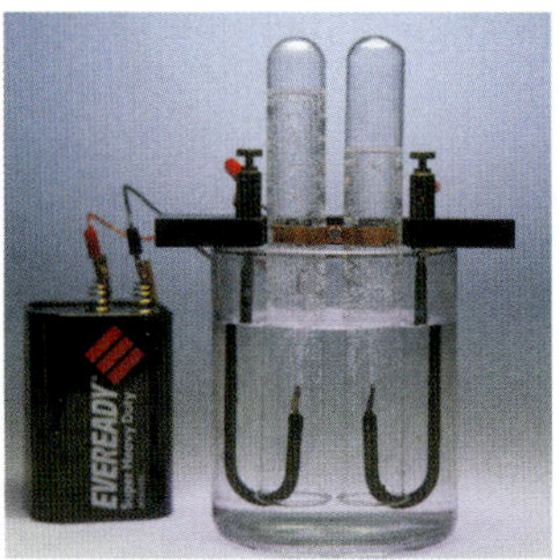

Figure 3: *Electrolysis of water being used to make hydrogen and oxygen.*

Pros and cons of fuel cells

The major advantage of using fuel cells in cars, rather than the internal combustion engine, is that fuel cells are more efficient — they convert more of their available energy into kinetic energy to get the car moving. Internal combustion engines waste a lot of their energy producing heat.

There are other benefits too. For example, the only waste product is water, so there are no nasty toxic chemicals to dispose of and no CO_2 emissions from the cell itself. Fuel cells don't need to be recharged like batteries. As long as hydrogen and oxygen are supplied, the cell will continue to produce electricity.

The downside is that you need energy to produce a supply of hydrogen. This can be produced from the electrolysis of water — i.e. by reusing the waste product from the fuel cell, but this requires electricity — and this electricity is normally generated by burning fossil fuels. So the whole process isn't usually carbon neutral. Hydrogen is also highly flammable so it needs to be handled carefully when it is stored or transported. The infrastructure to provide hydrogen fuel for cars doesn't exist on a large scale yet, so refuelling stations are very rare at the moment.

Practice Questions — Application

Q1 Nickel/cadmium batteries are a common type of rechargeable battery. The two half-equations for the reactions happening in this type of battery are shown below:

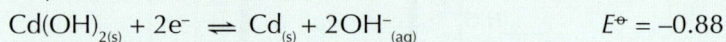

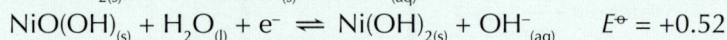

$Cd(OH)_{2(s)} + 2e^- \rightleftharpoons Cd_{(s)} + 2OH^-_{(aq)}$ $E^\ominus = -0.88$

$NiO(OH)_{(s)} + H_2O_{(l)} + e^- \rightleftharpoons Ni(OH)_{2(s)} + OH^-_{(aq)}$ $E^\ominus = +0.52$

a) Calculate the EMF of this cell.

b) Write an equation for the overall reaction occurring in this cell.

Q2 An electric car is being powered by an alkaline hydrogen-oxygen fuel cell. The half-equations for the reactions occurring in this cell are:

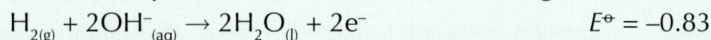

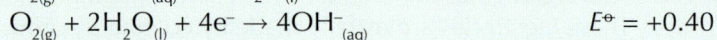

$H_{2(g)} + 2OH^-_{(aq)} \rightarrow 2H_2O_{(l)} + 2e^-$ $E^\ominus = -0.83$

$O_{2(g)} + 2H_2O_{(l)} + 4e^- \rightarrow 4OH^-_{(aq)}$ $E^\ominus = +0.40$

a) Write an equation for the overall reaction occurring in this cell.

b) Draw the conventional representation of this cell.

Tip: See pages 325-327 for a recap on how to draw cell diagrams and calculate EMF.

Tip: Don't forget that platinum electrodes need to be included in cell diagrams too.

Practice Questions — Fact Recall

Q1 Why can you recharge a rechargeable battery?

Q2 Describe how an alkaline hydrogen-oxygen fuel cell works.

Q3 Explain why hydrogen-oxygen fuel cells are not necessarily carbon neutral.

Q4 Give an advantage and a disadvantage of using fuel cells.

Section Summary

Make sure you know...

- That electrochemical cells can be made by dipping two different metals in salt solutions of their own ions and connecting them by a wire and a salt bridge.
- That the direction of the reaction in each half-cell depends on their relative electrode potentials.
- That the reaction with the more negative electrode potential goes in the direction of oxidation and the reaction with the more positive electrode potential goes in the direction of reduction.
- The conventions for drawing conventional diagrams of electrochemical cells.
- That the half-cell on the left of a conventional cell diagram has the more negative electrode potential and undergoes oxidation — the one on the right is more positive and is reduced.
- How to write oxidation and reduction half-equations from cell diagrams.
- How to calculate standard cell potentials and related quantities using the equation $E^{\ominus}_{cell} = E^{\ominus}_{right\text{-}hand\ side} - E^{\ominus}_{left\text{-}hand\ side}$.
- That temperature, pressure and concentration can all influence the cell potential.
- That standard electrode potentials are measured against the standard hydrogen electrode.
- That in the standard hydrogen electrode, hydrogen gas is bubbled into a solution of aqueous H^+ ions and platinum is used as the electrode.
- That standard conditions are 298 K (25 °C), 100 kPa and 1.00 mol dm^{-3}.
- What an electrochemical series is and how it can be used to predict the direction or feasibility of a reaction.
- That batteries are electrochemical cells.
- How to draw cell diagrams and calculate the EMF for the electrochemical cells in batteries.
- That non-rechargeable batteries use non-reversible reactions and rechargeable batteries use reversible reactions.
- How to write the electrode reactions in a lithium cell.
- How an alkaline hydrogen-oxygen fuel cell works.
- The advantages and disadvantages of using fuel cells instead of conventional batteries.

Exam-style Questions

1 The table below shows a short electrochemical series:

Half-reaction	$E^{\ominus}$ / V
$Mg^{2+}_{(aq)} + 2e^- \rightleftharpoons Mg_{(s)}$	–2.38
$V^{2+}_{(aq)} + 2e^- \rightleftharpoons V_{(s)}$	–1.18
$V^{3+}_{(aq)} + e^- \rightleftharpoons V^{2+}_{(aq)}$	–0.26
$Sn^{4+}_{(aq)} + 2e^- \rightleftharpoons Sn^{2+}_{(aq)}$	+0.15
$VO^{2+}_{(aq)} + 2H^+_{(aq)} + e^- \rightleftharpoons V^{3+}_{(aq)} + H_2O_{(l)}$	+0.34
$Fe^{3+}_{(aq)} + e^- \rightleftharpoons Fe^{2+}_{(aq)}$	+0.77
$VO_2^+{}_{(aq)} + 2H^+_{(aq)} + e^- \rightleftharpoons VO^{2+}_{(aq)} + H_2O_{(l)}$	+1.00

An electrochemical cell can be made by connecting an Mg^{2+}/Mg half-cell to an Fe^{2+}/Fe^{3+} half-cell with a platinum electrode.

1.1 Write half-equations for the oxidation and reduction reactions occurring in this cell.

(2 marks)

1.2 Draw a cell diagram for this cell using the conventional representation.

(2 marks)

1.3 Calculate the EMF of this cell under standard conditions.

(1 mark)

1.4 Using the information in the table, give the equations for any reactions which occur when aqueous Sn^{2+} ions are mixed with an acidified solution of VO_2^+ ions. Explain your answer.

(4 marks)

Standard electrode potentials are measured relative to the standard hydrogen electrode.

1.5 Explain what is meant by 'standard conditions'.

(1 mark)

1.6 The electrode potential of the standard hydrogen electrode is 0.00 V. Explain why.

(1 mark)

1.7 The standard hydrogen electrode is constructed using a platinum electrode. Suggest why platinum is a suitable metal to use for this purpose.

(1 mark)

When a standard hydrogen electrode was connected to an Ag^+/Ag half-cell, the reading on the voltmeter was +0.80 V.

1.8 What is the standard electrode potential for the reaction $Ag^+_{(aq)} + e^- \rightleftharpoons Ag_{(s)}$?

(1 mark)

2 Lithium-ion batteries are a type of rechargeable battery commonly used in mobile phones. The half-equations for the reactions that occur in a lithium-ion battery are shown below:

$$Li^+_{(aq)} + CoO_{2(s)} + e^- \rightleftharpoons Li^+[CoO_2]^-_{(s)}$$

$$Li_{(s)} \rightleftharpoons Li^+_{(aq)} + e^-$$

2.1 Combine the two half-equations to give the equation for the overall reaction occurring in this cell.

(1 mark)

2.2 Explain why rechargeable batteries can be recharged.

(1 mark)

Alkaline hydrogen-oxygen fuel cells can be used as an alternative source of commercially available electrical energy. They are commonly used to power electric vehicles.

2.3 Explain why the electrodes of a hydrogen-oxygen fuel cell are typically made by coating a porous ceramic material with a thin layer of platinum, rather than using solid platinum rods.

(2 marks)

2.4 Give two advantages of using hydrogen-oxygen fuel cells over the internal combustion engine.

(2 marks)

3 The table below shows a short electrochemical series:

Half-reaction	$E^{\ominus}$ / V
$Al^{3+}_{(aq)} + 3e^- \rightleftharpoons Al_{(s)}$	–1.66
$Zn^{2+}_{(aq)} + 2e^- \rightleftharpoons Zn_{(s)}$	–0.76
$Ni^{2+}_{(aq)} + 2e^- \rightleftharpoons Ni_{(s)}$	–0.25
$Cu^{2+}_{(aq)} + 2e^- \rightleftharpoons Cu_{(s)}$	+0.34
$Fe^{3+}_{(aq)} + e^- \rightleftharpoons Fe^{2+}_{(aq)}$	+0.77

Using the information in the table, determine which of the following reactions is not feasible.

A $Zn_{(s)} + 2Fe^{3+}_{(aq)} \rightarrow Zn^{2+}_{(aq)} + 2Fe^{2+}_{(aq)}$

B $3Zn_{(s)} + 2Al^{3+}_{(aq)} \rightarrow 3Zn^{2+}_{(aq)} + 2Al_{(s)}$

C $Ni_{(s)} + Cu^{2+}_{(aq)} \rightarrow Ni^{2+}_{(aq)} + Cu_{(s)}$

D $2Al_{(s)} + 3Ni^{2+}_{(aq)} \rightarrow 2Al^{3+}_{(aq)} + 3Ni_{(s)}$

(1 mark)

4 The conventional representation of an electrochemical cell is shown below.

$$Pt \mid Sn^{2+}_{(aq)}, Sn^{4+}_{(aq)} \parallel Tl^{3+}_{(aq)} \mid Tl_{(s)}$$

4.1 Write a half-equation for the reaction occurring at the positive electrode of this electrochemical cell.

(1 mark)

4.2 Calculate the electrode potential of the $Tl^{3+}_{(aq)}/Tl_{(s)}$ half-cell given that the overall cell potential is +0.57 V and the electrode potential of the $Sn^{4+}_{(aq)}/Sn^{2+}_{(aq)}$ half-cell is +0.15 V.

(1 mark)

4.3 Draw the conventional representation of an electrochemical cell that could be used to measure the standard electrode potential of the $Tl^{3+}_{(aq)}/Tl_{(s)}$ half-cell.

(1 mark)

1. Acids, Bases and K_w

There are a few different theories about what makes an acid and what makes a base. One of those theories is the Brønsted–Lowry theory. Here it is...

Brønsted–Lowry acids and bases

Brønsted–Lowry acids are proton donors — they release hydrogen ions (H^+) when they're mixed with water. You never get H^+ ions by themselves in water though — they're always combined with H_2O to form **hydroxonium ions**, H_3O^+. For example, for the general acid HA:

$$HA_{(aq)} + H_2O_{(l)} \rightarrow H_3O^+_{(aq)} + A^-_{(aq)}$$

Brønsted–Lowry bases do the opposite — they're proton acceptors. When they're in solution, they grab hydrogen ions from water molecules. For example, for the general base B:

$$B_{(aq)} + H_2O_{(l)} \rightarrow BH^+_{(aq)} + OH^-_{(aq)}$$

Dissociation in water

Acids and bases dissociate in water. This just means they break up into positively and negatively charged ions. The amount of dissociation depends on how weak or strong the acid or base is. **Strong acids** dissociate (or ionise) almost completely in water — nearly all the H^+ ions will be released. **Strong bases** (like sodium hydroxide) ionise almost completely in water too.

Examples

Hydrochloric acid is a strong acid: $HCl_{(g)} \rightarrow H^+_{(aq)} + Cl^-_{(aq)}$

Sodium hydroxide is a strong base: $NaOH_{(s)} \rightarrow Na^+_{(aq)} + OH^-_{(aq)}$

These reactions are actually reversible reactions but the equilibrium lies extremely far to the right, so only the forward reaction is shown in the equation.

Weak acids (e.g. ethanoic or citric) dissociate only very slightly in water — so only small numbers of H^+ ions are formed. An equilibrium is set up which lies well over to the left. **Weak bases** (such as ammonia) only slightly dissociate in water too. Just like with weak acids, the equilibrium lies well over to the left.

Examples

Ethanoic acid is a weak acid: $CH_3COOH_{(aq)} \rightleftharpoons CH_3COO^-_{(aq)} + H^+_{(aq)}$

Ammonia is a weak base: $NH_{3(aq)} + H_2O_{(l)} \rightleftharpoons NH_4^+{}_{(aq)} + OH^-_{(aq)}$

Acid and base reactions

Acids can't just throw away their protons — they can only get rid of them if there's a base to accept them. In this reaction the acid, HA, transfers a proton to the base, B:

$$HA_{(aq)} + B_{(aq)} \rightleftharpoons BH^+_{(aq)} + A^-_{(aq)}$$

Learning Objectives:

- Know that an acid is a proton donor.
- Know that a base is a proton acceptor.
- Know that acid–base equilibria involve the transfer of protons.
- Know that water is slightly dissociated.
- Know that $K_w = [H^+][OH^-]$ is derived from the equilibrium constant of the dissociation of water.
- Know that the value of K_w varies with temperature.
- Know that weak acids and weak bases dissociate only slightly in aqueous solution.

Specification Reference 3.1.12.1, 3.1.12.3, 3.1.12.4

Tip: The strength of an acid has nothing to do with its concentration. How strong an acid is depends on how much it dissociates in water. The concentration is just how diluted the acid is.

Tip: Whenever you see $H^+_{(aq)}$, it should really be $H_3O^+_{(aq)}$ but in practice it's usually okay to use just H^+, unless you need to show water acting as a base.

Tip: The equilibrium is far to the left for weak acids, and far to the right for strong acids.

It's an equilibrium, so if you add more HA or B, the position of equilibrium moves to the right. But if you add more BH^+ or A^-, the equilibrium will move to the left, according to Le Chatelier's principle. When an acid is added to water, water acts as the base and accepts the proton:

$$HA_{(aq)} + H_2O_{(l)} \rightleftharpoons H_3O^+_{(aq)} + A^-_{(aq)}$$

The ionic product of water, K_w

Tip: Le Chatelier's principle was covered on pages 133-135, so check back if it's a little hazy in your mind.

Water dissociates into hydroxonium ions and hydroxide ions. So this equilibrium exists in water:

$$H_2O_{(l)} + H_2O_{(l)} \rightleftharpoons H_3O^+_{(aq)} + OH^-_{(aq)}$$

If you remove an H_2O from both sides, this simplifies to:

$$H_2O_{(l)} \rightleftharpoons H^+_{(aq)} + OH^-_{(aq)}$$

And, just like for any other equilibrium reaction, you can apply the equilibrium law and write an expression for the equilibrium constant:

$$K_c = \frac{[H^+][OH^-]}{[H_2O]}$$

Water only dissociates a tiny amount, so the equilibrium lies well over to the left. There's so much water compared to the amounts of H^+ and OH^- ions that the concentration of water is considered to have a constant value. So if you multiply the expression you wrote for K_c (which is a constant) by $[H_2O]$ (another constant), you get a constant. This new constant is called the **ionic product of water** and it is given the symbol K_w.

Tip: You'll be using K_w to calculate pH on page 345, so make sure you understand what it means.

$$K_w = K_c \times [H_2O] = \frac{[H^+][OH^-]}{\cancel{[H_2O]}} \times \cancel{[H_2O]}$$

So... $K_w = [H^+][OH^-]$

K_w always has the same value for an aqueous solution at a given temperature. For example, at 298 K (25 °C), K_w has a value of 1.00×10^{-14} mol² dm⁻⁶. The value of K_w changes as the temperature changes. In pure water, there is always one H^+ ion for each OH^- ion. So $[H^+] = [OH^-]$. That means if you are dealing with pure water, then you can say that:

Tip: The units of K_w are always $mol^2\ dm^{-6}$ because $mol\ dm^{-3} \times mol\ dm^{-3} = mol^2\ dm^{-6}$.

$$K_w = [H^+]^2$$

Exam Tip
Remember — K_w only equals $[H^+]^2$ in pure water. If you're asked to write an expression for K_w, always give $K_w = [H^+][OH^-]$.

Practice Questions — Fact Recall

Q1 What is the definition of:
a) A Brønsted–Lowry acid? b) A Brønsted–Lowry base?

Q2 Write out the equations for:
a) A general acid (HA) mixed with water.
b) A general base (B) mixed with water.
c) A general acid (HA) reacting with a general base (B)

Q3 Write out the equation for the dissociation of water.

Q4 a) What is K_w?
b) Write out the equation for K_w.
c) What are the units of K_w?
d) Why can the equation be simplified to $K_w = [H^+]^2$ for water?

2. pH Calculations

There are lots of pH calculations that you'll need to be able to do in your exam. The next few pages tell you everything you need to know to do them.

The pH scale

The pH scale is a measure of the hydrogen ion concentration in a solution. The concentration of hydrogen ions in a solution can vary enormously, so those wise chemists of old decided to express the concentration on a logarithmic scale. pH can be calculated using the following equation:

$$pH = -\log_{10}[H^+]$$

$[H^+]$ is the concentration of hydrogen ions in a solution, measured in mol dm^{-3}. So, if you know the hydrogen ion concentration of a solution, you can calculate its pH by sticking the numbers into the formula.

Example — Maths Skills

A solution of hydrochloric acid has a hydrogen ion concentration of 0.010 mol dm^{-3}. What is the pH of the solution?

$$\begin{aligned} pH &= -\log_{10}[H^+] \\ &= -\log_{10} 0.010 \\ &= 2.00 \end{aligned}$$

Just substitute the [H⁺] value into the pH formula and solve.

The pH scale normally goes from 0 (very acidic) to 14 (very alkaline). pH 7 is regarded as being neutral. Solutions that have a very low pH include strong acids such as HCl and H_2SO_4. Strong bases such as NaOH and KOH have a very high pH. Pure water has a pH of 7 and is neutral.

Calculating $[H^+]$ from pH

If you've got the pH of a solution, and you want to know its hydrogen ion concentration, then you need the inverse of the pH formula:

$$[H^+] = 10^{-pH}$$

Now you can use this formula to find $[H^+]$.

Example — Maths Skills

A solution of sulfuric acid has a pH of 1.52. What is the hydrogen ion concentration of this solution?

$$\begin{aligned} [H^+] &= 10^{-pH} \\ &= 10^{-1.52} \\ &= 0.030 \text{ mol dm}^{-3} \\ &= 3.0 \times 10^{-2} \text{ mol dm}^{-3} \end{aligned}$$

Just substitute the pH value into the inverse pH formula and solve.

Practice Questions — Application

Q1 A solution of sulfuric acid (H_2SO_4) has a hydrogen ion concentration of 0.050 mol dm^{-3}. Calculate the pH of this solution.

Q2 A solution of nitric acid (HNO_3) has a pH of 2.86. Calculate the concentration of hydrogen ions in this solution.

Q3 Calculate the pH of a solution of hydrochloric acid (HCl) with a hydrogen ion concentration of 0.020 mol dm^{-3}.

Learning Objectives:

- Know that the concentration of hydrogen ions in aqueous solution covers a very wide range. Therefore a logarithmic scale, the pH scale, is used as a measure of hydrogen ion concentration.
- Know that $pH = -\log_{10}[H^+]$.
- Be able to convert concentration of hydrogen ions into pH and vice versa.
- Be able to calculate the pH of a strong acid from its concentration.
- Be able to use K_w to calculate the pH of a strong base from its concentration.

Specification Reference 3.1.12.2, 3.1.12.3

Tip: Look at page 529 for more on how to use logs.

Tip: To calculate logarithms you need to use the 'log' button on your calculator. Different calculators work differently so make sure you know how to calculate logs on yours.

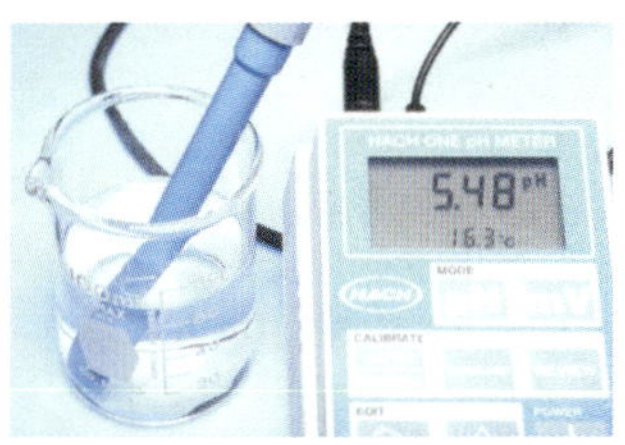

Figure 1: *pH can be measured using a pH meter like this one.*

Tip: There are details on the different ways that you can measure pH in the Practical Skills section — see page 6.

pH of strong monoprotic acids

Monoprotic means that each molecule of an acid will release one proton when it dissociates. Hydrochloric acid (HCl) and nitric acid (HNO_3) are strong acids so they ionise fully:

$$HCl_{(aq)} \rightarrow H^+_{(aq)} + Cl^-_{(aq)}$$

HCl and HNO_3 are also monoprotic, so each mole of acid produces one mole of hydrogen ions. This means the H^+ concentration is the same as the acid concentration. So, if you know the concentration of the acid you know the H^+ concentration and you can calculate the pH.

Exam Tip
Calculating the pH of acids from their concentrations comes up in loads of calculations so you need to be really confident that you know how to do it.

Examples — Maths Skills

Calculate the pH of 0.10 mol dm⁻³ hydrochloric acid:

$$[HCl] = [H^+] = 0.10 \text{ mol dm}^{-3}.$$

$$\text{So: } pH = -\log_{10}[H^+] = -\log_{10} 0.10 = 1.00$$

Calculate the pH of 0.050 mol dm⁻³ nitric acid:

$$[HNO_3] = [H^+] = 0.050 \text{ mol dm}^{-3}.$$

$$\text{So: } pH = -\log_{10} 0.050 = 1.30$$

If a solution of hydrochloric acid has a pH of 2.45, what is the concentration of the acid?

$$[HCl] = [H^+] = 10^{-pH}.$$

$$\text{So: } [HCl] = 10^{-2.45} = 3.5 \times 10^{-3} \text{ mol dm}^{-3}$$

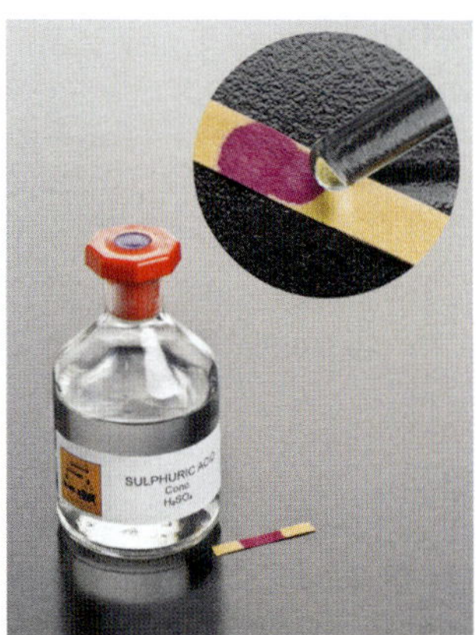

Figure 2: Concentrated H_2SO_4 is a strong acid. It turns this universal indicator paper purple, which shows it has a pH between 0 and 1.

pH of strong diprotic acids

Diprotic means that each molecule of an acid will release two protons when it dissociates. Sulfuric acid is an example of a strong diprotic acid:

$$H_2SO_{4(l)} \rightarrow 2H^+_{(aq)} + SO_4^{2-}{}_{(aq)}$$

So, diprotic acids produce two moles of hydrogen ions for each mole of acid, meaning that the H^+ concentration is twice the concentration of the acid.

Examples — Maths Skills

Calculate the pH of 0.100 mol dm⁻³ sulfuric acid:

$$[H^+] = 2 \times [H_2SO_4] = 0.200 \text{ mol dm}^{-3}.$$

$$\text{So: } pH = -\log_{10}[H^+] = -\log_{10} 0.200 = 0.699$$

Calculate the pH of 0.25 mol dm⁻³ sulfuric acid:

$$[H^+] = 2 \times 0.25 = 0.50.$$

$$\text{So: } pH = -\log_{10}[H^+] = -\log_{10}[0.50] = 0.30$$

Tip: The rule that $[H^+] = 2 \times [HA]$ is only true for <u>strong</u> diprotic acids — you can't apply the same rule to weak diprotic acids because they don't fully ionise.

pH of strong bases

Sodium hydroxide (NaOH) and potassium hydroxide (KOH) are strong bases that fully ionise in water — they donate one mole of OH^- ions per mole of base. This means that the concentration of OH^- ions is the same as the concentration of the base. So for 0.02 mol dm^{-3} sodium hydroxide solution, $[OH^-]$ is also 0.02 mol dm^{-3}. But to work out the pH you need to know $[H^+]$ — luckily this is linked to $[OH^-]$ through the ionic product of water, K_w:

$$K_w = [H^+][OH^-]$$

Tip: See page 342 for more information on K_w.

So if you know $[OH^-]$ for a strong aqueous base and K_w at a certain temperature, you can work out $[H^+]$ and then the pH. Just follow these steps:

Step 1: Find the values of K_w and $[OH^-]$. You may be told these in the question or you may have to work them out.

Step 2: Rearrange the equation, substitute the values for K_w and $[OH^-]$ into the equation, and solve it to find $[H^+]$.

Step 3: Once you know $[H^+]$, substitute this into the pH equation ($pH = -\log_{10}[H^+]$) and solve it to find out the pH.

Figure 3: Sodium hydroxide turns this universal indicator paper dark blue, which shows it has a high pH.

Example — Maths Skills

The value of K_w at 298 K is 1.0×10^{-14} $mol^2\,dm^{-6}$. Find the pH of 0.100 mol dm^{-3} NaOH at 298 K.

1. Find the values of K_w and $[OH^-]$:
   - The value of K_w is given in the question as 1.0×10^{-14} $mol^2\,dm^{-6}$
   - NaOH is a strong base so will donate one mole of OH^- ions per mole of base. The concentration of NaOH is 0.100 mol dm^{-3} so $[OH^-]$ must be 0.100 mol dm^{-3}.
2. Substitute the values of K_w and $[OH^-]$ into the K_w equation:

$$K_w = [H^+][OH^-]$$

$$\text{So } [H^+] = \frac{K_w}{[OH^-]} = \frac{1.0 \times 10^{-14}}{0.100} = 1.0 \times 10^{-13}\ \text{mol dm}^{-3}$$

3. Substitute the value of $[H^+]$ into the pH equation:

$$pH = -\log_{10}(1.0 \times 10^{-13}) = 13.00$$

Tip: You can use $K_w = [H^+]^2$ (p. 342) to calculate the pH of water. pH will vary with temperature, but if you know K_w at that temperature you can work out $[H^+]$ and then pH.

Practice Questions — Application

Q1 Hydrochloric acid (HCl) is a strong monoprotic acid. Calculate the pH of a 0.080 mol dm^{-3} solution of HCl.

Q2 Sulfuric acid (H_2SO_4) is a strong diprotic acid. Calculate the pH of a 0.025 mol dm^{-3} solution of H_2SO_4.

Q3 Potassium hydroxide (KOH) is a strong base. A 0.200 mol dm^{-3} solution of KOH was prepared at 50 °C. The value of K_w is 5.48×10^{-14} $mol^2\,dm^{-6}$ at this temperature. Calculate the pH of this solution.

Practice Questions — Fact Recall

Q1 Write out the equation that defines pH.

Q2 Explain what is meant by the terms monoprotic and diprotic.

Exam Tip

Before diving in to answer a question, you might find it useful to highlight the key bits of information in the question that you're likely to use.

3. The Acid Dissociation Constant

Calculating the pH of strong acids isn't too bad, but for weak acids things are a bit more complicated — you need to use the acid dissociation constant, K_a.

Learning Objectives:

- Be able to construct an expression for the dissociation constant K_a for a weak acid.
- Be able to perform calculations relating the pH of a weak acid to the dissociation constant K_a and the concentration.
- Know that $pK_a = -\log_{10}K_a$.
- Be able to convert K_a into pK_a, and vice versa.

Specification Reference 3.1.12.4

What is the acid dissociation constant?

Weak acids (like CH_3COOH) and weak bases dissociate only slightly in solution, so the $[H^+]$ isn't the same as the acid concentration. This makes it a bit trickier to find their pH. You have to use yet another equilibrium constant — the acid dissociation constant, K_a. The units of K_a are mol dm^{-3} and the equation for K_a is derived as follows:

For a weak aqueous acid, HA, you get the equilibrium $HA_{(aq)} \rightleftharpoons H^+_{(aq)} + A^-_{(aq)}$.

As only a tiny amount of HA dissociates, you can assume that $[HA_{(aq)}] >> [H^+_{(aq)}]$, so $[HA_{(aq)}]_{start} \approx [HA_{(aq)}]_{equilibrium}$.

So if you apply the equilibrium law, you get:

$$K_a = \frac{[H^+][A^-]}{[HA]}$$

When dealing with weak acids, you can assume that all the H^+ ions come from the acid, so $[H^+_{(aq)}] \approx [A^-_{(aq)}]$. So the formula for K_a can be simplified to:

$$K_a = \frac{[H^+]^2}{[HA]}$$

Tip: K_a is a type of equilibrium constant, so the formula for K_a is based on K_c.

Finding the pH of weak acids

You can use K_a to find the pH of a weak acid. Just follow these steps.

Step 1: Write an expression for K_a for the weak acid.

Step 2: Rearrange the equation and substitute in the values you know to find $[H^+]^2$.

Step 3: Take the square root of the number to find $[H^+]$.

Step 4: Substitute $[H^+]$ into the pH equation to find the pH.

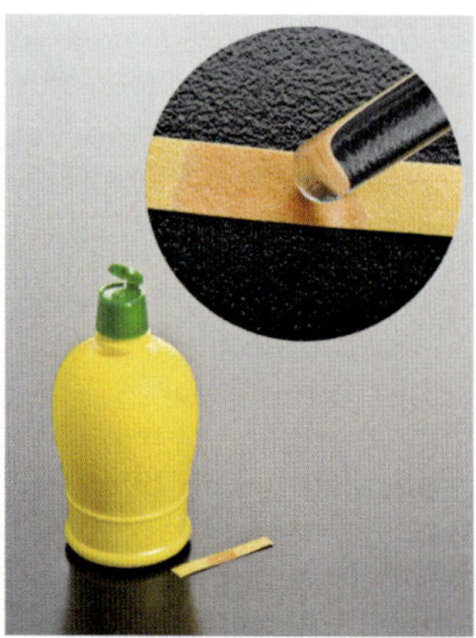

***Figure 1**: Lemon juice contains citric acid. It turns this universal indicator paper orange, which shows that it's a weak acid.*

Example — Maths Skills

Find the pH of a 0.020 mol dm^{-3} solution of propanoic acid (CH_3CH_2COOH) at 298K. K_a for propanoic acid at this temperature is 1.30×10^{-5} mol dm^{-3}.

1. Write an expression for K_a for the weak acid:

$$K_a = \frac{[H^+][CH_3CH_2COO^-]}{[CH_3CH_2COOH]} = \frac{[H^+]^2}{[CH_3CH_2COOH]}$$

2. Rearrange it to find $[H^+]^2$:

$$[H^+]^2 = K_a[CH_3CH_2COOH] = (1.30 \times 10^{-5}) \times 0.020 = 2.60 \times 10^{-7}$$

3. Take the square root of this number to find $[H^+]$:

$$[H^+] = \sqrt{2.60 \times 10^{-7}} = 5.10 \times 10^{-4} \text{ mol dm}^{-3}$$

4. Use $[H^+]$ to find the pH of the acid:

$$pH = -\log_{10}[H^+] = -\log_{10} 5.10 \times 10^{-4} = 3.292$$

Exam Tip

It's important to show all the steps of your working out when you're answering an exam question — that way, if your final answer is wrong, you can still pick up some marks.

Finding the concentration of weak acids

If you already know the pH you can use K_a to find the concentration of the acid. You don't need to know anything new for this type of calculation — you use the same formulas you used to find the pH.

Step 1: Substitute the pH into the inverse pH equation to calculate $[H^+]$.

Step 2: Write an expression for K_a.

Step 3: Rearrange the equation to give the concentration of the acid.

Step 4: Substitute the values for K_a and $[H^+]$ into the equation and solve it.

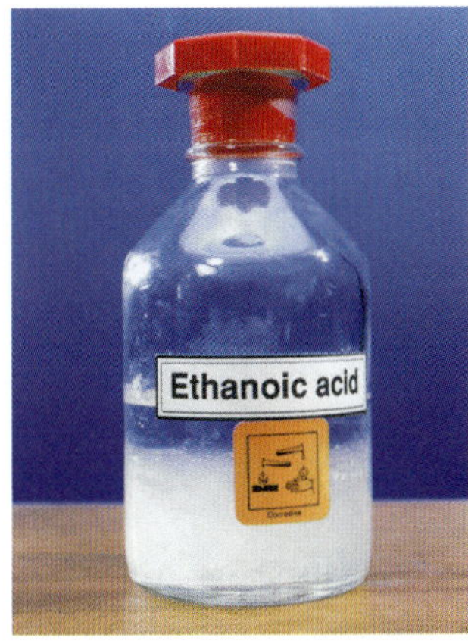

***Figure 2**: Ethanoic acid is a weak acid.*

Example — Maths Skills

The pH of an ethanoic acid (CH_3COOH) solution is 3.02 at 298 K. Calculate the molar concentration of this solution. The K_a of ethanoic acid is 1.75×10^{-5} mol dm^{-3} at 298 K.

1. Use the pH of the acid to find $[H^+]$:

$$[H^+] = 10^{-pH} = 10^{-3.02} = 9.5 \times 10^{-4} \text{ mol dm}^{-3}$$

2. Write an expression for K_a:

$$K_a = \frac{[H^+][CH_3COO^-]}{[CH_3COOH]} = \frac{[H^+]^2}{[CH_3COOH]}$$

3. Rearrange it to give $[CH_3COOH]$:

$$[CH_3COOH] = \frac{[H^+]^2}{K_a}$$

4. Substitute in K_a and $[H^+]$ and solve the equation to find $[CH_3COOH]$:

$$[CH_3COOH] = \frac{(9.5 \times 10^{-4})^2}{1.75 \times 10^{-5}} = 0.052 \text{ mol dm}^{-3}$$

Exam Tip

In the exam you might not be given the chemical formula of the acid you're looking at — you may have to work it out for yourself. See page 243 if you need a recap on the nomenclature for carboxylic acids.

Finding the K_a of weak acids

If you know both the concentration and the pH, you can use them to find the K_a of the weak acid. Just find $[H^+]$ (as shown in the example above), substitute the values you know into the expression for K_a and solve.

Exam Tip

If you can't work out the chemical formula of an acid, just use HA instead and you might still get the marks.

Example — Maths Skills

A solution of 0.162 mol dm^{-3} HCN has a pH of 5.050 at 298 K. What is the value of K_a for HCN at 298 K?

1. Use the pH of the acid to find $[H^+]$:

$$[H^+] = 10^{-pH} = 10^{-5.050} = 8.91 \times 10^{-6} \text{ mol dm}^{-3}$$

2. Write an expression for K_a:

$$K_a = \frac{[H^+][CN^-]}{[HCN]} = \frac{[H^+]^2}{[HCN]}$$

3. Substitute in the values for $[H^+]$ and $[HCN]$:

$$K_a = \frac{[H^+]^2}{[HCN]} = \frac{(8.91 \times 10^{-6})^2}{0.162} = 4.90 \times 10^{-10} \text{ mol dm}^{-3}$$

Exam Tip

You need to be able to do all three types of calculation in the exam — any one of them could crop up so make sure you understand these examples.

The logarithmic constant pK_a

The value of K_a varies massively from one acid to the next. This can sometimes make the numbers difficult to manage so to make life easier, scientists often use the pK_a instead. pK_a is calculated from K_a in exactly the same way as pH is calculated from $[H^+]$ — and vice versa:

$$pK_a = -\log_{10}(K_a) \qquad K_a = 10^{-pK_a}$$

Tip: Notice how pK_a values aren't annoyingly tiny like K_a values.

So if an acid has a K_a value of 1.50×10^{-7} mol dm^{-3}, then:

$$\begin{aligned} pK_a &= -\log_{10}(K_a) \\ &= -\log_{10}(1.50 \times 10^{-7}) \\ &= 6.824 \end{aligned}$$

Tip: The larger the pK_a, the weaker the acid. Strong acids have very small pK_a values.

And if an acid has a pK_a value of 4.320, then:

$$\begin{aligned} K_a &= 10^{-pK_a} \\ &= 10^{-4.320} \\ &= 4.79 \times 10^{-5} \text{ mol dm}^{-3} \end{aligned}$$

Just to make things that bit more complicated, there might be a pK_a value in a 'find the pH' type of question. If so, you need to convert it to K_a so that you can use the K_a expression.

Tip: These are the same steps as you followed on page 346 but with an extra step (converting the pK_a into K_a) at the beginning.

Example — Maths Skills

Calculate the pH of 0.050 mol dm^{-3} methanoic acid (HCOOH). Methanoic acid has a pK_a of 3.750 at 298 K.

1. Convert the pK_a value to a K_a value:

$$\begin{aligned} K_a &= 10^{-pK_a} \\ &= 10^{-3.750} \\ &= 1.78 \times 10^{-4} \text{ mol dm}^{-3} \end{aligned}$$

2. Write out an expression for K_a:

$$K_a = \frac{[H^+][HCOO^-]}{[HCOOH]} = \frac{[H^+]^2}{[HCOOH]}$$

3. Rearrange it to give $[H^+]^2$:

$$\begin{aligned} [H^+]^2 &= K_a[HCOOH] \\ &= 1.78 \times 10^{-4} \times 0.050 \\ &= 8.9 \times 10^{-6} \end{aligned}$$

4. Take the square root to get $[H^+]$:

$$\begin{aligned} [H^+] &= \sqrt{8.9 \times 10^{-6}} \\ &= 2.98 \times 10^{-3} \text{ mol dm}^{-3} \end{aligned}$$

5. Substitute $[H^+]$ into the pH equation and solve:

$$\begin{aligned} pH &= -\log(2.98 \times 10^{-3}) \\ &= 2.526 \end{aligned}$$

Tip: '$\log_{10}$' is often written as just 'log'.

Exam Tip
pH values for weak acids are usually between 2 and 5. If you get an answer much bigger or smaller than this in your exam, double-check your calculation — you may have gone wrong somewhere.

It works the other way around too. Sometimes you are asked to calculate K_a but have to give your answer as a pK_a value. In this case, you just work out the K_a value as usual and then convert it to pK_a — and Bob's your pet hamster.

Practice Questions — Application

Q1 A solution of the weak acid hydrocyanic acid (HCN) has a concentration of 2.0 mol dm^{-3}. The K_a of this acid at 25 °C is 4.9×10^{-10} mol dm^{-3}.

a) Write down an expression for the K_a of this acid.

b) Calculate the pH of this solution at 25 °C

Q2 A solution of the weak acid nitrous acid (HNO_2) has a pH of 3.80 at 25 °C. The K_a of this acid at 25 °C is 4.0×10^{-4} mol dm^{-3}. Determine the concentration of this solution at 25 °C.

Q3 The K_a of lactic acid (a weak acid) at 25 °C is 1.38×10^{-4} mol dm^{-3}. Calculate the pH of a 0.48 mol dm^{-3} solution of lactic acid at 25 °C.

Q4 A 0.28 mol dm^{-3} solution of a weak acid (HA) has a pH of 4.11 at 25 °C. Calculate the K_a of this acid at 25 °C.

Q5 Methanoic acid (HCOOH) is a weak acid. It has a K_a of 1.8×10^{-4} mol dm^{-3} at 298 K. The pH of a solution of methanoic acid was measured to be 3.67 at 298 K. Determine the concentration of this solution of methanoic acid.

Q6 The weak acid ethanoic acid has a pK_a of 4.78 at 298 K. A 0.25 mol dm^{-3} solution of ethanoic acid was prepared.

a) Determine the K_a of ethanoic acid at 298 K.

b) Calculate the pH of this solution at 298 K.

Q7 A 0.154 mol dm^{-3} weak acid solution has a pH of 4.50 at 45 °C. Calculate the pK_a of this acid at 45 °C.

Q8 The pK_a of hydrofluoric acid, HF (a weak acid), is 3.14 at a certain temperature. Calculate the concentration of a solution of hydrofluoric acid that has a pH of 3.20 at this temperature.

Q9 A weak acid (HX) has a pK_a of 4.50 at 25 °C. Calculate the pH of a 0.6 mol dm^{-3} solution of this acid at 25 °C.

Exam Tip

You'll almost certainly have to use expressions for terms like pH and pK_a in the exam so you really, really need to make sure that you know the formulas. Have a look at pages 523-524 for a summary of all the formulas in this section.

Practice Questions — Fact Recall

Q1 What are the units for K_a?

Q2 a) Write the general equation for the acid dissociation constant K_a for a general acid HA.

b) How can you simplify this formula for weak acid?

c) Rearrange your answer to b) to give an expression for calculating [HA].

Q3 Give two things that K_a can be used for.

Q4 a) Write out the expression that defines pK_a.

b) Rearrange this equation to give an expression for K_a.

4. Titrations and pH Curves

When acids and bases are mixed together a neutralisation reaction occurs — H^+ ions from the acid join with OH^- ions from the base to create water. If there are equal numbers of H^+ and OH^- ions the mixture will be neutral (pH 7). What do titrations and pH curves have to do with this? Read on...

Learning Objectives:

- Be able to carry out titrations of acids with bases.
- Understand the typical shape of pH curves for acid–base titrations in all combinations of weak and strong monoprotic acids and bases.
- Be able to sketch and explain the shapes of typical pH curves.
- Be able to use pH curves to select an appropriate indicator.
- Be able to make up a volumetric solution and carry out a simple acid-base titration (Required Practical 1).
- Be able to investigate how pH changes when a weak acid reacts with a strong base and when a strong acid reacts with a weak base (Required Practical 9).

Specification Reference 3.1.12.5

Titrations

REQUIRED PRACTICAL 1

Titrations allow you to find out exactly how much alkali is needed to neutralise a quantity of acid. Here's how you do one...

Step 1: You measure out some acid of known concentration using a pipette and put it in a flask, along with some appropriate **indicator** (indicators change colour at a certain pH — see page 352 for more).

Step 2: Do a rough titration — add the alkali to the acid using a burette fairly quickly to get an approximate idea where the solution changes colour. This is the **end point** — the point at which all of the acid is just neutralised. Give the flask a regular swirl to make sure the acid and alkali are mixed properly.

Step 3: Repeat step one and then do an accurate titration. Run the alkali in to within 2 cm^3 of the end point, then add it drop by drop. If you don't notice exactly when the solution changes colour you've overshot and your result won't be accurate.

Step 4: Record the amount of alkali needed to neutralise the acid.

It's best to repeat this process a few times, making sure you get very similar answers each time (within about 0.1 cm^3 of each other).

You can also find out how much acid is needed to neutralise a quantity of alkali. It's exactly the same process as above, but you add acid to alkali instead. The equipment you'll need to do a titration is illustrated in Figure 1.

Tip: Sometimes you'll hear about alkalis being used in titrations, sometimes bases. Don't get yourself in a muddle over it — an alkali is just a type of base, it's a base that's soluble in water.

Tip: Before you carry out any experiment, make sure you have considered all safety precautions that you need to take.

Tip: You need to repeat titrations to make sure your results are reliable.

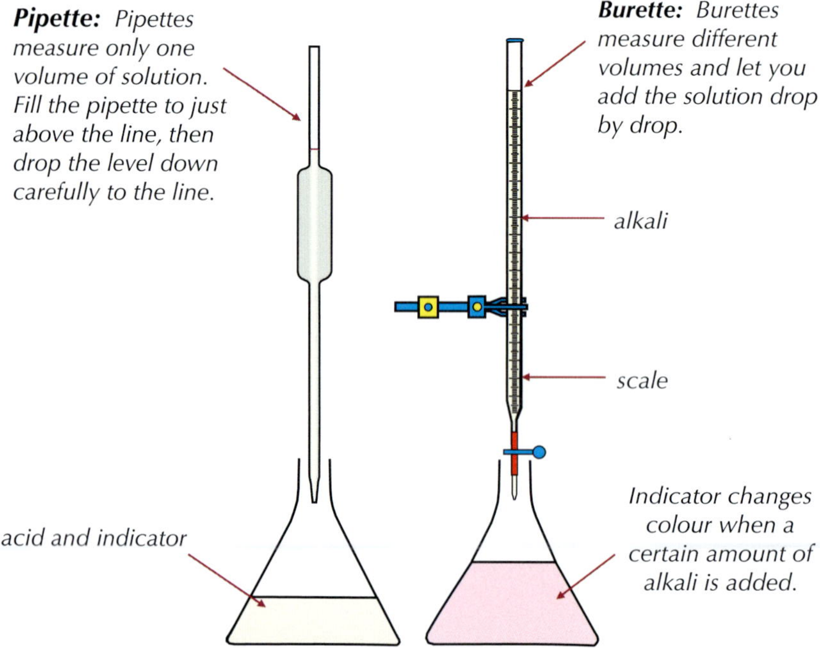

Pipette: Pipettes measure only one volume of solution. Fill the pipette to just above the line, then drop the level down carefully to the line.
Burette: Burettes measure different volumes and let you add the solution drop by drop.
alkali
scale
acid and indicator
Indicator changes colour when a certain amount of alkali is added.

Figure 1: *Equipment needed to perform a titration.*

pH curves

REQUIRED PRACTICAL 9

pH curves show the results of titration experiments. They can be made by plotting the pH of the titration mixture against the amount of base added as the titration goes on. The pH of the mixture can be measured using a pH meter and the scale on the burette can be used to see how much base has been added.

The shape of the curve looks a bit different depending on the strengths of the acid and base that were used. The graphs below show the pH curves for the different combinations of strong and weak monoprotic acids and bases:

Figure 2: A student using a pH meter to measure pH during an acid–base titration.

Strong acid/strong base

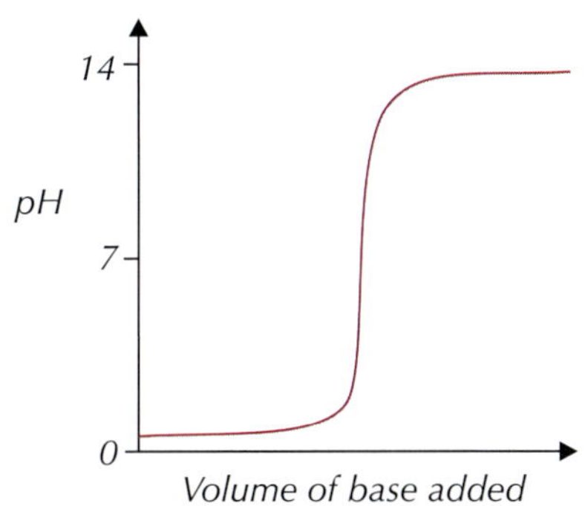

The pH starts around 1, as there's an excess of strong acid.
It finishes up around pH 13, when you have an excess of strong base.

Strong acid/weak base

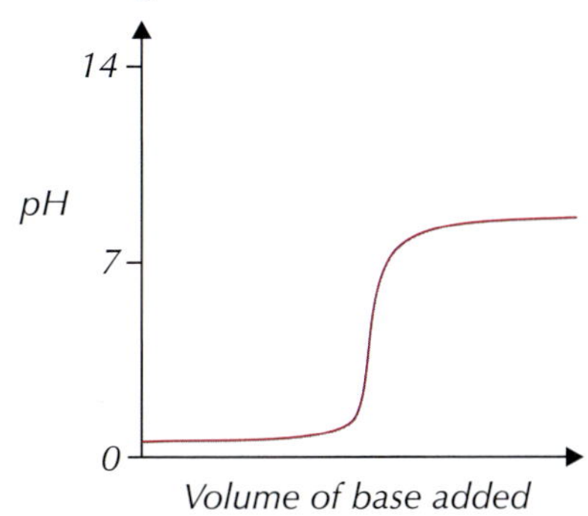

The pH starts around 1, as there's an excess of strong acid.
It finishes up around pH 9, when you have an excess of weak base.

Weak acid/strong base

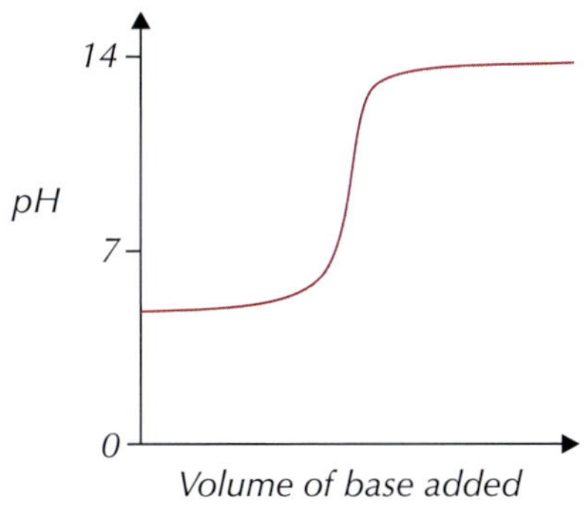

The pH starts around 5, as there's an excess of weak acid.
It finishes up around pH 13, when you have an excess of strong base.

Weak acid/weak base

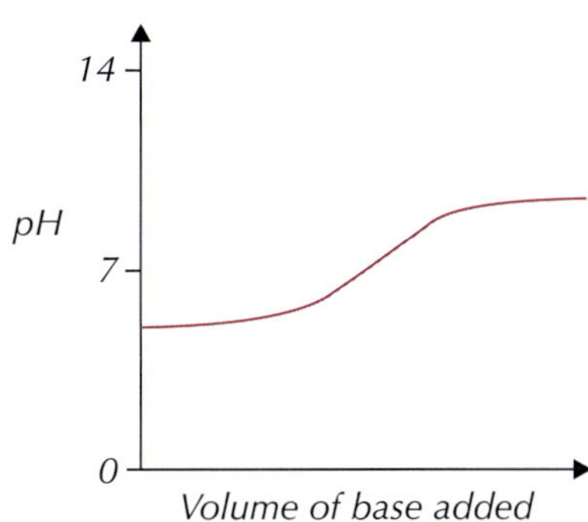

The pH starts around 5, as there's an excess of weak acid.
It finishes up around pH 9, when you have an excess of weak base.

All the graphs apart from the weak acid/weak base graph have a bit that's almost vertical — the mid-point of this vertical section is the equivalence point or end point. At this point, a tiny amount of base causes a sudden, big change in pH — it's here that all the acid is just neutralised.

You don't get such a sharp change in a weak acid/weak base titration. If you used an indicator for this type of titration, its colour would change very gradually, and it would be very tricky to see the exact end point. So you're usually better using a pH meter to find the end point for this type of titration.

Tip: If you titrate a base with an acid instead, the shapes of the curves stay the same, but they're reversed. For example:

<u>Strong base/strong acid</u>

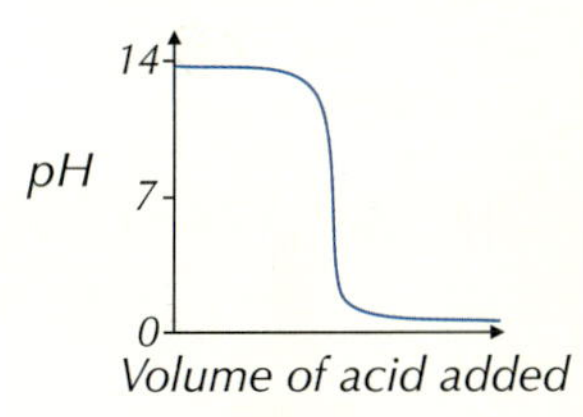

<u>Strong base/weak acid</u>

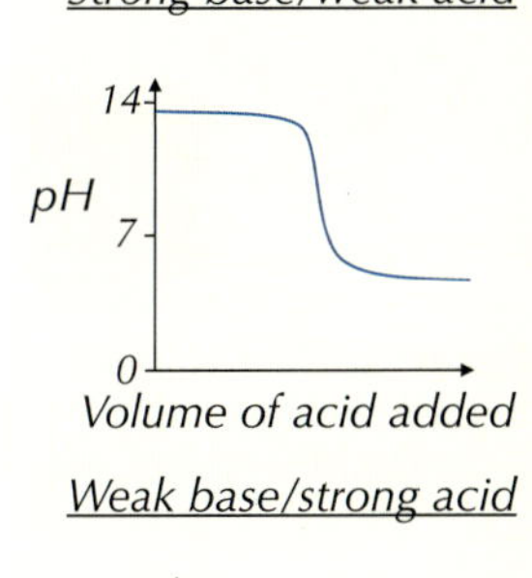

<u>Weak base/strong acid</u>

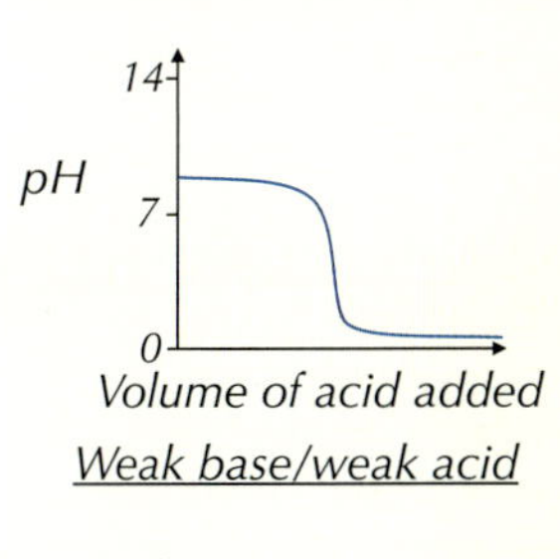

<u>Weak base/weak acid</u>

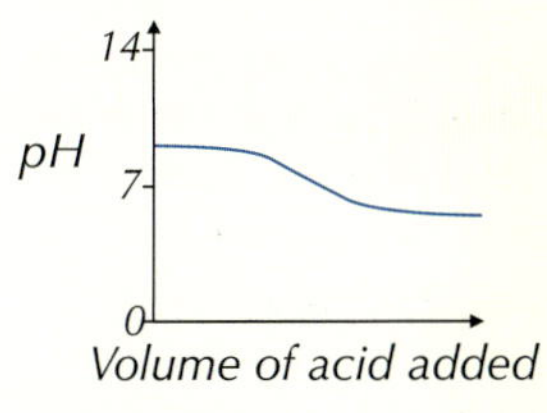

Indicators

REQUIRED PRACTICAL 9

When you use an indicator, you need it to change colour exactly at the end point of your titration. So you need to pick one that changes colour over a narrow pH range that lies entirely on the vertical part of the pH curve. So for the titration shown in Figure 3 (below) you'd want an indicator that changed colour somewhere between pH 8 and pH 11:

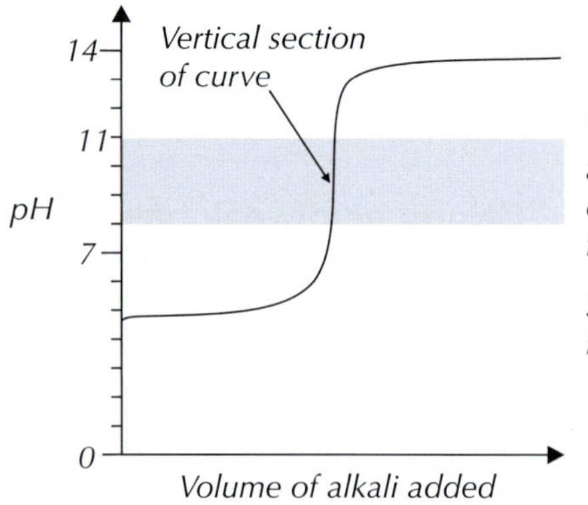

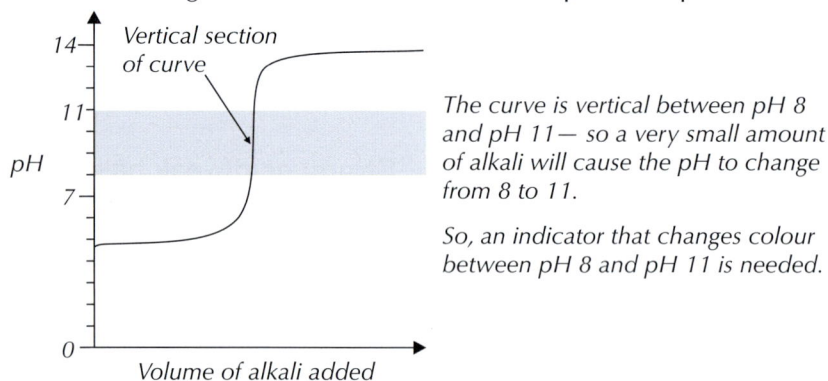

The curve is vertical between pH 8 and pH 11 — so a very small amount of alkali will cause the pH to change from 8 to 11.

So, an indicator that changes colour between pH 8 and pH 11 is needed.

Figure 3: *Graph showing how to select an indicator.*

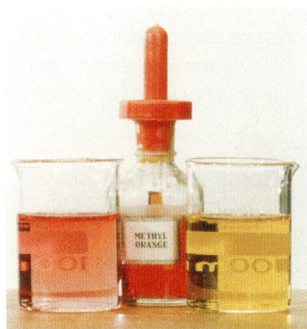

Figure 4: *The red to yellow colour change of methyl orange.*

Methyl orange and **phenolphthalein** are indicators that are often used for acid–base titrations. They each change colour over a different pH range:

Name of indicator	Colour at low pH	Approx. pH of colour change	Colour at high pH
Methyl orange	red	3.1 – 4.4	yellow
Phenolphthalein	colourless	8.3 – 10	pink

Figure 5: *The colourless to pink colour change of phenolphthalein*

- For a strong acid/strong base titration, you can use either of these indicators — there's a rapid pH change over the range for both indicators.
- For a strong acid/weak base only methyl orange will do. The pH changes rapidly across the range for methyl orange, but not for phenolphthalein.
- For a weak acid/strong base, phenolphthalein is the stuff to use. The pH changes rapidly over phenolphthalein's range, but not over methyl orange's.
- For weak acid/weak base titrations there's no sharp pH change, so no indicator will work — you should just use a pH meter.

You need to be able to use a pH curve to select an appropriate indicator:

Example

The graph to the right shows the pH curve produced when a strong acid is added to a weak base. From the table below, select an indicator that you could use for this titration.

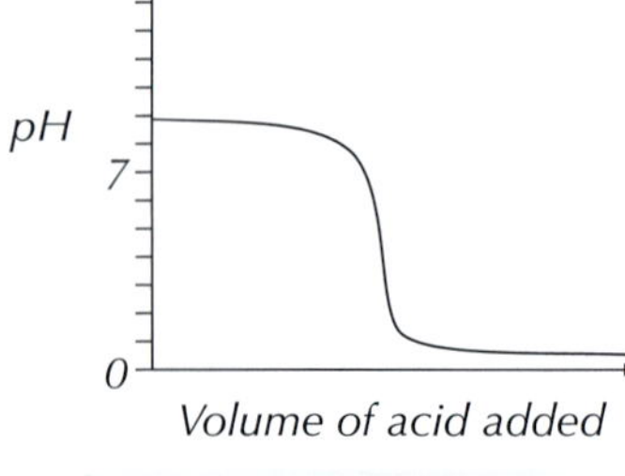

Indicator	pH range
Bromophenol blue	3.0 – 4.6
Litmus	5.0 – 8.0
Cresol purple	7.6 – 9.2

The graph shows that the vertical part of the pH curve is between about pH 2 and pH 6. So you need an indicator with a pH range between 2 and 6. The only indicator that changes colour within this range is bromophenol blue. So in this example, bromophenol blue is the right indicator to choose.

Exam Tip

In your exam, you'll usually be given a table of indicators to choose from, like in this example — so don't worry, you don't have to learn all these indicators and their pH ranges.

Practice Questions — Application

Q1 The graphs below show the pH curves for four different acid–base titrations. For each reaction state what type of acid and base (i.e. strong/weak) were used and select an appropriate indicator from the table below.

Indicator	pH range
Orange IV	1.4 – 2.6
Methyl orange	3.1 – 4.4
Litmus	5.0 – 8.0
Cresol purple	7.6 – 9.2
Phenolphthalein	8.3 – 10

a)

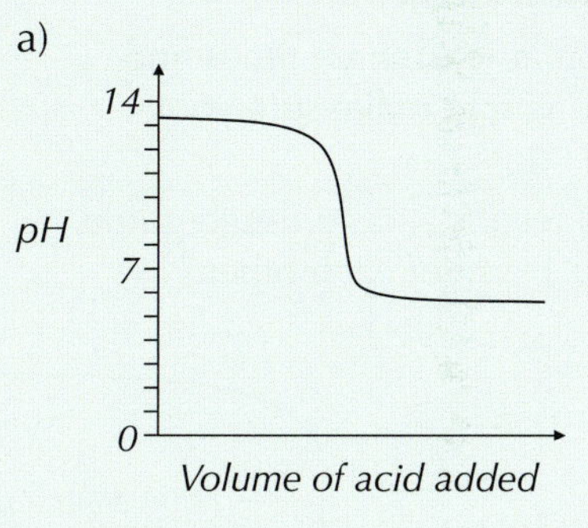

b)

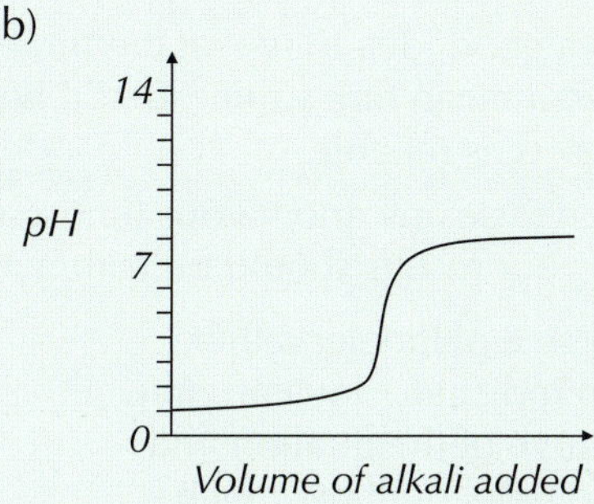

c)

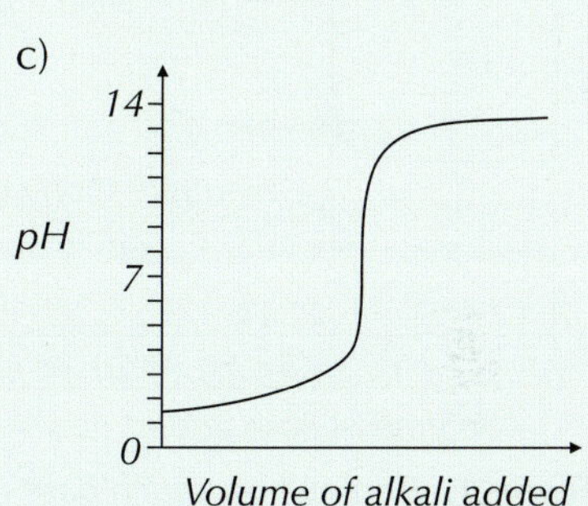

d)

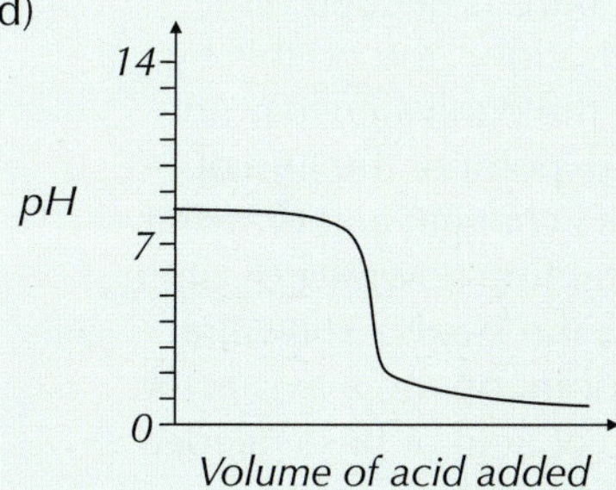

Q2 Neutral red is an indicator which changes colour from red to yellow between pH 6.8 and pH 8.0. Sketch the pH curve for a titration reaction that this indicator could be used for.

Practice Questions — Fact Recall

Q1 Sketch the pH curve produced when:

a) A strong acid neutralises a weak base.

b) A strong base neutralises a strong acid.

c) A weak acid neutralises a strong base.

Q2 a) What happens at the end point of a titration reaction?

b) How can you see that the end point has been reached when you're carrying out a titration?

c) How can you tell the end point has been reached using a pH curve?

Q3 How would you know if an indicator is suitable for a particular titration reaction?

5. Titration Calculations

Learning Objective:

- Be able to perform calculations for the titrations of acids with bases, based on experimental results.

Specification Reference 3.1.12.5

Now that you've learnt all about titrations it's time to find out what you can do with the results. There are a few calculations you'll need to be able to do — the next few pages tell you how.

Titration results

When you've done a titration you can use your results to calculate the concentration of your acid or base. There are a few things you can do to make sure your titration results are as accurate as possible:

- Measure the neutralisation volume as precisely as you possibly can. This will usually be to the nearest 0.05 cm^3.
- It's a good idea to repeat the titration at least three times and take a mean titre value. That'll help you to make sure your answer is reliable.
- Don't use any anomalous (unusual) results — as a rough guide, all your results should be within ±0.1 cm^3 of each other.

If you use a pH meter rather than an indicator, you can draw a pH curve of the titration and use it to work out how much acid or base is needed for neutralisation. You do this by finding the equivalence point (the mid–point of the line of rapid pH change) and drawing a vertical line downwards until it meets the *x*-axis. The value at this point on the *x*-axis is the volume of acid or base needed — see Figure 1.

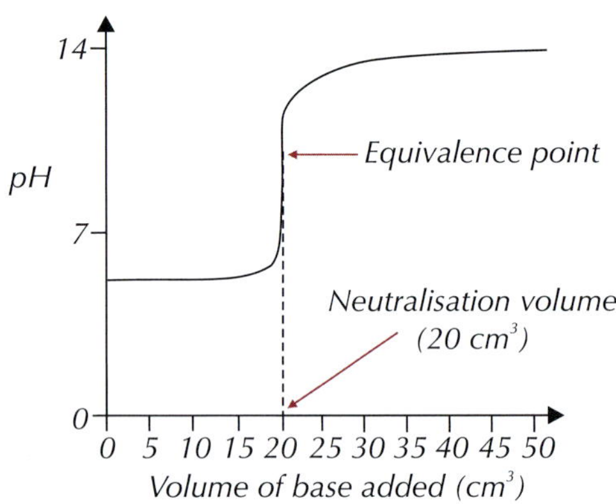

***Figure 1:** Finding how much base is needed for neutralisation.*

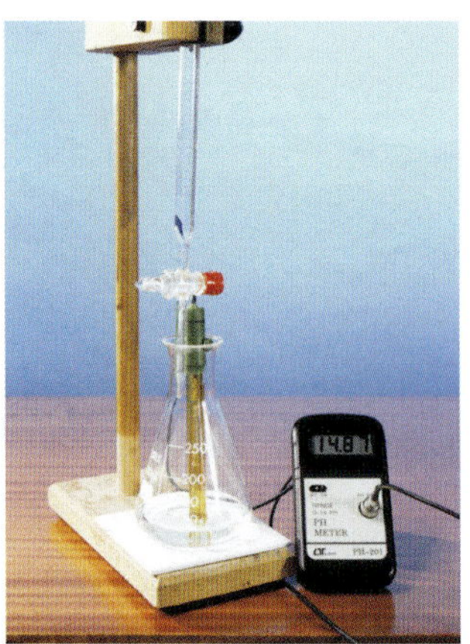

***Figure 2:** A pH meter being used to monitor the pH during a titration.*

Calculating concentrations

Monoprotic acids

Once you know the neutralisation volume you can use it to calculate the concentration of the acid or base. To do this:

Step 1: Write out a balanced equation for the titration reaction.

Step 2: Decide what you know already and what you need to know — usually you'll be given the two volumes and a concentration and you'll have to work out the other concentration.

Step 3: For one reagent you'll know both the concentration and the volume. Calculate the number of moles of this reagent using the equation:

$$\text{moles} = \frac{\text{concentration (mol dm}^{-3}) \times \text{volume (cm}^3)}{1000}$$

Step 4: Use the molar ratios in the balanced equation to find out how many moles of the other reagent reacted.

Step 5: Calculate the unknown concentration using the equation:

$$\text{concentration} = \frac{\text{moles} \times 1000}{\text{volume}}$$

in mol dm^{-3} (concentration) — *in cm^3* (volume)

Tip: You came across the formula:

$$\textbf{moles} = \frac{\textbf{conc.} \times \textbf{vol.}}{1000}$$

on page 48. The dividing by 1000 bit is to get the volume from cm^3 to dm^3 — if your volume is already in dm^3 then it's just:

moles = conc. × vol.

Example — Maths Skills

The graph on the right shows the results when 0.500 mol dm^{-3} HCl was titrated against 35.0 cm^3 of NaOH. Calculate the concentration of the NaOH solution.

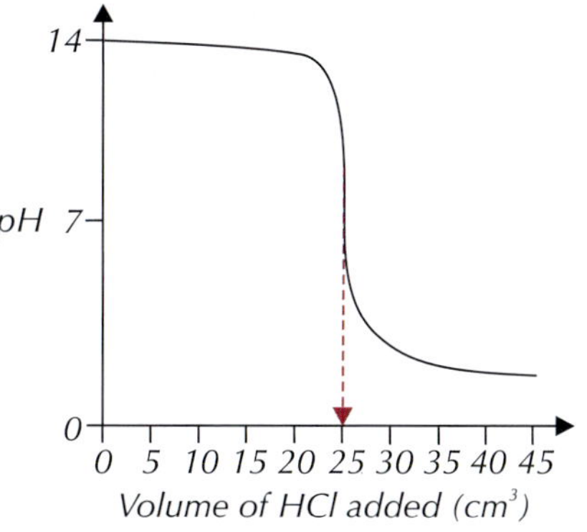

From the graph you can see that 25 cm^3 of HCl was required to neutralise the NaOH. You can use this information to work out the concentration of the NaOH solution by following the steps on the previous page.

1. The balanced equation for this titration reaction is:

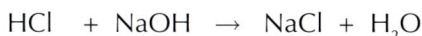

$$HCl + NaOH \rightarrow NaCl + H_2O$$

2. You know the concentration of HCl (0.500 mol dm^{-3}), the volume of HCl (25.0 cm^3) and the volume of NaOH (35.0 cm^3). You need to know the concentration of NaOH.

3. Calculate the moles of HCl:

$$\text{moles HCl} = \frac{\text{concentration} \times \text{volume}}{1000} = \frac{0.500 \times 25.0}{1000} = 0.0125 \text{ moles}$$

4. From the equation, you know 1 mole of HCl neutralises 1 mole of NaOH. So 0.0125 moles of HCl must neutralise 0.0125 moles of NaOH.

5. Calculate the concentration of NaOH:

$$\text{Conc. NaOH} = \frac{\text{moles} \times 1000}{\text{volume}} = \frac{0.0125 \times 1000}{35.0} = 0.357 \text{ mol dm}^{-3}$$

Exam Tip
You won't always have to read the volume of acid off a pH curve in the exam — you might just be told the neutralisation volume, in which case you'll have one less job to do.

Tip: In these calculations the units of concentration should always be mol dm^{-3}.

Diprotic acids

A diprotic acid is one that can release two protons when it's in solution. Ethanedioic acid (HOOC–COOH) is diprotic. When ethanedioic acid reacts with a base like sodium hydroxide, it's neutralised. But the reaction happens in two stages, because the two protons are removed from the acid separately. This means that when you titrate ethanedioic acid with a strong base you get a pH curve with two equivalence points:

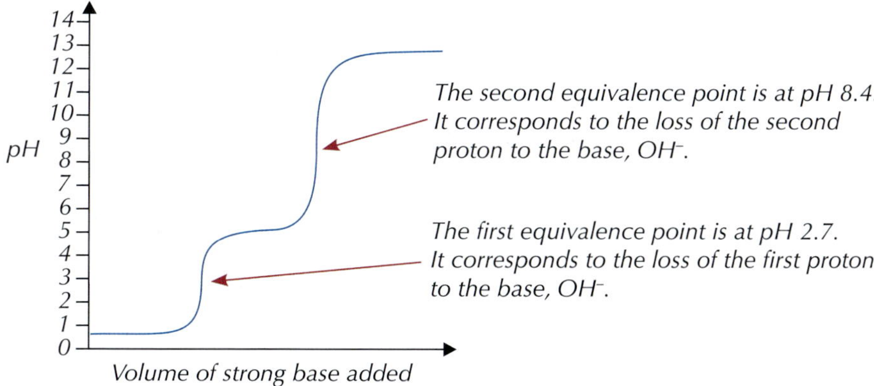

First equivalence point: $HOOC–COOH_{(aq)} + OH^-_{(aq)} \rightarrow HOOC–COO^-_{(aq)} + H_2O_{(l)}$

Second equivalence point: $HOOC–COO^-_{(aq)} + OH^-_{(aq)} \rightarrow {}^-OOC–COO^-_{(aq)} + H_2O_{(l)}$

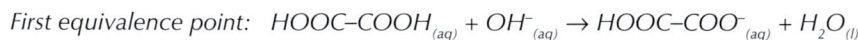

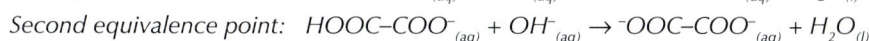

Tip: pH curves for diprotic acids look a bit different to those for monoprotic acids. For more on pH curves, see page 351.

Tip: The curve won't always look exactly like this. The pH at the start, end and equivalence points will vary depending on the concentrations of the acid and the base used.

You can calculate the concentration of a diprotic acid from titration data in the same way as you did for a monoprotic acid. Just remember that the acid is diprotic so you'll need twice as many moles of base as moles of acid.

Example — Maths Skills

25.0 cm^3 of ethanedioic acid, $C_2H_2O_4$, was completely neutralised by 20.0 cm^3 of 0.100 mol dm^{-3} KOH solution. Calculate the concentration of the ethanedioic acid solution.

1. The balanced equation for this titration reaction is:

$$C_2H_2O_4 + 2KOH \rightarrow K_2C_2O_4 + 2H_2O$$

2. You know the volume of $C_2H_2O_4$ (25.0 cm^3), the volume of KOH (20.0 cm^3) and the concentration of KOH (0.100 mol dm^{-3}). You need to know the concentration of $C_2H_2O_4$.

3. Calculate the moles of KOH:

$$\text{moles KOH} = \frac{\text{concentration} \times \text{volume}}{1000} = \frac{0.100 \times 20.0}{1000} = 0.002 \text{ moles}$$

4. You know from the equation that you need 2 moles of KOH to neutralise 1 mole of $C_2H_2O_4$. So 0.002 moles of KOH must neutralise (0.002 ÷ 2) = 0.001 moles of $C_2H_2O_4$.

5. Calculate the concentration of $C_2H_2O_4$:

$$\text{Conc. } C_2H_2O_4 = \frac{\text{moles} \times 1000}{\text{volume}} = \frac{0.001 \times 1000}{25} = 0.0400 \text{ mol dm}^{-3}$$

Exam Tip
You need to be able to do calculations for both monoprotic and diprotic acids so make sure you understand these two examples.

Tip: Because it's a diprotic acid, you need twice as many moles of base as moles of acid.

Practice Questions — Application

Q1 In a titration, the equivalence point was reached after 13.8 cm^3 of a 1.50 mol dm^{-3} solution of HCl had been added to 20.0 cm^3 of NaOH. Calculate the concentration of the NaOH solution.

Q2 Nitric acid (HNO_3) was added to 30.0 cm^3 of a 0.250 mol dm^{-3} solution of NaOH in the presence of methyl orange. A colour change was observed after 17.8 cm^3 of the acid had been added. Calculate the concentration of the nitric acid solution.

Q3 A 0.250 mol dm^{-3} KOH solution was titrated against 24.0 cm^3 of HCl. The experiment was repeated three times and the amount of KOH required to neutralise the HCl each time is given in the table below.

	Titration 1	Titration 2	Titration 3
Titre volume (cm^3 KOH)	22.40	22.50	22.40

a) Calculate the average titre volume for this titration.

b) Calculate the concentration of the HCl solution.

Q4 The graph below shows the pH curve produced when 0.850 mol dm^{-3} HCl was titrated against 30.0 cm^3 of an NaOH solution of unknown concentration.

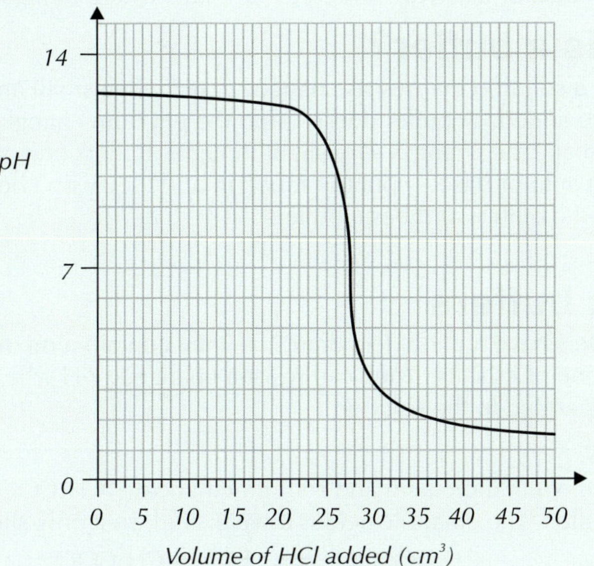

Calculate the concentration of the NaOH solution.

Q5 32.0 cm^3 of a KOH solution is fully neutralised when 18.0 cm^3 of a 0.400 mol dm^{-3} solution of H_2SO_4 is added to it. Calculate the concentration of the KOH solution.

Q6 When a 1.20 mol dm^{-3} solution of KOH was titrated against 20.0 cm^3 of carbonic acid (H_2CO_3), 26.2 cm^3 of KOH was required to fully neutralise the acid. Calculate the concentration of the carbonic acid solution.

Tip: When you're writing the equations for a titration remember that it's a neutralisation reaction so the products will be a salt and water.

Practice Questions — Fact Recall

Q1 Write out the equation that links the number of moles, the volume and the concentration of a solution.

Q2 Sketch the pH curve produced when a strong base is added to a strong diprotic acid in a titration.

Q3 Explain why the pH curve for the titration of a diprotic acid is different to a pH curve for the titration of a monoprotic acid.

Q4 How many moles of a diprotic acid does it take to neutralise 1 mole of NaOH?

6. Buffer Action

Sometimes, it's useful to have a solution that doesn't change pH when small amounts of acid or alkali are added to it. That's where buffers come in handy.

Learning Objectives:

- Know that a buffer solution maintains an approximately constant pH, despite dilution or addition of small amounts of acid or base.
- Know that acidic buffer solutions contain a weak acid and the salt of that weak acid.
- Know that basic buffer solutions contain a weak base and the salt of that weak base.
- Be able to explain qualitatively the action of acidic and basic buffers.
- Know some applications of buffer solutions.

Specification Reference 3.1.12.6

What is a buffer?

A **buffer** is a solution that resists changes in pH when small amounts of acid or alkali are added. A buffer doesn't stop the pH from changing completely — it does make the changes very slight though. Buffers only work for small amounts of acid or base — put too much in and they won't be able to cope. You get acidic buffers and basic buffers.

Acidic buffers

Acidic buffers have a pH of less than 7 — they contain a mixture of a weak acid with one of its salts. They can resist a change in pH when either an acid or a base is added to the solution.

Tip: The acid has to be a weak acid — you can't make an acidic buffer with a strong acid.

Tip: Acidic buffers work slightly differently to basic buffers, which are covered on the next page — but the principle of how buffers work is the same for both types.

Example

A mixture of ethanoic acid and sodium ethanoate ($CH_3COO^-Na^+$) is an acidic buffer. The ethanoic acid is a weak acid, so it only slightly dissociates:

$$CH_3COOH_{(aq)} \rightleftharpoons H^+_{(aq)} + CH_3COO^-_{(aq)}$$

But the salt fully dissociates into its ions when it dissolves:

$$CH_3COONa_{(s)} \xrightarrow{water} CH_3COO^-_{(aq)} + Na^+_{(aq)}$$

So in the solution you've got heaps of undissociated ethanoic acid molecules ($CH_3COOH_{(aq)}$), and heaps of ethanoate ions ($CH_3COO^-_{(aq)}$) from the salt.

When you alter the concentration of H^+ or OH^- ions in the buffer solution the equilibrium position moves to counteract the change (this is down to Le Chatelier's principle). Here's how it all works:

Resisting an acid

The large number of CH_3COO^- ions make sure that the buffer can cope with the addition of acid. If you add a small amount of acid the H^+ concentration increases. Most of the extra H^+ ions combine with CH_3COO^- ions to form CH_3COOH. This shifts the equilibrium to the left, reducing the H^+ concentration to close to its original value. So the pH doesn't change much.

Addition of H⁺ (acid) ←

$$CH_3COOH_{(aq)} \rightleftharpoons H^+_{(aq)} + CH_3COO^-_{(aq)}$$

Resisting a base

If a small amount of base (e.g. NaOH) is added, the OH^- concentration increases. Most of the extra OH^- ions react with H^+ ions to form water — removing H^+ ions from the solution. This causes more CH_3COOH to dissociate to form H^+ ions — shifting the equilibrium to the right. There's no problem doing this as there's loads of spare CH_3COOH molecules. The H^+ concentration increases until it's close to its original value, so the pH doesn't change much.

Addition of OH⁻ (base) →

$$CH_3COOH_{(aq)} \rightleftharpoons H^+_{(aq)} + CH_3COO^-_{(aq)}$$

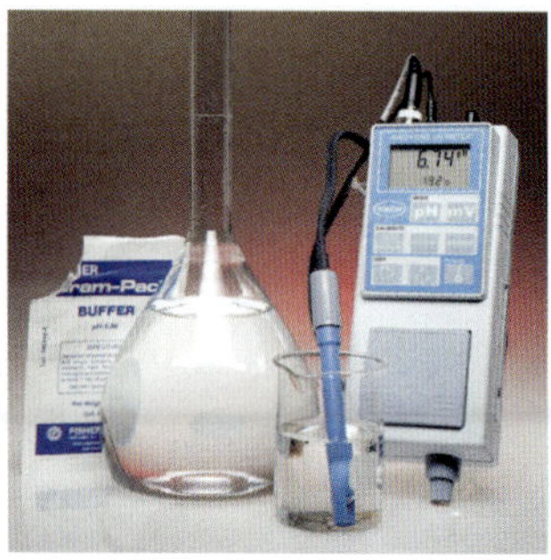

Figure 1: *An acidic buffer solution.*

Basic buffers

Basic buffers have a pH greater than 7 — and they contain a mixture of a weak base with one of its salts. They can resist changes in pH when acid or base is added, just like acidic buffers can.

Tip: You can't make basic buffers with strong bases — they have to be weak bases.

Example

A solution of ammonia (NH_3, a weak base) and ammonium chloride (NH_4Cl, a salt of ammonia) acts as a basic buffer.

The salt fully dissociates in solution:

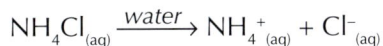

$$NH_4Cl_{(aq)} \xrightarrow{water} NH_4^+{}_{(aq)} + Cl^-{}_{(aq)}$$

Some of the NH_3 molecules will also react with water molecules:

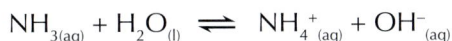

$$NH_{3(aq)} + H_2O_{(l)} \rightleftharpoons NH_4^+{}_{(aq)} + OH^-{}_{(aq)}$$

So the solution will contain loads of ammonium ions (NH_4^+), and lots of ammonia molecules too. The equilibrium position of this reaction can move to counteract changes in pH:

Resisting an acid

If a small amount of acid is added, the H^+ concentration increases, making the solution more acidic. Some of the H^+ ions react with OH^- ions to make H_2O. When this happens the equilibrium position moves to the right to replace the OH^- ions that have been used up. This reaction will remove most of the extra H^+ ions that were added — so the pH won't change much.

Addition of H⁺ (acid) →

$$NH_{3(aq)} + H_2O_{(l)} \rightleftharpoons NH_4^+{}_{(aq)} + OH^-{}_{(aq)}$$

Resisting a base

If a small amount of base is added, the OH^- concentration increases, making the solution more alkaline. Most of the extra OH^- ions will react with the NH_4^+ ions, to form NH_3 and H_2O. So the equilibrium will shift to the left, removing OH^- ions from the solution, and stopping the pH from changing much.

Addition of OH⁻ (base) ←

$$NH_{3(aq)} + H_2O_{(l)} \rightleftharpoons NH_4^+{}_{(aq)} + OH^-{}_{(aq)}$$

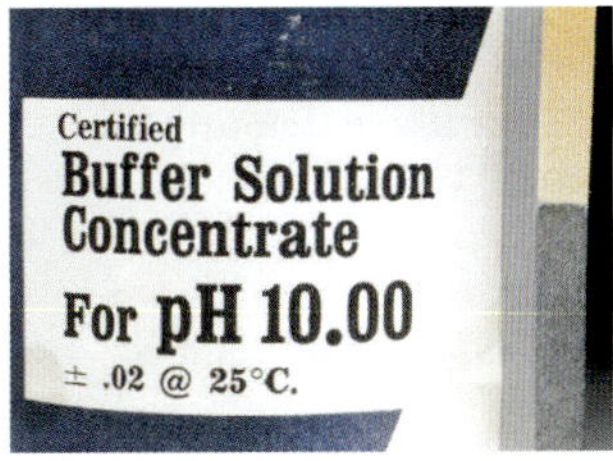

Figure 2: *This basic buffer will maintain a pH of 10.00 when small amounts of acid or alkali are added.*

Exam Tip
You don't need to memorise these specific examples but you do need to be able to explain how both acidic and basic buffers work.

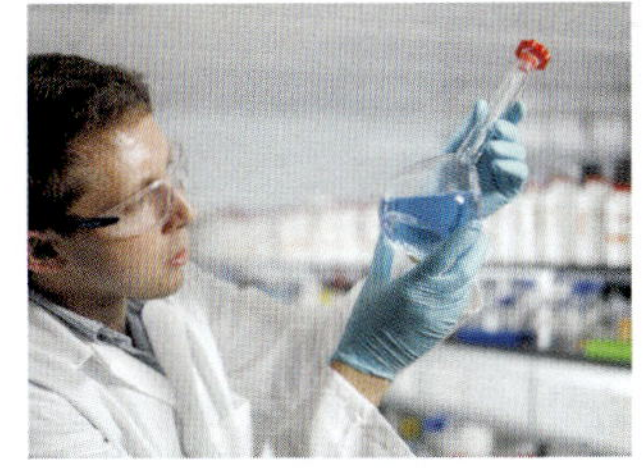

Figure 3: *A lovely blue buffer solution being prepared by a lovely blue-gloved scientist.*

Dilution

Acidic and basic buffers can also resist changes in pH when diluted by water. If a small amount of water is added to a buffer, the water slightly dissociates. So the extra H^+ and OH^- ions push the equilibrium the same amount in both directions, leaving it unchanged.

Acidic and basic buffers — summary

How buffers work can be a bit tricky to get your head round, so here's a nice summary diagram to help you out:

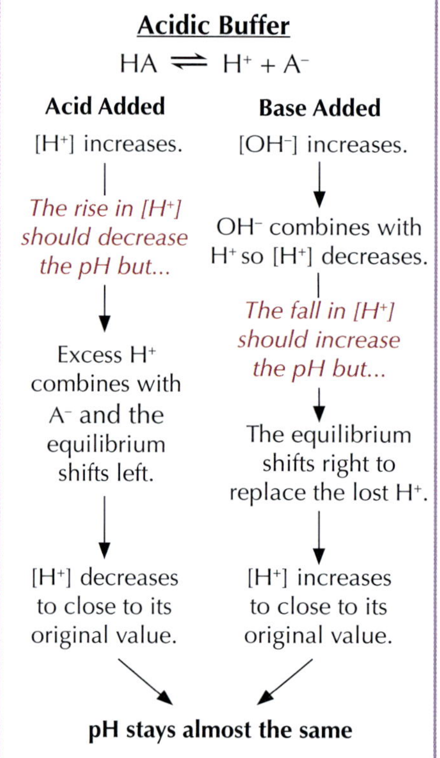

Acidic Buffer
$HA \rightleftharpoons H^+ + A^-$
Acid Added
$[H^+]$ increases.
The rise in $[H^+]$ should decrease the pH but...
Excess H^+ combines with A^- and the equilibrium shifts left.
$[H^+]$ decreases to close to its original value.
Base Added
$[OH^-]$ increases.
OH^- combines with H^+ so $[H^+]$ decreases.
The fall in $[H^+]$ should increase the pH but...
The equilibrium shifts right to replace the lost H^+.
$[H^+]$ increases to close to its original value.
pH stays almost the same

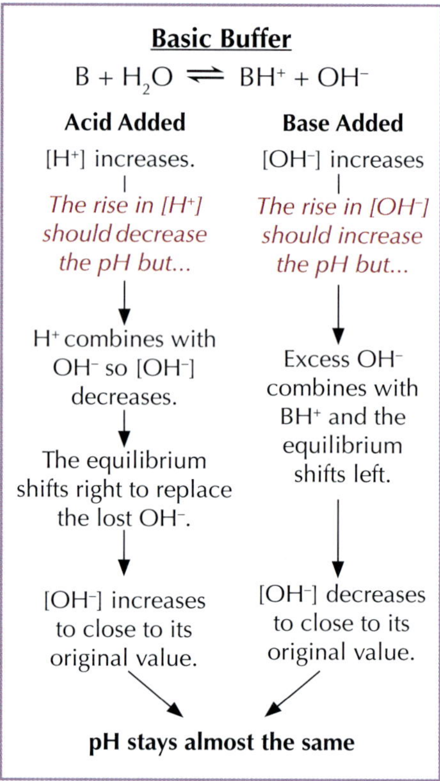

Basic Buffer
$B + H_2O \rightleftharpoons BH^+ + OH^-$
Acid Added
$[H^+]$ increases.
The rise in $[H^+]$ should decrease the pH but...
H^+ combines with OH^- so $[OH^-]$ decreases.
The equilibrium shifts right to replace the lost OH^-.
$[OH^-]$ increases to close to its original value.
Base Added
$[OH^-]$ increases
The rise in $[OH^-]$ should increase the pH but...
Excess OH^- combines with BH^+ and the equilibrium shifts left.
$[OH^-]$ decreases to close to its original value.
pH stays almost the same

Tip: pH is determined by $[H^+]$, but because OH^- will combine with H^+ and remove it from solution, $[OH^-]$ affects $[H^+]$ and so affects pH. A rise in $[OH^-]$ will increase the pH, a fall in $[OH^-]$ will decrease the pH. Unless there's a buffer present, that is.

Applications of buffers

Alkaline conditions make the surface of individual hairs rougher, so most shampoos have a pH of around 5.5, which keeps hair smooth and shiny. Buffers are added to shampoo to maintain this pH while you wash your hair. Biological washing powders contain buffers too. They keep the pH at the right level for the enzymes to work best. There are also lots of biological buffer systems in our bodies, making sure all our tissues are kept at the right pH. For example, it's vital that blood stays at a pH very near to 7.4, so it contains a buffer system.

Figure 4: *Shampoos contain buffers to stop the pH of your hair from changing too much.*

Practice Questions — Fact Recall

Q1 What is a buffer?

Q2 What do acidic buffers contain?

Q3 Explain how an acidic buffer resists changes in pH when:
- a) a small amount of acid is added.
- b) a small amount of base is added.

Q4 What do basic buffers contain?

Q5 Explain how a basic buffer resists changes in pH when:
- a) a small amount of acid is added.
- b) a small amount of base is added.

Q6 Give two real-life applications of buffers.

7. Calculating the pH of Buffers

You need to be able to calculate the pH of buffer solutions. These calculations look scary, but don't worry — they're not nearly as bad as they look.

Learning Objective:
- Be able to calculate the pH of acidic buffer solutions.

Specification Reference 3.1.12.6

Calculations using known concentrations

If you know the K_a of the weak acid and the concentrations of the weak acid and its salt, calculating the pH of an acidic buffer isn't too tricky. Here's how to go about it:

Step 1: Write out the expression for the K_a of the weak acid.

Step 2: Rearrange the equation to give an expression for $[H^+]$.

Step 3: Substitute the value for K_a and the concentrations of the acid and salt into the equation.

Step 4: Solve the equation to find a value for $[H^+]$.

Step 5: Substitute your value for $[H^+]$ into the pH equation ($pH = -\log[H^+]$) and solve it to calculate the pH.

Tip: Writing expressions for K_a was covered on page 346. Have a look back if you need a recap.

Tip: See page 343 for a reminder on how to calculate pH.

Example — Maths Skills

A buffer solution contains 0.400 mol dm^{-3} methanoic acid, HCOOH, and 0.600 mol dm^{-3} sodium methanoate, $HCOO^-Na^+$. For methanoic acid, $K_a = 1.6 \times 10^{-4}$ mol dm^{-3}. What is the pH of this buffer?

1. Write the expression for K_a of the weak acid:

$$HCOOH_{(aq)} \rightleftharpoons H^+_{(aq)} + HCOO^-_{(aq)} \quad \text{so} \quad K_a = \frac{[H^+][HCOO^-]}{[HCOOH]}$$

2. Rearrange the equation to get $[H^+]$:

$$[H^+] = \frac{K_a \times [HCOOH]}{[HCOO^-]}$$

3. Substitute in the value of K_a and the concentrations given in the question. You have to make a few assumptions here:
   - $HCOO^- Na^+$ is fully dissociated, so assume that the equilibrium concentration of $HCOO^-$ is the same as the initial concentration of $HCOO^- Na^+$.
   - HCOOH is only slightly dissociated, so assume that its equilibrium concentration is the same as its initial concentration.

$$[H^+] = \frac{K_a \times [HCOOH]}{[HCOO^-]} = \frac{(1.6 \times 10^{-4}) \times 0.400}{0.600}$$

4. Solve to find $[H^+]$:

$$[H^+] = \frac{(1.6 \times 10^{-4}) \times 0.400}{0.600} = 1.07 \times 10^{-4} \text{ mol dm}^{-3}$$

5. Use your value of $[H^+]$ to calculate the pH:

$$\begin{aligned} pH &= -\log[H^+] \\ &= -\log(1.07 \times 10^{-4}) \\ &= \mathbf{3.97} \end{aligned}$$

Tip: Remember — the concentrations in the expression for K_a all have to be equilibrium concentrations.

Another way to make acidic buffers

Mixing a weak acid with its salt is not the only way to make an acidic buffer. You could also take a weak acid and add a small amount of alkali, so that some of the acid is neutralised to make a salt, but some is left un-neutralised. The reaction mixture would then contain a weak acid and its salt, so would act as an acidic buffer. You can calculate the pH of an acidic buffer that has been made this way by following the steps below:

Step 1: Write out the equation for the neutralisation reaction — remember acid + base → salt + water.

Step 2: Calculate the number of moles of acid and base at the start of the reaction using the volumes and concentrations given in the question.

Step 3: Use the molar ratios in the equation to work out the moles of acid and salt left at the end of the reaction.

Step 4: Calculate the concentration of the acid and salt in the buffer solution by dividing by the volume of the solution — this is the volume of the acid and the base added together.

Step 5: Then you're ready to calculate the pH.

Tip: The molar ratios tell you how many moles of acid will react with a certain number of moles of base, and how many moles of salt are produced. These numbers will always be in the same ratio for a given reaction.

Tip: It's really important that you know equations like:

$$\text{moles} = \frac{\text{conc.} \times \text{vol.}}{1000}$$

You won't be able to do the harder calculations if you haven't got your head around the basics. Check out pages 522-524 for a summary of the equations that you need to know.

Example — Maths Skills

A buffer is formed by mixing 15 cm³ of 0.1 mol dm⁻³ sodium hydroxide (NaOH) and 30 cm³ of 0.6 mol dm⁻³ propanoic acid (CH_3CH_2COOH). Calculate the pH of this buffer solution ($K_a = 1.35 \times 10^{-5}$ mol dm⁻³).

1. Write out the equation for the reaction:

$$CH_3CH_2COOH + NaOH \rightarrow CH_3CH_2COO^-Na^+ + H_2O$$

2. Calculate the number of moles of acid and base:

$$\text{Moles } CH_3CH_2COOH = \frac{\text{Conc.} \times \text{Vol.}}{1000} = \frac{0.6 \times 30}{1000} = 0.018 \text{ moles}$$

$$\text{Moles NaOH} = \frac{\text{Conc.} \times \text{Vol.}}{1000} = \frac{0.1 \times 15}{1000} = 0.0015 \text{ moles}$$

3. The acid is in excess, so all the base reacts. There's 0.0015 moles of NaOH at the start of the reaction. If it all reacts there will be 0.0015 moles of salt at the end of the reaction.

   The equation shows us that 1 mole of base will react with 1 mole of acid to give 1 mole of salt. So if there are 0.0015 moles of salt, 0.0015 moles of acid must have been used up.
   This leaves 0.018 – 0.0015 = 0.0165 moles of acid.

   So, the buffer solution contains 0.0015 moles of $CH_3CH_2COO^-Na^+$ and 0.0165 moles of CH_3CH_2COOH.

4. Calculate the concentration of acid and salt in the buffer solution:

   The total volume of the solution is 15 + 30 = 45 cm³

$$\text{Conc. } CH_3CH_2COOH = \frac{\text{moles} \times 1000}{\text{volume}} = \frac{0.0165 \times 1000}{45}$$

$$= 0.366... \text{ mol dm}^{-3}$$

$$\text{Conc. } CH_3CH_2COO^-Na^+ = \frac{\text{moles} \times 1000}{\text{volume}} = \frac{0.0015 \times 1000}{45}$$

$$= 0.0333... \text{ mol dm}^{-3}$$

5. Work out the pH as before:

$$CH_3CH_2COOH \rightleftharpoons H^+ + CH_3CH_2COO^- \text{ so } K_a = \frac{[H^+][CH_3CH_2COO^-]}{[CH_3CH_2COOH]}$$

$$[H^+] = \frac{K_a \times [CH_3CH_2COOH]}{[CH_3CH_2COO^-]} = \frac{(1.35 \times 10^{-5}) \times 0.366...}{0.0333...}$$

$$= 1.485 \times 10^{-4} \text{ mol dm}^{-3}$$

$$pH = -\log[H^+] = -\log(1.485 \times 10^{-4}) = 3.828$$

Tip: This step was covered in more detail on page 346 — have a look back if you need a quick reminder of what's going on.

Practice Questions — Application

Q1 An acidic buffer solution contains 0.200 mol dm^{-3} propanoic acid (CH_3CH_2COOH) and 0.350 mol dm^{-3} potassium propanoate. For propanoic acid, $K_a = 1.35 \times 10^{-5}$ mol dm^{-3}.

a) Write an expression for the K_a of propanoic acid.

b) Calculate the concentration of H^+ ions in this buffer solution.

c) Calculate the pH of this buffer solution.

Q2 A buffer solution contains 0.150 mol dm^{-3} ethanoic acid and 0.250 mol dm^{-3} potassium ethanoate. For ethanoic acid $K_a = 1.74 \times 10^{-5}$ mol dm^{-3}. Calculate the pH of this buffer.

Q3 A buffer is made by mixing 30.0 cm^3 of 0.500 mol dm^{-3} propanoic acid (CH_3CH_2COOH) with 20.0 cm^3 of 0.250 mol dm^{-3} potassium hydroxide (KOH). For propanoic acid, $K_a = 1.35 \times 10^{-5}$ mol dm^{-3}.

a) Write an equation to show the reaction of propanoic acid with potassium hydroxide.

b) Calculate the number of moles of propanoic acid and potassium hydroxide at the beginning of the reaction.

c) Calculate the concentration of propanoic acid and potassium propanoate in the buffer solution.

d) Calculate the concentration of H^+ ions in this buffer solution.

e) Calculate the pH of the buffer solution.

Q4 A buffer is formed by mixing together 25.0 cm^3 of 0.200 mol dm^{-3} methanoic acid (HCOOH) and 15.0 cm^3 of 0.100 mol dm^{-3} sodium hydroxide (NaOH). For methanoic acid, $K_a = 1.6 \times 10^{-4}$ mol dm^{-3}. Calculate the pH of this buffer.

Section Summary

Make sure you know...

- That a Brønsted–Lowry acid is a proton donor and a Brønsted–Lowry base is a proton acceptor.
- That strong acids/bases dissociate fully in water while weak acids/bases only partially dissociate.
- That acid–base reactions involve the transfer of protons — $HA + B \rightleftharpoons BH^+ + A^-$.
- That water is weakly dissociated and the ionic product of water is $K_w = [H^+][OH^-]$ which is equivalent to $[H^+]^2$ for pure water.
- That the value of K_w varies with temperature.
- That $pH = -\log[H^+]$ where $[H^+]$ is the concentration of H^+ ions in mol dm^{-3}.
- How to convert pH into $[H^+]$ and vice versa.
- How to calculate the pH of a strong acid from its concentration.
- How to calculate the pH of a strong base from its concentration, using K_w.
- That K_a is the dissociation constant for a weak acid, and how to write expressions for K_a.
- How to calculate the pH of a weak acid from its concentration and K_a.
- How to calculate the concentration of a weak acid from its pH and K_a.
- How to calculate K_a for a weak acid from its pH and concentration.
- That $pK_a = -\log(K_a)$ and how to convert K_a to pK_a and vice versa.
- What the pH curves for all the different combinations of weak and strong acids and bases look like.
- How to use pH curves to select an appropriate pH indicator to use in a titration.
- How to calculate the concentration of monoprotic and diprotic acids using the results of a titration.
- What a buffer is and how both acidic and basic buffers can resist changes in pH and dilution.
- That buffer solutions contain a weak acid/base and its salt.
- That buffers are used in shampoos, biological washing powders and biological systems.
- How to calculate the pH of an acidic buffer.

Exam-style Questions

1 A scientist has a 0.20 mol dm^{-3} solution of aqueous potassium hydroxide, KOH. She knows that K_w has a value of 1.00×10^{-14} mol^2 dm^{-6} at 298 K. What is the pH of this solution, to 2 decimal places?

A 13.30

B 12.50

C 0.69

D 28.78

2 Ammonia, a weak base, is titrated with nitric acid, a strong acid. The change in pH is recorded in a graph. Which of the following graphs would you expect to represent the titration?

A

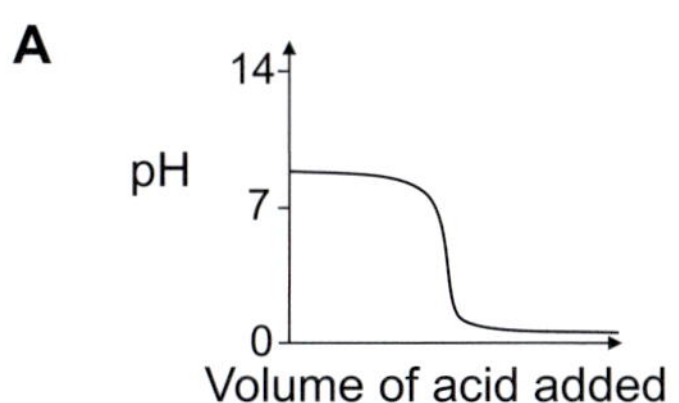

B

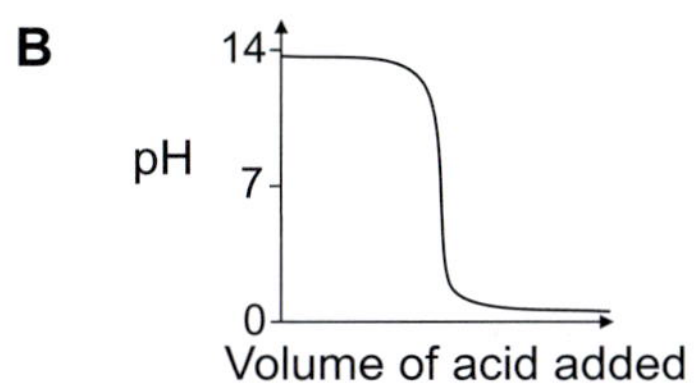

C

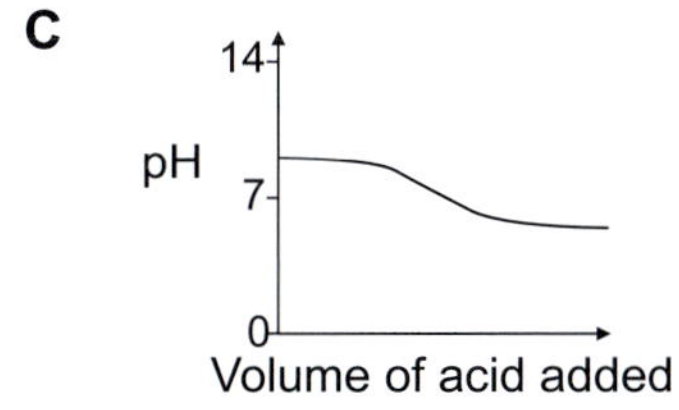

D

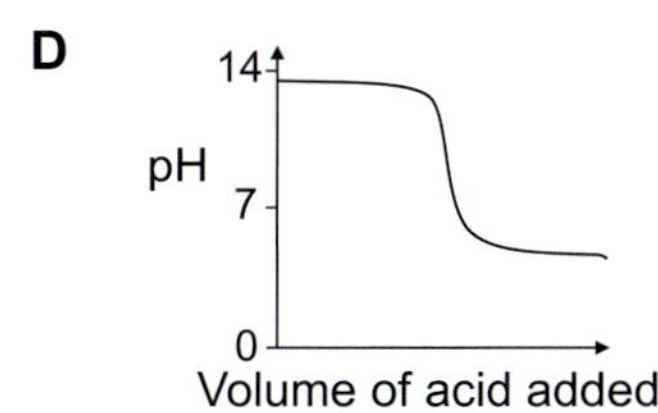

3 Thymol blue is an indicator that has a colour change at two different pH ranges.

Indicator	pH range	colour change
Thymol blue	1.2 – 2.8	red to yellow
	8.0 – 9.6	yellow to red

When titrating certain acids, it is useful to have an indicator with two colour changes. The titration of which of the following acids would benefit from using thymol blue, rather than an indicator with just one colour change?

A Nitric acid

B Methanoic acid

C Ethanedioic acid

D Hydrochloric acid

4 The concentrations of strong acids and strong bases can be found by carrying out titrations. Titrations are usually done at room temperature (25 °C). The value of K_w at 25 °C is 1.0×10^{-14} mol^2 dm^{-6}.

4.1 Give the expression for K_w.

(1 mark)

4.2 Give the expression for pH.

(1 mark)

4.3 Calculate the pH of a 0.15 mol dm^{-3} solution of NaOH at 25 °C.

(3 marks)

In a titration reaction at 25 °C, 25 cm^3 of this 0.15 mol dm^{-3} solution of NaOH was neutralised by 18.5 cm^3 of a HCl solution of unknown concentration.

4.4 Which of the graphs below (A, B and C) shows the pH curve for this reaction?

A

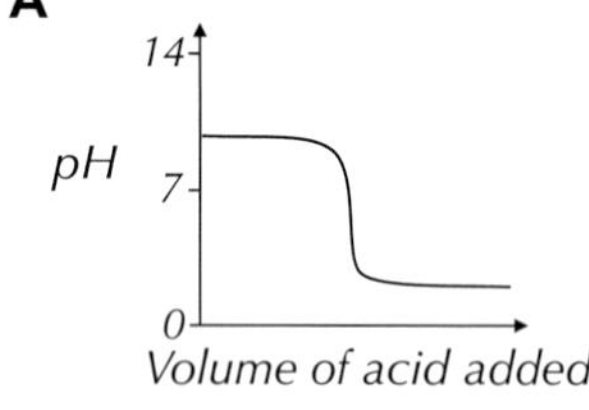

B

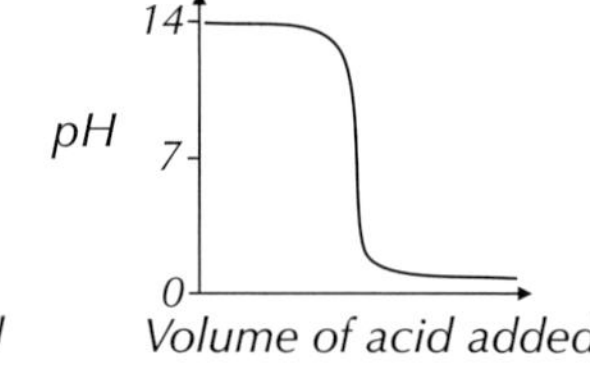

C

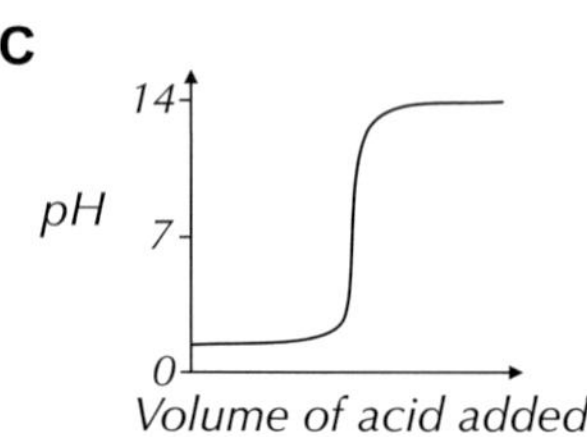

(1 mark)

4.5 Calculate the concentration of the unknown HCl solution.

(3 marks)

4.6 Calculate the pH of the HCl solution.

(2 marks)

The pH curve for another titration is shown below.

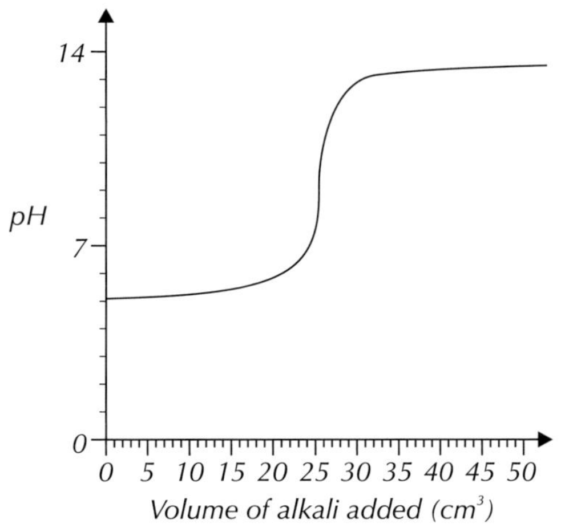

Indicator	pH range
Orange IV	1.4 – 2.6
Bromophenol blue	3.0 – 4.6
Litmus	5.0 – 8.0
Phenolphthalein	8.3 – 10

4.7 Suggest an acid and a base that could have been used in this titration.

(2 marks)

4.8 From the table above, suggest an indicator that would be suitable for this titration.

(1 mark)

5 Methanoic acid (HCOOH) is a weak acid. A 0.24 mol dm^{-3} solution of HCOOH has a pH of 2.2 at 25 °C.

5.1 Write out the equation for the dissociation of HCOOH.

(1 mark)

5.2 Write out an expression for K_a for this acid.

(1 mark)

5.3 Calculate the pK_a of methanoic acid at 25 °C.

(3 marks)

A buffer solution can be formed by mixing 30 cm^3 of this 0.24 mol dm^{-3} methanoic acid with 20 cm^3 of 0.15 mol dm^{-3} sodium hydroxide (NaOH).

5.4 Calculate the pH of this buffer.

(6 marks)

5.5 Explain how this buffer resists changes in pH when an acid is added.

(3 marks)

5.6 Suggest three substances where you could find a pH buffer.

(3 marks)

Learning Objectives:

- Know the reactions of Na and Mg with water.
- Be able to describe the trends in the reactions of the elements Na, Mg, Al, Si, P and S with oxygen, limited to the formation of Na_2O, MgO, Al_2O_3, SiO_2, P_4O_{10}, SO_2 and SO_3.

Specification Reference 3.2.4

1. Period 3 Elements

Periodicity means the trends that occur across a period of the periodic table. You've done a little bit on periodicity already (see pages 153-158), but now you need to know more about the trends which occur across Period 3.

Period 3

Period 3 is the third row in the periodic table. The elements of Period 3 are sodium (Na), magnesium (Mg), aluminium (Al), silicon (Si), phosphorus (P), sulfur (S), chlorine (Cl) and argon (Ar) — see Figure 1.

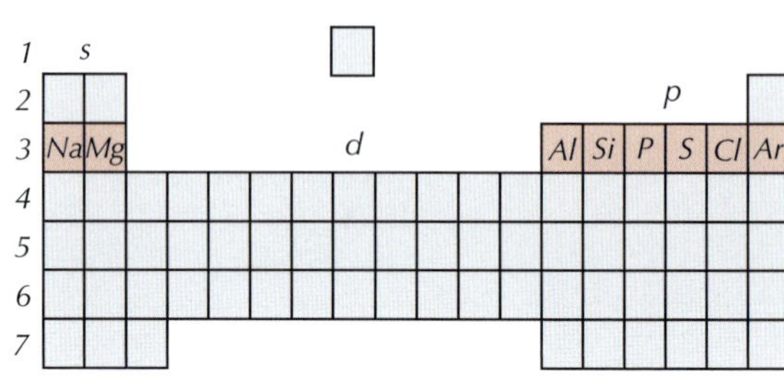

Figure 1: *Period 3 of the periodic table.*

Reactions with water

You need to know how two of the Period 3 elements — sodium and magnesium — react with water. Sodium and magnesium are the first two elements in Period 3. Sodium is in Group 1, and magnesium is in Group 2. When they react, sodium loses one electron to form an Na^+ ion, while magnesium loses two electrons to form Mg^{2+}. Sodium is more reactive than magnesium because it takes less energy to lose one electron than it does to lose two. So more energy (usually heat) is needed for magnesium to react. This is shown in their reactions with water.

Figure 2: *The reaction of sodium with water.*

Sodium

Sodium reacts vigorously with cold water, forming a molten ball on the surface, fizzing and producing H_2 gas:

$$2Na_{(s)} + 2H_2O_{(l)} \rightarrow 2NaOH_{(aq)} + H_{2(g)}$$

The reaction produces sodium hydroxide, so forms a strongly alkaline solution (pH 12 – 14).

Magnesium

Magnesium reacts very slowly with cold water. You can't see any reaction, but it forms a weakly alkaline solution (pH 9 – 10), which shows that a reaction has occurred:

$$Mg_{(s)} + 2H_2O_{(l)} \rightarrow Mg(OH)_{2(aq)} + H_{2(g)}$$

The solution is only weakly alkaline because magnesium hydroxide is not very soluble in water, so relatively few hydroxide ions are produced. Magnesium reacts much faster with steam (i.e. when there is more energy), to form magnesium oxide.

$$Mg_{(s)} + H_2O_{(g)} \rightarrow MgO_{(s)} + H_{2(g)}$$

Tip: Don't worry about the reactions of the other Period 3 elements with water. You only need to know about sodium and magnesium.

Reactions with oxygen

Period 3 elements form oxides when they react with oxygen. They're usually oxidised to their highest oxidation states — the same as their group numbers. So, sodium, which is in Group 1, has an oxidation state of +1 in sodium oxide. Magnesium (Group 2) has an oxidation state of +2 in magnesium oxide. Sulfur is the exception to this — it forms SO_2, in which it's only got a +4 oxidation state (a high temperature and a catalyst are needed to make SO_3, where sulfur has an oxidation state of +6).

The equations are all really similar — element + oxygen → oxide:

$2Na_{(s)} + \frac{1}{2}O_{2(g)} \rightarrow Na_2O_{(s)}$ sodium oxide

$Mg_{(s)} + \frac{1}{2}O_{2(g)} \rightarrow MgO_{(s)}$ magnesium oxide

$2Al_{(s)} + 1\frac{1}{2}O_{2(g)} \rightarrow Al_2O_{3(s)}$ aluminium oxide

$Si_{(s)} + O_{2(g)} \rightarrow SiO_{2(s)}$ silicon dioxide

$P_{4(s)} + 5O_{2(g)} \rightarrow P_4O_{10(s)}$ phosphorus(V) oxide

$S_{(s)} + O_{2(g)} \rightarrow SO_{2(g)}$ sulfur dioxide

The more reactive metals (Na, Mg) and the non-metals (P, S) react readily in air, while Al and Si react slowly — see the table below:

Element	Oxide	Reaction in air	Flame
Na	Na_2O	Vigorous	Yellow
Mg	MgO	Vigorous	Brilliant white
Al	Al_2O_3	Slow	N/A
Si	SiO_2	Slow	N/A
P	P_4O_{10}	Spontaneously combusts	Brilliant white
S	SO_2	Burns steadily	Blue

You can use the colours of the flames produced when the Period 3 elements react with air to identify them (e.g. if you burn a Period 3 element and it produces a blue flame, you know it's sulfur). This is known as a flame test.

Tip: Have a peek back at pages 143-145 for more on oxidation states.

Tip: SO_2 reacts with excess oxygen and a vanadium catalyst to form SO_3 (see p.399):

$SO_{2\,(g)} + \frac{1}{2}O_{2(g)} \xrightleftharpoons[\text{(cat.)}]{V_2O_5} SO_{3(g)}$

Tip: All of these oxides are white solids except for sulfur dioxide, which is a colourless gas.

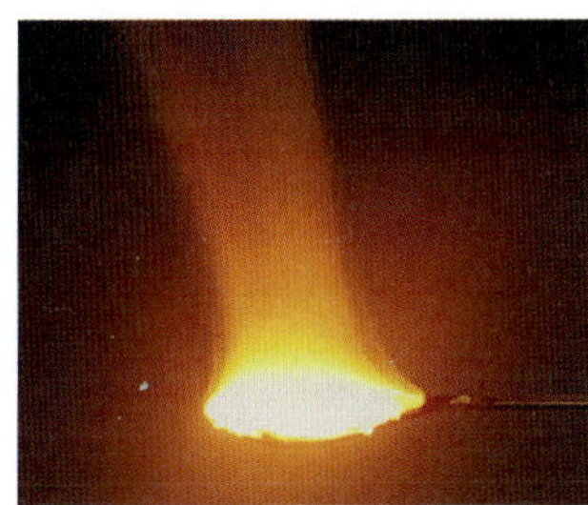

Figure 3: The yellow flame produced when sodium burns.

Figure 4: The blue flame produced when sulfur burns.

Practice Question — Application

Q1 The table shows the reactions of three Period 3 elements with oxygen:

Element	Reaction in air	Flame
A	Vigorous	Yellow
B	Vigorous	Brilliant white
C	Burns steadily	Blue

a) Identify the three elements A, B and C.

b) Write an equation for each of these reactions.

Practice Questions — Fact Recall

Q1 a) Write an equation to show the reaction of water with:

(i) sodium (ii) magnesium

b) Which of these elements reacts more vigorously with water? Why?

Q2 Write an equation to show the reaction of:

a) aluminium with oxygen. b) phosphorus with oxygen.

Learning Objectives:

- Be able to explain the trend in the melting point of the highest oxides of the elements Na-S in terms of their structure and bonding.
- Be able to describe the reactions of the oxides of the elements Na–S with water, limited to Na_2O, MgO, Al_2O_3, SiO_2, P_4O_{10}, SO_2 and SO_3, and know the pH of the solutions formed.
- Be able to explain the trends in the reactions of the oxides with water in terms of the type of bonding present in each oxide.
- Know the structures of the acids and anions formed when P_4O_{10}, SO_2 and SO_3 react with water.
- Be able to write equations for the reactions that occur between the oxides of the elements Na-S and given acids and bases.

Specification Reference 3.2.4

Exam Tip

If you're asked to explain the trend in melting points of the Period 3 oxides, make sure you mention their structure and the type of bonding.

2. Period 3 Oxides

Knowing the trends in the Period 3 elements isn't enough for A Level. You also need to know about the trends in the properties of the Period 3 oxides.

Melting points

The differences in the melting points of the Period 3 oxides are all down to differences in their structure and bonding. The trend in melting points across Period 3 is shown in Figure 1.

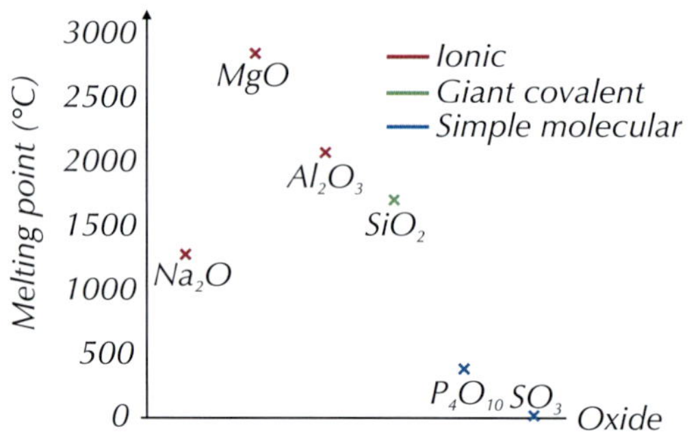

Figure 1: *The trend in melting points of the Period 3 oxides as you move across Period 3.*

Na_2O, MgO and Al_2O_3 — the metal oxides — all have high melting points because they form **giant ionic lattices**. The strong forces of attraction between each ion mean it takes a lot of heat energy to break the bonds and melt them. MgO has a higher melting point than Na_2O because magnesium forms 2+ ions, which attract O^{2-} ions more strongly than the 1+ sodium ions in Na_2O. Al_2O_3 has a lower melting point than you might expect because the difference in electronegativity between Al and O isn't as large as between Mg and O. This means that the O^{2-} ions in Al_2O_3 can't attract the electrons in the metal-oxygen bond as strongly as in MgO. This makes the bonds in Al_2O_3 partially covalent.

SiO_2 has a higher melting point than the other non-metal oxides because it has a giant **macromolecular** structure. Strong covalent bonds hold the structure together so lots of energy is needed to break the bonds and the melting temperature is high.

P_4O_{10} and SO_3 are covalent molecules. They have relatively low melting points because they form simple molecular structures. The molecules are attracted to each other by weak intermolecular forces (dipole-dipole and van der Waals), which take little energy to overcome.

Reactions with water

The ionic oxides of the metals Na and Mg both contain oxide ions (O^{2-}). When they dissolve in water, the O^{2-} ions accept protons from the water molecules to form hydroxide ions. The solutions are both alkaline, but NaOH is more soluble in water, so it forms a more alkaline solution than $Mg(OH)_2$:

$$Na_2O_{(s)} + H_2O_{(l)} \rightarrow 2NaOH_{(aq)} \qquad \text{pH 12 - 14}$$
$$MgO_{(s)} + H_2O_{(l)} \rightarrow Mg(OH)_{2(aq)} \qquad \text{pH 9 - 10}$$

The simple covalent oxides of the non-metals phosphorus and sulfur form acidic solutions. All of the acids are strong and so the pH of their solutions is about 0-1 (for solutions with a concentration of at least 1 mol dm^{-3}). They will dissociate in solution to form a conjugate base.

The reactions of some of the Period 3 oxides with water are shown in Figure 3.

Reaction of non-metal Period 3 oxide with water	Acid formed	Dissociation of Period 3 acid in water
$P_4O_{10(s)} + 6H_2O_{(l)} \rightarrow 4H_3PO_{4(aq)}$	phosphoric(V) acid	$H_3PO_{4(aq)} \rightarrow 3H^+_{(aq)} + PO_4{}^{3-}{}_{(aq)}$
$SO_{2(g)} + H_2O_{(l)} \rightarrow H_2SO_{3(aq)}$	sulfurous acid (or sulfuric(IV) acid)	$H_2SO_{3(aq)} \rightarrow 2H^+_{(aq)} + SO_3{}^{2-}{}_{(aq)}$
$SO_{3(g)} + H_2O_{(l)} \rightarrow H_2SO_{4(aq)}$	sulfuric(VI) acid	$H_2SO_{4(aq)} \rightarrow 2H^+_{(aq)} + SO_4{}^{2-}{}_{(aq)}$

Figure 3: *The acids and their conjugate bases that form when some of the Period 3 oxides react with water.*

The giant covalent structure of silicon dioxide means that it is insoluble in water. However, it will react with bases to form salts so it's classed as acidic.

Aluminium oxide, which is partially ionic and partially covalently bonded, is also insoluble in water. But, it will react with acids and bases to form salts — i.e. it can act as an acid or a base, so it is classed as **amphoteric**.

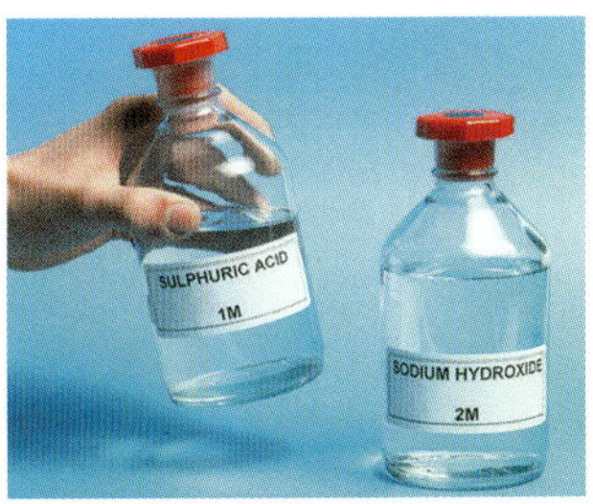

Figure 2: *NaOH and H_2SO_4 can both be made by reacting Period 3 oxides with water.*

Tip: If you're struggling to remember this, just think — SiO_2 is the main component of sand and sand definitely doesn't dissolve in water.

Tip: Amphoteric means it has the properties of an acid and a base.

Reactions with acids and bases

The equation for neutralising an acid with a base is a classic:

Acid + Base → Salt + Water

And it's no different for reactions of the Period 3 oxides. You may be asked to write equations for these reactions. Here are some examples:

Sodium and magnesium oxides are basic so will neutralise acids:

Examples

$$Na_2O_{(s)} + 2HCl_{(aq)} \rightarrow 2NaCl_{(aq)} + H_2O_{(l)}$$
$$MgO_{(s)} + H_2SO_{4(aq)} \rightarrow MgSO_{4(aq)} + H_2O_{(l)}$$

Silicon, phosphorus and sulfur oxides are acidic so will neutralise bases:

Examples

$$SiO_{2(s)} + 2NaOH_{(aq)} \rightarrow Na_2SiO_{3(aq)} + H_2O_{(l)}$$
$$P_4O_{10(s)} + 12NaOH_{(aq)} \rightarrow 4Na_3PO_{4(aq)} + 6H_2O_{(l)}$$
$$SO_{2(g)} + 2NaOH_{(aq)} \rightarrow Na_2SO_{3(aq)} + H_2O_{(l)}$$
$$SO_{3(g)} + 2NaOH_{(aq)} \rightarrow Na_2SO_{4(aq)} + H_2O_{(l)}$$

Aluminium oxides are amphoteric so can neutralise acids or bases:

Examples

$$Al_2O_{3(s)} + 3H_2SO_{4(aq)} \rightarrow Al_2(SO_4)_{3(aq)} + 3H_2O_{(l)}$$
$$Al_2O_{3(s)} + 2NaOH_{(aq)} + 3H_2O_{(l)} \rightarrow 2NaAl(OH)_{4(aq)}$$

Exam Tip
In your exam you could be asked to write an equation for any Period 3 oxide with any simple acid or base, so make sure you understand these reactions — just learning the examples won't be enough.

Practice Question — Application

Q1 The properties of three Period 3 oxides are shown in the table below:

Oxide	Structure	Melting point (°C)	Acidic/basic
A	Giant ionic lattice	2852	Basic
B	Giant ionic lattice	2045	Amphoteric
C	Macromolecular	1710	Acidic

a) Suggest the identities of the three oxides A, B and C.

b) Explain why A has a higher melting point than B.

Practice Questions — Fact Recall

Q1 a) Describe the structure and bonding of:

(i) MgO (ii) Al_2O_3 (iii) SiO_2 (iv) P_4O_{10}

b) Write equations for the reactions of oxides (i) and (iv) with water.

c) State whether the oxides (i) – (iv) are acidic, basic or amphoteric.

Q2 a) Write an equation to show the neutralisation of HCl by MgO.

b) Write an equation to show the neutralisation of NaOH by SO_2.

Section Summary

Make sure you know...

- That sodium is more reactive than magnesium because it takes less energy for sodium to lose one electron than it does for magnesium to lose two.
- How sodium and magnesium react with water (including equations and pH of the resulting solutions).
- How each of the elements in Period 3 reacts with oxygen to produce an oxide.
- The trend in melting points of the highest oxides of the Period 3 elements and that the differences in the melting points are due to differences in their structure and bonding.
- That the ionic Period 3 oxides (Na_2O/MgO) react with water to form basic hydroxides.
- That the simple covalent oxides (P_4O_{10}/SO_2/SO_3) react with water to form strong acids.
- The structures of the acids and anions formed when P_4O_{10}, SO_2 and SO_3 react with water.
- That SiO_2 is macromolecular and so is insoluble in water, but that it will react with bases — so is classified as acidic.
- That Al_2O_3 has ionic and covalent character so is insoluble in water, but that it will react with acids and bases — so is classified as amphoteric.
- How to write equations for the reactions between Period 3 oxides and simple acids and bases.

Exam-style Questions

1 Sodium and magnesium are both elements in Period 3 of the periodic table.

Sodium reacts vigorously with water but the reaction of magnesium with water is very slow.

1.1 Write an equation to show the reaction of magnesium with water.

(1 mark)

1.2 Explain why magnesium reacts less vigorously with water than sodium.

(2 marks)

Sodium and magnesium both react vigorously with oxygen to produce oxides. These oxides are structurally similar but they have very different melting temperatures.

1.3 Write an equation to show the reaction of sodium with oxygen.

(1 mark)

1.4 What colour is the flame that is produced when sodium is burnt in air?

(1 mark)

1.5 Describe the structure of sodium oxide and magnesium oxide.

(2 marks)

1.6 State which oxide has the higher melting point and explain why their melting temperatures are different.

(2 marks)

Non-metal Period 3 elements such as silicon, phosphorus and sulfur also react with oxygen to produce oxides.

1.7 Explain why sulfur dioxide has a relatively low melting point.

(2 marks)

1.8 Explain why the melting point of silicon dioxide is unusually high.

(2 marks)

2 The Period 3 oxides react differently with water.

2.1 Write equations showing the reactions of magnesium oxide and sulfur dioxide with water. Give the approximate pHs of the solutions formed when the concentration of the product of each reaction is greater than 1 mol dm^3.

(4 marks)

2.2 Name one Period 3 oxide that is insoluble in water.

(1 mark)

2.3 State which Period 3 oxide is amphoteric.

(1 mark)

2.4 Write two equations showing the reaction of this amphoteric oxide with an acid and a base.

(2 marks)

1. Transition Metals — The Basics

Learning Objectives:
- Know that transition metal characteristics of elements Ti – Cu arise from an incomplete d sub-level in atoms or ions.
- Know that these characteristics include complex formation, formation of coloured ions, catalytic activity and variable oxidation state.

Specification Reference 3.2.5.1

Some of the most precious materials in the world are transition metals. They're also responsible for some pretty interesting and important chemistry.

The d block

The **d block** is the block of elements in the middle of the periodic table. Most of the elements in the d block are **transition metals** (or transition elements). You only need to know about the transition elements in the first row of the d block (Period 4). These are the elements from titanium to copper — see Figure 1.

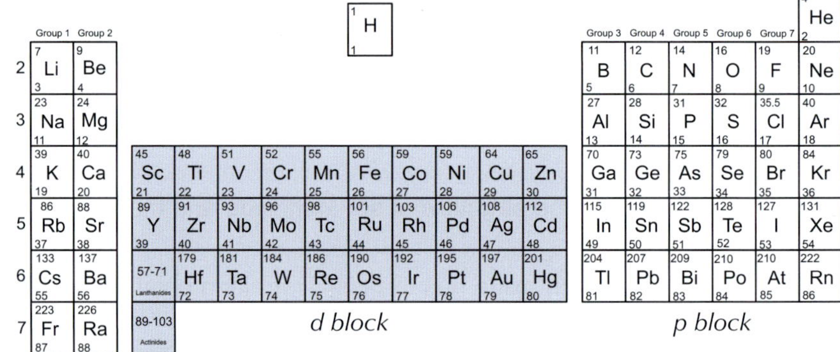

Figure 1: The three main blocks of the periodic table. The transition elements are in the d block.

Figure 2: A variety of transition metals.

What is a transition metal?

Here's the definition of a transition metal:

> A transition metal is a metal that can form one or more stable ions with an incomplete d sub-level.

A **d sub-level** can contain up to 10 electrons. So transition metals must form at least one ion that has between 1 and 9 electrons in the d sub-level. All the Period 4 d block elements are transition metals apart from scandium and zinc (p. 375).

Exam Tip
Make sure you learn the definition of a transition metal — there's a good chance you'll be asked for it in your exam.

Electron configurations

The electron configurations of elements can be figured out by following a few simple rules:

- Electrons fill up the lowest energy **sub-levels** first.
- Electrons fill **orbitals** singly before they start sharing.

The transition metals generally follow the same rules — see Figure 3. The 4s sub-level usually fills up first because it has lower energy than the 3d sub-level. Once the 4s sub-level is full, the 3d sub-level starts to fill up. The 3d orbitals are occupied singly at first. They only double up when they have to. But, there are a couple of exceptions...

Tip: Electron orbitals were covered on page 33, so if you've forgotten what a d orbital is — have a quick skim back.

- Chromium prefers to have one electron in each orbital of the 3d sub-level and just one in the 4s sub-level — this gives it more stability.
- Copper prefers to have a full 3d sub-level and just one electron in the 4s sub-level — it's more stable that way.

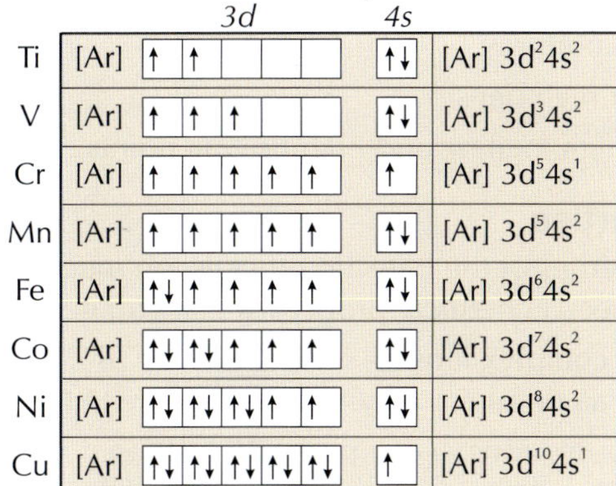

		3d	4s	
Ti	[Ar]	↑ ↑	↑↓	[Ar] $3d^2 4s^2$
V	[Ar]	↑ ↑ ↑	↑↓	[Ar] $3d^3 4s^2$
Cr	[Ar]	↑ ↑ ↑ ↑ ↑	↑	[Ar] $3d^5 4s^1$
Mn	[Ar]	↑ ↑ ↑ ↑ ↑	↑↓	[Ar] $3d^5 4s^2$
Fe	[Ar]	↑↓ ↑ ↑ ↑ ↑	↑↓	[Ar] $3d^6 4s^2$
Co	[Ar]	↑↓ ↑↓ ↑ ↑ ↑	↑↓	[Ar] $3d^7 4s^2$
Ni	[Ar]	↑↓ ↑↓ ↑↓ ↑ ↑	↑↓	[Ar] $3d^8 4s^2$
Cu	[Ar]	↑↓ ↑↓ ↑↓ ↑↓ ↑↓	↑	[Ar] $3d^{10} 4s^1$

Figure 3: *The electron configurations of the Period 4 d block transition metals.*

Transition metal ions

Transition metal atoms form positive ions. When this happens, the 4s electrons are removed first, then the 3d electrons.

Example

Iron forms Fe^{2+} ions and Fe^{3+} ions.

When it forms 2+ ions, it loses both its 4s electrons:

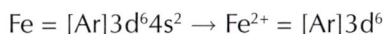

$$Fe = [Ar]3d^6 4s^2 \rightarrow Fe^{2+} = [Ar]3d^6$$

Only once the 4s electrons are removed can a 3d electron be removed.

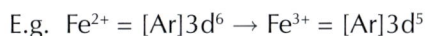

$$\text{E.g. } Fe^{2+} = [Ar]3d^6 \rightarrow Fe^{3+} = [Ar]3d^5$$

You might be asked to write the electron configuration of a transition metal ion in the exam. To do this, just follow these steps:

- Write down the electron configuration of the atom.
- Work out how many electrons have been removed to make the ion by looking at the charge on the ion.
- Remove that number of electrons from the electron configuration, taking them out of the 4s orbital first and then the 3d orbitals.

Example

Write out the electron configuration of Mn^{2+} ions.

- The electron configuration of Mn atoms is $[Ar]3d^5 4s^2$.
- Two electrons are removed to convert Mn atoms into Mn^{2+} ions.
- Removing the electrons starting from the 4s orbital gives $[Ar]3d^5 4s^0$.
- So the electron configuration of Mn^{2+} ions is $[Ar]3d^5$.

Scandium and zinc

Sc and Zn aren't transition metals as their stable ions don't have incomplete d sub-levels. Scandium only forms one ion, Sc^{3+}, which has an empty d sub-level. Scandium has the electron configuration $[Ar]3d^1 4s^2$, so when it loses three electrons to form Sc^{3+}, it ends up with the electron configuration [Ar]. Zinc only forms one stable ion, Zn^{2+}, which has a full d sub-level. Zinc has the electron configuration $[Ar]3d^{10} 4s^2$. When it forms Zn^{2+} it loses two electrons, both from the 4s sub-level. This means it keeps its full 3d sub-level.

Tip: Sub-levels are made of orbitals — for example, s sub-levels contain one orbital, p sub-levels contain three orbitals and d sub-levels contain five orbitals. Each orbital can hold two electrons.

Tip: The electron configurations of all the transition metals start in the same way — $1s^2 2s^2 2p^6 3s^2 3p^6$. This is the same as the electron configuration of the element argon, so [Ar] is used as a short way of writing it.

Exam Tip
The atomic number in the periodic table tells you how many electrons each atom has — you should be able to work the electron configuration out from there.

Tip: You can work out how many electrons have been lost using oxidation states — see page 376.

Figure 4: *A lump of zinc.*

Properties of the transition metals

Physical properties

The properties of the transition elements don't gradually change across the periodic table like you might expect. They're all typical metals and have similar physical properties.

- They all have a high density.
- They all have high melting and high boiling points.

Tip: Look back at pages 153-158 for more on periodicity.

Chemical properties

The chemical properties of the transition metals are much more interesting. They have a few special chemical properties that you need to know about:

- They can form **complex ions** — see pages 378-380.
  E.g. iron forms a complex ion with water — $[Fe(H_2O)_6]^{2+}$.
- They form coloured ions — see pages 384-385.
  E.g. Fe^{2+} ions are pale green and Fe^{3+} ions are yellow/purple.
- They're good **catalysts** — see pages 399-402.
  E.g. iron is the catalyst used in the Haber process.
- They can exist in variable **oxidation states** — see page 393.
  E.g. iron can exist in the +2 oxidation state as Fe^{2+} ions and in the +3 oxidation state as Fe^{3+} ions.

Tip: Aqueous Fe^{3+} ions are purple, but appear yellow due to a hydrolysis reaction (see pages 403-405).

Tip: All of the chemical properties of the transition elements are covered in more detail later in the section.

Some common coloured ions and oxidation states are shown below. The colours refer to the aqueous ions.

Element	Ion	Oxidation state	Colour
Ti	Ti^{2+}	+2	violet
	Ti^{3+}	+3	purple
V	V^{2+}	+2	violet
	V^{3+}	+3	green
	VO^{2+}	+4	blue
	VO_2^+	+5	yellow
Cr	Cr^{3+}	+3	green
	$Cr_2O_7^{2-}$	+6	orange
Mn	Mn^{2+}	+2	very pale pink/ colourless
	MnO_4^{2-}	+6	green
	MnO_4^-	+7	purple
Fe	Fe^{2+}	+2	pale green
	Fe^{3+}	+3	yellow/purple
Co	Co^{2+}	+2	pink
Ni	Ni^{2+}	+2	green
Cu	Cu^{2+}	+2	pale blue

Figure 6: A table showing some common stable ions of the Period 4 transition metals, their oxidation states and their colours.

Figure 5: The different colours of the aqueous transition metal ions.

These elements show variable oxidation states because the energies of the 4s and the 3d sub-levels are very similar. So different numbers of electrons can be gained or lost using fairly similar amounts of energy.

The incomplete d sub-level

It's the incomplete d sub-level that causes the special chemical properties of transition metals. d block elements without an incomplete d sub-level don't have these properties.

Example

Scandium and zinc don't form ions with incomplete d sub-levels. As a result, they don't have the same chemical properties as transition metals.

For example, they can't form complex ions, they don't form coloured ions and they can't exist in variable oxidation states.

Tip: Look back at pages 143-145 for more on oxidation states and how to find them.

Practice Questions — Application

Q1 Write out the electron configurations for the following transition metal elements:

a) V b) Co c) Mn d) Ni

Q2 Write out the electron configurations for the following transition metal ions:

a) V^{3+} b) Co^{2+} c) Mn^{2+} d) Ni^{2+}

Q3 Using electron configurations, explain why zinc is not a transition metal, despite being in the d block of the periodic table.

Exam Tip

If you're asked for a full electron configuration, you need to write it all out starting from $1s^2$ — don't use [Ar].

Practice Questions — Fact Recall

Q1 Where in the periodic table are transition elements found?

Q2 What is the definition of a transition metal?

Q3 How many electrons can a d sub-level hold?

Q4 Give two rules that are usually followed when working out electron configurations.

Q5 a) Explain why chromium has the electron configuration $[Ar]3d^54s^1$ and not $[Ar]3d^44s^2$ as you would expect.

b) Explain why copper has the electron configuration $[Ar]3d^{10}4s^1$ and not $[Ar]3d^94s^2$ as you would expect.

Q6 Give four chemical properties that all of the transition metals have in common.

Q7 What feature of transition metals causes their chemical properties?

2. Complex Ions

The ability to form complex ions is an important property of transition metals. You probably haven't come across complex ions before but the next few pages should tell you everything you need to know.

Learning Objectives:

- Know that a complex is a central metal ion surrounded by ligands.
- Know that a ligand is a molecule or ion that forms a co-ordinate bond with a transition metal by donating a pair of electrons.
- Understand that ligands can be monodentate (e.g. H_2O, NH_3 and Cl^-), bidentate (e.g. $NH_2CH_2CH_2NH_2$ and $C_2O_4^{2-}$) or multidentate (e.g. $EDTA^{4-}$).
- Know that co-ordination number is the number of co-ordinate bonds to the central metal atom or ion.
- Know that transition metal ions commonly form octahedral complexes with small ligands (e.g. H_2O and NH_3).
- Know that transition metal ions commonly form tetrahedral complexes with larger ligands (e.g. Cl^-).
- Know that square planar complexes are also formed.
- Know that Ag^+ forms the linear complex $[Ag(NH_3)_2]^+$ as used in Tollens' reagent.

Specification Reference 3.2.5.1, 3.2.5.2, 3.2.5.3

What are complex ions?

A **complex ion** is a metal ion surrounded by co-ordinately bonded ligands. A **co-ordinate bond** (or dative covalent bond) is a covalent bond in which both electrons in the shared pair come from the same atom. In a complex, they come from the ligands. So, a **ligand** is an atom, ion or molecule that donates a pair of electrons to a central metal ion.

Example

$[Cu(H_2O)_6]^{2+}$

The central metal ion is a Cu^{2+} ion and water molecules are acting as ligands. There are six water molecules that each form a co-ordinate bond with the Cu^{2+} ion:

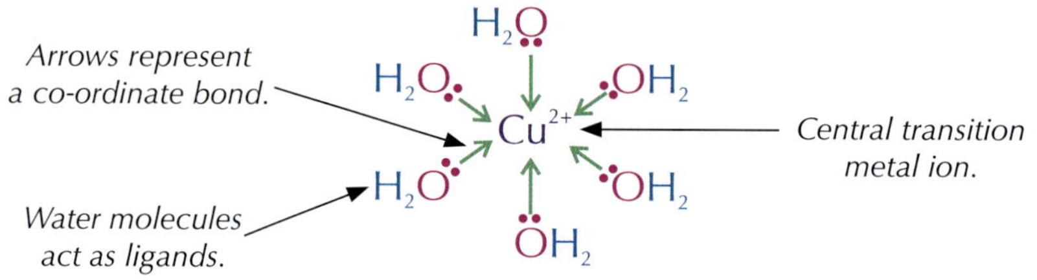

The different types of ligand

A ligand must have at least one lone pair of electrons, or it won't have anything to use to form a co-ordinate bond. But different ligands can have different numbers of lone pairs and can form different numbers of co-ordinate bonds. Ligands that can only form one co-ordinate bond are called **monodentate**.

Examples

Here are some examples of monodentate ligands:

Ammonia and chloride ions only have one lone pair of electrons to donate to form a co-ordinate bond. Water has two lone pairs of electrons but because they are so close together, it can only form one co-ordinate bond at a time.

Ligands that can form more than one co-ordinate bond are called **multidentate**.

Example

$EDTA^{4-}$ is a multidentate ligand:

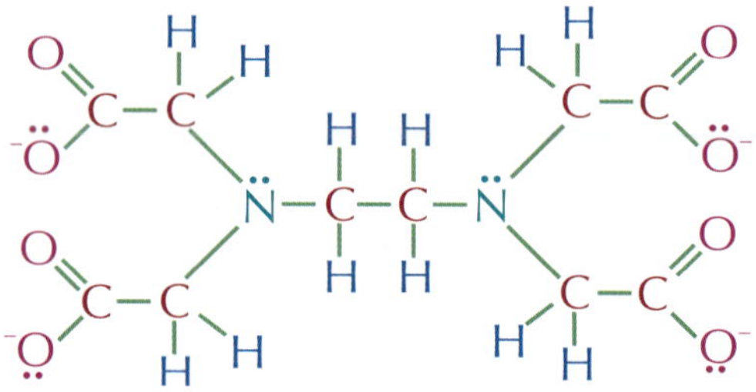

Tip: EDTA stands for ethylenediaminetetra-acetic acid.

$EDTA^{4-}$ has six lone pairs (two on nitrogen atoms and four on oxygen atoms) so it can form six co-ordinate bonds with a metal ion.

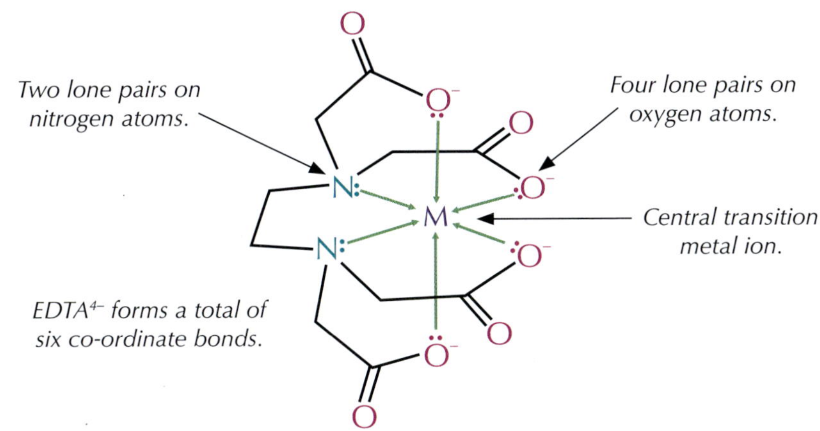
Two lone pairs on nitrogen atoms.
Four lone pairs on oxygen atoms.
Central transition metal ion.
EDTA4– forms a total of six co-ordinate bonds.
M
N:
O

Tip: $EDTA^{4-}$ is hexadentate — hexa meaning six.

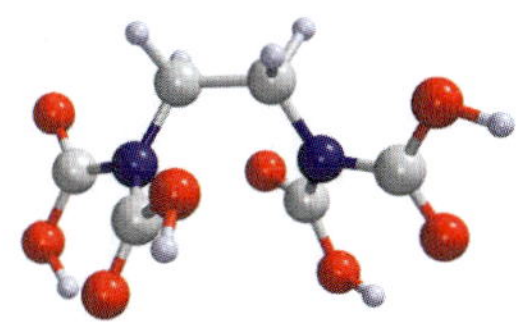

Figure 1: Molecular model of EDTA.

Multidentate ligands that can form two co-ordinate bonds are called **bidentate**.

Example

Ethane-1,2-diamine ($\ddot{N}H_2CH_2CH_2\ddot{N}H_2$) is a bidentate ligand. It has two amine groups, each of which has a lone pair of electrons that it can donate to form a co-ordinate bond. In complex ions, each ethane-1,2-diamine molecule forms two co-ordinate bonds with the metal ion. In the complex ion below, there are three ethane-1,2-diamine molecules forming six co-ordinate bonds:

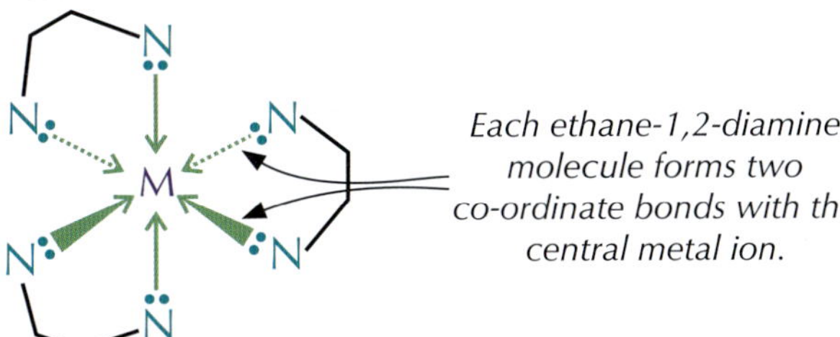
Each ethane-1,2-diamine molecule forms two co-ordinate bonds with the central metal ion.
M
N

Exam Tip
Don't worry, you'll never be expected to draw the shape of EDTA or complex ions containing EDTA in the exam.

Tip: Ethane-1,2-diamine ligands can be abbreviated to 'en'. For example, $[Cu(en)_3]^{2+}$.

Oxidation states of metals in complex ions

The overall charge on a complex ion is put outside the square brackets. For example, $[Cu(H_2O)_6]^{2+}$ has a total charge of +2. You can work out the oxidation state of the metal ion within a complex using this equation:

The oxidation state of the metal ion = The total charge of the complex – The sum of the charges of the ligands

Example

What is the oxidation state of Co in $[CoCl_4]^{2-}_{(aq)}$?

The total charge of $[CoCl_4]^{2-}_{(aq)}$ is –2 and each Cl^- ligand has an charge of –1. So in this complex:

$$\begin{aligned} \text{cobalt's oxidation state} &= \text{total charge} - \text{sum of } Cl^- \text{ charges} \\ &= (-2) - (4 \times -1) \\ &= \mathbf{+2} \end{aligned}$$

Tip: You should be familiar with oxidation states by now — but if not, have a look back at pages 143-145.

Exam Tip
You might have to work out the oxidation state of a transition metal in a complex ion in the exam, so make sure you know this equation.

Shapes of complex ions

The shape of a complex ion depends on its **co-ordination number**. This is the number of co-ordinate bonds that are formed with the central metal ion. The usual co-ordination numbers are 6 and 4. If the ligands are small, like H_2O or NH_3, 6 can fit around the central metal ion. But if the ligands are larger, like Cl^-, only 4 can fit around the central metal ion.

> **Tip:** A complex ion's co-ordination number depends on the number of bonds formed with the ligands — not the number of ligands. So a complex ion that has 3 bidentate ligands will have a co-ordination number of 6 (3 ligands each forming 2 bonds).

Six co-ordinate bonds

Complex ions that contain six co-ordinate bonds have an octahedral shape.

> **Tip:** In complex ions, the ligands don't all have to be the same — you can have mixtures of different ligands in the same complex.

Examples — Maths Skills

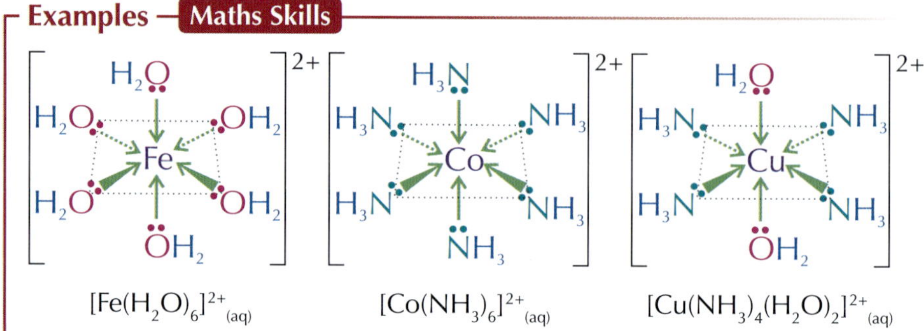

Many octahedral complex ions are hexaaqua complexes. This means there are six water ligands around the central ion.

The different types of bond arrow show that the complex is 3-D. The wedge-shaped arrows represent bonds coming towards you and the dashed arrows represent bonds sticking out behind the molecule.

Figure 2: The flask on the left shows the pink colour of hexaaqua cobalt(II) ions. The flask on the right shows the blue colour of cobalt(II) ions complexed as $[CoCl_4]^{2-}$.

Four co-ordinate bonds

Complex ions with four co-ordinate bonds usually have a tetrahedral shape.

Examples — Maths Skills

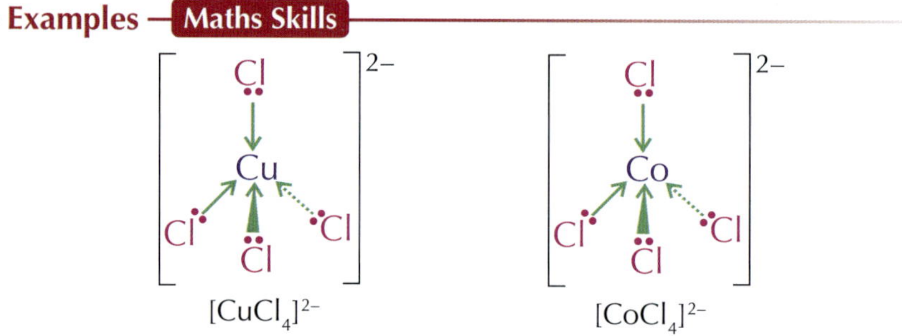

But in a few complexes four co-ordinate bonds form a square planar shape.

Example — Maths Skills

Cisplatin ($PtCl_2(NH_3)_2$) has a square planar shape:

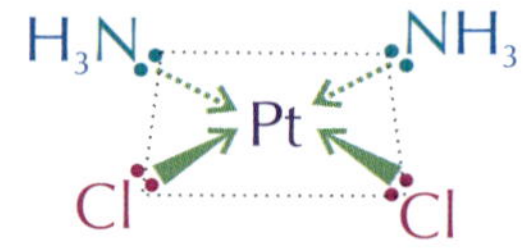

> **Tip:** Cisplatin is used as an anti-cancer drug. See page 489 for more information on this.

Two co-ordinate bonds

Some silver complexes have 2 co-ordinate bonds and form a linear shape.

Example — Maths Skills

$[Ag(NH_3)_2]^+$ forms a linear shape, as shown: $[H_3N{:}\rightarrow Ag \leftarrow{:}NH_3]^+$

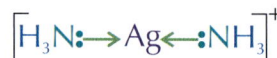

> **Tip:** $[Ag(NH_3)_2]^+$ is used in Tollens' reagent. See pages 251-252 and 394 for more.

Practice Questions — Application

Q1 What is the co-ordination number of the transition metal in each of the following complex ions:

a) $[CuF_6]^{4-}$ b) $[Ag(S_2O_3)_2]^{3-}$ c) $[CuCl_4]^{2-}$

Q2 Deduce the oxidation state of the transition metal in each of the complex ions in Q1.

Q3 Describe and draw the shape of each complex ion in Q1.

Q4 $C_2O_4^{2-}$ is a bidentate ligand. Its structure is shown below:

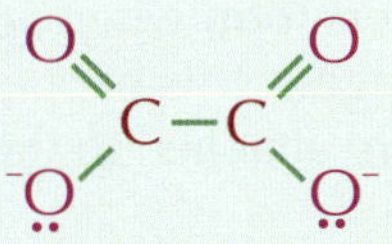

a) Copy the diagram and circle the atoms that could form a co-ordinate bond with a metal ion.

b) How many $C_2O_4^{2-}$ ligands would you expect to bind to a single Cu^{2+} ion?

Q5 $[Ni(CN)_4]^{2-}$ is a complex ion. It does not have a tetrahedral shape. What shape could it have instead? Draw the structure of $[Ni(CN)_4]^{2-}$.

Practice Questions — Fact Recall

Q1 Define the following terms:

a) Complex ion b) Co-ordinate bond c) Ligand

Q2 a) Explain what is meant by the terms monodentate, bidentate and multidentate.

b) Give an example of each type of ligand mentioned in a).

Q3 What is meant by the term co-ordination number?

3. Isomerism in Complex Ions

Complex ions can show optical and cis-trans isomerism because the ligands can be arranged in various different ways around the central metal ion.

Learning Objectives:

- Know that octahedral complexes can display optical isomerism with bidentate ligands.
- Know that octahedral and square planar complexes can display *cis-trans* isomerism (a special case of *E-Z* isomerism) with monodentate ligands.
- Know that cisplatin is a *cis* isomer.

Specification Reference 3.2.5.3

Optical isomerism in complex ions

Optical isomerism is a type of **stereoisomerism**. Stereoisomers have the same molecular formula but a different orientation of their bonds in space. For complex ions, optical isomers form when an ion can exist as two non-superimposable mirror images. This happens in octahedral complexes when three bidentate ligands are attached to the central ion. Optical isomers are also known as **enantiomers**.

Example — Maths Skills

When three ethane-1,2-diamine molecules ($H_2NCH_2CH_2NH_2$) use the lone pairs on both nitrogen atoms to co-ordinately bond with nickel, two optical isomers are formed.

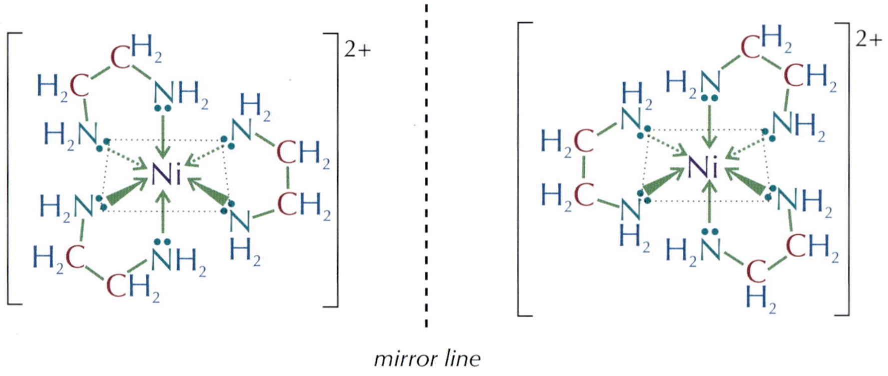

mirror line

Tip: If you're finding it difficult to see that the complexes in the example are non-superimposable mirror images of each other try building their 3-D structures using a molecular model kit (or matchsticks and modelling clay).

Cis-trans isomerism in complex ions

Cis-trans isomerism is another type of stereoisomerism.

Cis-trans isomerism in square planar complex ions

Square planar complex ions that have two pairs of ligands show *cis-trans* isomerism. When two paired ligands are directly opposite each other it's the *trans* isomer and when they're next to each other it's the *cis* isomer.

Example — Maths Skills

$NiCl_2(NH_3)_2$ has *cis* and *trans* isomers.

cis-$NiCl_2(NH_3)_2$

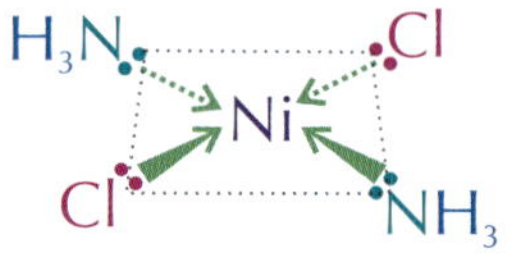

trans-$NiCl_2(NH_3)_2$

Exam Tip

If you're asked to draw a complex ion in the exam, make sure you use dashed bonds and wedged bonds to show that the structure is 3-D — you'll lose marks if you don't.

Cisplatin

Cisplatin is a complex of platinum(II) with two chloride ions and two ammonia molecules in a square planar shape (see Figure 1).

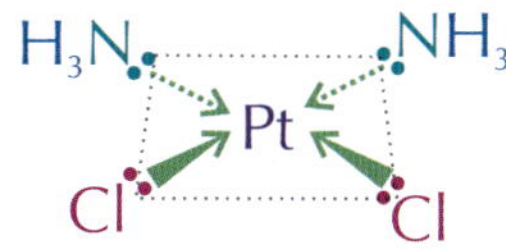

Figure 1: *The structure of cisplatin.*

Tip: If the two chloride ions were on opposite sides of the complex to each other the molecule would be transplatin. Transplatin has different biological effects to cisplatin.

Cisplatin can be used to treat some types of cancer. Cancer is caused by cells in the body dividing uncontrollably and forming tumours. Cisplatin is active against a variety of cancers, including lung and bladder cancer, because it prevents cancer cells from reproducing.

Tip: The downside is that cisplatin also prevents normal cells from reproducing — including blood and hair cells. This can cause hair loss and suppress the immune system, increasing the risk of infection. Cisplatin may also cause damage to the kidneys.

Cis-trans isomerism in octahedral complex ions

Octahedral complexes with four ligands of one type and two ligands of another type can also exhibit *cis-trans* isomerism. If the two odd ligands are opposite each other you've got the *trans* isomer, if they're next to each other then you've got the *cis* isomer.

Example — Maths Skills

$NiCl_2(H_2O)_4$ has a *trans* and a *cis* isomer.

trans-$NiCl_2(H_2O)_4$

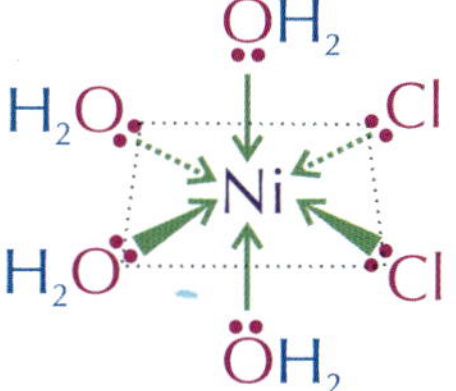

cis-$NiCl_2(H_2O)_4$

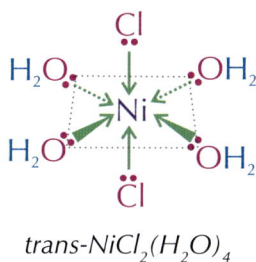

Tip: If a complex has an overall charge of zero, you don't need to put the formula in square brackets.

Practice Questions — Application

Q1 State what type of stereoisomerism the following complex ions will exhibit:

a) $PtCl_2(H_2O)_2$

b) $[Co(en)_3]^{2+}$

c) $Cu(OH)_2(H_2O)_4$

Q2 Draw the *cis* and *trans* isomers of $NiCl_2(H_2O)_2$.

Tip: If you're asked about the isomerism shown in a complex ion, sometimes it helps to draw out the structure of the ion first — then you'll get a better idea of the possible isomers.

Practice Questions — Fact Recall

Q1 Name two types of stereoisomerism that complex ions can exhibit.

Q2 Draw the structure of cisplatin.

Learning Objectives:

- Know that d electrons move from the ground state to an excited state when light is absorbed and that the energy difference between the ground state and the excited state is given by: $\Delta E = h\nu = hc/\lambda$.
- Know that transition metal ions can be identified by their colour.
- Know that colour arises when some of the wavelengths of visible light are absorbed and the remaining

e⁻ jump to higher orbitals because, when they absorb E from an external source its E and momentum changes to one that isn't the same as the E level it is in so it is forced to jump orbitals

- Know that absorption of visible light is used in spectroscopy.
- Know that a simple colorimeter can be used to determine the concentration of coloured ions in a solution.

Specification Reference 3.2.5.4

4. Formation of Coloured Ions

This section explains why different complex ions are different colours and how this can be used to identify them.

Energy levels

Normally the 3d orbitals of transition metal ions have the same energy. But when ligands bond to the ions, some orbitals are given more energy than others. This splits the 3d orbitals into different energy levels — see Figure 1.

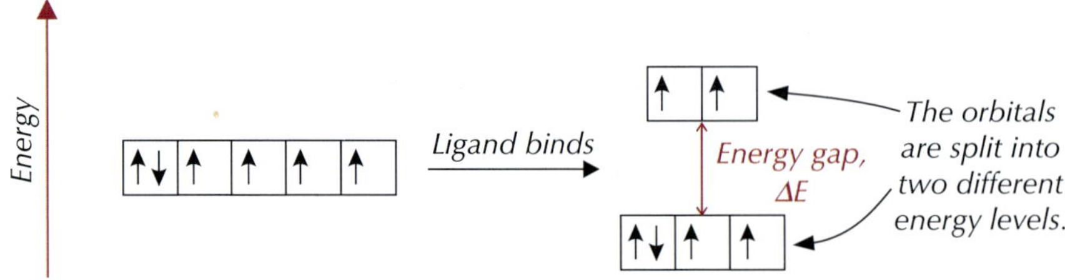

Figure 1: *The different orbital energy levels in complex ions.*

Electrons tend to occupy the lower orbitals (the ground state). To jump up to the higher orbitals (excited states) they need energy equal to the **energy gap, ΔE**. They get this energy from visible light — see Figure 2.

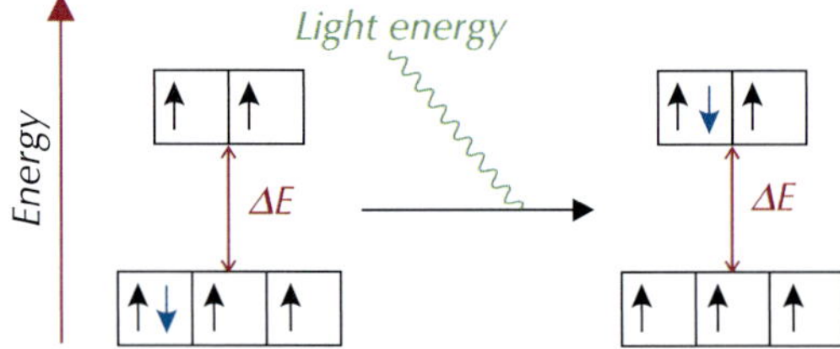

Figure 2: *The transition of an electron from the ground state to the excited state.*

The energy absorbed when electrons jump up can be worked out using this equation:

$$\Delta E = h\nu = hc/\lambda$$

Where ΔE is the energy absorbed (J), h = Planck's constant (6.63×10^{-34} Js), ν = frequency of light absorbed (hertz/Hz), c = the speed of light (3.00×10^{8} m s^{-1}) and λ = wavelength of light absorbed (m)

The amount of energy needed to make electrons jump depends upon the central metal ion and its oxidation state, the ligands, and the co-ordination number, as these affect the size of the energy gap.

The colours of compounds

When visible light hits a transition metal ion, some frequencies are absorbed as electrons jump up to the higher orbitals. The frequencies absorbed depend on the size of the energy gap (ΔE). The rest of the frequencies are reflected. These reflected frequencies combine to make the complement of the colour of the absorbed frequencies — this is the colour you see.

Example

$[Cu(H_2O)_6]^{2+}$ ions absorb yellow light. The remaining frequencies combine to produce the complementary colour — in this case that's blue. So $[Cu(H_2O)_6]^{2+}$ solution appears blue.

If there are no 3d electrons or the 3d sub-level is full, then no electrons will jump, so no energy will be absorbed. If there's no energy absorbed, the compound will look white or colourless because all the light will be reflected.

Tip: It's the incomplete d sub-level that means transition metals can absorb light. If it were empty/full there would be no electrons to jump between energy levels.

Identifying transition metal ions

It'd be nice if each transition metal formed ions or complexes with just one colour, but sadly it's not that simple. The colour of a complex can be altered by any of the factors that can affect the size of the energy gap (ΔE).

Changes in oxidation state

If the oxidation state of a transition metal in a complex ion changes, then the colour of the complex ion may also change.

Examples

Complex ions containing iron change colour when the oxidation state of iron increases from +2 to +3.

Complex:	$[Fe(H_2O)_6]^{2+}$	→	$[Fe(H_2O)_6]^{3+}$
Oxidation state:	+2	→	+3
Colour:	pale green	→	yellow

Complex ions containing vanadium change colour when the oxidation state of vanadium increases from +2 to +3.

Complex:	$[V(H_2O)_6]^{2+}$	→	$[V(H_2O)_6]^{3+}$
Oxidation state:	+2	→	+3
Colour:	violet	→	green

Tip: See page 376 for more examples of transition metals with multiple oxidation states.

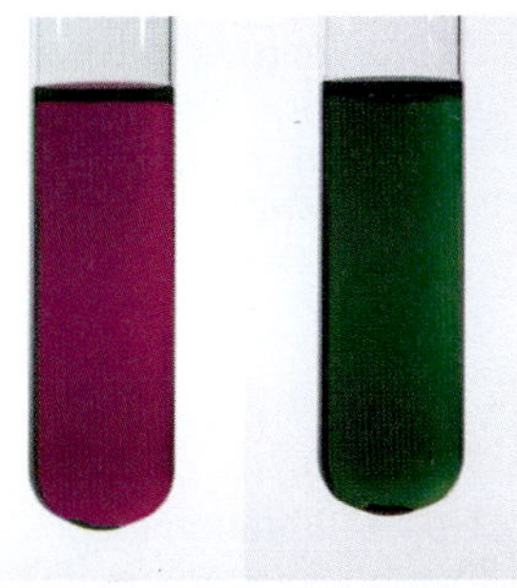

Figure 3: *The test tube on the left shows the violet colour of a V(II) solution. The test tube on the right shows the green colour of a V(III) solution.*

Changes in co-ordination number

Changes in co-ordination number may also result in a colour change. This always involves a change of ligand too.

Example

Complex ions containing Cu change colour when the co-ordination number of Cu decreases from 6 to 4.

Complex:	$[Cu(H_2O)_6]^{2+} + 4Cl^-$	→	$[CuCl_4]^{2-} + 6H_2O$
Co-ordination number:	6	→	4
Colour:	blue	→	yellow

Tip: All the colours on this page refer to aqueous solutions.

Changes in ligand

Changing the ligand can cause a colour change, even if the oxidation state and co-ordination number remain the same.

Example

If $[Co(H_2O)_6]^{2+}$ is converted to $[Co(NH_3)_6]^{2+}$ the colour will change from pink to straw coloured.

Complex:	$[Co(H_2O)_6]^{2+} + 6NH_3$	→	$[Co(NH_3)_6]^{2+} + 6H_2O$
Oxidation state:	+2	→	+2
Colour:	pink	→	straw coloured

Spectroscopy

Spectroscopy can be used to determine the concentration of a solution by measuring how much light it absorbs. White light is shone through a filter, which is chosen to only let the colour of light through that is absorbed by the sample. The light then passes through the sample to a **colorimeter**, which shows how much light was absorbed by the sample. The more concentrated a coloured solution is, the more light it will absorb.

Tip: Spectroscopy is the study of what happens when radiation interacts with matter. There are loads of different types — NMR spectroscopy is covered on pages 498-508.

So you can use this measurement to work out the concentration of a solution of transition metal ions — see Figure 5.

Figure 4: A student using a colorimeter to measure the absorbance of a solution of Cu^{2+} ions.

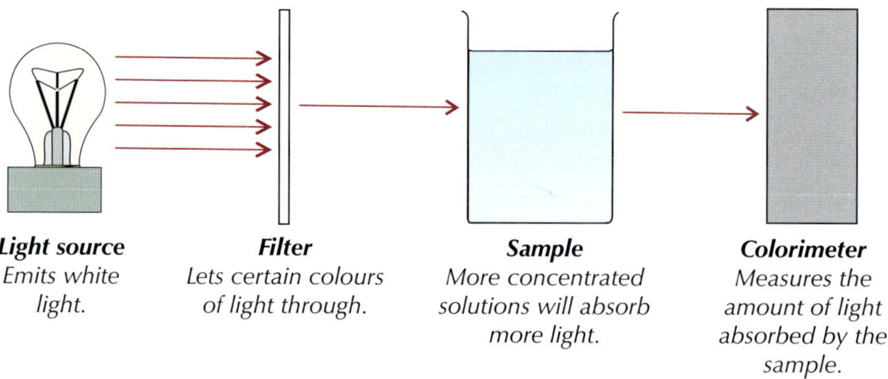
Light source
Emits white light.
Filter
Lets certain colours of light through.
Sample
More concentrated solutions will absorb more light.
Colorimeter
Measures the amount of light absorbed by the sample.

Figure 5: Measuring the concentration of a solution using spectrometry.

Before you can find the unknown concentration of a sample, you have to produce a **calibration graph** (or calibration curve) — like the lovely one below.

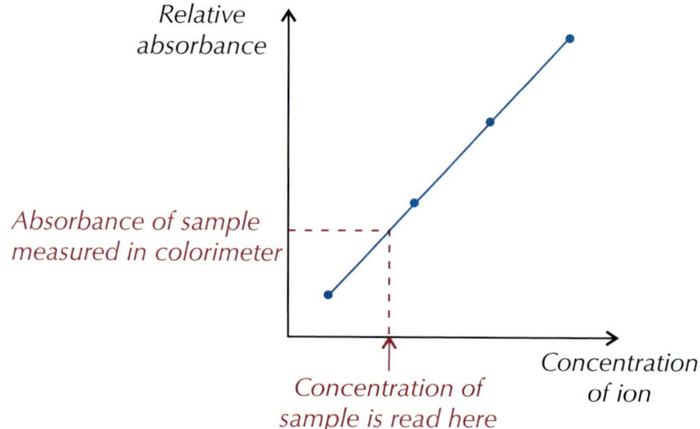
Relative absorbance
Absorbance of sample measured in colorimeter
Concentration of sample is read here
Concentration of ion

This involves measuring the absorbencies of known concentrations of solutions and plotting the results on a graph. Once you've done this, you can measure the absorbance of your sample and read its concentration off the graph.

Tip: Calibration graphs like this one are usually a straight line because, at low concentrations, the amount of light absorbed is directly proportional to the number of complex ions in the solution.

Example — Maths Skills

Say you want to find the concentration of a solution of $[Cu(H_2O)_6]^{2+}$ ions. These ions absorb yellow light so you'll need a filter that only lets yellow light through. You then need to measure the absorbance of some $[Cu(H_2O)_6]^{2+}$ ion solutions that you know the concentration of. Your results will look something like this:

Concentration (mol dm^{-3})	Relative Absorbance
0.2	0.050
0.3	0.055
0.4	0.100
0.6	0.150
1.0	0.250
1.6	0.400

Plotting these results on a graph gives you a calibration graph. Once you have your calibration graph, you measure the absorbance of the unknown solution, then read its concentration off the graph.

Tip: Different complex ions absorb different wavelengths of light so you'll need a different filter to measure the absorbance of each one.

For example, if its relative absorbance is 0.225, its concentration is around 0.9 mol dm^{-3}.

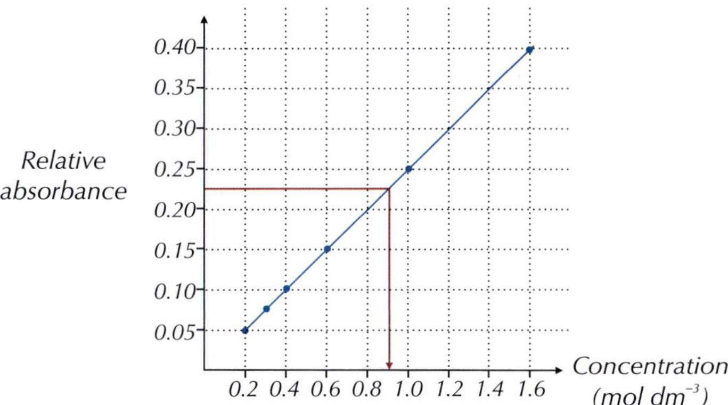

Tip: When you plot a calibration curve, concentration goes on the *x*-axis and absorbance goes on the *y*-axis.

Spectroscopy is a useful method for measuring the concentration of coloured ions because it's easy to get loads of readings and you can work out the concentration quite quickly. Plus, you can measure very low concentrations and it doesn't use up any of the substance or interfere with any reactions.

Exam Tip
For the exam you'll need to know how concentrations are determined using spectroscopy and why spectroscopy is a good method to use, so read these pages carefully.

Practice Questions — Application

Q1 When the following reaction occurs, the colour of the solution changes from violet to green.

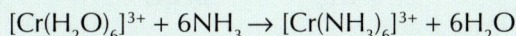

$$[Cr(H_2O)_6]^{3+} + 6NH_3 \rightarrow [Cr(NH_3)_6]^{3+} + 6H_2O$$

What causes the change in colour of the solution?

Q2 This reaction is accompanied by a green to yellow colour change:

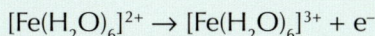

$$[Fe(H_2O)_6]^{2+} \rightarrow [Fe(H_2O)_6]^{3+} + e^-$$

What has caused the change in colour of the solution?

Q3 $[Cu(H_2O)_6]^{2+}$ ions absorb light with a wavelength of 580 nm (580×10^{-9} m). What is the energy gap (ΔE) between the split 3d orbitals of the Cu^{2+} ion?

Practice Questions — Fact Recall

Q1 What happens to the 3d orbitals of a transition metal when a ligand binds to it?

Q2 a) What type of energy is needed for an electron to jump from a lower energy orbital to a higher energy orbital?

b) Write the equation for calculating the energy absorbed (ΔE) when an electron jumps between orbitals.

Q3 Explain why transition metal ions have colour.

Q4 List three things that could cause the colour of a complex ion to change.

Q5 a) Describe how spectroscopy can be used to measure the amount of light absorbed by a solution of complex ions.

b) Explain how you would produce a calibration graph to measure the concentration of a solution of complex ions.

5. Ligand Substitution Reactions

Learning Objectives:

- Know that the exchange of similarly sized ligands (e.g. NH_3 and H_2O) occurs without change of co-ordination number.
- Know that the Cl^- ligand is larger than the uncharged ligands NH_3 and H_2O.
- Know that the exchange of the ligand H_2O by Cl^- can involve a change in co-ordination number.
- Know that substitution may be incomplete (e.g. the formation of $[Cu(NH_3)_4(H_2O)_2]^{2+}$.
- Know that haem is an iron(II) complex with a multidentate ligand.
- Know that oxygen forms a co-ordinate bond to Fe(II) in haemoglobin, enabling oxygen to be transported in the blood.
- Know that carbon monoxide is toxic because it replaces oxygen co-ordinately bonded to Fe(II) in haemoglobin.
- Know that bidentate and multidentate ligands replace monodentate ligands from complexes and that this is called the chelate effect.
- Be able to explain the chelate effect, in terms of the balance between the entropy and enthalpy changes in these reactions.

Specification Reference 3.2.5.2

Ligands around a central metal ion can switch places with other ligands in ligand substitution reactions.

Ligand substitution reactions

One ligand can be swapped for another ligand — this is **ligand substitution** (or ligand exchange). It pretty much always causes a colour change.

Substitution of similarly sized ligands

If the ligands are of similar size (e.g. H_2O and NH_3) then the co-ordination number of the complex ion doesn't change, and neither does the shape.

Example

H_2O and NH_3 ligands are similarly sized and are both uncharged. This means that H_2O ligands can be exchanged with NH_3 ligands without any change in co-ordination number or shape. There will still be a colour change due to the change of ligand.

$$[Co(H_2O)_6]^{2+}_{(aq)} + 6NH_{3(aq)} \rightarrow 6H_2O_{(l)} + [Co(NH_3)_6]^{2+}_{(aq)}$$

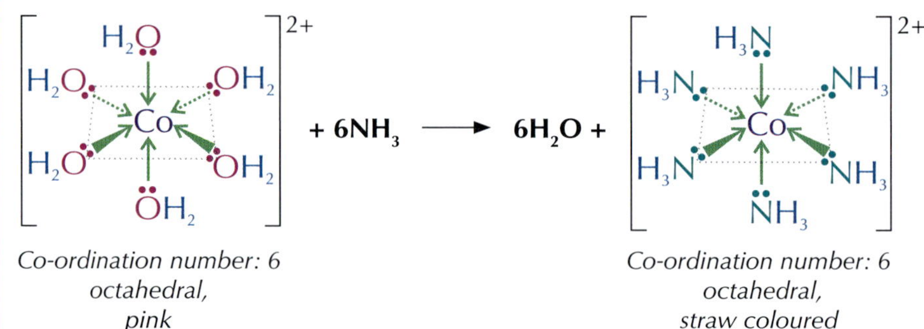

Co-ordination number: 6
octahedral,
pink

Co-ordination number: 6
octahedral,
straw coloured

Substitution of different sized ligands

If the ligands are different sizes (e.g. Cl^- is larger than H_2O and NH_3) there's a change of co-ordination number and a change of shape.

Example

In a copper-aqua complex, the H_2O ligands can be exchanged with Cl^- ligands. The shape of the complex changes from octahedral to tetrahedral because fewer of the larger Cl^- ligands can fit around the central metal ion. There is also a colour change during this reaction.

$$[Cu(H_2O)_6]^{2+}_{(aq)} + 4Cl^-_{(aq)} \rightarrow 6H_2O_{(l)} + [CuCl_4]^{2-}_{(aq)}$$

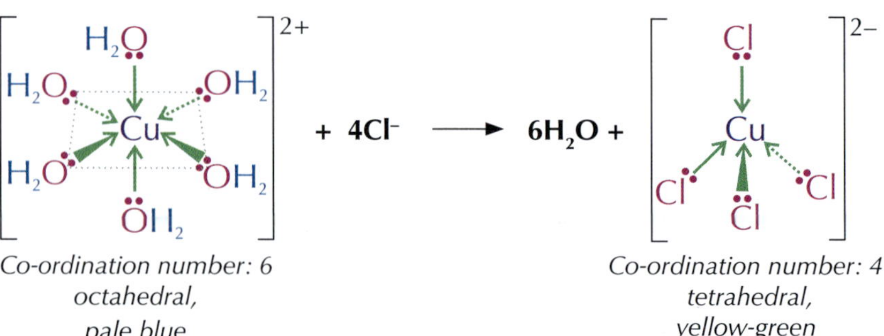

Co-ordination number: 6
octahedral,
pale blue

Co-ordination number: 4
tetrahedral,
yellow-green

Partial substitution of ligands

Sometimes the substitution is only partial — not all of the six H_2O ligands are substituted.

Example

In a copper-aqua complex, some of the H_2O ligands can be exchanged with NH_3 ligands whilst some H_2O ligands remain where they are.

In this example, four of the H_2O ligands are substituted with NH_3 ligands. The shape of the complex changes from octahedral to elongated octahedral and there is also a colour change.

$$[Cu(H_2O)_6]^{2+}_{(aq)} + 4NH_{3(aq)} \rightarrow [Cu(NH_3)_4(H_2O)_2]^{2+}_{(aq)} + 4H_2O_{(l)}$$

octahedral *pale blue* — *elongated octahedral* *deep blue*

Tip: In the example on the left you only get *trans*-$[Cu(NH_3)_4(H_2O)_2]^{2+}$ when an excess of NH_3 is added. Otherwise you get $[Cu(H_2O)_4(OH)_2]$ (a blue precipitate) instead.

Haem and haemoglobin

Haemoglobin is a protein found in blood that helps to transport oxygen around the body. It contains Fe^{2+} ions, which are hexa-co-ordinated — six lone pairs are donated to them to form six co-ordinate bonds. Four of the lone pairs come from nitrogen atoms, which form a circle around the Fe^{2+}. This part of the molecule is called **haem**. The molecule that the four nitrogen atoms are part of is a multidentate ligand called a **porphyrin**. A protein called a globin and either an oxygen or a water molecule also bind to the Fe^{2+} ion to form an octahedral structure — see Figure 2.

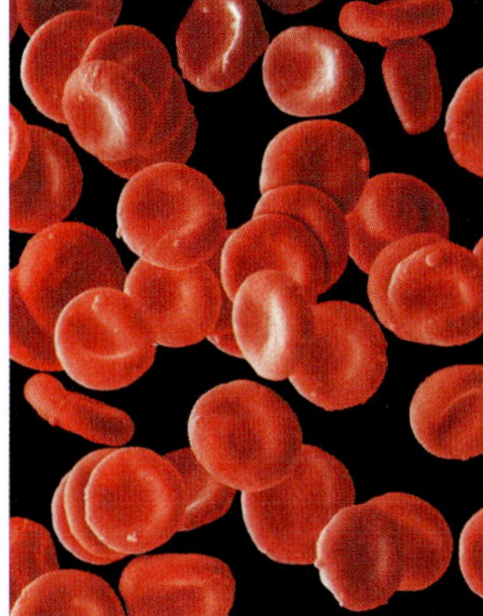

Figure 1: Red blood cells are packed full of haemoglobin.

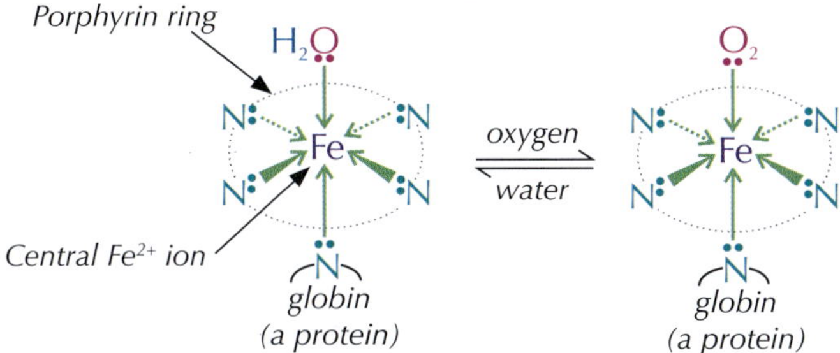

Figure 2: *The structure of haemoglobin when it is bound to oxygen or water.*

In the body, both water and oxygen will bind to the Fe^{2+} ions as ligands, so the complex can transport oxygen to where it's needed, and then swap it for a water molecule. In the lungs, where the oxygen concentration is high, water ligands are substituted for oxygen molecules to form oxyhaemoglobin, which is carried around the body in the blood. When the oxyhaemoglobin gets to a place where oxygen is needed, the oxygen molecules are exchanged for water molecules. The haemoglobin then returns to the lungs and the whole process starts again. The process of oxygen transport is summarised in Figure 4 (on the next page).

Tip: When the Fe^{2+} ion is bound to water the complex is called deoxyhaemoglobin and when it's bound to oxygen it's called oxyhaemoglobin.

Carbon monoxide poisoning

When carbon monoxide is inhaled, the haemoglobin can substitute its water ligands for carbon monoxide ligands, forming carboxyhaemoglobin (Figure 3). This is bad news because carbon monoxide forms a very strong bond with the Fe^{2+} ion and doesn't readily exchange with oxygen or water ligands, meaning the haemoglobin can't transport oxygen any more. Carbon monoxide poisoning starves the organs of oxygen — it can cause headaches, dizziness, unconsciousness and even death if it's not treated.

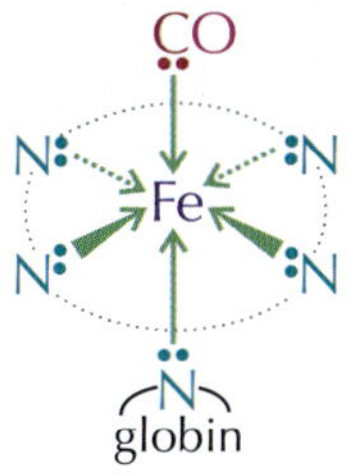

Figure 3: *Carboxyhaemoglobin.*

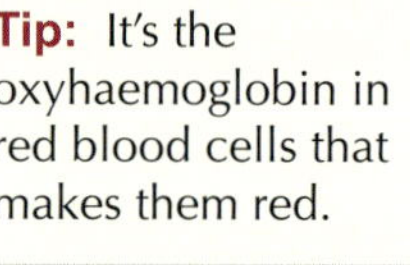
Tip: It's the oxyhaemoglobin in red blood cells that makes them red.

Summary of the oxygen transport process

Here's an overview of how haemoglobin transports oxygen round the body:

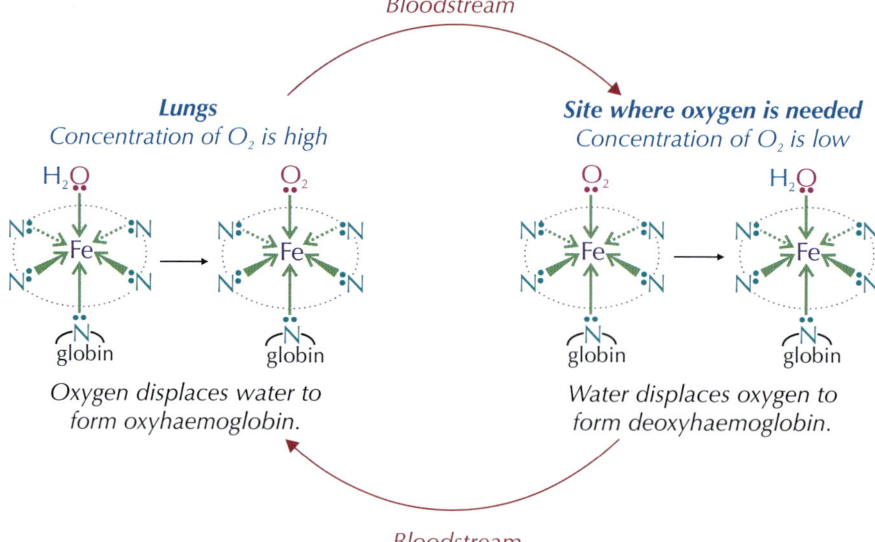

Figure 4: *The role of haemoglobin in the transport of oxygen around the body.*

Complex ion stability

Ligand exchange reactions can be easily reversed, unless the new complex ion is much more stable than the old one. If the new ligands form stronger bonds with the central metal ion than the old ligands did, the change is less easy to reverse.

Example

CN^- ions form stronger co-ordinate bonds than H_2O molecules with Fe^{3+} ions. So, the complex formed with CN^- ions will be more stable than the complex formed with H_2O molecules. This means that the substitution of water molecules by CN^- ions in an iron(III) complex is hard to reverse.

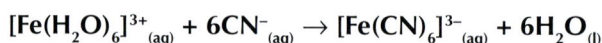

$$[Fe(H_2O)_6]^{3+}_{(aq)} + 6CN^-_{(aq)} \rightarrow [Fe(CN)_6]^{3-}_{(aq)} + 6H_2O_{(l)}$$

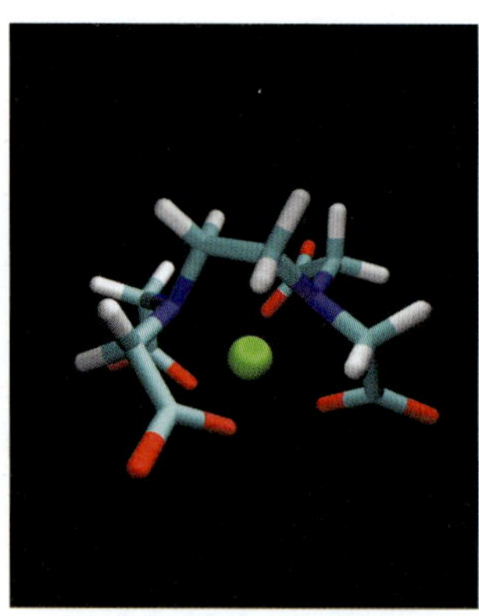

Figure 5: *$EDTA^{4-}$ bonding to a central metal ion. $EDTA^{4-}$ can bind with six bonds to the metal ion via nitrogen and oxygen atoms.*

Multidentate ligands, such as $EDTA^{4-}$, form more stable complexes than monodentate ligands, so ligand exchange reactions involving bidentate and multidentate ligands are hard to reverse.

Example

Complexes that contain the bidentate ligand ethane-1,2-diamine are more stable than those that contain water molecule ligands. So, reactions where water molecules are substituted by ethane-1,2-diamine are hard to reverse.

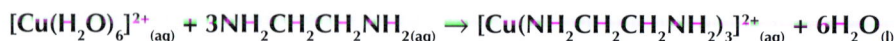

$$[Cu(H_2O)_6]^{2+}_{(aq)} + 3NH_2CH_2CH_2NH_{2(aq)} \rightarrow [Cu(NH_2CH_2CH_2NH_2)_3]^{2+}_{(aq)} + 6H_2O_{(l)}$$

Enthalpy change

When a ligand exchange reaction occurs, bonds are broken and formed. The strength of the bonds being broken is often very similar to the strength of the new bonds being made. So the enthalpy change for a ligand exchange reaction is usually very small.

Example

Substituting ammonia with ethane-1,2-diamine in a nickel complex

In this reaction, six co-ordinate bonds break between Ni and N in $[Ni(NH_3)_6]^{2+}$ and six co-ordinate bonds are formed between Ni and N in $[Ni(NH_2CH_2CH_2NH_2)_3]^{2+}$. This means the enthalpy change for the reaction is very small (only –13 kJ mol^{-1}).

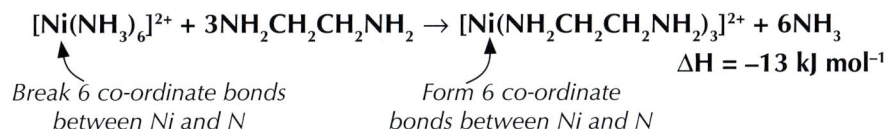

$$[Ni(NH_3)_6]^{2+} + 3NH_2CH_2CH_2NH_2 \rightarrow [Ni(NH_2CH_2CH_2NH_2)_3]^{2+} + 6NH_3 \quad \Delta H = -13 \text{ kJ mol}^{-1}$$

Break 6 co-ordinate bonds between Ni and N

Form 6 co-ordinate bonds between Ni and N

This is actually a reversible reaction, but the equilibrium lies so far to the right that it is thought of as being irreversible.

Entropy change

When monodentate ligands are substituted with bidentate or multidentate ligands, the number of particles increases — the more particles, the greater the entropy. Reactions that result in an increase in entropy are more likely to occur. So that's why multidentate ligands always form much more stable complexes than monodentate ligands. This is known as the **chelate effect**.

Exam Tip
When you're asked about the chelate effect in an exam make sure you talk about enthalpy and entropy otherwise you could lose out on valuable marks.

Example

In the example above, $[Ni(NH_2CH_2CH_2NH_2)_3]^{2+}$ is much more stable than $[Ni(NH_3)_6]^{2+}$. This isn't accounted for by an enthalpy change, but an increase in entropy can explain it.

The number of particles in the reaction increases from 4 to 7. This means that there is a large increase in entropy for the reaction and therefore the $[Ni(NH_2CH_2CH_2NH_2)_3]^{2+}$ formed in the reaction is more stable than the $[Ni(NH_3)_6]^{2+}$ complex.

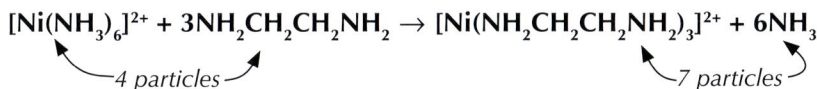

$$[Ni(NH_3)_6]^{2+} + 3NH_2CH_2CH_2NH_2 \rightarrow [Ni(NH_2CH_2CH_2NH_2)_3]^{2+} + 6NH_3$$

4 particles — *7 particles*

Tip: Entropy is covered in more detail on pages 274-275.

When the hexadentate ligand $EDTA^{4-}$ replaces monodentate or bidentate ligands, the complex formed is loads more stable. It's difficult to reverse these reactions, because reversing them would cause a decrease in the system's entropy.

Tip: There's more information on the $EDTA^{4-}$ ligand on pages 378 and 379.

Example

In the reaction below, the $EDTA^{4-}$ ligand is replacing six NH_3 ligands. This increases the number of particles in the reaction from 2 to 7.

$$[Cr(NH_3)_6]^{3+} + EDTA^{4-} \rightarrow [Cr(EDTA)]^- + 6NH_3$$

This means that the entropy of the system has greatly increased and so the $[Cr(EDTA)]^-$ complex formed will be more stable than the $[Cr(NH_3)_6]^{3+}$ complex.

Tip: The enthalpy change for this reaction is almost zero, and the entropy change is large and positive. This makes the free energy change negative ($\Delta G = \Delta H - T\Delta S$), so the reaction is feasible (see page 277).

Practice Questions — Application

Q1 The H_2O ligands in $[Fe(H_2O)_6]^{3+}$ can be exchanged for other ligands. Predict the shape of the complex ions formed after the following substitutions:

a) All the H_2O ligands exchanged for OH^- ligands.

b) The six H_2O ligands exchanged for four Cl^- ligands.

Q2 Write an equation for the formation of $[MnCl_4]^{2-}$ from a hexaaqua manganese(II) ion and chloride ions.

Q3 The metal-aqua ion $[Cu(H_2O)_6]^{2+}$ will undergo a ligand substitution reaction with excess ammonia to form $[Cu(NH_3)_4(H_2O)_2]^{2+}$.

a) Write an equation for this reaction.

b) State the shape of $[Cu(NH_3)_4(H_2O)_2]^{2+}$.

Q4 Ethane-1,2-diamine is a bidentate ligand.

a) Write an equation for the substitution of all the ammonia ligands in $[Cr(NH_3)_6]^{3+}$ with ethane-1,2-diamine ($NH_2CH_2CH_2NH_2$) ligands.

b) Explain why the new complex formed is more stable than $[Cr(NH_3)_6]^{3+}$.

Figure 6: *The reaction between $Cu^{2+}_{(aq)}$ ions and excess ammonia. A $[Cu(NH_3)_4(H_2O)_2]^{2+}$ complex ion is formed which has a deep blue colour.*

Practice Questions — Fact Recall

Q1 State whether the colour, co-ordination number and/or the shape of the complex ion changes in the following situations.

a) Ligand exchange of similarly sized ligands.

b) Ligand exchange of differently sized ligands.

Q2 a) What is the role of haemoglobin in the body?

b) Where do the six co-ordinate bonds come from in haemoglobin?

Q3 Haemoglobin is found in red blood cells.

a) Explain what happens to haemoglobin in the lungs.

b) Explain what happens to haemoglobin at sites where oxygen is needed.

Q4 Explain why carbon monoxide is toxic.

Q5 What is the chelate effect?

6. Variable Oxidation States

Transition metals can usually form more than one type of ion and, in different ions, transition metals can have different oxidation states. These variable oxidation states are another important feature of the transition metals.

Learning Objectives:

- Know that transition elements show variable oxidation states.
- Know that vanadium species in oxidation states IV, III and II are formed by the reduction of vanadate(V) ions by zinc in acidic solution.
- Know that the redox potential for a transition metal ion changing from a higher to a lower oxidation state is influenced by pH and by the ligand.
- Know that the reduction of $[Ag(NH_3)_2]^+$ (Tollens' reagent) to metallic silver is used to distinguish between aldehydes and ketones.

Specification Reference 3.2.5.5

Oxidation states of vanadium

You learnt on page 376 that one of the properties of transition metals is that they can exist in variable oxidation states. When you switch between oxidation states, it's a redox reaction — the metal ions are either oxidised or reduced. For example, vanadium can exist in four oxidation states in solution — the +2, +3, +4 and +5 states.

You can tell them apart by their different colours:

Oxidation state of vanadium	Formula of ion	Colour of ion
+5	$VO_2^+{}_{(aq)}$	Yellow
+4	$VO^{2+}{}_{(aq)}$	Blue
+3	$V^{3+}{}_{(aq)}$	Green
+2	$V^{2+}{}_{(aq)}$	Violet

Vanadium(V) can be reduced by adding it to zinc metal in an acidic solution.

Example

To begin with, the solution turns from yellow to blue as vanadium(V) is reduced to vanadium(IV).

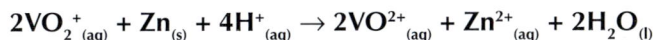

$$2VO_2^+{}_{(aq)} + Zn_{(s)} + 4H^+{}_{(aq)} \rightarrow 2VO^{2+}{}_{(aq)} + Zn^{2+}{}_{(aq)} + 2H_2O_{(l)}$$

The solution then changes colour from blue to green as vanadium(IV) is reduced to vanadium(III).

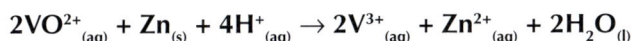

$$2VO^{2+}{}_{(aq)} + Zn_{(s)} + 4H^+{}_{(aq)} \rightarrow 2V^{3+}{}_{(aq)} + Zn^{2+}{}_{(aq)} + 2H_2O_{(l)}$$

Finally, vanadium(III) is reduced to vanadium(II), and so the solution changes from green to violet.

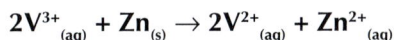

$$2V^{3+}{}_{(aq)} + Zn_{(s)} \rightarrow 2V^{2+}{}_{(aq)} + Zn^{2+}{}_{(aq)}$$

Figure 1: *From left to right — $VO_2^+{}_{(aq)}$, $VO^{2+}{}_{(aq)}$, $V^{3+}{}_{(aq)}$ and $V^{2+}{}_{(aq)}$.*

Tip: VO_2^+ is sometimes referred to as a 'vanadate(V) ion'.

Redox potentials

The **redox potential** of an ion or atom tells you how easily it is reduced to a lower oxidation state. They're the same as **electrode potentials** (see page 324). The more positive the redox potential, the less stable the ion will be and so the more likely it is to be reduced. For example, in the table below, copper(II) has a redox potential of +0.15 V, so is less stable and more likely to be reduced than chromium(III) which has a redox potential of –0.74 V.

Half equation	Standard electrode potential (V)
$Cr^{3+}{}_{(aq)} + e^- \rightleftharpoons Cr^{2+}{}_{(aq)}$	–0.74
$Cu^{2+}{}_{(aq)} + e^- \rightleftharpoons Cu^+{}_{(aq)}$	+0.15

The redox potentials in the table are standard electrode potentials — they've been measured with the reactants at a concentration of 1 mol dm^3 against a standard hydrogen electrode, under standard conditions (see page 328). The redox potential of an ion won't always be the same as its standard electrode potential. It can vary depending on the environment that the ion is in.

Ligands affect redox potential

Standard electrode potentials are measured in aqueous solution, so any ions will be surrounded by water ligands. Different ligands may make the redox potential larger or smaller, depending on how well they bind to the metal ion in a particular oxidation state.

pH affects redox potential

Some ions need H^+ ions to be present in order to be reduced.

Example

$$2VO_2^+{}_{(aq)} + 4H^+_{(aq)} + 2e^- \rightleftharpoons 2VO^{2+}_{(aq)} + 2H_2O_{(l)}$$

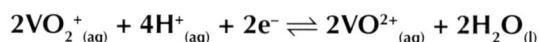

Others release OH^- ions into solution when they are reduced.

Example

$$CrO_4^{2-}{}_{(aq)} + 4H_2O_{(l)} + 3e^- \rightleftharpoons Cr(OH)_{3(s)} + 5OH^-_{(aq)}$$

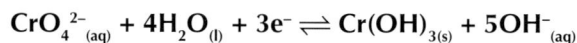

For reactions such as these, the pH of the solution affects the size of the redox potential. In general, redox potentials will be more positive in more acidic solutions, because the ion is more easily reduced.

Tollens' reagent

Silver is a transition metal that is most commonly found in the +1 oxidation state (Ag^+). It is easily reduced to silver metal.

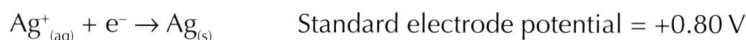

$Ag^+_{(aq)} + e^- \rightarrow Ag_{(s)}$ Standard electrode potential = +0.80 V

Tip: The standard electrode potential for this reaction is large, so Ag^+ is easily reduced.

Tollens' reagent uses this reduction reaction to distinguish between aldehydes and ketones. It's prepared by adding just enough ammonia solution to silver nitrate solution to form a colourless solution containing the complex ion $[Ag(NH_3)_2]^+$.

When added to aldehydes, Tollens' reagent reacts to give a silver mirror on the inside of the test tube. The aldehyde is oxidised to a carboxylate anion, and the Ag^+ ions are reduced to silver metal.

$$RCHO_{(aq)} + 2[Ag(NH_3)_2]^+_{(aq)} + 3OH^-_{(aq)} \rightarrow RCOO^-_{(aq)} + 2Ag_{(s)} + 4NH_{3(aq)} + 2H_2O_{(l)}$$

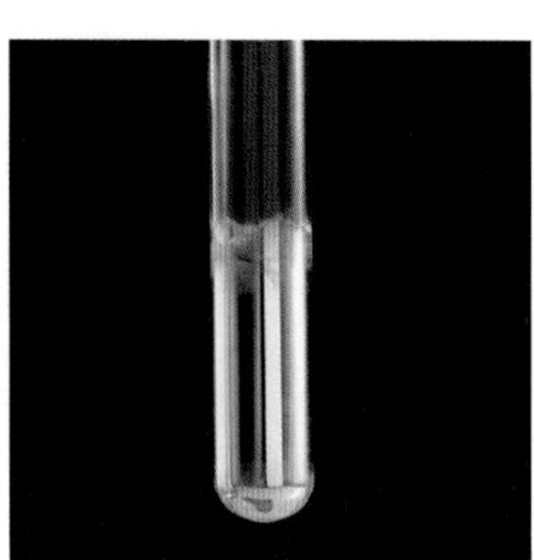

Figure 2: Tollens' reagent forms a 'silver mirror' when added to aldehydes.

Tollens' reagent can't oxidise ketones, so it won't react with them, and no silver mirror will form.

Practice Question — Application

Q1 The table below shows the redox potentials of iron(II) and manganate(VII).

Half equation	Standard electrode potential (V)
$Fe^{3+}_{(aq)} + e^- \rightleftharpoons Fe^{2+}_{(aq)}$	+0.77
$MnO^-_{4\,(aq)} + 5e^- \rightleftharpoons Mn^{2+}_{(aq)}$	+1.55

Which ion is more easily reduced? Explain your answer.

Practice Questions — Fact Recall

Q1 Write an equation for the reaction in which vanadium(III) is reduced to vanadium(II) by zinc.

Q2 Name two things that can influence the size of the redox potential for a transition metal ion.

Q3 What would you observe if you were to react Tollens' reagent with

a) an aldehyde,

b) a ketone?

Q4 What is the formula of the complex ion found in Tollens' reagent?

Learning Objectives:

- Know the redox titrations of Fe^{2+} and $C_2O_4^{2-}$ with MnO_4^-.
- Be able to perform calculations for these titrations and similar redox reactions.

Specification Reference 3.2.5.5

Tip: Titrations usually involve hazardous substances, so make sure you carry out a risk assessment first.

Tip: Titrations are covered in loads more detail on page 350.

Tip: You can also do titrations the other way round — adding the reducing agent to the oxidising agent.

Figure 1: *A student reading a burette during a titration experiment.*

7. Transition Metal Titrations

You might remember covering titrations in Unit 4. Well... you can do titrations with transition metals too. And the best bit is, you don't even need an indicator because the transition metal ions change colour all on their own.

Performing titrations

You can use titrations to find out how much oxidising agent is needed to exactly react with a quantity of reducing agent. If you know the concentration of either the oxidising agent or the reducing agent, you can use the titration results to work out the concentration of the other.

Transition metals have variable oxidation states which means they are often present in either the oxidising or reducing agent. Their colour changes also make them useful in titrations as it's easy to spot the end point.

- First you measure out a quantity of reducing agent (e.g. aqueous Fe^{2+} ions or aqueous $C_2O_4^{2-}$ ions) using a pipette, and put it in a conical flask.
- Using a measuring cylinder, you add about 20 cm^3 of dilute sulfuric acid to the flask — this is in excess, so you don't have to be too exact. The acid is added to make sure there are plenty of H^+ ions to allow the oxidising agent to be reduced.
- Now you add the oxidising agent, e.g. aqueous potassium manganate(VII), to the reducing agent using a burette, swirling the conical flask as you go.
- The oxidising agent that you add reacts with the reducing agent. This reaction will continue until all of the reducing agent is used up.
- The very next drop you add to the flask will give the mixture the colour of the oxidising agent. (You could use a coloured reducing agent and a colourless oxidising agent instead — then you'd be watching for the moment that the colour in the flask disappears.)
- Stop when the mixture in the flask just becomes tainted with the colour of the oxidising agent (the end point) and record the volume of the oxidising agent added. This is the rough titration.
- Now you do some accurate titrations. You need to do a few until you get two or more readings that are within 0.10 cm^3 of each other.

The equipment you'll need to do a titration is shown in Figure 2.

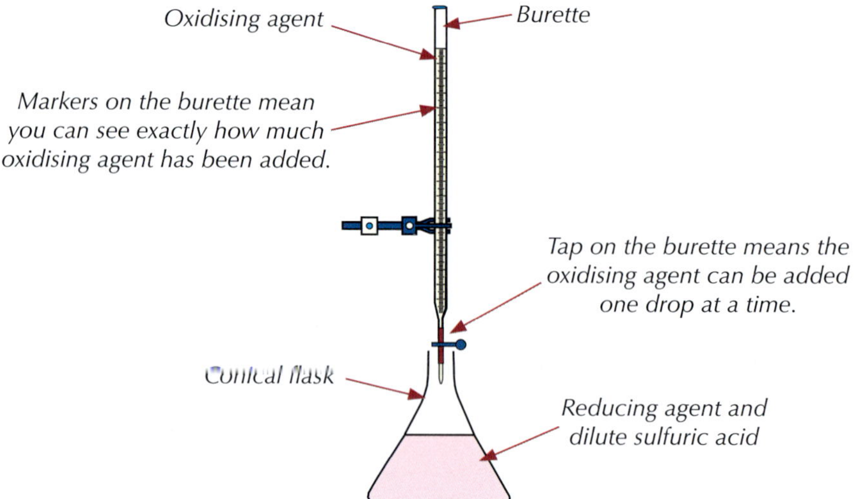

Oxidising agent
Burette
Markers on the burette mean you can see exactly how much oxidising agent has been added.
Tap on the burette means the oxidising agent can be added one drop at a time.
Conical flask
Reducing agent and dilute sulfuric acid

Figure 2: *The equipment needed to perform a titration.*

Titration reactions

You need to know the titration reactions of the reducing agents Fe^{2+} and $C_2O_4^{2-}$ ions with the oxidising agent manganate(VII), MnO_4^-, in aqueous potassium manganate(VII):

- Titration of Fe^{2+} with MnO_4^-:

$$\mathbf{MnO_4^-{}_{(aq)} + 8H^+{}_{(aq)} + 5Fe^{2+}{}_{(aq)} \rightarrow Mn^{2+}{}_{(aq)} + 4H_2O_{(l)} + 5Fe^{3+}{}_{(aq)}}$$

- Titration of $C_2O_4^{2-}$ with MnO_4^-:

$$\mathbf{2MnO_4^-{}_{(aq)} + 16H^+{}_{(aq)} + 5C_2O_4^{2-}{}_{(aq)} \rightarrow 2Mn^{2+}{}_{(aq)} + 8H_2O_{(l)} + 10CO_{2(g)}}$$

The solution turns purple at the end point of these reactions (see Figure 3).

Figure 3: *$KMnO_4$ being added to Fe^{2+} ions. The solution turns purple at the end point.*

Balancing half-equations

Before you do a titration calculation, you might need to work out the reaction equation first. You can do this by balancing half-equations.

You can always add electrons and water (H_2O) to your half-equations to balance them. If the reaction is taking place under acidic conditions, then you can also add H^+ ions. If the reaction is taking place under alkaline conditions, then you can add OH^- ions to balance your half-equation. But that's it — those are the only things that you're allowed to add.

Example — Maths Skills

Acidified manganate(VII) ions (MnO_4^-) can be reduced to Mn^{2+} by Fe^{2+} ions. Write the full equation for this reaction.

1. Iron is being oxidised. The half equation for this is: $Fe^{2+} \rightarrow Fe^{3+} + e^-$
2. Manganate is being reduced. Start by writing this down: $MnO_4^- \rightarrow Mn^{2+}$
3. To balance the oxygens, add water to the right-hand side of the equation:
$$MnO_4^- \rightarrow Mn^{2+} + 4H_2O$$
4. Add some H^+ ions to the left-hand side to balance the hydrogens:
$$MnO_4^- + 8H^+ \rightarrow Mn^{2+} + 4H_2O$$
5. Balance the charges by adding electrons:
$$MnO_4^- + 8H^+ + 5e^- \rightarrow Mn^{2+} + 4H_2O$$
6. Make sure the number of electrons released in the iron half-equation equal the number of electrons added in the manganate half-equation:
$$5Fe^{2+} \rightarrow 5Fe^{3+} + 5e^-$$
7. Finally, combine the half-equations:
$$MnO_4^- + 8H^+ + 5Fe^{2+} \rightarrow Mn^{2+} + 4H_2O + 5Fe^{3+}$$

Tip: Remember to finish by checking the charges balance — that way you know you haven't made a mistake.

Calculating the concentration of a reagent

Once you've done a titration you can use your results to calculate the concentration of either the oxidising agent or the reducing agent. To do this:

Step 1: Write out a balanced equation for the redox reaction that's happening in the conical flask.

Step 2: Decide what you know already and what you need to know — usually you'll know the two volumes and the concentration of one of the reagents and want to find the concentration of the other reagent.

Step 3: For the reagent you know both the concentration and the volume for, calculate the number of moles of this reagent using the equation:

$$\text{moles} = \frac{\text{concentration (mol dm}^{-3}) \times \text{volume (cm}^3)}{1000}$$

Step 4: Use the molar ratios in the balanced equation to find out how many moles of the other reagent were present in the solution.

Step 5: Calculate the unknown concentration using the equation:

$$\text{concentration (mol dm}^{-3}) = \frac{\text{moles} \times 1000}{\text{volume (cm}^3)}$$

Tip: $C_2O_4^{2-}$ is called ethanedioate.

Tip: These calculations are the same as those for acid-base titrations covered on page 354.

Tip: You can also work out the volume of one reagent if you know the volume of the other one and both concentrations. You follow the same steps as if you were trying to find the concentration — but at the end, you rearrange the equation to find the volume instead.

Tip: In these calculations the units of concentration should always be mol dm^{-3}.

Exam Tip
These questions are quite wordy. You might find it helpful to highlight the key bits of information in the question that you're going to use.

Example — Maths Skills

27.5 cm³ of 0.0200 mol dm⁻³ aqueous potassium manganate(VII) reacted with 12.5 cm³ of acidified sodium ethanedioate solution. Calculate the concentration of $C_2O_4^{2-}$ ions in the solution.

1. The balanced equation for this titration reaction is:

$$2MnO_4^-{}_{(aq)} + 16H^+{}_{(aq)} + 5C_2O_4^{2-}{}_{(aq)} \rightarrow 2Mn^{2+}{}_{(aq)} + 8H_2O_{(l)} + 10CO_{2(g)}$$

2. Work out the number of moles of MnO_4^- ions.

$$\text{moles } MnO_4^- = \frac{\text{conc.} \times \text{volume}}{1000} = \frac{0.0200 \times 27.5}{1000} = 5.50 \times 10^{-4} \text{ moles}$$

3. From the molar ratios in the equation, you know 2 moles of MnO_4^- reacts with 5 moles of $C_2O_4^{2-}$. So 5.50×10^{-4} moles of MnO_4^- must react with $(5.50 \times 10^{-4} \times 5) \div 2 = 1.375 \times 10^{-3}$ moles of $C_2O_4^{2-}$.

4. Calculate the concentration of $C_2O_4^{2-}$:

$$\text{conc } C_2O_4^{2-} = \frac{\text{moles} \times 1000}{\text{volume}} = \frac{(1.375 \times 10^{-3}) \times 1000}{12.5} = \mathbf{0.110 \text{ mol dm}^{-3}}$$

Practice Questions — Application

Q1 28.3 cm³ of a 0.0500 mol dm⁻³ acidified iron(II) sulfate solution reacted exactly with 30.0 cm³ of aqueous potassium manganate(VII). Calculate the concentration of the potassium manganate(VII) solution.

Q2 Aqueous potassium manganate(VII) with a concentration of 0.0750 mol dm⁻³ was used to completely react with 28.0 cm³ of a 0.600 mol dm⁻³ solution of acidified sodium ethanedioate. Calculate the volume of potassium manganate(VII) solution used.

Q3 A 0.450 mol dm⁻³ solution of acidified iron(II) sulfate completely reacted with 24.0 cm³ of a 0.0550 mol dm⁻³ solution of aqueous potassium manganate(VII). Calculate the volume of iron(II) sulfate solution used.

Practice Questions — Fact Recall

Q1 Why is acid added to the reducing agent in redox titrations?

Q2 a) Name an oxidising agent used in redox titrations.

b) What colour change would be observed if the oxidising agent from your answer to part a) was added to a colourless reducing agent?

8. Transition Metal Catalysts

A catalyst is something that speeds up the rate of a reaction by providing an alternative reaction pathway with a lower activation energy and doesn't get used up — you learnt all about them on pages 125-126. Transition metals make really good catalysts.

Why transition metals make good catalysts

Transition metals and their compounds make good **catalysts** because they can change oxidation states by gaining or losing electrons within their d orbitals. This means they can transfer electrons to speed up reactions.

Example

The **Contact Process** is used industrially to make sulfuric acid. One of the key steps in this process is:

$$SO_{2(g)} + \frac{1}{2}O_{2(g)} \xrightarrow{V_2O_{5(s)}\text{ catalyst}} SO_{3(g)}$$

Vanadium(V) oxide (V_2O_5) catalyses this reaction in two steps.
First, vanadium(V) oxide oxidises SO_2 to SO_3 and is itself reduced to vanadium(IV) oxide:

$$V_2O_5 + SO_2 \rightarrow V_2O_4 + SO_3$$

The reduced catalyst is then oxidised by oxygen gas back to its original state:

$$V_2O_4 + \frac{1}{2}O_2 \rightarrow V_2O_5$$

Vanadium(V) oxide is able to oxidise SO_2 to SO_3 because it can be reduced to vanadium(IV) oxide. It's then oxidised back to vanadium(V) oxide by oxygen ready to start all over again. If vanadium didn't have a variable oxidation state, it wouldn't be able to catalyse this reaction.

Elements in the other blocks of the periodic table (e.g. the s block) generally aren't used to catalyse redox reactions because they don't have an incomplete 3d sub-level and don't have variable oxidation states.

Heterogeneous catalysts

A **heterogeneous catalyst** is one that is in a different phase from the reactants — i.e. in a different physical state.

Examples

There are two examples of heterogeneous catalysts that you need to know about.

1. Iron, which is used in the **Haber Process** for making ammonia:

$$N_{2(g)} + 3H_{2(g)} \xrightarrow{Fe_{(s)}\text{ catalyst}} 2NH_{3(g)}$$

2. Vanadium(V) oxide that's used in the Contact Process (see above):

$$SO_{2(g)} + \frac{1}{2}O_{2(g)} \xrightarrow{V_2O_{5(s)}\text{ catalyst}} SO_{3(g)}$$

In all of these reactions, the catalyst is a solid and the reactants are gases. The gases are passed over the solid catalyst.

Learning Objectives:

- Know that transition metals and their compounds can act as catalysts.
- Understand the importance of variable oxidation states in catalysis.
- Explain, with equations, how V_2O_5 acts as a heterogeneous catalyst in the Contact Process.
- Know that a heterogeneous catalyst is in a different phase from the reactants and that the reaction occurs on the surface.
- Know that Fe is used as a heterogeneous catalyst in the Haber Process.
- Understand the use of a support medium to maximise the surface area of a catalyst and minimise the cost.
- Know that catalysts can become poisoned by impurities and consequently have reduced efficiency, and that this has a cost implication.
- Know that a homogeneous catalyst is in the same phase as the reactants.
- Know that in homogeneous catalysis, the reaction proceeds through an intermediate species.
- Explain, with equations, how Fe^{2+} ions catalyse the reaction between I^- and $S_2O_8^{2-}$, and how Mn^{2+} ions autocatalyse the reaction between $C_2O_4^{2-}$ and MnO_4^-.

Specification Reference 3.2.5.6

Use of a support medium

When a heterogeneous catalyst is used, the reaction happens on the surface of the catalyst. So increasing the surface area of the catalyst increases the number of molecules that can react at the same time, increasing the rate of the reaction. A support medium is often used to make the area of a catalyst as large as possible.

Example

Catalytic converters (which 'clean up' emissions from car engines) contain a ceramic lattice coated with a thin layer of rhodium. The rhodium acts as a catalyst helping to convert the really nasty waste gases to less harmful products. Here's one of the reactions that occurs in a catalytic converter:

$$2CO_{(g)} + 2NO_{(g)} \xrightarrow{Rh_{(s)} \text{ catalyst}} 2CO_{2(g)} + N_{2(g)}$$

The lattice structure maximises the surface area of the catalyst, making it more effective. And it minimises the cost of the catalyst because only a thin coating is needed.

Figure 1: *A catalytic converter fitted to a car.*

Catalyst poisoning

During a reaction, reactants are adsorbed onto active sites on the surfaces of heterogeneous catalysts. Impurities in the reaction mixture may also bind to the catalyst's surface and block reactants from being adsorbed. This process is called **catalyst poisoning**. Catalyst poisoning reduces the surface area of the catalyst available to the reactants, slowing down the reaction. It also increases the cost of a chemical process because less product can be made in a certain time or with a certain amount of energy. The catalyst may even need replacing or regenerating, which also costs money.

Tip: Adsorbed means stuck to the surface.

Tip: Catalytic poisoning is only a problem for heterogeneous catalysts — homogeneous catalysts aren't affected because the reaction doesn't occur on their surface.

Examples

Lead poisons the catalyst in catalytic converters:
Catalytic converters reduce harmful emissions from car engines. Lead can coat the surface of the catalyst in a catalytic converter, so vehicles that have them fitted must only be run on unleaded petrol.

Sulfur poisons the iron catalyst in the Haber process:
The hydrogen in the Haber process is produced from methane. The methane is obtained from natural gas, which contains impurities, including sulfur compounds. Any sulfur that is not removed is adsorbed onto the iron, forming iron sulfide, and stopping the iron from catalysing the reaction efficiently.

Tip: A lead compound called TEL used to be added to petrol to help engines run smoothly. It was phased out in the 80s and 90s so catalytic converters could be used (and also because lead is really poisonous).

Catalyst poisoning can be reduced by purifying the reactants. This removes many of the impurities which would otherwise poison the catalyst.

Homogeneous catalysts

Homogeneous catalysts are in the same physical state as the reactants. Usually a homogeneous catalyst is an aqueous catalyst for a reaction between two aqueous solutions.

A homogeneous catalyst works by forming an intermediate species. The reactants combine with the catalyst to make an intermediate species, which then reacts to form the products and reform the catalyst. This causes the enthalpy profile for a homogeneously catalysed reaction to have two humps in it, corresponding to the two reactions. The activation energy needed

Tip: Don't get confused between homogeneous and heterogeneous catalysts. Just remember — homo- means 'same' and homogeneous catalysts are in the same phase as the reactants.

to form the intermediates (and to form the products from the intermediates) is lower than that needed to make the products directly from the reactants (see Figure 2).

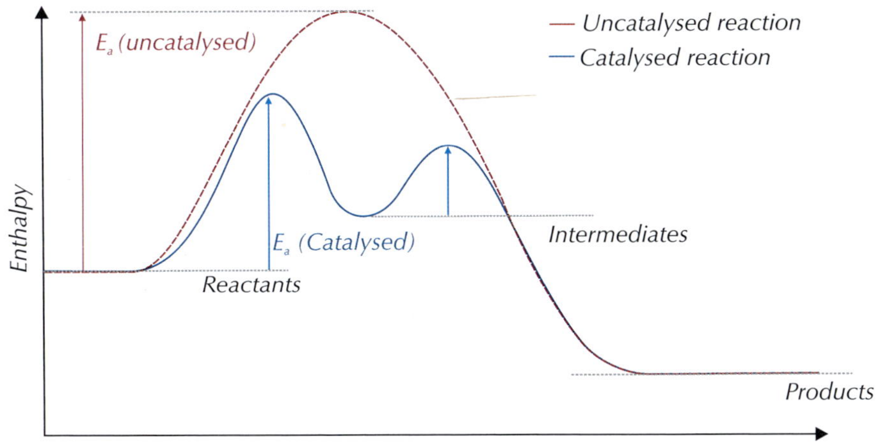

Figure 2: *Enthalpy profile for a reaction with a homogeneous catalyst. E_a represents the activation energy.*

Tip: You met enthalpy profile diagrams on page 122, so if you need a quick reminder about what it all means — have a quick look back.

The catalyst is always reformed so it can carry on catalysing the reaction. There are two reactions catalysed by homogeneous catalysts that you need to know about:

Fe^{2+} catalysing the reaction between $S_2O_8^{2-}$ and I^-

The redox reaction between iodide ions and peroxodisulfate ($S_2O_8^{2-}$) ions is shown below:

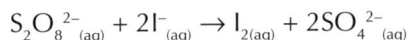

$$S_2O_8^{2-}{}_{(aq)} + 2I^-{}_{(aq)} \rightarrow I_{2(aq)} + 2SO_4^{2-}{}_{(aq)}$$

This reaction takes place annoyingly slowly because both ions are negatively charged. The ions repel each other, so it's unlikely they'll collide and react. But if Fe^{2+} ions are added, things are really speeded up because each stage of the reaction involves a positive and a negative ion, so there's no repulsion. First, the Fe^{2+} ions are oxidised to Fe^{3+} ions by the $S_2O_8^{2-}$ ions:

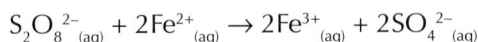

$$S_2O_8^{2-}{}_{(aq)} + 2Fe^{2+}{}_{(aq)} \rightarrow 2Fe^{3+}{}_{(aq)} + 2SO_4^{2-}{}_{(aq)}$$

The newly formed intermediate Fe^{3+} ions now easily oxidise the I^- ions to iodine, and the catalyst is regenerated:

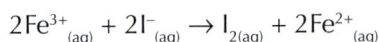

$$2Fe^{3+}{}_{(aq)} + 2I^-{}_{(aq)} \rightarrow I_{2(aq)} + 2Fe^{2+}{}_{(aq)}$$

You can test for iodine by adding starch solution — it'll turn blue-black if iodine is present — see Figure 3.

Tip: Fe^{2+}, $S_2O_8^{2-}$ and I^- are all in the aqueous phase so Fe^{2+} is a homogeneous catalyst.

Tip: The negative charges of the two ions is one reason why this reaction has high activation energy.

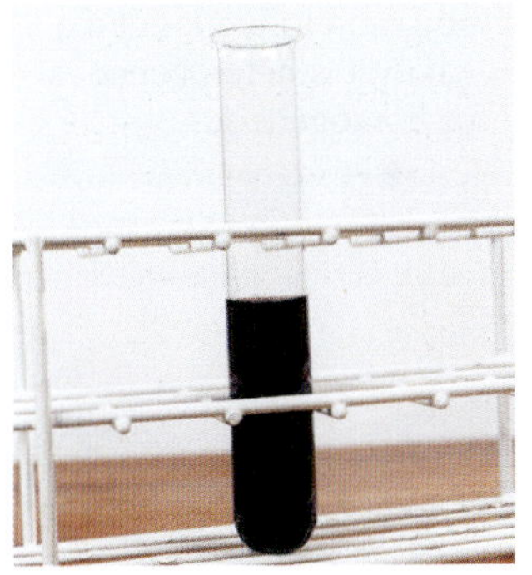

Figure 3: *A starch test for iodine. Starch solution is added and the solution turns black if iodine is present.*

Mn^{2+} autocatalysing the reaction between MnO_4^- and $C_2O_4^{2-}$

Another example of a homogeneous catalyst is Mn^{2+} in the reaction between $C_2O_4^{2-}$ and MnO_4^-. It's an **autocatalysis** reaction because Mn^{2+} is a product of the reaction and acts as a catalyst for the reaction. This means that as the reaction progresses and the amount of the product increases, the reaction speeds up. The equation for this reaction is shown below:

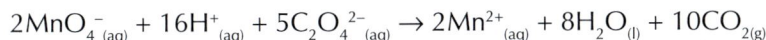

$$2MnO_4^-{}_{(aq)} + 16H^+{}_{(aq)} + 5C_2O_4^{2-}{}_{(aq)} \rightarrow 2Mn^{2+}{}_{(aq)} + 8H_2O_{(l)} + 10CO_{2(g)}$$

There isn't any Mn^{2+} present at the beginning of the reaction to catalyse it, so at first the rate of reaction is very slow. During this uncatalysed part of the reaction, the activation energy is very high. This is because the reaction proceeds via the collision of negative ions, which requires a lot of energy to achieve. But once a little bit of the Mn^{2+} catalyst has been made it reacts with the MnO_4^- ions to make Mn^{3+} ions:

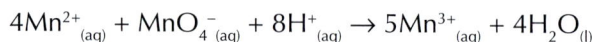

$$4Mn^{2+}_{(aq)} + MnO_{4\,(aq)}^{-} + 8H^{+}_{(aq)} \rightarrow 5Mn^{3+}_{(aq)} + 4H_2O_{(l)}$$

The Mn^{3+} ions are the intermediate. They then react with the $C_2O_4^{2-}$ ions to make CO_2 and reform the Mn^{2+} catalyst:

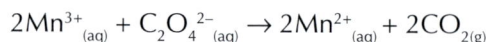

$$2Mn^{3+}_{(aq)} + C_2O_{4\,(aq)}^{2-} \rightarrow 2Mn^{2+}_{(aq)} + 2CO_{2(g)}$$

Because Mn^{2+} autocatalyses the reaction, the rate of reaction increases with time as more catalyst is made. This means a concentration-time graph for this reaction looks a bit unusual — see Figure 4.

Exam Tip
You could be asked to sketch this curve and explain its shape in the exam — so make sure you understand it.

Tip: This concentration-time graph shows the change in concentration of a reactant. If it showed the change in concentration of a product instead the shape of the graph would be reversed.

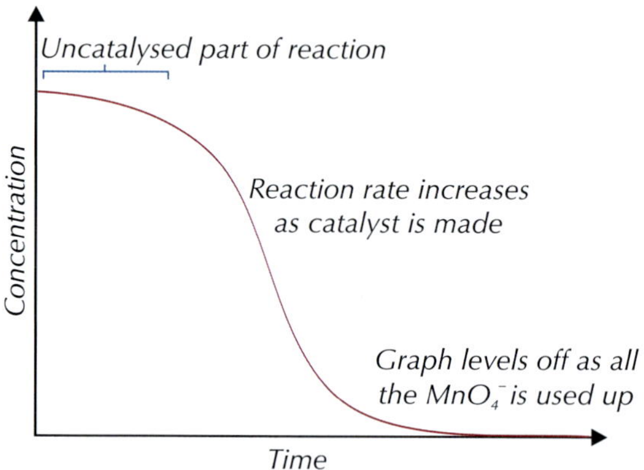

Uncatalysed part of reaction
Concentration
Reaction rate increases as catalyst is made
Graph levels off as all the MnO_4^- is used up
Time

Figure 4: *Concentration-time graph for a reaction where autocatalysis is taking place.*

Practice Question — Application

Q1 Manganese dioxide ($MnO_{2(s)}$) is used as a catalyst for the following reaction: $2H_2O_{2(l)} \rightarrow 2H_2O_{(l)} + O_{2(g)}$

a) What type of catalyst is MnO_2 in this reaction?

b) Explain why MnO_2 can act as a catalyst.

Tip: You can use the state symbols in the equation to work out if a catalyst is heterogeneous or homogeneous.

Practice Questions — Fact Recall

Q1 What are heterogeneous catalysts and homogeneous catalysts?

Q2 Give two examples of industrial processes which use a heterogeneous catalyst. For each example, name the catalyst and write an equation for the reaction it catalyses.

Q3 Why is a heterogeneous catalyst often spread over a support medium?

Q4 What is catalytic poisoning and how can it be reduced?

Q5 Explain how Fe^{2+} speeds up the rate of reaction between $S_2O_8^{2-}$ and I^-.

Q6 What is autocatalysis? Give an example of a reaction where it occurs.

9. Metal-Aqua Ions

This topic's all about complex ions and their reactions. If you've forgotten what complex ions are have a quick look back at pages 378-380.

Hydration of metal ions

Co-ordinate bonds always involve one substance donating an electron pair to another. When metal compounds dissolve in water, the water molecules form co-ordinate bonds with the metal ions. This forms **metal-aqua complex ions**. In general, six water molecules form co-ordinate bonds with each metal ion. The water molecules do this by donating a non-bonding pair of electrons from their oxygen.

Examples

Copper can form the metal-aqua ion $[Cu(H_2O)_6]^{2+}$.
Each H_2O ligand donates a lone pair of electrons to the copper ion, forming a co-ordinate bond.

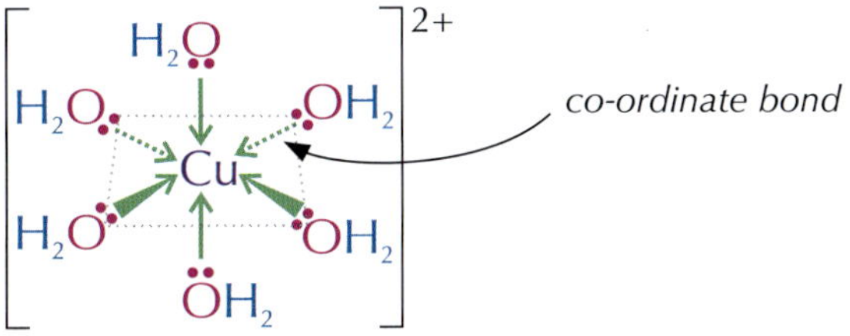

H_2O, H_2O, OH_2, H_2O, OH_2, OH_2, Cu, 2+, co-ordinate bond

$[Cu(H_2O)_6]^{2+}$

The diagrams show the metal-aqua ions formed by iron(II) — $[Fe(H_2O)_6]^{2+}$ and by aluminium — $[Al(H_2O)_6]^{3+}$.

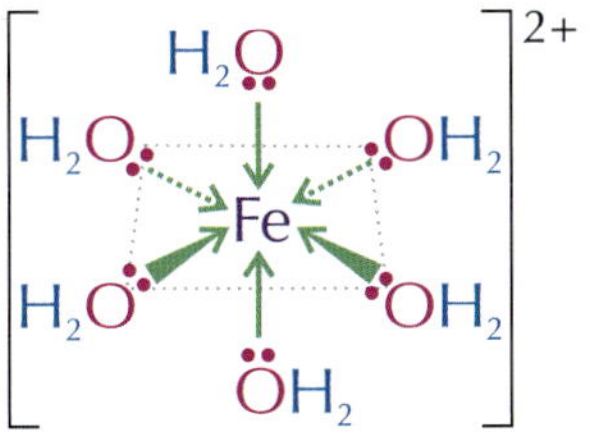

H_2O, H_2O, OH_2, H_2O, OH_2, OH_2, Fe, 2+

$[Fe(H_2O)_6]^{2+}$

H_2O, H_2O, OH_2, H_2O, OH_2, OH_2, Al, 3+

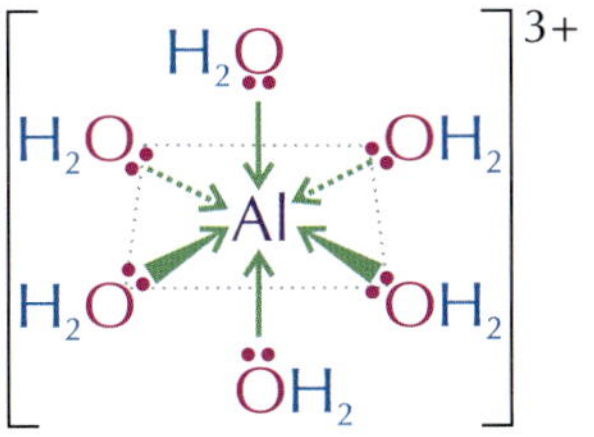

$[Al(H_2O)_6]^{3+}$

Water molecules are neutral, so the charge on the complex ion must also be the charge on the metal ion.

The acidity of metal-aqua ion solutions

In a solution containing metal-aqua 2+ ions, there's a reaction between the metal-aqua ion and the water — this is a hydrolysis or acidity reaction. The metal-aqua 2+ ions release H^+ ions, so an acidic solution is formed. There's only slight dissociation though, so the solution is only weakly acidic.

Learning Objectives:

- Know that metal-aqua ions are formed in aqueous solution: $[M(H_2O)_6]^{2+}$, limited to M = Fe and Cu, $[M(H_2O)_6]^{3+}$, limited to M = Al and Fe.
- Understand that the acidity of $[M(H_2O)_6]^{3+}$ is greater than that of $[M(H_2O)_6]^{2+}$.
- Be able to explain, in terms of the charge/size ratio of the metal ion, why the acidity of $[M(H_2O)_6]^{3+}$ is greater than that of $[M(H_2O)_6]^{2+}$.
- Know that some metal hydroxides show amphoteric character by dissolving in both acids and bases (e.g. hydroxides of Al^{3+}).
- Be able to describe and explain the simple test tube reactions of $M^{2+}_{(aq)}$ ions, limited to M = Fe and Cu, and of $M^{3+}_{(aq)}$ ions, limited to M = Al and Fe, with the bases OH^-, NH_3 and CO_3^{2-}.
- Know how to carry out simple test tube reactions to identify transition metal ions in aqueous solution. (Required Practical 11).

Specification Reference 3.2.6

Tip: Iron can form metal-aqua complexes in the oxidation states +2 and +3, so it can form the complexes $[Fe(H_2O)_6]^{2+}$ and $[Fe(H_2O)_6]^{3+}$.

Tip: In the forward reaction, the metal-aqua ion is acting as a Brønsted-Lowry acid. It donates a proton from one of its water ligands to a free water molecule.

Example

$[Fe(H_2O)_6]^{2+}$ will dissociate in water to form $[Fe(OH)(H_2O)_5]^{+}$:

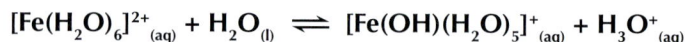

$$[Fe(H_2O)_6]^{2+}_{(aq)} + H_2O_{(l)} \rightleftharpoons [Fe(OH)(H_2O)_5]^{+}_{(aq)} + H_3O^{+}_{(aq)}$$

Metal-aqua 3+ ions react in the same way. They form more acidic solutions though.

Example

$[Al(H_2O)_6]^{3+}$ will dissociate in water to form $[Al(OH)(H_2O)_5]^{2+}$:

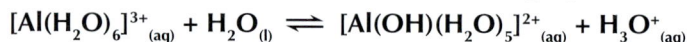

$$[Al(H_2O)_6]^{3+}_{(aq)} + H_2O_{(l)} \rightleftharpoons [Al(OH)(H_2O)_5]^{2+}_{(aq)} + H_3O^{+}_{(aq)}$$

Exam Tip

Make sure you understand why metal 3+ ions are more acidic than metal 2+ ions — it's all because of their polarising power.

Relative acidity of 2+ and 3+ metal-aqua ion solutions

Metal 3+ ions are pretty small but have a big charge — so they've got a high charge density (otherwise known as charge/size ratio). The metal 2+ ions have a much lower charge density. This makes the 3+ ions much more polarising than the 2+ ions. More polarising power means that they attract electrons from the oxygen atoms of the co-ordinated water molecules more strongly, weakening the O–H bond. So it's more likely that a hydrogen ion will be released. And more hydrogen ions means a more acidic solution — so metal 3+ ions are more acidic than metal 2+ ions.

Tip: You could also write out the equations using OH^- instead of water. This would leave you with $H_2O_{(l)}$ on the right hand side of the equation instead of $H_3O^+_{(aq)}$.

Further hydrolysis of metal-aqua ions

Adding OH^- ions to solutions of metal-aqua 3+ ions produces insoluble precipitates of metal hydroxides. Here's why:

- In water, metal-aqua 3+ ions form the equilibrium:

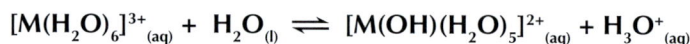

$$[M(H_2O)_6]^{3+}_{(aq)} + H_2O_{(l)} \rightleftharpoons [M(OH)(H_2O)_5]^{2+}_{(aq)} + H_3O^{+}_{(aq)}$$

  If you add OH^- ions to the equilibrium H_3O^+ ions are removed — this shifts the equilibrium to the right.

- Now another equilibrium is set up in the solution:

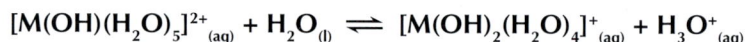

$$[M(OH)(H_2O)_5]^{2+}_{(aq)} + H_2O_{(l)} \rightleftharpoons [M(OH)_2(H_2O)_4]^{+}_{(aq)} + H_3O^{+}_{(aq)}$$

  Again the OH^- ions remove H_3O^+ ions from the solution, pulling the equilibrium to the right.

- This happens one last time — now you're left with an insoluble uncharged metal hydroxide:

$$[M(OH)_2(H_2O)_4]^{+}_{(aq)} + H_2O_{(l)} \rightleftharpoons M(OH)_3(H_2O)_{3(s)} + H_3O^{+}_{(aq)}$$

- The overall equation for this reaction is:

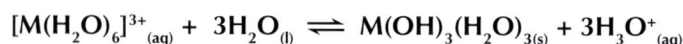

$$[M(H_2O)_6]^{3+}_{(aq)} + 3H_2O_{(l)} \rightleftharpoons M(OH)_3(H_2O)_{3(s)} + 3H_3O^{+}_{(aq)}$$

Tip: In this reaction M could be Fe or Al.

Figure 1: Metal-aqua ion precipitates. From left to right — copper(II) hydroxide, iron(II) hydroxide iron(III) hydroxide, and aluminium(III) hydroxide.

The same thing happens with metal-aqua 2+ ions (e.g. Fe or Cu), except this time there are only two steps:

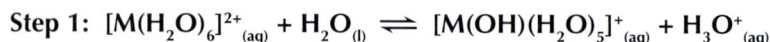

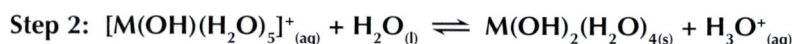

Step 1: $[M(H_2O)_6]^{2+}_{(aq)} + H_2O_{(l)} \rightleftharpoons [M(OH)(H_2O)_5]^{+}_{(aq)} + H_3O^{+}_{(aq)}$

Step 2: $[M(OH)(H_2O)_5]^{+}_{(aq)} + H_2O_{(l)} \rightleftharpoons M(OH)_2(H_2O)_{4(s)} + H_3O^{+}_{(aq)}$

Example

Adding NaOH to a solution of iron 3+ ions will produce a brown precipitate.

$$[Fe(H_2O)_6]^{3+}_{(aq)} + H_2O_{(l)} \rightleftharpoons [Fe(OH)(H_2O)_5]^{2+}_{(aq)} + H_3O^+_{(aq)}$$

$$[Fe(OH)(H_2O)_5]^{2+}_{(aq)} + H_2O_{(l)} \rightleftharpoons [Fe(OH)_2(H_2O)_4]^{+}_{(aq)} + H_3O^+_{(aq)}$$

$$[Fe(OH)_2(H_2O)_4]^{+}_{(aq)} + H_2O_{(l)} \rightleftharpoons Fe(OH)_3(H_2O)_{3(s)} + H_3O^+_{(aq)}$$

brown precipitate

Figure 2: *Brown precipitate of iron(III) hydroxide.*

All the metal hydroxide precipitates will dissolve in acid. They act as Brønsted-Lowry bases and accept H^+ ions. This reverses the hydrolysis reactions above.

Example

Adding an acid to iron(III) hydroxide will reform the soluble metal-aqua ion.

$$Fe(OH)_3(H_2O)_{3(s)} + H^+_{(aq)} \rightleftharpoons [Fe(OH)_2(H_2O)_4]^{+}_{(aq)}$$

$$[Fe(OH)_2(H_2O)_4]^{+}_{(aq)} + H^+_{(aq)} \rightleftharpoons [Fe(OH)(H_2O)_5]^{2+}_{(aq)}$$

$$[Fe(OH)(H_2O)_5]^{2+}_{(aq)} + H^+_{(aq)} \rightleftharpoons [Fe(H_2O)_6]^{3+}_{(aq)}$$

Figure 3: *Limonitic rocks. The rocks contain a large amount of hydrated iron(III) hydroxide, which gives them their distinctive reddish brown colour.*

Some metal hydroxides are **amphoteric** — they can act as both acids and bases. This means they'll dissolve in an excess of base as well as in acids.

Example

Aluminium hydroxide is amphoteric. It acts as a Brønsted-Lowry acid and donates H^+ ions to the OH^- ions:

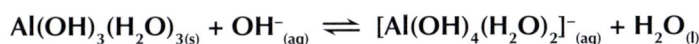

$$Al(OH)_3(H_2O)_{3(s)} + OH^-_{(aq)} \rightleftharpoons [Al(OH)_4(H_2O)_2]^-_{(aq)} + H_2O_{(l)}$$

It can also act as a Brønsted-Lowry base and accept H^+ ions:

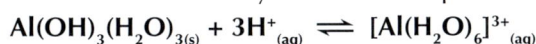

$$Al(OH)_3(H_2O)_{3(s)} + 3H^+_{(aq)} \rightleftharpoons [Al(H_2O)_6]^{3+}_{(aq)}$$

Hydrolysis of metal-aqua ions using NH_3

The obvious way of adding hydroxide ions is to use a strong alkali, like sodium hydroxide solution — but you can use ammonia solution too. When ammonia dissolves in water it can accept protons from the water molecules to form NH_4^+ ions and OH^- ligands. This gives the same results as adding NaOH.

Example

$[Fe(H_2O)_6]^{3+}_{(aq)}$ will react with ammonia to form a $Fe(OH)_3(H_2O)_3$ complex.

$$[Fe(H_2O)_6]^{3+}_{(aq)} + 3NH_{3(aq)} \rightleftharpoons Fe(OH)_3(H_2O)_{3(s)} + 3NH_4^+{}_{(aq)}$$

yellow solution — *brown precipitate*

Tip: When Fe^{3+} ions are surrounded by 6 water ligands, they're purple. But the solution may appear yellow because one of the water ligands hydrolyses, forming some $[Fe(OH)(H_2O)_5]^{2+}$ (see previous page).

In some cases, a further reaction happens if you add an excess of ammonia solution — the H_2O and OH^- ligands are displaced by NH_3 ligands. This will happen with Cu^{2+} complexes.

Example

Copper(II) hydroxide will react with excess ammonia to form a $[Cu(NH_3)_4(H_2O)_2]^{2+}$ complex.

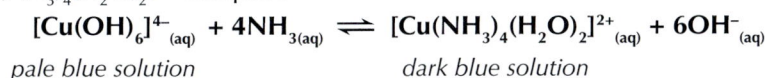

$$[Cu(OH)_6]^{4-}_{(aq)} + 4NH_{3(aq)} \rightleftharpoons [Cu(NH_3)_4(H_2O)_2]^{2+}_{(aq)} + 6OH^-_{(aq)}$$

pale blue solution — *dark blue solution*

Exam Tip

Make sure you remember that it's Cu(II) ions that will react with excess ammonia — Fe(II), Fe(III), and Al(III) ions won't.

Figure 4: Copper(II) carbonate. A green-blue precipitate is formed when carbonate ions react with aqueous copper(II) ions.

Hydrolysis of metal-aqua ions using Na_2CO_3

Metal 2+ ions react with sodium carbonate to form insoluble metal carbonates, like this:

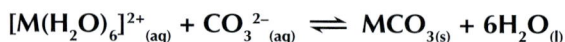

$$[M(H_2O)_6]^{2+}_{(aq)} + CO_3^{2-}{}_{(aq)} \rightleftharpoons MCO_{3(s)} + 6H_2O_{(l)}$$

But, metal 3+ ions don't form $M_2(CO_3)_3$ species when you react them with sodium carbonate. They are stronger acids (see page 404) so they always form hydroxide precipitates instead. The carbonate ions react with the H_3O^+ ions, removing them from the solution and forming bubbles of carbon dioxide gas.

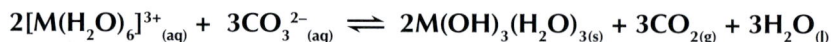

$$2[M(H_2O)_6]^{3+}_{(aq)} + 3CO_3^{2-}{}_{(aq)} \rightleftharpoons 2M(OH)_3(H_2O)_{3(s)} + 3CO_{2(g)} + 3H_2O_{(l)}$$

Tests to identify metal ions

REQUIRED PRACTICAL 11

Test tube reactions provide a qualitative way of working out the identity of unknown metal ions in solution. Adding different reagents, such as sodium hydroxide, ammonia and sodium carbonate, to separate samples of a metal ion solution and recording what you see should help you identify what metal ion is present. Here's what you should do:

- Measure out samples of the unknown metal ion solution into three separate test tubes.
- To the first test tube, add sodium hydroxide solution dropwise using a dropping pipette and record any changes you see. Then add more NaOH dropwise so that it is in excess. Record any changes.
- To the second test tube, add ammonia solution dropwise using a dropping pipette and record any changes you see. Keep adding ammonia so that it's in excess. Record any changes.
- To the third test tube, add sodium carbonate solution dropwise. Record your observations.

Tip: If you're carrying out this practical, don't forget to do a risk assessment. Some of the solutions may irritate your skin and eyes, so make sure you wear gloves, a lab coat and goggles. Ammonia is poisonous if you breathe it in, so it's best to use it in a fume cupboard.

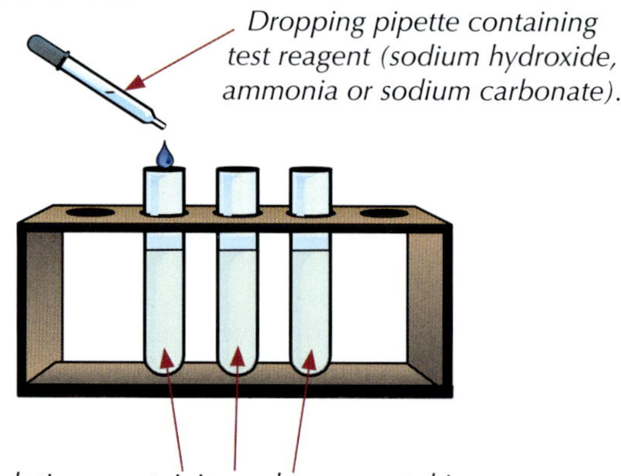

Figure 5: The equipment needed to perform metal ion tests.

On pages 403-406, you learnt about some of the reactions of copper(II), iron(II), iron(III) and aluminium(III) aqua ions. If you had four unknown solutions, each of which contained one of these metal aqua ions, you could use the method above and the differences in their reactions to distinguish between them. Here's how:

Tip: In real life, the colours of the solutions would give you a big clue. In the exam, they might tell you the observations from various reactions, but not the initial colours.

Reactions with sodium hydroxide

All four metal-aqua ions will form precipitates with sodium hydroxide, but only the aluminium hydroxide precipitate will dissolve in an excess of sodium hydroxide. This is because it's amphoteric (see page 405).

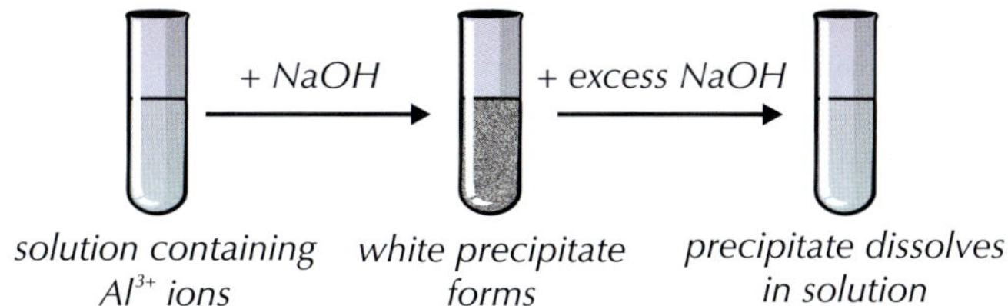

Reactions with ammonia

All four metal-aqua ions will form precipitates with ammonia, but only the copper hydroxide precipitate will dissolve in excess ammonia. This is because it undergoes a ligand exchange reaction with excess ammonia (see page 388).

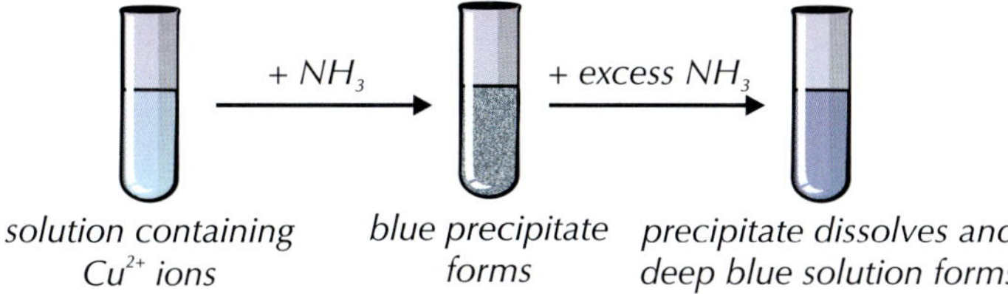

Reactions with sodium carbonate

All four metal-aqua ions will form precipitates with sodium carbonate. The solutions containing Al^{3+} or Fe^{3+} will also form bubbles as CO_2 is formed (see page 406). So if you're not sure whether a sample contains Fe^{2+} or Fe^{3+} ions, (which behave identically in the other two tests), you can use this test to decide — a sample containing Fe^{3+} ions will give off a gas (which you'll see as bubbles forming), whereas a sample containing Fe^{2+} ions won't.

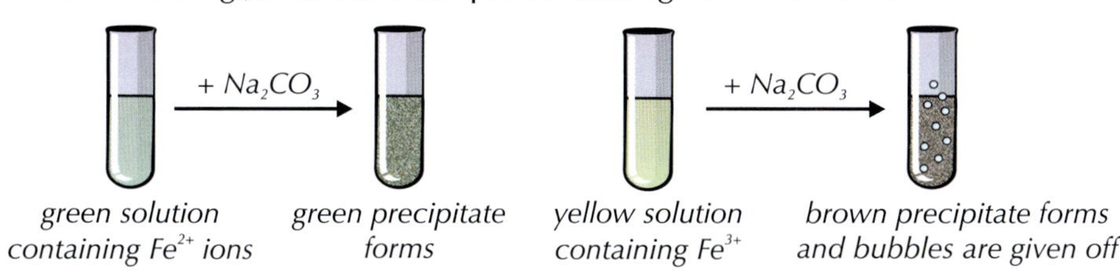

Tip: The solutions should be freshly made — if Fe^{2+} ions are left too long in contact with air, they will oxidise to Fe^{3+} ions.

Tip: In these tests with sodium hydroxide, ammonia and sodium carbonate, you could just look at the colour of the solution or precipitate to identify the ion. However, you need to know the other things that you'd expect to see, so you can't just learn the colours.

Complex ion solutions and precipitates

This handy table summarises all the compounds that are formed in the reactions on these pages. You need to know the formulas of all the complex ions, and their colours.

Metal-aqua ion	With $OH^-_{(aq)}$ or $NH_{3(aq)}$	With excess $OH^-_{(aq)}$	With excess $NH_{3(aq)}$	With $Na_2CO_{3(aq)}$
$[Cu(H_2O)_6]^{2+}$ blue solution	$Cu(OH)_2(H_2O)_4$ blue precipitate	no change	$[Cu(NH_3)_4(H_2O)_2]^{2+}$ deep blue solution	$CuCO_3$ green-blue precipitate
$[Fe(H_2O)_6]^{2+}$ green solution	$Fe(OH)_2(H_2O)_4$ green precipitate (goes brown on standing in air)	no change	no change	$FeCO_3$ green precipitate
$[Al(H_2O)_6]^{3+}$ colourless solution	$Al(OH)_3(H_2O)_3$ white precipitate	$[Al(OH)_4]^-$ colourless solution	no change	$Al(OH)_3(H_2O)_3$ white precipitate and bubbles
$[Fe(H_2O)_6]^{3+}$ yellow solution	$Fe(OH)_3(H_2O)_3$ brown precipitate	no change	no change	$Fe(OH)_3(H_2O)_3$ brown precipitate and bubbles

The $[Fe(H2O)_6]^{3+}$ complex is actually purple, but the solution appears yellow because one of the H_2O ligands hydrolyses (see pages 404-405).

Exam Tip

You need to learn the colours of all of these solutions and precipitates for the exam. Gutted. Any of them could come up at any time and they're easy marks if you can remember them.

Practice Question — Application

Q1 A student has an unknown metal-aqua ion. When she adds sodium carbonate to the aqueous metal-aqua ion a green-blue precipitate forms. What metal ion could be present in the solution?

Practice Questions — Fact Recall

Q1 What type of bonds hold metal-aqua ion complexes together?

Q2 Explain why metal-aqua 3+ ions are more acidic than metal-aqua 2+ ions.

Q3 Write equations to show what happens when OH^- ions are added to aqueous iron(III) ions.

Q4 What would you observe if you added a small amount of OH^- ions to aqueous aluminium(III) ions?

Q5 What would you observe if excess NH_3 was added to aqueous copper(II) ions?

Section Summary

Make sure you know...

- That the chemical properties of the transition metals are due to the incomplete d sub-level.
- That transition metal characteristics include complex ion formation, formation of coloured ions, catalytic activity, and variable oxidation states.
- That a complex ion is a metal ion surrounded by co-ordinately bonded ligands.
- That a ligand is an atom, ion or molecule that donates a pair of electrons to a central metal ion.
- The difference between monodentate (e.g. NH_3, H_2O, Cl^-), bidentate (e.g. $NH_2CH_2CH_2NH_2$, $C_2O_4^-$) and multidentate (e.g. $EDTA^{4-}$) ligands.
- That the co-ordination number is the number of co-ordinate bonds formed by the central metal ion.
- That complex ions with six co-ordinate bonds usually form octahedral complexes.
- That complex ions with four co-ordinate bonds can form tetrahedral or square planar complexes.
- That silver complexes can have two co-ordinate bonds and form linear complexes.
- That octahedral complexes can display optical isomerism with three bidentate ligands.
- That octahedral and square planar complexes can display *cis-trans* isomerism with monodentate ligands.
- That the anti-cancer drug cisplatin is a *cis* isomer.
- That energy from visible light can cause d electrons to jump from a lower energy orbital (ground state) to a higher energy orbital (excited state).
- That the energy difference between the ground state and excited state is given by $\Delta E = h\nu = hc/\lambda$.
- That transition metal ions can be identified by their colour.
- That frequencies of light which aren't absorbed are reflected and this gives complex ions colour.
- That changes in oxidation state, co-ordination number and ligand can cause changes in colour.
- How the concentration of a solution of complex ions can be determined using spectroscopy.
- That the exchange of similarly sized ligands occurs without change of co-ordination number.
- That the Cl^- ligand is larger than the uncharged ligands NH_3 and H_2O.

- That the exchange of the ligand H_2O by Cl^- can involve a change in co-ordination number because Cl^- is larger than H_2O.
- That substitution reactions may be incomplete.
- That haem is an iron(II) complex with a multidentate ligand.
- That oxygen forms a co-ordinate bond to Fe(II) in haemoglobin, enabling oxygen to be transported in the blood.
- That carbon monoxide is toxic because it replaces oxygen co-ordinately bonded to Fe(II) in haemoglobin.
- That the chelate effect is when complexes become more stable after monodentate ligands are replaced by bidentate and multidentate ligands, because there is a small change in enthalpy and a large, positive change in entropy.
- That vanadium species in oxidation states IV, III, and II are formed by the reduction of vanadate(V) ions by zinc in acidic solution
- That the redox potential for a transition metal ion changing from a higher to a lower oxidation state is influenced by pH and by the ligand.
- That the reduction of $[Ag(NH_3)_2]^+$ (Tollens' reagent) to metallic silver is used to distinguish between aldehydes and ketones.
- That transition metals can be used in titrations — specifically Fe^{2+} and $C_2O_4^{2-}$ with MnO_4^-.
- How to calculate the concentration or volume of a reagent using titration results.
- That the variable oxidation states of transition metals means they make good catalysts.
- That a heterogeneous catalyst is in a different phase from the reactants (and that the reaction occurs on its surface) and that a homogeneous catalyst is in the same phase as the reactants (and that the reaction proceeds through an intermediate species).
- That V_2O_5 acts as a heterogeneous catalyst in the Contact Process and that Fe is used as a heterogeneous catalyst in the Haber process.
- That a support medium can maximise the surface area of heterogeneous catalysts and minimise the costs involved in reactions.
- That heterogeneous catalysts can become poisoned by impurities, leading to reduced efficiency, which increases the cost involved in reactions.
- How Fe^{2+} ions catalyse the reaction between $S_2O_8^-$ and I^-, and how Mn^{2+} ions autocatalyse the reaction between MnO_4^- and $C_2O_4^{2-}$.
- That Fe(II) and Cu(II) ions form metal-aqua ions $[M(H_2O)_6]^{2+}$ in solution.
- That Al(III) and Fe(III) ions form metal-aqua ions $[M(H_2O)_6]^{3+}$ in solution.
- That $[M(H_2O)_6]^{3+}$ is more acidic than $[M(H_2O)_6]^{2+}$ due to the difference in the charge/size ratio of the metal ions.
- That some metal hydroxides (e.g. hydroxides of Al^{3+}) show amphoteric character by dissolving in both acids and bases.
- The reactions of $Fe^{2+}_{(aq)}$, $Cu^{2+}_{(aq)}$, $Al^{3+}_{(aq)}$ and $Fe^{3+}_{(aq)}$ ions with the bases OH^-, NH_3 and CO_3^{2-} and be able to explain them.
- How to carry out simple test tube reactions to identify transition metal ions in aqueous solution.

Exam-style Questions

1 Which of the following shows an incorrect electron configuration for the given transition element?

A vanadium: $1s^2\ 2s^2\ 2p^6\ 3s^2\ 3p^6\ 3d^3\ 4s^2$

B chromium: $1s^2\ 2s^2\ 2p^6\ 3s^2\ 3p^6\ 3d^5\ 4s^1$

C manganese: $1s^2\ 2s^2\ 2p^6\ 3s^2\ 3p^6\ 3d^6\ 4s^1$

D iron: $1s^2\ 2s^2\ 2p^6\ 3s^2\ 3p^6\ 3d^6\ 4s^2$

(1 mark)

2 Which of the following statements about the transition metal complex $[Cu(H_2O)_6]^{2+}_{(aq)}$ is untrue?

A When chloride ions are added, a square planar complex is formed.

B When $NaOH_{(aq)}$ is added, a blue precipitate is formed.

C When excess $NH_{3(aq)}$ is added, a deep blue solution is formed.

D $[Cu(H_2O)_6]^{2+}_{(aq)}$ is octahedral and blue in solution.

(1 mark)

3 A student is given an unknown solution containing a transition metal-aqua complex and asked to identify the ion present. The student carries out three tests on three separate samples, the results of which are shown below.

Test 1: A few drops of $NaOH_{(aq)}$ are added to the unknown solution and a precipitate forms. More $NaOH_{(aq)}$ is added until it is in excess and no change is observed.

Test 2: A few drops of $NH_{3(aq)}$ are added to the unknown solution and a precipitate forms. More $NH_{3(aq)}$ is added until it is in excess and no change is observed.

Test 3: A few drops of $Na_2CO_{3(aq)}$ are added to the unknown solution and no gas bubbles were observed.

What is the unknown transition metal ion?

A Al(III)

B Fe(II)

C Cu(II)

D Fe(III)

(1 mark)

4 Brass is an alloy of mainly copper and zinc.

Dissolving brass in acid produces a solution which contains aqueous copper ions.

4.1 What colour is the aqueous Cu^{2+} ion?

(1 mark)

4.2 Write down the full electron configurations of the copper atom and the Cu^{2+} ion.

(2 marks)

4.3 Explain why the Cu^{2+} ion has the chemical properties associated with transition metal elements.

(1 mark)

In the solution, the $[Cu(H_2O)_6]^{2+}$ complex ion forms.
In this complex ion, water acts as a ligand.

(1 mark)

4.4 Draw the 3D structure of $[Cu(H_2O)_6]^{2+}$ and state its shape.

(2 marks)

4.5 What is the co-ordination number of this ion?

(1 mark)

4.6 Explain why transition metal complex ions have a colour.

(2 marks)

4.7 Give three factors which can affect the colour of a complex ion.

(3 marks)

5 Many complex ions exist as stereoisomers.

$CoCl_2(NH_3)_2$ exists as two stereoisomers.

5.1 Give the shape of the $CoCl_2(NH_3)_2$ ion.

(1 mark)

5.2 What type of stereoisomerism does $CoCl_2(NH_3)_2$ exhibit?

(1 mark)

5.3 Draw the two possible stereoisomers of $CoCl_2(NH_3)_2$.

(2 marks)

The ethanedioate ion is a bidentate ligand. When it binds to chromium(III) ions it forms a complex ion with stereoisomers. The structure of the ethanedioate ion is shown below.

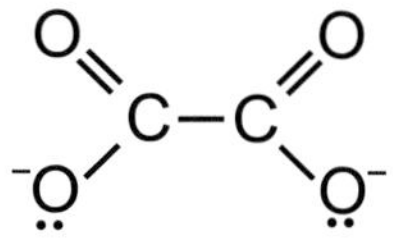

O=C(O⁻)–C(=O)O⁻ Ethanedioate ion

5.4 What type of stereoisomerism does $[Cr(C_2O_4)_3]^{3-}$ exhibit?

(1 mark)

5.5 Describe how the ethanedioate ion bonds to the chromium ion.

(2 marks)

5.6 Draw two stereoisomers of $[Cr(C_2O_4)_3]^{3-}$.
You can simplify the structure of the ethanedioate ion if you wish.

(2 marks)

6 Transition metals and their oxides make good catalysts. Rhodium (Rh) is used in catalytic converters to catalyse the reaction shown below:

$$2CO_{(g)} + 2NO_{(g)} \rightarrow 2CO_{2(g)} + N_{2(g)}$$

6.1 Explain why transition metals make good catalysts.

(2 marks)

6.2 What type of catalyst is rhodium in this reaction?

(1 mark)

6.3 Explain why catalytic converters contain a ceramic lattice coated with a thin layer of rhodium rather than a solid rhodium block.

(3 marks)

6.4 The rhodium used in catalytic converters is vulnerable to catalytic poisoning. Explain what catalytic poisoning is.

(3 marks)

Fe^{2+} catalyses this reaction between $S_2O_8^{2-}$ and I^-:

$$S_2O_8^{2-}{}_{(aq)} + 2I^-{}_{(aq)} \rightarrow I_{2(aq)} + 2SO_4^{2-}{}_{(aq)}$$

6.5 Explain why this reaction is extremely slow in the absence of a catalyst.

(2 marks)

6.6 Write two equations to show how Fe^{2+} catalyses this reaction.

(2 marks)

7 Metal-aqua complex ions form when transition metal compounds dissolve in water.

7.1 Describe the bonding in a metal-aqua complex ion.

(2 marks)

7.2 Give the formula of the species formed when Fe^{2+} ions dissolve in water.

(1 mark)

7.3 Write an equation for the reaction of the iron-aqua complex ion and water.

(1 mark)

7.4 Explain why this solution will be less acidic than a solution of Al^{3+} ions dissolved in water.

(4 marks)

7.5 Describe the observation you would expect when a small amount of NaOH is added to the solution of Fe^{2+} ions and water.

(1 mark)

7.6 Identify the species formed when Fe^{3+} ions dissolve in water.

(1 mark)

7.7 Write an equation for the addition of carbonate ions to this species.

(1 mark)

7.8 Describe what you would observe when this reaction takes place.

(2 marks)

When aluminium dissolves in water, $[Al(H_2O)_6]^{3+}$ is formed.

7.9 Write an equation for the ligand substitution reaction between this species and an excess of ethane-1,2-diamine ($NH_2CH_2CH_2NH_2$).

(1 mark)

7.10 Explain why the metal complex formed in this reaction is more stable than $[Al(H_2O)_6]^{3+}$.

(2 marks)

8 Aluminium hydroxide ($Al(OH)_3(H_2O)_{3(aq)}$) is amphoteric.

8.1 Define the term amphoteric.

(1 mark)

8.2 Describe the appearance of aluminium hydroxide.

(1 mark)

8.3 Write an equation for the reaction of aluminium hydroxide with OH^- ions.

(2 marks)

8.4 What type of acid is aluminium hydroxide acting as in this reaction? Explain your answer.

(2 marks)

9 This question is all about the uses of transition metals.

The key component of Tollens' reagent is a complex ion.

9.1 Identify the complex ion present in Tollens' reagent, sketch it and describe its structure.

(3 marks)

9.2 State what Tollens' reagent is used for and explain why it can be used for this purpose.

(3 marks)

9.3 Transition metals can also be used in titrations. In one titration, 15.8 cm^3 of acidified potassium manganate(VII) was needed to completely react with 30.0 cm^3 of a 0.150 mol dm^{-3} solution of acidified iron(II) sulfate. Calculate the concentration of the potassium manganate (VII) solution used.

(4 marks)

9.4 What colour will the solution be at the end point?

(1 mark)

Learning Objectives:

- Know that optical isomerism is a form of stereoisomerism, and occurs as a result of chirality in molecules, limited to molecules with a single chiral centre.
- Know that an asymmetric carbon atom is chiral and gives rise to optical isomers (enantiomers) which exist as non superimposable mirror images and differ in their effect on plane polarised light.
- Be able to draw the structural formulas and displayed formulas of enantiomers.
- Know that a mixture of equal amounts of enantiomers is called a racemic mixture (racemate).
- Understand how racemic mixtures (racemates) are formed and why they are optically inactive.

Specification Reference 3.3.7

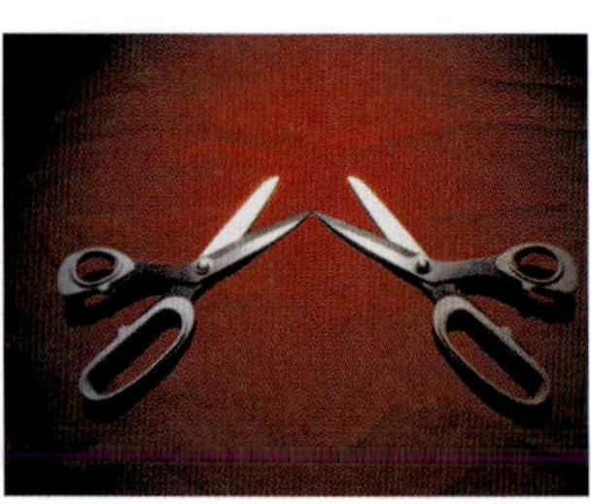

Figure 1: Left handed and right handed scissors are mirror images of each other — they can't be superimposed on each other.

1. Optical Isomerism

Optical isomerism, as with all other isomerism, is to do with how molecules arrange in space. So, if you're struggling to picture what they look like, get your hands on some molecular models to play around with.

What is optical isomerism?

Optical isomerism is a type of **stereoisomerism**. Stereoisomers have the same structural formula, but have their atoms arranged differently in space. Optical isomers have a **chiral** carbon atom. A chiral (or asymmetric) carbon atom is one that has four different groups attached to it.

Example

The molecule below, 1-aminoethan-1-ol, has a chiral carbon atom.

H
C
H_2N CH_3
OH

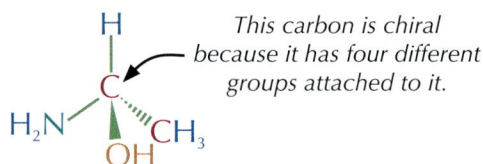

This carbon is chiral because it has four different groups attached to it.

It's possible to arrange the groups in two different ways around chiral carbon atoms so that two different molecules are made — these molecules are called **enantiomers** or **optical isomers**. The enantiomers are mirror images and, no matter which way you turn them, they can't be superimposed. If molecules can be superimposed, they're **achiral** — and there's no optical isomerism.

Example

Here are the two enantiomers of 1-aminoethan-1-ol. It doesn't matter how many times you turn and twist them, they can't be superimposed.

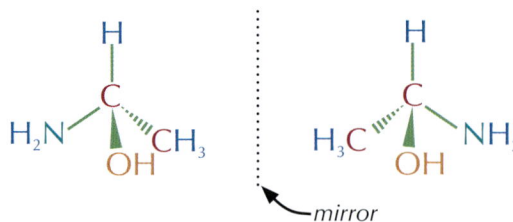

H | H
C | C
H_2N CH_3 | H_3C NH_2
OH | OH

mirror

You have to be able to draw optical isomers. Just follow these steps each time:

1. Locate the chiral centre — look for the carbon atom with four different groups attached.
2. Draw one enantiomer in a tetrahedral shape — put the chiral carbon atom at the centre and the four different groups in a tetrahedral shape around it. Don't try to draw the full structure of each group — it gets confusing. Just use the structural formulas.
3. Draw the mirror image of the enantiomer — put in a mirror line next to the enantiomer and then draw the mirror image of the enantiomer on the other side of it.

Don't panic — here are some examples to help you get to grips with this.

Examples

Draw the two enantiomers of 2-hydroxypropanoic acid.
The structure of 2-hydroxypropanoic acid is shown below.

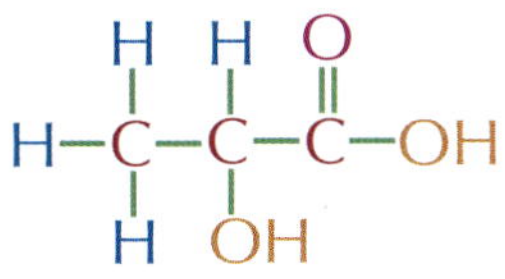

H H O
H–C–C–C–OH
H OH

1. Locate the chiral centre — the chiral carbon in this molecule is the carbon with the groups H, OH, COOH and CH_3 attached.

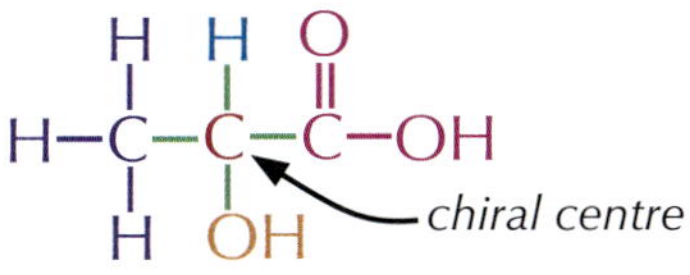

H H O
H–C–C–C–OH
H OH
chiral centre

2. Draw one enantiomer in a tetrahedral shape — put the chiral carbon atom at the centre and the groups H, OH, COOH and CH_3 in a tetrahedral shape around it.

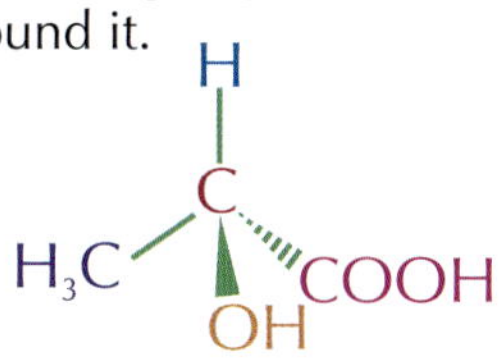

H
C
H_3C COOH
OH

3. Then draw its mirror image beside it.

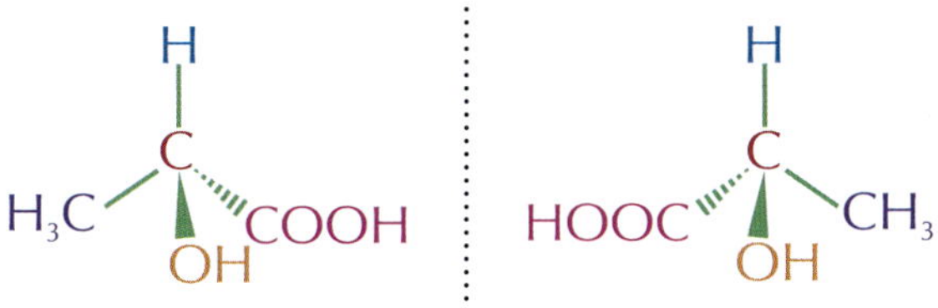

H
C
H_3C COOH
OH

H
C
HOOC CH_3
OH

Draw the two enantiomers of butan-2-ol.

1. Draw the structure of butan-2-ol and locate the chiral centre.

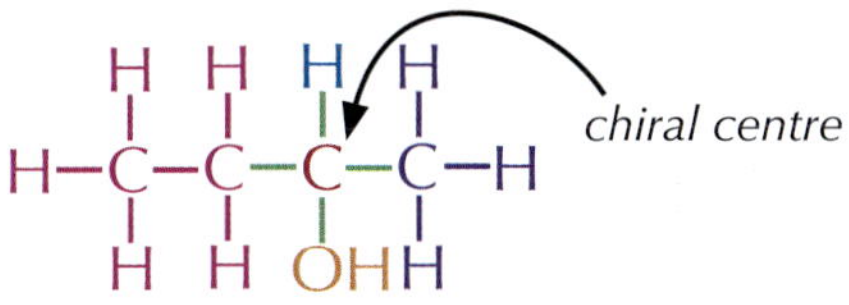

H H H H
H–C–C–C–C–H
H H OH H
chiral centre
butan-2-ol

2. Draw one enantiomer in a tetrahedral shape.

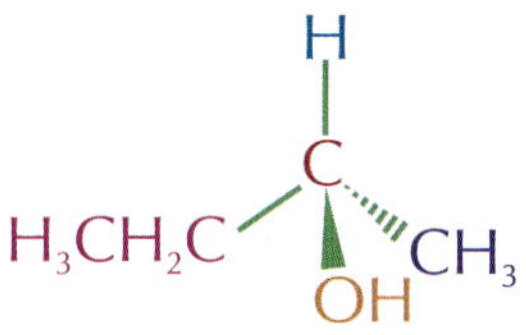

H
C
H_3CH_2C CH_3
OH

3. Then draw its mirror image beside it.

H
C
H_3CH_2C CH_3
OH

H
C
H_3C CH_2CH_3
OH

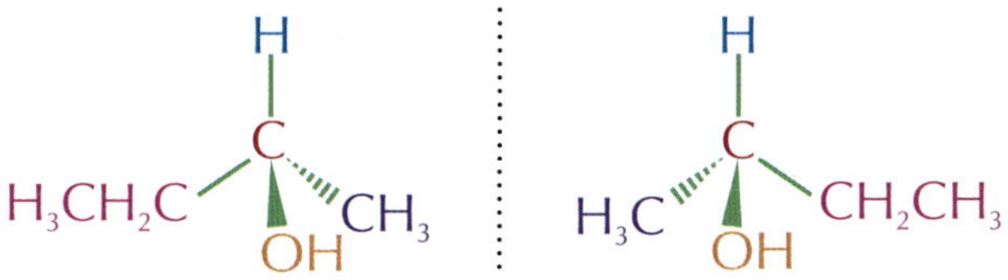

Tip: You can get molecules that contain more than one chiral carbon, but in your exam you'll only get asked about molecules with just one chiral centre.

Tip: The chiral centre and the chiral carbon atom are the same thing.

Tip: Remember, those dashed lines mean the bond is pointing into the page, and the wedged lines mean the bond is pointing out of the page towards you.

Tip: Remember the chiral carbon is the one with four different groups attached to it.

Tip: Sometimes a chiral carbon is shown in a diagram by placing a star next to the chiral carbon atom, like this:

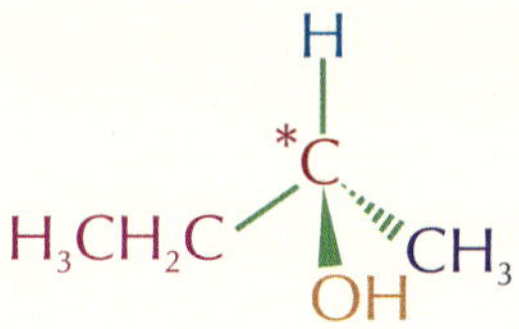

H
*C
H_3CH_2C CH_3
OH

Optical activity

Optical isomers are optically active — they rotate **plane polarised light**.

Plane polarised light

Normal light vibrates in all directions — some of it vibrates up and down, some of it vibrates side to side and some wiggles all over the place.
If normal light is passed through a polarising filter it becomes plane polarised. This means all the light is vibrating in the same plane (for example, only up and down or only side to side).

Exam Tip
Don't worry, you won't need to give all this detail about vibrations in the exam. All you need to know is that optical isomers are optically active and will rotate plane polarised light.

Rotation of plane polarised light

When you pass plane polarised light through an optically active mixture, the molecules interact with the light and rotate the plane of the vibration of the light. The two enantiomers of an optically active molecule will rotate the plane polarised light in opposite directions. One enantiomer rotates it in a clockwise direction, and the other rotates it in an anticlockwise direction.

Racemates

A **racemate** (or racemic mixture) contains equal quantities of each enantiomer of an optically active compound. Racemates don't show any optical activity — the two enantiomers cancel each other's light-rotating effect.
Chemists often react two achiral things together and get a racemic mixture of a chiral product. This is because when two molecules react, there's normally an equal chance of forming each of the enantiomers.

Examples

Here's the reaction between butane and chlorine:

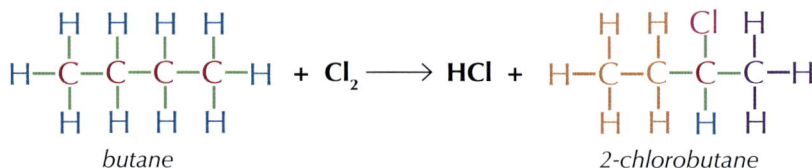

$$CH_3CH_2CH_2CH_3 + Cl_2 \longrightarrow HCl + CH_3CHClCH_2CH_3$$

butane — *2-chlorobutane*

A chlorine atom replaces one of the H groups on carbon 2, to give 2-chlorobutane. Either of the H's directly attached to carbon 2 can be replaced, so the reaction produces a mixture of the two possible enantiomers.

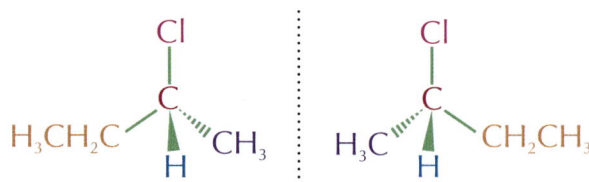

Cl / C / H_3CH_2C — CH_3 / H ⋮ Cl / C / H_3C — CH_2CH_3 / H

Each hydrogen has an equal chance of being replaced, so the two optical isomers are formed in equal amounts — you get a racemic mixture (a racemate).

Here's the reaction between propanal and acidified KCN. The nitrile group adds to the carbonyl group to form 2-hydroxybutanenitrile.

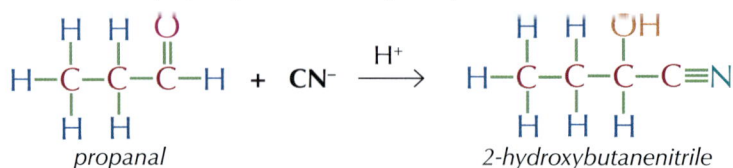

$$CH_3CH_2CHO + CN^- \xrightarrow{H^+} CH_3CH_2CH(OH)C{\equiv}N$$

propanal — *2-hydroxybutanenitrile*

Tip: Butane is achiral because it doesn't have four different groups attached to the central carbon atom — two of the groups are hydrogen atoms.

Tip: The reaction between butane and chlorine will also produce 1-chlorobutane:

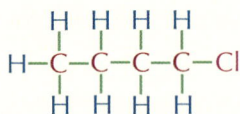

$$CH_3CH_2CH_2CH_2Cl$$

But, 1-chlorobutane is made via a less stable intermediate, so it is the minor product. It also doesn't contain a chiral centre, so it doesn't produce an optical isomer.

2-hydroxybutanenitrile has two enantiomers (shown below). They are equally likely to form via this reaction and so the two enantiomers are formed in equal amounts — you get a racemic mixture.

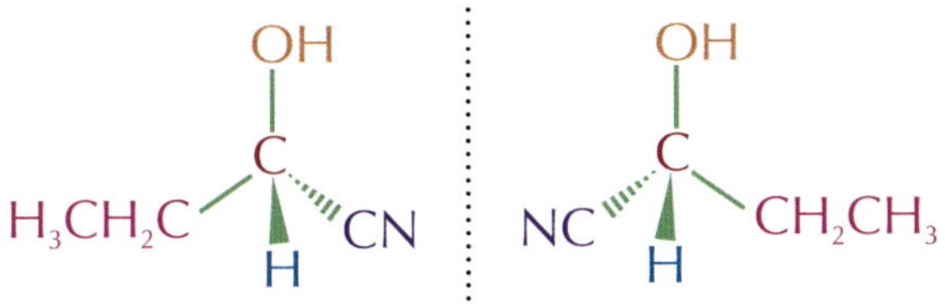

You can modify a reaction to produce a single enantiomer using chemical methods, but it's difficult and expensive.

Tip: See pages 422-424 for more detail on hydroxynitriles.

Tip: Not all reactions that produce chiral compounds will produce racemates. Sometimes, one enantiomer will be favoured over another enantiomer and you'll get an enantiomerically pure product — a product that contains only one enantiomer.

Practice Questions — Application

Q1 Circle the chiral carbon in each of these molecules.

a) b) 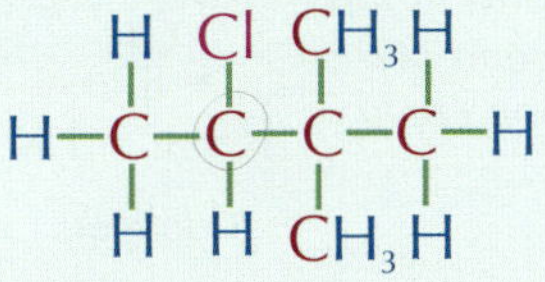 c) 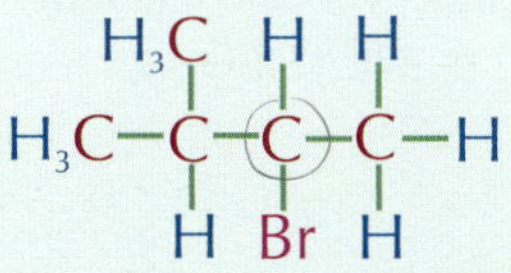 d)

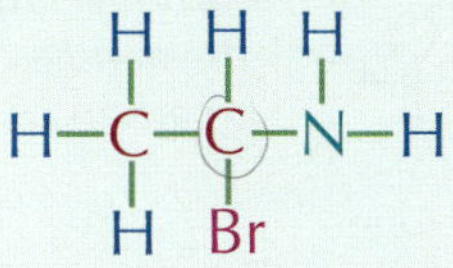

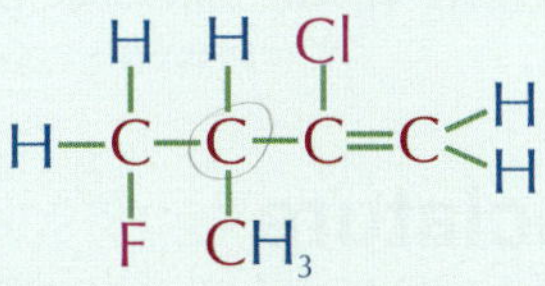

Q2 Draw the two enantiomers of the following molecules.

a) b)

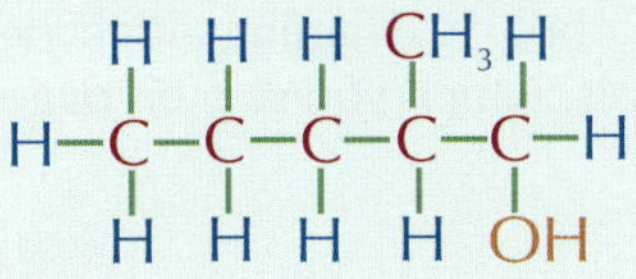

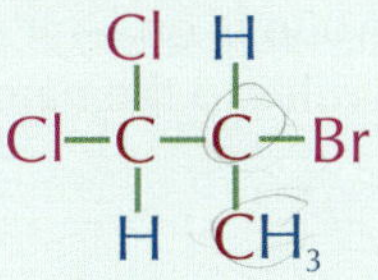

Q3 Here is an equation for the reaction between butanone and acidified KCN.

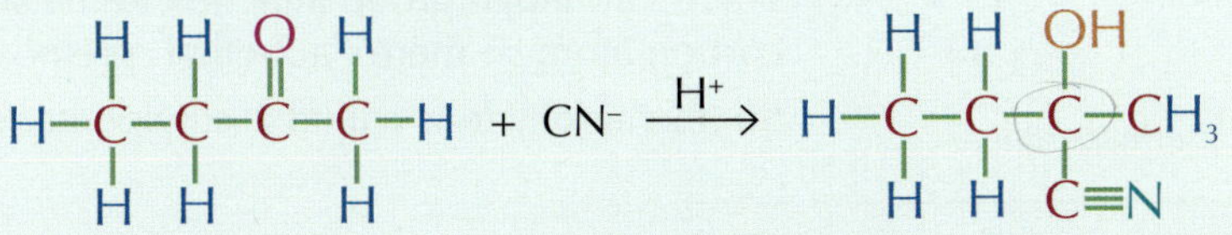

$$\text{butanone} + CN^- \xrightarrow{H^+} \text{2-hydroxy-2-methylbutanenitrile}$$

butanone *2-hydroxy-2-methylbutanenitrile*

Draw the two optical isomers that can be formed via this reaction.

Tip: When you're drawing out the enantiomers for molecules just write out the structural formula of each group — don't try and draw the displayed formula for each one or your diagram will become too confusing.

Practice Questions — Fact Recall

Q1 What is optical isomerism?

Q2 What is a chiral carbon?

Q3 What does it mean if a molecule is 'optically active'?

Q4 Explain why racemic mixtures are not optically active.

2. Aldehydes and Ketones

Learning Objectives:

- Be able to apply IUPAC rules for naming aldehydes and ketones.
- Know that aldehydes are readily oxidised to carboxylic acids
- Know the chemical tests to distinguish between aldehydes and ketones, including Fehling's solution and Tollens' reagent.
- Know that aldehydes can be reduced to primary alcohols and ketones to secondary alcohols, using $NaBH_4$ in aqueous solution, and that these reactions are examples of nucleophilic addition.
- Be able to write overall equations for reduction reactions using [H] as the reductant.
- Be able to outline the nucleophilic addition mechanism for reduction reactions with $NaBH_4$ (the nucleophile should be shown as H^-).

Specification Reference 3.3.1.1, 3.3.8

Aldehydes and ketones are compounds that are made by oxidising alcohols. Aldehydes are made by oxidising primary alcohols and ketones are made by oxidising secondary alcohols. You met aldehydes and ketones on pages 242-245, but now it's time to cover them in a bit more detail.

What are aldehydes and ketones?

Aldehydes and **ketones** are both **carbonyl compounds** as they both contain the carbonyl functional group, C=O. The difference between aldehydes and ketones is that they've got their carbonyl groups in different positions. Aldehydes have their carbonyl group at the end of the carbon chain. Ketones have their carbonyl group in the middle of the carbon chain, see Figure 1.

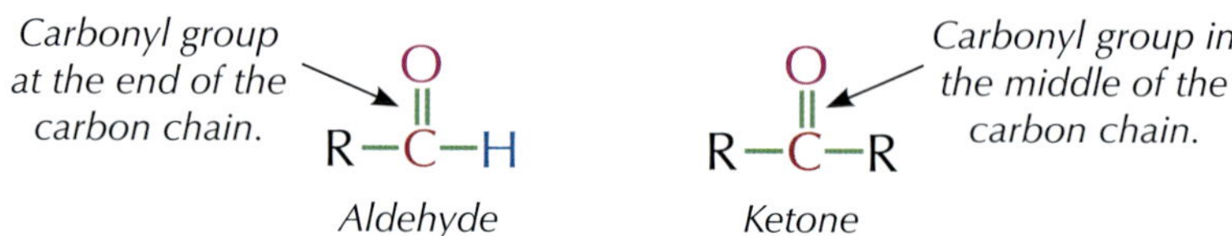

Figure 1: *The difference between an aldehyde and a ketone. 'R' represents a carbon chain of any length.*

Nomenclature

Before you can study aldehydes and ketones you need to know how to name them. Aldehydes have the suffix -al. You don't have to say which carbon the functional group is on — it's always on carbon-1. Naming aldehydes follows very similar rules to the naming of alcohols, which you saw on page 233.

Tip: This is a recap of stuff you learnt in Unit 3, so if it's not looking familiar, have a look back at pages 242-243.

Example

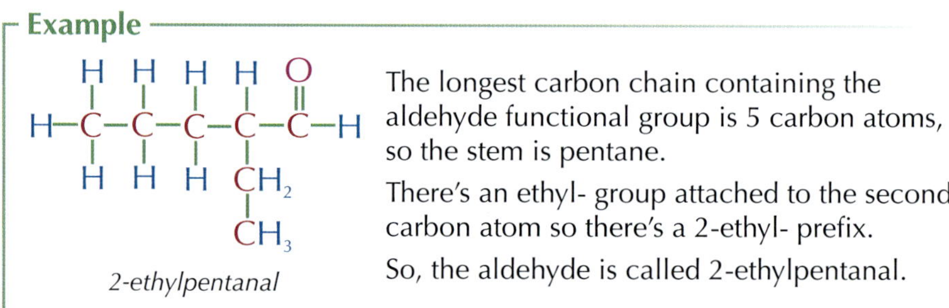

2-ethylpentanal

The longest carbon chain containing the aldehyde functional group is 5 carbon atoms, so the stem is pentane.

There's an ethyl- group attached to the second carbon atom so there's a 2-ethyl- prefix.

So, the aldehyde is called 2-ethylpentanal.

The suffix for ketones is -one. For ketones with five or more carbons (or four-carbon ketones which are branched), you always have to say which carbon the functional group is on.

Tip: When naming ketones, the carbonyl group is the highest priority, so you always number the carbons from the end that means the carbonyl group carbon has the lowest number.

Example

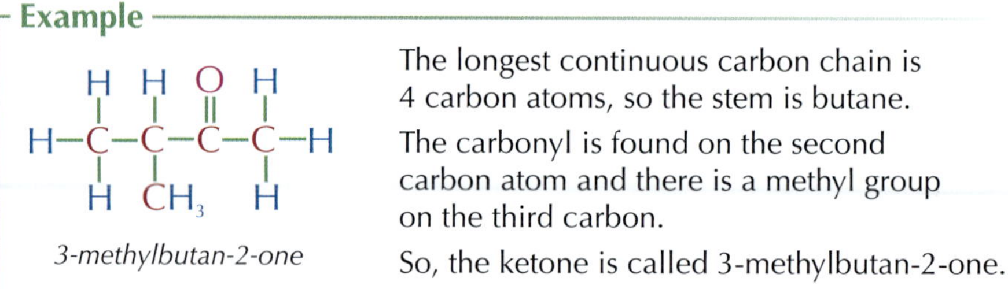

3-methylbutan-2-one

The longest continuous carbon chain is 4 carbon atoms, so the stem is butane.

The carbonyl is found on the second carbon atom and there is a methyl group on the third carbon.

So, the ketone is called 3-methylbutan-2-one.

Testing for aldehydes and ketones

There are a few tests to distinguish between aldehydes and ketones. They all work on the idea that an aldehyde can be easily oxidised to a carboxylic acid, but a ketone can't. Aldehydes can be easily oxidised to carboxylic acids because there's a hydrogen attached to the carbonyl group:

$$R-\overset{\overset{\displaystyle O}{\|}}{C}-H + [O] \longrightarrow R-\overset{\overset{\displaystyle O}{\|}}{C}-OH$$

The only way to oxidise a ketone would be to break a carbon-carbon bond so ketones are not easily oxidised:

$$R-\overset{\overset{\displaystyle O}{\|}}{C}-R + [O] \longrightarrow \textit{Nothing happens}$$

As an aldehyde is oxidised, another compound is reduced — so a reagent is used that changes colour as it's reduced. Two reagents that can be used to distinguish between aldehydes and ketones are Tollens' reagent and Fehling's solution.

Tip: In these equations [O] is used to represent an oxidising agent.

Figure 2: *Tollens' reagent. The test tube on the left shows the unreacted Tollens' reagent. The test tube on the right shows the result of a reaction of Tollens' reagent with an aldehyde.*

Tollens' reagent

Tollens' reagent is a colourless solution of silver nitrate dissolved in aqueous ammonia. If it's heated in a test tube with an aldehyde, a silver mirror forms after a few minutes (see Figure 2). As the aldehyde is oxidised, the diaminesilver ions in the Tollens' reagent are reduced, producing silver and ammonia:

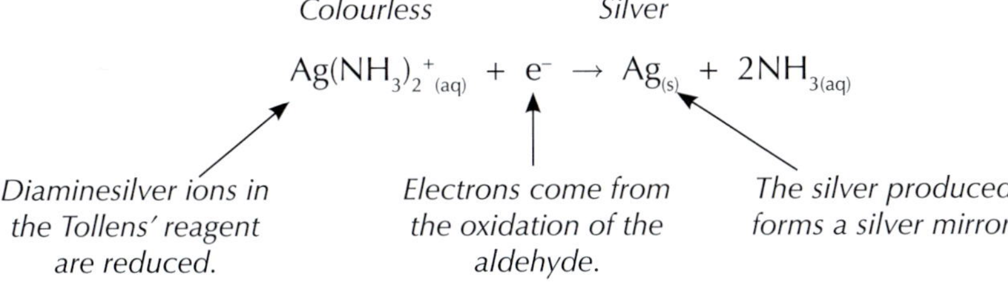
Colourless
Silver
$Ag(NH_3)_2{}^+{}_{(aq)} + e^- \rightarrow Ag_{(s)} + 2NH_{3(aq)}$
Diaminesilver ions in the Tollens' reagent are reduced.
Electrons come from the oxidation of the aldehyde.
The silver produced forms a silver mirror.

Ketones don't react with Tollens' reagent.

Tip: Aldehydes and ketones are flammable, so they must be heated in a water bath rather than over a flame. (See page 7 for methods of heating that don't involve a naked flame.) You also need to consider any other safety precautions before carrying out these tests.

Fehling's solution

Fehling's solution is a blue solution of complexed copper(II) ions dissolved in sodium hydroxide. If it's heated with an aldehyde the copper(II) ions are reduced to a brick-red precipitate of copper(I) oxide (see Figure 3):

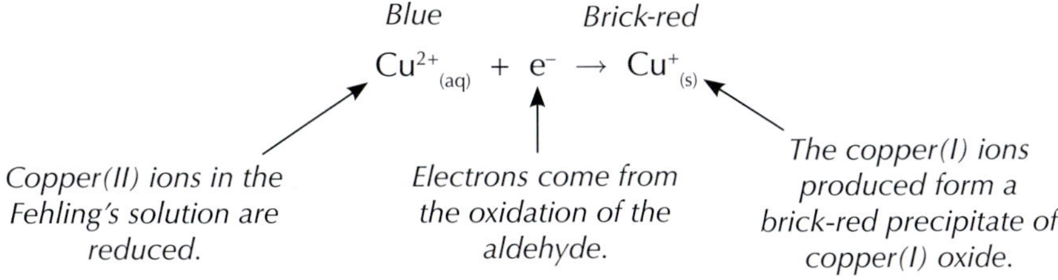
Blue
Brick-red
$Cu^{2+}{}_{(aq)} + e^- \rightarrow Cu^+{}_{(s)}$
Copper(II) ions in the Fehling's solution are reduced.
Electrons come from the oxidation of the aldehyde.
The copper(I) ions produced form a brick-red precipitate of copper(I) oxide.

Ketones also don't react with Fehling's solution.

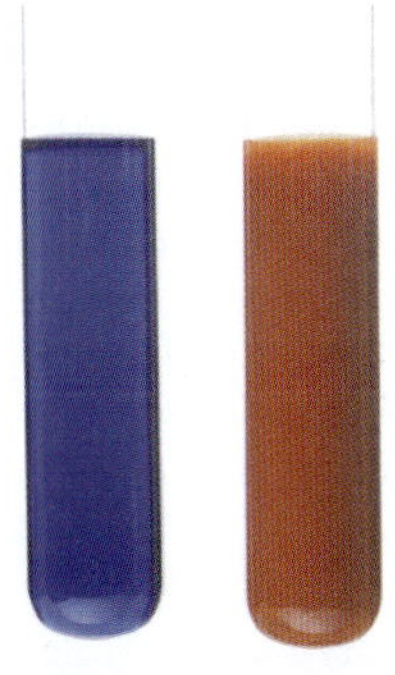

Figure 3: *Fehling's solution. The test-tube on the left shows the unreacted Fehling's solution. The test tube on the right shows the result of the reaction of Fehling's solution with an aldehyde.*

Figure 4: *$NaBH_4$ powder. This is dissolved in water with methanol to produce a commonly used reducing agent.*

Reducing aldehydes and ketones

You saw on pages 242-245 how primary alcohols can be oxidised to produce aldehydes and carboxylic acids, and how secondary alcohols can be oxidised to make ketones. Using a reducing agent you can reverse these reactions. $NaBH_4$ (sodium tetrahydridoborate(III) or sodium borohydride) is usually the reducing agent used (see Figure 4). But in equations, [H] is often used to indicate a hydrogen from a reducing agent. The equation below shows the reduction of an aldehyde to a primary alcohol:

$$R{-}C({=}O){-}H + 2[H] \longrightarrow R{-}CH(OH){-}H$$

And here's the reduction of a ketone to a secondary alcohol:

$$R{-}C({=}O){-}R + 2[H] \longrightarrow R{-}CH(OH){-}R$$

Exam Tip
When you're writing equations like this in an exam, make sure you balance the [H]'s as well as the molecules.

Nucleophilic addition reactions

You need to understand the reaction mechanisms for the reduction of aldehydes and ketones back to alcohols. These are **nucleophilic addition** reactions — an H^- ion from the reducing agent acts as a nucleophile and adds on to the δ^+ carbon atom of a carbonyl group. Here's the mechanism...

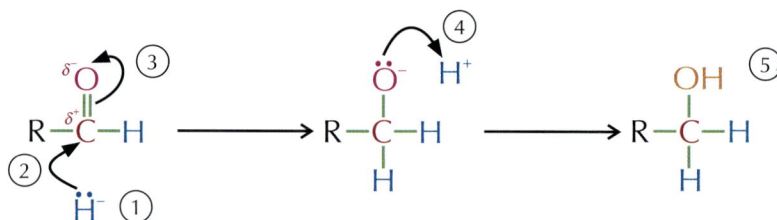

$$R{-}\overset{\delta+}{C}(={}^{\delta-}O){-}H \;\xrightarrow{:H^-}\; R{-}CH(O^-){-}H \;\xrightarrow{H^+}\; R{-}CH(OH){-}H$$

Tip: In reaction mechanisms, the curly arrows show the movement of an electron pair.

1. The C=O bond is polar, meaning the carbon in the carbonyl group is slightly positive. So, the $C^{\delta+}$ attracts the negatively charged lone pair of electrons on the H^- ion.
2. The H^- ion attacks the slightly positive carbon atom and donates its lone pair of electrons forming a bond with the carbon.
3. As carbon can only have 4 bonds, the addition of the H^- ion causes one of the carbon-oxygen bonds to break. This forces a lone pair of electrons from the C=O double bond onto the oxygen.
4. The oxygen donates its lone pair of electrons to a H^+ ion (this H^+ ion usually comes from water but sometimes a weak acid is added as a source of H^+).
5. A primary alcohol is produced.

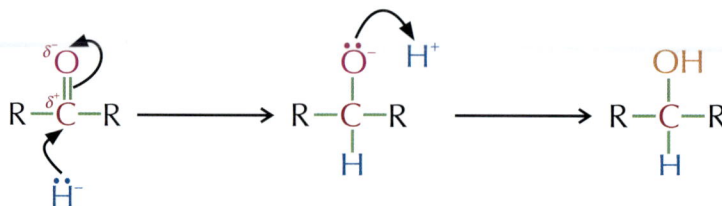

Exam Tip
You need to be able to draw the mechanism for the reduction of any aldehyde or ketone — so make sure you understand all the steps of these nucleophilic addition reactions.

The mechanism for the reduction of a ketone is the same as for an aldehyde — you just get a secondary alcohol at the end instead of a primary alcohol:

$$R{-}\overset{\delta+}{C}(={}^{\delta-}O){-}R \;\xrightarrow{:H^-}\; R{-}CH(O^-){-}R \;\xrightarrow{H^+}\; R{-}CH(OH){-}R$$

This reaction mechanism can be applied to any aldehyde or ketone.

Examples

Propanal can be reduced to propan-1-ol:

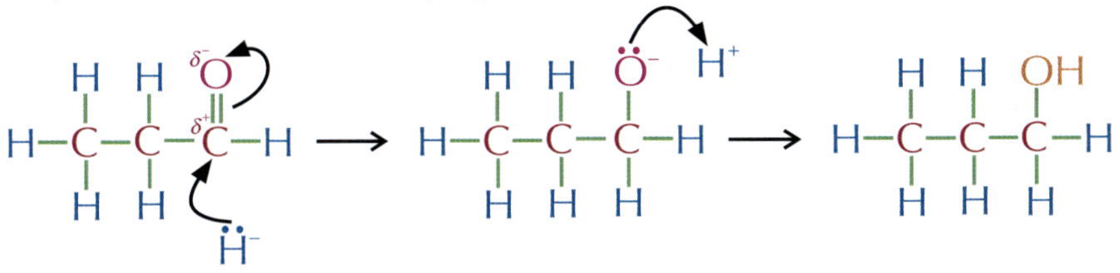

Butanone can be reduced to butan-2-ol:

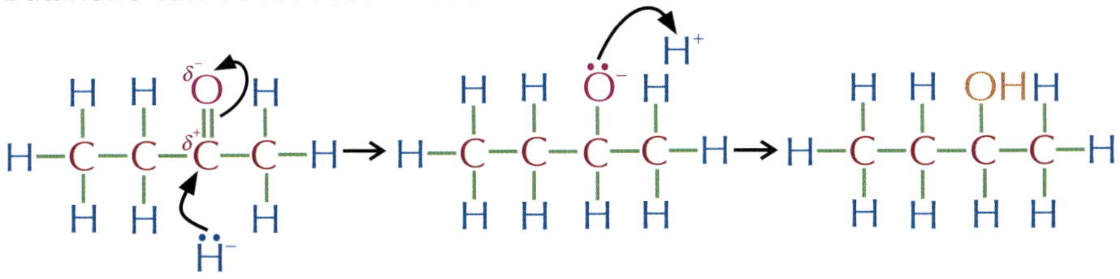

Exam Tip
You <u>must</u> draw the curly arrows coming from the lone pair of electrons. If you don't — you won't get the marks for the mechanism in the exam.

Practice Questions — Application

Q1 Name these molecules:

a) 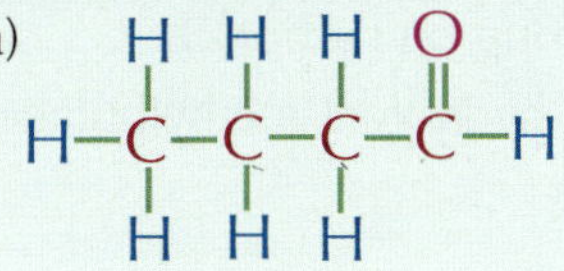
```
    H   H   H   O
    |   |   |   ||
H - C - C - C - C - H
    |   |   |
    H   H   H
```

b)
```
    H   O   H   H   H   H
    |   ||  |   |   |   |
H - C - C - C - C - C - C - H
    |       |   |   |   |
    H       H   H   H   H
```

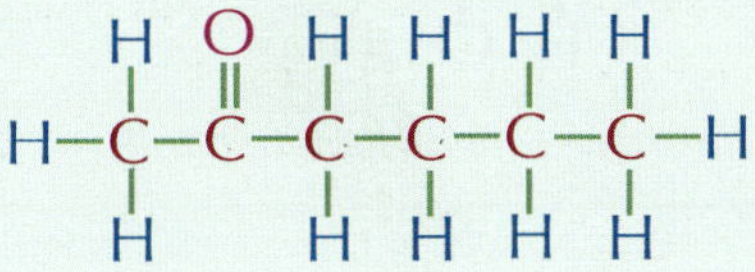

c)
```
    H   H   O   H   H
    |   |   ||  |   |
H - C - C - C - C - C - H
    |   |       |   |
    H   H       H   H
```

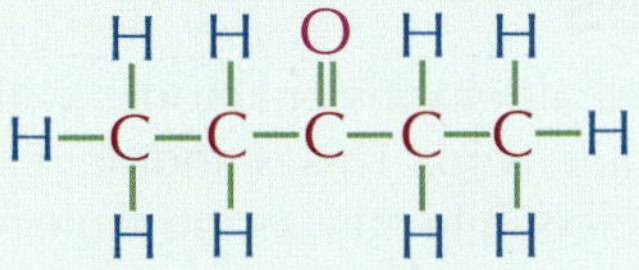

d) 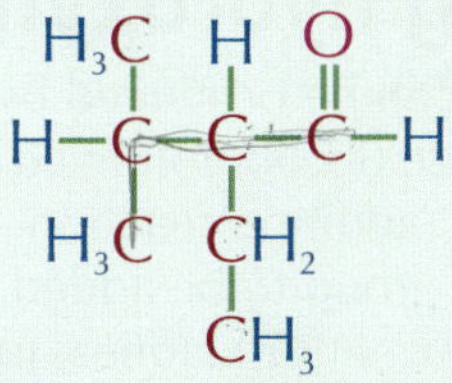
```
   H3C  H   O
    |   |   ||
H - C - C - C - H
    |   |
   H3C  CH2
        |
        CH3
```

Q2 Draw the mechanisms for the reduction of each molecule in Q1 to an alcohol. Use H^- as the nucleophile.

Q3 Write an equation, using structural formulas, for the reduction of pentan-2-one to pentan-2-ol. Use [H] to represent a reducing agent.

Tip: Don't forget — aldehydes are reduced to primary alcohols and ketones are reduced to secondary alcohols.

Practice Questions — Fact Recall

Q1 What is the difference between an aldehyde and a ketone?

Q2 What type of compound is produced when an aldehyde is oxidised?

Q3 Explain why ketones cannot be easily oxidised.

Q4 Describe what you would observe if you used the following to distinguish between an aldehyde and a ketone:

a) Tollens' reagent.

b) Fehling's solution.

Q5 Identify a reducing agent which could be used to reduce an aldehyde to a primary alcohol.

Tip: Sometimes you might see primary and secondary alcohols written as 1° and 2° alcohols.

3. Hydroxynitriles

Aldehydes and ketones can be used to produce another set of molecules known as hydroxynitriles.

What are hydroxynitriles?

Hydroxynitriles are molecules which contain a hydroxyl group (OH) and a nitrile group (CN) — see Figure 1.

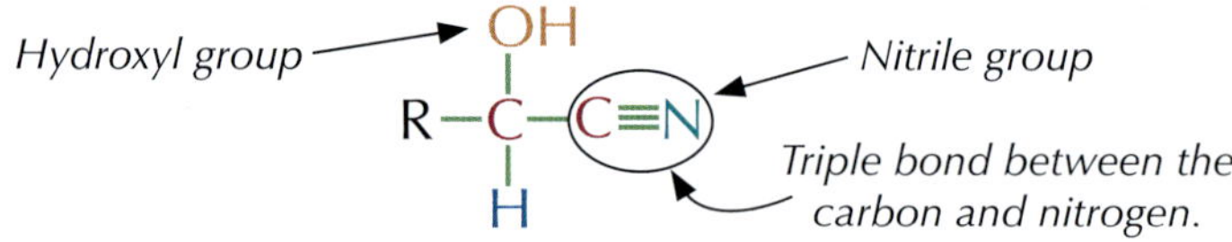

Figure 1: *The general structure of a hydroxynitrile.*

When naming hydroxynitriles the nitrile group is the most important so the suffix is -nitrile and the carbon that's attached to the nitrogen is always carbon-1. There's also a hydroxy- prefix because of the OH group. After that, naming hydroxynitriles is just the same as naming any other compound.

Example

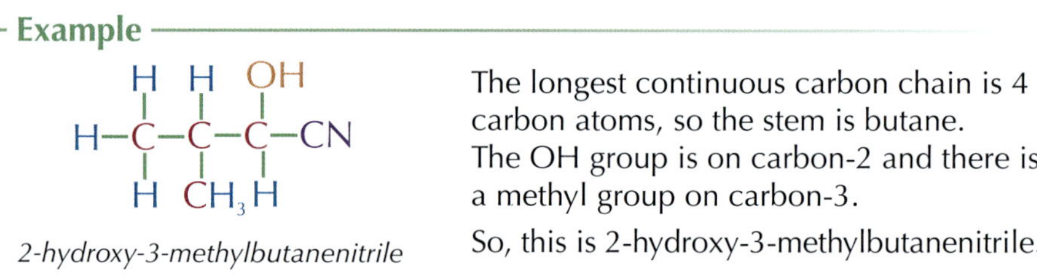

2-hydroxy-3-methylbutanenitrile

The longest continuous carbon chain is 4 carbon atoms, so the stem is butane. The OH group is on carbon-2 and there is a methyl group on carbon-3.
So, this is 2-hydroxy-3-methylbutanenitrile.

Producing hydroxynitriles

Hydroxynitriles can be produced by reacting aldehydes or ketones with potassium cyanide (KCN), followed by a dilute acid. This is another example of a nucleophilic addition reaction — a nucleophile attacks the molecule, causing an extra group to be added. Here's the mechanism for the reaction of aldehydes with KCN (for ketones, just replace the H with an R′ group):

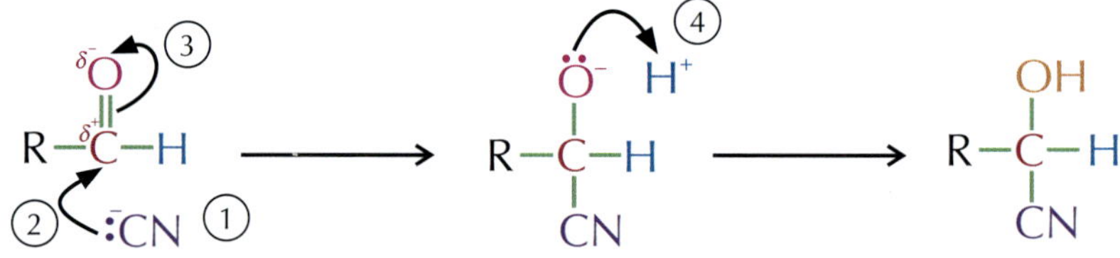

1. Potassium cyanide's an ionic compound.
   It dissociates in water to form K^+ ions and CN^- ions: $KCN \rightleftharpoons K^+ + CN^-$.
2. The CN^- ion from the KCN attacks the partially positive carbon atom and donates a pair of electrons, forming a bond with the carbon.
3. A pair of electrons from the C=O double bond is pushed onto the oxygen.
4. The oxygen bonds to a H^+ ion (from the dilute acid) to form the hydroxyl group (OH) and a hydroxynitrile is produced.

The overall reaction for an aldehyde is:

$$RCHO_{(aq)} + KCN_{(aq)} \xrightarrow{H^+_{(aq)}} RCH(OH)CN_{(aq)} + K^+_{(aq)}$$

Or for a ketone: $RCOR'_{(aq)} + KCN_{(aq)} \xrightarrow{H^+_{(aq)}} RC(OH)R'CN_{(aq)} + K^+_{(aq)}$

Learning Objectives:

- Be able to apply IUPAC rules for naming hydroxynitriles.
- Understand the nucleophilic addition reactions of carbonyl compounds with KCN, followed by dilute acid, to produce hydroxynitriles.
- Be able to outline the nucleophilic addition mechanism for the reaction of carbonyl compounds with KCN, followed by dilute acid.
- Be able to write overall equations for the formation of hydroxynitriles using HCN.
- Be able to explain why aldehydes and unsymmetrical ketones form mixtures of enantiomers when they react with KCN followed by dilute acid in nucleophilic addition reactions.
- Know the hazards associated with using KCN.

Specification Reference 3.3.1.1, 3.3.8

Tip: Don't forget to add a negative sign to the oxygen after step three.

Tip: This mechanism is exactly the same as the nucleophilic addition mechanism on page 420, except we're using CN^- instead of H^-.

Example

Propanone and acidified KCN react to form 2-hydroxy-2-methylpropanenitrile. The mechanism for this reaction is shown below:

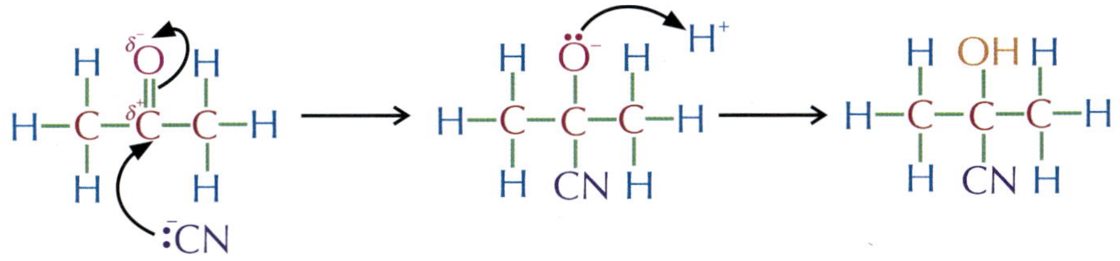

You can also make hydroxynitriles by reacting aldehydes or ketones with hydrogen cyanide. The overall equations for these reactions are:

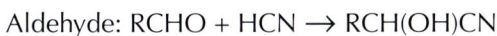

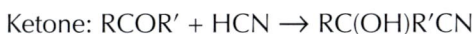

Aldehyde: $RCHO + HCN \rightarrow RCH(OH)CN$

Ketone: $RCOR' + HCN \rightarrow RC(OH)R'CN$

Racemic mixtures of hydroxynitriles

Double bonds, such as C=O and C=C bonds, are planar. So, when nucleophiles attack atoms at double bonds, they can attack from either above or below the plane of the bond. Nucleophilic attack from each side of the bond will produce a different enantiomer.

The nucleophilic attack of cyanide ions (from e.g. KCN) on aldehydes or unsymmetric ketones can lead to a racemic mixture of hydroxynitriles. This is because nucleophilic attack of the $C^{\delta+}$, by $CN:^-$, can occur from either above or below the plane of the C=O bond. Depending on which way the nucleophile attacks from, a different enantiomer will form. Usually, attack from both sides is equally likely, so a racemic mixture is formed.

Example

The products of the reaction of propanal with potassium cyanide, under weakly acidic conditions, are present as a racemic mixture. Explain this observation.

The reaction of acidified KCN with propanal involves the nucleophilic attack of CN^- at the positive carbon centre. The CN^- nucleophile can attack from two directions — from above the plane of the molecule, or from below it.

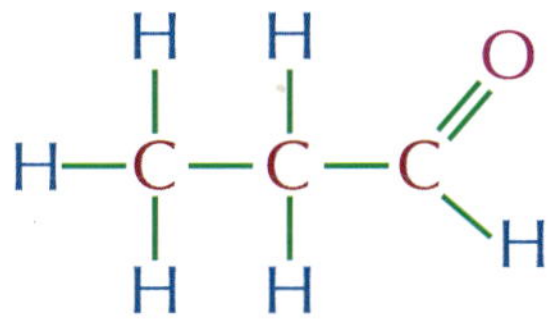

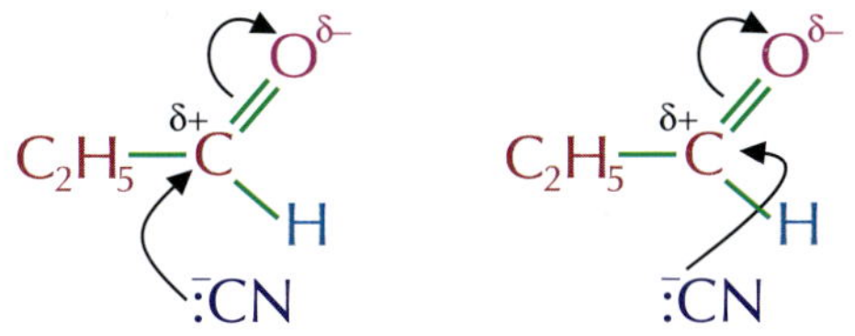

Depending on the direction from which the nucleophilic attack happens, one of two enantiomers is formed.

Because the C=O bond is planar, there is an equal chance that the nucleophile will attack from either of these directions. Therefore, equal amounts (a racemate) of the products of these reactions are formed.

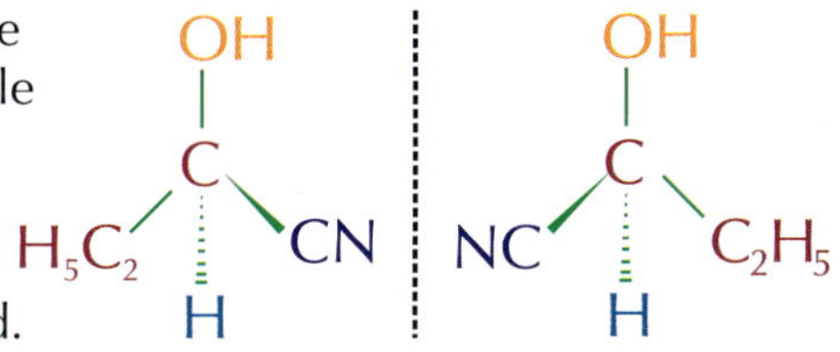

Exam Tip
In the exam, you could be asked to apply this mechanism to show how any aldehyde or ketone reacts with KCN.

Tip: The mechanism for the reactions of aldehydes/ketones is exactly the same with HCN as with KCN. The only difference is that you don't need to add an acid as HCN is a source of H^+ ions (since $HCN \rightleftharpoons H^+ + CN^-$).

Exam Tip
Unless a question asks otherwise, you can show overall equations using displayed formulas, if you find structural formulas hard to use.

Tip: Look back at pages 414-417 for more on enantiomers and racemic mixtures.

Tip: Try building these molecules with a molecular modelling kit if you're struggling to visualise the planar carbonyl group. It will make it so much clearer which directions the nucleophile can attack from.

Tip: This is just the same as the mechanism that you saw on the last page. Look back there for full details on all the steps.

Figure 2: *Warning labels on the bottles of chemicals warn you if a chemical is dangerous.*

Risk assessments

A **risk assessment** involves reviewing the hazards of the reacting chemicals, the products and any conditions needed, such as heat. You don't have to wrap yourself in cotton wool, but you do have to take all reasonable precautions to reduce the risk of an accident.

Example

Here's part of a risk assessment for reacting potassium cyanide with a carbonyl compound:

KCN is toxic but it can be stored safely. Acidified potassium cyanide is used for the reaction to supply the CN^- ions and also the H^+ ions needed. However, the reaction should be done in a fume cupboard as there is a risk of some HCN gas being released from the solution. Also, the organic compounds are flammable, so if you need to heat them use a water bath or electric mantle.

Practice Questions — Application

Q1 a) Draw the structure of the hydroxynitrile that would be produced if 2-methylpropanal was reacted with acidified KCN.
The structure of 2-methylpropanal is shown below:

```
    H  H  O
    |  |  ||
 H—C—C—C—H
    |  |
    H  CH₃
```

b) Draw the mechanism for this reaction.

Q2 The diagram below shows a hydroxynitrile:

```
    H  H  OH H
    |  |  |  |
 H—C—C—C—C—H
    |  |  |  |
    H  H  CN H
```

a) Name this hydroxynitrile.

b) Draw the structure of the carbonyl compound that could be used to produce this hydroxynitrile.

c) Name this carbonyl compound.

d) Draw the mechanism for the production of this hydroxynitrile.

Q3 Butanone reacts with acidified potassium cyanide to produce a mixture of products.

a) Draw the structures of the products from this reaction.

b) Explain why this reaction doesn't produce a single product.

Practice Questions — Fact Recall

Q1 What is a hydroxynitrile?

Q2 Name the mechanism for the production of hydroxynitriles from carbonyl compounds and acidified KCN.

Q3 a) What are the hazards associated with KCN?

b) What safety precautions should be taken when using KCN?

Tip: Remember — when drawing reaction mechanisms, curly arrows must go from a lone pair and to the atom the lone pair is going to.

Tip: Look back at page 418 if you need a reminder on how to name certain carbonyl compounds.

Tip: If a question asks you what a certain molecule is, you can always draw a picture to help you show its features.

4. Carboxylic Acids and Esters

Carboxylic acids and esters are two more types of carbonyl compound that you need to know about. Read on to learn all about them.

What are carboxylic acids?

Carboxylic acids contain the carboxyl functional group –COOH.

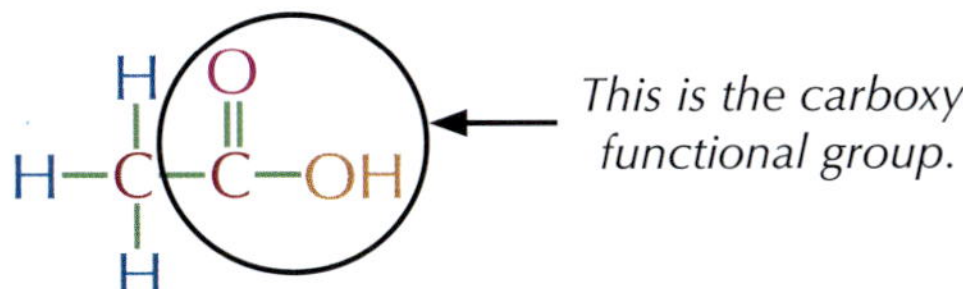

H
O
H–C–C–OH
H
This is the carboxyl functional group.

To name them, you find and name the longest alkane chain, take off the 'e' and add '–oic acid'. The carboxyl group is always at the end of the molecule and when naming it's more important than other functional groups — so all the other functional groups in the molecule are numbered starting from this carbon.

Example

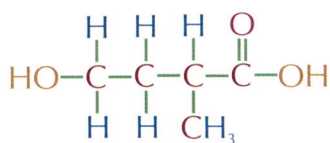

HO–CH₂–CH₂–CH(CH₃)–COOH

4-hydroxy-2-methylbutanoic acid

The longest continuous carbon chain containing the carboxylic acid functional group is 4 carbon atoms, so the stem is butane.

Numbering of the carbons starts at the COOH group so there's a COOH group on carbon-1, a methyl group on carbon-2 and a hydroxyl group on carbon-4.

So, this is 4-hydroxy-2-methylbutanoic acid.

Dissociation of carboxylic acids

Carboxylic acids are weak acids — in water they partially dissociate into a carboxylate ion and an H^+ ion.

$$R\text{–}COOH \rightleftharpoons R\text{–}COO^- + H^+$$

Carboxylic acid — *Carboxylate ion*

This reaction is reversible but the equilibrium lies to the left because most of the molecules don't dissociate.

Reaction with carbonates

Carboxylic acids react with carbonates (CO_3^{2-}) or hydrogen carbonates (HCO_3^-) to form a salt, carbon dioxide and water.

Examples

$$2CH_3COOH_{(aq)} + Na_2CO_{3(s)} \rightarrow 2CH_3COONa_{(aq)} + H_2O_{(l)} + CO_{2(g)}$$

Ethanoic acid — *Sodium carbonate* — *Sodium ethanoate*

$$HCOOH_{(aq)} + NaHCO_{3(s)} \rightarrow HCOONa_{(aq)} + H_2O_{(l)} + CO_{2(g)}$$

Methanoic acid — *Sodium hydrogen carbonate* — *Sodium methanoate*

In these reactions, carbon dioxide fizzes out of the solution.

Learning Objectives:

- Know the structures of carboxylic acids and esters.
- Be able to apply IUPAC rules for naming carboxylic acids and esters.
- Know that carboxylic acids are weak acids but will liberate CO_2 from carbonates.
- Know that carboxylic acids and alcohols react, in the presence of an acid catalyst, to give esters.

Specification Reference 3.3.1.1, 3.3.9.1

Tip: A carboxyl group contains a carbonyl group and a hydroxyl group on the same carbon atom.

Tip: See page 341 for more on the dissociation of weak acids.

Tip: When you're writing equations for the reaction of carboxylic acids with carbonates, always remember to balance the equation at the end.

Tip: Making esters in the lab, by reacting an alcohol and a carboxylic acid under reflux, won't result in a pure product. You'll need to purify your ester using a separating funnel and distillation (see pages 437-438).

Esterification reactions

If you heat a carboxylic acid with an alcohol in the presence of a strong acid catalyst, you get an **ester**. It's called an **esterification** reaction.

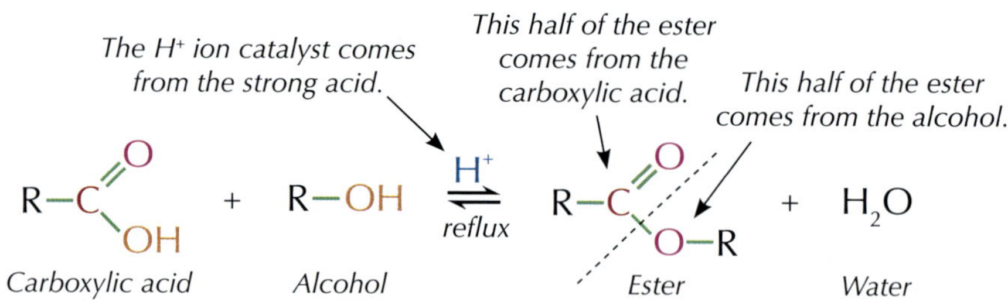
The H^+ ion catalyst comes from the strong acid.
This half of the ester comes from the carboxylic acid.
This half of the ester comes from the alcohol.
R–C(=O)–OH + R–OH ⇌ R–C(=O)–O–R + H_2O
H^+
reflux
Carboxylic acid
Alcohol
Ester
Water

Concentrated sulfuric acid (H_2SO_4) is usually used as the acid catalyst, but other strong acids such as HCl or H_3PO_4 can also be used.

Tip: Esterification reactions are also condensation reactions because they release a small molecule (water) when the carboxylic acid and alcohol join.

Naming esters

You've just seen that an ester is formed by reacting an alcohol with a carboxylic acid. Well, the name of an ester is made up of two parts — the first bit comes from the alcohol, and the second bit from the carboxylic acid.

To name an ester, just follow these steps:

1. Look at the alkyl group that came from the alcohol. This is the first bit of the ester's name.
2. Now look at the part that came from the carboxylic acid. Swap its '-oic acid' ending for 'oate' to get the second bit of the name.
3. Put the two parts together.

Figure 1: *Model showing the structure of the ester ethyl ethanoate.*

Example

Ethanoic acid reacts with methanol to produce the ester shown below:

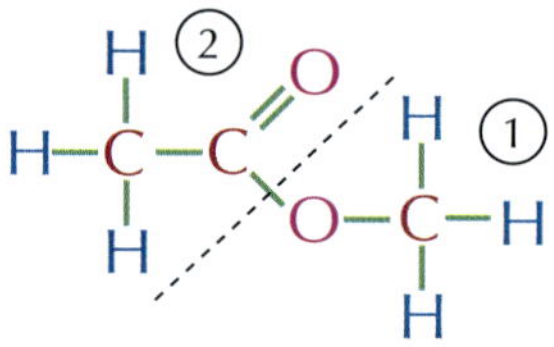

1. This part of the ester came from the alcohol. It's a methyl group so the first part of the ester's name is methyl-.
2. This part of the ester came from the carboxylic acid. It came from ethanoic acid so the second part of the ester's name is - ethanoate.

So this ester is methyl ethanoate.

Tip: Be careful — the name of the ester is generally written the opposite way round to the formula.

The same rules apply even if the carbon chains are branched or if the molecule has a benzene ring attached. Always number the carbons starting from the carbon atoms in the C–O–C bond.

Examples

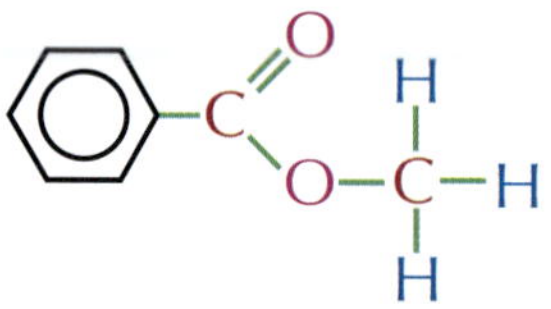

This ester has a methyl group that came from the alcohol so the name begins with methyl-.

There is a benzene ring that came from benzoic acid so the name ends in -benzoate.

So this is methyl benzoate.

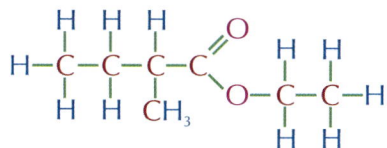

```
 H  H  H    O
 |  |  |   //    H  H
H-C--C--C--C      |  |
 |  |  |    \O--C--C-H
 H  H  CH3      |  |
                H  H
```

This ester has an ethyl group that came from the alcohol and a 2-methyl butanyl group that came from the acid so is called ethyl 2-methylbutanoate.

```
    O
   //    H  H  H
H-C      |  |  |
   \O--C--C--C-H
       |  |  |
      H3C H  H
```

This ester has a 1-methylpropyl group that came from the alcohol and a methyl group that came from the acid so is called 1-methylpropyl methanoate.

Sometimes you may be asked to predict which alcohol and which carboxylic acid are needed to form a particular ester.

Example

```
 H  H   O
 |  |  //    H
H-C--C--C     |
 |  |   \O--C-H
 H  H        |
             H
```

There are 3 carbons in the part of the molecule that came from the acid so the stem is propane. This part came from propanoic acid.

There is one carbon in the section that came from the alcohol so the stem is methane. This part of the molecule came from methanol.

Tip: The rules about naming esters only apply if the alcohol is a primary alcohol — if it's a secondary alcohol the first part of the name will be different. E.g. The alcohol that reacts with methanoic acid to form 1-methylpropyl methanoate is called butan-2-ol, not 1-methylpropan-1-ol.

Practice Questions — Application

Q1 Below are two carboxylic acids:

(i)

```
   O
   ||
H--C--OH
```

(ii)

```
 H  H  H  H  O
 |  |  |  |  ||
H-C--C--C--C--C-OH
 |  |  |  |
 H  OH H  C2H5
```

a) Name these carboxylic acids.

b) Write a balanced equation for the reaction of carboxylic acid (i) with sodium carbonate (Na_2CO_3).

Q2 Below are two esters:

(i)

```
    O
   //    H  H  H
H-C      |  |  |
   \O--C--C--C-H
       |  |  |
       H  H  H
```

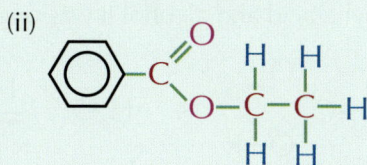

(ii)

```
          O
         //    H  H
(C6H5)--C      |  |
         \O--C--C-H
             |  |
             H  H
```

a) Name these esters

b) State which carboxylic acid and which alcohol have reacted to form each of these esters.

Q3 1-methylethyl methanoate is an ester.

a) Draw the structure of this ester.

b) Write an equation, using structural formulas, to show the formation of this ester from an acid and an alcohol.

Exam Tip

Sometimes examiners will try and throw you by drawing esters the opposite way round, e.g.

```
 H      O  H  H
 |      ||  |  |
H-C--O--C--C--C-H
 |          |  |
 H          H  H
```

Try not to get confused when naming esters — always think about the position of the O and C=O groups rather than just thinking about left and right. This ester is methyl propanoate.

Practice Questions — Fact Recall

Q1 Name the three products that are produced when a carboxylic acid reacts with a carbonate or a hydrogen carbonate.

Q2 Carboxylic acids react with alcohols to form esters.

a) What is the name given to this type of reaction?

b) What type of catalyst is used for this reaction? Give an example.

5. Reactions and Uses of Esters

Learning Objectives:

- Know the common uses of esters (e.g. in solvents, plasticisers, perfumes and food flavourings).
- Know that esters can be hydrolysed in acid or alkaline conditions to form alcohols and carboxylic acids or salts of carboxylic acids.
- Know that vegetable oils and animal fats are esters of propane-1,2,3-triol (glycerol).
- Know that vegetable oils and animal fats can be hydrolysed in alkaline conditions to give soap (salts of long-chain carboxylic acids) and glycerol.
- Know that biodiesel is a mixture of methyl esters of long-chain carboxylic acids.
- Know that biodiesel can be produced by reacting vegetable oils with methanol in the presence of a catalyst.

Specification Reference 3.3.9.1

Esters are extremely useful molecules. In fact — you've probably used an ester at some point today without even realising it. Read on to find out more.

Useful properties of esters

Esters have a number of properties that make them very useful.

- Esters have a sweet smell — it varies from gluey sweet for smaller esters to a fruity 'pear drop' smell for the larger ones. This makes them useful in perfumes. The food industry uses esters to flavour things like drinks and sweets too.
- Esters are polar liquids so lots of polar organic compounds will dissolve in them. They've also got quite low boiling points, so they evaporate easily from mixtures. This makes them good solvents in glues and printing inks.
- Esters are used as plasticisers — they're added to plastics during polymerisation to make the plastic more flexible. Over time, the plasticiser molecules escape though, and the plastic becomes brittle and stiff.

Tip: The nice fragrances and flavours of lots of flowers and fruits come from esters.

Hydrolysis of esters

Hydrolysis is when a substance is split up by water — but using just water is often really slow, so an acid or an alkali is often used to speed it up. There are two types of hydrolysis of esters — acid hydrolysis and base hydrolysis. With both types you get an alcohol, but the second product in each case is different.

Acid hydrolysis

Acid hydrolysis splits the ester into an acid and an alcohol — it's the reverse of the reaction on page 426. You have to reflux the ester with a dilute acid, such as hydrochloric or sulfuric acid. The ester will then split back into the carboxylic acid and alcohol it was originally made from.

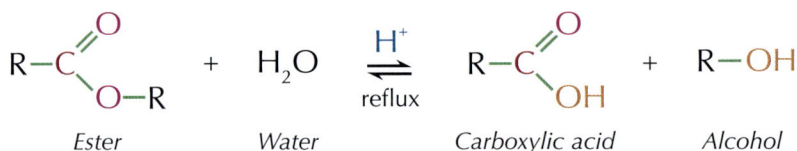

$$\mathrm{RCOOR} + H_2O \underset{\text{reflux}}{\overset{H^+}{\rightleftharpoons}} \mathrm{RCOOH} + \mathrm{R{-}OH}$$

Ester + *Water* ⇌ *Carboxylic acid* + *Alcohol*

Example

Acid hydrolysis of methyl ethanoate produces ethanoic acid and methanol:

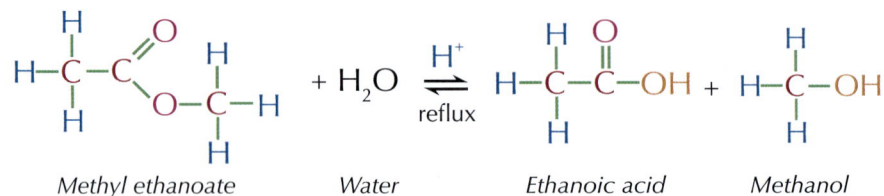

$$CH_3COOCH_3 + H_2O \underset{\text{reflux}}{\overset{H^+}{\rightleftharpoons}} CH_3COOH + CH_3OH$$

Methyl ethanoate + *Water* ⇌ *Ethanoic acid* + *Methanol*

As these acid hydrolysis reactions are reversible you need to use lots of water to push the equilibrium over to the right so you get lots of product.

Exam Tip

Don't worry, you don't need to know the mechanism for these reactions — just knowing the equations is enough.

Base hydrolysis

For a base hydrolysis reaction you have to reflux the ester with a dilute alkali, such as sodium hydroxide. OH^- ions from the base react with the ester and you get a carboxylate ion and an alcohol.

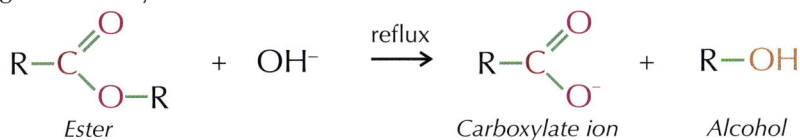

$$\underset{\text{Ester}}{R-C(=O)-O-R} + OH^- \xrightarrow{\text{reflux}} \underset{\text{Carboxylate ion}}{R-C(=O)-O^-} + \underset{\text{Alcohol}}{R-OH}$$

Example

Base hydrolysis of methyl ethanoate produces ethanoate ions and methanol:

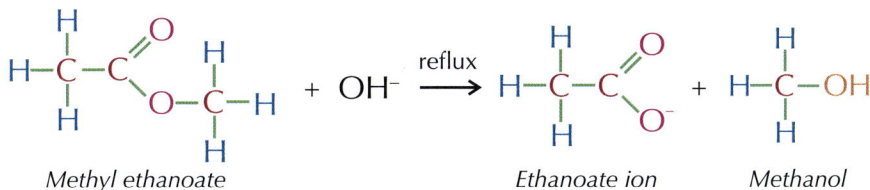

$$\underset{\text{Methyl ethanoate}}{CH_3-C(=O)-O-CH_3} + OH^- \xrightarrow{\text{reflux}} \underset{\text{Ethanoate ion}}{CH_3-C(=O)-O^-} + \underset{\text{Methanol}}{CH_3-OH}$$

Tip: Make sure you know the difference between the two different types of hydrolysis — acid hydrolysis produces a carboxylic acid and base hydrolysis produces a carboxylate ion.

Tip: The negatively charged carboxylate ions bond with the positively charged ions from the base (e.g. Na^+ ions if the base is NaOH) to form salts like the one shown below:

$$CH_3-C(=O)-O^-Na^+$$

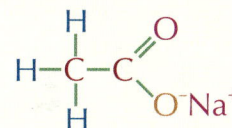

This is sodium ethanoate.

Fats and oils

Animal fats and vegetable oils are esters of glycerol and **fatty acids**. Fatty acids are long chain carboxylic acids. They combine with glycerol (propane-1,2,3-triol) to make fats and oils — see Figure 1. The fatty acids can be **saturated** (no double bonds) or **unsaturated** (with C=C double bonds). Most of a fat or oil is made from fatty acid chains — so it's these that give them many of their properties.

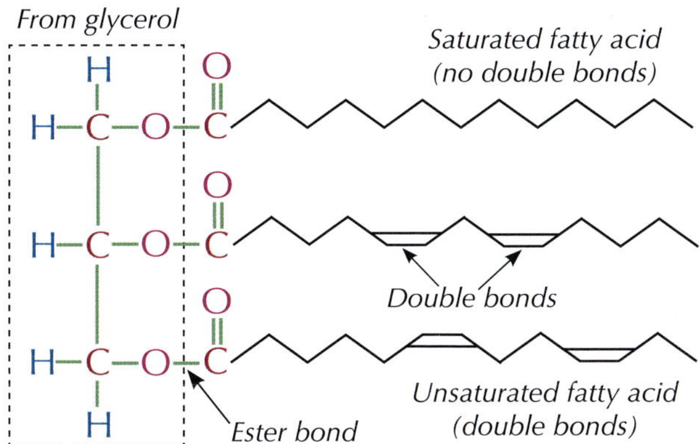

Figure 1: *Diagram showing the structure of a fat/oil.*

Figure 2: *Fats and oils are esters formed from glycerol (propane-1,2,3-triol) and long-chain carboxylic acids.*

'Fats' have mainly saturated hydrocarbon chains — they fit neatly together, increasing the van der Waals forces between them (see Figure 3). This means higher temperatures are needed to melt them and they're solid at room temperature.

Figure 3: *The arrangement of hydrocarbon chains in a fat.*

'Oils' have unsaturated hydrocarbon chains — the double bonds mean the chains are bent and don't pack together well, decreasing the effect of the van der Waals forces (see Figure 4). So they're easier to melt and are liquids at room temperature.

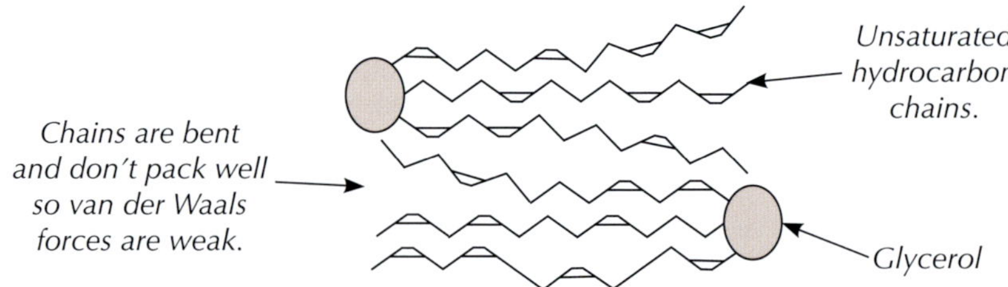
Unsaturated hydrocarbon chains.
Chains are bent and don't pack well so van der Waals forces are weak.
Glycerol

Figure 4: *The arrangement of hydrocarbon chains in an oil.*

Tip: See the previous page for more on the base hydrolysis of esters.

Hydrolysis of fats and oils

Like any ester, you can hydrolyse oils and fats by heating them with sodium hydroxide. This is a type of base hydrolysis. OH^- ions from the sodium hydroxide react with the fat/oil to form a carboxylate ion and an alcohol. The alcohol that is formed is glycerol (propane-1,2,3-triol) and the carboxylate ions combine with Na^+ ions from the sodium hydroxide to form a sodium salt. And you'll never guess what the sodium salt produced is — a soap.

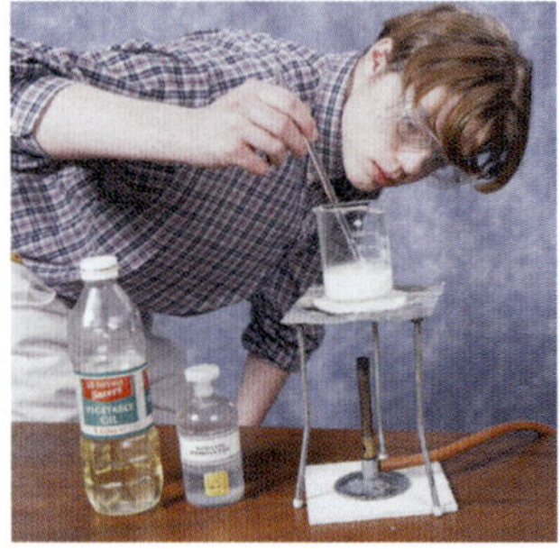

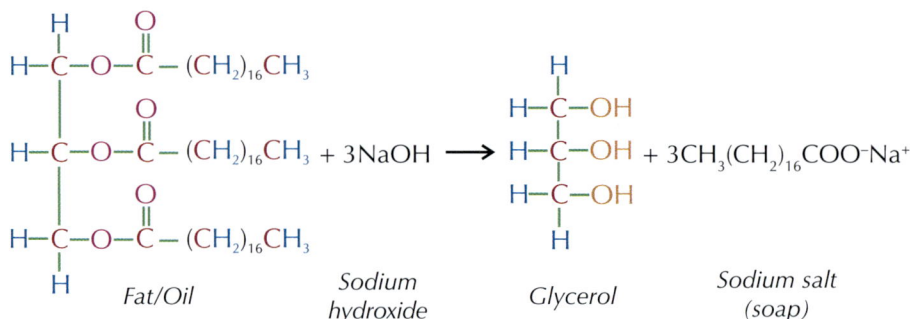

Figure 5: *Student making soap by heating vegetable oil with sodium hydroxide.*

$$\text{Fat/Oil (triglyceride with three } -O-CO-(CH_2)_{16}CH_3 \text{ groups)} + 3NaOH \longrightarrow \text{Glycerol } (CH_2OH-CHOH-CH_2OH) + 3CH_3(CH_2)_{16}COO^-Na^+$$

Fat/Oil — Sodium hydroxide — Glycerol — Sodium salt (soap)

If you want to, you can then convert the sodium salt back into a long chain carboxylic acid (fatty acid) by adding an acid such as HCl. H^+ ions from the acid displace the Na^+ ions in the salt to form a carboxylic acid, which releases a free Na^+ ion. This reaction is shown below:

$$CH_3(CH_2)_{16}COO^-Na^+ + H^+ \rightarrow CH_3(CH_2)_{16}COOH + Na^+$$

sodium salt (soap) — *fatty acid*

Tip: You may get the chance to make soap and biodiesel in the lab. If you do, make sure you carry out a risk assessment before you start your experiment, and take any necessary safety precautions.

Biodiesel

Vegetable oils, e.g. rapeseed oil, make good vehicle fuels, but you can't burn them directly in engines. The oils must be converted into **biodiesel** first. This involves reacting them with methanol, using a strong alkali (e.g. potassium hydroxide or sodium hydroxide) as a catalyst. You get a mixture of methyl esters of long-chain fatty acids — this is biodiesel.

$$\text{Fat/Oil (triglyceride with three } -O-CO-R \text{ groups)} + 3CH_3OH \xrightarrow[\text{(catalyst)}]{KOH} \text{Glycerol } (CH_2OH-CHOH-CH_2OH) + 3R-COO-CH_3$$

R = long carbon chain

Fat/Oil — Methanol — Glycerol (propane-1,2,3-triol) — Methyl ester (biodiesel is a mixture of methyl esters)

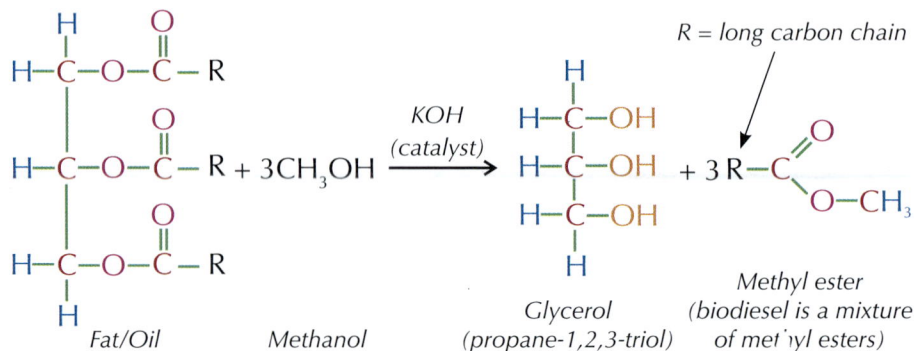

Exam Tip
Make sure you learn the equation for the formation of biodiesel — it could come up in your exam.

Is biodiesel 100% carbon neutral?

Biodiesel can be thought of as carbon neutral, because when crops grow they absorb the same amount of CO_2 as they produce when they're burned. But it's not quite as simple as that — energy is used to make the fertilizer to grow the crops, and it's used in planting, harvesting and converting the oil. If this energy comes from fossil fuels, then the process won't be carbon neutral overall.

Figure 6: *A biodiesel fuel pump.*

Practice Questions — Application

Q1 Below is the ester methyl propanoate:

```
    H   H    O
    |   |   //       H
H — C — C — C        |
    |   |    \ O  —  C — H
    H   H            |
                     H
```

Write an equation to show:

a) the acid hydrolysis of this ester.

b) the base hydrolysis of this ester.

Q2 Stearic acid ($CH_3(CH_2)_{16}COOH$) is a long chain carboxylic acid found in most animal fats.

a) Draw the triester that would be produced if three molecules of stearic acid reacted with glycerol.

b) Write an equation for the production of soap from this triester.

c) Explain how soap could be converted back into glycerol and stearic acid.

Q3 Below is an ester commonly found in vegetable oils.

$$\begin{array}{l} CH_2OOC(CH_2)_{14}CH_3 \\ | \\ CH_2OOC(CH_2)_{14}CH_3 \\ | \\ CH_2OOC(CH_2)_{14}CH_3 \end{array}$$

a) Write an equation for the conversion of this ester into biofuel.

b) Suggest a suitable catalyst for this reaction.

Exam Tip

If you don't recognise the structure of a molecule in your exam, try re-drawing it in a different way — it might start to look more familiar.

Practice Questions — Fact Recall

Q1 Give three common uses of esters.

Q2 What two products are produced when an ester is broken down by:

a) acid hydrolysis?

b) base hydrolysis?

Q3 What two things are fats and oils made of?

Q4 Explain why the properties of fats and oils are different.

Q5 Fats can be hydrolysed by heating them with NaOH. What are the products of this reaction?

Q6 How is biodiesel made?

Q7 Explain why biodiesel usually isn't 100% carbon neutral.

Learning Objectives:

- Know the structure of acyl chlorides.
- Be able to apply IUPAC rules for naming acyl chlorides.
- Be able to outline the nucleophilic addition-elimination reactions of water, alcohols, ammonia and primary amines with acyl chlorides.
- Be able to outline the mechanism of nucleophilic addition-elimination reactions of acyl chlorides with water, alcohols, ammonia and primary amines.

Specification Reference 3.3.1.1, 3.3.9.2

6. Acyl Chlorides

Acyl chlorides are a particularly useful type of carbonyl compound because they are good starting points for making lots of different types of molecule.

What are acyl chlorides?

Acyl (or **acid**) **chlorides** have the functional group COCl — their general formula is $C_nH_{2n-1}OCl$. Naming acyl chlorides is similar to naming carboxylic acids. All their names end in –oyl chloride and the carbon atoms are numbered from the end with the acyl functional group.

Example

```
    H  H  CH3 H  O
    |  |  |   |  ||
 H—C—C—C—C—C—Cl
    |  |  |   |
    H  H  H   CH3
```

2,3-dimethylpentanoyl chloride

The longest continuous carbon chain containing the acyl chloride functional group is 5 carbon atoms, so the stem is pentane.

There are methyl groups on carbon-2 and carbon-3.

So, it's 2,3-dimethylpentanoyl chloride.

Reactions of acyl chlorides

Acyl chlorides can react with a wide range of different molecules. In each of these reactions, Cl is substituted by an oxygen or nitrogen group and misty fumes of hydrogen chloride are given off. The key reactions involving acyl chlorides that you need to know are:

Figure 1: The reaction of an acyl chloride with water. You can see the misty fumes of HCl that are being given off.

Reaction with water

Acyl chlorides react vigorously with cold water, producing a carboxylic acid.

```
    H  O                          H  O
    |  ||                         |  ||
 H—C—C—Cl  +  H2O  ———→  H—C—C—OH  +  HCl
    |                             |
    H                             H
 Ethanoyl chloride             Ethanoic acid
```

Reaction with alcohols

Acyl chlorides react vigorously with alcohols at room temperature, producing an ester.

```
                                   H     O
    H  O                           |    //        H
    |  ||                       H—C—C           |
 H—C—C—Cl  +  CH3OH  ———→        |    \O—C—H   +  HCl
    |                              H             |
    H                                            H
 Ethanoyl chloride   Methanol        Methyl ethanoate
```

This irreversible reaction is a much easier, faster way to produce an ester than esterification (see page 426).

Reaction with ammonia

Acyl chlorides react vigorously with ammonia at room temperature, producing an amide.

```
    H  O                          H  O
    |  ||                         |  ||
 H—C—C—Cl  +  NH3  ———→  H—C—C—NH2  +  HCl
    |                             |
    H                             H
 Ethanoyl chloride             Ethanamide
```

Tip: Amides are organic compounds with a C=O double bond and an $-NH_2$ group attached to the same carbon.

Reaction with primary amines

Acyl chlorides react vigorously with **primary amines** at room temperature, producing an **N-substituted amide**.

$$H{-}\underset{H}{\overset{H}{C}}{-}\overset{O}{\overset{\|}{C}}{-}Cl \; + \; CH_3NH_2 \longrightarrow H{-}\underset{H}{\overset{H}{C}}{-}\overset{O}{\overset{\|}{C}}{-}\overset{H}{N}{-}\underset{H}{\overset{H}{C}}{-}H \; + \; HCl$$

Ethanoyl chloride *Methylamine* *N-methylethanamide*

Tip: See pages 453-460 for more on amines and amides.

Nucleophilic addition-elimination

All of the reactions involving acyl chlorides shown on this page and the previous page follow the same mechanism — they are all **nucleophilic addition-elimination** reactions. Acyl chloride nucleophilic addition-elimination reactions have two steps:

1. The nucleophile adds onto the acyl chloride, displacing a Cl^- ion.
2. The hydrogen leaves to create an acyl chloride derivative.

Tip: A nucleophile is a molecule that can donate a lone pair of electrons.

Step 1:

In acyl chlorides, both the chlorine and the oxygen atoms draw electrons towards themselves, so the carbon has a slight positive charge — meaning it's easily attacked by nucleophiles.

2. A pair of electrons is transferred from the C=O bond onto the oxygen.

3. The pair of electrons on the oxygen reform the double bond and the chlorine is kicked off.

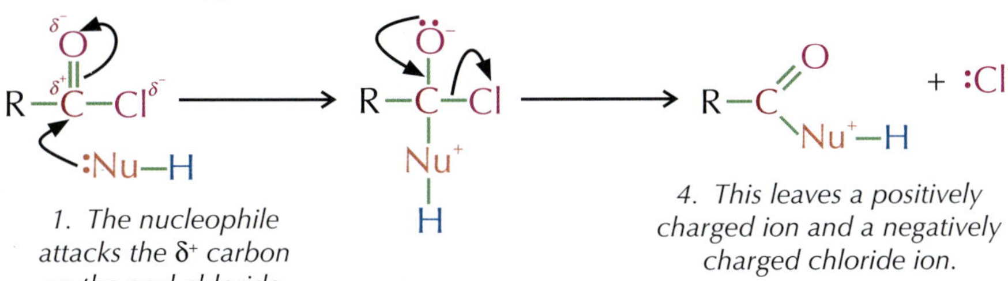

1. The nucleophile attacks the δ⁺ carbon on the acyl chloride.

4. This leaves a positively charged ion and a negatively charged chloride ion.

Exam Tip
Sometimes you will be asked to apply this mechanism to a nucleophilic reaction that you haven't studied before, so make sure you really understand it — it won't be enough to just learn the examples.

Step 2:

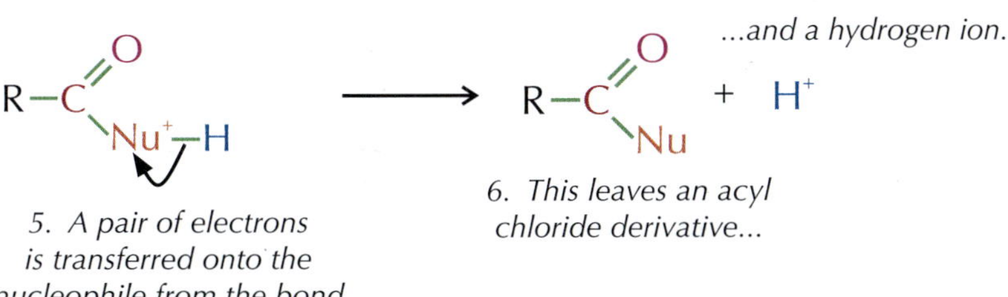

5. A pair of electrons is transferred onto the nucleophile from the bond

6. This leaves an acyl chloride derivative...

...and a hydrogen ion.

Tip: An acyl chloride derivative is just something that is made from an acyl chloride.

All the reactions that you need to know involving acyl chlorides work in exactly the same way. You just need to change the nucleophile (Nu) to whichever nucleophile is present in the reaction, for example water (H_2O:), an alcohol (e.g. $CH_3\ddot{O}H$), ammonia ($\dot{N}H_3$) or an amine (e.g. $CH_3\dot{N}H_2$).

Exam Tip
You don't have to show the two steps separately in your exam — drawing them one after the other is fine.

Tip: Any nucleophile could be involved in a nucleophilic addition-elimination reaction — not just the ones shown here.

Example

Below is the mechanism for the reaction of ethanoyl chloride with methanol:

Step 1:

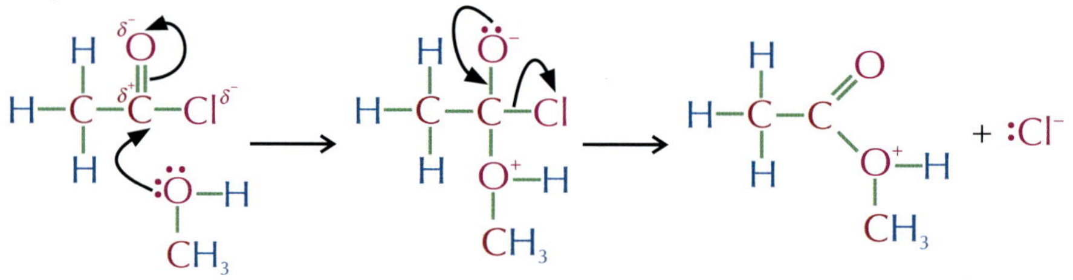

$$CH_3C^{\delta+}(=O^{\delta-})Cl^{\delta-} + :\ddot{O}H(CH_3) \longrightarrow CH_3C(O^-)(Cl)(O^+H\,CH_3) \longrightarrow CH_3C(=O)O^+H(CH_3) + :Cl^-$$

Methanol is the nucleophile here. It attacks the partially positive carbon on the ethanoyl chloride, and a pair of electrons from the C=O bond are transferred to the oxygen. Then the pair of electrons on the oxygen reform the double bond and the chlorine's kicked off.

Step 2:

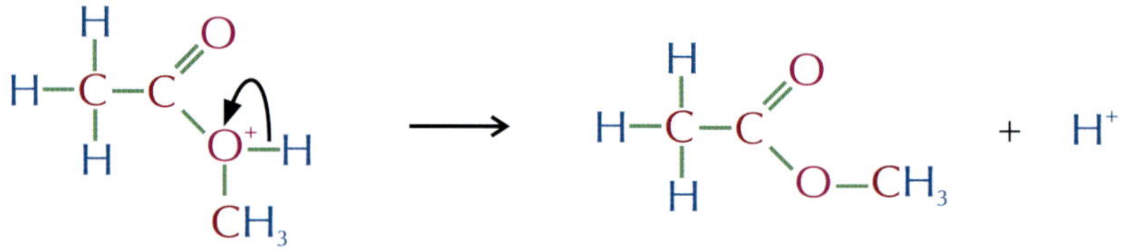

$$CH_3C(=O)O^+H(CH_3) \longrightarrow CH_3C(=O)O{-}CH_3 + H^+$$

The hydrogen in the hydroxyl group of the methanol leaves, leaving methyl ethanoate and a hydrogen ion.

Practice Questions — Application

Q1 Name the acyl chlorides that are shown below:

(a) $HCOCl$ (H–C(=O)–Cl)

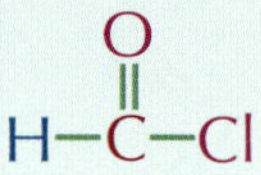

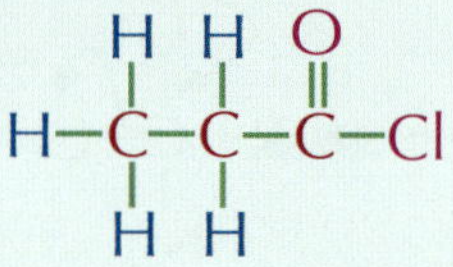

(b) CH_3CH_2COCl

(c) $CH_3CH(CH_3)COCl$

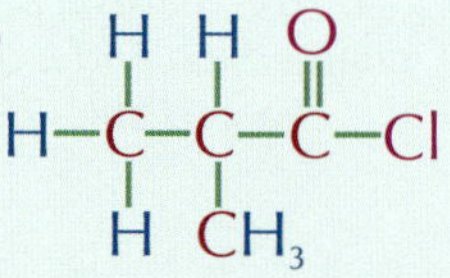

(d) $CH_2(OH)CH_2CH(CH_3)COCl$

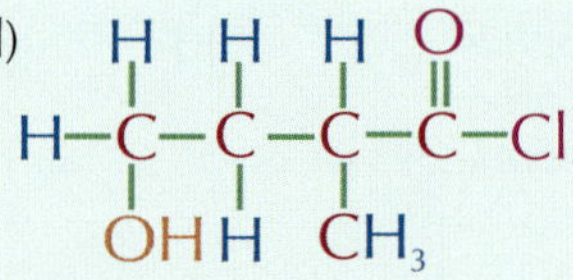

Q2 Draw the mechanisms for the following reactions:

a) acyl chloride (a) with methanol.

b) acyl chloride (b) with water.

c) acyl chloride (c) with ammonia.

d) acyl chloride (d) with methylamine.

Practice Questions — Fact Recall

Q1 What is the general formula for an acyl chloride?

Q2 What are the products when an acyl chloride is reacted with:

a) water? b) an alcohol? c) ammonia? d) an amine?

Q3 What is the name given to the mechanism for the reactions of acyl chlorides with water, alcohols, ammonia and amines?

7. Acid Anhydrides

Acid anhydrides are compounds that react in a similar way to acyl chlorides.

What are acid anhydrides?

An **acid anhydride** is made from two identical carboxylic acid molecules. The two carboxylic acid molecules are joined together via an oxygen with the carbonyl groups on either side. This oxygen has come from the OH group of one of the carboxylic acids. The other OH group and the spare hydrogen are released as water. The formation of an acid anhydride is shown below.

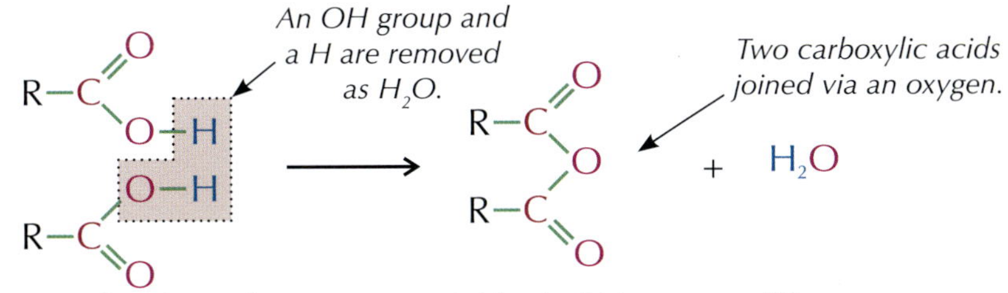

An OH group and a H are removed as H_2O.

Two carboxylic acids joined via an oxygen.

$$2\ R{-}COOH \longrightarrow (RCO)_2O + H_2O$$

2 × Carboxylic acid — *Acid anhydride* — *Water*

If you know the name of the carboxylic acid, acid anhydrides are easy to name — just take away 'acid' and add 'anhydride'. So methanoic acid gives methanoic anhydride, ethanoic acid gives ethanoic anhydride, etc...

Example

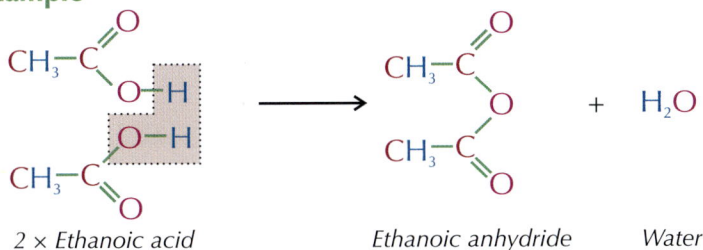

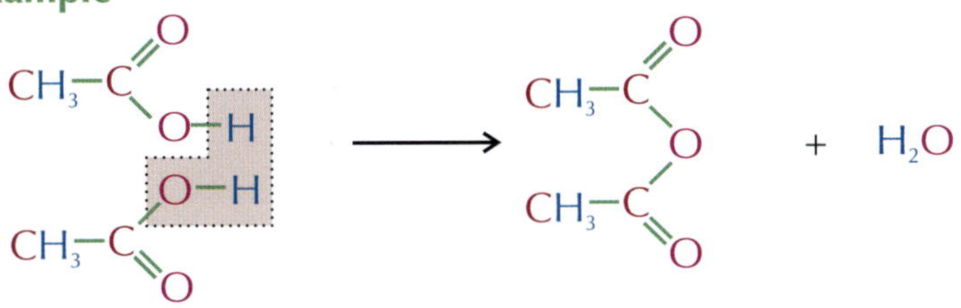

$$2\ CH_3{-}COOH \longrightarrow (CH_3CO)_2O + H_2O$$

2 × Ethanoic acid — *Ethanoic anhydride* — *Water*

Reactions of acid anhydrides

You need to know the reactions of water, alcohol, ammonia and primary amines with acid anhydrides. Luckily, they're almost the same as those of acyl chlorides — the reactions are just less vigorous and you get a carboxylic acid formed instead of HCl.

Examples

Acid anhydrides react with water, producing a carboxylic acid.

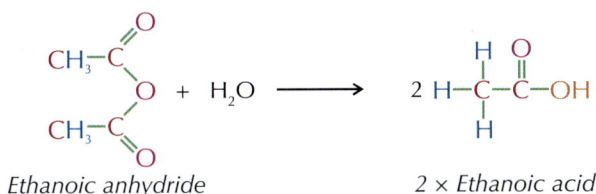

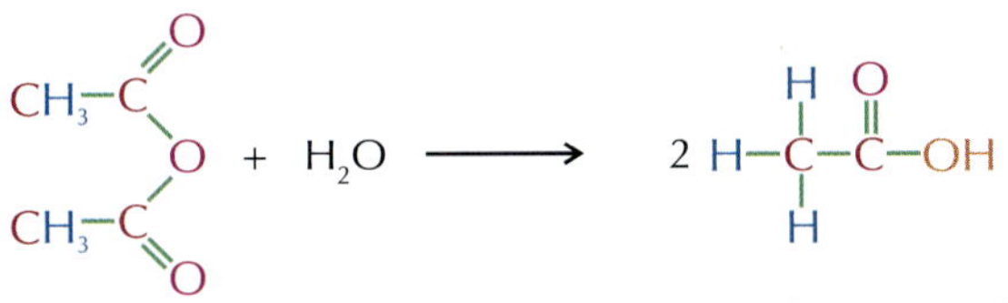

$$(CH_3CO)_2O + H_2O \longrightarrow 2\ CH_3COOH$$

Ethanoic anhydride — *2 × Ethanoic acid*

Acid anhydrides react with alcohols, producing an ester.

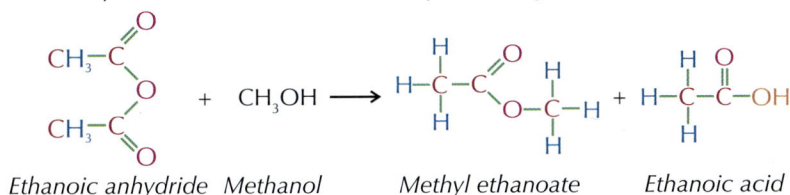

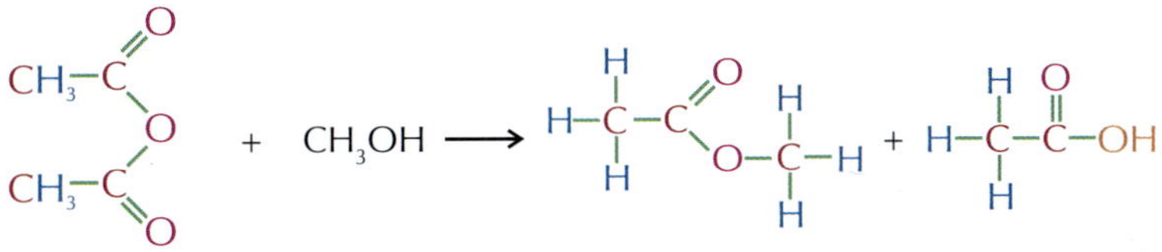

$$(CH_3CO)_2O + CH_3OH \longrightarrow CH_3COOCH_3 + CH_3COOH$$

Ethanoic anhydride — *Methanol* — *Methyl ethanoate* — *Ethanoic acid*

Learning Objectives:

- Know the structure of acid anhydrides.
- Be able to apply IUPAC rules for naming acid anhydrides.
- Be able to outline the nucleophilic addition-elimination reactions of water, alcohols, ammonia and primary amines with acid anhydrides.
- Understand the industrial advantages of ethanoic anhydride over ethanoyl chloride in the manufacture of the drug aspirin.

Specification Reference 3.3.1.1, 3.3.9.2

Tip: See page 425 for how to name carboxylic acids.

Tip: See pages 432-434 for more on the reactions of acyl chlorides.

Tip: All of these reactions are nucleophilic addition-elimination reactions — the same as with the acyl chlorides.

Acid anhydrides react with ammonia, producing an amide.

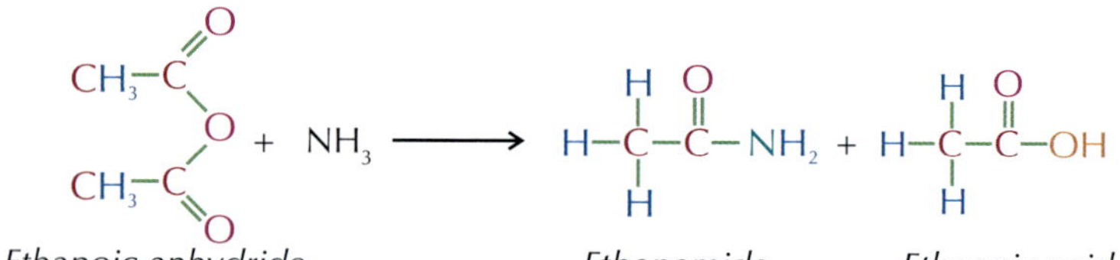

$$(CH_3CO)_2O + NH_3 \longrightarrow CH_3CONH_2 + CH_3COOH$$

Ethanoic anhydride → *Ethanamide* + *Ethanoic acid*

Acid anhydrides react with amines, producing an N-substituted amide.

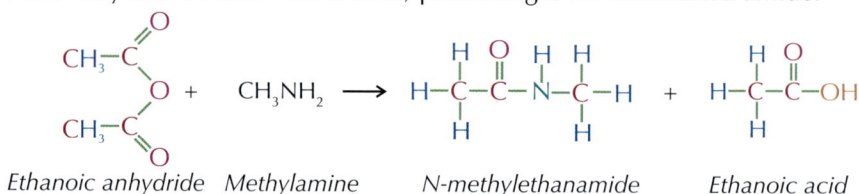

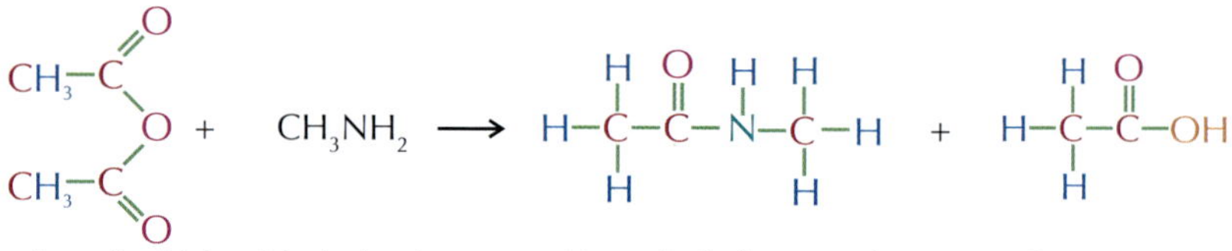

$$(CH_3CO)_2O + CH_3NH_2 \longrightarrow CH_3CONHCH_3 + CH_3COOH$$

Ethanoic anhydride + *Methylamine* → *N-methylethanamide* + *Ethanoic acid*

Tip: The carboxylic acid formed at the end of all these reactions is the same as the carboxylic acid that would have made the anhydride.

Tip: Aspirin is a drug used to relieve symptoms such as minor aches and pains, fever and swelling.

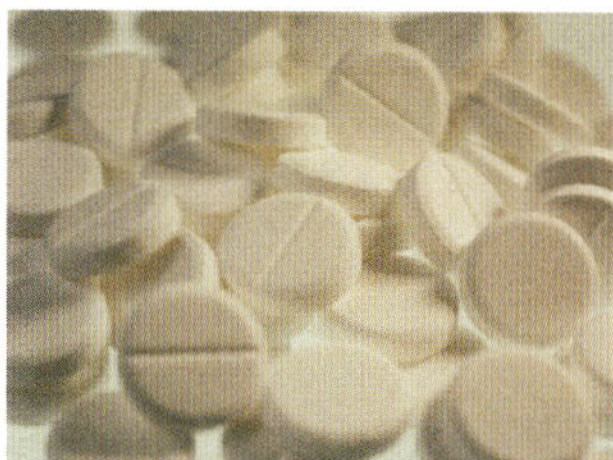

Figure 1: *Aspirin tablets.*

Tip: In these reactions salicylic acid is behaving like an alcohol so everything you've just learnt about the reactions of acid anhydrides/acyl chlorides with alcohols applies here too.

Manufacture of aspirin

Aspirin is an ester — it can be made by reacting salicylic acid (which has an alcohol group) with either ethanoic anhydride (see Figure 2) or ethanoyl chloride. Ethanoic anhydride is used in industry because it's cheaper than ethanoyl chloride. It's also safer to use than ethanoyl chloride as it's less corrosive, reacts more slowly with water, and doesn't produce dangerous hydrogen chloride fumes.

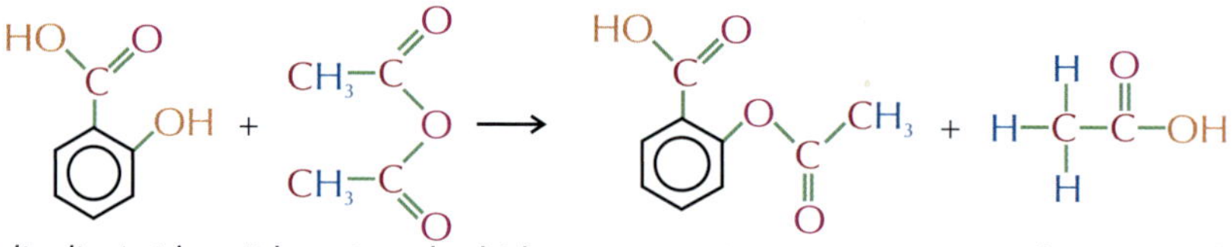

$$HOOC\text{-}C_6H_4\text{-}OH + (CH_3CO)_2O \longrightarrow HOOC\text{-}C_6H_4\text{-}OCOCH_3 + CH_3COOH$$

Salicylic Acid + *Ethanoic anhydride* → *Aspirin* + *Ethanoic acid*

Figure 2: *Using ethanoic anhydride to produce aspirin.*

Practice Questions — Application

Q1 Draw the following anhydrides:

a) propanoic anhydride

b) 2-ethylpentanoic anhydride

c) 4-hydroxy-2,3-dimethylbutanoic anhydride

Q2 Write an equation for the following reactions:

a) propanoic anhydride with water

b) 2-ethylpentanoic anhydride with methanol

c) 4-hydroxy-2,3-dimethylbutanoic anhydride with ammonia

Q3 Write the equation for the production of aspirin from salicylic acid and ethanoyl chloride.

Practice Questions — Fact Recall

Q1 What is produced when propanoic anhydride reacts with propanol?

Q2 Give two reason why ethanoic anhydride is used instead of ethanoyl chloride when producing aspirin.

8. Purifying Organic Compounds

Synthesising organic compounds is hardly ever as simple as it sounds. The products of organic reactions are almost always riddled with impurities. It's a good thing there are many ways of purifying them, outlined on these pages.

REQUIRED PRACTICAL 10

Learning Objectives:
- Be able to prepare a pure organic solid and test its purity (Required Practical 10).
- Be able to prepare a pure organic liquid (Required Practical 10).

Specification Reference 3.3.9.2

Separation

If an organic product is insoluble in water, then you can use separation to remove any impurities that do dissolve in water, such as salts or water soluble organic compounds (e.g. alcohols).

The organic layer and the aqueous layer (which contains any water soluble impurities) are immiscible, i.e. they don't mix, so separate out into two distinct layers (see Figure 1). You can then open the tap and run each layer off into a separate container.

The organic layer is normally less dense than the aqueous layer so should float on top. Any water soluble impurities should have dissolved in the lower aqueous layer. You can then open the stopper on the separating funnel, run off the aqueous layer and collect your product.

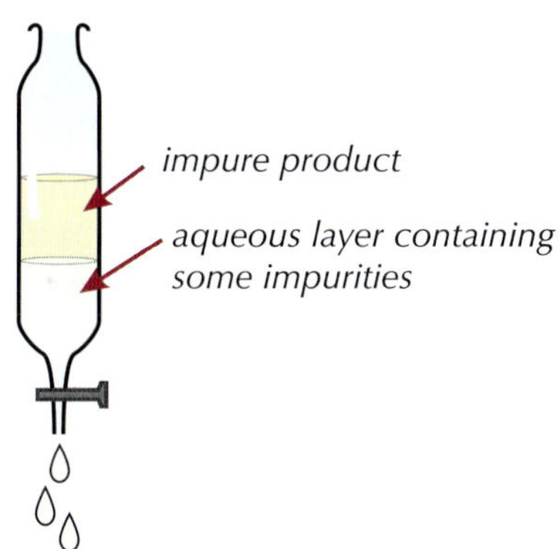

Figure 1: *Separating apparatus.*

Tip: Water soluble impurities might include salts (such as NaCl) or short-chain organic compounds that are capable of forming hydrogen bonds (such as alcohols, carbonyls and carboxylic acids).

Tip: Your product won't necessarily be pure after separation — any organic impurities that don't dissolve in water will still be in the organic layer alongside your product. You'll probably need to remove them by redistillation (see page 439).

Solvent extraction

If your product and impurities are all dissolved in a solution together, you can use another form of separation known as **solvent extraction**. This involves vigorously shaking your impure product with an immiscible solvent, so they temporarily mix. Your product needs to be more soluble in the added immiscible solvent than the one it was initially dissolved in. If this is the case, the product will dissolve in the added solvent, and separate from the solution containing the impurities. The solvent containing the product can then be run off using a separating funnel, as shown above.

Figure 2: *A student using a separating funnel to separate an organic and an aqueous layer.*

Drying agents

If you use separation to purify a product, the organic layer will end up containing trace amounts of water, so it has to be dried. To do this you can add an anhydrous salt such as magnesium sulfate ($MgSO_4$) or calcium chloride ($CaCl_2$). The salt is used as a **drying agent** — it binds to any water present to become hydrated.

When you first add the salt to the organic layer it will form clumps. This means you need to add more. You know that all the water has been removed when you can swirl the mixture and it looks like a snow globe. You can filter the mixture to remove the solid drying agent.

Tip: For all of these methods of purification, you must take all necessary safety precautions before and during the processes.

Tip: Adding water to an impure organic product before you separate it using a separating funnel is another example of washing — you're removing water soluble impurities,

Tip: Organic compounds are often flammable, so heating should never be done using an open flame — use a water bath, sand bath or electric heater instead.

Washing

The product of a reaction can be contaminated with unreacted reagents or unwanted side products. You can remove some of these by washing the product.

Example

An excess of aqueous sodium hydrogen carbonate solution can be used to remove acid from an organic product.

Any acid is reacted with the sodium hydrogen carbonate to give CO_2 gas, a salt and water. This leaves an aqueous solution of the excess aqueous sodium hydrogen carbonate, salt and water. The organic product (assuming it's insoluble in the aqueous layer) can be easily removed using a separating funnel.

Distillation

Distillation separates liquids with different boiling points. It works by gently heating a mixture in distillation apparatus. The substances will evaporate out of the mixture in order of increasing boiling point.

A thermometer is placed at the neck of the condenser and shows the boiling point of the substance that is evaporating at any given time. If you know the boiling point of your pure product, you can use the thermometer to tell you when it's evaporating and therefore when it's condensing.

If the product of a reaction has a lower boiling point than the starting materials, then the reaction mixture can be heated in distillation apparatus so that the product evaporates from the reaction mixture as it forms. If the starting materials have a higher boiling point than the product, so as long as the temperature is controlled, they won't evaporate out from the reaction mixture.

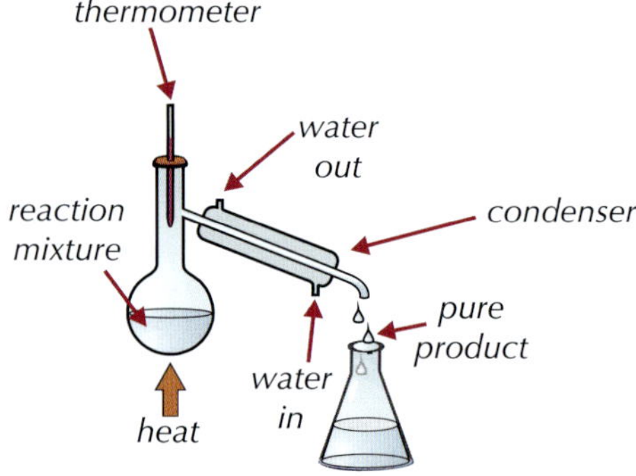

Figure 3: *Distillation apparatus.*

Tip: Secondary alcohols are oxidised to ketones by acidified potassium dichromate(VI) solution. Primary alcohols are oxidised first to aldehydes and then to carboxylic acids.

Example

Hexan-1-ol reacts with acidified dichromate(VI) solution to form hexanal. However, hexanal can be further oxidised by acidified dichromate(VI) to form hexanoic acid.

By heating the reaction in distillation apparatus at a temperature of around 131 °C (the boiling point of hexanal), any hexanal that is produced will immediately evaporate out of the reaction mixture. Hexan-1-ol, which has a boiling point of about 155 °C, won't evaporate so stays in the reaction mixture and can react fully to form hexanal. The hexanal that is distilled out of the reaction mixture can be collected in a separate vessel.
This prevents it from being oxidised further to hexanoic acid.

Redistillation

Mixtures that contain volatile liquids can be purified using a technique called **redistillation**. If a product and its impurities have different boiling points, then redistillation can be used to separate them. You just use the same distillation apparatus as shown on the previous page, but this time you're heating an impure product, instead of the reaction mixture.

When the liquid you want boils (this is when the thermometer is at the boiling point of the liquid), you place a flask at the open end of the condenser ready to collect your product. When the thermometer shows the temperature is changing, put another flask at the end of the condenser because a different liquid is about to be delivered.

Recrystallisation

If the product of an organic reaction is a solid, then the simplest way of purifying it is a process called **recrystallisation**. First you dissolve your solid in a hot solvent to make a saturated solution. Then you let it cool. As the solution cools, the solubility of the product falls. When it reaches the point where it can't stay in solution, it starts to form crystals. Here's how it's done:

1. Add very hot solvent to the impure solid until it just dissolves — it's really important not to add too much solvent.
2. This should give a saturated solution of the impure product.
3. Leave the solution to cool down slowly. Crystals of the product form as it cools.
4. The impurities stay in solution. They're present in much smaller amounts than the product, so they'd take much longer to crystallise out.
5. Remove the crystals by filtration and wash them with ice-cold solvent. The crystals then need to be dried — leaving you with crystals of your product that are much purer than the original solid.

Tip: In a saturated solution, the maximum possible amount of solid is dissolved in the solvent.

Tip: When you recrystallise a product, you must use an appropriate solvent for that particular substance. It will only work if the solid is very soluble in the hot solvent, but nearly insoluble when the solvent is cold.

Tip: You can separate out the crystals using filtration under reduced pressure — see page 8 for details on this technique.

Testing purity

You can use melting point apparatus to accurately determine the melting point of an organic solid.

Pack a small sample of the solid into a glass capillary tube and place it inside the heating element. Increase the temperature until the sample turns from solid to liquid. You usually measure a melting range, which is the range of temperatures from where the solid begins to melt to where it has melted completely.

You can look up the melting point of a substance in data books and compare it to your measurements. Impurities in the sample will lower the melting point and increase the melting range.

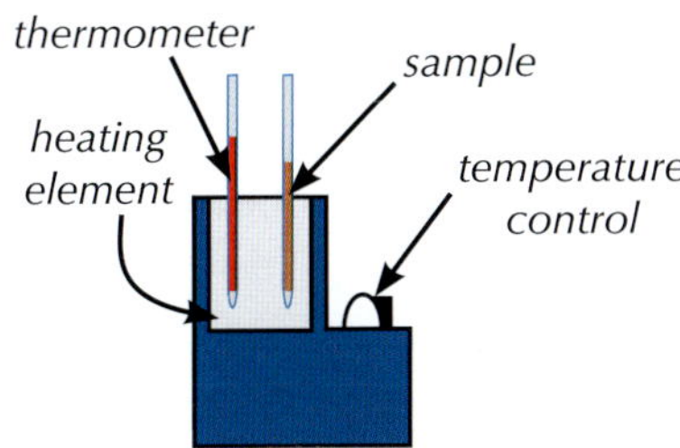

Figure 4: *Melting point apparatus.*

Tip: You can also use melting point analysis to identify an unknown substance. For example, if you have an unknown ester, you could hydrolyse it and purify the carboxylic acid using recrystallisation. You could then use melting point analysis to work out the identity of the carboxylic acid. Working out the identity of the alcohol as well would let you identify the ester.

Tip: If you're asked about how you could go about separating a mixture, you need to think about the properties of the different compounds.

Practice Questions — Application

Q1 Hexane is insoluble in water and has a boiling point of 68 °C. Ethanol is soluble in water and has a boiling point of 78 °C. Suggest how you could separate a mixture of hexane and ethanol into its component parts.

Q2 A student wants to make the compound 4-hydroxypentan-2-one by heating pentan-2,4-diol with acidified potassium dichromate. If the reaction is left for too long, the starting material ends up being oxidised twice to form a compound with two carbonyl groups. Suggest a method the student could use to prevent the starting material from being over-oxidised.

Exam Tip
Always remember to label diagrams if you have to draw them in the exam.

Practice Questions — Fact Recall

Q1 Name two compounds that can be used as drying agents.

Q2 Draw a labelled diagram of the apparatus used to carry out a distillation.

Q3 What physical property is used to separate compounds in redistillation?

Q4 Will a pure product have a wide or a narrow melting point range?

Section Summary

Make sure you know...

- That optical isomerism is a form of stereoisomerism.
- That chiral carbon atoms are carbon atoms with four different groups attached, and that optical isomers have chiral carbon atoms.
- That enantiomers (optical isomers) are molecules with a chiral carbon atom and the same structural formula that exist as non superimposable mirror images.
- How to draw optical isomers.
- That enantiomers rotate plane polarised light in opposite directions.
- That a racemic mixture contains equal quantities of each enantiomer of an optically active compound.
- That racemic mixtures are optically inactive.
- The difference between aldehydes and ketones — aldehydes have their carbonyl group at the end of the carbon chain and ketones have their carbonyl group in the middle.
- How to name aldehydes and ketones.
- That aldehydes are easily oxidised to form carboxylic acids but ketones aren't.
- How Tollens' reagent and Fehling's solution can be used to distinguish between aldehydes and ketones.
- That aldehydes and ketones can be reduced to primary and secondary alcohols respectively, using reducing agents such as $NaBH_4$.
- The nucleophilic addition reaction mechanism for the reduction of aldehydes and ketones.
- What hydroxynitriles are and how to name them.
- That aldehydes and ketones react with KCN to produce hydroxynitriles.

- How to write the overall eqautions to show the reaction of aldehydes and ketones with HCN and acidified KCN.
- That the production of hydroxynitriles using aldehydes/ketones and KCN is an example of nucleophilic addition.
- That the reaction between KCN and aldehydes or unsymmetrical ketones leads to enantiomers.
- The hazards of using KCN in a reaction.
- What carboxylic acids are and how to name them.
- That carboxylic acids are weak acids that partially dissociate in water.
- That carboxylic acids react with carbonates and hydrogen carbonates to form a salt, CO_2 and H_2O.
- That carboxylic acids react with alcohols in the presence of a strong acid catalyst (e.g. HCl, H_2SO_4 or H_3PO_4) to form esters — this is called an esterification reaction.
- What esters are and how to name them.
- That esters can have pleasant smells so are often used in perfumes and as food flavourings.
- That esters can also be used as plasticisers and solvents.
- That esters can undergo acid hydrolysis to give carboxylic acids and alcohols.
- That esters can undergo base hydrolysis to give carboxylate ions and alcohols.
- That fats/oils are esters of glycerol (propane-1,2,3-triol) and long chain carboxylic acids (fatty acids).
- The difference between fats and oils — fats contain mainly saturated fatty acids while oils contain mostly unsaturated fatty acids.
- Why fats are solid at room temperature when oils are liquid.
- That fats and oils can be hydrolysed to form glycerol and sodium salts which can be used as soap.
- That vegetable oils react with methanol in the presence of a KOH catalyst to form methyl esters which are used in biodiesel.
- That biodiesel is sometimes referred to as carbon neutral, but it actually is not.
- What acyl chlorides are and how to name them.
- The reactions of acyl chlorides with water, alcohols, ammonia and primary amines — including the nucleophilic addition-elimination mechanism for these reactions.
- What acid anhydrides are and how to name them.
- The reactions of acid anhydrides with water, alcohols, ammonia and primary amines.
- That aspirin can be made by reacting salicylic acid with ethanoic anhydride or ethanoyl chloride.
- That ethanoic anhydride is used in industry to produce aspirin because it is cheaper and safer to work with than ethanoyl chloride.
- That you can use a separating funnel to separate an organic product from impurities that dissolve in a solvent that's immiscible with the organic product.
- That solvent extraction is a form of separation that involves the vigorous shaking of a product with an immiscible solvent.
- That drying agents remove water from an organic product.
- That washing is a technique used to purify organic products.
- That distillation is a technique that uses the difference in boiling points of different substances to isolate a product from a reaction as it forms.
- That redistillation uses differences in boiling points to purify volatile liquids.
- That organic solids can be purified by recrystallisation.
- That you can test the purity of a product using melting point apparatus.

Exam-style Questions

1 Which of the following molecules have optical isomers?

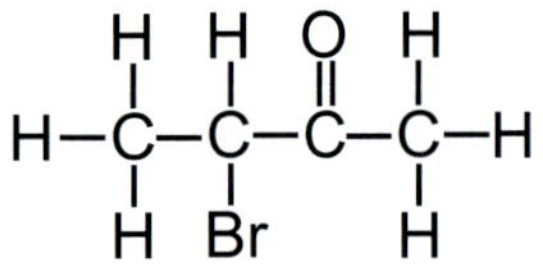

molecule **x**

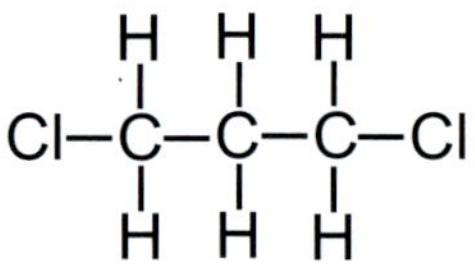

molecule **y**

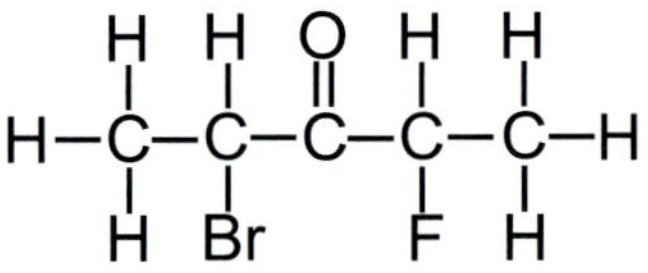

molecule **z**

A **x** and **z** only
B **y** only
C **x**, **y** and **z**
D **y** and **z** only

(1 mark)

2 Which of the following methods could be used to purify a solid organic product?

A Redistillation
B Washing
C Reflux
D Recrystallisation

(1 mark)

3 Which of the options shows the molecule with the correctly labelled chiral carbon atom?

A

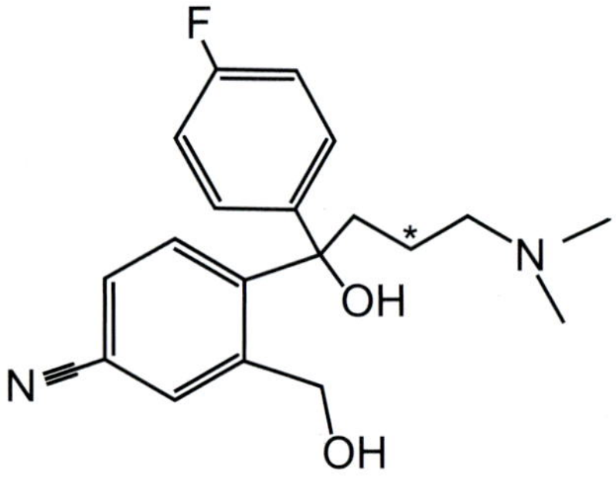

B

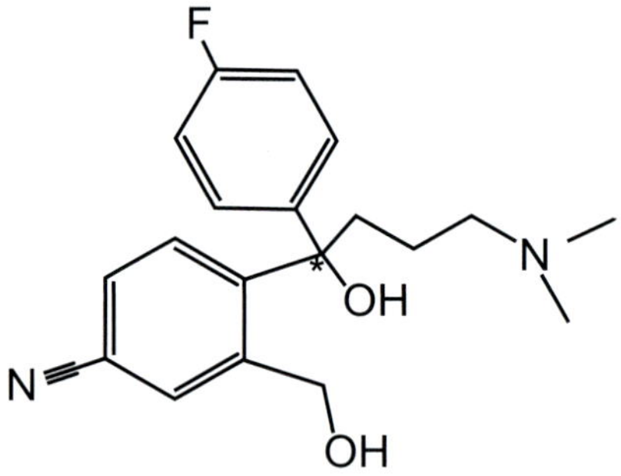

C

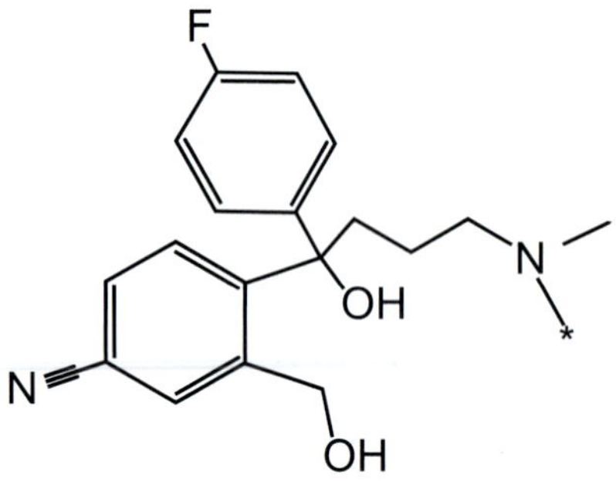

D

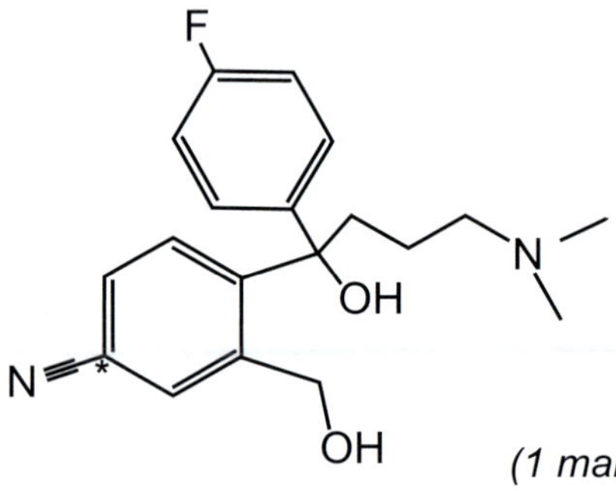

(1 mark)

4 The two compounds, A and B, shown below are structurally very similar, but a simple test can be used to distinguish between them.

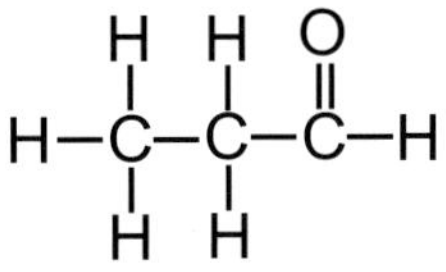

```
     H  H  O                    H  O  H
     |  |  ‖                    |  ‖  |
  H—C—C—C—H                  H—C—C—C—H
     |  |                       |     |
     H  H                       H     H
```

Compound A

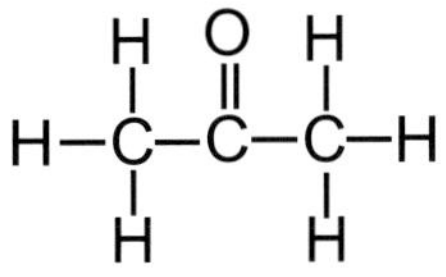

Compound B

4.1 Give the IUPAC names for these two compounds.

(2 marks)

4.2 Describe a test that could be used to differentiate between these compounds.

(2 marks)

Compound A can be reduced to form a primary alcohol.

4.3 Suggest a suitable reducing agent for this reaction.

(1 mark)

4.4 Outline the mechanism for this reduction reaction.

(4 marks)

Compound B can react with acidified potassium cyanide (KCN).

4.5 Name and outline the mechanism for this reaction.

(5 marks)

4.6 Give the IUPAC name for the product formed.

(1 mark)

4.7 Compound A can also react with acidified KCN. What effect will the product of this reaction have on plane polarised light? Explain your answer.

(3 marks)

5 2-methylbutanoic acid is a carboxylic acid.

5.1 Draw the structure of this carboxylic acid.

(1 mark)

5.2 2-methylbutanoic acid is optically active. Draw the two enantiomers of 2-methylbutanoic acid, and mark the chiral centre on each molecule with a *.

(2 marks)

5.3 Write an equation for the reaction of 2-methylbutanoic acid with sodium carbonate (Na_2CO_3).

(2 marks)

Carboxylic acids react with alcohols to form esters.

5.4 Draw the structure of the ester that would be formed if 2-methylbutanoic acid reacted with methanol.

(1 mark)

5.5 Give the IUPAC name for this ester.

(1 mark)

5.6 Suggest a suitable catalyst for this reaction.

(1 mark)

5.7 Write an equation for the hydrolysis of this ester with dilute alkali. Show the structure of the reactants and products in your answer.

(2 marks)

5.8 Give one common use of esters.

(1 mark)

6 Fats and oils are esters formed from glycerol (propane-1,2,3-triol) and fatty acids (long chain carboxylic acids).

6.1 Explain why fats are usually solid at room temperature while oils are normally liquid.

(3 marks)

6.2 The structure below shows a naturally occurring triester commonly found in fats and oils.

$$\begin{array}{l} CH_2OOC(CH_2)_{16}CH_3 \\ | \\ CH_2OOC(CH_2)_{16}CH_3 \\ | \\ CH_2OOC(CH_2)_{16}CH_3 \end{array}$$

Write an equation to show how this ester could be converted to a sodium salt.

(3 marks)

6.3 Suggest a use for this sodium salt.

(1 mark)

Biodiesel can be produced from fats and oils.

6.4 What is biodiesel?

(1 mark)

6.5 Write an equation to show how biodiesel could be produced from the ester shown in part **6.2**.

(3 marks)

6.6 What is the role of KOH in this reaction?

(1 mark)

7 This question is about acyl chlorides and acid anhydrides.

The structure below shows the acyl chloride 2,3-dimethylbutanoyl chloride.

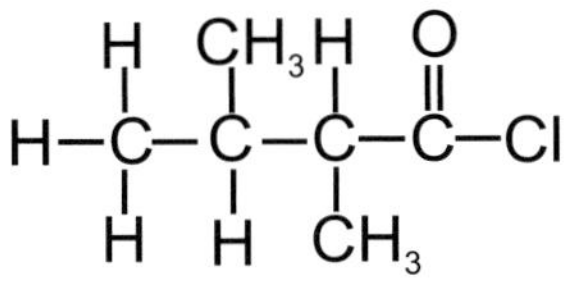

7.1 Write an equation to show the reaction of this acyl chloride with ammonia.

(2 marks)

7.2 Name and outline the mechanism for this reaction.

(6 marks)

Acid anhydrides have very similar properties to acyl chlorides but they are structurally very different.

7.3 Draw the structure of ethanoic anhydride.

(1 mark)

7.4 Write an equation to show the reaction of this acid anhydride with water.

(2 marks)

Ethanoic anhydride can be reacted with salicylic acid to produce aspirin. The structure of salicylic acid is shown below.

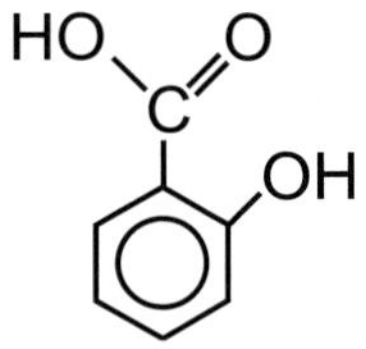

Salicylic acid has an OH group and behaves like an alcohol when reacting with ethanoic anhydride.

7.5 Using your knowledge of the reactions of acid anhydrides with alcohols, write an equation showing the formation of aspirin from ethanoic anhydride and salicylic acid.

(2 marks)

7.6 Aspirin can also be produced by reacting ethanoyl chloride with salicylic acid.

Which method of aspirin production is favoured in industry? Explain your answer.

(3 marks)

7.7 In the lab, aspirin forms as a solid organic product which contains trace impurities. Outline a method that could be used to purify aspirin made in the lab.

(3 marks)

Learning Objectives:

- Be able to describe the nature of the bonding in a benzene ring, limited to its planar structure and bond length intermediate between single and double.
- Understand that the delocalisation of p electrons makes benzene more stable than the theoretical molecule cyclohexa-1,3,5-triene.
- Be able to use thermochemical evidence from enthalpies of hydrogenation to account for the extra stability of the benzene ring.
- Be able to apply IUPAC rules for nomenclature to name organic compounds limited to chains and rings with up to six carbon atoms each.
- Be able to apply IUPAC rules for nomenclature to draw the structure of an organic compound from the IUPAC name limited to chains and rings with up to six carbon atoms each.

Specification Reference
3.3.1.1, 3.3.10.1

Tip: Have a look back at page 33 for a recap on p-orbitals.

1. Aromatic Compounds

Aromatic compounds all contain a benzene ring. Benzene's a bit of a funny thing — it contains a very stable ring of delocalised electrons, and this affects its properties. Read on and all shall be revealed...

The structure of benzene

Benzene has the formula C_6H_6. It has a planar cyclic structure (which is just a complicated way of saying that its six carbon atoms are joined together in a flat ring). Each carbon atom forms single covalent bonds to the carbons on either side of it and to one hydrogen atom. The final unpaired electron on each carbon atom is located in a p-orbital that sticks out above and below the plane of the ring. The p-orbitals on each carbon atom combine to form a ring of delocalised electrons (see Figure 1). The electrons in the ring are said to be delocalised because they don't belong to a specific carbon atom. All the bonds in the ring are the same, so they are the same length — 140 pm. This lies in between the length of a single C–C bond (147 pm) and a double C=C bond (135 pm).

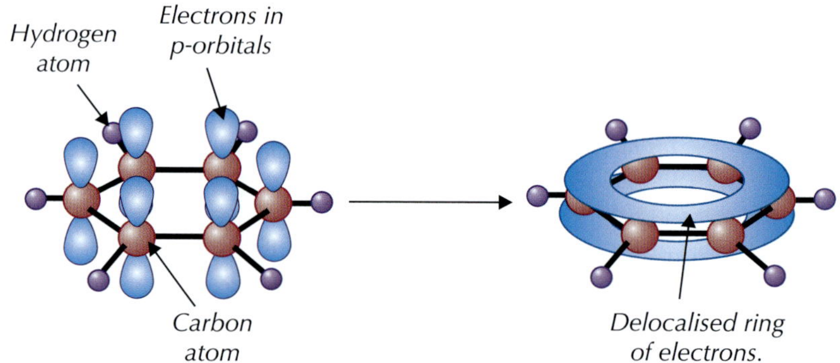

Figure 1: *The formation of a delocalised ring of electrons in benzene.*

The delocalised ring of electrons can be represented by a circle in the ring of carbon-carbon single bonds rather than as alternating double and single bonds (see Figure 3).

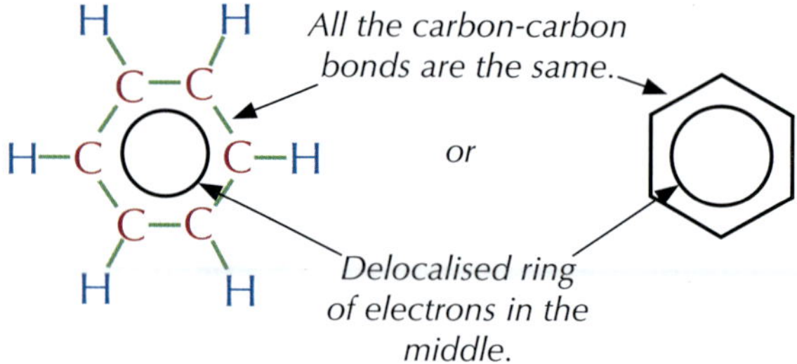

Figure 2: *The structure of benzene according to the delocalised model.*

Figure 4 shows another representation of benzene that you may come across. Don't get confused by this drawing — those aren't really alternating single and double bonds between the carbon atoms. Scientists used to think this was the structure, and their way of drawing it has just stuck around.

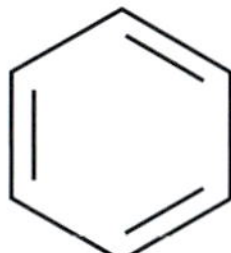

Figure 3: *Another representation of a benzene ring.*

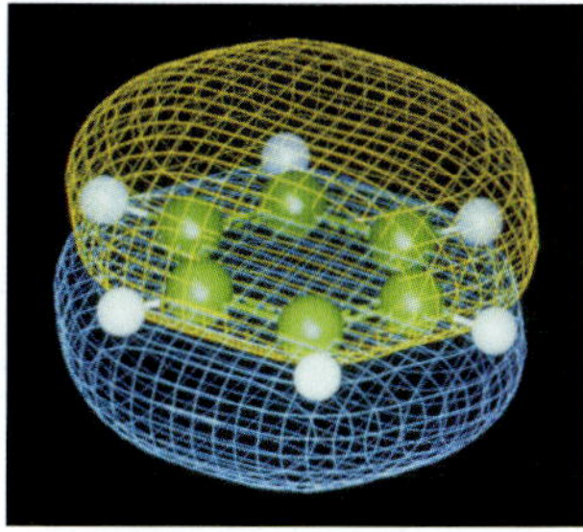

Figure 4: *A computer graphic showing the structure of benzene.*

Stability of benzene

Benzene is far more stable than the theoretical compound cyclohexa-1,3,5-triene would be (where the ring would be made up of alternating single and double bonds). You can see this by comparing the enthalpy change of hydrogenation for benzene with the enthalpy change of hydrogenation for cyclohexene. Cyclohexene has one double bond. When it's hydrogenated, the enthalpy change is –120 kJ mol^{-1}. If benzene had three double bonds, you'd expect it to have an enthalpy of hydrogenation of –360 kJ mol^{-1}. But the experimental enthalpy of hydrogenation of benzene is –208 kJ mol^{-1} — 152 kJ mol^{-1} less exothermic than expected (see Figure 5).

Tip: See page 264 for more on enthalpies.

Tip: Cyclohexa-1,3,5-triene (a six-carbon ring with three double bonds) is a theoretical molecule (it doesn't actually exist), so we can't measure its enthalpy of hydrogenation directly. We can only predict it from the enthalpy of hydrogenation for cyclohexene.

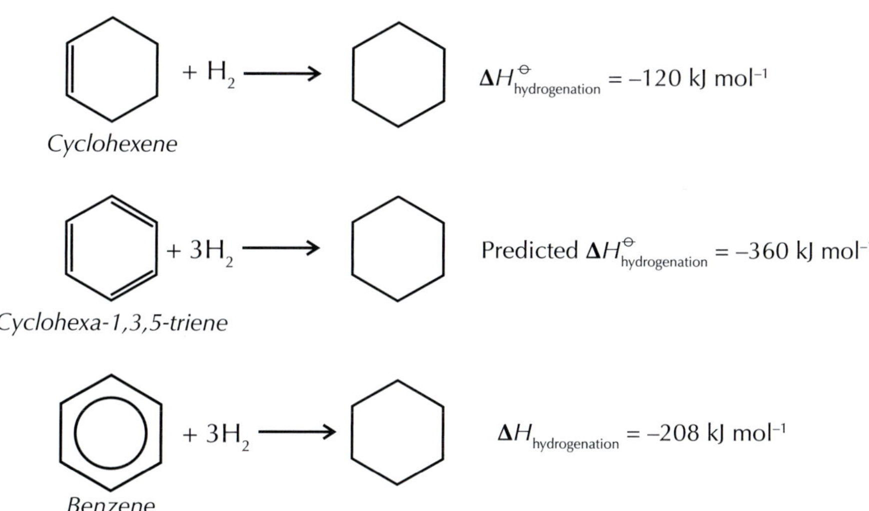

Figure 5: *Enthalpies of hydrogenation providing further evidence for delocalisation.*

Energy is put in to break bonds and released when bonds are made. So more energy must have been put in to break the bonds in benzene than would be needed to break the bonds in a theoretical cyclohexa-1,3,5-triene molecule. This difference indicates that benzene is more stable than cyclohexa-1,3,5-triene would be. This is thought to be due to the delocalised ring of electrons. In a delocalised ring the electron density is shared over more atoms, which means the energy of the molecule is lowered and it becomes more stable.

Tip: Aromatic compounds are also known as arenes, so don't get confused if you see this term around.

Practice Question — Application

Q1 The graph below shows the enthalpies of hydrogenation of cyclohexene and benzene.

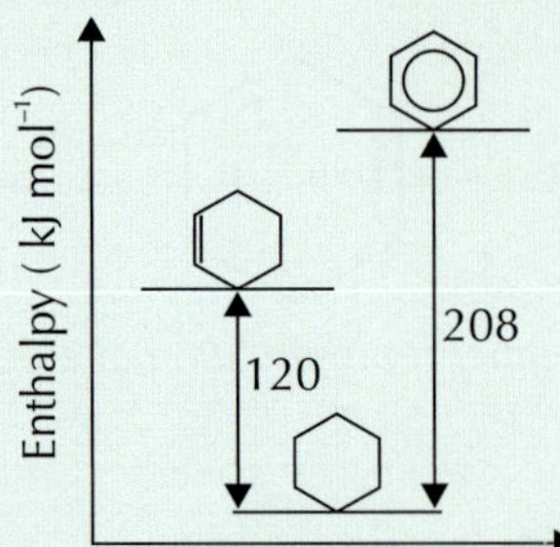

Use the graph to explain why benzene does not have the structure of the theoretical molecule cyclohexa-1,3,5-triene.

Naming aromatic compounds

Aromatic compounds all contain a benzene ring, but naming them isn't that simple. They're named in two ways:

1. In some cases, the benzene ring is the main functional group and the molecule is named as a substituted benzene ring — the suffix is -benzene and there are prefixes to represent any other functional groups.

Examples

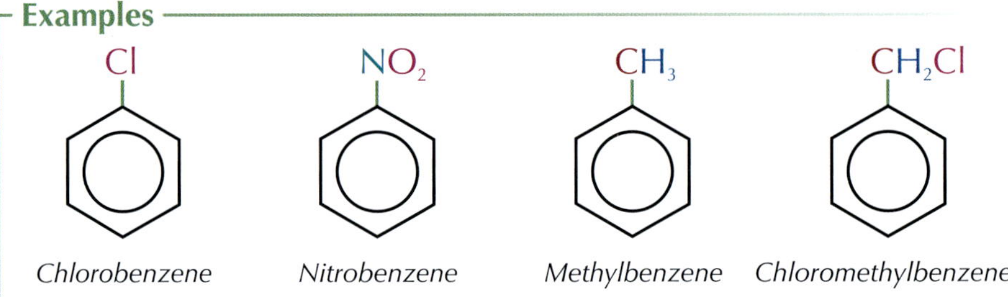

Tip: Non-aromatic molecules are called aliphatic molecules — don't get the terms aromatic and aliphatic mixed up.

2. In other cases, the benzene ring is not the main functional group and the molecule is named as having a phenyl group (C_6H_5) attached. Phenyl- or phen- are used as a prefixes to show the molecule has a benzene ring and the suffix comes from other functional groups on the molecule (e.g. -ol if it's an alcohol, -amine if it's an amine).

Examples

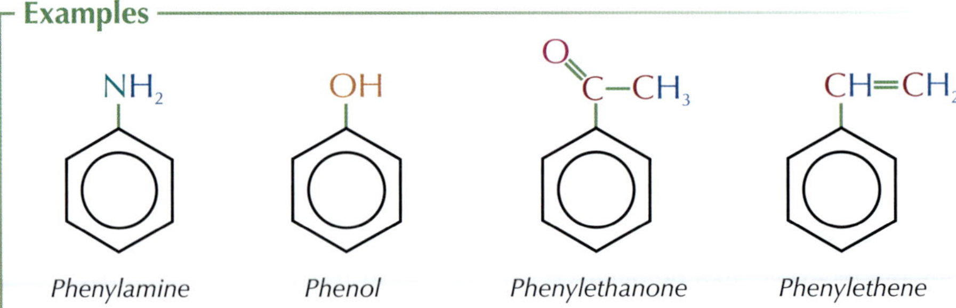

Unfortunately, there's no simple rule to help remember which name to use — you just have to learn which are '-benzene' and which are 'phenyl-'

Numbering the benzene ring

If there is more than one functional group attached to the benzene ring you have to number the carbons to show where the groups are. If all the functional groups are the same, pick any group to start from and count round either clockwise or anticlockwise — whichever way gives the smallest numbers. If the functional groups are different, start from whichever functional group gives the molecule its suffix (e.g. the –OH group for a phenol) and continue counting round whichever way gives the smallest numbers. If there's no suffix from a functional group, start numbering from whichever group is first alphabetically.

Tip: If there is only one functional group you don't have to number the carbons because all the positions around the carbon ring are the same.

Examples

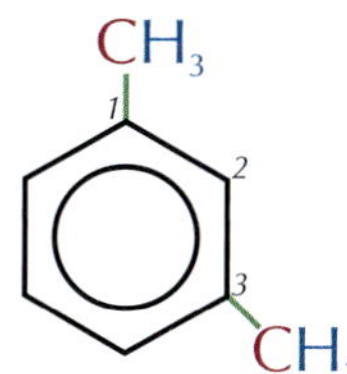

This benzene ring only has methyl groups attached so it will be named as a substituted benzene ring.

Starting from the methyl group at the top and counting clockwise there is another methyl group on carbon-3.

So this is 1,3-dimethylbenzene.

This benzene ring has an OH group attached so the stem is phenol.

Starting from the OH group (the group which gives the molecule its name) and counting anticlockwise there are chlorines on carbon-2 and carbon-4.

So this is 2,4-dichlorophenol.

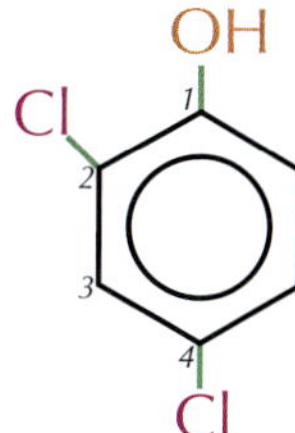

Exam Tip

You're unlikely to be asked to name anything too complicated in the exams. If you need to give the product of a reaction, don't try and name it unless they specifically ask for the name — giving the structure should be enough to get the marks.

Practice Question — Application

Q1 Name these aromatic compounds:

a) NO_2, NO_2

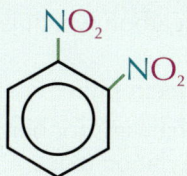

b)

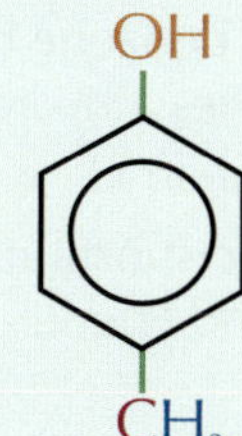

c) NH_2, Cl, Cl, Cl

Practice Questions — Fact Recall

Q1 What are aromatic compounds?

Q2 Describe the structure of benzene.

2. Reactions of Aromatics

Learning Objectives:

- Know that electrophilic attack on benzene rings results in substitution, limited to monosubstitutions.
- Be able to explain why substitution reactions occur at a benzene ring in preference to addition reactions.
- Be able to outline the electrophilic substitution mechanism of nitration, including the generation of the nitronium ion.
- Understand that nitration is an important step in synthesis, including for the manufacture of explosives and formation of amines.
- Be able to outline the electrophilic substitution mechanism of acylation using $AlCl_3$ as a catalyst.
- Understand that Friedel-Crafts acylation reactions are important steps in synthesis.

Specification Reference
3.3.10.1, 3.3.10.2

Tip: Don't get confused between electrophiles and nucleophiles. Just remember, electrophiles love electrons.

Exam Tip
In your exams, you might be asked to write out the equation for the formation of the nitronium ion before giving the rest of the mechanism for nitration — so make sure you learn it.

You need to know some of the main reactions of aromatic compounds — plus, there's another reaction mechanism for you to learn.

Electrophilic substitution

The benzene ring is a region of high electron density, so it attracts **electrophiles**. Electrophiles are electron pair acceptors — they are usually a bit short of electrons so are attracted to areas of high electron density. Common electrophiles include positively charged ions (e.g. H^+ or NO_2^+), and polar molecules (e.g. carbonyl compounds), which have a partial positive charge.

As the benzene ring's so stable, it doesn't undergo electrophilic addition reactions, which would destroy the delocalised ring of electrons. Instead, it undergoes **electrophilic substitution** reactions where one of the hydrogen atoms (or another group on the ring) is substituted for the electrophile.

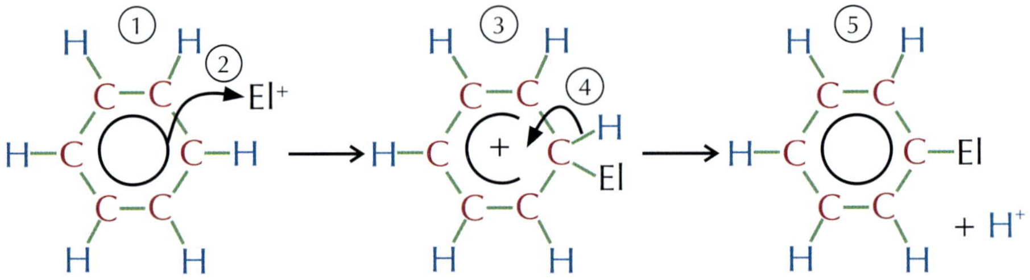

1. The electron-dense region at the centre of the benzene ring attracts an electrophile (El^+).
2. The electrophile steals a pair of electrons from the centre of the benzene ring and forms a bond with one of the carbons.
3. This partially breaks the delocalised electron ring and gives the molecule a positive charge.
4. To regain the stability of the benzene ring, the carbon which is now bound to the electrophile loses a hydrogen.
5. So you get the substitution of an H^+ with the electrophile.

You need to know two electrophilic substitution mechanisms for benzene — the nitration reaction (below) and Friedel-Crafts acylation (page 451).

Nitration

When you warm benzene with concentrated nitric acid and concentrated sulfuric acid, you get nitrobenzene. The overall equation for this reaction is:

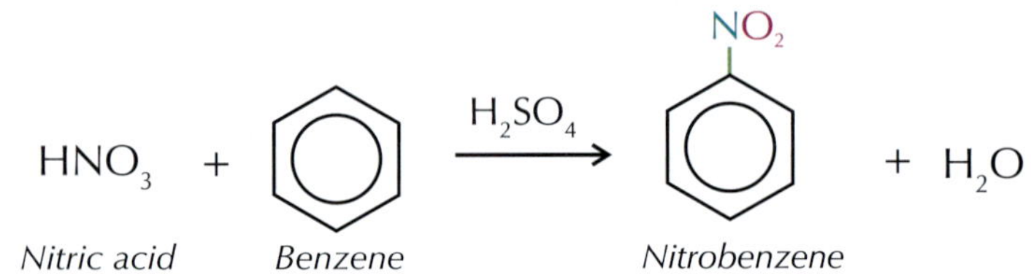

$$HNO_3 + C_6H_6 \xrightarrow{H_2SO_4} C_6H_5NO_2 + H_2O$$

Nitric acid, Benzene, Nitrobenzene

Sulfuric acid acts as a catalyst — it helps to make the nitronium ion, NO_2^+, which is the electrophile. The formation of the nitronium ion is the first step of the reaction mechanism. The equation for this step is:

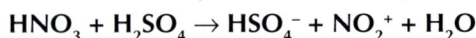

$$\mathbf{HNO_3 + H_2SO_4 \rightarrow HSO_4^- + NO_2^+ + H_2O}$$

Once the nitronium ion has been formed, it can react with benzene to form nitrobenzene. This is the electrophilic substitution step in the reaction.

Here's the mechanism for the electrophilic substitution part of the reaction:

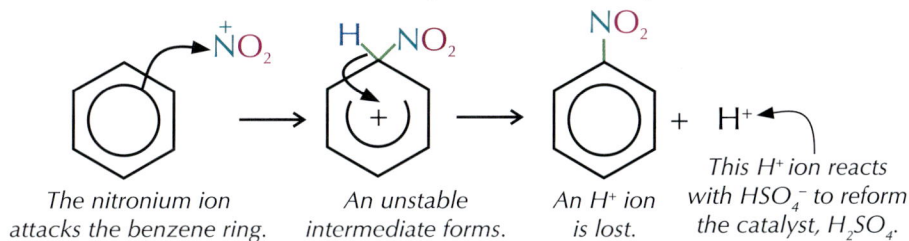

$\overset{+}{N}O_2$ → (H, NO_2, +) → (NO_2) + H^+

The nitronium ion attacks the benzene ring. *An unstable intermediate forms.* *An H^+ ion is lost.* *This H^+ ion reacts with HSO_4^- to reform the catalyst, H_2SO_4.*

Exam Tip
If you're asked how the catalyst reforms, here's the reaction:
$H^+ + HSO_4^- \rightarrow H_2SO_4$

If you only want one NO_2 group added (mononitration), you need to keep the temperature below 55 °C. Above this temperature you'll get lots of substitutions.

Uses of nitration reactions

Nitro compounds can be reduced to form aromatic amines (see page 460). These are used to manufacture dyes and pharmaceuticals. Nitro compounds decompose violently when heated, so they are used as explosives — such as 2,4,6-trinitromethylbenzene (trinitrotoluene (TNT) — see Figures 1 and 2).

CH_3, O_2N, NO_2, NO_2

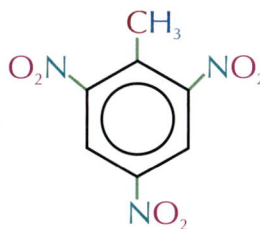

Figure 1: *TNT.*

Figure 2: *Blocks of TNT explosive.*

Friedel-Crafts acylation

Many useful chemicals such as dyes and pharmaceuticals contain benzene rings, but because benzene is so stable, it's fairly unreactive. Friedel-Crafts acylation reactions are important in the synthesis of these compounds, as they are used to add an acyl group (–C(=O)–R) to the benzene ring. The products of the reaction are HCl and a phenylketone. The reactants need to be heated under reflux in a non-aqueous solvent (like dry ether) for the reaction to occur. Here's the equation for the reaction:

$$R{-}C({=}O){-}Cl + C_6H_6 \xrightarrow[\text{reflux}]{AlCl_3} C_6H_5{-}C({=}O){-}R + HCl$$

Acyl chloride *Benzene* *Phenylketone*

Figure 3: *Charles Friedel — co-developer of Friedel-Crafts acylation.*

Once an acyl group has been added, the side chains can be modified by further reactions to produce useful products.

An electrophile has to have a pretty strong positive charge to be able to attack the stable benzene ring — most just aren't polarised enough. But some can be made into stronger electrophiles using a catalyst called a **halogen carrier**.

Friedel-Crafts acylation uses an acyl chloride to provide the electrophile and a halogen carrier such as $AlCl_3$. $AlCl_3$ accepts a lone pair of electrons from the acyl chloride. As the lone pair of electrons is pulled away, the polarisation in the acyl chloride increases and it forms a **carbocation**. This makes it a much, much stronger electrophile, and gives it a strong enough charge to react with the benzene ring. The formation of the carbocation is the first step in the reaction mechanism and is shown below:

Tip: There are other types of halogen carrier, but $AlCl_3$ is the only one you need to know about.

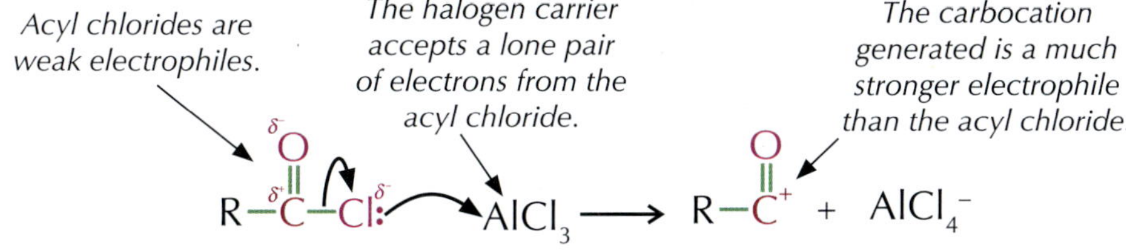

Acyl chlorides are weak electrophiles.
The halogen carrier accepts a lone pair of electrons from the acyl chloride.
The carbocation generated is a much stronger electrophile than the acyl chloride.
δ−
O
δ+
R—C—Cl:
δ−
AlCl3
R—C+
O
+ AlCl4−

Exam Tip
The formation of the carbocation is part of the reaction mechanism so make sure you include it in the exams.

And here's the second step in the reaction mechanism, the electrophilic substitution bit:

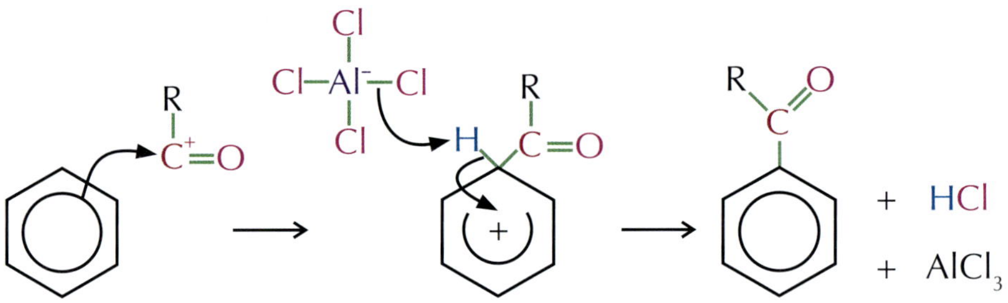

Electrons in the benzene ring are attracted to the positively charged carbocation. Two electrons from the benzene bond with the carbocation. This partially breaks the delocalised ring and gives it a positive charge.

The negatively charged $AlCl_4^-$ ion is attracted to the positively charged ring. One chloride ion breaks away from the aluminium chloride ion and bonds with the hydrogen. This removes the hydrogen from the ring, forming HCl. It also allows the catalyst to reform. Any acyl chloride can react with benzene in this way.

Example

Ethanoyl chloride reacts with benzene to produce phenyl ethanone:

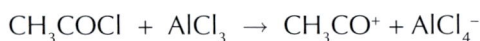

$$CH_3COCl + AlCl_3 \rightarrow CH_3CO^+ + AlCl_4^-$$

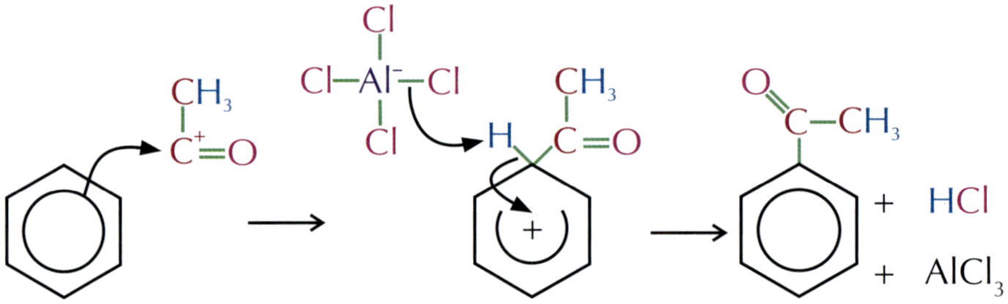

Exam Tip
You could draw out the full structure of benzene when writing out this equation — but these simplified diagrams are a lot quicker and could save you valuable time in your exams.

Exam Tip
When drawing the intermediates of these reactions, make sure the horseshoe in the middle of the benzene ring only extends from carbon-2 to carbon-6.

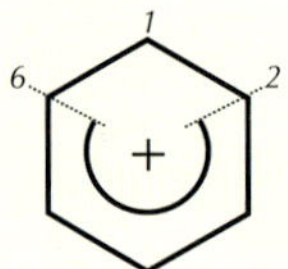

Exam Tip
When drawing these mechanisms in your exams, make sure your curly arrows clearly go from the delocalised ring of electrons to the carbocation and from the C-H bond back to the delocalised electron ring. Just drawing your curly arrows to/from the centre of benzene won't be enough to get you all the marks.

Practice Questions — Application

Q1 Phenylmethanal can be produced using Friedel-Crafts acylation.
- a) Suggest a suitable catalyst for this reaction.
- b) Write out an equation for this reaction.
- c) Name and outline the mechanism for this reaction.

Q2 a) Draw the mechanism for the formation of nitrobenzene from benzene and concentrated nitric acid.
- b) What two conditions are needed to produce nitrobenzene with this reaction?

Practice Questions — Fact Recall

Q1 Explain why electrophiles are attracted to aromatic compounds.

Q2 Give two uses of nitro compounds produced by nitration of benzene.

Q3 Explain how $AlCl_3$ acts as a catalyst in Friedel-Crafts acylation.

3. Amines and Amides

Amines and amides are produced from ammonia (NH_3). You met amines on page 214, where you saw that they can be formed when halogenoalkanes react with ammonia. Now you need to know a bit more about them. The next few pages are all about amines, amides and why they're useful.

Learning Objectives:

- Be able to apply IUPAC rules for nomenclature to name organic compounds limited to chains and rings with up to six carbon atoms each.
- Be able to apply IUPAC rules for nomenclature to draw the structure of an organic compound from the IUPAC name limited to chains and rings with up to six carbon atoms each.
- Be able to describe the use of quaternary ammonium salts as cationic surfactants.
- Know that amines are weak bases.
- Be able to describe and explain the difference in base strength between ammonia, primary aliphatic and primary aromatic amines in terms of the availability of the lone pair of electrons on the N atom.
- Be able to recognise the structures of amides.

Specification Reference
3.3.1.1, 3.3.9.2, 3.3.11.2, 3.3.11.3

What are amines?

If one or more of the hydrogens in ammonia (NH_3) is replaced with an organic group such as an alkyl or aromatic group, you get an **amine**. If one hydrogen is replaced, you get a primary amine, if two are replaced it's a secondary amine, and three means it's a tertiary amine. If a fourth organic group is added you get a quaternary ammonium ion (see Figure 1).

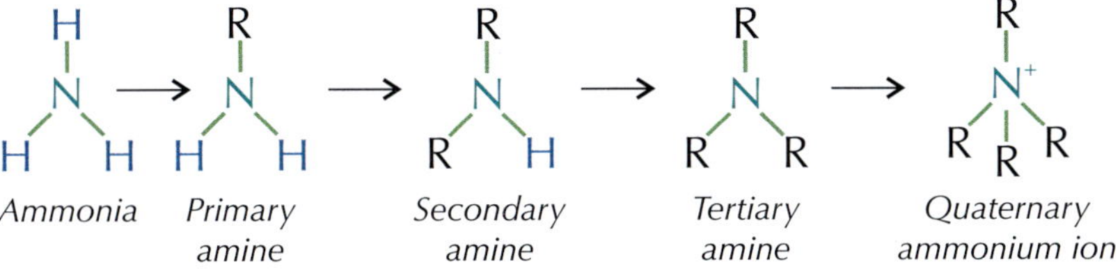

Figure 1: *Diagram showing the different types of amine.*

Because quaternary ammonium ions are positively charged, they will hang around with any negative ions that are near. The complexes formed are called quaternary ammonium salts — like tetramethylammonium chloride ($(CH_3)_4N^+Cl^-$). Small amines smell similar to ammonia, with a slightly 'fishy' twist. Larger amines smell very 'fishy'. (Nice.)

Naming amines

Naming amines is similar to naming other organic compounds. The suffix is -amine (or -amine ion if it's a quaternary ammonium ion). The prefix depends on what organic groups are attached. If the organic groups are all the same you also need to add di- for secondary amines, tri- for tertiary amines and tetra- for quaternary ammonium ions.

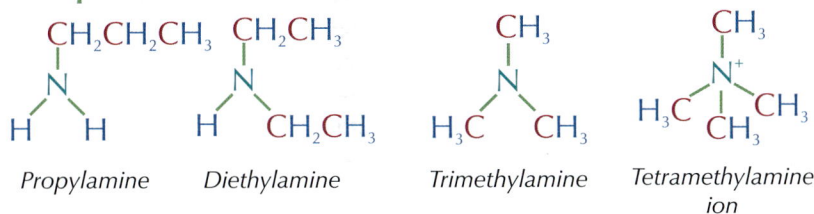

Tip: Have a look at pages 188-191 for the general rules for naming organic compounds.

Examples

$CH_3CH_2CH_2NH_2$ — *Propylamine*

$(CH_3CH_2)_2NH$ — *Diethylamine*

$(CH_3)_3N$ — *Trimethylamine*

$(CH_3)_4N^+$ — *Tetramethylamine ion*

If the amine has more than one type of organic group attached, you list the different groups in alphabetical order.

Example

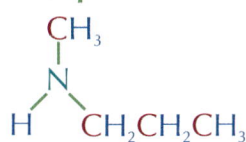

$CH_3NHCH_2CH_2CH_3$

This is a secondary amine. It has a methyl group and a propyl group attached.
So this is methylpropylamine.

You can also get aromatic amines. See page 448 for how to name these.

Tip: Aliphatic amines are amines without a benzene ring. Aromatic amines are amines where the nitrogen atom is directly bonded to a benzene ring.

Cationic surfactants

Surfactants are compounds which are partly soluble and partly insoluble in water. Some quaternary ammonium compounds can be used as **cationic surfactants** (surfactants which are positively charged). These compounds are quaternary ammonium salts with at least one long hydrocarbon chain (see Figure 2).

Long hydrocarbon chain: insoluble

Positively charged head group: soluble

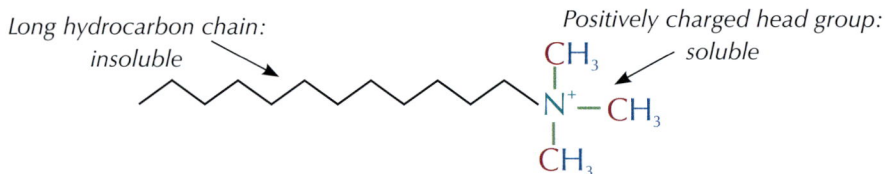

Figure 2: *A cationic surfactant.*

> **Tip:** Cationic surfactants form induced dipole-dipole interactions with non-polar substances and hydrogen bonds with water.

The long hydrocarbon chain is insoluble in water, but will bind to non-polar substances such as grease, whilst the positively charged head group is soluble in water. This makes cationic surfactants useful in detergents, as the non-polar end will bind to grease, and the polar head group will dissolve in water, allowing spots of grease to mix with water and be washed away (see Figure 3).

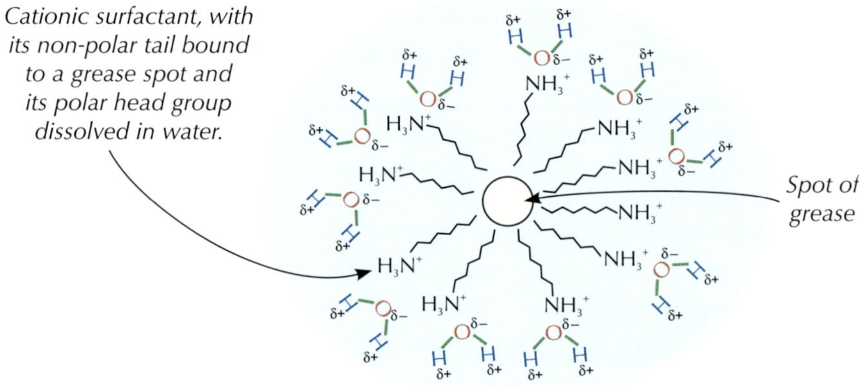
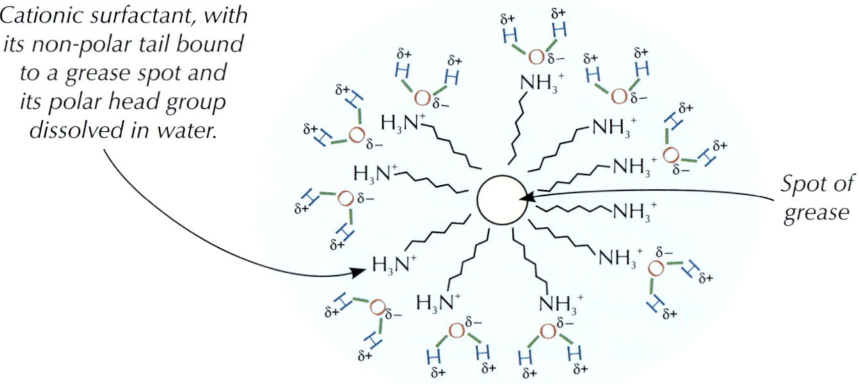

Cationic surfactant, with its non-polar tail bound to a grease spot and its polar head group dissolved in water.
Spot of grease

Figure 3: *How cationic surfactants can help non-polar substances such as grease mix with water.*

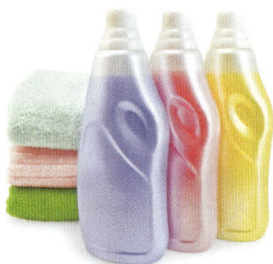

Figure 4: *Fabric softeners contain cationic surfactants.*

Cationic surfactants are used in things like fabric conditioners and hair products. When hair or fabric get wet they pick up negative charges on their surface. The positively charged part of the surfactant is attracted to these negatively charged surfaces and forms a coating over the surface.
This coating prevents the build-up of static electricity. In fabric conditioners this is important to keep the fabric smooth and in hair products it helps to keep hair flat.

> **Tip:** See page 341 for more on acids and bases.

> **Tip:** A co-ordinate bond is a type of covalent bond where both of the electrons have been provided by the same atom (in this case the nitrogen atom).

Amines as bases

Amines act as weak bases because they accept protons. There's a lone pair of electrons on the nitrogen atom that forms a **co-ordinate** (dative covalent) bond with an H^+ ion (see Figure 5).

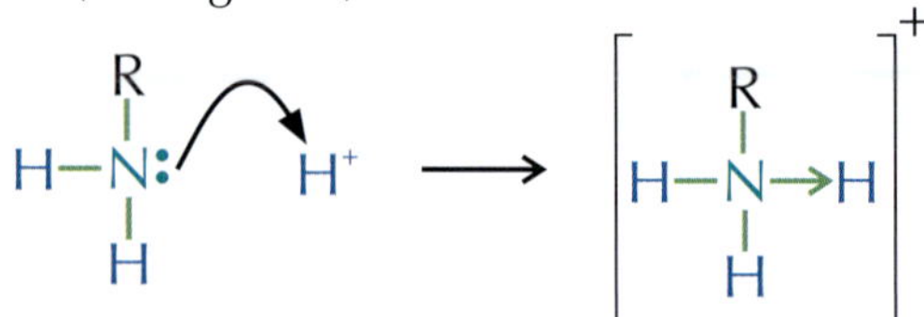

Figure 5: *Formation of a dative bond between a primary amine and an H^+ ion.*

The strength of the base depends on how available the nitrogen's lone pair of electrons is. A lone pair of electrons will be more available if its electron density is higher. The more available the lone pair is, the more likely the amine is to accept a proton, and the stronger a base it will be.

Primary aromatic amines are weak bases. This is because the benzene ring draws electrons towards itself and the nitrogen's lone pair gets partially delocalised onto the ring. So the electron density on the nitrogen decreases. This makes the lone pair much less available.

Primary aliphatic amines are strong bases. This is because alkyl groups push electrons onto attached groups. So the electron density on the nitrogen atom increases. This makes the lone pair more available.

The strength of ammonia as a base lies somewhere in between. This is because ammonia doesn't have an aromatic group to pull the lone pair of electrons away, or an alkyl group to push the lone pair of electrons forward — see Figure 6.

Exam Tip
You need to be able to explain the difference in base strength between ammonia, primary aromatic amines and primary aliphatic amines — so make sure you understand these explanations.

Tip: The effect of an alkyl group on a lone pair of electrons is sometimes called the inductive effect.

Areas of highest electron density

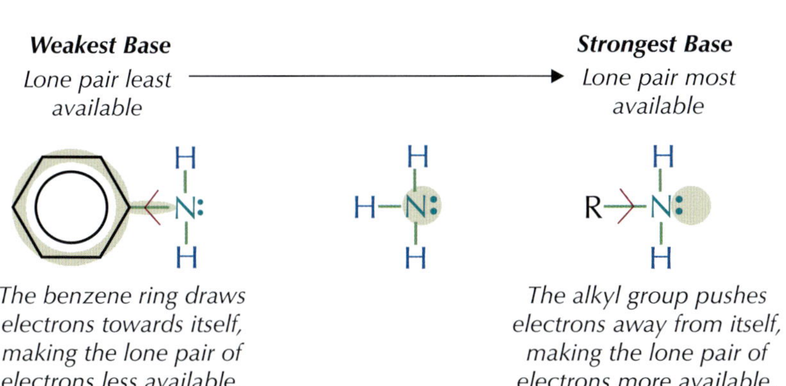

Figure 6: *Diagram showing why primary aromatic amines are weaker bases than ammonia and why primary aliphatic amines are stronger bases than ammonia.*

Tip: Don't get confused between amides and amines. Just remember, ami<u>d</u>es contain nitrogen bonded directly to a carbon with a C=O <u>d</u>ouble bond — ami<u>n</u>es contain <u>n</u>o double bonds.

Amides

Amides are derivatives of carboxylic acids. They contain the functional group $-CONH_2$. The carbonyl group pulls electrons away from the NH_2 group, so amides behave differently from amines. Here's the general structure of an amine:

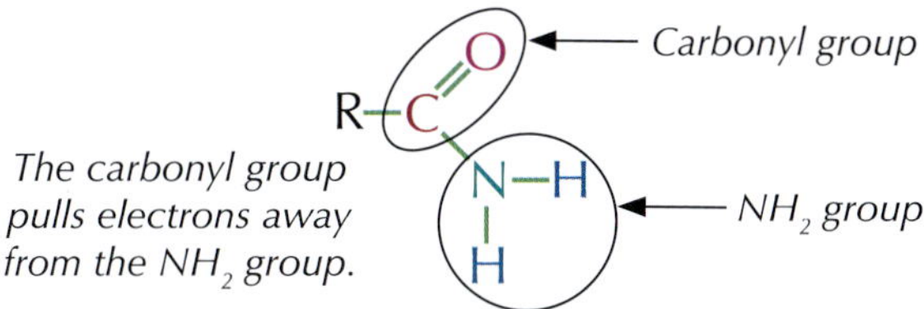

Tip: Amides can be made by reacting acyl chlorides with ammonia See page 432 for more.

Naming amides is pretty easy — the suffix is always -amide and the prefix comes from the acyl group (R–C=O).

Examples

$H-C(=O)-NH_2$ — *Methanamide*

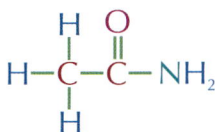

$H-CH_2-C(=O)-NH_2$ — *Ethanamide*

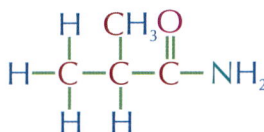

$H-CH_2-CH(CH_3)-C(=O)-NH_2$ — *Methylpropanamide*

Tip: N-substituted amides can be made by reacting acyl chlorides with amines. There's more about this on page 433.

Exam Tip
You don't need to know much about amides. As long as you can name them and know how they can be produced from acyl chlorides (see pages 432-433) or acid anhydrides (see page 436) you should be fine in your exams.

You can also get **N-substituted amides**. These are amides where one of the hydrogens attached to the nitrogen has been substituted with an alkyl group. When naming N-substituted amides, you have to add an extra bit onto the beginning of the name. The name starts with N-something (e.g. N-methyl, N-ethyl, N-propyl) — whatever the alkyl group that's been added onto the nitrogen is.

Examples

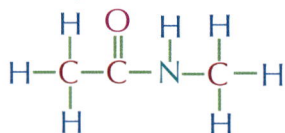

```
   H  O  H  H
   |  ‖  |  |
H—C—C—N—C—H
   |        |
   H        H
```

This molecule is based on ethanamide.

One of the hydrogens attached to the nitrogen has been substituted with a methyl group.

So this is N-methylethanamide.

This is based on propanamide.

One of the hydrogens attached to the nitrogen has been substituted with an ethyl group.

So this is N-ethylpropanamide.

```
   H  H  O  H  H  H
   |  |  ‖  |  |  |
H—C—C—C—N—C—C—H
   |  |        |  |
   H  H        H  H
```

Practice Questions — Application

Q1 Name these amines:

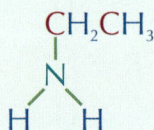

a) CH_2CH_3 bonded to N, with N also bonded to H and H

b) $CH_2CH_2CH_3$ bonded to N, with N also bonded to H and $CH_2CH_2CH_3$

c) CH_3 bonded to N, with N also bonded to H_3C and CH_2CH_3

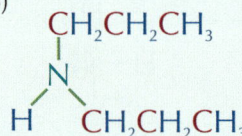

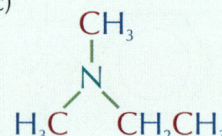

Q2 Ammonia, phenylamine and ethylamine, shown below, can all act as bases:

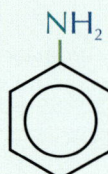

```
   H                NH2              H  H  H
   |                 |               |  |  |
   N             (benzene ring)   H—C—C—N—H
  / \                                |  |
 H   H                               H  H
```

Ammonia *Phenylamine* *Ethylamine*

a) Which of these is the strongest base? Explain your answer.

b) Which of these is the weakest base? Explain your answer.

Q3 Name these amides:

a)
```
   H  H  O
   |  |  ‖
H—C—C—C—NH2
   |  |
   H  H
```

b)
```
   O  H  H  H  H
   ‖  |  |  |  |
H—C—N—C—C—C—H
         |  |  |
         H  H  H
```

Practice Questions — Fact Recall

Q1 What is a quaternary ammonium ion?

Q2 a) What type of ammonium compound can be used as a cationic surfactant?

b) Give one use of cationic surfactants.

Q3 Amines can act as bases. What determines the strength of the base?

4. Reactions of Amines

The last few pages talked about amines and why they are useful. This section covers how amines are made from other organic compounds, and how they can react.

Formation of amines from halogenoalkanes

There are two ways to produce aliphatic amines — the first uses halogenoalkanes in a nucleophilic substitution reaction you met on page 214. Amines can be made by heating a halogenoalkane with excess ethanolic ammonia (ammonia dissolved in ethanol).

Example

Ethylamine can be produced by heating bromoethane with ammonia:

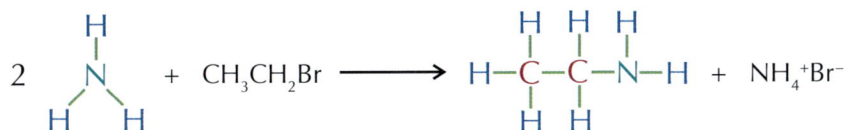

$$2\,NH_3 + CH_3CH_2Br \longrightarrow CH_3CH_2NH_2 + NH_4^+Br^-$$

But, things aren't that simple. You'll actually get a mixture of primary, secondary and tertiary amines, and quaternary ammonium salts, as more than one hydrogen is likely to be substituted.

Example

When producing ethylamine you'll actually get a mixture of ethylamine, diethylamine, triethylamine and tetraethylamine ions:

$$NH_3 + CH_3CH_2Br \longrightarrow NH_2CH_2CH_3 \text{ / } NH(CH_2CH_3)_2 \text{ / } N(CH_2CH_3)_3 \text{ / } N(CH_2CH_3)_4^+ \; + NH_4^+Br^-$$

The formation of amines from halogenoalkanes is a **nucleophilic substitution** reaction. Here's the general mechanism for this reaction:

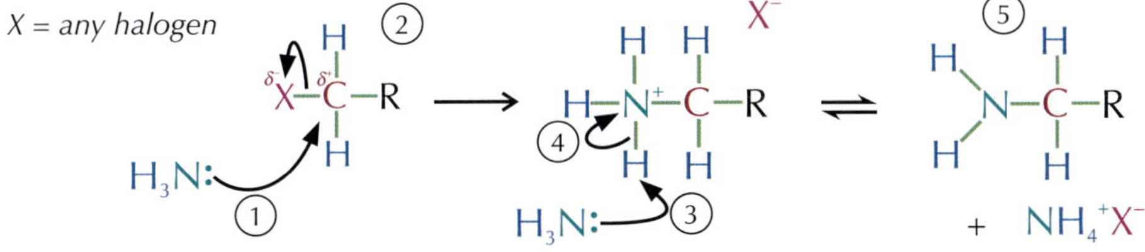

***Figure 1:** The nucleophilic substitution mechanism for the reaction between ammonia and a halogenoalkane.*

1. NH_3 (the nucleophile) attacks the δ^+ carbon on the halogenoalkane and donates its lone pair of electrons, forming a bond with the carbon.
2. As carbon can only have four bonds, the addition of ammonia causes the C–X bond to break, releasing a negatively charged halide ion and leaving an alkylammonium salt.
3. A second ammonia molecule then donates its lone pair of electrons to one of the hydrogens attached to the nitrogen.
4. This hydrogen breaks off from the alkyl ammonium salt and joins ammonia to form an NH_4^+ ion.
5. This leaves a primary amine. The NH_4^+ ion forms an ionic bond with the X^- ion to produce $NH_4^+X^-$.

Learning Objectives:

- Know that primary aliphatic amines can be prepared by the reaction of ammonia with halogenoalkanes and by the reduction of nitriles.
- Be able to describe and outline the mechanisms of the nucleophilic substitution reactions of ammonia and amines with halogenoalkanes to form primary, secondary and tertiary amines and quaternary ammonium salts.
- Know that amines are nucleophiles.
- Be able to describe and outline the mechanism of the nucleophilic addition-elimination reactions of ammonia and primary amines with acyl chlorides.
- Be able to describe the nucleophilic addition-elimination reactions of ammonia and primary amines with acid anhydrides.
- Know that aromatic amines can be prepared by the reduction of nitro compounds, and are used in the manufacture of dyes.

Specification Reference 3.3.11.1, 3.3.11.3

Tip: The nitrogen in the amine has a lone pair of electrons on it. This means it can accept a hydrogen and form an alkylammonium salt — so the conversion of an alkylammonium salt to an amine is reversible.

Exam Tip
You could be asked to name and outline the mechanism for the production of any amine from a halogenoalkane, so make sure you understand this mechanism — just learning the example won't be enough.

Example

Here's the mechanism for the formation of ethylamine from ammonia and bromoethane:

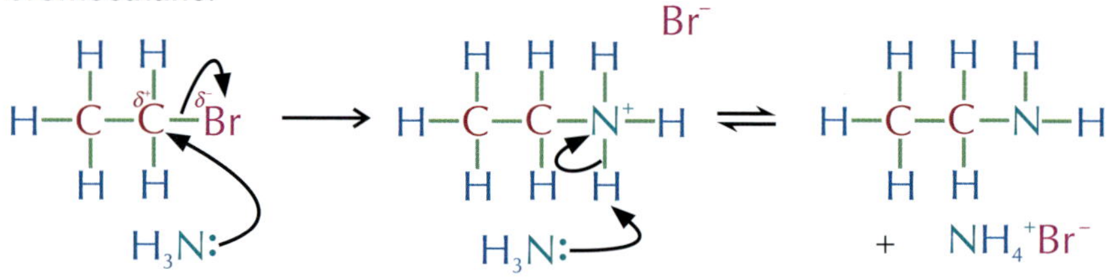

$$CH_3CH_2Br + H_3N: \rightarrow CH_3CH_2NH_3^+Br^- \rightleftharpoons CH_3CH_2NH_2 + NH_4^+Br^-$$

The amine product still has a lone pair of electrons on the nitrogen, so it is still a nucleophile. This means that further substitutions can take place. They keep happening until you get a quaternary ammonium salt, which can't react any further as it doesn't have a lone pair of electrons (see Figure 2).

Primary Amine $\xrightarrow{\text{Halogenoalkane}}$ Secondary Amine $\xrightarrow{\text{Halogenoalkane}}$ Tertiary Amine $\xrightarrow{\text{Halogenoalkane}}$ Quaternary ammonium salt

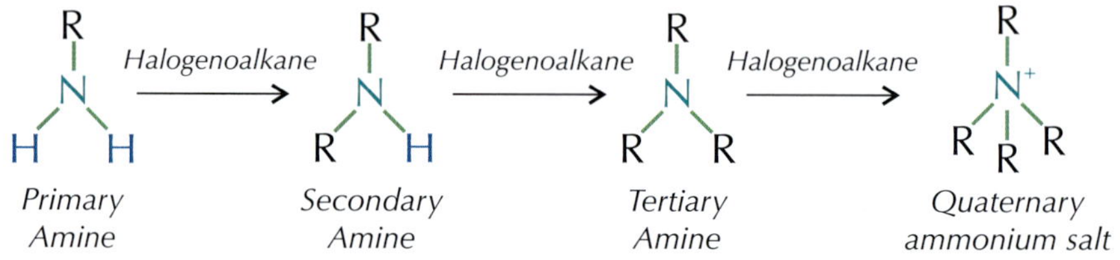

Figure 2: Further substitutions lead to the formation of secondary, tertiary and quaternary amines.

Tip: The number of further substitutions that happen can be reduced by increasing the concentration of ammonia relative to the halogenoalkane — the higher the concentration of ammonia, the less likely it is that a halogenoalkane will meet and react with an amine rather than ammonia.

The mechanism for these further substitutions is the same — you just use an amine instead of ammonia.

Example

Here's the mechanism for the formation of dimethylamine from methylamine and bromomethane:

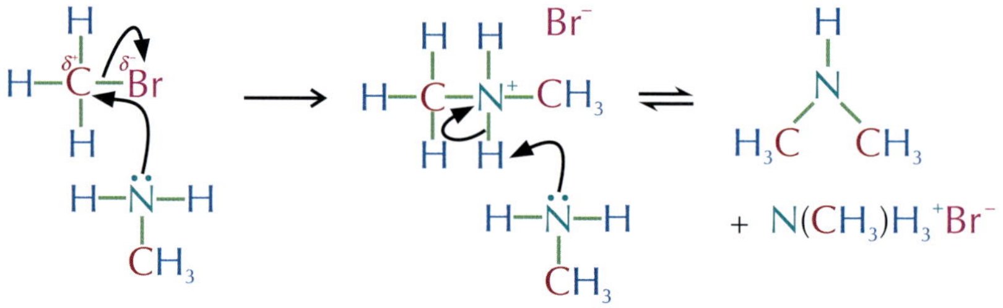

$$CH_3Br + CH_3NH_2 \rightarrow CH_3NH_2^+CH_3\ Br^- \rightleftharpoons (CH_3)_2NH + N(CH_3)H_3^+Br^-$$

Tip: Amines and ammonia can both act as nucleophiles in nucleophilic substitution reactions because they both have a lone pair of electrons.

Formation of amines from nitriles

The second way to produce aliphatic amines uses nitriles. You can reduce a nitrile to an amine using a strong reducing agent such as lithium aluminium hydride ($LiAlH_4$) in dry ether, followed by some dilute acid.

Tip: Forming amines by reducing nitriles is useful if you just want the primary amine because you can't get further substitutions — the primary amine is the only product.

Example

Here's the equation for the formation of ethylamine from ethanenitrile:

$$\underset{\text{Ethanenitrile}}{CH_3C{\equiv}N} + 4[H] \xrightarrow[\text{2. Dilute acid}]{\text{1. } LiAlH_4} \underset{\text{Ethylamine}}{CH_3CH_2NH_2}$$

Tip: [H] means any reducing agent.

This is great in the lab, but $LiAlH_4$ is too expensive for industrial use. Industry uses a metal catalyst such as platinum or nickel at a high temperature and pressure — it's called **catalytic hydrogenation**.

Example

Here's the equation for the formation of ethylamine from ethanenitrile using catalytic hydrogenation:

$$\underset{\textit{Ethanenitrile}}{CH_3C{\equiv}N} + 2H_2 \xrightarrow[\textit{High temp \& pressure}]{Ni} \underset{\textit{Ethylamine}}{CH_3CH_2NH_2}$$

Amines as nucleophiles

Acyl chlorides react with ammonia to form primary amides, and with primary amines to form N-substituted amides. These reactions occur via a **nucleophilic addition-elimination** mechanism. The general mechanism is:

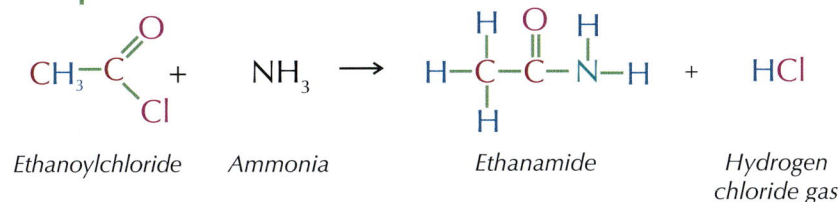

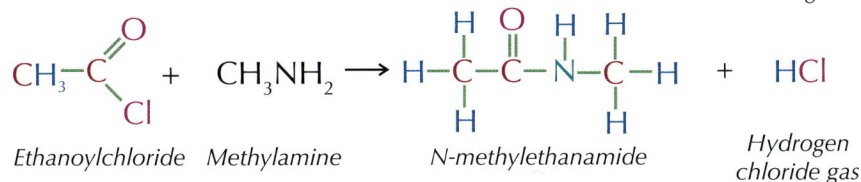

Step 1:

1. The lone pair of electrons on the nitrogen atom of the amine attacks the δ⁺ carbon on the acyl chloride.

4. This leaves a positively charged amide ion and a negatively charged chloride ion.

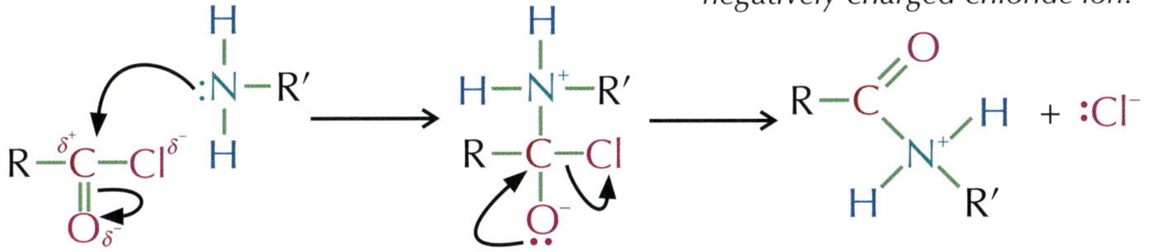

2. A pair of electrons is transferred from the C=O bond onto the oxygen.

3. The pair of electrons on the oxygen reform the double bond and the chlorine is kicked off.

Step 2:

5. The Cl⁻ ion bonds with a hydrogen on the nitrogen atom.

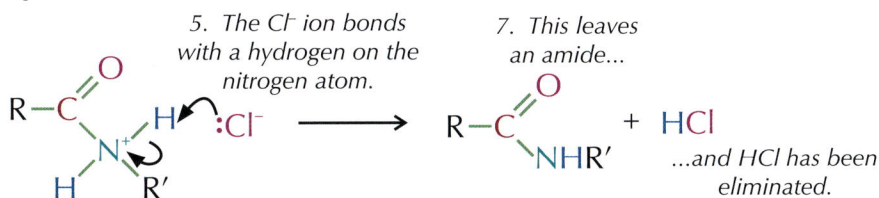

6. A pair of electrons is transferred onto the nitrogen atom from the bond.

7. This leaves an amide...

$$R{-}C(=O){-}N^+H_2R' + :Cl^- \longrightarrow R{-}C(=O){-}NHR' + HCl$$

...and HCl has been eliminated.

Examples

$$\underset{\textit{Ethanoylchloride}}{CH_3COCl} + \underset{\textit{Ammonia}}{NH_3} \longrightarrow \underset{\textit{Ethanamide}}{CH_3CONH_2} + \underset{\textit{Hydrogen chloride gas}}{HCl}$$

$$\underset{\textit{Ethanoylchloride}}{CH_3COCl} + \underset{\textit{Methylamine}}{CH_3NH_2} \longrightarrow \underset{\textit{N-methylethanamide}}{CH_3CONHCH_3} + \underset{\textit{Hydrogen chloride gas}}{HCl}$$

Tip: Remember, nucleophiles are lone pair donors. The lone pair on the nitrogen atom in aliphatic amines lets them act as nucleophiles.

Tip: The mechanism for the reaction with ammonia is exactly the same, but an amide containing a $-C(O)NH_2$ group is formed

Tip: There's lots more detail about acyl chlorides and their reactions on pages 432-433.

***Figure 3:** The reaction of an amine (the orange liquid in the pipette) with an acyl chloride, to produce an amide and hydrogen chloride gas.*

Tip: Acid anhydrides have the general formula RC(O)OC(O)R′. There's more about them and their reactions on pages 435-436.

Tip: See page 450 for more on nitrobenzene and how it is made.

Acid anhydrides also react with primary amines or ammonia to form amides and a carboxylic acid. The reaction with primary amines produces and N-substituted amide, whilst the reaction with ammonia produces a primary amide. The general reactions is:

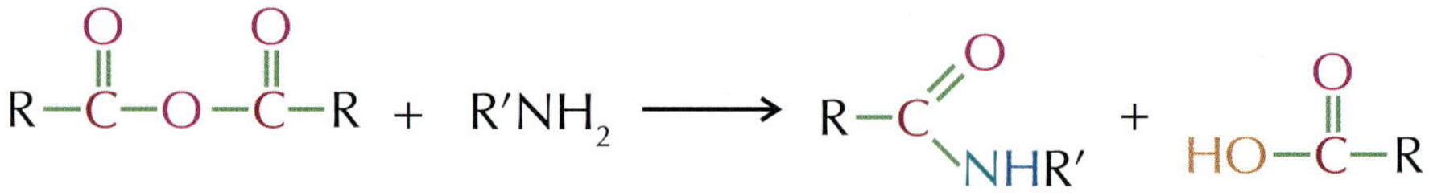

$$R-C(=O)-O-C(=O)-R + R'NH_2 \longrightarrow R-C(=O)-NHR' + HO-C(=O)-R$$

Figure 4: The general reaction of an amine (if R′ is an organic group) or ammonia (if R′ is an H atom) with an acid anhydride.

Formation of aromatic amines

Aromatic amines are produced by reducing a nitro compound, such as nitrobenzene. There are two steps to the method:

- First, a mixture of a nitro compound, tin metal and concentrated hydrochloric acid is heated under reflux — this makes a salt.
- Then to turn the salt into an aromatic amine, an alkali, such as sodium hydroxide solution, is added to the mixture.

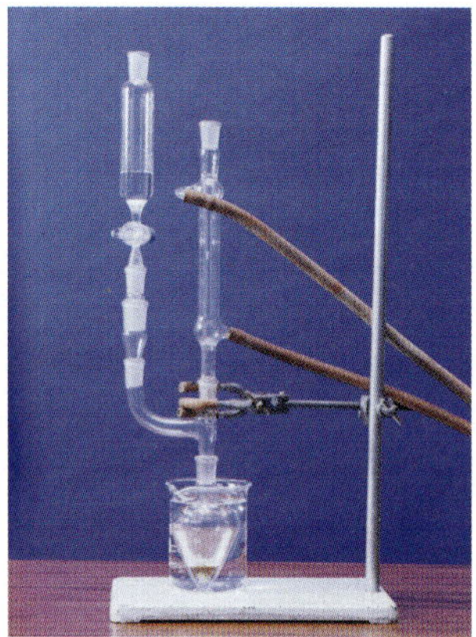

Figure 5: The reflux equipment used to make phenylamine.

Example

Phenylamine can be made by reducing nitrobenzene:

- Mixing nitrobenzene with tin and concentrated HCl and heating under reflux produces the salt $C_6H_5NH_3^+ Cl^-$.
- Adding NaOH to this salt then releases phenylamine.

Here's the overall equation for the reaction:

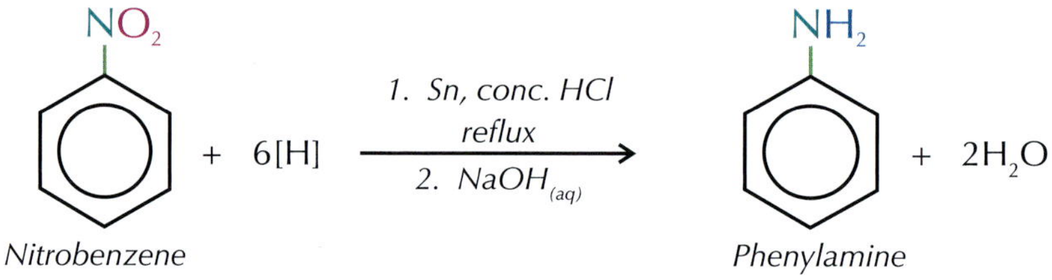

$$C_6H_5NO_2 + 6[H] \xrightarrow[\text{2. NaOH}_{(aq)}]{\text{1. Sn, conc. HCl reflux}} C_6H_5NH_2 + 2H_2O$$

Nitrobenzene → Phenylamine

Producing aromatic amines is important in industry — for example, they're used in the manufacture of dyes.

Practice Questions — Application

Q1 a) Write an equation to show the formation of methylamine from chloromethane.

b) Name and outline the mechanism for this reaction.

Q2 a) Write an equation to show the formation of propylamine from propanenitrile by catalytic hydrogenation.

b) What conditions are required for this reaction?

Q3 Write an equation to show the formation of 3-methylphenylamine from 3-methylnitrobenzene.

Practice Questions — Fact Recall

Q1 a) How can amines be produced from halogenoalkanes?

b) Explain why a mixture of primary, secondary and tertiary amines, and quaternary ammonium salts are produced in these reactions.

Q2 a) Give two ways that amines can be produced from nitriles.

b) Which of these methods is used in industry, and why?

Q3 Name the type of organic product(s) formed when ammonia reacts with:

a) a chloroalkane b) an acyl chloride c) an acid anhydride

Q4 Describe how aromatic amines can be produced.

Section Summary

Make sure you know...

- That benzene has the chemical formula C_6H_6 with six carbons joined together in a planar (flat) ring.
- That the carbon-carbon bonds in benzene are all the same length (intermediate between double and single bonds).
- That the delocalisation of p electrons makes benzene more stable than a theoretical cyclohexa-1,3,5-triene molecule.
- That enthalpies of hydrogenation provide evidence for the extra stability of benzene compared to a theoretical cyclohexa-1,3,5-triene molecule.
- What aromatic compounds are and how to name them.
- That the benzene ring is a region of high electron density so it attracts electrophiles.
- That electrophilic attack on benzene rings usually results in substitution because the ring is so stable.
- That Friedel-Crafts acylation and nitration are both electrophilic substitution reactions, and how to draw the mechanism for these reactions.
- That nitration is used to add nitro groups onto benzene rings, using HNO_3 and an H_2SO_4 catalyst.
- That nitro compounds are used to make explosives and aromatic amines (which are used to manufacture dyes and pharmaceuticals).
- That Friedel-Crafts acylation adds an acyl group onto benzene rings, producing a phenylketone.
- The role of halogen carriers such as $AlCl_3$ as catalysts in Friedel-Crafts acylation.
- What amines are and how to name them.
- That quaternary ammonium salts can be used as cationic surfactants, and the uses of these surfactants.
- That amines can accept protons, so are able to act as weak bases.
- Why aromatic amines are weaker bases than ammonia and why aliphatic amines are stronger bases.
- What amides are and how to name them.
- That amines can be formed by heating a halogenoalkane with an excess of ethanolic ammonia.
- That nucleophilic substitution reactions produce a mixture of primary, secondary and tertiary amines and quaternary ammonium salts because more than one hydrogen is likely to be substituted.
- The mechanism for nucleophilic substitution reactions of amines and ammonia with halogenoalkanes.
- That amines can also be formed by reducing nitriles.
- The mechanism for the nucleophilic addition-elimination reactions of ammonia and primary amines with acyl chlorides and acid anhydrides to produce amides.
- That aromatic amines are formed by reducing nitro compounds such as nitrobenzene.

Exam-style Questions

1 Which of the following statements about cationic surfactants is correct?

A Cationic surfactants can be made by reacting tertiary amines with halogenoalkanes.

B Cationic surfactants can cause a build-up of static electricity on hair and clothing.

C Cationic surfactants contain an amine group, so can act as bases.

D The long hydrocarbon chain in cationic surfactants allows the molecule to dissolve in water.

(1 mark)

2 Benzene is an extremely important molecule in the synthesis of aromatic compounds. The benzene ring is very stable and as a result benzene is fairly unreactive. Friedel-Crafts acylation reactions are used to add acyl side chains onto the benzene ring which can then be modified to make useful products.

2.1 Write an equation to show the production of phenylpropanone by Friedel-Crafts acylation. Include all the reagents and conditions required for this reaction.

(2 marks)

2.2 Friedel-Crafts acylation uses an $AlCl_3$ catalyst. Explain its role.

(2 marks)

2.3 Outline the mechanism for this reaction.

(4 marks)

Nitro groups can be added to the benzene ring via nitration reactions, where an aromatic compound is heated with concentrated nitric and sulfuric acids.

2.4 Outline the mechanism for the addition of one nitro group onto the benzene ring of methylbenzene. Your mechanism should include an equation to show the formation of a nitronium ion from concentrated nitric and sulfuric acids.

(4 marks)

Nitro compounds can be reduced to give aromatic amines.

2.5 Write an equation to show the reduction of nitrobenzene to phenylamine. Describe the conditions needed for this reaction.

(5 marks)

3 What type of reactions will typically take place at a benzene ring?

A Electrophilic addition

B Electrophilic substitution

C Nucleophilic addition-elimination

D Nucleophilic substitution

(1 mark)

4 The diagrams below show two amines (**X** and **Y**).

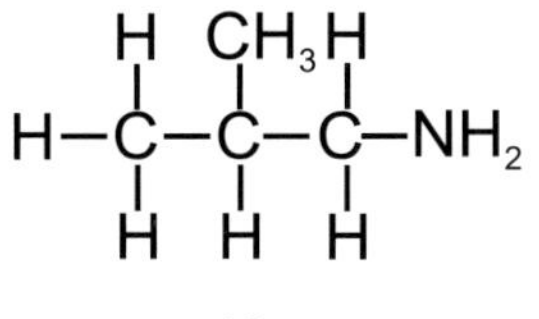

H CH$_3$H
H–C–C–C–NH$_2$
H H H

X

NH$_2$
CH$_3$
CH$_3$

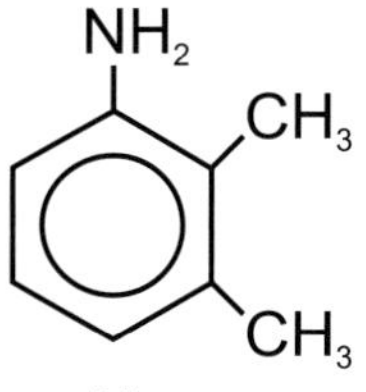

Y

4.1 Give the IUPAC names for these two amines.

(2 marks)

Aliphatic amines such as amine **X** can be produced by heating halogenoalkanes with an excess of ethanolic ammonia.

4.2 Write an equation to show the formation of amine **X** from a bromoalkane.

Name and outline the mechanism for this reaction.

(7 marks)

Preparations of amine **X** from a halogenoalkane are usually impure and contain a number of different products.

4.3 State why a mixture of different products can be produced in these reactions.

Draw the structure of one of the other products likely to be in the mixture with amine **X**.

(2 marks)

4.4 Suggest how the amount of amine **X** in the mixture could be increased.

(1 mark)

Amine **X** can also be made by reducing a nitrile with a strong reducing agent such as $LiAlH_4$.

4.5 Write an equation to show the reduction of a nitrile to amine **X**.

Explain why the amine **X** produced via this reaction is purer than the amine **X** made from a halogenoalkane.

(2 marks)

Amines **X** and **Y** can both act as bases.

4.6 State why amines are able to act as bases.
Which of these two amines is the stronger base? Explain your answer.

(5 marks)

1. Condensation Polymerisation

On page 229, you were introduced to addition polymerisation, which is where lots of alkenes join up together to form one long molecule. Now it's time for another type of polymerisation — condensation polymerisation.

Learning Objectives:

- Know that condensation polymers are formed by reactions between dicarboxylic acids and diols, dicarboxylic acids and diamines or between amino acids.
- Be able to describe the repeating units in polyesters (e.g. Terylene™) and polyamides (e.g. nylon 6,6 and Kevlar®) and the linkages between these repeating units.
- Be able to describe some typical uses of polyesters (e.g. Terylene™) and polyamides (e.g. nylon 6,6 and Kevlar®).
- Know that polyesters and polyamides can be broken down by hydrolysis.
- Be able to explain the nature of the intermolecular forces between molecules of condensation polymers.

Specification Reference 3.3.12.1, 3.3.12.2

Condensation polymers

Condensation polymerisation usually involves two different types of monomer. Each monomer has at least two functional groups. Each functional group reacts with a group on another monomer to form a link, creating polymer chains. Each time a link is formed, a small molecule is lost (usually water) — that's why it's called condensation polymerisation.

Examples of condensation polymers include polyamides, polyesters and polypeptides (or proteins). In polyesters, an ester link (–COO–) is formed between the monomers. In polyamides and polypeptides, amide links (–CONH–) are formed between the monomers. In polypeptides, these amide links are usually called peptide bonds.

Polyamides

Reactions between dicarboxylic acids and diamines make **polyamides**. The carboxyl groups of dicarboxylic acids react with the amino groups of diamines to form **amide links**. A water molecule is lost each time an amide link is formed — it's a condensation reaction (see Figure 1).

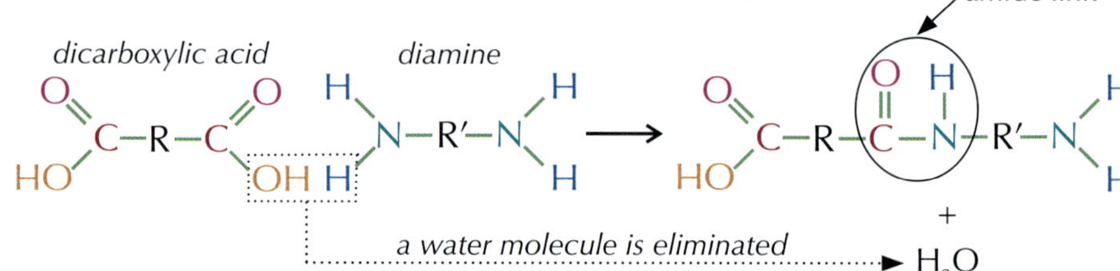

Figure 1: *The formation of an amide link.*

Dicarboxylic acids and diamines have functional groups at both ends, which means that they can each form two amide links and long chains can form. There are two polyamides you need to learn the reactions for:

- Nylon 6,6 — made from 1,6-diaminohexane and hexanedioic acid:

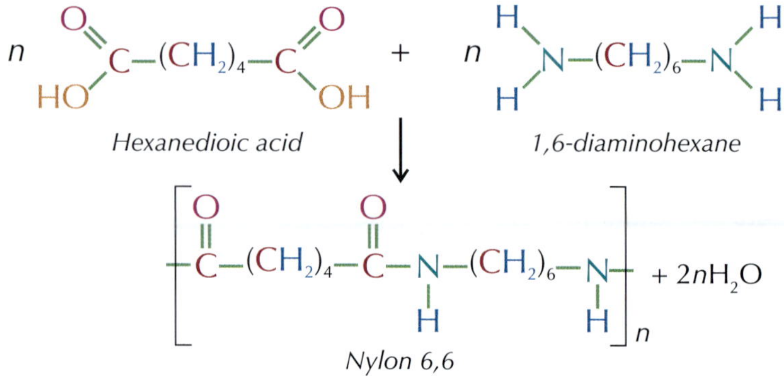

$$n\ HOOC{-}(CH_2)_4{-}COOH + n\ H_2N{-}(CH_2)_6{-}NH_2 \rightarrow \left[{-}CO{-}(CH_2)_4{-}CO{-}NH{-}(CH_2)_6{-}NH{-} \right]_n + 2nH_2O$$

Hexanedioic acid — 1,6-diaminohexane — Nylon 6,6

Figure 2: *The formation of nylon 6,6.*

Tip: Figure 2 shows the formation of nylon 6,6. The product shown has brackets around it and an '*n*' after it — this is the formula of the polymer. The repeating unit is this formula without the brackets, and without the '*n*'.

Nylon 6,6 is strong and resistant to abrasion so is used to make clothing, carpet, rope, airbags and parachutes.

- Kevlar® — made from benzene-1,4-dicarboxylic acid and 1,4-diaminobenzene:

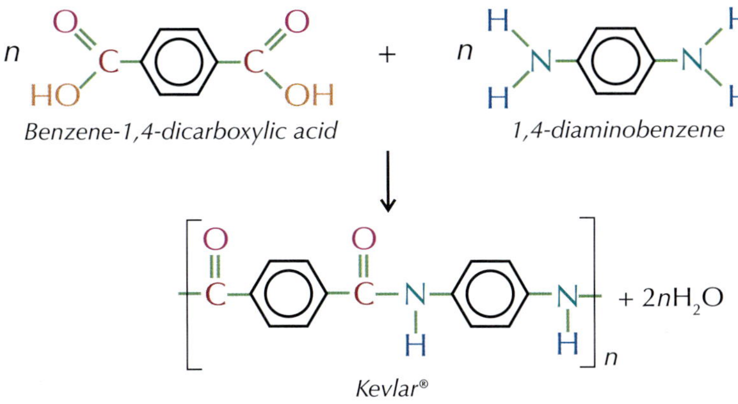

Benzene-1,4-dicarboxylic acid

1,4-diaminobenzene

$+ 2nH_2O$

Kevlar®

Figure 4: *The formation of Kevlar®.*

Kevlar® is both light and very strong, so it's used in bulletproof vests, boat construction, car tyres and lightweight sports equipment.

Figure 3: *Kevlar® is used to make bulletproof clothing.*

Exam Tip
If you're asked to draw the repeating unit of Kevlar® you should leave off the square brackets and the '*n*'.

Peptides

Amino acids contain an amine functional group and a carboxylic acid functional group. These groups can react in condensation polymerisation reactions to form polyamides. Polymers between amino acids are more commonly called **peptides**, though — see page 479 for more information.

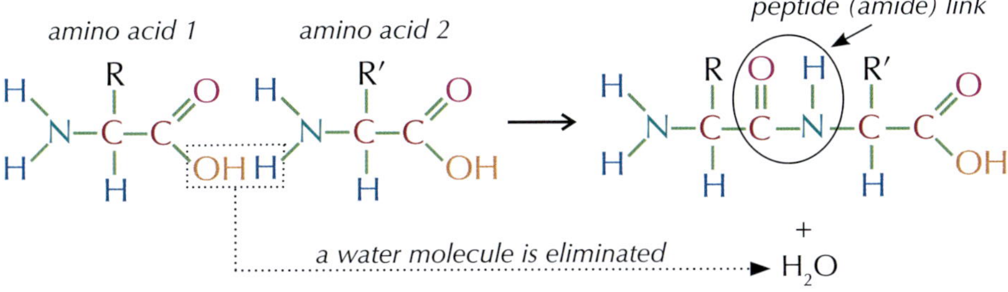

Figure 5: *The formation of a peptide link.*

Polyesters

The carboxyl groups of dicarboxylic acids can react with the hydroxyl groups of diols to form **ester links** — it's another condensation reaction (see Figure 6).

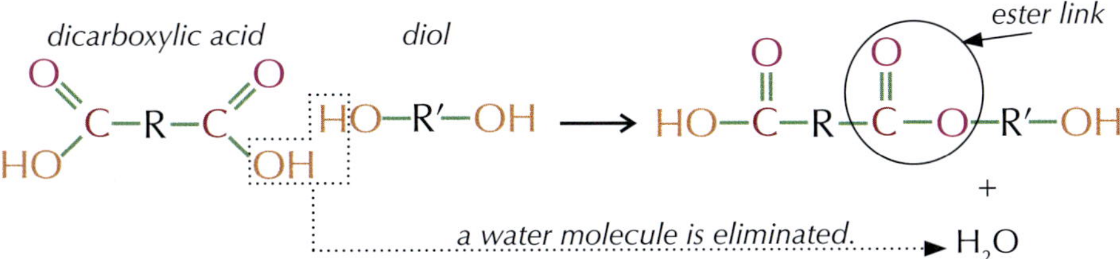

Figure 6: *The formation of an ester link.*

Tip: The formation of polyesters is very similar to the formation of polyamides and polypeptides.

Figure 7: *The containers that microwave meals come in are made of Terylene™.*

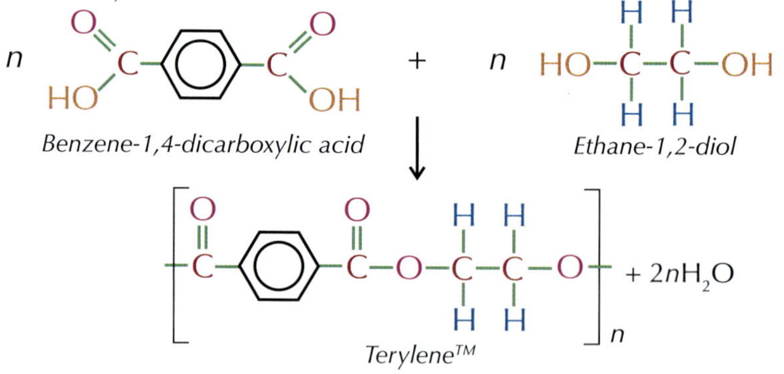

Polymers joined by ester links are called polyesters — an example is Terylene™ (PET). Terylene™ is formed from benzene-1,4-dicarboxylic acid and ethane-1,2-diol.

$$n\ HOOC{-}C_6H_4{-}COOH + n\ HO{-}CH_2{-}CH_2{-}OH \rightarrow [{-}CO{-}C_6H_4{-}CO{-}O{-}CH_2{-}CH_2{-}O{-}]_n + 2nH_2O$$

Benzene-1,4-dicarboxylic acid | *Ethane-1,2-diol* | *Terylene™*

Figure 8: *The formation Terylene™.*

Some forms of Terylene™ are stable at both cold and hot temperatures, making them useful for creating containers for ready meals. Other forms are used to make plastic bottles, clothing, sheets and sails.

Hydrolysis of polyesters and polyamides

Tip: In hydrolysis reactions, a molecule of water breaks down a large molecule into two smaller molecules.

The ester or amide link in polyesters and polyamides can be broken down by **hydrolysis** — water molecules are added back in and the links are broken. The products of the hydrolysis are the monomers that were used to make the polymer — you're basically reversing the condensation reaction that was used to make them:

$$n(\text{monomers}) \underset{\textit{hydrolysis}}{\overset{\textit{condensation}}{\rightleftharpoons}} \text{polymer} + \text{water}$$

Water hydrolyses the amide links in polyamides, resulting in the formation of diamine and dicarboxylic acid molecules:

$$[{-}CO{-}R{-}CO{-}NH{-}R'{-}NH{-}]_n + 2nH_2O \longrightarrow n\ HOOC{-}R{-}COOH + n\ H_2N{-}R'{-}NH_2$$

amide link | *polyamide* | *dicarboxylic acid* | *diamine*

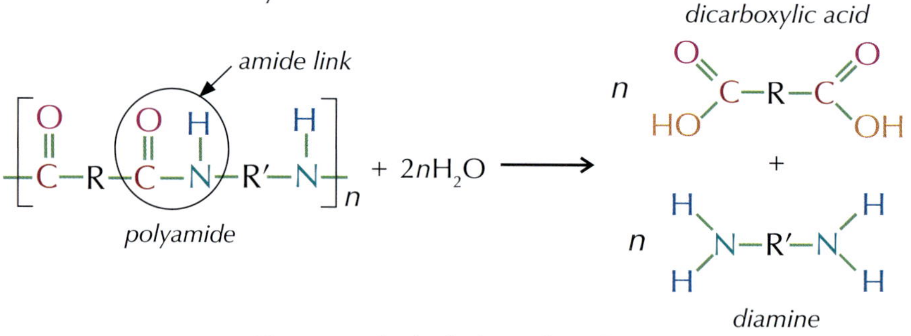

Figure 9: *The hydrolysis of a polyamide.*

Water hydrolyses the ester links in polyesters, resulting in the formation of diol and dicarboxylic acid molecules:

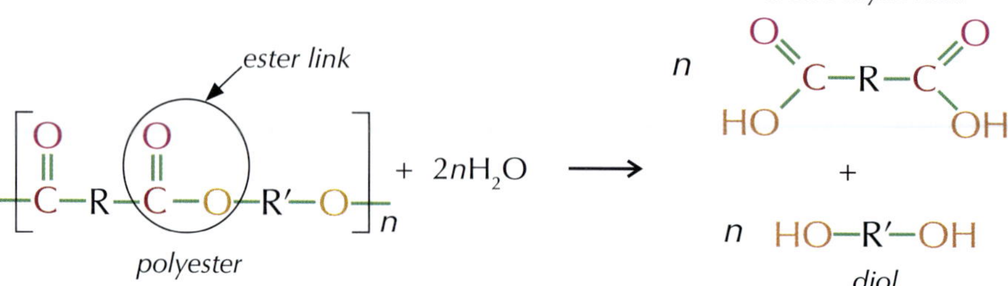

Figure 10: *The hydrolysis of a polyester.*

In practice, hydrolysis with just water is far too slow, so in the lab, the reaction is done with an acid or alkali. Polyamides are hydrolysed more easily in acidic conditions, and polyesters are hydrolysed more easily in basic conditions.

Bonding between polymer chains

Condensation polymers are generally stronger and more rigid than addition polymers. This is because condensation polymers are made up of chains containing polar bonds, e.g. C=O and C–N or C–O.
So, as well as induced dipole-dipole forces, there are permanent dipole-dipole forces and very strong hydrogen bonds between the polymer chains (Figure 11).

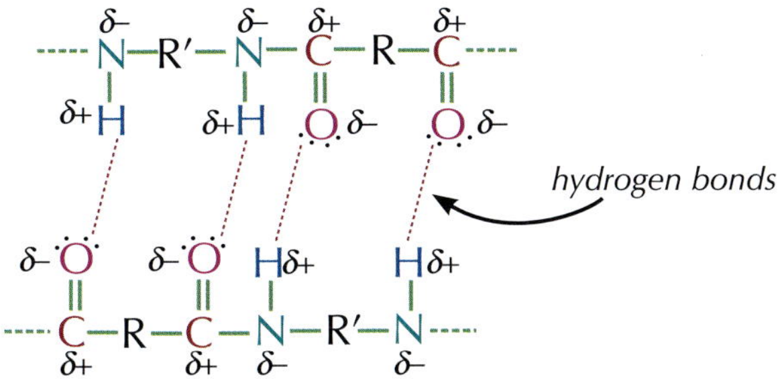

Figure 11: *Intermolecular forces between condensation polymer chains.*

Tip: You met all the different types of intermolecular force on pages 94-97.

Tip: Hydrogen bonds are the strongest type of intermolecular force, so need lots of energy to be broken. This makes condensation polymers more rigid than addition polymers.

Practice Questions — Fact Recall

Q1 Suggest two types of molecule that can react together to form:
- a) polyamides
- b) polyesters

Q2 What type of reaction gives rise to polyamides and polyesters?

Q3 Draw the repeating units of:
- a) Nylon 6,6
- b) Kevlar®
- c) Terylene™

Q4 What are the products when you hydrolyse:
- a) a polyamide,
- b) a polyester?

Q5 a) What is the strongest type of intermolecular force that can form between the chains of a condensation polymer?
b) Draw a diagram to show the formation of these bonds between two sections of polymer chains.

2. Monomers & Repeating Units

Learning Objectives:

- Be able to draw the repeating unit of a condensation polymer from monomer structure(s) or from a section of the polymer chain.
- Be able to draw the structure(s) of the monomer(s) of a condensation polymer from a section of the polymer chain.

Specification Reference 3.3.12.1

You need to be able to look at a polymer and work out its repeating unit and what monomers it's made from, or look at some monomers and figure out the repeating unit of the polymer they'll form. Luckily for you, help is here...

Repeating units

Condensation polymers are made up of **repeating units** (a bit of molecule that repeats over and over again). If you know the formulas of a pair of monomers that react together in a condensation polymerisation reaction, you can work out the repeating unit of the condensation polymer that they would form. Here are the steps you should take to find the repeating unit of the polyamide that forms when a diamine and a dicarboxylic acid react together:

- Draw out the two monomer molecules next to each other.
- Remove an OH from the dicarboxylic acid, and an H from one of the nitrogen atoms in the diamine — that gives you a water molecule.
- Join the C=O and the N together to make an amide link.
- Take an H off the other nitrogen atom, and an OH off the other -COOH group at the ends of your molecule.

Example

Draw the repeating unit of the condensation polymer that is made from 1,4-diaminobutane, $H_2N(CH_2)_4NH_2$, and decanedioic acid, $HOOC(CH_2)_8COOH$.

If you draw out the monomers next to each other, this makes it easier to see which groups will be lost to form the water molecule:

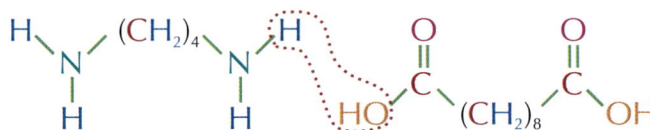

$$H_2N-(CH_2)_4-N(H)-\boxed{H \quad HO}-C(=O)-(CH_2)_8-C(=O)-OH$$

Draw a bond between the C and the N that have lost groups to the water molecule:

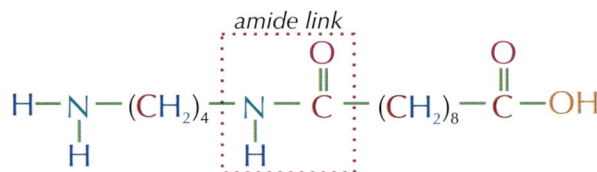

$$H-N(H)-(CH_2)_4-\underbrace{N(H)-C(=O)}_{\text{amide link}}-(CH_2)_8-C(=O)-OH$$

To get the repeating unit, all that needs to be done now is to remove another -OH and -H from the groups at the ends of the molecule.

$$-N(H)-(CH_2)_4-N(H)-C(=O)-(CH_2)_8-C(=O)-$$

If you start by drawing the monomers the other way round, your repeating unit may end up looking slightly different:

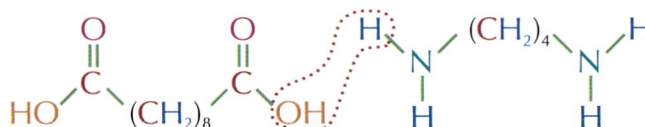

$$HO-C(=O)-(CH_2)_8-C(=O)-\boxed{OH \quad H}-N(H)-(CH_2)_4-NH_2$$

Tip: If your monomers are a dicarboxylic acid and a diol, you follow the same steps, but in step two you take an H from one of the oxygen atoms of your diol, and in the third step, you join the C and the O together to make an ester link.

Tip: Even if you've got a dicarboxylic acid and a diol, the -OH group is always lost from the carboxylic acid.

Tip: The formula of a polymer is just its repeating unit with square brackets round it and an *n* at the end.

Tip: If you've got a diol rather than a diamine, the process is exactly the same except you end up with a polyester rather than a polyamide.

This would cause your repeating unit to look like this:

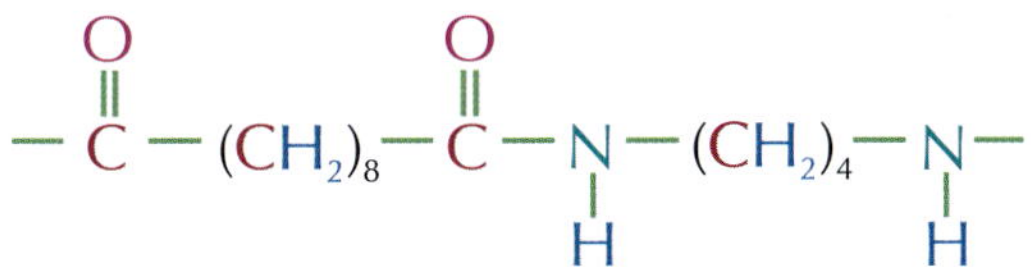

$$-\overset{O}{\overset{\|}{C}}-(CH_2)_8-\overset{O}{\overset{\|}{C}}-\underset{H}{\underset{|}{N}}-(CH_2)_4-\underset{H}{\underset{|}{N}}-$$

Tip: It doesn't matter which way round you draw the repeating unit — both are correct.

You might also have to find the repeating unit from a longer section of a polyester or polyamide chain. Here's how you do it:

- Look along the chain and find the repeating pattern. Starting from one end, write out the chain until you get to the end of the repeating section.
- The repeating unit should have a C=O from an ester or amide link at one end, and the -O- or -NH- part of the link at the other. If the section you've written out doesn't, just find an ester or amide link, break it, and move the bit you've broken off from one end of the unit to the other.

Example

A section of a polymer chain is shown below. Draw its repeating unit.

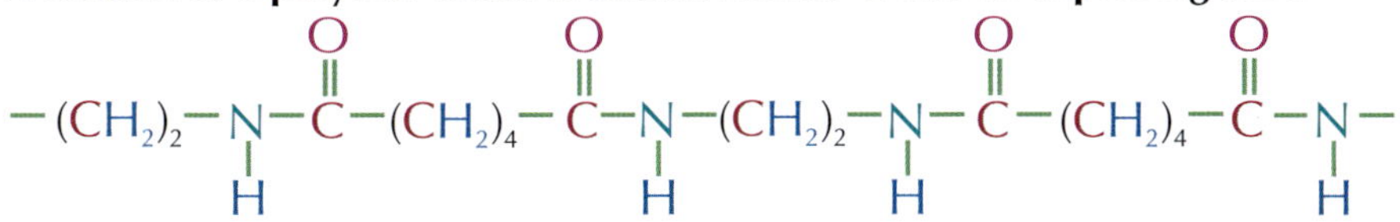

$$-(CH_2)_2-\underset{H}{\underset{|}{N}}-\overset{O}{\overset{\|}{C}}-(CH_2)_4-\overset{O}{\overset{\|}{C}}-\underset{H}{\underset{|}{N}}-(CH_2)_2-\underset{H}{\underset{|}{N}}-\overset{O}{\overset{\|}{C}}-(CH_2)_4-\overset{O}{\overset{\|}{C}}-\underset{H}{\underset{|}{N}}-$$

Starting from the first $(CH_2)_2$, the repeating pattern goes to the second -NH-:

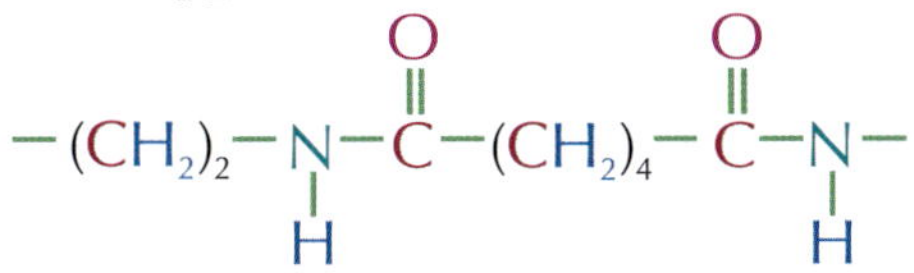

$$-(CH_2)_2-\underset{H}{\underset{|}{N}}-\overset{O}{\overset{\|}{C}}-(CH_2)_4-\overset{O}{\overset{\|}{C}}-\underset{H}{\underset{|}{N}}-$$

Move the $-(CH_2)_2$–NH- section from the start to the end, to get a C=O from an amide link at one end and the corresponding -NH- at the other:

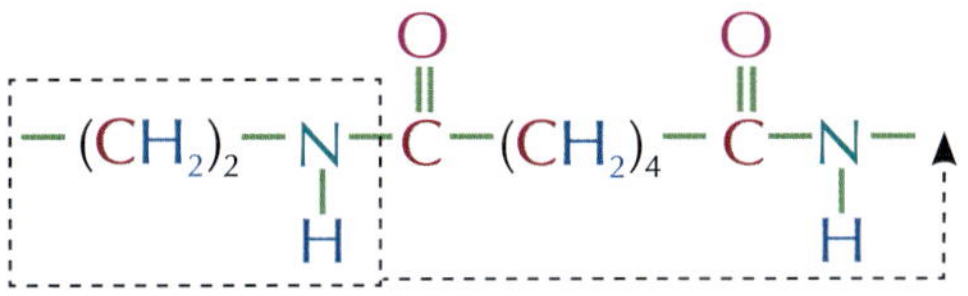

$$\boxed{-(CH_2)_2-\underset{H}{\underset{|}{N}}-}\overset{O}{\overset{\|}{C}}-(CH_2)_4-\overset{O}{\overset{\|}{C}}-\underset{H}{\underset{|}{N}}-\uparrow$$

So the repeating unit looks like this:

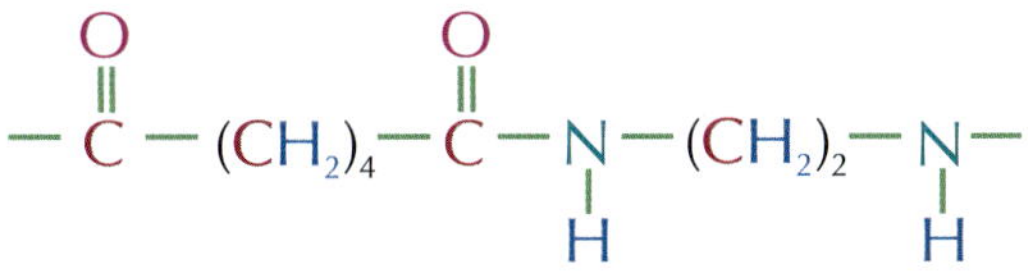

$$-\overset{O}{\overset{\|}{C}}-(CH_2)_4-\overset{O}{\overset{\|}{C}}-\underset{H}{\underset{|}{N}}-(CH_2)_2-\underset{H}{\underset{|}{N}}-$$

Tip: As before, you could write the repeating unit the other way round — you'd get the alternative version by moving the final -NH- from the end of the repeating section to the beginning instead of moving the $-(CH_2)_2$–NH- section.

Identifying monomers

In your exam, you might be asked to identify the monomers that formed a particular condensation polymer. To do this just follow these steps:

- If you're given a section of the polymer chain, first find the repeating unit, as shown above.
- Remove the bond in the middle of the central amide or ester link — that's the bond between the -C=O group and the -NH group in polyamides or the bond between the -C=O group and the oxygen in polyesters.
- Add –OH groups on to the -C=O groups to make carboxyl groups.
- For polyamides, add hydrogens to the -NH groups to make $-NH_2$ groups.
- For polyesters, add hydrogens on to the oxygens to make -OH groups and add -OH groups on to any terminal carbon atoms.

Exam Tip
Don't be caught out — there won't necessarily be two different monomers. There could be one monomer with a COOH and an NH_2 group on it. See the next page for an example of this.

Tip: This example is a polyamide. You do exactly the same for a polyester but you end up with a dicarboxylic acid and a diol.

Tip: The monomers are just the products of hydrolysing the polymer with water. This means it's easy to check that you've added enough -H and -OH groups. For a repeating unit containing two monomers, you should add in two H atoms and two -OH groups (which makes up two H_2O molecules).

Example

Draw the monomers that formed this condensation polymer:

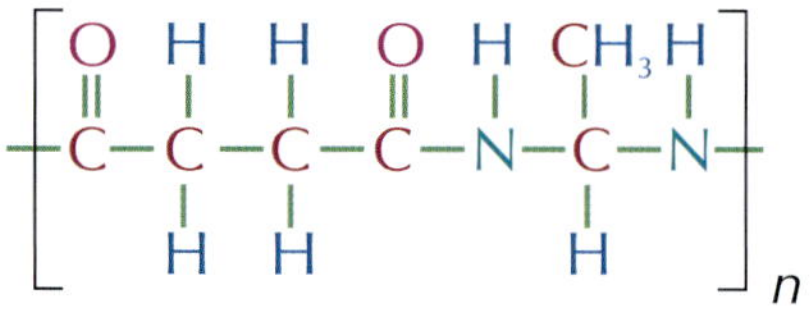

$$\left[-C(=O)-CH_2-CH_2-C(=O)-NH-CH(CH_3)-NH- \right]_n$$

This is a polyamide so if you remove the bond in the middle of the amide link you get:

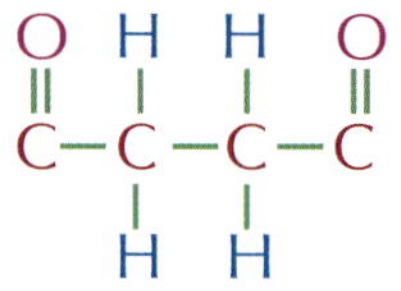

$$O{=}C-CH_2-CH_2-C{=}O \quad \text{and} \quad HN-CH(CH_3)-NH$$

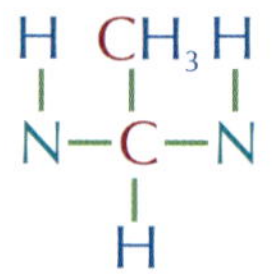

Adding an OH group on to the C=O groups gives you:

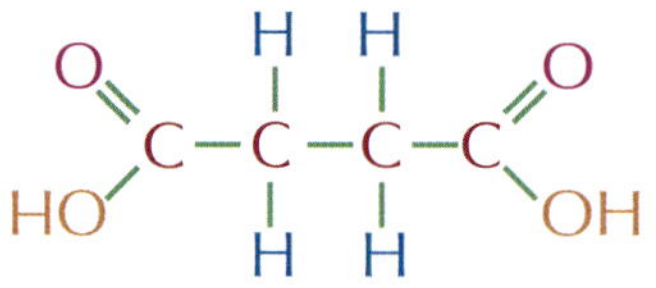

$$HOOC-CH_2-CH_2-COOH$$

And adding a hydrogen on to each of the NH groups gives you:

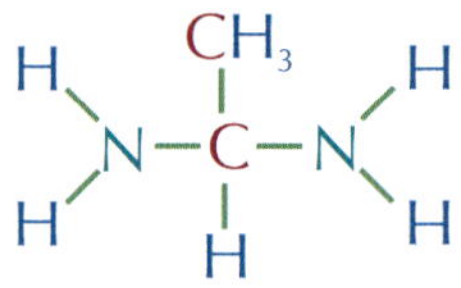

$$H_2N-CH(CH_3)-NH_2$$

So these must be the monomers that joined to form the polymer.

More complex condensation polymers

Not all condensation polymers are made from a dicarboxylic acid and a diamine or diol.

For example, if a molecule contains a carboxylic acid group and either an alcohol or an amine group, it can polymerise with itself to form a condensation polymer with only one monomer. This is the case for amino acids, which contain an amine and a carboxylic acid group, and polymerise to form peptides (see page 479).

Tip: When a molecule can form a condensation polymer with itself, it's called 'self-polymerisation'.

Tip: Amino acids contain a carboxylic acid group and an amine group, so they can react with themselves to form polymers containing only one monomer. More usually, they react with other amino acids to form a protein with a long sequence of different amino acids all joined together by peptide bonds (p.479).

Example

3-aminopropanoic acid contains both an amine and a carboxylic acid group. It can react with itself to form a condensation polymer:

$$n\,H_2N-(CH_2)_2-C(=O)-OH \longrightarrow \left[-N(H)-(CH_2)_2-C(=O)- \right]_n + nH_2O$$

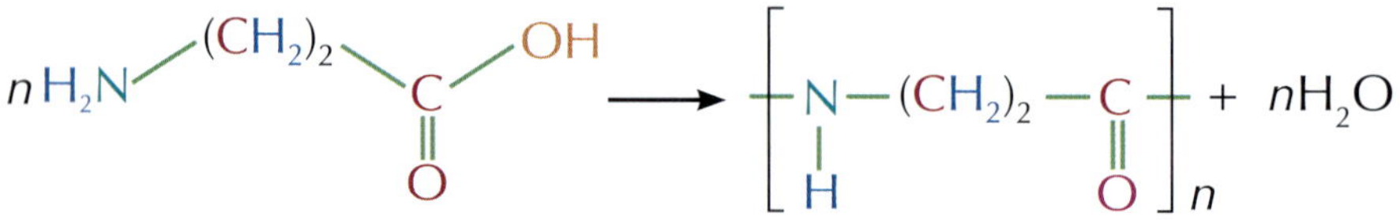

3-aminopropanoic acid

Molecules that contain both an amine and an alcohol group can react with dicarboxylic acids in a condensation polymerisation reaction. The polymers they form contain both amide and ester links.

Example

4-aminobutan-1-ol contains an amine group and an alcohol group. It can react with dicarboxylic acids, such as butanedioic acid, in a condensation polymerisation reaction:

4-aminobutan-1-ol

butanedioic acid

Amide link.

The ester link goes between the repeating units.

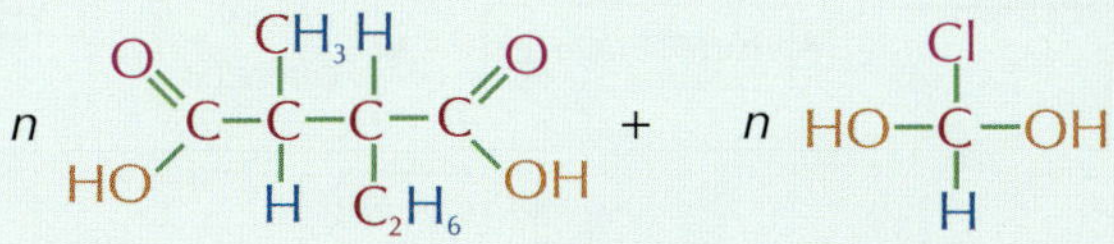

Tip: You could also draw the repeat unit so that the ester link is in the middle and the amide link is split between the repeat units — both ways are correct.

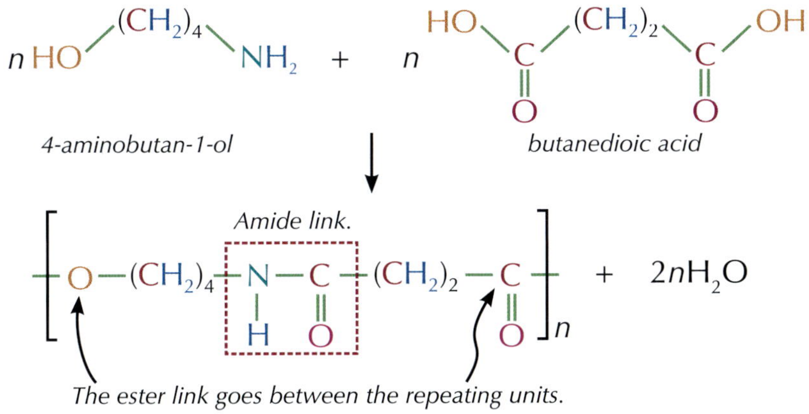

Practice Questions — Application

Q1 Identify whether the following are formulas of polyesters or polyamides and draw the monomer(s) they were formed from.

a)

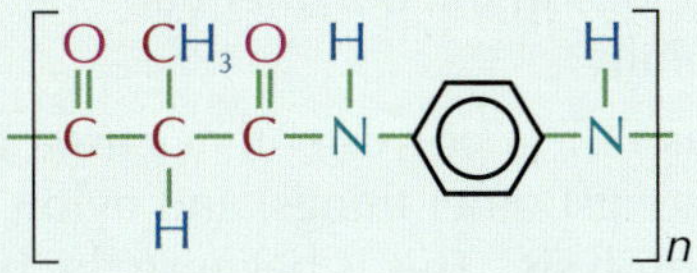

b)

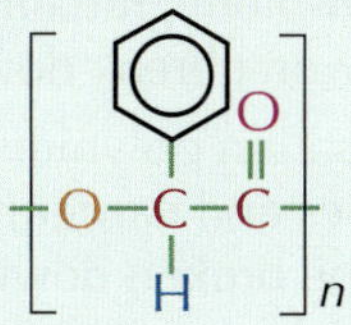

c)

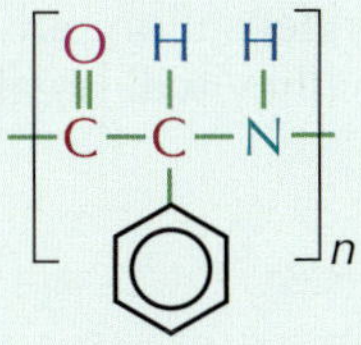

d)

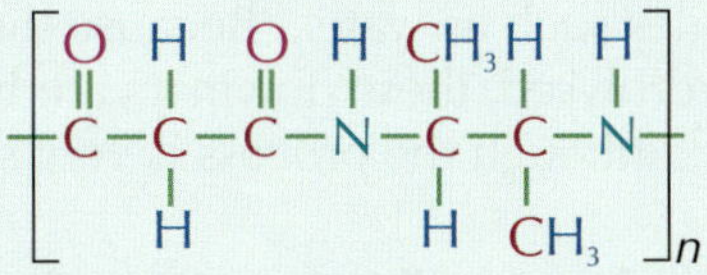

Q2 Draw the repeating units of the polymers that would be formed from these monomers:

a)

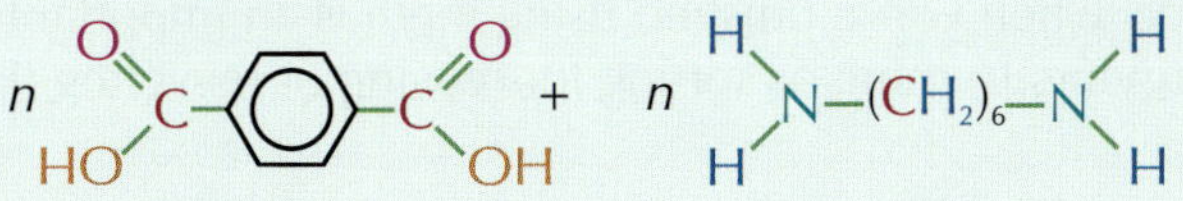

b)

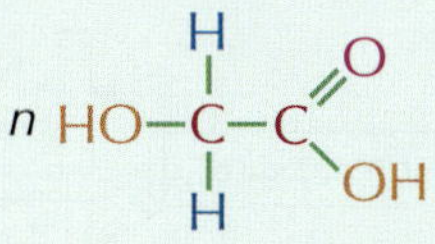

c)

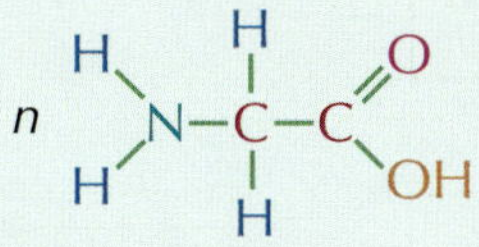

d)

Exam Tip

It's easy to get in a muddle when you draw repeating units of condensation polymers, so make sure you practise these types of questions.

3. Disposing of Polymers

Learning Objectives:

- Know that polyalkenes are chemically inert and non-biodegradable.
- Know that polyesters and polyamides can be broken down by hydrolysis and are biodegradable.
- Be able to explain why polyesters and polyamides can be hydrolysed but polyalkenes cannot.
- Be able to describe the advantages and disadvantages of different methods of disposal of polymers, including recycling.

Specification Reference 3.3.12.2

Polymers are really useful — loads of things are made from them. But unfortunately, disposing of polymers once they've been used can be a tad tricky. Read on to find out why.

The widespread use of polymers

Synthetic polymers have loads of advantages, so they're incredibly widespread these days — we take them pretty much for granted. Just imagine what you'd have to live without if there were no polymers (see Figure 1).

Figure 1: *Some of the many items that are made out of synthetic polymers.*

Biodegradability of polymers

Polyalkenes such as poly(ethene) and polystyrene are chemically inert (very unreactive). This is because the bonds between the repeating units are non-polar, so they aren't susceptible to attack by nucleophiles. Being chemically inert is an advantage when polymers are being used (e.g. a polystyrene cup won't react with your coffee), but it has the disadvantage of making them **non-biodegradable** (the bonds in the polymer can't be hydrolysed and won't break down naturally).

Condensation polymers such as PET (a polyester that's used to make fizzy drinks bottles and carrier bags amongst other things) and nylon (a polyamide) can be broken down by hydrolysis. This is because the bonds between the repeating units are polar and so are susceptible to attack by nucleophiles such as water. Because the bonds in condensation polymers can be hydrolysed, these polymers are **biodegradable** (they will break down naturally), although the process is very slow.

Tip: You met addition polymers (polyalkenes) back on pages 229-231. They form when the double bonds in alkenes open up and join together.

Disposing of waste plastics

It's estimated that in the UK we throw away over 3 million tons of plastic (i.e. synthetic polymers) every year. Because plastics either take a very long time to biodegrade or are non-biodegradable, the question of what to do with all those plastic objects when we've finished using them is an important one. The options are burying, burning or sorting for reusing or recycling (Figure 2).

Tip: A ton is just over 1000 kg (1016.047 kg to be exact) so 3 million tons is an awful lot of plastic to dispose of.

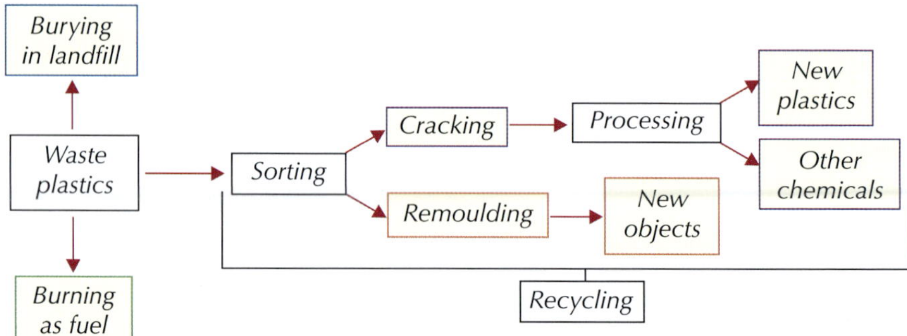

Figure 2: *The different methods for disposing of waste plastic.*

None of these methods is an ideal solution — they all have advantages and disadvantages associated with them.

Figure 3: A landfill site — burying waste plastic is one way of disposing of it.

Burying waste plastic

Landfill is one option for dealing with waste plastics. It is generally used when the plastic is difficult to separate from other waste, not in sufficient quantities to make separation financially worthwhile or too difficult technically to recycle. Landfill is a relatively cheap and easy method of waste disposal, but there are disadvantages to burying waste plastic too:

- It requires areas of land.
- As the waste decomposes it can release methane — a greenhouse gas.
- As the waste decomposes it can release toxins which can be washed away and contaminate water supplies.

The amount of waste we generate is becoming more and more of a problem, so there's a need to reduce landfill as much as possible.

Burning waste plastic

Waste plastics can be burned and the heat used to generate electricity. This process needs to be carefully controlled to reduce toxic gases. For example, polymers that contain chlorine (such as PVC) produce HCl when they're burned — this has to be removed. So, waste gases from the combustion are passed through scrubbers which can neutralise gases such as HCl by allowing them to react with a base. But, the waste gases, e.g. carbon dioxide, will still contribute to the greenhouse effect.

Recycling plastics

Because many plastics are made from non-renewable oil fractions, it makes sense to recycle plastics as much as possible. There's more than one way to recycle plastics. After sorting into different types some plastics (poly(propene), for example) can be melted and remoulded, while others can be cracked into monomers which can be used to make more plastics or other chemicals. Plastic products are usually marked to make sorting easier. The different numbers show different polymers.

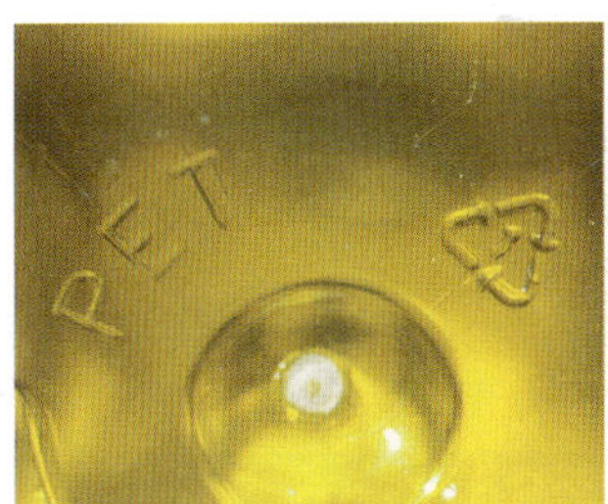

Figure 4: The recycling symbol on the bottom of a plastic bottle.

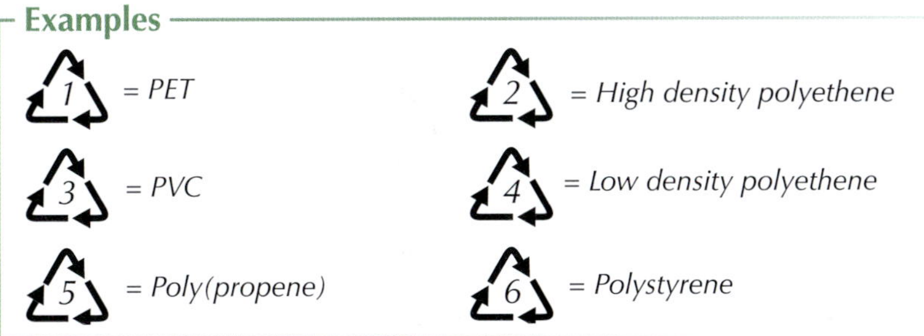

Like other disposal methods, there are advantages and disadvantages to recycling plastics:

Advantages	Disadvantages
It reduces the amount of waste going into landfill.	It is technically difficult to recycle plastics.
It saves raw materials — which is important because oil is non-renewable.	Collecting, sorting and processing the plastic is more expensive than burning/landfill.
The cost of recycling plastics is lower than making the plastics from scratch.	You often can't remake the plastic you started with — you have to make something else.
It produces less CO_2 emissions than burning the plastic.	The plastic can be easily contaminated during the recycling process.

Exam Tip
When answering a question on the advantages and disadvantages of recycling plastics, don't just say cost — you need to justify your answer because the cost can be an advantage and a disadvantage.

Practice Question — Application

Q1 The structures of two polymers (**A** and **B**) are shown below:

A

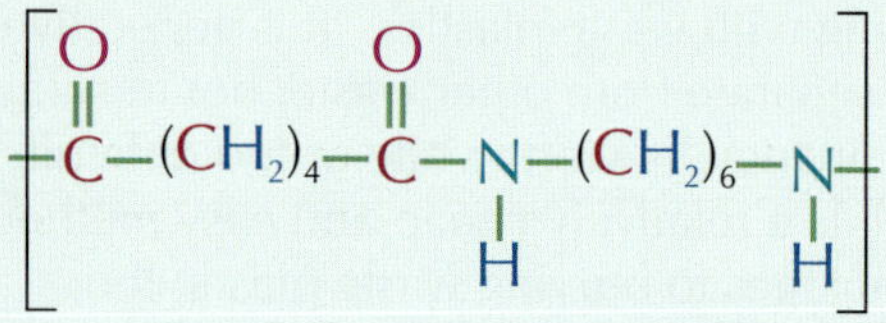

$$\left[\begin{array}{c} O \\ \| \\ C \end{array} - (CH_2)_4 - \begin{array}{c} O \\ \| \\ C \end{array} - \begin{array}{c} N \\ | \\ H \end{array} - (CH_2)_6 - \begin{array}{c} N \\ | \\ H \end{array} \right]$$

B

$$\left[\begin{array}{c} H_3C \\ | \\ C \\ | \\ H \end{array} - \begin{array}{c} H \\ | \\ C \\ | \\ CH_3 \end{array} \right]$$

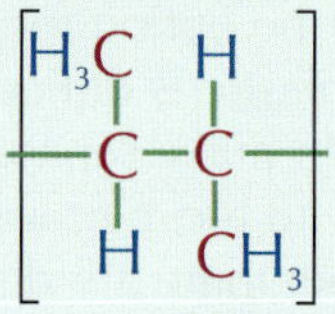

Predict, with reasoning, which of these polymers is biodegradable.

Practice Questions — Fact Recall

Q1 a) Explain why addition polymers are chemically inert.

b) Give an advantage of addition polymers being chemically inert.

c) What does being chemically inert mean for the biodegradability of the plastic?

Q2 a) Are condensation polymers biodegradable or non-biodegradable?

b) Explain why the biodegradabilities of addition and condensation polymers are different.

Q3 Disposing of waste plastic in landfill sites is cheap and easy. Give three disadvantages of using landfill sites.

Q4 Burning plastic produces toxic gases. How can these toxic gases be removed?

Q5 a) Give two ways in which plastics can be recycled.

b) How is the sorting of plastics made easier?

c) Give two advantages and two disadvantages of recycling waste plastics.

Section Summary

Make sure you know...

- What condensation polymers are.
- How polyamides, polyesters and polypeptides are formed.
- The typical uses of polyesters and polyamides.
- How polyamides and polyesters can be hydrolysed back to their constituent monomers using acid or base hydrolysis.
- That condensation polymers can form strong hydrogen bonds between the polymer chains.
- How to draw the repeating unit(s) of condensation polymers from monomer structures or from a section of a polymer chain.
- How to draw the monomer structure(s) from a section of a condensation polymer chain.
- Why addition polymers are non-biodegradable and condensation polymers are biodegradable.
- The advantages and disadvantages of disposing of waste plastic in landfill and burning waste plastic.
- The advantages and disadvantages of recycling waste plastics.

Exam-style Questions

1 Which of the options below shows the correct monomers that are formed when the general polymer shown below is hydrolysed by water?

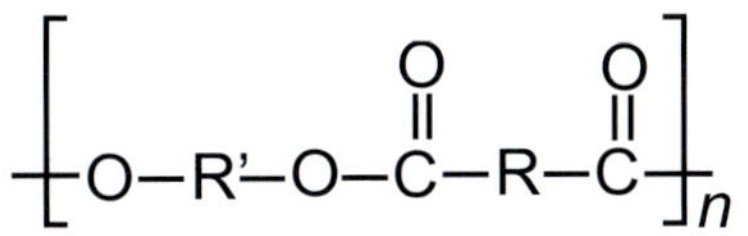

$$\left[-O-R'-O-\overset{\overset{O}{\|}}{C}-R-\overset{\overset{O}{\|}}{C}- \right]_n$$

	Monomer 1	Monomer 2
A	HO–R'–OH	HOOC–R–COOH
B	HO–R–OH	HOOC–R'–COOH
C	HO–R'–COOH	HOOC–R–OH
D	HO–R'–$CONH_2$	H_2NOC–R–OH

(1 mark)

2 The diagram below shows the repeating unit formed when two amino acids polymerise to form a peptide:

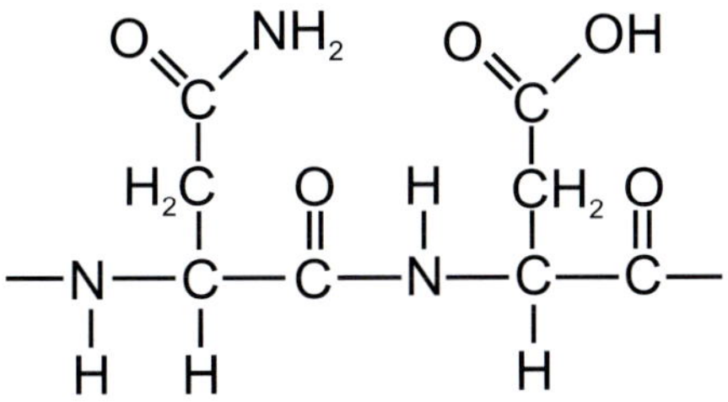

$$-N(H)-C(H)(CH_2CONH_2)-C(=O)-N(H)-C(H)(CH_2COOH)-C(=O)-$$

2.1 Draw the two amino acids which have joined together to make this repeating unit. *(2 marks)*

2.2 What type of polymerisation occurs between the amino acids to form this polymer? *(1 mark)*

2.3 What is the name of the link that forms between two amino acids? *(1 mark)*

3 Nylon 6,6 is made from hexanedioic acid and 1,6-diaminohexane.

3.1 What type of condensation polymer is nylon 6,6? *(1 mark)*

3.2 Draw the repeating unit of nylon 6,6. *(1 mark)*

3.3 Is nylon 6,6 biodegradable or non-biodegradable? Explain your answer. *(4 marks)*

3.4 Discuss the advantages and disadvantages of recycling objects made from nylon 6,6. *(4 marks)*

1. Amino Acids

This is where chemistry and biology start to overlap. Amino acids are the building blocks of all living organisms. Read on to find out all about them.

Learning Objectives:
- Understand that amino acids have both acidic and basic properties.
- Be able to apply IUPAC rules for naming amino acids.
- Understand that amino acids can form zwitterions.
- Be able to draw the structures of amino acids as zwitterions, and the ions formed from amino acids in acid and alkaline solution.
- Know that mixtures of amino acids can be separated and identified by thin-layer chromatography.
- Know that amino acids can be located on a chromatogram using developing agents such as ninhydrin or ultraviolet light and identified by their R_f values.
- Be able to calculate R_f values from a chromatogram.

Specification Reference 3.3.1.1, 3.3.13.1, 3.3.13.2

What are amino acids?

An **amino acid** has two functional groups — an amine (or amino) group (NH_2) and a carboxyl group (COOH). The structure of all amino acids is the same apart from an organic side chain (R) that is different in each. The general structure of an amino acid is given in Figure 1.

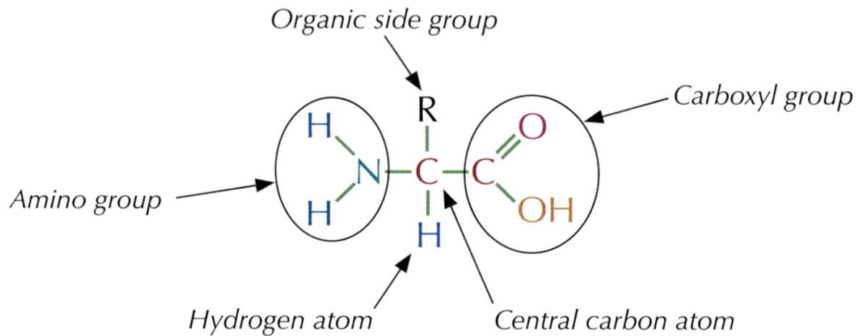

Figure 1: *The general structure of an amino acid.*

Amino acids are interesting molecules because they are **amphoteric** — this means they've got both acidic and basic properties. They can act as acids because the carboxyl group is acidic — it can donate a proton:

$$\text{-COOH} \rightleftharpoons \text{-COO}^- + \text{H}^+$$

They can act as bases because the amino group is basic — it can accept a proton:

$$\text{-NH}_2 + \text{H}^+ \rightleftharpoons \text{-NH}_3^+$$

Amino acids are chiral molecules because the carbon usually has four different groups attached. So a solution of a single amino acid enantiomer will rotate polarised light — see Figure 2.

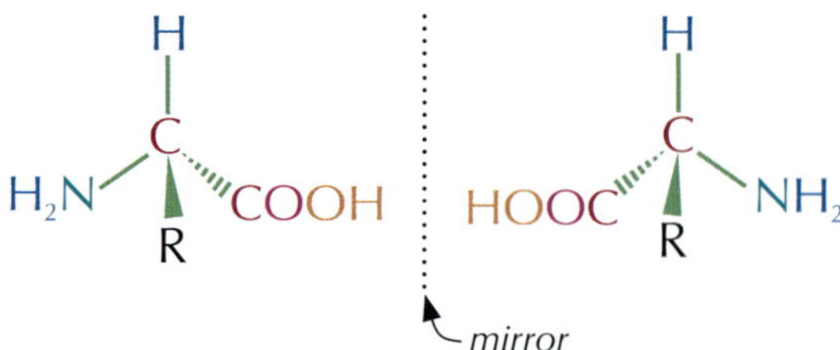

Figure 2: *The two different amino acid enantiomers.*

Tip: The only amino acid that isn't chiral is glycine because its organic side chain is just H — so there are two hydrogens attached to the central carbon.

Tip: See pages 414-417 for more on optical isomerism and chirality.

Naming amino acids

Amino acids often have a common and a systematic name. There's no way of working out the common names, but you can work out the systematic names. To name an amino acid systematically just follow these steps:

1. Find the longest carbon chain that includes the carboxylic acid group and write down its name.
2. Number the carbons in the chain starting with the carbon in the carboxylic acid group as number 1.
3. Write down the positions of any NH_2 groups and show that they are NH_2 groups with the word 'amino'.
4. Write down the names of any other functional groups and say which carbon they are on.

Examples

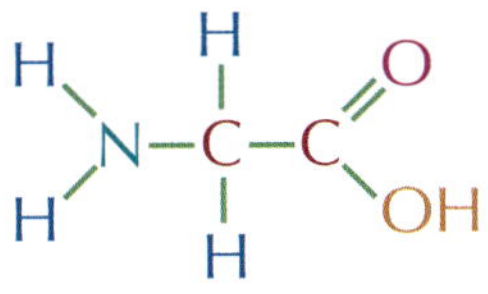

The common name for this amino acid is glycine.

The longest carbon chain is two carbons long so the stem is ethane and the amino group is on carbon-2.

So the systematic name is 2-aminoethanoic acid.

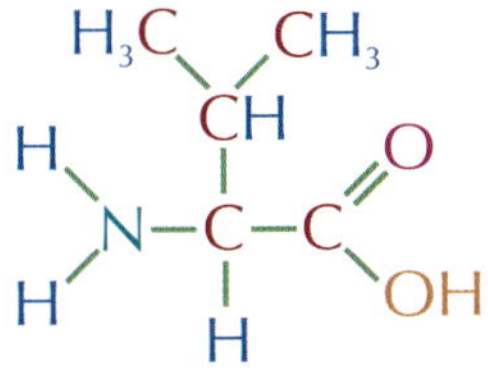

The common name for this amino acid is valine.

The longest carbon chain is four carbons long so the stem is butane. The amino group is on carbon-2 and there is a methyl group on carbon-3.

So this is 2-amino-3-methylbutanoic acid.

Zwitterions

Amino acids can exist as **zwitterions**. A zwitterion is a dipolar ion — it has both a positive and a negative charge in different parts of the molecule. Zwitterions only exist near an amino acid's **isoelectric point**. This is the pH where the overall charge on the amino acid is zero. It's different for different amino acids — it depends on their R-group.

An amino acid becomes a zwitterion when its amino group is protonated to NH_3^+ and its COOH group is deprotonated to COO^-.

- In conditions more acidic than the isoelectric point, the $-NH_2$ group is likely to be protonated but the –COOH group will be unchanged — so the amino acid will carry a positive charge but not a negative charge.
- In conditions more basic than the isoelectric point, the –COOH group is likely to lose its proton but the $-NH_2$ group will be unchanged — so the amino acid will carry a negative charge but not a positive charge.
- Only at or near the isoelectric point are both the carboxyl group and the amino group likely to be ionised — forming a zwitterion (see Figure 3).

Zwitterion

Low pH *Isoelectric point* *High pH*

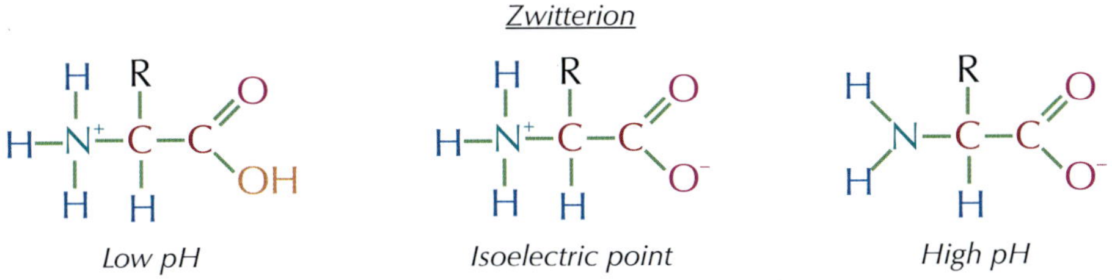

Figure 3: *Formation of a zwitterion.*

Tip: 'Carboxylic acid groups' and 'carboxyl groups' are the same thing.

Exam Tip
Don't worry about learning the common names of the amino acids. If you're asked to name an amino acid in your exam just give the systematic name — you can work this out without having to memorise them all.

Exam Tip
The amino acid side chains can get really complicated, but don't worry — you won't be asked to name anything too complicated in your exam.

Tip: 'Protonated' just means 'has gained a proton' or 'has gained an H^+'. 'Deprotonated' means the opposite.

Exam Tip
You need to be able to draw amino acids at different pHs in your exam, so make sure you understand why amino acids have different charges at different pHs.

Separating mixtures of amino acids

Since different amino acids have different 'R' groups, they will all have different solubilities in the same solvent. This means you can easily separate and identify the different amino acids in a mixture using **thin-layer chromatography** (see Figure 4). Thin-layer chromatography (TLC) is covered in more detail on pages 509-511.

Tip: The thin-layer chromatography plate is usually a piece of plastic or glass covered with a thin layer of silica (silicon dioxide) gel or alumina (aluminium oxide) powder.

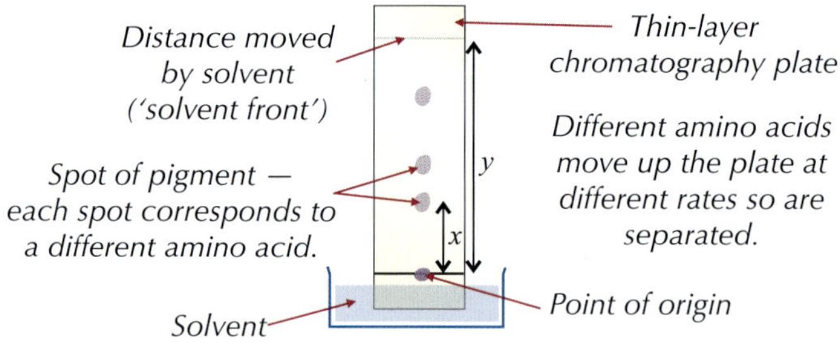

Distance moved by solvent ('solvent front')
Thin-layer chromatography plate
Different amino acids move up the plate at different rates so are separated.
Spot of pigment — each spot corresponds to a different amino acid.
y
x
Solvent
Point of origin

Figure 4: *Thin-layer chromatography.*

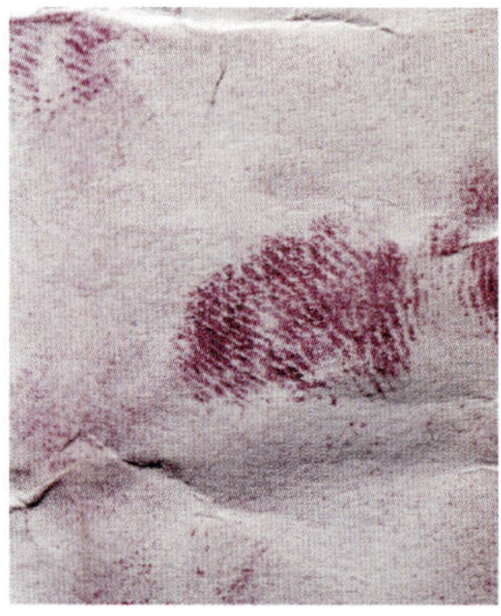

Figure 5: *Ninhydrin can also be used to detect fingerprints as it reacts with the amino acids in the fingerprint, turning them purple and making the print visible.*

Amino acids aren't coloured — you need to do something to make them visible. You could spray ninhydrin solution on the plate, which causes the amino acids to turn purple. Alternatively, you could use a special plate that has a fluorescent dye added to it. The dye glows when UV light shines on it. Where there are spots of chemical on the plate, they cover the fluorescent dye — so the spots appear dark. You can put the plate under a UV lamp and draw around the dark patches to show where the spots are.

Identifying amino acids

You can identify what amino acids you've got in the mixture by using the chromatogram to calculate R_f values. You work out the R_f value of each amino acid spot using this formula:

$$R_f = \frac{x}{y} = \frac{\text{distance travelled by spot}}{\text{distance travelled by solvent}}$$

Using a table of known amino acid R_f values, you can identify the amino acids in the mixture.

Tip: When finding the value of x, measure from the point of origin to the middle of the spot of pigment.

Tip: There's more on calculating R_f values on page 510.

Practice Question — Application

Q1 a) Below are two amino acids (**A** and **B**). Name these amino acids.

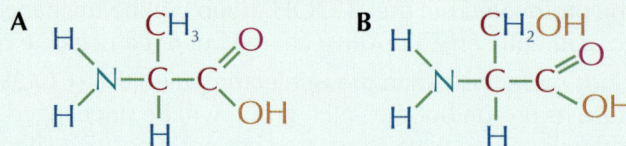

A: $H_2N-C(H)(CH_3)-COOH$ (drawn as H, H on N; N–C–C; CH_3 above and H below the central C; =O and OH on the carboxyl C)

B: $H_2N-C(H)(CH_2OH)-COOH$ (drawn as H, H on N; N–C–C; CH_2OH above and H below the central C; =O and OH on the carboxyl C)

b) Draw species **A** at:

(i) low pH (ii) its isoelectric point (iii) high pH

Practice Questions — Fact Recall

Q1 Explain why amino acids are said to be amphoteric.

Q2 What is a zwitterion?

Q3 What is the formula used to calculate the R_f value of an amino acid?

Q4 Why would ninhydrin solution, or a fluorescent dye with UV light be used in thin-layer chromatography involving amino acids?

2. Proteins

Amino acids are just the starting point for making living things — next come proteins. They're used in loads of different ways, and make up loads of different things, in your body and other animals and plants.

What are proteins?

Proteins are condensation polymers of amino acids — they are made up of lots of amino acids joined together by peptide links. The chain is put together by condensation reactions and broken apart by hydrolysis reactions. Here's how two amino acids join together to make a dipeptide:

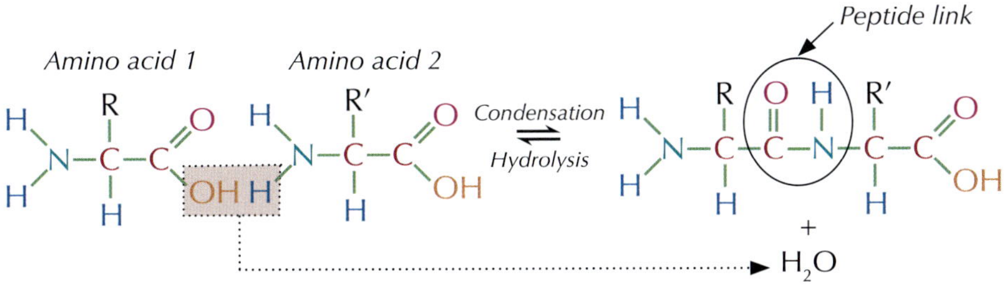

If the two amino acids that are combining are different, then two different dipeptides will be formed because the amino acids can join either way round.

Example

A molecule of alanine and a molecule of glycine could join to form either of these dipeptides:

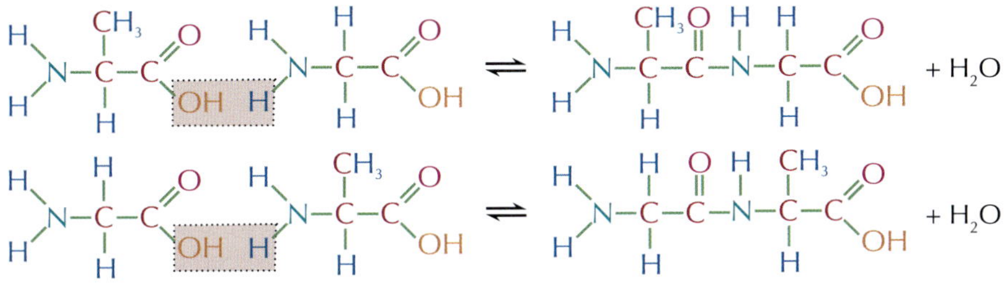

The dipeptide still has an NH_2 group at one end and a COOH group at the other. This means the dipeptides can undergo further condensation reactions with amino acids or other peptides to make longer peptide chains.

To break up the protein into its individual amino acids (hydrolysis) you need to use pretty harsh conditions. One way of hydrolysing proteins is to add aqueous 6 mol dm^{-3} hydrochloric acid, and then heat the mixture under reflux for 24 hours. Proteins are condensation polymers, so to work out the formulas of the amino acids that a protein chain was made from, you can use the same method as on page 469 — just break each of the peptide links down the middle, then add either an H atom or an OH group to each of the broken ends to get the amino acids back.

Protein structure

Proteins are big, complicated molecules. They're easier to explain if you describe their structure in four 'levels' — primary, secondary, tertiary and quaternary. You don't need to know about the quaternary structure though, so only the first three are explained here.

Learning Objectives:

- Know that proteins are sequences of amino acids joined by peptide links.
- Be able to draw the structure of a peptide formed from up to three amino acids.
- Know that hydrolysis of the peptide link produces the constituent amino acids.
- Be able to draw the structure of the amino acids formed by hydrolysis of a peptide.
- Know the primary, secondary (α-helix and β-pleated sheets) and tertiary structure of proteins.
- Be able to identify primary, secondary and tertiary structures in diagrams, and explain how these structures are maintained by hydrogen bonding and S–S bonds.
- Understand the importance of hydrogen bonding and sulfur-sulfur bonding in proteins.

Specification Reference 3.3.13.2

Tip: A condensation reaction is a reaction which joins two molecules together and releases a small molecule (often water). A hydrolysis reaction is a reaction that splits a larger molecule into two smaller molecules using water.

> **Tip:** Proteins are really just polyamides — the monomers are joined by amide groups. But in proteins these are called peptide links.

Primary structure

The primary structure is the sequence of amino acids in the long chain that makes up the protein (the polypeptide chain).

Example

The diagram below shows the primary structure of a polypeptide chain, showing individual amino acids.

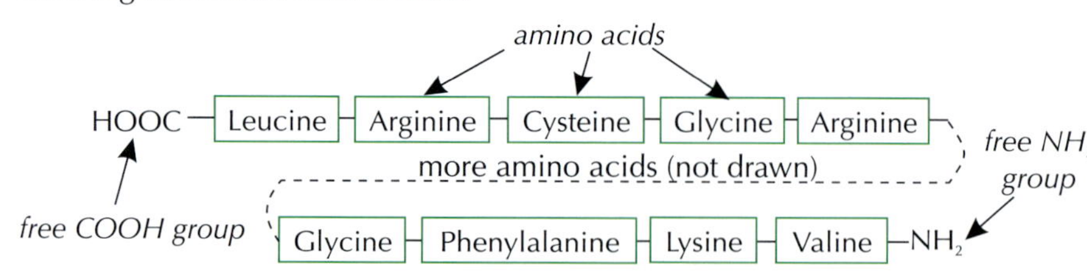

> **Tip:** Proteins often have a mixture of different types of secondary structure — see Figure 1.

> **Tip:** The letter 'α' is read as 'alpha', and the letter 'β' is read as 'beta'

Secondary structure

The peptide links can form hydrogen bonds with each other (see the next page), meaning the chain isn't a straight line. The shape of the chain is called its secondary structure. The most common secondary structure is a spiral called α-helix. Another common type of secondary structure is a β-pleated sheet. This is a layer of protein folded like a concertina.

Examples

These are the two main forms of secondary structure of proteins.

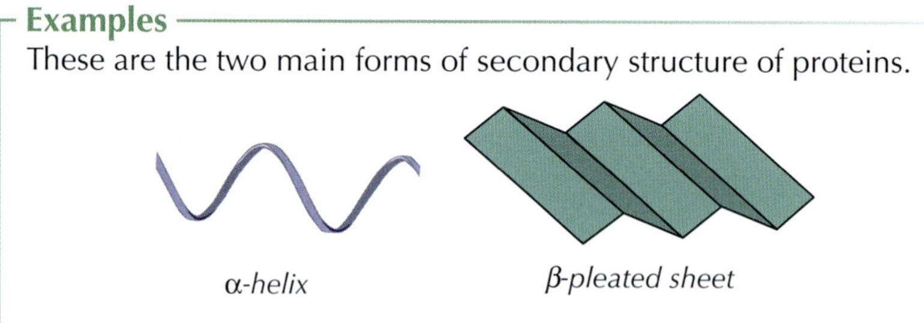

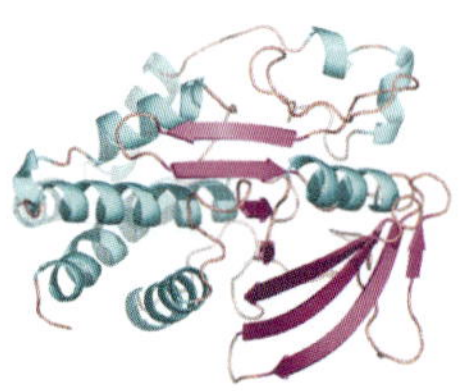

Figure 1: A computer model showing the secondary structure of the protein tyrosine phosphatase. The α-helices are shown in blue and the β-pleated sheets are shown in purple.

Tertiary structure

The chain of amino acids is itself often coiled and folded in a characteristic way that identifies the protein. Extra bonds can form between different parts of the polypeptide chain, which gives the protein a kind of three-dimensional shape (see Figure 2). This is its tertiary structure.

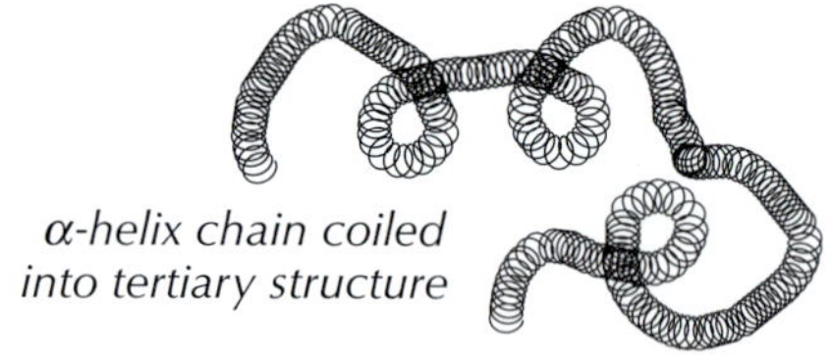

Figure 2: An example of tertiary structure.

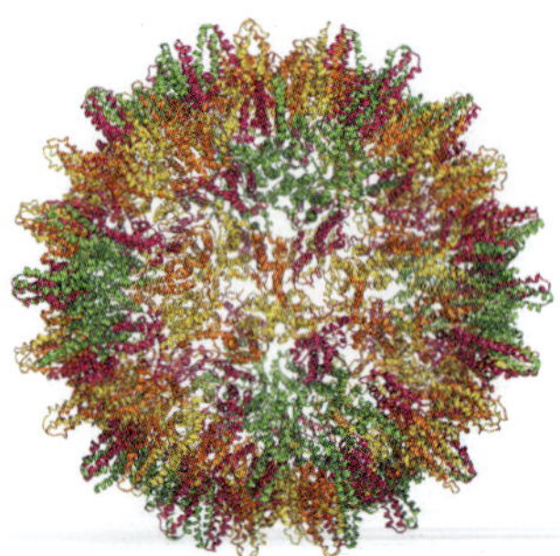

Figure 3: A molecular model showing the tertiary structure of the protein of the hepatitis B virus.

Intermolecular forces in proteins

The secondary and tertiary structures of proteins are formed by intermolecular forces causing the amino acid chains to fold or twist. These intermolecular forces are really important, because the three-dimensional shape of a protein is vital to how it functions. For example, changing the shape of an enzyme can stop it working (see page 483). There are two main types of bond that hold proteins in shape — hydrogen bonds and disulfide bonds.

1. Hydrogen bonding

Hydrogen bonding is one type of force that holds proteins in shape. Hydrogen bonds only exist between polar groups such as –OH and $-NH_2$. These groups contain electronegative atoms which induce a partial positive charge on the hydrogen atom. The hydrogen is then attracted to lone pairs of electrons on adjacent polar groups and a hydrogen bond is formed — see Figure 3.

Tip: Hydrogen bonding was covered on pages 96-97, so look back now if you need a bit of a reminder.

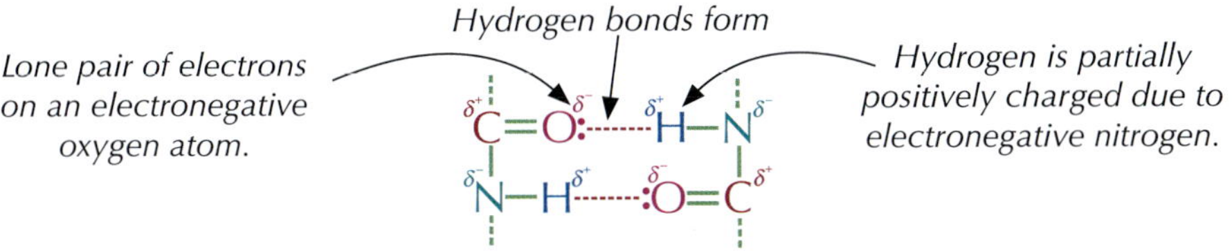

Figure 4: *Hydrogen bonding.*

2. Disulfide bonding

An amino acid that's part of a protein is called a residue. Disulfide bonding (or sulfur-sulfur bonding) occurs between residues of the amino acid cysteine. Cysteine contains a thiol group (–SH) and this thiol group can lose its H atom and join together to form a disulfide –S–S– bond with another thiol group. These disulfide bonds link together different parts of the protein chain, and help to stabilise the tertiary structure — see Figure 5.

Tip: Don't be put off by the term 'residue' — it's just the term for an amino acid that's part of a protein. So cysteine, when it's part of a protein, is a 'cysteine residue'. You may also hear it called an 'amino acid residue'.

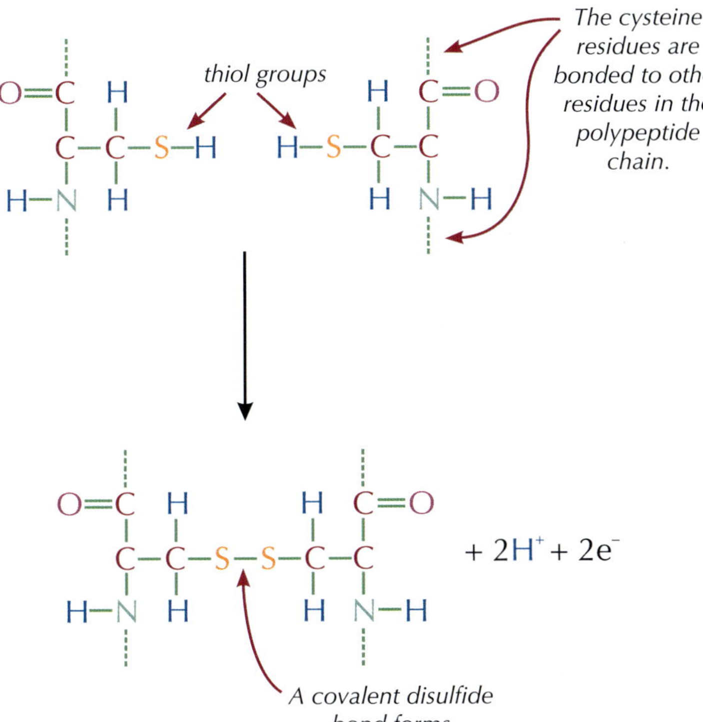

Figure 5: *Disulfide bonding between two cysteine residues.*

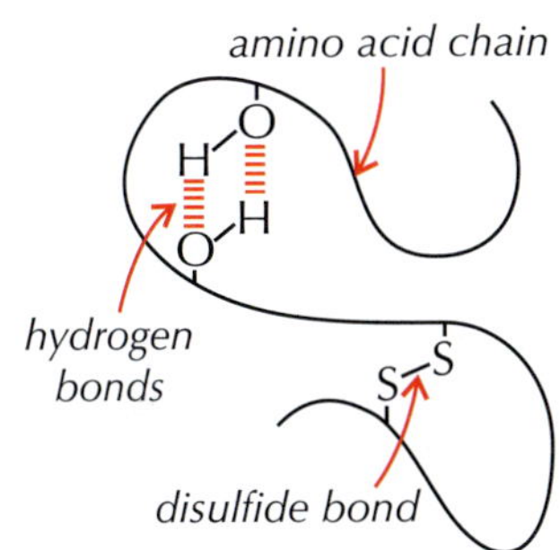

Figure 6: *An amino acid chain folds and can form both hydrogen and disulfide bonds.*

Factors such as temperature and pH can affect hydrogen bonding and the formation of disulfide bonds and so can change the shape of proteins.

Tip: Remember — the primary structure of a protein is held together by covalent bonds. The secondary structure is formed by hydrogen bonds. The tertiary structure is the result of hydrogen bonds <u>and</u> disulfide bonds.

Practice Question — Application

Q1 The amino acids cysteine and serine are shown below. There are two possible dipeptides that could form when one cysteine molecule reacts with one serine molecules. Draw the structures of both of these dipeptides.

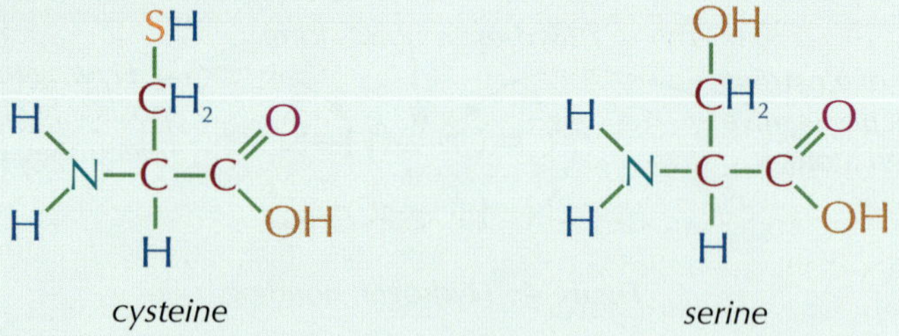

cysteine *serine*

Q2 The diagram below shows a tripeptide. Draw the structure of the molecules that would form when the tripeptide is fully hydrolysed.

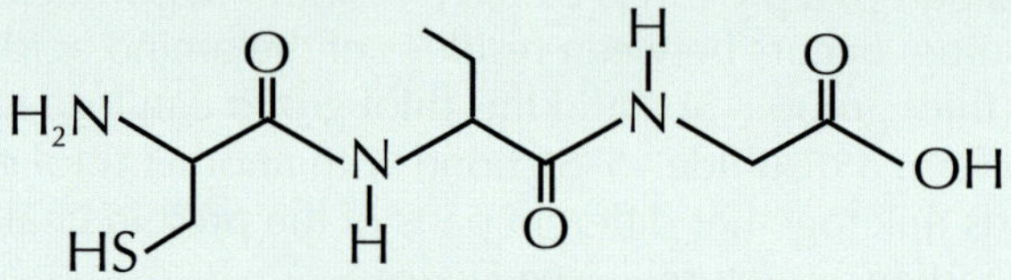

Q3 Below is a simplified diagram of part of a protein. The side groups are represented by the letter R.

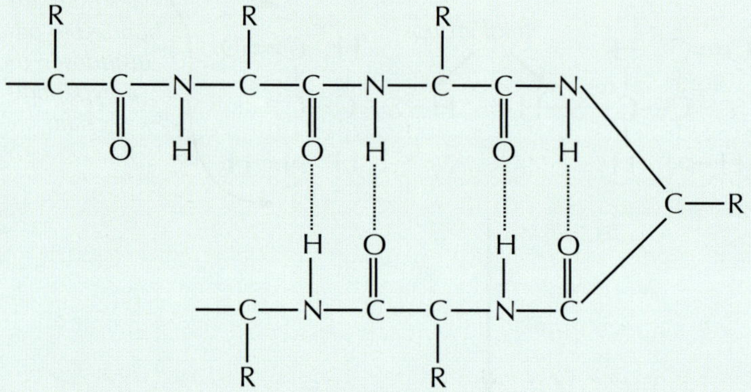

a) What types of protein structure are shown in this diagram?

b) What type of bond is represented by the dashed lines in the diagram?

c) i) Name a type of bonding that helps maintain the structure of proteins that isn't shown in the diagram above.

ii) What type of structure would these bonds result in?

Practice Questions — Fact Recall

Q1 What is a protein?

Q2 Give two common secondary structure of proteins.

Q3 What is a cysteine residue?

Q4 Explain how disulfide bonds and hydrogen bonds maintain the three-dimensional structure of proteins.

3. Enzymes

Enzymes are a type of protein that work wonders in your body all the time. They help speed up chemical reactions, and each has a very specific purpose.

Learning Objectives:

- Know that enzymes are proteins.
- Understand the action of enzymes as catalysts, including the concept of a stereospecific active site that binds to a substrate molecule.
- Be able to explain why a stereospecific active site can only bond to one enantiomeric form of a substrate or drug.
- Understand the principle of a drug acting as an enzyme inhibitor by blocking the active site.
- Know that computers can be used to help design such drugs.

Specification Reference 3.3.13.3

Biological catalysts

Enzymes speed up chemical reactions by acting as biological **catalysts**. They catalyse every metabolic reaction in the bodies of living organisms. They are proteins, but some also have non-protein components. The molecules that enzymes act upon are known as **substrates**.

Every enzyme has an area called its active site. This is the part that the substrate fits into so that it can interact with the enzyme. The active site is three-dimensional — it's part of the tertiary structure of the enzyme protein (see page 480).

Specificity

Enzymes are a bit picky. They only work with specific substrates — usually only one. This can be explained by the 'lock-and-key' model. This states that, for an enzyme to work, the substrate has to fit into the active site. If the substrate's shape doesn't match the active site's shape then the reaction won't be catalysed. The lock-and-key model is a simplified model of what actually happens, but it does a good job of explaining an enzyme's specificity.

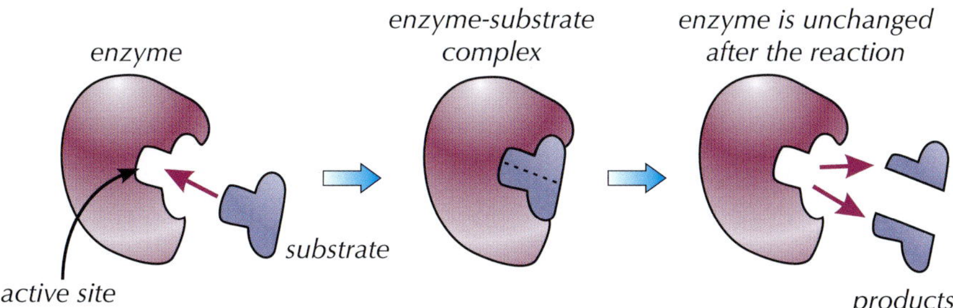

Figure 1: *An example of how an enzyme interacts with a substrate.*

Tip: Figure 1 shows an enzyme breaking up a substrate, but they can do lots of other things as well. For example, they can be used to combine two substrates, or add a functional group to a substrate.

The active sites of enzymes are **stereospecific** — they'll only work on one enantiomer of a substrate. The other enantiomer won't fit properly in the active site, so the enzyme can't work on it — it's a bit like how your left shoe doesn't fit on your right foot properly.

Inhibitors

Molecules that have a similar shape to the substrate act can as enzyme inhibitors. They compete with the substrate to bond with the active site, but no reaction follows. Instead they block the active site, so no substrate can fit in it (see Figure 3).

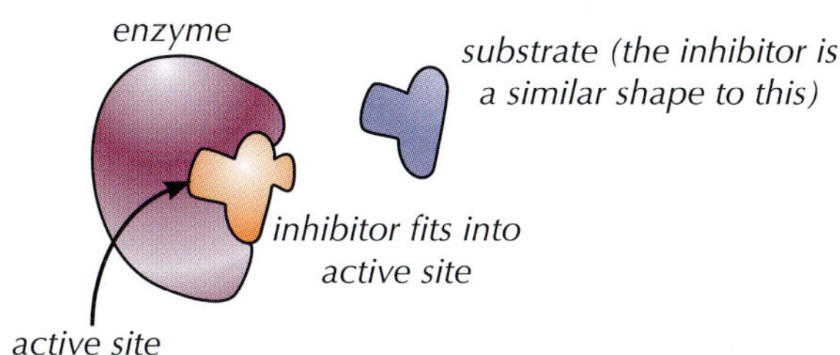

Figure 3: *Diagram showing how an inhibitor affects an enzyme.*

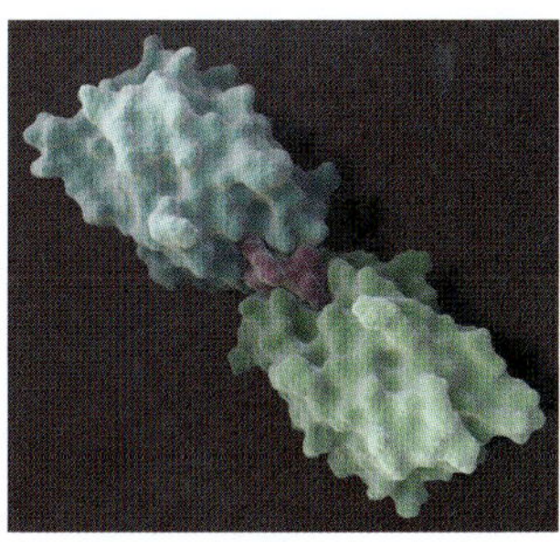

Figure 2: *A computer model showing an inhibitor (purple) binding to the active sites of two enzyme molecules (green), preventing them from binding to a substrate molecule.*

The amount of inhibition that happens depends on the relative concentrations of inhibitor and substrate — if there's a lot more of the inhibitor, it'll take up most of the active sites and very little substrate will be able to get to the enzyme. The amount of inhibition is also affected by how strongly the inhibitor bonds to the active site.

Use of inhibitors as drugs

Some drugs are inhibitors that block the active site of an enzyme and stop it from working. For example, some antibiotics work by blocking the active site of an enzyme in bacteria that helps to make their cell walls. This causes their cell walls to weaken over time, so the bacteria eventually burst.

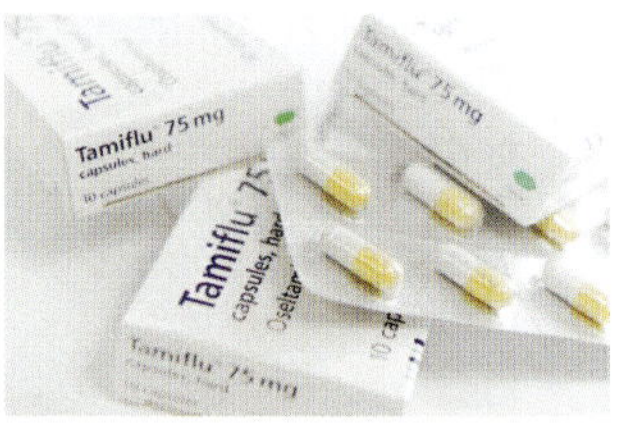

Figure 4: *Tamiflu® is an example of an enzyme inhibitor drug — it is used to treat people with the flu.*

The active site of an enzyme is very specific, so it takes a lot of effort to find a drug molecule that will fit into the active site. It's even trickier if the drug molecule is chiral — then only one enantiomer will fit into the active site, because the active sites of enzymes are usually stereospecific. Often, new drug molecules are found by trial and error. Scientists will carry out experiments using lots of compounds to see if they work as inhibitors for a particular enzyme. They'll then adapt any that work to try and improve them. This process takes a long time. One way that scientists are speeding this process up is by using computers to model the shape of the enzyme's active site and predict how well potential drug molecules will interact with it. They can quickly test hundreds of molecules to look for ones that might be the right shape before they start testing anything in the laboratory.

Practice Questions — Fact Recall

Q1 Describe the 'lock and key' model.

Q2 What is a stereospecific active site?

Q3 a) How does an enzyme inhibitor work?

b) Give one use of inhibitors.

4. DNA

DNA is the stuff that makes you, you. It's pretty complicated stuff, but after everything it's done for you, the least you could do is learn the basics...

Nucleotides

DNA stands for **d**eoxyribo**n**ucleic **a**cid. It contains all the genetic information of an organism. DNA is made up of lots of monomers called **nucleotides**, which are in turn made up of three components — a phosphate group, the pentose sugar 2-deoxyribose, and a base.

1. Phosphate group

The phosphate group is an ion, due to the negative charge on one of the oxygens.

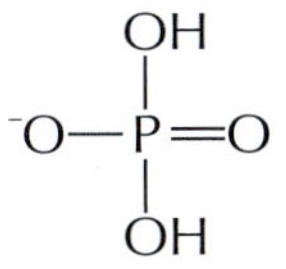

Figure 1: *A phosphate ion.*

2. Pentose sugar

A pentose sugar is a sugar made up of five carbons. All pentose sugars in DNA are 2-deoxyribose.

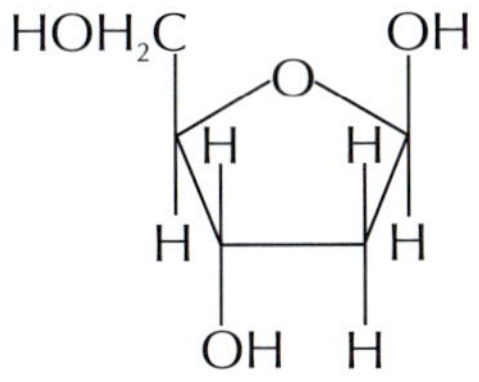

Figure 2: *2-deoxyribose*

3. Base

Each nucleotide in DNA contains one of four different bases: adenine (A), cytosine (C), guanine (G) and thymine (T).

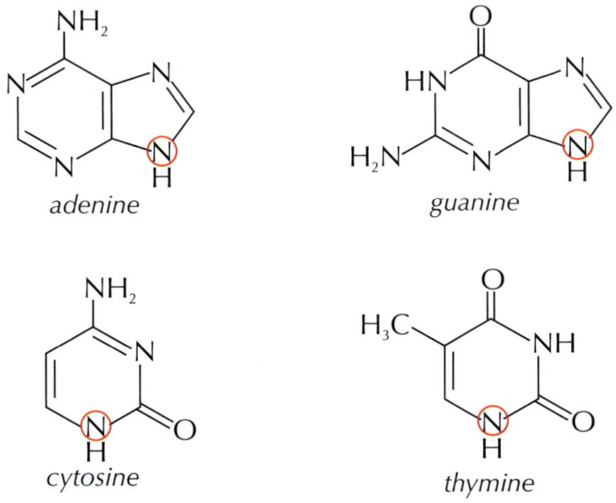

Figure 3: *The four bases found in DNA.*

The circled nitrogens in Figure 3 are the atoms that bond with the 2-deoxyribose molecule (see the next page).

Learning Objectives:

- Know that a nucleotide is made up from a phosphate ion bonded to 2-deoxyribose which is in turn bonded to one of the four bases adenine, cytosine, guanine and thymine.
- Understand that a single strand of DNA (deoxyribonucleic acid) is a polymer of nucleotides linked by covalent bonds between the phosphate group of one nucleotide and the 2-deoxyribose of another nucleotide, and that this results in a sugar-phosphate-sugar-phosphate polymer chain with bases attached to the sugars in the chain.
- Know that DNA exists as two complementary strands arranged in the form of a double helix.
- Be able to explain how hydrogen bonding between base pairs leads to the two complementary strands of DNA.

Specification Reference 3.3.13.4

Tip: When the phosphate and the sugar bond, a molecule of water is lost. Similarly, when the sugar and the base bond, a molecule of water is lost. Remembering this can help you work out what atoms bonds are formed between when the phosphate, sugar and base join together to form a nucleotide.

Luckily, you don't need to learn the structures of the phosphate ion and 2-deoxyribose, or of the bases — they'll all be in your data booklet. But you do need to know how they are arranged to form nucleotides (see Figure 4).

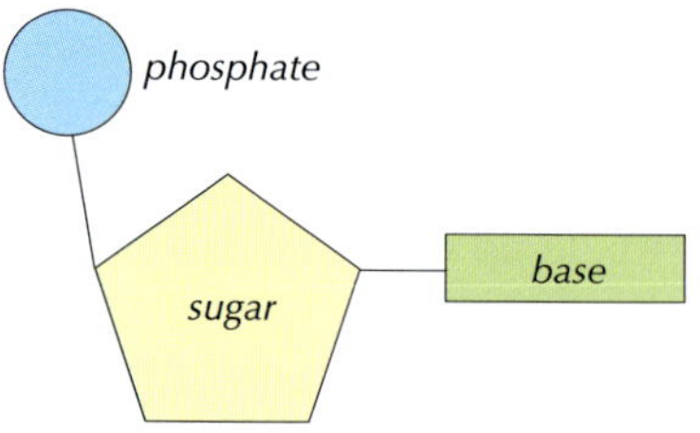

Figure 4: *Simplified structure of a nucleotide.*

Once you know the basic structure of a nucleotide (a phosphate ion bonded to a pentose sugar bonded to a base), you can use the structures in the data booklet to work out the structures of the four nucleotides in DNA. And here they are:

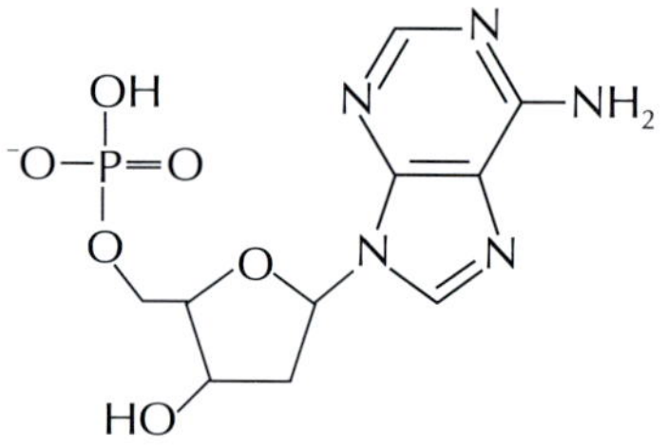

adenine nucleotide

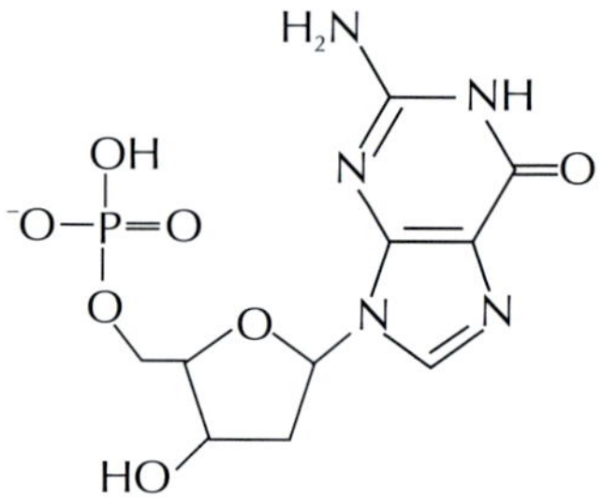

guanine nucleotide

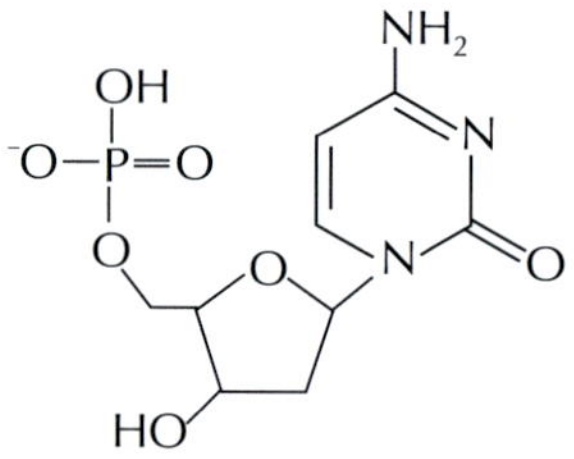

cytosine nucleotide

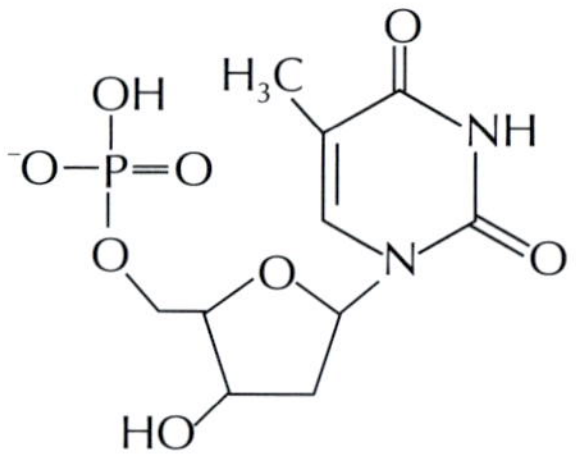

thymine nucleotide

Figure 5: *The four nucleotides that are found in DNA.*

Exam Tip
You should be able to work out these structures from the info in the exam data booklet if you need to, so you don't have to learn them. But you do need to know where the phosphate group and the bases join to the sugar.

Polynucleotides

The nucleotides join together to form a **polynucleotide chain.** Covalent bonds form between the phosphate group of one nucleotide and the sugar of another — this makes what's called the sugar-phosphate backbone of the chain (see Figure 6).

Tip: The polynucleotide chain formed by nucleotides joining together is a single strand of DNA.

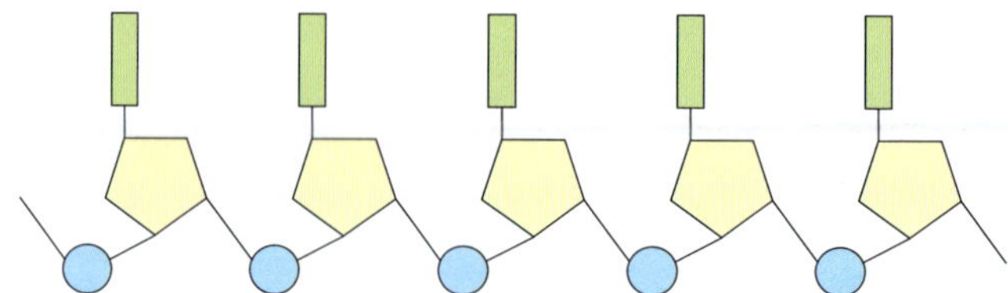

Figure 6: *The sugar-phosphate polymer chain that makes up DNA.*

The sugar-phosphate backbone of DNA is formed by condensation polymerisation — a molecule of water is lost and a covalent phosphodiester bond is formed. There are still OH groups at either end of the chain, so further links can be made — see Figure 7. This allows the nucleotides to form a polymer made up of an alternating sugar-phosphate-sugar-phosphate chain.

Tip: In this polymerisation reaction the monomer units are the nucleotides. Hence the name 'polynucleotides'.

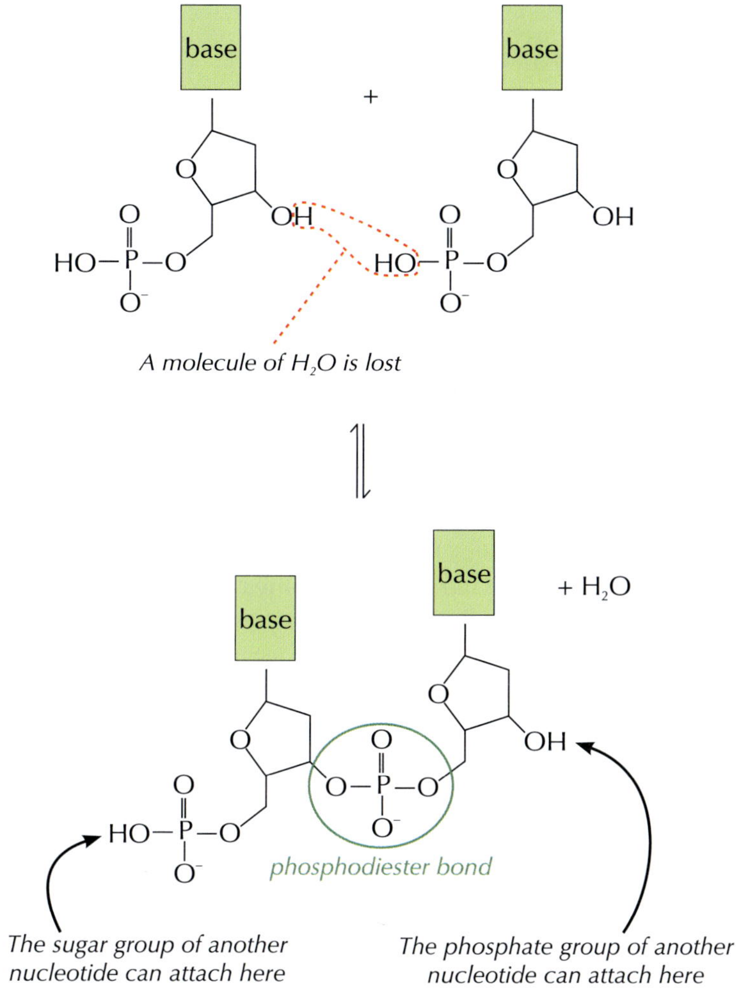

A molecule of H_2O is lost

phosphodiester bond

The sugar group of another nucleotide can attach here

The phosphate group of another nucleotide can attach here

Figure 7: *The condensation polymerisation of two nucleotides.*

DNA structure

DNA is formed of two polynucleotide strands. The two strands spiral together to form a **double helix** structure, which is held together by hydrogen bonds between the bases. Each base can only join with one particular partner — adenine always pairs with thymine (A – T) and cytosine always pairs with guanine (C – G). So the two strands of DNA are complementary — whenever there is an adenine base on one strand, there'll be a thymine base on the other, and whenever there's a guanine base on one strand, there's a cytosine base on the other.

Tip: To help remember which bases go together, just think of us folk **at** **CG**P. We're always here to help...

Figure 8: *James Watson and Francis Crick discovered the double helix structure of DNA in 1953.*

Tip: A double helix looks a bit like a twisted ladder. Imagine holding a ladder at one end and twisting the other end so that the two sides start coiling around each other — that's a bit like the shape of a DNA molecule.

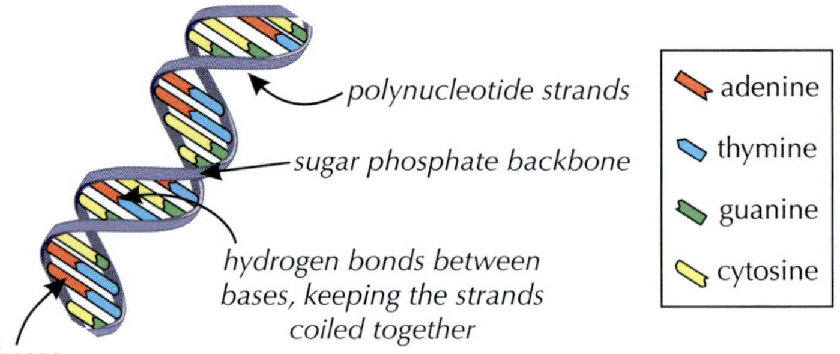

Figure 9: *Double helix structure of DNA.*

The complementary base pairing is due to the arrangement of atoms in the base molecules that are capable of forming hydrogen bonds. A hydrogen bond forms between a hydrogen atom with partial positive charge (a hydrogen attached to anything highly electronegative, such as a nitrogen atom) and a lone pair of electrons on a nearby oxygen, nitrogen or fluorine atom. To bond, the two atoms have to be the right distance apart.

Adenine and thymine have the right atoms in the right places to each form two hydrogen bonds, so they can pair up. Cytosine and guanine can each form three hydrogen bonds, so they can pair up too (see Figure 10). These are the only possible base combinations. Other base pairings would put the partially charged atoms too close together (they'd repel each other), or too far apart, or the bonding atoms just wouldn't line up properly. The DNA double helix has to twist so that the bases are in the right alignment and at the right distance apart for the complementary base pairs to form.

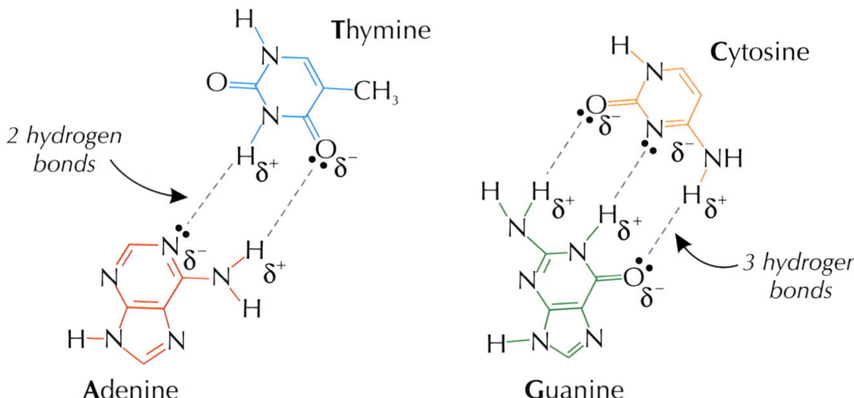

Figure 10: *Two hydrogen bonds join adenine and thymine together, three hydrogen bonds join cytosine and guanine together.*

Practice Questions — Fact Recall

Q1 a) Name the components of a nucleotide of DNA,

b) Sketch a labelled diagram to show how the components are arranged within a nucleotide.

Q2 Nucleotides join together to form polynucleotides.

a) What type of bond is formed between two nucleotides?

b) What parts of the nucleotide react to form these bonds?

Q3 a) Name the four bases found in DNA.

b) State which bases form complementary pairs, and how many hydrogen bonds form between the bases in each pair.

5. Cisplatin

Cancer is caused by cells in the body dividing and reproducing uncontrollably. Cisplatin is an anticancer drug that works by combating this reproduction.

Learning Objectives:

- Know that the Pt(II) complex cisplatin is used as an anticancer drug.
- Be able to explain that cisplatin prevents DNA replication in cancer cells by a ligand replacement reaction with DNA in which a bond is formed between platinum and a nitrogen atom on guanine.
- Be able to explain why drugs such as cisplatin can have adverse effects.
- Be able to appreciate that society needs to assess the balance between the benefits and the adverse effects of drugs, such as the anticancer drug cisplatin.

Specification Reference 3.3.13.5

What is cisplatin?

Cisplatin is a complex of platinum(II) with two chloride ion ligands and two ammonia ligands in a square planar shape. The two chloride ions are next to each other, making it cisplatin. If they were opposite each other, that would be transplatin, which has different biological effects (see pages 382-383 for more on *cis/trans* isomerism).

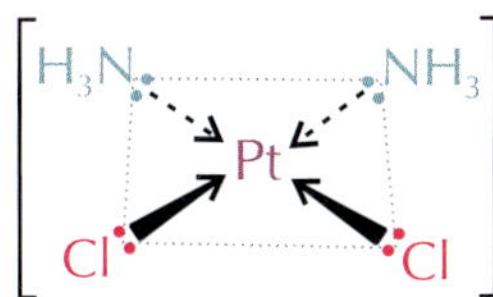

***Figure 1:** The molecular structure of cisplatin.*

Cisplatin as an anticancer drug

Cancer is caused when cells mutate and start dividing uncontrollably to form tumours. In order for a cell to divide it has to replicate its DNA. The two strands of the DNA double helix have to unwind so that they can be copied. Cisplatin stops this happening properly, so the tumour cells stop reproducing. Here's how it works:

- A nitrogen atom on a guanine base in DNA forms a co-ordinate bond with cisplatin's platinum ion, replacing one of the chloride ion ligands (see Figure 2). This is a ligand replacement reaction.

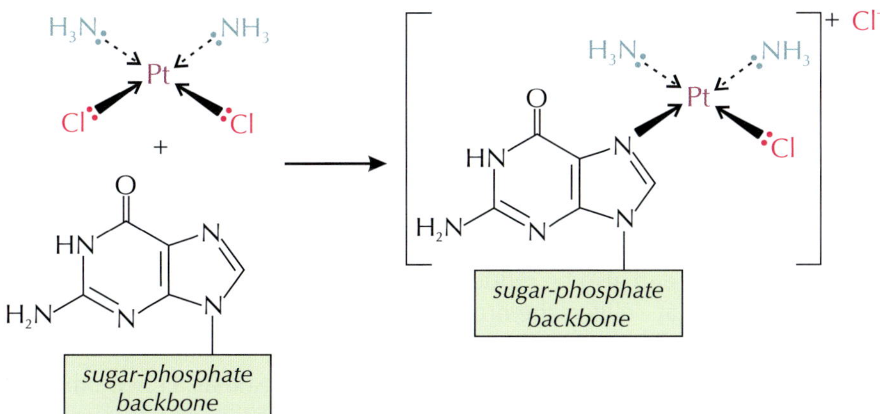

***Figure 2:** A guanine molecule replaces a chloride ion in cisplatin*

Tip: See pages 378-380 for more on ligands and the shapes of complex ions and pages 388-391 for more on ligand replacement reactions.

- A second nitrogen atom from a nearby guanine molecule (either on the same strand of the DNA, or the opposite strand) can bond to the platinum and replace the second chloride ion too (see Figure 4).

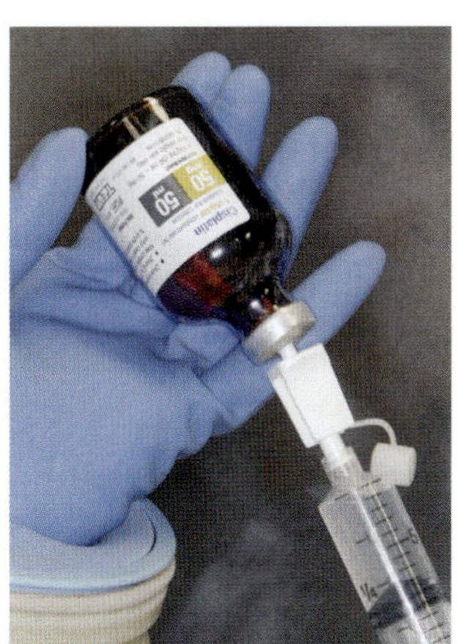

***Figure 3:** Cisplatin is used as part of chemotherapy treatment of some cancers.*

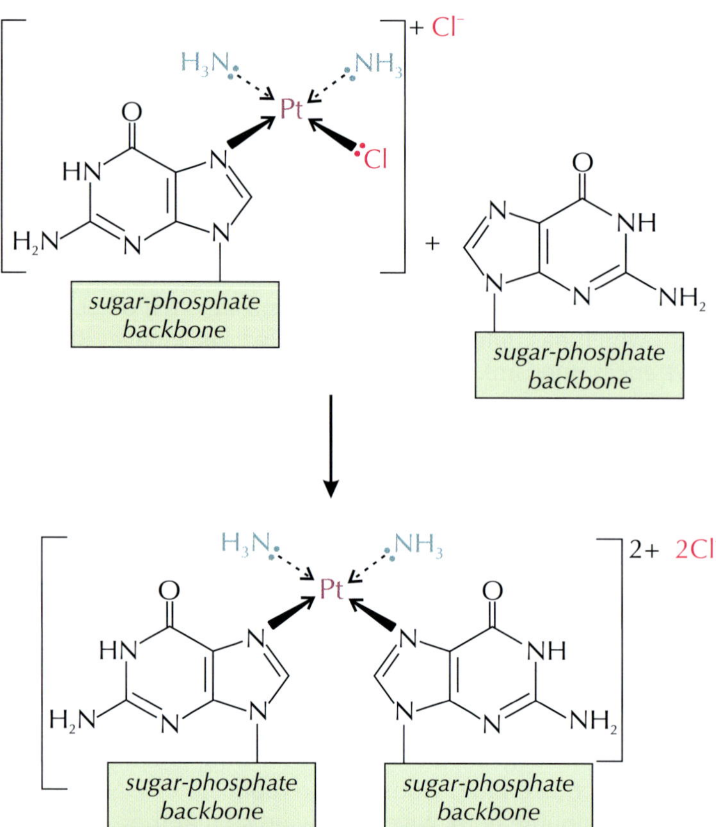

Figure 4: *A second guanine molecule replaces the other chloride ion in cisplatin.*

- The presence of the cisplatin complex bound to the DNA strands causes the strands to kink. This means that the DNA strands can't unwind and be copied properly — so the cell can't replicate properly.

Tip: The sugar-phosphate backbones in this reaction could be on separate polynucleotide strands, or they could be two parts of the same polynucleotide strand.

Tip: The damage to the DNA caused by the cisplatin binding to it also triggers mechanisms that lead to the death of the cell.

Adverse effects

Unfortunately, cisplatin can bind to DNA in normal cells as well as cancer cells. This is a problem for any healthy cells that replicate frequently, such as hair cells and blood cells, because cisplatin stops them from replicating in the same way as it does the cancer cells. This means that cisplatin can cause hair loss and suppress the immune system (which is controlled by white blood cells). It can also cause kidney damage.

These side effects can be lessened by giving patients very low dosages of cisplatin. They can also be reduced by targeting the tumour directly — i.e. using a method that delivers the drug only to the cancer cells, so it doesn't get the chance to attack healthy cells.

Despite the side effects of cisplatin, it is still used as a chemotherapy drug. This is because the balance of the long-term positive effects (curing cancer) outweigh the negative short-term effects.

Tip: The balance of the benefits and the adverse effects must be assessed before deciding whether any drug should be used, not just cisplatin.

Practice Questions — Fact Recall

Q1 Draw the structure of cisplatin.

Q2 a) Outline how cisplatin binds to DNA strands and prevents DNA replication in cancer cells.

b) What are the disadvantages of using cisplatin as an anticancer drug?

Section Summary

Make sure you know...

- What amino acids are and how to name them.
- That amino acids are amphoteric — they have both acidic and basic properties.
- That amino acids can form zwitterions and what a zwitterion is.
- How to draw the structures of amino acids as zwitterions.
- How to draw the structures formed when amino acids are in acid and alkaline conditions.
- That mixtures of amino acids can be separated using thin-layer chromatography.
- That developing agents, such as ninhydrin, and ultraviolet light can be used to locate amino acids on a chromatogram.
- How to calculate R_f values and how to identify amino acids by their R_f values.
- That proteins are sequences of amino acids joined by peptide links.
- That proteins can be hydrolysed and this produces the constituent amino acids.
- How to draw the peptide formed from specified amino acids and vice versa.
- That protein structure can be described by looking at the primary, secondary and tertiary structures.
- That the primary structure of a protein is the chain of amino acids.
- That the secondary structure of proteins is often an α-helix or a β-pleated sheet.
- That the tertiary structure of proteins describes how the chain folds to form its 3D shape.
- That hydrogen bonding and disulfide bonding are important in maintaining the 3-dimensional structure of proteins.
- That enzymes are proteins that can be used as biological catalysts.
- That enzymes recognise specific substrate molecules using the 'lock and key' model.
- That enzymes are stereospecific — only one enantiomer of a chiral substrate will fit in the active site.
- That molecules with a similar shape to an enzyme's substrate can work as inhibitors by blocking the active site of the enzyme.
- That many drugs are enzyme inhibitors.
- That a nucleotide is made up of a phosphate ion, 2-deoxyribose and a base.
- That there are four bases found in DNA — adenine (A), thymine (T), cytosine (C) and guanine (G).
- That a polynucleotide is a polymer made up of nucleotide monomers, joined together by covalent bonds, in a sugar-phosphate-sugar-phosphate backbone.
- That DNA is formed from two complementary polynucleotide strands linked by hydrogen bonds and arranged in a double helix shape.
- That adenine and thymine are complementary bases that form 2 hydrogen bonds, and cytosine and guanine are complementary bases that form 3 hydrogen bonds.
- That cisplatin is a platinum(II) complex with two ammonia and two chloride ion ligands.
- That cisplatin can be used as an anticancer drug.
- How cisplatin works as an anticancer drug.
- What the adverse effects of cisplatin are, and why society needs to assess the benefits and adverse effects of any drug before making it available.

Exam-style Questions

1 What is the skeletal formula of the amino acid, 2-amino-3-hydroxybutanoic acid?

A

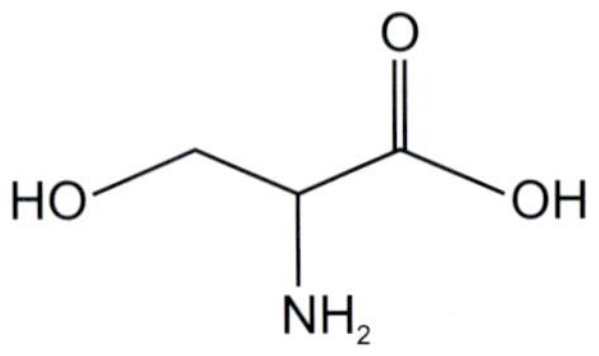

HO, O, OH, NH_2

B

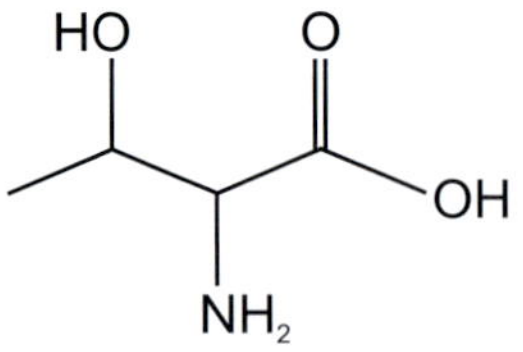

HO, O, OH, NH_2

C

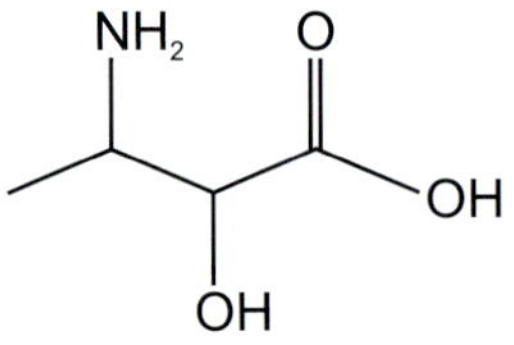

NH_2, O, OH, OH

D

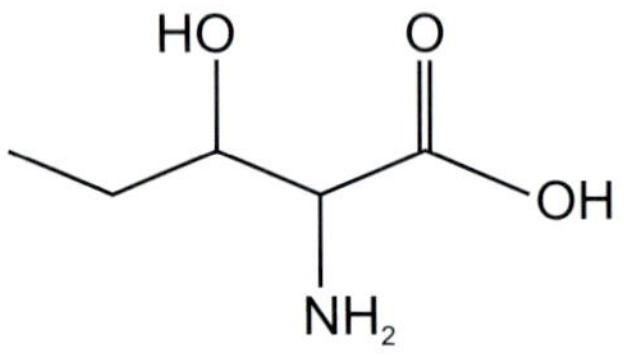

HO, O, OH, NH_2

(1 mark)

2 Adenine (A), cytosine (C), guanine (G) and thymine (T) are the four bases found in DNA. Which pairing is formed by two hydrogen bonds?

A A-G

B A-T

C C-G

D C-T

(1 mark)

3 A group of researchers are trying to find out which amino acids are present in a mixture by using thin-layer chromatography. Their observations are recorded in Table 1 below. The R_f values of some amino acids are shown in Table 2.

Mark on plate	Distance from point of origin (in cm)
Solvent front	6.5
Spot A	3.6
Spot B	1.0

Table 1

Amino acid	R_f value
Arginine	0.32
Cystine	0.15
Isoleucine	0.73
Proline	0.48
Tyrosine	0.55
Valine	0.66

Table 2

3.1 Which of the amino acids in Table 2 could be present in this mixture?

(1 mark)

The diagrams below shows the amino acids methionine and histidine.

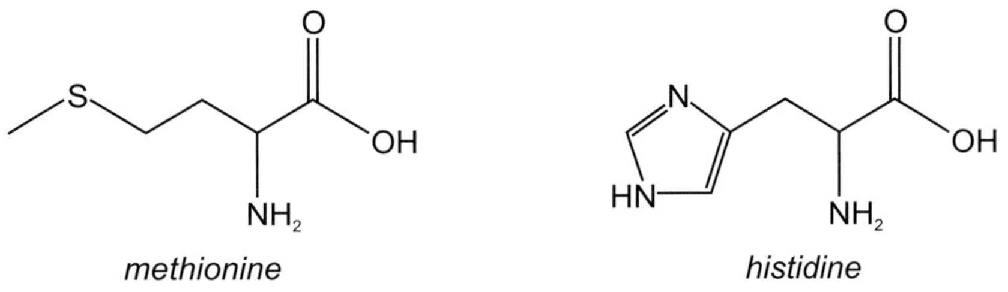

methionine *histidine*

3.2 They react to form a dipeptide known as Met-His. In Met-His, the free $–NH_2$ group is from the methionine. Draw Met-His.

(1 mark)

The dipeptide Met-His has an isoelectric point of around 6.5.

3.3 What is the isoelectric point of an amino acid?

(1 mark)

3.4 Draw the structure of Met-His when in a solution of pH 13.

(1 mark)

4 The diagram below shows a dipeptide formed from two different amino acids.

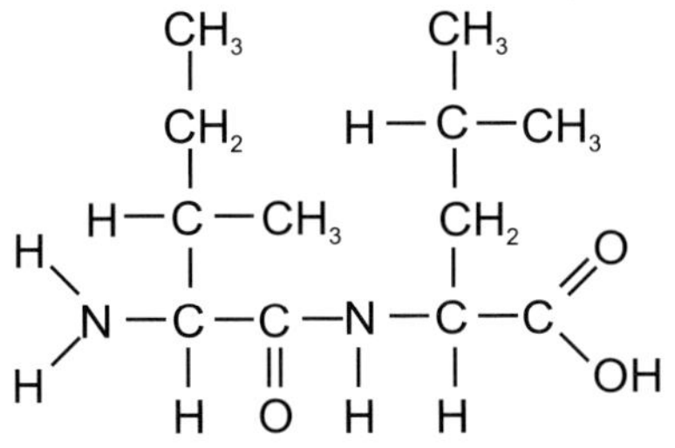

4.1 What type of reaction could be used to break down the dipeptide into its component amino acids?

(1 mark)

4.2 Draw the structures of the two amino acids that the dipeptide breaks down into.

(2 marks)

Proteins are made up of many amino acids put together.

4.3 This dipeptide forms part of the protein FIH. The structure of FIH can be described in terms of its primary, secondary and tertiary structure. Explain what is meant by each of these terms. Include details on the shapes resulting from each level of structure and the bonding present in each level.

(7 marks)

1. Organic Synthesis

Organic synthesis deals with how to make organic compounds using different chemical reactions. Organic synthesis is important as it provides ways to create materials and chemicals like pharmaceutical drugs, fertilisers and plastics.

Synthetic routes

Chemists have got to be able to make one compound from another. It's vital for things like designing medicines. It's also good for making imitations of useful natural substances when the real things are hard to extract. Chemists use **synthetic routes** to show the reagents, conditions and any special procedures needed to get from one compound to another.

Designing synthetic routes

Chemists try to design synthetic routes that use non-hazardous starting materials — this limits the potential for accidents and environmental damage. Chemists are also concerned with designing processes that are not too wasteful. Processes with high atom economies and high percentage yields are preferred because they convert more of the starting materials into useful products.

Waste can be reduced by designing synthesis routes that have as few steps as possible. However, it's not always possible to synthesise one chemical straight from another — a synthetic route can involve multiple steps. For example, you can't synthesise an aldehyde directly from a halogenoalkane. You'd have to carry out a two step reaction and first synthesise an alcohol.

Example

Avoiding using solvents wherever possible is one way of reducing both the hazards associated with a process and the amount of waste created by a synthetic route. Solvents are often flammable and toxic so can pose safety risks. Having to separate solvents from products or dispose of solvents after the reaction is complete can create a lot of waste too.

Organic synthesis in the exam

In your exam you could be asked to provide a step-wise synthesis for the production of one chemical from another. You might have to use any of the reactions you've learnt in A-Level chemistry, so it's important that you learn them all really well. If you're asked how to make one compound from another in the exam, make sure you include:

- Any special procedures, such as refluxing.
- The conditions needed, e.g. high temperature or pressure, or the presence of a catalyst.
- Any safety precautions, e.g. do it in a fume cupboard. (If there are toxic gases like hydrogen chloride or hydrogen cyanide around, you really don't want to go breathing them in. Stuff like bromine and strong acids and alkalis are corrosive, so you don't want to splash them on your bare skin.)

The reaction diagrams on the next two pages summarise the organic reactions you've come across in the A-Level chemistry course.

Learning Objectives:

- Know that the synthesis of an organic compound can involve several steps.
- Be able to explain why chemists aim to design processes that do not require a solvent and that use non-hazardous starting materials.
- Be able to explain why chemists aim to design production methods with fewer steps that have a high percentage atom economy.
- Be able to use reactions in this specification to devise a synthesis, with up to four steps, for an organic compound.

Specification Reference 3.3.14

Tip: You'll almost always have to use a solvent in organic synthesis as most organic reactions only happen in solution.

Exam Tip

You will be asked to recall some of the synthetic routes in your exam, so it's important that you know all the reactions on the next page. Don't panic, though — there's nothing new to learn there, it's just a summary of what you've already learnt.

Synthetic routes for making organic compounds

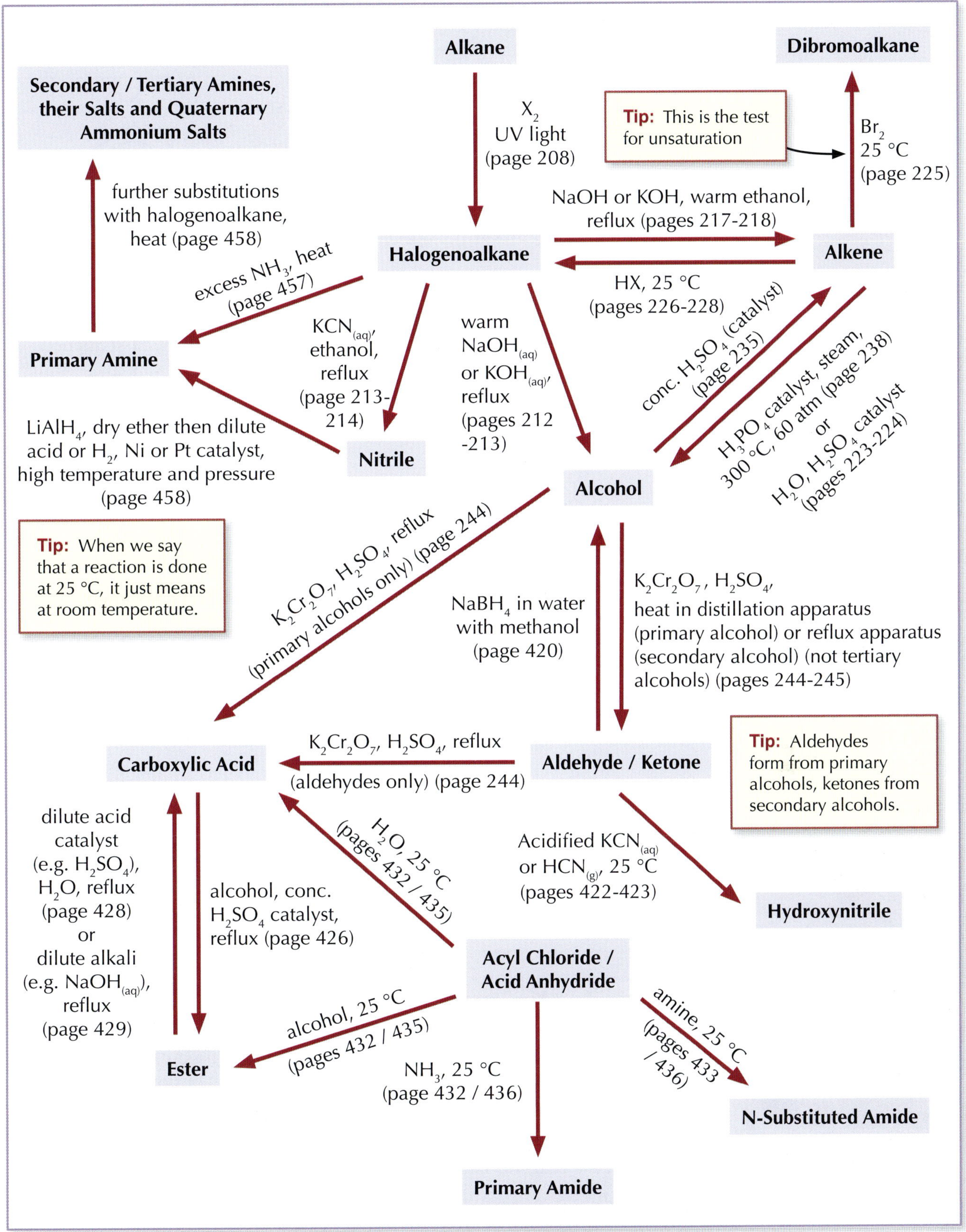

Alkane
Dibromoalkane
Secondary / Tertiary Amines, their Salts and Quaternary Ammonium Salts
X_2 UV light (page 208)
Tip: This is the test for unsaturation
Br_2 25 °C (page 225)
further substitutions with halogenoalkane, heat (page 458)
NaOH or KOH, warm ethanol, reflux (pages 217-218)
excess NH_3, heat (page 457)
Halogenoalkane
Alkene
HX, 25 °C (pages 226-228)
$KCN_{(aq)}$, ethanol, reflux (page 213-214)
warm $NaOH_{(aq)}$ or $KOH_{(aq)}$, reflux (pages 212 -213)
conc. H_2SO_4 (catalyst) (page 235)
H_3PO_4 catalyst, steam, 300 °C, 60 atm (page 238) or H_2O, H_2SO_4 catalyst (pages 223-224)
Primary Amine
LiAlH$_4$, dry ether then dilute acid or H_2, Ni or Pt catalyst, high temperature and pressure (page 458)
Nitrile
Alcohol
Tip: When we say that a reaction is done at 25 °C, it just means at room temperature.
$K_2Cr_2O_7$, H_2SO_4, reflux (primary alcohols only) (page 244)
NaBH$_4$ in water with methanol (page 420)
$K_2Cr_2O_7$, H_2SO_4, heat in distillation apparatus (primary alcohol) or reflux apparatus (secondary alcohol) (not tertiary alcohols) (pages 244-245)
$K_2Cr_2O_7$, H_2SO_4, reflux (aldehydes only) (page 244)
Carboxylic Acid
Aldehyde / Ketone
Tip: Aldehydes form from primary alcohols, ketones from secondary alcohols.
dilute acid catalyst (e.g. H_2SO_4), H_2O, reflux (page 428) or dilute alkali (e.g. $NaOH_{(aq)}$), reflux (page 429)
alcohol, conc. H_2SO_4 catalyst, reflux (page 426)
H_2O, 25 °C (pages 432 / 435)
Acidified $KCN_{(aq)}$ or $HCN_{(g)}$, 25 °C (pages 422-423)
Hydroxynitrile
Acyl Chloride / Acid Anhydride
alcohol, 25 °C (pages 432 / 435)
amine, 25 °C (pages 433 / 436)
Ester
NH_3, 25 °C (page 432 / 436)
N-Substituted Amide
Primary Amide

Synthetic routes for aromatic compounds

There are not so many of these reactions to learn, but it's still just as important that you know all the itty-bitty details. If you can't remember any of the reactions, look back to the relevant pages and take a quick peek over them.

> **Exam Tip**
> When you're writing down a step-wise synthesis in the exam, don't forget to put the conditions as well as the reagents. You might lose marks if you don't.

> **Exam Tip**
> For acylation questions in the exam, don't just put RCOCl — you need to specify what the R group is.

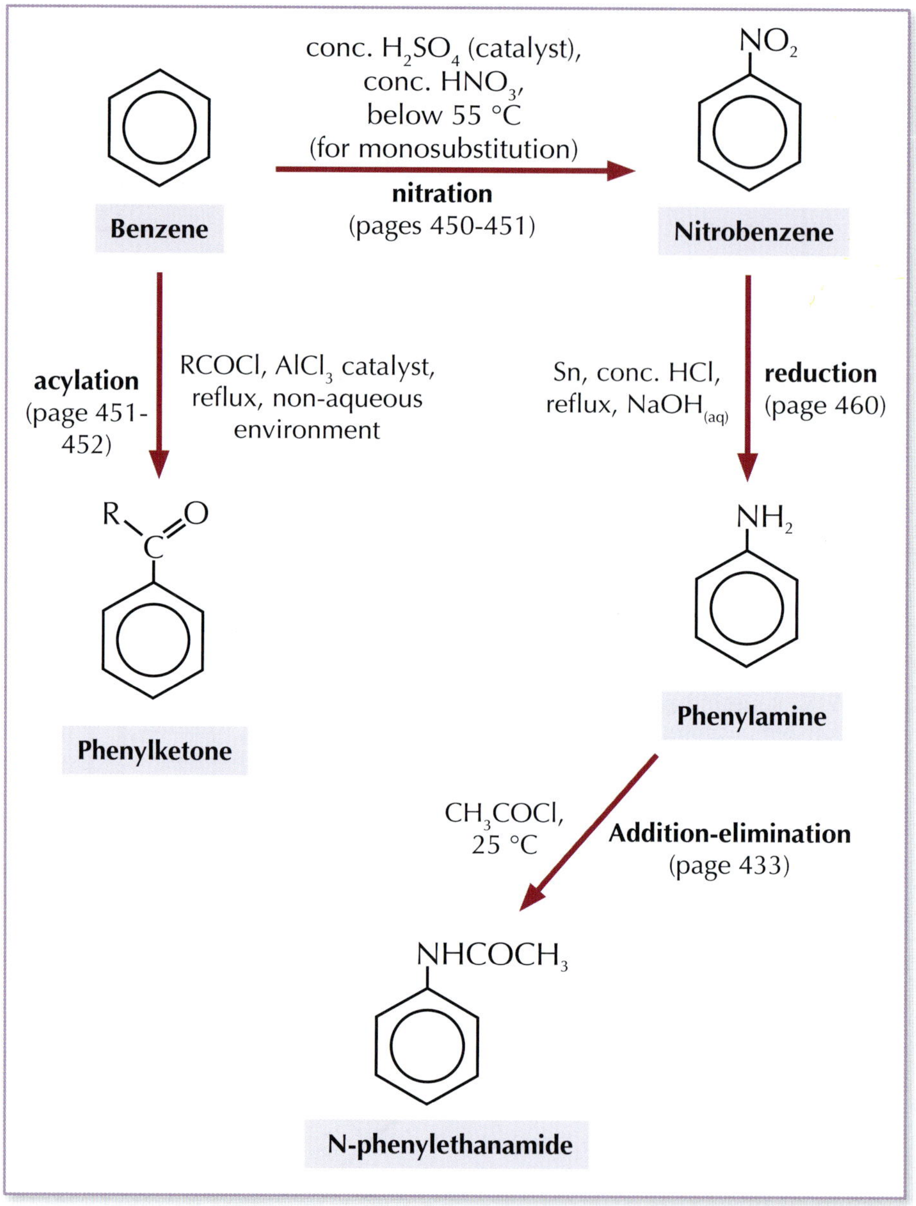

Benzene → (conc. H_2SO_4 (catalyst), conc. HNO_3, below 55 °C (for monosubstitution); **nitration** (pages 450-451)) → **Nitrobenzene** (NO_2)

Benzene → (RCOCl, $AlCl_3$ catalyst, reflux, non-aqueous environment; **acylation** (page 451-452)) → **Phenylketone** (R–C=O)

Nitrobenzene → (Sn, conc. HCl, reflux, $NaOH_{(aq)}$; **reduction** (page 460)) → **Phenylamine** (NH_2)

Phenylamine → (CH_3COCl, 25 °C; **Addition-elimination** (page 433)) → **N-phenylethanamide** ($NHCOCH_3$)

Exam technique

In the exam you may be asked to 'identify' the name of a reagent used in a synthesis step. A reagent is just a chemical that can be used straight out of a bottle or container, for example, hydrochloric acid (HCl) or potassium cyanide (KCN). You must be really careful about this...

Example

Suggest the reagents used to transform a halogenoalkane into a nitrile.

Giving the reagent CN^- in this case would be incorrect as you can't just take it out of a bottle. A correct answer could be KCN and ethanol.

Figure 1: Reagents are chemicals that can be taken from bottles or containers. For example, NaOH can come as a solution or a solid.

Practice Questions — Application

Q1 Write down the reagents and conditions you would use to carry out the following organic syntheses.

a) Making ethanol from ethene.

b) Forming chloromethane from methane.

c) Creating phenylethanone from benzene.

Q2 The following syntheses require a two-step organic synthesis. Write down the reagents and conditions you would use to carry out each step in the following organic syntheses, and state the organic product formed after the first step.

a) Forming phenylamine from benzene.

b) Creating propanal from 1-bromopropane.

c) Turning ethane into propanenitrile.

d) Making methyl butanoate from butanal.

Q3 Give a three-step synthesis of N-phenylethanamide starting from benzene. In your second synthesis step you should form phenylamine. For each step, give the reagents and conditions you would use to carry out the reaction, and state the organic product formed.

Q4 Give a three-step synthesis of propyl ethanoate starting from chloroethane. For each step, give the reagents and conditions you would use to carry out the reaction, and state the organic product formed.

Exam Tip
For questions like these you don't have to give the mechanisms or equations for the reactions.

Exam Tip
You need to be specific when you're talking about reagents in the exam. Make sure you write down exactly what you need to use — writing 'an alcohol' when you really mean 'ethanol' just won't get you all of the marks.

Practice Questions — Fact Recall

Q1 Give one reason why chemists try to avoid using hazardous starting materials when designing organic synthesis processes.

Q2 Give two ways a chemist can reduce the amount of waste produced in organic synthesis.

Q3 Explain why chemists try to design organic synthesis processes that not require a solvent.

Q4 Write down the reagents and conditions needed to carry out the following organic syntheses.

a) Making an alkene from a halogenoalkane.

b) Making a hydroxynitrile from an aldehyde.

c) Making a carboxylic acid from an ester.

d) Making a primary amine from a nitrile.

Q5 What reagents and conditions are needed to turn an acyl chloride into:

a) an N-substituted amide?

b) an ester?

c) a primary amide?

Learning Objectives:

- Appreciate that scientists have developed a range of analytical techniques which together enable the structures of new compounds to be confirmed.
- Know that nuclear magnetic resonance (NMR) gives information about the position of ^{13}C or ^{1}H atoms in a molecule.
- Know that chemical shift depends on the molecular environment.
- Know the use of the δ scale for recording chemical shift.
- Know that tetramethylsilane (TMS) is used as a standard and be able to explain why it is a suitable substance to use as a standard.

Specification Reference 3.3.15

2. NMR Spectroscopy

NMR spectroscopy is just one of several techniques that scientists have come up with to help determine the structure of a molecule. It provides information on the different environments of atoms in molecules.

Types of NMR spectroscopy

There are two types of nuclear magnetic resonance (NMR) spectroscopy that you need to know about — ^{13}C NMR, which gives you information about how the carbon atoms in a molecule are arranged, and ^{1}H (or proton) NMR, which tells you how the hydrogen atoms in a molecule are arranged.

Nuclear environments

A nucleus is partly shielded from the effects of an external magnetic field by its surrounding electrons. Any other atoms and groups of atoms that are around a nucleus will also affect the amount of electron shielding. So the overall effect of the external magnetic field on each nucleus is different, depending on its environment within the molecule.

The size of the energy gap between a nucleus being aligned with or against an external magnetic field is determined by the strength of the magnetic field. Nuclei in different environments will therefore absorb at different frequencies. It's these differences in absorption of energy between environments that you're looking for in NMR spectroscopy.

Example

If a carbon atom bonds to a more electronegative atom (like oxygen) the amount of electron shielding around its nucleus will decrease.

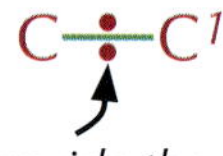

These electrons provide the carbon atoms with shielding from a magnetic field.

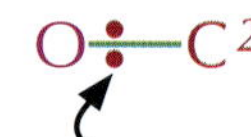

These electrons are pulled further away from the carbon atom by the electronegative oxygen atom. The carbon atom is less shielded.

This means that carbon 1 and carbon 2 are in different environments.

An atom's environment depends on all the groups that it's connected to, going right along the molecule — not just the atoms it's actually bonded to.

Examples

Chloroethane has 2 carbon environments — each of its two carbons are bonded to different atoms.

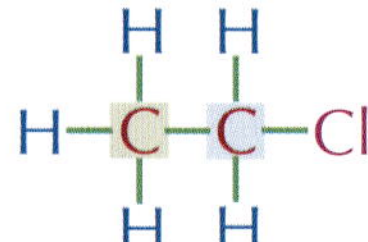

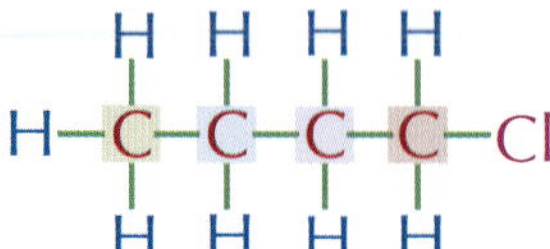

1-chlorobutane has 4 carbon environments — the carbons in the CH_2 groups are different distances from the electronegative Cl atom, so their environments are different.

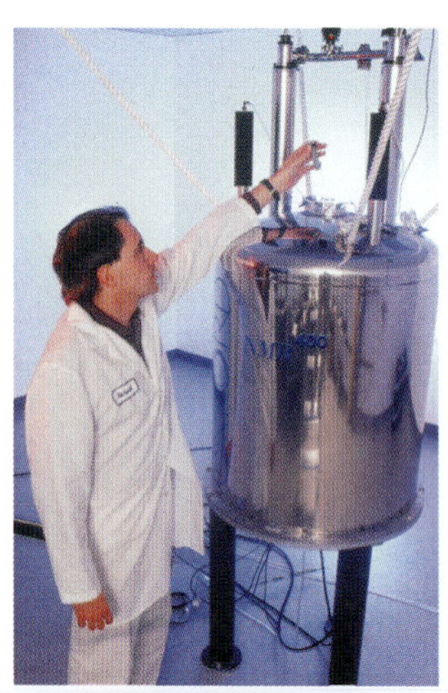

Figure 1: *An NMR spectrometer.*

2-chloropropane has 2 carbon environments — it's symmetrical, so the carbons in CH_3 groups on each end are in the same environment.

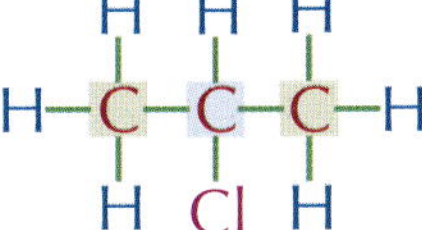

Chemical shift

NMR spectroscopy measures differences in the energy absorbed by nuclei in different environments relative to a standard substance — the difference is called the **chemical shift** (δ).
Chemical shift is measured in parts per million (or ppm) relative to a standard. The standard substance is tetramethylsilane (TMS).

$$H_3C-\underset{\underset{CH_3}{|}}{\overset{\overset{CH_3}{|}}{Si}}-CH_3$$

tetramethylsilane (TMS)

Tip: Here's the ^{13}C NMR spectrum for TMS:

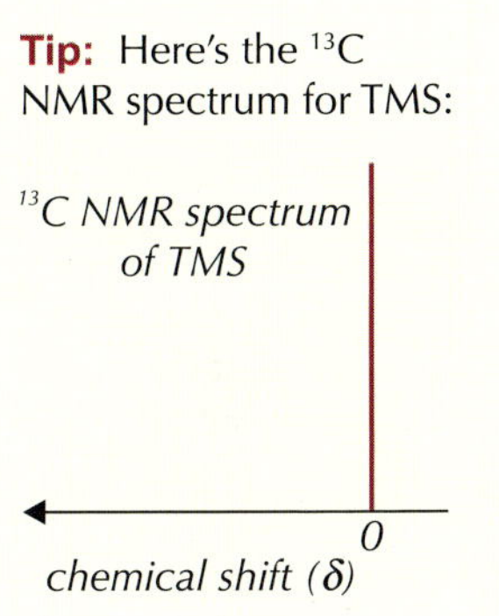

TMS has 12 hydrogen atoms all in identical environments and 4 carbon atoms all in identical environments. This means that, in both ^{1}H NMR and ^{13}C NMR, it will produce a single absorption peak, well away from most other absorption peaks. The single peak produced by TMS is given a chemical shift value of 0. You'll often see a peak at $\delta = 0$ on spectra because TMS is added to the test compound for calibration purposes. Tetramethylsilane is inert (so it doesn't react with the sample), non-toxic, and volatile (so it's easy to remove from the sample).

Practice Questions — Fact Recall

Q1 What is chemical shift (δ) measured in?

Q2 Explain why TMS is used as the standard substance used in NMR.

Learning Objective:

- Be able to use ^{13}C NMR spectra and chemical shift data from the Chemistry Data Booklet to suggest possible structures or part structures for molecules.

Specification Reference 3.3.15

3. ^{13}C NMR Spectroscopy

Time for some more information about carbon-13 NMR. This topic's all about working out where the carbon atoms in an organic molecule are.

^{13}C NMR

The number of peaks on a ^{13}C NMR spectrum tells you how many different carbon environments are present in a particular molecule. The spectrum will have one peak on it for each carbon environment in the molecule. The carbon atoms which are closer to more electronegative atoms (for example, oxygen, nitrogen or chlorine) will be less shielded and have a higher chemical shift.

Examples

Ethanol

There are two carbon atoms in a molecule of ethanol:

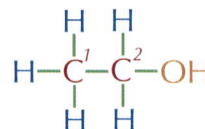

Because they are bonded to different atoms, each has a different amount of electron shielding — so there are two carbon environments in the ethanol molecule and two peaks on its ^{13}C NMR spectrum.

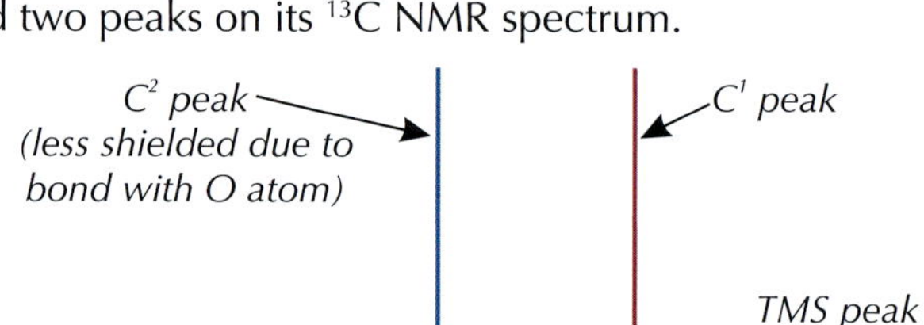

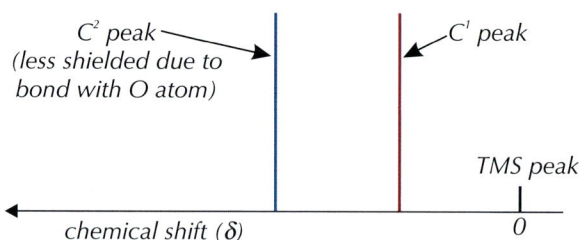

Tip: You'll often see a peak at $\delta = 0$ on spectra because TMS is added to the test compound for calibration purposes. You can ignore this peak when interpreting NMR spectra.

Ethane

Both carbon nuclei in ethane have the same environment. Each C has 3 Hs and a CH_3 group attached. So ethane has one peak on its ^{13}C NMR spectrum.

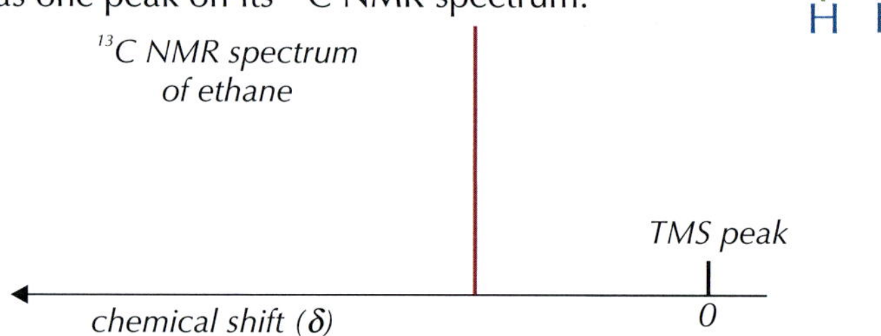

Exam Tip

Watch out for the scale on ^{13}C spectra in the exam — the chemical shift increases from right to left.

Propanone

There are two different carbon environments. The end carbons both have the same environment — they each have 3 Hs and a $COCH_3$ attached. The centre carbon has 2 CH_3 groups and an O attached by a double bond. So propanone's ^{13}C NMR spectrum has 2 peaks.

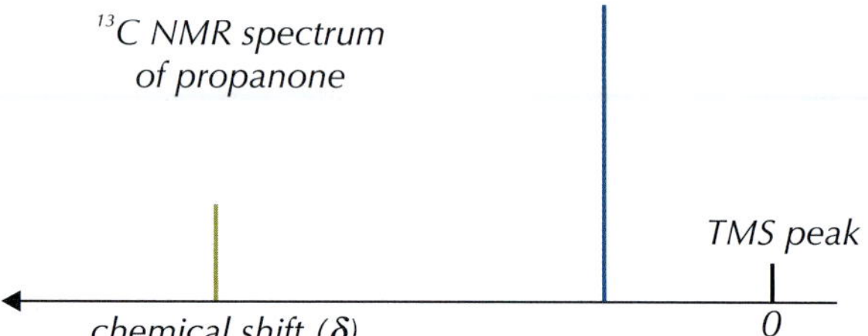

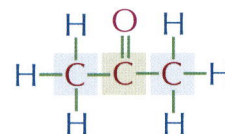

Tip: Remember, carbons next to more electronegative atoms will have a higher chemical shift and will be further to the left on the spectrum.

Cyclohexane-1,3-diol

In cyclohexane-1,3-diol there are four different carbon environments — each different carbon environment is shown in a different colour on the diagram below. If you think about the symmetry of the molecule you can see why this is.

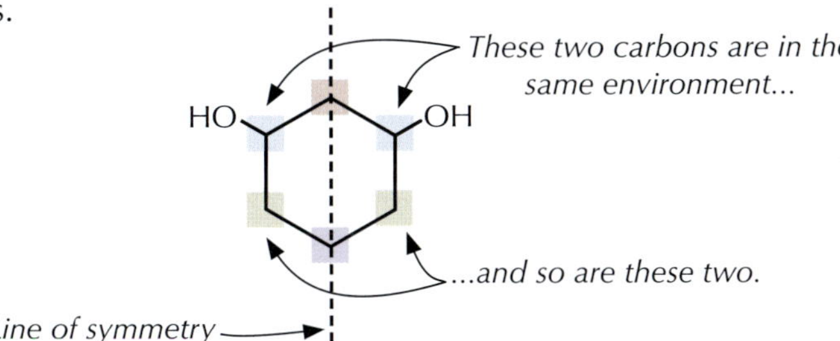

So cyclohexane-1,3-diol's ^{13}C NMR spectrum will have 4 peaks.

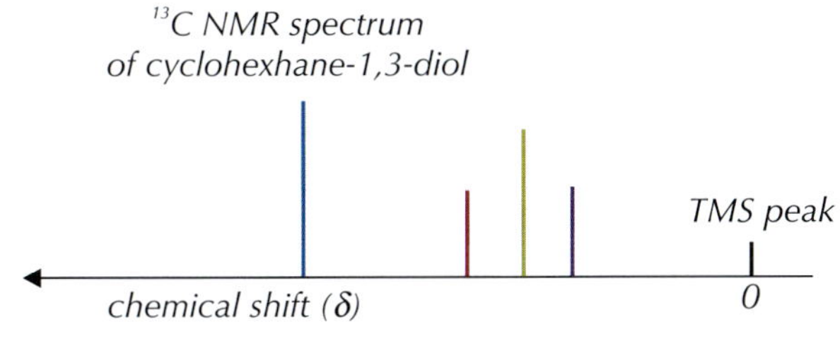

Chemical shifts on ^{13}C NMR spectra

In your exam you'll get a data sheet that will include a table like the one in Figure 1. The table shows the chemical shifts experienced by carbon–13 nuclei in different environments. You can match up the peaks in a ^{13}C NMR spectrum with the chemical shifts in the table to work out which carbon environments they could represent.

^{13}C NMR Chemical Shifts Relative to TMS	
Chemical shift, δ (ppm)	**Type of Carbon**
5 – 40	C–C
10 – 70	R–C–Cl or R–C–Br
20 – 50	R–C(=O)–C–
25 – 60	R–C–N (amines)
50 – 90	C–O (alcohols, ethers or esters)
90 – 150	C=C (alkenes)
110 – 125	R–C≡N
110 – 160	aromatic
160 – 185	carbonyl (ester or carboxylic acid) R–C(=O)–
190 – 220	carbonyl (ketone or aldehyde) R–C(=O)–

Figure 1: *Chemical shifts of different carbon environments.*

Tip: When you're thinking about the chemical shift of different carbon atoms you have to take into account all the atoms they're near — not just the ones right next to them. For example, in 1-chlorobutane all of the carbons are in different environments:

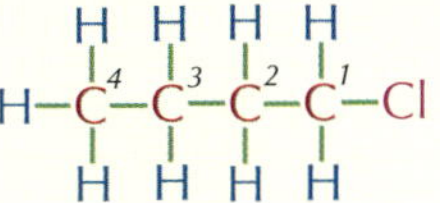

Carbon-2 and carbon-3 look like they might be in the same environment. But carbon-2 is much closer to the electronegative chlorine atom — so it will have a higher chemical shift.

Exam Tip

You'll get a table like this in the exam, so you don't need to learn it (phew). Just read up from any chemical shift value to find what groups could be causing a peak there.

Tip: If you're given a spectrum containing a peak that could be caused by a carbon atom in a number of different environments, don't panic. You'll often need to refer to information you're given in the question to work out which environment the peak corresponds to.

Tip: The best way to figure out the structure of a molecule is often to sketch out some possibilities. Then you can rule out any that don't have the right number of carbon environments or chemical shift values.

Tip: Don't forget that if there's a peak at δ = 0 it's a TMS peak, so you can ignore it.

Tip: There are actually three ketones that have the formula $C_5H_{10}O$ (you could try sketching them all out) but pentan-3-one is the only one that has three carbon environments.

Tip: Benzene (C_6H_6) is completely symmetrical. It only has one peak on its ^{13}C NMR spectrum as all of its carbon atoms are in the same environment.

Matching peaks to the groups that cause them isn't always straightforward, because the chemical shifts can overlap. For example, a peak at δ ≈ 30 might be caused by C–C, C–Cl or C–Br. A peak at δ ≈ 210, is due to a C=O group in an aldehyde or a ketone — but you don't know which.

Interpreting ^{13}C NMR spectra

If you have a sample of a chemical that contains carbon atoms you can use a ^{13}C NMR spectrum of the molecule to help work out what it is. The spectrum gives you information about the number of carbon atoms that are in a molecule, and the environments that they are in.

Here are the three steps to follow to interpret a ^{13}C NMR spectrum:

1. Count the number of peaks in the spectrum (excluding the TMS peak) — this tells you the number of carbon environments in the molecule.
2. Use a table of chemical shift data to work out what kind of carbon environment is causing each peak. A table of typical chemical shift data for carbon-13 NMR is shown in Figure 1.
3. Use this information to figure out the structure of the molecule.

Example

The carbon-13 NMR spectrum of a straight-chain molecule with the molecular formula $C_5H_{10}O$ is shown below. Use the spectrum, along with the data in Figure 1 to identify the molecule.

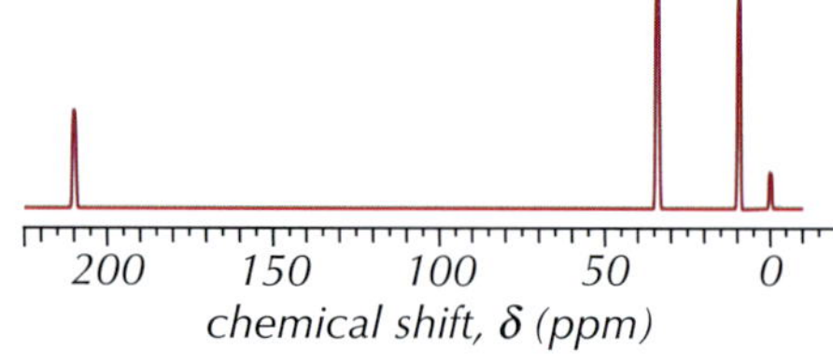

1. The spectrum has three peaks (and a TMS peak at δ = 0), so the molecule must have three carbon environments.
2. The peak at δ = 10 ppm represents carbon atoms in C–C bonds. Looking at the molecular formula, the peak at δ = 35 ppm must also be due to carbons in C–C bonds (rather than C–Cl or C–Br bonds), as the molecule only contains C, H, and O atoms. The carbons causing this peak have a different chemical shift to those causing the first peak, so they must be in a slightly different environment.
   The peak at δ = 210 ppm is due to a carbon atom in a C=O group.
3. You know you're looking for a straight-chain carbonyl with the formula $C_5H_{10}O$ that has three different carbon environments. The only one that fits the bill is pentan-3-one:

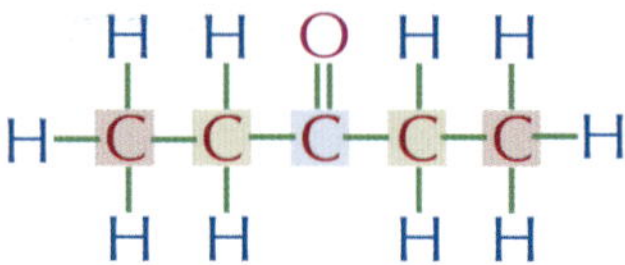

^{13}C NMR spectra of cyclic molecules

The number of peaks on the ^{13}C spectrum of a cyclic compound depends on the symmetry of the molecule. There's an example on the next page to show how it works.

Example

The ^{13}C NMR spectrum of a cyclic molecule with the formula $C_6H_4Cl_2$ is shown below. Identify the molecule that produced this spectrum.

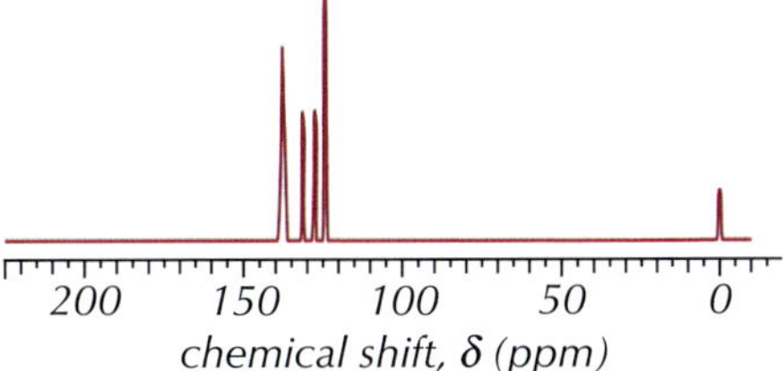

1. The spectrum has four peaks, so it must have four carbon environments.
2. All four peaks are between $\delta = 125$ ppm and $\delta = 140$ ppm. Looking at the chemical shift table these can only be due to alkene groups or carbons in a benzene ring. The question tells you that the molecule is cyclic with six carbons, so these carbons must be in a benzene ring, as a six-carbon cyclic molecule with three double bonds does not exist.
3. There are only three aromatic molecules with the formula $C_6H_4Cl_2$ — they're all isomers of dichlorobenzene (see below).

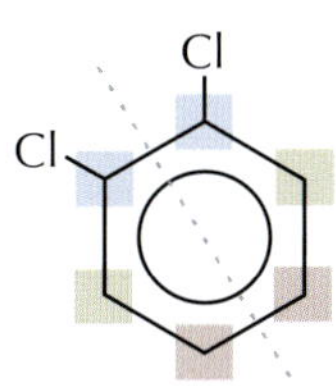

1,2-dichlorobenzene

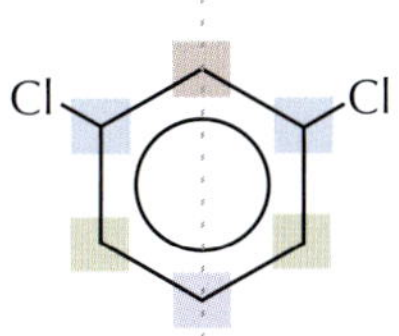

1,3-dichlorobenzene

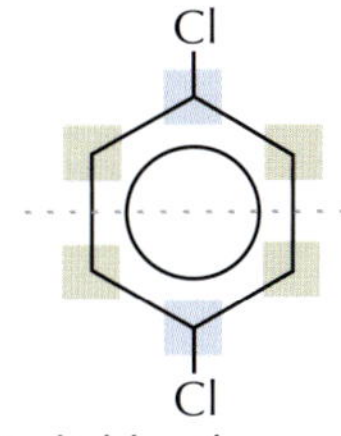

1,4-dichlorobenzene

If you look at the symmetry of the molecules you can see that 1,2-dichlorobenzene has three carbon environments, while 1,3-dichlorobenzene has four and 1,4-dichlorobenzene only has two. So the spectrum must have been produced by 1,3-dichlorobenzene.

Tip: If a cyclic molecule is symmetrical, you can draw a mirror line (like the dashed lines on the molecules on the left) across it. Atoms or groups that are in the same place on opposite sides of the line will be in the same environment.

Practice Question — Application

Q1 How many peaks would you see on a ^{13}C NMR spectrum of:

a) butanal?
b) pentane?
c) 2-methylpropane?
d) 4-chlorocyclohexanone?

Q2 The carbon-13 NMR spectrum of a compound with the molecular formula C_5H_{12} is shown below.

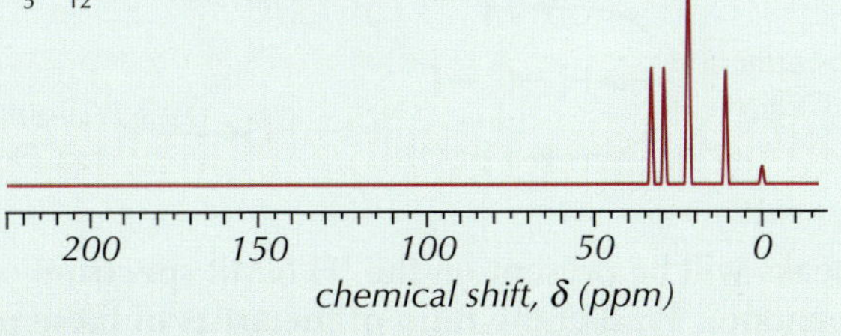

Use the spectrum to identify the structure of the molecule. Explain your reasoning.

Q3 The carbon-13 NMR spectrum of a molecule with the formula $C_4H_{10}O$ is shown on the right. Identify the structure of the molecule.

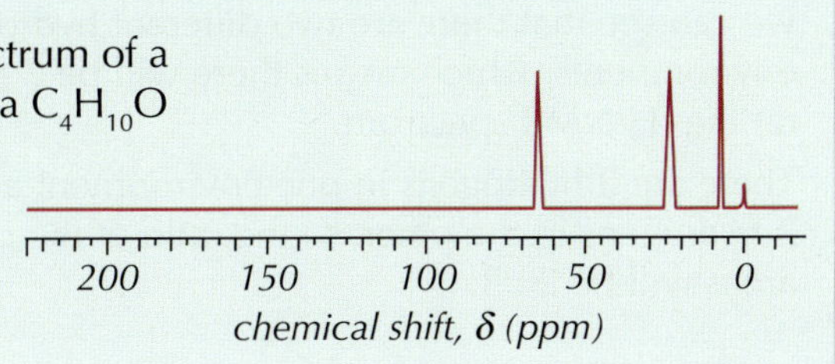

Tip: Use the chemical shift data table from page 501 to help you with these questions.

Tip: Make sure you use any information that you're given about what atoms are in the molecule — for instance if you're told that the molecule's a hydrocarbon, then it will only have carbon and hydrogen in it. This will help you narrow down the options for what type of environment each shift could represent.

4. ¹H NMR Spectroscopy

¹H NMR works in pretty much the same way as ¹³C NMR except this time the spectra tell you about the different hydrogen environments in a molecule.

Learning Objectives:

- Know that integrated spectra indicate the relative numbers of ¹H atoms in different environments.
- Be able to use integration data from ¹H NMR spectra to determine the relative numbers of equivalent protons in the molecule.
- Be able to use the n+1 rule to deduce the spin-spin splitting patterns of adjacent, non-equivalent protons, limited to doublet, triplet and quartet formation in aliphatic compounds.
- Know that ¹³C NMR gives simpler spectra than ¹H NMR.
- Know that ¹H NMR spectra are obtained using samples dissolved in deuterated solvents or CCl_4.
- Be able to use ¹H NMR spectra and chemical shift data from the Chemistry Data Booklet to suggest possible structures or part structures for molecules.

Specification Reference 3.3.15

Hydrogen environments

¹H NMR is all about how hydrogen nuclei react to a magnetic field. The nucleus of a hydrogen atom is a single proton. So ¹H NMR is also known as 'proton NMR' — and you might see the hydrogen atoms involved being called 'protons'. Each peak on a ¹H NMR spectrum is due to one or more hydrogen nuclei (protons) in a particular environment — this is similar to a ¹³C NMR spectrum (which tells you the number of different carbon environments). But, with ¹H NMR, the relative area under each peak also tells you the relative number of H atoms in each environment.

Examples

The spectrum below is the ¹H NMR spectrum of ethanoic acid (CH_3COOH).

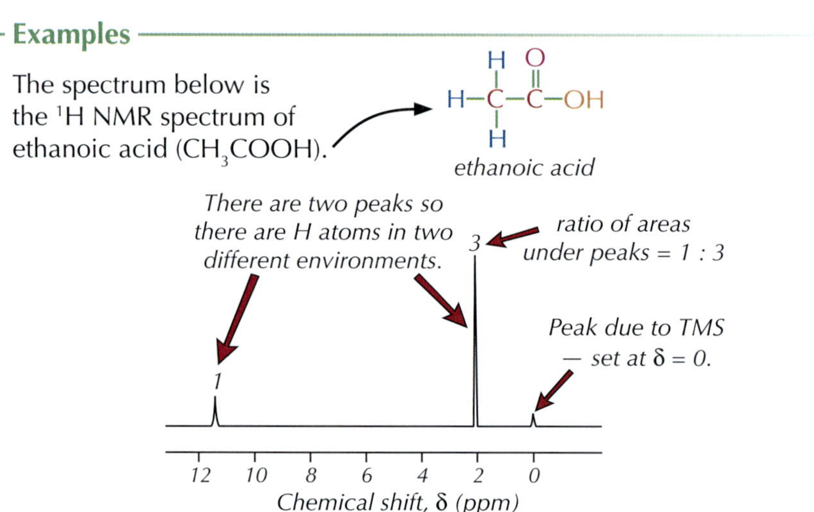

There are two peaks — so there are two hydrogen environments. The numbers above the peaks tell you the area ratio between the hydrogens in each environment, which here is 1 : 3 — so there's 1 H atom in the environment at δ = 11.5 ppm to every 3 H atoms in the other environment. If you look at the structure of ethanoic acid, this makes sense:

3 H atoms attached to CH_2COOH.

1 H atom attached to $COOCH_3$.

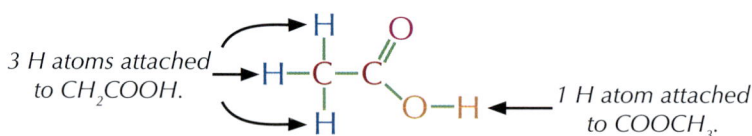

How many peaks will be present on the ¹H NMR spectrum of 1-chloropropanone? Predict the ratio of the areas of these peaks.

By looking at the structure of 1-chloropropanone we can see that there are two different hydrogen environments, which means there will be 2 peaks on the ¹H NMR spectrum.

There are 2 hydrogens in one environment and 3 hydrogens in the other, so the ratio of the peak areas will be 2 : 3.

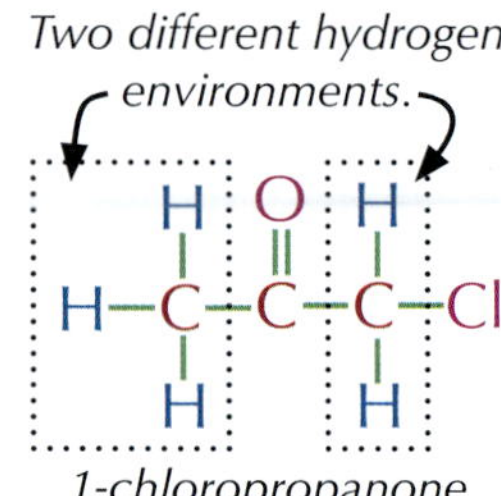

1-chloropropanone

Exam Tip

It can be difficult to work out how many hydrogen environments you have in a molecule. Sometimes it helps if you highlight all the hydrogens that are in the same environment the same colour — then you can go back and count up how many environments there are.

Integration traces

^{1}H NMR spectra can get quite cramped and sometimes it's not easy to see the ratio of the areas — so an **integration trace** is often shown. The height increases shown on the integration trace are proportional to the areas of the peaks.

Example

Here's the spectrum for ethanoic acid again:

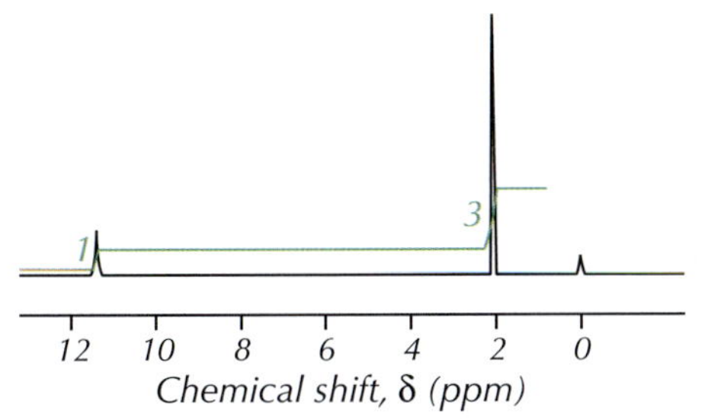

The integration trace (shown in green on the diagram) has a peak around 11.5 ppm and one around 2 ppm.

The heights of the vertical lines are in the ratio 1 : 3 — this means that for every one hydrogen in the first environment there, are three in the second environment.

Tip: If the ratio of areas isn't given to you, you can use a ruler to measure the ratios off the integration trace. Measure the heights of the trace peaks and divide the measurements by the smallest peak height, et voilà — you've got your area ratio.

Tip: Don't worry if the relative areas are not always whole numbers — they are ratios and not exact numbers.

Chemical shift

You use a table like the one in Figure 1 to identify which functional group each peak in a ^{1}H NMR spectrum is due to. Don't worry — you'll be given one in your exam, so you don't need to memorise it, you just need to learn how to use it.

^{1}H NMR Chemical Shifts Relative to TMS	
Chemical shift, δ (ppm)	**Type of H atom**
0.5 – 5.0	ROH
0.7 – 1.2	RCH_3
1.0 – 4.5	RNH_2
1.2 – 1.4	R_2CH_2
1.4 – 1.6	R_3CH
2.1 – 2.6	R–C(=O)–C(H)–
3.1 – 3.9	R–O–C(H)–
3.1 – 4.2	RCH_2Br or RCH_2Cl
3.7 – 4.1	R–C(=O)–O–C(H)–
4.5 – 6.0	R(–)C=C(–)H
9.0 – 10.0	R–C(=O)H
10.0 – 12.0	R–C(=O)–O–H

The hydrogen atoms that cause the shift are highlighted in red.

R stands for the rest of the organic molecule.

Empty bonds just mean that there are other things attached to the carbon too. (So this group could be $COCH_3$, $COCH_2R$, or $COCHR_2$.)

Figure 1: *Chemical shifts of different hydrogen environments.*

Exam Tip
The copy of this table you get in your exam may look a little different, and have different values — they depend on the solvent, temperature and concentration.

Exam Tip
Try and get your hands on a copy of the data sheet that you'll get in the exam before the day (you can download one from AQA's website). That way you'll know exactly what's given to you and what you'll be expected to know.

Example

According to the table, ethanoic acid (CH_3COOH) should have a peak at 10.0 – 12.0 ppm due to R-COOH, and a peak at 2.1 – 2.6 ppm due to R-$COCH_3$. You can see these peaks on ethanoic acid's spectrum below.

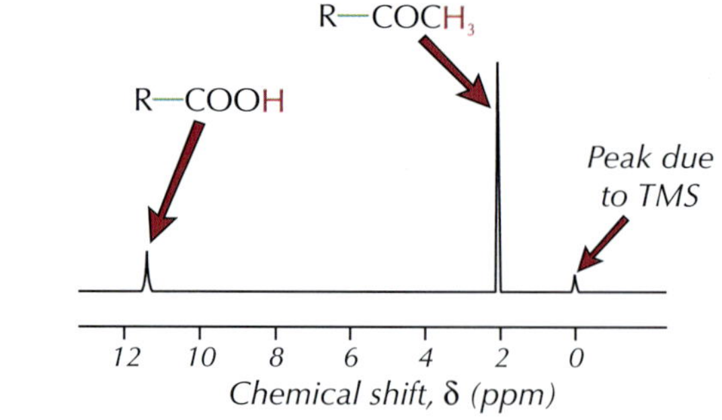

Tip: Neighbouring carbons are carbons one along in the carbon chain from the carbon the hydrogen's attached to. In the diagram below C^2 is the neighbouring carbon to C^1, so the peak for the red H atom is split into the number of purple H atoms, plus one.

Splitting patterns

The peaks on a ^1H NMR spectrum may be split into smaller peaks (this is called spin-spin splitting). These split peaks are called **multiplets**. Peaks always split into the number of hydrogens on the neighbouring carbon(s), plus one. It's called the **n+1 rule**. Some of the different **splitting patterns** you'll find in ^1H spectra are shown in Figure 2.

Type of peak	Structure of peak	Number of hydrogens on adjacent carbon(s)
Singlet (not split)		0
Doublet (split into two)		1
Triplet (split into three)		2
Quartet (split into four)		3

Figure 2: *Splitting patterns in ^1H NMR.*

Tip: Hydrogens in -OH groups are a special case when it comes to splitting. They are always singlets, and they don't affect the splitting patterns of neighbouring hydrogens.

Example

Here's the ^1H NMR spectrum for 1,1,2-trichloroethane:

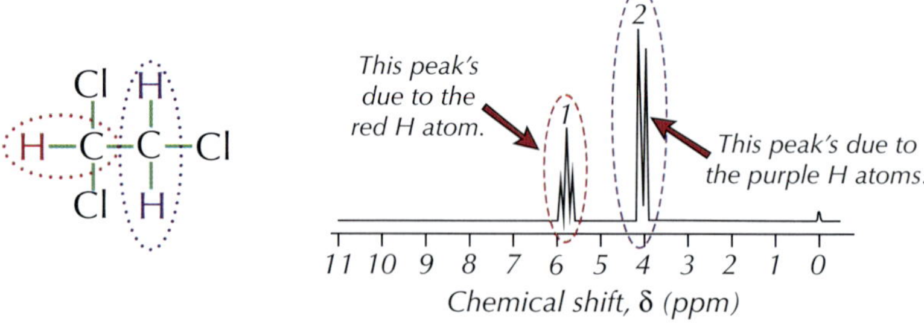

The peak due to the purple hydrogens is split into two because there's one hydrogen on the adjacent carbon atom. The peak due to the red hydrogen is split into three because there are two hydrogens on the adjacent carbon atom. The numbers above the peaks confirm that the ratio of hydrogens in the red environment to those in the purple environment is 1 : 2.

^1H NMR spectra are usually more complicated than ^{13}C NMR spectra because they have more, unclear split peaks to worry about.

Tip: For your exam, you don't need to know why ^1H NMR peaks split. It's all down to how the tiny magnetic fields of the hydrogen nuclei interact — it's pretty tricky stuff...

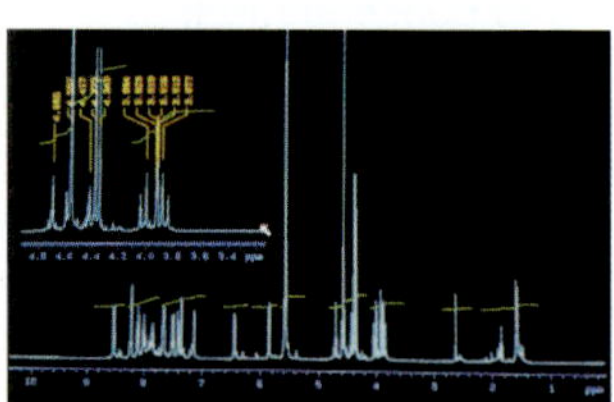

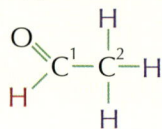

Figure 3: *^1H NMR spectrum. In the ^1H NMR spectra of large organic molecules the splitting patterns can be very complicated. Thankfully you'll never have to deal with one like this in the exam.*

Proton-free solvents

If a sample has to be dissolved, then a solvent is needed that doesn't contain any ^1H atoms — because these would show up on the spectrum and confuse things. **Deuterated solvents** are often used — their ^1H atoms have been replaced by deuterium (D or ^2H) (see Figure 4). Deuterium's an isotope of hydrogen that's got two nucleons (a proton and a neutron).
Unlike ^1H, deuterium don't interact with the external magnetic field, and don't produce peaks on the ^1H NMR spectrum. Here are some examples of deuterated solvents:

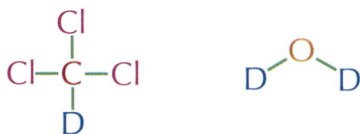

Cl–C(Cl)(Cl)–D   D–O–D

CCl_4 can also be used as a solvent — it doesn't contain any ^1H atoms either.

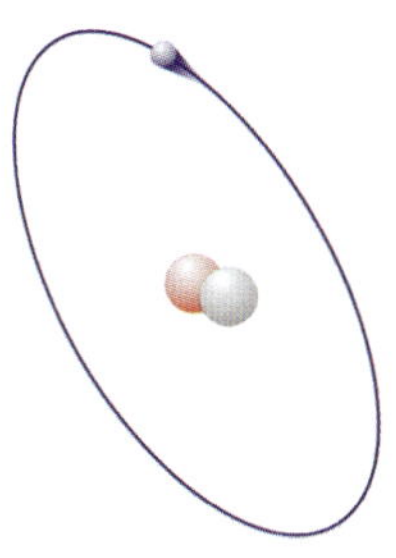

Figure 4: Atomic model of deuterium. Deuterium has one neutron and one proton in its nucleus, whereas ^1H has no neutrons. The atom also has one electron orbiting the nucleus.

Predicting structure from ^1H NMR spectra

^1H NMR spectra provide you with an awful lot of information to analyse. Here's a run down of the things to look out for:

- The number of peaks tells you how many different hydrogen environments there are in your compound.
- The ratio of the peak areas tells you about the relative number of hydrogens in each environment. Sometimes these ratios are written above the peaks on the spectrum for you, other times you may have to use the integration traces.
- You can use the chemical shift of each peak to work out what type of environment the hydrogen is in. You can use a table like the one in Figure 1 to help you.
- The splitting pattern of each peak tells you the number of hydrogens on the adjacent carbon. You can use the n+1 rule to work this out.

Tip: If you want to work out the number of hydrogens on an adjacent carbon from a peak you have to take 1 away from the number of peaks. For example, a doublet (two peaks) on a spectrum means there's 1 hydrogen on the adjacent carbon.

And here's one last example to help you on your way...

Example

Using the spectrum below, and the table of chemical shift data on page 505, predict the structure of the compound.

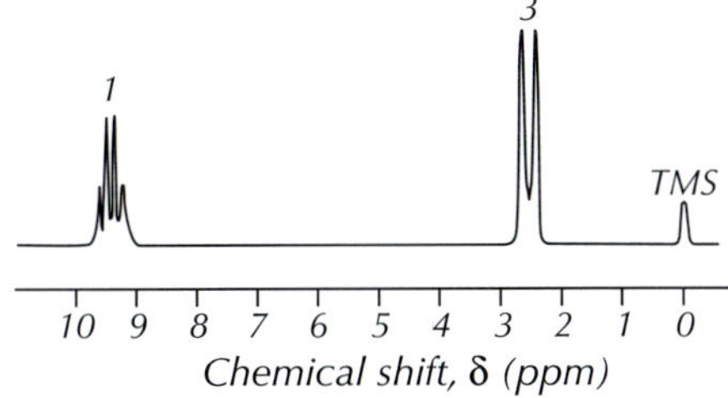

- There are two peaks so there are two different hydrogen environments.
- From the area ratios, there's one proton in the environment at $\delta = 9.5$ ppm for every three in the environment at $\delta = 2.5$ ppm.
- Now using the chemical shift data, the peak at $\delta = 2.5$ ppm is likely to be due to an R–$COCH_3$ group, and the peak at $\delta = 9.5$ ppm is likely to be due to an R–CHO group. This fits nicely with the area ratio data.
- The peak at $\delta = 9.5$ ppm is a quartet so this proton has got three neighbouring hydrogens. The peak at $\delta = 2.5$ ppm is a doublet so these protons have got one neighbouring hydrogen. So, it's likely these two groups are next to each other.

Exam Tip
When you're analysing a spectrum in an exam write down all the information you're given and can work out first. Then try and work out the structure — that way you're less likely to miss out an important detail.

You know the molecule has to contain these groups:

```
   O            H  O
   ‖            |  ‖
 —C—H        H—C—C—
                |
                H
```

Exam Tip
Always go back and check to see if your structure matches the spectrum — you need to make sure you haven't overlooked an important piece of information.

So all you have to do is fit them together so that each environment has the correct number of neighbouring hydrogens:

```
   H  O
   |  ‖
H—C—C—H
   |
   H
```

So this is the ^{1}H NMR spectrum for ethanal.

Practice Questions — Application

Q1 The spectrum below is the ^{1}H NMR spectrum for propanoic acid.

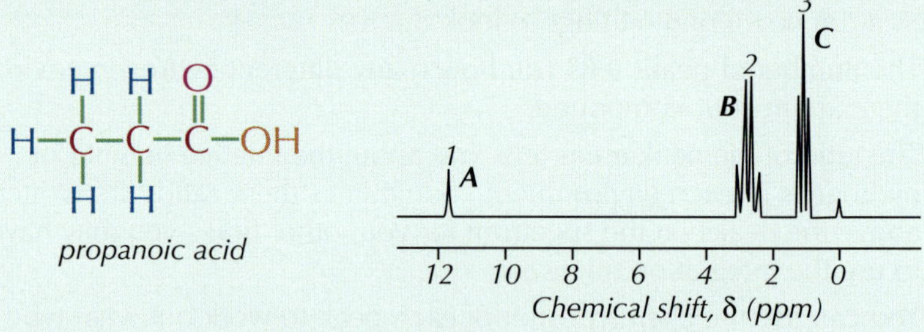

Assign the peaks **A** – **C** to the hydrogen environments found in propanoic acid.

Q2 Use the ^{1}H spectrum of compound **X** below, along with the chemical shift data on page 505 to work out the structure of compound **X**. HINT: Compound **X** has the molecular formula C_4H_8O.

Tip: Remember to use all the different types of information that the spectrum gives you...

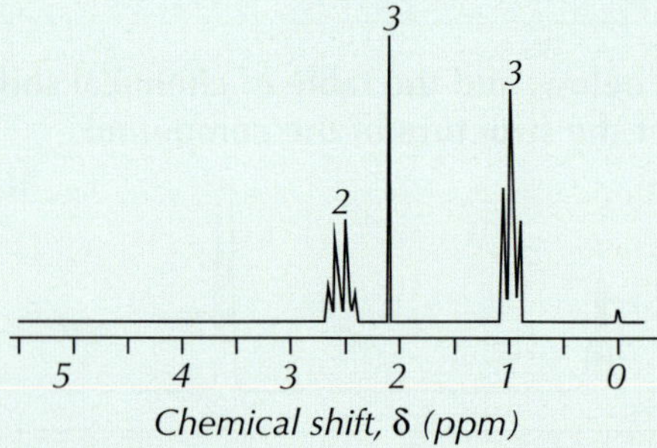

Practice Questions — Fact Recall

Q1 What does the number of peaks on the ^{1}H NMR spectrum of a compound correspond to?

Q2 What does the area under the peaks on an ^{1}H NMR spectrum of a compound correspond to?

Q3 What data about a peak in a ^{1}H NMR spectrum tells you about the functional group that hydrogen is in?

Q4 Suggest one solvent that could be used to dissolve a sample for analysis by ^{1}H NMR that wouldn't interfere with the spectrum.

5. Chromatography

Chromatography is used to separate and identify chemicals in mixtures.

The basics

Chromatography is used to separate stuff in a mixture — once it's separated out, you can often identify the components. There are quite a few different types of chromatography — but they all have the same basic set up:

- A **mobile phase** — where the molecules can move. This is always a liquid or a gas.
- A **stationary phase** — where the molecules can't move. This must be a solid, or a liquid on a solid support.

They also use the same basic principle. The mobile phase moves through or over the stationary phase. The distance each substance moves depends on its solubility in the mobile phase and its retention by the stationary phase — components that are more soluble in the mobile phase will travel further than components that are more strongly **adsorbed** (attracted) to the stationary phase. It's these differences in solubility and retention that separate out the different substances.

Thin-layer chromatography

REQUIRED PRACTICAL 12

In **thin-layer chromatography** (TLC):

- The mobile phase is a liquid solvent, such as ethanol.
- The stationary phase is a thin layer of silica (silicon dioxide) or alumina (aluminium oxide) fixed to a glass or metal plate.

Here's how you separate a mixture using TLC:

1. Draw a line in pencil near the bottom of the TLC plate (the baseline) and put a very small drop of each mixture to be separated on the line.
2. Allow the spots on the plate to dry.
3. Place the plate in a beaker with a small volume of solvent (this is the mobile phase). The solvent level must be below the baseline, so it doesn't dissolve your samples away. Cover the top of the beaker with a watch glass.
4. The solvent will start to move up the plate. As it moves, the solvent will carry the substances in the mixture with it — some chemicals will be carried faster than others and so travel further up the plate.
5. Leave the beaker until the solvent has moved almost to the top of the plate. Then remove the plate from the beaker. Before it's evaporated, use a pencil to mark how far the solvent travelled up the plate (this line is called the solvent front).
6. Place the plate in a fume cupboard and leave it to dry. The fume cupboard will prevent any toxic or flammable fumes from escaping into the room.
7. The result is called a chromatogram (see Figure 2). You can use the positions of the chemicals on the chromatogram to identify what the chemicals are.

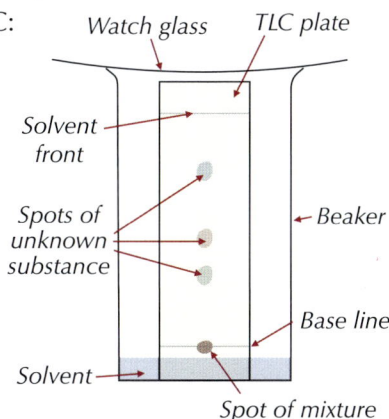

Figure 1: *A thin-layer chromatography plate.*

Learning Objectives:

- Know that chromatography can be used to separate and identify the components in a mixture.
- Know that separation depends on the balance between solubility in the moving phase and retention by the stationary phase.
- Know that in thin-layer chromatography (TLC) a plate is coated with a solid and a solvent moves up the plate.
- Be able to separate species by thin-layer chromatography (Required practical 12).
- Know that R_f values are used to identify different substances.
- Be able to calculate R_f values from a chromatogram and compare them with standards to identify different substances.
- Know that in column chromatography (CC) a column is packed with a solid and a solvent moves down the column.

Specification Reference 3.3.16

Tip: Placing a watch glass over the top of the beaker stops the solvent from evaporating.

Tip: It's important that you carry out a risk assessment before beginning your experiment.

Tip: It's a good idea to wear gloves when handling the plate, to avoid any contamination by substances on your hands.

Revealing colourless chemicals

If the chemicals in the mixture are coloured (such as the dyes that make up an ink) then you'll see them as a set of coloured dots at different heights on the TLC plate. But if there are colourless chemicals such as amino acids (see page 476-478) in the mixture, you need to find a way of making them visible.

Here are two ways:

- Many TLC plates have a special fluorescent dye added to the silica or alumina layer that glows when UV light shines on it. You can put the plate under a UV lamp and draw around the dark patches to show where the spots of chemical are.
- Expose the chromatogram to iodine vapour (leaving the plate in a sealed jar with a couple of iodine crystals does the trick). Iodine vapour is a locating agent — it sticks to the chemicals on the plate and they'll show up as purple spots.

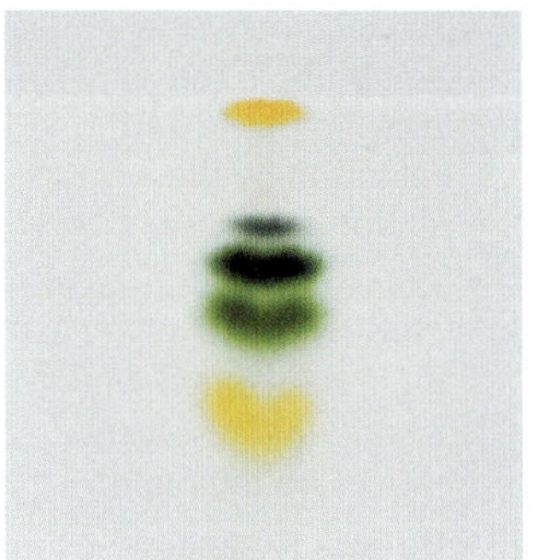

Figure 2: *A TLC chromatogram.*

R_f values

If you just want to know how many chemicals are present in a mixture, all you have to do is count the number of spots that form on the plate. But if you want to find out what each chemical is, you can calculate something called an **R_f value**. The formula for this is:

$$R_f \text{ value} = \frac{\text{distance travelled by spot}}{\text{distance travelled by solvent}}$$

Tip: R_f stands for 'retardation factor'.

When you're measuring how far a spot has travelled, you just measure from the base line (point of origin) to the vertical centre of the spot. When you're measuring how far the solvent has travelled, you measure from the base line to the solvent front (see Figure 3).

Tip: Small R_f values arise from substances that are not very soluble in the mobile phase, and are strongly adsorbed onto the plate. A large R_f value shows a substance is very soluble in the mobile phase and weakly adsorbed.

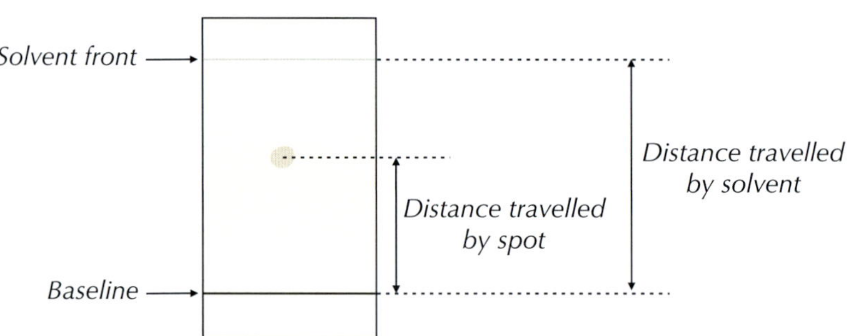

Figure 3: *The distances needed for calculating an R_f value.*

R_f values aren't dependent how big the plate is or how far the solvent travels — they're properties of the chemicals in the mixture and so can be used to identify those chemicals. This means you can look your R_f value up in a table of standard R_f values to identify what that substance is. But if the composition of the TLC plate, the solvent, or the temperature at which you carry out your chromatography experiment change even slightly, you'll get different R_f values. It's hard to keep the conditions at which you carry out your chromatography experiment identical. So, if you suspect that a mixture contains, say, chlorophyll, it's best to put a spot of chlorophyll on the baseline of the same plate as the mixture and run them both at the same time.

Tip: The stationary phase and solvent that you use will affect the R_f value. If you're looking up R_f values, you need to check that they were recorded under the same conditions as your experiment.

Example

A sugar solution containing a mixture of three sugars is separated using TLC. The chromatogram is shown on the right.

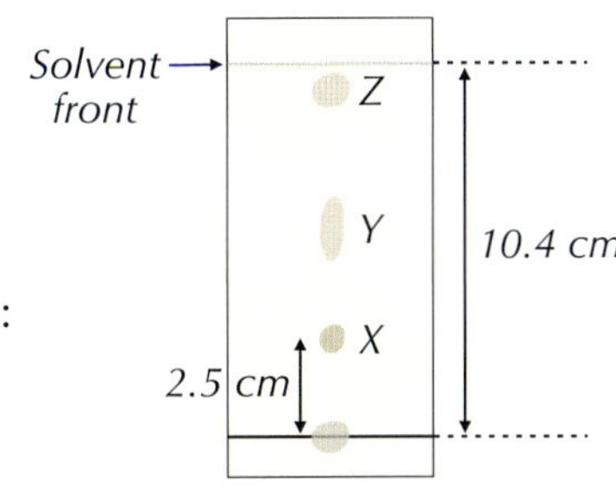

a) Calculate the R_f value of spot X.

To find the R_f value of spot X all you have to do is stick the numbers into the formula:

$$R_f \text{ value} = \frac{\text{distance travelled by spot}}{\text{distance travelled by solvent}}$$

$= 2.5 \text{ cm} \div 10.4 \text{ cm} = \mathbf{0.24}$

b) Figure 4 shows the R_f values of three sugars under the conditions used in the experiment. Use the table to suggest the sugar present in spot X.

Spot X has an R_f value of 0.24. Fructose also has an R_f value of 0.24. So the sugar present in spot X could be fructose.

Tip: R_f values are always between 0 and 1.

Sugar	R_f value
Glucose	0.20
Fructose	0.24
Xylose	0.30

Figure 4: *R_f values of sugars.*

Column chromatography

Column chromatography (CC) is mostly used for purifying an organic product. This needs to be done to separate it from unreacted chemicals or side products. It involves packing a glass column with a solid, absorbent material such as aluminium oxide coated with water — this is called a slurry. This is the stationary phase. The mixture to be separated is added to the top of the column and allowed to drain down into the slurry. A solvent is then run slowly and continually through the column. This solvent is the mobile phase.

Figure 5: *Column chromatography equipment.*

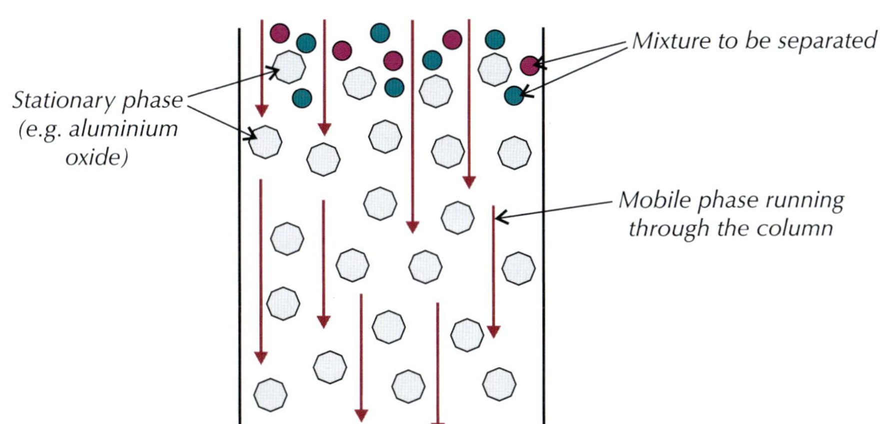

Figure 6: *The initial set-up of a column for column chromatography.*

As the mixture is washed through the column, its components separate out according to how soluble they are in the mobile phase and how strongly they are adsorbed onto the stationary phase (**retention**). Each different component will spend some time adsorbed onto (stuck to) the stationary phase and some time dissolved in the mobile phase.

The longer a component spends dissolved in the mobile phase, the quicker it travels down the column. If a component spends a long time adsorbed onto the stationary phase, it will take a long time to travel down the column. So, the more soluble a component is in the mobile phase, the quicker it'll pass through the column (see Figure 7).

Tip: This chromatography stuff isn't as difficult as it first seems. You just have to remember that the mixture separates depending on how soluble components are in the mobile phase. Components that are most soluble will travel through the column, or up the plate, quickest.

Exam Tip
If you're asked to explain how column chromatography works in the exam, don't forget to say both that some components are more soluble in the mobile phase and that some are more strongly adsorbed by the stationary phase.

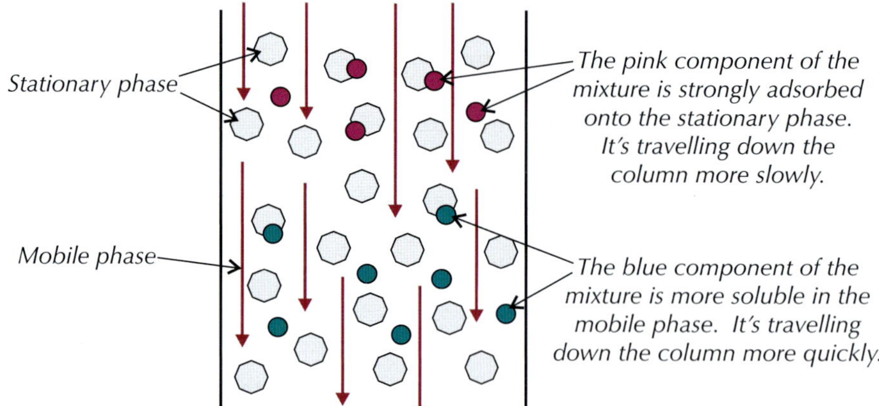

Figure 7: *The passage of a two component mixture through a chromatography column.*

As a component of the mixture reaches the end of the column it is collected. It can then be identified using the time taken to pass through the column (**retention time**) or another technique (e.g. mass spectrometry, which you met on pages 27-32).

Practice Questions — Application

Amino acid	R_f value
Glycine	0.26
Alanine	0.39
Tyrosine	0.46
Valine	0.61
Leucine	0.74

Figure 8: *R_f values of amino acids.*

Q1 A student used thin-layer chromatography to separate out a mixture of three amino acids. The chromatogram that she produced is shown on the right.

R
Q
P
5.9 cm
8.0 cm
3.7 cm
2.1 cm

a) Calculate the R_f values of the three spots, P, Q and R.

b) P,Q and R are each known to be one of the amino acids shown in Figure 8. Use the table in Figure 8 to identify the amino acid present in each spot.

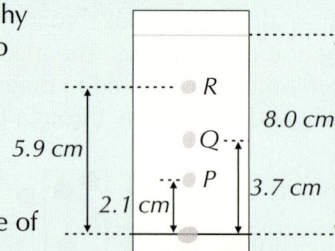

Q2 A scientist is using column chromatography to purify an organic product. She knows that the pure product is more strongly adsorbed to the stationary phase than the impurities. Will the pure product leave the column before or after the impurities? Explain your answer.

Practice Questions — Fact Recall

Q1 What is chromatography used for?

Q2 Describe a method you could use to separate a mixture using thin-layer chromatography.

Q3 a) What are R_f values used for?

b) Give the formula for calculating R_f values.

Q4 Briefly explain why the components in a mixture separate out during column chromatography.

6. Gas Chromatography

Another type of chromatography you need to know about is gas chromatography. It's more high-tech than thin-layer chromatography, but the idea's just the same — a mobile phase, a stationary phase and things separating.

Learning Objectives:

- Know that in gas chromatography (GC) a column is packed with a solid or with a solid coated by a liquid, and a gas is passed through the column under pressure at high temperature.
- Know that retention times are used to identify different substances.
- Be able to compare retention times with standards to identify different substances.
- Understand the use of mass spectrometry to analyse the components separated by GC.

Specification Reference 3.3.16

What is gas chromatography?

If you've got a mixture of volatile liquids (ones that turn into gases easily), then **gas chromatography** (GC) is the way to separate them out so that you can identify them. The stationary phase is a solid or a solid coated by a viscous liquid, such as an oil, packed into a long column. The column is coiled to save space, and built into an oven. The mobile phase is an unreactive carrier gas such as nitrogen. The sample is vaporised and passes through the oven as a gas.

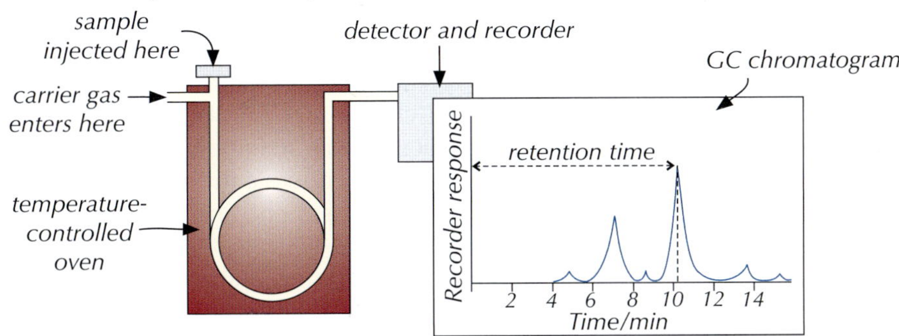

Figure 1: *GC equipment and chromatogram.*

Each component takes a different amount of time from being injected into the column to being recorded at the other end. This is the retention time. The retention times for the components in a mixture are shown on a chromatogram (see Figure 2). The retention time depends on how much time the component spends moving along with the carrier gas, and how much time it spends adsorbed to the viscous liquid. Under standard conditions, each separate substance will have a unique retention time — so you can use the retention time to identify the components of the mixture. (You have to run a known sample under the same conditions for comparison).

Tip: You can also use retention times to identify the components of a mixture in column chromatography.

Example

If you wanted to know if a mixture contained octane, you could run a sample of the mixture through the system, then run a sample of pure octane through, and see if there's a peak at the same retention time on both spectra.

Tip: You have to run the known sample under the same conditions as the unknown to ensure the variables of the experiment are kept the same.

The area under each peak tells you the relative amount of each component that's present in the mixture (see Figure 2).

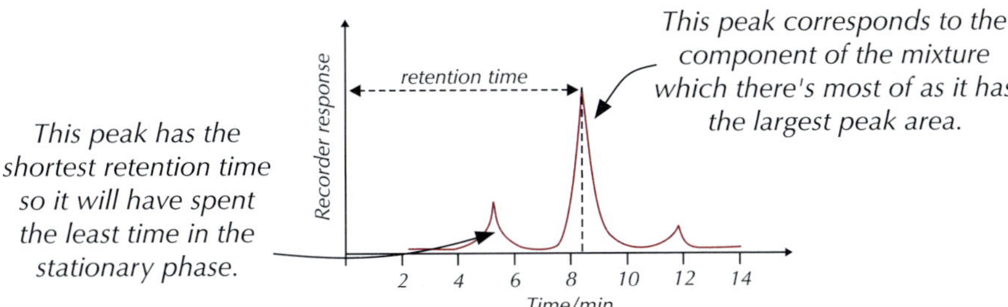

Figure 2: *A chromatogram showing the separation of three components in a mixture.*

GC can be used to find the level of alcohol in blood or urine — the results are accurate enough to be used as evidence in court. It's also used to find the proportions of various esters in oils used in paints — this lets picture restorers know exactly what paint was originally used.

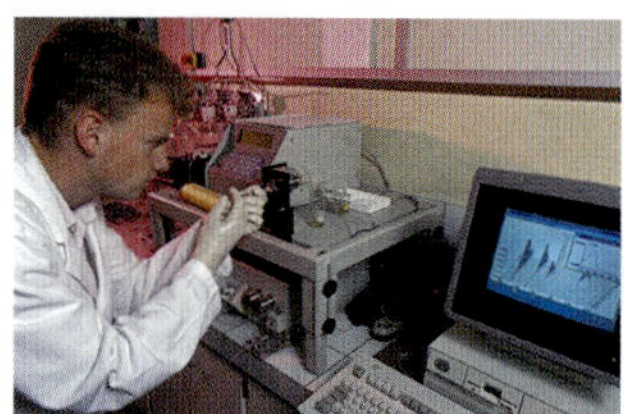

Figure 3: *GC-MS being used in a police lab to analyse evidence found at a crime scene.*

GC-MS

Mass spectrometry is a technique used to identify substances from their mass/charge ratio. It's very good at identifying unknown compounds, but would give confusing results from a mixture of substances. Gas chromatography (see previous page), on the other hand, is very good at separating a mixture into its individual components, but not so good at identifying those components. If you put these two techniques together, you get an extremely useful analytical tool. Gas chromatography-mass spectrometry (or GC-MS for short) combines the benefits of gas chromatography and mass spectrometry to make a super analysis tool.

The sample is separated using gas chromatography, but instead of going to a detector, the separated components are fed into a mass spectrometer (see Figure 4). The spectrometer produces a mass spectrum for each component, which can be used to identify each one and show what the original sample consisted of. Computers can be used to match up the mass spectrum for each component of the mixture against a database, so the whole process can be automated.

Tip: Mass spectrometry was covered back on pages 27-32, so have a look back at those notes if you need a reminder on how it works, and the sort of spectra it produces.

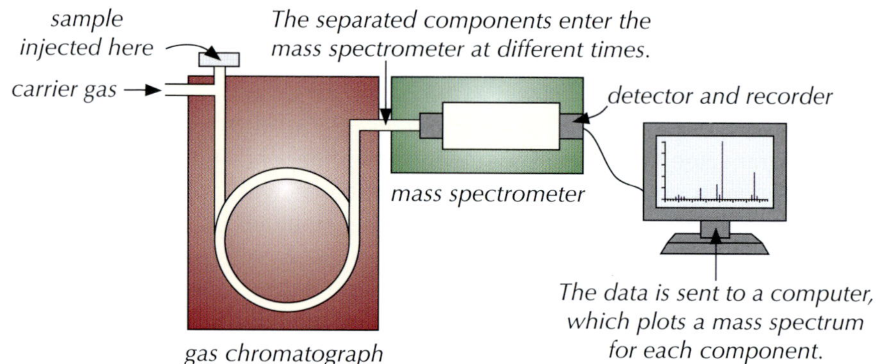

Figure 4: *GC-MS equipment.*

The advantage of this method over normal GC is that the components separated out by the chromatography can be positively identified, which can be impossible from a chromatogram alone (because compounds which are similar often have very similar retention times).

Tip: Combining different analytical techniques creates really powerful tools for working out the structures of unknown chemicals.

Practice Questions — Application

Q1 An environmental scientist ran a water sample through a GC-MS machine. He was looking for traces of the herbicide glyphosate.

a) What function does the gas chromatograph (GC) part of the scientist's GC-MS machine perform?

b) Explain how the scientist can confirm that the water is contaminated with glyphosate once he has the results from the GC-MS machine.

Q2 The GC chromatogram below shows the retention times of three components in a mixture.

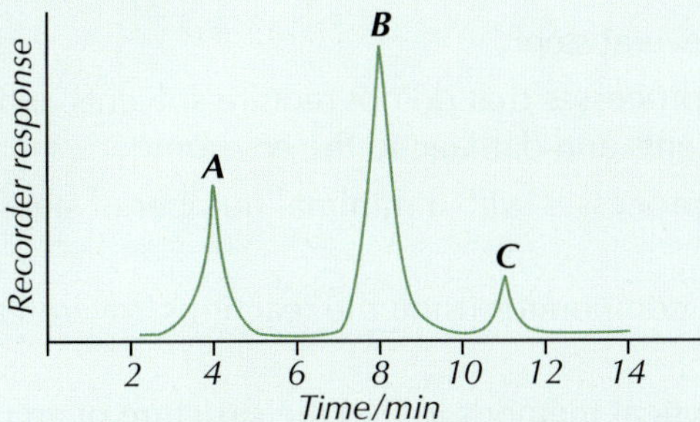

a) State which peak corresponds to the component that spends the highest proportion of its time in the tube in the mobile phase.

b) One component of the mixture is hexene. When pure hexene is run through the machine it has a retention time of 8 minutes. Which of the components, A, B or C, is hexene?

Q3 A student runs an unknown mixture through gas chromatography machine. The GC chromatogram below shows the retention times of the two components in the mixture.

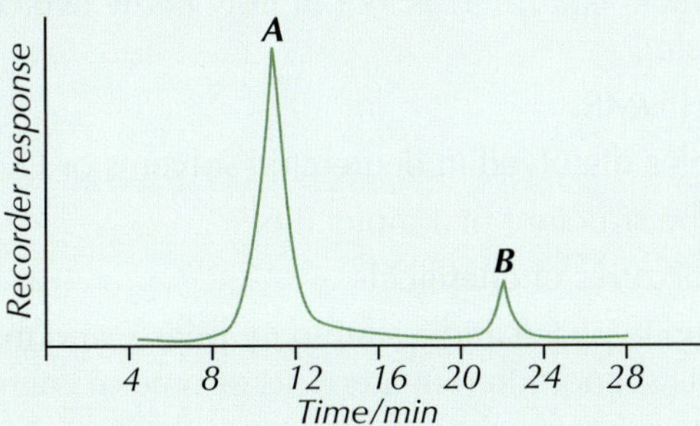

a) Estimate the retention time of component A.

b) Which component was present in the largest amount? Explain your answer.

c) The table in Figure 6 gives the standard retention times of some alcohols. Given that component A is an alcohol, use the table to suggest its identity.

d) How else could the student confirm that component A has been identified correctly?

Practice Questions — Fact Recall

Q1 Give an example of a mobile phase used in gas chromatography.

Q2 What is retention time?

Q3 Outline how GC-MS works.

Q4 Why is GC-MS a better technique for identifying compounds in a mixture than GC alone?

Exam Tip

You might get a question about GC-MS where you have to analyse a gas chromatogram or a mass spectrum — examiners do tend to like mixing and matching topics...

Alcohol	Retention time (min)
2,3-dimethyl butan-2-ol	7.51
butan-1-ol	11.27
pentan-1-ol	21.62
hexan-1-ol	40.30

Figure 5: *Retention times of alcohols.*

Section Summary

Make sure you know...

- That organic synthesis processes can involve several steps.
- That chemists aim to design organic synthesis processes that do not require solvents and that use non-hazardous starting materials to limit accidents and damage to the environment.
- That chemists aim to design organic synthesis processes with a minimal number of steps and a high atom economy to reduce waste.
- How to deduce the synthesis routes of organic compounds using the reactions covered across the A Level Chemistry course.
- That you can combine data from different analytical methods to find the structure of an unknown molecule.
- That nuclear magnetic resonance (NMR) spectroscopy gives information about the position of ^{13}C or ^{1}H atoms in a molecule.
- That chemical shift (δ) depends on the molecular environment of a nucleus.
- Why tetramethylsilane (TMS) is used as a standard in NMR spectroscopy.
- How to use a ^{13}C NMR spectrum to work out the structure of a molecule.
- That integration traces can tell you the relative numbers of ^{1}H atoms in different environments.
- How to use peak splitting patterns on a ^{1}H NMR spectrum to work out how many hydrogens there are on neighbouring carbons (using the n+1 rule).
- That ^{13}C NMR gives a simpler spectrum that ^{1}H NMR.
- That ^{1}H NMR spectra are obtained using samples dissolved in deuterated solvents or CCl_4.
- How to use a ^{1}H NMR spectrum to work out the structure of a molecule.
- That chromatography is used to separate out mixtures of chemicals.
- That in chromatography you always have a mobile phase and a stationary phase, and that the mixture separates out because the different components spend different amounts of time in each phase.
- That how long a component spends in each phase depends on how soluble it is in the mobile phase and how strongly adsorbed it is to the stationary phase.
- That in thin-layer chromatography (TLC) the stationary phase is a plate coated with a solid, and the mobile phase is a solvent that moves up the plate.
- How to separate a mixture by thin-layer chromatography.
- What an R_f value is and how to calculate one using a TLC chromatogram.
- That you can compare R_f values with standard values to identify different substances.
- That in column chromatography (CC) the stationary phase is a column packed with a solid, and the mobile phase is a solvent that moves down the column.
- That in gas chromatography (GC) the stationary phase is a column packed with a solid or a solid coated by a liquid, and the mobile phase is a gas that passes through the column.
- What retention times are and how they can be used to identify different substances.
- How you can compare retention times to standards to identify different substances.
- That chromatography can be combined with mass spectrometry (GC-MS) to give a way of separating mixtures and identifying chemicals that works better than chromatography on its own.

Exam-style Questions

1 A scientist is trying to synthesise phenylethanone. The structure of phenylethanone is shown on the right.

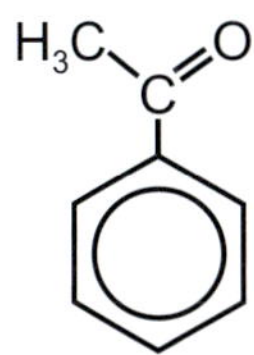

1.1 Give a one step synthesis of this molecule from benzene.

(3 marks)

1.2 The scientist purifies his compound using column chromatography. Suggest a stationary phase that the scientist could use in the column.

(1 mark)

1.3 Briefly explain why the pure phenylethanone separates out from the impurities as the mixture passes through the column.

(2 marks)

1.4 The table below gives details of the δ values of the peaks seen on the ^{13}C NMR spectra of three different compounds. Which compound, **A, B** or **C**, is phenylethanone? Explain your answer.

Peak number	Spectrum A — δ / ppm	Spectrum B — δ / ppm	Spectrum C — δ / ppm
1	26.5	26.5	26.5
2	128.3	128.3	128.3
3	137.2	128.6	128.6
4	197.9	133.0	133.0
5		137.2	137.2
6			197.9

(2 marks)

2 The ^1H NMR spectrum of an alcohol is shown below.

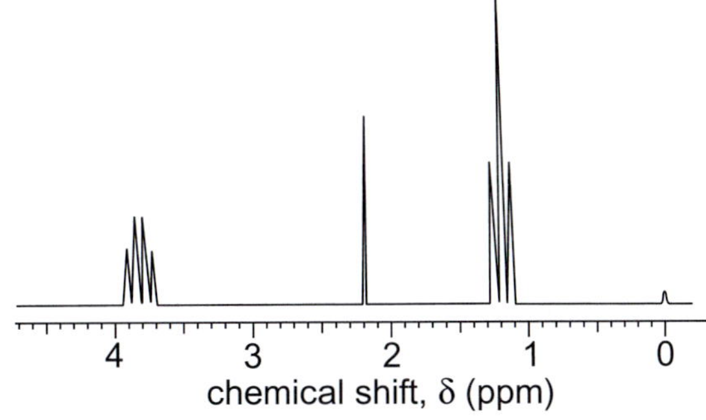

2.1 Suggest a solvent that you could use to dissolve the sample for this spectrum.

(1 mark)

2.2 Use the chemical shift values and splitting patterns to identify the chemical. Explain your reasoning. You may use the table on page 505 to help with this question.

(6 marks)

3 A forensic scientist is analysing a sample of chemicals taken from the scene of a fire. The scientist uses GC-MS to identify the compounds in the mixture.

3.1 Explain how GC-MS is used to identify the different components in a mixture.

(2 marks)

3.2 The sample contains several short-chain alkanes.
Suggest why the scientist could not use gas chromatography alone to identify the components of the mixture.

(2 marks)

3.3 Several of the substances that the scientist finds in the mixture are common components of petrol. The structures of one of these molecules, methylbenzene, is shown below.

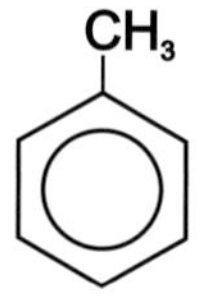

CH_3

The scientist runs a sample of methylbenzene through an NMR spectrometer.
How many peaks will appear on the ^{13}C NMR spectrum of this molecule?

(1 mark)

4 A scientist has a sample of a fruit flavouring that is used in soft drinks.
He uses thin-layer chromatography to separate out the components of the mixture.
The thin-layer chromatogram that he produces is shown below.

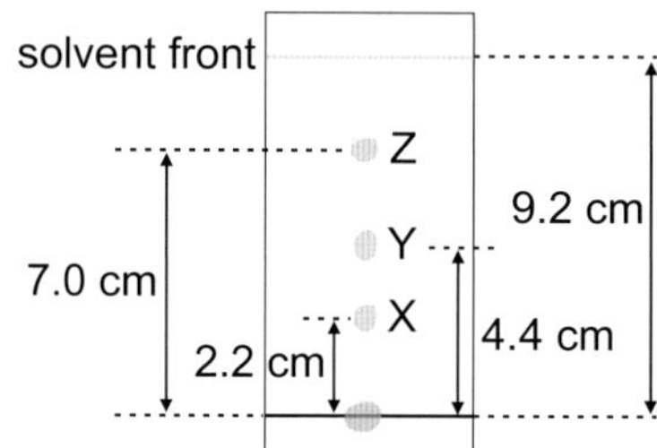

4.1 Explain why the components of the mixture separate as they travel up the plate.

(2 marks)

4.2 Calculate the R_f value of spot X.

(1 mark)

4.3 Spot Y is made up of the ester methyl 2-methylpropanoate.
The structure of this ester is shown below, with its carbon atoms labelled.

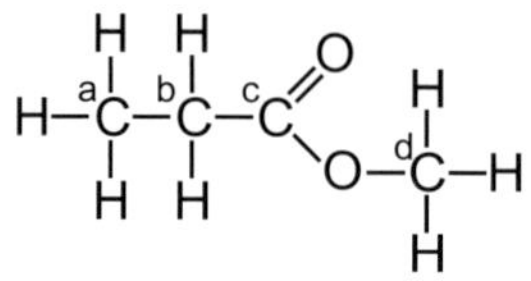

The ^1H NMR spectrum of this ester has three peaks. Describe how the peaks will be split. State which hydrogen atoms are responsible for each splitting pattern.

(3 marks)

Maths Skills

Maths skills for A-Level Chemistry

Maths crops up quite a lot in A-Level Chemistry so it's really important that you've mastered all the maths skills you'll need before sitting your exams. Maths skills are covered throughout this book but here's an extra little section, just on maths, to help you out.

Exam Tip
At least 20% of the marks in the exams will depend on your maths skills. That's a lot of marks so it's definitely worth making sure you're up to speed.

1. Exam Technique

The way you answer calculation questions in the exams is important — you should always show your working, and remember all those pesky things like units and significant figures when giving your final answer.

Tip: All the examples in this book that include the kind of maths you're expected to know for your exams are clearly marked. You can spot them by the big label that says...

Maths Skills

Showing your working

When you're doing calculations the most important thing to remember is to show your working. You've probably heard it a million times before but it makes perfect sense. You won't get a mark for a wrong answer but you could get marks for the method you used to work out the answer.

Units

Make sure you always give the correct units for your answer (see pages 526-527 for more on units).

Exam Tip
You'll need to know what units your figures need to be in for different formulas — see pages 522-524 for the units used in different formulas and pages 526-527 for how to convert between units.

Example — Maths Skills

Here's an example of a question where you need to change the units so they match the answer the examiner wants.

1.1 Calculate the free energy change for the following reaction at 298 K, giving your answer in kJ mol^{-1}:

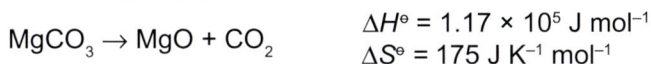

$MgCO_3 \rightarrow MgO + CO_2$

$\Delta H^{\ominus} = 1.17 \times 10^5$ J mol^{-1}
$\Delta S^{\ominus} = 175$ J K^{-1} mol^{-1}

Here you use the equation: $\Delta G = \Delta H - T\Delta S$. Plugging the numbers in gives you an answer of +64 900 J mol^{-1} to three significant figures. But the question asks for the answer in kJ mol^{-1}, so you have to divide by 1000, giving an answer of 64.9 kJ mol^{-1}. If you left your answer as +64 900 J mol^{-1} you'd lose a mark.

Standard form

You might need to use numbers written in standard form in calculations. Standard form is used for writing very big or very small numbers in a more convenient way. Standard form must always look like this:

$$A \times 10^n$$

This number must always be between 1 and 10. (A)

This number is an integer and tells you the number of places and the direction in which the decimal point moves. (n)

Tip: *'A'* can be 1 or any number up to 10 but it can't be 10 — there can only be a single digit before the decimal point, and this digit can't be 0.

You need to be able to convert between numbers written in ordinary form and standard form.

Tip: If you're converting a number from standard form back to ordinary form, you just move the decimal point the other way. If *n* is positive, that means the number you're converting is a large number, so you move the decimal point to the right (you need more digits in front of the decimal point). If *n* is negative then you're dealing with a very small number, so the decimal point needs to move to the left.

Examples — Maths Skills

Here's how to write 3 500 000 in standard form.

- First write all the digits up to the last non-zero digit. Put a decimal point after the first digit and a '× 10' at the end:

$$3.5 \times 10$$

- Then count how many places the decimal point has moved to the left. This number sits to the top right of the 10, as a superscript.

$$3\,5\,0\,0\,0\,0\,0 = 3.5 \times 10^6$$

(6 5 4 3 2 1 — places counted as the decimal point moves left)

- Et voilà... that's 3 500 000 written in standard form.

Here are some more examples.

- You can write 450 000 as 4.5×10^5.
- The number 0.000056 is 5.6×10^{-5} in standard form — the *n* is negative because the decimal point has moved to the right instead of the left.
- You can write 0.003456 as 3.456×10^{-3}.

Significant figures

Use the number of significant figures given in the data as a guide for how many you need to give in your answer. Whether you're doing calculations with the results of an experiment or in an exam, the rule is the same — round your answer to the lowest number of significant figures in the data you're using.

You should always write down the number of significant figures you've rounded to after your answer, so that other people can see what rounding you've done.

Tip: The fewer significant figures a measurement has, the less accurate it is. Your answer can only be as accurate as the least accurate measurement in the calculation — that's why you round your answer to the lowest number of significant figures in the data.

Exam Tip
You might get told in the question how many significant figures to give your answer to. If you are, make sure that you follow the instructions — you'll lose marks if you don't.

Tip: Even if the last significant figure is a zero, you still need to include it in your answer.

Examples — Maths Skills

In this question the data given to you is a good indication of how many significant figures you should give your answer to.

1.2 Calculate the free energy change of a reaction carried out at 298 K, given that $\Delta H = -2.43$ kJ mol^{-1} and $\Delta S = 94.5$ J mol^{-1} K^{-1}.

All the data values in the question are given to 3 s.f. so it makes sense to give your answer to 3 s.f. too. But sometimes it isn't as clear as that.

3.2 At 298 K, K_p for the following reaction is 1.15×10^{-1} kPa: $N_2O_4 \rightleftharpoons 2NO_2$. Given that, at equilibrium, the partial pressure of N_2O_4 is 0.42 kPa, calculate the partial pressure of NO_2.

There are two types of data that you need to do the calculation in this question — K_p data and partial pressure data. K_p is given to 3 s.f. and the partial pressure data is given to 2 s.f. You should always give your answer to the lowest number of significant figures given — in this case that's to 2 s.f. The answer in full is 0.219772... kPa, so the answer rounded correctly would be 0.22 kPa (2 s.f.).

Remember that when you're converting between ordinary and standard form, you need to make sure the number you've converted to has the same number of significant figures as the number you've converted from.

Diagrams of molecules

When you're asked to draw a diagram of a molecule in an exam it's important that you draw everything correctly. There's more about how to draw the displayed and skeletal formulas of molecules on pages 180 and 182, but you're also going to have to tackle how to represent the 3D shapes of molecules on the page. Here's a reminder of the main things to remember.

Tip: See pages 86-90 for more on shapes of molecules.

Examples — **Maths Skills**

Here's how to draw a 2D representation of the 3D shape of an ammonia molecule (NH_3):

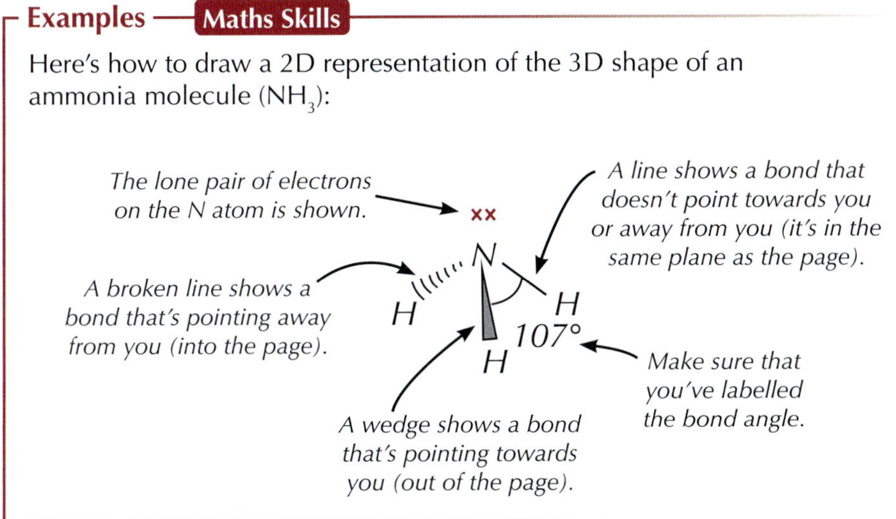

The lone pair of electrons on the N atom is shown.
xx
N
H
H
H
107°
A broken line shows a bond that's pointing away from you (into the page).
A line shows a bond that doesn't point towards you or away from you (it's in the same plane as the page).
Make sure that you've labelled the bond angle.
A wedge shows a bond that's pointing towards you (out of the page).

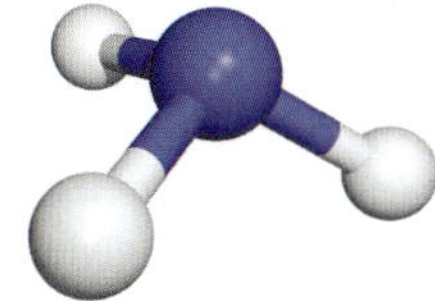

Figure 1: *An ammonia molecule. This is the 3D shape you are trying to represent in your drawing.*

When you're drawing any diagram make sure it's really clear what you're drawing. Draw diagrams nice and big, so you can see all of the details clearly. But do stay within the answer space — you won't get marks for anything that's drawn in the margin.

2. Formulas and Equations

A big part of the maths you need to do in Chemistry involves using formulas. Here are all of the ones you'll need for this course neatly summarised for you.

Tip: M_r is relative molecular mass (or relative formula mass). You work it out by adding up the A_r (atomic mass) of each atom in the molecule or compound.

Amounts and identities of substances

First up is perhaps the most useful formula of all:

$$\text{Number of moles} = \frac{\text{Mass of substance}}{M_r} \qquad \text{also written as...} \quad n = \frac{m}{M_r}$$

Then there's the one that uses the Avogadro constant (6.02×10^{23}):

$$\text{Number of particles} = \text{Number of moles} \times \text{the Avogadro constant}$$

You'll need these ones when you're dealing with solutions...

$$\text{Number of moles} = \frac{\text{Concentration (mol dm}^{-3}) \times \text{Volume (cm}^3)}{1000}$$

$$\text{Number of moles} = \text{Concentration (mol dm}^{-3}) \times \text{Volume (dm}^3)$$

Exam Tip
You don't need to learn the value of the Avogadro constant, or of the gas constant. They'll both be given to you if you need them in the exams.

Here's the ideal gas equation:

$$pV = nRT$$

- p — pressure (Pa)
- V — volume (m^3)
- n — number of moles
- R — the gas constant ($= 8.31\ J\ K^{-1}\ mol^{-1}$)
- T — temperature (K)

These two are handy when you're working out how much stuff you've made...

$$\%\ \text{atom economy} = \frac{\text{Molecular mass of desired product}}{\text{Sum of molecular masses of reactants}} \times 100$$

$$\%\ \text{yield} = \frac{\text{Actual yield}}{\text{Theoretical yield}} \times 100$$

Exam Tip
All these formulas are really important — you have to learn them because they won't be given to you in the exam (with the exception of the Arrhenius equation — see next page).

R_f values let you identify unknown substances based on how far they travel up the plate in thin-layer chromatography. Here's how you calculate them:

$$R_f\ \text{value} = \frac{\text{distance travelled by spot}}{\text{distance travelled by solvent}}$$

Reaction rates

First up, here's a simple formula that you can use to find the average rate of a reaction over a set period of time from experimental data:

$$\text{rate of reaction} = \frac{\text{amount of reactant used or product formed}}{\text{time}}$$

The rate equation links the rate of a reaction to the concentration of the reactants, the orders of these reactants, and the rate constant:

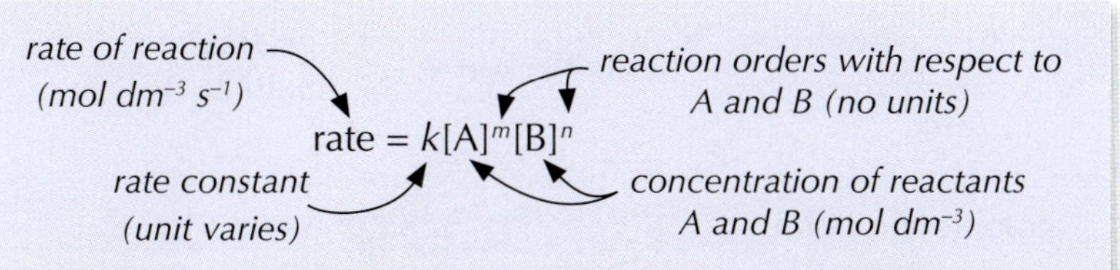

Tip: In this equation there are only two reactants (A and B), but you can have more. Here's the equation if you have three reactants: rate $= k[A]^m[B]^n[C]^x$

You can find the rate constant, k, for a reaction using the Arrhenius equation:

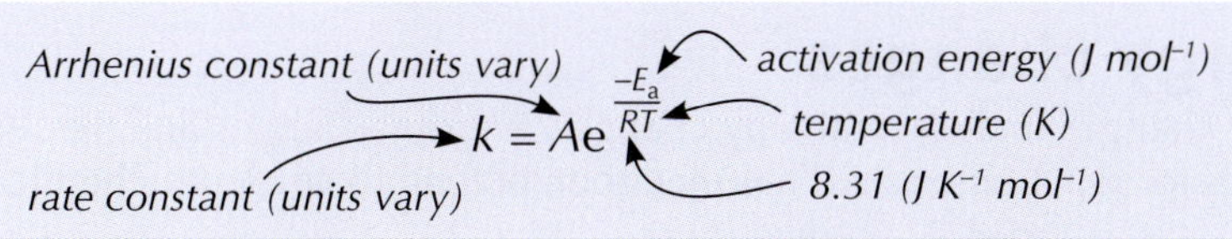

Tip: The Arrhenius equation is the only equation that you don't have to learn — both its forms will be given to you if you need them in the exams.

This rearranges into a logarithmic form: $\ln k = \frac{-E_a}{RT} + \ln A$

pH

You need to be able to convert between pH and the concentration of H^+ ions in a solution, and vice versa:

$$\text{pH} = -\log_{10}[H^+] \qquad [H^+] = 10^{-\text{pH}}$$

Tip: Make sure you know how to use the log button on your calculator.

You also need to be able to convert between pK_a and K_a:

$$pK_a = -\log_{10} K_a \qquad K_a = 10^{-pK_a}$$

Tip: You may also see '$\log_{10} x$' written as '$\log x$'.

Equilibria

You need to be able to work out equilibrium constants for reactions. Here's the formula for the equilibrium constant (K_c).

For the general reaction $aA + bB \rightleftharpoons dD + eE$: $K_c = \frac{[D]^d[E]^e}{[A]^a[B]^b}$ *where [X] is the concentration of X (mol dm⁻³)*

Tip: The lower-case letters a, b, d and e are the number of moles of each substance in the equation. The square brackets, [ ], mean concentration in mol dm⁻³.

For acids, the equilibrium constant is called the acid dissociation constant, K_a, has units of mol dm⁻³ and is found using the formula:

$$K_a = \frac{[H^+][A^-]}{[HA]}$$

Tip: For calculations involving K_a for a weak acid, you can assume that $[H^+] = [A^-]$, so the formula for K_a becomes: $K_a = \frac{[H^+]^2}{[HA]}$. If you're asked to give the formula of K_a, however, always write the other version.

Water has a special dissociation constant — the ionic product of water, K_w. The units of K_w are mol² dm⁻⁶, and it's found using the formula:

$$K_w = [H^+][OH^-]$$

Tip: In pure water, $[H^+] = [OH^-]$ so $K_w = [H^+]^2$.

Equilibrium reactions between gases have the equilibrium constant K_p.

For the general reaction $aA + bB \rightleftharpoons dD + eE$: $K_p = \frac{p(D)^d p(E)^e}{p(A)^a p(B)^b}$ *where p(X) is the partial pressure of X in Pa*

You can work out the partial pressures and mole fractions of the gases in a mixture using the following formulas:

$$\text{partial pressure of a gas in a mixture (Pa)} = \text{mole fraction} \times \text{total pressure of the mixture (Pa)}$$

Tip: The mole fraction of a substance doesn't have any units.

$$\text{mole fraction of a gas in a mixture} = \frac{\text{number of moles of gas}}{\text{total number of moles of all gases in the mixture}}$$

Energy changes

Exam Tip
Make sure you can rearrange all these formulas and give the units of each quantity as well.

Energy changes include enthalpy changes, entropy changes and free energy changes for reactions, as well as electrode potentials of electrochemical cells.

Here's the formula that you'll need for calculating the energy change of a reaction from experimental data:

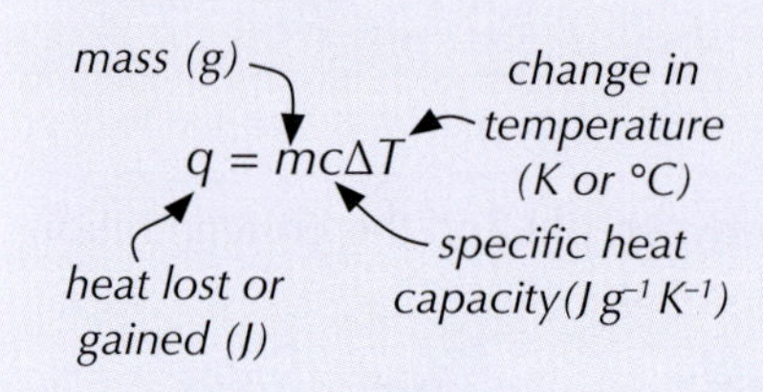

It doesn't matter whether the temperature is in K or °C — it's the change in temperature that goes into the formula, and that will be the same whether the units are K or °C.

To calculate the entropy change of a system, you use:

$\Delta S = S_{\text{products}} - S_{\text{reactants}}$ where ΔS is the entropy change in J K^{-1} mol^{-1}

If you know the enthalpy change, the entropy change and temperature of a reaction, you can calculate the free energy change (ΔG) using:

$$\Delta G = \Delta H - T\Delta S$$

Free energy change ($J\ mol^{-1}$) → ΔG; *Enthalpy change ($J\ mol^{-1}$)* → ΔH; *Temperature (K)* → T; *Entropy change ($J\ K^{-1}\ mol^{-1}$)* → ΔS

Tip: Electrode potentials are always measured in volts (V).

Here's the formula for calculating the cell potential of an electrochemical cell: $E^{\ominus}_{\text{cell}} = E^{\ominus}_{\text{reduced}} - E^{\ominus}_{\text{oxidised}}$

Or, if you're given a cell diagram to work from, you can use this formula instead: $E^{\ominus}_{\text{cell}} = E^{\ominus}_{\text{right-hand side}} - E^{\ominus}_{\text{left-hand side}}$

Finally, you need to know how to calculate the change in energy when a particle absorbs light of a certain frequency:

$$\Delta E = h\nu = \frac{hc}{\lambda}$$

Planck's constant = 6.63×10^{-34} Js → h; *energy absorbed (J)* → ΔE; *speed of light = 3.00×10^{8} m s^{-1}* → c; *frequency of light absorbed (Hz)* → ν; *wavelength of light absorbed (m)* → λ

Rearranging formulas

Being able to rearrange formulas is a must in chemistry, since you'll often need to make a different quantity the subject of a formula. Just remember the golden rule — whatever you do to one side of the formula, you must do to the other side of the formula.

Example — Maths Skills

The ideal gas equation is $pV = nRT$.
Rearrange the equation to make T the subject.

$pV = nRT$

$T = \frac{pV}{nR}$ Divide both sides by nR

Exam Tip
Once you've rearranged your formula, all you need to do is to pop the values from the question into the right places — making sure that they're in the right units first, of course.

Formula triangles

Formula triangles are useful tools for changing the subject of a formula.

If three things are related by a formula like this: $a = b \times c$ or like this: $b = \frac{a}{c}$

then you can put them into a formula triangle:

Example — Maths Skills

If you have this formula:

number of moles (n) = mass (m) ÷ M_r

then you can turn it into a formula triangle like this:

- As m is divided by M_r, m goes on top, leaving $n \times M_r$ on the bottom:

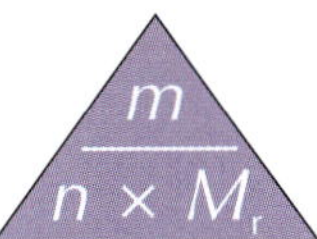

- If you wanted to find the number of moles (n) you would just cover up n, which leaves $m \div M_r$.
- So, number of moles = mass ÷ M_r.

Tip: You don't have to use formula triangles to rearrange formulas. If you're happy rearranging formulas without them, that's fine.

Symbols

You'll need to know these symbols which might be used in equations in the exam. You'll have seen many of them before but here's a quick refresher:

Symbol	Meaning
$=$	equal to
$<$	less than
$<<$	much less than
$>$	greater than
$>>$	much greater than
$\propto$	proportional to
$\sim$	approximately
$\rightleftharpoons$	reversible

3. Units

Units aren't the most exciting bit of chemistry but you need to be able to use them. Here's how to convert between units and work them out from scratch.

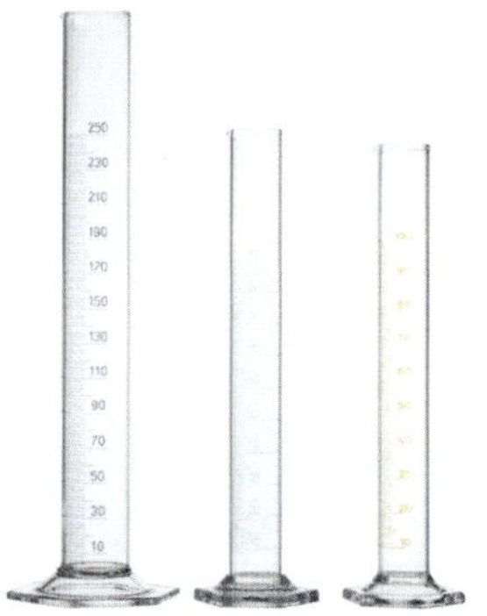

Figure 1: *Measuring cylinders like these measure volumes in cm³.*

Converting between units

Volume

Volume can be measured in m^3, dm^3 and cm^3.

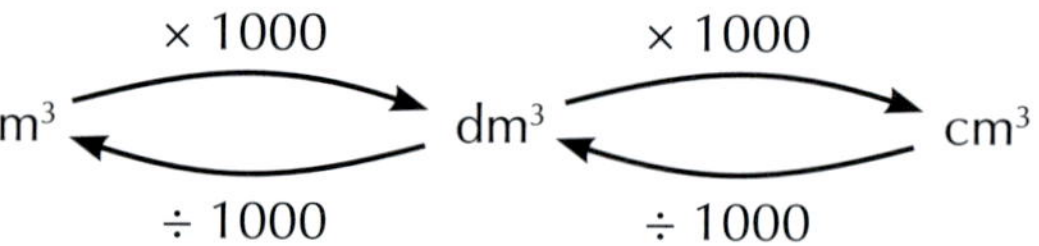

Example — Maths Skills

Write 0.3 cm³ in dm³ and m³.

First, to convert 0.3 cm³ into dm³ you need to divide by 1000.

$$0.3 \div 1000 = 0.0003\ dm^3 = 3 \times 10^{-4}\ dm^3$$

Then, to convert 0.0003 dm³ into m³ you need to divide by 1000.

$$0.0003 \div 1000 = 0.0000003\ m^3 = 3 \times 10^{-7}\ m^3$$

Tip: Standard form (that's showing numbers as, for example, 3×10^{-7}) is covered on pages 519-520.

Exam Tip
Make sure you practise these conversions. It could save you valuable time in the exam if you can change between units confidently.

Temperature

Temperature can be measured in K and °C.

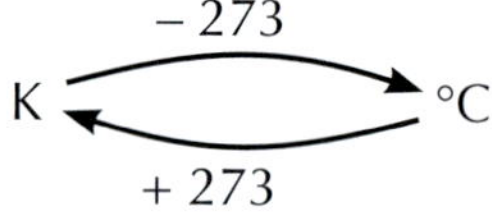

Example — Maths Skills

Write 25 °C in kelvins.

To convert 25 °C into K you need to add 273: $25 + 273 = 298$ K

Tip: A kPa is bigger than a Pa, so you'd expect the number to get smaller when you convert from Pa to kPa — each unit is worth more so you'll have fewer of them.

Pressure

Pressure can be measured in Pa and kPa.

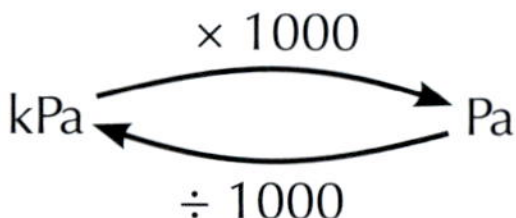

Example — Maths Skills

Write 3200 Pa in kPa.

To convert 3200 Pa into kPa you need to divide by 1000.

$$3200 \div 1000 = 3.2\ kPa$$

Mass

Mass can be measured in kg and g.

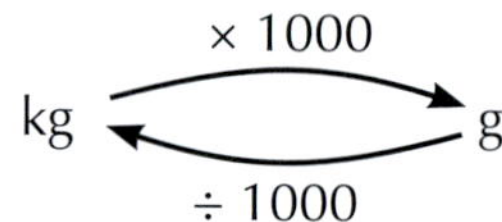

Example — Maths Skills

Write 5.2 kg in g.

To convert 5.2 kg into g you need to multiply by 1000.

$$5.2 \times 1000 = 5200\ g$$

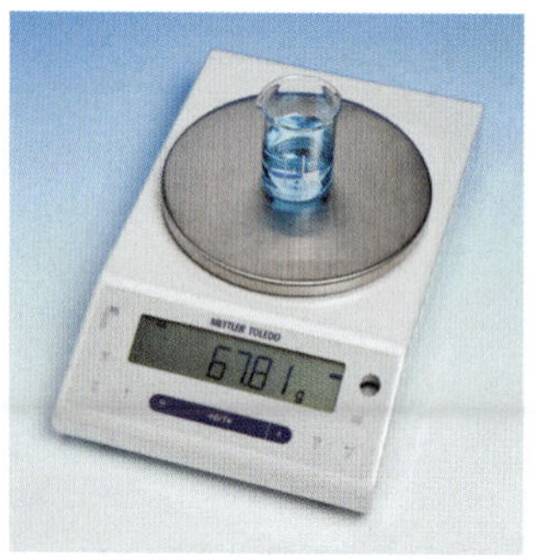

Figure 2: *This balance measures mass in g.*

Energy

Energy can be measured in kJ and J.

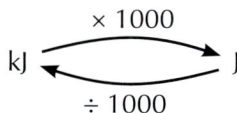

$$\text{kJ} \underset{\div 1000}{\overset{\times 1000}{\rightleftarrows}} \text{J}$$

Example — Maths Skills

Write 78 kJ in J.

To convert 78 kJ into J you need to multiply by 1000.

$$78 \times 1000 = 78\,000 \text{ J} = 7.8 \times 10^4 \text{ J}$$

Life gets a bit confusing if you have to do lots of calculations one after the other — sometimes it can be difficult to keep track of your units. To avoid this, write down the units you're using with each line of the calculation. Then when you get to the end you know what units to give with your answer.

Tip: If you're struggling with a unit conversion, try thinking about an example for that unit that you're certain of — e.g. if you know that 1 kJ is 1000 J, you know that to get from kJ to J you must have to multiply by 1000. Simple.

Figure 3: *When you're under pressure in an exam it's easy to make mistakes, so don't be afraid to use your calculator, even for simple calculations.*

Calculating units

In chemistry, some quantities, like the equilibrium constant (K_c) and the rate constant (k), have variable units. This means you'll need to work the units out — you can't just learn them. To work out the units, you just follow these steps:

- Substitute the units that you know into the equation you're using.
- Cancel out units wherever possible — if the same unit is present on the top and the bottom of a fraction, you can cancel them out.
- Get rid of any fractions by using the rule $\frac{1}{a^n} = a^{-n}$ (see page 528) — any positive powers on the bottom of the fraction become negative and any negative powers become positive.

Example — Maths Skills

The rate equation for the reaction $CH_3COCH_3 + I_2 \rightarrow CH_3COCH_2I + H^+ + I^-$ is Rate = $k[CH_3COCH_3][H^+]$. The rate of reaction is in mol dm^{-3} s^{-1} and the concentrations are in mol dm^{-3}. Find the units of k.

$$\text{Rate} = k[CH_3COCH_3][H^+] \quad \text{so} \quad k = \frac{\text{Rate}}{[CH_3COCH_3][H^+]}$$

First substitute in the units you know:

$$\text{units of } k = \frac{\text{mol dm}^{-3}\text{s}^{-1}}{(\text{mol dm}^{-3})(\text{mol dm}^{-3})}$$

Cancel out units where you can. In this case you can cancel a mol dm^{-3} from the top and the bottom of the fraction:

$$\text{units of } k = \frac{\cancel{\text{mol dm}^{-3}}\text{s}^{-1}}{(\cancel{\text{mol dm}^{-3}})(\text{mol dm}^{-3})} = \frac{\text{s}^{-1}}{\text{mol dm}^{-3}}$$

Then get rid of the fraction by using the rule $\frac{1}{a^n} = a^{-n}$:

$$\text{units of } k = \text{s}^{-1}\text{ mol}^{-1}\text{ dm}^{3}$$

Exam Tip
Always, always, give units with your answers. It's really important that the examiner knows what units you're working in — 10 g is very different from 10 kg. (Unless, of course, it's a quantity without units, for example pH.)

Tip: If you have more than one of a particular unit multiplied together, you add the powers together (see p.528), e.g. (mol dm^{-3})(mol dm^{-3}) is (mol^2 dm^{-6}).

Tip: Writing mol is the same as writing mol^1, so 1/mol = mol^{-1}.

4. Powers and Logarithms

A power is like an instruction that tells you how many times to multiply a number by itself. By contrast, a logarithm tells you how many times a number has been multiplied by itself in order to get to another number.

Powers

When you see a term with a small number beside it, the small number is telling you what power the term has been raised to:

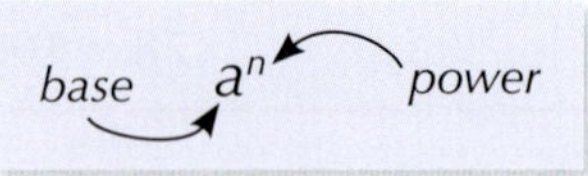

Powers can be positive or negative numbers.

A positive power tells you how many times to multiply the term by itself.

Examples — Maths Skills

$2^5 = 2 \times 2 \times 2 \times 2 \times 2 = \mathbf{32}$

$10^3 = 10 \times 10 \times 10 = \mathbf{1000}$

This is how to calculate a negative power: $a^{-n} = \frac{1}{a^n}$

Examples — Maths Skills

$4^{-2} = \frac{1}{4^2} = \frac{1}{4 \times 4} = \mathbf{0.0625}$

$5^{-3} = \frac{1}{5^3} = \frac{1}{5 \times 5 \times 5} = \mathbf{0.008}$

There are some rules for using powers in calculations that are useful to know:

- If you raise a number to the power 1, the result is just that number:
  $$\boldsymbol{a^1 = a}$$
- If you raise a number to the power 0, the result is always 1:
  $$\boldsymbol{a^0 = 1}$$
- If you multiply a number raised to a power by the same number raised to a power, the result is the same as the number raised to the sum of the two powers: $\boldsymbol{a^n \times a^m = a^{(n+m)}}$
- If you divide a number raised to a power by the same number raised to a power, the result is the same as the number raised to the value of the first power minus the value of the second power:
  $$\boldsymbol{a^n \div a^m = a^{(n-m)}}$$
- A number raised to a power in the denominator of a fraction is the same as that number raised to the negative of the power and then multiplied by the numerator: $\boldsymbol{\frac{1}{a^n} = 1 \times a^{-n} = a^{-n}}$

Examples — Maths Skills

$7^1 = \mathbf{7}$

$11^0 = \mathbf{1}$

$3^2 \times 3^4 = 3^6 = \mathbf{729}$

$4^7 \div 4^5 = 4^2 = \mathbf{16}$

$\frac{1}{\text{mol dm}^{-3}} = \textbf{mol}^{-1}\,\textbf{dm}^{3}$

The number 'e'

The number 'e' crops up occasionally in chemistry. Like 'π', it is a fixed, irrational number with an infinite number of decimal places.

$$e = 2.71828182845904523536\ldots$$

Tip: The small superscript number is sometimes called an exponent or an index.

Tip: Powers are used a lot in rate equations. Have a look at page 294 to see them in action.

Tip: These rules all apply to powers you find in units as well as in numerical equations.

Tip: An irrational number is one that you <u>can't</u> write as a fraction. Irrational numbers have an infinite number of non-recurring decimal places (they go on forever and never repeat).

In most of the calculations involving e that you will come across, it is raised to a power, so it looks like this: e^x. The function e^x is known as the 'exponential function'. To stop you having to type in a long string of numbers every time you carry out calculations involving exponentials, there should be a button on your calculator that looks a bit like this: $e^{■}$.

> **Tip:** The power rules on the previous page all apply to calculations involving e raised to a power.

Examples — Maths Skills

$e^9 = \mathbf{8103.1}$ $e^{-4} = \mathbf{0.01832}$ $e^0 = \mathbf{1}$

Logarithms

Logarithms tell you how many times a number (the **base**) has been multiplied by itself to get to another number. They are the opposite of powers.

$$\text{If } a^n = x \quad \text{then} \quad \log_a x = n$$

It's important to specify the base of your logarithm, but some numbers are so common they have their own symbols:

- You often want to know how many times the number 10 has been multiplied by itself. Sometimes $\log_{10} x$ is just written as $\log x$.
- The **natural logarithm** is the logarithm with a base of e. You write $\log_e x$ as $\ln x$.

These two functions are used so often that you'll probably have a button for each of them on your calculator — handy.

> **Tip:** Your calculator may well have buttons that look like: $\log_{■}\square$, log and ln. The first button lets you specify your base. The second is $\log_{10}$ and the third is $\log_e$.

> **Tip:** Logarithms tell you the order of magnitude a number is in, so they're really useful for comparing numbers that fall across a very large range.

Examples — Maths Skills

$\log_2 16 = \mathbf{4}$ $\log 100 = \mathbf{2}$ $\ln 14 = \mathbf{2.6}$

Just like for powers, there are some rules for working with logarithms in calculations that are useful to know:

- The logarithm of two things multiplied together is the same as the sum of their individual logarithms:

$$\mathbf{\log_a (xy) = \log_a x + \log_a y}$$

- The logarithm of the value of the base raised to a power is the value of the power:

$$\mathbf{\log_a a^x = x}$$

> **Tip:** The pH scale is a logarithmic scale in base 10. Have a look at page 343 to see more about it.

Example — Maths Skills

You can use these rules to rearrange the Arrhenius equation into its logarithmic form. Here's the ordinary version: $k = Ae^{\frac{-E_a}{RT}}$

Taking the natural logarithm of both sides gives you:

$$\ln (k) = \ln (Ae^{\frac{-E_a}{RT}})$$

Using the first rule, you know that $\ln (xy) = \ln x + \ln y$, so:

$$\ln k = \ln (A) + \ln(e^{\frac{-E_a}{RT}})$$

Using the second rule, you know that $\ln e^x = \log_e (e^x) = x$, so:

$$\ln k = \ln A + \frac{-E_a}{RT}$$

This is the logarithmic form of the Arrhenius equation.

> **Tip:** To make your calculations clear, always put brackets around the thing you're taking the log of. For example, $\log (a + b)$ and $\log(a) + b$ both clearly show what you're doing in a calculation. If you just write $\log a + b$, it's not obvious what's being done.

> **Tip:** See pages 308-311 for lots more about the Arrhenius equation.

5. Graphs

Being able to interpret graphs can be a really important skill in Chemistry, so here's a recap of some of the trickier bits.

Straight line graphs

All straight line graphs have the general equation:

$$y = mx + c$$

dependent variable (y), independent variable (x), gradient (m), y-intercept (c)

If you have an equation in this form and you plot a graph of the dependent variable against the independent variable, then the gradient of the graph will be equal to m, and the y-intercept will be equal to c.

Tip: Straight line graphs you're likely to come across include rate-concentration graphs for first order reactions (page 304), Arrhenius plots (page 310) and free-energy graphs (page 279).

Example — Maths Skills

The logarithmic form of the Arrhenius equation is: $\ln k = \frac{-E_a}{RT} + \ln A$

This is an equation in the form $y = mx + c$ where:

$\ln k = y$, $\frac{-E_a}{R} = m$, $\frac{1}{T} = x$, $\ln A = c$

So if you plot a graph of $\ln k$ against $\frac{1}{T}$, it will have a gradient equal to $\frac{-E_a}{R}$ and a y-intercept of $\ln A$.

Finding the gradient of a straight line graph

The gradient of a graph tells you how much the y value changes as you change the x value — it's a measure of the steepness of your graph.

gradient = change in y ÷ change in x

To measure the gradient of a straight line, start by picking two points on the line that are easy to read. Draw a vertical line down from one point and a horizontal line across from the other to make a triangle. The length of the vertical side of the triangle is the change in y and the length of the horizontal side is the change in x.

Tip: You might also see the gradient of a graph referred to as its 'slope'.

Tip: The change in y can be written as Δy, and the change in x can be written as Δx.

Tip: A line that slopes downwards from left to right has a negative gradient, because y decreases as x increases.

Example — Maths Skills

First order reactions have the rate equation: Rate = k[A].

This is in the form $y = mx + c$, where Rate = y, $k = m$, [A] = x and $0 = c$.

So a graph of rate against [A] will have a gradient equal to k.

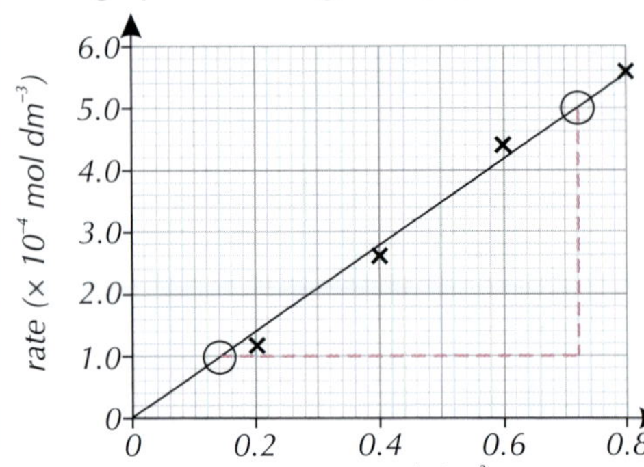

$\Delta x = 0.72 - 0.14 = 0.58$

$\Delta y = (5.0 \times 10^{-4}) - (1.0 \times 10^{-4}) = 4.0 \times 10^{-4}$

Gradient $= (4.0 \times 10^{-4}) \div 0.58 = 6.9 \times 10^{-4}$

$k =$ **6.9×10^{-4} s^{-1}**

Exam Tip
Always check the labels of the axes. Here, the y axis units are 10^{-4} mol dm^{-3}, so each of the values on the y axis is multiplied by 10^{-4} to find its true value.

Finding the *y*-intercept of a graph

The *y*-intercept of a graph is the point at which the line of best fit crosses the *y* axis. You may have to extrapolate your line of best fit beyond your data points to find the *y*-intercept.

> **Tip:** Finding a value by extending your line of best fit beyond your data points is called extrapolation. Finding a value between two of your data points using a line of best fit is called interpolation.

Example — Maths Skills

From the logarithmic form of the Arrhenius equation, $\ln k = \frac{-E_a}{RT} + \ln A$, you know that the y-intercept of a graph of ln *k* against $\frac{1}{T}$ is equal to ln *A*.

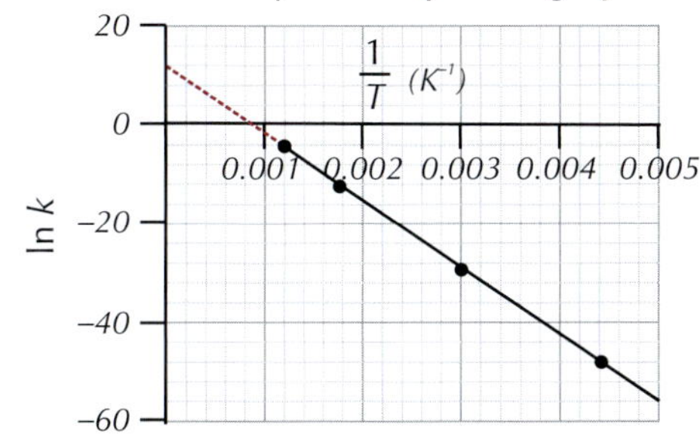

By extrapolating the line of best fit for this graph to the point where x = 0, you can see that *y* = 12.

So ln *A* = **12**

> **Tip:** You can't find the *y*-intercept from a graph if the axes have been contracted. Both axes need to run continuously from zero for it to work.

Finding the gradient of a curve

The gradient is different at different points along a curve. To find the gradient at a particular point, you first have to draw a tangent to the curve at that point, and then find the gradient of that.

Example — Maths Skills

The gradient of a graph of concentration against time at any point is equal to the rate of the reaction at that time.

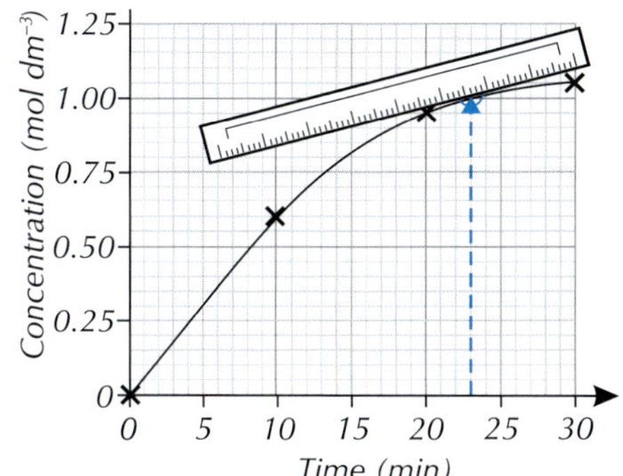

For example, if you want to find the rate after 23 minutes, place a ruler at the point on the curve at 23 minutes so it's just touching the line. Position the ruler so that you can see the whole curve.

Adjust the ruler until the space between the ruler and the curve is equal on both sides of the point.

> **Tip:** There's more about finding gradients on pages 291-292.

Now draw a line along the ruler to make the tangent.

Extend the line right across the graph — it'll help to make your gradient calculation easier as you'll have more points to choose from.

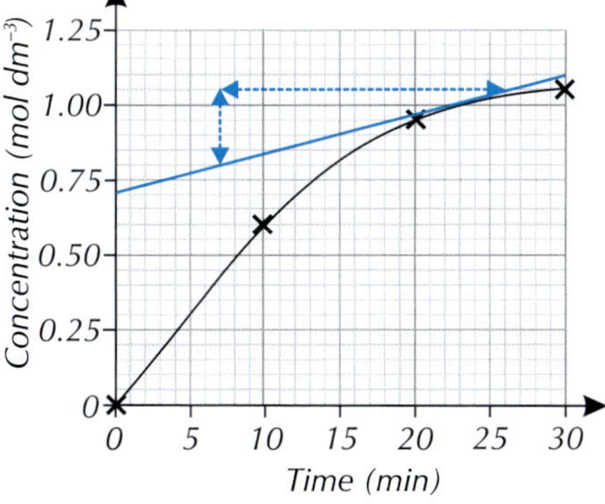

Finally, calculate the gradient of the tangent to find the rate:

- gradient = change in *y* ÷ change in *x*
  = (1.05 – 0.80) ÷ (26.0 – 7.0) = 0.013... $\text{mol dm}^{-3}\,\text{min}^{-1}$
- So, the rate of reaction at 23 mins was **0.013 $\text{mol dm}^{-3}\,\text{min}^{-1}$**.

1. Exam Structure and Technique

Passing exams isn't all about revision — it really helps if you know how the exam is structured and have got your exam technique nailed so that you pick up every mark you can.

***Figure 1:** The Room of Doom awaits you. But don't panic — prepare properly and there's no reason you can't ace the exam.*

Course structure

AQA A-Level Chemistry is split into three parts:

Physical chemistry — Units 1 and 4 of this book

Inorganic chemistry — Units 2 and 5 of this book

Organic chemistry — Units 3 and 6 of this book

Exam structure

For AQA A-Level Chemistry you're going to have to sit three exams — Paper 1, Paper 2 and Paper 3.

- Paper 1 will test you on most of Physical chemistry and all of Inorganic chemistry.
- Paper 2 will test you on some of Physical chemistry and all of Organic chemistry.
- Paper 3 can test you on any of the material covered in the course.

All three papers could include questions on practical skills. If you want more detail on what could come up in each paper, look at the table below:

Paper	Content Assessed
1	▪ Physical chemistry Specification references 3.1.1.1 to 3.1.4.4 and 3.1.6.1 to 3.1.7 (Unit 1 in this book). Specification references 3.1.8.1 to 3.1.8.2 and 3.1.10 to 3.1.12.6 (Unit 4 in this book). ▪ Inorganic chemistry Specification references 3.2.1.1 to 3.2.3.2 (Unit 2 in this book). Specification references 3.2.4 to 3.2.6 (Unit 5 in this book). ▪ Relevant practical skills (see next page).
2	▪ Physical chemistry Specification references 3.1.2.1 to 3.1.6.2 (Unit 1 in this book). Specification references 3.1.9.1 to 3.1.9.2 (Unit 4 in this book). ▪ Organic chemistry Specification references 3.3.1.1 to 3.3.6.3 (Unit 3 in this book). Specification references 3.3.7 to 3.3.16 (Unit 6 in this book). ▪ Relevant practical skills (see next page).
3	▪ Any content from across the course — all Units in this book. ▪ Relevant practical skills (see next page).

Exam Tip
Make sure you have a good read through of this exam structure. It might not seem important now but you don't want to get any nasty surprises just before an exam.

Exam Tip
Short answer questions are broken down into parts, but they can still be worth lots of marks overall.

All the exams are 2 hours long. Papers 1 and 2 are both worth 105 marks, and are each worth 35% of your total mark. Paper 3 is worth 90 marks and is worth 30% of your total mark.

Paper 1 and Paper 2 are made up of short and long answer questions. Paper 3 will be made up of 40 marks on questions relating to practical techniques and data analysis, 20 marks on general questions from all parts of the specification, and 30 marks' worth of multiple choice questions.

Exam Tip
Make sure you know which units and sections will be assessed in each exam. That way, you can make sure that you're properly prepared when you actually come to sit the papers.

Relevant practical skills

At least 15% of the marks on your A-Level Chemistry exams will focus on practical skills. This means you will be given questions where you're asked to do things like comment on the design of experiments, make predictions, draw graphs, calculate means and percentage uncertainties — basically, anything related to planning experiments or analysing results. Handily, the Practical Skills section of this book covers these skills, so make sure you've read through it.

Quality of written communication

Your A-Level Chemistry exam papers will have some extended response questions — in these questions you'll be tested on your ability to write clearly as well as on your chemistry knowledge. So for extended response questions, you need to make doubly sure that:

- your writing is legible.
- your spelling, punctuation and grammar are accurate.
- you organise your answer clearly and coherently — make sure everything you say is relevant and is in a logical order — it should be clear how your points are linked.
- you use specialist scientific vocabulary where it's appropriate.

Tip: Don't worry too much about extended response questions — they sound a bit scary, but as long as you write full and coherent answers whenever you tackle a long answer question, you'll be fine.

Time management

This is one of the most important exam skills to have. How long you spend on each question is really important in an exam — it could make all the difference to your grade. Some questions will require lots of work for only a few marks but other questions will be much quicker. Don't spend ages struggling with questions that are only worth a couple of marks — move on. You can come back to them later when you've bagged loads of other marks elsewhere.

Example

The questions below are both worth the same number of marks but require different amounts of work.

1.1 Define the term 'lattice enthalpy'.

(2 marks)

2.1 Draw the structures of the two monomers that react to form the condensation polymer shown above.

(2 marks)

Question 1.1 only requires you to write down a definition — if you can remember it this shouldn't take you too long.

Question 2.1 requires you to work out and draw the structures of two monomers used to make a condensation polymer — this may take longer than writing down a definition.

So, if you're running out of time it makes sense to do questions like 1.1 first and come back to 2.1 if you've got time at the end.

Exam Tip
Don't forget to go back and do any questions that you left the first time round — you don't want to miss out on marks because you forgot to do the question.

Exam Tip
It's a really good idea to do as many practice papers as you can before you actually sit the exams — that way you'll get a good feel for how to divide up your time. You could ask your teacher if they have any practice papers you can do, or look on the AQA website to see if there are any specimen papers or past papers for you to download.

Exam Tip
Don't leave any questions blank if you can avoid it. If you're getting short on time towards the end of the exam it only takes a moment to answer a multiple choice question (even if it is just an educated guess).

It's worth keeping in mind that the multiple choice questions are only worth 1 mark each and some of them can be quite time-consuming to answer. If you're pressed for time, it's usually a good idea to focus on the written answer questions first, where there are more marks available, and then go back to the harder multiple choice questions later.

Command words

Command words are just the bit of the question that tell you what to do. You'll find answering exam questions much easier if you understand exactly what they mean, so here's a summary of the most common command words.

Command word	What to do
Give / State	Write a concise answer, from fact recall or from information that you've been given in the question.
Name	Give the correct technical term for a chemical or a process.
Identify	Say what something is.
Describe	Write about what something is like or how it happens.
Explain	Give reasons for.
Suggest / Predict	Use your scientific knowledge to work out what the answer might be.
Outline	Give a brief description of the main characteristics of a process or an issue.
Calculate	Work out the solution to a mathematical problem.
Estimate	Give an approximate answer.
Determine	Use the information given in the question to work something out.
Compare	Give the similarities and differences between two things.
Draw	Produce a diagram or graph.
Sketch	Draw something approximately — for example, draw a rough line graph to show the main trend of some data.
Justify	Give the case for or against, supported by evidence.

Not all of the questions will have a command word like this — instead they may just be a which / what / how type of question.

Exam Tip
Some questions might contain more than one command word. For example, a question could start 'state and explain'. In this case, you should treat the question a little bit like two questions rolled into one — start by stating what the questions asks for and then explain it.

Exam Tip
Some questions will also ask you to answer 'using the information (or data) provided' (e.g. a graph, table, equation, etc.). If so, make sure that you do use it in your answer, or you may not get all the marks.

Exam data booklet

When you sit your exams, you'll be given a data booklet with the exam paper. In it you'll find some useful information to help you, including...

- the characteristic infrared absorptions, ^{13}C NMR shifts and ^{1}H NMR shifts of some common functional groups.
- the structures of some biologically important molecules, such as the DNA bases, some amino acids, common phosphates and sugars and Heme B.
- a copy of the periodic table.

You might have seen a few slightly different versions of the periodic table — in the exams you should use the information from the one in the data booklet, even if it's slightly different to something you've seen elsewhere. The examiners will use the information in the data booklet when they're marking.

Exam Tip
It's a good idea to have a look at the data booklet before the exams — you should be able to find it on the AQA website.

2. Diagrams

When you're asked to draw diagrams or mechanisms in an exam it's important that you draw everything correctly and include all the details that are needed.

Organic reaction mechanisms

Organic reaction mechanisms are used to show what happens during a chemical reaction. One of the most common mistakes people make when drawing these is to get the curly arrows wrong.

Example

When you're drawing organic reaction mechanisms the curly arrows must come from either a lone pair of electrons or from a bond, like this:

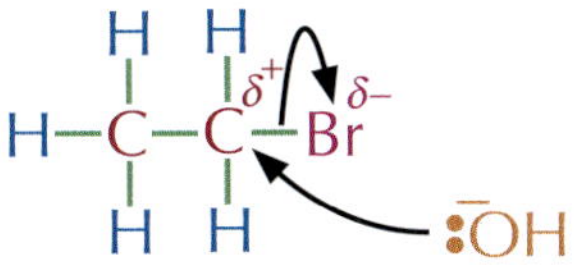

The examples below are incorrect — you wouldn't get marks for them:

You won't get marks if the curly arrows come from atoms, like this...

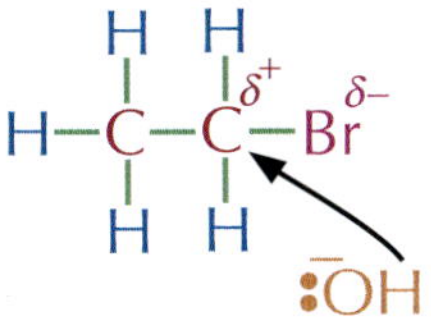

...or if they come from a charge or from thin air like these.

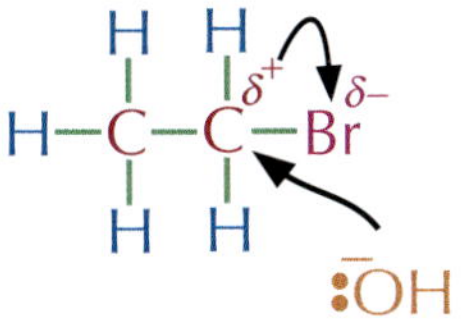

Tip: It's important that the curly arrows come from a lone pair or a bond because curly arrows show the movement of electrons and that's where the electrons are found.

Exam Tip

Make sure you draw full charges (+ and –) or dipoles (partial charges shown using δ+ and δ–) clearly — you could lose marks if it's not clear which type of charge you mean.

Displayed and skeletal formulas

Displayed formulas show how all the atoms are connected in a molecule.

Examples

If a question asks you for a displayed formula you have to show all of the bonds and all of the atoms in the molecule. That means you have to draw displayed formulas like this:

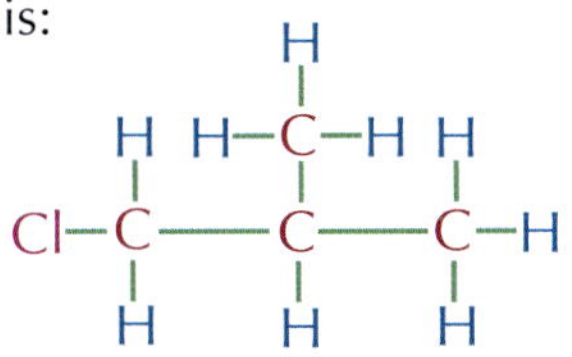

And not like this:

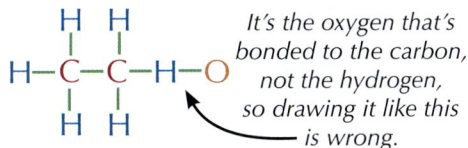

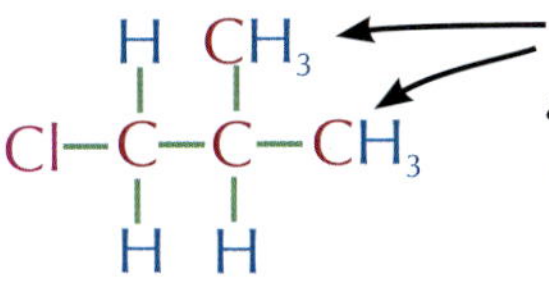

Some of the bonds between the carbon atoms and the hydrogen atoms haven't been shown, so it's not a displayed formula and you wouldn't get the marks.

If you're not asked specifically for a displayed formula then either of the diagrams above will do. Just make sure that the bonds are always drawn between the right atoms. For example, ethanol should be drawn like this:

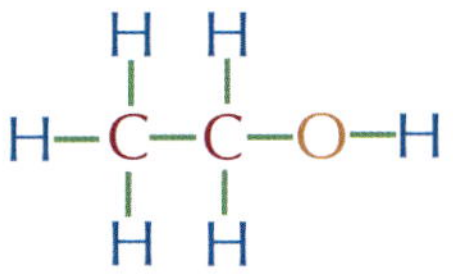

And not like this:

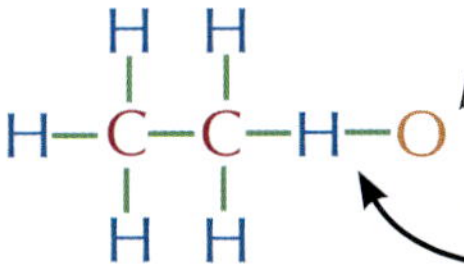

It's the oxygen that's bonded to the carbon, not the hydrogen, so drawing it like this is wrong.

Skeletal formulas are handy when you're drawing large organic molecules. You might have to either draw or interpret skeletal formulas in your exam, so make sure that you're happy drawing and using them. Remember — bonds between carbon atoms are shown by a line and carbon atoms are found at each end. Atoms that aren't carbon or hydrogen have to be drawn on:

Tip: If you're not totally comfortable drawing skeletal formulas, it might help to draw the displayed formula first. That will help you to see what atoms need to be drawn in the skeletal formula.

Example

1,5-difluoropentane ($FCH_2CH_2CH_2CH_2CH_2F$)

The carbon-carbon bonds are shown by lines.

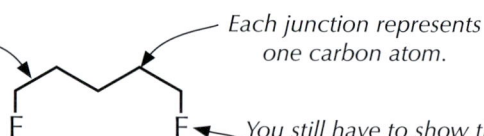

Each junction represents one carbon atom.

F F

You still have to show the atoms that aren't carbon or hydrogen.

You don't draw any carbon or hydrogen atoms from the main carbon chain when you're drawing a skeletal formula, so both of the diagrams below are wrong.

C C C F F

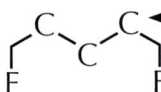

You don't show the carbon atoms or the hydrogen atoms.

H F F

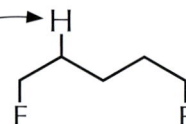

Hydrogen bonds

Drawing a diagram showing hydrogen bonding is a fairly common exam question. You need to know how to draw hydrogen bonds properly to pick up all the marks you can.

Exam Tip
Make sure you include everything the question asks for in your answer. For example, if the question asks you to show lone pairs and dipoles in a diagram, don't forget to add them.

Example

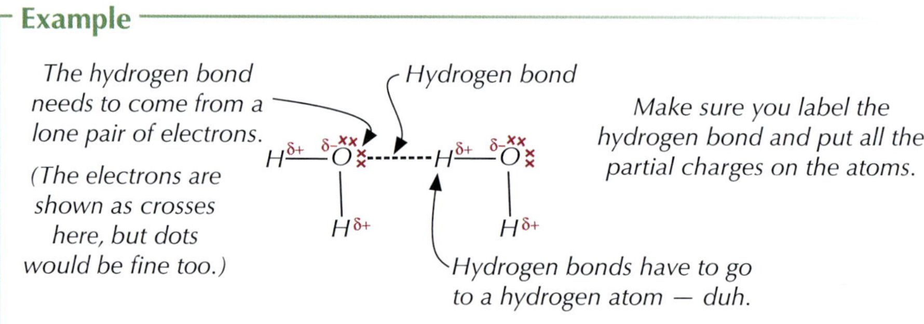
The hydrogen bond needs to come from a lone pair of electrons.
(The electrons are shown as crosses here, but dots would be fine too.)
Hydrogen bond
Make sure you label the hydrogen bond and put all the partial charges on the atoms.
Hydrogen bonds have to go to a hydrogen atom — duh.

General advice on diagrams

These pages cover some of the types of diagram that are likely to come up in your exams. But you could be asked to draw other diagrams. Whatever diagram you're drawing, make sure it's really clear. A small scribble in the bottom corner of a page isn't going to show enough detail to get you the marks. Draw the diagrams nice and big, but make sure that you stay within the space given for that answer. It may also help to add clear, concise labels to your diagram, particularly for things like drawings of the apparatus needed for an experiment.

If you've drawn a diagram incorrectly don't scribble part of it out and try to fix it — it'll look messy and be hard for the examiner to figure out what you're trying to show. Cross the whole thing out and start again. And always double check that you've included all the things that you should have done.

3. The Periodic Table — Facts and Trends

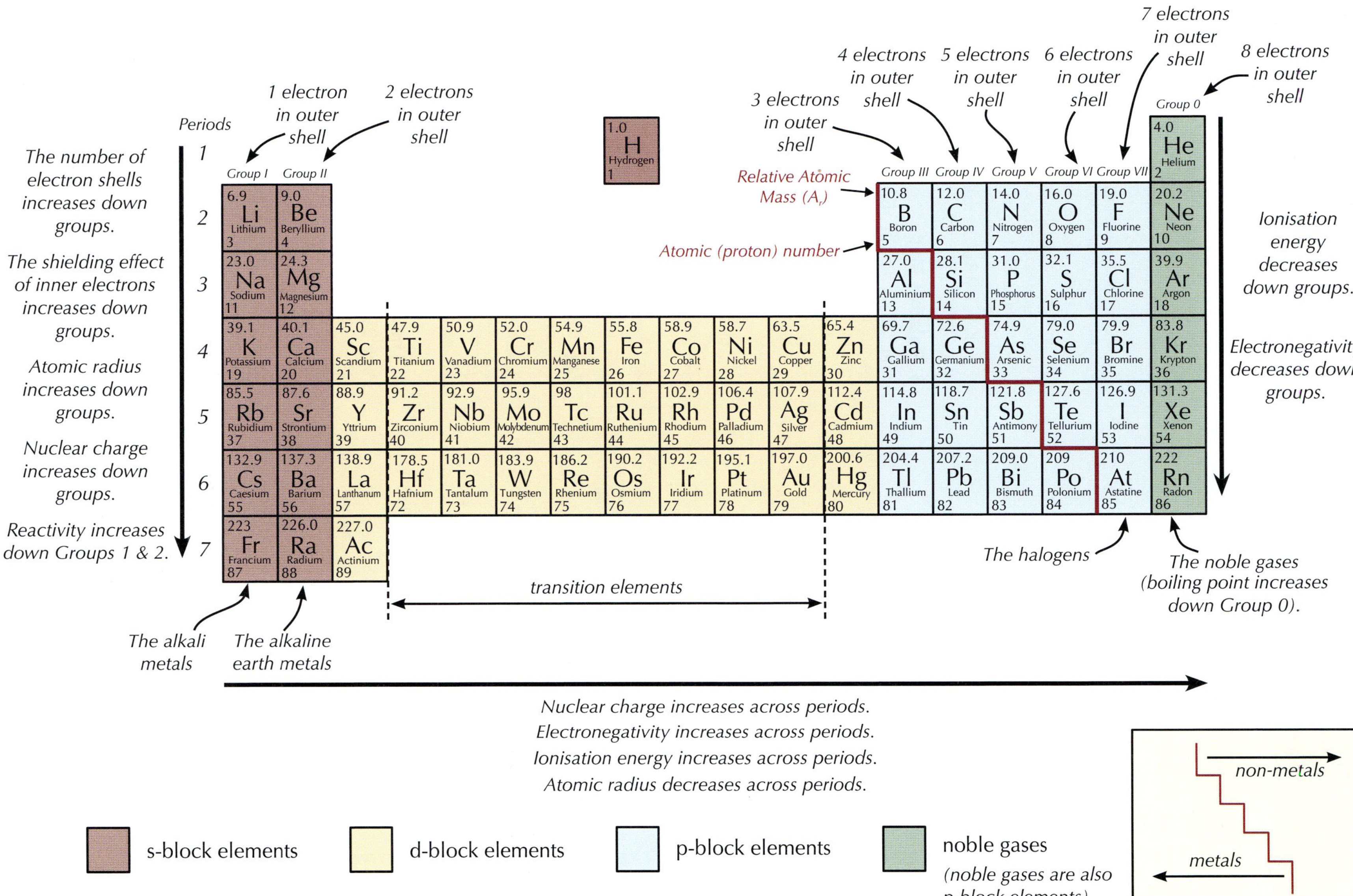

Periods
1
2
3
4
5
6
7
1 electron in outer shell
2 electrons in outer shell
3 electrons in outer shell
4 electrons in outer shell
5 electrons in outer shell
6 electrons in outer shell
7 electrons in outer shell
8 electrons in outer shell
Group I
Group II
Group III
Group IV
Group V
Group VI
Group VII
Group 0
Relative Atomic Mass (A_r)
Atomic (proton) number
1.0 H Hydrogen 1
4.0 He Helium 2
6.9 Li Lithium 3
9.0 Be Beryllium 4
10.8 B Boron 5
12.0 C Carbon 6
14.0 N Nitrogen 7
16.0 O Oxygen 8
19.0 F Fluorine 9
20.2 Ne Neon 10
23.0 Na Sodium 11
24.3 Mg Magnesium 12
27.0 Al Aluminium 13
28.1 Si Silicon 14
31.0 P Phosphorus 15
32.1 S Sulphur 16
35.5 Cl Chlorine 17
39.9 Ar Argon 18
39.1 K Potassium 19
40.1 Ca Calcium 20
45.0 Sc Scandium 21
47.9 Ti Titanium 22
50.9 V Vanadium 23
52.0 Cr Chromium 24
54.9 Mn Manganese 25
55.8 Fe Iron 26
58.9 Co Cobalt 27
58.7 Ni Nickel 28
63.5 Cu Copper 29
65.4 Zn Zinc 30
69.7 Ga Gallium 31
72.6 Ge Germanium 32
74.9 As Arsenic 33
79.0 Se Selenium 34
79.9 Br Bromine 35
83.8 Kr Krypton 36
85.5 Rb Rubidium 37
87.6 Sr Strontium 38
88.9 Y Yttrium 39
91.2 Zr Zirconium 40
92.9 Nb Niobium 41
95.9 Mo Molybdenum 42
98 Tc Technetium 43
101.1 Ru Ruthenium 44
102.9 Rh Rhodium 45
106.4 Pd Palladium 46
107.9 Ag Silver 47
112.4 Cd Cadmium 48
114.8 In Indium 49
118.7 Sn Tin 50
121.8 Sb Antimony 51
127.6 Te Tellurium 52
126.9 I Iodine 53
131.3 Xe Xenon 54
132.9 Cs Caesium 55
137.3 Ba Barium 56
138.9 La Lanthanum 57
178.5 Hf Hafnium 72
181.0 Ta Tantalum 73
183.9 W Tungsten 74
186.2 Re Rhenium 75
190.2 Os Osmium 76
192.2 Ir Iridium 77
195.1 Pt Platinum 78
197.0 Au Gold 79
200.6 Hg Mercury 80
204.4 Tl Thallium 81
207.2 Pb Lead 82
209.0 Bi Bismuth 83
209 Po Polonium 84
210 At Astatine 85
222 Rn Radon 86
223 Fr Francium 87
226.0 Ra Radium 88
227.0 Ac Actinium 89
The number of electron shells increases down groups.
The shielding effect of inner electrons increases down groups.
Atomic radius increases down groups.
Nuclear charge increases down groups.
Reactivity increases down Groups 1 & 2.
Ionisation energy decreases down groups.
Electronegativity decreases down groups.
The alkali metals
The alkaline earth metals
transition elements
The halogens
The noble gases (boiling point increases down Group 0).
Nuclear charge increases across periods.
Electronegativity increases across periods.
Ionisation energy increases across periods.
Atomic radius decreases across periods.
s-block elements
d-block elements
p-block elements
noble gases
(noble gases are also p-block elements)
non-metals
metals

Answers

Unit 1

Section 1: Atomic Structure

1. The Atom

Page 22 — Application Questions

Q1 a) 13
b) 13
c) 27 – 13 = **14**

Q2 a) 19
b) 19 + 20 = **39**
c) $^{39}_{19}\mathrm{K}$
d) 19 – 1 = **18**

Q3 a) 20 – 2 = **18**
b) 40 – 20 = **20**

Q4 Isotope X is an isotope of niobium. Its nuclear symbol is:
$^{93}_{41}\mathrm{Nb}$
Looking at the periodic table, niobium's atomic number is 41. Isotope X has 41 protons, so it must be an isotope of niobium. Then, mass number = protons + neutrons = 41 + 52 = 93.

Q5 a) A and C both have 10 electrons.
b) A and D both have 8 protons.
c) B and C both have 10 neutrons.
(17 – 7 = 10 and 20 – 10 = 10)
d) B and D both have 10 neutrons.
(17 – 7 = 10 and 18 – 8 = 10)
e) A and D are isotopes of each other because they have the same number of protons (8) but different numbers of neutrons.
(A has 16 – 8 = 8 neutrons and D has 18 – 8 = 10 neutrons.)

Page 22 — Fact Recall Questions

Q1 proton, neutron, electron
Q2 Protons and neutrons are found in the nucleus. Electrons are found in orbitals around the nucleus.
Q3 proton: 1, neutron: 1, electron: 1/2000
Q4 The total number of protons and neutrons in the nucleus of an atom.
Q5 The number of protons in the nucleus of an atom.
Q6 By subtracting the atomic number from the mass number.
Q7 Atoms with the same number of protons but different numbers of neutrons.
Q8 The chemical properties of an element are decided by its electron configuration. Isotopes have the same configuration of electrons, so have the same chemical properties.
Q9 Physical properties depend on the mass of an atom. Isotopes have different masses, so can have different physical properties.

2. Atomic Models

Page 24 — Fact Recall Questions

Q1 Dalton described atoms as solid spheres. J. J. Thomson suggested that atoms were not solid spheres — he thought they contained small negatively charged particles (electrons) in a positively charged "pudding".
Q2 If Thomson's model was correct the alpha particles fired at the sheet of gold should have been deflected very slightly by the positive "pudding" that made up most of the atom. Instead, most of the alpha particles passed straight through the gold atoms, and a very small number were deflected backwards. So the plum pudding model couldn't be right.
Q3 Rutherford's model has a tiny positively charged nucleus at the centre surrounded by a "cloud" of negative electrons. Most of the atom is empty space.
Q4 In Bohr's model the electrons only exist in fixed shells and not anywhere in between. Each shell has a fixed energy. When an electron moves between shells electromagnetic radiation is emitted or absorbed. Because the energy of the shells is fixed, the radiation will have a fixed frequency.

3. Relative Mass

Page 26 — Application Questions

Q1 a) 85.5
b) 200.6
c) 65.4

Q2 a) 14.0 + (3 × 1.0) = **17.0**
b) 12.0 + (16.0 × 2) = **44.0**
c) (12.0 × 2) + (1.0 × 4) + (16.0 × 6) + (14.0 × 2) = **152.0**

Q3 a) 40.1 + (35.5 × 2) = **111.1**
b) 24.3 + 32.1 + (16.0 × 4) = **120.4**
c) 23.0 + 16.0 + 1.0 = **40.0**

Q4 A_r = ((0.1 × 180) + (26.5 × 182) + (14.3 × 183) + (30.7 × 184) + (28.4 × 186)) ÷ 100 = **183.9 (to 1 d.p.)**

Q5 A_r = ((51.5 × 90) + (11.2 × 91) + (17.1 × 92) + (17.4 × 94) + (2.8 × 96)) ÷ 100 = **91.3 (to 1 d.p.)**

Page 26 — Fact Recall Questions

Q1 The average mass of an atom of an element on a scale where an atom of carbon-12 is exactly 12.
Q2 The average mass of a molecule on a scale where an atom of carbon-12 is exactly 12.
Q3 The average mass of a formula unit on a scale where an atom of carbon-12 is exactly 12.

4. The Mass Spectrometer

Page 29 — Application Question

Q1 a) 2
b) 63, 65
c) ^{63}Cu = 69.1%, ^{65}Cu = 30.9%

Page 29 — Application Question

Q2 E.g. It's unlikely to be magnesium, as the abundance of the particle with the smallest mass/charge is about 90%, which is too high for magnesium. It's unlikely to be silicon, as you'd expect the second peak to be higher than the third peak in the mass spectrum for silicon. It can't be indium, as the mass spectrum for indium would only have two peaks.

5. Using Mass Spectra

Page 31 — Application Questions

Q1 $69.1 \times 63 = 4353.3$ $30.9 \times 65 = 2008.5$
$4353.3 + 2008.5 = 6361.8$
$6361.8 \div 100 =$ **63.6 (to 3 s.f.)**

Q2 $20 \times 10 = 200$ $80 \times 11 = 880$
$200 + 880 = 1080$
$1080 \div 100 =$ **10.8**

Q3 $8.0 \times 6 = 48$ $100 \times 7 = 700$
$700 + 48 = 748$
$748 \div (100 + 8) =$ **6.9 (to 2 s.f.)**

Q4 $100 \times 69 = 6900$ $65.5 \times 71 = 4650.5$
$6900 + 4650.5 = 11550.5$
$11550.5 \div (100 + 65.5) =$ **69.8 (to 3 s.f.)**

Page 32 — Application Questions

Q1 a) $(3 \times 12.0) + (6 \times 1.0) + 16.0 =$ **58.0**
b) $(4 \times 12.0) + (8 \times 1.0) =$ **56.0**
c) $(3 \times 12.0) + (6 \times 1.0) + (2 \times 16.0) =$ **74.0**
The mass/charge value is equal to the M_r of the compound.

Q2 C
M_r of $N_2 = 28.0$, M_r of $CH_2CH_2 = 28.0$, M_r of $CH_3NH_2 = 31.0$, and M_r of $CO = 28.0$. So the gas could be N_2, CH_2CH_2 or CO, but not CH_3NH_2.

Q3 B
M_r of $CH_3CH_3 = 30.0$, M_r of $CH_3F = 34.0$, M_r of $CO_2 = 44.0$, and M_r of $HCN = 27.0$. The only one of these with a molecular ion of $m/z = 34$ is CH_3F.

Q4 The spectrum has two peaks, one at $m/z = 18$ and one at $m/z = 46$.
M_r of $NH_3 = 14.0 + (3 \times 1.0) = 17.0$
M_r of $H_2O = (2 \times 1.0) + 16.0 = 18.0$
M_r of $CH_4 = 12.0 + (4 \times 1.0) = 16.0$
M_r of $CH_3CH_2OH = (2 \times 12.0) + (6 \times 1.0) + 16.0 = 46.0$
M_r of $CH_3CH_2CH_3 = (3 \times 12.0) + (8 \times 1.0) = 44.0$
The compounds in the mixture must be **water and ethanol**.

6. Electronic Structure

Page 35 — Application Questions

Q1 a) $1s^2\ 2s^1$
b) $1s^2\ 2s^2\ 2p^6\ 3s^2\ 3p^6\ 3d^2\ 4s^2$
(or $1s^2\ 2s^2\ 2p^6\ 3s^2\ 3p^6\ 4s^2\ 3d^2$)
c) $1s^2\ 2s^2\ 2p^6\ 3s^2\ 3p^6\ 3d^{10}\ 4s^2\ 4p^1$
d) $1s^2\ 2s^2\ 2p^3$

Q2

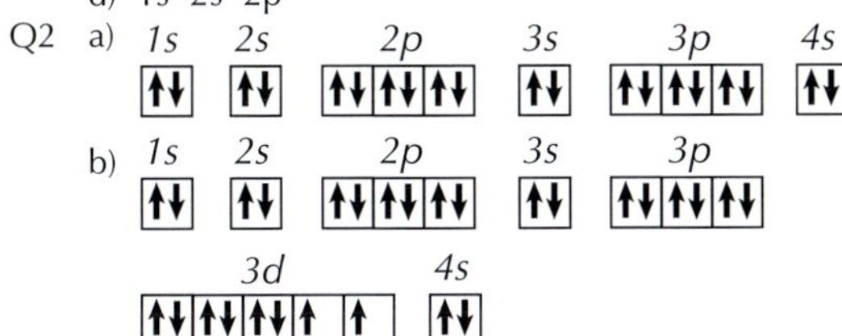

Remember that for each sub-shell you add, you should fill up each orbital singly before they start to share.

c) 1s 2s 2p 3s
[↑↓] [↑↓] [↑↓|↑↓|↑↓] [↑]

d) 1s 2s 2p
[↑↓] [↑↓] [↑↓|↑|↑]

Q3

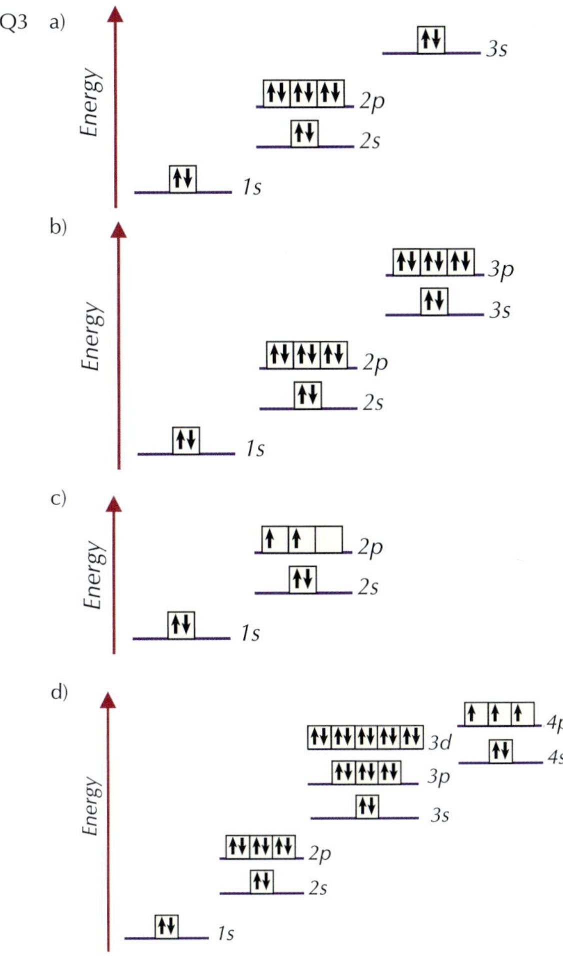

Q4 a) $1s^2\ 2s^2\ 2p^6$
b) $1s^2\ 2s^2\ 2p^6$
c) $1s^2\ 2s^2\ 2p^6$
d) $1s^2\ 2s^2\ 2p^6\ 3s^2\ 3p^6$

Q5 a) bromine
b) phosphorus
c) vanadium

Page 36 — Application Questions

Q1 $1s^2\ 2s^2\ 2p^6\ 3s^2\ 3p^6\ 3d^4$
Q2 $1s^2\ 2s^2\ 2p^6\ 3s^2\ 3p^6\ 3d^8$
Q3 $1s^2\ 2s^2\ 2p^6\ 3s^2\ 3p^6\ 3d^2$

Page 36 — Fact Recall Questions

Q1 3
Q2 6 (it can hold two electrons in each orbital)
Q3 18
Q4 The number of electrons that an atom or ion has and how they are arranged.
Q5 Electrons fill orbitals singly before they start sharing, so the electrons in the 2p sub-shell should be in separate orbitals.
Q6 $1s^2\ 2s^2\ 2p^6\ 3s^2\ 3p^6\ 3d^5\ 4s^1$
Q7 Copper donates one of its 4s electrons to the 3d sub-shell so that the 3d sub-shell is full, making it more stable.
Q8 They form negative ions with an inert gas electron configuration.

7. Ionisation Energies

Pages 40-41 — Application Questions

Q1 a) $Cl_{(g)} \rightarrow Cl^{+}_{(g)} + e^{-}$

b) $Cl^{+}_{(g)} \rightarrow Cl^{2+}_{(g)} + e^{-}$

Q2

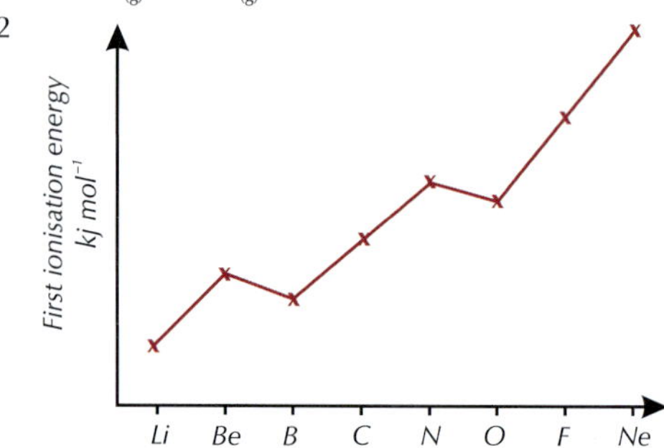

Q3 In both nitrogen and oxygen, the first electron to be removed is in the same sub-shell and the shielding is identical in the two atoms. However, the electron being removed from oxygen is in an orbital where there are two electrons. The repulsion between these two electrons means that it's easier to remove the electron than it would be if it was unpaired (like in nitrogen), so the first ionisation energy of oxygen is lower than that of nitrogen.

Q4 Boron's outer electron is in a 2p orbital rather than a 2s (like beryllium's), which means it has a higher energy and is located further from the nucleus. The 2p orbital also has additional shielding provided by the 2s electrons. These two factors override the effect of the increased nuclear charge of the boron atom, and result in the first ionisation energy of beryllium being higher than the first ionisation energy of boron.

Q5 The first electron in lithium is removed from the 2s orbital, whereas the second and third electrons are removed from the 1s orbital. The 2s orbital is further from the nucleus and is shielded by the inner electrons, so it takes much less energy to remove the first electron than the second. The second and third electrons have no shielding and are the same distance from the nucleus, so there's less difference between their ionisation energies.

Q6 a) Group 6

b) 2 electrons in the first shell and 6 electrons in the second shell.

c) Oxygen

Q7 a)

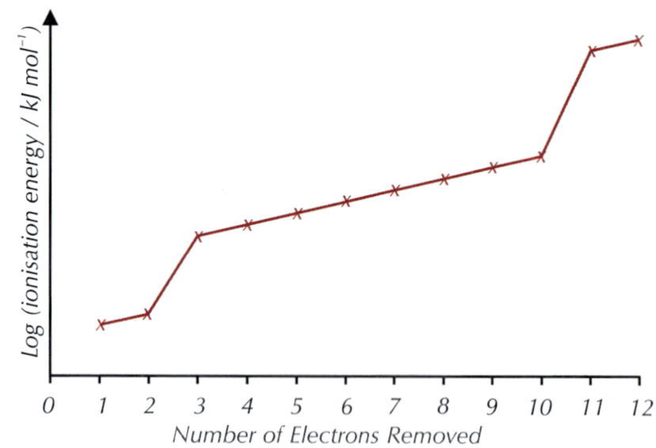

b) Within each shell, successive ionisation energies increase. This is because electrons are being removed from an increasingly positive ion. There's less repulsion amongst the remaining electrons, so they're held more strongly by the nucleus. The big jumps in ionisation energy happen when a new shell that's closer to the nucleus is broken into — the first shell that electrons are removed from has 2 electrons in it, the second shell has 8 electrons and the third shell has 2 electrons.

Page 41 — Fact Recall Questions

Q1 The first ionisation energy is the energy needed to remove 1 electron from each atom in 1 mole of gaseous atoms to form one mole of gaseous 1+ ions.

Q2 The more protons there are, the more positively charged the nucleus is, the stronger the attraction for the electrons and the higher the first ionisation energy.

Q3 The distance between the outer electron and the nucleus, and the shielding effect of inner electrons.

Q4 The second ionisation energy is the energy needed to remove 1 electron from each ion in 1 mole of gaseous 1+ ions to form one mole of gaseous 2+ ions.

Q5 Within each shell, successive ionisation energies increase because electrons are being removed from an increasingly positive ion — there's less repulsion amongst the remaining electrons, so they're held more strongly by the nucleus.

Q6 It decreases.

Q7 The first ionisation energy increases.

Exam-style Questions — pages 43-45

1.1 silicon ***(1 mark)***

1.2 The p block ***(1 mark)***

1.3 $1s^2\ 2s^2\ 2p^6\ 3s^2\ 3p^1$ ***(1 mark)***

1.4 three ***(1 mark)***

1.5 $^{28}_{14}\mathbf{Si}$

(2 marks for correct nuclear symbol, otherwise 1 mark for a correct mass number of 28.)

1.6 The isotopes would have the same chemical properties, as they have the same electron arrangement and it's this that determines the chemical properties of an element ***(1 mark)***.

2.1 electrospray ionisation ***(1 mark)***

2.2 The ions are accelerated to constant kinetic energy ***(1 mark)*** by an electric field ***(1 mark)***. Lighter ions are accelerated more than heavier ions ***(1 mark)***, so ions with different masses take different amounts of time to travel through the mass spectrometer and reach the detector ***(1 mark)***.

3.1 The first ionisation energy increases across Period 3 ***(1 mark)***.

3.2 $Na_{(g)} \rightarrow Na^{+}_{(g)} + e^{-}$ ***(1 mark)***

3.3 Magnesium's outer electron is in a 3s orbital and aluminium's outer electron is in a 3p orbital ***(1 mark)***. The 3p orbital has a higher energy than the 3s orbital, so the electron is further from the nucleus ***(1 mark)***. The 3p orbital also has additional shielding provided by the 3s electrons ***(1 mark)***.

3.4 B ***(1 mark)***

3.5 Aluminium, silicon and phosphorus all have their outer electron in the same/3p sub-shell ***(1 mark)***. So the shielding effect and the distance between the nucleus and the outer electron is very similar for all three elements ***(1 mark)***. But the nuclear charge of silicon is higher than aluminium and lower than phosphorus ***(1 mark)***. So the energy needed to remove the outer electron/first ionisation energy of silicon should be greater than aluminium, but lower than phosphorus ***(1 mark)***.

3.6 In phosphorus, the electron is being removed from a singly-occupied orbital, but in sulfur the electron is being removed from an orbital containing two electrons ***(1 mark)***. The repulsion between the two electrons means that electrons are easier to remove from shared orbitals and so the first ionisation energy of sulfur is lower than that of phosphorus ***(1 mark)***.

4.1 $1s^2\ 2s^2\ 2p^6\ 3s^2\ 3p^6\ 3d^3\ 4s^2$ ***(1 mark)***

4.2 V^{2+}: $1s^2\ 2s^2\ 2p^6\ 3s^2\ 3p^6\ 3d^3$ ***(1 mark)***
V^{3+}: $1s^2\ 2s^2\ 2p^6\ 3s^2\ 3p^6\ 3d^2$ ***(1 mark)***

4.3 $1s^2\ 2s^2\ 2p^6\ 3s^2\ 3p^6\ 3d^{10}\ 4s^1$ ***(1 mark)***
Copper donates one of its 4s electrons to the 3d sub-shell to make a more stable full 3d sub-shell ***(1 mark)***.

4.4 iron ***(1 mark)***

4.5 manganese ***(1 mark)***

5.1 $\dfrac{(20 \times 90.48) + (21 \times 0.27) + (22 \times 9.25)}{90.48 + 0.27 + 9.25} = \mathbf{20.2}$
(1 mark for correct calculation, 1 mark for correct answer.)
Element X is neon ***(1 mark)***.

5.2 ^{20}X has 10 protons, 10 electrons and 20 – 10 = **10** neutrons ***(1 mark)***.
^{21}X has 10 protons, 10 electrons and 21 – 10 = **11** neutrons ***(1 mark)***.
^{22}X has 10 protons, 10 electrons and 22 – 10 = **12** neutrons ***(1 mark)***.

5.3 The mass number of an atom is the total number of protons and neutrons in the nucleus of the atom ***(1 mark)***.

5.4 The proton number of an atom is the number of protons in the nucleus of the atom ***(1 mark)***.

5.5 The first ionisation energy is the energy needed to remove 1 electron from each atom in 1 mole of gaseous atoms ***(1 mark)*** to form 1 mole of gaseous 1+ ions ***(1 mark)***.

5.6 The first ionisation energy of element A would be lower than the first ionisation energy of element X. Element A has more electrons than element X, so the distance between the outer electrons and the nucleus is greater ***(1 mark)***. The shielding effect of the inner electrons is greater in element A than in element X ***(1 mark)***.

5.7 The eighth electron is being removed from the second electron shell, but the ninth electron is being removed from the first electron shell ***(1 mark)***. The ninth electron is much closer to the nucleus than the eighth and experiences less shielding ***(1 mark)***, so it takes much more energy to remove it ***(1 mark)***.

Section 2: Amount of Substance

1. The Mole

Page 47 — Application Questions

Q1 Number of molecules = 0.360 × (6.02 × 10^{23}) = **2.17 × 10^{23}**

Q2 Number of ions = 0.0550 × (6.02 × 10^{23}) = **3.31 × 10^{22}**

Q3 M_r = 66 ÷ 1.5 = **44**

Q4 M_r = 23.0 + 14.0 + (16.0 × 3) = 85.0
number of moles = 212.5 ÷ 85.0 = **2.50 moles**

Q5 M_r = 65.4 + (35.5 × 2) = 136.4
number of moles = 15.5 ÷ 136.4 = **0.114 moles**

Q6 M_r = 23.0 + 35.5 = 58.5
Mass = 58.5 × 2 = **117 g**

Page 49 — Application Questions

Q1 Number of moles = (2 × 50) ÷ 1000 = **0.1 moles**

Q2 Number of moles = 0.08 × 0.5 = **0.04 moles**

Q3 Number of moles = (0.70 × 30) ÷ 1000 = **0.021 moles**

Q4 Concentration = 0.25 ÷ 0.50 = **0.50 mol dm^{-3}**

Q5 Concentration = 0.080 ÷ 0.75 = **0.11 mol dm^{-3}**

Q6 Concentration = 0.10 ÷ (36 ÷ 1000) = **2.8 mol dm^{-3}**

Q7 Volume = 0.46 ÷ 1.8 = **0.26 dm^3**

Q8 Volume = 0.010 ÷ 0.55 = **0.018 dm^3**

Q9 Number of moles = concentration × volume (dm^3)
= 0.80 × (75 ÷ 1000) = 0.060
M_r of Na_2O = (23.0 × 2) + 16.0 = 62.0
Mass = moles × molar mass = 0.060 × 62.0 = **3.7 g**

Q10 Number of moles = concentration × volume (dm^3)
= 0.50 × (30 ÷ 1000) = 0.015
M_r of $CoBr_2$ = 58.9 + (79.9 × 2) = 218.7
Mass = number of moles × molar mass
= 0.015 × 218.7 = **3.3 g**

Q11 Number of moles = concentration × volume (dm^3)
= 1.20 × (100 ÷ 1000) = 0.120
M_r = mass ÷ number of moles
= 4.08 ÷ 0.120 = **34.0**

Page 49 — Fact Recall Questions

Q1 a) 6.02 × 10^{23}
b) The Avogadro constant

Q2 number of particles = number of moles × Avogadro constant

Q3 Number of moles = mass of substance ÷ M_r

Q4 1000

Q5 E.g mol dm^{-3}

Q6 Number of moles = $\dfrac{\text{concentration} \times \text{volume (in cm}^3)}{1000}$
Number of moles = concentration × volume (dm^3)

2. Gases and the Mole

Page 51 — Application Questions

Q1 $n = pV \div RT$
= (70 000 × 0.040) ÷ (8.31 × 350) = **0.96 moles**

Q2 $V = nRT \div p$
= (0.65 × 8.31 × 280) ÷ 100 000 = **0.015 m^3**

Q3 0.55 dm^3 = 5.5 × 10^{-4} m^3 35 °C = 308 K
$n = pV \div RT$
= (90 000 × (5.5 × 10^{-4})) ÷ (8.31 × 308)
= **0.019 moles**

Q4 1200 cm^3 = 1.2 × 10^{-3} m^3
$T = pV \div nR$
= (110 000 × (1.2 × 10^{-3})) ÷ (0.0500 × 8.31) = 318 K
318 K = (318 – 273) °C = **45 °C**

Q5 75 kPa = 75 000 Pa 22 °C = 295 K
$V = nRT \div p$
= (0.75 × 8.31 × 295) ÷ 75 000 = **0.025 m^3**

Q6 80 kPa = 80 000 Pa 1.5 dm^3 = 1.5 × 10^{-3} m^3
$n = pV \div RT$
= (80 000 × (1.5 × 10^{-3})) ÷ (8.31 × 300)
= 0.048... moles
M_r = mass ÷ moles = 2.6 ÷ 0.048... = 54
So the relative molecular mass is **54**.

Q7 44 °C = 317 K 100 kPa = 100 000 Pa
$n = pV \div RT$
= (100 000 × 0.00300) ÷ (8.31 × 317)
= 0.113... moles
M_r of neon = 20.2
mass = number of moles × M_r
= 0.113... × 20.2 = **2.30 g**

Page 51 — Fact Recall Question

Q1 $pV = nRT$
p = pressure measured in pascals (Pa)
V = volume measured in m^3
n = number of moles
R = 8.31 J K^{-1} mol^{-1}. R is the gas constant.
T = temperature measured in kelvin (K)

3. Chemical Equations

Page 53 — Application Questions

Q1 a) $Mg + \mathbf{2}HCl \rightarrow MgCl_2 + H_2$

b) $S_8 + \mathbf{24}F_2 \rightarrow \mathbf{8}SF_6$

c) $Ca(OH)_2 + H_2SO_4 \rightarrow CaSO_4 + \mathbf{2}H_2O$

d) $Na_2CO_3 + \mathbf{2}HCl \rightarrow \mathbf{2}NaCl + CO_2 + H_2O$

e) $C_4H_{10} + \mathbf{6\frac{1}{2}}O_2 \rightarrow \mathbf{4}CO_2 + \mathbf{5}H_2O$

Q2 a) $Fe + Cu^{2+} \rightarrow Fe^{2+} + Cu$

b) $Ba^{2+} + SO_4^{2-} \rightarrow BaSO_4$

This reaction looks like it has the same ions on each side of the equation, so wouldn't have an ionic equation. But the state symbols show you that $BaSO_4$ is a solid, so you shouldn't split it up into ions when writing your ionic equation.

c) $CO_3^{2-} + 2H^+ \rightarrow H_2O + CO_2$

4. Equations and Calculations

Page 55 — Application Questions

Q1 a) $Zn + 2HCl \rightarrow ZnCl_2 + H_2$

b) A_r of Zn = 65.4

number of moles = mass ÷ A_r

= 3.4 ÷ 65.4 = **0.052 moles**

c) The molar ratio of Zn : $ZnCl_2$ is 1 : 1. So 0.052 moles of Zn will give **0.052 moles** of $ZnCl_2$.

d) M_r of $ZnCl_2$ = 65.4 + (2 × 35.5) = 136.4

mass = number of moles × M_r = 0.052 × 136.4 = **7.1 g**

Q2 a) $C_2H_4 + 3O_2 \rightarrow 2CO_2 + 2H_2O$

b) M_r of H_2O = (2 × 1.0) + 16.0 = 18.0

number of moles = mass ÷ M_r = 15 ÷ 18.0 = **0.83 moles**

c) The molar ratio of H_2O : C_2H_4 is 2 : 1.

So 0.83 moles of H_2O must be made from (0.83 ÷ 2) = **0.42 moles** of C_2H_4.

d) M_r of C_2H_4 = (2 × 12.0) + (4 × 1.0) = 28.0

mass = number of moles × M_r = 0.42 × 28.0 = **12 g**

Q3 $Na_2CO_3 + BaCl_2 \rightarrow 2NaCl + BaCO_3$

M_r of $BaCl_2$ = 137.3 + (2 × 35.5) = 208.3

number of moles = mass ÷ M_r

= 4.58 ÷ 208.3 = 0.0219... moles

The molar ratio of $BaCl_2$: $BaCO_3$ is 1 : 1.

So 0.0219... moles of $BaCO_3$ must be made from 0.0219... moles of $BaCl_2$.

M_r of $BaCO_3$ = 137.3 + 12.0 + (16.0 × 3) = 197.3

mass = number of moles × M_r = 0.0219... × 197.3 = **4.34 g**

Page 56 — Application Questions

Q1 a) aq

b) s

c) l

d) aq

e) g

f) s

Q2 a) $2H_2O_{(l)} \rightarrow 2H_{2\,(g)} + O_{2\,(g)}$

b) M_r of H_2O = (2 × 1.0) + 16.0 = 18.0

number of moles = mass ÷ M_r = 9.00 ÷ 18.0 = **0.500**

c) The molar ratio of H_2O to O_2 is 2 : 1.

So 0.500 moles of H_2O will produce (0.500 ÷ 2) = **0.250 moles** of O_2.

d) Using the ideal gas equation:

$V = nRT \div p$

= (0.0250 × 8.31 × 298) ÷ 100 000

= **0.00619 m³**

Q3 a) $ZnS_{(s)} + 1\frac{1}{2}O_{2\,(g)} \rightarrow ZnO_{(s)} + SO_{2\,(g)}$

b) M_r of ZnS = 65.4 + 32.1 = 97.5

number of moles = mass ÷ M_r

= 7.0 ÷ 97.5 = **0.072 moles**

c) The molar ratio of ZnS to SO_2 is 1 : 1.

So 0.072 moles of ZnS will give **0.072 moles** of SO_2.

d) Using the ideal gas equation:

$V = nRT \div p$

= (0.072 × 8.31 × 298) ÷ 100 000

= **0.0018 m³**

Q4 a) $C_6H_{14\,(g)} \rightarrow C_4H_{10\,(g)} + C_2H_{4\,(g)}$

b) M_r of C_4H_{10} = (4 × 12.0) + (10 × 1.0) = 58.0

number of moles = mass ÷ M_r

= 3.0 ÷ 58.0 = **0.052 moles**

c) The molar ratio of C_4H_{10} to C_6H_{14} is 1 : 1.

So 0.052 moles of C_4H_{10} must be made from **0.052 moles** of C_6H_{14}.

d) Using the ideal gas equation:

$V = nRT \div p$

= (0.052 × 8.31 × 308) ÷ 100 000

= **0.0013 m³**

Q5 $Mg_{(s)} + H_2O_{(g)} \rightarrow MgO_{(s)} + H_{2\,(g)}$

M_r of MgO = 24.3 + 16.0 = 40.3

number of moles = mass ÷ M_r = 10 ÷ 40.3 = 0.248... moles

The molar ratio of MgO : H_2O is 1 : 1. So 0.248... moles of MgO is made from 0.248... moles of H_2O.

Using the ideal gas equation:

$V = nRT \div p$

= (0.248... × 8.31 × (100 + 273)) ÷ 101 325

= **0.0076 m³**

5. Titrations

Page 60 — Application Questions

Q1 a) $HCl_{(aq)} + KOH_{(aq)} \rightarrow KCl_{(aq)} + H_2O_{(l)}$

b) moles HCl = (conc. × volume (cm³)) ÷ 1000

= (0.75 × 28) ÷ 1000 = **0.021 moles**

c) 1 mole of HCl reacts with 1 mole of KOH.

So 0.021 moles of HCl must react with **0.021 moles** of KOH.

d) concentration = (moles KOH × 1000) ÷ vol. (cm³)

= (0.021 × 1000) ÷ 40 = **0.53 mol dm⁻³**

Q2 a) $NaOH_{(aq)} + HNO_{3\,(aq)} \rightarrow NaNO_{3\,(aq)} + H_2O_{(l)}$

b) moles NaOH = (conc. × volume (cm³)) ÷ 1000

= (1.5 × 15.3) ÷ 1000 = **0.023 moles**

c) 1 mole of NaOH reacts with 1 mole of HNO_3.

So 0.023 moles of NaOH must react with **0.023 moles** of HNO_3.

d) concentration = (moles HNO_3 × 1000) ÷ vol. (cm³)

= (0.023 × 1000) ÷ 35 = **0.66 mol dm⁻³**

Q3 $KOH_{(aq)} + HCl_{(aq)} \rightarrow KCl_{(aq)} + H_2O_{(l)}$

moles HCl = (conc. × volume (cm³)) ÷ 1000

= (0.50 × 12) ÷ 1000 = 0.0060 moles

1 mole of HCl reacts with 1 mole of KOH, so 0.0060 moles of HCl must react with 0.0060 moles of KOH.

concentration = (moles KOH × 1000) ÷ vol. (cm³)

= (0.0060 × 1000) ÷ 24 = **0.25 mol dm⁻³**

Page 61 — Application Questions

Q1 a) $HNO_{3\,(aq)} + KOH_{(aq)} \rightarrow KNO_{3\,(aq)} + H_2O_{(l)}$

b) moles HNO_3 = (conc. × volume (cm³)) ÷ 1000

= (0.20 × 18.8) ÷ 1000 = **0.0038 moles**

c) 1 mole of HNO_3 reacts with 1 mole of KOH.

So 0.0038 moles of HNO_3 must react with **0.0038 moles** of KOH.

d) volume = (moles KOH × 1000) ÷ concentration

= (0.0038 × 1000) ÷ 0.45 = **8.4 cm³**

Q2 a) $KOH_{(aq)} + CH_3COOH_{(aq)} \rightarrow CH_3COOK_{(aq)} + H_2O_{(l)}$
b) moles KOH = (conc. × volume (cm^3)) ÷ 1000
= (0.420 × 37.3) ÷ 1000 = **0.0157 moles**
c) 1 mole of KOH reacts with 1 mole of CH_3COOH.
So 0.0157 moles of KOH must react with **0.0157 moles** of CH_3COOH.
d) volume = (moles CH_3COOH × 1000) ÷ conc.
= (0.0157 × 1000) ÷ 1.10 = **14.3 cm³ (3 s.f.)**

Q3 $2NaOH_{(aq)} + H_2SO_{4\,(aq)} \rightarrow Na_2SO_{4\,(aq)} + 2H_2O_{(l)}$
2 moles of NaOH = (conc. × volume (cm^3)) ÷ 1000
= (14 × 1.5) ÷ 1000 = 0.021 moles
2 moles of NaOH react with 1 mole of H_2SO_4.
So, 0.021 moles of NaOH must react with 0.0105 moles of H_2SO_4.
volume = (moles H_2SO_4 × 1000) ÷ conc.
= (0.0105 × 1000) ÷ 0.60 = **18 cm³**

Page 61 — Fact Recall Questions

Q1 Any solution that you know the exact concentration of.
Q2 pipette
Q3 burette
Q4 The exact point at which the indicator changes colour (at this point the amount of acid added is just enough to neutralise the alkali).

6. Formulas

Page 63 — Application Questions

Q1 empirical mass = (4 × 12.0) + (9 × 1.0) = 57.0
M_r = 171, so there are (171 ÷ 57.0) = 3 empirical units in the molecule.
molecular formula = $\mathbf{C_{12}H_{27}}$

Q2 empirical mass = (3 × 12.0) + (5 × 1.0) + (2 × 16.0) = 73.0
M_r = 146, so there are (146 ÷ 73.0) = 2 empirical units in the molecule.
molecular formula = $\mathbf{C_6H_{10}O_4}$

Q3 empirical mass = (2 × 12.0) + (6 × 1.0) + (1 × 16.0) = 46.0
M_r = 46, so there is (46 ÷ 46.0) = 1 empirical unit in the molecule.
molecular formula = $\mathbf{C_2H_6O}$

Q4 empirical mass = (4 × 12.0) + (6 × 1.0) + (2 × 35.5) + (1 × 16.0) = 141.0
M_r = 423, so there are (423 ÷ 141.0) = 3 empirical units in the molecule.
molecular formula = $\mathbf{C_{12}H_{18}Cl_6O_3}$

Page 65 — Application Questions

Q1 Mass of each element:
H = 5.9 g   O = 94.1 g
Moles of each element:
H = (5.9 ÷ 1.0) = 5.9 moles
O = (94.1 ÷ 16.0) = 5.9 moles
Divide each by 5.9:
H = (5.9 ÷ 5.9) = 1   O = (5.9 ÷ 5.9) = 1
The ratio of H : O is 1 : 1.
So the empirical formula is **HO**.

Q2 Mass of each element:
Al = 20.2 g   Cl = 79.8 g
Moles of each element:
Al = (20.2 ÷ 27.0) = 0.748 moles
Cl = (79.8 ÷ 35.5) = 2.25 moles
Divide each by 0.748:
Al = (0.748 ÷ 0.748) = 1   Cl = (2.25 ÷ 0.748) = 3
The ratio of Al : Cl is 1 : 3.
So the empirical formula is $\mathbf{AlCl_3}$.

Q3 Mass of each element:
C = 8.5 g   H = 1.4 g   I = 90.1 g
Moles of each element:
C = (8.5 ÷ 12.0) = 0.71 moles
H= (1.4 ÷ 1.0) = 1.4 moles
I = (90.1 ÷ 126.9) = 0.71 moles
Divide each by 0.71:
C = (0.71 ÷ 0.71) = 1
H = (1.4 ÷ 0.71) = 2
I = (0.71 ÷ 0.71) = 1
The ratio of C : H : I is 1 : 2 : 1.
So the empirical formula is $\mathbf{CH_2I}$.

Q4 Mass of each element:
Cu = 50.1 g   P = 16.3 g   O = 33.6 g
Moles of each element:
Cu = (50.1 ÷ 63.5) = 0.789 moles
P = (16.3 ÷ 31.0) = 0.526 moles
O = (33.6 ÷ 16.0) = 2.10 moles
Divide each by 0.526:
Cu = (0.789 ÷ 0.526) = 1.5
P = (0.526 ÷ 0.526) = 1.0
O = (2.10 ÷ 0.526) = 4.0
The ratio of Cu : P : O is 1.5 : 1.0 : 4.0.
Multiply by 2 — 2 × (1.5 : 1.0 : 4.0) = 3 : 2 : 8.
So the empirical formula is $\mathbf{Cu_3P_2O_8}$.

Q5 % V = 32.3   % Cl = 100 – 32.3 = 67.7
Mass of each element:
V = 32.3 g   Cl = 67.7 g
Moles of each element:
V = (32.3 ÷ 50.9) = 0.635 moles
Cl = (67.7 ÷ 35.5) = 1.91 moles
Divide each by 0.635:
V = (0.635 ÷ 0.635) = 1   Cl = (1.91 ÷ 0.635) = 3
The ratio of V : Cl is 1 : 3.
So the empirical formula is $\mathbf{VCl_3}$.

Q6 % O = 31.58   % Cr = 100 – 31.58 = 68.42
Mass of each element:
O = 31.58 g   Cr = 68.42 g
Moles of each element:
O = (31.58 ÷ 16.0) = 1.97 moles
Cr = (68.42 ÷ 52.0) = 1.32 moles
Divide each by 1.32:
O = (1.97 ÷ 1.32) = 1.5   Cr = (1.32 ÷ 1.32) = 1.0
The ratio of Cr : O is 1.0 : 1.5.
Multiply by 2 — 2 × (1.0 : 1.5) = 2 : 3.
So the empirical formula is $\mathbf{Cr_2O_3}$.

Q7 Mass of each reactant:
P = 2.00 g   O = 4.58 – 2.00 = 2.58 g
Moles of each reactant:
P = (2.00 ÷ 31.0) = 0.0645 moles
O = (2.58 ÷ 16.0) = 0.161 moles
Divide each by 0.0645:
P = (0.0645 ÷ 0.0645) = 1.0
O = (0.161 ÷ 0.0645) = 2.5
The ratio of P : O is 1.0 : 2.5.
Multiply by 2 — 2 × (1.0 : 2.5) = 2 : 5.
So the empirical formula is $\mathbf{P_2O_5}$.

Q8 Mass of each reactant:
Ag = 0.503 g   Cl = 0.669 – 0.503 = 0.166 g
Moles of each reactant:
Ag = (0.503 ÷ 107.9) = 0.00466 moles
Cl = (0.166 ÷ 35.5) = 0.00468 moles
Divide each by 0.00466:
Ag = (0.00466 ÷ 0.00466) = 1
Cl = (0.00468 ÷ 0.00466) = 1
The ratio of Ag : Cl is 1 : 1.
So the empirical formula is **AgCl**.

Q9 M_r of CO_2 = (12.0 + (2 × 16.0)) = 44.0
M_r of H_2O = (16.0 + (2 × 1.0) = 18.0
moles of CO_2 = (9.70 ÷ 44.0) = 0.220 moles
moles of H_2O = (7.92 ÷ 18.0) = 0.440 moles
So the hydrocarbon must contain 0.220 moles of carbon and 2 × 0.440 = 0.880 moles of hydrogen.
Divide each by 0.220:
C = (0.220 ÷ 0.220) = 1
H = (0.880 ÷ 0.220) = 4
The ratio of C : H is 1 : 4.
So the empirical formula is **CH_4**.

Page 65 — Fact Recall Questions

Q1 The empirical formula gives the smallest whole number ratio of atoms present in a compound.

Q2 The molecular formula gives the actual numbers of atoms in a molecule.

7. Chemical Yield

Page 68 — Application Questions

Q1 % yield = (1.76 ÷ 3.24) × 100 = **54.3%**

Q2 % yield = (3.7 ÷ 6.1) × 100 = **61%**

Q3 % yield = (138 ÷ 143) × 100 = **96.5%**

Q4 a) M_r of Fe = 55.8
moles of Fe = mass ÷ M_r
= 3.0 ÷ 55.8 = **0.054 moles**

b) From the equation: 4 moles of Fe produces 2 moles of Fe_2O_3, so 0.054 moles of Fe will produce (0.054 ÷ 2) = 0.027 moles of Fe_2O_3.
M_r of Fe_2O_3 = (2 × 55.8) + (3 × 16.0) = 159.6
theoretical yield = moles Fe_2O_3 × M_r
= 0.027 × 159.6 = **4.3 g**

c) % yield = (actual yield ÷ theoretical yield) × 100
= (3.6 ÷ 4.3) × 100 = **84%**

Q5 M_r of Al_2O_3 = (2 × 27.0) + (3 × 16.0) = 102.0
Number of moles Al_2O_3 = mass ÷ M_r
= 1000 ÷ 102.0 = 9.80 moles
From the equation: 2 moles of Al_2O_3 produce 4 moles of Al, so 9.80 moles of Al_2O_3 will produce (9.80 × 2) = 19.6 moles of Al.
M_r of Al = 27.0
theoretical yield = moles Al × M_r
= 19.6 × 27.0 = **529 g**

Q6 M_r of NaOH = 23.0 + 16.0 + 1.0 = 40.0
Number of moles NaOH = mass ÷ M_r
= 4.70 ÷ 40.0 = 0.118 moles
From the equation: 2 moles of NaOH produce 1 mole of Na_2SO_4, so 0.118 moles of NaOH will produce (0.118 ÷ 2) = 0.0590 moles of Na_2SO_4.
M_r of Na_2SO_4 = (2 × 23.0) + 32.1 + (4 × 16.0) = 142.1
theoretical yield = moles Na_2SO_4 × M_r
= 0.0590 × 142.1 = 8.38... g
% yield = (actual yield ÷ theoretical yield) × 100
= (6.04 ÷ 8.38...) × 100 = **72.0%**

Q7 M_r of Mg = 24.3
Number of moles Mg = mass ÷ M_r
= 40.0 ÷ 24.3 = 1.64... moles
From the equation: 1 mole of Mg produces 1 mole of H_2, so 1.64... moles of Mg will produce 1.64... moles of H_2.
M_r of H_2 = 2 × 1.0 = 2.0
theoretical yield = moles H_2 × M_r
= 1.64... × 2.0 = 3.29... g
% yield = (actual yield ÷ theoretical yield) × 100
= (1.70 ÷ 3.29...) × 100 = **51.6%**

Page 68 — Fact Recall Questions

Q1 The theoretical yield is the mass of product that should be formed in a chemical reaction.

Q2 $$\text{percentage yield} = \frac{\text{actual yield}}{\text{theoretical yield}} \times 100$$

8. Atom Economy

Page 71 — Application Questions

Q1 a) mass of reactants = (12.0 + (4 × 1.0)) + (2 × 35.5) = **87.0**

b) mass of CH_3Cl = 12.0 + (3 × 1.0) + 35.5 = **50.5**

c) $$\text{\% atom economy} = \frac{\text{molecular mass of desired product}}{\text{sum of molecular masses of all reactants}} \times 100$$
= (50.5 ÷ 87.0) × 100 = **58.0%**

d) E.g. sell the HCl so it can be used in other chemical reactions / use the HCl as a reactant in another reaction.

Q2 % atom economy = **100%**
Any reaction where there's only one product will have 100% atom economy. If you don't notice this, you can just do the calculations as usual:
mass of reactants = (2 × 27.0) + (3 × (2 × 35.5)) = 267
mass of 2$AlCl_3$ = 2 × (27.0 + (3 × 35.5)) = 267
% atom economy = (267 ÷ 267) × 100 = 100%

Q3 mass of reactants =
[2 × ((2 × 55.8) + (3 × 16.0))] + [3 × 12.0] = 355.2
mass of 4Fe = 4 × 55.8 = 223.2
$$\text{\% atom economy} = \frac{\text{molecular mass of desired product}}{\text{sum of molecular masses of all reactants}} \times 100$$
= (223.2 ÷ 355.2) × 100 = **62.8%**

Q4 a) Reaction 1:
% atom economy = **100%**
There's only one product so the atom economy has to be 100%. If you didn't spot this, the calculations would have been:
mass of reactants = (2 × 14.0) + (3 × (2 × 1.0)) = 34.0
mass of 2NH_3 = 2 × (14.0 + (3 × 1.0)) = 34.0
% atom economy = (34.0 ÷ 34.0) × 100 = 100%
Reaction 2:
mass of reactants = (2 × (14.0 + (4 × 1.0) + 35.5)) + (40.1 + ((16.0 + 1.0) × 2)) = 107 + 74.1 = 181.1
mass of 2NH_3 = 2 × (14.0 + (3 × 1.0)) = 34.0
$$\text{\% atom economy} = \frac{\text{molecular mass of desired product}}{\text{sum of molecular masses of all reactants}} \times 100$$
= (34.0 ÷ 181.1) × 100 = **18.8%**

b) E.g. reaction 1 has a much higher atom economy / produces no waste.

Page 71 — Fact Recall Questions

Q1 Atom economy is a measure of the proportion of reactant atoms that become part of the desired product (rather than by-products) in the balanced chemical equation.

Q2 E.g. Processes with high atom economies are better for the environment because they use fewer raw materials and produce less waste. They're also less expensive. A company using a process with a high atom economy will spend less on raw materials, and also less on treating waste.

Q3 $$\text{\% atom economy} = \frac{\text{molecular mass of desired product}}{\text{sum of molecular masses of all reactants}} \times 100$$

Exam-style Questions — pages 73-75

1 B ***(1 mark)***

One mole of any substance contains 6.02×10^{23} particles, so the substance with the largest number of particles will be the one with the most number of moles.

2 C ***(1 mark)***

M_r of CO = 12.0 + 16.0 = 28.0

Moles of CO = 5.00 ÷ 28.0 = 0.178... moles

From the equation, 0.178... moles of CO will produce 0.178... moles of CO_2.

Theoretical yield = moles × M_r = 0.178... × 44.0 = 7.85... g

Percentage yield = 6.40 ÷ 7.85... = 81.5%

3.1 M_r = 23.0 + 35.5 = 58.5 ***(1 mark)***

Number of moles = 20.0 ÷ 58.5 = **0.342 moles** ***(1 mark)***

3.2 2 moles of NaCl react to form 1 mole of Cl_2, so the number of moles of Cl_2 produced from 0.342 moles NaCl is:

0.342 ÷ 2 = **0.171 moles**

(1 mark)

Even if you got the first part wrong you can still get the marks for the second part as long as you've used the correct method. This will happen for all of the calculation questions.

3.3 98 kPa = 98 000 Pa

$pV = nRT$

$V = nRT \div p$

= (0.65 × 8.31 × 330) ÷ 98 000 = **0.018 m³**

(3 marks for correct answer, otherwise 1 mark for converting 98 kPa into Pa and 1 mark for correctly rearranging the equation to find V.)

3.4 mass of reactants = (2 × (23.0 + 35.5)) + (2 × ((2 × 1.0) + 16.0)) = 153

mass of Cl_2 = 2 × 35.5 = 71.0

% atom economy = (71.0 ÷ 153) × 100 = **46.4%**

(2 marks for correct answer, otherwise 1 mark for correct masses.)

3.5 E.g. the other products (H_2 and NaOH) are useful starting chemicals for other reactions/can be sold to make money ***(1 mark)***.

4.1 O = 43.6   P = 100 – 43.6 = 56.4 ***(1 mark)***

Moles of each element:

P = (56.4 ÷ 31.0) = 1.82 moles

O = (43.6 ÷ 16.0) = 2.73 moles

Divide each by 1.82:

P = (1.82 ÷ 1.82) = 1.0   O = (2.73 ÷ 1.82) = 1.5

The ratio of P : O is 1.0 : 1.5 ***(1 mark)***.

Multiply by 2: 2 × (1.0 : 1.5) = 2 : 3.

So the empirical formula is $\mathbf{P_2O_3}$ ***(1 mark)***.

All the numbers in an empirical formula must be whole numbers — that's why you need to multiply the ratio by two here.

4.2 empirical mass = (2 × 31.0) + (3 × 16.0) = 110 g

molecular mass = 220 g

(220 ÷ 110) = 2 empirical units in the molecular formula ***(1 mark)***

Molecular formula = $\mathbf{P_4O_6}$ ***(1 mark)***

4.3 M_r of NH_3 = 14.0 + (3 × 1.0) = 17.0 ***(1 mark)***

number of moles = mass ÷ M_r

= 2.50 ÷ 17 = **0.147** ***(1 mark)***

4.4 Number of moles = 0.147 ÷ 2 = **0.0735** ***(1 mark)***

The equation tells you that the molar ratio of $NH_3 : (NH_4)_2HPO_4$ is 2 : 1. So to find the number of moles of $(NH_4)_2HPO_4$ you divide the number of moles of NH_3 by 2.

4.5 M_r of $(NH_4)_2HPO_4$ = (2 × (14.0 + (4 × 1.0))) + 1.0 + 31.0 + (4 × 16.0)

= 132 ***(1 mark)***

mass = number of moles × M_r

= 0.0735 × 132 = **9.70 g** ***(1 mark)***

5.1 The atom economy is 100% ***(1 mark)*** because the reaction only has one product ***(1 mark)***.

If the reaction only has one product, all of the reactants end up in the desired product.

5.2 M_r of Ca = 40.1 ***(1 mark)***

number of moles = mass ÷ M_r

= 3.40 ÷ 40.1 = **0.0848** ***(1 mark)***

5.3 Number of moles = **0.0848** ***(1 mark)***

You can see from the equation that the molar ratio of Ca : CaO is 2 : 2. So the number of moles of CaO must be the same as the number of moles of Ca.

5.4 M_r of CaO = 40.1 + 16.0 = 56.1 ***(1 mark)***

mass = number of moles × M_r

= 0.0848 × 56.1 = **4.76 g** ***(1 mark)***

5.5 % yield = (actual yield ÷ theoretical yield) × 100

= (3.70 ÷ 4.76) × 100 = **77.7%** ***(1 mark)***

6.1 $C_8H_{18\,(l)} + 12\frac{1}{2}O_{2\,(g)} \rightarrow 8CO_{2\,(g)} + 9H_2O_{(l)}$

(1 mark for balanced equation, 1 mark for state symbols. Allow any correct multiple of the balanced equation. Allow (g) as the state symbol for water.)

6.2 $pV = nRT$

$n = pV \div RT$ ***(1 mark)***

= (101 000 × 0.020) ÷ (8.31 × 308) = **0.79 moles** ***(1 mark)***

6.3 Number of moles = 0.79 ÷ 8 = 0.099 ***(1 mark)***

From the balanced equation you wrote in part 6.1, you know that the molar ratio of $CO_2 : C_8H_{18}$ is 8 : 1. So, to find how many moles of octane were burnt, you can divide the number of moles of CO_2 by 8.

6.4 C = 85.7   H = 100 – 85.7 = 14.3 ***(1 mark)***

Moles of each element:

C = (85.7 ÷ 12) = 7.14 moles

H = (14.3 ÷ 1) = 14.3 moles

Divide each by 7.14:

C = (7.14 ÷ 7.14) = 1   H = (14.3 ÷ 7.14) = 2

The ratio of C : H is 1 : 2 ***(1 mark)***.

So the empirical formula is $\mathbf{CH_2}$ ***(1 mark)***.

7.1 moles KOH = concentration × volume (dm³)

= 0.5 × (150 ÷ 1000) = 0.075 ***(1 mark)***

M_r of KOH = 39.1 + 16.0 + 1.0 = 56.1 ***(1 mark)***

Mass = number of moles × M_r

= 0.075 × 56.1 = **4.2 g** ***(1 mark)***

7.2 Average titre = (26.00 + 26.05 + 26.00) ÷ 3

= 26.02 cm³ ***(1 mark)***

Titration 3 is an anomalous result so you should ignore it when calculating the average titre.

moles KOH = concentration × volume (dm³)

= 0.5 × (26.02 ÷ 1000) = 0.013 ***(1 mark)***

moles HCl = 0.013 ***(1 mark)***

concentration of HCl = moles ÷ volume (dm³)

= 0.013 ÷ (20 ÷ 1000)

= **0.65 mol dm⁻³** ***(1 mark)***

7.3 moles HX = concentration × volume (dm³)

= 0.12 × 0.20 ***(1 mark)***

= 0.024 ***(1 mark)***

M_r of HX = mass ÷ number of moles

= 3.07 ÷ 0.024 ***(1 mark)***

= 127.9 ***(1 mark)***

A_r of X = M_r of HX – A_r of H

= 127.9 – 1.0 = 126.9 ***(1 mark)***

So the halogen, X, is iodine (I) ***(1 mark)***.

Section 3: Bonding

1. Ionic Bonding

Page 80 — Application Questions

Q1 a) −1
b) +1
c) +2

Q2 a) +2
b) −1
c) CaI_2

Q3 a) LiF
b) A lithium atom (Li) loses 1 electron to form a lithium ion (Li^+). The fluorine atom (F) gains 1 electron to form a fluoride ion (F^-). Electrostatic attraction holds the positive and negative ions together — this is an ionic bond.

Q4 Magnesium sulfate is an ionic compound, so will have a giant ionic lattice structure where the ions are held together by very strong electrostatic forces. These bonds require a lot of energy to break, so magnesium sulfate will have a very high melting point.

Page 80 — Fact Recall Questions

Q1 a) SO_4^{2-}
b) NH_4^+

Q2 It holds positive and negative ions together.

Q3 A regular structure made up of ions.

Q4 E.g

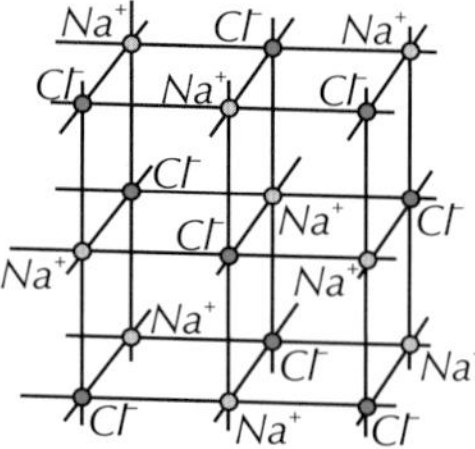

Q5 The ions in a liquid are free to move and carry a charge.

Q6 It will have a high melting point. It will dissolve in water.

2. Covalent Bonding

Page 84 — Application Questions

Q1 a) 5
b) 8
c) phosphorus
d)

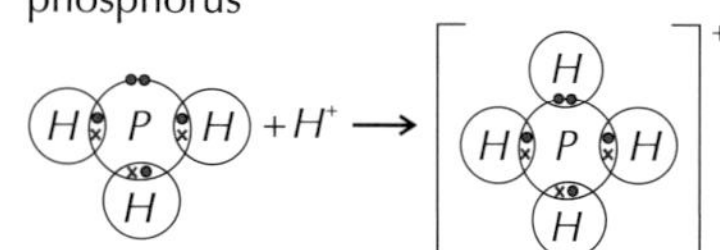

You could also draw PH_4^+ like this: $\left[\begin{array}{c} H \\ \uparrow \\ H-P-H \\ | \\ H \end{array}\right]^+$

Q2 SiO_2 is a giant covalent compound so the bonds between the atoms are very strong covalent bonds which require a lot of energy to break, so it will have a very high melting point. All the outer electrons in SiO_2 are held in localised bonds, so it won't conduct electricity. As the covalent bonds are very difficult to break, SiO_2 will be insoluble.

Page 84 — Fact Recall Questions

Q1 It forms when two atoms share electrons so that they've both got full outer shells of electrons.

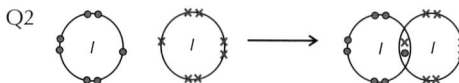

Q2 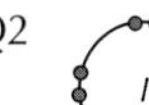  → 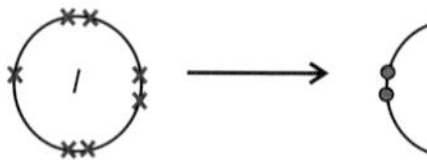

Q3 It is a covalent bond formed when two atoms share three pairs of electrons.

Q4 The weak bonds between layers are easily broken, so the sheets can slide over each other.

Q5 Diamond is a giant covalent structure made up of carbon atoms. Each carbon atom is covalently bonded to four other carbon atoms and the atoms arrange themselves in a tetrahedral shape.

Q6 A bond formed between two atoms where one of the atoms provides both of the shared electrons.

Q7 The arrow shows a pair of electrons from one atom shared between 2 atoms in a co-ordinate bond. The direction of the arrow shows which atom is the donor atom.

3. Charge Clouds

Page 85 — Fact Recall Questions

Q1 Electrons in an atom that are unshared.

Q2 A charge cloud is an area where you have a really big chance of finding an electron pair.

Q3 Lone-pair charge clouds repel more than bonding-pair charge clouds. So, the greatest angles are between lone pairs of electrons, and bond angles between bonding pairs are often reduced because they are pushed together by lone-pair repulsion.

4. Shapes of Molecules

Page 91 — Application Questions

Q1 a) Sulfur has 6 outer electrons and 2 hydrogen atoms donate one electron each. So there are 8 electrons on the S atom, which is **4** electron pairs.
b) 2 electron pairs are involved in bonding, so there are **2** lone pairs.
c)

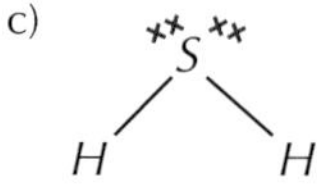

d) non-linear/bent
e) 104.5°

Q2 a) Oxygen has 6 outer electrons and 3 hydrogen atoms donate one electron each. It's a positive ion so one electron has been removed. So there are 8 electrons on the O atom, which is **4** electron pairs.
b) 3 electron pairs are involved in bonding, so there is **1** lone pair.
c)

The shape is trigonal pyramidal.

d) 107°

Q3 a)

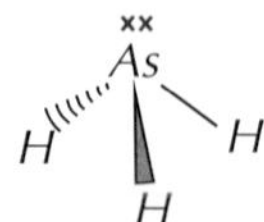

The shape is trigonal pyramidal.

Arsenic has 5 outer electrons and 3 hydrogen atoms donate one electron each. So there are 8 electrons on the As atom, which is 4 electron pairs. 3 electron pairs are involved in bonding, so there is 1 lone pair.

b) 107°

Q4

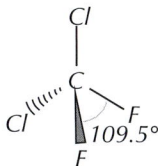

The shape is tetrahedral.

Carbon has 4 outer electrons and 2 chlorine and 2 fluorine atoms donate one electron each. So there are 8 electrons on the C atom, which is 4 electron pairs. All the electron pairs are involved in bonding.

Page 91 — Fact Recall Questions

Q1 120°

Q2 4

Q3 seesaw

5. Polarisation

Page 93 — Application Questions

Q1 a)

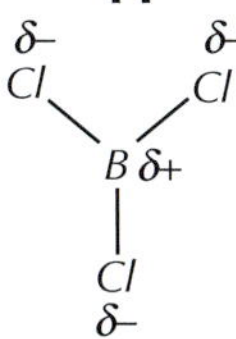

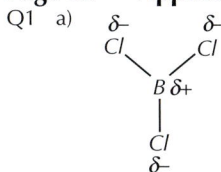

The polar B-Cl bonds cancel each other out, so the molecule has no permanent dipole (it's non-polar).

b)

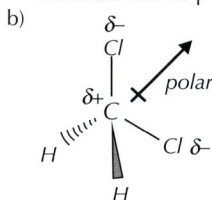

The two C–Cl bonds point in roughly the same direction, so the molecule has a permanent dipole (it's polar).

c)

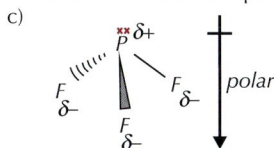

The P–F bonds all point in roughly the same direction, so the molecule has a permanent dipole (it's polar).

Q2 Chlorine has a much higher electronegativity than silicon, so $SiCl_4$ will have polar bonds — the Si–Cl bonds will be polarised towards chlorine. The shape of the molecule is tetrahedral, so the four polar Si–Cl will balance each other out, and the molecule won't have a permanent dipole.

Page 93 — Fact Recall Questions

Q1 Electronegativity is the ability to attract the bonding electrons in a covalent bond. So a chlorine atom is better able to attract the bonding electrons than a hydrogen atom.

Q2 Fluorine is more electronegative than hydrogen so attracts the electrons in the H—F covalent bond more than hydrogen. The bonding electrons are pulled towards the fluorine atom. This makes the bond polar.

Q3 A dipole is a difference in charge between two atoms caused by a shift in the electron density in the bond between them.

6. Intermolecular Forces

Page 99 — Applications Questions

Q1 van der Waals forces / induced dipole-dipole forces

Q2 a) The weak van der Waals forces between the chlorine molecules are easily broken.

b) No — chlorine wouldn't conduct electricity because there are no free ions to carry the charge.

Q3 Decane is a larger molecule/contains more atoms than methane, therefore it has a larger electron cloud and stronger van der Waals forces between the molecules. This means more energy is needed to break the forces between the molecules.

Q4 a) Oxygen is more electronegative than chlorine so it has a greater ability to pull the bonding electrons away from hydrogen atoms. So the bonds are more polarised in H_2O than in HCl, which means that hydrogen bonds form in H_2O but not in HCl.

b) fluorine/F

c) carbon/C

Q5 a) NH_3 has hydrogen bonds between molecules whereas PH_3 only has van der Waals forces, as the electronegativity values of P and H are very similar. It takes less energy to break van der Waals forces than hydrogen bonds so the boiling point of PH_3 is lower.

b) lower

Page 99 — Fact Recall Questions

Q1 There are covalent bonds within iodine molecules and van der Waals forces between iodine molecules.

Q2 Permanent dipole-dipole forces are weak electrostatic forces of attraction between polar molecules.

Q3 hydrogen bonding

Q4 a) hydrogen bonding

b) E.g.

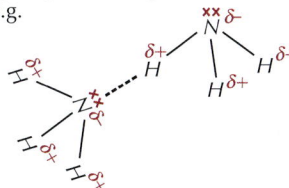

Make sure you draw the hydrogen bonds coming from the lone pair of electrons on the nitrogen atoms.

Q5 Ice has more hydrogen bonds than liquid water, and hydrogen bonds are relatively long. So the H_2O molecules in ice are further apart on average than in liquid water, making ice less dense than liquid water.

7. Metallic Bonding

Page 100 — Fact Recall Questions

Q1 Magnesium exists as a giant metallic lattice structure. The outermost shell of electrons of a magnesium atom is delocalised — the electrons are free to move about the metal. This leaves positive metal ions, Mg^{2+}, which are attracted to the delocalised negative electrons. They form a lattice of closely packed positive ions in a sea of delocalised electrons.

Q2 metallic bonding

Q3 a) As there are no bonds holding specific ions together, the copper ions can slide over each other when the structure is pulled, so it can be drawn into a wire.

b) Copper has delocalised electrons which can pass kinetic energy to each other, making copper a good thermal conductor.

8. Properties of Materials

Page 103 — Application Questions

Q1 a) Giant covalent/macromolecular.
b) Silicon dioxide has a high melting point because the strong covalent forces between all the atoms have to be broken to turn silicon dioxide from a solid to a liquid. This takes a lot of energy.

Q2 Simple covalent/molecular
The sample won't conduct electricity when solid, so it isn't metallic. It won't dissolve in water, so it isn't ionic. It has a low melting point, so it isn't giant covalent. That just leaves simple covalent (molecular), which fits all three properties.

Page 103 — Fact Recall Questions

Q1 The particles go from vibrating about a fixed point and being unable to move about freely, to being able to move about freely and randomly.

Q2 To melt a simple covalent substance you only have to overcome the intermolecular forces that hold the molecules together. To melt a giant covalent substance you need to break the much stronger covalent bonds that hold the structure together. This takes a lot more energy, so the melting points of giant covalent substances are higher.

Q3 Three from: e.g. high melting and boiling points / typically solid at room temperature / conducts electricity as a solid and a liquid / insoluble in water / malleable.

Exam-style Questions — pages 105-107

1 C ***(1 mark)***

2 B ***(1 mark)***

3.1 Covalent bonding ***(1 mark)***. A germanium atom and a hydrogen atom share a pair of electrons, with each atom donating one electron ***(1 mark)***.

3.2

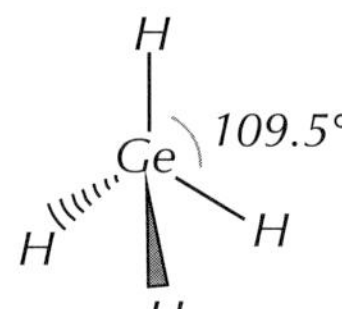

(2 marks — 1 mark for correct shape, 1 mark for correct bond angle)

The shape is tetrahedral ***(1 mark)***.
Ge has 4 outer electrons and each H atom donates 1 electron, so there are 8 electrons on the Ge atom, which is 4 electron pairs. There are 4 bonding pairs, so there are no lone pairs.

3.3 It will not conduct electricity ***(1 mark)*** as there are no free ions or electrons to carry the charge ***(1 mark)***.

3.4

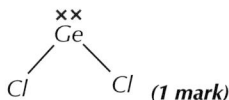

××
Ge
Cl Cl ***(1 mark)***

The shape is non-linear/bent ***(1 mark)***
Ge has 4 outer electrons and each Cl atom donates 1 electron, so there are 6 outer electrons on the Ge atom, which is 3 electron pairs. There are 2 bonding pairs, so there is 1 lone pair.

3.5 Accept 112°-119° ***(1 mark)*** There are 3 pairs of electrons on the central atom, but lone pair/bonding pair repulsion is greater than bonding pair/bonding pair repulsion, so the bond angle will be less than 120° ***(1 mark)***.

4.1 As you go down Group 5 the atoms have more electrons, so the van der Waals forces between molecules increase ***(1 mark)***. It takes more energy to break the stronger van der Waals forces so the boiling points increase from PH_3 to SbH_3 ***(1 mark)***.

4.2 hydrogen bonding ***(1 mark)***

4.3 It is a dative/co-ordinate bond ***(1 mark)***. The ammonia molecule has a lone pair of electrons ***(1 mark)*** which it donates to the hydrogen ion to form a co-ordinate bond ***(1 mark)***.

4.4 NH_3 has one lone pair of electrons and three bonding pairs whereas NH_4^+ has four bonding pairs ***(1 mark)***.
As lone-pair/bonding pair repulsion is greater than bonding-pair/bonding-pair repulsion ***(1 mark)***, the bond angle is pushed smaller in the NH_3 molecule than in NH_4^+ ***(1 mark)***.

5.1 Sodium metal has a giant metallic lattice structure ***(1 mark)*** made up of Na^+ ions surrounded by delocalised electrons ***(1 mark)***. The atoms in sodium are held together by metallic bonds formed due to the electrostatic attraction between the positively charged metal ions and the delocalised sea of electrons ***(1 mark)***.

5.2 As there are no bonds holding specific ions together, the sodium ions can slide over each other when the structure is pulled ***(1 mark)***.

5.3 Oppositely charged ions held together by electrostatic attraction ***(1 mark)***.

5.4 Sodium chloride is a giant ionic lattice ***(1 mark)*** with a cube shape made up of alternating sodium and chloride ions ***(1 mark)***.

6.1 Electronegativity is the ability to attract the bonding electrons in a covalent bond ***(1 mark)***.
If there's a difference in the electronegativities of two covalently bonded atoms, there's a shift in the electron density towards the more electronegative atom, giving the bond a dipole and causing it to be polar ***(1 mark)***. If the shape of the molecule means the dipoles aren't cancelled out then the molecule will have a permanent dipole ***(1 mark)*** and there will be weak electrostatic forces of attraction, known as permanent dipole-dipole interactions, between molecules ***(1 mark)***.

6.2 HCl: van der Waals forces / induced dipole-dipole forces and permanent dipole-dipole forces ***(1 mark)***
CH_4: van der Waals forces / induced dipole-dipole forces ***(1 mark)***
H_2O: van der Waals forces / induced dipole-dipole forces, permanent dipole-dipole forces and hydrogen bonds ***(1 mark)***

6.3

$\delta+$ $\delta-$ $\delta+$ $\delta-$
H — F ---- H — F (with lone pairs ×× on each F)

(3 marks — 1 mark for showing all lone pairs, 1 mark for showing the partial charges and 1 mark for showing the hydrogen bond going from a lone pair on an F atom to an H atom.)

6.4 The Cl atoms in Cl_2 have equal electronegativities so the covalent bond is non-polar so only van der Waals forces will form between the molecules (permanent dipole-dipole forces and hydrogen bonds can't form) ***(1 mark)***.

7.1 Giant covalent/macromolecular ***(1 mark)***

7.2 Graphite consists of flat sheets of carbon atoms covalently bonded to three other carbon atoms and arranged in hexagons ***(1 mark)***. The sheets of hexagons are bonded together by weak van der Waals forces ***(1 mark)***. Each carbon atom in diamond is covalently bonded to four other carbon atoms ***(1 mark)***. The atoms arrange themselves in a tetrahedral shape ***(1 mark)***.

7.3 The delocalised electrons in graphite are free to move along the sheets, so an electric current can flow, unlike in diamond where there are no free electrons ***(1 mark)***.

7.4 Van der Waals forces / induced dipole-dipole forces ***(1 mark)*** The forces between the C_3H_8 molecules would be stronger because each molecule contains more atoms so has larger electron clouds ***(1 mark)***.

7.5 For diamond to boil the covalent bonds between carbon atoms have to be broken ***(1 mark)***. This would need a lot more energy than breaking the van der Waals forces between methane molecules ***(1 mark)***.

Section 4: Energetics

1. Enthalpy

Page 108 — Fact Recall Questions

Q1 $\Delta H_{298}^{\ominus}$

Q2 Exothermic reactions give out energy, and endothermic reactions absorb energy. For exothermic reactions, the enthalpy change (ΔH) is negative. For endothermic reactions it is positive.

2. Bond Enthalpies

Page 111 — Application Questions

Q1 a) Bonds broken = (1 × C=C) + (1 × H–H)
So total energy absorbed = 612 + 436
= 1048 kJ mol^{-1}.
Bonds formed = (2 × C–H) + (1 × C–C)
So total energy released = (2 × 413) + 347
= 1173 kJ mol^{-1}.
Enthalpy change of reaction
= total energy absorbed – total energy released
= 1048 – 1173 = **–125 kJ mol^{-1}**.

b) Bonds broken = (1 × C–O) + (1 × H–Cl)
So total energy absorbed = 358 + 432
= 790 kJ mol^{-1}.
Bonds formed = (1 × C–Cl) + (1 × O–H)
So total energy released = 346 + 464
= 810 kJ mol^{-1}.
Enthalpy change of reaction
= total energy absorbed – total energy released
= 790 – 810 = **–20 kJ mol^{-1}**.

c) Bonds broken =
(2 × C–C) + (8 × C–H) + (5 × O=O)
So total energy absorbed =
(2 × 347) + (8 × 413) + (5 × 498) = 6488 kJ mol^{-1}.
Bonds formed = (6 × C=O) + (8 × O–H)
So total energy released = (6 × 805) + (8 × 464)
= 8542 kJ mol^{-1}.
Enthalpy change of reaction/combustion
= total energy absorbed – total energy released
= 6488 – 8542 = **–2054 kJ mol^{-1}**.

It really helps if you've drawn a sketch for this question.

d) Bonds broken = (1 × C–Cl) + (1 × N–H)
So total energy absorbed = 346 + 391
= 737 kJ mol^{-1}.
Bonds formed = (1 × C–N) + (1 × H–Cl)
So total energy released = 286 + 432
= 718 kJ mol^{-1}.
Enthalpy change of reaction
= total energy absorbed – total energy released
= 737 – 718 = **+19 kJ mol^{-1}**.

Q2 The balanced equation for the combustion of ethene is:
$C_2H_4 + 3O_2 \rightarrow 2CO_2 + 2H_2O$.
Bonds broken =
(1 × C=C) + (4 × C–H) + (3 × O=O)
So total energy absorbed =
(1 × 612) + (4 × 413) + (3 × 498) = 3758 kJ mol^{-1}.
Bonds formed = (4 × C=O) + (4 × O–H)
So total energy released = (4 × 805) + (4 × 464)
= 5076 kJ mol^{-1}.
Enthalpy change of combustion
= total energy absorbed – total energy released
= 3758 – 5076 = **–1318 kJ mol^{-1}**.

Q3 The balanced equation for the formation of 1 mole of HCl is: $\frac{1}{2}H_2 + \frac{1}{2}Cl_2 \rightarrow HCl$.
The enthalpy change of formation is the enthalpy change when 1 mole of a compound is formed, so your equation needs to have 1 mole of HCl on the RHS, which means you need half a mole of H_2 and half a mole of Cl_2 on the LHS of the equation.
Bonds broken = ½(1 × H–H) + ½(1 × Cl–Cl)
So total energy absorbed = (½ × 436) + (½ × 243)
= 339.5 kJ mol^{-1}.
Bonds formed = 1 × H–Cl
So total energy released = 432 kJ mol^{-1}.
Enthalpy change of formation
= total energy absorbed – total energy released
= 339.7 – 432 = **–92.5 kJ mol^{-1}**.

Q4 Call the unknown bond enthalpy between N and O 'X'.
Bonds broken = 2 × X
So total energy absorbed = 2X kJ mol^{-1}.
Bonds formed = (1 × N≡N) + (1 × O=O)
So total energy released = 945 + 498
= 1443 kJ mol^{-1}.
Enthalpy change of reaction = –181 kJ mol^{-1}
= total energy absorbed – total energy released.
So: –181 = 2X – 1443
2X = –181 + 1443 = 1262
X = 1262 ÷ 2 = **+631 kJ mol^{-1}**.

Page 111 — Fact Recall Questions

Q1 a) $\Delta_f H^{\ominus}$
b) $\Delta_c H^{\ominus}$

Q2 Standard enthalpy change of formation, $\Delta_f H^{\ominus}$, is the enthalpy change when 1 mole of a compound is formed from its elements in their standard states under standard conditions.

3. Measuring Enthalpy Changes

Page 115 — Application Questions

Q1 $q = mc\Delta T$ = 220 × 4.18 × (301 – 298) = 2758.8 J
= 2.7588 kJ

$$\Delta H = \frac{q}{n} = -\frac{2.7588\text{ kJ}}{0.0500\text{ mol}} = \mathbf{-55.2\ kJ\ mol^{-1}}\text{ (to 3 s.f.)}.$$

Don't forget — the enthalpy change must be negative because it's an exothermic reaction (you can tell because the temperature increased).

Q2 a) $q = mc\Delta T$ = 200 × 4.18 × 29.0 = 24244 J
= 24.244 kJ

$$n = \frac{\text{mass}}{M} = \frac{0.500\text{ g}}{72\text{ g mol}^{-1}} = 0.00694\text{ moles of fuel}$$

$$\Delta H = \frac{q}{n} = -\frac{24.244\text{ kJ}}{0.00694\text{ mol}}$$

= **–3490 kJ mol^{-1}** (to 3 s.f.).

b) E.g. some heat from the combustion will have been lost to the surroundings (rather than being transferred to the water). / Some of the combustion may have been incomplete combustion. / If the fuel was volatile, some of it may have been lost to evaporation. / The experiment may not have taken place under standard conditions.

Q3 $q = mc\Delta T = 300 \times 4.18 \times 55 = 68\,970$ J
$= 68.97$ kJ

$\Delta_c H^\circ$ octane $= -5470$ kJ mol^{-1} $= \frac{q}{n}$

$n = \frac{q}{\Delta H} = \frac{68.97\text{ kJ}}{5470\text{ kJ mol}^{-1}} = 0.0126...$ mol.

$n = \frac{\text{mass}}{M}$, so mass $= n \times M = 0.0125... \times 114$

$=$ **1.44 g** of octane (to 3 s.f.).

Page 115 — Fact Recall Questions

Q1 The temperature change due to the reaction, the mass of the reactant (which is used to calculate the number of moles that has reacted), and the mass of the stuff that's being heated.

Q2 E.g.

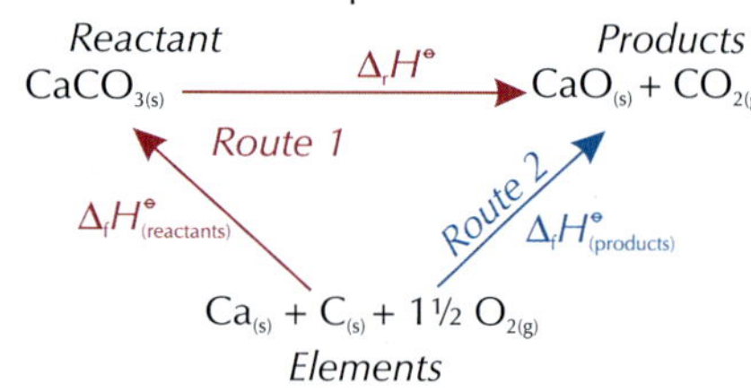

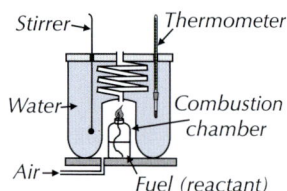

4. Hess's Law

Page 119 — Application Questions

Q1 a) First draw out a reaction scheme with an alternative reaction route that includes balanced equations for the formation of each compound:

Reactant — Products

$CaCO_{3(s)} \xrightarrow{\Delta_r H^\circ} CaO_{(s)} + CO_{2(g)}$

Route 1: $\Delta_f H^\circ_{(reactants)}$ — Route 2: $\Delta_f H^\circ_{(products)}$

$Ca_{(s)} + C_{(s)} + 1\frac{1}{2} O_{2(g)}$

Elements

$\Delta_f H^\circ_{(reactants)} = \Delta_f H^\circ_{[CaCO_3]} = -1207$ kJ mol^{-1}.

$\Delta_f H^\circ_{(products)} = \Delta_f H^\circ_{[CaO]} + \Delta_f H^\circ_{[CO_2]}$

$\Delta_f H^\circ_{(products)} = -635 + -394 = -1029$ kJ mol^{-1}.

Using Hess's Law: Route 1 = Route 2, so:

$\Delta_f H^\circ_{(reactants)} + \Delta_r H^\circ = \Delta_f H^\circ_{(products)}$

$-1207 + \Delta_r H^\circ = -1029$

$\Delta_r H^\circ = -1029 + 1207 =$ **+178 kJ mol^{-1}**.

b) Reactants — Products

$2N_2O_{5(s)} \xrightarrow{\Delta_r H^\circ} 4NO_{2(g)} + O_{2(g)}$

Route 1: $\Delta_f H^\circ_{(reactants)}$ — Route 2: $\Delta_f H^\circ_{(products)}$

$2N_{2(g)} + 5O_{2(g)}$

Elements

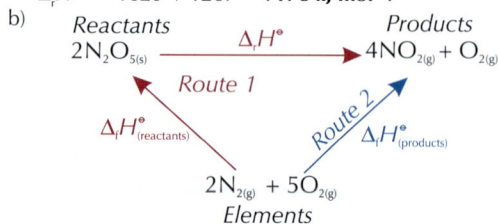

$\Delta_f H^\circ_{(reactants)} = 2 \times \Delta_f H^\circ_{[N_2O_5]}$

$\Delta_f H^\circ_{(reactants)} = 2 \times -41 = -82$ kJ mol^{-1}.

$\Delta_f H^\circ_{(products)} = (4 \times \Delta_f H^\circ_{[NO_2]}) + \Delta_f H^\circ_{[O_2]}$

$\Delta_f H^\circ_{(products)} = (4 \times 33) + 0 = 132$ kJ mol^{-1}.

Remember that the enthalpy change of formation of O_2 is zero because it's an element.

Using Hess's Law: Route 1 = Route 2, so:

$\Delta_f H^\circ_{(reactants)} + \Delta_r H^\circ = \Delta_f H^\circ_{(products)}$

$-82 + \Delta_r H^\circ = 132$

$\Delta_r H^\circ = 132 + 82 =$ **+214 kJ mol^{-1}**.

Q2 a) *First draw out balanced reactions for the formation of the compound, and the combustion of the reactants and product:*

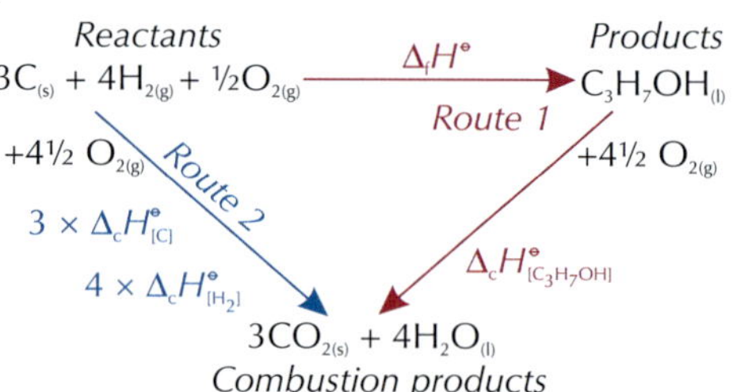

Using Hess's Law: Route 1 = Route 2, so:

$\Delta_f H^\circ + \Delta_c H^\circ_{[C_3H_7OH]} = (3 \times \Delta_c H^\circ_{[C]}) + (4 \times \Delta_c H^\circ_{[H_2]})$

$\Delta_f H^\circ + (-2021) = (3 \times -394) + (4 \times -286)$

$\Delta_f H^\circ = -1182 - 1144 + 2021 =$ **−305 kJ mol^{-1}**.

b)

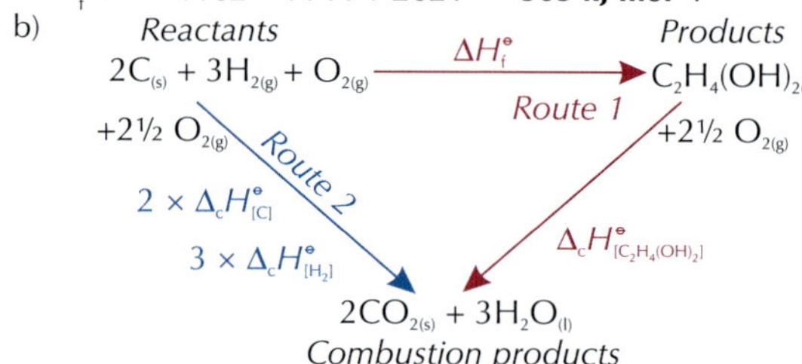

Using Hess's Law: Route 1 = Route 2, so:

$\Delta_f H^\circ + \Delta_c H^\circ_{[C_2H_4(OH)_2]} = (2 \times \Delta_c H^\circ_{[C]}) + (3 \times \Delta_c H^\circ_{[H_2]})$

$\Delta_f H^\circ + (-1180) = (2 \times -394) + (3 \times -286)$

$\Delta_f H^\circ = -788 - 858 + 1180 =$ **−466 kJ mol^{-1}**.

c)

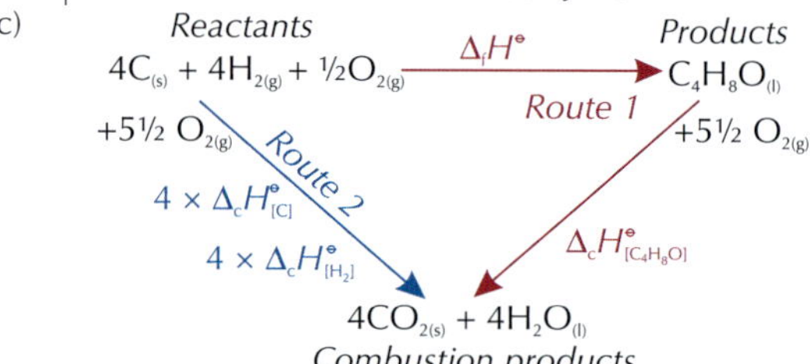

Using Hess's Law: Route 1 = Route 2, so:

$\Delta_f H^\circ + \Delta_c H^\circ_{[C_4H_8O]} = (4 \times \Delta_c H^\circ_{[C]}) + (4 \times \Delta_c H^\circ_{[H_2]})$

$\Delta_f H^\circ + (-2442) = (4 \times -394) + (4 \times -286)$

$\Delta_f H^\circ = -1576 - 1144 + 2442 =$ **−278 kJ mol^{-1}**.

Exam-style Questions — pages 120-121

1.1 $q = mc\Delta T = 50.0 \times 2.46 \times 1.00 = 123$ J $= 0.123$ kJ

$\Delta H = \frac{q}{n} = -\frac{0.123\text{ kJ}}{0.0500\text{ mol}} =$ **−2.46 kJ mol^{-1}**.

(3 marks for correct answer, otherwise 1 mark for q = mcΔT, 1 mark for 235 J or 0.235 kJ)

You've got to remember the negative sign on your answer — the temperature rises so it's an exothermic reaction.

1.2 The reaction must be carried out at a pressure of 100 kPa ***(1 mark)*** with all reactants and products in their standard states at that pressure ***(1 mark)***.

2.1 Hess's Law says that the total enthalpy change for a reaction is independent of the route taken ***(1 mark)***.

2.2 $C_8H_{18(l)} + 12\frac{1}{2}O_{2(g)} \rightarrow 8CO_{2(g)} + 9H_2O_{(l)}$ ***(1 mark)***

2.3

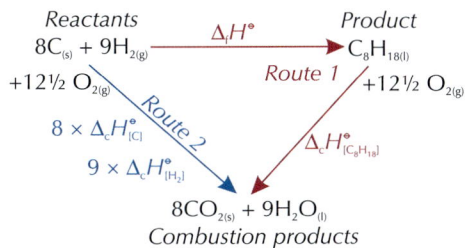

Reactants: $8C_{(s)} + 9H_{2(g)} + 12\frac{1}{2}\,O_{2(g)}$ — $\Delta_f H^\circ$ → Product: $C_8H_{18(l)} + 12\frac{1}{2}\,O_{2(g)}$

Route 1: $\Delta_c H^\circ_{[C_8H_{18}]}$ → Combustion products

Route 2: $8 \times \Delta_c H^\circ_{[C]}$, $9 \times \Delta_c H^\circ_{[H_2]}$ → Combustion products

Combustion products: $8CO_{2(s)} + 9H_2O_{(l)}$

Using Hess's Law: Route 1 = Route 2, so:

$\Delta_f H^\circ + \Delta_c H^\circ_{[C_8H_{18}]} = (8 \times \Delta_c H^\circ_{[C]}) + (9 \times \Delta_c H^\circ_{[H_2]})$

$\Delta_f H^\circ + (-5470) = (8 \times -394) + (9 \times -286)$

$\Delta_f H^\circ = -3152 - 2574 + 5470 =$ $\mathbf{-256\ kJ\ mol^{-1}}$.

(3 marks for correct answer, otherwise 1 mark for correct equation using Hess's Law and 1 mark for correct molar quantities.)

2.4 Exothermic ***(1 mark)***. The enthalpy change for the formation of octane is negative ***(1 mark)***.

3.1 Standard enthalpy of combustion, $\Delta_c H^\circ$, is the enthalpy change when 1 mole of a substance ***(1 mark)*** is completely burned in oxygen ***(1 mark)*** under standard conditions with all reactants and products in their standard states ***(1 mark)***.

3.2 The balanced equation for the combustion of but-1-ene is: $C_4H_8 + 6O_2 \rightarrow 4CO_2 + 4H_2O$.
Bonds broken =
(1 × C=C) + (2 × C–C) + (8 × C–H) + (6 × O=O)
So total energy absorbed =
(1 × 612) + (2 × 347) + (8 × 413) + (6 × 498)
= 7598 kJ mol^{-1}.
Bonds formed = (8 × C=O) + (8 × O–H)
So total energy released = (8 × 805) + (8 × 464)
= 10 152 kJ mol^{-1}.
Enthalpy change of combustion
= total energy absorbed – total energy released
= 7598 – 10 152 = **–2554 kJ** $\mathbf{mol^{-1}}$.
(3 marks for correct answer, otherwise 1 mark for 'total energy absorbed – total energy released', and 1 mark for correct value for either energy released or energy absorbed.)

3.3 Some of the mean bond enthalpies are average values for the bonds in many different compounds, so they are not accurate for the specific molecules involved in this combustion ***(1 mark)***.

4.1 Standard enthalpy change of formation, $\Delta_f H^\circ$, is the enthalpy change when 1 mole ***(1 mark)*** of a compound is formed from its elements ***(1 mark)*** in their standard states under standard conditions ***(1 mark)***.

4.2

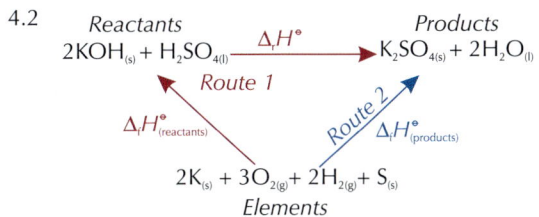

Reactants: $2KOH_{(s)} + H_2SO_{4(l)}$ — $\Delta_r H^\circ$ → Products: $K_2SO_{4(s)} + 2H_2O_{(l)}$

Route 1: Elements → Reactants: $\Delta_f H^\circ_{(reactants)}$

Route 2: Elements → Products: $\Delta_f H^\circ_{(products)}$

Elements: $2K_{(s)} + 3O_{2(g)} + 2H_{2(g)} + S_{(s)}$

$\Delta_f H^\circ_{(reactants)} = (2 \times \Delta_f H^\circ_{[KOH]}) + \Delta_f H^\circ_{[H_2SO_4]}$

$\Delta_f H^\circ_{(reactants)} = (2 \times -425) + -814 = -1664\ kJ\ mol^{-1}$.

$\Delta_f H^\circ_{(products)} = \Delta_f H^\circ_{[K_2SO_4]} + (2 \times \Delta_f H^\circ_{[H_2O]})$

$\Delta_f H^\circ_{(products)} = -1438 + (2 \times -286) = -2010\ kJ\ mol^{-1}$.

Using Hess's Law: Route 1 = Route 2, so:

$\Delta_f H^\circ_{(reactants)} + \Delta_r H^\circ = \Delta_f H^\circ_{(products)}$

$-1664 + \Delta_r H^\circ = -2010$

$\Delta_r H^\circ = -2010 + 1664 =$ $\mathbf{-346\ kJ\ mol^{-1}}$.

(3 marks for correct answer, otherwise 1 mark for correct equation using Hess's Law and 1 mark for correct molar quantities.)

Section 5: Kinetics

1. Reaction Rates

Page 124 — Application Question

Q1 B is the curve for the gas at a higher temperature because it is shifted over to the right showing that more molecules have more energy.

Page 124 — Fact Recall Questions

Q1 The particles must collide in the right direction (facing each other the right way) and with at least a certain minimum amount of kinetic energy.

Q2 The minimum amount of kinetic energy that particles need to have in order to react when they collide.

Q3 A small increase in temperature gives all molecules more energy, so a greater number of them have at least the activation energy to react when they collide. They will also collide more frequently because they will be moving about faster.

Q4 If you increase the concentration of reactants in a solution, the particles will be closer together in a given volume and so collide more frequently, increasing the reaction rate.

2. Catalysts

Page 126 — Application Question

Q1 a) E.g.

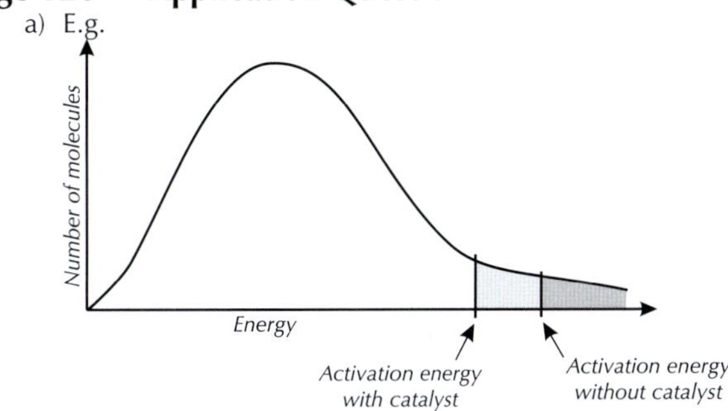

You can draw the activation energy line anywhere as long as it is lower than the activation energy for the uncatalysed reaction. The shape of the graph should stay the same.

b) Adding a catalyst would lower the activation energy for the reaction so that it would not need such a high temperature in order to take place. Being able to carry out the reaction at a lower temperature would save energy and money.

Page 126 — Fact Recall Questions

Q1 A substance that increases the rate of a reaction by providing an alternative reaction pathway with a lower activation energy. The catalyst is chemically unchanged at the end of the reaction.

Q2 A catalyst increases the rate of a reaction by providing an alternative reaction pathway with a lower activation energy. This means that more of the molecules collide with energies above the activation energy, and so can react.

3. Measuring Reaction Rates

Page 129 — Application Question

Q1 a) One of the products is a gas so the student could measure the change in mass of the open reaction vessel over time / use a gas syringe to measure the volume of gas produced by the reaction mixture over time.

b) If the student is measuring the change in mass, then the reaction is finished when the mass of the reaction vessel stops decreasing. If the student is measuring the change in gas volume, then the reaction is finished when the volume stops increasing.

c) 188-196 s.

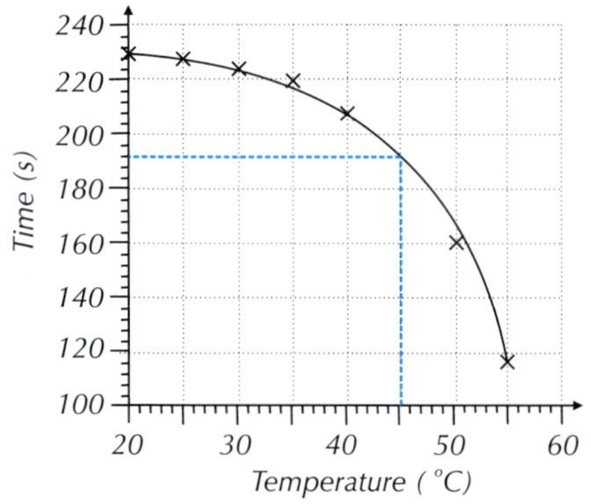

The best way to get an estimate is to draw a graph and plot the data points from the table. You can then draw a curve and read off the value on the time axis (y-axis in the example above) at 45 °C. For the curve drawn above, the estimate of time taken is 192 s. Everyone will draw a slightly different curve though so any answer from 188 s to 196 s is fine.

Page 129 — Fact Recall Questions

Q1 The rate of reaction is the change in the amount of a reactant or product over time.

Q2 Mix the reactants in a conical flask placed on a black cross. Using a stop clock, measure the time taken for enough precipitate to form so that the cross is no longer visible through the reaction mixture.

Exam-style Questions — pages 131-132

1 B ***(1 mark)***

2 C ***(1 mark)***

3 A ***(1 mark)***

4 D ***(1 mark)***

5.1 The minimum amount of kinetic energy that particles need to have in order to react when they collide ***(1 mark)***.

5.2 E.g.

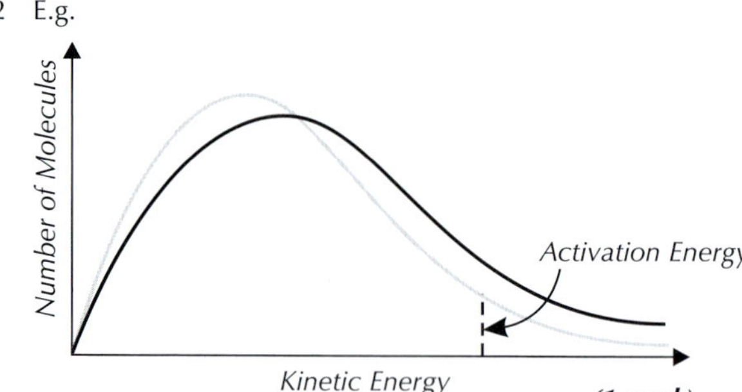

(1 mark)

There isn't just one right answer here. Your curve just needs to be pushed over to the right, with a greater number of molecules having at least the activation energy.

5.3 A catalyst is a substance that increases the rate of a reaction by providing an alternative reaction pathway with a lower activation energy. The catalyst is chemically unchanged at the end of the reaction ***(1 mark)***.

5.4 The shape doesn't change ***(1 mark)*** because the molecules still have the same amount of energy ***(1 mark)***.

5.5 The rate of reaction will increase ***(1 mark)***. Heating the reactants gives the molecules on average more kinetic energy and they will move faster ***(1 mark)***. A greater proportion of molecules will have at least the activation energy, and because the molecules are moving about faster, the frequency of collisions increases ***(1 mark)***.

5.6 Lowering the pressure will reduce the rate of reaction ***(1 mark)***. This is because there will be fewer gas molecules in a given volume/the concentration will be reduced/the molecules will be further apart ***(1 mark)***, so the frequency of collisions between molecules that result in a reaction decreases ***(1 mark)***.

Section 6: Equilibria and Redox Reactions

1. Reversible Reactions

Page 136 — Application Questions

Q1 a) Increasing the concentration of A will shift the equilibrium to the right (favouring the forwards reaction) in order to get rid of the excess A.

b) There are 3 moles on the left and only 2 on the right. Increasing the pressure will shift the equilibrium to the right (favouring the forwards reaction) in order to reduce the number of moles to reduce the pressure again.

c) The forwards reaction is exothermic, so the backwards reaction must be endothermic. Increasing the temperature will shift the equilibrium to the left (favouring the endothermic backwards reaction) in order to remove the extra heat.

d) The reaction should ideally be performed with a high concentration of A and B, at a high pressure and low temperature.

Q2 There are the same number of moles on either side of the reaction. Increasing the pressure favours the reaction producing the fewest moles, but since both reactions are equal in this respect, increasing the pressure will not shift the position of equilibrium.

Page 136 — Fact Recall Questions

Q1 At dynamic equilibrium, the concentrations of reactants and products are constant, and the forwards reaction and the backwards reaction are going at the same rate.

Q2 For a dynamic equilibrium to be established it must be a closed system and the temperature must be constant.

Q3 Le Chatelier's principle states that if there's a change in concentration, pressure or temperature, the equilibrium will move to help counteract the change.

Q4 The addition of a catalyst has no effect on the position of equilibrium in a reversible reaction.

2. Industrial Processes

Page 137 — Fact Recall Question

Q1 The temperature and pressure used in the industrial production of ethanol are a compromise. Using a low temperature increases the yield of the reaction, but at low temperatures the reaction rate becomes much slower. So a moderate temperature is used as a compromise between a good yield and a fast reaction. Using a high pressure also increases the yield of the reaction, but high pressures are expensive to produce. You need stronger pipes and containers to withstand high pressure. So a moderate pressure is used as a compromise between maximum yield and minimum expense.

3. The Equilibrium Constant

Page 140 — Application Questions

Q1 a) $K_c = \frac{[C_2H_5OH]}{[C_2H_4][H_2O]}$

b) The equation tells you that if 1 mole of C_2H_5OH decomposes, 1 mole of C_2H_4 and 1 mole of H_2O are formed. So if 1.85 moles of C_2H_4 are produced at equilibrium, there will also be **1.85** moles of H_2O. 1.85 moles of C_2H_5OH has decomposed so there must be 5.00 – 1.85 = **3.15** moles of C_2H_5OH remaining.

c) The volume of the reaction is 15.0 dm³. So the molar concentrations are:
For H_2O and C_2H_4: 1.85 ÷ 15.0 = **0.123 mol dm⁻³**.
For C_2H_5OH: 3.15 ÷ 15.0 = **0.210 mol dm⁻³**.

d) $K_c = \frac{[C_2H_5OH]}{[C_2H_4][H_2O]} = \frac{0.210}{(0.123)(0.123)} = 13.9$

Units of $K_c = \frac{\text{mol dm}^{-3}}{(\text{mol dm}^{-3})(\text{mol dm}^{-3})} = \frac{1}{(\text{mol dm}^{-3})}$

$K_c = \mathbf{13.9\ mol^{-1}\ dm^{3}}$

For the units here, you've ended up with mol dm⁻³ on its own on the bottom of the fraction. So you simplify it by swapping the signs of the powers — it becomes mol⁻¹ dm³.

e) $K_c = \frac{[C_2H_5OH]}{[C_2H_4][H_2O]}$ so $3.8 = \frac{0.80}{[C_2H_4][H_2O]}$

$[C_2H_4][H_2O] = 0.8 \div 3.8 = 0.21$
$[C_2H_4]$ and $[H_2O] = \sqrt{0.21} = \mathbf{0.46\ mol\ dm^{-3}}$

Q2 a) $K_c = \frac{[SO_3]^2}{[SO_2]^2[O_2]}$

b) $K_c = \frac{[SO_3]^2}{[SO_2]^2[O_2]} = \frac{0.36^2}{(0.25)^2(0.18)} = 11.5$

Units of $K_c = \frac{(\text{mol dm}^{-3})^2}{(\text{mol dm}^{-3})^2(\text{mol dm}^{-3})^2} = \frac{1}{\text{mol dm}^{-3}}$

$K_c = \mathbf{11.5\ mol^{-1}\ dm^{3}}$

c) $[SO_2]^2 = \frac{[SO_3]^2}{K_c \times [O_2]} = \frac{0.360^2}{15.0 \times 0.180} = 0.0480$

$[SO_2] = \sqrt{0.0480} = \mathbf{0.219\ mol\ dm^{-3}}$

4. Factors Affecting the Equilibrium Constant

Page 142 — Application Questions

Q1 a) No effect — if the concentration of C_2F_4 increases the equilibrium will shift to counteract the change and K_c will stay the same.

b) The reaction is endothermic in the forward direction so increasing the temperature will shift the equilibrium to the right. As a result more product will be produced, so K_c will increase.

c) No effect — catalysts only affect the time taken to reach equilibrium and not the position of the equilibrium itself.
Remember, it's only temperature that affects the value of K_c.

Q2 Exothermic. If decreasing the temperature increases K_c then it must increase the amount of product formed. The equilibrium must have shifted to the right, so the forward reaction must be exothermic.

Page 142 — Fact Recall Questions

Q1 a) It will increase K_c.
b) It will decrease K_c.

Q2 If the concentration of a reagent is changed the equilibrium will shift and the concentrations of other reagents will also change. So K_c will stay the same.

Q3 Adding a catalyst doesn't change K_c, but it decreases the time taken to reach equilibrium.

5. Redox Reactions

Pages 145-146 — Application Questions

Q1 a) +1
b) –1
c) +2

Q2 a) –1
b) –2
c) –1

Q3 a) H: +1, Cl: –1
b) S: +4, O: –2
c) C: +4, O: –2
Oxygen has an oxidation state of -2. There are 3 oxygen atoms here so the total is –6. The overall oxidation state of the ion is –2. So, carbon must have an oxidation state of +4 (as –6 + 4 = –2).
d) Cl: +7, O: –2
e) Cu: +1, O: –2
f) H: +1, S: +6, O: –2
Oxygen has an oxidation state of –2. There are 4 oxygen atoms here so the total is –8. The overall oxidation state of the ion is –1. Hydrogen has an oxidation state of +1, so sulfur must have an oxidation state of +6 (as –8 + 1 + 6 = –1).

Q4 a) +2
b) +4
c) +4
d) 0
e) +4
Hydrogen has an oxidation state of +1, so in C_3H_6 carbon must have an oxidation state of –2 (as (6 x +1) + (3 x –2) = 0).

Q5 a) 0
b) –3
c) +5
d) +2
Fluorine is the most electronegative element so its oxidation state is equal to its ionic charge, –1. There are 4 fluorine atoms here so the total is –4. So, phosphorus must have an oxidation state of +2 (as (4 x –1) + (2 x +2) = 0).
e) +5

f) –2

Q6 a) beginning: 0, end: +2
b) beginning: +6, end: +6
c) beginning: +1, end: 0
d) beginning: –2, end: –2

Page 146 — Fact Recall Questions

Q1 Oxidation is a loss of electrons.
Q2 Reduction is a gain of electrons.
Q3 An oxidising agent accepts electrons from another reactant and is reduced.
Q4 A reducing agent donates electrons to another reactant and is oxidised.
Q5 0
Q6 0
Q7 –1
Q8 –1

6. Redox Equations

Page 148 — Application Questions

Q1 $Zn + 2Ag^+ \rightarrow Zn^{2+} + 2Ag$
You need to multiply everything in the silver half-equation by 2 so that the e^- will cancel when you combine the two half-equations.
Q2 Oxidation equation: $Ca \rightarrow Ca^{2+} + 2e^-$
Reduction equation: $Cl_2 + 2e^- \rightarrow 2Cl^-$
Q3 $2NO_3^- + 12H^+ + 10e^- \rightarrow N_2 + 6H_2O$
Q4 $Cr_2O_7^{2-} + 14H^+ + 6e^- \rightarrow 2Cr^{3+} + 7H_2O$
Q5 $H_2SO_4 + 8H^+ + 8e^- \rightarrow H_2S + 4H_2O$

Page 148 — Fact Recall Questions

Q1 An ionic half-equation shows oxidation or reduction.
Q2 **B**

Exam-style Questions — pages 150-152

1 C ***(1 mark)***
2 D ***(1 mark)***
3 A ***(1 mark)***
4 B ***(1 mark)***
5.1 In a reversible reaction, dynamic equilibrium is reached when the concentrations of reactants and products are constant ***(1 mark)***, and the forwards reaction and the backwards reaction are going at the same rate ***(1 mark)***.
5.2 If there's a change in concentration, pressure or temperature, the equilibrium will move to help counteract the change ***(1 mark)***.
5.3 The higher the pressure, the faster the reaction rate ***(1 mark)***. A high pressure also favours the forwards reaction (which produces fewer moles) so the higher the pressure, the greater the yield of ethanol ***(1 mark)***. However, high pressures are very expensive/produce side reactions/require strong and expensive equipment, so the pressure used is limited by these factors ***(1 mark)***.
5.4 Reducing the amount of H_2O will shift the position of equilibrium to the left ***(1 mark)*** in order to increase the amount of H_2O present ***(1 mark)***. This shift will reduce the maximum yield of ethanol from the forwards reaction ***(1 mark)***.
6.1 It has no effect on the position of equilibrium ***(1 mark)***.
Catalysts increase the rate at which equilibrium is reached but don't affect the position of equilibrium.
6.2 exothermic
6.3 The forwards reaction is exothermic, so the backwards reaction is endothermic. Increasing the temperature will shift the position of equilibrium to the left ***(1 mark)***, favouring the endothermic backwards reaction ***(1 mark)*** in order to remove the excess heat ***(1 mark)***.
6.4 Any two from: e.g. increasing the pressure/increasing the concentration of N_2 and H_2/reducing the concentration of NH_3 (e.g. by removing it) ***(1 mark for each — maximum of 2 marks overall)***.
7.1 $K_c = \frac{[CO_2][H_2]^4}{[CH_4][H_2O]^2}$ ***(1 mark)***.
7.2 $K_c = (0.200 \times (0.280)^4) \div (0.0800 \times (0.320)^2) = 0.150$
$K_c = (\text{mol dm}^{-3} \times (\text{mol dm}^{-3})^4) \div (\text{mol dm}^{-3} \times (\text{mol dm}^{-3})^2) = \text{mol}^2\,\text{dm}^{-6}$ so $K_c =$ **0.150 mol² dm⁻⁶**
(3 marks for correct answer, otherwise 1 mark for correct method and 1 mark for correct units.)
7.3 $[CH_4] = \frac{[CO_2][H_2]^4}{K_c \times [H_2O]^2}$
$= (0.420 \times 0.480^4) \div (0.0800 \times 0.560^2)$
$=$ **0.889 mol dm⁻³** ***(2 marks for correct answer, otherwise 1 mark for correct equation.)***
7.4 Lower ***(1 mark)***. K_c is lower at temperature Z than at Y. This means at Z the equilibrium is being shifted to the left ***(1 mark)***. As the reaction is endothermic in the forward direction, the temperature must be lower to make it shift in the exothermic direction ***(1 mark)***.
7.5 The value of K_c would not change ***(1 mark)***. Catalysts do not affect the position of the equilibrium, only the time taken to reach equilibrium ***(1 mark)***.
8.1 $3Cl_2 + 6e^- \rightarrow 6Cl^-$ ***(1 mark)***
$2Al \rightarrow 2Al^{3+} + 6e^-$ ***(1 mark)***
8.2 Chlorine is the oxidising agent ***(1 mark)***.
8.3 +6 ***(1 mark)***

Unit 2

Section 1: Periodicity

1. The Periodic Table

Page 155 — Application Questions

Q1 a) 3
b) 2
c) 4
d) 2
e) 5
Q2 a) 6
b) 1
c) 7
d) 3
e) 2
Q3 a) $1s^2\ 2s^2\ 2p^6\ 3s^1$
b) $1s^2\ 2s^2\ 2p^6\ 3s^2\ 3p^6\ 4s^2$
c) $1s^2\ 2s^2\ 2p^6\ 3s^2\ 3p^5$
d) $1s^2\ 2s^2\ 2p^6\ 3s^2\ 3p^6\ 3d^{10}\ 4s^2\ 4p^3$
e) $1s^2\ 2s^2\ 2p^6\ 3s^2\ 3p^6\ 3d^3\ 4s^2$
f) $1s^2\ 2s^2\ 2p^6\ 3s^2\ 3p^6\ 3d^1\ 4s^2$

Page 155 — Fact Recall Questions

Q1 The periodic table is arranged into periods (rows) and groups (columns) by atomic (proton) number.
Q2 a) Z
b) Y
c) X

2. Periodicity

Page 158 — Application Questions

Q1 a) Aluminium has 13 protons and sulfur has 16 protons. So the positive charge of the nucleus of sulfur is greater. This means electrons are pulled closer to the nucleus, making the atomic radius of sulfur smaller than the atomic radius of aluminium.
b) sodium/magnesium

Q2 Sulfur has more protons than aluminium, so its electrons are attracted more strongly towards the nucleus. This means it takes more energy to remove an electron from each atom, so it has a higher first ionisation energy.

Q3 a) Silicon is macromolecular so has strong covalent bonds linking all its atoms together. Phosphorus is a molecular substance with van der Waals forces between its molecules. It takes much less energy to break van der Waals forces than covalent bonds so the melting point of phosphorus is much lower than the melting point of silicon.
b) chlorine / argon

Page 158 — Fact Recall Questions

Q1 The atomic radius decreases across Period 3.

Q2 The melting points increase from sodium to silicon, but then generally decrease from silicon to argon.

Q3 There's a general increase in the first ionisation energy as you go across Period 3.

Exam-style Questions — page 159

1 B ***(1 mark)***

2 C ***(1 mark)***
Sulfur is in the p block of the periodic table.

3 B ***(1 mark)***
First ionisation energy generally increases across Period 3.

4.1 The atomic radius of aluminium is smaller than that of sodium because an aluminium atom contains more protons, which increases the positive charge of the nucleus ***(1 mark)***. This means the electrons are pulled closer to the nucleus, making the atomic radius smaller ***(1 mark)***.

4.2 Magnesium ions have a 2+ charge, whereas sodium ions only have a 1+ charge ***(1 mark)***. Magnesium ions also have more delocalised electrons and a smaller radius than sodium ions ***(1 mark)***. So the metal-metal bonds are stronger in magnesium than in sodium ***(1 mark)***.

4.3 Sodium, magnesium and aluminium are all metals with strong metallic bonds holding the atoms together, so they have quite high melting points ***(1 mark)***. However, argon exists as individual atoms with only weak van der Waals forces between the atoms, so it has a very low melting point ***(1 mark)***.

Section 2: Group 2 and Group 7 Elements

1. Group 2 — The Alkaline Earth Metals

Page 162 — Application Questions

Q1 Strontium with dilute hydrochloric acid.

Q2 a) Element X
b) Element Y

Page 162 — Fact Recall Questions

Q1 Atomic radius increases down the group.

Q2 First ionisation energy decreases down the group. This is because each element down Group 2 has an extra electron shell compared to the one above. The extra inner shells shield the outer electrons from the attraction of the nucleus. Also, the extra shell means that the outer electrons are further away from the nucleus, which greatly reduces the nucleus's attraction. Both of these factors make it easier to remove outer electrons, resulting in a lower first ionisation energy.

Q3 a) Barium has a smaller atomic radius than radium, so its two delocalised electrons will be closer to the positive ion/nucleus. This means they will be more strongly attracted to the positive ion/nucleus than the delocalised electrons in radium. So it will take more energy to break the bonds in barium, which means it has a higher melting point.
b) Magnesium — the crystal structure changes.

2. Group 2 Compounds

Page 165 — Application Questions

Q1 The element with the lower ionisation energy will react more readily because it will be oxidised more easily.

Q2 There are no sulfate ions in the solution.
If there were sulfate ions in the sample, a white precipitate of barium sulfate would form.

Page 165 — Fact Recall Questions

Q1 The elements react more readily down the group.

Q2 They become more soluble down the group.

Q3 They become less soluble down the group.

Q4 To get rid of any sulfites or carbonates that might affect the result of the test.

Q5 It is a suspension of barium sulfate that a patient swallows before an X-ray. It is opaque to X-rays so will show up the structure of the oesophagus, stomach or intestines.

Q6 Titanium(IV) oxide is first converted to titanium(IV) chloride by heating it with carbon in a stream of chlorine gas. The titanium chloride is then purified by fractional distillation, before being reduced by magnesium in a furnace at almost 1000 °C.

Q7 Calcium oxide and calcium carbonate.

Q8 Calcium hydroxide is used in agriculture to neutralise acidic soils.

3. Group 7 — The Halogens

Page 169 — Application Questions

Q1 a) A — bromide, B — chloride/fluoride, C — iodide
b) There would be no change in the colours of the halide solutions. Iodine is below bromine and chlorine in Group 7 so it is less oxidising than them and cannot displace them from a halide solution.

Q2 When chlorine reacts with sodium hydroxide solution it produces sodium chlorate solution. This is bleach, so it will turn the litmus paper white by bleaching it.

Page 169 — Fact Recall Questions

Q1 The boiling points increase down the group.

Q2 fluorine

Q3 bromide and iodide

Q4 a) The solution turns from colourless to brown.
b) $Br_{2(aq)} + 2KI_{(aq)} \rightarrow 2KBr_{(aq)} + I_{2(aq)}$
c) $Br_{2(aq)} + 2I^-_{(aq)} \rightarrow 2Br^-_{(aq)} + I_{2(aq)}$

Q5 a) sodium chlorate(I) solution, sodium chloride, water
b) $2NaOH_{(aq)} + Cl_{2(g)} \rightarrow NaClO_{(aq)} + NaCl_{(aq)} + H_2O_{(l)}$

Q6 When you mix chlorine with water, it undergoes disproportionation. You end up with a mixture of chloride ions and chlorate(I) ions. In sunlight, chlorine can also decompose water to form chloride ions and oxygen.

Q7 a) Chlorine kills disease-causing microorganisms. It also prevents the growth of algae, eliminating bad tastes and smells, and removes discolouration caused by organic compounds.

b) Chlorine gas is very harmful if it's breathed in — it irritates the respiratory system. Liquid chlorine on the skin or eyes causes severe chemical burns. Accidents involving chlorine could be really serious, even fatal. Water contains a variety of organic compounds, e.g. from the decomposition of plants. Chlorine reacts with these compounds to form chlorinated hydrocarbons, e.g. chloromethane (CH_3Cl) — and many of these chlorinated hydrocarbons are carcinogenic (cancer-causing).

4. Halide Ions

Page 172 — Application Questions

Q1 He is correct for sodium chloride — it will only produce hydrogen chloride. However, the hydrogen bromide produced in the sodium bromide reaction is a strong reducing agent, so it will reduce the sulfuric acid further to sodium dioxide gas and bromine gas.

Q2 A — iodide
B — fluoride
C — bromide

Page 172 — Fact Recall Questions

Q1 How easy it is for a halide ion to lose an electron depends on the attraction between the nucleus and the outer electrons. As you go down the group, the attraction gets weaker because the ions get bigger, so the electrons are further away from the positive nucleus. There are extra inner electron shells, so there's a greater shielding effect too. Therefore, the reducing power of the halides increases down the group.

Q2 a) $NaF_{(s)} + H_2SO_{4(l)} \rightarrow NaHSO_{4(s)} + HF_{(g)}$

b) $NaI_{(s)} + H_2SO_{4(l)} \rightarrow NaHSO_{4(s)} + HI_{(g)}$
$2HI_{(g)} + H_2SO_{4(l)} \rightarrow I_{2(s)} + SO_{2(g)} + 2H_2O_{(l)}$
$6HI_{(g)} + SO_{2(g)} \rightarrow H_2S_{(g)} + 3I_{2(s)} + 2H_2O_{(l)}$

Q3 a) You can use the silver nitrate test. First add dilute nitric acid (to remove ions which might interfere with the test). Then add silver nitrate solution ($AgNO_{3\,(aq)}$). Chloride ions give a white precipitate, whereas fluoride ions give no precipitate.

b) A precipitate of silver chloride will dissolve in dilute ammonia solution.

5. Tests for Ions

Page 175 — Fact Recall Questions

Q1 a) pale green
b) brick red
c) red

Q2 sodium hydroxide

Q3 Add a little dilute hydrochloric acid, followed by barium chloride solution. If a white precipitate of barium sulfate forms, it means the original compound contained a sulfate.

Q4 Dip a piece of red litmus paper into the solution and if hydroxide ions are present, the paper will turn blue.

Q5 Dilute nitric acid and silver nitrate solution.

Q6 If you add dilute hydrochloric acid to a solution containing carbonate ions, carbon dioxide will be given off. Carbon dioxide turns limewater cloudy so if you bubble the gas through a test tube of limewater and it goes cloudy, your solution contains carbonate ions.

Exam-style Questions — pages 176-178

1 D ***(1 mark)***
When you add hydrochloric acid and barium chloride to a solution, a white precipitate means that sulfate ions are present. In a flame test, a brick red flame means that calcium ions are present. So the substance must be calcium sulfate.

2 B ***(1 mark)***
Hydroxide ions react with ammonium ions to release ammonia gas, which turns damp red litmus paper blue.

3.1 $Cl_{2(g)} + H_2O_{(l)} \rightleftharpoons 2H^+_{(aq)} + Cl^-_{(aq)} + ClO^-_{(aq)}$ ***(1 mark)***

3.2 It kills disease-causing microorganisms ***(1 mark)***. It persists in the water and prevents reinfection further down the supply ***(1 mark)***.

3.3 E.g. chlorine reacts with organic compounds in the water to form chlorinated hydrocarbons, e.g. chloromethane, many of which are carcinogenic ***(1 mark)***.

3.4 E.g. the increased risk of cancer is small compared to the risk of thousands of people dying from untreated water ***(1 mark)***.

4.1 Chlorine has a lower boiling point than bromine ***(1 mark)***. This is because chlorine molecules are smaller than bromine molecules ***(1 mark)***. So chlorine molecules have weaker van der Waals forces holding them together ***(1 mark)***.

4.2 Potassium bromide ***(1 mark)***. Chlorine displaces the halide ions, so the halogen must be below it in Group 7 ***(1 mark)***. However, there's no reaction with bromine water so the halide ion can't be less reactive than bromine ***(1 mark)***.

4.3 Potassium iodide ***(1 mark)***.
$Cl_2 + 2I^- \rightarrow 2Cl^- + I_2$ ***(1 mark)***
You must make sure you balance ionic equations or you won't get the marks in the exam.

4.4 NaClO is used as bleach ***(1 mark)***.

4.5 Dilute nitric acid is added to the solution to remove ions which might interfere with the test ***(1 mark)***.

4.6 A cream precipitate would form ***(1 mark)***.

4.7 The precipitate of silver bromide would dissolve in concentrated (but not dilute) ammonia solution ***(1 mark)***.

5.1 D ***(1 mark)***
Of the three elements listed, magnesium is the first in Period 2, so it will have the smallest atomic radius, the highest 1st ionisation energy and the lowest melting point.

5.2 Metal E has an extra electron shell compared to metal D. The extra inner shells shield the outer electrons from the attraction of the nucleus ***(1 mark)***. The outer electrons in E are further away from the nucleus, which greatly reduces the nucleus's attraction ***(1 mark)***. This makes it easier to remove outer electrons, resulting in a lower ionisation energy in metal E ***(1 mark)***.

5.3 Metal F will react more quickly with water ***(1 mark)*** because it has a lower first ionisation energy ***(1 mark)***.

5.4 Barium has a larger atomic radius than metal F ***(1 mark)***. So its delocalised electrons will be further away from the positive ion/nucleus than those in metal F ***(1 mark)***. This means they will be less strongly attracted to the positive ion/nucleus than in metal F so the metallic bonding is weaker ***(1 mark)***.

5.5 It is sparingly soluble / has very low solubility ***(1 mark)***.

5.6 It is used in indigestion tablets ***(1 mark)*** to neutralise acid in the stomach ***(1 mark)***.

5.7 $TiCl_{4(g)} + 2Mg_{(l)} \rightarrow Ti_{(s)} + 2MgCl_{2(l)}$ ***(1 mark)***

Unit 3

Section 1: Introduction to Organic Chemistry

1. Formulas

Pages 182-183 — Application Questions

Q1

```
    H  Br  H
    |  |   |
 H—C—C—C—H
    |  |   |
    H  H   H
```

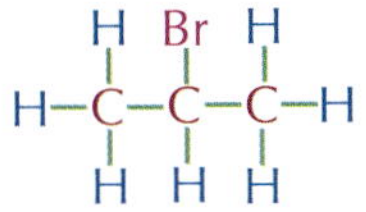

It doesn't matter if you draw the bromine atom above or below the carbon atom — it means the same thing.

Q2 C_8H_{18}

Q3 a) CH_2

b) C_4H_7Br

c) $C_9H_{17}Cl_3$

Q4 a) C_4H_8

b) Heptene contains 14 H atoms.

Q5 a) $C_3H_6Br_2$

b)

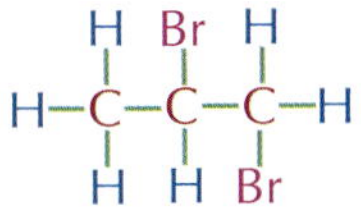

```
    H  Br  H
    |  |   |
 H—C—C—C—H
    |  |   |
    H  H   Br
```

c) $C_3H_6Br_2$

For this molecule the empirical formula is the same as the molecular formula because you can't cancel the numbers of atoms down and still have whole numbers.

Q6 a) C_5H_{10}

b) $CH_3CH_2CH_2CHCH_2$

c) CH_2

Q7 a)

Skeletal formulas have a carbon atom at each end and at each junction.

b)

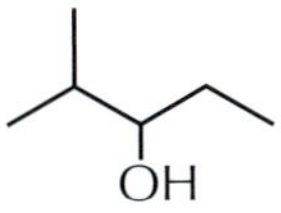

OH

c) e.g.

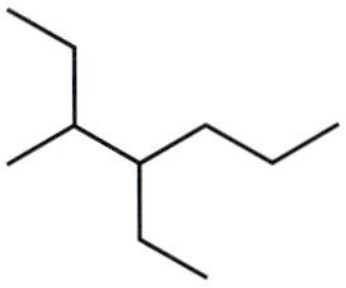

d)

Cl

Q8 a) $CH_3CH(CH_3)CH(CH_3)CH_2CH_2CH_3$

b) $CH_3CH_2C(CH_2)CH_2CH_2CH_3$

c) $CH_3CH_2CClCHCH_2CH_3$

d) $CH_3CHBrCH(OH)CH_2CHCH_2$

Page 183 — Fact Recall Questions

Q1 A molecular formula is a formula which gives the actual number of atoms of each element in a molecule.

Q2 A displayed formula shows how all the atoms are arranged, and all the bonds between them.

Q3 To find the empirical formula you have to divide the molecular formula by the highest number that goes into each number given in the molecular formula.

Q4 A homologous series is a family of organic compounds which have the same general formula and similar chemical properties.

2. Functional Groups

Page 187 — Fact Recall Questions

Q1 C_nH_{2n+2}

Q2 hex-

Q3 A cycloalkane is a ring of carbon atoms with two hydrogens attached to each carbon (assuming there's only one ring).

Q4 ethyl group

Q5 cycloalkanes and alkenes

Q6 An alkyl group or a hydrogen atom.

Q7 An aldehyde has a C=O at the end of a carbon chain, while a ketone has a C=O between two alkyl groups.

3. Nomenclature

Page 191 — Application Questions

Q1 a) 3-methylpentane

b) 3-ethyl-3-methylpentane

c) 3,3-diethylhexane

d) 3,3-diethyl-2-methylhexane

Q2 a) 2,3,4-trimethylpentanoic acid

b) 2-methlypentan-3-one

Q3 a)

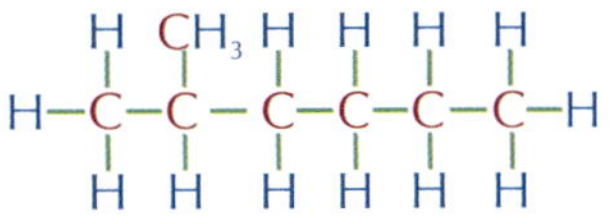

```
    H  CH3 H  H  H  H
    |  |   |  |  |  |
 H—C—C — C—C—C—C—H
    |  |   |  |  |  |
    H  H   H  H  H  H
```

b)

```
    H  H   H   H
    |  |   |   |
 H—C—C—C—C—H
    |  |   |   |
    H  Br  OH  H
```

c)

```
    H  H  H   O  H
    |  |  |   ‖  |
 H—C—C—C—C—C—H
    |  |  |      |
    H  H  CH2    H
          |
          CH3
```

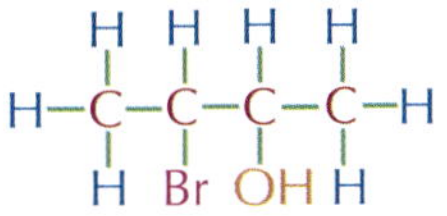

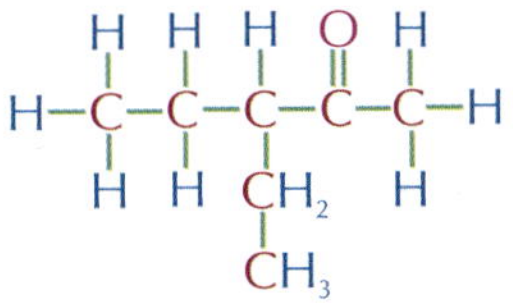

d)

```
    H       F  H
    |       |  |
 H—C—C=C—C—H
    |  |       |
    F  H       H
```

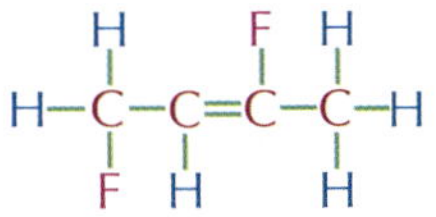

These could all be drawn the other way round (i.e. with the 1-carbon on the other side).

Page 191 — Fact Recall Questions

Q1 di-

Q2 Alcohol, Alkene, Alkyl, Halogen

4. Mechanisms

Page 192 — Fact Recall Questions

Q1 The arrow points from where the electrons are moving from, and points to where they move to.

Q2 Potassium hydroxide dissociates in aqueous solution, giving K^+ and OH^- ions. It is the OH^- ion that interacts with the chloroethane. In the reaction, ethanol and potassium chloride, KCl, are formed. So K^+ ions are unchanged in the reaction, so they play no part in the mechanism.

5. Isomers

Page 195 — Application Questions

Q1

```
        H  CH3
        |  |
    Cl—C—C—CH3
        |  |
        H  H
```

You could draw the chlorine atom attached to any other carbon atom apart from the one it was on originally.

Q2 There are three chain isomers of C_5H_{12}.

```
     H  H  H  H  H
     |  |  |  |  |
  H—C—C—C—C—C—H
     |  |  |  |  |
     H  H  H  H  H

     H CH3 H  H
     |  |  |  |
  H—C—C—C—C—H
     |  |  |  |
     H  H  H  H

        CH3
        |
  H3C—C—CH3
        |
        CH3
```

Q3

```
     H  O  H
     |  ‖  |
  H—C—C—C—H
     |     |
     H     H
```

Q4 a) i) Yes
ii) Yes
iii) No
iv) No
b) chain isomerism and positional isomerism

Page 195 — Fact Recall Questions

Q1 A chain isomer is a molecule that has the same molecular formula but a different arrangement of the carbon skeleton to another molecule. Some are straight chains and others branched in different ways.

Q2 A position isomer has the same skeleton and the same atoms or groups of atoms attached as another molecule. The difference is that the atoms or groups of atoms are attached to different carbon atoms.

Q3 A functional group isomer has the same atoms as another molecule but the atoms are arranged into different functional groups.

6. *E/Z* Isomers

Page 199 — Application Questions

Q1 a) *Z*-isomer
b) *E*-isomer

Q2

```
  H3C         CH2CH3
     \       /
      C=C                 E-isomer
     /       \
  H3CH2C      CH3

  H3C         CH3         Z-isomer
     \       /
      C=C
     /       \
  H3CH2C      CH2CH3
```

Page 199 — Fact Recall Questions

Q1 A stereoisomer is one of two or more forms of a molecule having the same structural formula but a different arrangement in space.

Q2 a) A *Z*-isomer is an isomer which has the higher priority groups both above or both below the double bond.
b) An *E*-isomer is an isomer that has the higher priority groups across the double bond from each other.

Q3 a) CH_3
b) Cl
c) OH
d) $CH_2CH_2CH_2CH_2COOH$

Exam-style Questions — pages 200-202

1 B ***(1 mark)***

2 D ***(1 mark)***

3 C ***(1 mark)***

4.1 3,4-dichlorohex-1-ene ***(1 mark)***

4.2 $C_6H_{10}Cl_2$ ***(1 mark)***

4.3 C_3H_5Cl ***(1 mark)***

4.4 functional group isomer ***(1 mark)***

4.5 E.g.

```
       H    Cl
        \  /
    H—C—C—H
      /     \
 Cl—C        C—H
 H /          \ H
      \     /
   H—C—C—H
       |   \
       H    H
```

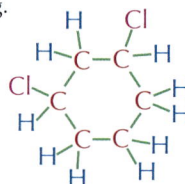

(1 mark)

Any position isomer where the two chlorine atoms aren't on adjacent carbons and the carbon skeleton is the same as the other isomer will get the marks.

4.6 position isomer ***(1 mark)***

4.7 It does not show *E/Z*-isomerism ***(1 mark)***. This is because one of the carbons in the C=C double bond has two identical groups (the carbon on the right of the diagram has two Hs) ***(1 mark)***.

5.1 C_nH_{2n+2} ***(1 mark)***

5.2 E.g.

```
     H  H  H  H  H
     |  |  |  |  |
  H—C—C—C—C—C—H
     |  |  |  |  |
     H  H  |  H  H
        H—C—H
           |
        H—C—H
           |
           H
```

(1 mark)

There are loads of possible answers for this question. As long as you've drawn a molecule with seven carbon atoms, sixteen hydrogen atoms and have arranged the carbon skeleton so there is at least one branch from the main carbon chain you'll pick up the mark.

5.3 pentane ***(1 mark)***

5.4 2,3-dimethylbutane ***(1 mark)***

6.1 A stereoisomer is a molecule that has the same structural formula as another molecule but its atoms are arranged differently in space ***(1 mark)***.

6.2 E.g.

CH_3 CH_3
C=C
H_3CH_2C H

(1 mark)

CH_3 H
C=C
H_3CH_2C CH_3

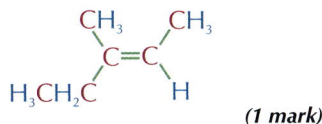

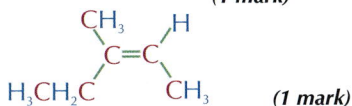

(1 mark)

The two isomers shown above would still be correct if they were rotated and reflected in any direction.

6.3 The top isomer is *E*-3-methylpent-2-ene ***(1 mark)***.
The bottom isomer is *Z*-3-methylpent-2-ene ***(1 mark)***.

Section 2: Alkanes and Halogenoalkanes

1. Alkanes and Petroleum

Page 205 — Fact Recall Questions

Q1 A mixture that consists mainly of alkane hydrocarbons.
Q2 They only contain carbon and hydrogen atoms, and each of their carbon atoms forms four single bonds — there are no C=C double bonds present.
Q3 They are separated by boiling point.
Q4 Some fractions have lower boiling points than others. This means they condense further up the column, so are drawn off higher up.
Q5 Cracking is breaking long-chain alkanes into smaller hydrocarbons.
Q6 There is more demand for lighter petroleum fractions so, to meet the demand, the heavier fractions are cracked into lighter fractions.
Q7 Using a catalyst cuts costs, because the reaction can be done at a lower temperature and pressure, and at an increased speed.

2. Alkanes as Fuels

Page 207 — Application Questions

Q1 $C_5H_{12} + 8O_2 \rightarrow 5CO_2 + 6H_2O$
Q2 $C_5H_{12} + 5\frac{1}{2}O_2 \rightarrow 5CO + 6H_2O$

Page 207 — Fact Recall Questions

Q1 Any three from: e.g. nitrogen oxides / carbon monoxide / carbon (soot) / unburned hydrocarbons / carbon dioxide.
Q2 A catalytic converter.
Q3 The sulfur burns to produce sulfur dioxide gas. If sulfur dioxide enters the atmosphere, it can dissolve in the moisture, and form sulfuric acid.
Q4 Powdered calcium carbonate or calcium oxide can be mixed with water to make an alkaline slurry. When the flue gases mix with the alkaline slurry, the acidic sulfur dioxide gas reacts with the calcium compounds to form a harmless salt (calcium sulfite).

3. Synthesis of Chloroalkanes

Page 210 — Fact Recall Questions

Q1 Reactions that are started by light.
Q2 $Cl_2 \xrightarrow{UV} 2Cl\bullet$
Q3 Two free radicals join together to make a stable molecule. This terminates the chain reaction.

Q4 It acts as a chemical sunscreen by absorbing a lot of ultraviolet radiation from the Sun, and stopping it from reaching us. This helps to prevent sunburn and even skin cancer.
Q5 Chlorine free radicals, Cl•, are formed in the upper atmosphere when the C–Cl bonds in CFCs are broken down by ultraviolet radiation.
Q6 $Cl\bullet_{(g)} + O_{3(g)} \rightarrow O_{2(g)} + ClO\bullet_{(g)}$
$ClO\bullet_{(g)} + O_{3(g)} \rightarrow 2O_{2(g)} + Cl\bullet_{(g)}$
Q7 Research by several different scientific groups demonstrated that CFCs were causing damage to the ozone layer. The advantages of CFCs couldn't outweigh the environmental problems they were causing, so they were banned.

4. Halogenoalkanes

Page 211 — Fact Recall Questions

Q1 An alkane with at least one halogen atom in place of a hydrogen atom.
Q2 Halogen atoms are generally more electronegative than carbon atoms, and so they withdraw electron density from carbon atoms. This leaves the carbon atoms with a partial positive charge and the halogen atoms with a partial negative charge — resulting in a polar bond.
Q3 An electron-pair donor.
Q4 Any two from: e.g. CN^- / NH_3 / OH^-.
Q5 A pair of dots represents a lone pair of electrons.

5. Nucleophilic Substitution

Page 216 — Application Questions

Q1

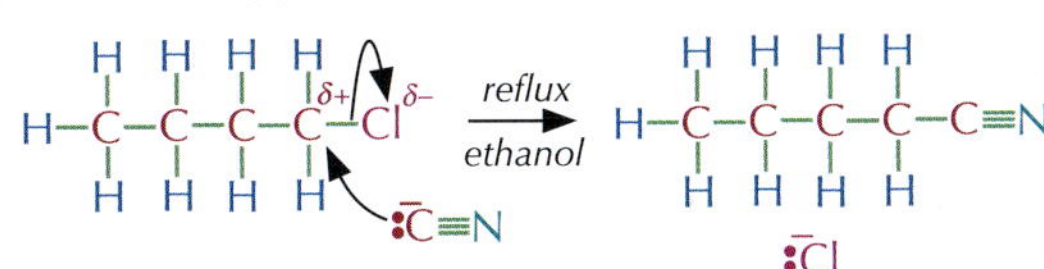

Q2 Reaction C would happen the quickest because the C–I bond has the lowest bond enthalpy of all the carbon-halogen bonds. This means that the C–I bond is the easiest to break, and therefore this reaction will happen the quickest.

Q3

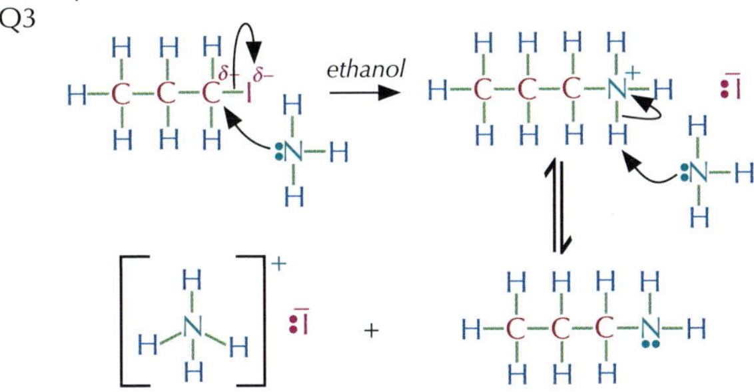

Q4

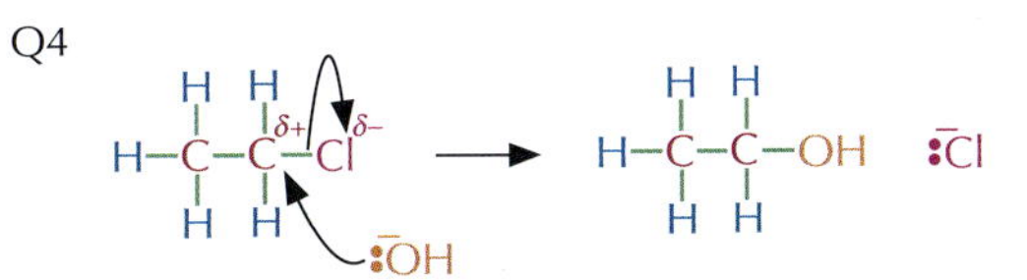

Page 216 — Fact Recall Questions

Q1 In nucleophilic substitution, one functional group is substituted for another.
Q2 Aqueous sodium hydroxide or potassium hydroxide.
Q3 e.g. ^-CN / cyanide ion
Q4 Warm in ethanol in a sealed tube.

Q5 The C–F bond is the strongest — it has the highest bond enthalpy. So fluoroalkanes are substituted more slowly than other halogenoalkanes.

6. Elimination Reactions

Page 218 — Application Question

Q1

reflux ethanol

H_2O :Br^-

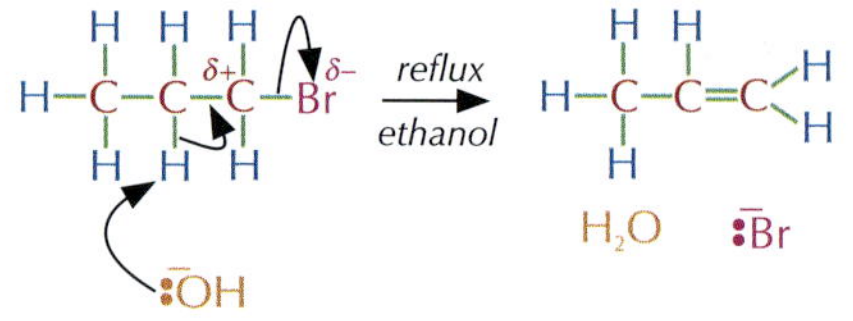

Page 218 — Fact Recall Questions

Q1 An elimination reaction happens when a molecule loses atoms or groups of atoms from two neighbouring carbon atoms and forms a carbon-carbon double bond.

Q2 a) The ^-OH acts as a nucleophile.
b) This is a nucleophilic substitution reaction.

Q3 a) The ^-OH acts as a base.
b) This is an elimination reaction.

Exam-style Questions — pages 220-221

1 C ***(1 mark)***

In aqueous conditions, the OH^- acts as a nucleophile, so a nucleophilic substitution reaction will dominate.

2 B ***(1 mark)***

3.1 $Cl\bullet + CH_4 \rightarrow \bullet CH_3 + HCl$ ***(1 mark)***
$\bullet CH_3 + Cl_2 \rightarrow CH_3Cl + Cl\bullet$ ***(1 mark)***

3.2 $\bullet CH_3 + Cl\bullet \rightarrow CH_3Cl$ ***(1 mark)***

3.3 The chlorine free radicals act as catalysts. They react with ozone to form an intermediate (ClO·), and an oxygen molecule in the following reaction:
$Cl\bullet_{(g)} + O_{3(g)} \rightarrow O_{2(g)} + ClO\bullet_{(g)}$ ***(1 mark)***
The ClO• then reacts with more ozone to regenerate the chlorine free radical in this reaction:
$ClO\bullet_{(g)} + O_{3(g)} \rightarrow 2O_{2(g)} + Cl\bullet_{(g)}$ ***(1 mark)***
Because the chlorine free radical is regenerated, just one chlorine free radical can destroy lots of ozone molecules ***(1 mark)***. This is a problem as ozone absorbs ultraviolet radiation, and acts as a chemical sunscreen. When ozone is destroyed, it will increase people's risk of getting sunburn and skin cancer ***(1 mark)***.

4.1 Because the boiling points of alkanes increase as the molecules get bigger, each fraction condenses at a different temperature ***(1 mark)***. As the crude oil vapour goes up the fractionating column, it gets cooler ***(1 mark)***. So the fractions are drawn off at different levels in the column depending on where they condense ***(1 mark)***.

4.2 E.g. $C_7H_{16(g)} + 7\frac{1}{2}O_{2(g)} \rightarrow 7CO_{(g)} + 8H_2O_{(g)}$
(1 mark for the correct products and 1 mark for a correctly balanced equation)
There's more than one correct answer here. You get both marks if you've shown carbon monoxide as one of the products (as well as water) and your equation is balanced. The other possible products are carbon and carbon dioxide.

4.3 It can be dangerous because carbon monoxide gas is poisonous ***(1 mark)***. Carbon monoxide molecules bind to the same sites on haemoglobin molecules in red blood cells as oxygen molecules ***(1 mark)***. So oxygen can't be carried around the body ***(1 mark)***.

4.4 There is more demand for light fractions. / Light fractions are more useful***(1 mark)***.

4.5 E.g. $C_7H_{16} \rightarrow C_4H_8 + C_3H_8$ ***(1 mark)***
Any balanced equation including an alkane and alkene gets a mark here.

4.6 Using a catalyst cuts costs, because the reaction can be done at a lower temperature and pressure ***(1 mark)***. The catalyst also speeds up the rate of reaction ***(1 mark)***.

5.1 $CH_3CH_2Br + OH^- \rightarrow CH_3CH_2OH + Br^-$ ***(1 mark)***

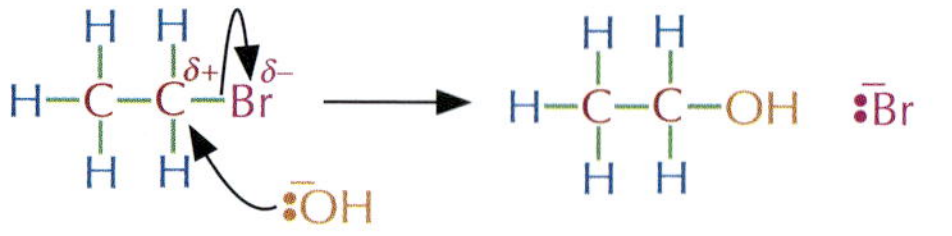

(1 mark for each curly arrow, up to a maximum of 2 marks)

5.2 The reaction of water with iodoethane would be quicker than the reaction of bromoethane with water ***(1 mark)***. This is because the reaction involves the breaking of a carbon-halogen bond and the C–I bond has a lower bond enthalpy than the C–Br bond ***(1 mark)***, which means it is more easily broken ***(1 mark)***.

5.3 It will predominantly undergo an elimination reaction, instead of a nucleophilic reaction ***(1 mark)***. This is because the OH– acts as a base and removes a hydrogen atom from bromoethane ***(1 mark)***.

Section 3: Alkenes and Alcohols

1. Alkenes

Page 225 — Application Question

Q1

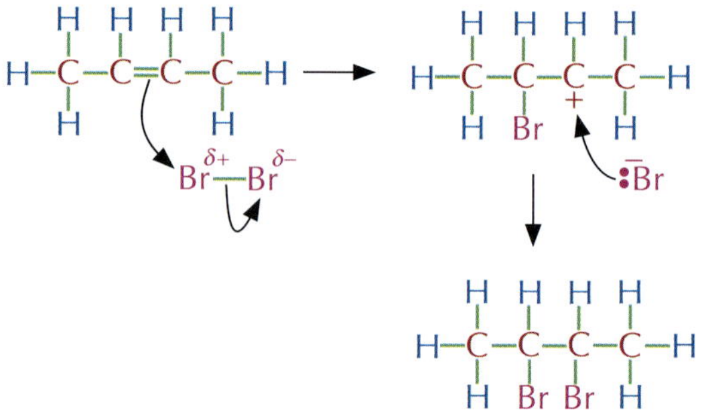

Page 225 — Fact Recall Questions

Q1 C_nH_{2n}

Q2 Alkenes are unsaturated because they can make more bonds with extra atoms in addition reactions across the carbon-carbon double bond.

Q3 Alkenes can undergo electrophilic addition reactions because they have a double bond which has a high electron density and is easily attacked by electrophiles.

Q4 Electrophiles are electron-pair acceptors.

Q5 Any two from: e.g. NO_2^+ / H^+ / CH_3CH_2Br.

Q6 Carbon-carbon double bonds/unsaturation/alkenes.

2. Reactions of Alkenes

Page 228 — Application Questions

Q1 a)

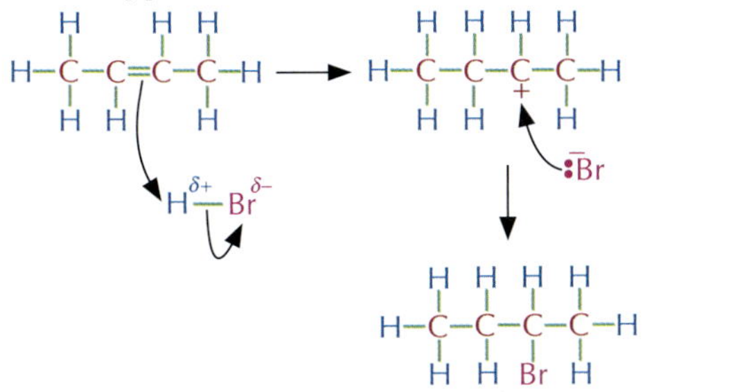

It doesn't matter which carbon you add the bromine to as you'll always end up with the same product (2-bromobutane).

b) $C_4H_8 + HBr \rightarrow C_4H_9Br$

Q2 The reaction mechanism for the production of 2-bromobutane contains a secondary carbocation, which is more stable than the primary carbocation formed in the reaction mechanism for 1-bromobutane. More stable carbocations are more likely to form, so 2-bromobutane is the major product of this reaction.

Page 228 — Fact Recall Questions

Q1 hydrogen bromide

Q2 Tertiary carbocations are more stable than secondary carbocations, which are more stable than primary carbocations.

3. Addition Polymers

Page 232 — Application Questions

Q1 a)

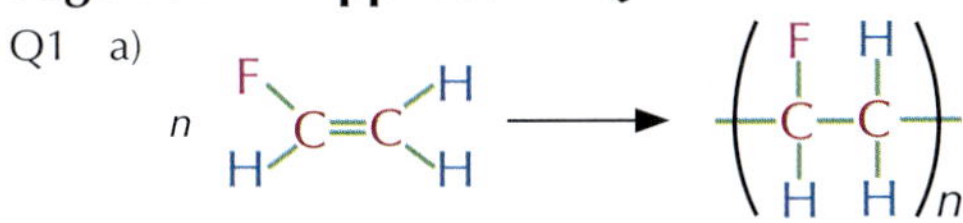

b) poly(fluoroethene) / polyfluoroethene

Q2

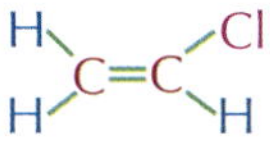

Q3

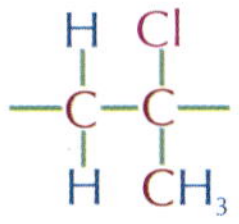

Page 232 — Fact Recall Questions

Q1 The double bonds in alkenes (monomers) open up and join together to make long chains called polymers.

Q2 Poly(alkenes) are unreactive because they have lost their carbon-carbon double bonds/they are saturated. Also, the carbon chain is non-polar.

Q3 a) The chlorine-carbon bonds are polar, with the chlorine atoms being δ–. The chlorine and carbon atoms of different polymer chains have permanent dipole-dipole forces bonding them.

b) E.g. window frames, drain pipes.

Q4 a) Plasticisers are molecules that get between polymer chains. This pushes them apart, which reduces the strength of the intermolecular forces between the chains, which makes the polymer more flexible.

b) E.g. electric cable insulation, clothing.

4. Alcohols

Page 234 — Application Questions

Q1 a) pentan-1-ol

b) 2,2-dimethylpropan-1-ol

c) 2,3-dimethylbutan-2-ol

d) 3-ethyl-2-methylpentane-1,5-diol

That last one is a tricky one... Make sure you have the side chains in alphabetical order and the name has the lowest possible numbers in it.

Q2 a) primary

b) primary

c) tertiary

d) primary

Page 234 — Fact Recall Questions

Q1 $C_nH_{2n+1}OH$

Q2 A secondary alcohol is an alcohol with the –OH group attached to a carbon with two alkyl groups attached.

5. Dehydrating Alcohols

Page 237 — Application Question

Q1 a) $C_3H_7OH \rightarrow C_3H_6 + H_2O$

b)

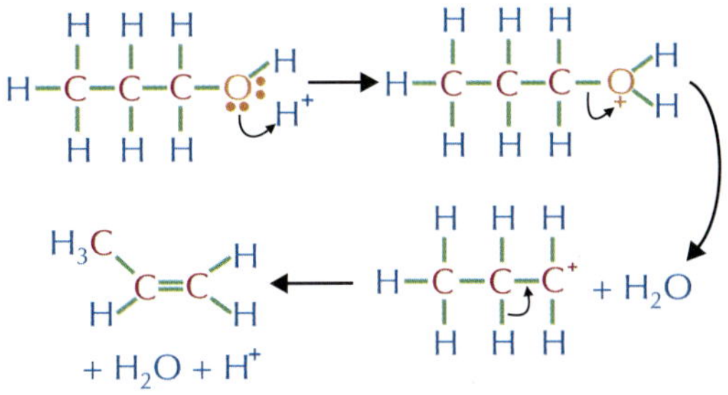

Page 237 — Fact Recall Questions

Q1 $C_nH_{2n+1}OH \rightarrow C_nH_{2n} + H_2O$

Q2 E.g.

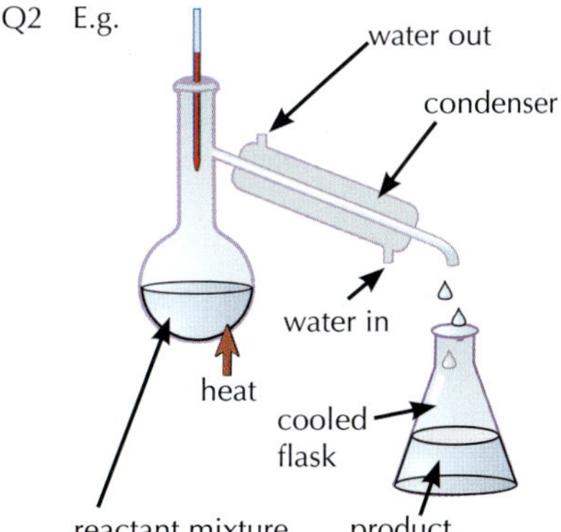

Q3 Put impure cyclohexene in a round-bottomed flask. Add anhydrous $CaCl_2$ and stopper the flask. Let the mixture dry for at least 20 minutes with occasional swirling. Distil the resulting mixture. Collect the product that is released when the mixture is at around 83 °C — this will be the pure cyclohexene.

6. Ethanol Production

Page 241 — Application Question

Q1

```
H       H                  H H
 \     /                   | |
  C = C     ——————→      H—C—C⁺   :O—H
 /  (  \                   | |        \
H    H⁺  H                 H H         H
                                   ↓
   H H                       H H
   | |                       | |
 H—C—C—O:—H    ←———————    H—C—C—O⁺—H
   | |   ··                  | |    \
   H H        + H⁺           H H     H
```

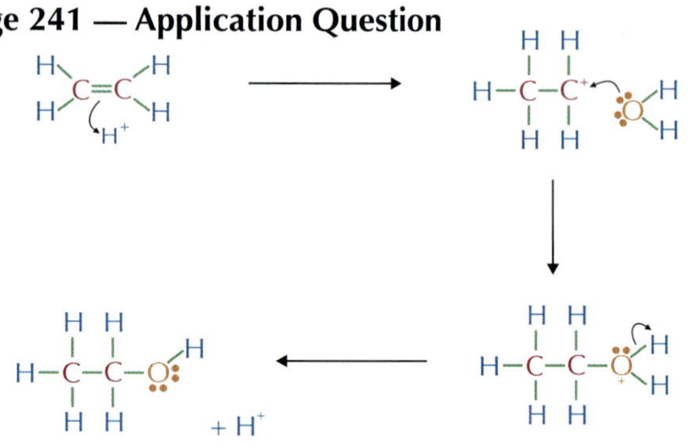

Page 241 — Fact Recall Questions

Q1 $C_6H_{12}O_{6(aq)} \xrightarrow[\text{yeast}]{30\text{-}40°C} 2C_2H_5OH_{(aq)} + 2CO_{2(g)}$

Q2 If it's too cold, the reaction is slow — if it's too hot, the enzyme is denatured (damaged).

Q3 A biofuel is a fuel that's made from biological material that's recently died.

Q4 a) $6CO_2 + 6H_2O \rightarrow C_6H_{12}O_6 + 6O_2$
$C_6H_{12}O_6 \rightarrow 2C_2H_5OH + 2CO_2$
$2C_2H_5OH + 6O_2 \rightarrow 4CO_2 + 6H_2O$

b) These equations don't take into account the carbon dioxide produced by other stages in the process, e.g. making fertilisers, powering agricultural machinery, and transporting the fuel.

Q5 a) E.g. unlike fossil fuels, biofuels are sustainable/renewable. / Burning a biofuel releases the same amount of carbon dioxide that the plant took in as it grew, so biofuels are nearly carbon neutral/don't contribute to global warming as much as fossil fuels.

b) Any three from: e.g. land which is being used to grow crops for fuel, can't be used to grow food. / Trees may be cut down to create more land to grow crops for biofuels, which destroys habitats/removes trees/releases carbon dioxide if they are burnt. / Fertilisers, which are added to soils to increase crop production, can pollute waterways/release nitrous oxide. / Most current car engines couldn't run on biofuels without being modified.

7. Oxidising Alcohols

Page 245 — Application Question

Q1 a)

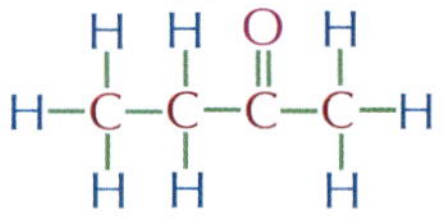

```
  H H O H
  | | ‖ |
H—C—C—C—C—H
  | |   |
  H H   H
```

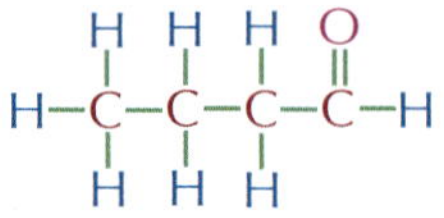

b)

```
  H H H O
  | | | ‖
H—C—C—C—C—H
  | | |
  H H H
```

c)

```
  H H H O
  | | | ‖
H—C—C—C—C—OH
  | | |
  H H H
```

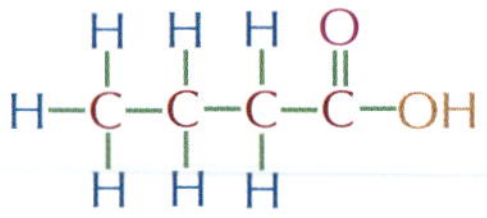

Page 245 — Fact Recall Questions

Q1 Aldehydes have the functional group C=O and have one hydrogen atom and one R group attached to the carbon atom. Ketones have the functional group C=O and have two R groups attached either side of the carbon atom. Carboxylic acids have the functional group COOH.

Q2 $R{-}CH_2{-}OH + 2[O] \xrightarrow{\text{reflux}} R{-}C(=O){-}OH + H_2O$
primary alcohol → *carboxylic acid*

Q3 Any two from: e.g. Fehling's solution / Benedict's solution / Tollens' reagent.

Exam-style Questions — pages 247-249

1 B ***(1 mark)***

2 C ***(1 mark)***

3.1 $C_6H_{12}O_{6(aq)} \rightarrow 2C_2H_5OH_{(aq)} + 2CO_{2(g)}$ ***(1 mark)***
The reaction needs to be carried out in the presence of yeast and at 30–40 °C ***(1 mark)***.

3.2 Fermentation uses renewable resources to produce ethanol and so will become more important as the amount of crude oil decreases ***(1 mark)***.

3.3 The hydration of ethene by steam is carried out at 300 °C and at a pressure of 60 atm ***(1 mark)***. It also needs a solid phosphoric(V) acid catalyst ***(1 mark)***.

3.4

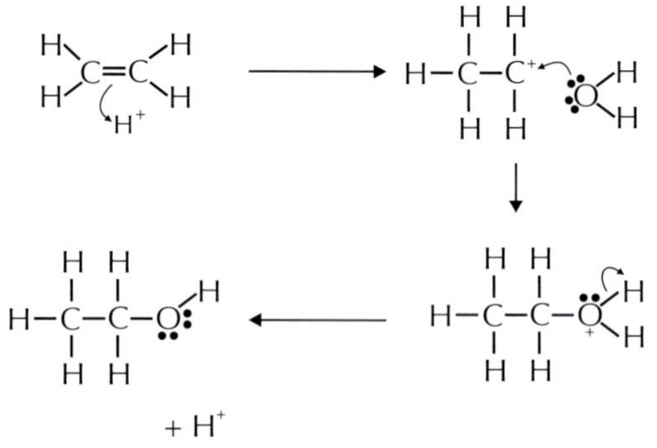

```
H       H                  H H
 \     /                   | |
  C = C     ——————→      H—C—C⁺   :O—H
 /  (  \                   | |        \
H    H⁺  H                 H H         H
                                   ↓
   H H                       H H
   | |                       | |
 H—C—C—O:—H    ←———————    H—C—C—O⁺—H
   | |   ··                  | |    \
   H H                       H H     H
      + H⁺
```

(4 marks, 1 mark for each correct curly arrow and 1 mark for the structure of the final product.)

3.5 A biofuel is a fuel that's made from biological material that's recently died ***(1 mark)***.

3.6 Carbon dioxide goes into crops during photosynthesis:
$6CO_2 + 6H_2O \rightarrow C_6H_{12}O_6 + 6O_2$ ***(1 mark)***
Taking the glucose that is produced in photosynthesis and fermenting it produces ethanol and carbon dioxide:
$C_6H_{12}O_6 \rightarrow 2C_2H_5OH + 2CO_2$ ***(1 mark)***
Burning the ethanol produced gives off water and carbon dioxide:
$2C_2H_5OH + 6O_2 \rightarrow 4CO_2 + 6H_2O$ ***(1 mark)***
In these three balanced equations, 6 moles of CO_2 go into the system, and 6 moles of CO_2 come out.
So ethanol is thought of as a carbon neutral fuel ***(1 mark)***.

3.7 There are still carbon emissions if you consider the whole ethanol production process ***(1 mark)***. For example, the machinery used to produce the ethanol fuel may be powered by fossil fuels which release CO_2 into the atmosphere when they're burnt ***(1 mark)***.

3.8 Any two of: e.g. they aren't carbon neutral and the manufacturing of them could still use fossil fuels / fertilisers used to grow the crops needed to make biofuels produce nitrous oxide which contributes to the greenhouse effect / deforestation takes place to make room for crops ***(1 mark for each correct statement up to 2 marks)***.

3.9 E.g. it uses farmable land to produce fuel instead of food, and where food is scarce this could be a problem ***(1 mark)***.

4.1 The student could shake the alkene with bromine water ***(1 mark)***. The solution will turn from orange to colourless if a carbon-carbon double bond is present ***(1 mark)***.

4.2 $C_6H_{12} + HBr \rightarrow C_6H_{13}Br$ ***(1 mark)***

4.3
```
    H   H   Br  H   H
    |   |   |   |   |
H - C - C - C - C - C - H
    |   |   |   |   |
    H   H  CH3  H   H
```
(1 mark)

4.4 This isomer is the major product of the reaction because it's formed when the reaction proceeds via the most stable carbocation intermediate ***(1 mark)***. For the major product the reaction goes via a tertiary (3°) carbocation ***(1 mark)*** and for the minor product the reaction goes via a secondary (2°) carbocation ***(1 mark)***.

4.5
```
   H   CH2CH3
   |   |
 — C — C —
   |   |
 H3C   CH3
```
(1 mark)

5.1 A monomer is an alkene molecule that is used to make a polymer ***(1 mark)***.

5.2 tetrafluoroethene ***(1 mark)***

5.3 Poly(tetrafluoroethene) is saturated (it only has single bonds in the carbon chain) ***(1 mark)*** and its main carbon chain is non-polar ***(1 mark)***.

5.4 It makes the polymer more bendy/flexible ***(1 mark)***.

6.1 butan-1-ol ***(1 mark)***

6.2 potassium dichromate(VI) ***(1 mark)***

6.3 Gently heat ***(1 mark)*** excess butan-1-ol with the oxidising agent in distillation apparatus ***(1 mark)***. The aldehyde, butanal, is distilled off as soon as it forms as it passes up a fractionating column and through a condenser ***(1 mark)***.

6.4 $C_4H_9OH + [O] \longrightarrow C_4H_8O + H_2O$ ***(1 mark)***

You can also have the mark if you used the full structural formulas here.

6.5 She could test the unknown product with Fehling's/ Benedict's solution ***(1 mark)***. As it is an aldehyde, the solution will go from a deep blue colour to a brick-red colour ***(1 mark)***. She could also test the product with Tollens' reagent ***(1 mark)***. The aldehyde would make a silver mirror coating form on the inside of the testing apparatus ***(1 mark)***.

6.6
```
    H   OH  H
    |   |   |
H - C - C - C - H
    |   |   |
    H  CH3  H
```
(1 mark)

Section 4: Organic Analysis

1. Tests for Functional Groups

Page 253 — Application Questions

Q1 Tollens' reagent is reduced to silver when warmed with an aldehyde, so the compound must be an aldehyde.

Q2 No reaction is seen between ketones and Benedict's solution. Therefore the test tube containing the ketone must be the one that doesn't produce a precipitate with Benedict's solution. So the ketone is compound B.

Q3 a) The potassium dichromate(VI) solution would change from orange to green.

b) The limewater would turn cloudy.

Q4 E.g. first warm a small sample of solution from each test tube with a few drops of Tollens' reagent/Fehling's solution/ Benedict's solution. Propanal will give a silver mirror/ brick-red precipitate. No reaction will be seen with propanone or propanoic acid.
Next add a small spatula of solid sodium carbonate to samples of the two remaining unknown solutions. The propanoic acid will fizz, producing carbon dioxide gas, which will turn limewater cloudy when bubbled through it. Propanone will not react with calcium carbonate.
You could do these two steps the other way around if you wanted to — identifying the carboxylic acid first using the carbonate test, then using Tollens' reagent, Fehling's solution or Benedict's solution to identify the aldehyde and the ketone.

Q5 a) Potassium dichromate(VI) shows the same colour change (orange to green) when it reacts with both primary and secondary alcohols.

b) Either: Add acidified potassium dichromate(VI) to both samples and heat them in distillation apparatus. Collect the products of both reactions. Add Fehling's/ Benedict's solution / Tollens' reagent to both and warm them. The primary alcohol will have been oxidised to an aldehyde and will produce a brick red precipitate / silver mirror. The secondary alcohol will have been oxidised to a ketone and will not react with Fehling's/ Benedict's solution / Tollens' reagent.
Or: Add acidified potassium dichromate(VI) to both samples and heat them under reflux. Collect the products of both reactions. Add sodium carbonate to both. The primary alcohol will have been oxidised to a carboxylic acid and so it will begin to fizz.
(If you collect the gas produced and bubble it through limewater, the limewater will turn cloudy.)
The secondary alcohol will have been oxidised to a ketone, which will not react with the sodium carbonate (so it will not fizz).

Page 253 — Fact Recall Questions

Q1 Brick-red.

Q2 Put 2 cm^3 of 0.10 mol dm^{-3} silver nitrate solution in a test tube. Add a few drops of dilute sodium hydroxide solution. A light brown precipitate should form. Then, add drops of dilute ammonia solution until the brown precipitate dissolves completely.

Q3 The orange/brown bromine water would decolourise.

2. Mass Spectrometry

Page 255 — Application Questions

Q1 a) At lower resolution, all three of these molecules would appear to have the same molecular mass.
$M_r\ (C_3H_6O) = (3 \times 12) + (6 \times 1) + 16 = 58$
$M_r\ (C_3H_8N) = (3 \times 12) + (8 \times 1) + 14 = 58$
$M_r\ (C_4H_{10}) = (4 \times 12) + (10 \times 1) = 58$

b) $M_r\ (C_3H_6O)$
$= (3 \times 12.0000) + (6 \times 1.0078) + 15.9949 = 58.0417$
$M_r\ (C_3H_8N)$
$= (3 \times 12.0000) + (8 \times 1.0078) + 14.0031 = 58.0655$
$M_r\ (C_4H_{10}) = (4 \times 12.0000) + (10 \times 1.0078) = 58.0780$
The molecular formula is C_3H_8N.

Q2 Find the precise molecular mass of each compound.
M_r of butanoic acid
= (4 × 12.0000) + (8 × 1.0078) + (2 × 15.9949) = 88.0522
M_r of pentan-1-ol
= (5 × 12.0000) + (12 × 1.0078) + 15.9949 = 88.0885
M_r of pentan-3-one
= (5 × 12.0000) + (10 × 1.0078) + 15.9949 = 86.0729
M_r of hexane = (6 × 12.0000) + (14 × 1.0078) = 86.1092
So the compounds in the mixture are pentan-1-ol and hexane.

3. Infrared Spectroscopy

Page 258 — Application Questions

Q1 a) A (~3000 cm⁻¹) — O–H (carboxylic acid)
B (~1700 cm⁻¹) — C=O

b) CH_3CH_2COOH (displayed formula: H–C(H)(H)–C(H)(H)–C(=O)–O–H)

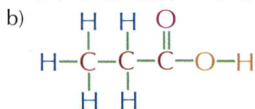

The mass of the carboxylic acid group (COOH) is 45 (12 + 16 + 16 + 1). The M_r of the molecule is 74, so the rest of the molecule has a mass of 74 – 45 = 29. This corresponds to an ethyl group (CH_3CH_2), so the molecule must be propanoic acid.

Q2 E.g. There is a strong, sharp peak at about 1700 cm^{-1}, this indicates that a C=O bond is present in the molecule.

Page 258 — Fact Recall Questions

Q1 A beam of IR radiation is passed through a sample of a chemical. The IR radiation is absorbed by the covalent bonds in the molecules, increasing their vibrational energy. Bonds between different atoms absorb different frequencies of IR radiation. Bonds in different places in a molecule absorb different frequencies too. The frequencies where they absorb IR radiation are plotted to give an IR spectrum.

Q2 1000 cm^{-1} – 1550 cm^{-1}

Exam-style Questions — pages 260-263

1.1 The IR spectrum of molecule A has a peak at around 1650 cm^{-1}, which corresponds to a C=C double bond in an alkene / doesn't contain a peak corresponding to O–H bonds in an alcohol ***(1 mark)***.
So molecule **A** is:
$H_2C=CH–CH_3$ (displayed formula) ***(1 mark)***

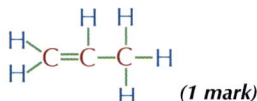

The IR spectrum of molecule B shows a strong, broad peak at around 3400 cm^{-1}. This corresponds to O–H bonds in an alcohol ***(1 mark)***.
So molecule **B** is:
$CH_3CH_2CH_2OH$ (displayed formula: H–C(H)(H)–C(H)(H)–C(H)(H)–OH) ***(1 mark)***

The first thing you need to here is to work out what different molecules you could make by reacting 1-bromopropane with OH^- ions. Once you've drawn the two possible products (an alcohol and an alkene), you get the rest of the marks for working out which spectrum matches which product.

1.2 Molecule A is propene ***(1 mark)***.
Molecule B is propan-1-ol ***(1 mark)***.

1.3 Molecule A is produced by reacting 1-bromopropane with ethanol and potassium/sodium hydroxide under reflux ***(1 mark)***.
Molecule B is produced by reacting 1-bromopropane with water and potassium/sodium hydroxide under reflux ***(1 mark)***.

2 Water vapour is the most effective greenhouse gas ***(1 mark)***. The infrared spectrum for water vapour shows that, out of the three gases, water vapour absorbs the most energy in the infrared region of the electromagnetic spectrum ***(1 mark)***.

3 Compound A must be ethanol (a primary alcohol) since it is oxidised by potassium dichromate(VI) solution, but does not react with calcium carbonate or Tollens' reagent ***(1 mark)***. Compound B must be butanal (an aldehyde) since it reacts with Tollens' reagent, a test for aldehydes, to produce a silver mirror ***(1 mark)***. Therefore, compound **C** must be propanoic acid. This fits with the observations — propanoic acid is a carboxylic acid so reacts with calcium carbonate to produce carbon dioxide, which turns limewater cloudy ***(1 mark)***.

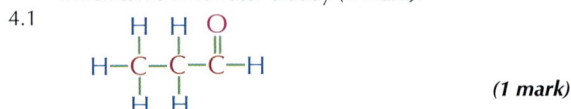

4.1 CH_3CH_2CHO (displayed formula: H–C(H)(H)–C(H)(H)–C(=O)–H) ***(1 mark)***

4.2 CH_3COCH_3 (displayed formula: H–C(H)(H)–C(=O)–C(H)(H)–H) ***(1 mark)***

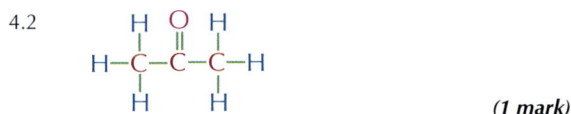

4.3 The IR spectra of propanal and propanone would be very similar/it would be hard to tell the difference between the spectra of propanal and propanone as they both only contain C=O, C–C and C–H bonds ***(1 mark)***.

4.4 The relative molecular masses of propanal and propanone are the same, so high resolution mass spectroscopy could not distinguish between them ***(1 mark)***.

4.5 E.g. Use Benedict's/Fehling's solution ***(1 mark)***. Benedict's/Fehling's solution will produce a brick-red precipitate when warmed with propanal, but not propanone ***(1 mark)***.
OR Use Tollens' reagent ***(1 mark)***. Tollens' reagent will produce a silver mirror when warmed with propanal, but not propanone ***(1 mark)***.
Care should be taken when heating aldehydes and ketones as they are flammable — a water bath should be used instead of a naked flame. You should also always wear safety goggles and a lab coat to prevent any irritant reagents splashing into your eyes or onto your skin ***(1 mark)***.

5.1 E.g. Spectrum X shows a peak just below 3000 cm^{-1}, probably for C–H bonds, but no other peaks corresponding to specific bonds. This suggests that one of the compounds is an alkane.
Alkanes with molecular formula C_5H_{12} have M_r = (5 × 12.0) + (12 × 1.0) = 72, which could be A or B. Find the precise molecular mass of C_5H_{12}: (5 × 12.0000) + (12 × 1.0078) = 72.0936 — i.e. compound B.
Spectrum Y has a peak at just over 1700 cm^{-1}, which suggests a C=O bond. There's no O–H/C–O peak for a carboxylic acid, but it could be an aldehyde or a ketone. The molecular formula could be C_4H_8O, which has M_r = (4 × 12.0) + (8 × 1.0) + 16.0 = 72, so check the precise molecular mass of C_4H_8O:
(4 × 12.0000) + (8 × 1.0078) + 15.9949 = 72.0573 — i.e. compound A.
Spectrum Z has a peak at about 1650 cm^{-1}. This could be caused by the C=C bond in an alkene.
Alkenes with molecular formula C_5H_{10} have M_r = (5 × 12.0) + (10 × 1.0) = 70, which could be C.

Check by finding the precise molecular mass of C_5H_{10}:
(5 × 12.0000) + (10 × 1.0078) = 70.0780 — i.e. compound C.
So compound A is $\mathbf{C_4H_8O}$, which gives spectrum **Y**.
Compound B is $\mathbf{C_5H_{12}}$, which gives spectrum **X**.
Compound C is $\mathbf{C_5H_{10}}$, which gives spectrum **Z**.
(8 marks available in total: 1 mark each for a correct analysis of the three spectra, 1 mark for each correct molecular formula and 2 marks for matching all three compounds to the correct spectra, otherwise 1 mark for matching one compound to the correct spectrum.)
You might have done things in a different order here — e.g. if you had a hunch about what the molecular formulas of the compounds were, you could have used the precise molecular masses to confirm them first, then matched them up to the IR spectra.

5.2 **A** is either an aldehyde or a ketone with molecular formula C_4H_8O. It doesn't have a branched carbon chain, so it must be either butanal or butanone ***(1 mark for either)***.
B is a straight-chain alkane with molecular formula C_5H_{12}, so it must be pentane ***(1 mark)***.
C is a straight-chain alkene with molecular formula C_5H_{10}. That means it's pentene. You don't know where the double bond is, so it could be pent-1-ene or pent-2-ene ***(1 mark for either, allow 1 mark for pentene)***.

5.3 The infrared spectrum of every compound has a unique 'fingerprint region' between 1000 cm^{-1} and 1550 cm^{-1}. You can compare spectra X, Y and Z against spectra for the suggested identities. If the suggestions are correct, the fingerprint regions will match ***(1 mark)***.
In case you're wondering, the spectra here are actually for pentane (X), butanone (Y) and pent-1-ene (Z).

Unit 4

Section 1 — Thermodynamics

1. Enthalpy Changes

Page 266 — Fact Recall Questions

Q1 Enthalpy change is the heat energy transferred in a reaction at constant pressure.

Q2 ΔH

Q3 Lattice formation enthalpy is the enthalpy change when 1 mole of a solid ionic compound is formed from its gaseous ions.
Lattice dissociation enthalpy is the enthalpy change when 1 mole of a solid ionic compound is completely dissociated into its gaseous ions.

Q4 a) Enthalpy change of formation is the enthalpy change when 1 mole of a compound is formed from its elements in their standard states under standard conditions.
b) Second electron affinity is the enthalpy change when 1 mole of gaseous 2– ions is made from 1 mole of gaseous 1– ions.
c) The enthalpy change of hydration is the enthalpy change when 1 mole of aqueous ions is formed from gaseous ions.

Q5 a) Enthalpy change of atomisation of an element.
b) The first ionisation energy.

Q6 a) ΔH_{hyd}
b) ΔH_{diss}

2. Born-Haber Cycles

Pages 270-271 — Application Questions

Q1
$Ca^{2+}_{(g)} + 2e^- + 2Br_{(g)}$
$Ca^{+}_{(g)} + e^- + 2Br_{(g)}$
$Ca_{(g)} + 2Br_{(g)}$
$Ca^{2+}_{(g)} + 2Br^-_{(g)}$
$Ca_{(s)} + 2Br_{(g)}$
$Ca_{(s)} + Br_{2(g)}$
$CaBr_{2(s)}$

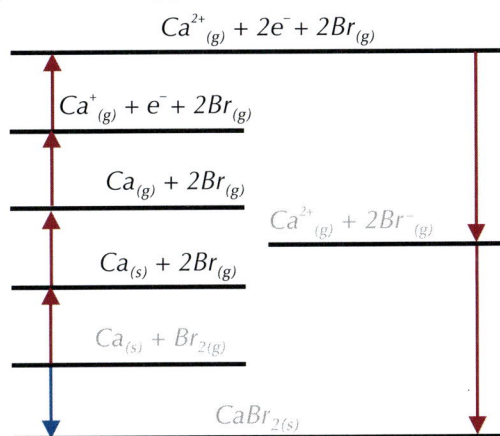

Q2 a)

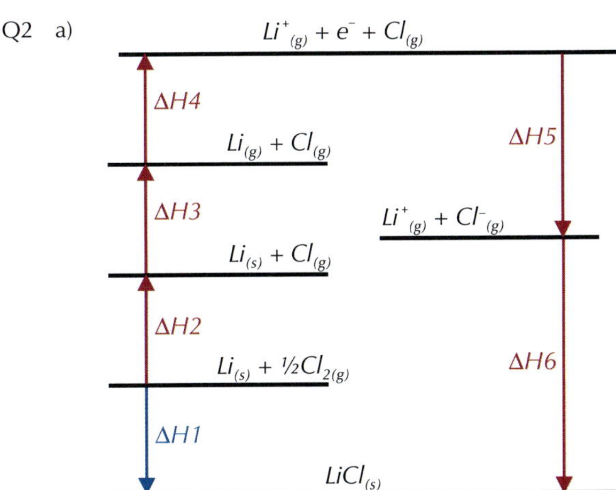

b) $\Delta H1 = \Delta H2 + \Delta H3 + \Delta H4 + \Delta H5 + \Delta H6$
$= (+122) + (+159) + (+520) + (-349) + (-861)$
$= \mathbf{-409\ kJ\ mol^{-1}}$

Q3 a)

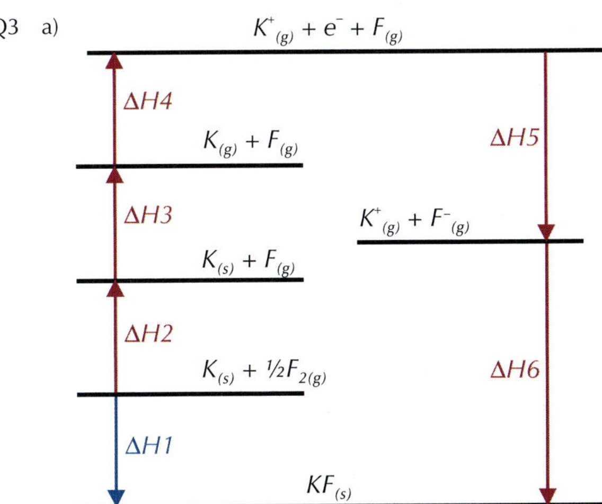

b) $\Delta H6 = -\Delta H5 - \Delta H4 - \Delta H3 - \Delta H2 + \Delta H1$
$= -(-328) - (+419) - (+89) - (+79) + (-563)$
$= \mathbf{-822\ kJ\ mol^{-1}}$

c) The purely ionic model of a lattice, which theoretical lattice enthalpies are based on, assumes that all the ions are spherical, and have their charge evenly distributed around them. But the experimental lattice enthalpy from the Born-Haber cycle is usually different. This is because ionic compounds usually have some covalent character. The positive and negative ions in a lattice aren't usually exactly spherical. Positive ions polarise neighbouring negative ions to different extents, and the more polarisation there is, the more covalent the bonding will be.

3. Enthalpies of Solution

Page 273 — Application Questions

Q1 $\Delta H3 = \Delta H1 + \Delta H2 = +826 + (-520 + -364) =$ **-58 kJ mol^{-1}**

Q2 a)

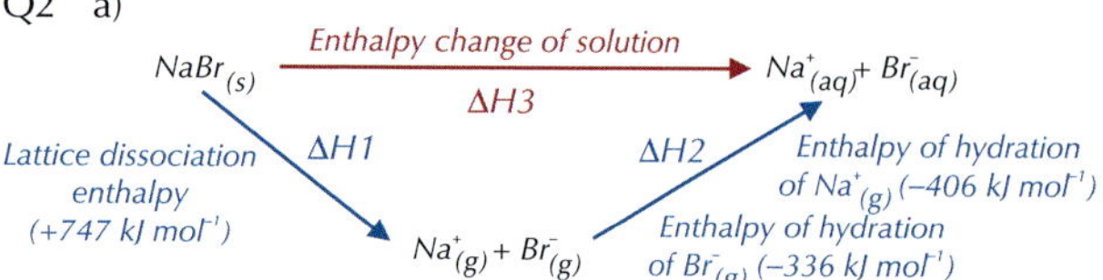

b) $\Delta H3 = \Delta H1 + \Delta H2 = +747 + (-406 + -336)$
= **+5 kJ mol^{-1}**

Page 273 — Fact Recall Question

Q1 a) The bonds between the ions break (the ions dissociate), and bonds between the ions and water are made (the ions become hydrated).

b) Dissociation is endothermic, and hydration is exothermic.

4. Entropy

Page 276 — Application Questions

Q1 The entropy will increase when the solid sodium hydroxide dissolves in the aqueous hydrogen chloride. The entropy of the water will increase when it turns to a gas and the entropy of the sodium chloride will decrease when it turns to a solid.

Q2 $\Delta S = S_{products} - S_{reactants}$
$= (214 + (2 \times 69.9)) - (186 + (2 \times 205))$
= **-242.2 J K^{-1} mol^{-1}**

Q3 $\Delta S = S_{products} - S_{reactants}$
$= ((3 \times 31.6) + (2 \times 69.9)) - (248 + (2 \times 206))$
= **-425.4 J K^{-1} mol^{-1}**

Page 276 — Fact Recall Questions

Q1 a) It is a measure of the amount of disorder in a system (e.g. the number of ways that particles can be arranged and the number of ways that the energy can be shared out between particles).

b) ΔS

Q2 a) The entropy increases because particles move around more in gases than in liquids and their arrangement is more disordered.

b) The entropy increases because particles move around more in solution than in solids and their arrangement is more disordered.

c) The entropy increases because the more gaseous particles you've got, the more ways they and their energy can be arranged.

5. Free-Energy Change

Page 280 — Application Questions

Q1 a) $\Delta S = S_{products} - S_{reactants}$
$= ((2 \times 28.3) + (3 \times 27.0)) - (51.0 + (3 \times 32.5))$
= **-10.9 J K^{-1} mol^{-1}**

b) $\Delta H^\circ = -130$ kJ mol^{-1} $= -130 \times 10^3$ J mol^{-1}
$\Delta G = \Delta H - T\Delta S$
$= -130 \times 10^3 - (298 \times -10.9)$
= **$-127\ 000$ J mol^{-1}** (3 s.f.)

c) The reaction is feasible at 298 K because ΔG is negative.

Q2 $\Delta H^\circ = 178.0$ kJ mol^{-1} $= 178.0 \times 10^3$ J mol^{-1}
$T = \Delta H \div \Delta S = 178.0 \times 10^3 \div 165.0 =$ **1079 K**

Q3 a) $\Delta H = -20\ 000$ J mol^{-1} = **-20 kJ mol^{-1}**
The enthalpy change is the y-intercept of the straight line.

b) $\Delta S = -\text{gradient}$
$= -(\Delta y \div \Delta x) = -((30\ 000 - (-20\ 000)) \div (600 - 0))$
$= -(50\ 000 \div 600) =$ **-83.3 J K^{-1} mol^{-1}**

c) The reaction is not feasible at 400 K. From the graph, ΔG is positive at 400 K.

Page 280 — Fact Recall Questions

Q1 a) Free-energy change is a measure used to predict whether a reaction is feasible. It is 0 or negative for a feasible reaction and positive for a non-feasible reaction.

b) ΔG

c) J mol^{-1} OR kJ mol^{-1}

Q2 $\Delta G = \Delta H - T\Delta S$

Q3 no

Q4 $T = \Delta H \div \Delta S$

Exam-style Questions — pages 282-285

1 D ***(1 mark).***
The equation $\Delta G = \Delta H - T\Delta S$ is the equation of a straight line, $y = mx + c$. So the enthalpy change is the y-intercept, which is negative. The gradient is equal to the entropy change with the opposite sign. So, since the gradient is positive, entropy change must be negative.

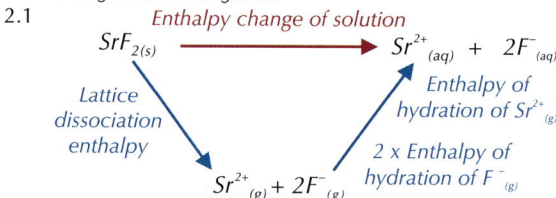

2.1

Enthalpy change of solution
$SrF_{2(s)}$
$Sr^{2+}_{(aq)} + 2F^-_{(aq)}$
Lattice dissociation enthalpy
Enthalpy of hydration of $Sr^{2+}_{(g)}$
$Sr^{2+}_{(g)} + 2F^-_{(g)}$
2 x Enthalpy of hydration of $F^-_{(g)}$

(1 mark for a complete correct cycle, 1 mark for correctly labelled arrows)

2.2 Enthalpy change of solution ($SrF_{2(s)}$)
= lattice dissociation enthalpy ($SrF_{2(s)}$)
+ enthalpy of hydration ($Sr_2{}^+{}_{(g)}$)
+ [2 × enthalpy of hydration ($F^-_{(g)}$)] ***(1 mark)***
$= 2492 + (-1480) + (2 \times -506) =$ **0 kJ mol^{-1}** ***(1 mark)***
You have to double the enthalpy of hydration for F^- because there are two in SrF_2.

3.1 It is the enthalpy change when 1 mole of gaseous atoms is formed from an element in its standard state ***(1 mark)***.

3.2

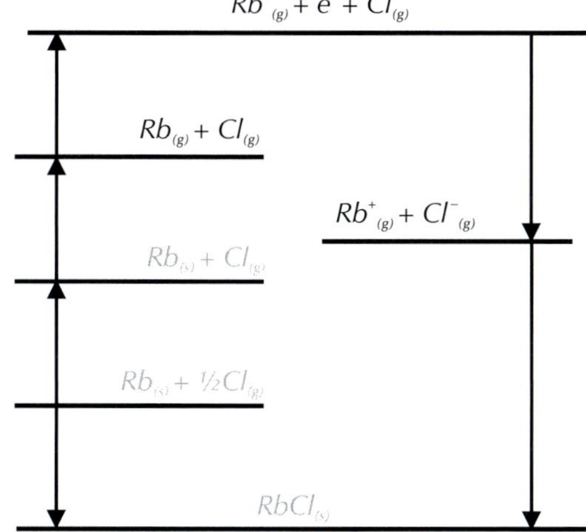

(1 mark for each correct label).

3.3 $\Delta H6 = -\Delta H5 - \Delta H4 - \Delta H3 - \Delta H2 + \Delta H1$
$= -(-349) - (+403) - (+81) - (+122) + (-435)$
$= \mathbf{-692\ kJ\ mol^{-1}}$
(3 marks for correct answer, otherwise 1 mark for correct equation, 1 mark for correct substitution)

3.4 When experimental and theoretical lattice enthalpies are very similar, it shows that a compound has very little covalent character ***(1 mark)***. This means that rubidium chloride fits the purely ionic model very well and has almost 100% ionic character. ***(1 mark)***.

3.5 +692 kJ mol^{-1} ***(1 mark)***

3.6 Enthalpy change of solution ***(1 mark)***.

3.7 $\Delta H3 = \Delta H1 + \Delta H2 = +692 + (-296 + -364)$
$= \mathbf{+32\ kJ\ mol^{-1}}$ ***(1 mark)***
If you used the value of +300 kJ mol^{-1} for the enthalpy change of dissociation you'll get an answer of −360 kJ mol^{-1}. If you got this answer, give yourself the mark.

4.1 Increase. Every 5 molecules of reactants produces 6 molecules of products ***(1 mark)*** so the products will have higher entropy because there are more ways of arranging 6 molecules than 5 ***(1 mark)***.
You generally don't get marks for 50:50 guesses — so you won't get any marks just for saying it increases. The marks come for the explanation.

4.2 It would decrease. Particles move around less in liquids than in gases and their arrangement is less disordered, so liquids have lower entropy ***(1 mark)***.

4.3 $\Delta S = S_{products} - S_{reactants}$
$= ((2 \times 198) + (4 \times 189)) - ((2 \times 186) + (3 \times 205))$
$= \mathbf{+165\ J\ K^{-1}\ mol^{-1}}$
(3 marks for correct answer, otherwise 1 mark for correct equation, 1 mark for correct substitution).

4.4 $\Delta G = \Delta H - T\Delta S$ ***(1 mark)***.

4.5 ΔH is negative and ΔS_{system} is positive ***(1 mark)***. This means that the value of ΔG will always be negative irrespective of the temperature ***(1 mark)***.

5.1 $\Delta S = S_{products} - S_{reactants}$
$= (32.0 + 214) - (53.0 + 5.70)$
$= +187.3\ J\ K^{-1}\ mol^{-1}$ ***(1 mark)***.
$\Delta H^{\circ} = +127\ kJ\ mol^{-1} = +127 \times 10^3\ J\ mol^{-1}$
$\Delta G = \Delta H - T\Delta S$ ***(1 mark)***.
$= (+127 \times 10^3) - (1473 \times +187.3)$ ***(1 mark)***.
$= \mathbf{-149000\ J\ mol^{-1}}$ or $\mathbf{-149\ kJ\ mol^{-1}}$ (3 s.f.) ***(1 mark)***.

5.2 If the free-energy change is negative or equal to zero the reaction is feasible ***(1 mark)***.

5.3 $T = \Delta H \div \Delta S$ ***(1 mark)***.

5.4 $T = (+127 \times 10^3) \div 187.3$
$= 678\ K$
(1 mark for correct substitution, 1 mark for correct answer).
Don't forget, reactions become feasible when ΔG is 0.

5.5 The standard enthalpy change of formation is the enthalpy change when 1 mole of a compound is formed from its elements in their standard states and under standard conditions ***(1 mark)***.

5.6 The formation of manganese(IV) oxide would be exothermic ***(1 mark)***.
It's an exothermic reaction because ΔH is negative.

Section 2 — Rate Equations and K_p

1. Monitoring Reactions

Page 290 — Application Questions

Q1 a) E.g. use colorimetry to measure the change in absorbance, measure the change in pH.
b) E.g. measure the loss of mass, measure the volume of gas produced, measure the change in pH.
c) E.g. use colorimetry to measure the change in absorbance, measure the loss of mass, measure the volume of gas produced.

Q2 The calculated rate would be less than the true rate. Not all of the gas produced would be collected, as some would escape from the system. This means that the reaction would appear to have produced less gas than it really had, and so the rate would appear lower.

Q3 Initial number of moles HCl = conc × vol
= 1.5 × (250 ÷ 1000) = 0.375

Time / seconds	Mass lost / g	Concentration of HCl / mol dm^{-3}
0	0.00	**1.50**
10	0.60	**1.36**
20	1.18	**1.22**
30	1.77	**1.08**
40	2.37	**0.944**
50	2.94	**0.810**
60	3.51	**0.677**

See page 289 for a full method for this type of question.

Q4 $Na_2CO_3 + H_2SO_4 \rightarrow Na_2SO_4 + H_2O + CO_2$
Initial number of moles H_2SO_4 = conc × vol
= 1.00 × (100 ÷ 1000)
= 0.100

Time / seconds	Concentration of H_2SO_4 / mol dm^{-3}
0	**1.00**
15	**0.999**
30	**0.997**
45	**0.996**
60	**0.995**
75	**0.994**

See page 287 for a full method for this type of question.

Page 290 — Fact Recall Questions

Q1 Continuous monitoring is a method for following the progress of a reaction. It involves taking measurements at regular intervals, over the course of a reaction, to measure the loss of a reactant or the formation of a product.

Q2

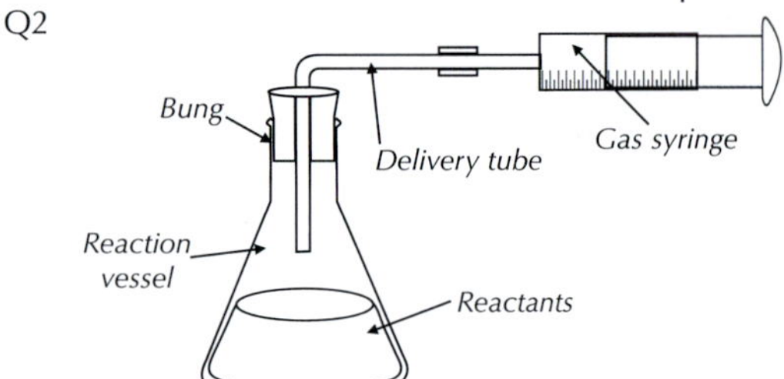

Q3 Colorimeters measure the absorbance of a solution.

Q4 E.g. you could measure the loss of mass, at regular intervals, as a gaseous product forms and escapes. By carrying out the reaction on a balance, you can follow how the mass changes throughout the experiment. / You could collect the gas in a gas syringe and record the volume of gas produced, at regular intervals, over the course of the reaction.

2. Reaction Rates and Graphs

Page 293 — Application Questions

Q1 a)

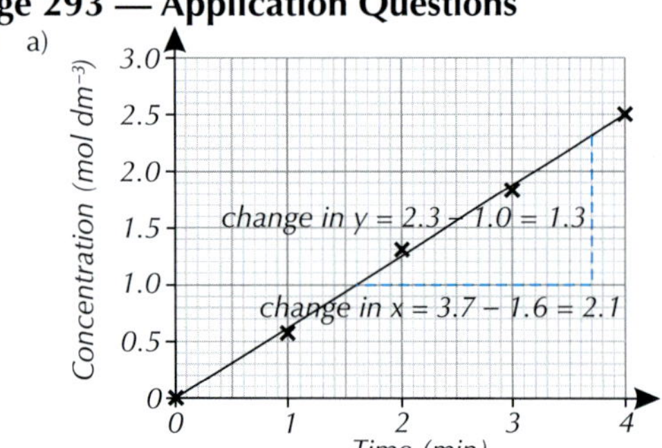

E.g. gradient = change in y ÷ change in x

= 1.3 ÷ 2.1 = **0.62 mol dm^{-3} min^{-1} (2 s.f.)**

In questions where you have to find a gradient, it's okay if you get a slightly different answer — your rate will vary if you use a slightly different line of best fit, tangent or pair of points to calculate the gradient from.

b) The rate of reaction is the same throughout the 4 minutes.

Q2 a)

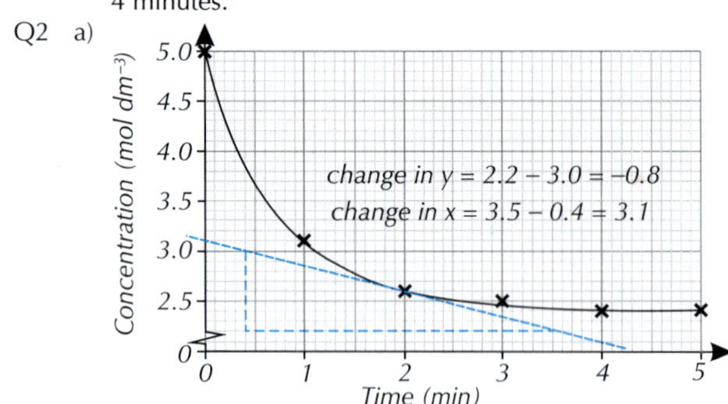

b) see above

c) E.g. gradient = change in y ÷ change in x

= –0.8 ÷ 3.1 = –0.258...

So the rate is **0.26 mol dm^{-3} min^{-1} (2 s.f.)**

Q3

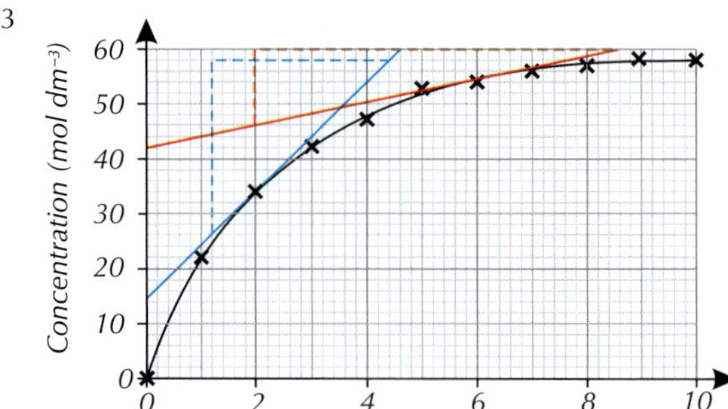

a) i) At 2 minutes, e.g.

change in y = 58 – 26 = 32

change in x = 4.4 – 1.2 = 3.2

gradient = change in y ÷ change in x

= 32 ÷ 3.2 = **10 mol dm^{-3} min^{-1} (2 s.f.)**

ii) At 6 minutes, e.g.

change in y = 60 – 46 = 14

change in x = 8.6 – 2 = 6.6

gradient = change in y ÷ change in x

= 14 ÷ 6.6 = **2.1 mol dm^{-3} min^{-1} (2 s.f.)**

b) E.g. when the reaction is complete (after about 9 minutes), the student is correct in their statement. However, the rate of the reaction changes/decreases throughout the reaction so, before this point, the student is incorrect.

3. Rate Equations

Page 297 — Application Questions

Q1 a) No. You can't predict the orders of reactions, and therefore the rate equation, by looking at the balanced chemical equation. You can only work out reaction orders and the rate equation by carrying out experiments.

b) The student's rate equation shows that the reaction is second order with respect to iodide ions. Therefore, halving the concentration of iodide would cause the rate of reaction to be 4 times slower.

Q2 a) Rate = $k[H_2][NO]^2$

b) 3

c) If the concentration of NO were doubled the rate of the reaction would be four times faster.

d) Rate = 221 × (1.54 × 10^{-3}) × (1.54 × 10^{-3})^2

= 8.07 × 10^{-7} mol dm^{-3} s^{-1} (3 s.f.)

Q3 a) Rate = $k[NO]^2[Cl_2]$

b) 5.85 × 10^{-6} = k(0.400)^2(0.400) so

$$k = \frac{5.85 \times 10^{-6}}{[0.400]^2[0.400]} = 9.14 \times 10^{-5}$$

$$\text{units of } k = \frac{\text{mol dm}^{-3}\text{ s}^{-1}}{(\text{mol dm}^{-3})^2(\text{mol dm}^{-3})} = \text{mol}^{-2}\text{ dm}^6\text{ s}^{-1}$$

k = **9.14 × 10^{-5} mol^{-2} dm^6 s^{-1}**

When calculating k in exams, don't forget to work out the units as well.

c) If the temperature was increased the value of k would increase.

d) Rate = $k[NO]^2[Cl_2]$

Rate = (9.14 × 10^{-5}) × 0.500^2 × 0.200

= **4.57 × 10^{-6} mol dm^{-3} s^{-1}**

Page 297 — Fact Recall Questions

Q1 The rate equation tells you how the rate of a reaction is affected by the concentration of the reactants.

Q2 Rate = $k[A]^m[B]^n$

Q3 a) The rate constant.
b) The concentration of reactant Y.
c) The order of the reaction with respect to X.

Q4 Input all of the units for the different values into the rate equation. Cancel down as much as possible.

4. The Initial Rates Method

Page 301 — Application Questions

Q1 E.g.

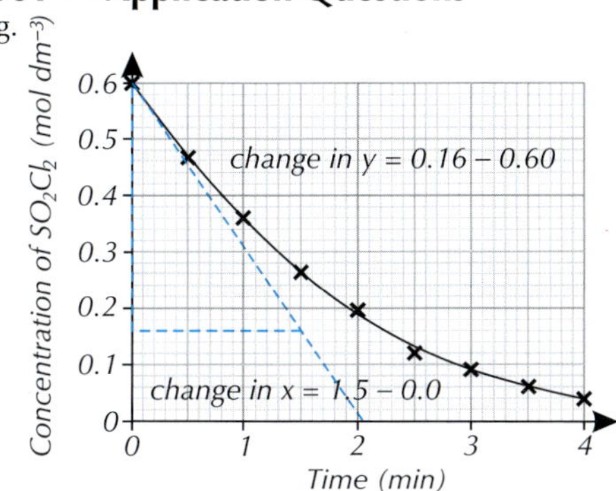

Gradient = change in y ÷ change in x
= –0.44 ÷ 1.5 = –0.293...
So the rate is **0.29 mol dm^{-3} min^{-1} (2 s.f.)**

Q2 Start by calculating the 1/reaction time for each trial number.

Trial no.	$[I_2]$ / mol dm^{-3}	$[S_2O_3^{2-}]$ / mol dm^{-3}	Reaction time / s	1/time / s^{-1}
1	0.040	0.040	312	3.21×10^{-3}
2	0.080	0.040	156	6.41×10^{-3}
3	0.040	0.020	624	1.60×10^{-3}

$1/t$ can be used as a measure of the rate, so from the 1/t values, you can see that if the concentration of I_2 is doubled, the rate also doubles. If the concentration of $S_2O_3^{2-}$ is halved, then the rate also halves. Therefore the reaction must be first order with respect to both reactants.
Rate = $k[I_2]\ [S_2O_3^{2-}]$

Q3 a) Looking at experiments 1 and 2: doubling [B] quadruples the rate so the reaction is order 2 with respect to [B]. Looking at experiments 2 and 3: tripling [C] triples the rate so the reaction is order 1 with respect to [C]. Looking at experiments 2 and 4: halving [A] halves the rate so the reaction is order 1 with respect to [A].
So, Rate = $k[A][B]^2[C]$

b) Rate = $k[A][B]^2[C]$ so k = rate ÷ $[A][B]^2[C]$
For experiment 1:
$k = 0.25 \div [(1.2) \times (1.2)^2 \times (1.2)] = 0.12$
units of k = mol dm^{-3} s^{-1} ÷ [(mol dm^{-3}) × (mol dm^{-3})^2 × (mol dm^{-3})]
k = **0.12 mol^{-3} dm^9 s^{-1}**

Page 301 — Fact Recall Questions

Q1 Take the gradient of the tangent to the curve, right at the start of the reaction/at time = 0.

Q2 E.g. set up a reaction where you are monitoring the loss of a reactant or the production of a product at the start of a reaction. Use this data to draw a concentration-time graph and calculate the initial rate of the reaction by working out the gradient at time = 0. Then, repeat the same experiment multiple times, but altering the concentration of one of the reactants each time. After each reaction, draw a concentration-time graph. Continue this until you have varied the concentration of each reactant in isolation. Compare the initial rates from each of your reactions to obtain information on how the initial rate varies with the concentration with respect to each reactant.

5. Clock Reactions

Page 303 — Fact Recall Questions

Q1 In a clock reaction, you measure how the time taken for a set amount of product to form changes as you vary the concentration of one of the reactants.

Q2 You have to assume that the concentration of each reactant doesn't change significantly over the time period of your clock reaction, the temperature stays constant and that when the endpoint is seen, the reaction has not proceeded too far.

Q3 The solution turns from colourless to blue-black.

6. Rate-concentration Graphs

Page 305 — Application Questions

Q1 a) A = first order
B = zero order
C = second order

b) (i) If [A] was halved the rate of reaction would also halve.
(ii) If [B] was tripled the rate of reaction would stay the same.
(iii) If [C] was doubled, the reaction rate would increase by a factor of four (quadruple).

Q2 a)

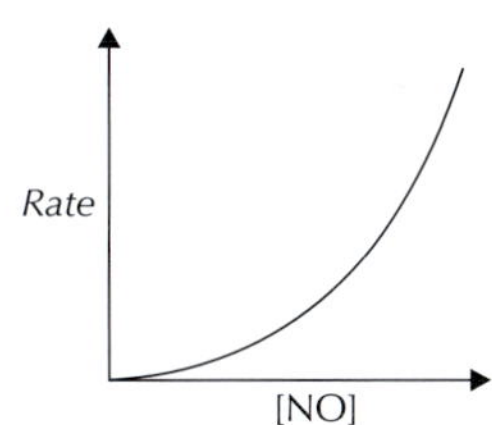

If you're asked to draw or sketch a graph in your exam, make sure you label the axes — if you don't your graph could be showing anything and you won't get the marks.

b) If the concentration of O_2 was doubled the reaction rate would also double.

Pages 305 — Fact Recall Questions

Q1 Find the gradient at various points along the concentration-time graph. Plot rate of reaction against concentration to produce a rate-concentration graph.

Q2 0

Q3

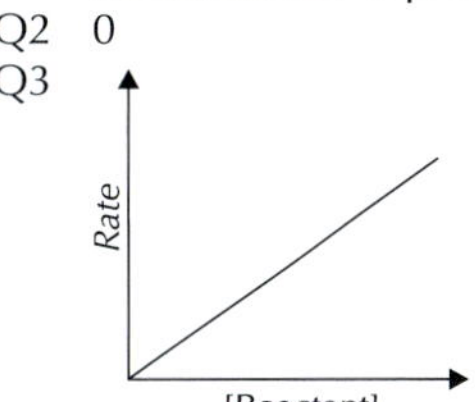

7. The Rate-determining Step

Page 307 — Application Questions

Q1 a) Step 1. The rate equation shows that the rate-determining step involves two molecules of NO_2, and there are two NO_2 molecules in step 1.

Don't forget — if it's in the rate equation it must be involved in the rate-determining step.

b) If the reaction had a one-step mechanism, that step would have to involve CO. CO is not in the rate equation so can't be involved in the rate-determining step. So a one-step mechanism isn't possible.

Q2 Rate = $k[A]^2[B][C]$

Page 307 — Fact Recall Questions

Q1 The rate-determining step is the slowest step in a reaction mechanism, so it's the step which determines the overall rate of the reaction.

Q2 2

8. The Arrhenius Equation

Page 312 — Application Questions

Q1 a) The rate constant of the catalysed reaction would be greater than that of the uncatalysed reaction. This is because introducing a catalyst decreases the activation energy of a reaction. Since the rate constant is proportional to $e^{-E_a/RT}$, decreasing the value of E_a makes the number that the exponent is raised to less negative. This increases the value of the exponent, and therefore increases the value of the rate constant.

b) Decreasing the temperature would decrease the rate constant of the reaction. Since the rate constant is proportional to $e^{-E_a/RT}$, decreasing the value of T makes the number that the exponent is raised to more negative. This decreases the value of the exponent, and therefore decreases the value of the rate constant.

Q2 a) $k = Ae^{\frac{-E_a}{RT}}$

$A = k \div e^{-E_a/RT}$

$= 1.10 \times 10^{-5} \div e^{-180\,000/(8.31 \times 600)}$

$= 1.10 \times 10^{-5} \div 2.09... \times 10^{-16}$

$= \mathbf{5.25 \times 10^{10}\ mol^{-1}\ dm^3\ s^{-1}}$ **(3 s.f.)**

b) $k = Ae^{\frac{-E_a}{RT}}$

$\ln (k) = \ln (A) - E_a/RT$

$\ln (k) - \ln (A) = -E_a/RT$

$E_a = (\ln (A) - \ln(k)) \times RT$

$E_a = (\ln(5.25 \times 10^{10}) - \ln (2.02 \times 10^{-5})) \times (8.31 \times 600)$

$= 176\,972...\ J\ mol^{-1} = \mathbf{177\ kJ\ mol^{-1}}$ **(3 s.f.)**

Q3 a)

$1/T$ / K^{-1}	ln k
0.00180	−14.9
0.00159	−10.4
0.00150	−8.42
0.00143	−6.77
0.00127	−3.23

Plotting these values on a graph of $1/T$ vs $\ln k$ gives:

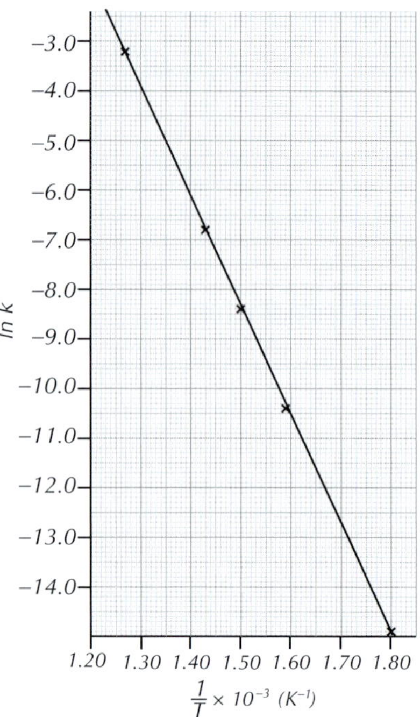

b) E.g. gradient = $-E_a/R$

gradient = change in y ÷ change in x

gradient = $[-14.5 - (-3.5)] \div (1.78 \times 10^{-3} - 1.28 \times 10^{-3})$

gradient = $-11 \div (5.0 \times 10^{-4}) = -22\,000$

$-E_a/R = -22\,000$

$E_a = 22000 \times 8.31 = 182820\ J\ mol^{-1} = \mathbf{183\ kJ\ mol^{-1}}$

c) E.g. by substituting E_a and a data point for T and k into the equation $k = Ae^{-E_a/RT}$.

when $k = 3.50 \times 10^{-7}$ and $T = 555$,

$3.50 \times 10^{-7} = Ae^{(-182\,820 \div (8.31 \times 555))}$

$A = \mathbf{5.7 \times 10^{10}}$

Page 312 — Fact Recall Questions

Q1 $k = Ae^{\frac{-E_a}{RT}}$, where k = the rate constant, A = the Arrhenius constant, E_a = the activation energy, R = the gas constant and T = the temperature.

Q2 When the temperature of a reaction is increased, the rate increases.

Q3 $\ln k = -E_a/RT + \ln A$

Q4 $1/T$ and $\ln k$

9. Gas Equilibria

Page 316 — Application Questions

Q1 a) $2NH_{3\,(g)} + 1\frac{1}{2}O_{2(g)} \rightleftharpoons N_{2\,(g)} + 3H_2O_{(g)}$

b) total pressure = $p_{NH_3} + p_{O_2} + p_{N_2} + p_{H_2O}$

c) total pressure = 42 + 85 + 21 + 12 = **160 kPa**

d) partial pressure of a gas in a mixture = mole fraction of gas × total pressure of the mixture

So mole fraction of H_2O = 12 ÷ 160 = **0.075**

Q2 No. moles of He = 4.00 ÷ 4 = 1.00

No. moles of O_2 = 2.81 ÷ (16 × 2) = 0.0878...

Total moles of gas = 1.00 + 0.0878... = 1.0878...

Mole fraction of O_2 = 0.0878... ÷ 1.0878... = 0.0807...

So partial pressure of O_2 = 0.0807... × 8.12

= **0.655 kPa (3 s.f.)**

Q3 a) $K_p = \frac{p_{C_6H_{12}}}{p_{C_6H_6} \times (p_{H_2})^3}$

b) Since the partial pressures of all three gases are equal, $p_{C_6H_{12}} = p_{C_6H_6} = p_{H_2} = x$

$4.80 \times 10^{-13} = \frac{x}{x \times (x)^3} = 1/x^3$

$x^3 = 1 \div 4.80 \times 10^{-13}$

$x = \sqrt[3]{2.08... \times 10^{12}}$

$x =$ **12 800 kPa**

Q4 a) $2HF \rightleftharpoons H_2 + F_2$

Equal amounts of H_2 and F_2 are produced, so if there are 9.34 mol of F_2 present at equilibrium, there are also 9.34 mol of H_2.

There are the same number of moles of gas on each side of the equation, so the total number of moles of gas at equilibrium must still be 24.32.

mole fraction of H_2 = 9.34 ÷ 24.32 = 0.384...

partial pressure of H_2 = 0.384... × 2313 = **888.3 kPa**

b) partial pressure of H_2 = partial pressure of F_2, so partial pressure of F_2 = 888.3 kPa.

p_{HF} = 2313 – 888.3 – 888.3 = 536.4 kPa

$K_p = \frac{p_{H_2} \times p_{F_2}}{(p_{HF})^2}$ = 888.3 × 888.3 ÷ (536.4)² = **2.742** (no units)

Page 316 — Fact Recall Questions

Q1 The partial pressure of a gas is the pressure that an individual gas in a mixture exerts on a system.

Q2 mole fraction of a gas = number of moles of that gas ÷ total number of moles of gas in the mixture

Q3 $K_p = \frac{(p_D)^d (p_E)^e}{(p_A)^a (p_B)^b}$

10. Changing Gas Equilibria

Page 318 — Application Questions

Q1 The student is incorrect. Because there are the same number of moles of gas on both sides of the equation, this particular reaction would not be affected by changes in pressure.

Q2 a) There are two moles of reactants, and only one mole of product. Decreasing pressure would shift equilibrium to the side of the reaction where there are more moles of gas. Therefore, equilibrium would move to the left.

b) Since the forward reaction is exothermic, a decrease in temperature would favour the forward reaction. This would increase the moles of product present at equilibrium, and there would be comparatively fewer moles of reactant. This would increase the value of K_p.

Page 318 — Fact Recall Questions

Q1 Le Chatelier's principle states that if a reversible reaction at equilibrium is subjected to a change in temperature, pressure or concentration, then the equilibrium will shift to try and counteract the change.

Q2 Increasing the temperature of a reaction that is exothermic in the forward direction would shift equilibrium to the left. The endothermic reaction is favoured as it can absorb some of the extra heat.

Q3 K_p is not affected by changes in pressure.

Q4 Catalysts do not affect the position of equilibrium or K_p.

Exam-style Questions — pages 320-322

1 B ***(1 mark)***

2 B ***(1 mark)***

3 C ***(1 mark)***

4 D ***(1 mark)***

5.1 E.g. measure the change in pH / measure the volume of gas produced using a gas syringe ***(1 mark)***.

5.2

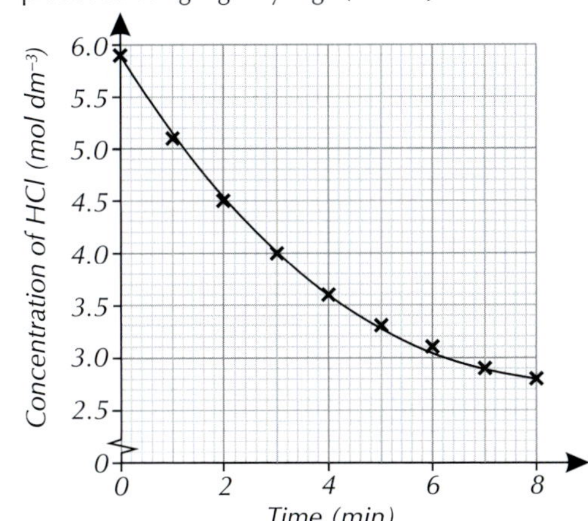

(4 marks in total — 1 mark for appropriate scale chosen on each axis, 1 mark for axes correctly labelled, 1 mark for points correctly plotted, 1 mark for correctly drawn line of best fit.)

5.3 At $t = 0$: E.g. gradient = change in y ÷ change in x

= (2.7 – 6.0) ÷ (3.6 – 0.0) ***(1 mark)***

= –3.3 ÷ 3.6 = 0.916...

So rate is **0.92 mol dm⁻³ min⁻¹** ***(1 mark)***

6.1 $k = Ae^{\frac{-E_a}{RT}}$

$\ln k = -E_a/RT + \ln A$

$\ln k = \ln A - E_a/RT$

$RT \times (\ln A - \ln k) = E_a$ ***(1 mark)***

$(8.31 \times 600) \times (\ln (2.21) - \ln (1.28 \times 10^{-5})) = E_a$

E_a = 60 126.46... J mol⁻¹ = **60.1 kJ mol⁻¹** ***(1 mark)***

6.2 A catalyst decreases the activation energy of a reaction ***(1 mark)***. The Arrhenius equation shows that the rate of reaction is proportional to e^{-Ea}, so as E_a decrease, the value of the exponential increases, therefore the rate constant increases ***(1 mark)***.

6.3 First, you need to convert your values of T and k into $1/T$ and $\ln k$ values.

$1/T$ / K⁻¹	ln k
3.50×10^{-3}	–9.50
3.40×10^{-3}	–8.50
3.34×10^{-3}	–7.90
3.19×10^{-3}	–6.50
3.11×10^{-3}	–5.65
2.99×10^{-3}	–4.50

(2 marks — 1 mark for correct values of 1/T, 1 mark for correct values of ln k)

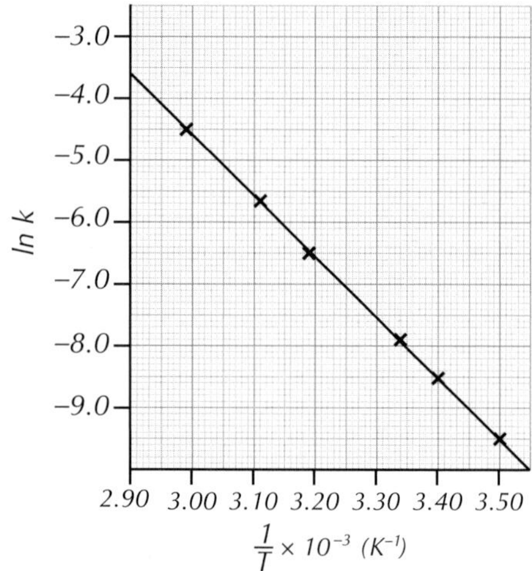

(2 marks — 1 mark for correct plot, 1 mark for correctly drawn line of best fit.)

e.g. gradient = $-E_a/R$ ***(1 mark)***

gradient = change in y ÷ change in x

gradient = $[-8.5 - (-4.0)] \div (3.40 \times 10^{-3} - 2.94 \times 10^{-3})$

gradient = $-4.5 \div (4.6 \times 10^{-4})$

$-E_a/R = 9782.60...$

$E_a = 81\ 293.47...$ J mol^{-1} = **81.3 kJ mol^{-1}** ***(1 mark)***

By substituting E_a and a data point for T and k into the Arrhenius equation, $k = Ae^{-Ea/RT}$.

When $k = 7.49 \times 10^{-5}$ and $T = 286$,

$7.49 \times 10^{-5} = Ae^{(-81\ 293.47... \div (8.31 \times 286))}$ ***(1 mark)***

A = **5.36 × 10^{10} mol dm^{-3} s^{-1}** ***(1 mark)***

Remember, the units of A are the same as the units of k.

7.1 $K_p = (p_{NH_3})^2 \div [p_{N_2} \times (p_{H_2})^3]$ ***(1 mark)***

$p_{N_2} = (p_{NH_3})^2 \div [K_p \times (p_{H_2})^3]$

$= 8920^2 \div (52.7 \times 19^3)$

= **220 kPa** ***(1 mark)***

7.2 The introduction of a catalyst would have no effect on K_p ***(1 mark)***.

7.3 Increasing the temperature would shift the equilibrium position over to the left to favour the endothermic, reverse reaction and absorb the extra heat ***(1 mark)***. This would decrease the value of K_p ***(1 mark)*** (as there would be more reactants and less product present in the reaction container).

8.1 The reaction is first order with respect to H_2 ***(1 mark)*** and second order with respect to NO ***(1 mark)***.

Finding the reaction order with respect to NO is tricky. You know the reaction is first order with respect to H_2 from experiments 1 and 2, so if only $[H_2]$ changed from experiment 2 to 3 you would expect the rate of reaction to halve. But the rate of reaction is four times greater than this, so the reaction must be second order with respect to NO.

8.2 rate = $k[H_2][NO]^2$ ***(1 mark)***

8.3 rate = $k[H_2][NO]^2$ so k = rate ÷ $[H_2][NO]^2$

E.g. Using experiment 1:

$k = (4.5 \times 10^{-3}) \div (6 \times 10^{-3})(3 \times 10^{-3})^2 = 8.3 \times 10^4$

k = mol dm^{-3} s^{-1} ÷ (mol dm^{-3})(mol dm^{-3})^2 = mol^{-2} dm^6 s^{-1}

k = **8.3 × 10^4 mol^{-2} dm^6 s^{-1}**

(3 marks for correct answer, otherwise 1 mark for correct method and 1 mark for correct units.)

8.4 rate = $k[H_2][NO]^2$

$= (8.3 \times 10^4) \times (2.5 \times 10^{-3}) \times (4.5 \times 10^{-3})^2$

= **4.2 × 10^{-3} mol dm^{-3} s^{-1}**

(1 mark for correct value, 1 mark for correct units — full marks if the method is correct but error carried forward from (c).)

8.5 The rate equation shows that the rate-determining step involves 2 molecules of NO and 1 molecule of H_2 ***(1 mark)***. There are 2 molecules of H_2 in the overall equation so there must be another step involving another molecule of H_2 ***(1 mark)***.

8.6 A possible mechanism for this reaction would be:

Step 1: $2NO + H_2 \rightarrow N_2O + H_2O$

Step 2: $N_2O + H_2 \rightarrow H_2O + N_2$

(1 mark for the left-hand side of step 1, 1 mark for rest of step 1 and step 2)

Other mechanisms are possible and will gain credit as long as:
The reactants in the first equation are 2NO and H_2.
The steps add together to give the overall reaction.
Both equations are balanced.

Section 3 — Electrode Potentials and Cells

1. Electrode Potentials

Page 327 — Application Questions

Q1 a) (i) $Ag^+_{(aq)} + e^- \rightarrow Ag_{(s)}$

(ii) $Ca_{(s)} \rightarrow Ca^{2+}_{(aq)} + 2e^-$

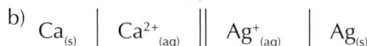

b) $Ca_{(s)}$ | $Ca^{2+}_{(aq)}$ || $Ag^+_{(aq)}$ | $Ag_{(s)}$

c) $E_{cell} = E_{RHS} - E_{LHS}$ = 0.80 − (−2.87) = **+3.67 V**

Q2 a) At the positive electrode: $Tl^{3+}_{(aq)} + 2e^- \rightarrow Tl^+_{(aq)}$

At the negative electrode: $Fe^{2+}_{(aq)} \rightarrow Fe^{3+}_{(aq)} + e^-$

b) $E_{RHS} = E_{cell} + E_{LHS}$ = 0.48 + 0.77 = **+1.25 V**

Q3 a) The Mg^{2+}/Mg half-cell.

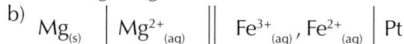

b) $Mg_{(s)}$ | $Mg^{2+}_{(aq)}$ || $Fe^{3+}_{(aq)}$, $Fe^{2+}_{(aq)}$ | Pt

Page 327 — Fact Recall Questions

Q1 a) platinum

b) Platinum is inert, so it won't react with the solution, but it's a solid that conducts electricity.

Q2 reduction

Q3 Any two from, e.g.: the half-cell with the more negative potential goes on the left / the oxidised forms go in the centre of the cell diagram / the reduced forms go at the edge of the cell diagram / double vertical lines are used to show a salt bridge / things in different phases are separated by a vertical line / things in the same phase are separated by a comma.

Any two facts about how to draw a cell diagram will do here — just flick back to page 325 if you need a quick reminder.

2. Standard Electrode Potentials

Page 329 — Application Question

Q1 a) Pt | $H_{2(g)}$ | $H^+_{(aq)}$ || $Pb^{2+}_{(aq)}$ | $Pb_{(s)}$

b) −0.13 V

c) oxidation

The electrode potential of the standard hydrogen electrode is 0.00 V, so the Pb^{2+}/Pb half-cell has the more negative electrode potential and oxidation occurs in this half-cell.

Page 329 — Fact Recall Questions

Q1 E.g. temperature, pressure and concentrations of reactants.

Q2 Hydrogen gas is bubbled into a solution of aqueous H^+ ions. The electrode is made of platinum. The standard conditions used are a temperature of 298 K (25 °C), a pressure of 100 kPa and all solutions of ions have a concentration of 1.00 mol dm^{-3}.

Q3 a) 0.00 V
b) It's zero by definition (scientists decided it would have that value).

Q4 The voltage measured under standard conditions when a half-cell is connected to a standard hydrogen electrode.

3. Electrochemical Series

Page 333 — Application Questions

Q1 a) $E^\circ_{cell} = E^\circ_{reduced} - E^\circ_{oxidised} = 0.80 - (-1.66) = \mathbf{+2.46\ V}$
b) $E^\circ_{cell} = E^\circ_{reduced} - E^\circ_{oxidised} = 1.36 - 0.34 = \mathbf{+1.02\ V}$

Q2 a) $Mg^{2+}_{(aq)} + 2e^- \rightarrow Mg_{(s)}$ $E^\circ = -2.38$ V
$Ni^{2+}_{(aq)} + 2e^- \rightarrow Ni_{(s)}$ $E^\circ = -0.25$ V
The electrode potential of the magnesium half-cell is more negative than the nickel half-cell, so $Mg_{(s)}$ will be oxidised by $Ni^{2+}_{(aq)}$.
So the full reaction is: $Mg_{(s)} + Ni^{2+}_{(aq)} \rightarrow Mg^{2+}_{(aq)} + Ni_{(s)}$
b) $Fe^{3+}_{(aq)} + e^- \rightarrow Fe^{2+}_{(aq)}$ $E^\circ = +0.77$ V
$Br_{2(aq)} + 2e^- \rightarrow 2Br^-_{(aq)}$ $E^\circ = +1.07$ V
The electrode potential of the iron half-cell is more negative than the bromine half-cell, so $Br_{2(aq)}$ will be oxidised by $Fe^{2+}_{(aq)}$.
So the full reaction is: $2Fe^{2+}_{(aq)} + Br_{2(aq)} \rightarrow 2Fe^{3+}_{(aq)} + 2Br^-_{(aq)}$
Don't forget to balance the charges when you combine redox equations — that's why you need $2Fe^{2+}$ ions here.

Q3 a) $Sn^{4+}_{(aq)} + 2e^- \rightarrow Sn^{2+}_{(aq)}$ $E^\circ = +0.15$ V
$Cu^{2+}_{(aq)} + 2e^- \rightarrow Cu_{(s)}$ $E^\circ = +0.34$ V
So if Cu^{2+} is reduced by Sn^{2+},
$E^\circ_{cell} = E^\circ_{reduced} - E^\circ_{oxidised} = 0.34 - 0.15 = +0.19$ V
E°_{cell} is positive, so the reaction is feasible.
b) $Sn^{4+}_{(aq)} + 2e^- \rightarrow Sn^{2+}_{(aq)}$ $E^\circ = +0.15$ V
$Zn^{2+}_{(aq)} + 2e^- \rightarrow Zn_{(s)}$ $E^\circ = -0.76$ V
So if Zn^{2+} is reduced by Sn^{2+},
$E^\circ_{cell} = E^\circ_{reduced} - E^\circ_{oxidised} = -0.76 - 0.15 = -0.91$ V
E°_{cell} is negative, so the reaction is not feasible.

Q4 The proposed reaction is $Sn^{2+}_{(aq)} + 2Ag^+_{(aq)} \rightarrow Sn^{4+}_{(aq)} + 2Ag_{(s)}$
$Sn^{4+}_{(aq)} + 2e^- \rightarrow Sn^{2+}_{(aq)}$ $E^\circ = +0.15$
$Ag^+_{(aq)} + e^- \rightarrow Ag_{(s)}$ $E^\circ = +0.80$
So if Ag^+ is reduced by Sn^{2+},
$E^\circ_{cell} = E^\circ_{reduced} - E^\circ_{oxidised} = 0.80 - 0.15 = +0.65$ V
E°_{cell} is positive, so the reaction is feasible, so Sn^{2+} ions will react with Ag^+ ions in solution.

Page 333 — Fact Recall Questions

Q1 A list of electrode potentials for different electrochemical half-cells, written in order from the most negative to the most positive.

Q2 In the direction of reduction.

Q3 reduction

4. Electrochemical Cells

Page 336 — Application Questions

Q1 a) $E^\circ_{cell} = (E^\circ_{R.H.S.} - E^\circ_{L.H.S}) = 0.52 - (-0.88) = \mathbf{+1.40\ V}$
b) $2NiO(OH)_{(s)} + Cd_{(s)} + 2H_2O_{(l)} \rightarrow 2Ni(OH)_{2(s)} + Cd(OH)_{2(s)}$
Don't forget, you need to cancel out the OH^- ions.

Q2 a) $2H_{2(g)} + O_{2(g)} \rightarrow 2H_2O_{(l)}$
b) $Pt \mid H_{2(g)} \mid OH^-_{(aq)}, H_2O_{(l)} \parallel O_{2(g)} \mid H_2O_{(l)}, OH^-_{(aq)} \mid Pt$

Page 336 — Fact Recall Questions

Q1 You can recharge a rechargeable battery because the reaction that occurs within it can be reversed if a current is supplied to force the electrons to flow in the opposite direction around the circuit.

Q2 There are two platinum electrodes separated by anion exchange membranes. Oxygen is fed into the positive electrode, where it reacts with water and electrons to produce OH^- anions. The OH^- anions can cross the anion exchange membrane and travel through the electrolyte to the negative electrode. Hydrogen is fed into the negative electrode where it reacts with the OH^- anions to produce water and electrons. The electrons travel round the external circuit, towards the positive electrode, generating electricity.

Q3 You need energy to produce the supply of hydrogen. This is normally generated by burning fossil fuels.

Q4 Advantages: E.g. they don't need electrically recharging / the only waste product is water / they don't produce CO_2 emissions / they are more efficient than the internal combustion engine.
Disadvantages: E.g. they are not usually carbon neutral / hydrogen is highly flammable so needs to be handled carefully.

Exam-style Questions — pages 338-340

1.1 Oxidation: $Mg_{(s)} \rightarrow Mg^{2+}_{(aq)} + 2e^-$ ***(1 mark)***
Reduction: $Fe^{3+}_{(aq)} + e^- \rightarrow Fe^{2+}_{(aq)}$ ***(1 mark)***

1.2 $Mg_{(s)} \mid Mg^{2+}_{(aq)} \parallel Fe^{3+}_{(aq)}, Fe^{2+}_{(aq)} \mid Pt$
(1 mark for left-hand side correct, 1 mark for right-hand side correct)
Don't forget to include the platinum electrode...

1.3 $E_{cell} = E_{RHS} - E_{LHS} = 0.77 - (-2.38) = \mathbf{+3.15\ V}$ ***(1 mark)***

1.4 E° for Sn^{4+}/Sn^{2+} is more negative than E° for VO_2^+/VO^{2+}, so Sn^{2+} ions will reduce VO_2^+ ions ***(1 mark)***
$2VO_2^+{}_{(aq)} + Sn^{2+}_{(aq)} + 4H^+_{(aq)} \rightarrow 2VO^{2+}_{(aq)} + Sn^{4+}_{(aq)} + 2H_2O_{(l)}$
(1 mark)
E° for Sn^{4+} / Sn^{2+} is more negative than E° for VO^{2+} / V^{3+}, so Sn^{2+} ions will reduce VO^{2+} ions created in the previous reaction ***(1 mark)***.
$2VO^{2+}_{(aq)} + Sn^{2+}_{(aq)} + 4H^+_{(aq)} \rightarrow 2V^{3+}_{(aq)} + Sn^{4+}_{(aq)} + 2H_2O_{(l)}$
(1 mark)
The V^{3+} ions made in this reaction don't react with the Sn^{2+} ions because the electrode potential for V^{3+}/V^{2+} is more negative than that for Sn^{4+}/Sn^{2+}.

1.5 A temperature of 298 K. A pressure of 100 kPa. Any solutions of ions have a concentration of 1.00 mol dm^{-3} ***(1 mark)***.

1.6 It is zero by definition/scientists decided it would be that value ***(1 mark)***.

1.7 Platinum conducts electricity and is inert (so it doesn't react with the ions formed) ***(1 mark)***.

1.8 +0.80 V ***(1 mark)***

2.1 $CoO_{2(s)} + Li_{(s)} \rightleftharpoons Li^+[CoO_2]^-_{(s)}$ ***(1 mark)***

2.2 The reactions that occur within them can be reversed if a current is supplied to force the electrons to flow in the opposite direction around the circuit ***(1 mark)***.

2.3 It is cheaper ***(1 mark)*** and it provides a larger surface area so the reaction goes faster ***(1 mark)***.

2.4 Any two from, e.g.: they're more efficient / they don't need to be electrically recharged / the only waste product is water / they don't produce CO_2 emissions ***(1 mark for each, up to a maximum of 2 marks)***.

3 B ***(1 mark)***

Aluminium is above zinc in the electrochemical series so the Al^{3+}/Al half-reaction has a more negative electrode potential than Zn^{2+}/Zn half-reaction. So $Al^{3+}_{(aq)}$ cannot oxidise $Zn_{(s)}$.

4.1 $Tl^{3+}_{(aq)} + 3e^- \rightleftharpoons Tl_{(s)}$ ***(1 mark)***

4.2 $E_{RHS} = E_{LHS} + E_{cell} = 0.15 + 0.57 =$ **+0.72 V *(1 mark)***

4.3 $Pt \mid H_{2(g)} \mid H^+_{(aq)} \| Tl^{3+}_{(aq)} \mid Tl_{(s)}$ ***(1 mark)***

Section 4 — Acids, Bases and pH

1. Acids, Bases and K_w

Page 342 — Fact Recall Questions

Q1 a) Brønsted-Lowry acids are proton donors.
b) Brønsted-Lowry bases are proton acceptors.

Q2 a) $HA_{(aq)} + H_2O_{(l)} \rightarrow H_3O^+_{(aq)} + A^-_{(aq)}$
(or $HA_{(aq)} \rightarrow H^+_{(aq)} + A^-_{(aq)}$)
b) $B_{(aq)} + H_2O_{(l)} \rightarrow BH^+_{(aq)} + OH^-_{(aq)}$
c) $HA_{(aq)} + B_{(aq)} \rightleftharpoons BH^+_{(aq)} + A^-_{(aq)}$
When you're writing equations for reversible reactions don't forget to use the funky double arrows. If you just draw a normal arrow it won't be correct.

Q3 $H_2O_{(l)} \rightleftharpoons H^+_{(aq)} + OH^-_{(aq)}$

Q4 a) K_w is a constant called the ionic product of water.
b) $K_w = [H^+][OH^-]$
c) mol^2 dm^{-6}
d) In pure water there is always one H^+ ion per OH^- ion so $[H^+]=[OH^-]$ and $[H^+][OH^-]$ is the same as $[H^+]^2$.

2. pH Calculations

Page 343 — Application Questions

Q1 pH = −log[H$^+$] = −log[0.050] = **1.30**

Q2 [H$^+$] = $10^{-pH} = 10^{-2.86}$ = **1.4 × 10^{-3} mol dm^{-3}**

Q3 pH = −log[H$^+$] = −log[0.020] = **1.70**

Page 345 — Application Questions

Q1 HCl is monoprotic so [H$^+$] = [HCl] = 0.080 mol dm^{-3}
pH = −log[H$^+$] = −log[0.080] = **1.1**

Q2 H_2SO_4 is diprotic so [H$^+$] = 2[H_2SO_4] = 2 × 0.025 = 0.050 mol dm^{-3}
pH= −log[H$^+$] = −log[0.050] = **1.3**

Q3 KOH is a strong base so [OH$^-$] = [KOH] = 0.200 mol dm^{-3}
$K_w = [H^+][OH^-]$ so $[H^+] = K_w \div [OH^-]$
= (5.48 × 10^{-14}) ÷ 0.200 = 2.74 × 10^{-13}
pH = −log[H$^+$] = −log[2.74 × 10^{-13}] = **12.562**

Page 345 — Fact Recall Questions

Q1 pH = $-\log_{10}[H^+]$

Q2 Monoprotic means that each molecule of acid releases one proton. Diprotic means that each molecule of acid releases two protons.

3. The Acid Dissociation Constant

Page 349 — Application Questions

Q1 a) $K_a = [H^+][CN^-] \div [HCN]$
You can use $[H^+]^2$ in your calculations, but not as the expression for K_a — this isn't correct, so you wouldn't get the marks for it.
b) $K_a = [H^+]^2 \div [HCN]$ so $[H^+]^2 = K_a \times [HCN]$
= (4.9 × 10^{-10}) × 2.0 = 9.8 × 10^{-10}
[H$^+$] = $\sqrt{9.8 \times 10^{-10}}$ = 3.13 × 10^{-5} mol dm^{-3}
pH = −log[H$^+$] = −log[3.13 × 10^{-5}] = **4.504**

Q2 [H$^+$] = $10^{-pH} = 10^{-3.8}$ = 1.584... × 10^{-4} mol dm^{-3}
$K_a = [H^+]^2 \div [HNO_2]$ so $[HNO_2] = [H^+]^2 \div K_a$
= (1.584... × 10^{-4})^2 ÷ 4.0 × 10^{-4} = **6.28 × 10^{-5} mol dm^{-3}**

Q3 $K_a = [H^+]^2 \div [HA]$ so $[H^+]^2 = K_a \times [HA]$
= (1.38 × 10^{-4}) × 0.48 = 6.624 × 10^{-5}
[H$^+$] = $\sqrt{6.624 \times 10^{-5}}$ = 8.14 × 10^{-3} mol dm^{-3}
pH = −log[8.14 × 10^{-3}] = **2.09**

Q4 [H$^+$] = $10^{-pH} = 10^{-4.11}$ = 7.76 × 10^{-5} mol dm^{-3}
$K_a = [H^+]^2 \div [HA]$ = (7.76 × 10^{-5})^2 ÷ 0.28
= **2.15 × 10^{-8} mol dm^{-3}**

Q5 [H$^+$] = $10^{-pH} = 10^{-3.67}$ = 2.14 × 10^{-4} mol dm^{-3}
$K_a = [H^+]^2 \div [HCOOH]$ so $[HCOOH] = [H^+]^2 \div K_a$
= (2.14 × 10^{-4})^2 ÷ (1.8 ×10^{-4}) = **2.54 × 10^{-4} mol dm^{-3}**

Q6 a) $K_a = 10^{-pK_a} = 10^{-4.78}$ = **1.7 × 10^{-5} mol dm^{-3}**
b) $K_a = [H^+]^2 \div [CH_3COOH]$ so
$[H^+]^2 = K_a \times [CH_3COOH]$ = (1.7 × 10^{-5}) × 0.25
= 4.25 × 10^{-6} mol^2 dm^{-6}
[H$^+$] = $\sqrt{4.25 \times 10^{-6}}$ = 2.06 × 10^{-3} mol dm^{-3}
pH = −log[H$^+$] = −log[2.06 × 10^{-3}] = **2.69**

Q7 [H$^+$] = $10^{-pH} = 10^{-4.5}$ = 3.16 × 10^{-5} mol dm^{-3}
$K_a = [H^+]^2 \div [HA]$ = (3.16 × 10^{-5})^2 ÷ 0.154
= 6.49 × 10^{-9} mol dm^{-3}
$pK_a = -\log(K_a)$ = −log(6.49 × 10^{-9}) = **8.188**

Q8 $K_a = 10^{-pK_a} = 10^{-3.14}$ = 7.24 × 10^{-4} mol dm^{-3}
[H$^+$] = $10^{-pH} = 10^{-3.2}$ = 6.31 × 10^{-4} mol dm^{-3}
$K_a = [H^+]^2 \div [HF]$ so $[HF] = [H^+]^2 \div K_a$
= (6.31 × 10^{-4})^2 ÷ 7.24 × 10^{-4} = **5.50 × 10^{-4} mol dm^{-3}**

Q9 $K_a = 10^{-pK_a} = 10^{-4.5}$ = 3.16 × 10^{-5} mol dm^{-3}
$K_a = [H^+]^2 \div [HX]$ so $[H^+]^2 = K_a \times [HX]$
= (3.16 × 10^{-5}) × 0.6 = 1.90 × 10^{-5} mol^2 dm^{-6}
so [H$^+$] = $\sqrt{1.9 \times 10^{-5}}$ = 4.36 ×10^{-3} mol dm^{-3}
pH = −log[H$^+$] = −log[4.36 × 10^{-3}] = **2.36**

Page 349 — Fact Recall Questions

Q1 mol dm^{-3}

Q2 a) $K_a = [H^+][A^-] \div [HA]$
b) $K_a = [H^+]^2 \div [HA]$
c) $[HA] = [H^+]^2 \div K_a$

Q3 E.g. Calculating the pH of a weak acid and calculating the concentration of a weak acid.

Q4 a) $pK_a = -\log_{10}(K_a)$
b) $K_a = 10^{-pK_a}$

4. Titrations and pH Curves

Page 353 — Application Questions

Q1 a) Strong base/weak acid, phenolphthalein
b) Strong acid/weak base, methyl orange
c) Strong acid/strong base, phenolphthalein/cresol purple/ litmus
d) Weak base/strong acid, methyl orange

Q2 Any curve with the vertical section covering pH 6.8 and pH 8.0. E.g.

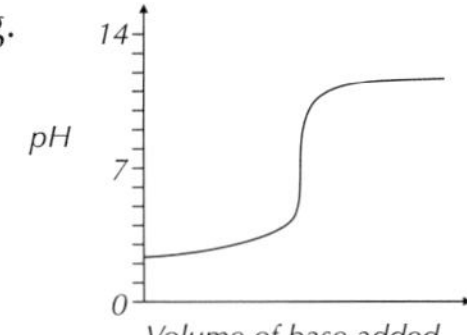

You could also have drawn a curve showing an acid being added to a base.

Page 353 — Fact Recall Questions

Q1 a)

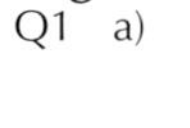

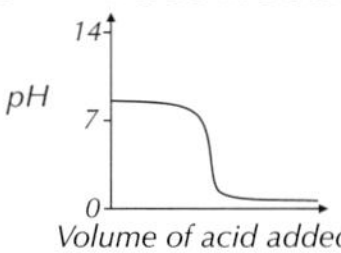

b)

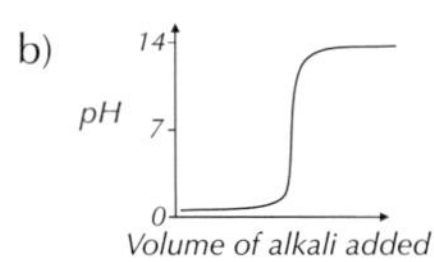

c)

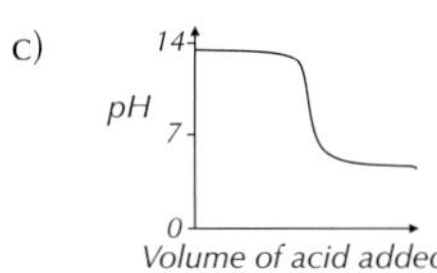

If the question says that a strong acid neutralises a strong base, the base is being neutralised so it's the acid that's added.

Q2 a) In most titrations, at the end point a tiny amount of acid/ base causes a sudden big change in pH and the base/ acid is just neutralised. But in weak acid/weak base titrations, the change in pH is more gradual, and the end point is when the base/acid is just neutralised.
b) The pH indicator changes colour/the pH meter shows a sudden big change/the pH meter shows a reading of 7.
c) In most titrations, the pH curve becomes close to vertical. In weak acid/weak base titrations, the end point is at the point on pH curve where pH = 7.

Q3 For an indicator to be suitable it must change colour over a narrow pH range that lies entirely within the vertical part of the pH curve for the titration.

5. Titration Calculations

Pages 356-357 — Application Questions

Q1 Moles HCl = (conc. × volume) ÷ 1000
= (1.5 × 13.8) ÷ 1000 = 0.0207 moles.
$HCl + NaOH \rightarrow NaCl + H_2O$ so 1 mole of HCl neutralises 1 mole of NaOH and 0.0207 moles of HCl must neutralise 0.0207 moles of NaOH.
Conc. NaOH = (moles × 1000) ÷ volume
= (0.0207 × 1000) ÷ 20 = **1.04 mol dm^{-3}**

Q2 Moles NaOH = (conc. × volume) ÷ 1000
= (0.25 × 30) ÷ 1000 = 7.5 × 10^{-3} moles.
$HNO_3 + NaOH \rightarrow NaNO_3 + H_2O$ so 1 mole of NaOH neutralises 1 mole of HCl and 7.5 ×10^{-3} moles of NaOH must neutralise 7.5 × 10^{-3} moles of HCl.
Conc. HNO_3 = (moles × 1000) ÷ volume
= (7.5 × 10^{-3} × 1000) ÷ 17.8 = **0.421 mol dm^{-3}**

Q3 a) average = (22.4 + 22.5 + 22.4) ÷ 3
= 67.3 ÷ 3 = **22.4 cm^3**
b) Moles KOH = (conc. × volume) ÷ 1000
= (0.25 × 22.4) ÷ 1000 = 5.6 × 10^{-3} moles.
$HCl + KOH \rightarrow KCl + H_2O$ so 1 mole of KOH neutralises 1 mole of HCl and 5.6 × 10^{-3} moles of KOH must neutralise 5.6 × 10^{-3} moles of HCl.
Conc. HCl = (moles × 1000) ÷ volume
= (5.6 × 10^{-3} × 1000) ÷ 24 = **0.233 mol dm^{-3}**

Q4 From the graph, the volume of HCl required to neutralise the NaOH solution was 27.5 cm^3
Allow any answer between 27 cm^3 and 28 cm^3. This is the volume of HCl at the equivalence point.
Moles HCl = (conc. × volume) ÷ 1000
= (0.85 × 27.5) ÷ 1000 = 0.0234 moles.
$HCl + NaOH \rightarrow NaCl + H_2O$ so 1 mole of HCl neutralises 1 mole of NaOH and 0.0234 moles of HCl must neutralise 0.0234 moles of NaOH.
Conc. NaOH = (moles × 1000) ÷ volume
= (0.0234 × 1000) ÷ 30 = **0.780 mol dm^{-3}**
Allow any answer between 0.77 mol dm^{-3} and 0.79 mol dm^{-3}, depending on the value from reading the graph.

Q5 Moles H_2SO_4 = (conc. × volume) ÷ 1000
= (0.4 × 18) ÷ 1000 = 7.2 × 10^{-3} moles.
$H_2SO_4 + 2KOH \rightarrow K_2SO_4 + 2H_2O$ so 1 mole of H_2SO_4 neutralises 2 moles of KOH and 7.2 × 10^{-3} moles of H_2SO_4 must neutralise (7.2 × 10^{-3}) × 2 = 0.0144 moles of KOH.
Conc. KOH = (moles × 1000) ÷ volume
= (0.0144 × 1000) ÷ 32 = **0.450 mol dm^{-3}**

Q6 Moles KOH = (conc. × volume) ÷ 1000
= (1.2 × 26.2) ÷ 1000 = 0.0314 moles.
$H_2CO_3 + 2KOH \rightarrow K_2CO_3 + 2H_2O$ so 2 moles of KOH neutralises 1 mole of H_2CO_3 and 0.03144 moles of KOH must neutralise 0.03144 ÷ 2 = 0.01572 moles of H_2CO_3.
Conc. H_2CO_3 = (moles × 1000) ÷ volume
= (0.01572 × 1000) ÷ 20 = **0.786 mol dm^{-3}**

Page 357 — Fact Recall Questions

Q1 moles = (concentration × volume) ÷ 1000 if the units are cm^3 or moles = concentration × volume if the units are dm^3.

Q2

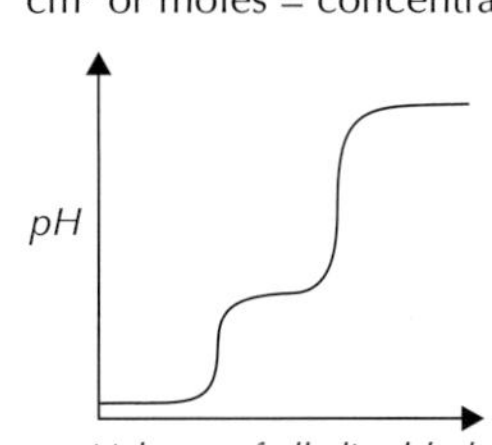

Q3 Diprotic acids release two protons when they react with a base. So, the pH curve for a diprotic acid has two equivalence points (one corresponding to the loss of the first proton and one corresponding to the loss of the second proton). Monoprotic acids only release one proton so pH curves for monoprotic acids only have one equivalence point.

Q4 0.5 moles

6. Buffer Action

Page 360 — Fact Recall Questions

Q1 A buffer is a solution that resists changes in pH when small amounts of acid or alkali are added.

Q2 Acidic buffers contain a weak acid with one of its salts.

Q3 a) When acid is added $[H^+]$ increases. Most of the extra H^+ ions combine with A^- ions to form HA. This shifts the equilibrium to the left so $[H^+]$ is reduced to close to its original value and the pH doesn't change much.

b) When a base is added $[OH^-]$ increases. Most of the extra OH^- ions react with H^+ ions to form water. This removes H^+ ions from the solution so the equilibrium shifts to the right to compensate by more HA dissociating. So more H^+ ions are formed and the pH doesn't change much.

Q4 Basic buffers contain a weak base with one of its salts.

Q5 a) When acid is added the H^+ concentration increases. These H^+ ions react with OH^- ions so $[OH^-]$ goes down. The equilibrium then shifts to the right to replace the lost OH^- ions, so the pH doesn't change much.

b) When a base is added, the OH^- concentration increases. These OH^- ions react with the salt to form more base and H_2O. So, the equilibrium shifts to the left to remove the excess OH^- ions in the solution and the pH doesn't change much.

Q6 Any two from, e.g. in shampoo/in biological washing powders/in biological systems (such as blood).

7. Calculating the pH of Buffers

Page 363 — Application Questions

Q1 a) $CH_3CH_2COOH \rightleftharpoons H^+ + CH_3CH_2COO^-$ so

$$K_a = \frac{[H^+][CH_3CH_2COO^-]}{[CH_3CH_2COOH]}$$

b) $K_a = [H^+][CH_3CH_2COO^-] \div [CH_3CH_2COOH]$ so
$[H^+] = (K_a \times [CH_3CH_2COOH]) \div [CH_3CH_2COO^-]$
$= ((1.35 \times 10^{-5}) \times 0.2) \div 0.35 =$ **7.7×10^{-6} mol dm^{-3}**

c) $pH = -\log[H^+] = -\log(7.7 \times 10^{-6}) =$ **5.11**

Q2 $CH_3COOH \rightleftharpoons H^+ + CH_3COO^-$ so
$K_a = [H^+][CH_3COO^-] \div [CH_3COOH]$ so
$[H^+] = (K_a [CH_3COOH]) \div [CH_3COO^-]$
$= ((1.74 \times 10^{-5}) \times 0.15) \div 0.25 = 1.04 \times 10^{-5}$ mol dm^{-3}
$pH = -\log[H^+] = -\log(1.04 \times 10^{-5}) =$ **4.98**

Q3 a) $CH_3CH_2COOH + KOH \rightarrow$
$CH_3CH_2COO^-K^+ + H_2O$

b) initial moles CH_3CH_2COOH = (conc. × vol.) ÷ 1000
= (0.500 × 30) ÷ 1000 = **0.0150 moles.**
initial moles KOH = (conc. × vol.) ÷ 1000
= (0.250 × 20) ÷ 1000 = **5.00×10^{-3} moles.**

c) From the equation, moles salt = moles base = 5.00×10^{-3} moles. Also, 1 mole of base neutralises 1 mole of acid so 5.00×10^{-3} moles of base neutralises 5.00×10^{-3} moles of acid. So $0.0150 - (0.500 \times 10^{-3}) =$ 0.0100 moles of acid remain.
Total volume = 30.0 + 20.0 = 50.0 cm^3
conc. acid in buffer = (moles × 1000) ÷ vol.
= (0.0100 × 1000) ÷ 50.0 = **0.200 mol dm^{-3}**
conc. salt in buffer = (moles × 1000) ÷ vol.
$= ((5.00 \times 10^{-3}) \times 1000) \div 50.0 =$ **0.100 mol dm^{-3}**

d) $K_a = [H^+][CH_3CH_2COO^-] \div [CH_3CH_2COOH]$ so
$[H^+] = (K_a \times [CH_3CH_2COOH]) \div [CH_3CH_2COO^-]$
$= ((1.35 \times 10^{-5}) \times 0.200) \div 0.100$
= **2.70×10^{-5} mol dm^{-3}**

e) $pH = -\log[H^+] = -\log(2.70 \times 10^{-5}) =$ **4.569**

Q4 initial moles HCOOH = (conc. × vol.) ÷ 1000
$= (0.2 \times 25) \div 1000 = 5 \times 10^{-3}$ moles.
initial moles NaOH = (conc. × vol.) ÷ 1000
$= (0.1 \times 15) \div 1000 = 1.5 \times 10^{-3}$ moles.
$HCOOH + NaOH \rightarrow HCOO^-Na^+ + H_2O$ so moles salt = moles base = 1.5×10^{-3} moles. Also, 1 mole of base neutralises 1 mole of acid and 1.5×10^{-3} moles of base neutralises 1.5×10^{-3} moles of acid.
So $(5 \times 10^{-3}) - (1.5 \times 10^{-3}) = 3.5 \times 10^{-3}$ moles of acid remain.
Total volume = 15 + 25 = 40 cm^3
final conc. acid = (moles × 1000) ÷ vol.
$= ((3.5 \times 10^{-3}) \times 1000) \div 40 = 0.0875$ mol dm^{-3}
final conc. salt = (moles × 1000) ÷ vol.
$= ((1.5 \times 10^{-3}) \times 1000) \div 40 = 0.0375$ mol dm^{-3}
$K_a = [H^+][HCOO^-] \div [HCOOH]$ so
$[H^+] = (K_a \times [HCOOH]) \div [HCOO^-]$
$= ((1.6 \times 10^{-4}) \times 0.0875) \div 0.0375$
$= 3.73 \times 10^{-4}$ mol dm^{-3}
$pH = -\log[H^+] = -\log(3.73 \times 10^{-4}) =$ **3.43**

Pages 365-367 — Exam-style Questions

1 A ***(1 mark)***
$K_w = [H^+][OH^-]$, so $[H^+] = K_w \div [OH^-] = 1.00 \times 10^{-14} \div 0.20$
$= 5.0 \times 10^{-14}$.
$pH = -\log[H^+] = -\log(5.0 \times 10^{-14}) = 13.30$.

2 A ***(1 mark)***
Ammonia is a weak base, so the graph starts off at around pH 9. It gets titrated with nitric acid, a strong acid, so it has a steep curve that ends at around pH 1.

3 C ***(1 mark)***
A titration of a diprotic acid would benefit from using an indicator with two colour changes. The only option that's diprotic is ethanedioic acid.

4.1 $K_w = [H^+][OH^-]$ ***(1 mark)***

4.2 $pH = -\log[H^+]$ ***(1 mark)***

4.3 NaOH is a strong base so
$[OH^-] = [NaOH] = 0.15$ mol dm^{-3}
$K_w = [H^+][OH^-]$ so $[H^+] = K_w \div [OH^-]$
$= 1 \times 10^{-14} \div 0.15 = 6.67 \times 10^{-14}$ mol dm^{-3}
$pH = -\log[H^+] = -\log[6.67 \times 10^{-14}] =$ **13.18**
(3 marks for correct answer, otherwise 1 mark for $[OH^-]$ = 0.15 and 1 mark for $[H^+] = 6.67 \times 10^{-4}$.)

4.4 B ***(1 mark)***
This titration was a strong acid against a strong base so the curve should start at around pH 14 and fall to around pH 1.

4.5 Moles NaOH = (conc. × volume) ÷ 1000
$= (0.15 \times 25) \div 1000 = 3.75 \times 10^{-3}$ moles.
$HCl + NaOH \rightarrow NaCl + H_2O$ so 1 mole of NaOH neutralises 1 mole of HCl and
3.75×10^{-3} moles of NaOH must neutralise 3.75×10^{-3} moles of HCl.
Conc. HCl = (moles × 1000) ÷ volume
$= ((3.75 \times 10^{-3}) \times 1000) \div 18.5 =$ **0.20 mol dm^{-3}**
(3 marks for correct answer, otherwise 1 mark for moles NaOH = 3.75×10^{-3} and 1 mark for moles HCl = 3.75×10^{-3}.)

4.6 HCl is a strong acid and fully dissociates so
$[H^+] = [HCl] = 0.20$ mol dm^{-3}
$pH = -\log[H^+] = -\log(0.2) =$ **0.70**
(2 marks for correct answer, otherwise 1 mark for $[H^+]$ = 0.20 mol dm^{-3}.)

4.7 Any weak acid (e.g. methanoic acid/ethanoic acid/hydrogen cyanide) ***(1 mark)***. Any strong base (e.g. potassium hydroxide/sodium hydroxide) ***(1 mark)***.

4.8 Phenolphthalein ***(1 mark)***

5.1 $HCOOH \rightleftharpoons H^+ + HCOO^-$ ***(1 mark)***

5.2 $K_a = [H^+][HCOO^-] \div [HCOOH]$ ***(1 mark)***

5.3 $[H^+] = 10^{-pH} = 10^{-2.2} = 6.31 \times 10^{-3}$
$K_a = [H^+]^2 \div [HCOOH] = (6.31 \times 10^{-3})^2 \div 0.24$
$= 1.66 \times 10^{-4}$ mol dm^{-3}
$pK_a = -\log(K_a) = -\log(1.66 \times 10^{-4}) = \mathbf{3.78}$
(3 marks for correct answer, otherwise 1 mark for $[H^+] = 6.31 \times 10^{-3}$ and 1 mark for $K_a = 1.66 \times 10^{-4}$.)

5.4 initial moles HCOOH = (conc. × vol.) ÷ 1000
= (0.24 × 30) ÷ 1000 = 7.2×10^{-3} moles ***(1 mark)***
initial moles NaOH = (conc. × vol.) ÷ 1000
= (0.15 × 20) ÷ 1000 = 3.0×10^{-3} moles ***(1 mark)***
moles salt = moles base = 3.0×10^{-3} moles ***(1 mark)***
$HCOOH + NaOH \rightarrow HCOO^-Na^+ + H_2O$ so 1 mole of base neutralises 1 mole of acid and 3.0×10^{-3} moles of base neutralise 3.0×10^{-3} moles of acid. So $(7.2 \times 10^{-3}) - (3.0 \times 10^{-3}) = 4.2 \times 10^{-3}$ moles of acid remain ***(1 mark)***.
Total volume = 20 + 30 = 50
final conc. acid = (moles × 1000) ÷ vol.
= $((4.2 \times 10^{-3}) \times 1000) \div 50 = 0.084$ mol dm^{-3}
final conc. salt = (moles × 1000) ÷ vol.
= $((3.0 \times 10^{-3}) \times 1000) \div 50 = 0.06$ mol dm^{-3}
$HCOOH \rightleftharpoons H^+ + HCOO^-$ so
$K_a = [H^+][HCOO^-] \div [HCOOH]$ so
$[H^+] = (K_a \times [HCOOH]) \div [HCOO^-]$
$= ((1.66 \times 10^{-4}) \times 0.084) \div (0.06)$
$= 2.32 \times 10^{-4}$ mol dm^{-3} ***(1 mark)***
$pH = -\log[H^+] = -\log(2.32 \times 10^{-4}) = \mathbf{3.63}$
(1 mark) (Maximum of 6 marks for correct answer.)
You calculated K_a for this acid in the previous part of the question.

5.5 $HCOOH \rightleftharpoons H^+ + HCOO^-$. Adding an acid increases $[H^+]$ ***(1 mark)*** so the equilibrium shifts to the left to remove the excess H^+ ***(1 mark)***. The excess H^+ combines with $HCOO^-$ to form HCOOH ***(1 mark)***.

5.6 E.g. biological washing powder/shampoo/human blood ***(1 mark for each correct answer, maximum of 3 marks)***.

Unit 5

Section 1 — Period 3 Elements

1. Period 3 Elements

Page 369 — Application Question

Q1 a) A is sodium, B is magnesium, C is sulfur.
b) A: $2Na_{(s)} + \frac{1}{2}O_{2(g)} \rightarrow Na_2O_{(s)}$
B: $Mg_{(s)} + \frac{1}{2}O_{2(g)} \rightarrow MgO_{(s)}$
C: $S_{(s)} + O_{2(g)} \rightarrow SO_{2(g)}$

Page 369 — Fact Recall Questions

Q1 a) (i) $2Na_{(s)} + 2H_2O_{(l)} \rightarrow 2NaOH_{(aq)} + H_{2(g)}$
(ii) $Mg_{(s)} + 2H_2O_{(l)} \rightarrow Mg(OH)_{2(aq)} + H_{2(g)}$
b) Sodium reacts more vigorously. Sodium is in Group 1 and magnesium is in Group 2. So when they react sodium loses one electron and magnesium loses two. It takes less energy to lose one electron than it does to lose two, so sodium is more reactive.

Q2 a) $2Al_{(s)} + 1\frac{1}{2}O_{2(g)} \rightarrow Al_2O_{3(s)}$
b) $P_{4(s)} + 5O_{2(g)} \rightarrow P_4O_{10(s)}$

2. Period 3 Oxides

Page 372 — Application Question

Q1 a) E.g. A is magnesium oxide (MgO), B is aluminium oxide (Al_2O_3), C is silicon dioxide (SiO_2).
b) Magnesium oxide has a higher melting point than aluminium oxide because the difference in electronegativity between Al and O isn't as large as between Mg and O. This means that the O^{2-} ions in Al_2O_3 can't attract the electrons in the metal-oxygen bond as strongly as in MgO. This makes the bonds in Al_2O_3 partially covalent.

Page 372 — Fact Recall Questions

Q1 a) (i) MgO forms a giant ionic lattice with strong ionic bonds between the ions.
(ii) Al_2O_3 forms a giant ionic lattice. Bonding is ionic with partial covalent character because there is a relatively small difference in electronegativity between Al and O.
(iii) SiO_2 has a giant macromolecular structure with strong covalent bonds between the atoms.
(iv) P_4O_{10} has a simple molecular structure with weak intermolecular forces (e.g. dipole-dipole/van der Waals) between the molecules.
b) $MgO_{(s)} + H_2O_{(l)} \rightarrow Mg(OH)_{2(aq)}$
$P_4O_{10(s)} + 6H_2O_{(l)} \rightarrow 4H_3PO_{4(aq)}$
c) (i) Basic
(ii) Amphoteric
(iii) Acidic
(iv) Acidic

Q2 a) $MgO_{(s)} + 2HCl_{(aq)} \rightarrow MgCl_{2(aq)} + H_2O_{(l)}$
b) $SO_{2(g)} + 2NaOH_{(aq)} \rightarrow Na_2SO_{3(aq)} + H_2O_{(l)}$

Exam-style Questions — page 373

1.1 $Mg_{(s)} + 2H_2O_{(l)} \rightarrow Mg(OH)_{2(aq)} + H_{2(g)}$ ***(1 mark)***.

1.2 Magnesium loses two electrons to form an Mg^{2+} ion but sodium only loses one electron to form an Na^+ ion ***(1 mark)***. It takes less energy for sodium to lose one electron than it does for magnesium to lose two, so more energy is needed for magnesium to react ***(1 mark)***.

1.3 $2Na_{(s)} + \frac{1}{2}O_{2(g)} \rightarrow Na_2O_{(s)}$ ***(1 mark)***

1.4 Yellow ***(1 mark)***

1.5 Both form giant ionic lattices ***(1 mark)***, with strong ionic bonds between each ion ***(1 mark)***.

1.6 MgO has the higher melting temperature ***(1 mark)***. Magnesium forms Mg^{2+} ions, which attract O^{2-} ions more strongly than the Na^+ ions in Na_2O, so the ionic bonds in MgO are stronger and more heat energy is required to break the bonds ***(1 mark)***.

1.7 Sulfur dioxide has a simple molecular structure ***(1 mark)***. Only weak intermolecular forces (e.g. dipole-dipole/van der Waals) hold the molecules together so little heat energy is needed to overcome these forces ***(1 mark)***.

1.8 Silicon dioxide has a giant macromolecular structure ***(1 mark)***, with strong covalent bonds holding the molecules together, so lots of energy is required to break these bonds ***(1 mark)***.

2.1 $MgO_{(s)} + H_2O_{(l)} \rightarrow Mg(OH)_{2(aq)}$ ***(1 mark)***
pH 9-10 ***(1 mark)***
$SO_{2(g)} + H_2O_{(l)} \rightarrow H_2SO_{3(aq)}$ ***(1 mark)***
pH 0-1 ***(1 mark)***

2.2 Aluminium oxide/silicon dioxide ***(1 mark)***.

2.3 Aluminium oxide ***(1 mark)***.

2.4 $Al_2O_{3(s)} + 3H_2SO_{4(aq)} \rightarrow Al_2(SO_4)_{3(aq)} + 3H_2O_{(l)}$ ***(1 mark)***
$Al_2O_{3(s)} + 2NaOH_{(aq)} + 3H_2O_{(l)} \rightarrow 2NaAl(OH)_{4(aq)}$ ***(1 mark)***

Section 2 — Transition Metals

1. Transition Metals — The Basics

Page 377 — Application Questions

Q1 a) $[Ar]3d^34s^2$
b) $[Ar]3d^74s^2$
c) $[Ar]3d^54s^2$
d) $[Ar]3d^84s^2$
For these questions you don't have to write [Ar]. Writing out the whole electron configuration starting from $1s^2$ is fine too.

Q2 a) $[Ar]3d^2$
b) $[Ar]3d^7$
c) $[Ar]3d^5$
d) $[Ar]3d^8$
Don't forget, the s electrons are always removed first, then the d electrons.

Q3 Zinc has the electron configuration $1s^22s^22p^63s^23p^63d^{10}4s^2$. It can only form Zn^{2+} ions which have an electron configuration of $1s^22s^22p^63s^23p^63d^{10}$. This ion has a full d sub-level, so zinc cannot form a stable ion with an incomplete d sub-level and it therefore can't be a transition metal.

Page 377 — Fact Recall Questions

Q1 In the d block.
Q2 A transition metal is a metal that can form one or more stable ions with an incomplete d sub-level.
Q3 10
Q4 Electrons fill up the lowest energy sub-levels first and electrons fill orbitals singly before they start sharing.
Q5 a) Chromium prefers to have one electron in each orbital of the 3d sub-level and just one in the 4s sub-level because this gives it more stability.
b) Copper prefers to have a full 3d sub-level and one electron in the 4s sub-level because it's more stable that way.
Q6 They can form complex ions, form coloured ions, act as catalysts and exist in variable oxidation states.
Q7 Their incomplete d sub-level.

2. Complex Ions

Page 381 — Application Questions

Q1 a) 6
b) 2
c) 4

Q2 a) Overall charge of $[CuF_6]^{4-}$ is –4. Each F^- ion has a charge of –1 so the oxidation state of copper must be $-4 - (6 \times -1) = \mathbf{+2}$.
b) Overall charge of $[Ag(S_2O_3)_2]^{3-}$ is –3. Each $S_2O_3^{2-}$ ion has charge of –2 so the oxidation state of silver must be $-3 - (2 \times -2) = \mathbf{+1}$.
c) Overall charge of $[CuCl_4]^{2-}$ is –2. Each Cl^- ion has a charge of –1 so the oxidation state of copper must be $-2 - (4 \times -1) = \mathbf{+2}$

Q3 a) Octahedral:

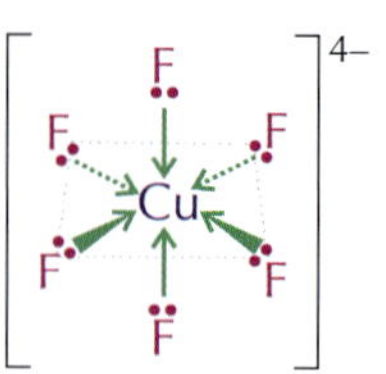

When you're drawing the shapes of ions and molecules make sure you include the dashed arrows and the wedged arrows to show that it's 3-D.

b) Linear: $[S_2O_3 \rightarrow Ag \leftarrow O_3S_2]^{3-}$

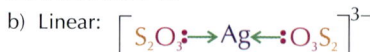

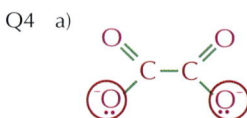

c) Tetrahedral:

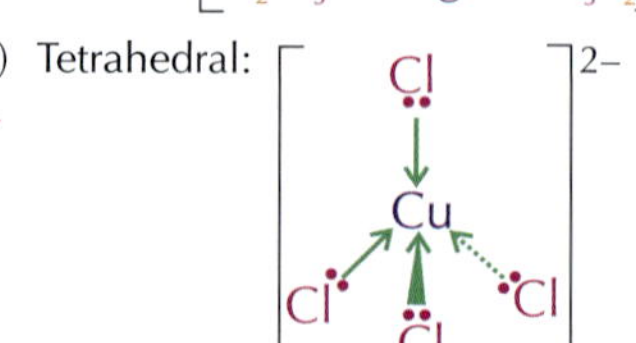

Q4 a)

b) 3
Oxygen atoms are small so six of them will fit around a Cu^{2+} ion and form co-ordinate bonds with it. This means that three molecules of $C_2O_4^{2-}$ will each form two co-ordinate bonds with the ion.

Q5 Square planar:

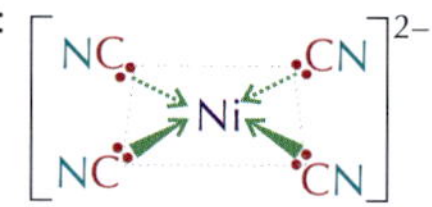

Page 381 — Fact Recall Questions

Q1 a) A metal ion surrounded by co-ordinately bonded ligands.
b) A covalent bond where both of the electrons in the shared pair come from the same atom.
c) An atom, ion or molecule that donates a pair of electrons to a central metal ion.

Q2 a) Monodentate means that a ligand can form one co-ordinate bond, bidentate means that a ligand can form two co-ordinate bonds and multidentate means that a ligand can form two or more co-ordinate bonds.
b) monodentate: e.g. $H_2O/NH_3/Cl^-$
bidentate: e.g. $NH_2CH_2CH_2NH_2/C_2O_4^{2-}$
multidentate: e.g. $EDTA^{4-}$

Q3 The number of co-ordinate bonds that are formed with the central metal ion in a complex ion.

3. Isomerism in Complex Ions

Page 383 — Application Questions

Q1 a) *cis-trans* isomerism
b) optical isomerism
c) *cis-trans* isomerism

Q2 *cis*: *trans*:

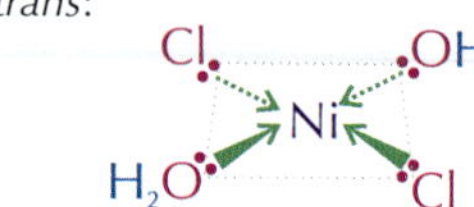

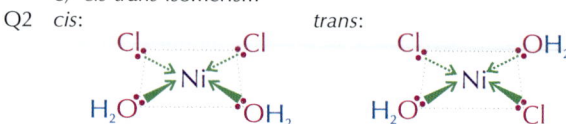

Page 383 — Fact Recall Questions

Q1 optical isomerism and *cis-trans* isomerism

Q2

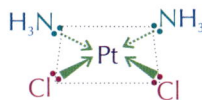

H_3N NH_3
Pt
Cl Cl

4. Formation of Coloured Ions

Page 387 — Application Questions

Q1 The change in the ligand.

Q2 The change in the oxidation state of iron.

Q3 $\Delta E = hc/\lambda$
$= (6.63 \times 10^{-34} \times 3.00 \times 10^{8}) \div 580 \times 10^{-9}$
$= \mathbf{3.43 \times 10^{-19}\ J}$

Page 387 — Fact Recall Questions

Q1 They split into two different energy levels.

Q2 a) Light energy.
b) $\Delta E = h\nu = hc/\lambda$

Q3 Transition metal ions have incomplete 3d sub-levels. When visible light hits transition metal ions, some of the frequencies are absorbed by electrons which jump up to the higher energy orbitals. The frequencies that remain are reflected and these make up the colour you see.

Q4 A change in oxidation state, a change in co-ordination number and a change in ligand.

Q5 a) White light is shone through a filter, which is chosen to only let the colour of light through that is absorbed by the sample. The light then passes through the sample to a colorimeter, which measures how much light is absorbed by the sample.
b) Measure the absorbances of known concentrations of solutions and plot the results on a graph with absorbance on the *y*-axis and concentration on the *x*-axis.

5. Ligand Substitution Reactions

Page 392 — Application Questions

Q1 a) octahedral
b) tetrahedral

Q2 $[Mn(H_2O)_6]^{2+}{}_{(aq)} + 4Cl^-{}_{(aq)} \rightleftharpoons [MnCl_4]^{2-}{}_{(aq)} + 6H_2O_{(l)}$

Q3 a) $[Cu(H_2O)_6]^{2+}{}_{(aq)} + 4NH_{3(aq)} \rightarrow [Cu(NH_3)_4(H_2O)_2]^{2+}{}_{(aq)} + 4H_2O_{(l)}$
b) (elongated) octahedral

Q4 a) $[Cr(NH_3)_6]^{3+}{}_{(aq)} + 3NH_2CH_2CH_2NH_{2\,(aq)} \rightarrow [Cr(NH_2CH_2CH_2NH_2)_3]^{3+}{}_{(aq)} + 6NH_{3\,(aq)}$
b) The enthalpy change (ΔH) for this reaction will be very small because six Cr–N bonds are broken and six Cr–N bonds are formed. The entropy change (ΔS) will be positive and large because the number of particles increases during the reaction.

Page 392 — Fact Recall Questions

Q1 a) There will be no change in co-ordination number or shape but the colour may change.
b) There will be a change in co-ordination number and shape and the colour may change.

Q2 a) It helps transport oxygen around the body.
b) Four come from nitrogen atoms in the porphyrin ring, one comes from a nitrogen atom in a globin protein and one comes from either water or oxygen.

Q3 a) In the lungs, the concentration of oxygen is high. So water ligands that were bound to the haemoglobin are substituted for oxygen ligands, forming oxyhaemoglobin.
b) At sites where oxygen is needed the concentration of oxygen is low. So oxygen ligands that were bound to haemoglobin are substituted for water ligands, forming deoxyhaemoglobin.

Q4 Carbon monoxide is a very strong ligand for haemoglobin. When it is inhaled, it binds to the central Fe^{2+} ion and prevents it from binding to oxygen. As a result, oxygen can no longer be transported around the body.

Q5 When monodentate ligands are substituted with bidentate or multidentate ligands, the number of particles and the entropy increases. The enthalpy change (ΔH) for these reaction will be very small because six metal-ligand bonds are broken and six metal-ligand bonds are formed. Reactions that result in an large increase in entropy and a small change in enthalpy are more likely to occur and so multidentate ligands form much more stable complexes than monodentate ligands. This is the chelate effect.

6. Variable Oxidation States

Page 395 — Application Question

Q1 The manganate(VII) ion has a higher redox potential so it is more unstable than the iron(III) ion and therefore, more easily reduced.

Page 395 — Fact Recall Questions

Q1 $2V^{3+}{}_{(aq)} + Zn_{(s)} \rightarrow 2V^{2+}{}_{(aq)} + Zn^{2+}{}_{(aq)}$

Q2 E.g. type of ligand and pH.

Q3 a) A silver mirror would form on the inside of the test tube.
b) No change.

Q4 $[Ag(NH_3)_2]^+$

7. Transition Metal Titrations

Page 398 — Application Questions

Q1 $MnO_4^-{}_{(aq)} + 8H^+{}_{(aq)} + 5Fe^{2+}{}_{(aq)} \rightarrow Mn^{2+}{}_{(aq)} + 4H_2O_{(l)} + 5Fe^{3+}{}_{(aq)}$
Moles Fe^{2+} = (conc. × volume) ÷ 1000
= (0.0500 × 28.3) ÷ 1000 = 1.415×10^{-3} moles.
5 moles of Fe^{2+} reacts with 1 mole of MnO_4^- so 1.415×10^{-3} moles of Fe^{2+} must react with $(1.415 \times 10^{-3}) \div 5 = 2.83 \times 10^{-4}$ moles of MnO_4^-.
Conc. MnO_4^- = (moles × 1000) ÷ volume
= $((2.83 \times 10^{-4}) \times 1000) \div 30.0 = \mathbf{0.00943\ mol\ dm^{-3}}$

Q2 $2MnO_4^-{}_{(aq)} + 16H^+{}_{(aq)} + 5C_2O_4^{2-}{}_{(aq)} \rightarrow 2Mn^{2+}{}_{(aq)} + 8H_2O_{(l)} + 10CO_{2(g)}$
Moles $C_2O_4^{2-}$ = (conc. × volume) ÷ 1000
= (0.600 × 28.0) ÷ 1000 = 0.0168 moles.
5 moles of $C_2O_4^{2-}$ reacts with 2 moles of MnO_4^- so 0.0168 moles of $C_2O_4^{2-}$ must react with $(0.0168 \div 5) \times 2 = 6.72 \times 10^{-3}$ moles of MnO_4^-.
Volume MnO_4^- = (moles × 1000) ÷ conc.
= $((6.72 \times 10^{-3}) \times 1000) \div 0.0750 = \mathbf{89.6\ cm^3}$

Q3 $MnO_4^-{}_{(aq)} + 8H^+{}_{(aq)} + 5Fe^{2+}{}_{(aq)} \rightarrow Mn^{2+}{}_{(aq)} + 4H_2O_{(l)} + 5Fe^{3+}{}_{(aq)}$
Moles MnO_4^- = (conc. × volume) ÷ 1000
= (0.0550 × 24.0) ÷ 1000 = 1.32×10^{-3} moles.
1 mole of MnO_4^- reacts with 5 moles of Fe^{2+} so 1.32×10^{-3} moles of MnO_4^- must react with $(1.32 \times 10^{-3}) \times 5 = 6.60 \times 10^{-3}$ moles of Fe^{2+}.
Volume Fe^{2+} = (moles × 1000) ÷ conc.
= $((6.60 \times 10^{-3}) \times 1000) \div 0.450 = \mathbf{14.7\ cm^3}$

Page 398 — Fact Recall Questions

Q1 Acid is added to make sure there are plenty of H^+ ions to allow all the oxidising agent to be reduced.

Q2 a) e.g. manganate(VII) ions (MnO_4^-).
b) e.g. colourless to purple.

8. Transition Metal Catalysts

Page 402 — Application Question

Q1 a) A heterogeneous catalyst.
b) Mn is a transition metal. Transition metals have incomplete d sub-levels so can have multiple oxidation states. This means they can transfer electrons to speed up reactions.

Page 402 — Fact Recall Questions

Q1 A heterogeneous catalyst is a catalyst in a different phase from the reactants and a homogeneous catalyst is a catalyst in the same phase as the reactants.

Q2 E.g. the Haber Process. Iron is the catalyst. It catalyses the reaction $N_{2(g)} + 3H_{2(g)} \rightarrow 2NH_{3(g)}$.
The Contact Process. Vanadium(V) oxide (V_2O_5) is the catalyst. It catalyses the reaction $SO_{2(g)} + \frac{1}{2}O_{2(g)} \rightarrow SO_{3(g)}$.

Q3 The reaction happens on the surface of the heterogeneous catalyst. So increasing the surface area of the catalyst increases the rate of reaction. Using a support medium maximises the surface area for minimal extra cost.

Q4 Catalytic poisoning is when impurities bind to the surface of the catalyst and block reactants from being adsorbed. It can be reduced by purifying the reactants to remove as many impurities as possible.

Q5 The reaction between $S_2O_8^{2-}$ and I^- is very slow because both ions are negatively charged and repel each other. When Fe^{2+} ions are added, the reaction can proceed in two stages, both involving a positive and a negative ion, so there's no repulsion and the reaction happens a lot faster.

Q6 Autocatalysis is when one of the products of a reaction also catalyses it. E.g. the catalysis of the reaction between $C_2O_4^{2-}$ and MnO_4^- by Mn^{2+}.

9. Metal-Aqua Ions

Page 408 — Application Question

Q1 copper 2+/Cu^{2+}

Page 408 — Fact Recall Questions

Q1 co-ordinate/dative covalent bonds

Q2 Metal 3+ ions have a high charge density (charge/size ratio). The metal 2+ ions have a much lower charge density. This makes the 3+ ions much more polarising than the 2+ ions. More polarising power means that they attract electrons from the oxygen atoms of the co-ordinated water molecules more strongly, weakening the O–H bond. So it's more likely that a hydrogen ion will be released. And more hydrogen ions means a more acidic solution. This means that metal-aqua 3+ ions are more acidic than metal-aqua 2+ ions.

Q3 $[Fe(H_2O)_6]^{3+}_{(aq)} + H_2O_{(l)} \rightleftharpoons [Fe(OH)(H_2O)_5]^{2+}_{(aq)} + H_3O^+_{(aq)}$
$[Fe(OH)(H_2O)_5]^{2+}_{(aq)} + H_2O_{(l)} \rightleftharpoons [Fe(OH)_2(H_2O)_4]^+_{(aq)} + H_3O^+_{(aq)}$
$[Fe(OH)_2(H_2O)_4]^+_{(aq)} + H_2O_{(l)} \rightleftharpoons Fe(OH)_3(H_2O)_{3(s)} + H_3O^+_{(aq)}$
You could also write these equations using $OH^-_{(aq)}$ instead of $H_2O_{(l)}$ and replacing $H_3O^+_{(aq)}$ with $H_2O_{(l)}$. For example:
$[Fe(H_2O)_6]^{3+}_{(aq)} + OH^-_{(aq)} \rightleftharpoons [Fe(OH)(H_2O)_5]^{2+}_{(aq)} + H_2O_{(l)}$

Q4 A white precipitate would form in the colourless solution.

Q5 The solution would turn from blue to deep blue.

Exam-style Questions — pages 410-413

1 C **(1 mark)**.
The electron configuration of manganese is $1s^2\,2s^2\,2p^6\,3s^2\,3p^6\,3d^5\,4s^2$.

2 A **(1 mark)**.
When chloride ions are added to $[Cu(H_2O)_6]^{2+}_{(aq)}$, $[CuCl_4]^{2-}$ is formed, which is tetrahedral.

3 B **(1 mark)**.
Test 1 rules out Al(III) as its precipitate dissolves in excess $NaOH_{(aq)}$. Test 2 rules out Cu(II) as its precipitate dissolves in excess $NH_{3(aq)}$. Test 3 rules out Fe(III) as +3 ions produce carbon dioxide gas after reacting with sodium carbonate.

4.1 blue **(1 mark)**

4.2 Cu: $1s^2 2s^2 2p^6 3s^2 3p^6 3d^{10} 4s^1$ **(1 mark)**
Cu^{2+}: $1s^2 2s^2 2p^6 3s^2 3p^6 3d^9$ **(1 mark)**

4.3 Cu^{2+} has the characteristic chemical properties of transition element ions because it has an incomplete d sub-level **(1 mark)**.

4.4 $[Cu(H_2O)_6]^{2+}$ **(1 mark)**
Don't forget to include the charge on the complex ion when you're drawing its structure.
The shape of $[Cu(H_2O)_6]^{2+}$ is octahedral **(1 mark)**.

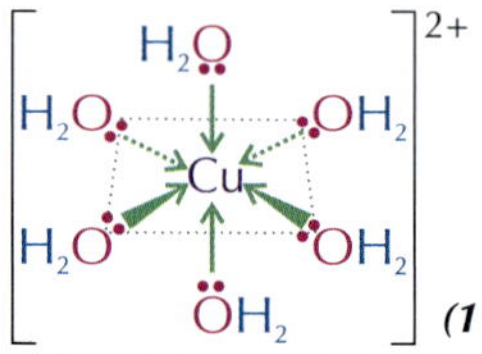

4.5 6 **(1 mark)**

4.6 When visible light hits a transition metal ion, some frequencies are absorbed but others are reflected **(1 mark)**. The reflected frequencies combine to make the complement of the colour of the absorbed frequencies and this is the colour you see **(1 mark)**.

4.7 Change in oxidation state **(1 mark)**, ligand **(1 mark)** or co-ordination number **(1 mark)**.

5.1 square planar **(1 mark)**

5.2 *cis-trans* isomerism **(1 mark)**

5.3 $[Co(NH_3)_2Cl_2]$ cis and trans isomers

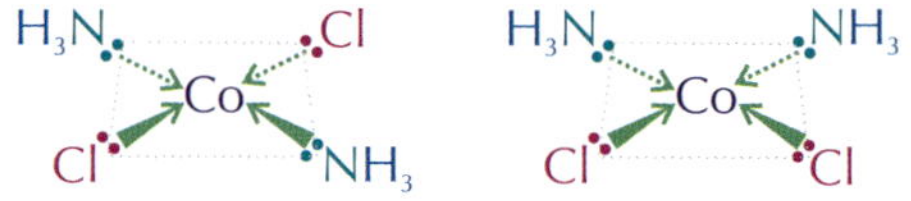

(1 mark for each correctly drawn isomer)

5.4 optical isomerism **(1 mark)**

5.5 The oxygen atoms have lone pairs of electrons **(1 mark)** which they can donate to the chromium(III) ion to form co-ordinate bonds **(1 mark)**.

5.6 $[Cr(C_2O_4)_3]^{3-}$

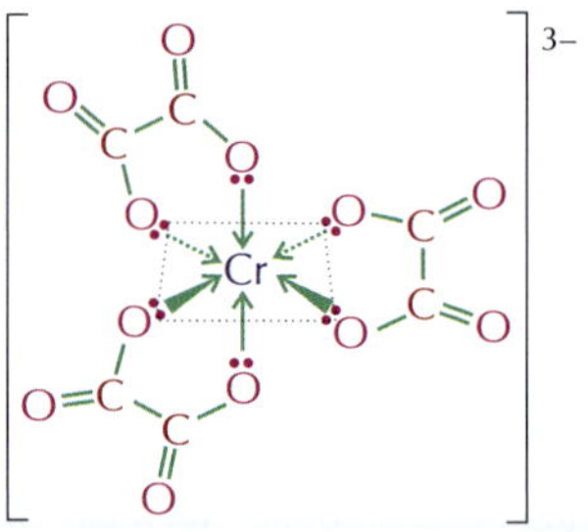

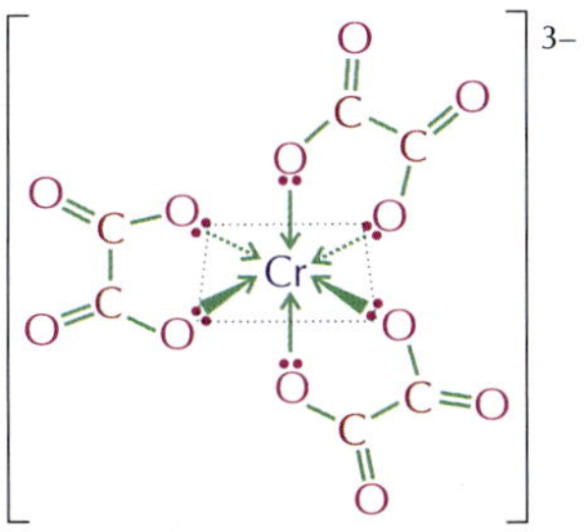

(1 mark for each correctly drawn isomer. The structures must be shown as 3-D, i.e. with wedged and dotted bonds)

These are fiendishly difficult isomers to draw so don't worry if you've messed up your first few attempts — you can always simplify the structure of the ligands to make the whole thing a bit easier. Make sure you get lots of practice at drawing them before the exam and you'll be fine.

6.1 Transition metals exist in variable oxidation states ***(1 mark)*** so they can transfer electrons to speed up reactions ***(1 mark)***.

6.2 A heterogeneous catalyst ***(1 mark)***.
It's a heterogeneous catalyst because it's in a different physical state than the reactants.

6.3 The reaction happens on the surface of the heterogeneous catalyst ***(1 mark)***. Using a ceramic lattice increases the surface area, making it more effective ***(1 mark)*** and reduces the cost because only a thin layer of catalyst is needed ***(1 mark)***.

6.4 Catalytic poisoning is when impurities bind to the surface of a catalyst ***(1 mark)*** and block reactants from being adsorbed ***(1 mark)***. This reduces the surface area of the catalyst available to the reactants, slowing down the reaction ***(1 mark)***.

6.5 The reacting ions are both negatively charged ***(1 mark)*** so they will repel each other and are unlikely to collide ***(1 mark)***.

6.6 $S_2O_8^{2-}{}_{(aq)} + 2Fe^{2+}{}_{(aq)} \rightarrow 2Fe^{3+}{}_{(aq)} + 2SO_4^{2-}{}_{(aq)}$ ***(1 mark)***.
$2Fe^{3+}{}_{(aq)} + 2I^-{}_{(aq)} \rightarrow I_{2(aq)} + 2Fe^{2+}{}_{(aq)}$ ***(1 mark)***.

7.1 Water molecules form co-ordinate/dative covalent bonds with the central metal ion ***(1 mark)***. Each water molecule donates a pair of electrons to the central metal ion ***(1 mark)***.

7.2 $[Fe(H_2O)_6]^{2+}{}_{(aq)}$ ***(1 mark)***

7.3 $[Fe(H_2O)_6]^{2+}{}_{(aq)} + H_2O_{(l)} \rightleftharpoons [Fe(OH)(H_2O)_5]^{+}{}_{(aq)} + H_3O^+{}_{(aq)}$ ***(1 mark)***

7.4 Al^{3+} has a higher charge density (charge/size ratio) than Fe^{2+} ***(1 mark)***. This makes the Al^{3+} ions much more polarising than the Fe^{2+} ions ***(1 mark)***. More polarising power means that they attract electrons from the oxygen atoms of the co-ordinated water molecules more strongly, weakening the O–H bond ***(1 mark)***. So it's more likely that a hydrogen ion will be released. And more hydrogen ions means a more acidic solution ***(1 mark)***.

7.5 The solution will remain green but a green precipitate will form in it ***(1 mark)***.

7.6 $[Fe(H_2O)_6]^{3+}{}_{(aq)}$ / $[Fe(OH)(H_2O)_5]^{2+}{}_{(aq)}$ ***(1 mark)***

7.7 $2[Fe(H_2O)_6]^{3+}{}_{(aq)} + 3CO_3^{2-}{}_{(aq)}$
$\rightleftharpoons 2Fe(OH)_3(H_2O)_{3(s)} + 3CO_{2(g)} + 3H_2O_{(l)}$ ***(1 mark)***

7.8 A brown precipitate would form ***(1 mark)*** and bubbles of CO_2 gas would be given off ***(1 mark)***.

7.9 $[Al(H_2O)_6]^{3+}{}_{(aq)} + 3NH_2CH_2CH_2NH_{2(aq)} \rightarrow$
$[Al(NH_2CH_2CH_2NH_2)_3]^{3+}{}_{(aq)} + 6H_2O_{(l)}$ ***(1 mark)***

7.10 The products have a lot more entropy than the reactants while the enthalpy change in both directions is small ***(1 mark)*** so the forward reaction happens easily while the reverse reaction doesn't ***(1 mark)***.

8.1 An amphoteric species can act as both a base and an acid ***(1 mark)***.

8.2 white precipitate ***(1 mark)***

8.3 $Al(OH)_3(H_2O)_{3(s)} + OH^-{}_{(aq)} \rightleftharpoons [Al(OH)_4]^-{}_{(aq)} + 3H_2O_{(l)}$
(2 marks, 1 mark for correct products, 1 mark for correct charge on product ion)

8.4 It is acting as a Brønsted-Lowry acid ***(1 mark)*** because it's donating H^+ ions in the reaction ***(1 mark)***.

9.1 The complex ion present is $[Ag(NH_3)_2]^+$ ***(1 mark)***.

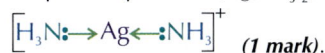

$[H_3N{:}\rightarrow Ag \leftarrow{:}NH_3]^+$ ***(1 mark)***.

The shape of the molecule is linear ***(1 mark)***.

9.2 Tollens' reagent is used to distinguish between aldehydes and ketones ***(1 mark)***. It reacts with aldehydes to form a silver mirror ***(1 mark)***. It does not react with ketones ***(1 mark)***.

9.3 $MnO_4^-{}_{(aq)} + 8H^+{}_{(aq)} + 5Fe^{2+}{}_{(aq)}$
$\rightarrow Mn^{2+}{}_{(aq)} + 4H_2O_{(l)} + 5Fe^{3+}{}_{(aq)}$ ***(1 mark)***
Moles Fe^{2+} = (conc. × volume) ÷ 1000
= (0.150 × 30.0) ÷ 1000 = 4.50×10^{-3} moles ***(1 mark)***.
5 moles of Fe^{2+} reacts with 1 mole of MnO_4^- so 4.50×10^{-3} moles of Fe^{2+} must react with (4.50×10^{-3}) ÷ 5 = 9.00×10^{-4} moles of MnO_4^- ***(1 mark)***.
Conc. MnO_4^- = (moles × 1000) ÷ volume
= ((9.00×10^{-4}) × 1000) ÷ 15.8 = **0.0570 mol dm^{-3}** ***(1 mark)***.

9.4 The solution will be purple at the end point ***(1 mark)***.

Unit 6

Section 1 — Isomerism and Carbonyl Compounds

1. Optical Isomerism

Page 417 — Application Questions

Q1 a)

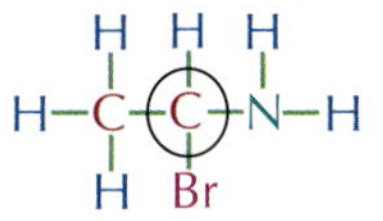

b)

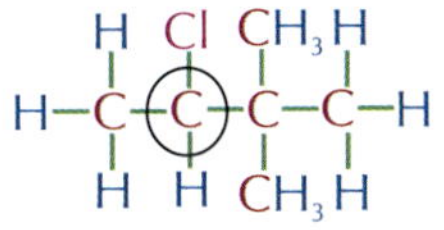

c)

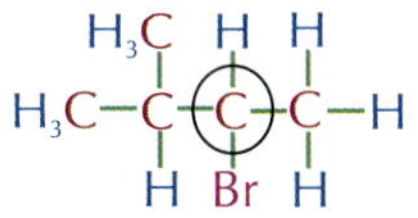

d)

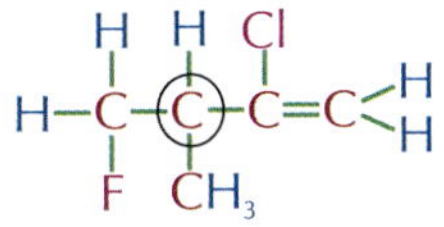

Q2 a)

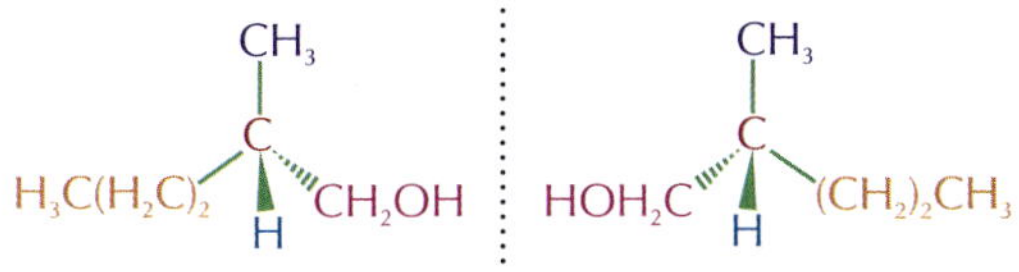

b) [Structural formulas: two enantiomers of a chiral carbon bonded to H, Cl_2HC, Br and CH_3 / Br, $CHCl_2$, H and CH_3]

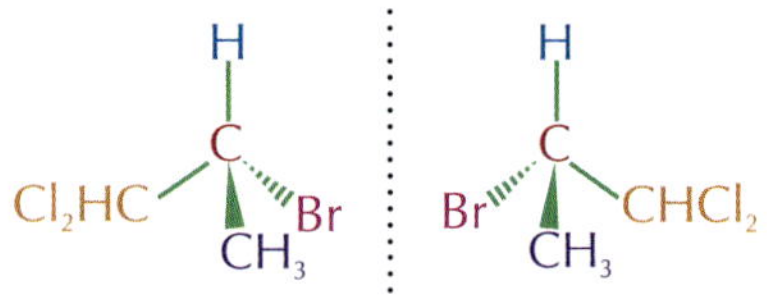

Q3 [Structural formulas: two enantiomers of a chiral carbon bonded to OH, H_3CH_2C, CH_3 and CN / H_3C, CH_2CH_3, OH and CN]

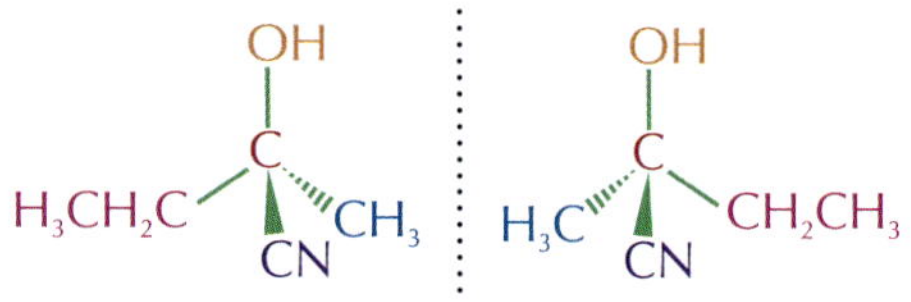

Page 417 — Fact Recall Questions

Q1 Optical isomerism is a type of stereoisomerism. In optical isomers four groups are arranged in two different ways around a central carbon atom so that two different molecules are made — these molecules are non-superimposable mirror images of each other and are called enantiomers or optical isomers. Optical isomers differ by their effect on plane polarised light.

Q2 A chiral carbon atom is a carbon atom that has four different groups attached to it.

Q3 Optically active molecules will rotate plane polarised light.

Q4 Racemic mixtures contain equal quantities of two enantiomers. The two enantiomers cancel each other's light-rotating effect so the mixture doesn't show any optical activity.

2. Aldehydes and Ketones

Page 421 — Application Questions

Q1 a) butanal
b) hexan-2-one
c) pentan-3-one
d) 2-ethyl-3-methylbutanal

Q2 a) [Reaction mechanism: H^- attacks the $\delta+$ carbon of the $C=O^{\delta-}$ group in butanal, giving the O^- intermediate, which is protonated by H^+ to give butan-1-ol]

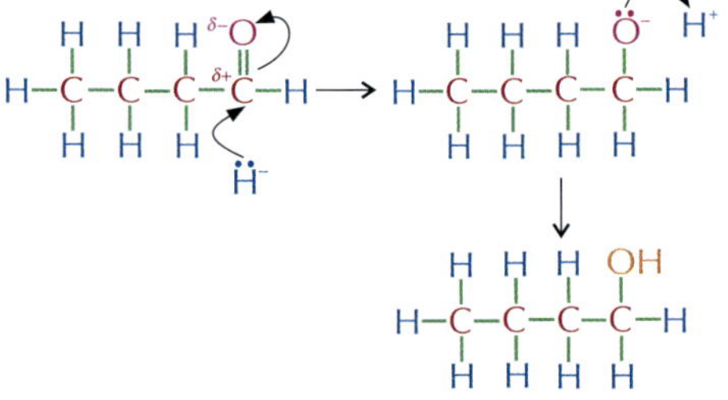

b) [Reaction mechanism: H^- attacks the $\delta+$ carbonyl carbon of hexan-2-one, giving the O^- intermediate, which is protonated by H^+ to give hexan-2-ol]

c) [Reaction mechanism: H^- attacks the $\delta+$ carbonyl carbon of pentan-3-one, giving the O^- intermediate, which is protonated by H^+ to give pentan-3-ol]

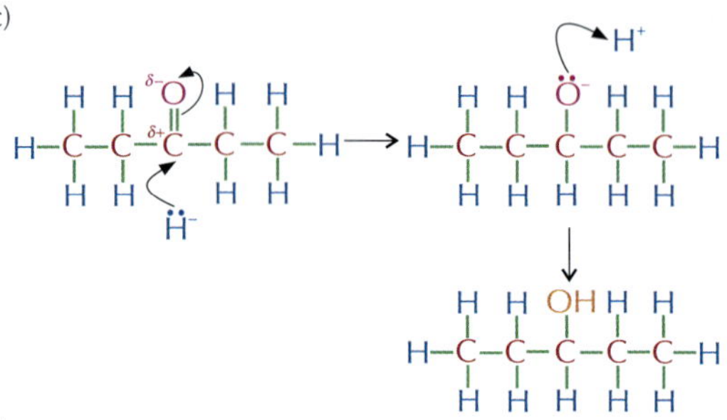

d) [Reaction mechanism: H^- attacks the $\delta+$ carbonyl carbon of 2-ethyl-3-methylbutanal, giving the O^- intermediate, which is protonated by H^+ to give 2-ethyl-3-methylbutan-1-ol]

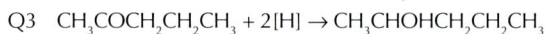

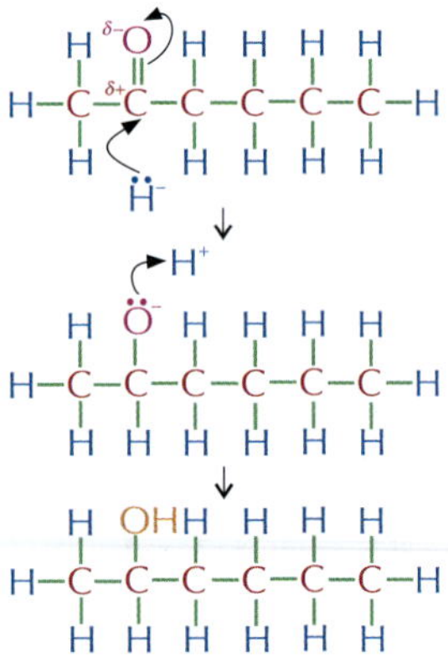

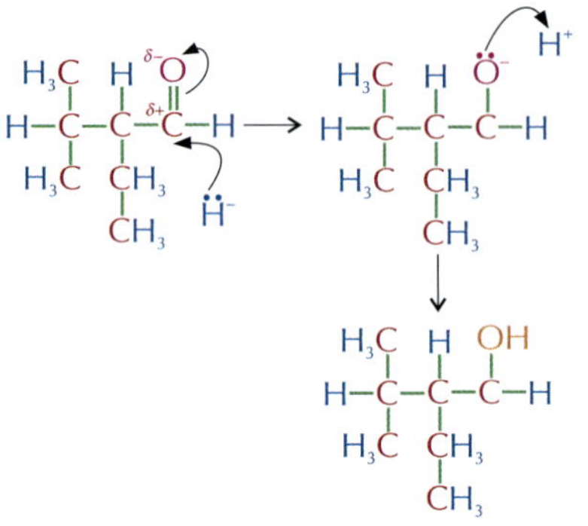

Q3 $CH_3COCH_2CH_2CH_3 + 2[H] \rightarrow CH_3CHOHCH_2CH_2CH_3$

Page 421 — Fact Recall Questions

Q1 Aldehydes have a carbonyl group at the end of the carbon chain, ketones have a carbonyl group in the middle of the carbon chain.

Q2 A carboxylic acid.

Q3 The carbonyl group is in the middle of the carbon chain so to oxidise it, a carbon-carbon bond would have to be broken. This requires a lot of energy, so ketones are not easily oxidised.

Q4 a) In the presence of an aldehyde a silver mirror is produced. In the presence of a ketone there is no change.
b) In the presence of an aldehyde a brick-red precipitate is formed. In the presence of a ketone there is no change.

Q5 e.g. $NaBH_4$

3. Hydroxynitriles

Page 424 — Application Questions

Q1 a) [Structural formula: $CH_3CH_2C(OH)(H)$ chain drawn as H–C–C–C–H with OH on the end carbon, CH_3 and CN substituents — 2-hydroxy-2-methylbutanenitrile-type product: H H OH / H–C–C–C–H / H CH_3 CN]

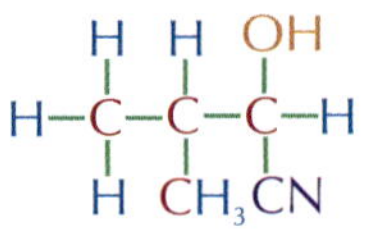

b) [Reaction mechanism: $:CN^-$ attacks the $\delta+$ carbonyl carbon of the $C=O^{\delta-}$ group, giving the O^- intermediate with CN attached, which is protonated by H^+ to give the hydroxynitrile]

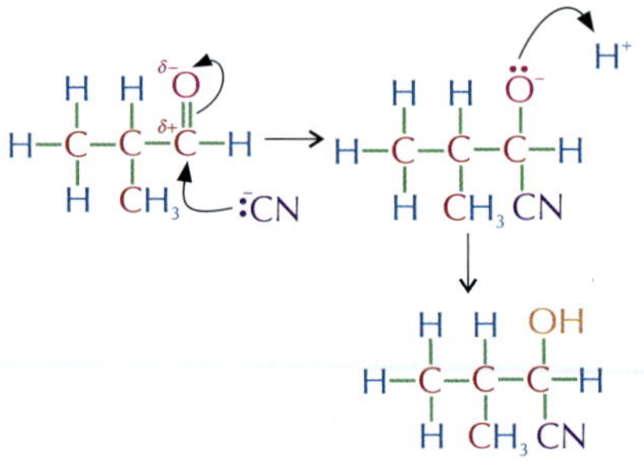

Q2 a) 2-hydroxy-2-methylbutanenitrile

b)
```
    H  H  O  H
    |  |  ‖  |
 H—C—C—C—C—H
    |  |     |
    H  H     H
```

c) butanone

d)
```
    H  H  δ-O  H                    H  H  :Ö⁻  H    ↷ H⁺
    |  |   ‖   |                    |  |   |   |
 H—C—C—δ+C—C—H   →   H—C—C—C—C—H
    |  |       |                    |  |   |   |
    H  H  ⤺   H                    H  H  CN  H
          :CN⁻                            ↓
                                    H  H  OH  H
                                    |  |   |   |
                                 H—C—C—C—C—H
                                    |  |   |   |
                                    H  H  CN  H
```

Q3 a)
```
       OH                  OH
       |                   |
       C                   C
     /  ⋮  \             /  ⋮  \
 H5C2  CH3  CN       NC   CH3  C2H5
```

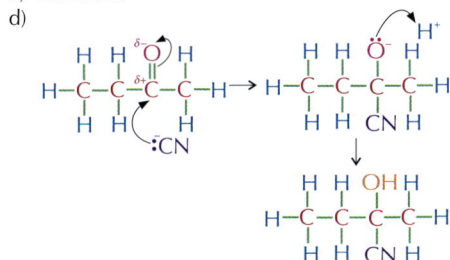

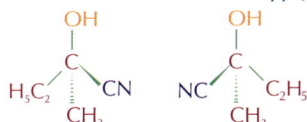

b) The C=O bond is planar, so nucleophilic attack of the δ+ carbon can occur from above or below the C=O bond. Depending on which side the nucleophile attacks from, a different optical isomer/enantiomer will form.

Page 424 — Fact Recall Questions

Q1 A hydroxynitrile is a molecule with a nitrile (CN) group and a hydroxyl (OH) group.

Q2 nucleophilic addition

Q3 a) KCN is toxic.

b) E.g. Any reaction involving KCN should be done in a fume cupboard If heating KCN, this should be done using a water bath or electric mantle, rather than an open flame.

4. Carboxylic Acids and Esters

Page 427 — Application Questions

Q1 a) (i) methanoic acid

(ii) 2-ethyl-4-hydroxypentanoic acid

b) $2HCOOH_{(aq)} + Na_2CO_{3(s)} \rightarrow 2HCOONa_{(aq)} + H_2O_{(l)} + CO_{2(g)}$

Q2 a) (i) propyl methanoate

(ii) ethyl benzoate

b) (i) propan-1-ol and methanoic acid

(ii) ethanol and benzoic acid

Q3 a)
```
      O
     //    H   H
 H—C       |   |
     \O—C—C—H
           |   |
         H3C   H
```

b) $HCOOH + CH_3CH(OH)CH_3 \rightarrow HCOOCH(CH_3)CH_3 + H_2O$

Page 427 — Fact Recall Questions

Q1 A salt, water and carbon dioxide.

Q2 a) Esterification reaction / condensation reaction.

b) A strong acid catalyst. E.g. H_2SO_4/HCl/H_3PO_4.

5. Reactions and Uses of Esters

Page 431 — Application Questions

Q1 a) $CH_3CH_2COOCH_3 + H_2O \rightarrow CH_3CH_2COOH + CH_3OH$

b) $CH_3CH_2COOCH_3 + OH^- \rightarrow CH_3CH_2COO^- + CH_3OH$

Q2 a)
```
    H        O
    |        ‖
 H—C—O—C—(CH2)16CH3
    |        O
    |        ‖
 H—C—O—C—(CH2)16CH3
    |        O
    |        ‖
 H—C—O—C—(CH2)16CH3
    |
    H
```

b)
```
    H        O
    |        ‖
 H—C—O—C—(CH2)16CH3
    |        O
    |        ‖
 H—C—O—C—(CH2)16CH3   + 3NaOH
    |        O
    |        ‖
 H—C—O—C—(CH2)16CH3
    |
    H
              ↓
    H
    |
 H—C—OH
    |
 H—C—OH        + 3CH3(CH2)16COO⁻Na⁺
    |
 H—C—OH
    |
    H
```

We've had to write the equation vertically here so it'll fit on the page. In the exam, write the equation horizontally like normal.

c) Mix the sodium salt produced with a strong acid such as HCl.

Q3 a)
```
 CH2OOC(CH2)14CH3
 |
 CH2OOC(CH2)14CH3   + 3CH3OH
 |
 CH2OOC(CH2)14CH3
          ↓
    H
    |
 H—C—OH
    |                              O
 H—C—OH   + 3 CH3(CH2)14—C//
    |                              \O—CH3
 H—C—OH
    |
    H
```

b) E.g. KOH/potassium hydroxide or NaOH/sodium hydroxide.

Page 431 — Fact Recall Questions

Q1 Any three from: e.g. in perfumes / as food flavourings / as solvents (e.g. in glue/printing ink) / as plasticisers / to produce methyl esters for biofuels.

Q2 a) A carboxylic acid and an alcohol.

b) A carboxylate ion/carboxylate salt and an alcohol.

Q3 Glycerol (propane-1,2,3-triol) and long chain carboxylic acids (fatty acids).

Q4 Fats contain mostly saturated hydrocarbon chains which fit neatly together, which increases the strong van der Waals forces. As a result a higher temperature is needed to melt them and they are solid at room temperature.
Oils contain mostly unsaturated hydrocarbon chains which don't fit together very closely, which decreases the effect of the van der Waals forces. As a result a lower temperature is needed to melt them and they are liquid at room temperature.

Q5 Glycerol and a sodium salt of a fatty acid (soap).

Q6 By reacting rapeseed/vegetable oils with methanol in the presence of a strong alkali catalyst (e.g. KOH or NaOH) to form methyl esters. Biodiesel is a mixture of methyl esters.

Q7 Biodiesel isn't always 100% carbon neutral because energy is used to make the fertilizer to grow the crop and is used in planting, harvesting and converting the plants to oil. This energy often comes from burning fossil fuels, which produces CO_2, so the process isn't carbon neutral overall.

6. Acyl Chlorides

Page 434 — Application Questions

Q1 a) methanoyl chloride

b) propanoyl chloride

c) 2-methylpropanoyl chloride

d) 4-hydroxy-2-methylbutanoyl chloride

Q2 a)

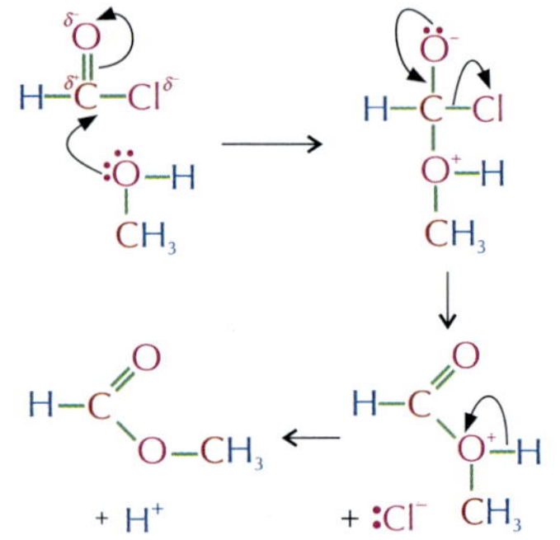

b)

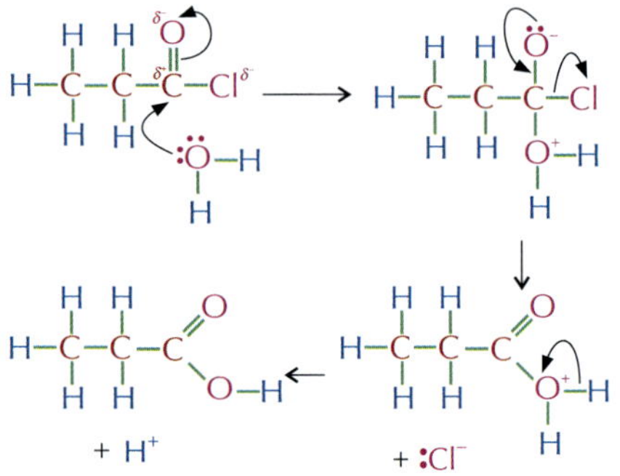

c)

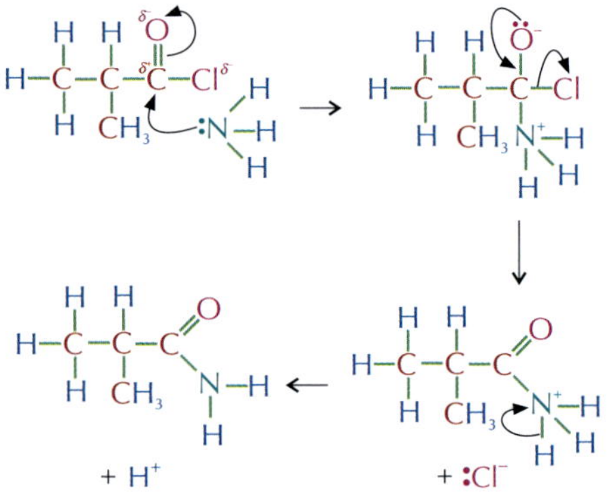

d)

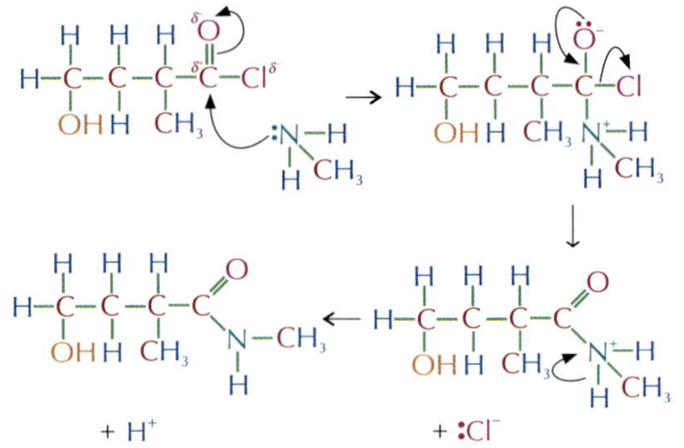

Page 434 — Fact Recall Questions

Q1 $C_nH_{2n-1}OCl$

Q2 a) A carboxylic acid and HCl.

b) An ester and HCl.

c) An amide and HCl.

d) An N-substituted amide and HCl.

Q3 nucleophilic addition-elimination

7. Acid Anhydrides

Page 436 — Application Questions

Q1 a)

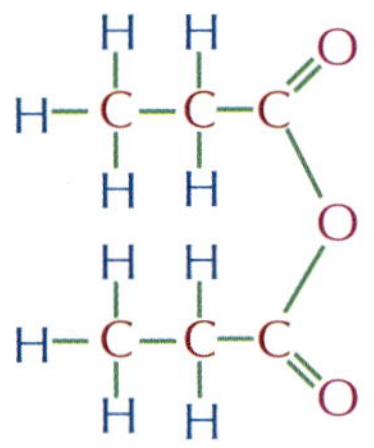

b)

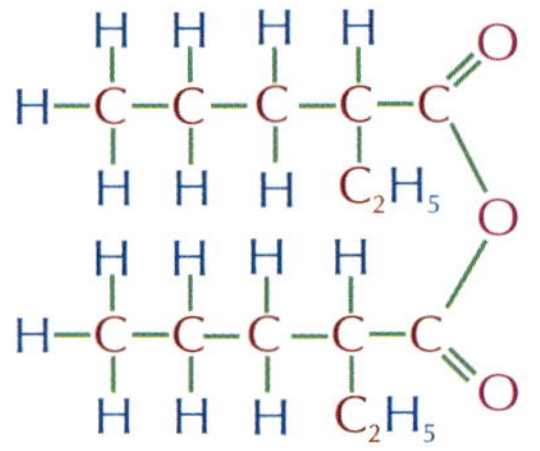

c)

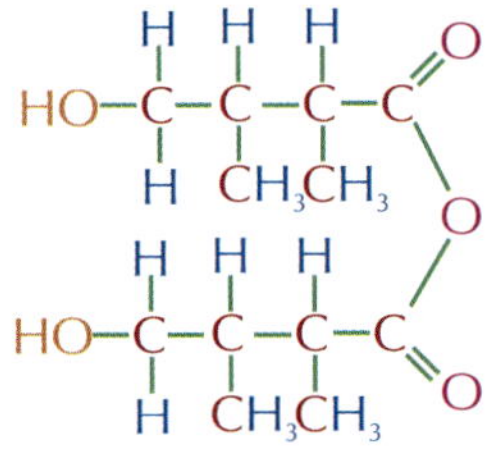

Q2 a) $+ H_2O \longrightarrow$ 2

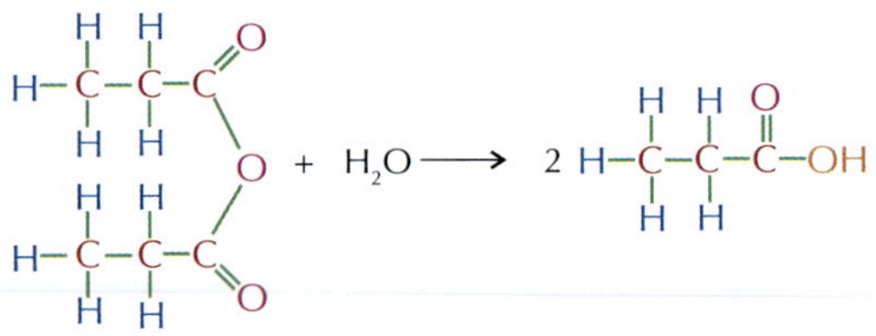

Unless you're asked for it, you don't have to draw out the full displayed formula for the molecule in the exam. You could just have written this equation as $CH_3CH_2COOCOCH_2CH_3 + H_2O \rightarrow 2CH_3CH_2COOH$. Either way of writing the equation will get you the marks.

b) $(CH_3CH_2CH_2CH(C_2H_5)CO)_2O + CH_3OH$

$\downarrow$

$CH_3CH_2CH_2CH(C_2H_5)COOCH_3$

$+$

$CH_3CH_2CH_2CH(C_2H_5)COOH$

c) $(HOCH_2CH(CH_3)CH(CH_3)CO)_2O + NH_3$

$\downarrow$

$HOCH_2CH(CH_3)CH(CH_3)CONH_2$

$+$

$HOCH_2CH(CH_3)CH(CH_3)COOH$

Q3 $C_6H_4(OH)COOH + CH_3COCl$

$\downarrow$

$C_6H_4(OCOCH_3)COOH + HCl$

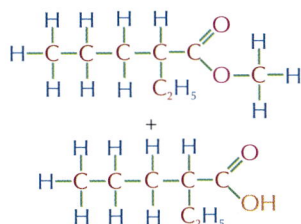

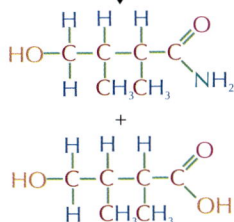

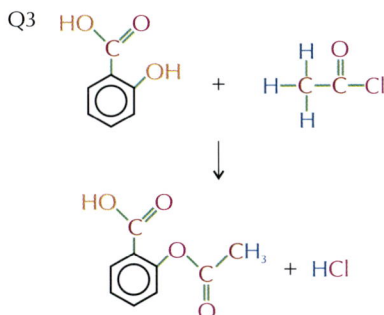

The equations in Q2 and Q3 are written vertically for space. In the exam, write them horizontally.

Page 436 — Fact Recall Questions

Q1 Propanyl propanoate and propanoic acid.

Q2 Ethanoic anhydride is cheaper than ethanoyl chloride. Ethanoic anhydride is safer than ethanoyl chloride because it's less corrosive, reacts more slowly with water and doesn't produce dangerous HCl fumes.

8. Purifying Organic Compounds

Page 440 — Application Questions

Q1 E.g. Redistill the mixture as hexane will distil out of the mixture at a lower temperature than ethanol, so the two liquids can be collected separately. / Use separation to separate hexane (insoluble in water) from ethanol (soluble in water). You'll be left will an organic layer containing hexane and an aqueous layer containing water and ethanol. The ethanol can be separated from the water by redistillation (as ethanol has a lower boiling point than water) or can be dried using an anhydrous salt, e.g. $MgSO_4$ or $CaCl_2$, which can then be removed by filtration.

Q2 The student could carry out the reaction in distillation apparatus, as the product would distil out of the reaction mixture as it formed, and so it wouldn't be over-oxidised.

Page 440 — Fact Recall Questions

Q1 E.g. anhydrous magnesium sulfate ($MgSO_4$) and anhydrous calcium chloride ($CaCl_2$).

Q2

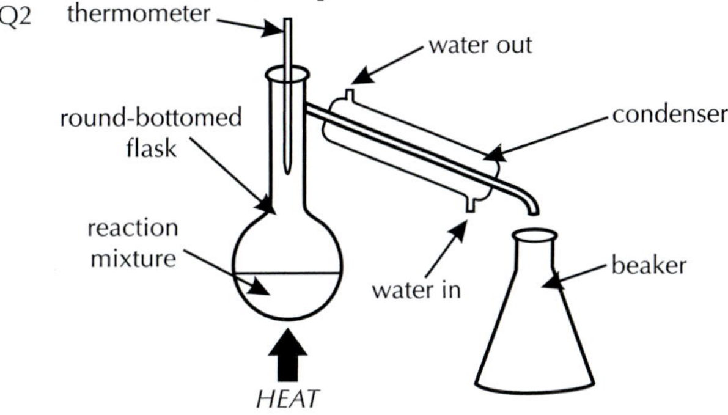

Q3 The different boiling points of the compounds.

Q4 A narrow range.

Exam-style Questions — pages 442-445

1 A ***(1 mark)***

2 D ***(1 mark)***

3 B ***(1 mark)***

4.1 A — propanal ***(1 mark)***
B — propanone ***(1 mark)***

4.2 E.g. Tollens' reagent ***(1 mark)*** produces a silver mirror with A but not with B ***(1 mark)***/Fehling's solution or Benedict's solution ***(1 mark)*** produces a brick-red precipitate with A but not B ***(1 mark)***.

4.3 $NaBH_4$/sodium borohydride/sodium tetrahydridoborate(III) ***(1 mark)***.

4.4 $CH_3CH_2CHO + :H^- \rightarrow CH_3CH_2CH_2O^- \xrightarrow{H^+} CH_3CH_2CH_2OH$

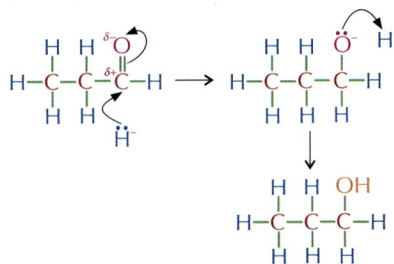

(1 mark for each correct curly arrow, 1 mark for correct structures.)

To get a mark for your curly arrow the end of the arrow must start at a lone pair of electrons or a bond and the point of the arrow must point to wherever the electrons are heading.

4.5 nucleophilic addition ***(1 mark)***

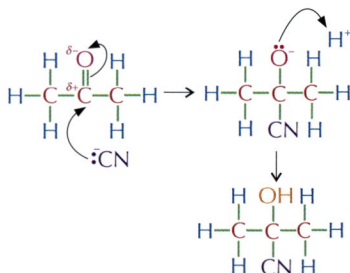

```
   H δ-O  H                 H  Ö⁻ H      H⁺
   |   ||  |                |   |  |
H—C—δ+C—C—H   →        H—C—C—C—H
   |      |                 |   |  |
   H      H                 H  CN  H
     :CN⁻
                                ↓
                           H  OH  H
                           |   |  |
                        H—C—C—C—H
                           |   |  |
                           H  CN  H
```

(1 mark for each correct curly arrow, 1 mark for correct structures.)

4.6 2-hydroxy-2-methylpropanenitrile ***(1 mark)***

4.7 The reaction of acidified KCN and compound A involves nucleophilic attack at the carbonyl carbon. The C=O bond is planar, and nucleophilic attack can occur from above or below the plane of the bond ***(1 mark)***. Depending on which direction the nucleophile attacks from, a different enantiomer will be formed. There is an equal chance that the nucleophilic attack will occur from each side of the bond, so the two enantiomers form in a racemate/racemic mixture ***(1 mark)***.

A racemate/racemic mixture has no effect on plane polarised light, as the two enantiomers cancel out each other's light-rotating effect ***(1 mark)***.

5.1

```
   H  H  H  O
   |  |  |  ||
H—C—C—C—C—OH
   |  |  |
   H  H  CH3
```

(1 mark)

5.2

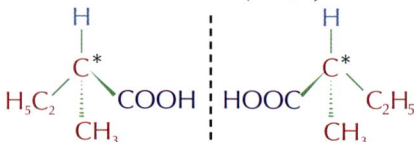

```
        H          |          H
        |          |          |
        C*         |          C*
      /  ⋮  \      |        /  ⋮  \
  H5C2  CH3  COOH  |  HOOC   CH3   C2H5
```

(1 mark for correct chiral centre marked, 1 mark for correctly drawn enantiomers.)

5.3 $2C_2H_5CH(CH_3)COOH + Na_2CO_3 \rightarrow$ $2C_2H_5CH(CH_3)COONa + H_2O + CO_2$

(1 mark for correct chemical formulas, 1 mark for balancing the equation).

5.4

```
   H  H  H     O         H
   |  |  |    //         |
H—C—C—C—C              
   |  |  |    \          |
   H  H  CH3   O————C—H
                         |
                         H
```

(1 mark)

5.5 methyl 2-methylbutanoate ***(1 mark)***

5.6 Any strong acid (e.g. $HCl/H_2SO_4/H_3PO_4$) ***(1 mark)***.

5.7

```
   H  H  H     O       H
   |  |  |    //       |
H—C—C—C—C             
   |  |  |    \        |
   H  H  CH3   O——C—H       +   OH⁻
                       |
                       H

            ↓ reflux

   H  H  H  O            H
   |  |  |  ||           |
H—C—C—C—C—O⁻   +   H—C—OH
   |  |  |               |
   H  H  CH3             H
```

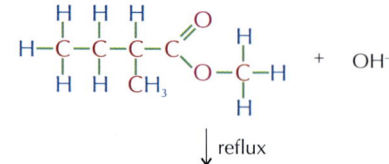

(1 mark for correct equation, 1 mark for correct structure of products).

5.8 E.g. food flavouring / perfume / plasticisers / solvent / biodiesel production / soap production ***(1 mark)***.

6.1 Fats contain mainly saturated hydrocarbon chains, whereas oils contain mainly unsaturated hydrocarbon chains ***(1 mark)***. Saturated hydrocarbon chains fit neatly together increasing the van der Waals forces between them, whereas unsaturated hydrocarbon chains are bent and don't pack well decreasing the van der Waals forces between them ***(1 mark)***. Stronger van der Waals forces mean a higher melting temperature so fats are solid at room temperature, whereas weaker van der Waals forces mean a lower melting temperature so oils are liquid at room temperature ***(1 mark)***.

6.2

```
CH2OOC(CH2)16CH3
|
CH2OOC(CH2)16CH3      + 3NaOH
|
CH2OOC(CH2)16CH3

        ↓

   H
   |
H—C—OH
   |
H—C—OH   + 3CH3(CH2)16COO⁻Na⁺
   |
H—C—OH
   |
   H
```

(1 mark for glycerol, 1 mark for correct sodium salt, 1 mark for balancing the equation).

6.3 soap ***(1 mark)***

6.4 Biodiesel is a mixture of methyl esters that can be used as a fuel ***(1 mark)***.

6.5

```
CH2OOC(CH2)16CH3
|
CH2OOC(CH2)16CH3   + 3CH3OH
|
CH2OOC(CH2)16CH3

        ↓

   H
   |
H—C—OH
   |
H—C—OH   + 3CH3(CH2)16COOCH3
   |
H—C—OH
   |
   H
```

(1 mark for methanol, 1 mark for correct methyl ester, 1 mark for balancing the equation).

These equations are written vertically for space. In the exam, write them horizontally.

6.6 KOH acts as a catalyst ***(1 mark)***.

7.1

```
   H  CH3 H  O
   |  |   |  ||
H—C—C—C—C—Cl  + NH3
   |  |   |
   H  H   CH3

        ↓

   H  CH3 H  O
   |  |   |  ||
H—C—C—C—C—NH2  + HCl
   |  |   |
   H  H   CH3
```

(1 mark for each correct product, maximum 2 marks).

This equation could also have been written as: $CH_3CH(CH_3)CH(CH_3)COCl + NH_3 \rightarrow$ $CH_3CH(CH_3)CH(CH_3)CONH_2 + HCl$

7.2 nucleophilic addition-elimination ***(1 mark)***

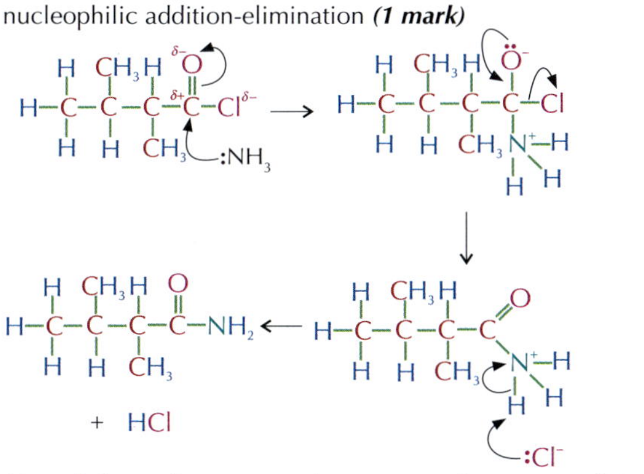

(1 mark for each correct curly arrow, maximum 5 marks)

Examiners often ask you to name and outline mechanisms. Don't forget to name it — it's easy to get carried away with the mechanism and lose a mark just for forgetting to put the name of the mechanism.

7.3 ***(1 mark)***

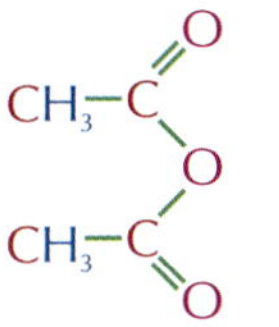

7.4 $CH_3COOCOCH_3 + H_2O \rightarrow 2CH_3COOH$
(1 mark for correct product, 1 mark for balancing the equation.)

7.5

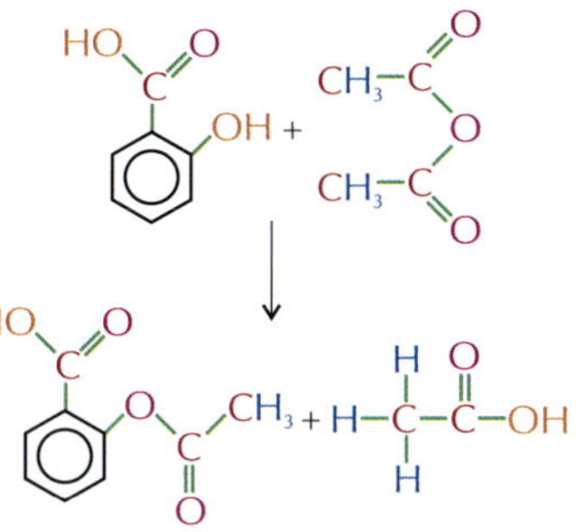

(1 mark for correct reactants, 1 mark for correct products.)

7.6 Ethanoic anhydride ***(1 mark)***. Any two from, e.g. ethanoic anhydride is cheaper / less corrosive / reacts less vigorously / doesn't produce toxic HCl fumes ***(1 mark for each)***.

7.7 E.g. you could use recrystallisation. To do this, first add just enough hot solvent to the impure aspirin so that it completely dissolves ***(1 mark)***. Then, leave the saturated solution of the impure product to cool slowly, allowing solid crystals of pure aspirin to form ***(1 mark)***. The aspirin crystals can be removed using filtration, washed with ice-cold solvent and left to dry ***(1 mark)***.

Section 2 — Aromatic Compounds and Amines

1. Aromatic Compounds

Page 448 — Application Question

Q1 The $\Delta H_{hydrogenation}$ for cyclohexene is -120 kJmol^{-1}. So if benzene had three double bonds, as in the theoretical molecule cyclohexa-1,3,5-triene, you would expect it to have a $\Delta H_{hydrogenation}$ of -360 kJmol^{-1}. But, $\Delta H_{hydrogenation}$ for benzene is only -208 kJmol^{-1}. This is 152 kJ mol^{-1} less exothermic than expected so benzene must be more stable than cyclohexa-1,3,5-triene would be, so must have a different structure.

Page 449 — Application Questions

Q1 a) 1,2-dinitrobenzene
b) 4-methylphenol
c) 2,4,6-trichlorophenylamine

Page 449 — Fact Recall Questions

Q1 Aromatic compounds are compounds that contain a benzene ring.

Q2 Benzene has a planar cyclic structure. Each carbon atom forms single covalent bonds to the carbons on either side of it and to one hydrogen atom. The final unpaired electron on each carbon atom is located in a p-orbital that sticks out above and below the plane of the ring. The p-orbitals on each carbon atom combine to form a ring of delocalised electrons. All the bonds in the ring are the same, so they are the same length which lies between the length of a single C–C bond and a double C=C bond.

2. Reactions of Aromatics

Page 452 — Application Questions

Q1 a) $AlCl_3$
b) $HCOCl + C_6H_6 \xrightarrow{AlCl_3} C_6H_5CHO + HCl$
c) Electrophilic substitution.

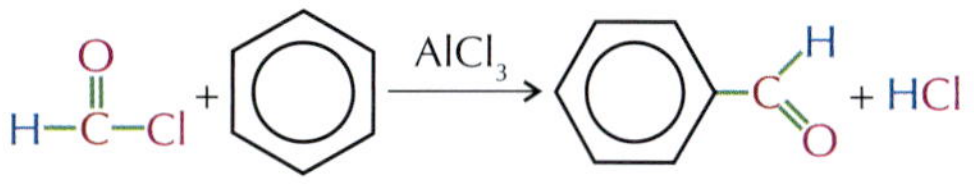

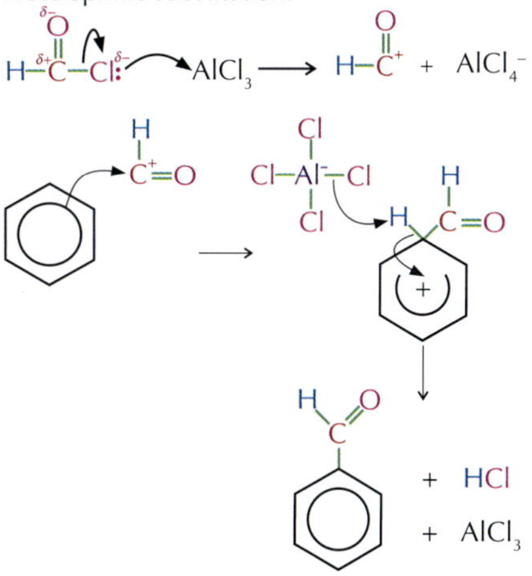

$HCOCl + AlCl_3 \rightarrow HCO^+ + AlCl_4^-$

+ HCl
+ $AlCl_3$

Q2 a) $HNO_3 + H_2SO_4 \rightarrow HSO_4^- + NO_2^+ + H_2O$

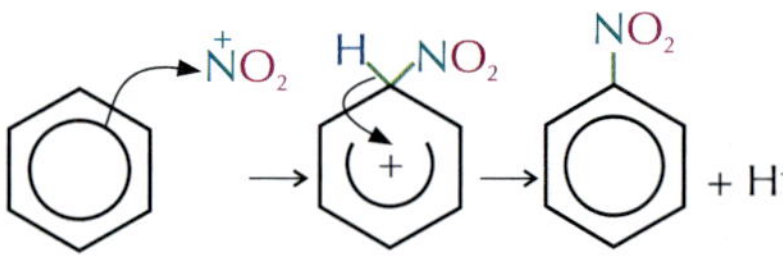

Don't forget that the formation of the nitronium ion is part of the mechanism too.

b) A concentrated H_2SO_4 catalyst and a temperature below 55 °C.

Page 452 — Fact Recall Questions

Q1 Aromatic compounds contain a benzene ring, which is a region of high electron density. Electrophiles are electron deficient so are attracted to these regions of high electron density.

Q2 E.g. nitro compounds can be reduced to form aromatic amines which are used to manufacture dyes and pharmaceuticals / nitro compounds decompose violently when heated so can be used as explosives (e.g. TNT).

Q3 $AlCl_3$ is a halogen carrier. It accepts a lone pair of electrons from the acyl chloride, forming a carbocation. This carbocation is a much stronger electrophile than the acyl chloride and is strong enough to react with the benzene ring.

3. Amines and Amides

Page 456 — Application Questions

Q1 a) ethylamine

b) dipropylamine

c) ethyldimethylamine

Q2 a) Ethylamine is the strongest base. The alkyl group on this amine pushes electrons away from itself so the electron density on the nitrogen atom increases. This makes the lone pair more available to form a co-ordinate bond with a hydrogen ion.

b) Phenylamine is the weakest base. The benzene ring draws electrons towards itself so the nitrogen's lone pair of electrons is partially delocalised onto the ring and the electron density on the nitrogen decreases. This makes the lone pair less available to bond with a hydrogen ion.

Q3 a) propanamide

b) N-propylmethanamide

Page 456 — Fact Recall Questions

Q1 A quaternary ammonium ion is a positively charged ion that consists of a nitrogen atom with four alkyl groups attached.

Q2 a) A quaternary ammonium salt with at least one long hydrocarbon chain.

b) E.g. in fabric softener / in hair conditioner / in detergent.

Q3 The availability of the lone pair of electrons on the nitrogen/ the electron density around the nitrogen.

4. Reactions of Amines

Page 460 — Application Questions

Q1 a) $CH_3Cl + 2NH_3 \rightarrow CH_3NH_2 + NH_4Cl$

b) nucleophilic substitution

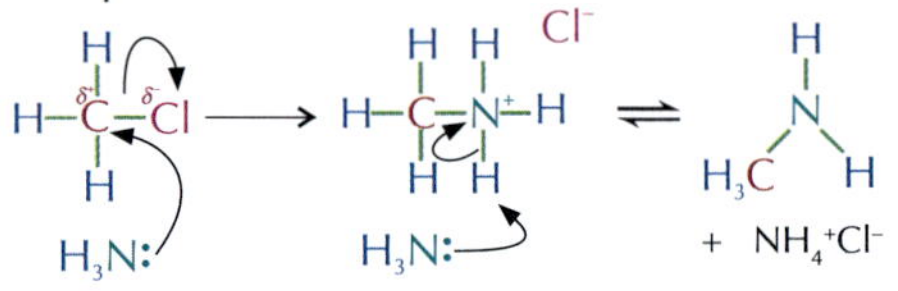

$+ NH_4^+Cl^-$

Q2 a) $CH_3CH_2CN + 2H_2 \rightarrow CH_3CH_2CH_2NH_2$

b) E.g. a platinum/nickel catalyst and high temperature and pressure.

Q3

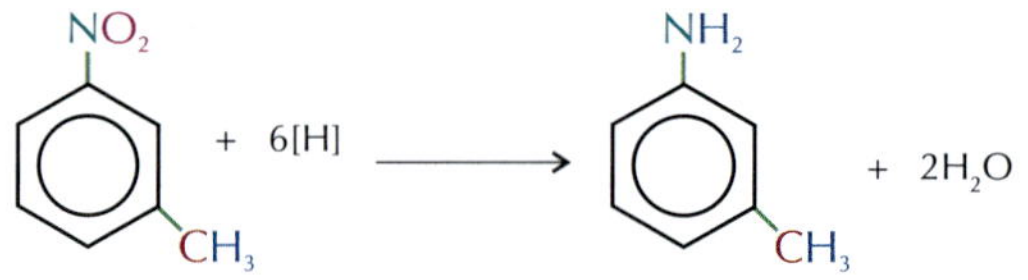

$+ 6[H] \longrightarrow$ $+ 2H_2O$

Page 461 — Fact Recall Questions

Q1 a) Heat the halogenoalkane with an excess of ethanolic ammonia.

b) The amine that is formed still has a lone pair of electrons on the nitrogen so it can still act as a nucleophile. This means that further substitutions can take place, and a mixture of primary, secondary and tertiary amines and quaternary ammonium salts can be produced.

Q2 a) E.g. by reducing the nitrile with $LiAlH_4$ and adding dilute acid. By catalytic hydrogenation (using e.g. a platinum/ nickel catalyst and a high temperature and pressure).

b) Catalytic hydrogenation. $LiAlH_4$ is too expensive for industrial use.

Q3 a) primary, secondary and tertiary amines and quaternary ammonium salts

b) an amide

c) an amide and a carboxylic acid

Q4 Take a nitro compound (e.g. nitrobenzene). Heat the nitro compound with tin metal and concentrated HCl under reflux to form a salt. Mix the salt with an alkali (e.g. NaOH) to produce an aromatic amine.

Exam-style Questions — Pages 462-463

1 A ***(1 mark)***

2.1

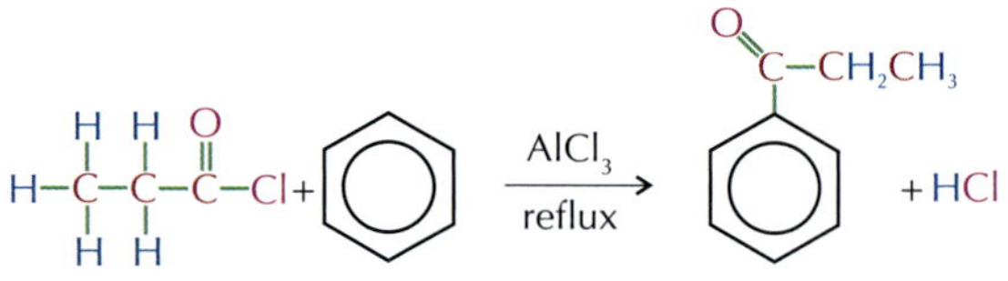

$\xrightarrow[\text{reflux}]{AlCl_3}$ $+ HCl$

(1 mark for $AlCl_3$ and reflux, 1 mark for correct products).

2.2 $AlCl_3$ accepts a lone pair of electrons from the propanoyl chloride ***(1 mark)***. The resulting carbocation is a strong enough electrophile to react with benzene ***(1 mark)***.

2.3 $CH_3CH_2COCl + AlCl_3 \rightarrow CH_3CH_2CO^+ + AlCl_4^-$

(1 mark, this mark is also awarded if the mechanism for this part of the reaction is correctly shown with curly arrows).

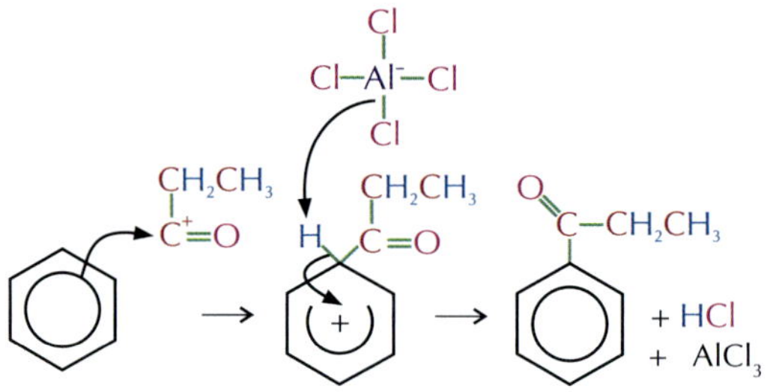

$+ HCl + AlCl_3$

(1 mark for each correct curly arrow, 1 mark for correct structure of intermediate, maximum 3 marks).

2.4 $HNO_3 + H_2SO_4 \rightarrow HSO_4^- + NO_2^+ + H_2O$ ***(1 mark)***
E.g.

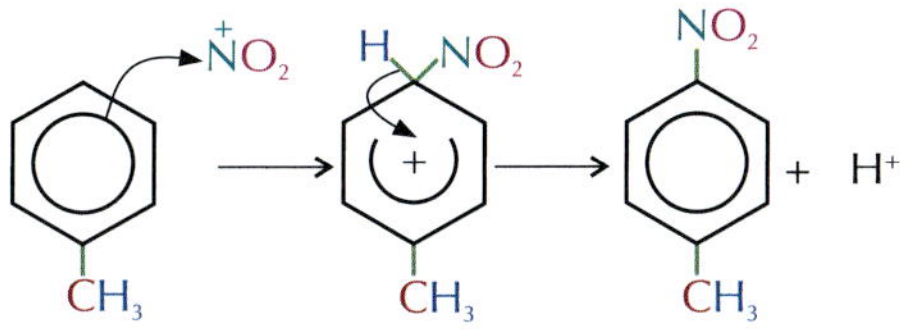

(1 mark for each correct curly arrow, 1 mark for correct structure of intermediate).

2.5

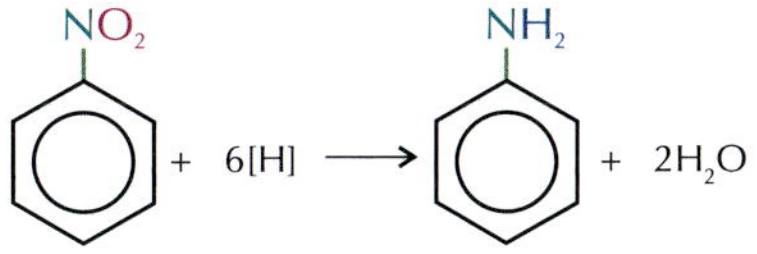

$C_6H_5NO_2 + 6[H] \rightarrow C_6H_5NH_2 + 2H_2O$

(1 mark for correct reactants and products, 1 mark for balanced equation).
The conditions are: reflux ***(1 mark)*** with tin (Sn) and concentrated HCl ***(1 mark)***, then add NaOH ***(1 mark)***.

3 B ***(1 mark)***

4.1 X is 2-methylpropylamine ***(1 mark)***.
Y is 2,3-dimethylphenylamine ***(1 mark)***.

4.2 $2NH_3 + CH_3CH(CH_3)CH_2Br \rightarrow CH_3CH(CH_3)CH_2NH_2 + NH_4^+Br^-$
(1 mark for correct reactants and products, 1 mark for balanced equation).
It's a nucleophilic substitution reaction ***(1 mark)***.

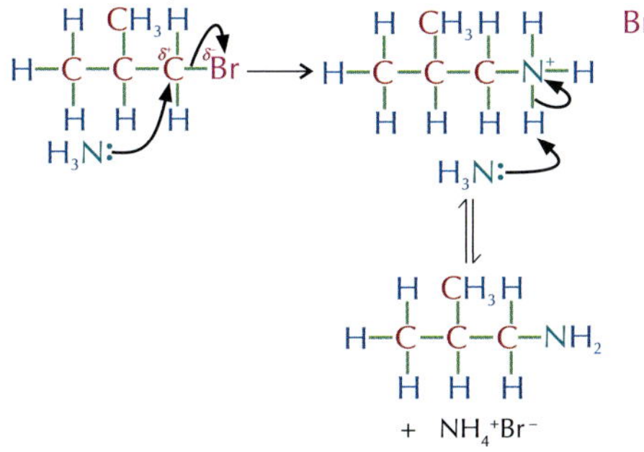

$+ \ NH_4^+Br^-$

(1 mark for each correct curly arrow, 1 mark for correct structure of intermediate, maximum 4 marks).

4.3 More than one of the hydrogens attached to the nitrogen may be substituted ***(1 mark)***.
E.g.

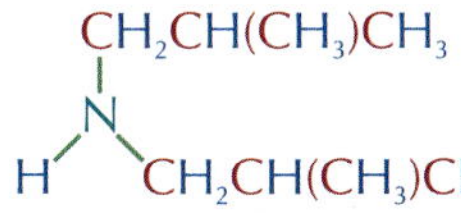

$HN(CH_2CH(CH_3)CH_3)_2$

Triamine and quaternary ammonium salt also acceptable ***(1 mark)***.
If you're asked to draw a structure like this and you're given a choice, always just draw the easiest one. You won't get any extra marks for giving harder answers and you're more likely to get it wrong. For example, if you drew the quaternary ammonium salt for this answer and forgot to give it a positive charge you wouldn't get the mark.

4.4 E.g. increase the concentration of ammonia ***(1 mark)***.

4.5 E.g. $CH_3CH(CH_3)CN + 4[H] \rightarrow CH_3CH(CH_3)CH_2NH_2$
(1 mark).
Further substitutions aren't possible, so amine X is the only possible product ***(1 mark)***.

4.6 There is a lone pair of electrons on the nitrogen which can be donated to form a co-ordinate bond with an H^+ ion ***(1 mark)***.
Amine X is the stronger base ***(1 mark)***. In amine X the alkyl group pushes electrons away from itself/in amine Y the benzene ring draws electrons towards itself ***(1 mark)***. This increases electron density on the nitrogen in amine X/ this decreases electron density on the nitrogen in amine Y ***(1 mark)***. So the lone pair of electrons is more available to form a bond with hydrogen in amine X/so the lone pair of electrons is less available to form a bond with hydrogen in amine Y ***(1 mark)***.

Section 3 — Polymers

1. Condensation Polymerisation

Page 467 — Fact Recall Questions

Q1 a) e.g dicarboxylic acids and diamines / amino acids
b) dicarboxylic acids and diols

Q2 condensation reactions

Q3 a) E.g. $-CO-(CH_2)_4-CO-NH-(CH_2)_6-NH-$

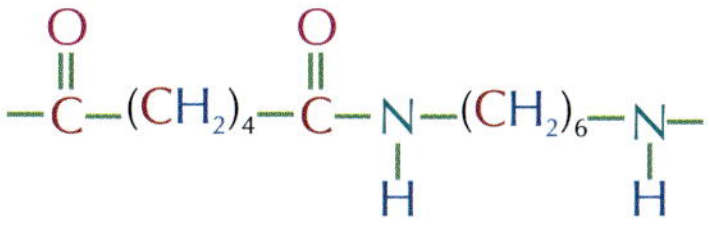

b) E.g. $-CO-C_6H_4-CO-NH-C_6H_4-NH-$

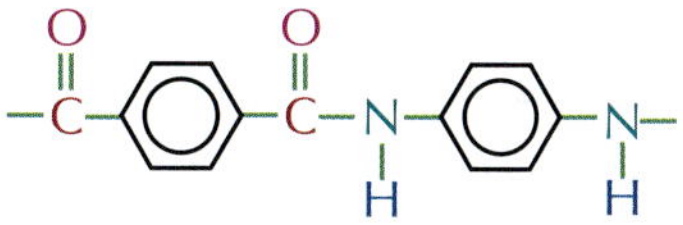

c) E.g. $-CO-C_6H_4-CO-O-CH_2-CH_2-O-$

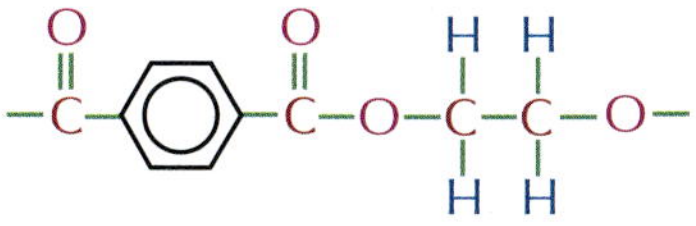

Q4 a) a dicarboxylic acid and a diamine
b) a dicarboxylic acid and a diol

Q5 a) hydrogen bonds
b) E.g.

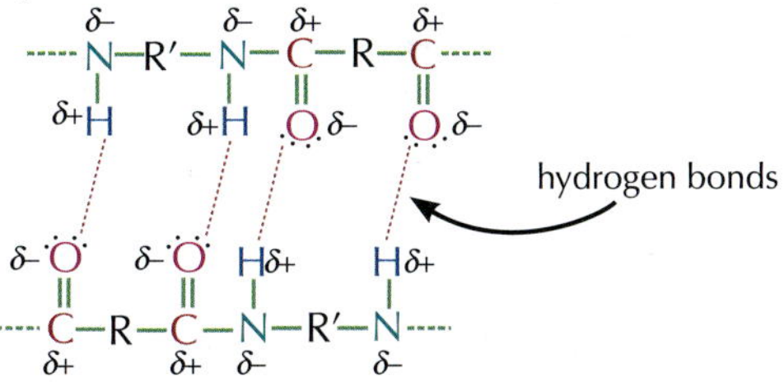

hydrogen bonds

2. Monomers & Repeating Units

Pages 471 — Application Questions

Q1 a) polyamide

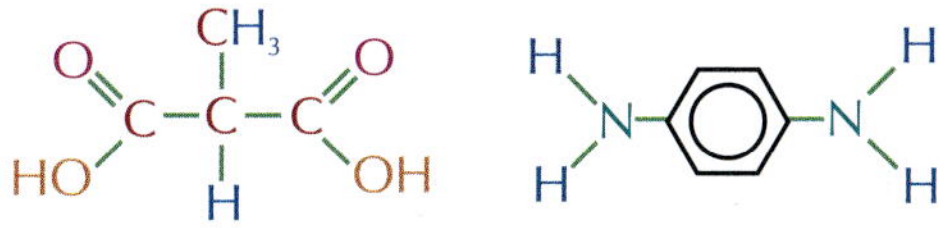

$HOOC-CH(CH_3)-COOH$ and $H_2N-C_6H_4-NH_2$

b) polyester

$HO-CH(C_6H_5)-COOH$

c) polyamide

$HOOC-CH(C_6H_5)-NH_2$

d) polyamide

$HOOC-CH_2-COOH$ $\quad H_2N-CH(CH_3)-CH(CH_3)-NH_2$

Q2 a) $-CO-C_6H_4-CO-NH-(CH_2)_6-NH-$

b) $-O-CH_2-CO-$

c) $-NH-CH_2-CO-$

d) $-CO-CH(CH_3)-CH(C_2H_6)-CO-O-CH(Cl)-O-$

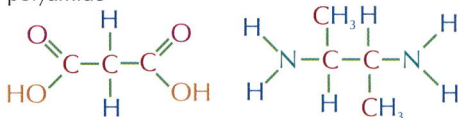

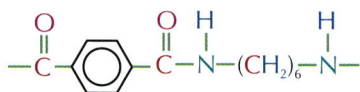

3. Disposing of Polymers

Page 474 — Application Question

Q1 Polymer A is biodegradable. It's a polyamide, so it can be broken down by hydrolysis. (Polymer B is a polyalkene, so it's chemically inert and non-biodegradable.)

Page 474 — Fact Recall Questions

Q1 a) The bonds in addition polymers are non-polar so they are not susceptible to attack by nucleophiles and won't react easily.

b) It means they don't react with things and degrade when they are being used.

c) It means that addition polymers are non-biodegradable.

Q2 a) biodegradable

b) The bonds in condensation polymers are polar. This means they are susceptible to attack by nucleophiles and can be broken by hydrolysis. This means the polymer will break down naturally. Addition polymers have non-polar bonds which can't be hydrolysed so won't break down naturally.

Q3 E.g. it requires areas of land / as the waste decomposes it releases methane (a greenhouse gas) / leaks from landfill sites can contaminate water supplies.

Q4 By passing the waste gases through scrubbers which can neutralise toxic gases by allowing them to react with a base.

Q5 a) E.g. by melting and remoulding / by cracking into monomers which can be used to make more plastic or other chemicals.

b) Plastic products are marked with numbers which show what type of plastic they are made from.

c) Advantages: any two from: e.g. it reduces the amount of waste going into landfill / it saves raw materials / the cost is lower than making plastics from scratch / it produces less CO_2 emissions than burning plastic.
Disadvantages: any two from: e.g. it is technically difficult / it is more expensive than burning/landfill / the product isn't usually suitable for its original purpose / the plastic can be easily contaminated.

Exam-style Questions — Page 475

1 A ***(1 mark)***

2.1

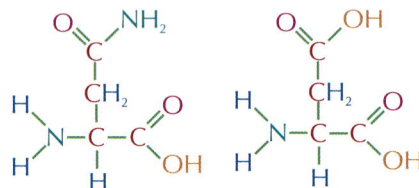

$H_2N-CH(CH_2CONH_2)-COOH$ $\quad H_2N-CH(CH_2COOH)-COOH$

(1 mark for each, maximum 2 marks).

2.2 Condensation polymerisation ***(1 mark)***

2.3 An amide / peptide link ***(1 mark)***

3.1 A polyamide ***(1 mark)***.

3.2 $-CO-(CH_2)_4-CO-NH-(CH_2)_6-NH-$ ***(1 mark)***.

When drawing repeating units make sure you include the trailing bonds which join the repeating units together. You'll lose a mark if you don't.

3.3 Biodegradable ***(1 mark)***. The amide bonds in nylon 6,6 are polar ***(1 mark)***, so they are susceptible to attack by nucleophiles ***(1 mark)*** and can be hydrolysed ***(1 mark)***.

3.4 Advantages: e.g. it reduces the amount of waste going into landfill / it saves raw materials / the cost is lower than making plastics from scratch / it produces less CO_2 emissions than burning the nylon ***(1 mark for each, maximum 2 marks)***.
Disadvantages: e.g. it is technically difficult / it is more expensive than burning/landfill / the recycled nylon may not be suitable for its original purpose / the nylon can easily be contaminated during recycling ***(1 mark for each, maximum 2 marks)***.

If you get an "advantages/disadvantages" question, make sure you discuss both. If you only give advantages or only give disadvantages you'll only get half marks, no matter how many correct points you've made.

Section 4 — Amino Acids, Proteins and DNA

1. Amino Acids

Page 478 — Application Question

Q1 a) A — 2-aminopropanoic acid
B — 2-amino-3-hydroxypropanoic acid

b) (i)

```
     H  CH3
     |  |    //O
  H—N+—C—C
     |  |    \OH
     H  H
```

(ii)

```
     H  CH3
     |  |    //O
  H—N+—C—C
     |  |    \O-
     H  H
```

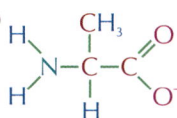

(iii)

```
  H       CH3
   \      |    //O
    N—C—C
   /      |    \O-
  H       H
```

Page 478 — Fact Recall Questions

Q1 Amino acids are said to be amphoteric because they have a basic amino group and an acidic carboxyl group and so have both acidic and basic properties.

Q2 A zwitterion is a dipolar ion, i.e. an ion that has both a positive and a negative charge in different parts of the molecule.

Q3 $R_f = \frac{\text{distance travelled by spot}}{\text{distance travelled by solvent front}}$.

Q4 Ninhydrin or a fluorescent dye and UV light would be used to make the colourless amino acids visible to the naked eye.

2. Proteins

Page 482 — Application Question

Q1

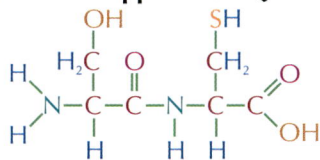

```
          OH             SH
          |              |
   H     H2C  O          CH2
    \     |   ||         |    //O
     N—C—C—N—C—C
    /     |        |     |    \OH
   H      H        H     H
```

```
          SH             OH
          |              |
   H     H2C  O          CH2
    \     |   ||         |    //O
     N—C—C—N—C—C
    /     |        |     |    \OH
   H      H        H     H
```

Q2

```
          O
  H2N     ||
     \/\/ \OH
       |
  HS/
```

```
              /
  H—N\  /\/   \OH
     |        ||
     H        O
```

```
    H      O
    |      ||
  H—N\  /\/ \
                OH
```

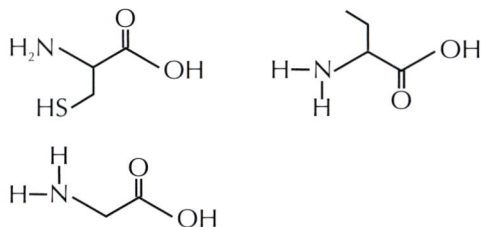

Q3 a) Primary and secondary.
b) hydrogen bonds
c) i) disulfide bonding
ii) tertiary structure

Page 482 — Fact Recall Questions

Q1 A polymer/sequence of amino acids joined together by peptide links.

Q2 E.g. α-helix and β-pleated sheet.

Q3 The amino acid cysteine, when it's part of a protein.

Q4 They bond different parts of the same protein molecule to help hold the 3D shape of the protein.

3. Enzymes

Page 484 — Fact Recall Questions

Q1 The lock and key model represents how an enzyme works with a substrate. The substrate fits neatly into the active site of an enzyme, forming an enzyme-substrate complex. The enzyme then catalyses the reaction of a substrate to form its products. The enzyme is left unchanged by the reaction.

Q2 An active site that only works on one enantiomer of a chiral substrate.

Q3 a) An inhibitor is roughly the same shape as a substrate of an enzyme. It fits into the active site of an enzyme, which blocks the substrate from getting to the active site.
b) E.g. Inhibitors can be used as drugs.

4. DNA

Page 488 — Fact Recall Questions

Q1 a) A phosphate ion, a 2-deoxyribose sugar molecule and a base.

b)

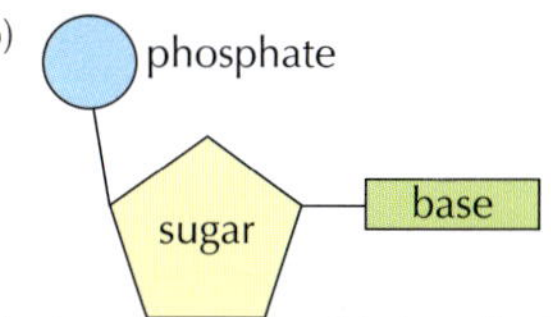

Q2 a) A covalent bond / phosphodiester bond
b) The phosphate group of one nucleotide and the 2-deoxyribose of another nucleotide.

Q3 a) adenine, cytosine, guanine, thymine
b) Adenine and thymine pair up and form two hydrogen bonds. Cytosine and guanine pair up and form three hydrogen bonds.

5. Cisplatin

Page 490 — Fact Recall Questions

Q1

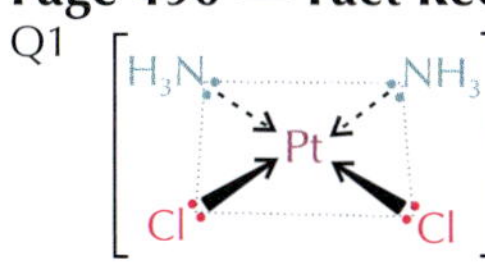

Q2 a) A nitrogen atom on guanine forms a coordinate bond with the platinum ion of cisplatin, replacing one of the chloride ion ligands. Another nearby guanine molecule undergoes the same reaction by replacing the other chloride ion ligand of cisplatin. This stops the DNA double helix from unwinding and being copied properly, which means it can't replicate.
b) Cisplatin can bind to healthy cells and stop them replicating. This is a problem for cells that replicate frequently such as hair and blood cells and can lead to hair loss and a suppressed immune system.

Exam-style Questions — Pages 492-493

1 B ***(1 mark)***

2 B ***(1 mark)***

3.1 Tyrosine and cystine ***(1 mark)***.

To find the R_f values, divide the distance of the spot from the origin by the distance of the solvent front from the origin. So for Spot A, 3.6 ÷ 6.5 ≈ 0.55, so Spot A could be tyrosine. For Spot B, 1.0 ÷ 6.5 = ≈ 0.15, so Spot B could be cytosine.

3.2

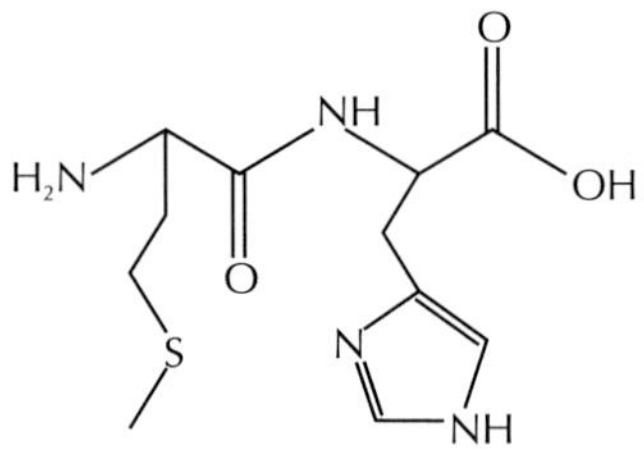

(1 mark)

3.3 The isoelectric point is the pH where the overall charge of an amino acid is zero ***(1 mark)***.

3.4

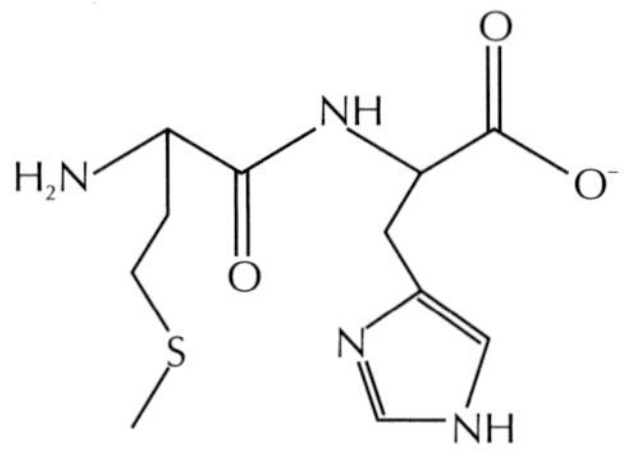

(1 mark)

4.1 hydrolysis ***(1 mark)***.

4.2

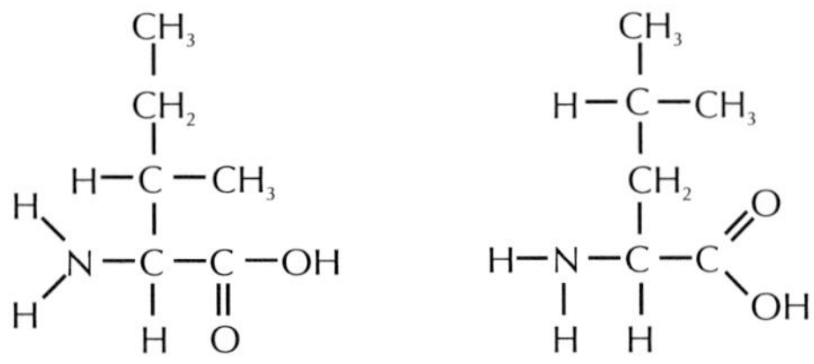

(2 marks available — 1 mark for each correct amino acid)

4.3 E.g. The primary structure of a protein is the chain of amino acids joined together by peptide links ***(1 mark)***. The secondary structure is a result of hydrogen bonds forming between peptide links ***(1 mark)***. This is often either an α-helix or a β-pleated sheet ***(1 mark)***. The tertiary structure is due to bonds forming between different parts of the polypeptide chain, folding the chain in a characteristic 3D shape ***(1 mark)***. These bonds may be hydrogen bonds or disulfide bonds ***(1 mark)***. Hydrogen bonds form between polar groups on different amino acid residues ***(1 mark)***. Disulfide bonds form between the sulfide groups on different cysteine residues ***(1 mark)***.

Section 5 — Further Synthesis and Analysis

1. Organic Synthesis

Page 497 — Application Questions

Q1 a) e.g. H_3PO_4 catalyst, water (steam), 300 °C, 60 atm
b) Cl_2, UV light
c) CH_3COCl, $AlCl_3$ catalyst, reflux, non-aqueous environment

Q2 a) step 1: conc. H_2SO_4, conc. HNO_3, below 55 °C
product: nitrobenzene
step 2: Sn, conc. HCl, reflux, $NaOH_{(aq)}$
b) step 1: e.g. warm $NaOH_{(aq)}$ or $KOH_{(aq)}$, reflux
product: propanol
step 2: $K_2Cr_2O_7$, H_2SO_4, heat in distillation apparatus
You have to do this reaction in distillation apparatus so that you don't form the carboxylic acid.
c) step 1: e.g. Cl_2, UV light
product: e.g. chloroethane
step 2: $KCN_{(aq)}$, ethanol, reflux
d) step 1: $K_2Cr_2O_7$, H_2SO_4, reflux
product: butanoic acid
step 2: methanol, conc. H_2SO_4 catalyst, heat

Q3 step 1: conc. H_2SO_4, conc. HNO_3, below 55 °C
product: nitrobenzene
step 2: Sn, concentrated HCl, reflux, $NaOH_{(aq)}$
product: phenylamine
step 3: CH_3COCl, 25 °C
product: N-phenylethanamide

Q4 step 1: e.g. warm $NaOH_{(aq)}$ or $KOH_{(aq)}$, reflux
product: ethanol
step 2: $K_2Cr_2O_7$, H_2SO_4, heat, reflux
product: ethanoic acid
step 3: propanol, conc. H_2SO_4 catalyst, reflux
product: propyl ethanoate

Page 497 — Fact Recall Questions

Q1 E.g. limit the potential for accidents / environmental damage.

Q2 E.g. use processes with high atom economies and high percentage yields.
Use synthesis routes that have as few steps as possible.

Q3 E.g. solvents are often flammable and toxic so can pose safety risks. If the solvent has to be separated from the product or disposed of after the reaction is complete that can create a lot of waste too.

Q4 a) NaOH or KOH, warm ethanol, reflux
b) $HCN_{(g)}$ or acidified $KCN_{(aq)}$ and H_2SO_4, 25 °C
c) dilute H_2SO_4 catalyst, H_2O, reflux / dilute $NaOH_{(aq)}$, reflux
d) $LiAlH_4$, dry ether then dilute H_2SO_4 or H_2, Ni or Pt catalyst, high temperature and pressure

Q5 a) amine, 25 °C
b) alcohol, 25 °C
c) NH_3, 25 °C

2. NMR Spectroscopy

Page 499 — Fact Recall Questions

Q1 parts per million/ppm

Q2 E.g. TMS produces a single absorption peak. / Its absorption peak is well away from most other absorption peaks. / TMS is inert (so it doesn't react with the sample). / TMS is volatile (so it's easy to remove from the sample).

3. ^{13}C NMR Spectroscopy

Page 503 — Application Questions

Q1 a) 4

The number of peaks on the ^{13}C NMR spectrum is the same as the number of carbon environments in the molecule...

b) 3

c) 2

This one's a bit tricky. If you draw the molecule, you should see that three of the carbons are CH_3 groups joined to the central carbon atom. These carbons are all in the same environment.

d) 4

Q2 There are four peaks on the spectrum, so the molecule must have four carbon environments. All the peaks lie between $\delta = 10$ ppm and $\delta = 40$ ppm. Since the molecule only contains carbon and hydrogen, these must all represent carbon atoms in C–C bonds. The formula of the molecule is C_5H_{12}, so it must be an isomer of pentane with four carbon environments. The only molecule that fits this description is 2-methylbutane:

```
    H CH3 H  H
    |  |  |  |
 H—C—C—C—C—H
    |  |  |  |
    H  H  H  H
```

Carbon 1 and the methyl group carbon are in the same environment. Carbons 2, 3 and 4 are all in different environments.

Q3 There are three peaks on the spectrum, so the molecule must have three carbon environments. The peaks at $\delta = 8$ ppm must represent carbons in C–C bonds. The peak at $\delta = 25$ ppm could represent a carbon in an R–C–Cl, R–C–Br, C–C group, or a carbon adjacent to a C=O group. From the molecular formula, you can see that it can't be ab R–C–Cl or R–C–Br group, and as there is no peak equivalent to a carbon in a C=O group, it must be due to a carbon in a C–C bond. The peak at $\delta = 65$ ppm must represent a carbon in a C–O bond. The formula of the molecule is $C_4H_{10}O$ and it must have three carbon environments. The only molecule that fits this description is 2-methylpropan-1-ol:

```
    H CH3 H
    |  |  |
 H—C—C—C—OH
    |  |  |
    H  H  H
```

4. ^{1}H NMR Spectroscopy

Page 508 — Application Questions

Q1 These hydrogen atoms correspond to peak **B**.

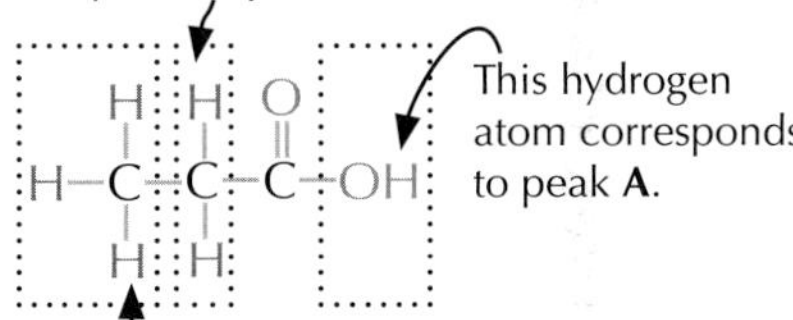

This hydrogen atom corresponds to peak **A**.

These hydrogen atoms correspond to peak **C**.

Q2 There are three peaks, so there must be three hydrogen environments in compound **X**.

The peak at $\delta = 2.5$ ppm is likely to be formed by hydrogens in a –COCH– environment.

It can't be an R–NH environment because that wouldn't fit with the molecular formula. And it can't be an R–OH group because the peak isn't a singlet.

The peak at $\delta = 2.1$ ppm is also likely to be formed by hydrogens in a –COCH– environment.

This isn't likely to be an R–OH group because the area ratio tells you that the environment contains more than one hydrogen.

The peak at $\delta \approx 1$ ppm is likely to be formed by hydrogens in an R–CH_3 environment.

It can't be an R–NH group because that wouldn't fit with the formula, and it isn't an R–OH group because it isn't a singlet.

From the area ratios, there are two protons in the environment at $\delta = 2.5$ ppm for every three protons in the environment at $\delta = 2.1$ ppm and for every three protons in the environment at $\delta \approx 1$ ppm. To fit this data, the groups contained in compound **X** must be –$COCH_2$–, –$COCH_3$ and –CH_3.

We know that the molecular formula is C_4H_8O and that the –CH_3 group must be next to the –$COCH_2$– (because their peaks are split into a triplet and a quartet).

So, the molecule must be **butan-2-one**:

```
    H  H  O  H
    |  |  ‖  |
 H—C—C—C—C—H
    |  |     |
    H  H     H
```

Page 508 — Fact Recall Questions

Q1 The number of hydrogen (proton) environments in a compound.

Q2 The relative number of hydrogen atoms in each environment.

Q3 Chemical shift data.

Q4 E.g. CCl_4 / D_2O / $CDCl_3$.

5. Chromatography

Page 512 — Application Questions

Q1 a) $R_f \text{ value} = \dfrac{\text{distance travelled by spot}}{\text{distance travelled by solvent}}$

Spot P: R_f value = 2.1 ÷ 8.0 = **0.26**

Spot Q: R_f value = 3.7 ÷ 8.0 = **0.46**

Spot R: R_f value = 5.9 ÷ 8.0 = **0.74**

b) Spot P contains glycine, spot Q contains tyrosine and spot R contains leucine.

Q2 Components that are more strongly adsorbed to the stationary phase take longer to pass down the column so the pure product will leave the column after the impurities.

Page 512 — Fact Recall Questions

Q1 Chromatography is used to separate out the components of a mixture.

Q2 Draw a line in pencil near the bottom of the TLC plate and put a very small drop of each mixture to be separated on the line. Allow the spots on the plate to dry. Place the plate in a beaker with a small volume of solvent and place a watch glass on top of the beaker. Leave the beaker until the solvent has moved almost to the top of the plate. Then remove the plate from the beaker. Use a pencil to mark how far the solvent travelled up the plate. Place the plate in a fume cupboard and leave it to dry.

Q3 a) R_f values are used to identify different substances.

b) $R_f \text{ value} = \frac{\text{distance travelled by spot}}{\text{distance travelled by solvent}}$

Q4 A chromatography column has a liquid mobile phase, and a solid stationary phase. When a mixture of components is added to the column, some will be more soluble in the mobile phase and some will be more strongly adsorbed to the stationary phase. This means that the components will spend different amounts of time adsorbed onto the stationary phase and dissolved in the mobile phase. So the different components will take different amounts of time to pass through the column and will be separated out.

6. Gas Chromatography

Page 514-515 — Application Questions

Q1 a) The gas chromatograph (GC) separates out the chemicals in the water sample.

b) E.g. the GC-MS machine will produce a mass spectrum for each individual chemical in the water sample. The scientist can compare these mass spectra to the mass spectrum of glyphosate looking for a match / use a computer to match up the mass spectra for each component of the mixture against a database.

Q2 a) peak A

b) component B

Q3 a) 11 minutes

b) Component A because it has the largest peak and the area under each peak tells you the relative amount of each component that's present in the mixture.

c) butan-1-ol

d) Run a sample of butan-1-ol under the same conditions as the sample of the mixture and see whether it has the same retention time as component A.

Page 515 — Fact Recall Questions

Q1 E.g. nitrogen gas.

Q2 The time taken from the injection of a sample to the detection of a substance.

Q3 The sample is separated out into its components using gas chromatography (GC). The separated components are fed into a mass spectrometer. This produces a mass spectrum for each component, which can be used to identify each one.

Q4 The components separated out by the GC can be positively identified by the MS. Identifying the components using GC alone can be impossible, because similar compounds often have similar retention times.

Pages 517-518 — Exam-style Questions

1.1 CH_3COCl ***(1 mark)***, reflux in a non-aqueous environment ***(1 mark)***, with an $AlCl_3$ catalyst ***(1 mark)***.

1.2 E.g. aluminium oxide (coated with water) ***(1 mark)***.

1.3 Some components of the mixture will be more soluble in the mobile phase and some will be more strongly adsorbed to the stationary phase ***(1 mark)***. This means that the phenylethanone will take a different amount of time passing through the column to the impurities ***(1 mark)***.

1.4 Spectrum C ***(1 mark)***. Phenylethanone has six different carbon environments and so will have six peaks on its ^{13}C NMR spectrum ***(1 mark)***.

2.1 E.g. $CDCl_3$ / D_2O / CCl_4 ***(1 mark)***.

Any solvent that contains no ^1H atoms would get the mark here.

2.2 The molecule is ethanol:

```
    H  H
    |  |
H—C—C—OH
    |  |
    H  H
```

(1 mark for correctly identifying the molecule)

Plus any five from: There are three peaks, so the molecule must have three hydrogen environments ***(1 mark)***. The peak at $\delta = 1.2$ ppm must be due to hydrogens in an R–CH_3 environment ***(1 mark)***. This peak is a triplet, so there must be two hydrogens on adjacent carbons ***(1 mark)***. The singlet peak at $\delta = 2.2$ ppm must be due to a hydrogen in the alcohol group / an R–OH environment ***(1 mark)***. The peak at $\delta = 3.8$ ppm must be due to a hydrogen in an R–O–CH environment ***(1 mark)***. It is a quartet, so there must be three hydrogens on adjacent carbons ***(1 mark)***.

3.1 The sample is separated using gas chromatography ***(1 mark)***. The separated components are fed straight into a mass spectrometer, which is used to identify the individual components ***(1 mark)***.

3.2 Gas chromatography uses retention times to identify the components of a mixture ***(1 mark)***. Similar compounds have similar retention times, so can't be identified accurately ***(1 mark)***.

3.3 5 ***(1 mark)***

If you had trouble with this, have a look back at the bit of the ^{13}C NMR topic about aromatic compounds and symmetry.

4.1 Some components of the mixture will be more strongly adsorbed to the stationary phase than others ***(1 mark)***. So the components will spend different amounts of time adsorbed to the stationary phase and will travel different distances in the same time ***(1 mark)***.

4.2 $R_f \text{ value} = \frac{\text{distance travelled by spot}}{\text{distance travelled by solvent}}$

$= 2.2 \text{ cm} \div 9.2 \text{ cm} = \mathbf{0.24}$ ***(1 mark)***

4.3 The peaks will be: a singlet caused by the hydrogens attached to carbon d ***(1 mark)***, a quartet caused by the hydrogens attached to carbon b ***(1 mark)*** and a triplet caused by the hydrogens attached to carbon a ***(1 mark)***.

Glossary

A

Accurate result
A result that's really close to the true answer.

Achiral molecule
A molecule that can be superimposed on its mirror image.

Acid anhydride
A molecule formed from two identical carboxylic acid molecules, joined via an oxygen atom with the carbonyl groups on either side.

Acid dissociation constant, K_a
An equilibrium constant specific to weak acids that relates the acid concentration to the concentration of $[H^+]$ ions. $K_a = [H^+][A^-] \div [HA]$.

Acidic buffer
A buffer with a pH of less than 7 containing a mixture of a weak acid with one of its salts.

Activation energy
The minimum amount of kinetic energy that particles need to have in order to react when they collide.

Acyl chloride
A molecule which contains the functional group COCl.

Acylation
When an acyl group (–COR) is added to a molecule.

Addition polymer
A type of polymer formed by joining small alkenes (monomers) together.

Adsorption
The attraction between a substance and the surface of the solid stationary phase in thin-layer chromatography, column chromatography and gas chromatography.

Aim
The question an experiment is trying to answer.

Alcohol
A substance with the general formula $C_nH_{2n+1}OH$.

Aldehyde
A substance with the general formula $C_nH_{2n}O$ which has a hydrogen and one alkyl group attached to a carbonyl carbon atom.

Alkaline earth metal
An element in Group 2 of the periodic table.

Alkane
A hydrocarbon with the general formula C_nH_{2n+2}.

Alkene
A hydrocarbon with the general formula C_nH_{2n}, containing at least one carbon-carbon double bond.

Amide
A carboxylic acid derivative which contains the functional group $CONH_2$.

Amide link
The -CONH- group which is found between monomers in a polyamide.

Amine
A molecule where one or more of the hydrogen atoms in ammonia have been replaced with an organic functional group, such as an alkyl or an aromatic group.

Amino acid
A molecule with an amino group (NH_2) and a carboxyl group (COOH).

Amphoteric
Having both acidic and basic properties.

Anomalous result
A result that doesn't fit in with the pattern of the other results in a set of data.

Aromatic compound
A compound that contains a benzene ring.

Arrhenius equation
An equation that links the rate constant, k, to temperature and activation energy.

Arrhenius plot
A graph where 1/temperature is plotted against $\ln k$, where k is the rate constant.

Atom
A neutral particle made up of protons and neutrons in a central nucleus and electrons orbiting the nucleus.

Atom economy
A measure of the proportion of reactant atoms that become part of the desired product in a balanced chemical reaction.

Atomic number
The number of protons in the nucleus of an atom. Also called proton number.

Autocatalysis
When a reaction is catalysed by one of its products.

Avogadro constant
6.02×10^{23} — the number of particles in 1 mole of a substance.

B

Barium chloride test
Test that uses acidified barium chloride to test for sulfate ions in solution.

Barium meal
A suspension of barium sulfate swallowed by a patient before an X-ray in order to show up the structure of their oesophagus, stomach or intestine.

Base (of DNA)
One of four molecules (adenine, thymine, cytosine, guanine) attached to the sugar-phosphate backbone in DNA.

Basic buffer
A buffer with a pH of more than 7 containing a mixture of a weak base with one of its salts.

Benedict's solution
A deep blue solution containing Cu^{2+} ions, which are reduced to a brick-red precipitate of Cu_2O when warmed with an aldehyde, but stays blue when warmed with a ketone.

Bidentate ligand
A ligand that can form two co-ordinate bonds in a complex ion.

Biodegradable
Will break down naturally.

Biodiesel
A mixture of methyl esters of fatty acids which can be used as a carbon neutral fuel.

Biofuel
A fuel that's made from biological material that's recently died.

Bond dissociation enthalpy, ΔH_{diss}
The enthalpy change when all the bonds of the same type in 1 mole of gaseous molecules are broken.

Bond enthalpy
The energy required to break a bond between two atoms.

Born-Haber cycle
An enthalpy cycle that allows you to calculate the lattice enthalpy change of formation for a system.

Brønsted-Lowry acid
A proton donor.

Brønsted-Lowry base
A proton acceptor.

Buffer
A solution that resists changes in pH when small amounts of acid or alkali are added.

C

Calibration curve/graph
A graph which shows the relationship between a measurement taken in an experiment and a property of a substance. It can be used to determine an unknown value.

Calorimetry
Method of finding out how much energy is given out or taken in by a reaction, by measuring the temperature change that takes place during the reaction.

Carbocation
An ion containing a positively charged carbon atom.

Carbonyl compound
A compound that contains a carbon-oxygen double bond.

Carboxylic acid
A molecule which contains a carboxyl group (COOH).

Catalyst
A substance that increases the rate of a reaction by providing an alternative reaction pathway with a lower activation energy. The catalyst is chemically unchanged at the end of the reaction.

Catalyst poisoning
When impurities in a reaction mixture bind to a catalyst's surface, blocking reactants from being adsorbed and reducing the effectiveness of the catalyst.

Catalytic hydrogenation
A chemical reaction in which hydrogen is added to an unsaturated molecule by reacting the molecule with molecular hydrogen (H_2) in the presence of a catalyst.

Categoric data
Data that can be sorted into categories.

Cationic surfactant
A surfactant which is positively charged.

Causal link
The relationship between two variables where a change in one variable causes a change in the other.

Cell potential, E_{cell}
The voltage between two half-cells in an electrochemical cell.

Chain isomer
An organic molecule that contains the same atoms and functional groups as another molecule but has a different arrangement of the carbon skeleton.

Charge cloud
An area in an atom or molecule where there's a high chance of finding an electron pair.

Chelate effect
When monodentate ligands are substituted with multidentate ligands, the number of particles and the entropy of the system increases without enthalpy changing significantly. Reactions that result in an increase in entropy and little change in enthalpy are more likely to occur, so multidentate ligands form much more stable complexes than monodentate ligands.

Chemical shift
Nuclei in different environments absorb energy of different frequencies. NMR spectroscopy measures these differences relative to a standard substance — the difference is called the chemical shift (δ).

Chiral carbon
A carbon atom that has four different groups attached to it.

Chloroalkane
An alkane with one or more hydrogen atoms substituted for chlorine atoms.

Chromatogram
A visual record (such as a pattern of spots or a graph) of the results of a chromatography experiment.

Chromatography
An analytical technique which uses a mobile phase and a stationary phase to separate out mixtures into their constituent components.

***Cis-trans* isomerism**
A special type of *E/Z* isomerism where two of the groups attached to the carbon atoms around the C=C double bond are the same.

Cisplatin
A platinum-containing complex ion with a square planar shape that can be used as an anti-cancer drug.

Clock reaction
A reaction where, after a period of time, there's a sudden increase in the concentration of a product.

Closed system
A system where nothing can get in or out.

Co-ordinate bond
A covalent bond in which both electrons in the shared pair come from the same atom (also called a dative covalent bond).

Co-ordination number
The number of co-ordinate bonds that are formed with the central metal ion in a complex ion.

Collision theory
The theory that a reaction will not take place between two particles unless they collide in the right direction and with at least a certain minimum amount of kinetic energy.

Colorimeter
An instrument for measuring how much light is absorbed by a sample.

Column chromatography
A type of chromatography where the stationary phase is a column packed with a solid, and the mobile phase is a liquid solvent.

Complete combustion
Burning a substance completely in oxygen to produce carbon dioxide and water only.

Complex ion
A metal ion surrounded by co-ordinately bonded ligands.

Condensation polymer
A type of polymer formed through a series of condensation reactions.

Condensation reaction
A chemical reaction in which two molecules are joined together and a small molecule is eliminated.

Contact Process
An industrially used method of producing sulfuric acid.

Continuous data
Data that can have any value on a scale.

Continuous monitoring
A method of following a reaction by monitoring the formation of a product or the loss of a reactant, over the course of a reaction.

Control variable
A variable that is kept constant in an experiment.

Correlation
The relationship between two variables.

Corrosive substance
A substance that may cause chemical burns.

Covalent bond
A pair of electrons shared between two atoms. Both nuclei are attracted electrostatically to the shared electrons.

Cracking
Breaking long-chain alkanes into smaller hydrocarbons.

Crude oil
A mixture consisting mainly of alkane hydrocarbons that can be separated into different fractions.

Curly arrow
An arrow used in reaction mechanism diagrams to show the movement of a pair of electrons.

Cycloalkane
A type of alkane where the carbon atoms form a ring, with two hydrogen atoms attached to each carbon.

d-block
The block of elements in the middle of the periodic table.

d sub-level
A type of sub-level. Each can hold ten electrons.

Data logger
A device that can record data readings automatically at set intervals and store them to be looked at later.

Dative covalent bond
A covalent bond in which both electrons in the shared pair come from the same atom (also called a co-ordinate bond).

Dehydration
A reaction where water is eliminated.

Dependent variable
The variable that you measure in an experiment.

Deuterated solvent
A solvent which has had all of its hydrogen atoms exchanged for deuterium atoms.

Deuterium
An isotope of hydrogen. It contains one neutron, one proton and one electron.

Dichromate(VI) ion
An oxidising agent, used to test for the presence of primary or secondary alcohols. Orange dichromate(VI) ions ($Cr_2O_7^{2-}$) turn to green chromium(III) ions (Cr^{3+}) ions in the presence of primary or secondary alcohols.

Dipole
A difference in charge between two atoms caused by a shift in the electron density in a bond.

Diprotic acid
An acid that releases two H^+ ions per molecule.

Discrete data
Data that can only take certain values.

Displacement reaction
A reaction where a more reactive element pushes out a less reactive element and takes its place.

Displayed formula
A way of representing a molecule that shows how all the atoms are arranged and all the bonds between them.

Disproportionation
When an element is both oxidised and reduced in a single chemical reaction.

Distillation
A method of separating liquids with different boiling points by gently heating them.

Disulfide bond
A covalent bond between sulfur atoms in two different thiol groups.

DNA
Deoxyribonucleic acid — two complementary polynucleotide chains joined together in a double helix structure.

Double helix
A shape formed by two helices twisted around each other.

Drying agent
An anhydrous salt, such as magnesium sulfate or calcium chloride, that can be used to remove water from an organic product.

Dynamic equilibrium
In a reversible reaction dynamic equilibrium is reached when the concentrations of the reactants and products are constant, and the forward and backward reactions are going at the same rate.

***E*-isomer**
A stereoisomer of an alkene that has the two highest priority groups on opposite sides of the carbon-carbon double bond.

***E/Z* isomerism**
A type of stereoisomerism that is caused by the restricted rotation about a carbon-carbon double bond. Each of the carbon atoms must have two different groups attached.

Electric heater
A piece of equipment used to heat a reaction mixture, consisting of a plate of metal that is heated to a set temperature.

Electrochemical cell
An electrical circuit made from two metal electrodes dipped in salt solutions and connected by a wire.

Electrochemical series
A list of electrode potentials written in order from most negative to most positive.

Electrode potential
The voltage measured when a half-cell is connected to a standard hydrogen electrode.

Electromotive force (EMF)
Another name for the cell potential.

Electron
A subatomic particle with a relative charge of –1 and a relative mass of 1/2000, located in orbitals around the nucleus.

Electron configuration
The number of electrons an atom or ion has and how they are arranged.

Electronegativity
The ability of an atom to attract the bonding electrons in a covalent bond.

Electrophile
An electron deficient (and usually positively charged) species which is attracted to regions of high electron density.

Electrophilic addition
A reaction mechanism where a C=C double bond in an alkene opens up and atoms are added to the carbon atoms.

Electrophilic substitution
A reaction mechanism where an electrophile substitutes for an atom (or group of atoms) in a molecule.

Electrospray ionisation
A method of producing ions for analysis in a mass spectrometer by applying high pressure and high voltage to a sample of a substance.

Elimination reaction
A reaction mechanism in which a molecule loses atoms or groups of atoms.

Empirical formula
A formula that gives the simplest whole number ratio of atoms of each element in a compound.

Enantiomer
A molecule that has the same structural formula as another molecule but with four groups arranged around a chiral carbon atom so that it is a non-superimposable mirror image of the other molecule.

End point
The point in a titration at which all the acid is just neutralised and the pH curve becomes vertical.

Endothermic reaction
A reaction that absorbs energy (ΔH is positive).

Energy gap, ΔE
The amount of energy needed for an electron to transfer to a higher orbital.

Energy level
A region of an atom with a fixed energy that contains electrons orbiting the nucleus. Also known as a shell.

Enthalpy change
The energy transferred in a reaction at constant pressure.

Enthalpy change of atomisation of a compound, ΔH_{at}
The enthalpy change when 1 mole of a compound in its standard state is converted to gaseous atoms.

Enthalpy change of atomisation of an element, ΔH_{at}
The enthalpy change when 1 mole of gaseous atoms is formed from an element in its standard state.

Enthalpy change of formation, ΔH°_f
The enthalpy change when 1 mole of a compound is formed from its elements in their standard states under standard conditions.

Enthalpy change of hydration, ΔH_{hyd}
The enthalpy change when 1 mole of aqueous ions is formed from gaseous ions.

Enthalpy change of solution $\Delta H_{solution}$
The enthalpy change when 1 mole of an ionic substance dissolves in enough solvent to form an infinitely dilute solution.

Entropy, S
A measure of the amount of disorder in a system (e.g. the number of ways that particles can be arranged and the number of ways that the energy can be shared out between the particles).

Enzyme
A protein that acts as a biological catalyst.

Equilibrium constant, K_c
A ratio worked out from the concentration of the products and reactants once a reversible reaction has reached equilibrium.

Equilibrium constant, K_p
A ratio worked out from the partial pressures of the gaseous products and reactants once a reversible reaction has reached equilibrium.

Ester
A molecule that contains the functional group RCOOR.

Ester link
The -COO- group which is found between monomers in a polyester.

Esterification
Forming an ester by heating a carboxylic acid and an alcohol in the presence of a strong acid catalyst.

Exothermic reaction
A reaction that gives out energy (ΔH is negative).

Fatty acid
A long chain carboxylic acid which can combine with glycerol to form a fat or an oil.

Feasible reaction
A reaction that, once started, will carry on to completion, without any energy being supplied to it.

Fehling's solution
A blue solution of complexed copper(II) ions dissolved in sodium hydroxide which can be used to distinguish between aldehydes and ketones.

Filtration
A technique used to separate solids from liquids.

Fingerprint region
The region between 1000 cm^{-1} and 1550 cm^{-1} on an infrared spectrum. It's unique to a particular compound.

First electron affinity, ΔH_{ea1}
The enthalpy change when 1 mole of gaseous 1– ions is made from 1 mole of gaseous atoms.

First ionisation energy, ΔH_{ie1}
The enthalpy change when 1 mole of gaseous 1+ ions is formed from 1 mole of gaseous atoms.

Flammable substance
A substance that catches fire easily.

Fractional distillation
A method of separating crude oil fractions using their boiling points.

Free-energy change, ΔG
A measure which links enthalpy and entropy changes to predict whether a reaction is feasible.
$\Delta G = \Delta H - T\,\Delta S$

Free radical
A particle with an unpaired electron, written like this: Cl• or •Cl.

Fuel cell
A device that converts the energy of a fuel into electricity through an oxidation reaction.

Functional group
The group of atoms that is responsible for the characteristic reactions of a molecule (e.g. OH for alcohols, COOH for carboxylic acids, C=C for alkenes).

Functional group isomer
A molecule with the same molecular formula as another molecule, but with the atoms arranged into different functional groups.

G

Gas chromatography (GC)
A type of chromatography where the stationary phase is a tube packed with a solid, and the mobile phase is an unreactive gas.

Gas constant
A constant used in the ideal gas equation. It has the symbol R, and its value is 8.31 J K^{-1} mol^{-1}.

General formula
An algebraic formula that can describe any member of a family of compounds.

Giant covalent structure
A structure consisting of a huge network of covalently bonded atoms. Sometimes called macromolecular structures.

Giant ionic lattice structure
A regular repeated structure made up of oppositely charged ions held together by electrostatic attraction.

Giant metallic lattice structure
A regular structure consisting of closely packed positive metal ions in a sea of delocalised electrons.

Greenhouse effect
The trapping of energy from the Sun that has been absorbed and re-emitted by the Earth, by certain gases in the Earth's atmosphere.

Greenhouse gas
A gas that contributes to the greenhouse effect by absorbing energy in the infrared region of the electromagnetic spectrum.

Group
A column in the periodic table.

Haber Process
An industrially used method of producing ammonia.

Haem
A part of the haemoglobin molecule, consisting of a central Fe(II) ion bonded to four nitrogen atoms from a porphyrin ring.

Haemoglobin
A protein found in blood that helps to transport oxygen around the body.

Half-cell
One half of an electrochemical cell.

Half-equation
An ionic equation that shows oxidation or reduction — one half of a full redox equation.

Halide
A negative ion of a halogen.

Halogen
An element in Group 7 of the periodic table.

Halogen carrier
A molecule which can accept a halogen atom (e.g. $AlCl_3$). Used as a catalyst in Friedel-Crafts acylation reactions.

Halogenoalkane
An alkane with at least one halogen atom in place of a hydrogen atom.

Hess's Law
The total enthalpy change for a reaction is independent of the route taken.

Heterogeneous catalyst
A catalyst which is in a different physical state to the reactants.

Homogeneous catalyst
A catalyst which is in the same physical state as the reactants.

Homologous series
A family of organic compounds that have the same general formula and similar chemical properties.

Hydrocarbon
A molecule that only contains hydrogen and carbon atoms.

Hydrogen bonding
A type of weak bonding which occurs between hydrogen atoms and electronegative atoms in polar groups (e.g. $-NH_2$ and $-OH$).

Hydrolysis
A reaction where molecules are split apart by water molecules.

Hydroxynitrile
A molecule which contains a hydroxyl group (OH) and a nitrile group (CN).

Hypothesis
A suggested explanation for a fact or observation.

Ideal gas equation
The ideal gas equation is $pV = nRT$. It allows you to find the number of moles in a volume of gas at any temperature and pressure.

Incomplete combustion
Burning a substance in a poor supply of oxygen to produce carbon monoxide, water and sometimes carbon and carbon dioxide.

Independent variable
The variable that you change in an experiment.

Indicator
A substance that changes colour over a particular pH range.

Infrared (IR) spectroscopy
An analytical technique used to identify the functional groups present in a molecule by measuring the vibrational frequencies of its bonds.

Inhibitor
A molecule that is a similar shape to a substrate of an enzyme, that blocks the active site of the enzyme and stops it interacting with the substrate.

Initial rates method
An experimental technique that can be used to work out the orders of a reaction.

Integration trace
A line on a ^{1}H NMR spectrum that has a change in height that is proportional to the area of the peak it's next to.

Intermediate
A short-lived, reactive molecule that occurs in the middle of a step-wise reaction mechanism.

Intermolecular forces
Forces between molecules, e.g. van der Waals forces, permanent dipole-dipole forces and hydrogen bonding.

Ion
A charged particle formed when one or more electrons are lost or gained by an atom or molecule.

Ionic bond
An electrostatic attraction between positive and negative ions in a lattice.

Ionic equation
An equation which only shows the reacting particles of a reaction involving ions in solution.

Ionic product of water, K_w
A constant generated by multiplying the K_c for the dissociation of water by $[H_2O]$. $K_w = [H^+][OH^-]$.

Ionisation
The removal (or addition) of one or more electrons from an atom or molecule, resulting in an ion forming.

Ionisation energy
The energy needed to remove 1 electron from each atom or ion in 1 mole of gaseous atoms or ions.

Irritant substance
A substance that may cause inflammation or discomfort.

Isoelectric point
The pH at which the average overall charge on a molecule is zero.

Isomer
A molecule with the same molecular formula as another molecule, but with the atoms arranged in a different way.

Isotope
An atom with the same number of protons as another atom but a different number of neutrons.

K

Ketone
A substance with the general formula $C_nH_{2n}O$ which has two alkyl groups attached to a carbonyl carbon atom.

L

Lattice dissociation enthalpy
The enthalpy change when 1 mole of a solid ionic compound is completely dissociated into its gaseous ions.

Lattice formation enthalpy
The enthalpy change when 1 mole of a solid ionic compound is formed from its gaseous ions.

Le Chatelier's principle
A theory that states that if there's a change in concentration, pressure or temperature, the equilibrium position will move to help counteract the change.

Ligand
An atom, ion or molecule that donates a pair of electrons to a central metal ion in a complex ion.

Ligand substitution/exchange reaction
A reaction where one or more ligands are changed for one or more other ligands in a metal complex ion.

Lone pair (of electrons)
An unshared pair of electrons in the outer shell of an atom. Also called a non-bonding pair of electrons.

M

Mass number
The total number of protons and neutrons in the nucleus of an atom.

Mass spectrometry
An analytical technique that gives information on relative mass and relative abundance of atoms or molecules in a sample.

Mass spectrum
A chart produced by a mass spectrometer which can give you information about the relative masses and relative abundances of particles in a sample.

Maxwell-Boltzmann distribution
A theoretical model that describes the distribution of kinetic energies of molecules in a gas.

Mean bond enthalpy
An average value for the bond enthalpy of a particular bond over the range of compounds it is found in.

Melting point apparatus
Equipment that can be used to test the purity of an organic product by measuring its melting point.

Metal-aqua complex ion
A species formed when metal ions dissolve in water. The water molecules form co-ordinate bonds with the metal ions.

Metallic bond
The attraction between delocalised electrons and a lattice of positive metal ions.

Method
A set of instructions detailing how to carry out an experiment safely.

Methyl orange
A pH indicator that changes colour between pH 3.1 and 4.4.

Mobile phase
A liquid or a gas used in chromatography which contains molecules that can move.

Model
A simplified picture or representation of a real physical situation.

Molar ratio
The ratio of the number of moles of each reactant and product in a balanced chemical equation.

Mole
The unit of amount of substance. One mole is equal to 6.02×10^{23} particles (the Avogadro constant).

Mole fraction
A measure of the proportion of a mixture that is made up of a particular substance.

Molecular formula
A way of representing molecules that shows the actual number of atoms of each element in a molecule.

Molecule
A group of two or more atoms bonded together with covalent bonds.

Monodentate ligand
A ligand that can only form one co-ordinate bond in a complex ion.

Monomer
A small molecule which can join together with other monomers to form a polymer.

Monoprotic acid
An acid that releases one H^+ ion per molecule.

Multidentate ligand
A ligand that can form two or more co-ordinate bonds in a complex ion.

Multiple covalent bond
A covalent bond that contains more than one shared pair of electrons.

Multiplet
A split peak on a ^1H NMR spectrum. (General term for a doublet, triplet, quartet, etc.)

N

n+1 rule
Peaks on a ^1H NMR spectrum always split into the number of hydrogens on the neighbouring carbon(s), plus one.

N-substituted amide
An amide where one of the hydrogens attached to the nitrogen has been substituted with an alkyl group.

Neutron
A subatomic particle with a relative charge of 0 and a relative mass of 1, located in the nucleus of an atom.

Nitration
A reaction in which a nitro group (NO_2) is added to a molecule.

Nomenclature
A fancy word for naming things, in particular organic compounds.

Nuclear magnetic resonance (NMR) spectroscopy
An analytical technique used to determine the relative environment of the nuclei in a compound.

Nucleophile
A species that forms a bond with an electrophile by donating a pair of electrons.

Nucleophilic addition-elimination
A reaction mechanism where a nucleophile adds on to the δ^+ carbon atom of a carbonyl group and another molecule is eliminated.

Nucleophilic substitution
A reaction mechanism where a nucleophile substitutes for an atom (or group of atoms) in a molecule.

Nucleotide
A molecule that is made up of a phosphate ion, a pentose sugar, and a (genetic) base.

Nucleus
The central part of an atom or ion, made up of protons and neutrons.

O

Optical isomer
A molecule that has the same structural formula as another molecule but with four groups arranged around a chiral carbon atom so that it is a non-superimposable mirror image of the other molecule.

Orbital
A region of a sub-level that contains a maximum of 2 electrons.

Order of reaction (reaction order)
A number that tells you how the concentration of a particular reactant affects the reaction rate.

Ordered (ordinal) data
Categoric data where the categories can be put in order.

Oxidation
The loss of electrons.

Oxidation state
The total number of electrons an element has donated or accepted. Also called an oxidation number.

Oxidising agent
Something that accepts electrons and gets reduced.

Oxidising substance
A substance that reacts to form oxygen, meaning that other substances burn more easily in its presence.

Ozone layer
A layer of ozone (O_3) found in the Earth's upper atmosphere which protects the Earth from ultraviolet radiation.

P

Partial pressure
The pressure that an individual gas, in a mixture of gases, exerts on a system.

Pentose sugar
A sugar with five carbon atoms.

Peptide
A polymer formed from reactions between amino acids.

Peptide link (bond)
The bonds which hold amino acids together in a protein.

Percentage yield
A comparison between the amount of product that should form during a reaction and the amount that actually forms.

Period
A row in the periodic table.

Periodicity
The trends in the physical and chemical properties of elements as you go across the periodic table.

Permanent dipole-dipole forces
Intermolecular forces that exist because the difference in electronegativities in a polar bond causes weak electrostatic forces of attraction between molecules.

Petrochemical
Any compound that is made from crude oil (or any of its fractions).

Petroleum
A mixture consisting mainly of alkane hydrocarbons that can be separated into different fractions.

pH
A measure of the hydrogen ion concentration in a solution.
$pH = -\log_{10}[H^+]$

pH chart
A chart that shows the colour of an indicator at different pHs.

pH curve
A graph of pH against volume of acid/alkali added.

pH meter
An electronic device used to measure pH, made up of a probe connected to a digital display.

Phenolphthalein
A pH indicator that changes colour between pH 8.3 and 10.

Photochemical reaction
A reaction started by ultraviolet light.

Plane polarised light
Light in which all the waves are vibrating in the same plane.

Polar bond
A covalent bond where a difference in electronegativity has caused a shift in electron density in the bond.

Polar molecule
A molecule containing polar bonds that are arranged so that the dipoles don't cancel each other out.
This causes a permanent dipole across the whole molecule.

Polyamide
A polymer containing amide links between the monomers. Can be formed from reactions between dicarboxylic acids and diamines.

Polyester
A polymer containing ester links between the monomers. Can be formed from reactions between dicarboxylic acids and diols.

Polymer
A long molecule formed from lots of repeating units (called monomers).

Polynucleotide
A polymer formed of nucleotides.

Polypeptide
A polymer formed from reactions between amino acids.

Porphyrin
A multidentate ligand found in a number of biological molecules including haemoglobin.

Position isomer
A molecule with the same skeleton and molecular formula as another molecule but with the functional group attached to a different carbon.

Precise result
Results where the data have a very small spread around the mean.

Prediction
A specific testable statement about what will happen in an experiment, based on observation, experience or a hypothesis.

Primary structure (of proteins)
The first 'level' that describes the structure of protein — i.e. the sequence of amino acids in the polypeptide chain.

Protein
One or more polypeptides folded into a structure which has a biological function.

Proton
A subatomic particle with a relative charge of +1 and a relative mass of 1, located in the nucleus of an atom.

Proton number
The number of protons in the nucleus of an atom. Also called atomic number.

Purely ionic model of a lattice
A model which assumes that all the ions in a lattice are spherical and have their charge evenly distributed around them. This model is also known as the perfect ionic model.

R

Racemate (or racemic mixture)
A mixture that contains equal quantities of each enantiomer of an optically active compound.

Random error
An error introduced by a factor that you cannot control.

Rate constant, *k*
A constant in the rate equation for a reaction at a certain temperature. The larger it is the faster the rate of reaction.

Rate-determining step
The slowest step in a reaction mechanism which determines the overall rate of a reaction.

Rate equation
An equation of the form rate = $k[A]^m[B]^n$ which tells you how the rate of a reaction is affected by the concentration of reactants.

Reaction rate
The change in the amount of reactants or products over time.

Recrystallisation
A process of purifying a solid organic compound.

Redistillation
A process of purification, by repeatedly distilling.

Redox potential
A measure of how easily an atom, molecule or ion is reduced.

Redox reaction
A reaction where reduction and oxidation happen simultaneously.

Redox titration
A titration which can be performed to determine how much reducing agent is needed to exactly react with a quantity of oxidising agent, or vice versa.

Reducing agent
Something that donates electrons and gets oxidised.

Reduction
The gain of electrons.

Refluxing
A method for heating a reaction so that you can increase the temperature of an organic reaction to boiling without losing volatile solvents, reactants or products. Any vaporised compounds are cooled, condense and drip back into the reaction mixture.

Relative atomic mass, A_r
The average mass of an atom of an element on a scale where an atom of carbon-12 is exactly 12.

Relative formula mass
The average mass of a formula unit on a scale where an atom of carbon-12 is exactly 12.

Relative isotopic abundance
The relative amount of each isotope present in a sample of an element.

Relative isotopic mass
The mass of an atom of an isotope of an element on a scale where an atom of carbon-12 is exactly 12.

Relative molecular mass, M_r
The average mass of a molecule on a scale where an atom of carbon-12 is exactly 12.

Repeatable result
A result is repeatable if you can repeat an experiment multiple times and get the same result.

Repeating unit
A part of a polymer that repeats over and over again.

Reproducible result
A result is reproducible if someone else can recreate your experiment and get the same result you do.

Retention
The process of being adsorbed onto the stationary phase in chromatography.

Retention time
The time taken for a component of a mixture to pass through a chromatography column to the detector at the other end.

R_f value
The ratio of the distance travelled by a spot to the distance travelled by the solvent in thin-layer chromatography.

Risk assessment
A procedure carried out to identify any hazards associated with an experiment and how to reduce the risks these hazards present.

S

Salt bridge
A connection between two half-cells that ions can flow through, used to complete the circuit. Usually a piece of filter paper soaked in a salt solution or a glass tube filled with a salt solution.

Sand bath
A piece of equipment used to heat a reaction mixture, consisting of a container filled with sand that can be heated to a set temperature.

Saturated fatty acid
A fatty acid that contains no double bonds — found in fats.

Saturated molecule
A molecule that only has single carbon-carbon bonds.

Second electron affinity, ΔH_{ea2}
The enthalpy change when 1 mole of gaseous 2– ions is made from 1 mole of gaseous 1– ions.

Second ionisation energy, ΔH_{ie2}
The enthalpy change when 1 mole of gaseous 2+ ions is formed from 1 mole of gaseous 1+ ions.

Secondary structure (of proteins)
The second 'level' that describes the structure of protein — e.g. how the primary structure folds and twists to form an a-helix or b-pleated sheet, held together by hydrogen bonds.

Separation
A technique to separate the water-soluble impurities out of an organic mixture. The aqueous and organic solutions can be separated as they are immiscible, and separate out into two distinct layers due to their different densities.

Shell
A region of an atom with a fixed energy that contains electrons orbiting the nucleus.

Silver nitrate test
Test that uses silver nitrate solution to identify halide ions in solution.

Skeletal formula
A simplified organic formula which only shows the carbon skeleton and associated functional groups.

Solvent extraction
A form of separation, where the product is shaken vigorously with an immiscible solvent.

Solvent front
The distance travelled by the solvent in thin-layer chromatography.

Spectroscopy
The study of what happens when radiation interacts with matter.

Splitting pattern
Peaks in ^{1}H NMR spectra may be split into further peaks. The resultant group of peaks is called a splitting pattern.

Standard conditions
A temperature of 298 K (25 °C), a pressure of 100 kPa and all ion solutions having a concentration of 1.00 mol dm^{-3}.

Standard electrode potential
The voltage measured under standard conditions when a half-cell is connected to a standard hydrogen electrode.

Standard enthalpy change of combustion, $\Delta_c H^\circ$
The enthalpy change when 1 mole of a substance is completely burned in oxygen under standard conditions with all reactants and products in their standard states.

Standard enthalpy change of formation, $\Delta_f H^\circ$
The enthalpy change when 1 mole of a compound is formed from its elements in their standard states under standard conditions.

Standard enthalpy change of reaction, $\Delta_r H^\circ$
The enthalpy change when a reaction occurs in the molar quantities shown in the chemical equation, under standard conditions with all reactants and products in their standard states.

Standard hydrogen electrode
An electrode where hydrogen gas is bubbled through a solution of aqueous H^+ ions under standard conditions.

Standard solution
A solution that you know the exact concentration of.

State symbol
A symbol placed after a chemical in an equation to tell you what state of matter the chemical is in.

Stationary phase
A solid, or a liquid held in a solid, used in chromatography which contains molecules that can't move.

Stereoisomer
A molecule that has the same structural formula as another molecule but with the atoms arranged differently in space.

Stereospecific active site
An active site that only reacts with one enantiomer of a chiral compound.

Strong acid/base
An acid or base that fully dissociates in water.

Structural formula
A way of representing molecules that shows the atoms carbon by carbon, with the attached hydrogens and functional groups.

Structural isomer
A molecule with the same molecular formula as, but a different structural formula to another molecule (i.e. the atoms are connected in different ways).

Sub-level (sub-shell)
A subdivision of an energy level. A sub-level may be an s, p, d or f sub-level.

Substituted alkene
An alkene where one of the hydrogen atoms has been swapped for another atom or group.

Substrate
A molecule that is acted on by an enzyme.

Surfactant
A compound which is partly soluble and partly insoluble in water.

Synthetic route (synthesis)
A method detailing how to create a chemical.

Systematic error
An error introduced by the apparatus or method you use in an experiment.

Tertiary structure (of proteins)
The third 'level' that describes the structure of protein — i.e. how the secondary structure folds and twists to form a 3D molecule, held together by hydrogen and disulfide bonds.

Theoretical yield
The mass of product that should be formed in a chemical reaction.

Thin-layer chromatography (TLC)
A type of chromatography where the stationary phase is a plate coated with a solid and the mobile phase is a liquid solvent.

Titration
An experimental technique that lets you work out exactly how much alkali is needed to neutralise a quantity of acid.

Tollens' reagent
A colourless solution of silver nitrate dissolved in aqueous ammonia which can be used to distinguish between aldehydes and ketones.

Toxic substance
A substance that can cause illness or even death.

Transition metal
A metal that can form one or more stable ions with a partially filled d sub-level.

Unsaturated fatty acid
A fatty acid that contains double bonds — found in oils.

Unsaturated molecule
A molecule with one or more carbon-carbon double bonds.

V

Valence shell electron pair repulsion theory
The theory that in a molecule lone pair/lone pair bond angles are the biggest, lone pair/bonding pair bond angles are the second biggest and bonding pair/bonding pair bond angles are the smallest.

Valid result
A result which answers the question it was intended to answer.

Van der Waals forces
A type of intermolecular force caused by temporary dipoles, which causes all atoms and molecules to be attracted to each other. Also called induced dipole-dipole forces or London forces.

Variable
A factor in an experiment or investigation that can change or be changed.

Washing
A method of purifying a product by washing it with chemicals, such as washing with sodium hydrogen carbonate solution to remove acids.

Water bath
A piece of equipment used to heat a reaction mixture, consisting of a container filled with water that can be heated to a set temperature.

Weak acid/base
An acid or base that only partially dissociates in water.

Yield
The amount of product you get from a reaction.

Z

***Z*-isomer**
A stereoisomer of an alkene that has the two highest priority groups on the same side of the carbon-carbon double bond.

Zwitterion
A dipolar ion which has both a negative and a positive charge in different parts of the molecule

Acknowledgements

Photograph Acknowledgements

Cover Image © **TomekD76**/iStock / Getty Images Plus/Getty Images, p 2 **Martyn F. Chillmaid**/Science Photo Library, p 4 © **Ecelop**/iStockphoto.com, p 5 **Andrew Lambert Photography**/Science Photo Library, p 7 (top) **Martin Shields**/Science Photo Library, p 7 (bottom) **Andrew Lambert Photography**/Science Photo Library, p 8 **Andrew Lambert Photography**/Science Photo Library, p 10 **Martyn F. Chillmaid**/Science Photo Library, p 15 **Gustoimages**/ Science Photo Library, p 23 Science Photo Library, p 24 **Prof. Peter Fowler**/Science Photo Library, p 27 **Andrew Brookes, National Physical Laboratory**/Science Photo Library, p 29 **NASA**/Science Photo Library, p 32 **Food & Drug Administration**/Science Photo Library, p 38 **Andrew Lambert Photography**/Science Photo Library, p 46 **Andrew Lambert Photography**/Science Photo Library, p 49 **GIPhotostock**/Science Photo Library, p 55 **Charles D. Winters**/Science Photo Library, p 57 **Martyn F. Chillmaid**/Science Photo Library, p 58 **Andrew Lambert Photography**/Science Photo Library, p 61 **Martyn F. Chillmaid**/Science Photo Library, p 66 **Martyn F. Chillmaid**/ Science Photo Library, p 77 **Charles D. Winters**/Science Photo Library, p 79 **Bill Beatty, Visuals Unlimited**/ Science Photo Library, p 83 © **sarbiewski**/iStockPhoto.com, p 86 © **JeffreyRasmussen**/iStockPhoto.com, p 88 © **StasKhom**/iStockPhoto.com, p 89 © **JeffreyRasmussen**/iStockPhoto.com, p 90 **Dr Tim Evans**/Science Photo Library, p 96 **Marytn F. Chillmaid**/Science Photo Library, p 97 (top) **Clive Freeman/Biosym Technologies**/ Science Photo Library, p 97 (bottom) **Clive Freeman/Biosym Technologies**/Science Photo Library, p 100 © **wertorer**/iStockPhoto.com, p 101 **Charles D. Winters**/Science Photo Library, p 102 (top) Science Photo Library, p 102 (bottom) **Andrew Lambert Photography**/Science Photo Library, p 103 **GIPhotostock**/Science Photo Library, p 112 **Charles D. Winters**/Science Photo Library, p 123 (top) **Sheila Terry**/Science Photo Library, p 123 (bottom) Science Photo Library, p 125 **Emilio Segre Visual Archives/American Institute Of Physics**/Science Photo Library, p 134 (top) **Martyn F. Chillmaid**/Science Photo Library, p 134 (bottom) **Martyn F. Chillmaid**/ Science Photo Library, p 136 **Martyn F. Chillmaid**/Science Photo Library, p 137 **Deloche**/Science Photo Library, p 145 **Martyn F. Chillmaid**/Science Photo Library, p 153 **Ria Novosti**/Science Photo Library, p 154 (top left) **Charles D. Winters**/Science Photo Library, p 154 (top right) **Charles D. Winters**/Science Photo Library, p 154 (bottom) **E. R. Degginger**/Science Photo Library, p 156 (top) **Martyn F. Chillmaid**/Science Photo Library, p 156 (bottom) **Andrew Lambert Photography**/Science Photo Library, p 157 (top) © **cerae**/iStockPhoto.com, p 157 (middle) **Charles D. Winters**/Science Photo Library, p 157 (bottom) **Charles D. Winters**/Science Photo Library, p 161 (top) **Charles D. Winters**/Science Photo Library, p 161 (bottom) **Martyn F. Chillmaid**/Science Photo Library, p 164 **Miriam Maslo**/Science Photo Library, p 167 (top) **Andrew Lambert Photography**/Science Photo Library, p 167 (middle) **Andrew Lambert Photography**/Science Photo Library, p 167 (bottom) **Charles D. Winters**/Science Photo Library, p 172 (top) **Andrew Lambert Photography**/Science Photo Library, p 172 (bottom) **Andrew Lambert Photography**/Science Photo Library, p 173 (top) **Andrew Lambert Photography**/ Science Photo Library, p 173 (bottom left) **Andrew Lambert Photography**/Science Photo Library, p 173 (bottom right) **Andrew Lambert Photography**/Science Photo Library, p 174 **Trevor Clifford Photography**/Science Photo Library, p 179 **Laguna Design**/Science Photo Library, p 181 **Martyn F. Chillmaid**/Science Photo Library, p 184 **US Geological Survey**/Science Photo Library, p 187 **Nicolas Reusens**/Science Photo Library, p 188 Science Photo Library, p 191 **Faye Norman**/Science Photo Library, p 198 Science Photo Library, p 204 **Paul Rapson**/ Science Photo Library, p 205 **Martyn F. Chillmaid**/Science Photo Library, p 207 (top) **Astrid & Hanns-Frieder Michler**/Science Photo Library, p 207 (bottom) **Simon Fraser**/Science Photo Library, p 210 **NASA/Goddard Space Flight Center**/Science Photo Library, p 218 **Andrew Lambert Photography**/Science Photo Library, p 224 **Andrew Lambert Photography**/Science Photo Library, p 231 **Astrid & Hanns-Frieder Michler**/Science Photo Library, p 236 **Andrew Lambert Photography**/Science Photo Library, p 239 (top) **Power and Syred**/Science Photo Library, p 239 (bottom) **Ed Young**/Science Photo Library, p 240 **Alex Bartel**/Science Photo Library, p 243 (top) © **ScantyNebula**/iStockPhoto.com, p 243 (bottom) **Andrew Lambert Photography**/Science Photo Library, p 244 **Martyn F. Chillmaid**/Science Photo Library, p 245 (top) **Andrew Lambert Photography**/Science Photo Library, p 245 (bottom) **Andrew Lambert Photography**/Science Photo Library, p 250 **Andrew Lambert Photography**/Science Photo Library, p 251 **Andrew Lambert Photography**/Science Photo Library, p 252 (top) **Andrew Lambert Photography**/Science Photo Library, p 252 (middle) Science Photo Library, p 252 (bottom) **Andrew Lambert Photography**/Science Photo Library, p 254 **Gustoimages**/Science Photo Library, p 258 **Science**

Source/Science Photo Library, p 265 **David Taylor**/Science Photo Library, p 267 Science Photo Library, p 268 **Andrew Lambert Photography**/Science Photo Library, p 272 **Andrew Lambert Photography**/Science Photo Library, p 274 **Adam Hart-Davis**/Science Photo Library, p 277 Science Photo Library, p 288 (top) © **photongpix**/ iStockphoto.com, p 288 (bottom) **Andrew Lambert Photography**/Science Photo Library, p 295 **Andrew Lambert Photography**/Science Photo Library, p 302 (top) **Andrew Lambert Photography**/Science Photo Library, p 302 (bottom) **Andrew Lambert Photography**/Science Photo Library, p 309 **Emilio Segre Visual Archives/American Institute of Physics**/Science Photo Library, p 312 **Martin Bond**/Science Photo Library, p 313 Science Photo Library, p 316 © **MichaelJay**/iStockphoto.com, p 318 **Charles D. Winters**/Science Photo Library, p 323 **Charles D. Winters**/ Science Photo Library, p 324 **Martyn F. Chillmaid**/Science Photo Library, p 328 **E. R. Degginger**/Science Photo Library, p 335 (top) **Cordelia Molloy**/Science Photo Library, p 335 (bottom) **Martin Bond**/Science Photo Library, p 336 **Charles D. Winters**/Science Photo Library, p 343 **Charles D. Winters**/Science Photo Library, p 344 **Martyn F. Chillmaid**/Science Photo Library, p 345 **Martyn F. Chillmaid**/Science Photo Library, p 346 **Martyn F. Chillmaid**/ Science Photo Library, p 347 **Andrew Lambert Photography**/Science Photo Library, p 351 Science Photo Library, p 352 (top) **Andrew Lambert Photography**/Science Photo Library, p 352 (bottom) **Andrew Lambert Photography**/ Science Photo Library, p 354 **Andrew Lambert Photography**/Science Photo Library, p 358 **Charles D. Winters**/ Science Photo Library, p 359 (top) **Charles D. Winters**/Science Photo Library, p 359 (bottom) **Andrew Brookes, National Physical Laboratory**/Science Photo Library, p 360 **Paul Whitehill**/Science Photo Library, p 368 **Charles D. Winters**/Science Photo Library, p 369 (top) **Andrew Lambert Photography**/Science Photo Library, p 369 (bottom) **Jerry Mason**/Science Photo Library, p 371 **Martyn F. Chillmaid**/Science Photo Library, p 374 **Klaus Guldbrandsen**/ Science Photo Library, p 375 **Astrid & Hanns-Frieder Michler**/Science Photo Library, p 376 **Andrew Lambert Photography**/Science Photo Library, p 379 **Dr Mark J. Winter**/Science Photo Library, p 380 **Andrew Lambert Photography**/Science Photo Library, p 385 (left) **Andrew Lambert Photography**/Science Photo Library, p 385 (right) **Andrew Lambert Photography**/Science Photo Library, p 386 **Andrew Lambert Photography**/Science Photo Library, p 389 **Power And Syred**/Science Photo Library, p 390 **Dr Tim Evans**/Science Photo Library, p 392 **Andrew Lambert Photography**/Science Photo Library, p 393 **Andrew Lambert Photography**/Science Photo Library, p 394 Science Photo Library, p 396 **Martyn F. Chillmaid**/Science Photo Library, p 397 **Andrew Lambert Photography**/Science Photo Library, p 400 **Sheila Terry**/Science Photo Library, p 401 **Andrew Lambert Photography**/Science Photo Library, p 404 **Andrew Lambert Photography**/Science Photo Library, p 405 (top) **Charles D. Winters**/Science Photo Library, p 405 (bottom) **Chris Hellier**/Science Photo Library, p 406 **Andrew Lambert Photography**/Science Photo Library, p 414 **Klaus Guldbrandsen**/Science Photo Library, p 419 (top) **Andrew Lambert Photography**/Science Photo Library, p 419 (bottom) **Andrew Lambert Photography**/Science Photo Library, p 420 **Andrew Lambert Photography**/ Science Photo Library, p 424 **Tek Image**/Science Photo Library, p 426 **Cristina Pedrazzini**/Science Photo Library, p 429 **Maximilian Stock Ltd**/Science Photo Library, p 430 **Andrew Lambert Photography**/Science Photo Library, p 431 **Ria Novosti**/Science Photo Library, p 432 **Andrew Lambert Photography**/Science Photo Library, p 436 **Cordelia Molloy**/Science Photo Library, p 437 Science Photo Library, p 447 **Clive Freeman, The Royal Institution**/ Science Photo Library, p 451 (top) © **rclassenlayouts**/iStockphoto.com, p 451 (bottom) Science Photo Library, p 454 © **rezkrr**/iStockphoto.com, p 459 **Martyn F. Chillmaid**/Science Photo Library, p 460 **Andrew Lambert Photography**/ Science Photo Library, p 465 © **fotokostic**/iStockphoto.com, p 466 © **monkeybusinessimages**/iStockphoto.com, p 473 **Andrew Lambert Photography**/Science Photo Library, p 478 **Mauro Fermariello**/Science Photo Library, p 480 (top) **Laguna Design**/Science Photo Library, p 480 (bottom) **Laguna Design**/Science Photo Library, p 483 **Kenneth Eward**/Science Photo Library, p 484 **Jan Van De Vel/Reporters**/Science Photo Library, p 488 **A. Barrington Brown**/ Science Photo Library, p 489 **Dr P. Marazzi**/Science Photo Library, p 496 **Martyn F. Chillmaid**/Science Photo Library, p 498 **Hank Morgan**/Science Photo Library, p 506 **Hank Morgan**/Science Photo Library, p 507 **Friedrich Saurer**/Science Photo Library, p 510 **Sinclair Stammers**/Science Photo Library, p 511 **Sinclair Stammers**/Science Photo Library, p 514 **Patrick Landmann**/Science Photo Library, p 521 **Dr Tim Evans**/Science Photo Library, p 526 (top) © **Achim Prill**/iStockphoto.com, p 526 (bottom) **Martyn F. Chillmaid**/Science Photo Library, p 532 © **AtomStudios**/iStockphoto.com

Index

D

E

F

G

T

U

V

W

Y

Z

The Periodic Table

Relative Atomic Mass (A_r) → top number in each box; Atomic (proton) number → bottom number.

Periods	Group 1	Group 2											Group 3	Group 4	Group 5	Group 6	Group 7	Group 0
1								1.0 H hydrogen 1										4.0 He helium 2
2	6.9 Li lithium 3	9.0 Be beryllium 4											10.8 B boron 5	12.0 C carbon 6	14.0 N nitrogen 7	16.0 O oxygen 8	19.0 F fluorine 9	20.2 Ne neon 10
3	23.0 Na sodium 11	24.3 Mg magnesium 12											27.0 Al aluminium 13	28.1 Si silicon 14	31.0 P phosphorus 15	32.1 S sulfur 16	35.5 Cl chlorine 17	39.9 Ar argon 18
4	39.1 K potassium 19	40.1 Ca calcium 20	45.0 Sc scandium 21	47.9 Ti titanium 22	50.9 V vanadium 23	52.0 Cr chromium 24	54.9 Mn manganese 25	55.8 Fe iron 26	58.9 Co cobalt 27	58.7 Ni nickel 28	63.5 Cu copper 29	65.4 Zn zinc 30	69.7 Ga gallium 31	72.6 Ge germanium 32	74.9 As arsenic 33	79.0 Se selenium 34	79.9 Br bromine 35	83.8 Kr krypton 36
5	85.5 Rb rubidium 37	87.6 Sr strontium 38	88.9 Y yttrium 39	91.2 Zr zirconium 40	92.9 Nb niobium 41	96.0 Mo molybdenum 42	[98] Tc technetium 43	101.1 Ru ruthenium 44	102.9 Rh rhodium 45	106.4 Pd palladium 46	107.9 Ag silver 47	112.4 Cd cadmium 48	114.8 In indium 49	118.7 Sn tin 50	121.8 Sb antimony 51	127.6 Te tellurium 52	126.9 I iodine 53	131.3 Xe xenon 54
6	132.9 Cs caesium 55	137.3 Ba barium 56	138.9 La lanthanum 57	178.5 Hf hafnium 72	180.9 Ta tantalum 73	183.8 W tungsten 74	186.2 Re rhenium 75	190.2 Os osmium 76	192.2 Ir iridium 77	195.1 Pt platinum 78	197.0 Au gold 79	200.6 Hg mercury 80	204.4 Tl thallium 81	207.2 Pb lead 82	209.0 Bi bismuth 83	[209] Po polonium 84	[210] At astatine 85	[222] Rn radon 86
7	[223] Fr francium 87	[226] Ra radium 88	[227] Ac actinium 89	[267] Rf rutherfordium 104	[268] Db dubnium 105	[271] Sg seaborgium 106	[272] Bh bohrium 107	[270] Hs hassium 108	[276] Mt meitnerium 109	[281] Ds darmstadtium 110	[280] Rg roentgenium 111	Elements with atomic numbers 112–116 have been reported but not fully authenticated						

58–71: Lanthanides	140.1 Ce cerium 58	140.9 Pr praseodymium 59	144.2 Nd neodymium 60	[145] Pm promethium 61	150.4 Sm samarium 62	152.0 Eu europium 63	157.3 Gd gadolinium 64	158.9 Tb terbium 65	162.5 Dy dysprosium 66	164.9 Ho holmium 67	167.3 Er erbium 68	168.9 Tm thulium 69	173.1 Yb ytterbium 70	175.0 Lu lutetium 71
90–103: Actinides	232.0 Th thorium 90	231.0 Pa protactinium 91	238.0 U uranium 92	[237] Np neptunium 93	[244] Pu plutonium 94	[243] Am americium 95	[247] Cm curium 96	[247] Bk berkelium 97	[251] Cf californium 98	[252] Es einsteinium 99	[257] Fm fermium 100	[258] Md mendelevium 101	[259] No nobelium 102	[262] Lr lawrencium 103